Vitamins		Minerals											
Vitamin E RDA (mg/day)[e]	Vitamin K AI (µg/day)	Calcium AI (mg/day)	Phosphorus RDA (mg/day)	Magnesium RDA (mg/day)	Iron RDA (mg/day)	Zinc RDA (mg/day)	Iodine RDA (µg/day)	Selenium RDA (µg/day)	Copper RDA (µg/day)	Manganese AI (mg/day)	Fluoride AI (mg/day)	Chromium AI (µg/day)	Molybdenum RDA (µg/day)
4	2.0	210	100	30	0.27	2	110	15	200	0.003	0.01	0.2	2
5	2.5	270	275	75	11	3	130	20	220	0.6	0.5	5.5	3
6	30	500	460	80	7	3	90	20	340	1.2	0.7	11	17
7	55	800	500	130	10	5	90	30	440	1.5	1.0	15	22
11	60	1300	1250	240	8	8	120	40	700	1.9	2	25	34
15	75	1300	1250	410	11	11	150	55	890	2.2	3	35	43
15	120	1000	700	400	8	11	150	55	900	2.3	4	35	45
15	120	1000	700	420	8	11	150	55	900	2.3	4	35	45
15	120	1200	700	420	8	11	150	55	900	2.3	4	30	45
15	120	1200	700	420	8	11	150	55	900	2.3	4	30	45
11	60	1300	1250	240	8	8	120	40	700	1.6	2	21	34
15	75	1300	1250	360	15	9	150	55	890	1.6	3	24	43
15	90	1000	700	310	18	8	150	55	900	1.8	3	25	45
15	90	1000	700	320	18	8	150	55	900	1.8	3	25	45
15	90	1200	700	320	8	8	150	55	900	1.8	3	20	45
15	90	1200	700	320	8	8	150	55	900	1.8	3	20	45
15	75	1300	1250	400	27	13	220	60	1000	2.0	3	29	50
15	90	1000	700	350	27	11	220	60	1000	2.0	3	30	50
15	90	1000	700	360	27	11	220	60	1000	2.0	3	30	50
19	75	1300	1250	360	10	14	290	70	1300	2.6	3	44	50
19	90	1000	700	310	9	12	290	70	1300	2.6	3	45	50
19	90	1000	700	320	9	12	290	70	1300	2.6	3	45	50

[e] Vitamin E recommendations are expressed as α-tocopherol.

SOURCE: Adapted with permission from the *Dietary Reference Intakes* series, National Academy Press. Copyright 1997, 1998, 2000, 2001, by the National Academy of Sciences. Courtesy of the National Academy Press, Washington, D.C.

Minerals									
Zinc (mg/day)	Iodine (µg/day)	Selenium (µg/day)	Copper (µg/day)	Manganese (mg/day)	Fluoride (mg/day)	Molybdenum (µg/day)	Boron (mg/day)	Nickel (mg/day)	Vanadium (mg/day)
4	—	45	—	—	0.7	—	—	—	—
5	—	60	—	—	0.9	—	—	—	—
7	200	90	1000	2	1.3	300	3	0.2	—
12	300	150	3000	3	2.2	600	6	0.3	—
23	600	280	5000	6	10	1100	11	0.6	—
34	900	400	8000	9	10	1700	17	1.0	—
40	1100	400	10,000	11	10	2000	20	1.0	1.8
40	1100	400	10,000	11	10	2000	20	1.0	1.8
34	900	400	8000	9	10	1700	17	1.0	—
40	1100	400	10,000	11	10	2000	20	1.0	—
34	900	400	8000	9	10	1700	17	1.0	—
40	1100	400	10,000	11	10	2000	20	1.0	—

NOTE: An Upper Limit was not established for vitamins and minerals not listed and for those age groups listed with a dash (—) because of a lack of data, not because these nutrients are safe to consume at any level of intake. All nutrients can have adverse effects when intakes are excessive.

SOURCE: Adapted with permission from the *Dietary Reference Intakes* series, National Academy Press. Copyright 1997, 1998, 2000, 2001, by the National Academy of Sciences. Courtesy of the National Academy Press, Washington, D.C.

Daily Values (DV, used on food labels)[a]

Daily Reference Values (DRVs)[b]		Reference Daily Intakes (RDI)			
Food Component	**Amount**	**Nutrient**	**Amount**	**Nutrient**	**Amount**
protein[c]	50 g	Thiamin	1.5 mg	Calcium	1,000 mg
fat	65 g[d]	Riboflavin	1.7 mg	Iron	18 mg
saturated fatty acids	20 g	Niacin	20 mg	Zinc	15 mg
cholesterol	300 mg[e]	Biotin	300 μg	Iodine	150 μg
total carbohydrate	300 g	Pantothenic Acid	10 mg	Copper	2 mg
fiber	25 g	Vitamin B_6	2 mg	Chromium	120 μg
sodium	2,400 mg	Folate	400 μg[f]	Selenium	70 μg
potassium	3,500 mg	Vitamin B_{12}	6 μg	Molybdenum	75 μg
		Vitamin C	60 mg	Manganese	2 mg
		Vitamin A	5,000 IU[g]	Chloride	3,400 mg
		Vitamin D	400 IU[g]	Magnesium	400 mg
		Vitamin E	30 IU[g]	Phosphorus	1 g
		Vitamin K	80 μg		

[a]Based on 2,000 calories a day for adults and children over 4 years old.

[b]Formerly the U.S. RDA, based on National Academy of Sciences' 1968 Recommended Dietary Allowances.

[c]DRV for protein does not apply to certain populations; Reference Daily Intake (RDI) for protein has been established for these groups: children 1 to 4 years: 16 g; infants under 1 year: 14 g; pregnant women: 60 g; nursing mothers: 65 g.

[d](g) grams

[e](mg) milligrams

[f](μg) micrograms

[g]Equivalent values for the three DV nutrients expressed as IU are: vitamin A, 900 RE (assumes a mixture of 40% retinol and 60% beta-carotene); vitamin D, 10 μg; vitamin E, 20 mg.

1989 Recommended Dietary Allowances (RDA) for Energy and Protein

Age (yr)	Energy (cal)	Protein (g)
Infants		
0.0–0.5	650	13
0.5–1.0	850	14
Children		
1–3	1300	16
4–6	1800	24
7–10	2000	28
Males		
11–14	2500	45
15–18	3000	59
19–24	2900	58
25–50	2900	63
51+	2300	63
Females		
11–14	2200	46
15–18	2200	44
19–24	2200	46
25–50	2200	50
51+	1900	50
Pregnancy	+300	60
Lactation		
1st 6 mo.	+500	65
2nd 6 mo.	+500	62

Petrov-Vodkin, Kuzma (1878–1930), *Still Life with Apple Tree Branch,* Tretyakov Gallery, Moscow, Russia © Scala/Art Resource, New York

NUTRITION

CONCEPTS AND CONTROVERSIES

NINTH EDITION

Frances Sienkiewicz Sizer

Eleanor Noss Whitney

Australia • Canada • Mexico • Singapore • Spain
United Kingdom • United States

PUBLISHER: Peter Marshall
DEVELOPMENT EDITOR: Elizabeth Howe
ASSISTANT EDITOR: John Boyd
EDITORIAL ASSISTANT: Madinah Chang
MARKETING MANAGER: Jennifer Somerville
MARKETING ASSISTANT: Mona Weltmer
PROJECT EDITORS: Sandra Craig, Teri Hyde
PRINT BUYER: Barbara Britton
PERMISSIONS EDITOR: Joohee Lee
PRODUCTION SERVICE: The Book Company
TEXT AND COVER DESIGNER: Baugher Design Inc.

PHOTO RESEARCHERS: Myrna Engler, Lindsay Kefauver
COPY EDITORS: Patricia Lewis, Pat Brewer
ILLUSTRATIONS: Impact Publications, McMahon Medical Art, Rolin Graphics, Dorothy Reinhardt
PHOTOGRAPHER AND FOOD STYLISTS: Polara Studios, Inc. and Carol Ladd
COVER IMAGE: Art Resource
COMPOSITOR: Parkwood Composition Services
PRINTER: Transcontinental Interglobe

Printed in Canada
3 4 5 6 7 06 05 04

For more information about our products, contact us at:
Thomson Learning Academic Resource Center
1-800-423-0563
For permission to use material from this text, contact us by:
Phone: 1-800-730-2214 **Fax:** 1-800-730-2215
Web: http://www.thomsonrights.com

Library of Congress Control Number: 2002105893

ISBN 0534-57799-7

WADSWORTH/THOMSON LEARNING
10 Davis Drive
Belmont, CA 94002-3098
USA

ASIA
Thomson Learning
5 Shenton Way #01-01
UIC Building
Singapore 068808

AUSTRALIA
Nelson Thomson Learning
102 Dodds Street
South Melbourne, Victoria 3205
Australia

CANADA
Nelson Thomson Learning
1120 Birchmount Road
Toronto, Ontario M1K 5G4
Canada

EUROPE/MIDDLE EAST/AFRICA
Thomson Learning
High Holborn House
50/51 Bedford Row
London WC1R 4LR
United Kingdom

LATIN AMERICA
Thomson Learning
Seneca, 53
Colonia Polanco
11560 Mexico D.F.
Mexico

SPAIN
Paraninfo Thomson Learning
Calle/Magallanes, 25
28015 Madrid, Spain

About the Authors

TO PHILIP AND ALL OF MY FAMILY, TO KITTY AND ALL OF MY FRIENDS (THANKS FOR WAITING), TO ECHO AND ALL WHO HELP OTHERS, AND TO ALL WHO RECEIVE HELP.

FRAN

FRANCES SIENKIEWICZ SIZER, M.S., R.D., F.A.D.A., attended Florida State University where, in 1980, she received her B.S., and in 1982, her M.S. in nutrition. She is certified as a charter Fellow of the American Dietetic Association. She is a founding member and vice president of Nutrition and Health Associates, an information and resource center in Tallahassee, Florida, that maintains an ongoing bibliographic database tracking research in more than 1,000 topic areas of nutrition. Her textbooks include *Life Choices: Health Concepts and Strategies; Making Life Choices; The Fitness Triad: Motivation, Training, and Nutrition*; and others. She is a primary author of *Nutrition Interactive*, an instructional college-level nutrition CD-ROM. In addition to writing, she lectures at universities and at national and regional conferences, and serves actively on the board of directors of ECHO, a local hunger and homelessness relief organization in her community.

TO ALL OF MY GRANDCHILDREN AND STEPGRANDCHILDREN: MAX, ZOEY, EMILY, REBECCA, SARAH, WILL, TOOTTOOT, AND JACOB.

ELLIE

ELEANOR NOSS WHITNEY, PH.D., received her B.A. in Biology from Radcliffe College in 1960 and her Ph.D. in Biology from Washington University, St. Louis, in 1970. Formerly on the faculty at Florida State University, and a dietitian registered with the American Dietetic Association, she now devotes full time to research, writing, and consulting in nutrition, health, and environmental issues. Her earlier publications include articles in *Science*, *Genetics*, and other journals. Her textbooks include *Understanding Nutrition*, *Understanding Normal and Clinical Nutrition*, *Nutrition and Diet Therapy*, and *Essential Life Choices* for college students and *Making Life Choices* for high-school students. Her most intense interests presently include energy conservation, solar energy uses, alternatively fueled vehicles, and ecosystem restoration.

Contents in Brief

Contents

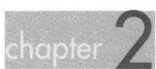

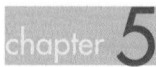

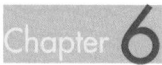

The Vitamins 211

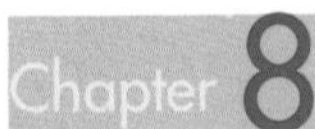

Water and Minerals 265

Preface

A BILLBOARD IN Louisiana reads, "Come as you are. Leave different," meaning that once you've seen, smelled, tasted, and listened to Louisiana, you'll never be the same. This book extends the same invitation to its readers: Come to nutrition science as you are, with all of the knowledge and enthusiasm you possess, with all of your unanswered questions and misconceptions, and with the habits and preferences that now dictate what you eat.

But leave different. Take with you from this study a more complete understanding of nutrition. Take a greater ability to discern between nutrition truth and fiction, to ask sophisticated questions and find the answers. Finally, leave with a better sense of how to feed yourself in ways that not only please you and soothe your spirit, but nourish your body as well.

For almost a quarter of a century, *Nutrition: Concepts and Controversies* has been a cornerstone in nutrition classes across North America, serving the needs of students and professors in building a healthier future. In keeping with the tradition of past editions, we continue to explore the ever-changing frontier of nutrition science, confronting its mysteries through its scientific roots. We maintain our sense of personal connection with instructors and learners alike, writing to them in the clear, informal style that has become our trademark, but with a fresh crispness that we hope you will enjoy.

Throughout these chapters, features both pique the reader's interest and inform. For both verbal and visual learners, our logical presentation and our direct, colorful figures keep interest high and understanding at a peak. New photos adorn many of our pages, adding pleasure to reading. Each chapter begins with *Frequently Asked Questions*, to arouse curiosity, with the answers sprinkled throughout the chapter. Little reminders, called *Think Fitness*, appear from time to time to alert readers to ways in which physical activity links with nutrition in supporting the health of the body. For example, the *Think Fitness* feature of Chapter 1 points out the health benefits that one can expect from regular physical activity. In remaining chapters, these features emphasize the roles of the nutrients in supporting physical activity and spell out the advantages of an active lifestyle throughout life.

Many tried-and-true features return in this edition: The practical, interactive activities called *Do It!* invite students to apply chapter contents in their everyday encounters with nutrition choices. The popular *Self Check* features provide review questions at the end of each chapter; answers in Appendix G provide immediate feedback to the learner. Many relevant Internet websites are listed in each chapter's *Nutrition on the Net*.

By popular demand, we have retained our *Snapshots* of vitamins and minerals. These capsules of information depict food sources of nutrients and reinforce concepts concerning the Food Guide Pyramid and the Daily Values found on food labels. In this edition, the food label format appears often in figures to familiarize readers with the process of comparing the nutrients in foods that bear labels.

New to this ninth edition, the *Nutrition Connections* CD-ROM serves as a springboard to nutrition investigation in an electronic environment. Authored by Michelle Grodner of William Patterson University, this interactive CD-ROM is fully integrated with the text, using many of the book's illustrations, tables, and key points in 75 learning assignments that encourage exploration and research using Internet websites. *Nutrition Connections* also includes a hot glossary and student quizzes as well as 35 dynamic interactions. These animated activities help to explain complex nutrition concepts through student interaction. This icon signifies that the *Nutrition Connections* CD-ROM further explores a concept discussed in the text.

Chapter 1 begins the text with a personal challenge to students. It asks the question so many people ask of nutrition educators—"Why should people care about nutrition?" We answer with a lesson in the ways nutritious foods affect diseases, and present a continuum of diseases from the purely genetic in origin to those almost totally preventable by nutrition. After offering some beginning facts about nutrients and foods, the chapter goes on to present the *Dietary Guidelines for Americans* and the *Healthy People 2010* goals for the nation. It concludes with a discussion of scientific research in nutrition to provide a perspective on the context in which study results may be rightly viewed. Chapter 2 brings together the concepts of diet planning through nutrient allowances, such as the new *Dietary Reference Intakes* and the *Daily Food Guide* with its *Food Guide Pyramid* of food choices. Chapter 3 presents a thorough, but brief, introduction to the workings of the human body with emphasis on the digestive system. Chapters 4–6 are devoted to the energy-yielding nutrients—carbohydrates, lipids, and protein. Chapters 7 and 8 present the vitamins, minerals, and water, with special emphasis on the emerging importance of antioxidant nutrients. Chapter 9 relates energy balance to body composition, obesity, and underweight and presents weight maintenance as a lifelong effort. Chapter 10 examines the relationships between physical activity, athletic performance, and nutrition. Chapter 11 applies the essence of the first ten chapters to two broad and rapidly changing areas within nutrition: immunity and disease prevention. Chapters 12 and 13 emphasize the importance of nutrition throughout the life span. Chapter 14 delivers urgently important concepts of food safety, and Chapter 15 touches on the vast problems of the global food supply and world and U.S. hunger and explains how readers are linked to the meaningful whole through the daily choices available to them.

The *Controversies* of this book's title invite you to explore beyond the safe boundaries of established nutrition knowledge. These optional readings, which appear at the end of each chapter and are printed with colored borders, delve into current scientific topics and emerging controversies. Some are new to this edition, and the others have been updated. Of special current interest is Controversy 2, which presents the struggle between scientific exploration of the phytochemicals in foods and consumers' need for guidance in the appropriate use of these substances. Controversy 7 sets up a lively competition between food and supplements as vitamin sources, while exploring the research to date on the antioxidant vitamins. Controversy 9 presents current thinking about eating disorders. Controversy 10 tackles some questions surrounding the safety and effectiveness of products sold to athletes. Controversy 12 delves into theories concerning the influence of early nutrition on later disease development. Controversy 14 examines issues surrounding organic foods and the products of biotechnology. Controversy 15 explores ways in which agriculture can ensure a high-quality food supply throughout this century and beyond.

The *Food Feature* sections that appear in most chapters act as bridges between theory and practice; they are practical applications of the chapter concepts that help readers to choose foods according to nutrition principles. *Consumer Corners* empower students to make informed decisions by presenting information on olestra and other fat replacers, amino acid supplements, vitamin C and the common cold, bottled water, irradiation of foods, and other nutrition-related marketplace issues.

New or major terms in chapters are defined in the margins of the pages where they are introduced and also in the Glossary at the end of the book. Terms in Controversy sections are grouped together and defined in tables within the sections and in the Glossary. The reader who wishes to locate any term can do so by consulting the index, which lists the page numbers of definitions in boldface type.

The appendixes have proved useful to past readers. Appendix A presents complete and accurate listings of the nutrient contents of more than 2,200 foods. For compatibility with the new RDA, many vitamin A values are now expressed in the DRI committee's preferred units, micrograms Retinol Activity Equivalents. Appendix B, *Canadiana*, supplies the RNI, the Guidelines, the Food Guide, Food Labels, and the Choice System for our Canadian readers. Appendix C demonstrates nutrition calculations, with special emphasis on finding the percentage of calories from fat in a diet and percentages of the Daily Values. Appendix D provides full coverage with applications of the U.S. Exchange System. Appendix E offers an invaluable list of current addresses, telephone numbers, and Internet websites for those interested in additional information. To save space, we have collected all reference notes into Appendix F. Older source notes have been removed but are easily available by consulting older editions of this book or by contacting the publisher.

As always, our purpose in writing this text is to enhance our readers' understanding of nutrition science and motivation to apply it. We hope the information on this book's pages will reach beyond the classroom into our readers' lives. Take the information you find inside this book home with you. Use it in your life: nourish yourself, educate your loved ones, and nurture others to be a part of our healthy new future. Stay up with the news, too. For despite all the conflicting messages, inflated claims, and even quackery that abound in media reports, true nutrition knowledge progresses with a genuine scientific spirit, and important new truths are constantly unfolding.

Acknowledgments

OUR THANKS to Linda Kelly DeBruyne for the newly updated Chapter 10 and Chapter 12 of this edition. Thanks to Lynn Earnest for her enthusiastic, competent, and dedicated assistance at each stage of this writing. Thanks also to Donna Lindquist for applying her creative mind and skilled hand to Controversy 10. Many thanks to Michelle Grodner for the Nutrition on the Net features that appear throughout this edition, and for creating our exciting new *Nutrition Connections* CD-ROM.

Thanks to Jayme Ebaugh for organizing our manuscript and for office help. Thanks, too, to our associate Lori Turner for much of the *Instructor's Manual;* thanks to Margaret Hedley, University of Guelph, who prepared the Canadian material for the manual and reviewed the Canadian resources listed in the text. For the special Instructor's Edition of the text, we thank Lori Turner, University of Arkansas-Fayetteville (lecture outlines); Judy Kaufman, Monroe Community College (margin references to overheads, *Nutrition Interactive* CD-ROM, and *NutriLink* CD-ROM). Thanks also to Judy Kaufman who developed the *Self Check* questions for this edition. Thanks to Pat Brewer and Pat Lewis for their creative and excellent editing help. Thank you, Myrna Engler, Dorothy Reinhardt, Polara Studios, Rolin Graphics, and associates for bringing our figures and photos to their full potential. For the creation of this edition's bright, new design, we thank Norman Baugher.

Special thanks to our publisher, Peter Marshall, and to our editors Elizabeth Howe, Dusty Friedman, and Sandra Craig, and to their staff, for their unflagging efforts to ensure the highest quality of all facets of this book. Our applause and gratitude goes to the marketing team. Thanks also to Jana Kicklighter of Georgia State University for preparing the *Student Study Guide* and the *Test Bank*. As always, we are grateful to Bob Geltz and Betty Hands and their staff at ESHA research for Appendix A and for the computerized diet analysis program that accompanies this book. To our reviewers, many heartfelt thanks for your many thoughtful ideas and suggestions:

Kwaku Addo	University of Kentucky
Janet Anderson	Utah State University
Jenna Anding	University of Houston
Raga Bakhit	Virginia Polytechnic University
Linda C. Barnes	Tidewater Community College
Dea Hanson Baxter	Georgia State University
Ethan A. Bergman	Central Washington University
Debra Boardley	University of Toledo
Mallory Boylan	Texas Technical University
Pat Brown	Cuesta College
Ardith Brunt	Tennessee Tech University
N. Joanne Caid	California State University–Fresno
Marjorie Caldwell	University of Rhode Island
Leah Carter	Bakersfield College
Tom W. Castonguay	University of Maryland
Janet Colson	Middle Tennessee State University
Stacy Coseio	University of Houston
Margaret Craig-Schmidt	Auburn University
Georgia Crews	South Dakota State University
Leslie Edwards Cummings	University of Nevada
Earlene Davis	Bakersfield College
Allan Davison	Simon Fraser University
Bonnie Dean	West Virginia State College
Robert DiSilvestro	Ohio State University
Carol Friesen	Ball State University
Sherrie Frye	University of Northern Colorado
Cindy J. Fuller	University of North Carolina
Leonard Gerber	University of Rhode Island
Jan Goodwin	University of North Dakota
Cynthia Gossage	Prince George's Community College
Artis Grady	Southern Utah University
Margaret Gunther	Palomar Community College
Deborah Gustafson	Utah State University
Evette Hackman	Seattle Pacific University
Charlene Harkins	University of Minnesota, Duluth
Karen Heller	University of New Mexico
Deloy G. Hendricks	Utah State University
Ann A. Hertzier	Virginia Polytechnic Institute
David Holben	Ohio University
Clarie Hollenbeck	San Jose State University
Wendy T. Hunt	American River College
Kendra K. Kattelmann	South Dakota State University
Kathy Keiver	University of British Columbia
Margaret Kessel	Ohio State University
Jeanne W. Lawless	Ithaca College
Ryna Levy-Milne	University of British Columbia
Karen Lieberman	Johnson and Wales University
Dalia Lima	Houston Community College
Heather Lynne	Mission College
Paula May	Eastern Kentucky University
Julie McCullough	University of Southern Indiana
Sharon L. McWhimey	Prairie View A & M University
Janis Mena	University of Florida
Stella Miller	Mt. San Antonio College
Ryna Levy Milne	University of British Columbia
Cherie L. Moore	Cuesta College
Mary Etta Moorachian	Johnson and Wales University
Sharon K. Morcos	Kansas State University
Carmen Nochera	Grand Valley State University
Emilia Papakonstantinou	University of Georgia
Leonard A. Piche	Brescia College
Susan Polasek	University of Texas, Austin

Jennifer Ricketts	University of Arizona
Clay Robinson	Lewis Clark State College
Judith Rodriguez	University of North Florida
Brian Luke Seaward	University of Colorado
Kathryn Silliman	California State University Chico
Samuel C. Smith	University of New Hampshire
Carol Stinson	University of Louisville
Wendy Stuhldreher	Slippery Rock University
Joan Thompson	Weber State University
Dellman Walker	Middle Tennessee State University
Hope Weiler	University of Manitoba
Doug White	Auburn University
Julian H. Williford, Jr.	Bowling Green State University
Bonny Burns Whitmore	California State Polytechnic University, Pomona
Fred H. Wolfe	University of Arizona
Cynthia Wright	Southern Utah University
Lisa Young	New York University

chapter 1

Henry Church, Jr. (1836–1908), *Still Life* (detail) date unknown

Food Choices and Human Health

Frequently Asked Questions

How powerful is a nutritious diet in preventing diseases? p. 3

Can I forgo food and live just on supplements? p. 5

How, exactly, can I recognize a nutritious diet? p. 7

If the scientists don't know, how can I? p. 14

Can I trust the media to deliver nutrition news? p. 17

Contents

food medically, any substance that the body can take in and assimilate that will enable it to stay alive and to grow; the carrier of nourishment; socially, a more limited number of such substances defined as acceptable by each culture.

nutrition the study of the nutrients in foods and in the body; sometimes also the study of human behaviors related to food.

diet the foods (including beverages) a person usually eats and drinks.

nutrients components of food that are indispensable to the body's functioning. They provide energy, serve as building material, help maintain or repair body parts, and support growth. The nutrients include water, carbohydrate, fat, protein, vitamins, and minerals.

malnutrition any condition caused by excess or deficient food energy or nutrient intake or by an imbalance of nutrients. Nutrient or energy deficiencies are classed as forms of undernutrition; nutrient or energy excesses are classed as forms of overnutrition.

If YOU CARE about your body, and if you have strong feelings about **food,** then you have much to gain from learning about **nutrition**—the study of how food nourishes the body. Nutrition is a fascinating, much talked-about subject. Each day, newspapers, radio, and television present stories of new findings on nutrition and heart health or nutrition and cancer prevention, and at the same time advertisements and commercials bombard us with multicolored pictures of tempting foods—pizza, burgers, cakes, and chips. If you are like most people, when you eat you sometimes wonder, "Is this food good for me?" or you berate yourself, "I probably shouldn't be eating this."

When you study nutrition, you learn which foods serve you best, and you can work out ways of choosing foods, planning meals, and designing your **diet** wisely. Knowing the facts can enhance your health and your enjoyment of eating while relieving your feelings of guilt or worry that you aren't eating well.

As a starting point, this chapter addresses these "why, what, and how" questions about nutrition:

- *Why* care about nutrition? Nutrients interact with body tissues, adding a little or subtracting a little, day by day, and thus change the very foundations upon which the health of the body is built.
- *What* are the nutrients in foods, and what roles do they play in the body? Meet the nutrients, and discover their general roles in building and maintaining body tissues.
- *What* constitutes a nutritious diet? Can you choose foods wisely, for nutrition's sake? And what motivates our choices?
- *How* do governments suggest that their citizens choose their diets to meet the national health objectives?
- *How* do we know what we know about nutrition? Scientific research reports provide an important foundation for understanding nutrition science.

The Controversy section demonstrates the differences between trustworthy sources of nutrition information and those that are less reliable.

A Lifetime of Nourishment

If you live for 65 years or longer, you will have consumed more than 70,000 meals, and your remarkable body will have disposed of 50 tons of food. The foods you choose have cumulative effects on your body. As you age, you will see and feel those effects, if you know what to look for.

Your body renews its structures continuously, and each day it builds a little muscle, bone, skin, and blood, replacing old tissues with new. It may also add a little fat, if you consume excess food energy (calories), or subtract a little, should you consume less than you require. In this way, some of the food you eat today becomes part of "you" tomorrow. The best food for you, then, is the kind that supports the growth and maintenance of strong muscles, sound bones, healthy skin, and sufficient blood to cleanse and nourish all parts of your body. This means you need food that provides not only energy but also sufficient **nutrients,** that is, enough water, carbohydrates, fat, protein, vitamins, and minerals. If the foods you eat provide too little or too much of any nutrient today, your health may suffer just a little. If the foods you eat provide too little or too much of one or more nutrients every day for years, then, by the time you are old, you may well suffer severe disease effects.

The point is that a well-chosen array of foods supplies enough energy and enough of each nutrient to prevent **malnutrition.** Malnutrition includes deficiencies, imbalances, and excesses of nutrients, any of which can take a toll on health over time.

When you choose foods with nutrition in mind, you can enhance your own well-being.

key point · · *The nutrients in food support growth, maintenance, and repair of the body. Deficiencies, excesses, and imbalances of nutrients bring on the diseases of malnutrition.*

How Powerful Is a Nutritious Diet in Preventing Diseases?

Your choice of diet profoundly influences your long-term health prospects.[1] Only two common lifestyle habits are more influential: smoking and other tobacco use, and excessive drinking of alcohol. Many older people suffer from debilitating conditions that could have been largely prevented had they known and applied the nutrition principles that we know today.

The poor health conveyed by a poor diet involves not only the various forms of malnutrition, but also other diseases, especially the **chronic diseases:** heart disease, diabetes, some kinds of cancer, dental disease, and adult bone loss. Of the leading causes of death listed in Table 1-1, five are related to nutrition, and others such as suicide, liver disease, and homicide are related to drinking alcohol. We should hasten to say that although diet can powerfully influence these diseases, they cannot be prevented by a good diet alone; they are to some extent determined by a person's genetic constitution, activities, and lifestyle.[2] Within the range set by your genetic inheritance, however, the likelihood that you will develop these diseases is strongly influenced by your food choices.

Some people overestimate and some underestimate the influence of diet in preventing diseases and poor health. Putting diet's exact role in perspective is difficult not only for ordinary people, but also for research scientists who spend their working lives trying to figure out precisely how diet relates to health and various diseases.

chronic diseases long-duration degenerative diseases characterized by deterioration of the body organs; examples include heart disease, cancer, and diabetes.

genome the complete set of chromosomes that comprises the entirety of an organism's genetic information. The *human genome* is the genetic map of a human being, revealing the location of hereditary information carried on the chromosomes. The map may be used to identify disease-causing genetic variations, allowing advances in disease prevention and treatment.

Nutrition profoundly affects health.

Genetics and Individuality

Consider the role of genetics. The recent completion of the human **genome** establishes the entire sequence of genetic material that defines the human being. This leap of knowledge promises improved understanding of the interactions between human genetic makeup and nutrition and brings hope for greater control over health and disease.[3]

The influence of genetics and nutrition varies for different diseases (see Figure 1-1 on the next page). The anemia caused by sickle-cell disease, for example, is purely hereditary and thus appears at the left of Figure 1-1 as a nutrition-unrelated, genetic disease. Nothing a person eats affects the person's chances of contracting this anemia, although nutrition therapy may help ease its course. At the other end of the spectrum in Figure 1-1, iron-deficiency anemia most often results from undernutrition. Diseases and conditions of poor health appear all along this continuum from purely genetic to purely nutritional; the more nutrition-related a disease or health condition is, the more successfully sound nutrition can prevent it.

Furthermore, some diseases, such as heart disease and cancer, are not one disease but many. Two people may both have heart disease, but not the same form. Whereas the human genome defines a general human pattern of genes, individual people differ genetically from each other in thousands of subtle ways. One person's heart disease or cancer may be nutrition-related, but another's may not be. The concept presented in Figure 1-1 is based on the experience of millions of people; in contrast, no simple statement can be made about the extent to which diet can help any one person avoid a disease or slow its progress.

key point · · *Choice of diet influences long-term health within the range set by genetic inheritance. Nutrition has no influence on some diseases but is closely linked to others.*

Table 1•1

LEADING CAUSES OF DEATH IN THE UNITED STATES, 1999

The five blue-shaded diseases are related to nutrition; those shaded in light yellow are related to alcohol.

1.	Heart disease
2.	Cancers
3.	Strokes
4.	Chronic lower respiratory diseases
5.	Motor vehicle and other accidents
6.	Diabetes mellitus
7.	Pneumonia and influenza
8.	Alzheimer's disease
9.	Kidney disease
10.	Infections of the blood
11.	Suicide
12.	Liver disease and cirrhosis
13.	Hypertension (high blood pressure)
14.	Homicide
15.	Aneurysm (internal hemorrhage)

SOURCE: National Center for Health Statistics, 2001.

Anemia is a blood condition in which red blood cells are inadequate or impaired, and so cannot meet the oxygen demands of the body. More about the anemia of sickle-cell disease in Chapter 6; iron-deficiency anemia is described in Chapter 8.

energy the capacity to do work. The energy in food is chemical energy; it can be converted to mechanical, electrical, heat, or other forms of energy in the body. Food energy is measured in calories, defined on page 5.

organic carbon containing. Four of the six classes of nutrients are organic: carbohydrate, fat, protein, and vitamins. Strictly speaking, organic compounds include only those made by living things and do not include carbon dioxide and a few carbon salts.

energy-yielding nutrients the nutrients the body can use for energy. They may also supply building blocks for body structures.

Lifestyle Choices

Besides people's food choices, other lifestyle choices also affect their health. Tobacco and alcohol use and other substance abuse can all be destructive of health. Other major health determinants include physical activity, sleep, stress, and home and job conditions, including environmental quality. In all of these contexts, healthful nutrition can help prevent or reduce the severity of some diseases.

Personal life choices, such as use of tobacco or alcohol or staying physically active, also affect health for the better or worse.

The Human Body and Its Food

As your body moves and works each day, it must use **energy.** The energy that fuels the body's work comes indirectly from the sun by way of plants. Plants capture and store the sun's energy in their tissues as they grow. When you eat plant-derived foods such as fruits, grains, or vegetables, you obtain and use the solar energy they have stored. Plant-eating animals obtain their energy in the same way, so when you eat animal tissues, you are eating compounds containing energy that came originally from the sun.

The body also requires six kinds of nutrients—families of molecules indispensable to its functioning—and foods deliver these. Table 1-2 lists the six classes of nutrients. Four of these six are **organic;** that is, the nutrients contain the element carbon derived from living things. The human body and foods are made of the same materials, arranged in different ways (see Figure 1-2).

The Nutrients in Foods

Foremost among the six classes of nutrients in foods is water, which is constantly lost from the body and must constantly be replaced. Of the four organic nutrients, three are **energy-yielding nutrients,** meaning that the body can use the energy they contain. The *carbohydrates* and *fats* (fats are properly called *lipids*) are especially important energy-yielding nutrients. As for *protein,* it does double duty: it can yield energy, but it also provides materials that form structures and working parts of body tissues. (Alcohol yields energy, too, but it is a toxin, not a nutrient—see the note to Table 1-3, page 6).

Alcohol use and abuse and their effects on body tissues are topics of Controversy 3.

Figure 1•1

NUTRITION AND DISEASE

Not all diseases are equally influenced by diet. Some are purely genetic, like the anemia of sickle-cell disease. Some may be inherited (or the tendency to develop them may be inherited) but may be influenced by diet, like some forms of diabetes. Some are purely dietary, like the vitamin and mineral deficiency diseases.

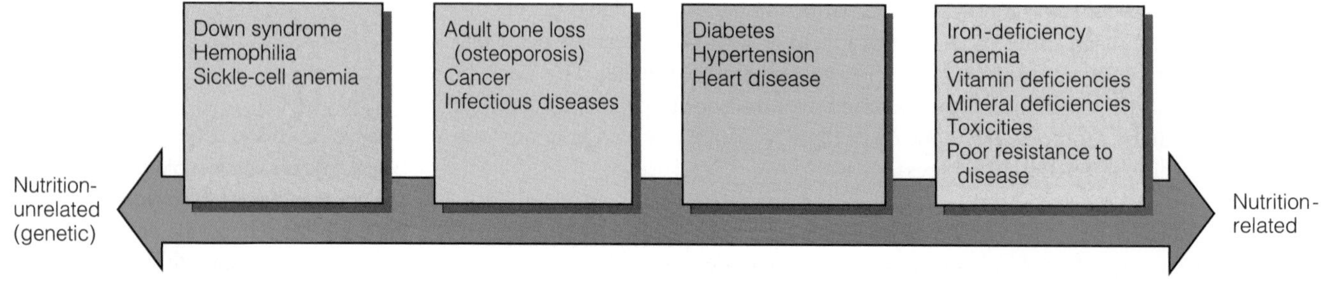

Table 1•2

ELEMENTS IN THE SIX CLASSES OF NUTRIENTS

The nutrients that contain carbon are organic.

	Carbon	Oxygen	Hydrogen	Nitrogen	Minerals
Water		✔	✔		
Carbohydrate	✔	✔	✔		
Fat	✔	✔	✔		
Protein	✔	✔	✔	✔	b
Vitamins	✔	✔	✔	✔[a]	b
Minerals					✔

[a] All of the B vitamins contain nitrogen; *amine* means nitrogen.
[b] Protein and some vitamins contain the mineral sulfur; vitamin B_{12} contains the mineral cobalt.

The fifth and sixth classes of nutrients are the *vitamins* and the *minerals*. These provide no energy to the body. A few minerals serve as parts of body structures (calcium and phosphorus, for example, are major constituents of bone), but all vitamins and minerals act as regulators. As regulators, the vitamins and minerals assist in all body processes: digesting food; moving muscles; disposing of wastes; growing new tissues; healing wounds; obtaining energy from carbohydrate, fat, and protein; and participating in every other process necessary to maintain life. Later chapters are devoted to these six classes of nutrients.

When you eat food, then, you are providing your body with energy and nutrients. Furthermore, some of the nutrients are **essential nutrients,** meaning that if you do not receive them from food, you will develop deficiencies; the body cannot make these nutrients for itself. Essential nutrients are found in all six classes of nutrients. Water is essential; so is a form of carbohydrate; so are some lipids, some parts of protein, all of the vitamins, and the minerals important in human nutrition.

To support understanding of many of the discussions that follow, two definitions and a set of numbers are needed. Food scientists measure food energy in **calories,** units of heat. Food and nutrient quantities are often measured in **grams,** units of weight. The most energy-rich of the nutrients is fat, which contains 9 calories in each gram. Carbohydrate and protein each contain only 4 calories in a gram (see Table 1-3).

Scientists have worked out ways to measure the energy and nutrient contents of foods. They have also calculated the amounts of energy and nutrients various types of people need—people of both genders, of different ages, and of different walks of life. Thus, after studying human nutrient requirements (the subject of Chapter 2 of this book), you can state with some accuracy just what your own body needs—this much water, that much carbohydrate and fat, so much protein, and so forth. Might it be possible, then, to simply take pills or **supplements** in place of food? No, because, as it turns out, food offers more than just the six basic nutrients.

key point *Food supplies energy and nutrients. The most vital nutrient is water. The energy-yielding nutrients are carbohydrates, fats (lipids), and protein. The helper nutrients are vitamins and minerals. Food energy is measured in calories; food and nutrient quantities are often measured in grams.*

Can I Forgo Food and Live Just on Supplements?

Nutrition science can state what nutrients human beings need to survive—at least for a time. Scientists are becoming skilled at making **elemental diets**—diets with a precise chemical composition that are life-saving for people in the hospital who cannot

essential nutrients the nutrients the body cannot make for itself (or cannot make fast enough) from other raw materials; nutrients that must be obtained from food to prevent deficiencies.

calories units of energy. Strictly speaking, the unit used to measure the energy in foods is a kilocalorie (*kcalorie* or *Calorie*): it is the amount of heat energy necessary to raise the temperature of a kilogram (a liter) of water 1 degree Celsius. This book follows the common practice of using the lowercase term *calorie* (abbreviated *cal*) to mean the same thing.

grams units of weight. A gram (g) is the weight of a cubic centimeter (cc) or milliliter (ml) of water under defined conditions of temperature and pressure. About 28 grams equal an ounce.

supplements pills, liquids, or powders that contain purified nutrients or other ingredients (see Chapter 7).

elemental diets diets composed of purified ingredients of known chemical composition; intended to supply all essential nutrients to people who cannot eat foods.

Figure 1•2

MATERIALS OF FOOD AND THE HUMAN BODY

Foods and the human body are made of the same materials.

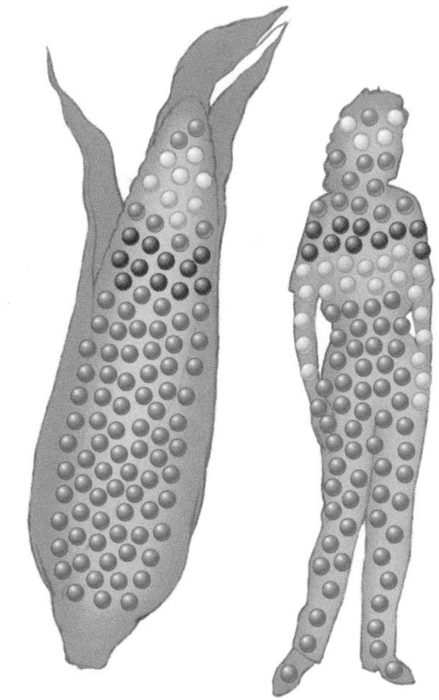

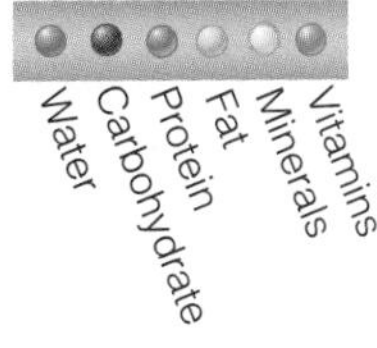

nonnutrients a term used in this book to mean compounds other than the six nutrients that are present in foods and that have biological activity in the body.

phytochemicals nonnutrient compounds in plant-derived foods that have biological activity in the body.

eat ordinary food. These formulas, administered to severely ill people for days or weeks, support not only continued life but also recovery from nutrient deficiencies and infections and the healing of wounds.

Lately, marketers have taken these formulas out of the medical setting and have advertised them heavily to healthy people as "insurance" against malnutrition. The truth is that such products are not superior to a sound diet of real foods. Formula diets are essential to help sick people to survive, but they are not sufficient to enable people to thrive over long periods. Elemental diet formulas do not support optimal growth and health, and they often lead to medical complications.[4] Although these problems are rare and can be detected and corrected, they show that the composition of these diets is not yet perfect for all people in all settings. Healthy people who eat a healthy diet do not need such formulas or, in fact, any sort of supplements.[5]

Even if a person's basic nutrient needs are perfectly understood and met, concoctions of nutrients still lack something that foods provide. Hospitalized clients who are fed nutrient mixtures through a vein often improve dramatically when they can finally eat food. Something in real food is important to health—but what is it? What does food offer that cannot be provided through a needle? Science has some partial explanations, some physical and some psychological.

In the digestive tract, the stomach and intestine are dynamic living organs, changing constantly in response to the foods they receive—even to just the sight, aroma, and taste of food.[6] When a person is fed through a vein, the digestive organs, like unused muscles, weaken and grow smaller. Medical wisdom now dictates that a person should be fed through a vein for as short a time as possible and that real food taken by mouth should be reintroduced as early as possible.[7] The intestine also releases hormones in response to food, and these send messages to the brain that bring the eater a feeling of satisfaction: "There, that was good. Now I'm full." Eating offers both physical and emotional comfort: after a good meal, you can relax, enjoy entertainment, rest, or sleep.

Food does still more than maintain the intestine and convey messages of comfort to the brain. Foods are chemically complex. In addition to nutrients, they contain **nonnutrients,** including the **phytochemicals.** These compounds confer color, taste, and other characteristics on foods and are believed to affect health (see Controversy 2). Even an ordinary baked potato contains hundreds of different compounds. In view of all this, it is not surprising that food gives us more than just nutrients. If it were otherwise, that would be surprising.

Table 1•3

CALORIE VALUES OF ENERGY NUTRIENTS

Energy Nutrient	Energy
Carbohydrate	4 cal/g
Fat (lipid)	9 cal/g
Protein	4 cal/g

NOTE: Alcohol contributes 7 calories/gram that the human body can use for energy. Alcohol is not classed as a nutrient, however, because it interferes with growth, maintenance, and repair of body tissues.

"One cannot think well, love well, sleep well, if one has not dined well."—V. Woolf, *A Room of One's Own* (New York: Harcourt Brace, 1929), p. 30.

When you eat foods, you are receiving more than just nutrients.

© Bob Daemmrich/Stock, Boston

key point *In addition to nutrients, food conveys emotional satisfaction and hormonal stimuli that contribute to health. Foods also contain phytochemicals that give them their tastes, aromas, colors, and other characteristics. Some phytochemicals are believed to play roles in disease prevention.*

adequacy the dietary characteristic of providing all of the essential nutrients, fiber, and energy in amounts sufficient to maintain health and body weight.

balance the dietary characteristic of providing foods of a number of types in proportion to each other, such that foods rich in some nutrients do not crowd out of the diet foods that are rich in other nutrients. Also called *proportionality*.

calorie control control of energy intake; a feature of a sound diet plan.

The Challenge of Choosing Foods

At their best, well-planned meals convey pleasure and are nutritious, too, fitting your tastes, personality, family and cultural traditions, lifestyle, and budget. Given the astounding numbers and varieties available, consumers can lose track of what individual foods contain and how to put them together into health-promoting diets. A few guidelines can help.

The Variety of Foods to Choose From

A list of the variety of foods available a hundred years ago would be relatively short. It would consist of **basic foods**—foods that have been around for a long time such as vegetables, fruits, meats, milk, and grains. These foods have been called unprocessed, natural, whole, or farm foods. An easy way to obtain a nutritious diet is to consume a variety of selections from among these foods each day. On a given day, however, almost half of our population consume no fruits or fruit juices. Also, although people generally consume a few servings of vegetables, the vegetable they most often choose is potatoes, usually prepared as french fries.[8] Such diet patterns are believed to be predictive of chronic disease risks.[9]

In 1900, Americans chose from among 500 or so different foods; today, they choose from more than 50,000.

The food industry today offers thousands of foods—many are processed mixtures of the basic ones, and some are even constructed mostly from artificial ingredients. Ironically, this variety may make it more difficult, rather than easier, to plan a nutritious diet.

Table 1-4 on the next page presents a glossary of terms related to foods. The terms reveal that all types of food—including **fast foods** and **processed foods**—offer various constituents to the eater. You may also hear about **functional foods,** a term coined in an attempt to identify those foods that might lend protection against chronic diseases by way of the nutrients or nonnutrients they contain.[10] The trouble is, scientists trying to single out the most health-promoting foods find that almost every naturally occurring food—even chocolate—is functional in some way with regard to human health. Controversy 2 provides more information about functional foods.

The extent to which foods support good health depends on the calories, nutrients, and nonnutrients they contain. In short, to select well among foods, as among people, you need to know more than their names; you need to know the foods' inner qualities.

Even more importantly, you need to know how to combine foods into nutritious diets. Foods are not nutritious by themselves; each is of value only in so far as it contributes to a nutritious diet. A key to wise diet planning is to make sure that the foods you eat daily, your **staple foods,** are especially nutritious.

Some foods offer beneficial nonnutrients called phytochemicals.

© Richard Fukuhara/CORBIS

key point *Foods come in a bewildering variety in the marketplace, but the foods that form the basis of a nutritious diet are staple foods, such as ordinary milk and milk products; meats, fish, and poultry; vegetables and dried peas and beans; fruits; and grains.*

How, Exactly, Can I Recognize a Nutritious Diet?

A nutritious diet has five characteristics. First is **adequacy:** the foods provide enough of each essential nutrient, fiber, and energy. Second is **balance:** the choices do not overemphasize one nutrient or food type at the expense of another. Third is **calorie control:** the

moderation the dietary characteristic of providing constituents within set limits, not to excess.

variety the dietary characteristic of providing a wide selection of foods—the opposite of monotony.

legumes (leg-GOOMS, LEG-yooms) beans, peas, and lentils valued as inexpensive sources of protein, vitamins, minerals, and fiber that contribute little fat to the diet. Also defined in Chapter 6.

Table 1•4

GLOSSARY OF FOOD TYPES

The purpose of this little glossary is to show that good-sounding food names don't necessarily signify that foods are nutritious. Read the comment at the end of each definition.

- **basic foods** milk and milk products; meats and similar foods such as fish and poultry; vegetables, including dried beans and peas; fruits; and grains. These foods are generally considered to form the basis of a nutritious diet. Also called *whole foods*.
- **enriched foods** and **fortified foods** foods to which nutrients have been added. If the starting material is a whole, basic food such as milk or whole grain, the result may be highly nutritious. If the starting material is a concentrated form of sugar or fat, the result may be less nutritious.
- **fast foods** restaurant foods that are available within minutes after customers order them—traditionally, hamburgers, french fries, and milkshakes; more recently, salads and other vegetable dishes as well. These foods may or may not meet people's nutrient needs well, depending on the selections made and on the energy allowances and nutrient needs of the eaters.
- **functional foods** a term that reflects an attempt to define as a group the foods known to possess nutrients or nonnutrients that might lend protection against diseases. However, all nutritious foods can support health in some ways; the Controversy section of Chapter 2 provides details.
- **natural foods** a term that has no legal definition.
- **nutraceutical** a term with no legal or scientific meaning, but sometimes used to refer to foods, nutrients, or dietary supplements believed to have medicinal effects (see Chapter 11). Often used to sell unnecessary or unproved supplements.
- **organic foods** understood to mean foods grown without synthetic pesticides or fertilizers. In chemistry, however, all foods are made mostly of organic (carbon-containing) compounds. (See Chapter 14's Controversy section for details.)
- **partitioned foods** foods composed of parts of whole foods, such as butter (from milk), sugar (from beets or cane), or corn oil (from corn). Partitioned foods are generally overused and provide few nutrients with many calories.
- **processed foods** foods subjected to any process, such as milling, alteration of texture, addition of additives, cooking, or others. Depending on the starting material and the process, a processed food may or may not be nutritious.
- **staple foods** foods used frequently or daily, for example, rice (in East and Southeast Asia) or potatoes (in Ireland). If well chosen, these foods are nutritious; certainly, they should be.

foods provide the amount of energy you need to maintain appropriate weight—not more, not less. Fourth is **moderation:** the foods do not provide excess fat, salt, sugar, or other unwanted constituents. Fifth is **variety:** the foods chosen differ from one day to the next. In addition, meals should occur with regular timing throughout the day.

A nutritious diet follows the a, b, c, m, v principles:

- **a**dequacy
- **b**alance
- **c**alorie control
- **m**oderation
- **v**ariety

Adequacy Any nutrient could be used to demonstrate the importance of dietary *adequacy*. Iron provides a familiar example. It is an essential nutrient: you lose some every day, so you have to keep replacing it; and you can get it into your body only by eating foods that contain it.* If you eat too few of the iron-containing foods, you can develop iron-deficiency anemia: with anemia you may feel weak, tired, cold, sad, and unenthusiastic; you may have frequent headaches; and you can do very little muscular work without disabling fatigue. If you add iron-rich foods to your diet, you soon feel more energetic. Some foods are rich in iron; others are notoriously poor. Meat, fish, poultry, and **legumes** are in the iron-rich category, and an easy way to obtain the needed iron is to include these foods in your diet regularly.

Balance To appreciate the importance of dietary *balance*, consider a second essential nutrient, calcium. Most foods that are rich in iron are poor in calcium. Calcium's best food sources are milk and milk products, which happen to be extraordinarily poor iron sources. A diet lacking calcium causes poor bone development during the growing years and increases a person's susceptibility to disabling bone loss in

*A person can also take supplements of iron, but as later discussions demonstrate, this is not as effective as eating iron-rich foods.

Foods once looked like this . . .

© David R. Frazier/Photolibrary

. . . but now foods often look like this.

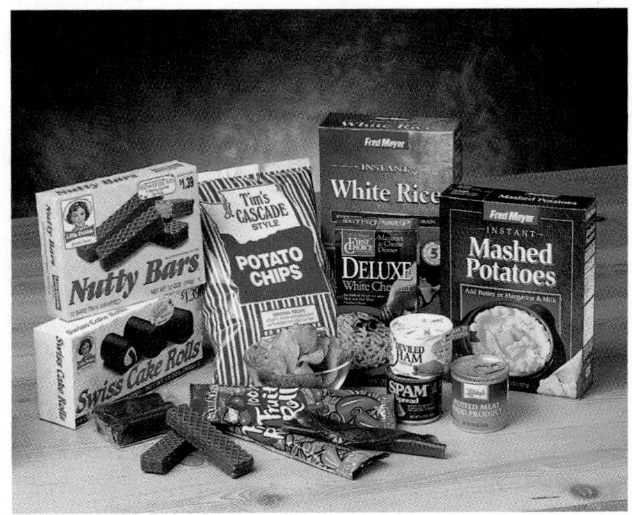

© Polara Studios Inc.

adult life. Clearly, to obtain enough of both iron and calcium, which seldom appear together in the same foods, people have to balance their food choices. Balancing the whole diet to provide enough but not too much of every one of the 40-odd nutrients the body needs for health requires considerable juggling. As you will see in Chapter 2, food group plans can help you achieve dietary adequacy and balance because they recommend specific amounts of foods of each type.

Calorie Control Energy intakes should not exceed energy needs. Nicknamed *calorie control,* this diet characteristic ensures that energy intakes from food balance energy expenditures in activity. Eating such a diet gives you control of body fat content and weight. The many strategies that promote this goal appear in Chapter 9.

Moderation Intakes of certain food constituents such as fat, cholesterol, sugar, and salt should be limited for health's sake (more on health effects in later chapters). A major guideline for healthy people is to keep fat intake below 30 percent of total calories.[11] Some people take this to mean that they must never indulge in a delicious beefsteak or hot-fudge sundae, but they are misinformed: *moderation,* not total abstinence, is the key. A steady diet of steak and ice cream might be harmful, but once a week as part of an otherwise moderate diet plan, these foods may have little impact; as once-a-month treats, these foods would have practically no effect at all. Moderation also means that limits are necessary, even for desirable food constituents. For example, a certain amount of fiber in foods contributes to the health of the digestive system, but too much fiber leads to nutrient losses.

Variety As for *variety,* nutrition scientists agree that people should not eat the same foods, even highly nutritious ones, day after day. One reason is that some less well-known nutrients and some nonnutrient food components could be important to health; some foods may be better sources of these than others. Another reason is that a monotonous diet may deliver large amounts of toxins or contaminants. Each such undesirable item in a food is diluted by all the other foods eaten with it and is even further diluted if the food is not eaten again for several days. Last, variety adds interest—trying new foods can be a source of pleasure. Table 1-5 on the next page takes a whimsical look at obstacles to eating well.

A caution is in order. Any one of these dietary principles alone cannot ensure a healthful diet. For example, the most likely outcome of relying solely on variety could easily be a low-nutrient, high-calorie diet consisting of a variety of snack foods and nutrient-poor sweets.[12] If you establish the habit of using all of the principles just described, you will find that choosing a healthful diet becomes as automatic as brushing your teeth or falling asleep.

All of these factors help to build a nutritious diet.

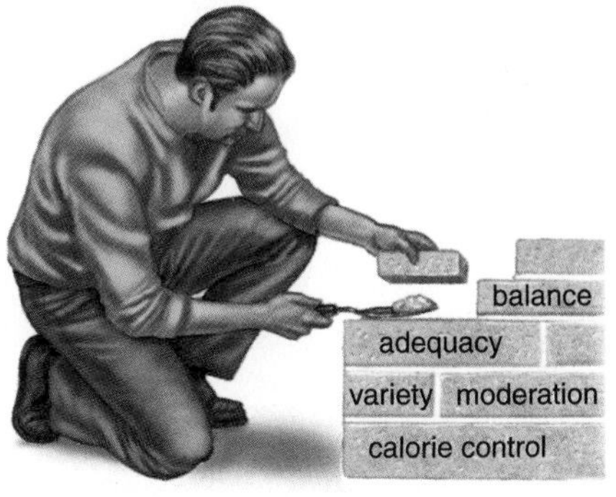

cuisines styles of cooking.

foodways the sum of a culture's habits, customs, beliefs, and preferences concerning food.

ethnic foods foods associated with particular cultural subgroups within a population.

omnivores people who eat foods of both plant and animal origin, including animal flesh.

vegetarians people who exclude from their diets animal flesh and possibly other animal products such as milk, cheese, and eggs.

Table 1•5

WHAT'S TODAY'S EXCUSE FOR NOT EATING WELL?

If you find yourself saying, "I know I should eat well, but I'm too busy" (or too fond of fast food, or have too little money, or a dozen other excuses), take note:

- *No time*. Everyone is busy. In truth, eating well takes little time. Convenience packages of frozen vegetables, jars of pasta sauce, and prepared meats and salads are abundant in markets today and take no longer to pick up than snack chips and colas. Priorities change drastically and instantly when illness strikes—better to spend a little time now nourishing your body's defenses than to spend time later treating illness.
- *Crave fast food*. Occasional fast-food meals can support health, if you choose wisely (see Chapter 5).
- *Too little money*. Eating right costs no more than eating poorly. Chips, colas, fast food, doughnuts, and premium ice cream cost *more* than foods such as fruits, vegetables, legumes, cereal, and milk. And serious illness costs more than a well person can imagine.
- *Like to eat large portions*. An occasional splurge, say, once a month, is a healthy part of moderation.
- *Take vitamins instead*. Vitamin pills cannot make up for consistently poor food choices. Food constituents such as fiber and phytochemicals are also important to good health.
- *Love sweets*. Sweets in moderation are an acceptable, and even desirable, part of a balanced diet.

SOURCE: Ideas adapted from "Seven excuses for not eating better," *Tufts University Healthletter*, December 1998, p. 8.

key point *A well-planned diet is adequate in nutrients, is balanced with regard to food types, offers food energy that matches energy expended in activity, is moderate in unwanted constituents, and offers a variety of nutritious foods.*

Why People Choose Foods

Eating is an intentional act. Each day, people choose from the available foods, prepare the foods, decide where to eat, which rules to follow, and with whom to dine. Among the many factors influencing food-related choices is the culture to which a person is accustomed.

Cultural and Social Meanings Attached to Food Like wearing traditional clothing or speaking a native language, enjoying traditional **cuisines** and **foodways** can be a celebration of your own or a friend's heritage. Sharing **ethnic food** can be symbolic: people offering foods are expressing a willingness to share cherished values with others. People accepting those foods are symbolically accepting not only the person doing the offering but the person's culture.

Cultural traditions regarding food are not inflexible; they keep evolving as people move about, learn about new foods, and teach each other.[13] Today, some people are ceasing to be **omnivores** and are becoming **vegetarians.** Vegetarians often choose this lifestyle because they honor the lives of animals or because they have discovered the health and other advantages associated with diets rich in beans, whole grains, fruits, nuts, and vegetables. The Controversy of Chapter 6 explores the pros and the cons of both the vegetarian's and the meat eater's diets.

Factors That Drive Food Choices Consumers today value convenience so highly that they are willing to spend over half of their food budget on meals that require little or no preparation.[14] They frequently eat out, bring home ready-to-eat meals, or have food delivered. In their own kitchens, they want to prepare a meal in 15 to 20 minutes, using only four to six ingredients.[15] Such convenience limits food choices but doesn't necessarily mean that nutrition is out the window. This chapter's Food Feature addresses the time and nutrition tradeoff.

Convenience is only one consideration. Physical, psychological, social, and philosophical factors all influence how you choose the foods you generally eat:

- *Advertising*. The media have persuaded you to eat these foods.
- *Availability*. There are no others to choose from.

© Michael Newman/PhotoEdit

Sharing ethnic food is a way of sharing culture.

- *Economy.* They are within your means.
- *Emotional comfort.* They make you feel better for a while.
- *Habit.* They are familiar; you always eat them.
- *Personal preference and genetic inheritance.* You like the way these foods taste, with some preferences possibly determined by the genes.[16]
- *Positive associations.* They are eaten by people you admire, or they indicate status, or they remind you of fun.
- *Region of the country.* They are foods favored in your area.
- *Social pressure.* They are offered; you feel you can't refuse them.
- *Values or beliefs.* They fit your religious tradition, square with your political views, or honor the environmental ethic.
- *Weight.* You think they will help to control body weight.
- *Nutritional value.* You think they are good for you.

Figure 2-5 of Chapter 2 depicts some ethnic foods that have become an integral part of the "American diet."

Just the last two of these reasons for choosing foods assign a high priority to nutritional health. Similarly, the choice of where, as well as what, to eat is often based more on social considerations than on nutrition judgments. College students often choose to eat at fast-food and other restaurants to socialize, to get out, to save time, or to date; they are not always conscious of the need to obtain healthful food.[17] For those who wish to choose intentionally for health's sake, the next sections aim to set the table with some broad guidelines for choosing an adequate diet and some larger goals for the nutrition of the nation.

Cultural traditions and social values revolve around food. Some values are expressed through foodways. Many factors other than nutrition drive food choices.

Dietary Guidelines and Nutrition Objectives

Many countries set forth dietary recommendations for their citizens, striving to answer the question, "What should an individual eat to stay healthy?" Governments also set health objectives for the nation. The guidelines and objectives are related: if

The Dietary Guidelines *suggest that physical activity should be part of a healthy lifestyle.*

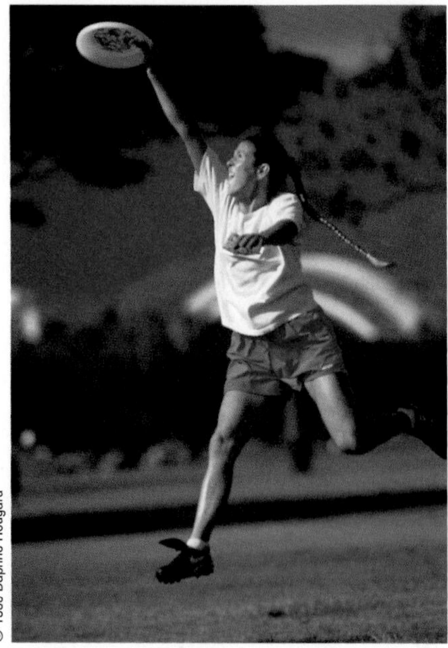

everyone followed the guidelines for individuals, many of the goals for the nation would fall into place.

Dietary Guidelines

The *Dietary Guidelines for Americans* (Figure 1-3) consist of ten guidelines clustered into three groupings.[18] The first cluster encourages people to combine a healthy body weight with regular physical activity for the sake of achieving fitness. In fact, physical activity is so closely linked with nutrition in supporting health that this book asks you to focus on fitness in features called *"Think Fitness,"* such as the one near here.

The second cluster of the *Guidelines* urges people to build a healthy diet base by using the diet-planning guide, the Food Guide Pyramid, explained in Chapter 2. The diet should provide a variety of whole grains, fruits, and vegetables (the reasons will become clear as you move through this book). Also, foods should be kept safe from spoilage or contamination (see Chapter 14).

The final cluster of the *Guidelines* focuses on limiting potentially harmful dietary constituents. According to the *Guidelines*, the most healthful diet is moderate in sugar and total fat, and low in saturated fat and cholesterol (Chapters 4 and 5 explain why). Finally, people are asked to consume less salt and choose sensibly if they use alcohol.

Other dietary recommendations include the *Nutrition Recommendations for Canadians* (Table 1-6) and *Canada's Guidelines for Healthy Eating* (Table 1-7). Numerous others have been published, and all offer similar advice on which nutrients people should emphasize and which they most often need to control for health's sake.[19]

Notice that the *Dietary Guidelines* do not require that you give up your favorite foods or eat strange, unappealing foods. Almost anyone's diet, with some adjustments, can fit most of these recommendations. Making these changes may sound simple, but do not be deceived. Even nutrition experts struggle to design diets of appealing, nutrient-dense foods that meet nutrient needs and follow the *Dietary Guidelines*.

If the experts who develop such guidelines were to ask us, we would add one more recommendation to their lists: choose foods that you enjoy. The joys of eating are physically beneficial to the body because they trigger health-promoting changes in the nervous, hormonal, and immune systems. Pleasure from food ensures that people will eat and thus obtain the nutrients needed for healthy body systems, as well as for

Figure 1•3

DIETARY GUIDELINES FOR AMERICANS: THE ABC'S OF GOOD HEALTH

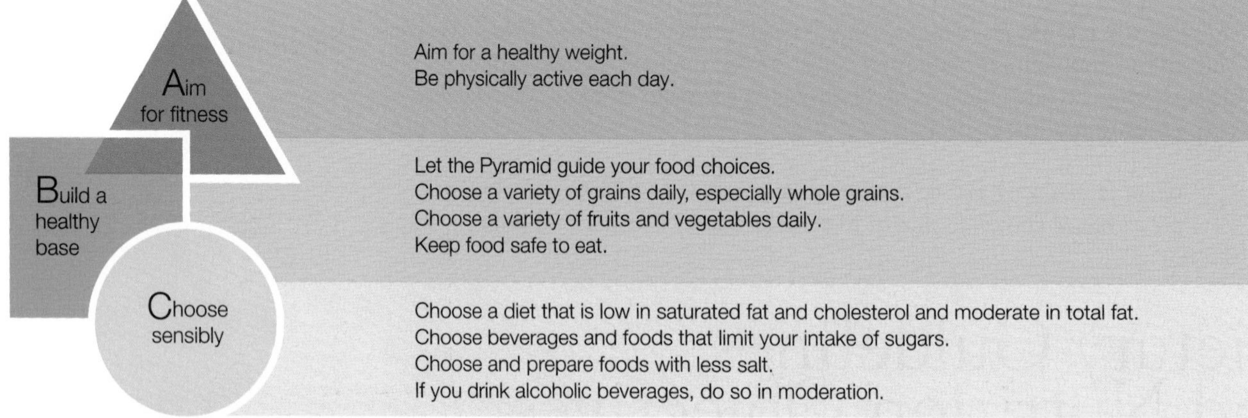

Aim for fitness
Aim for a healthy weight.
Be physically active each day.
Build a healthy base
Let the Pyramid guide your food choices.
Choose a variety of grains daily, especially whole grains.
Choose a variety of fruits and vegetables daily.
Keep food safe to eat.
Choose sensibly
Choose a diet that is low in saturated fat and cholesterol and moderate in total fat.
Choose beverages and foods that limit your intake of sugars.
Choose and prepare foods with less salt.
If you drink alcoholic beverages, do so in moderation.

NOTE: These guidelines are intended for adults and healthy children ages 2 and older.

SOURCE: U.S. Department of Agriculture and U.S. Department of Health and Human Services, *Nutrition and Your Health: Dietary Guidelines for Americans*, Home and Garden Bulletin no. 232 (Washington, D.C.: 2000).

Table 1•6

NUTRITION RECOMMENDATIONS FOR CANADIANS

- The Canadian diet should provide energy consistent with the maintenance of *body weight* within the recommended range.
- The Canadian diet should include *essential nutrients* in amounts recommended.
- The Canadian diet should include no more than 30% of energy as *fat* (33 grams/1,000 calories or 39 grams/5,000 kilojoules) and no more than 10% as saturated fat (11 grams/1,000 calories or 13 grams/5,000 kilojoules).
- The Canadian diet should provide 55% of energy as *carbohydrate* (138 grams/1,000 calories or 165 grams/5,000 kilojoules) from a variety of sources.
- The *sodium* content of the Canadian diet should be reduced.
- The Canadian diet should include no more than 5% of total energy as *alcohol*, or two drinks daily, whichever is less.
- The Canadian diet should contain no more *caffeine* than the equivalent of four regular cups of coffee per day.
- Community water supplies containing less than 1 milligram per liter should be *fluoridated* to that level.

NOTE: Italics added to highlight areas of concern.
SOURCE: Health and Welfare Canada, *Nutrition Recommendations: The Report of the Scientific Review Committee* (Ottawa: Canadian Government Publishing Centre, 1990).

the healthy skin, glossy hair, and natural good looks that accompany health. People tend to repeat what brings them pleasure, so they are most likely to stay with foods they like. Remember to enjoy your foods.

key point *The* Dietary Guidelines for Americans, Nutrition Recommendations for Canadians, *and other recommendations address the problems of overnutrition and undernutrition. To implement them requires exercising regularly; following the Food Guide Pyramid; amplifying servings of low-fat grains, fruits, and vegetables; limiting intakes of fats, sugar, and salt; and moderating alcohol intake.*

Why Be Physically Active?

Why should people bother to be physically active? While a person's daily food choices can powerfully affect health, the combination of nutrition and physical activity is more powerful still. People who are physically active can expect to receive at least some of the benefits listed in the margin. If even half of these benefits were yours for the asking, wouldn't you step up to claim them? In truth, they are yours to claim, at the price of including physical activity in your day.

THINK FITNESS

Potential benefits of physical activity:

- Reduced risk of cardiovascular diseases.
- Reduced risk of some types of cancer (especially colon and breast).
- Improved mental outlook and lessened likelihood of depression.
- Improved mental functioning.
- Feeling of vigor.
- Feeling of belonging—the companionship of sports.
- Strong self-image and belief in one's abilities.
- Reduced body fatness, increased lean tissue.
- Greater bone density and lessened risk of adult bone loss in later life.
- Sound, beneficial sleep.
- A more youthful appearance, healthy skin, and improved muscle tone.
- Faster wound healing.
- Lessening or elimination of menstrual pain.
- Improved resistance to infection.

Table 1•7

CANADA'S GUIDELINES FOR HEALTHY EATING

- Enjoy a variety of foods.
- Emphasize cereals, breads, other grain products, vegetables, and fruits.
- Choose lower-fat dairy products, leaner meats, and foods prepared with little or no fat.
- Achieve and maintain a healthy body weight by enjoying regular physical activity and healthy eating.
- Limit salt, alcohol, and caffeine.

SOURCE: These guidelines derive from *Action Towards Healthy Eating: The Report of the Communications/Implementation Committee and Nutrition Recommendations. . . . A Call for Action: Summary Report of the Scientific Review Committee and the Communications/Implementation Committee*, which are available from Branch Publications Unit, Health Services and Promotion Branch, Department of Health and Welfare, 5th Floor, Jeanne Manice Building, Ottawa, Ontario K1A 1B4.

Healthy People 2010: Nutrition Objectives for the Nation

The U.S. Department of Health and Human Services (DHHS) sets ten-year health objectives for the nation in its document *Healthy People*.[20] The objectives for the year 2010, listed in Table 1-8, provide a quick scan of the nutrition-related objectives set for this decade. The inclusion of nutrition objectives and those pertaining to food safety reveal that public health officials consider these areas to be top national priorities.[21]

At the close of the twentieth century, the nation's progress toward meeting its *Healthy People 2000* goals was mixed.[22] For well over half of the objectives, the population either met the target or was moving in the right direction. Successes included reductions in the incidence of food- and water-borne infections, cancers of the mouth and breast, and infant mortality. Overall cancer death rates were below the year 2000 target. Deaths from heart disease and stroke had also declined substantially, but on the negative side, heart disease was still the leading cause of death among adults. Also, the number of overweight people had jumped significantly.

All of these government efforts reflect ambitious goals for our national nutritional health. Without a firm base of scientific knowledge, however, such efforts would not be possible. The next section addresses the final "how" question of this chapter: How do we know what we know about nutrition?

The U.S. Department of Health and Human Services sets nutrition objectives for the nation each decade.

The Science of Nutrition

Nutrition is a science—a field of knowledge composed of organized facts. Unlike sciences such as astronomy and physics, nutrition is a relatively young science. Most nutrition research has been conducted since 1900. The first vitamin was identified in 1897, and the first protein structure was not fully described until 1945. Much remains to be learned about the effects of foods, nutrients, and nonnutrients on the body. Because nutrition science is an active, changing, growing body of knowledge, scientific findings often seem to contradict one another or are subject to conflicting interpretations.

For this reason, people may despair as they try to decipher current reports to learn what is really going on: "When the scientists themselves can't agree on what is true, how am I supposed to know?" Yet, beyond the theories now being investigated, many facts in nutrition are known with great certainty. And where there are conflicts and contradictions, researchers are energetically attempting to resolve them. The next section can help consumers understand why apparent contradictions sometimes arise in nutrition science.

If the Scientists Don't Know, How Can I?

Everyone stampedes for oat bran, red wine, or fish oil based on today's news that these products are good for health. Then tomorrow's news reports, "It isn't true after all," and everyone drops oat bran, red wine, or fish oil and takes up the next craze. Meanwhile, bewildered consumers complain in frustration, "Those scientists don't know anything."

In truth, though, it is a scientist's business not to know. Scientists obtain facts by systematically asking questions—that's their job. Then they conduct experiments designed to test for various possible answers (see Figure 1-4 on page 16 and Table 1-9 on page 17). When they have ruled out some possibilities and found evidence for others, they submit their findings, not to the news media, but to boards of reviewers composed of other scientists who try to pick the findings apart. If these reviewers consider

Table 1•8

HEALTHY PEOPLE 2010 NUTRITION-RELATED OBJECTIVES

- Increase *nutrition education* among consumers and in educational settings at all levels.
- Increase the proportion of children, adolescents, and adults who are at a *healthy weight*.
- Reduce *growth retardation* among low-income children under age 5 years.
- Increase the proportion of persons aged 2 years and older who consume at least two daily servings of *fruit*.
- Increase the proportion of persons aged 2 years and older who consume at least three daily servings of *vegetables*, with at least one-third being dark green or orange vegetables.
- Increase the proportion of persons aged 2 years and older who consume at least six daily servings of *grain products*, with at least three being whole grains.
- Increase the proportion of persons aged 2 years and older who consume less than 10% of calories from *saturated fat*.
- Increase the proportion of persons aged 2 years and older who consume no more than 30% of calories from *total fat*.
- Increase the proportion of persons aged 2 years and older who consume 2,400 milligrams or less of *sodium*.
- Increase the proportion of adults with *high blood pressure* who are taking action to control their blood pressure.
- Increase the proportion of persons aged 2 years and older who meet dietary recommendations for *calcium*.
- Reduce *iron deficiency* among young children, females of childbearing age, and pregnant females.
- Reduce *anemia* among low-income pregnant females in their third trimester.
- Reduce key *vitamins and mineral deficiencies* in pregnant women.
- Increase the proportion of children and adolescents aged 6 to 19 years whose intake of *meals and snacks at school* contributes to good overall dietary quality.
- Increase the proportion of worksites that offer *nutrition or weight management classes or counseling*.
- Increase the proportion of physician office visits made by patients with a diagnosis of cardiovascular disease, diabetes, or hyperlipidemia that include *counseling or education related to diet and nutrition*.
- Reduce deaths from anaphylaxis caused by *food allergies*.
- Increase the number of consumers and retail establishments who follow key *food safety* practices and reduce key food-borne illnesses.
- Increase *food security* among U.S. households and in so doing reduce hunger.

Details about these and hundreds of other objectives are available from the U.S. Department of Health and Human Services, *Healthy People 2010*, 2nd ed. (Washington, D.C.: Government Printing Office, 2000) online at www.health.gov/healthypeople or call (800) 367-4725.

the conclusions to be well supported by the evidence, they endorse the work for publication, not in the news media, but in scientific journals where still other scientists can read it. You can read it, too; Table 1-10 on page 17 explains what you can expect to find in a journal article.

As you study nutrition, you are likely to hear of findings based on two ongoing national scientific research projects. The National Health and Nutrition Examination Surveys (NHANES) is a nationwide project that gathers information from about 50,000 people using diet histories, physical examinations and measurements, and laboratory tests.[23] Boiled down to its essence, NHANES involves:

- Asking people what they have eaten.
- Recording measures of their health status.

The Continuing Survey of Food Intakes by Individuals (CSFII) involves:

- Recording what people have actually eaten for two days.
- Comparing the foods they have chosen with recommended food selections.

Nutrition monitoring makes it possible for researchers to assess the nutrient status, health indicators, and dietary intakes of the U.S. population. The agencies involved with these efforts are listed in the margin.

Agencies active in nutrition policy, research, and monitoring:

- Department of Health and Human Services (DHHS).
- United States Department of Agriculture (USDA).
- Centers for Disease Control and Prevention (CDC).

Ongoing national nutrition research projects:

- National Health and Nutrition Examination Surveys (NHANES).
- Continuing Survey of Food Intakes by Individuals (CSFII).

Figure 1•4

RESEARCH DESIGN

The source of valid nutrition information is scientific research. Nutrition is a science, an organized body of knowledge composed of facts that we believe to be true because they have been supported, time and again, in experiments designed to rule out all other possibilities. Each fact has been established by many different kinds of experiments. For example, we know that eyesight depends partly on vitamin A because animals deprived of that vitamin and only that vitamin begin to go blind; when it is restored soon enough to their diet, they regain their sight. The same fact holds true in observations of human beings.

*The type of study chosen for research depends upon what sort of information the researchers require. Studies of individuals (**case studies**) yield observations that may lead to possible avenues of research. A study of a man who ate gumdrops and became a famous dancer might suggest that an experiment be done to see if gumdrops contain dance-enhancing power.*

*Studies of whole populations (**epidemiological studies**) provide another sort of information. Such a study can reveal a **correlation.** For example, an epidemiological study might find no worldwide correlation of gumdrop eating with fancy footwork but, unexpectedly, might reveal a correlation with tooth decay.*

*Studies in which researchers actively intervene to alter people's eating habits (**intervention studies**) go a step further. In such a study, one set of subjects (the **experimental group**) receive a treatment, and another set (the **control group**) go untreated or receive a **placebo** or sham treatment. If the study is a **blind experiment,** the subjects do not know who among the members receives the treatment or who receives the sham. If the two groups experience different effects, then the treatment's effect can be pinpointed. For example, an intervention study might show that withholding gumdrops, together with other candies and confections, reduced the incidence of tooth decay in an experimental population compared to that in a control population.*

*Finally, **laboratory studies** can pinpoint the mechanisms by which nutrition acts. What is it about gumdrops that contributes to tooth decay: their size, shape, temperature, color, ingredients? Feeding various forms of gumdrops to rats might yield the information that sugar, in a gummy carrier, promotes tooth decay. In the laboratory, using animals or plants or cells, scientists can inoculate with diseases, induce deficiencies, and experiment with variations on treatments to obtain in-depth knowledge of the process under study. Intervention studies and laboratory experiments are among the most powerful tools in nutrition research because they show the effects of treatments.*

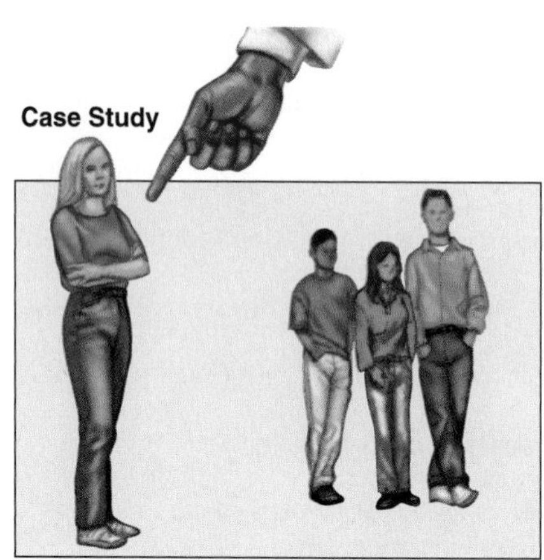

"This person eats too little of nutrient X and has illness Y."

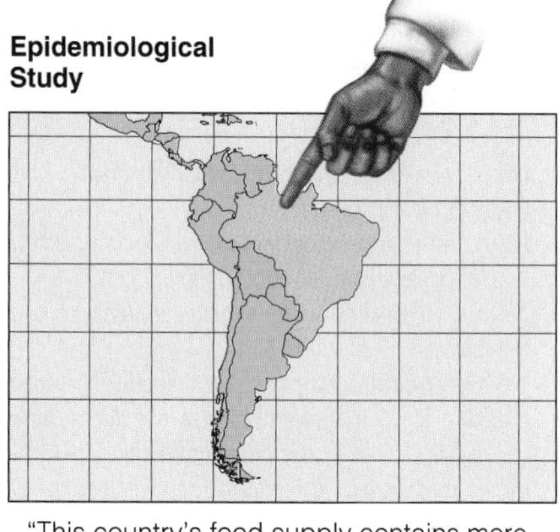

"This country's food supply contains more nutrient X, and these people suffer less illness Y."

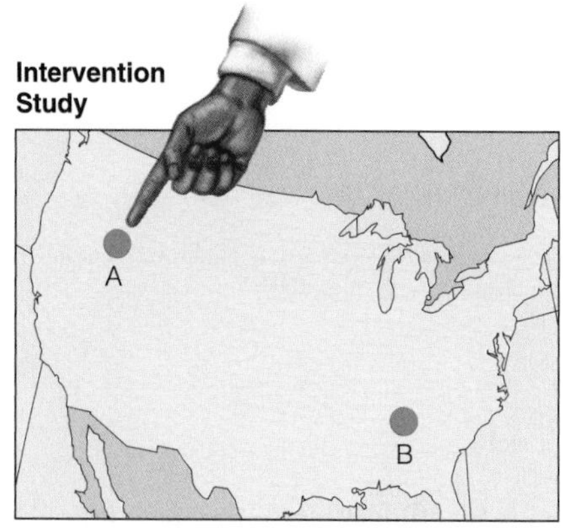

"Let's add foods containing nutrient X to city A's food supply and compare illness Y rates of city A with those of city B."

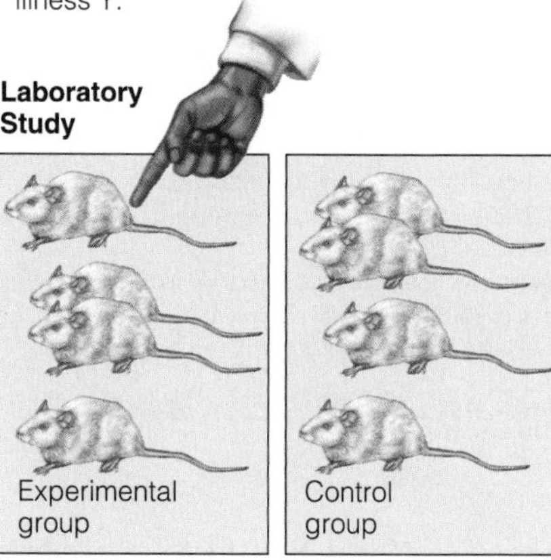

"Now let's prove that a nutrient X deficiency causes illness Y by inducing a deficiency in these rats."

Scientific Challenge

Once a new finding is published, it is still only preliminary. One experiment does not "prove" or "disprove" anything. The next step is for other scientists to attempt to duplicate and support the work of the first researchers or to challenge the finding by designing experiments to refute it.

Only when a finding has stood up to rigorous, repeated testing in several kinds of experiments performed by several different researchers is it finally considered confirmed. Even then, strictly speaking, science consists not of facts that are set in stone but of *hypotheses* that can always be challenged and revised. Some, though, like the hypothesis that the earth revolves about the sun, are so well supported by observations and experimental findings that they are generally accepted as facts. What we "know" in nutrition is confirmed in the same way—through years of replicating study

Table 1•9

RESEARCH DESIGN TERMS

- **blind experiment** an experiment in which the subjects do not know whether they are members of the experimental group or the control group. In a *double-blind experiment*, neither the subjects nor the researchers know to which group the members belong until the end of the experiment.
- **case studies** studies of individuals. In clinical settings, researchers can observe treatments and their apparent effects. To prove that a treatment has produced an effect requires simultaneous observation of an untreated similar subject (a *case control*).
- **control group** a group of individuals who are similar in all possible respects to the group being treated in an experiment but who receive a sham treatment instead of the real one. Also called *control subjects*. See also *experimental group* and *intervention studies*.
- **correlation** the simultaneous change of two factors, such as the increase of weight with increasing height (a *direct* or *positive* correlation) or the decrease of cancer incidence with increasing fiber intake (an *inverse* or *negative* correlation). A correlation between two factors suggests that one may cause the other, but does not rule out the possibility that both may be caused by chance or by a third factor.
- **epidemiological studies** studies of populations; often used in nutrition to search for correlations between dietary habits and disease incidence; a first step in seeking nutrition-related causes of diseases.
- **experimental group** the people or animals participating in an experiment who receive the treatment under investigation. Also called *experimental subjects*. See also *control group and intervention studies*.
- **intervention studies** studies of populations in which observation is accompanied by experimental manipulation of some population members—for example, a study in which half of the subjects (the *experimental subjects*) follow diet advice to reduce fat intakes while the other half (the *control subjects*) do not, and both groups' heart health is monitored.
- **laboratory studies** studies that are performed under tightly controlled conditions and are designed to pinpoint causes and effects. Such studies often use animals as subjects.
- **placebo** a sham treatment often used in scientific studies; an inert harmless medication. The *placebo effect* is the healing effect that the act of treatment, rather than the treatment itself, often has.

Table 1•10

THE ANATOMY OF A RESEARCH ARTICLE

Here's what you can expect to find inside a research article:

- *Abstract*. The abstract provides a brief overview of the article.
- *Introduction*. The introduction clearly states the purpose of the current study.
- *Review of literature*. A review of the literature reveals all that science has uncovered on the subject to date.
- *Methodology*. The methodology section defines key terms and describes the procedures used in the study.
- *Results*. The results report the findings and may include summary tables and figures.
- *Conclusions*. The conclusions drawn are those supported by the data and reflect the original purpose as stated in the introduction. Usually, they answer a few questions and raise several more.
- *References*. The references list relevant studies (including key studies several years old as well as current ones).

findings. This slow path of repeated studies stands in sharp contrast to the media's desire for today's latest news, described next.

Can I Trust the Media to Deliver Nutrition News?

The news media are hungry for new findings, and reporters often latch onto ideas from the scientific laboratories before they have been fully tested. Also, a reporter who lacks a strong understanding of science may misunderstand complex scientific principles. To tell the truth, sometimes scientists get excited about their findings, too, and leak them to the press before they have been through a rigorous review by the scientists' peers. As a result, the public is often exposed to late-breaking nutrition news stories before the findings are fully confirmed.[24] Then, when the hypothesis being tested fails to hold up to a later challenge, consumers feel betrayed by what is simply the normal course of science at work. The Consumer Corner offers some tips for evaluating news stories about nutrition.

Some newspapers, magazines, talk shows, Internet websites, and other media strive for accuracy in reporting but others specialize in sensationalism that borders on quackery—see this chapter's Controversy for details.

It also follows that people who take action based on single studies are almost always acting impulsively, not scientifically. The real scientists are trend watchers. They evaluate the methods used in each study, assess each study in light of all the evidence gleaned from other studies, and, little by little, modify their picture of what is true. As evidence accumulates, the scientists become more and more confident about their ability to make recommendations that apply to people's health and lives. Single studies are interesting, perhaps even exciting, but experienced observers learn to withhold judgment about the application of a study's findings until they have been repeated and confirmed.

Reading Nutrition News with an Educated Eye

A NEWS READER, who had sworn off butter years ago for his heart's sake, bemoaned this headline: *Margarine Fat as Bad as Butter for Heart Health*. "Do you mean to say that I could have been eating butter all these years? That's it. I quit. No more diet changes for me." His response is understandable—diet changes, after all, take effort to make and commitment to sustain. Those who do make changes may feel betrayed when, years later, science appears to have turned its advice upside down.

It bears repeating that the findings of a single study never prove or disprove anything. Study results may constitute strong supporting evidence for one view or another, but they rarely merit the sort of finality implied by journalistic phrases such as "Now we know . . ." or "The answer has been found." Misinformed readers who look for simple answers to complex nutrition problems often take such phrases literally.

To read news stories with an educated eye, keep these points in mind:

- The study being described should be published in a peer-reviewed journal such as the *American Journal of Clinical Nutrition*. An unpublished study or one from a less credible source may or may not be valid; the reader has no way of knowing because the study has not been challenged or reviewed by other experts in the field.
- The report should follow a standard format, stating the purpose of the study and describing the research methods used to obtain the data; it should also note their limitations (in the Methodology section—look again at Table 1-10 on page 17). For example, it matters whether the study participants numbered eight or eight thousand, or whether the researchers personally observed participants' behaviors or relied on self-reports collected over the telephone.
- The subjects of the study may have been single cells, animals, or human beings, and the report should clearly make this distinction. If the study subjects were human beings, the more you have in common with them (age and gender, for example), the more applicable the findings may be for you.
- Valid reports also describe previous research and put the current research in proper context. Some reporters regularly follow developments in a research area and thus acquire the background knowledge they need to report meaningfully in that area.
- Useful for their broad perspective on a single topic are review articles appearing in journals such as *Nutrition Reviews*. Such articles allow judgment about a single study within the context of many other studies on the same topic.

A person wanting the whole story on a nutrition topic is wise to seek articles from peer-reviewed journals such as these. A review journal examines all available evidence on major topics. Other journals report details of the methods, results, and conclusions of single studies.

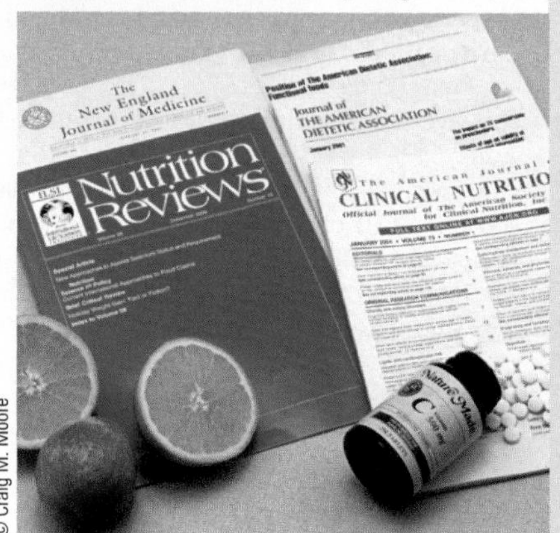

© Craig M. Moore

Finally, ask yourself if the study makes common sense. Even if it turns out that the fat of margarine is damaging to the heart, do you eat enough margarine to worry about its effects? Before making a decision, learn more about the effects of fats on the arteries in Chapters 5 and 11 and then ask the critical questions about yourself.

When a headline touts a shocking new "answer" to a nutrition question, read the story with a critical eye. It may indeed be a carefully researched report, but often it is a sensational story intended to catch the attention of newspaper and magazine buyers, not to offer useful nutrition information.

nutrient density a measure of nutrients provided per calorie of food.

Sometimes media sensationalism overrates the importance of even true, replicated findings. For example, a few years ago the media eagerly reported that oat bran lowers blood cholesterol, a lipid indicative of heart disease risk. Although the reports were true, oat bran is only one of several hundred factors that affect blood cholesterol. News reports on oat bran often failed to mention that cutting saturated fat intake is still the major step to take to lower blood cholesterol.

Also, new findings need refinements. Oat bran is truly a cholesterol reducer, but how much bran must a person eat to produce the desired effects? Do little oat bran pills or powders meet the need? Do oat bran cookies? If so, how many cookies must be eaten? As for oatmeal, it takes a bowl-and-a-half daily to affect blood lipids. A few cookies cannot provide nearly so much and certainly cannot undo all the damage from a high-fat meal.

Today, oat bran's cholesterol-lowering effect is considered to be established, and labels on food packages can proclaim that a diet high in oats may reduce the risk of heart disease. The whole process of discovery, challenge, and vindication took almost ten years of research; establishing some other effects has taken many years longer. In science, a single finding almost never makes a crucial difference to our knowledge as a whole, but like each individual frame in a movie, it contributes a little to the big picture. Many such frames are needed to tell the whole story.

key point *Scientists uncover nutrition facts by experimenting. Single studies must be replicated before their findings can be considered valid. New nutrition news is not always to be believed; established nutrition news has stood up to the test of time.*

Conclusion

According to the experts, adults in the United States are not very successful at meeting the nutrition goals set forth in this chapter.[25] In particular, only 1 percent of several thousand adults who were surveyed managed to achieve both adequacy and moderation.[26] People seem to behave as though they must choose between two alternatives—getting all their nutrients but overconsuming fat calories in the process, or keeping their fat in line but running short on nutrients.[27] That finding defines the challenge for the health-conscious eater: try to achieve adequacy and moderation at the same time. Because this challenge is the key to good nutrition, this chapter's Food Feature begins to piece together a plan by offering a tool to help make it easier—the concept of **nutrient density.** The next chapter then takes up the challenge again, offering details about diet planning for adequacy and moderation.

How Can I Get Enough Nutrients without Consuming Too Many Calories?

IN TRYING to control calories while balancing the diet and making it adequate, certain foods are especially useful. These foods are rich in nutrients relative to their energy contents: that is, they are foods with high nutrient density. Consider calcium sources, for example. Ice cream and fat-free milk both supply calcium, but the milk is "denser" in calcium per calorie. A cup of rich ice cream contributes more than 350 calories, a cup of fat-free milk only 85—and with almost double the calcium. Most people cannot, for their health's sake, afford to choose foods without regard to their energy contents. Those who do very often use up their calorie allowances while leaving nutrient needs unmet.

For busy people, being conscious of nutrient density is especially important. To save both time and money, you are well advised to *center* the meal on foods of high nutrient density. The foods that offer the most nutrients per calorie are the vegetables, especially the nonstarchy vegetables such as broccoli, carrots, mushrooms, peppers, and tomatoes. These foods are also rich in phytochemicals thought to protect against diseases. These inexpensive foods take time to prepare, but time invested this way pays off in nutritional health. Twenty minutes spent peeling and slicing vegetables for a salad is a better investment in nutrition than 20 minutes spent fixing a fancy, high-fat, high-sugar dessert. Besides, the dessert ingredients cost more money and strain the calorie budget, too.

In today's households, although both men and women spend some 70 hours a week sleeping and taking care of personal needs, women still do most of the cooking and food shopping. Few households can afford a stay-at-home spouse, so families have very little time for food preparation. Busy chefs should seek out convenience foods that are nutrient dense, such as bags of ready-to-serve salads, refrigerated prepared meats, and frozen vegetables. To round out the meal, fat-free milk is both nutritious and convenient. Other selections, such as most potpies, are less helpful because they contain too few vegetables and too much fat and salt.

Nutrient density is such a useful concept in diet planning that this book encourages you to think in those terms (see Figure 1-5). Watch for the tables and figures in later chapters that show

Figure 1•5

HOW THE EXPERTS JUDGE WHICH FOODS ARE MOST NUTRITIOUS

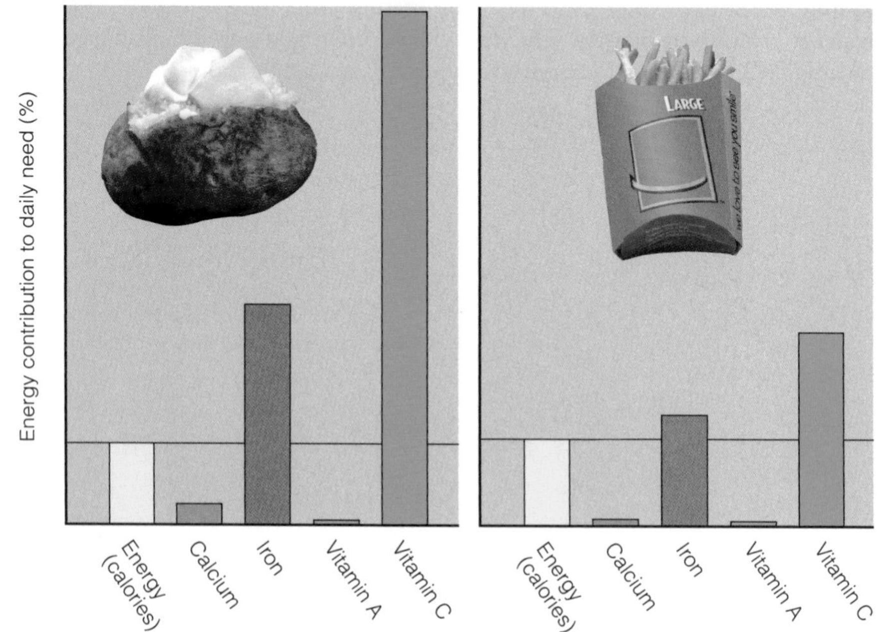

Would it take more time to prepare this dish than to prepare a batch of cookies? No, less time and less cleanup, too.

the best buys among foods, not necessarily in nutrients per dollar (although nutrient-dense foods are often among the least expensive), but in nutrients per calorie. This viewpoint can help you distinguish between more and less nutritious foods.

All of this discussion leads to the principle that is central to achieving nutritional health: It is not the individual foods you choose, but the way you combine them into meals and the way you arrange meals to follow one another over days and weeks that determine how well you are nourishing yourself. Nutrition is a science, not an art, but it can be used artfully to create a pleasing, nourishing diet. The remainder of this book is dedicated to helping you make informed choices and combine them artfully to meet all the body's needs.

ANALYZE NUTRITION NEWS

Here is a chance to practice your skills in reading nutrition news with an educated eye. Find an article in your local newspaper or any publication that carries nutrition news. On a copy of Form 1-1, answer these questions:

1. What sort of language does the writer use? Do the words imply sensationalism or conclusive findings? Phrases such as "startling revelation" or "now we know" or "the study proved" are clues to whether the report is a sensational one. Does the author take a tentative approach, using words such as *may, might,* or *could*? What do these words imply?
2. Is the finding placed in the context of previous nutrition findings? Does the article imply that the current finding wipes out all that has gone before it? Can you detect a broad understanding of nutrition on the writer's part? From what clues? For example, an article about folate and heart disease should say that saturated fat probably plays the major nutrition role in heart disease development.
3. Does the article mention whether the research results under discussion are published in a medical or nutrition journal? Where? The following Controversy section has more information about which types of journals publish valid scientific findings and which do not.
4. How were the results obtained? Can you tell from the article whether this was a case study, an epidemiological study, an intervention study, or a laboratory study? How does that information affect your understanding of what the results have contributed to nutrition science?
5. Does the finding apply to you? Should you change your eating patterns because of it? In what ways did the subjects resemble or differ from you? Were there enough subjects to make the study seem valid? (In a serious evaluation, a statistical analysis would be used to answer this question.)
6. Does the finding make sense to you in light of what you know about nutrition? You may not know enough to make this judgment yet, but by the end of this course, you should have developed a "feel" for identifying information that fits with reality.

This sort of assessment can guide you through the numerous nutrition articles appearing in newspapers and magazines, so you can avoid making nutrition decisions based on passing fads.

Form 1•1

CRITIQUING NUTRITION NEWS

The news report I am critiquing comes from ______________ (publication), dated ______________ (attach the report to this form).

1. I evaluate the language used in the publication as follows:

____

____

____

2. I believe the author's understanding of previously reported findings to be:

____

____

____

3. I judge the credibility of the item to be:

____

____

____

4. The methods used to obtain these results were:

____

____

____

5. The results of the study apply to the following populations:

____

____

____

6. To a reader without extensive nutritional background, the results of a study may be misleading. This report might mislead by:

____

____

____

Answers to these Self Check questions are in Appendix G.

1. Energy-yielding nutrients include all of the following *except:*
   a. vitamins
   b. carbohydrates
   c. fat
   d. protein
2. Organic nutrients include all of the following *except:*
   a. minerals
   b. fat
   c. carbohydrates
   d. protein
3. One of the characteristics of a nutritious diet is that the diet provides no constituent in excess. This principle of diet planning is called:
   a. adequacy
   b. balance
   c. moderation
   d. variety
4. A slice of peach pie supplies 357 calories with 48 units of vitamin A; one large peach provides 42 calories and 53 units of vitamin A. This is an example of:
   a. calorie control
   b. nutrient density
   c. variety
   d. essential nutrients
5. Which of the following adjustments in one's diet would agree with the *Dietary Guidelines for Americans?*
   a. eating baked potatoes rather than french fries
   b. eating fruits rather than cakes and pies
   c. drinking fat-free rather than 2% milk
   d. all of the above
6. Studies of populations in which observation is accompanied by experimental manipulation of some population members are referred to as:
   a. case studies
   b. intervention studies
   c. laboratory studies
   d. epidemiological studies
7. Both heart disease and cancer are due to genetic causes, and diet cannot influence whether they occur. T F
8. Both carbohydrates and protein have 4 calories per gram. T F
9. Once a new finding about nutrition is published, you can feel confident about changing your diet accordingly. T F
10. The Healthy People 2010 nutrition-related objectives recommend that the proportion of persons aged 2 years and older who eat no more than 30 percent of calories from total fat be increased. T F

nutrition on the net

For further study of the topics of this chapter, access these websites and search for the phrases or words in quotation marks:

1. Find updates and quick links to these and other nutrition-related sites at our website: www.wadsworth.com/nutrition
2. Search for "nutrition" at the U.S. Government health information site: www.healthfinder.gov
3. View *Healthy People 2010:* www.health.gov/healthypeople
4. Review the Canadian *National Plan of Action for Nutrition:* www.hc-sc.gc.ca/datahpsb/npu
5. Learn about NHANES: www.cdc.gov/nchs/nhanes.htm
6. Get information from the Food Surveys Research Group: www.barc.usda.gov/bhnrc/foodsurvey
7. Visit the nutrition center of the Mayo Clinic: www.mayohealth.org
8. Visit the National Council for Reliable Health Information: www.ncrhi.org
9. Check the ratings and reviews of websites by Tufts University Nutrition Navigator: navigator.tufts.edu
10. Find a registered dietitian in your area through the American Dietetic Association: www.eatright.org
11. Find a nutrition professional in Canada through the Dietitians of Canada: www.dietitians.ca

INTERNET ACTIVITY

Understanding and applying nutrition concepts also means having knowledge of other factors that affect our nutritional status. Complete this health assessment to determine aspects of your lifestyle that influence your own nutritional status.

1. Go to: www.healthyroads.com
2. Click on "My Profile."
3. Click on "My Health Assessment."
4. Follow instructions and then select "You and Your Lifestyle" and complete the quiz.

Discuss your results. Include what you were already aware of and what was new. From the results, identify a few health behaviors you would like to change and explain why it is important to address these.

CONTROVERSY 1

SORTING THE IMPOSTORS FROM THE REAL NUTRITION EXPERTS

THERE'S an old saying that goes, "If it sounds too good to be true, it probably is." This stems from a time when a shady "snake oil salesman" stood on the back of an open-air wagon, assuring every man, woman, and child in each tumbleweed town that a bottle of his Magic Elixir could cure whatever ailed them. Fast-forward at Internet speed to today's global marketplace, and you'll find that such dubious promises are far from dead and buried in ghost towns. Despite rapid advances in information access, familiar scams with new names and slick appearances still plague this nation: innumerable products are sold on late-night television **infomercials**, in magazine **advertorials,** and as **urban legends** on the Internet with promises of astonishing results requiring minimal effort, and all at bargain prices. When scam products are things like garden tools or stain removers, hoodwinked consumers may lose a few dollars and some pride. But when lapses in judgment lead to consumption of ineffective, untested, or even hazardous "dietary supplements," a person stands to lose much more. Ironically, the sham products and procedures of quacks not only rob people of the very health they are seeking, but also delay their use of legitimate strategies that could be of real help.[1]

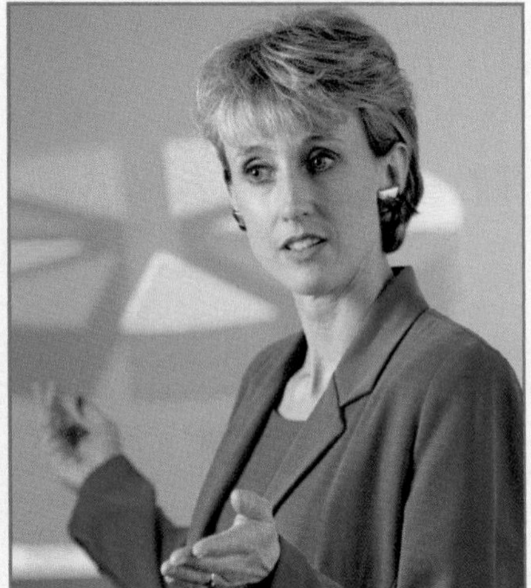

© 2002 Larry Bray/FPG/Getty Images

People want to know what nutrition news they can believe and safely use. This heightened interest in nutrition translates into a deluge of dollars spent on services and products peddled by both legitimate and fraudulent businesses. Unfortunately, nutrition and other health **fraud** (or **quackery,** defined in Table C1-1) rings cash registers to the tune of $27 billion annually.[2] Consumers with questions about fraud or suspicions about a product or individual can contact one of the consumer fraud organizations listed in Appendix E.

Table C1 • 1

MISINFORMATION TERMS

- **advertorials** lengthy advertisements in newspapers and magazines that read like feature articles but are written for the purpose of touting the virtues of products.
- **anecdotal evidence** information based on interesting and entertaining, but not scientific, personal accounts of events.
- **fraud** or **quackery** the promotion, for financial gain, of devices, treatments, services, plans, or products (including diets and supplements) that alter or claim to alter a human condition without proof of safety or effectiveness. (The word *quackery* comes from the term *quacksalver,* meaning a person who quacks loudly about a miracle product—a lotion or a salve.)
- **infomercials** feature-length television commercials that follow the format of regular programs but are intended to convince viewers to buy products, and not to educate or entertain them.
- **urban legend** a story, usually false, that may travel rapidly throughout the world via the Internet gaining strength of conviction solely on the basis of repetition.

How can people distinguish valid nutrition information from misinformation? One excellent approach is to notice who is purveying the information: quacks or qualified sources. Some quackery may be easy to identify—like the salesman in Figure C1-1—but most fraudulent nutrition claims are not so blatant.

Between the extremes of quackery and scientific data lies an abundance of less easily recognized nutrition information and misinformation. An instructor at a gym, a physician, a health-store clerk, an author of a book (and seller of juice machines) all recommend nutrition regimens. A famous talk show host advises viewers about the safety of hamburgers. What qualifies these people to give advice? Would following the advice be helpful or harmful? To sift the meaningful nutrition information from the rubble, you must first learn to recognize quackery wherever it presents itself.

Identifying Valid Nutrition Information

Nutrition derives information from scientific research. Scientists must systematically conduct research studies and cautiously interpret the findings before they can provide practical nutrition information. The following are characteristics of scientific research:

- Scientists test their ideas by conducting properly designed scientific experiments. They report their methods and procedures in detail so that other scientists can verify the findings through replication.
- Scientists recognize the inadequacy of **anecdotal evidence** or testimonials.
- Scientists who use animals in their research do not apply their findings directly to human beings.
- Scientists may use specific segments of the population in their research. When they do, they are careful not to generalize the findings to all people.
- Scientists report their findings in respected scientific journals. Their work must survive a screening review by their peers before it is accepted for publication.

Figure C1•1

EARMARKS OF NUTRITION QUACKERY

The more of these claims you hear about nutrition information, the less likely it is to be valid.

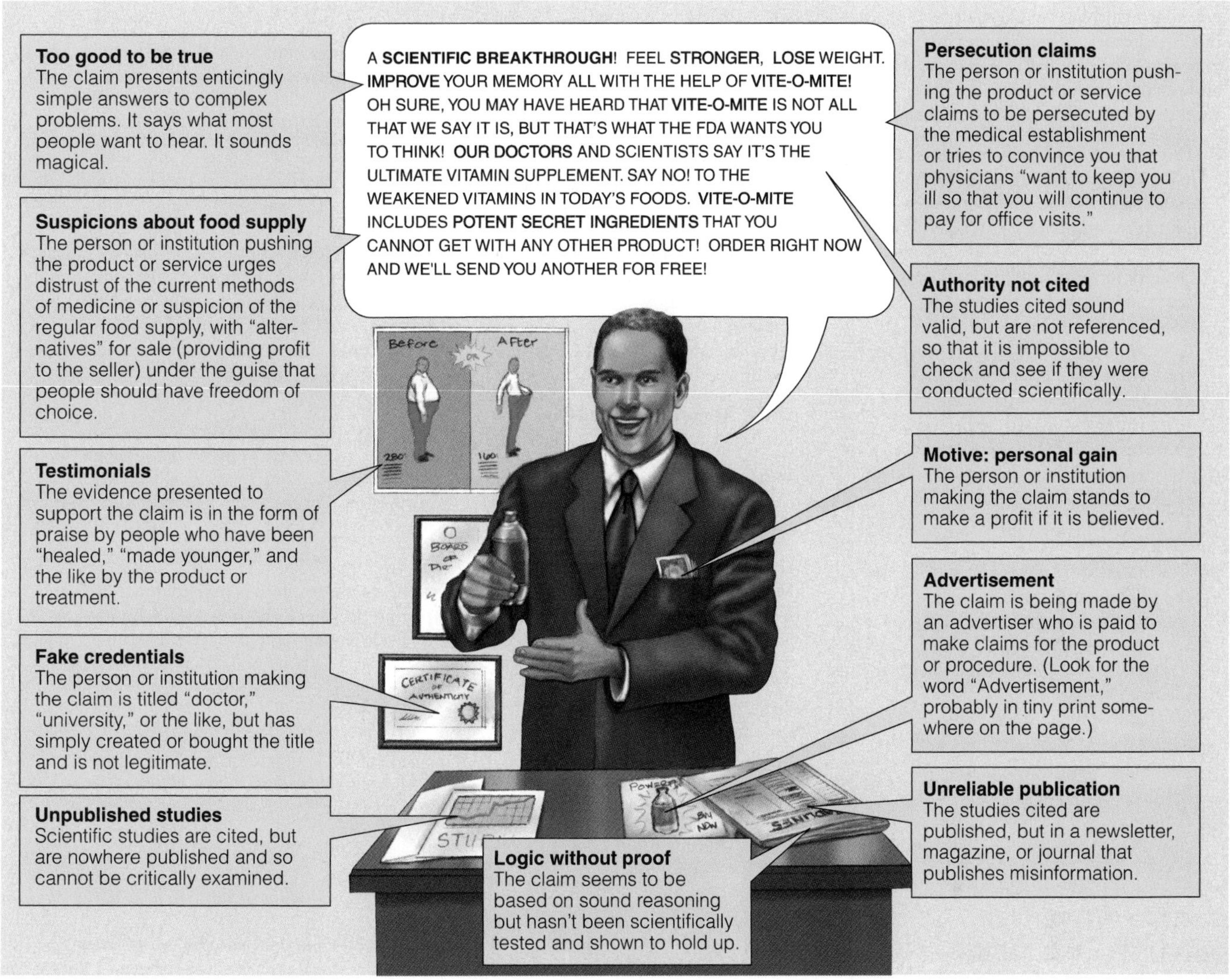

A **SCIENTIFIC BREAKTHROUGH**! FEEL **STRONGER**, **LOSE** WEIGHT. **IMPROVE** YOUR MEMORY ALL WITH THE HELP OF **VITE-O-MITE!** OH SURE, YOU MAY HAVE HEARD THAT **VITE-O-MITE** IS NOT ALL THAT WE SAY IT IS, BUT THAT'S WHAT THE FDA WANTS YOU TO THINK! **OUR DOCTORS** AND SCIENTISTS SAY IT'S THE ULTIMATE VITAMIN SUPPLEMENT. SAY NO! TO THE WEAKENED VITAMINS IN TODAY'S FOODS. **VITE-O-MITE** INCLUDES **POTENT SECRET INGREDIENTS** THAT YOU CANNOT GET WITH ANY OTHER PRODUCT! ORDER RIGHT NOW AND WE'LL SEND YOU ANOTHER FOR FREE!

Too good to be true
The claim presents enticingly simple answers to complex problems. It says what most people want to hear. It sounds magical.

Suspicions about food supply
The person or institution pushing the product or service urges distrust of the current methods of medicine or suspicion of the regular food supply, with "alternatives" for sale (providing profit to the seller) under the guise that people should have freedom of choice.

Testimonials
The evidence presented to support the claim is in the form of praise by people who have been "healed," "made younger," and the like by the product or treatment.

Fake credentials
The person or institution making the claim is titled "doctor," "university," or the like, but has simply created or bought the title and is not legitimate.

Unpublished studies
Scientific studies are cited, but are nowhere published and so cannot be critically examined.

Logic without proof
The claim seems to be based on sound reasoning but hasn't been scientifically tested and shown to hold up.

Persecution claims
The person or institution pushing the product or service claims to be persecuted by the medical establishment or tries to convince you that physicians "want to keep you ill so that you will continue to pay for office visits."

Authority not cited
The studies cited sound valid, but are not referenced, so that it is impossible to check and see if they were conducted scientifically.

Motive: personal gain
The person or institution making the claim stands to make a profit if it is believed.

Advertisement
The claim is being made by an advertiser who is paid to make claims for the product or procedure. (Look for the word "Advertisement," probably in tiny print somewhere on the page.)

Unreliable publication
The studies cited are published, but in a newsletter, magazine, or journal that publishes misinformation.

With each report from scientists, the field of nutrition changes a little—each finding contributes another piece to the whole body of knowledge. Table C1-2 on the next page lists some sources of credible nutrition information.

Nutrition on the Net

Hundreds of millions of websites await users of the Internet. Be forewarned: much of the nutrition "information" found on the Internet is pure fiction, aimed at separating consumers from their money. Sales of unproven and dangerous products over the Internet have reached huge proportions, partly because the Internet is difficult to regulate. The Internet also offers access to high-quality information, however, so it pays to learn to use it wisely. Table C1-3 (next page) provides some clues to reliable nutrition information websites.

Hoaxes and scare stories abound on unsound websites and in e-mails. Be suspicious of the content of an Internet source when:

- The contents were written by someone other than the sender or some authority you know.
- Something like the phrase, "Forward this to everyone you know" appears anywhere in the piece.
- The piece states something like, "This is not a hoax"; chances are the opposite is true.
- The information seems shocking or something that you've never heard from legitimate sources.
- The language is overly emphatic or sprinkled with capitalized words or exclamation marks.
- No references are offered.
- The message has been debunked on websites such as www.quackwatch.com or www.urbanlegends.com.[3]

Table C1•2

CREDIBLE SOURCES OF NUTRITION INFORMATION

Professional health organizations, government health agencies, volunteer health agencies, and consumer groups provide consumers with reliable health and nutrition information. Credible sources of nutrition information include:

- Professional health organizations, especially the American Dietetic Association's National Center for Nutrition and Dietetics (NCND) www.eatright.org/ncnd.html; also the Society for Nutrition Education www.sne.org and the American Medical Association www.ama-assn.org.
- Government health agencies such as the Federal Trade Commission (FTC) www.ftc.gov, the U.S. Department of Health and Human Services (DHHS) www.os.dhhs.gov, the Food and Drug Administration (FDA) www.fda.gov, and the U.S. Department of Agriculture (USDA) www.usda.gov.
- Volunteer health agencies such as the American Cancer Society www.cancer.org, the American Diabetes Association www.diabetes.org, and the American Heart Association www.americanheart.org.
- Reputable consumer groups such as the Better Business Bureau www.bbb.org, the Consumers Union www.consumersunion.org, the American Council on Science and Health www.acsh.org, and the National Council Against Health Fraud www.ncahf.org.

Appendix E provides addresses and websites for these and other organizations.

Table C1•3

IS THIS SITE RELIABLE?

To judge whether an Internet site offers reliable nutrition information, answer the following questions.

Who is responsible for the site?
Clues can be found in the three-letter "tag" that follows the dot in the site's name. For example, "gov" and "edu" indicate government and university sites, usually reliable sources of information.

Do the names and credentials of information providers appear? Is an editorial board identified?
Many legitimate sources provide e-mail addresses or other ways to obtain more information about the site and the information providers behind it.

Are links with other reliable information sites provided?
Reputable organizations almost always provide links with other similar sites because they want you to know of other experts in their area of knowledge. Caution is needed when you evaluate a site by its links, however. Anyone, even a quack, can link a Web page to a reputable site without the organization's permission. Doing so may give the quack's site the appearance of legitimacy, just the effect the quack is hoping for.

Is the site updated regularly?
Nutrition information changes rapidly, and sites should be updated often.

Is the site selling a product or service?
Commercial sites may provide accurate information, but they also may not, and their profit motive increases the risk of bias.

Does the site charge a fee to gain access to it?
Many academic and government sites offer the best information, usually for free. Some legitimate sites do charge fees, but before paying up, check the free sites. Chances are good you'll find what you are looking for without paying.

Some credible websites include:

National Council Against Health Fraud
www.ncahf.org

Stephen Barrett's Quackwatch
www.quackwatch.com

Centers for Disease Control and Prevention's Current Health Related Hoaxes and Rumors
www.cdc.gov/hoax_rumors.htm

Tufts University
www.navigator.tufts.edu

Federal Trade Commission's Operation Cure All
www.ftc.gov/opa/2001/06/cureall.htm

SOURCE: Adapted from M. Larkin, Health information on-line, *FDA Consumer*, June 1996. Available from http://vm.cfsan.fda.gov/list.html.

Of course, these hints alone are insufficient for judging material on the Net. The user must also scrutinize those posting materials even when they possess legitimate degrees, as described next.

Who Are the True Nutrition Experts?

Most people turn to their physicians for dietary advice. Physicians are expected to know all about health-related matters. But only about a quarter of all medical schools in the United States require students to take even one nutrition course, and less than half provide an elective nutrition course.[4] Students attending these classes receive an average of 20 hours of nutrition instruction—an amount most graduates consider inadequate. A decade ago, Congress passed a law mandating that "students enrolled in United States medical schools and physicians practicing in the United States [must] have access to adequate training in the field of nutrition and its relationship to human health." Plans are now in the works to make nutrition education a standard course in medical schools.

Enlarging on this idea, the **American Dietetic Association (ADA)** asserts that nutrition education should be part of the curriculum for all sorts of health-care professionals: physician's assistants, dental hygienists, physical and occupational therapists, social workers, and all others who provide services directly to clients.[5] This plan would bring access to reliable nutrition information to more people.

Some physicians are superbly qualified to speak on nutrition, particularly those who specialized in clinical nutrition in medical schools.

Table C1•4

TERMS ASSOCIATED WITH NUTRITION ADVICE

- **American Dietetic Association (ADA)** the professional organization of dietitians in the United States. The Canadian equivalent is the Dietitians of Canada (DC),[a] which operates similarly.
- **dietetic technician** a person who has completed a two-year acadmic degree from an accredited college or university and an approved dietetic technician program. A **dietetic technician, registered** (DTR) has also passed a national examination and maintains registration through continuing professional education.
- **dietitian** a person trained in nutrition, food science, and diet planning. See also *registered dietitian*.
- **license to practice** permission under state or federal law, granted on meeting specified criteria, to use a certain title (such as *dietitian*) and to offer certain services. Licensed dietitians may use the initials LD after their names.
- **medical nutrition therapy** nutrition services used in the treatment of injury, illness, or other conditions; includes assessment of nutrition status and dietary intake, and corrective applications of diet, counseling, and other nutrition services.
- **nutritionist** someone who engages in the study of nutrition. Some nutritionists are RDs, whereas others are self-described experts whose training is questionable and who are not qualified to give advice. In states with responsible legislation, the term applies only to people who have masters of science (MS) or doctor of philosophy (PhD) degrees from properly accredited institutions.
- **public health nutritionist** a dietitian or other person with an advanced degree in nutrition who specializes in public health nutrition.
- **registered dietitian (RD)** a dietitian who has graduated from a university or college after completing a program of dietetics. The program must be approved or accredited by the American Dietetic Association (or Dietitians of Canada). The dietitian must serve in an approved internship, coordinated program, or preprofessional practice program to practice the necessary skills; pass the five parts of the association's *registration* examination; and maintain competency through continuing education.[b] Many states also require licensing for practicing dietitians.
- **registration** listing with a professional organization that requires specific course work, experience, and passing of an examination.

[a]A new organization comprised of the former Canadian Dietetic Association (CDA) and ten provincial dietetic associations.
[b]The five content areas included on the registration examination for dietitians are nutrition services, foodservice systems, management, education and communication, and evaluation and standards.

Table C1•5

RESPONSIBILITIES OF A CLINICAL DIETITIAN

The first six items on this list play essential roles in **medical nutrition therapy** as part of a medical treatment plan.

- Assesses clients' nutrition status.
- Determines clients' nutrient requirements.
- Monitors clients' nutrient intakes.
- Develops, implements, and evaluates clients' medical nutrition therapy.
- Counsels clients to cope with unique diet plans.
- Teaches clients and their families about nutrition and diet plans.
- Provides training for other dietitians, nurses, interns, and dietetics students.
- Serves as liaison between clients and the foodservice department.
- Communicates with physicians, nurses, pharmacists, and other health-care professionals about clients' progress, needs, and treatments.
- Participates in professional activities to enhance knowledge and skill.

Membership in the American Society for Clinical Nutrition, whose journal is cited many times throughout this text, is another sign of nutrition knowledge. Still, few physicians have the knowledge, time, or experience to develop diet plans and provide detailed diet instruction for clients. Often physicians wisely refer their clients to nutrition specialists for diet advice. Table C1-4 lists the best specialists to choose.

Fortunately, the credential that indicates a qualified nutrition expert is easy to spot—you can confidently call on a **registered dietitian (RD).** Additionally, some states require that **nutritionists,** as well as **dietitians,** receive a **license to practice.** Meeting these established criteria certifies that an expert is the genuine article.

Dietitians are easy to find in most communities because they perform a multitude of duties in a variety of settings.[6] They work in foodservice operations, pharmaceutical companies, sports nutrition programs, corporate wellness programs, the food industry, home health agencies, long-term care institutions, private practice, community and public health settings, cooperative extension offices,* research centers, universities and other educational settings, and hospitals, health maintenance organizations (HMOs), and other health-care facilities.

Dietitians in hospitals have many subspecialties. Administrative dietitians manage the foodservice system; clinical dietitians provide client care (see Table C1-5); and nutrition support team dietitians coordinate nutrition care with the efforts of other health-care professionals. In the food industry, dietitians conduct research, develop products, and market services. In government, **public health nutritionists** play key roles in delivering nutrition services to people in the community. A public health nutritionist may plan, coordinate, administer, and evaluate food assistance programs; act as a consultant to other agencies; manage finances; and much more.

In some facilities, a **dietetic technician, registered (DTR)** assists registered dietitians in both administrative and clinical responsibilities.[7] A DTR has been educated and trained to work under the guidance of a registered dietitian.

Detecting Fake Credentials

In contrast to RDs, thousands of people possess fake nutrition degrees and claim to be nutrition counselors, nutritionists, or "dietists." These and other such titles may sound meaningful, but most of these people lack the

*Cooperative extension agencies are associated with land grant colleges and universities and may be found in the phone book's government listings.

established credentials of the ADA-sanctioned dietitian. If you look closely, you can see signs that their expertise is fake.

Take, for example, a nutrition expert's educational background. The minimum standards of education for a dietitian specify a bachelor of science (BS) degree in food science and human nutrition (or related fields) from an **accredited** college or university (Table C1-6 defines this and related terms). Such a degree generally requires four to five years of study. In contrast, a fake nutrition expert may display a degree from a six-month correspondence course; such a degree is simply not the same. In some cases, schools posing as legitimate **correspondence schools** offer even less. They are actually **diploma mills**—fraudulent businesses that sell certificates of competency to anyone who pays the fees, from under a thousand dollars for a bachelor's degree to several thousand for a doctorate. Buyers ordering multiple degrees are given discounts. To obtain these "degrees," a candidate need not read any books or pass any examinations.

Lack of proper accreditation is the identifying sign of a fake educational institution. To guard educational quality, an accrediting agency recognized by the U.S. Department of Education certifies that certain schools meet the criteria defining a complete and accurate schooling, but in the case of nutrition, quack accrediting agencies cloud the picture. Fake nutrition degrees are available from schools "accredited" by more than 30 phony accrediting agencies.*

Sassafras and Charlie display their professional credentials.

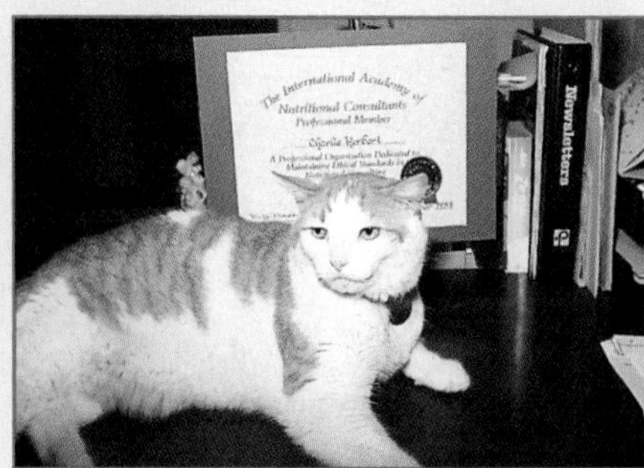

© Marilyn Herbert Photography (both)

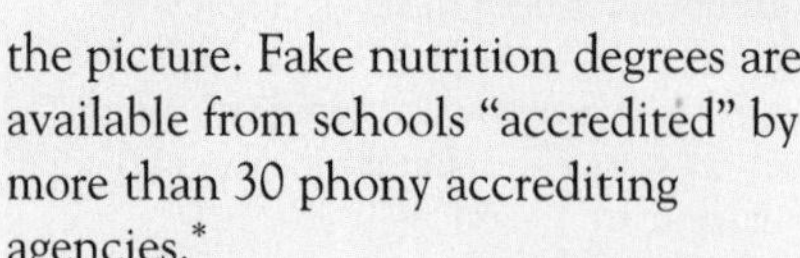

To dramatize the ease with which anyone can obtain a fake nutrition degree, one writer enrolled for $82 in a nutrition diploma mill that billed itself as a correspondence school. She made every attempt to fail, intentionally answering all the examination questions incorrectly. Even so, she received a "nutritionist" certificate at the end of the course, together with a letter from the "school" officials explaining that they were sure she must have misread the test.

In a similar stunt, Ms. Sassafras Herbert was named a "professional member" of a nutrition association. For her efforts, Sassafras received a wallet card and is listed in a fake "Who's Who in Nutrition" that is distributed at health fairs and trade shows nationwide. Sassafras is a poodle. Her master, Victor Herbert, MD, paid $50 to prove that she could be awarded these honors merely by sending in her name. Mr. Charlie Herbert is also a professional member of such an organization; Charlie is a cat.

State laws do not necessarily help consumers distinguish experts from fakes; some states allow anyone to use the title *dietitian* or *nutritionist*. But other states have responded to the need by allowing only RDs or people with certain graduate degrees and state licenses to call themselves dietitians. Licensing provides a way to identify people who have met minimum standards of education and experience.

In summary, to stay one step ahead of the nutrition quacks, check a provider's qualifications. First look for the degrees and credentials listed after the person's name (such as MD, RD, MS, PhD, or LD). Next find out what you can about the reputations of the institutions that awarded the degrees. Then call your state's health-licensing agency and ask if dietitians are licensed in your state. If they are, find out whether the person giving you dietary advice has a license—and if not, find someone better qualified. Your health is your most precious asset, and protecting it is well worth the time and effort it takes to do so.

Table C1•6

TERMS DESCRIBING INSTITUTIONS OF HIGHER LEARNING, LEGITIMATE AND FRAUDULENT

- **accredited** approved; in the case of medical centers or universities, certified by an agency recognized by the U.S. Department of Education.
- **correspondence school** a school that offers courses and degrees by mail. Some correspondence schools are accredited; others are *diploma mills*.
- **diploma mill** an organization that awards meaningless degrees without requiring its students to meet educational standards.

*To find out whether a correspondence school is accredited, write the Distance Education and Training Council, Accrediting Commission, 1601 Eighteenth Street NW, Washington, DC 20009; call (202) 234-5100; or visit their website (www.detc.org).

To find out whether a school is properly accredited for a dietetics degree, write the American Dietetic Association, Division of Education and Research, 216 West Jackson Boulevard, Chicago, IL 60606; call (312) 899-4870; or visit their website (www.eatright.org/caade).

The American Council on Education publishes a directory of accredited institutions, professionally accredited programs, and candidates for accreditation in *Accredited Institutions of Postsecondary Education Programs* (available at many libraries). For additional information, write the American Council on Education, One Dupont Circle NW, Suite 800, Washington, DC 20036; call (202) 939-9382; or visit their website (www.acenet.edu).

chapter 2

Paul Cézanne (1839–1906), *Four Peaches*, (detail) (1890–1894)

Nutrition Tools—Standards and Guidelines

Frequently Asked Questions

Why are Daily Values used on labels? p. 34

How can the Food Guide Pyramid help me to eat well? p. 37

How many servings do I need each day? p. 39

Which packaged foods and restaurant choices are best for my health? p. 47

Contents

Dietary Reference Intakes (DRI) a set of four lists of values for the dietary nutrient intakes of healthy people in the United States and Canada. The values include Estimated Average Requirements (EAR), Recommended Dietary Allowances (RDA), Adequate Intakes (AI), and Tolerable Upper Intake Levels (UL). Descriptions of the DRI values and other nutrient standards are found in Table 2-1 on page 32.

A directory of recommendations:

- DRI lists, inside front cover pages A and B.
- Daily Values, inside front cover page C.
- 1989 RDA for energy and protein, inside cover page C.

EATING WELL IS easy in theory—just choose a selection of foods that supplies appropriate amounts of the essential nutrients, fiber, phytochemicals, and energy without excess intakes of fat, sugar, and salt, and be sure to get enough exercise to balance the foods you eat. A few people do these things automatically, but most do not.[1] Many people are overweight, or undernourished, or suffer from nutrient excesses or deficiencies that impair their health—that is, they are malnourished. You may not think that this statement applies to you, but you may already have less-than-optimal nutrient intakes and activity without knowing it. Accumulated over years, the effects of your habits can seriously impair the quality of your life. Putting it positively, you can enjoy the best possible vim, vigor, and vitality if you learn now to nourish yourself optimally.

To master the task of meeting your nutrition needs, you may find it useful to learn the answers to several questions. How much energy and how much of each nutrient do you need? How much physical activity do you need to balance the energy you take in from foods? Which types of foods supply which nutrients? How much of each type of food do you have to eat to get enough? And how can you eat all these foods without gaining weight? This chapter begins by identifying some ideals for nutrient intakes and ends by showing how to achieve them.

Nutrient Recommendations

Nutrient recommendations are sets of yardsticks used as standards for measuring healthy people's energy and nutrient intakes. Nutrition experts use them to make nutrient recommendations and to assess nutrient intakes. Individuals may use them to decide how much of a nutrient they need to consume and how much is too much.

The standards in use in the United States and Canada are the **Dietary Reference Intakes (DRI).** A committee of qualified nutrition experts from the United States and Canada develops and publishes the DRI.* DRI values for the vitamins and most minerals are in place, and those for carbohydrates, lipids, protein, and water and the minerals sodium and potassium are forthcoming.

Another set of nutrient standards are practical for the person striving to make wise choices among packaged foods. These are the **Daily Values,** familiar to anyone who has stopped to read a food label. (Read about the Daily Values and other nutrient standards in Table 2-1 on page 32.) Nutrient standards—the DRI and Daily Values—are used and referred to so often that they are printed on the inside front covers of this book.

key point *The Dietary Reference Intakes are nutrient intake standards set for people living in the United States and Canada. The Daily Values are U.S. standards used on food labels.*

Goals of the DRI Committee

For each nutrient, the DRI establish two or three values, each serving a different purpose. Most people need to focus on only two kinds of DRI values: those that set nutrient intake goals for individuals (RDA and AI, described below) and those that define an upper limit of safety for nutrient intakes (UL, addressed later). The following sections address the different DRI values, arranged by the goals of the DRI committee.

GOAL #1. SETTING RECOMMENDED INTAKE VALUES—RDA AND AI One of the great advantages of the DRI values lies in their applicability to the diets of individuals.[2] The committee offers two sets of values setting intake goals for

* This is a committee of the Food and Nutrition Board, Institute of Medicine of the National Academy of Sciences, working in association with Health Canada.

individuals: one is known as the **Recommended Dietary Allowances (RDA),** and the other is referred to as **Adequate Intakes (AI).**

The RDA are indisputably the bedrock of the DRI recommended intakes, for they are based on solid experimental evidence and other reliable observations. The AI values are also as scientifically based as possible, but setting them requires some educated guesswork. The committee establishes an AI value whenever scientific evidence is insufficient to generate an RDA.[3] Both the RDA and AI values are intended to be used as nutrient goals in planning nutritious diets for individuals, so there is no practical need to distinguish between them.[4] This book refers to the RDA and AI values collectively as the DRI recommended intakes.

The DRI table on the inside front cover distinguishes the RDA and AI values, but both kinds of values are intended as nutrient intake goals for individuals. This book refers to both kinds of values as DRI recommended intakes.

See Table 2-1 on the next page for definitions of terms on this page.

Tolerable Upper Intake Levels (UL) are listed on pages A and B inside the front cover.

GOAL #2. FACILITATING NUTRITION RESEARCH AND POLICY—EAR Another set of values established by the DRI committee, the **Estimated Average Requirements (EAR),** establishes average nutrient intake requirements for given life stage and gender groups that researchers and nutrition policymakers use in their work. Public health officials may also use them to assess nutrient intakes of populations and make recommendations. The EAR values form the scientific basis upon which the RDA values are set (a later section explains how).

GOAL #3. ESTABLISHING SAFETY GUIDELINES—UL Beyond a certain point, it is unwise to consume large amounts of any nutrient. People need to know how much of a nutrient is too much, so the DRI committee sets the **Tolerable Upper Intake Levels (UL)** to identify potentially hazardous levels of nutrient intake. The UL are indispensable to consumers who take supplements or consume foods and beverages to which vitamins or minerals are added—a group that includes almost everyone. Public health officials also rely on UL values to set safe upper limits for nutrients added to our food and water supplies.

Nutrient needs fall within a range, and there is a danger zone both below and above that range. Figure 2-1 illustrates this point. Because people's tolerances for high doses of nutrients vary, the scientists who developed the UL values urge individuals to use caution when applying them. The UL values are listed on the inside front cover.

Some nutrients lack UL values. The absence of a UL for a nutrient does not imply that it is safe to consume it in any amount, however. It means only that insufficient data exist to establish a value.

GOAL #4. PREVENTING CHRONIC DISEASES The DRI also take into account chronic disease prevention, wherever appropriate. In the last decade, abundant new research has linked nutrients in the diet with the promotion of health and the prevention of chronic diseases, and the DRI committee uses this research in setting intake recommendations. For example, the committee sets lifelong intake goals for the mineral calcium at the levels believed to lessen the likelihood of osteoporosis-related fractures in the later years.

All in all, the DRI values are well designed to meet the diverse needs of individuals, the scientific and medical communities, and others.[5] Table 2-1 sums up the names and purposes of the nutrient intake standards just introduced. A later section comes back to the Daily Values.

key point *The DRI provide nutrient intake goals for individuals, provide a set of standards for researchers and makers of public policy, establish tolerable upper limits for nutrients that can be toxic in excess, and take into account new research on disease prevention. The DRI are composed of the RDA, AI, UL, and EAR lists of values. The Daily Values are nutrient intake standards used on food labels.*

Figure 2•1

THE NAIVE VIEW VERSUS THE ACCURATE VIEW OF OPTIMAL NUTRIENT INTAKES

Consuming too much of a nutrient endangers health, just as consuming too little does. The DRI recommended intake values fall within a safety range with the UL marking the tolerable upper levels.

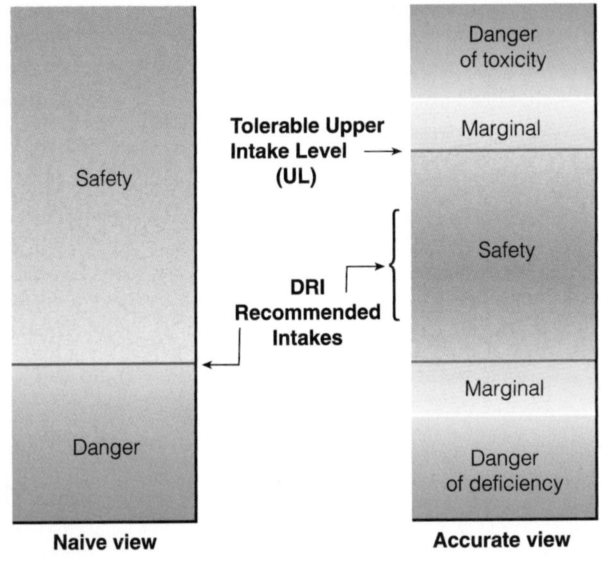

Don't let the "alphabet soup" of nutrient intake standards confuse you. Their names make sense when you learn their purposes.

Understanding the DRI Intake Recommendations

Nutrient recommendations have been much misunderstood. One young woman was outraged: "You mean that some bureaucrat says that I must eat exactly 5.0 micrograms of vitamin D every day?" This is not the DRI committee's intention. The DRI are recommendations, not commandments. The following facts will help put the DRI recommended intakes into perspective:

- The government funds the ongoing DRI creation, but the committee that determines the values is composed of scientists representing a variety of specialties.
- The values are based on available scientific research to the greatest extent possible and are updated periodically in light of new knowledge.
- The values are recommendations for optimal intakes, not minimum requirements. They include a generous margin of safety and meet the needs of virtually all healthy people in a specific age and gender group.
- The values are chosen in reference to specific indicators of nutrient adequacy, such as blood nutrient concentrations, normal growth, and reduction of certain chronic diseases or other disorders.
- The values reflect daily intakes to be achieved, on average, over time. They assume that intakes will vary from day to day, and they are set high enough to ensure that body nutrient stores will meet nutrient needs during periods of inadequate intakes lasting a day or two for some nutrients and up to a month or two for others.
- The recommendations apply to healthy persons only.

Separate recommendations are made for specific sets of people: men, women, pregnant women, children, and other life-stage groups have different needs. Children aged

Table 2•1

NUTRIENT STANDARDS

Standards from the DRI Committee

Dietary Reference Intakes (DRI) a set of four lists of nutrient intake values for healthy people in the United States and Canada. These values are used for planning and assessing diets:

1. **Recommended Dietary Allowances (RDA)**
   - Nutrient intake goals for individuals.[a] Derived from the Estimated Average Requirements (see below).
2. **Adequate Intakes (AI)**
   - Nutrient intake goals for individuals.[a] Set whenever scientific data are insufficient to allow establishment of an RDA value.
3. **Tolerable Upper Intake Levels (UL)**
   - Suggested upper limits of intake for potentially toxic nutrients. Intakes above the UL are likely to cause illness from toxicity.
4. **Estimated Average Requirements (EAR)**
   - Population-wide average nutrient requirements used in nutrition research and policymaking. The basis upon which RDA values are set.

Daily Values

Daily Values (DV)
- Nutrient standards used on food labels, in grocery stores, and on some restaurant menus. The DV allow comparisons among foods with regard to their nutrient content.

[a] For simplicity, this book combines the two sets of nutrient goals for individuals (AI and RDA) and refers to them as *DRI recommended intakes*.

four to eight years, for example, have their own DRI recommended intakes. Each individual can look up the recommendations for his or her own age and gender group.

The DRI recommended intakes are generous allowances. Even so, they do not necessarily cover every individual for every nutrient. On average, one should probably try to get 100 percent or more of the DRI recommended intake for every nutrient to ensure an adequate intake over time.

The DRI are designed for health maintenance and disease prevention in healthy people, not for the restoration of health. Under the stress of serious illness or malnutrition, a person may require a much higher intake of certain nutrients or may not be able to handle even the DRI amount. Therapeutic diets take into account the increased nutrient needs imposed by certain medical conditions, such as recovery from surgery, burns, fractures, illnesses, or addictions.

The DRI represent up-to-date, optimal, and safe nutrient intakes for healthy people in the United States and Canada.

balance study a laboratory study in which a person is fed a controlled diet and the intake and excretion of a nutrient are measured. Balance studies are valid only for nutrients like calcium (chemical elements) that do not change while they are in the body.

requirement the amount of a nutrient that will just prevent the development of specific deficiency signs; distinguished from the DRI recommended intake value, which is a generous allowance with a margin of safety.

How the Committee Establishes DRI Values—An RDA Example

A theoretical discussion will help to explain how the DRI committee goes about its work of setting the DRI values. Suppose we are the DRI committee members with the task of setting an RDA for nutrient X (an essential nutrient).* Ideally, our first step will be to find out how much of that nutrient various healthy individuals need. To do so, we review studies of deficiency states, nutrient stores and their depletion, and the factors influencing them. We then select the most valid data for use in our work. Of the DRI family of nutrient standards, the setting of an RDA value demands the most rigorous science and tolerates the least guesswork.

One experiment we might review or conduct is a **balance study.** In this type of study, scientists measure the body's intake and excretion of a nutrient to find out how much intake is required to balance excretion. For each individual subject, we can determine a **requirement** to achieve balance for nutrient X. With an intake below the requirement, a person will slip into negative balance or experience declining stores that could, over time, lead to deficiency of the nutrient.

With additional study, we find that different individuals, even of the same age and gender, have different requirements. Mr. A needs 40 units of the nutrient each day to maintain balance; Mr. B needs 35; Mr. C, 57. If we look at enough individuals, we find that their requirements are distributed as shown in Figure 2-2—with most requirements near the midpoint (here, 45), and only a few at the extremes.

To set the value, we have to decide what intake to recommend for everybody. Should we set it at the mean (45 units in Figure 2-2)? This is the Estimated Average Requirement (EAR) for nutrient X, mentioned earlier as valuable to scientists, but not appropriate as an individual's nutrient goal. The EAR value is probably close to everyone's minimum need, assuming the distribution shown in Figure 2-2. (Actually, the data for most nutrients indicate a distribution that is much less symmetrical.) But if people took us literally and consumed exactly this amount of nutrient X each day, half the population would begin to develop internal deficiencies and possibly even observable symptoms of deficiency diseases. Mr. C (at 57) would be one of those people.

Perhaps we should set the recommendation for nutrient X at or above the extreme, say, at 70 units a day so that everyone will be covered. (Actually, we didn't study everyone, so some individual we didn't happen to test might have an even higher requirement.) This might be a good idea in theory, but what about a person like Mr. B, who requires only 35 units a day? The recommendation would be twice his requirement, and to follow it, he might spend money needlessly on foods containing nutrient X to the exclusion of foods containing other nutrients he needs.

Figure 2•2

INDIVIDUALITY OF NUTRIENT REQUIREMENTS

Each square represents a person. A, B, and C are Mr. A, Mr. B, and Mr. C. Each has a different requirement.

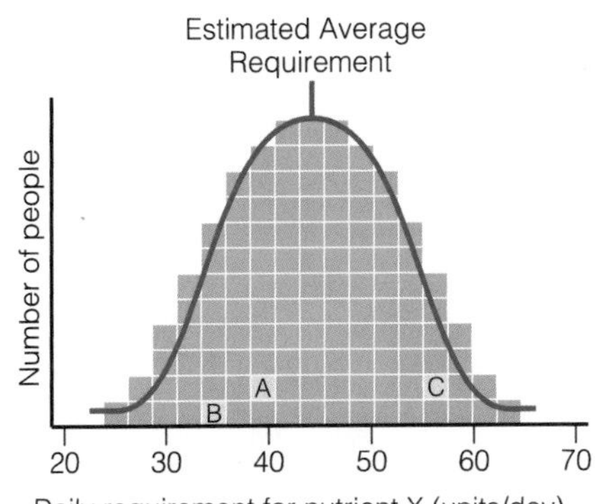

*This discussion describes how an RDA value is set; to set an AI value, the committee would use some educated guesswork as well as scientific research results to determine an approximate amount of the nutrient most likely to support health.

The decision we finally make is to set the value high enough so that the bulk of the population will be covered but not so high as to be excessive (the left-hand graph of Figure 2-3 illustrates such a value). In this example, a reasonable choice might be 63 units a day. Moving the DRI further toward the extreme would pick up a few additional people, but it would inflate the recommendation for most people, including Mr. A and Mr. B. The committee makes judgments of this kind when setting the DRI recommended intakes for nutrients. In theory, relatively few healthy people have requirements that are not covered by the DRI recommended intakes.

In contrast to the recommendations for nutrients, the value set for energy intake is not generous; instead it is set at the average of the population's estimated energy requirements (see the right-hand graph in Figure 2-3). Too much energy is as bad for health as too little because excess energy leads to obesity, whereas too little energy may cause undernutrition. The RDA for energy intakes are found on the inside back cover.

The DRI are based on scientific data and are designed to cover the needs of virtually all healthy people in the United States and Canada.

Why Are Daily Values Used on Labels?

Most careful diet planners are already familiar with the Daily Values because they are used on U.S. food labels. After learning about the DRI, many people wonder why yet another set of standards is needed for food labels. One answer is that the DRI values vary from group to group, whereas on a label, one set of values must apply to everyone. The Daily Values reflect the needs of an "average" person—someone eating 2,000 to 2,500 calories a day. Another answer is that a label must specify daily amounts of food constituents not yet covered by the DRI, such as carbohydrate, fat, and fiber, and Daily Values are set for those constituents.

The Daily Values are ideal for their intended purpose of allowing comparisons among *foods*. This strength is also their limitation, however. Because the Daily Values apply to all people, from children of age four through aging adults, they are much less useful as nutrient intake goals for individuals. Details about how to use the Daily Values appropriately in making comparisons among foods are offered in this chapter's Consumer Corner.

The Daily Values are standards used on food labels to enable consumers to compare the nutrient values among foods.

Figure 2•3

THE DIFFERENCES BETWEEN RECOMMENDED INTAKES OF NUTRIENTS AND ENERGY

The nutrient intake recommendations are set so that they will meet the requirements of nearly all people (boxes represent people). The recommended intake of energy is set at the average, or mean, so that half the population's requirements will fall below and half above the recommended level.

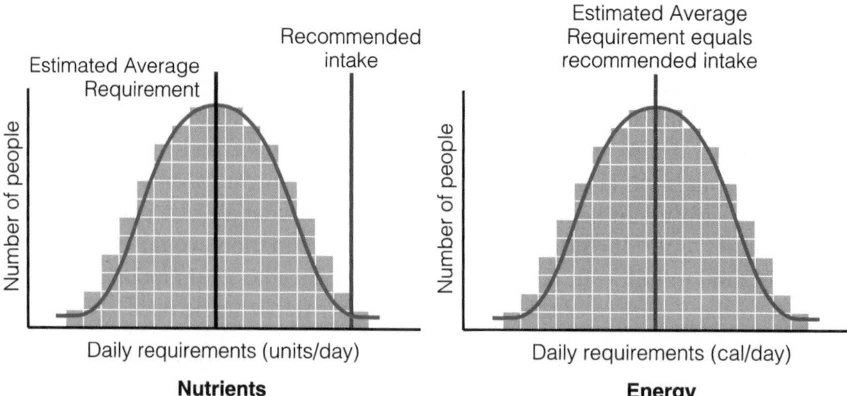

Other Nutrient Standards

Many nations and international groups have published sets of standards similar to the DRI. They differ from the DRI in some respects, though, partly because of different interpretations of the data from which they are derived and partly because people in different parts of the world have somewhat different food intakes and energy expenditures.

Many countries use recommendations developed by two international groups: the World Health Organization (WHO) and the Food and Agriculture Organization (FAO). The WHO/FAO recommendations are considered sufficient for the maintenance of health in nearly all healthy people worldwide. In addition to nutrient recommendations, recommendations for daily physical activity are intended to help people stay healthy and live long. See the Think Fitness box near here.

Many nations and groups issue recommendations for nutrient and energy intakes appropriate for specific groups of people.

food group plans diet-planning tools that sort foods into groups based on origin and nutrient content and then specify that people should eat certain minimum numbers of servings of foods from each group.

exchange system a diet-planning tool that organizes foods with respect to their nutrient contents and calorie amounts. Foods on any single exchange list can be used interchangeably. See the U.S. Exchange System, Appendix D, for details.

Appendix E, Nutrition Resources, provides addresses for WHO, FAO, and other agencies.

Recommendations for Daily Physical Activity

The American College of Sports Medicine (ACSM), a respected authority in exercise physiology, makes these minimum suggestions to maintain a healthy body:

- Engage in physical activity every day.
- Exercise at a comfortable effort level (this can be moderate, such as brisk walking).
- Exercise for a duration of at least 30 minutes total per day (this can be intermittent, a few minutes here and there throughout the day).

Other recommendations are found in Chapter 10.

THINK FITNESS

Diet Planning with the Daily Food Guide and the Food Guide Pyramid

Diet planning connects nutrition theory with the food on the table. To help people plan menus, **food group plans** describe food groups and dictate numbers and sizes of servings to choose each day. One example is the Daily Food Guide, explained next. Another is the Canadian food group plan, *Food Guide to Healthy Eating*, presented in full in Appendix B with a brief description in Table 2-2 in the margin. A different kind of planning tool, the **exchange system** (see Appendix D), focuses on controlling the carbohydrate, fat, protein, and energy (calories) in the diet.

The Food Guide Pyramid for Young Children is in Chapter 13. Other pyramids, such as one for vegetarians, also exist (see Controversy 6).

The Daily Food Guide

The Daily Food Guide of Figure 2-4 (pages 36–37) teaches people to recognize key nutrients provided by certain related groups of foods and recommends numbers of servings from each group to meet nutrient needs. The Food Guide Pyramid is a visual representation of the Daily Food Guide. This plan applies to adults only; children have their own food guide and pyramid, shown and explained in Chapter 13.

The foods in each group are well-known contributors of certain key nutrients, but you can count on them to supply many other nutrients as well. If you design your diet around this plan, it is assumed that you will obtain adequate amounts of not only the nutrients named in the figure but also the other two dozen or so essential nutrients because they are distributed among the same groups of foods. This is true in theory. In practice, however, diet planners must strive to choose mostly nutrient-dense foods in each group because some processes strip foods of some nutrients and add calories from sugar or fat. Color dots in Figure 2-4 identify a few foods in each group as having high,

Table 2•2

CANADA'S *FOOD GUIDE TO HEALTHY EATING*[a]

Food Group	Servings/Day
Grain products	5–12
Vegetables and fruits	5–10
Milk products	
Children aged 4–9 years	2–3
Youth aged 10–16 years	3–4
Adults	2–4
Pregnant and breast-feeding women	3–4
Meat and alternatives	2–3

[a]Canada's *Food Guide* and other Canadian guidelines are presented in full in the Canadiana appendix (Appendix B).

Figure 2•4

THE DAILY FOOD GUIDE AND THE FOOD GUIDE PYRAMID

KEY: Nutrient Density

- ■ *Foods generally highest in nutrient density (preferable first choice).*
- ■ *Foods moderate in nutrient density (reasonable second choice).*
- ■ *Foods lowest in nutrient density (limit selections).*

BREAD, CEREAL, RICE, AND PASTA

These foods contribute complex carbohydrates and fiber, plus riboflavin, thiamin, niacin, iron, protein, magnesium, and other nutrients.

6 to 11 servings per day.

Serving = 1 slice bread: ½ c cooked cereal, rice, or pasta: 1 oz ready-to-eat cereal; ½ bun, bagel, or English muffin; 1 small roll, biscuit, or muffin; 3 to 4 small or 2 large crackers.

- ■ Whole grains (wheat, oats, barley, millet, rye, bulgur), enriched breads, rolls, tortillas, cereals, bagels, rice, pastas (macaroni, spaghetti), air-popped corn.
- ■ Pancakes, muffins, cornbread, crackers, low-fat cookies, biscuits, presweetened cereals, granola.
- ■ Croissants, fried rice, doughnuts, pastries, sweet rolls.

VEGETABLES

These foods contribute fiber, vitamin A, vitamin C, folate, potassium, and magnesium.

3 to 5 servings per day (use dark green, leafy vegetables and legumes [dried beans] several times a week).

Serving = ½ c cooked or raw vegetables; 1 c leafy raw vegetables; ½ c cooked legumes;[a] ¾ c vegetable juice.

- ■ Bean sprouts, broccoli, brussels sprouts, cabbage, carrots, cauliflower, cucumbers, eggplant, green beans, green peas, bell peppers, leafy greens (spinach, mustard, and collard greens), legumes, lettuce, mushrooms, summer and winter squash, tomatoes.
- ■ Cassava, corn, potatoes, sweet potatoes, yams.
- ■ French fries, olives, tempura vegetables.

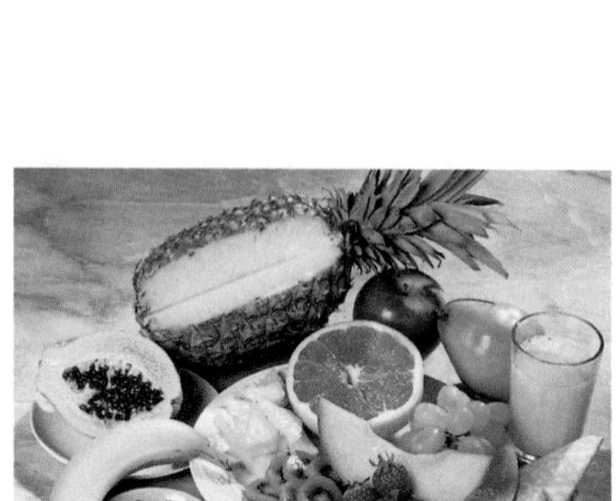

FRUITS

These foods contribute fiber, vitamin A, vitamin C, and potassium.

2 to 4 servings per day.

Serving = typical portion (such as 1 medium apple, banana, or orange, ½ grapefruit, 1 melon wedge; ¾ c juice; ½ c berries; ½ c diced, cooked, or canned fruit; ¼ c dried fruit).

- ■ Apples, apricots, bananas, cantaloupe, grapefruit, kiwi, oranges, orange juice, papaya, peaches, pears, pineapple, strawberries.
- ■ Canned or frozen fruit.
- ■ Avocados, dried fruit.

MEAT, POULTRY, FISH, DRY BEANS, EGGS, AND NUTS

These foods contribute protein, phosphorus, vitamin B_6, vitamin B_{12}, zinc, magnesium, iron, niacin, and thiamin.

2 to 3 servings per day.

Serving = 2 to 3 oz lean, cooked meat, poultry, or fish (total 5 to 7 oz per day); count 1 egg, ½ c cooked legumes,[a] ⅓ c nuts, or 2 tbs peanut butter as 1 oz meat (or about ⅓ serving).

- ■ Poultry, fish, lean meat (beef, lamb, pork, veal), legumes, egg whites.
- ■ Fat-trimmed beef, lamb, pork; refried beans; egg yolks, tofu, tempeh.
- ■ Hot dogs, luncheon meats, peanut butter, nuts (including coconut), sausage, bacon, fried fish or poultry, duck.

MILK, YOGURT, AND CHEESE

These foods contribute calcium, riboflavin, protein, vitamin B_{12}, and, when fortified, vitamin D and vitamin A.

2 servings per day.
3 servings per day for teenagers and young adults, pregnant/lactating women, women past menopause.
4 servings per day for pregnant/lactating teenagers.
Serving = 1 c milk or yogurt; 2 oz process cheese food; 1½ oz cheese.

- Fat-free and 1% low-fat milk (and fat-free products such as buttermilk, cottage cheese, cheese, yogurt); fortified soy drink.
- 2% reduced-fat milk (and low-fat products such as yogurt, cheese, cottage cheese); ice milk.
- Whole milk (and whole-milk products such as cheese, yogurt, cottage cheese[b]); custard; milkshakes; pudding; ice cream.

FATS, OILS, AND SWEETS

These foods contribute sugar, fat, alcohol, vitamin E, and food energy (calories). Their consumption should be limited because these foods provide few nutrients. Alcoholic beverages are not classed as foods on the pyramid; they contribute few nutrients, but they do contribute calories, and so are mentioned here.

- Foods high in fat include butter, margarine, lard, salad dressings, oils, mayonnaise, cream, sour cream, cream cheese, gravy, and sauces.
- Foods high in sugar include candy fruit rolls, other candies, soft drinks, fruit drinks, jelly, syrup, gelatin, desserts, sugar, and honey.
- Alcoholic beverages include wine, beer, and liquor.

[a]The Daily Food Guide lists legumes (dried beans, lentils, and peas) both under vegetables, for their starch, fiber, and vitamins, and under meats, for their protein and minerals.

[b]Cottage cheese is lower in calcium than most cheeses; 1 cup cottage cheese counts as ½ serving from the Milk, Yogurt, and Cheese group.

NOTE: Pregnant women may require additional servings of fruits, vegetables, meats, and breads to meet their higher needs for energy, vitamins, and minerals.

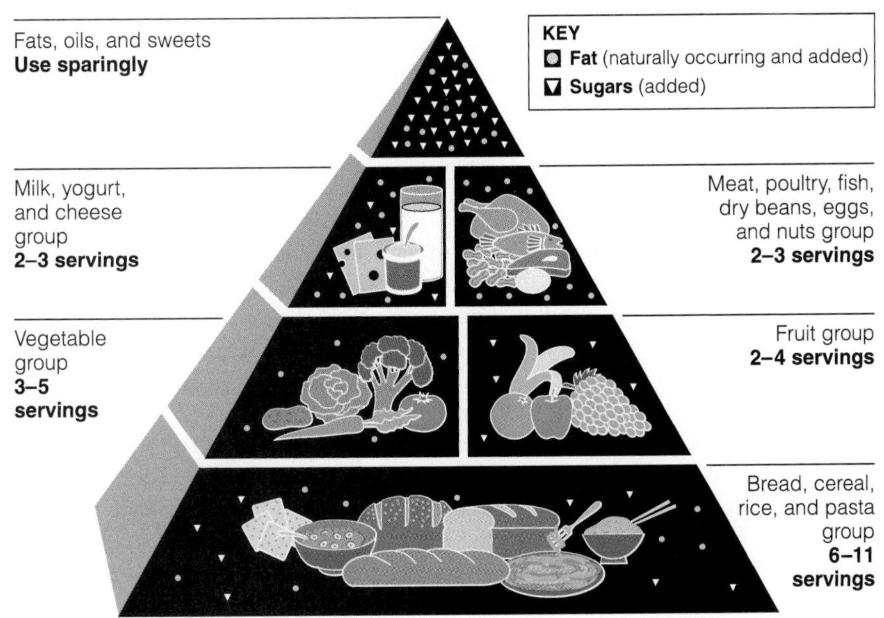

Food Guide Pyramid: A Guide to Daily Food Choices

The breadth of the base shows that grains (breads, cereals, rice, and pasta) deserve the most emphasis in the diet. The tip is smallest: use fats, oils, and sweets sparingly.

moderate, or low nutrient density to give you an idea of which are which. With this caution, the Daily Food Guide can provide a reasonable road map for diet planning.

key point *The Daily Food Guide sorts foods into groups based on their nutrients and origins. Then the plan suggests patterns of intake by group that will cover the nutrient needs of adults.*

How Can the Food Guide Pyramid Help Me to Eat Well?

Chapters to come give details about the energy nutrients in foods within the Pyramid.

The Food Guide Pyramid makes applying the Daily Food Guide easier to do. The Pyramid graphic assists you in planning a day's meals with the needed number of servings from each food group. The beauty of the Food Guide Pyramid lies in its simplicity. By using it wisely and by learning about the energy nutrients in various foods (as you will do in coming chapters), you can achieve the goals of a nutritious diet first mentioned in Chapter 1: adequacy, balance, calorie control, moderation, and variety.* A later section provides guidance in choosing an appropriate number of servings and points out the importance of holding serving sizes in line with recommendations.

Remember: 6, 3, 2, 2, and 2.

Adequacy To achieve adequacy, adults using the Food Guide Pyramid must choose at least 6 servings from the bread, cereal, rice, and pasta group; 3 from the vegetable group; 2 from the fruit group; 2 from the meat, poultry, fish, dry beans, eggs, and nuts group; and 2 from the milk, yogurt, and cheese group. To remember this pattern, think of the numbers: 6, 3, 2, 2, and 2.

These are the minimum numbers of servings. The plan's makers suggest that to meet additional energy needs, a person should choose proportionally more servings of foods from these very same groups.

Balance and Moderation The broad base of grains at the bottom of the Food Guide Pyramid conveys the idea that you should eat more grain foods than anything else; grains form the foundation of a healthful diet. Next in volume are the fruits and vegetables. Meats and milk products are dense in nutrients such as protein and are important sources of vitamins and minerals, but the number of servings must be limited because these foods can also be high in fat and calories.

The five groups are:
1. Bread, cereal, rice, and pasta.
2. Vegetables.
3. Fruits.
4. Meat, poultry, fish, dry beans, eggs, and nuts.
5. Milk, yogurt, and cheese.

The fats, oils, and sweets are extra and are not counted among the groups.

Fats, oils, and sweets occupy only a tiny triangle at the top of the Food Guide Pyramid, an indication that they should be used sparingly. These foods do not comprise a food group, because servings of them are optional; that is, they are not required to promote health. They do provide abundant energy. They also provide some essential lipids and vitamin E, but these nutrients are also amply provided by whole grains, fish, and other foods. Alcoholic beverages provide scant nutrients and are excluded from the Food Guide Pyramid altogether. They are high in calories, however, and must be counted in a day's tally, so a bottle of wine appears in Figure 2-4 as a reminder. Spices, coffee, tea, and diet soft drinks, also excluded from the Pyramid, provide few, if any, nutrients, but can add flavor and pleasure to meals as well as some potentially beneficial phytochemicals, such as those in tea or certain spices.

More on phytochemicals in foods in Controversy 2.

Variety within the Groups Although it may appear rigid, the Food Guide Pyramid can actually be very flexible once its intent is understood. For example, the user can substitute cheese for milk because both supply the key nutrients for the milk, yogurt, and cheese group. The user can choose legumes (beans) and nuts as alternatives to meats. One can adapt the plan to mixed dishes such as casseroles and to national and cultural cuisines as well, as Figure 2-5 on pages 40 and 41 shows.

Because the Food Guide Pyramid de-emphasizes meats and animal products such as milk, cheese, and eggs and emphasizes grains, fruits, and vegetables, it can assist vegetarians in their food choices, while encouraging others to choose foods from plants more often. The food group that includes the meats also includes *meat alternates*—foods such as legumes, nuts, and soybean products. As for the food group that includes milk and milk products, people who choose not to use dairy foods can substitute rice

*Balance is also called "proportionality."

or soy drinks—products made from rice or soybeans that fill some of the same nutrient needs, provided that they are fortified with calcium, riboflavin, vitamin A, vitamin D, and vitamin B_{12}. Thus, people who choose to eat no meats or products taken from animals can still use the Food Guide Pyramid to make their diets adequate.

Vegetarians will find more tips for choosing the right foods to supply the nutrients they need in the chapters to come.

Drawbacks to the Food Guide Pyramid The Food Guide Pyramid does have drawbacks, however. As mentioned, it does not limit food choices to foods low in calories. People who select the minimum number of servings from among the most nutrient-dense foods in each group and who strictly limit their use of fats, sweets, and alcoholic beverages can meet their nutrient needs and keep their energy intakes low. Even then, zinc and vitamin E are often lacking. However, people who use the higher-calorie foods in each group and who eat large servings, even without extra fats, sweets, or alcohol, can easily obtain too many calories.

In the cheese-for-milk substitution mentioned earlier, sufficient cheddar cheese to meet a day's calcium requirement would also provide almost 600 calories, over 70 percent of them from fat and most of those from saturated fat known to harm the heart and arteries. A day's calcium from fat-free milk comes with about 300 calories and with hardly any fat. Less obvious but significant over time are energy differences between sliced bread and biscuits, fish and hot dogs, nuts and legumes, or even green beans and sweet potatoes—all proper substitutions according to the Food Guide Pyramid. High-calorie choices may be just what some people, such as athletes, need to meet their large energy requirements, but for others such choices can dramatically boost energy intakes and, over time, add too much to body fat stores.

The Food Guide Pyramid attempts to caution consumers about fats and added sugars in food groups by sprinkling symbols for those two constituents across the pictures of the groups most likely to contain them. This warning may be difficult to put into practical use, however. The bread group, for example, is sprinkled with the symbols for both fat and added sugar because baked goods can be high in those constituents. From the symbols alone, though, the user cannot point to individual high-fat or high-sugar foods within the group.

Another criticism of the Food Guide Pyramid is that a person may choose the right number of servings from each group, yet make consistently nutrient-poor choices, and so fail to meet the day's needs for some nutrients. A diet can easily lack vitamin E or certain essential fatty acids, for example, because these nutrients are easily destroyed in processing or are refined out of foods.

key point *The Daily Food Guide and its visual image, the Food Guide Pyramid, convey the basics of planning a diet adequate in nutrients. The Food Guide Pyramid fails to show how to use nutrient-dense foods to form the bulk of food selections from each food group.*

A Note about Exchange Systems

Exchange systems can be useful to careful diet planners, especially those wishing to control calories (weight watchers), those who must control carbohydrate intakes (people with diabetes), and those who should control their intakes of fat and saturated fat (almost everyone). An exchange system presented in Appendix D (Appendix B for Canada) lists the estimated carbohydrate, fat, saturated fat, and protein contents of food portions, as well as their calorie values. The values in the exchange lists differ from the exacting values given for individual foods in Appendix A because exchange lists estimate values for whole groups of foods. With these estimates, exchange system users can make an educated approximation of the nutrients and calories in almost any food they might encounter.

The exchange system also highlights a fact that the Food Guide Pyramid overlooks: most foods provide more than just one energy nutrient. Meat, for example, is famous for protein, but meats like bacon and sausage deliver many more calories from fat than from protein. Pasta and bread contain significant protein with their carbohydrates and so on. This focus on nutrients in foods leads to some unexpected food groupings in the exchange lists. The high-fat meats mentioned above and also many cheeses are listed together as "high-fat meats" because fat constitutes the predominant form of energy in these foods, followed by protein. Potatoes and other vegetables high in starch are listed

Figure 2•5

ADDING VARIETY WITH ETHNIC AND REGIONAL FOODS

© Tony Freeman/PhotoEdit

CHINESE[a]

Bread, Cereal, Rice, and Pasta

- Millet; rice; rice noodles; steamed buns.
- Fried rice; fried noodles.

Vegetables

- Baby corn; bamboo shoots; bean sprouts; bok choy; cabbages; scallions; seaweed; snow peas; soybeans; water chestnuts.
- Fried vegetables.

Fruits

- Oranges, pears, plums, and other fresh fruit.

Milk, Yogurt, and Cheese

Not traditionally consumed.

Meat, Poultry, Fish, Dry Beans, Eggs, and Nuts

- Broiled or stir-fried fish and seafood; egg whites.
- Broiled or stir-fried beef or pork; egg yolks; tofu.
- Deep-fried meats and seafood; egg foo yung; pine nuts and cashews.

Fats, Oils, and Sweets

- Lard or oil for deep-frying.

Seasonings and Sauces[b]

- Bean sauce; garlic; ginger root; hoisin sauce;[c] oyster sauce;[c] plum sauce;[c] rice wine; scallions; soy sauce.[c]
- Sesame oil; other oils; oily gravies.

© Felicia Martinez/PhotoEdit

GREEK

Bread, Cereal, Rice, and Pasta

- Greek breads; flat bread; noodles.

Vegetables

- Cucumbers; eggplant; lentils and beans; onions; peppers; tomatoes.
- Olives.

Fruits

- Dates; figs; grapes; lemons; melons; raisins.

Milk, Yogurt, and Cheese

- Low-fat yogurt.
- Feta cheese; goat cheese.

Meat, Poultry, Fish, Dry Beans, Eggs, and Nuts

- Egg whites; fish and seafood; lentils and beans.
- Egg yolks; lamb; poultry; beef.
- Ground lamb; ground beef; gyros (spicy roasted meat and yogurt sauce, usually rolled in flat bread); almonds; walnuts.

Fats, Oils, and Sweets

- Olive oil; baklava (honey-soaked nut pastry); honey; cakes.

Seasonings and Sauces

- Garlic; herbs; lemons; egg and lemon sauce.
- Olive oil.

KEY: Nutrient Density

- *Foods generally highest in nutrient density (preferable first choice).*
- *Foods moderate in nutrient density (reasonable second choice).*
- *Foods lowest in nutrient density (limit selections).*

[a]Traditional cuisines of China and of West African influence exclude fluid milk as a beverage for adults and use few or no milk products in cooking. Calcium and certain other nutrients of milk are supplied by other foods, such as small fish eaten with the bones or large servings of leafy green vegetables.

[b]Many Chinese sauces are fat-free.

[c]May be high in sodium.

MEXICAN

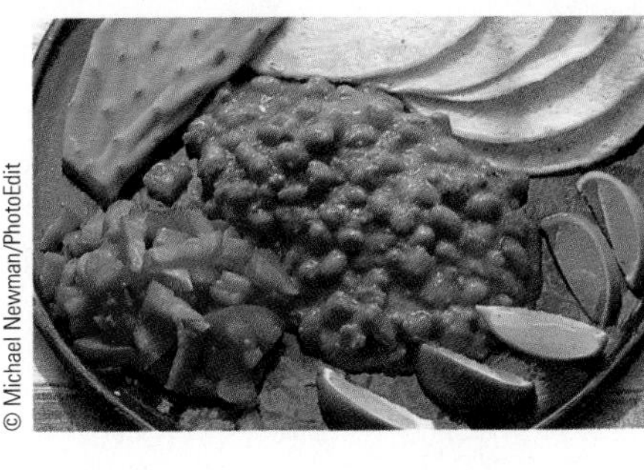
© Michael Newman/PhotoEdit

Bread, Cereal, Rice, and Pasta

- Cereal; corn or flour tortillas; macaroni; rice.
- Graham crackers.
- Churros (doughnuts); fried tortilla shells; tortilla chips.

Vegetables

- Cabbage; cactus; iceberg lettuce; legumes; squash; tomatoes.
- Corn; potatoes.
- Olives.

Fruits

- Bananas; guava; mango; oranges; papaya; pineapple.
- Avocados.

Milk, Yogurt, and Cheese

- Evaporated low-fat milk; powdered fat-free milk.
- Cheddar or jack cheese; flan (caramel custard); cocoa drink.

Meat, Poultry, Fish, Dry Beans, Eggs, and Nuts

- Fish; lean beef, poultry, lamb, and pork; many bean varieties.
- Egg yolks; refried beans.
- Bacon; fried fish, pork, or poultry; nuts; chorizo (sausages).

Fats, Oils, and Sweets

- Butter; candy; cream cheese; lard; margarine; pastries; soft drinks; vegetable oil; sour cream.

Seasonings and Sauces[d]

- Herbs; hot peppers; garlic; pico de gallo (finely chopped tomatoes, peppers, and onions with seasonings); salsas; spices.
- Guacamole; lard.

DEEP SOUTH (WEST AFRICAN INFLUENCE)[a]

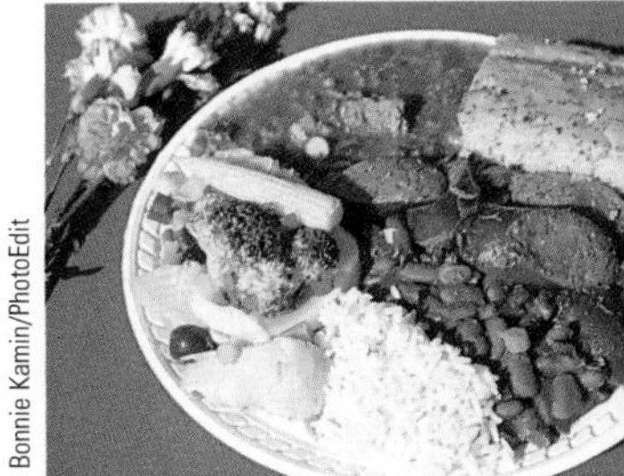
© Bonnie Kamin/PhotoEdit

Bread, Cereal, Rice, and Pasta

- Grits; macaroni; rice.
- Biscuits; cornbread; pastries.

Vegetables

- Beans; black-eyed peas; collards (other leafy greens); green beans; okra; tomatoes.
- Corn; sweet potatoes; hominy.
- Fried green tomatoes; fried okra.

Fruits

- Apples; bananas; berries; melons; peaches; pears.
- Fried pies; fruit pastries.

Milk, Yogurt, and Cheese

- Low-fat buttermilk; low-fat cheeses; low-fat milk.
- Full-fat American cheese; cheddar cheese.

Meat, Poultry, Fish, Dry Beans, Eggs, and Nuts

- Beans and peas; grilled or smoked poultry and fish.
- Braised or roasted meats (beef, lean ham, lean pork, poultry).
- Bacon; boiled peanuts; chitterlings; fat back; fried chicken, fish, or pork; ham hocks; peanut butter; pork rinds; salted pork; sausage; spareribs.

Fats, Oils, and Sweets

- Butter; lard; shortening; gravy.

[d]Many Mexican sauces are fat-free.

with the breads because one serving of bread and one serving of a starchy vegetable contain about the same amount of carbohydrate. To explore this powerful aid to diet planning, spend some time studying Appendix D or B.

Exchange lists facilitate calorie control by providing an understanding of how much carbohydrate, fat, and protein are in each food group.

How Many Servings Do I Need Each Day?

Note that for each food group the Food Guide Pyramid presents a range of numbers of servings. Find yourself among the people described at the top of Table 2-3 (below); then look at the column of numbers below for the approximate number of servings to take from each food group to meet your calorie goal.

Clearly, a sedentary person can meet the plan's requirements and still eat only about 1,600 calories (see Table 2-4). If you are only moderately active, you can probably eat an additional 600 to 1,200 calories without gaining weight. The more active you are, the higher the energy allowance you "earn." A wise choice is to invest many of these additional calories in additional nutrient-dense vegetables, legumes, fruits, and whole-grain foods and only a few in luxury items such as sweet desserts, butter, margarine, oil, or alcohol. If you make additions from the latter group, make them by conscious choice rather than through unintentional use. With judicious selections, the diet can supply all the necessary nutrients and provide some luxury items as well. Figure 2-6 demonstrates how the theory of the Pyramid translates to food on the plate.

The Daily Food Guide specifies how many servings of foods from each group people need to consume to meet their nutrient requirements.

Serving Sizes versus Helpings

To use the Food Guide Pyramid meaningfully, a person still must clarify what is meant by the word *serving*, for it can mean different things to different people.[6] Restaurants often deliver colossal helpings to ensure repeat business; a server on a cafeteria line may be instructed to deliver "about a spoonful"; fast-food burgers range from a one-ounce child-sized burger to a half-pound double deluxe; and so on. The trend in the United States has been toward consuming larger food portions, especially of foods rich in fat

Table 2•3

HOW MANY SERVINGS?

	Sedentary Women, Some Older Adults	Children, Teenage Girls, Active Women, Sedentary Men	Teenage Boys, Active Men
Calories[a]	About 1,600	About 2,000	About 2,800
Breads, cereals, rice, and pasta group	6	9	11
Vegetable group	3	4	5
Fruit group	2	3	4
Milk, yogurt, and cheese group[b]	2–3	2–3	2–3
Meat, poultry, fish, dry beans, eggs, and nuts group	2 (5 oz total)	2 (6 oz total)	3 (7 oz total)
Total fat (g)	53	73	93
Added sugar (tsp)	6	12	18

[a]Assumes mostly low-fat and low-calorie food choices.
[b]Women who are pregnant or lactating, teenagers, and young adults to age 24 need three servings. In fact, given the 1997 DRI, which raised the calcium recommendation, all individuals may need an additional milk serving to meet their calcium need.
SOURCE: U.S. Department of Agriculture.

Table 2•4

SAMPLE DIET PLANNED WITH THE FOOD GUIDE PYRAMID

Breakfast: Cornflakes with milk and sugar; toast; coffee; orange juice.
Lunch: Small cheeseburger, macaroni salad; banana; diet cola.
Supper: Chili with beans, beef, and rice; spinach salad with dressing; corn on the cob with margarine; water.

Pattern from the Pyramid	Example	Energy (cal)
Breads, cereals, rice, and pasta group—6 servings	½ c cooked white rice	103
	½ c low-fat macaroni salad	126
	1 oz cornflakes	100
	1 slice toast	119
	1 bun	123
Meat, poultry, fish, dry beans, eggs, and nuts group—2 servings (2 to 3 oz each)	½ c chili beans	126
	3 oz extra lean ground beef	225
Fruit group—2 servings	1 banana	109
	¾ c orange juice	84
Vegetable group—3 servings	½ c tomato sauce	37
	1 c spinach leaves	12
	1 medium corn on cob	72
Milk, yogurt, and cheese group—2 servings	1 c fat-free milk	85
	2 oz processed cheese	210
Added fat	2 pats reduced-fat margarine	32
	1 tbs low-calorie salad dressing	21
Added sugar—1 tsp	12 oz diet cola	0
	1 tsp sugar	16
		Total: 1,600

Figure 2•6

CREATING A MEAL WITH THE FOOD GUIDE PYRAMID

The meal shown at right provides 2 servings of grains, 1 serving of vegetables, 1 serving of meat, and 1 serving of fat. Other meals should include milk, fruit, vegetables, an additional meat serving, and whole grains to complete the Pyramid's requirements.

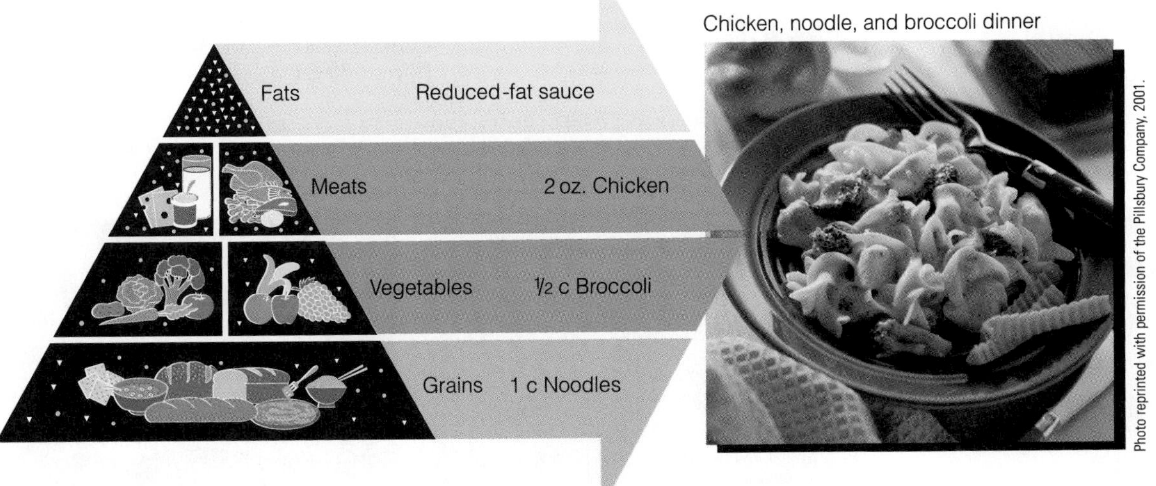

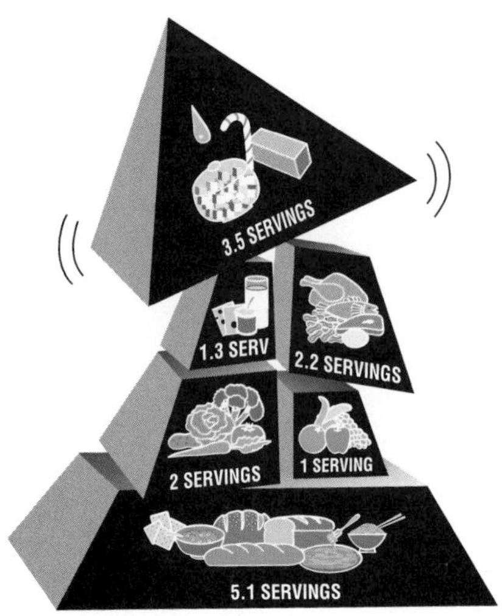

Here's how a typical U.S. diet stacks up.

SOURCE: National Livestock and Meat Board, courtesy of the National Cattlemen's Beef Association.

and sugar (see Figure 2-7), and, as the drawing in the margin shows, the result is an unbalanced diet. At the same time, body weights have been creeping upward, suggesting an increasing need to control portion sizes.

In contrast to the random-sized helpings found elsewhere, most serving sizes in the Food Guide Pyramid are specific and precise and can be relied upon to deliver certain amounts of key nutrients in foods.[7] They are based on four criteria:

1. An amount of food consumers typically consume at one sitting.
2. An amount of food that delivers a known quantity of key nutrients.
3. An amount of food easily recognized and multiplied or divided by most consumers.
4. An amount of food used in previous food guides to describe a serving.

Among volumetric measures, 1 "cup" refers to an 8-ounce measuring cup (not a teacup or drinking glass), filled to level (not heaped up, or shaken, or pressed down). Tablespoons and teaspoons refer to measuring spoons (not flatware), filled to level (not rounded). Ounces signify weight, not volume. Two ounces of meat, for example, refers to ⅛ pound of cooked meat. One ounce (weight) of granola cereal measures ¼ cup (volume), but take care: 1 ounce of crispy rice cereal measures a full cup. Also, some foods are specified as "medium," as in "one medium apple," but the word *medium* means different things to different people. When college students are asked to bring medium-sized foods to show the class, they bring bagels weighing anywhere from 2 to 5 ounces, muffins from about 2 to 8 ounces, baked potatoes from 4 to 9 ounces, and so forth.[8] The Food Guide Pyramid provides these standards for "medium": a bagel weighs 2 ounces; a muffin, 1.5 ounces; and a potato, 3.9 ounces. The Table of Food Composition, Appendix A, can help in determining serving sizes because it lists both weights and volumes of a wide variety of foods.

Wise diners also read labels on packaged foods to help them determine the foods' nutrient and energy contents and to decide how the foods may fit into their total eating plan. This chapter's Consumer Corner explains how to gain insight from the information on food labels.

A person wishing to avoid overconsuming calories must pay attention to serving sizes.

Figure 2•7

U.S TREND TOWARD COLOSSAL CUISINE

Chapter 9 discusses the consequences of increasing portion sizes in terms of body fitness.

Food	Food Guide Pyramid	Typical 1970s	Colossal 2002
Cola	—	10 oz bottle, 120 cal	40–60 oz fountain, 580 cal
French fries	10, 160 cal	about 30, 475 cal	about 50, 790 cal
Hamburger	2–3 oz meat, 240 cal	3–4 oz meat, 330 cal	6–8 oz meat, 650 cal
Bagel	½ bagel, 90 cal	2–3 oz, 230 cal	5–7 oz, 550 cal
Steak	2–3 oz, 170 cal	8–12 oz, 690 cal	16–22 oz, 1,260 cal
Pasta	½ cup, 100 cal	1 cup, 200 cal	2–3 cups, 600 cal
Baked potato	3–4 oz, 110 cal	5–7 oz, 180 cal	one pound, 420 cal
Candy bar	—	1½ oz, 220 cal	3–4 oz, 580 cal
Popcorn	—	1½ cups, 80 cal	8–16 cup tub, 880 cal

NOTE: Calories are rounded values for the largest portions in a given range.

SOURCE: Data for most entries from L. R. Young and M. Nestle, Portion sizes in dietary assessment: Issues and policy implications, *Nutrition Reviews* 53 (1995): 149–158.

© Tony Freeman/PhotoEdit (all)

1970s 2002

1970s 2002

1970s 2002

Checking Out Food Labels

NC CD ROM Assignment 5

NI • Menu > Labeling > Intro; Explore; Apply

WHAT CAN a package of potato chips tell you about its contents? Its label must list its ingredients: potato, fat, and salt; and its **Nutrition Facts** panel (see Table 2-5) must also reveal details about its nutrient composition. These requirements allow you to use packaged foods artfully in diet planning—if you can interpret food labels.

What Food Labels Must Include

The Nutrition Labeling and Education Act of 1990 set the requirements for label information to ensure that food labels truthfully inform consumers about the contents of the package. According to the law, every packaged food must state the following:

- The common or usual name of the product.
- The name and address of the manufacturer, packer, or distributor.
- The net contents in terms of weight, measure, or count.
- The nutrient contents of the product, presented in accordance with the rules governing the Nutrition Facts panel.

Then, the label must list the following in ordinary language:

- The ingredients, in descending order of predominance by weight.

Most food labels must conform with these requirements, but not every package need display information about every vitamin and mineral. The size of a package makes a difference. A large package, such as the box of cereal in Figure 2-8 on the next page, must provide all of the information listed above. A smaller label, such as the label on a can of tuna, provides some of the information in abbreviated form. A label on a roll of candy rings provides only a phone number, which is allowed for the tiniest labels. The Canadian version of a food label can be found in Appendix B.

The Nutrition Facts Panel

Most packaged foods have a Nutrition Facts panel, like the one shown in Figure 2-8. Many grocers also voluntarily post placards or offer handouts in fresh-food departments to provide consumers with similar sorts of nutrition information for the most popular types of fresh fruits, vegetables, meats, poultry, and seafoods.

When you read a Nutrition Facts panel, be aware that only the top portion of the panel conveys information specific to the food inside the package. The bottom portion is identical on every label—it lists the Daily Values standards.

Table 2•5

NUTRITION FACTS AND CLAIMS ON FOOD LABELS

- **health claims** claims linking food constituents with disease states; allowable on labels within the criteria established by the Food and Drug Administration.
- **nutrient claims** claims using approved wording to describe the nutrient values of foods, such as a claim that a food is "high" in a desirable constituent or "low" in an undesirable one.
- **Nutrition Facts** on a food label, the panel of nutrition information required to appear on almost every packaged food. Grocers may also provide the information for fresh produce, meats, poultry, and seafoods.

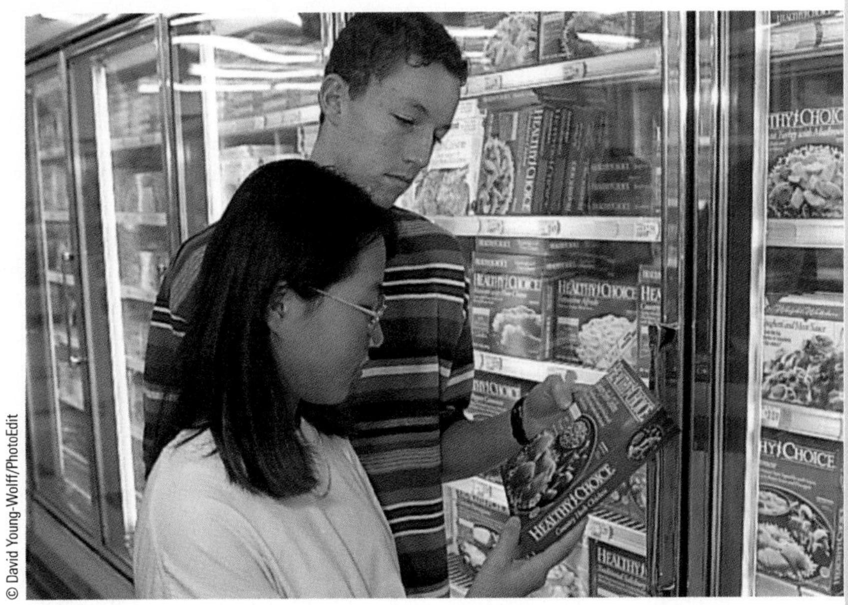

© David Young-Wolff/PhotoEdit

Modern-day entertainment: Reading food labels.

Figure 2•8

WHAT'S ON A FOOD LABEL?

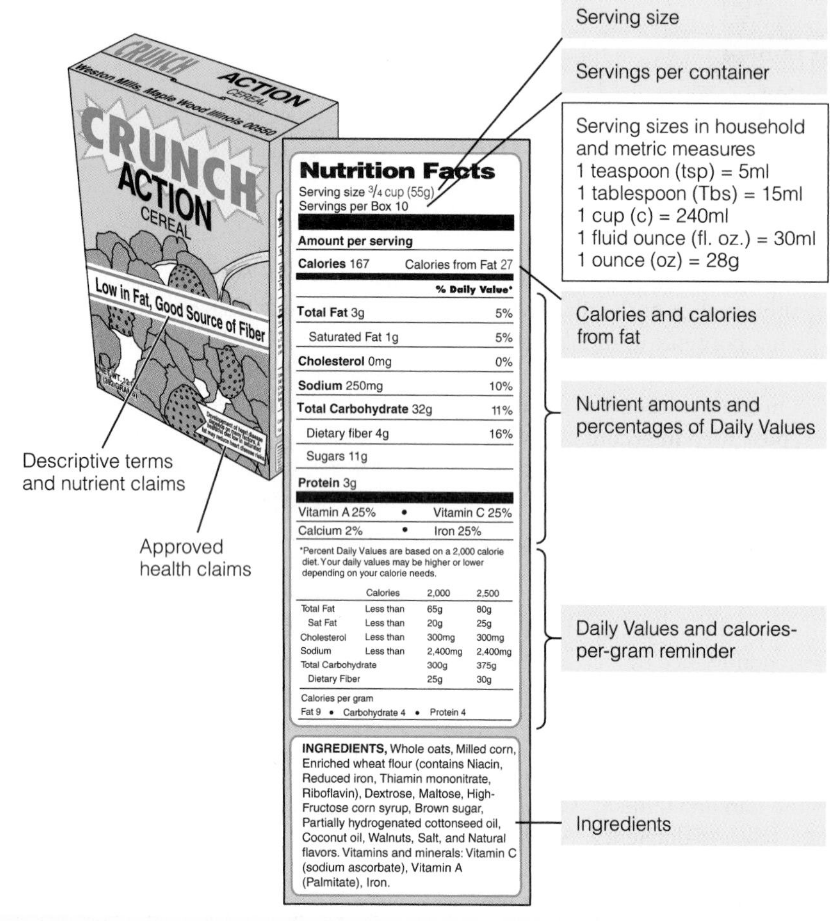

The highlighted items in this section correspond with those of Figure 2-8, which shows the location of the items listed below.

- Serving size. Common household and metric measures to allow comparison of foods within a food category. This is the amount of the food that constitutes a single serving and the portion that contains the nutrient amounts listed. A serving of chips may be 10 chips, so if you eat 50 chips, you will have consumed five times the nutrient amounts listed on the label.
- Servings per container. Number of servings per box, can, package, or other unit.
- Calories/calories from fat. Total food energy per serving, and energy from fat per serving.
- Nutrient amounts and percentages of Daily Values. This section provides the core information concerning these nutrients:
  - *Total fat*. Grams of fat per serving with a breakdown showing grams of *saturated fat* per serving.
  - *Cholesterol*. Milligrams of cholesterol per serving.
  - *Sodium*. Milligrams of sodium per serving.
  - *Total carbohydrate*. Grams of carbohydrate per serving, including starch, fiber, and sugars, with a breakdown showing grams of *dietary fiber* and *sugars*. The sugars include those that occur naturally in the food plus any added during processing.
  - *Protein*. Grams of protein per serving.

In addition, the label must state the contents of these nutrients, expressed as percentages of the Daily Values:

- *Vitamin A*.
- *Vitamin C*.
- *Calcium*.
- *Iron*.

Other nutrients present in significant amounts in the food may also be listed on the label. The percentages of the Daily Values (inside front cover, page C) are given in terms of a person requiring 2,000 calories each day.

- Daily Values and calories-per-gram reminder. This portion lists the Daily Values for a person needing 2,000 or 2,500 calories a day, and a calories-per-gram reminder as a handy reference for label readers.

An often neglected but highly valuable body of information is the list of:

- Ingredients. The product's ingredients must be listed in descending order of predominance by weight.

Knowing how to read an ingredient list puts you many steps ahead of the naive buyer. Consider the ingredient list on an orange drink powder whose first three ingredients are "sugar, citric acid, orange flavor." You can tell that sugar is the chief ingredient. Now consider a canned juice whose ingredient list begins with "water, orange juice concentrate, pineapple juice concentrate." This product is clearly made of **reconstituted** juice. Water is first on the label because it is the main constituent of juice. Sugar is nowhere to be found among the ingredients because sugar has not been added to the product. Sugar occurs naturally in juice, though, so the label does specify sugar grams; details are in Chapter 4.

Now consider a cereal whose entire list contains just one item: "100 percent shredded wheat." No question, this is a whole-grain food with nothing added. Finally, consider a cereal whose first three ingredients are "puffed milled corn, sweeteners (sugars: corn syrup, sucrose, honey, dextrose), salt." If you recognize that sugar, corn syrup, honey,

reconstituted a concentrate that has been brought back to its original strength by the addition of water.

and dextrose are all different versions of sugar (and you will, after Chapter 4), you might guess that this product contains close to half its weight as sugar.

More about Percentages of Daily Values As mentioned, some of the Daily Values are printed on each label in the Nutrition Facts panel. The entire list can be found on the inside cover of this text, page C. The calculations used to determine the "% Daily Value" figures for nutrient contributions from a serving of food are based on a 2,000-calorie diet. For example, if a food contributes 13 milligrams of cholesterol per serving, and the Daily Value is 300 milligrams, then a serving of that food provides about 4 percent of the Daily Value for cholesterol.

The Daily Values are of two types. Some, such as those for fiber, protein, vitamins, and most minerals, are akin to other nutrient intake recommendations. They suggest an intake goal to strive to reach; below that level, some people's needs may go unmet. Other Daily Values, such as those for cholesterol, total fat, saturated fat, and sodium, constitute healthy daily maximums.

Of course, though the Daily Values are based on a 2,000-calorie diet, people's actual calorie intakes vary widely; some people need fewer and some need many more. This makes the Daily Values most useful for comparing one food with another, and less useful as nutrient intake targets for individuals. Still, by examining a food's general nutrient profile, you can determine whether the food contributes "a little" or "a lot" of a nutrient, whether it contributes "more" or "less" than another food, and how well it fits into your overall diet. Chapter 5's Do It! provides some practice in comparing the percentages of the Daily Values for fat, saturated fat, and cholesterol on food labels.

What Food Labels May Include

If a food meets strict criteria, the label may display certain approved claims about the product:

- **Nutrient claims** concerning the product's nutritive value.
- **Health claims** concerning links between nutrients or food constituents and diseases.

One important distinction must be made before proceeding—the difference between labels of regular foods and labels of "functional foods" and other products that bear the words *dietary supplement*. The labels of regular foods are held to strict standards of accuracy in their meaning and constitute reliable sources of information. The labels of foods claiming to be dietary supplements are largely unregulated and can pose a quagmire of misleading, unsubstantiated (but legal) claims that can easily confuse health-conscious consumers. The information that follows applies to ordinary foods only. More about the labeling of dietary supplements and "functional foods" is found in this chapter's Controversy section.

Nutrient Claims on Food Labels The Daily Values are used on labels as an easily interpreted standard of comparison. They serve as the basis for claims that a food is "low" in cholesterol or a "good source" of vitamin A. Table 2-6 on the next page lists terms used on food labels along with their definitions. These definitions can help consumers in choosing foods. For example, any food providing 10 percent or more of the Daily Value for a nutrient is considered to be a good source of the nutrient; a food providing 20 percent is considered "high in" the nutrient. Additionally, as a rule of thumb, any food containing less than 5 percent of a Daily Value provides just a small amount of the nutrient per serving. For nutrients that must be limited, such as fat or sodium, foods providing less than 5 percent may be desirable. For hard-to-get nutrients such as iron or calcium, a reasonable goal might be to choose foods that are "good sources" or "high" in those nutrients several times each day. The vitamin and mineral Snapshot features in Chapters 7 and 8 point out "good sources" of the vitamins and minerals.

Q: Which Packaged Foods and Restaurant Choices Are Best for My Health? When chosen wisely, packages of convenience foods and restaurant meals can support nutritional health, but more often they carry too much fat, saturated fat, and salt and too little fiber, calcium, iron, and other nutrients to qualify as nutritious staple foods in the diet.[1] Luckily, health and nutrient claims on food labels and menus can be used as a sort of short cut to identifying the packaged or restaurant foods that provide the nutrients the body needs while limiting the constituents that are generally oversupplied. The trick lies in learning the meanings of the claims.

Approved Health Claims

Under guidelines established by the Food and Drug Administration (FDA), claims linking nutrients and food constituents to disease states are allowed on food labels in the United States when the claims are well supported by the available scientific evidence. Restaurant menus are held to the same standards set for labels of packaged foods.[2] An exception is made for packages with the words *dietary supplement* on their labels. Such products escape these regulations and therefore may make unsubstantiated claims, although they may not refer specifically to diseases (see this chapter's Controversy section).

Food labels can make statements concerning the following relationships (Table 2-6 defines the terms in bold type):

1. *Calcium and osteoporosis*. A food making this claim must be **high** in calcium.
2. *Sodium and hypertension (high blood pressure)*. The food must be **low sodium.**
3. *Dietary fat and cancer*. The food must be **low fat.**
4. *Dietary saturated fat and cholesterol and coronary heart disease*. The food must be **low saturated fat, low cholesterol,** and **low fat.**
5. *Fiber-containing grain products, fruits, vegetables, and cancer*. The food must be low in fat and have no added fiber. It must be a **good source** of dietary fiber.
6. *Fruits, vegetables, and grain products that contain fiber, especially soluble fiber, and reduced risk of*

coronary heart disease. The food must be **low saturated fat, low fat,** and **low cholesterol.** It must also contain at least 0.6 gram of soluble fiber (explained in Chapter 4) per serving.

7. *Fruits and vegetables and reduced risk of cancer*. The food must be **low fat,** and without added nutrients, it must be a **good source** of fiber, vitamin A, or vitamin C. (Fresh fruits and vegetables may make this claim whether or not they meet the criteria for vitamin A or vitamin C.)
8. *Diets high in oatmeal, oat bran, or soluble fiber from psyllium seed husk and the risk of coronary heart disease*. The food must be **low saturated fat, low cholesterol,** and **low fat;** provide at least 13 grams of oat bran, or 20 grams of oatmeal, or 1.7 grams of soluble fiber from psyllium seed husk.
9. *The vitamin folate and birth defects of the brain and spinal cord (neural tube defects)*. The food must be a **good source** of the vitamin folate and contain no more than 100 percent of the Daily Value for vitamin A or vitamin D.
10. *Sugar alcohols and tooth decay*. Sugar alcohols do not promote tooth decay; frequent between-meal snacks that are high in sugar and starch promote tooth decay.
11. *Soy protein and reduced risk of coronary heart disease*. The food must contain 6.25 grams of soy protein per serving and be **low fat, low saturated fat,** and **low cholesterol.**
12. *Whole grains and reduced risk of heart disease and certain cancers*. Whole grain must make up more than half of the food's ingredients, and the food must be **low fat, low saturated fat,** and **low cholesterol.**
13. *Potassium and reduced risk of hypertension and stroke*. The food must be a good source of potassium and **low sodium, low fat, low saturated fat,** and **low cholesterol.**
14. *Plant sterol or stanol esters and heart disease*. The food must provide the phytochemicals sterol esters or stanol esters in significant amounts and must also be **low saturated fat** and a **good source** of vitamin A, vitamin C, iron, calcium, protein, or fiber.[3] The Controversy section of this chapter presents information about sterol esters and stanol esters and many other phytochemicals.

Table 2•6

DESCRIPTIVE TERMS USED ON FOOD LABELS

Energy Terms

- **low calorie** 40 calories or fewer per serving.
- **reduced calorie** at least 25% lower in calories than a "regular," or reference, food.
- **calorie free** fewer than 5 calories per serving.

Fat Terms (Meat and Poultry Products)

- **extra lean**
  less than 5 g of fat *and*
  less than 2 g of saturated fat *and*
  less than 95 mg of cholesterol per serving.
- **lean**[a]
  less than 10 g of fat *and*
  less than 4 g of saturated fat *and*
  less than 95 mg cholesterol per serving.

Fat and Cholesterol Terms (All Products)

- **cholesterol free**
  less than 2 mg cholesterol *and*
  2 g or less saturated fat per serving.
- **fat free** less than 0.5 g of fat per serving.
- **low cholesterol**
  20 mg or less of cholesterol *and*
  2 g or less saturated fat per serving.
- **low fat** 3 g or less fat per serving.
- **low saturated fat** 1 g or less saturated fat per serving.
- **percent fat free** may be used only if the product meets the definition of *low fat* or *fat free*. Requires disclosure of g fat per 100 g food.
- **reduced** or **less cholesterol**
  at least 25% less cholesterol than a reference food *and*
  2 g or less saturated fat per serving.
- **reduced saturated fat**
  25% or less of saturated fat *and*
  reduced by more than 1 g saturated fat per serving compared with a reference food.
- **saturated fat free**
  less than 0.5 g of saturated fat *and*
  less than 0.5 g of *trans*-fatty acids.

Health claims may say only that a substance "may" or "might" reduce disease risks because science is still accumulating evidence concerning the roles of diet in diseases. Claims must also state that the development of a disease rests on many factors. A permissible health claim might look like this:

> Development of heart disease depends on many factors. A healthful diet low in saturated fat and cholesterol may lower blood cholesterol levels and may reduce the risk of heart disease.

When you choose a food with the word **healthy** as part of its name, you can rely on it to live up to its claim. To qualify, a serving of the food must contain at least 10 percent of at least one of these:

- Vitamin A.
- Vitamin C.
- Iron.
- Calcium.
- Protein.
- Fiber.

Some exceptions to these requirements exist. Fresh fruits and vegetables, and some canned and frozen varieties, can

Table 2•6

DESCRIPTIVE TERMS USED ON FOOD LABELS (continued)

Fiber Terms

- **high fiber** 5 g or more per serving. (Foods making high-fiber claims must fit the definition of low fat, or the level of total fat must appear next to the high-fiber claim.)
- **good source of fiber** 2.5 g to 4.9 g per serving.
- **more** or **added fiber** at least 2.5 g more per serving than a reference food.

Other Terms

- **free, without, no, zero** none or a trivial amount. *Calorie free* means containing fewer than 5 calories per serving; *sugar free* or *fat free* means containing less than half a gram per serving.
- **fresh** raw, unprocessed, or minimally processed with no added preservatives.
- **good source** 10 to 19% of the Daily Value per serving.
- **healthy** low in fat, saturated fat, cholesterol, and sodium and containing at least 10% of the Daily Value for vitamin A, vitamin C, iron, calcium, protein, or fiber.
- **high in** 20% or more of the Daily Value for a given nutrient per serving; synonyms include "rich in" or "excellent source."
- **less, fewer, reduced** containing at least 25% less of a nutrient or calories than a reference food. This may occur naturally or as a result of altering the food. For example, pretzels, which are usually low in fat, can claim to provide less fat than potato chips, a comparable food.
- **light** this descriptor has three meanings on labels:
  1. A serving provides one-third fewer calories or half the fat of the regular product.
  2. A serving of a low-calorie, low-fat food provides half the sodium normally present.
  3. The product is light in color and texture, so long as the label makes this intent clear, as in "light brown sugar."
- **more, extra** at least 10% more of the Daily Value than in a reference food. The nutrient may be added or may occur naturally.

Sodium Terms

- **low sodium** 140 mg or less sodium per serving.
- **sodium free** less than 5 mg per serving.
- **very low sodium** 35 mg or less sodium per serving.

[a]The word *lean* as part of the brand name (as in "Lean Supreme") indicates that the product contains fewer than 10 grams of fat per serving.

be labeled *healthy*, even if a serving falls short of the 10 percent mark for any of these nutrients. These foods support health, regardless.

Foods labeled as *healthy* or those making any kind of a health claim cannot contain any nutrient or food constituent in an amount known to increase disease risk. Specifically, a serving of the product may contain no more than 20 percent of the Daily Value for the following:

- Total fat.
- Saturated fat.
- Cholesterol.
- Sodium.

This means that whole milk, even though it is a rich source of calcium, may not make a claim about osteoporosis because whole milk contains too much saturated fat to qualify. Low-fat and fat-free milks, however, do qualify to bear the calcium and osteoporosis claim (milk's names and fat contents are clarified in the Food Feature of Chapter 5).

Conclusion

Health claims on labels and menus are so carefully controlled that consumers can rely on them instead of worrying about grams, percentages, and other mathematical speed bumps that would slow them down in grocery store aisles and at restaurants. Food manufacturers have now performed much of the math previously required of nutrition-conscious people. Between the honest and accurate numbers and the carefully defined words, consumers who take time to learn and use the lingo can choose among foods with confidence.

Getting a Feel for the Nutrients in Foods

FIGURE 2-9 ILLUSTRATES a playful contrast between two day's meals. "Monday's Meals" were selected following the recommendations of this chapter. "Tuesday's Meals" were chosen more for convenience and familiarity than out of concern for nutrition. The two sets of meals are similar in energy (calories) so that other differences will stand out.

Now, how can a person compare the nutrition that these sets of meals provide? One way is to look up each food in a table of food composition, write down the food's nutrient values, and compare each one to a standard such as the Daily Values, as we've done in Figure 2-9. The computer is a time saver—it performs nutrient calculations with lightning speed. This convenience may make working with paper, pencils, and erasers seem a bit old-fashioned, but computers are rarely available for diners in cafeterias or at fast-food counters where real-life decisions must be made. Those who can "see" the nutrients in their foods can make informed choices before eating meals, while others must wait until they visit their computers to find out how well they did in retrospect. By the time you reach Chapter 11 of this text, you will be ready to test your skill at "seeing" the nutrients in foods in that chapter's Do It! section. This chapter's Do It! provides an alternative method for judging the adequacy of diets that employs the Food Guide Pyramid as its standard.

Figure 2•9

TWO DAYS' MEALS COMPARED WITH FIVE DAILY VALUES

MONDAY'S MEALS

Monday's meals reflect nutrient-dense choices from the Food Guide Pyramid.

Foods	Energy (cal)	Fiber (g)	Total Fat (g)	Saturated Fat (g)	Vitamin C (mg)
Before heading off to class, a student eats breakfast:					
1 c sweetened cold cereal	166	1	—	—	—
1 c 1% low-fat milk	102	—	3	2	2
½ banana (sliced)	52	1	—	—	5
Then goes home for a quick lunch:					
1 turkey sandwich on roll with mayonnaise and mustard	294	1	12	2	—
1 c vegetable juice	46	2	—	—	67
While studying in the afternoon, the student eats a snack:					
4 whole-wheat crackers	80	2	3	—	—
1 oz low-fat cheddar cheese	49	—	2	1	—
1 apple	82	3	—	—	8
That night, the student makes dinner:					
A salad:					
1 c raw spinach leaves, shredded carrots, and sliced mushrooms	29	3	—	—	19
⅓ c garbanzo beans	135	4	2	—	1
5 lg olives and 1 tbs ranch salad dressing	80	1	8	1	—
A main course:					
1 c spaghetti with meat sauce	332	6	12	3	22
½ c green beans	18	2	—	—	6
2 tsp butter	68	—	8	5	—
And for dessert:					
Strawberry shortcake made with:	293	3	8	4	102
1¼ c strawberries					
1 piece spongecake					
2 tbs whipped cream					
Later that evening, the student enjoys a bedtime snack:					
3 graham crackers	89	—	2	—	—
1 c 1% low-fat milk	102	—	3	2	2
Totals:	2,017	29	63	20	234
Daily Values:[a]	2,000	25	65	20	60
Percentage of Daily Values:	101%	116%	97%	100%	390%
Percentage of calories from fat:	30%				

[a]Daily Values based on a 2,000-calorie diet.

© Polara Studios, Inc.

© Polara Studios, Inc.

© Polara Studios, Inc.

© Quest

© Quest

TWO DAYS' MEALS COMPARED WITH FIVE DAILY VALUES (continued)

© Quest

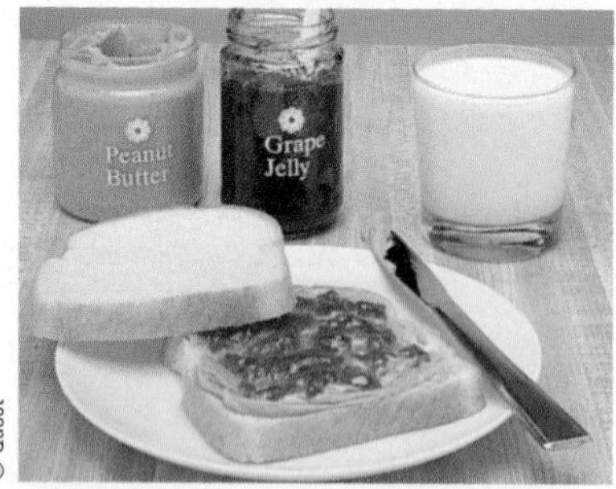

© Quest

© Polara Studios, Inc.

© Quest

© Michael Newman/PhotoEdit

TUESDAY'S MEALS

Tuesday's meals are lower on the nutrient density scale.

Foods	Energy (cal)	Fiber (g)	Total Fat (g)	Saturated Fat (g)	Vitamin C (mg)
Today, the student starts the day with a fast-food breakfast:					
1 c coffee	5	—	—	—	—
1 English muffin with egg, cheese, and bacon	383	2	20	9	1
Between classes, the student returns home for a quick lunch:					
1 peanut butter and jelly sandwich on white bread	350	3	14	3	—
1 c whole milk	150	—	8	5	2
While studying, the student has:					
12 oz diet cola	—	—	—	—	—
Bag of chips (14 chips)	228	2	15	5	13
That night for dinner, the student eats:					
A salad: 1 c lettuce 1 tbs blue cheese dressing	84	1	8	1	3
A main course:					
6 oz steak	343	—	14	5	—
½ baked potato (large)	110	2	—	—	13
1 tbs butter	102	—	12	7	—
1 tbs sour cream	31	—	3	2	—
12 oz diet cola	—	—	—	—	—
And for dessert:					
4 sandwich-type cookies	189	1	8	2	—
Later on, a bedtime snack:					
2 creme-filled snack cakes	214	—	8	2	—
1 c herbal tea	—	—	—	—	—
Totals:	2,189	11	110	41	32
Daily Values:[a]	2,000	25	65	20	60
Percentage of Daily Values:	109%	44%	169%	205%	53%
Percentage of calories from fat:	45%				

[a]Daily Values based on a 2,000-calorie diet.

SCORE YOUR DIET WITH THE FOOD GUIDE PYRAMID

This activity derives from a standard assessment tool, the *Healthy Eating Index*, used to assess diet quality in national food intake studies.* This exercise examines your diet for adherence to the Food Guide Pyramid and for variety. How can you tell if your diet meets the ideals of the Food Guide Pyramid? The number of servings you choose from each food group tells part of the story. More is revealed by the variety of foods selected *within* each group. The rest of the tale unfolds as you learn, in the chapters to come, about the carbohydrates, lipids, protein, vitamins, and minerals in foods.

In this activity, your diet will receive six scores (Form 2-2 on page 55). When added together, these six scores yield a single number. The first five scores measure how well your diet meets the ideals of the Food Guide Pyramid. The sixth score is awarded for variety among your food choices.

In this exercise, 60, not 100, is the highest score attainable, as a reminder that the Food Guide Pyramid gives only a partial accounting of a diet's attributes. The *Dietary Guidelines for Americans* (presented in Chapter 1) concerning moderation in fat, salt, sugar, and alcohol are important, too. Try to keep them in mind as you review your food choices.

Your study begins with the next section. Make several copies of Form 2-1, one to use now and others in case you want to repeat this activity later.

Preparing Your Food Record

Step 1. Record all of the foods you eat and all of the beverages you drink in one typical 24-hour period. Write them on the left-hand side of Form 2-1.

Step 2. As accurately as possible, record the numbers of servings or fractions of servings you obtained from each food group (see Figure 2-4 for serving sizes, pages 36–37). List the numbers in the appropriate columns and total each column. To do this accurately, as you eat the food, make careful note of the amount. Estimate the amount to the nearest ounce, quarter-cup, tablespoon, or other common measure. The Aids to Calculation Appendix (Appendix C) can help with conversion factors. Foods that belong in the tip of the Pyramid—the fats, oils, and sweets—are ignored in this exercise.

In guessing at serving sizes, use these rules of thumb:

- A 3-ounce serving of meat is about the size of the palm of a woman's hand or a deck of cards.
- A standard piece of fruit or potato is the size of a regular (60-watt) lightbulb.
- A 1½-ounce piece of cheese is about the size of a tiny (1½-ounce) "Snickers" candy bar.
- A standard slice of luncheon meat or American-type cheese weighs 1 ounce.
- A pat (1 teaspoon) of a quarter-pound stick of butter or margarine is about as thick as 250 pages of this book (pressed together).

You may have to break down mixed dishes into their ingredients to decide how many servings a food represents. For example, 1 cup of tuna noodle casserole may provide:

- An ounce of tuna (½ serving of meat, poultry, fish, dry beans, eggs, and nuts).
- A half-cup of noodles (1 serving of breads, cereals, rice, and pasta).
- A quarter-cup of a combination of peas, carrots, and onions (½ serving of vegetables).

Errors of up to 20 or 30 percent are expected and tolerated.

Step 3. On Form 2-1, write the suggested total number of servings from each food group for someone whose energy need is similar to yours (see Table 2-3 on page 42).

Scoring against the Food Guide Pyramid

Step 4. Compare your own totals on Form 2-1 with the suggested total number of servings for each food group and score yourself. Then transfer your scores to the pyramid of Form 2-2:

- 10 points for each food group in which you ate all or more of the recommended number of servings.
- 0 points for no servings from a group.
- For values between these extremes, find your score by dividing the number of servings you ate by the recommended number and multiplying by 10.

To help clarify the process, we use a student's diet as an example. Joe is an active young adult who needs 2,800 calories. After analyzing his diet on Form 2-1, Joe finds that he

Food Groups	Joe Needed This Many Servings	Joe Consumed
• Breads, cereals, rice, or pasta:	11	6
• Vegetables:	5	5
• Fruit:	4	0
• Milk, yogurt, or cheese:	2	3
• Meat, poultry, fish, dry beans, eggs, and nuts:	3	5

* This exercise represents six of the ten test categories of the *Healthy Eating Index (HEI)*, a diet assessment tool developed by the U.S. Department of Agriculture's (USDA's) Center for Nutrition Policy and Promotion for use in national studies. The full HEI also tests the diet for adherence to the *Dietary Guidelines for Americans*. For a copy of the entire HEI, contact the USDA (see the Nutrition Resources Appendix), or visit the HEI website at http://www.nalusda.gov/fnic/HEI/HEI.html.

Form 2•1

FOOD RECORD

	NUMBER OF SERVINGS FROM EACH FOOD GROUP				
FOOD/AMOUNT	**Breads, Cereals, Rice, and Pasta**	**Vegetables**	**Fruit**	**Milk, Yogurt, and Cheese**	**Meat, Poultry, Fish, Dry Beans, Eggs, and Nuts**
Breakfast:					
Snack:					
Lunch:					
Supper:					
Snack:					
Your total servings:					
Recommended servings (From Table 2-3, p. 42)					
Your *Score* (Transfer to Form 2-2)					

consumed 6 bread, cereal, rice, or pasta foods. Thus, he ate 6 of 11 recommended servings of breads, cereals, rice, or pasta.

$$(6 \div 11) \times 10 = 5.5 \text{ points}$$

He consumed 5 vegetable servings, right on target, so he gets 10 points for these. He ate no fruit, so here he scores a zero.

From the milk group, he ate more than the recommended 2 servings, but he gets no extra points. He receives the maximum 10 points. The same is true for the meat group: a maximum of 10. After adding the points earned for each of the five food groups, Joe finds that his total points were 35.5 out of a possible 50.

Add up your scores and write the total on the pyramid total line of Form 2-2, part A.

Scoring Variety

The final scoring component, variety, is subjective. It asks you to use your own judgment in your assessment.

Step 5. Look over your food intake record and jot down the approximate number of different types of foods listed. Note that the food types referred to are not the *groups* of foods specified in the Food Guide Pyramid. Here food types mean any of the individual foods that make up

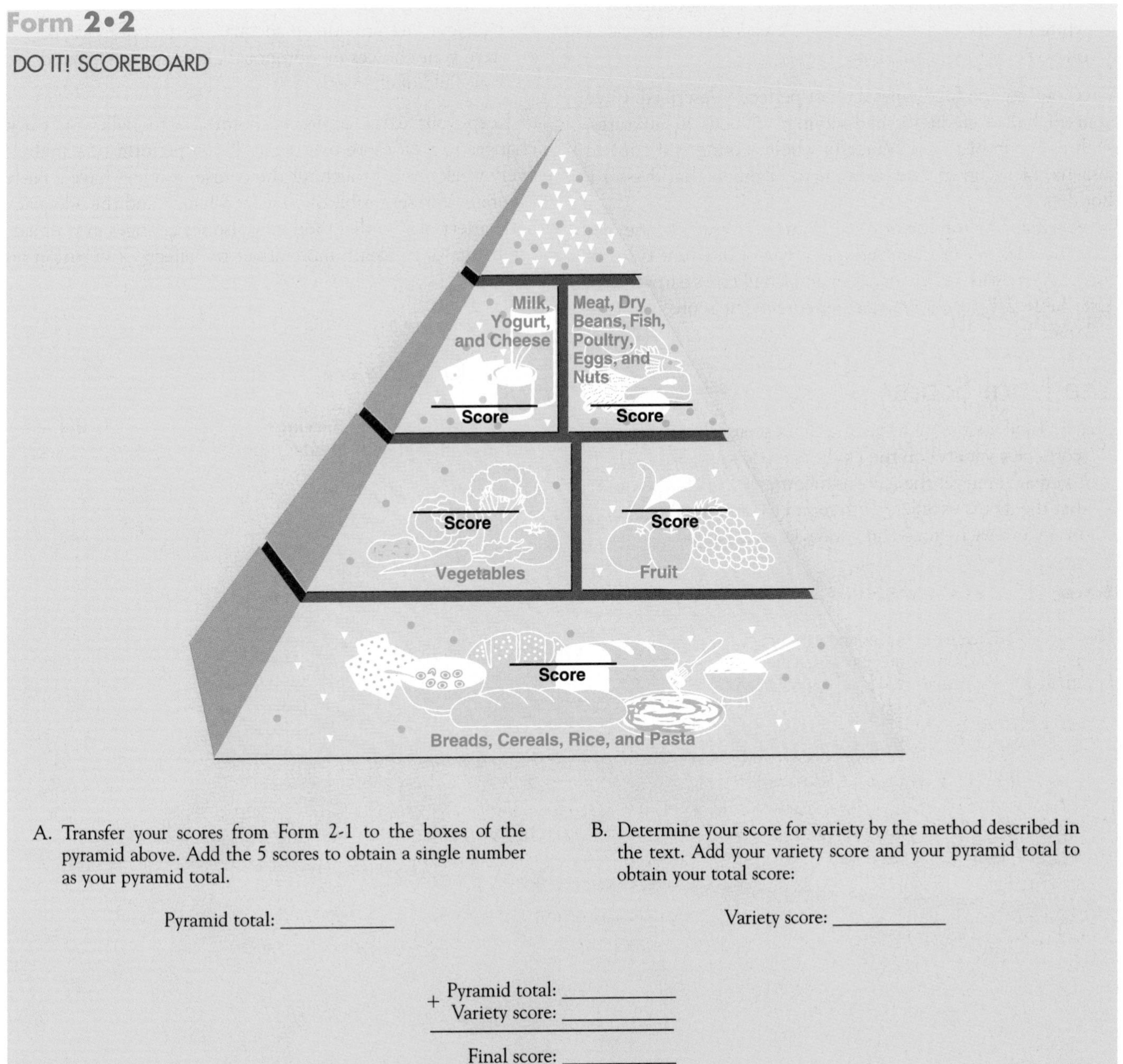

Form 2•2
DO IT! SCOREBOARD
Milk, Yogurt, and Cheese
Score
Meat, Dry Beans, Fish, Poultry, Eggs, and Nuts
Score
Score
Vegetables
Score
Fruit
Score
Breads, Cereals, Rice, and Pasta
A. Transfer your scores from Form 2-1 to the boxes of the pyramid above. Add the 5 scores to obtain a single number as your pyramid total.
Pyramid total: __________
B. Determine your score for variety by the method described in the text. Add your variety score and your pyramid total to obtain your total score:
Variety score: __________
Pyramid total: __________
+ Variety score: __________
Final score: __________

those groups. For example, a person choosing milk and cheese in a day counts them as two types of foods because milk and cheese differ significantly from each other. A person choosing milk and pudding, on the other hand, counts them both as the same food because pudding is made from fluid milk and so is counted as the same type of food as milk. Other examples:

- Beef and lamb differ from each other enough to count as two types of food, but hamburger and beefsteak are counted as the same type.
- Different fruits are counted as different foods, but apples, applesauce, and apple juice are all of one type.
- Whole-wheat bread and enriched white bread are different foods, but enriched white hamburger rolls, enriched white biscuits, and enriched white bread are counted as one type of food, and so on.

Divisions among food types are not perfectly described. Give yourself half a credit for half-servings of foods in mixtures, such as ¼ cup of tomato sauce (a whole serving is ½ cup) in lasagna, or ¼ cup of chopped tomato or onion that dresses a hot dog.

Scoring: The top honor of 10 points is awarded to the diet that includes 9.5 or more servings a day of different types of foods. A day with fewer than four and a half types earns a zero. Use Table 2-7 to obtain a score; enter your score on Form 2-2, part B.

The Final Score

Step 6. Total your score by adding all six scores (five pyramid scores plus variety) on the Do It! Scoreboard (Form 2-2). You may interpret the score as follows: a score of 60 means that the diet is excellent with regard to both number of servings in each group and variety in each group.

Score	Diet Characteristics
60	Excellent choices and variety
50–59	Fairly adequate and varied diet
Less than 50	Diet needs study and improvement

Analysis

Answer these questions:

1. How well did your diet's overall score compare with the perfect score of 60 points?
2. Did you consume the recommended minimum number of servings for each of the five groups of the Food Guide Pyramid?
3. Which groups of foods, if any, are underrepresented in your diet? Sometimes people fail to consume all of the servings from all of the food groups because of allergy or intolerance, but often people simply choose their diets impulsively with inadequate planning for nutrition. If you did not consume the minimum number of servings from each group, list some reasons why.
4. Did your diet provide an adequate variety of foods, or were your choices monotonous? How can you expand your field of choices?

Keep your extra copies of Form 2-1 to help you track changes in your score over time. If you perform this analysis every week or so throughout the course, you will have a fairly accurate representation of your food habits and the adequacy and variety of your diet. You may also see changes in your eating habits as you learn more about the effects of nutrition on health.

Table 2•7

SCORING VARIETY (see text instructions)

If You Ate This Many Servings of Different Types of Foods:	You Receive This Score:
9.5 or more	10
9.0	9
8.5	8
8.0	7
7.5	6
7.0	5
6.5	4
6.0	3
5.5	2
5.0	1
4.5 or fewer	0

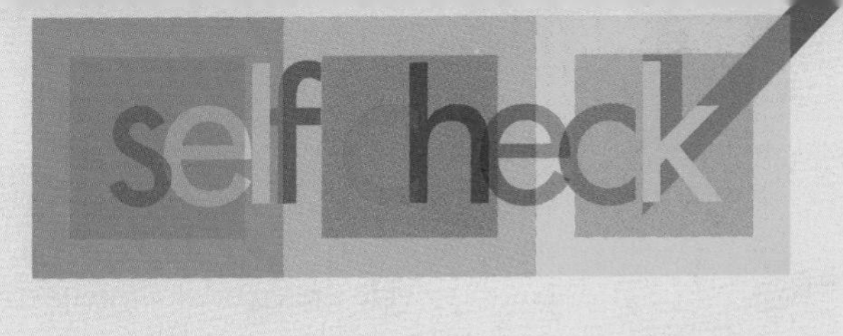

Answers to these Self Check questions are in Appendix G.

1. The nutrient standards in use today include all of the following *except:*
   a. Adequate Intakes (AI)
   b. Daily Minimum Requirements (DMR)
   c. Daily Values (DV)
   d. (a) and (c)
2. The Dietary Reference Intakes were devised for which of the following purposes?
   a. to set nutrient goals for individuals
   b. to suggest upper limits of intakes, above which toxicity is likely
   c. to set average nutrient requirements for use in research
   d. all of the above
3. According to the Food Guide Pyramid, which foods would form the foundation of a healthy diet?
   a. vegetables
   b. breads, cereals, and pasta
   c. fruits
   d. milk, yogurt, and cheese
4. Which of the following are considered *meat alternatives* according to the Food Guide Pyramid?
   a. legumes
   b. nuts
   c. soybean products
   d. all of the above
5. Which of the following values is found on food labels?
   a. Daily Values
   b. Dietary Reference Intakes
   c. Recommended Dietary Allowances
   d. Estimated Average Requirements
6. The energy intake recommendation is centered around the average requirements for each age and sex group. T F
7. The Dietary Reference Intakes (DRI) are for all people, regardless of their medical history. T F
8. People who choose not to eat animals or their products need to find an alternative food guide instead of the Food Guide Pyramid when planning their diets. T F
9. By law, food labels must state as a percentage of the Daily Values the amounts of vitamin C, vitamin A, niacin, and thiamin present in food. T F
10. To be labeled "low fat," a food must contain 3g of fat or less per serving. T F

nutrition on the net

For further study of the topics of this chapter, access these websites and search for the phrases or words in quotation marks:

1. Find updates and quick links to these and other nutrition-related sites at our website: www.wadsworth.com/nutrition
2. Search for "functional foods" at the International Food Information Council: ificinfo.health.org
3. Search for "functional foods" at the Center for Science in the Public Interest: www.cspinet.org
4. Review the Dietary Reference Intakes: www.nap.edu/readingroom or www2.nas.edu/fnb
5. Review nutrient recommendations from the Food and Agriculture Organization and the World Health Organization: www.fao.org and www.who.org
6. Visit the Food Guide Pyramid section (including its ethnic/cultural pyramids) of the U.S. Department of Agriculture: www.nal.usda.gov/fnic
7. Search for "exchange lists" at the American Diabetes Association: www.diabetes.org
8. Search for "diet" and "food labels" at the U.S. Government health information site: www.healthfinder.gov
9. Learn more about food labels from the Food and Drug Administration: www.cfsan.fda.gov or vm.cfsan.fda.gov/label.html
10. Search for "food labels" at the International Food Information Council: ificinfo.health.org
11. Check out the meal plans and recipe suggestions from Meals for You: www.mymenus.com
12. Visit the Healthy Lifestyle section of the American Dietetic Association: www.eatright.org

INTERNET ACTIVITY

The Food Guide Pyramid is a visual guide to the application of the Daily Food Guide. It is based on commonly chosen foods among most Americans. Our food selections, though, are often influenced by our cultural/ethnic origins. To meet the needs of diverse groups, additional pyramids have been created. Explore other pyramids.

1. Go to the website of the American Dietetic Association: http://www.eatright.org
2. Click "Knowledge Center" on the top of page.
3. Click "Nutrition Resources."
4. Under "For Consumers", click "Food Guide Pyramids."
5. Towards the bottom of the page will be a paragraph starting with "For more info...." Click the website address given to access other Pyramids.
6. Scroll to the bottom of the page and click "Ethnic/Cultural Food Guide Pyramids."

Choose a pyramid that is different than the traditional Food Guide Pyramid. (You may want to choose a pyramid for your own ethnic/cultural group.) Identify the site and compare it to the traditional Food Guide Pyramid. How is it the same? How is it different? Could you follow this pyramid?

PHYTOCHEMICALS AND FUNCTIONAL FOODS: WHAT DO THEY PROMISE? WHAT DO THEY DELIVER?

MANY health-conscious people are wondering about the **phytochemicals** these days. What are these mysterious "nonnutrients" in foods? Chapter 1 introduced nonnutrients as compounds in foods, other than the six nutrients, that have biological activity in the body. Phytochemicals are nonnutrients derived from plants (*phyto* means "plant"), and research indicates that some may support health in ways beyond the traditional roles of nutrients. Not all nonnutrients are plant derived. For example, yogurt, a milk product, is believed to support health by way of its bacterial cultures. As a group, foods that contain naturally occurring or added nonnutrients are broadly and imprecisely referred to as **functional foods.**

Exciting media headlines promise remarkable benefits from the phytochemicals and functional foods: "Miracle Tomato Chemical **Lycopene** Prevents Prostate Cancer," "Eat Soybeans and Beat Breast Cancer," "Drink Red Wine for **Flavonoids** to Cure Heart Disease," and so forth. Store shelves are overflowing with new products while consumers struggle to make wise decisions with little information about the potential risks and benefits of phytochemicals.

This Controversy invites you to weigh the evidence concerning both the effectiveness and, perhaps more importantly, the safety of a few selected phytochemicals and functional foods. Phytochemicals number in the thousands, so this discussion is necessarily limited to a few especially interesting ones. Table C2-1 begins by defining some terms that appear in this discussion. Table C2-2 introduces the names, possible physiological effects, and food sources of phytochemicals, not for memorization, but to provide a reference for interpreting media or research reports. Then functional foods take the spotlight, and the Controversy ends with some guidance about making choices.

Table C2•1

PHYTOCHEMICAL AND FUNCTIONAL FOOD TERMS

- **antioxidant** (anti-OX-ih-dant) a compound that protects other compounds from damaging reactions involving oxygen by itself reacting with oxygen (*anti* means "against"; *oxy* means "oxygen"). *Oxidation* is a potentially damaging effect of normal cell chemistry involving oxygen (more in Chapter 5 and Controversy 7).
- **broccoli sprouts** the sprouted seed of *Brassica italica*, or the common broccoli plant; believed to be a functional food by virtue of its high phytochemical content.
- **conjugated linoleic acid (CLA)** a type of fat in butter, milk, and other dairy products believed by some to have biological activity in the body. Not a phytochemical, but a biologically active chemical produced by animals.
- **drug** any substance that when taken into a living organism may modify one or more of its functions.
- **flavonoids** (FLAY-von-oyds) yellow pigments in foods; phytochemicals that may exert physiological effects on the body. *Flavus* means "yellow."
- **flaxseed** the small brown seed of the flax plant; used in baking, cereals, or other foods and valued by industry as a source of linseed oil and fiber.
- **functional foods** a general term for foods with beneficial physiological or psychological effects beyond providing essential nutrients. May also be referred to as *medical foods*, *foods for medical purposes*, or other terms with no legal or scientific definitions. Also defined in Chapter 1.
- **genistein** (GEN-ih-steen) a phytosterol found primarily in soybeans that both mimics and blocks the action of estrogen in the body.
- **lignans** phytochemicals present in flaxseed, but not in flax oil, that are converted to phytosterols by intestinal bacteria and are under study as possible anticancer agents.
- **lutein** (LOO-teen) a plant pigment of yellow hue; a phytochemical believed to play roles in eye functioning and health.
- **lycopene** (LYE-koh-peen) a pigment responsible for the red color of tomatoes and other red-hued vegetables; a phytochemical that may act as an antioxidant in the body.
- **miso** fermented soybean paste used in Japanese cooking. Soy products are considered to be functional foods.
- **organosulfur compounds** a large group of phytochemicals containing the mineral sulfur. Organosulfur phytochemicals are responsible for the pungent flavors and aromas of foods belonging to the onion, leek, chive, shallot, and garlic family and are thought to stimulate cancer defenses in the body.
- **phytochemicals** (FIGH-toe-CHEM-ih-cals) biologically active compounds of plants believed to confer resistance to diseases on the eater, also defined in Chapter 1. *Phyto* means "plant."
- **phytosterols** (FIGH-toe-STER-ols; figh-TOSS-ter-ols) phytochemicals structurally similar to mammalian steroid hormones, such as the female sex hormone estrogen. Phytosterols may or may not mimic hormone activity in the human body.
- **probiotics** consumable products containing live microorganisms in sufficient numbers to alter the bacterial colonies of the body in ways believed to benefit health. A *prebiotic* product is a substance that may not be digestible by the host, such as fiber, but serves as food for probiotic bacteria and thus promotes their growth.
- **soy drink** a milk-like beverage made from soybeans, claimed to be a functional food. Soy drink should be fortified with vitamin A, vitamin D, riboflavin, and calcium to approach the nutritional equivalency of milk. Also called *soy milk*.
- **stanol esters, sterol esters** compounds derived from vegetable oils or wood pulp that lower blood cholesterol in human beings by competing with cholesterol for absorption from the digestive tract. The term *sterol esters* often refers to both stanol and sterol esters.
- **tofu** a white curd made of soybeans, popular in Asian cuisines, and considered to be a functional food.

Table C2•2

A SAMPLING OF PHYTOCHEMICALS—POSSIBLE EFFECTS AND FOOD SOURCES

Name	Possible Effects	Food Sources
Capsaicin	Modulates blood clotting, possibly reducing the risk of fatal clots in heart and artery disease.	Hot peppers
Carotenoids (including beta-carotene, lutein, lycopene, and hundreds of related compounds)[a]	Act as antioxidants; possibly reduce risks of heart disease, age-related eye disease,[b] cancer, and other diseases.	Deeply pigmented fruits and vegetables (apricots, broccoli, cantaloupe, carrots, pumpkin, spinach, sweet potatoes, tomatoes)
Curcumin	May inhibit enzymes that activate carcinogens.	Turmeric, a yellow-colored spice
Flavonoids (including flavones, flavonols, isoflavones, catechin, and others)[c,d]	Act as antioxidants; scavenge carcinogens; bind to nitrates in the stomach, preventing conversion to nitrosamines; inhibit cell proliferation; flavonoids of blueberries may improve memory.	Berries, black tea, celery, chocolate, citrus fruits, green tea, olives, onions, oregano, purple grapes, purple grape juice, soybeans and soy products, vegetables, whole wheat, wine
Indoles	May trigger production of enzymes that block DNA damage from carcinogens; may inhibit estrogen action.	Broccoli and other cruciferous vegetables (brussels sprouts, cabbage, cauliflower), horseradish, mustard greens
Isothiocyanates (including sulforaphane)	Inhibit enzymes that activate carcinogens; trigger production of enzymes that detoxify carcinogens.	Broccoli and other cruciferous vegetables (brussels sprouts, cabbage, cauliflower), horseradish, mustard greens
Lignans[e]	Block estrogen activity in cells, possibly reducing the risk of cancer of the breast, colon, ovaries, and prostate.	Flaxseed, whole grains
Monoterpenes (including limonene)	May trigger enzyme production to detoxify carcinogens; inhibit cancer promotion and cell proliferation.	Citrus fruit peels and oils
Organosulfur compounds (including allicin)	May speed production of carcinogen-destroying enzymes; slow production of carcinogen-activating enzymes.	Chives, garlic, leeks, onions
Phenolic acids[d] (including ellagic acid)	May trigger enzyme production to make carcinogens water soluble, facilitating excretion.	Coffee beans, fruits (apples, blueberries, cherries, grapes, oranges, pears, prunes, strawberries), oats, potatoes, soybeans
Phytic acid	Binds to minerals, preventing free-radical formation, possibly reducing cancer risk.	Whole grains
Phytosterols (genistein and diadzein)	Estrogen inhibition may produce these actions: inhibit cell replication in GI tract; reduce risk of breast, colon, ovarian, prostate, and other estrogen-sensitive cancers; reduce cancer cell survival; may reduce risk of osteoporosis.	Soybeans, soy flour, soy milk, tofu, textured vegetable protein, other legume products
Protease inhibitors	May suppress enzyme production in cancer cells, slowing tumor growth; inhibit hormone binding; inhibit malignant changes in cells.	Broccoli sprouts, potatoes, soybeans and other legumes, soy products
Resveratrol[f]	Offsets artery-damaging effects of high-fat diets.	Red wine, peanuts
Saponins	May interfere with DNA replication, preventing cancer cells from multiplying; stimulate immune response.	Alfalfa sprouts, other sprouts, green vegetables, potatoes, tomatoes
Tannins[d]	May inhibit carcinogen activation and cancer promotion; act as antioxidants.	Black-eyed peas, grapes, lentils, red and white wine, tea

[a]Other carotenoids include alpha-carotene, beta-cryptoxanthin, and zeaxanthin.
[b]The age-related eye disease is macular degeneration.
[c]Other flavonoids of interest include ellagic acid and ferulic acid.
[d]A subset of the larger group *polyphenolic phytochemicals*.
[e]Lignans act as phytosterols, but their food sources are limited.
[f]A member of the chemical group stilbene, which is a subset of the larger group *polyphenolic phytochemicals*.

First, Some Cautions

A little caution is in order with regard to the phytochemicals. First, be aware that foods consist of dozens or hundreds of chemicals, and every chemical may be beneficial, neutral, or harmful to the body. Some may even be mixed: beneficial in some ways and harmful in others. To complicate matters further, some chemicals may exert different effects on different people and when taken in differing doses at different life stages.[1]

Research on phytochemicals is in its infancy and what is current today and written upon these pages may change in a year from now. A reliable truth is that consumption of naturally occurring foods is safe for healthy people, but virtually no safety studies exist to support the taking of any *purified* phytochemical. Consumers desiring the phytochemical benefits observed from consuming food may do themselves harm by taking phytochemical supplements. The conclusion of this Controversy and of every other valid assessment of phytochemicals is that *foods*, not supplements, are the best and safest source of these substances.

What Scientists Say about Phytochemicals

At one time, phytochemicals were viewed as conferring only sensory properties on foods such as taste, aroma, texture, or color. Thank phytochemicals for the burning sensation of hot peppers, the pungent flavor of onions and garlic, the bitter tang of chocolate, the aromatic qualities of herbs, and the deep red color of tomatoes.

Today, scientists recognize that phytochemicals have profound physiological effects on the body, including **antioxidant** activity, the mimicking of hormones, and the alterations of blood constituents in ways believed to be protective against some disease processes. Notably, cancer and heart disease are linked to processes involving oxygen compounds in the body, and antioxidants are thought to oppose these actions.

Evidence of a phytochemical's health effect may begin to surface when scientists notice that a group of people with a diet rich in a particular phytochemical have a low incidence of "disease X." The case for an association is strengthened if the diets of people who have contracted disease X are found to be low in the phytochemical. To follow up, laboratory scientists perform experiments in which animals or cell cultures are exposed to the phytochemical. If scientists detect a likely biological mechanism by which the phytochemical may reduce the incidence of disease X, then researchers deem it plausible that the phytochemical might help to prevent the disease.

Then science demands the answers to such questions as:

- What is the function of the phytochemical in the plant tissues that produce it, and might such a function also occur in the tissues of human beings?
- Can human beings absorb and metabolize enough of the phytochemical to have an effect?
- What safety issues pose concerns—is the phytochemical safe for consumption over many years, or in pregnancy, childhood, or old age?

Currently, the evidence is insufficient to say with certainty whether a particular phytochemical is effective in fighting diseases or whether it is safe to consume in concentrated doses. Research on a few of the likeliest candidates is discussed next.

Whole Foods and Flavonoids

By reputation whole grains, fruits, vegetables, herbs, spices, teas, and red wine, when consumed regularly, are health-promoting foods. An enormous body of epidemiological evidence spanning many countries reveals that deaths from cancer, heart disease, and heart attacks are less common wherever these foods are plentiful in the diet.[2] What these foods all have in common are phytochemicals known as flavonoids.

Because flavonoids impart a bitter taste to foods, manufacturers often refine away the natural flavonoids to please consumers who generally prefer milder flavors.[3] For example, the hearty taste of whole-wheat foods vanishes when whole wheat is refined into white flour by removing the tough brown parts that contain flavonoids (see Chapter 4 for details about refined grains). For white grape juice or white wine, makers remove the red, flavonoid-rich grape skins to lighten the flavor and the color of the product, while greatly reducing its beneficial flavonoid content. For example, one flavonoid of grapes and wine may have anticancer activity, although the flavonoid content of wine many be too low to affect human health.*[4] If preliminary research on the heart-defending and cancer-fighting nature of the flavonoids holds true, the best advice may be to seek out a *variety* of whole flavonoid-rich foods in preference to their more refined counterparts.[5]

Chocolate and Flavonoids

Imagine the delight of study subjects who were asked to consume three ounces of delicious dark (bittersweet) chocolate chips—a tough assignment but required for the sake of science. Upon analyzing blood from the chocolate eaters, researchers found a significant rise in the concentration of a flavonoid antioxidant and a substantial (40 percent) reduction in potentially harmful oxidizing substances in the blood.[6] It is estimated that chocolate may be as powerful as tea or red wine in reducing damaging oxidation in the body.[7] Besides reducing oxidation, chocolate may also function somewhat like aspirin in reducing the tendency of the blood to clot raising the possibility that chocolate phytochemicals may reduce the risk of heart attack and stroke.[8] However, much more evidence is needed to support this idea.

If eating chocolate daily is beginning to sound like a good idea, consider another centuries-old medicinal use of chocolate: to promote weight gain.[9] Three ounces of sweetened chocolate candy contain over 400 calories, a significant portion of most people's daily calorie allowance. For people concerned about body fatness, flavonoids are best obtained from nutrient-dense fruits and vegetables, or from calorie-

*The flavonoid is resveratrol.

free green or black tea. Chocolate is best enjoyed as an occasional treat.

Soybeans and Phytosterols

Compared with people in the West, Asians living in Asia suffer less frequently from osteoporosis (adult bone loss), cancer of the breast and prostate, and heart disease, and from symptoms related to menopause, the time when a woman's estrogen secretion sharply declines and menstruation ceases.[10] When Asians move to the United States and adopt Western diets and habits, however, they experience these problems at the same rate as native Westerners. Soybeans, a legume, and soy products such as **miso, soy drink,** and **tofu**—foods common to many Asian diets—contain **phytosterols.** Soy-containing Asian diets correlate with low rates of cancer, especially of the breast, prostate, and other hormone-sensitive organs, and low rates of heart attack.[11] Research even suggests a role for soy foods against lung cancer (in nonsmokers) and cancer of the thyroid gland.[12] No one yet knows whether the phytosterols of soy are responsible for these effects, but more research is warranted.

We can say with certainty that phytosterols are plant-derived relatives of steroid hormones of the human body. Phytosterols weakly mimic or modulate the effects of the hormones estrogen and progesterone—effects believed to alter the formation and growth of cancers of the breast, prostate, and uterus that grow when exposed to estrogen.

As for symptoms of menopause, phytosterols may alter a woman's monthly hormonal cycle in ways that may reduce her risk of breast cancer, osteoporosis, and the sensation of elevated body temperature known as "hot flashes," common in menopause.[13] Indeed, supplements of phytosterols are often sold to menopausal women as a "natural" alternative to hormone replacement therapy (HRT), the administration of female hormone drugs to prevent heart disease, bone loss, and symptoms of menopause. Research is lacking to support taking phytosterol supplements, however, and some researchers report no benefit at all to women who take them.[14]

Safety concerns often drive women's decision to forgo HRT and take phytosterols instead. While it is true that, like all drugs, HRT involves some health risks, it is still a reliable, effective, controlled medical treatment whereas phytochemical supplements have no proof of efficacy, and their health risks are only now beginning to emerge. While studying one soy phytosterol, **genistein,** researchers were astounded to find that, instead of suppressing cancer growth, genistein triggers the rapid division of breast cancer cells in laboratory cultures and in mice.[15] Also, the female offspring of mice treated with high doses of genistein during pregnancy seem prone to developing cancer of the uterus with an incidence even greater than that of a known cancer-causing drug.*[16]

The paired, opposing findings on phytosterols and cancer should send a red flag of warning against taking phytosterol supplements for relief from menopause or for any other purpose. Until more is known, a safer route to obtaining soy phytosterols may be to include moderate amounts of soy-based foods in the diet. Populations consuming soy foods have low rates of diseases.

Flaxseed and Lignans

Historically, people have used **flaxseed** for relieving constipation or digestive distress. Currently, flaxseed is under study as a functional food because it contains **lignans,** compounds that are converted into biologically active phytosterols by bacteria that normally reside in the intestine. Most of the flaxseed produced today is used by industry: ground for animal feed, pressed for linseed oil to make paint, or milled for fiber to make cloth. Consumers may soon take notice of flaxseed, however, if research bears out what some scientists suspect about its relationship to cancer development.

*The drug is DES, or diethylstilbestrol, once given to pregnant women before the discovery of greatly increased risk of uterine and breast cancer among their daughters.

A review of the literature offered this evidence:[17]

- Compared with rats fed an ordinary chow, rats fed chow high in flaxseed develop fewer cancerous changes and smaller tumors in mammary tissue after exposure to chemicals known to cause cancer.
- Cancerous tumors of the lung diminish in size, and new tumor development is significantly reduced in rats given flaxseed in the diet.
- Studies of population suggest that women who excrete more phytosterols in the urine (an indicator of phytosterol intake from flaxseed and other sources) have lower rates of breast cancer.

Studies of flaxseed in people are lacking, however, and some risks are possible with its use. Flaxseed contains compounds that may interfere with vitamin or mineral absorption and so high daily intakes may cause deficiency diseases; consuming large quantities of flaxseed can cause digestive distress; and workers in flaxseed processing plants have suffered asthma and allergy. While no clear role has been established for flaxseed in the prevention of human cancer, including a spoonful or two of flaxseed in the diet may not be a bad idea. Flaxseed richly supplies linolenic acid, a needed nutrient often lacking from the diet (see Chapter 5).

Tomatoes and Lycopene

People around the world who eat the most tomatoes, say, about five tomato-containing meals per week, are less likely to suffer from cancers of the esophagus, prostate, or stomach than those who avoid tomatoes.[18] Among the candidates for promoting this effect is lycopene, a red pigment with antioxidant activity found in guava, papaya, pink grapefruit, tomatoes (especially cooked tomatoes and tomato products), and watermelon.[19]

Lycopene may inhibit the reproduction of cancer cells.[20] In a case-controlled study, African American women with low lycopene intakes ran a three to four times greater risk of developing cancerous changes of the cervix than similar women with high lycopene intakes.[21] Studies also find an association between increased breast cancer risk and low intakes of lycopene and related compounds.[22] Further, researchers are investigating a tentative link between low lycopene in the blood and elevated incidence of heart disease, heart attack, and stroke.[23] Lycopene may also protect the skin from the carcinogenic effects of sunlight. When exposed to ultraviolet light in the laboratory, the skin loses more than a third of its lycopene concentration.[24]

Although cancer research favors eating foods high in lycopene, not a shred of evidence supports the consumption of purified supplements of lycopene. In fact, a lesson can be learned from experience with lycopene's chemical cousin, a normally beneficial vitamin A relative, beta-carotene. Diets high in beta-carotene-rich foods often correlate with low rates of lung cancer; in studies of smokers, however, concentrated beta-carotene supplements *increase* lung cancer rates (details in Controversy 7). The only safe option is to stick with food sources and avoid concentrated supplements of lycopene (and of beta-carotene) until safety studies are completed.

Garlic and Organosulfur Compounds

Garlic contains **organosulfur compounds** that appear to inhibit cancer development in laboratory animals.[25] These compounds may work by suppressing the formation of certain harmful compounds of nitrogen that damage DNA in animal cells and trigger cancerous changes. This evidence hints that eating garlic *may* affect cancer in human beings.[26]

Garlic also seems promising as a promoter of heart health. Garlic has been reported to temporarily improve measures of blood cholesterol in people whose cholesterol is too high for heart health. Considering the negative results associated with supplements of phytosterols and lycopene, perhaps it will be no surprise to learn that research on garlic *supplements*, such as powder and oil, has been disappointing.[27] No certainty exists about whether large doses of concentrated chemicals from garlic may improve a person's health or injure it.[28]

Many other foods and food constituents are credited with disease-fighting activity. Figure C2-1 provides a sampling.

What Supporters Say about Phytochemical Supplements

Users and sellers of supplements argue that the existing evidence is good enough to recommend that people take supplements of purified phytochemicals. Eager for the potential benefits, they discount the potential for harm and see no need to wait for studies to prove the efficacy and safety. Besides, they say, phytochemicals must be safe because people have been consuming foods containing them for tens of thousands of years. The body is clearly accustomed to handling them in foods, so supplements must be safe as well.

Such lines of thinking raise concerns among scientists. Although the body is equipped to handle phytochemicals in diluted form, mixed with all of the other constituents of natural foods, it is not adapted to phytochemicals in concentrated form. Five points about phytochemicals are important to consider in deciding whether to take phytochemical supplements for health's sake:

1. Phytochemicals can alter body functions, sometimes powerfully, in ways that are only partly understood.
2. Evidence for safety of isolated phytochemicals in human beings is lacking.[29]
3. No regulatory body oversees the safety of phytochemicals sold to consumers. No studies are required to prove that they are safe or effective before marketing them.
4. Phytochemical labels may make unsubstantiated claims about contributing to the body's structure or functioning, even though research supporting such claims is lacking (a later section tells more).
5. The best-known, most effective, and safest sources for phytochemicals are foods, not supplements.

Perhaps, then, it is wise to seek out the richest *food* sources of phytochemicals and consume them often. Maybe it would be even better to choose foods that have been enriched with extra doses of phytochemicals—functional foods. Many functional foods straddle the line between foods and medicines, as the next section makes clear.

Functional Foods—Groceries or Pharmaceuticals?

Today, the fastest-growing trend transforming the American food supply is the proliferation of functional foods—foods that supposedly provide health benefits beyond those of the traditional nutrients.[30] Not too long ago, most of us could agree on what was a food and what was a **drug.** Today, such distinctions are no longer clear.

Cholesterol-Reducing Margarine

Consider some options for lowering blood cholesterol, an indicator of heart disease risk. Replacing butter with ordinary liquid margarine in the diet may gradually lower blood cholesterol by a few percentage points over several months. Cholesterol-lowering medication, on the other hand, takes just weeks to dramatically lower cholesterol by as much as 40 percent—clearly, a drug action. But margarine enhanced with a phytosterol (**sterol esters** or **stanol esters** from vegetable oils) may achieve a sizable reduction in blood cholesterol, say, 10 to 15 percent over a relatively short time.[31] Achieving such results requires consuming the margarine three times a day for weeks.

Clearly this margarine, a food, acts like a drug in the body.[32] Also like drugs, these margarines may have an undesirable side effect: they may lower the blood concentration of a beneficial compound mentioned earlier, beta-carotene.[33]

Yogurt

Yogurt is a special case among functional foods because it contains living *Lactobacillus* or other bacteria that ferment milk into yogurt. Such microorganisms, or **probiotics,** are believed to alter the native bacterial colonies in the body in ways that may reduce diseases.[34] *Lactobacillus* organisms may be useful for relieving diarrhea caused by antibiotic use or infection.[35] Much more research is needed to verify suggestions that probiotic preparations may alleviate lactose intolerance, enhance immune function, protect against digestive tract cancers and ulcers, reduce urinary and vaginal infections in women, and lower blood cholesterol.[36] As information about the potential benefits from probiotics hits the newstands, *Lactobacillus* may soon turn up in all sorts of prepared foods and even snack foods proclaiming themselves to be "functional."

Is Every Food Functional?

Many functional foods occur in nature. A serving of **broccoli sprouts,** for example, provides a concentrated source of a phytochemical associated with cancer prevention. Broccoli itself contains this phytochemical in less impressive but still healthy doses, along with as many as 10,000 other phytochemicals with potential physiological activity.[37] Drinking a half-cup of cranberry juice daily may reduce the incidence of urinary tract infections in women because cranberries contain a phytochemical that dislodges bacteria from the tract.[38] (Conversely, concentrated cranberry tablets can injure the urinary tract, causing kidney stones.[39]) Tomatoes, as mentioned, contain lycopene, along with **lutein** (an antioxidant associated with healthy eye function), vitamin C (an antioxidant vitamin), and many other healthful attributes.[40] In fact, each kind of food possesses its own characteristic array of potentially healthful constituents.

Who can say whether even butter and cheese, foods that contain saturated fats known to be damaging to the heart and arteries, eaten in excess, may qualify as "functional foods" by virtue of their content of **conjugated linoleic acid (CLA)**? CLA is claimed to prevent cancer and heart disease, improve immune function, and nudge body composition toward leanness.[41] To date, there has been little human research on the effects of CLA.

Virtually all foods, even a chocolate bar, have some special value in supporting health.[42] How can a consumer choose foods wisely in the face of all of this? The final section of this Controversy provides some guidance, but first consider some concerns regarding functional foods.

Functional Food and Phytochemical Concerns

Problems exist concerning manufactured functional foods, and they are serious. As is true of concentrated phytochemicals

Figure C2•1

AN ARRAY OF PHYTOCHEMICALS IN A VARIETY OF FRUITS AND VEGETABLES

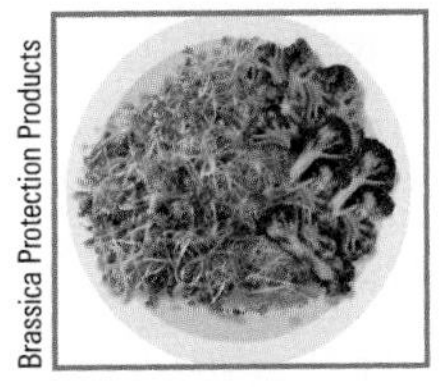

Broccoli sprouts contain an abundance of sulforaphane.

An apple a day—rich in flavonoids.

The phytosterols genistein and diadzein are found in soybeans.

Garlic, with its abundant allicin, may lower blood cholesterol and protect against stomach cancer.

The phytochemicals of grapes, red wine, and peanuts include resveratrol.

The ellagic acid of strawberries is a phenolic acid.

Citrus fruits provide limonene.

The flavonoids in black tea may protect against heart disease, whereas those in green tea may defend against cancer.

Tomatoes are famous for their abundant lycopene.

Flaxseed is the richest source of lignans.

Blueberries are a rich source of flavonoids.

in supplement form, large doses of purified phytochemicals added to foods may produce effects vastly different from those of the phytochemicals in whole foods.

Also, foods sold as functional foods often contain untested medicinal herbs. Such herbs have in recent years caused serious damage to health and even some deaths among consumers (see Table 11-9 of Chapter 11). Research has not matured nearly enough to identify which isolated phytochemicals, medicinal herbs, or other constituents may appropriately be added to foods to improve health, and consuming these substances may be risky in unimaginable ways.[43]

Another problem concerns the wisdom of dousing foods of low nutrient density, such as fried snack chips or candies, with phytochemicals and then labeling them "functional," implying that they will enhance health. In fact, such foods may be more deleterious to health than the original low-nutrient, high calorie product.[44] In defense, the manufacturers counter that if the public demands such foods, manufacturers must produce them or be swept away by the competition.

Until research determines more about functional foods, consumers are on their own to make sure that the products they use are safe and effective. A place to start is by finding answers to the following questions:[45]

- *Does it work?* Well-controlled, peer-reviewed research is generally lacking or inconclusive (see Chapter and Controversy 1 for guidelines to help identify such research).
- *Is it safe?* Check the research for well-controlled safety studies. The active ingredients of functional foods may cause allergies, drug interactions (see the Controversy of Chapter 11), dizziness, and other side effects.
- *Has the Food and Drug Administration (FDA) issued warnings about any of the ingredients?* Check for herbs and phytochemicals deemed harmful by the FDA (www.fda.gov, or call 1-888-INFO-FDA) and reject products that contain them.
- *How much of what does it contain?* Manufacturers are not required to list the quantities of added herbs and phytochemicals on labels, but must list their names. Beware, especially, of combinations of "functional" ingredients.
- *Is it in keeping with the* Dietary Guidelines for Americans? A candy bar or brownie or "smoothie" shake may be fortified with herbs and phytochemicals, but it is still made mostly of sugar and fat.

Regulation and Labeling of Functional Foods

Most consumers seeing the following statements on product labels would read similar meanings into them: "may reduce the risk of heart disease" and "promotes a healthy heart." Most would assume, reasonably, that the FDA approves all such label claims and that they would withstand scientific scrutiny, but those assumptions are false. What most consumers do not know is that currently the two claims are regulated differently. The first statement represents a reliable, well-researched FDA-approved health claim allowable on the labels of ordinary foods, as described by the Consumer Corner section of Chapter 2. The deceptively similar second example represents a far less regulated "structure/function" claim that may appear on functional food labels without scientific evidence. Some members of Congress are proposing to change the rules concerning supplement labels, but for now consumers must learn to distinguish between these claims.[46]

Many novel functional foods and ingredients advertised as health promoting cannot qualify to bear FDA health claims, but any food can bear a structure/function claim. Figure C2-2 depicts a functional food label and identifies the functional ingredients and the structure/function claim, described next.

Structure/Function Claims

Structure/function claims may appear without FDA approval. The foods need not be healthy; the ingredients need not be proved effective or safe; one requirement is that the label must also bear an FDA disclaimer similar to the one shown in Figure C2-2. The structure/function claim must be carefully worded and avoid any mention of a specific disease. Typical claims include "slows aging," "improves memory," and "builds strong bones" (see Table C2-3). To make a disease-related claim such as "prevents osteoporosis," however, the manufacturer would have to comply with all of the rigorous requirements for health claims made on food labels or meet the even stricter safety and efficacy standards applied to drugs.

Of course, some functional foods qualify to make a genuine FDA health claim, just as regular foods can. A functional food made of oats, for example, may state on its label that "A diet high in oats may reduce the risk of

Functional foods currently on the market promise to "enhance mood," "promote relaxation and good karma," "increase alertness," and "improve memory," among other claims.

heart disease." Know that when you see a claim that names a specific disease risk, it means that there is substantial scientific agreement that the food, in the context of a healthy diet, may help protect against the disease.

Health Claims in Advertisements

Even less controlled than claims made on labels are those made in magazine and television advertisements. Claims that would be banned from labels freely appear in these media. For example, the FDA approves no health claims for the phytochemical lycopene in tomatoes, yet magazine advertisements for tomato products make the health claim that lycopene "may help to reduce the risk of prostate and cervical cancer." Printed on a label, such a claim would invoke the FDA regulations for labeling of foods or drugs. Printed elsewhere, it stands under freedom of the press.

The Final Word

In light of all of the evidence for and against phytochemicals and functional foods, should consumers seek out each new phytochemical supplement to obtain potential benefits? Or, for safety's sake, should they steer clear of such supplements and even phytochemical-rich foods until research proves their safety?

Perhaps better than either of these extreme positions is a moderate approach based on the proven advice of the Food Guide Pyramid. People who eat the suggested five or more servings of a variety of fruits and vegetables in a day may cut their risk of many cancers by as much as half. Likewise, eating moderately of soy foods is associated with low cancer rates in populations. As for heart disease, in a recent study of 15,000 male physicians, those consuming at least two and a half servings of dark green, deep yellow, or red vegetables each day had a 23 percent lower risk of heart disease than men who ate less than one serving each day.[47] Choose freely among the fruits and vegetables shown in Figure C2-1 and many others, and rely on a variety of whole foods to supply health-promoting constituents to the diet.

Figure C2•2

FUNCTIONAL FOOD LABELS

A functional food may bear a structure/function claim but may not name specific diseases on the label unless approved by FDA. It must also supply standard label information such as a "Nutrition Facts Panel" and an ingredient list.

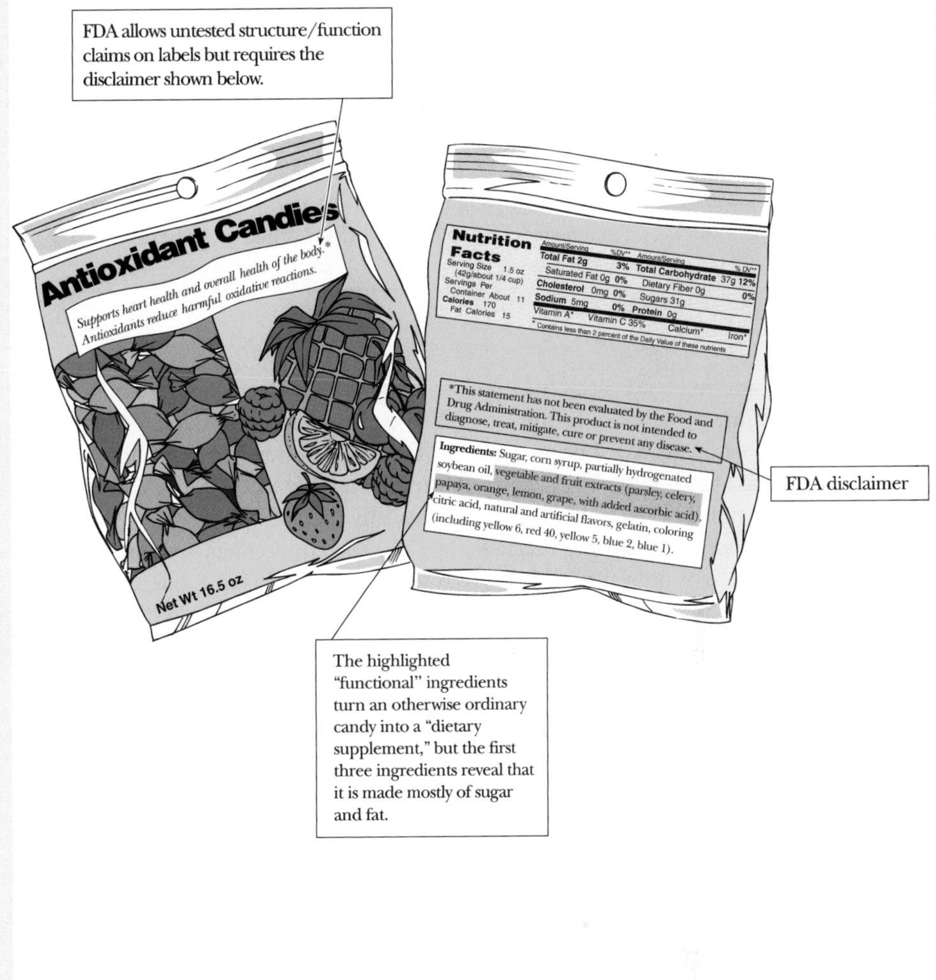

Table C2•3

EXAMPLES OF STRUCTURE/FUNCTION CLAIMS

Note that these claims mention body systems and organs but not diseases. Packages displaying these claims must also bear the FDA disclaimer shown in Figure C2-2.

- Boosts the immune system.
- Builds strong bones.
- Guards against colds.
- Improves memory or concentration.
- Improves mood.
- Lifts your spirits.
- Promotes relaxation.
- Provides a variety of health benefits.
- Slows aging.
- Supports heart health.

Table C2•4

TIPS FOR CONSUMING PHYTOCHEMICALS

- Eat more fruit. The average U.S. diet provides little more than one serving a day. Remember to choose juices and raw, dried , or cooked fruits and vegetables at mealtimes as well as for snacks. Choose dried fruit in place of candy.
- Double the normal serving size of vegetables to 1 cup.
- Use herbs and spices in cooking. Cookbooks offer ways to include parsley, basil, garlic, hot peppers, oregano, and other beneficial seasonings.
- Replace some of the meat in the diet with grains, legumes, and vegetables. Oatmeal, soy meat replacer, or grated carrots mixed with ground meat and seasonings make a luscious, nutritious meat loaf, for example.
- Add grated vegetables to other familiar foods: carrots in chili or meatballs, celery and squash in spaghetti sauce, etc.
- Try a new fruit, vegetable, or whole grain each week. Walk through vegetable aisles and visit farmers' markets. Read recipes. Try tofu, fortified soy drink, or soybeans in cooking.

National campaigns to "Eat 5 a Day" in the United States and "Reach for It" in Canada encourage consumers to eat enough fruits and vegetables daily.

Also, be sure the foods you choose most of the time are as close to the farm as possible. Diets based primarily on unprocessed foods appear to support health better than those founded on highly refined foods—probably because of the abundance of nutrients, fiber, *and* phytochemicals. The actions of other dietary constituents, such as the lipids of fish, also come into play in disease prevention (see Chapter 5).[48] Table C2-4 offers some tips for consuming phytochemicals in the context of a healthy diet.

In the end, remember that the beneficial constituents in foods are widespread among foods. Don't try to single out one phytochemical or functional food for its magical health effect, and don't trust supplements, for they are not trustworthy. Instead, take a no-nonsense approach where your health is concerned: choose a wide variety of fruits and vegetables in the context of an adequate, balanced, and varied diet and receive all of the health benefits offered by a nourishing diet.

chapter 3

Paul Gauguin (1848–1903), *Vahine No Te Vi* (Woman of the Mango), (detail), 1892

The Remarkable Body

Frequently Asked Questions

Contents

genes units of a cell's inheritance, made of the chemical DNA (deoxyribonucleic acid). Each gene directs the making of a protein to do the body's work. (Proteins are described fully in Chapter 6.)

cells the smallest units in which independent life can exist. All living things are single cells or organisms made of cells.

AT THE MOMENT of conception, you received from your mother and father the **genes** that determine how your body works. Many of these genes are thousands of centuries old and have not changed since the Stone Age, when your ancestors walked the earth clad in skins and carrying clubs. Your body has changed very little since then, but you are living with the food, the luxuries, the smog, the additives, and all the other pleasures and problems of the twenty-first century. Faced with a variety of cultures and traditions, you may not have learned any time-tested and proven way of patterning your food intake. There is no guarantee that your diet, haphazardly chosen, will meet the needs of your Stone Age body. Unlike your ancestors, you must learn how your body works and what it needs from food to serve it best.

The Body's Cells

The human body is composed of trillions of **cells,** and none of them knows anything about food. *You* may get hungry for fruit, milk, or bread, but each cell of your body needs nutrients—the vital components of foods. The ways in which the body's cells cooperate to obtain and use nutrients are the subjects of this chapter.

Each of the body's cells is a self-contained, living entity (see Figure 3-1), although each depends on the rest of the body to supply its needs. Among the cells' most basic needs are energy and the oxygen with which to burn it. Cells also need water to maintain the environment in which they live. They need building blocks and control systems. They especially need the nutrients they cannot make for themselves, the essential nutrients first described in Chapter 1, which must be supplied from food. The first principle of diet planning is that the foods we choose must provide energy and the essential nutrients, including water.

Interactions between vitamins and minerals and the genes are addressed in Chapters 7 and 8; other nutrition and gene interactions are addressed in the chapters on pregnancy and disease prevention.

As living things, cells also die off, although at varying rates. Some skin cells and red blood cells must replenish themselves every 10 to 120 days. Cells lining the digestive tract replace themselves every three days. Under ordinary conditions, many muscle

Figure 3•1

A TYPICAL CELL (SIMPLIFIED DIAGRAM)

A membrane encloses each cell's contents.
These fingerlike projections are typical of cells that absorb nutrients in the intestines.
A separate, inner membrane encloses the cell's nucleus.
Inside the nucleus is the hereditary material, which contains the genes. The genes control the inheritance of the cell's characteristics and its day-to-day workings. They are faithfully copied each time the cell duplicates itself.
On these membranes, instructions from the genes are translated into proteins that perform functions in the body.
Many other structures are present. This is a mitochondrion, a structure that takes in nutrients and releases energy from them.

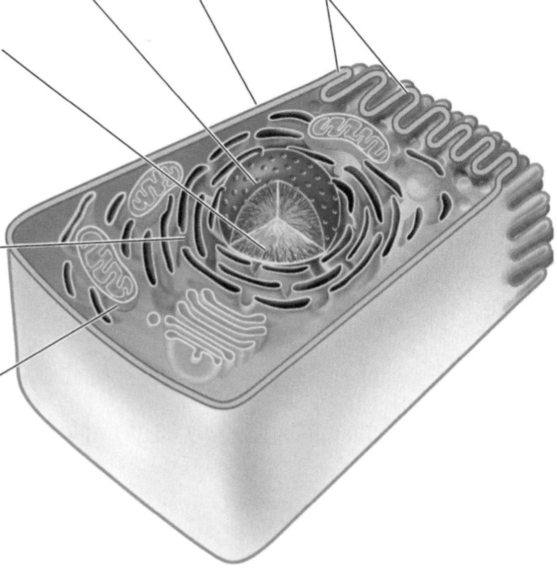

cells reproduce themselves only once every few years. Liver cells have the ability to reproduce quickly and do so whenever repairs to the organ are needed. Certain brain cells do not reproduce at all; if damaged by injury or disease, they are lost forever.

In the human body, every cell works in cooperation with every other cell to support the whole. A cell's genes determine the nature of that work. Each gene is a blueprint that directs the production of a piece of protein machinery, often an **enzyme,** that helps to do the cell's work. Each cell contains a complete set of genes, but different ones are active in different types of cells. For example, in some intestinal cells, the genes for making digestive enzymes are active; in some of the body's **fat cells,** the genes for making enzymes that metabolize fat are active.

Nutrients affect the genes' activities within the cells. For example, several vitamins are known to enter the cell nucleus where they help to direct the making of various proteins. Because proteins often function as working machinery within the body, nutrient and gene interactions have sweeping effects on body functioning, affecting everything from fetal development to the strength of a person's bones or the risk of developing a chronic disease.[1]

Cells are organized into **tissues** that perform specialized tasks. For example, individual muscle cells are joined together to form muscle tissue, which can contract. Tissues, in turn, are grouped together to form whole **organs.** In the organ we call the heart, for example, muscle tissues, nerve tissues, connective tissues, and other types all work together to pump blood. Some body functions are performed by several related organs working together as part of a **body system.** For example, the heart, lungs, and blood vessels cooperate as parts of the cardiovascular system to deliver oxygen to all the body cells. The next few sections present the body systems with special significance to nutrition.

key point *The body's cells need energy, oxygen, and nutrients, including water, to remain healthy and do their work. Genes direct the making of each cell's machinery, including enzymes. Specialized cells are grouped together to form tissues and organs; organs work together in body systems.*

The Body Fluids and the Cardiovascular System

Body fluids supply the tissues continuously with energy, oxygen, and nutrients, including water. The fluids constantly circulate to pick up fresh supplies and deliver wastes to points of disposal. Every cell continuously draws oxygen and nutrients from those fluids and releases carbon dioxide and other waste products into them.

The body's main fluids are the **blood** and **lymph.** Blood travels within the **arteries, veins,** and **capillaries,** as well as within the heart's chambers (see Figure 3-2 on the next page). Lymph travels in separate vessels of its own. Circulating around the cells are other fluids such as the **plasma** of the blood and the fluid surrounding muscle cells (see Figure 3-3 on page 71). Fluid surrounding cells (**extracellular fluid**) is derived from the blood in the capillaries; it squeezes out through the capillary walls and flows around the outsides of cells, permitting exchange of materials. Some of the extracellular fluid returns to the blood by reentering the capillaries. The fluid remaining outside the capillaries forms lymph, which travels around the body by way of lymph vessels. The lymph eventually returns to the bloodstream near the heart where large lymph and blood vessels join. In this way, all cells are served by the cardiovascular system.

The fluid inside cells (**intracellular fluid**) provides a medium in which all cell reactions take place. Its pressure also helps the cells to hold their shape. The intracellular fluid is drawn from the extracellular fluid that bathes the cells.

All the blood circulates to the **lungs** where it picks up oxygen and releases carbon dioxide wastes from the cells, as Figure 3-4 on page 72 shows. Then it returns to the

enzyme a protein that promotes a chemical reaction, described in Chapter 6.

fat cells cells that specialize in the storage of fat and form the fat tissue. Fat cells also produce hormones and enzymes involved in appetite and energy balance (details in Chapter 9).

tissues systems of cells working together to perform specialized tasks. Examples are muscles, nerves, blood, and bone.

organs discrete structural units made of tissues that perform specific jobs. Examples are the heart, liver, and brain.

body system a group of related organs that work together to perform a function. Examples are the circulatory system, respiratory system, and nervous system.

blood the fluid of the cardiovascular system; composed of water, red and white blood cells, other formed particles, nutrients, oxygen, and other constituents.

lymph (LIMF) the fluid that moves from the bloodstream into tissue spaces and then travels in its own vessels, which eventually drain back into the bloodstream.

arteries blood vessels that carry blood containing fresh oxygen supplies from the heart to the tissues (see Figure 3-2 on the next page).

veins blood vessels that carry blood, with the carbon dioxide it has collected, from the tissues back to the heart (see Figure 3-2).

capillaries minute, weblike blood vessels that connect arteries to veins and permit transfer of materials between blood and tissues (see Figures 3-2 and 3-3).

plasma the cell-free fluid part of blood and lymph.

extracellular fluid fluid residing outside the cells that transports materials to and from the cells.

intracellular fluid fluid residing inside the cells that provides the medium for cellular reactions.

lungs the body's organs of gas exchange. Blood circulating through the lungs releases its carbon dioxide and picks up fresh oxygen to carry to the tissues.

Figure 3•2

BLOOD FLOW IN THE CARDIOVASCULAR SYSTEM

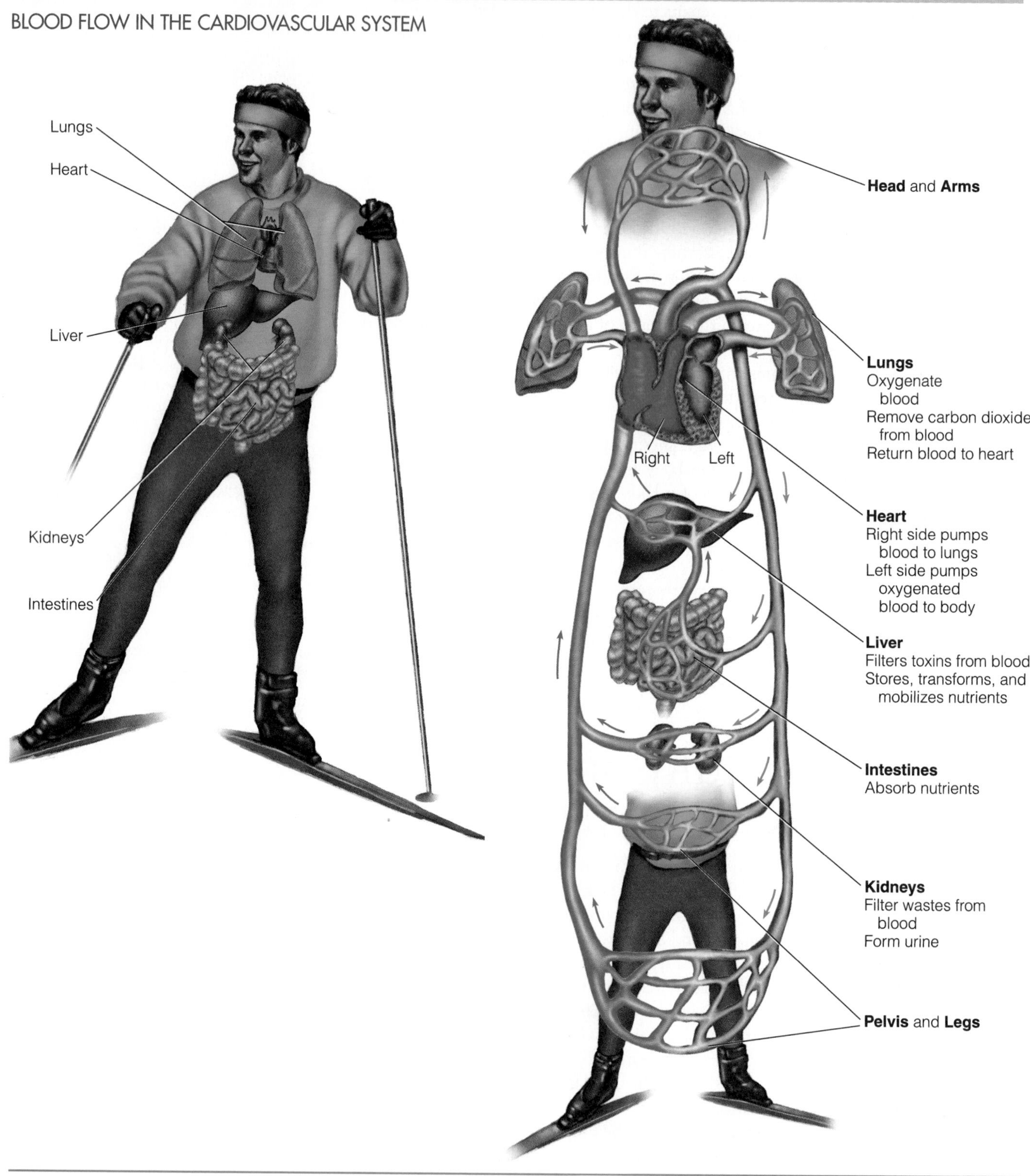

intestine the body's long, tubular organ of digestion and the site of nutrient absorption.

heart, where the pumping heartbeats push the fresh oxygenated blood from the lungs out to all body tissues. As the blood travels through the rest of the cardiovascular system, it delivers materials cells need and picks up their wastes.

As it passes through the digestive system, the blood delivers oxygen to the cells there and picks up most nutrients other than fats from the **intestine** for distribution elsewhere. Lymphatic vessels pick up most fats from the intestine and then transport

Figure 3•3

HOW THE BODY FLUIDS CIRCULATE AROUND CELLS

The upper box shows a tiny portion of tissue with blood flowing through its network or capillaries (greatly enlarged). The lower box illustrates the movement of the extracellular fluid. Exchange of materials also takes place between cell fluid and extracellular fluid.

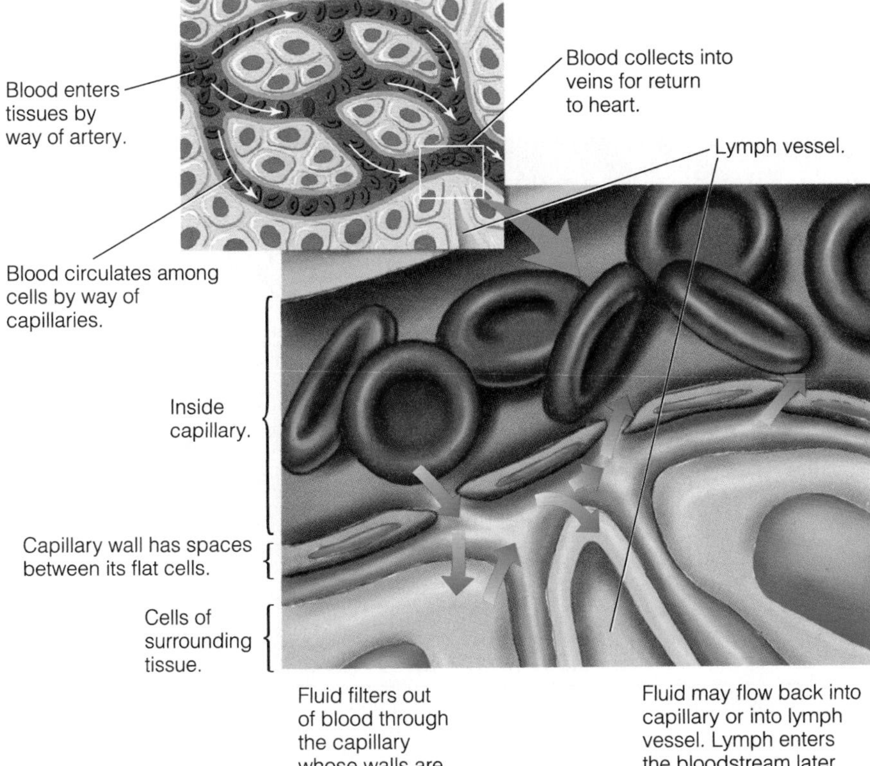

Blood enters tissues by way of artery.
Blood collects into veins for return to heart.
Lymph vessel.
Blood circulates among cells by way of capillaries.
Inside capillary.
Capillary wall has spaces between its flat cells.
Cells of surrounding tissue.
Fluid filters out of blood through the capillary whose walls are made of cells with small spaces between them.
Fluid may flow back into capillary or into lymph vessel. Lymph enters the bloodstream later through a large lymphatic vessel that empties into a large vein.

liver a large, lobed organ that lies just under the ribs. It filters the blood, removes and processes nutrients, manufactures materials for export to other parts of the body, and destroys toxins or stores them to keep them out of the circulation.

kidneys a pair of organs that filter wastes from the blood, make urine, and release it to the bladder for excretion from the body.

them to the blood. All blood leaving the digestive system is routed directly to the **liver,** which has the special task of chemically altering the absorbed materials to make them better suited for use by other tissues. Later, in passing through the **kidneys,** the blood is cleansed of wastes. In summary, the blood is routed as follows (look again at Figure 3-2):

- Heart to tissues to heart to lungs to heart (repeat).

The portion of the blood that flows by the intestine travels from:

- Heart to intestine to liver to heart.

To ensure efficient circulation of fluid to all your cells, you need an ample fluid intake. This means drinking sufficient water to replace the water lost each day. Cardiovascular fitness is essential, too, and constitutes an ongoing project that requires attention to both nutrition and physical activity. Healthy red blood cells also play a role, for they carry oxygen to all the other cells, enabling them to use fuels for energy. Since red blood cells arise, live, and die within about four months, your body replaces them constantly, a manufacturing process that requires many essential nutrients from food. Consequently, the blood is very sensitive to malnutrition and often serves as an indicator of disorders caused by dietary deficiencies or imbalances of vitamins or minerals.

key point *Blood and lymph deliver nutrients to all the body's cells and carry waste materials away from them. Blood also delivers oxygen to cells. The cardiovascular system ensures that these fluids circulate properly among all organs.*

Chapter 8 offers guidelines for water intake.

All the body's cells live in water.

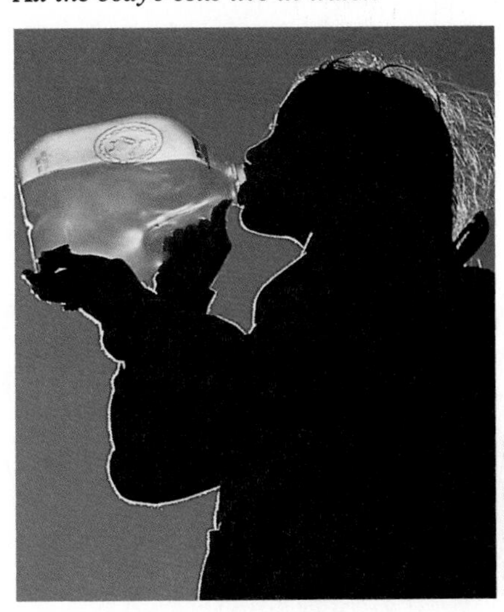

© Alan Oddie/PhotoEdit

hormones chemicals that are secreted by glands into the blood in response to conditions in the body that require regulation. These chemicals serve as messengers, acting on other organs to maintain constant conditions.

pancreas an organ with two main functions. One is an endocrine function—the making of hormones such as insulin, which it releases directly into the blood (*endo* means "into" the blood). The other is an exocrine function—the making of digestive enzymes, which it releases through a duct into the small intestine to assist in digestion (*exo* means "out" into a body cavity or onto the skin surface).

insulin a hormone from the pancreas that helps glucose enter cells from the blood (details in Chapter 4).

glucagon a hormone from the pancreas that stimulates the liver to release glucose into the bloodstream.

Figure 3•4

OXYGEN–CARBON DIOXIDE EXCHANGE IN THE LUNGS

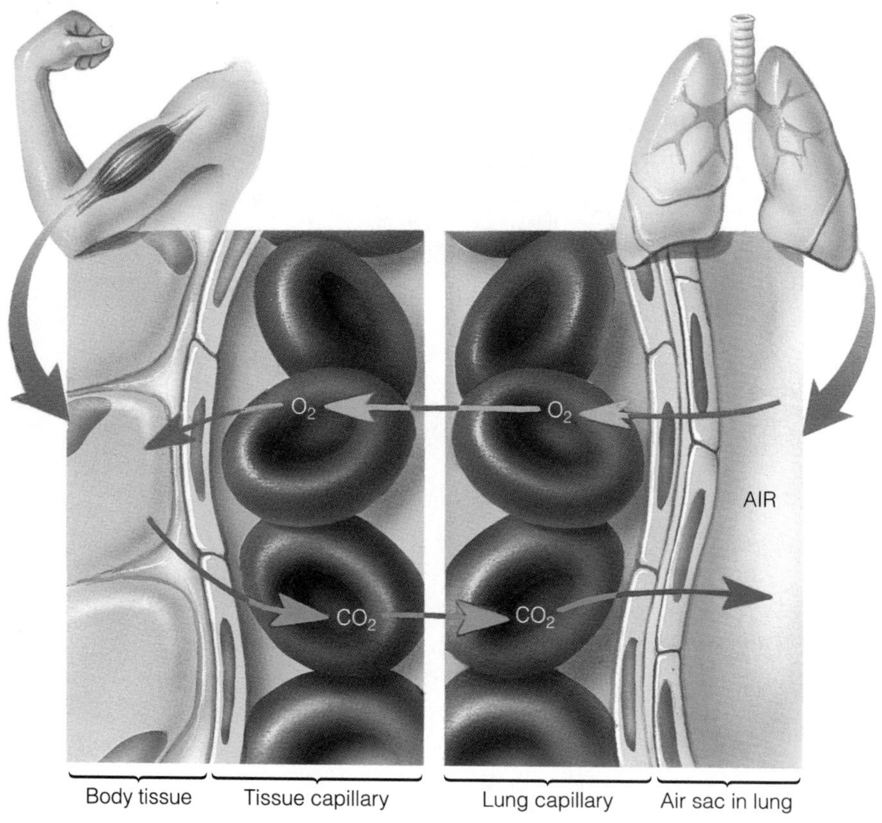

In body tissues, red blood cells give up their oxygen (O_2) and absorb carbon dioxide (CO_2).

In the air sacs of the lungs, the red blood cells give up their load of carbon dioxide (CO_2) and absorb oxygen (O_2) from air to supply to body tissues.

The Hormonal and Nervous Systems

In addition to nutrients, oxygen, and wastes, the blood also carries chemical messengers, **hormones,** from one system of cells to another. Hormones communicate changing conditions that demand responses from the body organs.

What Do Hormones Have to Do with Nutrition?

Hormones are secreted and released directly into the blood by organs known as glands. Glands and hormones abound in the body. Each gland monitors a condition and produces one or more hormones to regulate it. Each hormone acts as a messenger that stimulates various organs to take appropriate actions.

For example, when the **pancreas** (a gland) detects a high concentration of the blood's sugar, glucose, it releases **insulin,** a hormone. Insulin stimulates muscle and other cells to remove glucose from the blood and to store it. When the blood glucose level falls, the pancreas secretes another hormone, **glucagon,** to which the liver responds by releasing into the blood some of the glucose it stored earlier. Thus, normal blood glucose levels are maintained.

Nutrition affects the hormonal system. Fasting, feeding, and exercise alter hormonal balances. People who become very thin have an altered hormonal balance that may make them unable to maintain their bones. People who eat high-fat diets have hormone levels that may make them susceptible to certain cancers.

Hormones also affect nutrition. Along with the nervous system, they regulate hunger and affect appetite. They carry messages to regulate the digestive system, telling the digestive organs what kinds of foods have been eaten and how much of each digestive juice to secrete in response. Hormones regulate the menstrual cycle in women, and they affect the appetite changes many women experience during the cycle and in pregnancy. An altered hormonal state is thought to be at least partly responsible, too, for the loss of appetite that sick people experience. Hormones also regulate the body's reaction to stress, suppressing hunger and the digestion and absorption of nutrients. When there are questions about a person's nutrition or health, the state of that person's hormonal system is often part of the answer.

Glands secrete hormones that act as messengers to help regulate body processes.

How Does the Nervous System Interact with Nutrition?

The body's other major communication system is, of course, the nervous system. With the brain and spinal cord as central controllers, the nervous system receives and integrates information from sensory receptors all over the body—sight, hearing, touch, smell, taste, and others—which communicate to the brain the state of both the outer and inner worlds, including the availability of food and the need to eat. The nervous system also sends instructions to the muscles and glands, telling them what to do.

The nervous system's role in hunger regulation is coordinated by the brain. The sensations of hunger and appetite are perceived by the brain's **cortex,** the thinking, outer layer. Deep inside the brain, the **hypothalamus** (see Figure 3-5 on the next page) monitors many body conditions, including the availability of nutrients and water. To signal hunger, the physiological need for food, the digestive tract sends messages to the hypothalamus by way of hormones and nerves. The signals also stimulate the stomach to intensify its contractions and secretions, causing hunger pangs (and gurgling sounds). When your cerebral cortex becomes conscious of hunger, you want to eat. The conscious mind of the cortex, however, can override such signals and allow a person to choose to delay eating despite hunger or to eat when hunger is absent.

A marvelous adaptation of the human body, the ability to respond to physical danger, involves the workings of both the hormonal and nervous systems. Known as the **fight-or-flight reaction** or the *stress response,* this adaptation is present with only minor variations in all animals, showing how universally important it is to survival. It is a magnificently well-coordinated response. When danger is detected, nerves release **neurotransmitters,** and glands supply the compounds **epinephrine** and **norepinephrine.*** Every organ of the body responds and **metabolism** speeds up. The pupils of the eyes widen so that you can see better; the muscles tense up so that you can jump, run, or struggle with maximum strength; breathing quickens and deepens to provide more oxygen. The heart races to rush the oxygen to the muscles, and the blood pressure rises so that the fuel the muscles need for energy can be delivered efficiently. The liver pours forth glucose from its stores, and the fat cells release fat. The digestive system shuts down to permit all the body's systems to serve the muscles and nerves. With all action systems at peak efficiency, the body can respond with amazing speed and strength to whatever threatens it.

* Strictly speaking, norephinephrine is a neurotransmitter; see Controversy 13.

cortex the outermost layer of something. The brain's cortex is the part of the brain where conscious thought takes place.

hypothalamus (high-poh-THAL-uh-mus) a part of the brain that senses a variety of conditions in the blood, such as temperature, glucose content, salt content, and others. It signals other parts of the brain or body to adjust those conditions when necessary.

fight-or-flight reaction the body's instinctive hormone- and nerve-mediated reaction to danger. Also known as the *stress response*.

neurotransmitters chemicals that are released at the end of a nerve cell when a nerve impulse arrives there. They diffuse across the gap to the next cell and alter the membrane of that second cell to either inhibit or excite it.

epinephrine (EP-ih-NEFF-rin) the major hormone that elicits the stress response.

norepinephrine (NOR-EP-ih-NEFF-rin) a compound related to epinephrine that helps to elicit the stress response.

metabolism the sum of all physical and chemical changes taking place in living cells; including all reactions by which the body obtains and spends the energy from food.

Details about hormones, menstruation, and the bones appear in Controversy 8 and Controversy 9.

microbes bacteria, viruses, or other organisms invisible to the naked eye, some of which cause diseases. Also called *microorganisms*.

Figure 3•5

CUTAWAY SIDE VIEW OF THE BRAIN SHOWING THE HYPOTHALAMUS AND CORTEX

The hypothalamus monitors the body's conditions and sends signals to the brain's thinking portion, the cortex, which decides on actions. The pituitary gland is called the body's master gland, referring to its roles in regulating the activities of other glands and organs of the body.

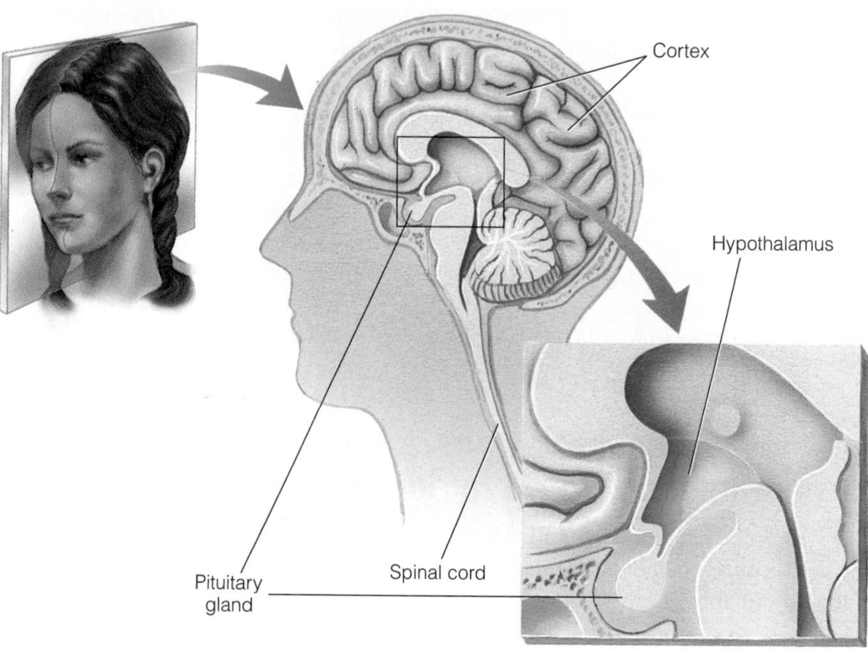

In ancient times, stress usually involved physical danger, and the response to it was violent physical exertion. In the modern world, stress is seldom physical, but the body reacts the same way. What stresses you today may be a checkbook out of control or a teacher who suddenly announces a pop quiz. Under these stresses, you are not supposed to fight or run as your Stone Age ancestor did. You smile at the "enemy" and suppress your fear. But your heart races, you feel it pounding, and hormones still flood your bloodstream with glucose and fat.

Your number-one enemy today is not a saber-toothed tiger prowling outside your cave, but a disease of modern civilization: atherosclerosis. Years of fat and other constituents accumulating in the arteries and stresses that strain the heart often lead to heart attacks, especially when a body accustomed to chronic underexertion experiences sudden high blood pressure. Daily exercise as part of a healthy lifestyle releases pent-up stress and helps to protect the heart against atherosclerosis.

key point *The nervous system joins the hormonal system to regulate body processes through communication among all the organs. Together, the hormonal and nervous systems respond to the need for food, govern the act of eating, regulate digestion, and call for the stress response.*

The Immune System

Many of the body's tissues cooperate to maintain defenses against infection. The skin presents a physical barrier, while the body's cavities (lungs, digestive tract, and others) are lined with membranes that resist penetration by invading **microbes** or unwanted substances. Vitamin and other nutrient deficiencies easily damage these linings, and health-care providers inspect both the skin and the inside of the mouth to detect signs of malnutrition. (Later chapters present details of the signs of deficiencies.) If an **antigen,** or foreign invader, penetrates the body's barriers, the **immune system** rushes in to defend the body against harm.

Of the 100 trillion cells that make up the human body, one in every hundred is a white blood cell. The actions of two types of white blood cells, the phagocytes and the **lymphocytes** known as T-cells and B-cells, are briefly described below:[2]

- **Phagocytes.** These scavenger cells travel throughout the body and are the first to defend body tissues against invaders. When a phagocyte recognizes a foreign particle, such as a bacterium, the phagocyte forms a pocket in its own outer membrane, engulfing the invader. Then, the phagocyte may attack the invader with oxidative chemicals in an "oxidative burst" or may otherwise digest or destroy them. Phagocytes also leave a chemical trail that helps other immune cells to join the defense against infection.
- **T-cells.** T-cells recognize chemical messages from phagocytes and "read" the identity of an invader from the messages. The T-cells then seek out and destroy all foreign particles having the same identity. T-cells defend against fungi, viruses, parasites, some bacteria, and some cancer cells. They also pose a formidable obstacle to a successful organ transplant—the physician must prescribe immunosuppressive drugs following surgery to hold down the T-cells' attack against the "foreign" organ. People suffering from the disease AIDS are rendered defenseless against other diseases because the human immunodeficiency virus (HIV) selectively attacks and destroys their T-cells.
- **B-cells.** B-cells respond rapidly to infection by dividing and releasing invader-fighting proteins, **antibodies,** into the bloodstream. Antibodies travel to the site of the infection and stick to the surface of the foreign particles, killing or inactivating them. The B-cells retain a chemical memory of each invader, and should the encounter recur, the response is swift. This is how immunizations work—a disabled or harmless form of a disease-causing organism is injected into the body so that the B-cells can learn to recognize it. Afterward, should the real, live infectious organism invade, the B-cells quickly release antibodies to destroy it.

Many other categories of white blood cells exist. Chapter 11 describes the roles of nutrition in supporting the body's defense system.

The immune system enables the body to resist disease.

antigen a microbe or substance that is foreign to the body.

immune system a system of tissues and organs that defend the body against antigens, foreign materials that have penetrated the skin or body linings.

lymphocytes (LIM-foe-sites) white blood cells that participate in the immune response; B-cells and T-cells.

phagocytes (FAG-oh-sites) white blood cells that can ingest and destroy antigens. The process by which phagocytes engulf materials is called *phagocytosis*. The Greek word *phagein* means "to eat."

T-cells lymphocytes that attack antigens. *T* stands for the thymus gland of the neck, where the T-cells are stored and matured.

B-cells lymphocytes that produce antibodies. *B* stands for bursa, an organ in the chicken where B-cells were first identified.

antibodies proteins, made by cells of the immune system, that are expressly designed to combine with and inactivate specific antigens.

More about oxidation in Chapter 5 and Controversy 7.

The Digestive System

When your body needs food, your brain and hormones alert your conscious mind to the sensation of hunger. Then, when you eat, your taste buds guide you in judging whether foods are acceptable.

On the surfaces of the taste buds are structures that detect four basic chemical tastes: sweet, sour, bitter, and salty. A fifth taste is sometimes included on this list: the taste of monosodium glutamate, sometimes called *savory* or by its Asian name, *umami* (ooh-MOM-ee). Aroma, texture, temperature, and other flavor elements present can also affect a food's flavor. In fact, the ability to detect a food's aroma is thousands of times more sensitive than the sense of taste. The nose can detect just a few molecules responsible for the aroma of frying bacon, for example, even when they are diluted in several rooms full of air.

Why Do People Like Sugar, Fat, and Salt?

Sweet, salty, and fatty foods seem to be universally desired, but most people have aversions to bitter and sour tastes in isolation (see Figure 3-6).[3] The enjoyment of sugars and fat encourages people to consume ample energy, especially in the form of foods containing sugars, which provide the energy fuel for the brain. Likewise, foods

digestive system the body system composed of organs that break down complex food particles into smaller, absorbable products. The *digestive tract* and *alimentary canal* are names for the tubular organs that extend from the mouth to the anus. The whole system, including the pancreas, liver, and gallbladder, is sometimes called the *gastrointestinal*, or *GI*, system.

digest to break molecules into smaller molecules; a main function of the digestive tract with respect to food.

absorb to take in, as nutrients are taken into the intestinal cells after digestion; the main function of the digestive tract with respect to nutrients.

containing fats provide energy and essential nutrients needed by all body tissues. The pleasure of a salty taste prompts eaters to consume sufficient amounts of two very important minerals—sodium and chloride. The aversion to bitterness discourages consumption of foods containing bitter toxins and also affects people's liking for foods in general. People born with great sensitivity to bitter tastes are apt to avoid foods with slightly bitter flavors, such as turnips and broccoli.[4]

The instinctive liking for sugar, fat, and salt can lead to drastic overeating of these substances. Sugar has become available in pure form only in the last hundred years, so it is relatively new to the human diet. Although fat and salt are much older, today all three substances are being added liberally to foods by manufacturers to tempt us to eat their products.

The preference for sweet, salty, and fatty tastes seems to be inborn and can lead to overconsumption of foods that offer them.

The Digestive Tract

Once you have eaten, your brain and hormones direct the many organs of the **digestive system** to **digest** and **absorb** the complex mixture of chewed and swallowed food. A diagram showing the digestive tract and its associated organs appears in Figure 3-7. The tract itself is a flexible, muscular tube extending from the mouth through the throat, esophagus, stomach, small intestine, large intestine, and rectum to the anus, for a total length of about 26 feet. The human body surrounds this digestive canal. When you swallow something, it still is not inside your body—it is only inside the inner bore of this tube. Only when a nutrient or other substance passes through the wall of the digestive tract does it actually enter the body's tissues. Many things pass into the digestive tract and out again, unabsorbed. A baby playing with beads may swallow one, but the bead will not really enter the body. It will emerge from the digestive tract within a day or two.

The digestive system's job is to digest food to its components and then to absorb the nutrients and some nonnutrients, leaving behind the substances, such as fiber, that are appropriate to excrete. To do this, the system works at two levels: one, mechanical; the other, chemical.

Figure 3•6

THE INNATE PREFERENCE FOR SWEET TASTE

This newborn baby is (a) resting, (b) tasting distilled water, (c) tasting sugar, (d) tasting something sour, and (e) tasting something bitter.

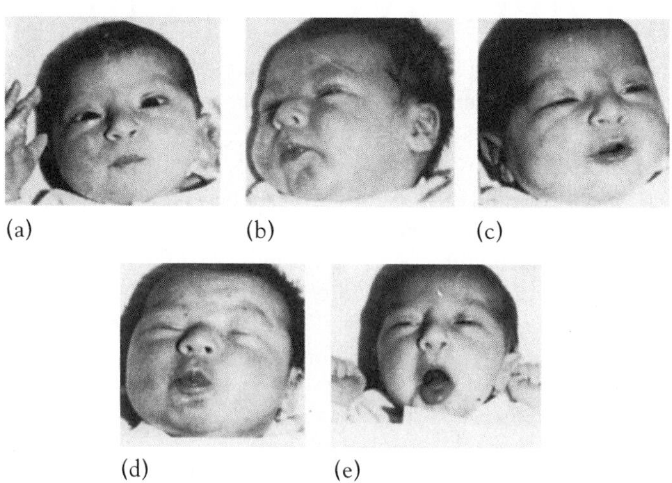

SOURCE: Taste-induced facial expressions of neonate infants from the classic studies of J. E. Steiner, in *Taste and Development: The Genesis of Sweet Preference*, ed. J. M. Weiffenbach, HHS publication no. NIH 77–1068 (Bethesda, Md.: U.S. Department of Health and Human Services, 1977), pp. 173–189, with permission of the author.

The digestive tract is a flexible, muscular tube that digests food and absorbs its nutrients and some nonnutrients.

The Mechanical Aspect of Digestion

The job of mechanical digestion begins in the mouth, where large, solid food pieces such as bites of meat are torn into shreds that can be swallowed without choking. Chewing also adds water in the form of saliva to soften rough or sharp foods, such as

Figure 3•7

THE DIGESTIVE SYSTEM

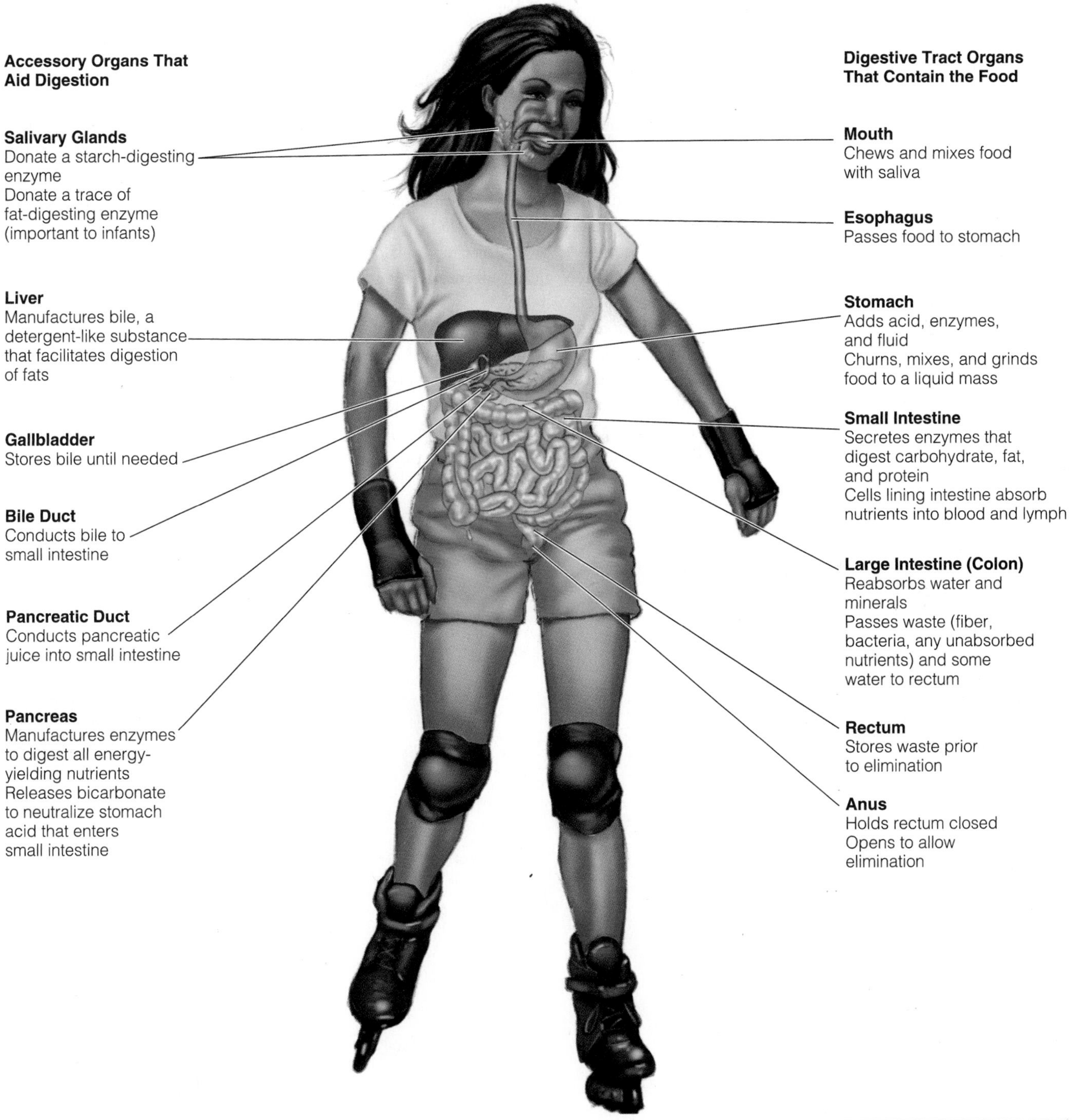

Accessory Organs That Aid Digestion

Salivary Glands
Donate a starch-digesting enzyme
Donate a trace of fat-digesting enzyme (important to infants)

Liver
Manufactures bile, a detergent-like substance that facilitates digestion of fats

Gallbladder
Stores bile until needed

Bile Duct
Conducts bile to small intestine

Pancreatic Duct
Conducts pancreatic juice into small intestine

Pancreas
Manufactures enzymes to digest all energy-yielding nutrients
Releases bicarbonate to neutralize stomach acid that enters small intestine

Digestive Tract Organs That Contain the Food

Mouth
Chews and mixes food with saliva

Esophagus
Passes food to stomach

Stomach
Adds acid, enzymes, and fluid
Churns, mixes, and grinds food to a liquid mass

Small Intestine
Secretes enzymes that digest carbohydrate, fat, and protein
Cells lining intestine absorb nutrients into blood and lymph

Large Intestine (Colon)
Reabsorbs water and minerals
Passes waste (fiber, bacteria, any unabsorbed nutrients) and some water to rectum

Rectum
Stores waste prior to elimination

Anus
Holds rectum closed
Opens to allow elimination

peristalsis (perri-STALL-sis) the wavelike muscular squeezing of the esophagus, stomach, and small intestine that pushes their contents along.

stomach a muscular, elastic, pouchlike organ of the digestive tract that grinds and churns swallowed food and mixes it with acid and enzymes, forming chyme.

sphincter (SFINK-ter) a circular muscle surrounding, and able to close, a body opening.

chyme (KIME) the fluid resulting from the actions of the stomach upon a meal.

pyloric (pye-LORE-ick) **valve** the circular muscle of the lower stomach that regulates the flow of partly digested food into the small intestine. Also called *pyloric sphincter.*

small intestine the 20-foot length of small-diameter intestine, below the stomach and above the large intestine, that is the major site of digestion of food and absorption of nutrients.

fried tortilla chips, to prevent them from tearing the esophagus. Saliva also moistens and coats each bite of food, making it slippery so that it can pass easily down the esophagus.

Nutrients trapped inside indigestible skins, such as seeds, must be liberated by breaking these skins before they can be digested. Chewing bursts open kernels of corn, for example, which would otherwise traverse the tract and exit undigested. Once food has been mashed and moistened for comfortable swallowing, longer chewing times provide no additional advantages to digestion. In fact, for digestion's sake, a relaxed, peaceful attitude during a meal aids digestion much more than chewing for an extended time.

The stomach and intestines then take up the task of liquefying foods through various mashing and squeezing actions. The best known of these actions is **peristalsis,** a series of squeezing waves that start with the tongue's movement during a swallow and pass all the way down the esophagus (see Figure 3-8). The stomach and the intestines also push food through the tract by waves of peristalsis. Besides these actions, the **stomach** holds swallowed food for a while and mashes it into a fine paste; the stomach and intestines also add water so that the paste becomes more fluid as it moves along.

Figure 3-9 shows the muscular stomach. Notice the circular **sphincter** muscle at the base of the esophagus. It squeezes the opening at the entrance to the stomach to narrow it and prevent the stomach's contents from creeping back up the esophagus as the stomach contracts. The stomach stores swallowed food in a lump in its upper portion and squeezes the food little by little to its lower portion. There the food is ground and mixed thoroughly, ensuring that digestive chemicals mix with the entire thick liquid mass, now called **chyme.** Chyme bears no resemblance to the original food. The starches have been partly split, proteins have been uncoiled and clipped, and fat has separated from the mass.

The stomach also acts as a holding tank. The muscular **pyloric valve** at the stomach's lower end (look again at Figure 3-9) controls the exit of the chyme, allowing only a little at a time to be squirted forcefully into the **small intestine.** Within a few hours after a meal, the stomach empties itself by means of these powerful squirts. The small intestine contracts rhythmically to move the contents along its length.

Figure 3•8

PERISTALTIC WAVE PASSING DOWN THE ESOPHAGUS

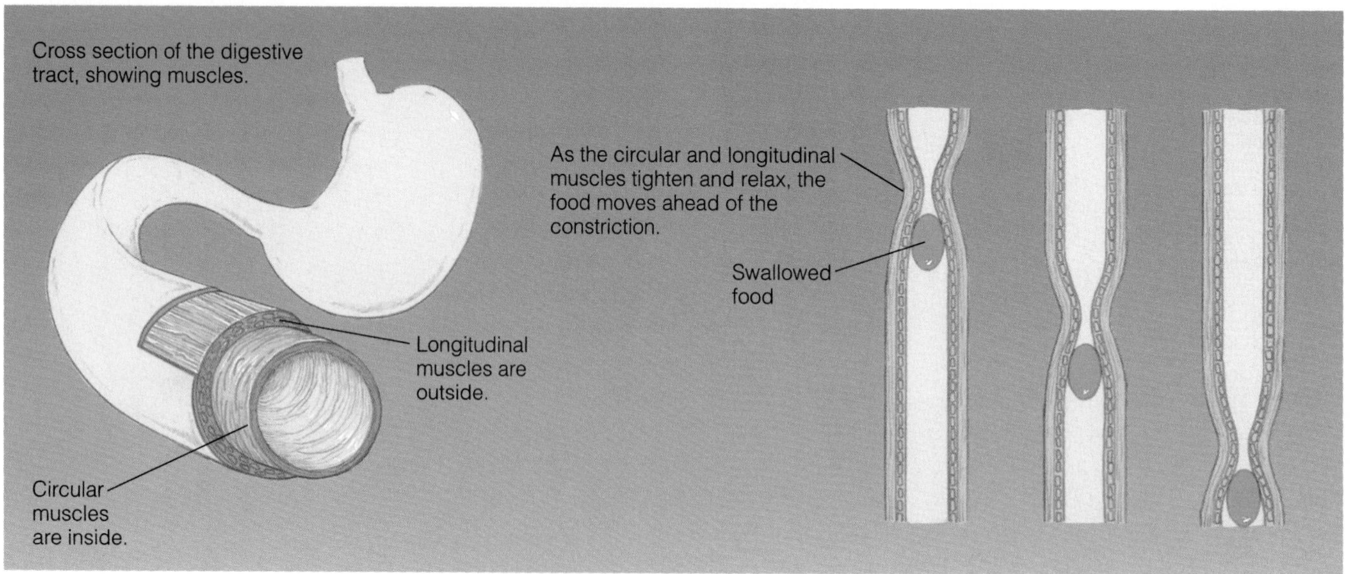

Cross section of the digestive tract, showing muscles.
Longitudinal muscles are outside.
Circular muscles are inside.
As the circular and longitudinal muscles tighten and relax, the food moves ahead of the constriction.
Swallowed food

large intestine the portion of the intestine that completes the absorption process.

colon the large intestine.

feces waste material remaining after digestion and absorption are complete; eventually discharged from the body.

Figure 3•9

THE MUSCULAR STOMACH

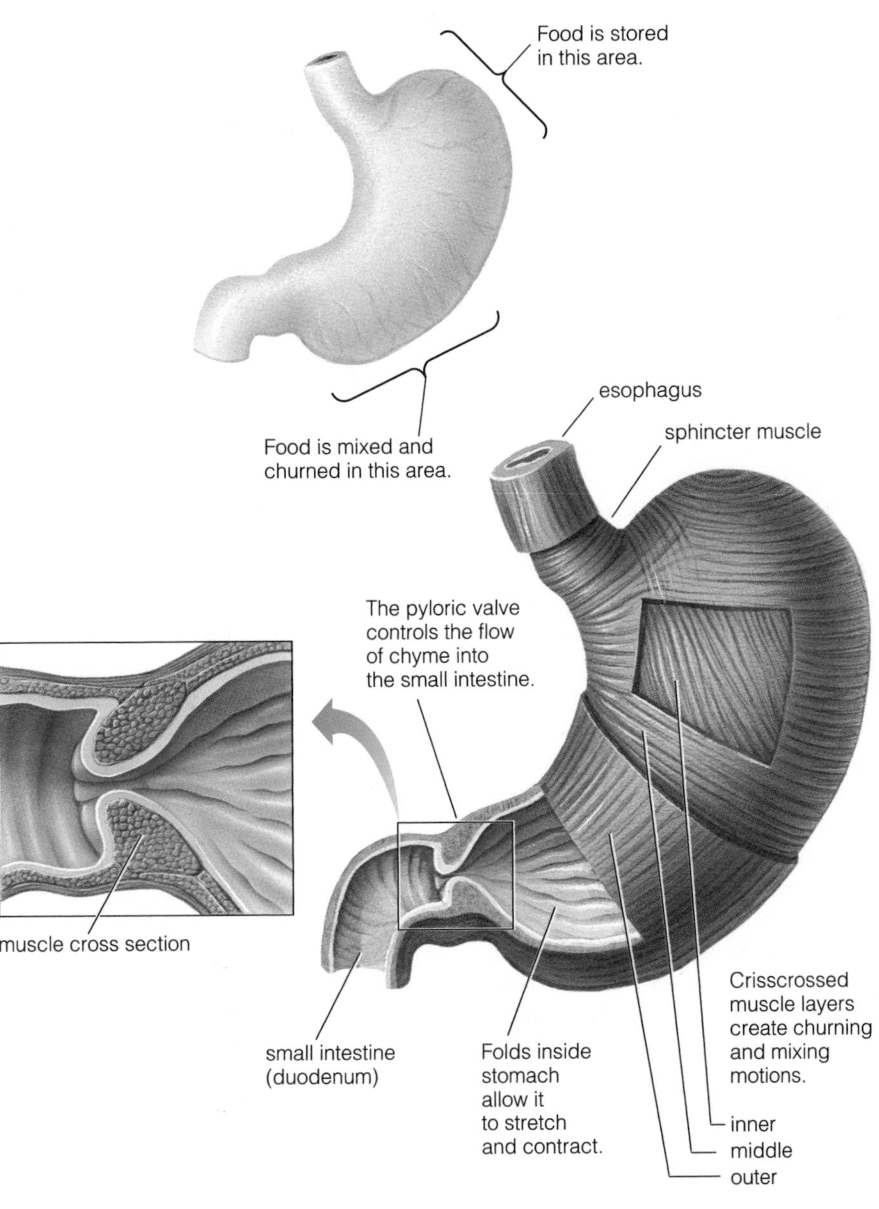

Food is stored in this area.
Food is mixed and churned in this area.
esophagus
sphincter muscle
The pyloric valve controls the flow of chyme into the small intestine.
muscle cross section
small intestine (duodenum)
Folds inside stomach allow it to stretch and contract.
Crisscrossed muscle layers create churning and mixing motions.
inner
middle
outer

By the time the intestinal contents have arrived in the **large intestine** (also called the **colon**), digestion and absorption are nearly complete. The colon's task is mostly to reabsorb the water donated earlier by digestive organs and to absorb minerals, leaving a paste of fiber and other undigested materials, the **feces,** suitable for excretion. The fiber provides bulk against which the muscles of the colon can work. The rectum stores this fecal material to be excreted at intervals. From mouth to rectum, the transit of a meal is accomplished in as short a time as a single day or as long as three days.

Some people wonder whether the digestive tract works best at some hours in the day and whether the timing of meals can affect how a person feels. Timing of meals is important to feeling well, not because the digestive tract is unable to digest food at certain times, but because the body requires nutrients to be replenished every few hours. Digestion is virtually continuous, being limited only during sleep and exercise. For some people, eating late may interfere with normal sleep. As for exercise, it is best pursued a few hours after eating because digestion can inhibit physical work (see Chapter 10 for details).

gastric juice the digestive secretion of the stomach.

pH a measure of acidity on a point scale. A solution with a pH of 1 is a strong acid; a solution with a pH of 7 is neutral; a solution with a pH of 14 is a strong base.

mucus (MYOO-cus) a slippery coating of the digestive tract lining (and other body linings) that protects the cells from exposure to digestive juices (and other destructive agents). The adjective form is *mucous* (same pronunciation). The digestive tract lining is a *mucous membrane*.

emulsifier (ee-MULL-sih-fire) a compound with both water-soluble and fat-soluble portions that can attract fats and oils into water to form an emulsion.

bile a compound made by the liver, stored in the gallbladder, and released into the small intestine when needed. It emulsifies fats and oils to ready them for enzymatic digestion (described in Chapter 5).

pancreatic juice fluid secreted by the pancreas that contains both enzymes to digest carbohydrate, fat, and protein and sodium bicarbonate, a neutralizing agent.

bicarbonate a common alkaline chemical; a secretion of the pancreas; also, the active ingredient of baking soda.

Alcohol needs no assistance from digestive juices to ready it for absorption; its handling by the body is described in this chapter's Controversy section.

key point *The digestive tract moves food through its various processing chambers by mechanical means. The mechanical actions include chewing, mixing by the stomach, adding fluid, and moving the tract's contents by peristalsis. After digestion and absorption, wastes are excreted.*

The Chemical Aspect of Digestion

Several organs of the digestive system secrete special digestive juices that perform the complex chemical processes of digestion. Digestive juices contain enzymes that break nutrients down into their component parts. The digestive organs that release digestive juices are the salivary glands, the stomach, the pancreas, the liver, and the small intestine. Their secretions are listed in Figure 3-7 (on page 77).

How Do "Digestive Juices" Work? Digestion begins in the mouth. An enzyme in saliva starts rapidly breaking down starch, and another enzyme initiates a little digestion of fat, especially the digestion of milk fat, important in infants. Saliva also helps maintain the health of the teeth in two ways: by washing away food particles that would otherwise foster decay and by neutralizing decay-promoting acids produced by bacteria in the mouth.

In the stomach, protein digestion begins. Cells in the stomach release **gastric juice,** a mixture of water, enzymes, and hydrochloric acid. A strong acid is needed to activate a protein-digesting enzyme and to initiate digestion of protein. As you might guess from the presence of acid and enzymes, protein digestion is the stomach's main function. The strength of an acid solution is expressed as its **pH.** As Figure 3-10 demonstrates, saliva is only weakly acidic, while the stomach's gastric juice is much more strongly acidic.

Upon learning of the powerful digestive juices and enzymes within the digestive tract, students often wonder how the tract's own cellular lining escapes being digested along with the food. The answer: Specialized cells secrete a thick, viscous substance known as **mucus,** which coats and protects the digestive tract lining.

In the small intestine, the digestive process gets under way in earnest. The small intestine is *the* organ of digestion and absorption, and it finishes what the mouth and stomach have started. The small intestine works with the precision of a laboratory chemist. As the thoroughly liquefied and partially digested nutrient mixture arrives there, hormonal messengers signal the gallbladder to contract and to squirt the right amount of the **emulsifier, bile,** into the intestine. Other hormones notify the pancreas to release **pancreatic juice** containing the alkaline compound **bicarbonate** in amounts precisely adjusted to neutralize the stomach acid that has reached the small intestine. All these actions alter the intestinal environment to perfectly support the work of the digestive enzymes.

Meanwhile, as the pancreatic and intestinal enzymes act on the chemical bonds that hold the large nutrients together, smaller and smaller pieces are released into the intestinal fluids. The cells of the intestinal wall also hold some digestive enzymes on their surfaces; these enzymes perform last-minute breakdown reactions required before nutrients can be absorbed. Finally, the digestive process releases pieces small enough for the cells to absorb and use. Digestion and absorption of carbohydrate, fat, and protein are essentially complete by the time the intestinal contents enter the colon. Water, fiber, and some minerals, however, remain in the tract. Table 3-1 on page 82 provides a summary of all the processes involved.

key point *Chemical digestion begins in the mouth, where food is mixed with an enzyme in saliva that acts on carbohydrates. Digestion continues in the stomach, where stomach enzymes and acid break down protein. Digestion then continues in the small intestine; there the liver and gallbladder contribute bile that emulsifies fat, and the pancreas and small intestine donate enzymes that continue digestion so that absorption can occur.*

Are Some Food Combinations More Easily Digested Than Others? People sometimes wonder if the digestive tract has trouble digesting certain foods in combination—for example, fruit and meat. Proponents of fad "food combining" diets claim that the digestive tract cannot perform certain digestive tasks at the same time, but this is a gross underestimation of the tract's capabilities. The digestive system can adjust to whatever mixture of foods is presented to it. The truth is that all foods, regardless of identity, are broken down by enzymes into the basic molecules that make them up.

Scientists who study digestion suggest that the tract analyzes the diet's nutrient contents and delivers juice and enzymes appropriate for digesting those nutrients. The pancreas is especially sensitive in this regard and has been observed to adjust its output of enzymes to digest carbohydrate, fat, or protein to an amazing degree. The pancreas of a person who suddenly consumes a meal unusually high in carbohydrate, for example, would begin increasing its output of carbohydrate-digesting enzymes within 24 hours, while reducing outputs of other types. This sensitive mechanism ensures that foods of all types are used fully by the body. The next section reviews the major processes of digestion by showing how the nutrients in a mixture of foods are handled.

The healthy digestive system is capable of adjusting to almost any diet and can handle any combination of foods with ease.

If "I Am What I Eat," Then How Does a Sandwich Become "Me"?

The process of rendering foods into nutrients and absorbing them into the body fluids is remarkably efficient. Within about 24 to 48 hours of eating, a healthy body digests and absorbs about 90 percent of the carbohydrate, fat, and protein in a meal. Here, we follow a peanut butter and banana sandwich on whole-wheat, sesame seed bread through the tract.

In the Mouth In each bite, food components are crushed, mashed, and mixed with saliva by the teeth and the tongue. The sesame seeds are crushed and torn open by the teeth, which break through the indigestible fiber coating so that digestive enzymes can reach the nutrients inside the seeds. The peanut butter is the "extra crunchy" type, but the teeth grind the chunks to a paste before the bite is swallowed. The carbohydrate-digesting enzyme of saliva begins to break down the starches of the bread, banana, and peanut butter to sugars. Each swallow triggers a peristaltic wave that travels the length of the esophagus and carries one bite of food to the stomach.

In the Stomach The stomach collects bite after swallowed bite in its upper storage area, where starch continues to be digested until the gastric juice mixes with the salivary enzymes and halts their action. Small portions of the mashed sandwich are pushed into the digesting area of the stomach where gastric juice mixes with the mass. Acid in gastric juice unwinds proteins from the bread, seeds, and peanut butter; then, an enzyme clips the protein strands into pieces. The sandwich has now become chyme. The watery carbohydrate- and protein-rich part of the chyme enters the small intestine first; a layer of fat follows closely behind.

In the Small Intestine Some of the sweet sugars in the banana require so little digesting that they begin to cross the linings of the small intestine immediately on contact. Nearby, the liver donates bile through a duct into the small intestine. The bile blends the fat from the peanut butter and seeds with the watery enzyme-containing digestive fluids. The nearby pancreas squirts enzymes into the

Figure 3•10

PH VALUES OF DIGESTIVE JUICES AND OTHER COMMON FLUIDS

A substance's acidity or alkalinity is measured in pH units. Each step down the scale indicates a tenfold increase in concentration of hydrogen particles, which determine acidity. For example, a pH of 2 is 1,000 times stronger than a pH of 5.

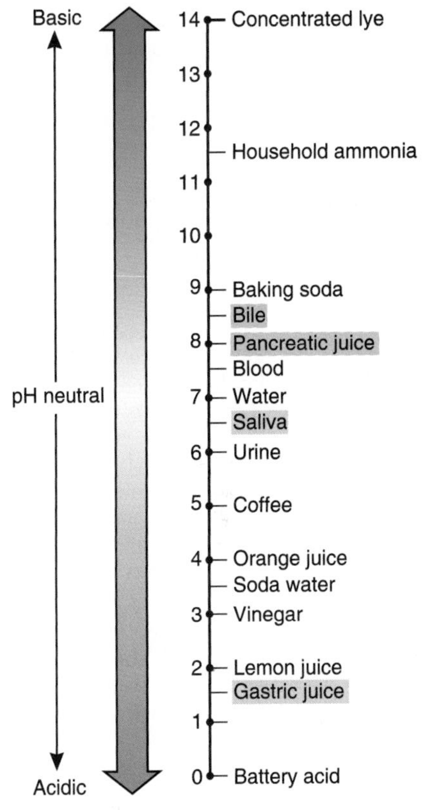

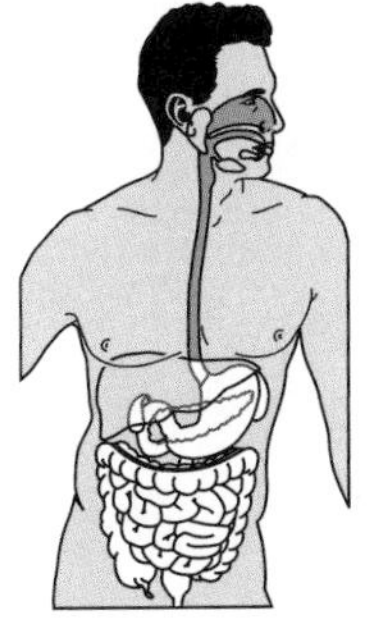

Time in mouth, less than a minute.

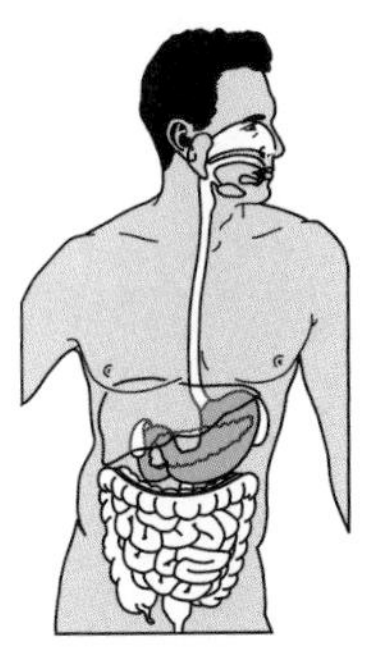

Time in stomach, about 1–2 hours.

Table 3•1

SUMMARY OF CHEMICAL DIGESTION

	Mouth	Stomach	Small Intestine, Pancreas, Liver, and Gallbladder	Large Intestine (Colon)
Sugar and Starch	The salivary glands secrete saliva to moisten and lubricate food; chewing crushes and mixes it with a salivary enzyme that initiates starch digestion.	Digestion of starch continues while food remains in the upper storage area of the stomach. In the lower digesting area of the stomach, hydrochloric acid and an enzyme of the stomach's juices halt starch digestion.	The pancreas produces a starch-digesting enzyme and releases it into the small intestine. Cells in the intestinal lining possess enzymes on their surfaces that break sugars and starch fragments into simple sugars, which then are absorbed.	Undigested carbohydrates reach the colon and are partly broken down by intestinal bacteria.
Fiber	The teeth crush fiber and mix it with saliva to moisten it for swallowing.	No action.	Fiber binds cholesterol and some minerals.	Most fiber excreted with feces; some fiber digested by bacteria in colon.
Fat	Fat-rich foods are mixed with saliva. The tongue produces traces of a fat-digesting enzyme that accomplishes some breakdown, especially of milk fats. The enzyme is stable at low pH and is important to digestion in nursing infants.	Fat tends to rise from the watery stomach fluid and foods and float on top of the mixture. Only a small amount of fat is digested. Fat is last to leave the stomach.	The liver secretes bile; the gallbladder stores it and releases it into the small intestine. Bile emulsifies the fat and readies it for enzyme action. The pancreas produces fat-digesting enzymes and releases them into the small intestine to split fats into their component parts (primarily fatty acids), which then are absorbed.	Some fatty materials escape absorption and are carried out of the body with other wastes.
Protein	Chewing crushes and softens protein-rich foods and mixes them with saliva.	Stomach acid works to uncoil protein strands and to activate the stomach's protein-digesting enzyme. Then the enzyme breaks the protein strands into smaller fragments.	Enzymes of the small intestine and pancreas split protein fragments into smaller fragments or free amino acids. Enzymes on the cells of the intestinal lining break some protein fragments into free amino acids, which then are absorbed. Some protein fragments are also absorbed.	The large intestine carries undigested protein residue out of the body. Normally, almost all food protein is digested and absorbed.
Water	The mouth donates watery, enzyme-containing saliva.	The stomach donates acidic, watery, enzyme-containing gastric juice.	The liver donates a watery juice containing bile. The pancreas and small intestine add watery, enzyme-containing juices; pancreatic juice is also alkaline.	The large intestine reabsorbs water and some minerals.

small intestine to break down the fat, protein, and starch in the chemical soup that just an hour ago was a sandwich. The cells of the small intestine itself produce enzymes to complete these processes. As the enzymes do their work, smaller and smaller chemical fragments are liberated from the chemical soup and are absorbed into the blood and lymph through the cells of the small intestine's wall. Vitamins and minerals are absorbed here, too. They all eventually enter the bloodstream to nourish the tissues.

In the Large Intestine (Colon) Only fiber fragments, fluid, and some minerals are absorbed in the large intestine. The fibers from the seeds, whole-wheat bread, peanut butter, and banana are partly digested by the bacteria living in the colon, and some of the products are absorbed.[5] Most fiber is not absorbed, however, and, along with some other components, it passes out of the colon, excreted as feces.

key point *The mechanical and chemical actions of the digestive tract break foods down to nutrients, and large nutrients to their smaller building blocks, with remarkable efficiency.*

villi (VILL-ee, VILL-eye) fingerlike projections of the sheets of cells that line the intestinal tract. The villi make the surface area much greater than it would otherwise be (singular: *villus*).

microvilli (MY-croh-VILL-ee, MY-croh-VILL-eye) tiny, hairlike projections on each cell of every villus that can trap nutrient particles and transport them into the cells (singular: *microvillus*).

Absorption and Transportation of Nutrients

Once the digestive system has broken food down to its nutrient components, the rest of the body awaits their delivery. First, though, every molecule of nutrient must traverse one of the cells of the intestinal lining. These cells absorb nutrients from the mixture within the intestine and deposit them in the blood and lymph. The cells are selective: they recognize some of the nutrients that may be in short supply in the body. The mineral calcium is an example. The less calcium in the body, the more calcium the intestinal cells absorb. The cells are also extraordinarily efficient: they absorb enough nutrients to nourish all the body's other cells.

The cells of the intestinal tract lining are arranged in sheets that poke out into millions of finger-shaped projections (**villi**). Every cell on every villus has a brushlike covering of tiny hairs (**microvilli**) that can trap the nutrient particles. Each villus (projection) has its own capillary network and a lymph vessel so that as nutrients move across the cells, they can immediately mingle with the body fluids. Figure 3-11 on the next page provides a close look at these details.

The small intestine's lining, villi and all, is wrinkled into thousands of folds, so that its absorbing surface is enormous. If the folds, and the villi that poke out from them, were spread out flat, they would cover a third of a football field. The billions of cells of that surface weigh only 4 to 5 pounds, yet they absorb enough nutrients to nourish the other 150 or so pounds of body tissues.

After the nutrients pass through the cells of the villi, the blood and lymph start transporting the nutrients to their ultimate consumers, the body's cells. The lymphatic vessels initially transport most of the products of fat digestion and a few vitamins, later delivering them to the bloodstream. The blood vessels carry the products of carbohydrate and protein digestion, most vitamins, and the minerals from the digestive tract to the liver. Thanks to these two transportation systems, every nutrient soon arrives at the place where it is needed.

The digestive system's millions of specialized cells are themselves sensitive to an undersupply of energy, nutrients, or dietary fiber. In cases of severe undernutrition of energy and nutrients, the absorptive surface of the small intestine shrinks. The surface may be reduced to a tenth of its normal area, preventing it from absorbing what few nutrients a limited food supply may provide. Without sufficient fiber to provide an undigested bulk for the tract's muscles to push against, the muscles become weak from lack of exercise. Malnutrition that impairs digestion is self-perpetuating because impaired digestion makes malnutrition worse. In fact, the digestive system's needs are few, but important. The body has much to say to the attentive listener, stated in a language of symptoms and feelings that you would be wise to study. The next section takes a lighthearted look at what your digestive tract might be trying to tell you.

key point *The digestive system feeds the rest of the body and is itself sensitive to malnutrition. The folds and villi of the small intestine enlarge its surface area to facilitate nutrient absorption through countless cells to the blood and lymph. These transport systems then deliver the nutrients to all the body cells.*

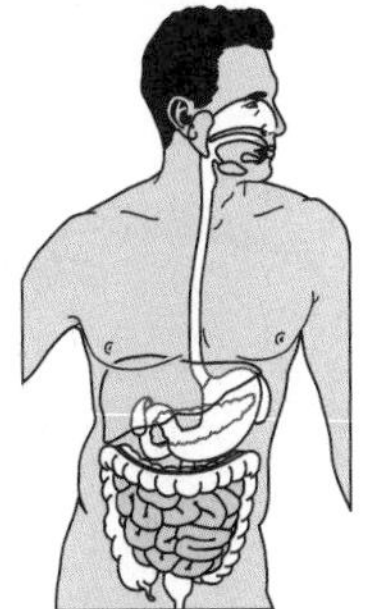

Time in small intestine, about 7–8 hours.*

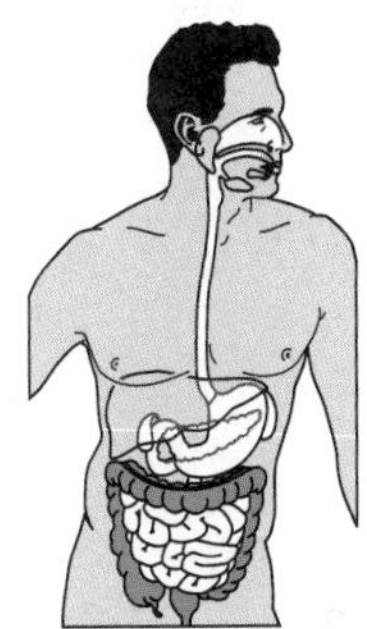

Time in colon, about 12–14 hours.*

*Based on a 24-hour transit time. Actual times vary widely.

What is your digestive tract trying to tell you?

Figure 3•11

DETAILS OF THE SMALL INTESTINAL LINING

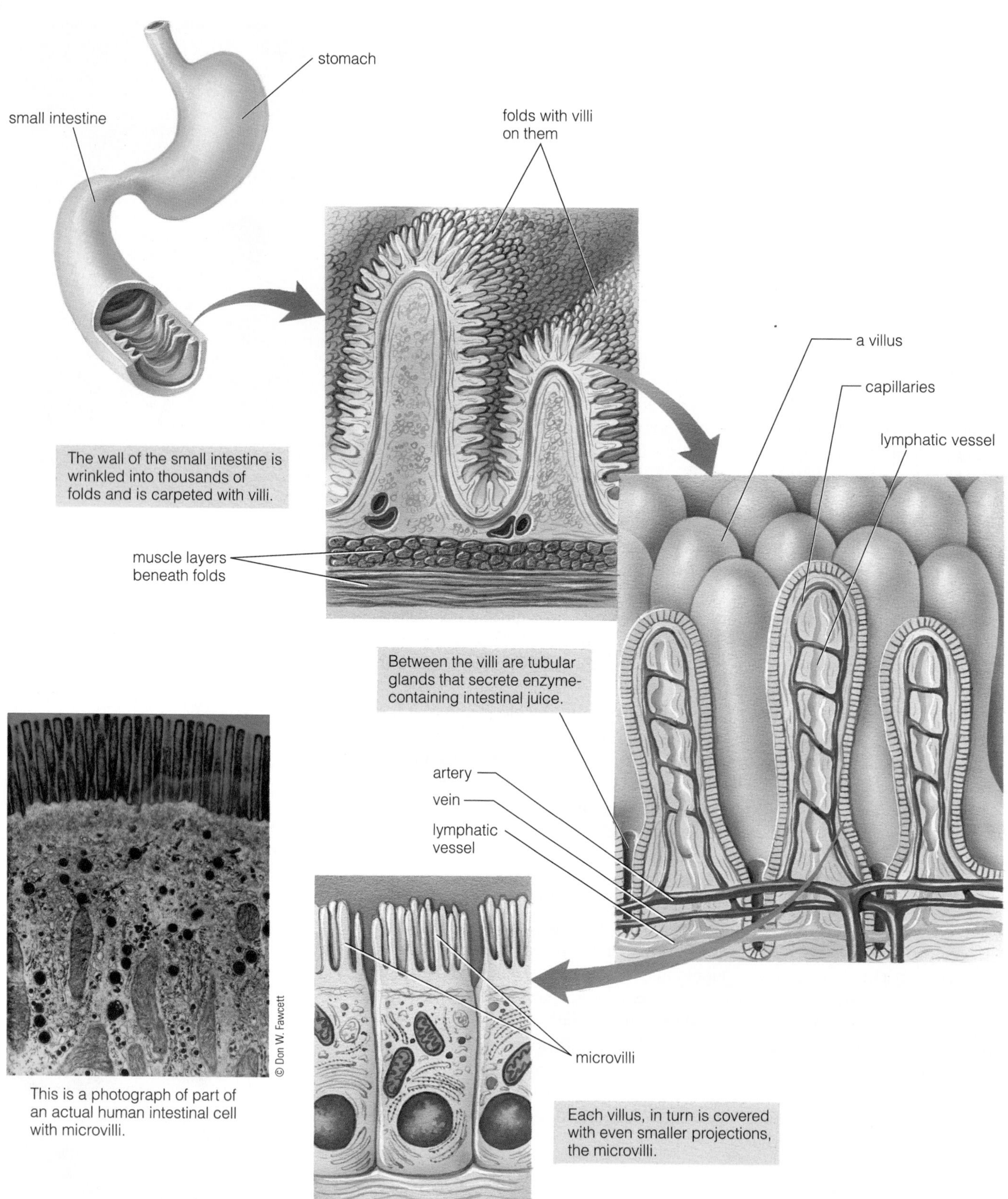

stomach
small intestine
folds with villi on them
a villus
capillaries
lymphatic vessel
The wall of the small intestine is wrinkled into thousands of folds and is carpeted with villi.
muscle layers beneath folds
Between the villi are tubular glands that secrete enzyme-containing intestinal juice.
artery
vein
lymphatic vessel
microvilli
© Don W. Fawcett
This is a photograph of part of an actual human intestinal cell with microvilli.
Each villus, in turn is covered with even smaller projections, the microvilli.

A Letter from Your Digestive Tract

To My Owner,

You and I are so close; I hope that I can speak frankly without offending you. I know that sometimes I *do* offend with my gurgling noises and belching at quiet times and, oh yes, the gas. But please understand that when you chew gum, drink carbonated beverages, or eat hastily, you gulp air with each swallow. I can't help making some noise as I move the air along my length or release it upward in a noisy belch. And if you eat or drink too fast, I can't help getting **hiccups.** Please sit and relax while you dine. You will ease my task, and we'll both be happier.

Also, when someone offers you a new food, you gobble away, trusting me to do my job. I try. It would make my life easier, and yours less gassy, if you would start with small amounts of new foods, especially those high in fiber. The bacteria that break down fiber produce gas in the process. I can handle just about anything if you introduce it slowly. But please: if you do notice more gas than normal from a specific food, avoid it. If the gas becomes excessive, check with a physician—the problem could be something simple, or it could be serious.

When you eat or drink too much, it just burns me up. Overeating causes **heartburn** because the acidic juice from my stomach backs up into my esophagus. Acid poses no problem to my healthy stomach, whose walls are coated with thick mucus to protect them. But when my too-full stomach squeezes some of its contents back up into the esophagus, the acid burns its unprotected surface. Also, those tight jeans you wear constrict my stomach, squeezing the contents upward into the esophagus. Just leaning over or lying down after a meal may allow the acid to escape up the esophagus because the muscular closure separating the two spaces is much looser than other such muscles. And if we need to lose a few pounds, let's get at it—excess body fat can also squeeze my stomach, causing acid to back up. When heartburn is a problem, do me a favor: try to eat smaller meals; drink liquids an hour before or after, but not during, meals; wear reasonably loose clothing; and relax after eating, but sit up (don't lie down).

Sometimes your food choices irritate me. Specifically, chemical irritants in foods, such as the "hot" component of chili peppers, chemicals in coffee, fat, chocolate, soda pop, and alcohol, may worsen heartburn in some people. Avoid the ones that cause trouble. Above all, do not smoke. Smoking makes my heartburn worse—and you should hear your lungs bellyache about it.

By the way, I can tell you've been taking heartburn medicines again. You must have been watching those misleading TV commercials. You need to know that **antacids** are designed only to temporarily relieve pain caused by heartburn by neutralizing stomach acid for a while. But when the antacids reduce my normal stomach acidity, I respond by producing *more* acid to restore the normal acid condition. Also, the ingredients in antacids can interfere with my ability to absorb nutrients. And don't decide that you need to take the heavily advertised **acid reducers** and **acid controllers** for my sake; these restrict my ability to produce acid so much that my job of digesting food becomes harder. In fact, the drugs often *cause* indigestion and diarrhea. Also, any heartburn medicine can mask the symptoms of **ulcer, hernia,** or the severe destructive form of chronic heartburn known as **gastro-esophageal reflux disease (GERD).**[6] This is serious business because evidence strongly suggests that, if not treated with antibiotic drugs, the bacterium that causes stomach ulcer may also cause stomach cancer.[7] A hernia can cause food to back up into the esophagus, so it can feel like heartburn, but many times hernias require corrective treatment by a physician. GERD can feel like heartburn, too, but may require surgery or drug therapy to prevent respiratory problems or serious damage to tissues.[8]

When you eat too quickly, I worry about choking (see Figure 3-12). Please take time to cut your food into small pieces, and chew it until it is crushed and moistened with saliva. Also, refrain from talking or laughing before swallowing, and never attempt to eat when you are breathing hard. Also, for our sake and the sake of others, learn the Heimlich maneuver as shown in Figure 3-13.

hiccups spasms of both the vocal cords and the diaphragm, causing periodic, audible, short, inhaled coughs. Can be caused by irritation of the diaphragm, indigestion, or other causes. Hiccups usually resolve in a few minutes, but can have serious effects if prolonged. Breathing into a paper bag (inhaling carbon dioxide) or dissolving a teaspoon of sugar in the mouth may stop them.

heartburn a burning sensation in the chest (in the area of the heart) area caused by backflow of stomach acid into the esophagus.

antacids medications that react directly and immediately with the acid of the stomach, neutralizing it. Antacids are most suitable for treating occasional heartburn. More about antacids appears in Controversy 8.

acid reducers and **acid controllers** drugs that reduce the acid output of the stomach. They are most suitable for treating severe, persistent forms of heartburn but are useless for neutralizing acid already present in the stomach. These drugs are now sold without prescription, but the packages bear warnings of side effects; some types interfere with the stomach's ability to destroy alcohol, so more of the alcohol in a drink enters the bloodstream.

ulcer an erosion in the topmost, and sometimes underlying, layers of cells that form a lining. Ulcers of the digestive tract commonly form in the esophagus, stomach, or upper small intestine.

hernia a protrusion of an organ or part of an organ through the wall of the body chamber that normally contains the organ. An example is a *hiatal* (high-AY-tal) *hernia*, in which part of the stomach protrudes up through the diaphragm into the chest cavity, which contains the esophagus, heart, and lungs.

gastro-esophageal reflux disease (GERD) a severe and chronic splashing of stomach acid and enzymes into the esophagus, throat, mouth, or airway that causes inflammation and injury to those organs. Untreated GERD may increase the risk of esophageal cancer; treatment may require surgery or management with medication.

For more information concerning ulcers and medication, call the Centers for Disease Control and Prevention at 888-MY-ULCER. The call is toll-free.

constipation infrequent, difficult bowel movements often caused by diet, inactivity, dehydration, or medication. (Also defined in Chapter 4.)

diarrhea frequent, watery bowel movements usually caused by diet, stress, or irritation of the colon. Severe, prolonged diarrhea robs the body of fluid and certain minerals, causing dehydration and imbalances that can be dangerous if left untreated.

irritable bowel syndrome intermittent disturbance of bowel function, especially diarrhea or alternating diarrhea and constipation; associated with diet, lack of physical activity, or psychological stress.

Figure 3•12

SWALLOWING AND CHOKING

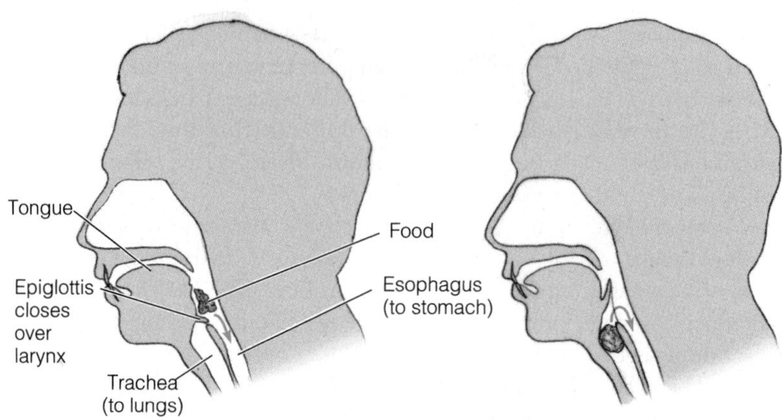

A normal swallow. The epiglottis acts as a flap to seal the entrance to the lungs (trachea) and direct food to the stomach via the esophagus.

Choking. A choking person cannot speak or gasp because food lodged in the trachea blocks the passage of air. The red arrow points to where the food should have gone to prevent choking.

Figure 3•13

THE HEIMLICH MANEUVER

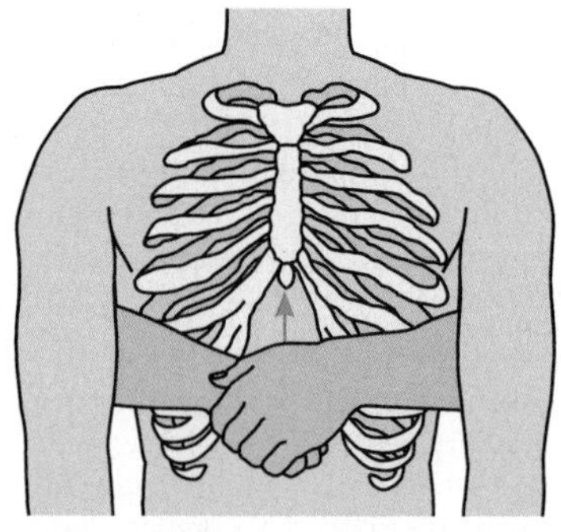

Rescuer positions fist directly against victim's abdomen as shown.

1. Rescuer stands behind victim and wraps her or his arms around victim's waist.
2. Rescuer makes a fist with one hand and places the thumb side of the fist against the victim's abdomen, slightly above the navel and below the rib cage.
3. Rescuer grasps fist with other hand and rapidly squeezes it inward and upward three or four times in rapid succession.
4. Rescuer repeats the process if necessary.

If the victim is alone, the victim positions himself or herself over edge of fixed horizontal object, such as a chair back, railing, or table edge, and presses abdomen into edge with quick movement.

When I'm suffering, you suffer, too. When **constipation** or **diarrhea** strikes, neither of us is having fun. Slow, hard, dry bowel movements can be painful, and failing to have a movement for too long brings on headaches and ill feelings. Don't rely on laxatives, though. They often contain stimulants that can cause side effects.* Instead of using laxatives, listen carefully for my signal that it is time to defecate, and make time for it even if you are busy. The longer you ignore my signal, the more time the colon has to extract water from the feces, hardening them. Also, please choose foods that provide enough fiber (Chapter 4 lists some of these foods). Fiber attracts water, creating softer, bulkier stools that stimulate my muscles to contract, pushing the contents along. Fiber helps my muscles to stay fit, too, making elimination easier. Be sure to drink enough water because dehydration causes the colon to absorb all the water it can get from the feces. And please make time to be physically active; exercise strengthens not just the muscles of your arms, legs, and torso, but those of the colon, too.

When I have the opposite problem, diarrhea, my system will rob you of water and salts. In diarrhea my intestinal contents have moved too quickly, drawing water and minerals from your tissues into the contents. When this happens, please rest a while and drink fluids. To avoid diarrhea, try not to change my diet too drastically or quickly. I'm willing to work with you and learn to digest new foods, but if you suddenly change your diet, we're both in for it. I hate even to think of it, but one likely cause of diarrhea is dangerous food poisoning (*please* read, and use, the tips in Chapter 14 to keep us safe). Also, if diarrhea lasts longer than a day or two, or if it alternates with constipation, the cause could be **irritable bowel syndrome,** and you should go see a physician.

Thank you for listening. I know we'll both benefit from communicating like this because you and I are in this together for the long haul.

Affectionately,
Your Digestive Tract

*One such stimulant, phenolphthalein, was recently banned because of an association with colon cancer.

 *The digestive tract has many ways to communicate its needs. By taking the time to listen, you will obtain a complete understanding of the mechanics of the digestive tract and its signals.*

The Excretory System

Cells generate a number of wastes, and all of them must be eliminated. Many of the body's organs play roles in removing wastes. Carbon dioxide waste from the cells travels in the blood to the lungs, where it is exchanged for oxygen. Other wastes are pulled out of the bloodstream by the liver. The liver processes these wastes and either tosses them out into the digestive tract with bile, to leave the body with the feces, or prepares them to be sent to the kidneys for disposal in the urine. Organ systems work together to dispose of the body's wastes, but the kidneys are waste- and water-removal specialists.

The kidneys straddle the cardiovascular system and filter the passing blood. Waste materials, dissolved in water, are collected by the kidneys' working units, the **nephrons.** These wastes become concentrated as urine, which travels through tubes to the urinary **bladder.** The bladder empties periodically, removing the wastes from the body. Thus, the blood is purified continuously throughout the day, and dissolved materials are excreted as necessary. One dissolved mineral, sodium, helps to regulate blood pressure, and its excretion or retention by the kidneys is a vital part of the body's blood pressure–controlling mechanism. As you might expect, the kidneys' work is regulated by hormones secreted by glands that respond to conditions in the blood (such as the sodium concentration).

Because the kidneys remove toxins that could otherwise damage body tissues, whatever supports the health of the kidneys supports the health of the whole body. A strong cardiovascular system and an abundant supply of water are important to keep blood flushing swiftly through the kidneys. In addition, the kidneys need sufficient energy to do their complex sifting and sorting job, and many vitamins and minerals serve as the cogs in their machinery. Exercise and nutrition are vital to healthy kidney function.

key point *The kidneys adjust the blood's composition in response to the body's needs, disposing of everyday wastes and helping remove toxins. Nutrients, including water, and exercise help keep the kidneys healthy.*

nephrons the working units in the kidneys, consisting of intermeshed blood vessels and tubules.

bladder the sac that holds urine until time for elimination.

glycogen a storage form of carbohydrate energy (glucose), described more fully in Chapter 4.

The lungs also excrete some small percentage of ingested alcohol—the basis for the "breathalyzer" test given to drivers to determine if they've been drinking.

Storage Systems

The human body is designed to eat at intervals of about four to six hours, but cells need nutrients around the clock. Providing the cells with a constant flow of the needed nutrients requires the cooperation of many body systems. These systems store and release nutrients to meet the cells' needs between meals. Among the major storage sites are the liver and muscles, which store carbohydrate, and the fat cells, which store fat as well as other things.

When I Eat More Than My Body Needs, What Happens to the Extra Nutrients?

Nutrients collected from the digestive system sooner or later all move through a vast network of capillaries that weave among the liver cells. This arrangement ensures that liver cells have access to the newly arriving nutrients for processing. Later chapters provide the details, but it is important to know now that the liver converts excess energy-containing nutrients into two forms. It makes some into **glycogen** (a carbohydrate) and some into fat. The liver stores the glycogen to meet the body's ongoing glucose needs. Liver glycogen can sustain cell activities when the intervals between

adipose tissue the body's fat tissue, consisting of masses of fat-storing cells and blood vessels to nourish them.

meals become long. Without glucose absorbed from food, the cells (including the muscle cells) draw on liver glycogen. Should no food be available, the liver's glycogen supply dwindles; it can be effectively depleted within as few as three to six hours. Muscle cells make and store glycogen, too, but selfishly reserve it for their own use.

Whereas the liver stores glycogen, it ships out fat in packages (see Chapter 5) to be picked up by cells that need it. All body cells may withdraw the fat they need from these packages, and the fat cells of the **adipose tissue** pick up the remainder and store it to meet long-term energy needs. Unlike the liver, fat tissue has virtually infinite storage capacity. It can continue to supply the body's cells with fat for days, weeks, or possibly even months when no food is eaten.

These storage systems for glucose and fat ensure that the body's cells will not go without energy even if the body is hungry for food. Body stores also exist for many other nutrients, each with a characteristic capacity. For example, liver and fat cells store many vitamins, and bones provide reserves of calcium, sodium, and other minerals. Stores of nutrients are available to keep the blood levels constant and to meet cellular demands.

Variations in Nutrient Stores

Some nutrients are stored in the body in much larger quantities than others are. For example, certain vitamins are stored without limit, even if they reach toxic levels within the body. Other nutrients are stored in only small amounts, regardless of the amount taken in, and these can readily be depleted. As you learn how the body handles various nutrients, pay particular attention to their storage so that you can know your tolerance limits. For example, you needn't eat fat at every meal because fat is stored abundantly. On the other hand, you normally do need to have a source of carbohydrate at intervals throughout the day because the liver stores less than one day's supply of glycogen.

key point *The body's energy stores are of two principal kinds: fat in fat cells (in potentially large quantities) and glycogen in muscle and liver cells (in smaller quantities). Other tissues store other nutrients.*

Conclusion

In addition to the systems just described, the body has many more: bones, muscles, reproductive organs, and others. All of these cooperate, enabling each cell to carry on its own life. For example, the skin and body linings defend other tissues against microbial invaders, while being nourished and cleansed by tissues specializing in these tasks. Each system needs a continuous supply of many specific nutrients to maintain itself and carry out its work. Calcium is particularly important for bones, for example; iron for muscles; glucose for the brain. But all systems need all nutrients, and every system is impaired by an undersupply or oversupply of them.

While external events clamor and vie for attention, the body quietly continues its life-sustaining work. Most of the body's work is directed automatically by the unconscious portions of the brain and nervous system, and this work is finely regulated to achieve a state of well-being. But you need to involve your cerebral cortex, your consciousness, to cultivate an understanding and appreciation of your body's needs. In doing so, attend to nutrition first. The rewards are liberating—ample energy to tackle life's tasks, a robust attitude, and the glowing appearance that comes from the best of health. Read on, and learn to let nutrition principles guide your choices.

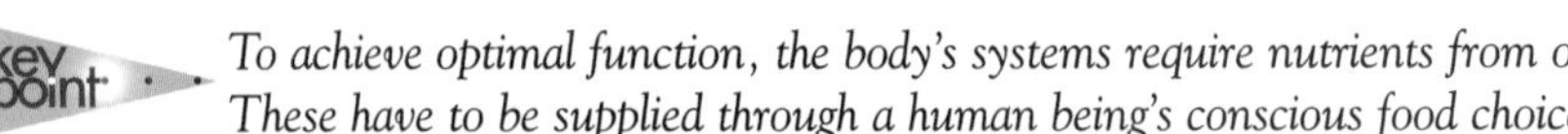

To achieve optimal function, the body's systems require nutrients from outside. These have to be supplied through a human being's conscious food choices.

Answers to these Self Check questions are in Appendix G.

1. All blood leaving the digestive system is routed directly to the:
   a. heart
   b. kidney
   c. liver
   d. lungs
2. Which of the following can affect the hormonal system?
   a. fasting
   b. feeding
   c. exercise
   d. all of the above
3. Chemical digestion of all nutrients mainly occurs in which organ?
   a. mouth
   b. stomach
   c. small intestine
   d. large intestine
4. Which chemical substance released by the pancreas neutralizes stomach acid that has reached the small intestine?
   a. mucus
   b. enzymes
   c. bicarbonate
   d. bile
5. Which nutrient passes through the large intestine mostly unabsorbed?
   a. starch
   b. vitamins
   c. minerals
   d. fiber
6. The immune system is so important that its tissues cannot be damaged by malnutrition. T F
7. Bile starts the process of protein digestion in the stomach. T F
8. To digest food efficiently, people should not combine certain foods, such as meat and fruit, at the same meal. T F
9. The gallbladder stores bile until it is needed to emulsify fat. T F
10. Absorption of the majority of nutrients takes place across the mucus-coated lining of the stomach. T F

nutrition on the net

For further study of the topics of this chapter, access these websites and search for the phrases or words in quotation marks:

1. Find updates and quick links to these and other nutrition-related sites at our website: www.wadsworth.com/nutrition
2. Visit the Center for Digestive Health and Nutrition: www.gihealth.com
3. Visit the Digest This! section of the American College of Gastroenterology: www.acg.gi.org
4. Search for "choking," "vomiting," "diarrhea," "constipation," "heartburn," "indigestion," and "ulcers" at the U.S. Government health information site: www.healthfinder.gov
5. Visit the Digestive Diseases section of the National Institute of Diabetes, Digestive, and Kidney Diseases: www.niddk.nih.gov/health/health.htm
6. Learn more about *H. pylori* (the bacterium that causes stomach cancer) from the Helicobacter Foundation: www.helico.com
7. Read more about food history at Oldways Preservation and Exchange Trust: www.oldwayspt.org

INTERNET ACTIVITY

Chronic heartburn seems to be a recurring theme of drug commercials on television. Heartburn can be due to simply overeating or the more serious gastro-esophageal reflux disease (GERD). Before using over-the-counter medications or asking physicians for prescriptions, we should be aware of these disorders. Take this test to determine your risk of GERD.

1. Go to the Intelihealth website at http://www.intelihealth.com
2. In box on left side of page, click on "Intelitools."
3. Scroll down the Interact box on right side of page. Under "Health Quizzes", click on "GERD Quiz."
4. Then click on "Have you heard of GERD?"

ALCOHOL AND NUTRITION: DO THE BENEFITS OUTWEIGH THE RISKS?

On average, people in the United States consume from 6 to 10 percent of their total daily energy intake as alcohol. Drinking habits span a wide spectrum: many adults drink no alcohol whatsoever, some take a glass of wine only with meals, others drink on social occasions, and others take in huge quantities of alcohol daily because of a life-shattering addiction. A third of U.S. college students are reported to be **binge drinkers,** and many pay a high price in terms of their health and safety as a result of episodes of heavy drinking.[1] People who are **moderate drinkers** usually consume the calories of alcohol in addition to their normal food intake, so the alcohol contributes to body fatness.[2] (Alcohol-related terms are defined in Table C3-1.) Alcohol is not just an energy source, however; it is also a psychoactive drug and a toxin to the body.

Despite its toxicity, many people want to know if there is an amount of alcohol they can drink safely or whether they may derive benefits, particularly for the health of the heart, by drinking. This Controversy first defines some terms and examines alcohol's actions within the body and its effects on the brain and other organs. It then summarizes the long-term effects of alcohol on the body and nutrition and concludes with research on moderate drinking.

Defining Drinks and Drinking

When people congregate to enjoy conversation and companionship, it is only natural to offer beverages. All beverages seem to ease conversation, whether they contain alcohol or not. Many people are **social drinkers** who choose alcohol over cola, juice, milk, tea, or coffee as a pleasant accompaniment to a meal, a drink of celebration, or a way to relax with friends. Taken in moderation, alcohol reduces inhibi-

Table C3•1

ALCOHOL AND DRINKING TERMS

- **acetaldehyde** (ass-et-AL-deh-hide) a substance to which ethanol is metabolized on its way to becoming harmless waste products that can be excreted.
- **alcohol dehydrogenase** (dee-high-DRAH-gen-ace) **(ADH)** an enzyme system that breaks down alcohol. The antidiuretic hormone listed below is also abbreviated ADH.
- **alcoholism** a dependency on alcohol marked by compulsive uncontrollable drinking with negative effects on physical health, family relationships, and social health.
- **antidiuretic** (AN-tee-dye-you-RET-ick) **hormone (ADH)** a hormone produced by the pituitary gland in response to dehydration (or a high sodium concentration in the blood). It stimulates the kidneys to reabsorb more water and so to excrete less. (This hormone should not be confused with the enzyme alcohol dehydrogenase, which is also abbreviated ADH.)
- **beer belly** central-body fatness associated with alcohol consumption.
- **binge drinkers** people who drink 4 or more drinks in a short period.
- **cirrhosis** (seer-OH-sis) advanced liver disease, often associated with alcoholism, in which liver cells have died, hardened, turned an orange color, and permanently lost their function.
- **congeners** (CON-jen-ers) chemical substances other than alcohol that account for some of the physiological effects of alcoholic beverages, such as appetite, taste, and aftereffects.
- **drink** a dose of any alcoholic beverage that delivers ½ ounce of pure ethanol.
- **ethanol** the alcohol of alcoholic beverages, produced by the action of microorganisms on the carbohydrates of grape juice or other carbohydrate-containing fluids.
- **euphoria** (you-FOR-ee-uh) an inflated sense of well-being and pleasure brought on by a moderate dose of alcohol and some other drugs.
- **fatty liver** an early stage of liver deterioration seen in several diseases, including kwashiorkor and alcoholic liver disease in which fat accumulates in the liver cells.
- **fibrosis** (fye-BROH-sis) an intermediate stage of alcoholic liver deterioration. Liver cells lose their function and assume the characteristics of connective tissue cells (fibers).
- **formaldehyde** a substance to which methanol is metabolized on the way to being converted to harmless waste products that can be excreted.
- **gout (GOWT)** a painful form of arthritis caused by the abnormal buildup of the waste product uric acid in the blood, with uric acid salt deposited as crystals in the joints.
- **MEOS** (microsomal ethanol oxidizing system) a system of enzymes in the liver that oxidize not only alcohol but also several classes of drugs.
- **methanol** an alcohol produced in the body continually by all cells.
- **moderate drinkers** people who do not drink excessively and do not behave inappropriately because of alcohol. A moderate drinker's health is not harmed by alcohol over the long term.
- **problem drinkers** or **alcohol abusers** people who suffer social, emotional, family, job-related, or other problems because of alcohol. A problem drinker is on the way to alcoholism.
- **proof** a statement of the percentage of alcohol in an alcoholic beverage. Liquor that is 100 proof is 50% alcohol, 90 proof is 45%, and so forth.
- **social drinkers** people who drink only on social occasions. Depending on how alcohol affects a social drinker's life, the person may be a moderate drinker or a problem drinker.
- **urethane** a carcinogenic compound that commonly forms in alcoholic beverages.
- **Wernicke-Korsakoff** (VER-nik-ee KOR-sah-koff) **syndrome** a cluster of symptoms involving nerve damage arising from a deficiency of the vitamin thiamin in alcoholism. Characterized by mental confusion, disorientation, memory loss, jerky eye movements, and staggering gait.

tions, encourages social interactions, and produces feelings of **euphoria,** a pleasant sensation that people seek. The term *moderation* is important because alcohol *worsens* social interactions at higher intakes. The nonalcoholic beers and wines now on the market also elevate mood and encourage social interaction, a testimony to the placebo effect at work.

In contrast to moderate social drinking, the effect of alcohol on **problem drinkers** or people with **alcoholism** is overwhelmingly negative. For these people, drinking alcohol brings irrational and often dangerous behavior, such as driving a car while intoxicated, and regrettable human interactions, such as arguments and violence. With continued drinking, such people face psychological depression, physical illness, severe malnutrition, and demoralizing erosion of self-esteem.

Moderation

Moderation is not easily defined, for no single upper limit of alcohol per day is appropriate for everyone. Tolerances to alcohol differ. In general, women cannot handle as much alcohol as men and should not try to match drinks with male companions. Health authorities have set limits at not more than two drinks a day for the average-sized, healthy man and not more than one drink a day for the average-sized, healthy woman. These amounts are supposed to be enough to elevate mood without incurring long-term harm to health—note that these are not average amounts, but daily maximums. In other words, a person who drinks no alcohol during the week but then takes seven drinks on Saturday night is not a moderate drinker—such alcohol intake patterns are characterized as binge drinking.

Doubtless some people can safely consume slightly more than the alcohol dose called moderate; others, especially those prone to alcohol addiction, definitely cannot handle that amount without significant risk. Table C3-2 lists those people advised by the *Dietary Guidelines for Americans* not to drink at all. If you think your own drinking might not be moderate or normal, if it has caused problems in your life, or if you feel guilty about your drinking, you may want to seek a professional evaluation.* Table C3-3 on the next page contrasts some behaviors of moderate drinkers with those of problem drinkers.

Table C3•2

WHO SHOULD NOT DRINK ALCOHOL?

The *Dietary Guidelines for Americans* suggest that these people not drink alcoholic beverages at all:

- *Children and adolescents.*
- *People of any age who cannot restrict their drinking to moderate levels.* Especially, people recovering from alcoholism, problem drinkers, and people whose family members have alcohol problems.
- *Women who may become pregnant or who are pregnant.* A safe level of alcohol intake has not been established for women during pregnancy (see Chapter 12), and alcohol may be especially hazardous during the first few weeks, before a woman knows she is pregnant.
- *People who plan to drive, operate machinery, or take part in other activities that require attention, skill, or coordination to remain safe.* Alcohol remains in the blood for several hours after taking even a single drink.
- *People taking prescription or over-the-counter medications that can interact with alcohol.* Alcohol alters the effectiveness or toxicity of many medications, and some medications may increase blood alcohol levels.

SOURCE: U.S. Department of Agriculture, Dietary Guidelines Advisory Committee, *Nutrition and Your Health: Dietary Guidelines for Americans, 5th ed.*, 2000, Home and Garden Bulletin no. 232, available online at www.usda.gov/cnpp or call (888) 878-3256.

Binge Drinking

Binge drinking (consuming four or more drinks in a short time) poses a serious health threat to college students, to others who engage in it, and to nearby nondrinkers.[3] A binge is most likely to occur at a party, a sporting event, or other social occasion. Compared with nondrinkers or moderate drinkers, binge drinkers are more likely to damage property, to assault other people, to cause fatal automobile accidents, and to engage in risky unprotected and unplanned sexual intercourse.

Binge drinkers skew the statistics on alcohol use on college campuses. The median number of drinks consumed by college students is 1.5 per week, but for binge drinkers, it is 14.5. Nationally, only 20 percent of all students are frequent binge drinkers; yet they account for two-thirds of all the alcohol students report consuming and most of the alcohol-related problems.[4]

Binge drinking is not limited to college campuses, of course, but the student environment seems most accepting of such behavior despite its problems. Social acceptance may make it difficult for binge drinkers to recognize themselves as problem drinkers (refer to Table C3-3) until their drinking behavior causes a crisis, such as a car crash, or until they've binged long enough to have caused substantial damage to their health. Those who start drinking at an early age suffer more often from alcoholism than people who start later.

What Is Alcohol?

In chemistry, the term *alcohol* refers to a class of chemical compounds whose names end in "-ol." The glycerol molecule of a triglyceride is an example. Alcohols affect living things profoundly, partly because they act as lipid solvents. Alcohols can easily penetrate a cell's outer lipid membrane, and once inside, they denature the cell's protein structures and kill the cell. Because some alcohols kill microbial cells, they make useful disinfectants and antiseptics.

The alcohol of alcoholic beverages, **ethanol,** is somewhat less toxic than others. Sufficiently diluted and taken in small enough doses, its action in the brain produces euphoria. Used in this way, alcohol is a drug, and like many drugs, alcohol presents both benefits and hazards to the taker. Its effects depend on the quantity of alcohol consumed.

* The U.S. center for facts on alcohol is the National Clearinghouse for Alcohol and Drug Information: (800) 729-6686.

Table C3•3

BEHAVIORS TYPICAL OF MODERATE DRINKERS AND PROBLEM DRINKERS

Moderate Drinkers Typically:	Problem Drinkers Typically:
• Drink slowly, casually. • Eat food while drinking or beforehand. • Don't binge drink; know when to stop. • Respect nondrinkers. • Avoid drinking when solving problems or making decisions. • Do not admire or encourage drunkenness. • Remain peaceful, calm, and unchanged by drinking. • Cause no problems to others or themselves by drinking.	• Gulp or "chug" drinks. • Drink on an empty stomach. • Binge drink; drink to get drunk. • Pressure others to drink. • Turn to alcohol when facing problems or decisions. • Consider drunks to be funny or admirable. • Become loud, angry, violent, or silent when drinking. • Physically or emotionally harm themselves, family members, or others when drinking.

What Is a "Drink"?

Alcoholic beverages contain a great deal of water and some other substances, as well as the alcohol ethanol. In wine, beer, and wine coolers, alcohol contributes a relatively low percentage of the beverage's volume. In contrast, whiskey, vodka, rum, and brandy may contain as much as 50 percent of their volume as alcohol. The percentage of alcohol is stated as **proof.** Proof equals twice the percentage of alcohol; for example, 100 proof liquor is 50 percent alcohol, 90 proof is 45 percent, and so forth.

A serving of alcoholic beverage, commonly called a **drink,** delivers ½ ounce of pure ethanol. Figure C3-1 depicts servings of alcoholic beverages that are considered to be one drink. These standard measures may have little in common with the drinks served by enthusiastic bartenders, however. Many wine glasses easily hold 6 to 8 ounces of wine; wine coolers may come packaged 12 ounces to a bottle; a large beer stein can hold 16, 20, or even more ounces; a strong liquor drink may contain 2 or 3 ounces of various liquors.

Figure C3•1

SERVINGS OF ALCOHOLIC BEVERAGES THAT EQUAL ONE DRINK

Alcohol Enters the Body

From the moment an alcoholic beverage is swallowed, the body pays special attention to it. Unlike food, which requires digestion before it can be absorbed, the tiny alcohol molecules can diffuse right through the stomach walls and reach the brain within a minute. Ethanol is a toxin, and a too-high dose of alcohol triggers one of the body's primary defenses against poison—vomiting. Many times, though, alcohol arrives gradually and in a beverage dilute enough that the vomiting reflex is delayed and the alcohol is absorbed.

A person can become intoxicated almost immediately when drinking, especially if the stomach is empty. When the stomach is full of food, molecules of alcohol have less chance of touching the stomach walls and diffusing through, so alcohol reaches the brain more gradually (see Figure C3-2). By the time the stomach contents are emptied into the small intestine, however, alcohol is absorbed rapidly whether food is present or not.

Figure C3•2

FOOD SLOWS ALCOHOL'S ABSORPTION

The alcohol in a stomach filled with food has a low probability of touching the walls and diffusing through. Food also holds alcohol in the stomach longer, slowing its entry into the highly absorptive small intestine.

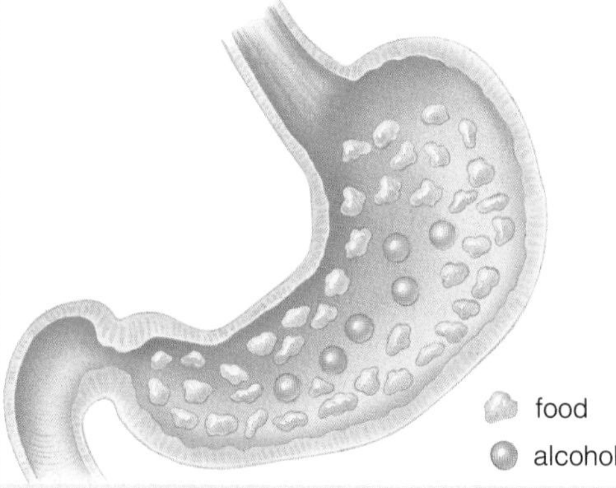

A person who wants to drink socially and not become intoxicated should eat the snacks provided by the host (avoid the salty ones; they make you thirstier). Carbohydrate snacks slow alcohol absorption, and high-fat snacks help too because they slow peristalsis, keeping the alcohol in the stomach longer. Other tips include adding ice or water to drinks to dilute them and choosing nonalcoholic beverages first and then every other round to quench thirst.

Anyone who has had an alcoholic drink has experienced one of alcohol's physical effects: alcohol increases urine output (because alcohol depresses the brain's production of **antidiuretic hormone**). Loss of body water leads to thirst. The only fluid that relieves dehydration is water, so alternating alcoholic beverages with nonalcoholic ones will quench thirst. Otherwise, each drink may worsen the thirst.

The water lost due to hormone depression takes with it important minerals, such as magnesium, potassium, calcium, and zinc, depleting the body's reserves. These minerals are vital to fluid balance and to nerve and muscle coordination. When drinking incurs mineral losses, the losses must be made up in subsequent meals to avoid deficiencies.

If a person drinks slowly enough, the alcohol, after absorption, will be collected by the liver and processed without much effect on other parts of the body. If a person drinks more rapidly, however, some of the alcohol bypasses the liver and flows for a while through the rest of the body and the brain.

Alcohol Arrives in the Brain

Some people use alcohol as a kind of social anesthetic to help them relax or to relieve anxiety. One drink relieves inhibitions, which gives people the impression that alcohol is a stimulant. Actually, it gives this impression by sedating the *inhibitory* nerves, allowing excitatory nerves to take over. This effect is temporary, and ultimately, alcohol acts as a depressant and sedates all the nerve cells. Figure C3-3 shows alcohol's effects on the brain.

It is lucky that the brain centers respond to rising blood alcohol in the order shown in the figure, because a person usually passes out before drinking a lethal dose. If a person drinks fast enough, though, the alcohol continues to be absorbed, and its effects continue to accelerate after the person has gone to sleep. Every year, deaths attributed to this effect take place during drinking contests. Before passing out, the drinker drinks fast enough to receive a lethal dose. Table C3-4 shows blood alcohol levels that correspond with progressively greater intoxication, and Table C3-5 on the next page shows brain responses that occur at these levels.

Brain cells are particularly sensitive to excessive exposure to alcohol. The brain shrinks, even in people who drink only moderately. The extent of the shrinkage is proportional to the amount drunk. Abstinence, together with good nutrition, reverses some of the brain damage, and possibly all of it, if heavy drinking has not continued for more than a few years. However, prolonged drinking beyond an individual's capacity to recover can do severe and irreversible harm to vision, memory, learning ability, and other brain functions.

Figure C3•3

ALCOHOL'S EFFECTS ON THE BRAIN

When alcohol flows to the brain, it first sedates the frontal lobe, the reasoning part. As the alcohol molecules diffuse into the cells of this lobe, they interfere with reasoning and judgment.

With continued drinking, the speech and vision centers of the brain become sedated, and the area that governs reasoning becomes more incapacitated.

Still more drinking affects the cells of the brain responsible for large-muscle control; at this point people under the influence stagger or weave when they try to walk.

Finally the conscious brain becomes completely subdued, and the person passes out. Now the person can drink no more. This is fortunate because a higher dose would anesthetize the deepest brain centers that control breathing and heartbeat, causing death.

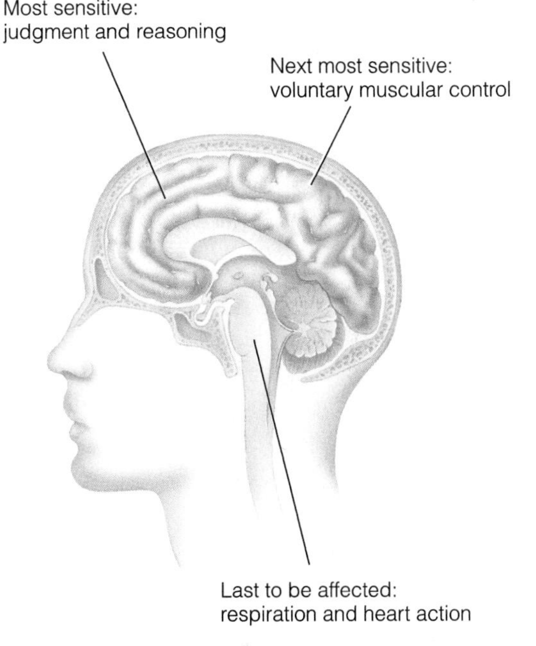

Alcohol Arrives in the Liver

The capillaries that surround the digestive tract merge into veins that carry the alcohol-laden blood to the liver. Here the veins branch and rebranch into capillaries that touch every liver cell. The liver cells make nearly all of the body's alcohol-processing machinery. The routing of blood through the liver allows the cells to go right to work detoxifying substances before they reach other body organs such as the heart and brain.

The Liver Metabolizes Alcohol

The liver makes and maintains two sets of equipment for metabolizing alcohol. One is an enzyme that removes hydrogens from alcohol to break it down; the name, **alcohol dehydrogenase (ADH)**, almost says what it does.* This enzyme

*There are actually two ADH enzymes; each performs a specific task in alcohol breakdown.

Table C3•4

ALCOHOL DOSES AND BLOOD LEVELS

Number of Drinks[a]	Average Percent Blood Alcohol by Body Weight 100 lb	120 lb	150 lb	180 lb	200 lb
2	0.08	0.06	0.05	0.04	0.04
4	0.15	0.13	0.10	0.08	0.08
6	0.23	0.19	0.15	0.13	0.11
8	0.30	0.25	0.20	0.17	0.15
12	0.45	0.36	0.30	0.25	0.23
14	0.52	0.42	0.35	0.34	0.27

[a]Taken within an hour or so; each drink equal to ½ ounce pure ethanol.

Table C3•5

BLOOD ALCOHOL LEVELS AND BRAIN RESPONSES

Blood Alcohol Level (%)	Brain Response
0.05[a]	Judgment impaired
0.10	Emotional control impaired
0.15	Muscle coordination and reflexes impaired
0.20	Vision impaired
0.30	Drunk, lacking control
0.35	In a stupor
0.50–0.60	Loss of consciousness; death.

[a]A 0.08 percent level is the legal limit for intoxication according to most states' highway safety ordinances; however, driving ability may be impaired at blood alcohol levels lower than 0.08 percent.

handles about 80 percent or more of the alcohol in the body. The other set of alcohol-metabolizing equipment is a chain of enzymes known as the **MEOS,** which is thought to handle about 10 percent of alcohol. The remaining 10 percent is excreted through the breath and in the urine. Because the alcohol in the breath is directly proportional to the alcohol in the blood, the breathalyzer test that law enforcement officers administer when someone may be driving under the influence of alcohol accurately reveals the person's degree of intoxication.

The amount of alcohol a person's body can process in a given time is limited by the amount of ADH enzymes residing in the liver. If more molecules of alcohol arrive at the liver cells than the enzymes can handle, the extra alcohol must wait. It circulates again and again through the brain, liver, and other organs until enzymes are available to degrade it.

Some ADH enzymes reside in the stomach and break down some alcohol before it enters the blood. Research shows that people with alcoholism make less stomach ADH than others and that women make less than men. Earlier, this Controversy warned that women should not try to keep up with male drinkers, and here is the reason why: women absorb about one-third more alcohol than men do, even when the women are the same size as the men and drink the same amount of alcohol.

The amount of ADH enzymes present is also affected by whether a person eats. Fasting for as little as a day causes degradation of body proteins, including the ADH enzymes, which can reduce the rate of alcohol metabolism by half.

It takes about an hour and a half to metabolize one drink, depending on a person's body size, previous drinking experience, how recently the person has eaten, and the person's current state of health. The liver is the only organ that can dispose of significant quantities of alcohol, and its maximum rate of alcohol clearance cannot be speeded up. This explains why only time restores sobriety. Walking will not because muscles cannot metabolize alcohol. Nor will drinking a cup of coffee help. Caffeine is a stimulant, but it won't speed up the metabolism of alcohol. The police say that a cup of coffee only makes a sleepy drunk into a wide-awake drunk. Table C3-6 presents other alcohol myths.

Alcohol Affects Body Functions

Upon exposure to alcohol, the liver speeds up its synthesis of fatty acids. Fat is known to accumulate in the livers of young men after a single night of heavy drinking and to remain there for more than a day. The first stage of liver deterioration seen in heavy drinkers is therefore known as **fatty liver;** it interferes with the distribution of nutrients and oxygen to the liver cells. If the condition lasts long enough, fibrous scar tissue invades the liver. This is the second stage of liver deterioration, called **fibrosis.** Fibrosis is reversible with good nutrition and abstinence from alcohol, but the next (last) stage, **cirrhosis,** is not. In cirrhosis, the liver cells harden, turn orange, and die, losing function forever. All of this points to the importance of moderation in the use of alcohol.

The presence of alcohol alters amino acid metabolism in the liver cells. Synthesis of some immune system proteins slows down, weakening the body's defenses against infection. Synthesis of blood lipids speeds up, increasing the concentration of triglycerides and high-density lipoproteins (see Chapter 5). In addition, excess alcohol adds to the body's acid burden and interferes with normal uric acid metabolism, causing symptoms like those of **gout.**

The reproductive system is also vulnerable to alcohol's effects. Heavy drinking in women may lead to infertility and spontaneous abortion.[5] Alcohol may also

Table C3•6

MYTHS AND TRUTHS CONCERNING ALCOHOL

Myth:	A shot of alcohol warms you up.
Truth:	Alcohol diverts blood flow to the skin making you *feel* warmer, but it actually cools the body.
Myth:	Wine and beer are mild; they do not lead to addiction.
Truth:	Wine and beer drinkers worldwide have high rates of death from alcohol-related illnesses. It's not what you drink, but how much, that makes the difference.
Myth:	Mixing drinks is what gives you a hangover.
Truth:	Too much alcohol in any form produces a hangover.
Myth:	Alcohol is a stimulant.
Truth:	Alcohol depresses the brain's activity.
Myth:	Alcohol is legal; therefore, it is not a drug.
Truth:	Alcohol is legal, but it alters body functions and is medically defined as a depressant drug.

Left, normal liver; center, fatty liver; right, cirrhosis.

© A. Glaubermann/Photo Researchers, Inc.

suppress the male reproductive hormone testosterone, leading to decreases in muscle and bone tissue, altered immunity, abnormal prostate gland, and decreased reproductive ability.[6]

The Fattening Power of Alcohol

Alcohol should probably be counted as fat in the diet because metabolic interactions occur between fat and alcohol in the body.[7] Presented with both fat and alcohol, the body stores the comparatively harmless fat and rids itself of the toxic alcohol by burning it off as fuel. Alcohol may promote fat storage particularly in the central abdominal area—the **"beer belly"** effect seen in moderate drinkers whose risks to the heart are described in Chapter 9.[8] Also, alcohol yields 7 calories of energy per gram to the body, so many alcoholic drinks can be much more fattening than their nonalcoholic counterparts. A general guideline states that each ounce of ethanol in a drink represents the same number of calories as about half an ounce of fat. An observant reader, knowing that, in the laboratory, a gram of fat and a gram of alcohol yield 9 and 7 calories, respectively, may wonder why alcohol in a drink is worth only half the calorie value of fat. The answer is that the body rids itself of a small but measurable amount of the alcohol by way of the breath and urine.

The Hangover

The hangover—the awful feeling of headache, unpleasant sensations in the mouth, and nausea that one has the morning after drinking too much—is a mild form of drug withdrawal. (The worst form is a delirium with severe tremors that presents a danger of death and demands medical management.) Hangovers are caused by several factors. One is the toxic effects of **congeners** that accompany the alcohol in alcoholic beverages. Mixing or switching drinks will not prevent hangover because congeners are only one factor. Dehydration of the brain is a second factor: alcohol reduces the water content of the brain cells. When they rehydrate the morning after and swell back to their normal size, nerve pain results.

Another contributor to the hangover is **formaldehyde,** the same chemical that medical laboratories use to preserve dead animals. Formaldehyde comes from **methanol,** an alcohol produced constantly by normal chemical processes in all the cells. Normally, a set of liver enzymes converts this methanol to formaldehyde, and then a second set immediately converts the formaldehyde to carbon dioxide and water, harmless waste products that can be excreted. But these same two sets of liver enzymes are also used to process ethanol to its own intermediate (also highly toxic) waste product, **acetaldehyde,** and then to carbon dioxide and water.[9] The enzymes prefer ethanol 20 times over methanol. Both alcohols are metabolized without delay until the excess acetaldehyde monopolizes the enzymes, leaving formaldehyde to wait for later detoxification. At that point, formaldehyde starts accumulating and the hangover begins.

Time alone is the cure for a hangover. Simple-minded remedies clearly will not work: vitamins, tranquilizers, aspirin, drinking more alcohol, breathing pure oxygen, exercising, eating, or drinking something awful are all useless. Fluid replacement can help to normalize the body's chemistry. The headache, unpleasantness in the mouth, and nausea of a hangover come simply from drinking too much.

Alcohol's Long-Term Effects

By far the longest-term effects of alcohol are those felt by the child of a woman who drinks during pregnancy. When a pregnant woman takes a drink, her fetus takes the same drink within minutes, and its body is defenseless against the effects. This topic is so important that it has its own section in Chapter 12, where the recommendation is made that pregnant women should not drink at all. For nonpregnant adults, however, what are the effects of alcohol over the long term?

A couple of drinks set in motion many destructive processes in the body. The next day's abstinence can reverse them only if the doses taken are moderate, the time between them is ample, and nutrition is adequate meanwhile.

If the doses of alcohol are heavy, however, and the time between them is short, complete recovery cannot take place, and repeated onslaughts of alcohol take a toll on the body. For example, alcohol is directly toxic to skeletal and cardiac muscle, causing weakness and deterioration that is greater, the larger the dose. Alcoholism makes heart disease likely, probably because chronic alcohol use raises blood pressure. At autopsy, the heart of a person with alcoholism appears bloated and weighs twice as much as a normal heart.

Alcohol attacks brain cells directly and can result in dementia. Cirrhosis also develops after 10 to 20 years from the cumulative effects of frequent episodes of heavy drinking.

Alcohol abuse also leads to cancers of the breast, mouth, throat, esophagus, rectum, and lungs. Daily human exposure to ethanol ranks high among carcinogenic hazards. Alcohol seems to promote the development of cancer once the disease has started.

A convincing body of evidence implicates alcohol intake by women in the causation of cancer of the breast—women who drink less than one drink per day elevate their risk slightly, and those who drink more increase their risk accordingly.[10] Cancer of the rectum increases with intakes of more than 15 ounces of beer each day. In the case of beer, alcohol may be acting together with other compounds formed during brewing to promote the cancer. One compound, **urethane,** is often found in alcoholic beverages. Urethane is known to cause cancer in animals, but the risk to human beings is unknown.

Other long-term effects of alcohol abuse include the following:

- Diabetes (type 2 or noninsulin-dependent).
- Ulcers and inflammation of the stomach and intestines.
- Nonviral hepatitis.
- Disease of the muscles of the heart.
- Severe psychological depression.
- Kidney, bladder, prostate, and pancreas damage.
- Skin rashes and sores.
- Impaired immune response.
- Deterioration of the testicles and adrenal glands.
- Feminization and sexual impotence in men.
- Brain disease and central nervous system damage.
- Impaired memory and balance.
- Malnutrition.
- Bone deterioration and osteoporosis.
- Increased risks of death from all causes.[11]

This list is by no means all-inclusive. Alcohol abuse exerts direct toxic effects on all body organs. Monetarily, alcoholism costs our society an estimated $166 *billion* every year in medical services, lost wages, criminal offenses, auto crashes, and other losses.[12]

Alcohol's Effect on Nutrition

Alcohol abuse also does damage indirectly via malnutrition. The more alcohol a person drinks, the less likely that he or she will eat enough food to obtain adequate nutrients. Like pure sugar and pure fat, alcohol is empty calories; it displaces nutrients. Table C3-7 shows the calorie amounts of typical alcoholic beverages.

Alcohol abuse also disrupts every tissue's metabolism of nutrients. Stomach cells oversecrete both acid and histamine, an agent of the immune system that produces inflammation. Intestinal cells fail to absorb thiamin, folate, vitamin B_6, and other vitamins. Liver cells lose efficiency in activating vitamin D and alter their production and excretion of bile. Rod cells in the retina, which normally process vitamin A alcohol (retinol) to the form needed in vision, find themselves processing drinking alcohol instead. Liver cells, too, suffer a reduced capacity to process and use vitamin A.[13] The kidneys excrete magnesium, calcium, potassium, and zinc.

The inadequate food intake and impaired nutrient absorption that accompany chronic alcohol abuse frequently lead to a deficiency of the B vitamin thiamin. In fact, the cluster of thiamin-deficiency symptoms commonly seen in chronic alcoholism has its own name—the **Wernicke-Korsakoff syndrome.** This syndrome is characterized by paralysis of the eye muscles, poor muscle coordination, impaired memory, and damaged nerves; the syndrome and other alcohol-related memory problems may respond to treatment with thiamin supplements.[14]

Most dramatic is alcohol's effect on folate. When an excess of alcohol is present, the body actively expels folate from all of its sites of action and storage. The liver, which normally contains enough folate to meet all needs, leaks its folate into the blood. As blood folate rises, the kidneys are deceived into excreting it, as if it were in excess. The intestine normally releases and retrieves folate continuously, but it becomes so damaged by folate deficiency and alcohol toxicity that it fails to absorb folate. Alcohol also interferes with the action of what little folate is left, causing a buildup in the blood of a compound suspected of involvement with many diseases, including heart disease, stroke, and birth defects.*[15] This interference inhibits the production of new cells, especially the rapidly dividing cells of the intestine and the blood.

Nutrient deficiencies are thus an inevitable consequence of alcohol abuse, not only because alcohol displaces food but also because alcohol interferes directly with the body's use of nutrients. People treated for alcohol addiction also need nutrition therapy to reverse deficiencies and even deficiency diseases rarely seen in others: night blindness, beriberi, pellagra, scurvy, and protein-energy malnutrition.

Does Moderate Alcohol Use Benefit Health?

What are the risks—and possible benefits to heart health—of moderate drinking? Does age matter?

Age does matter. Young people do not benefit their health by drinking; rather, they increase their risk of dying

*The compound is homocysteine; see Chapter 7.

Table C3•7

CALORIES IN ALCOHOLIC BEVERAGES AND MIXERS

Beverage	Amount (oz)	Energy (cal)
Beer	12	150
Dessert wine	3½	140
Fruit-flavored soda, Tom Collins mix	8	115
Gin, rum, vodka, whiskey (86 proof)	1½	105
Cola, root beer	8	100
Light beer	12	100
Table wine	3½	85
Tonic, ginger ale	8	80
Club soda, plain seltzer, diet drinks	8	1

from any cause.[16] Young nondrinkers are found to have a lower risk of dying than even light drinkers (fewer than 15 drinks per month) of the same age.[17] Alcohol is related to car crashes, homicides, and other violence that account for the great majority of deaths of people in this age group each year. Young women in particular should not drink alcohol for the sake of their heart. For women before menopause, the risk of heart disease is low, but the risk of breast cancer is substantial; and alcohol in the amounts that have been said to benefit the hearts of older people raises the risk of breast cancer in young women.[18]

Alcohol and Heart Disease

One to two standard drinks of alcoholic beverages a day are credited with reducing the risk of death from heart disease in people over 60 years old who have an increased risk of heart disease.[19] Increasing alcohol beyond this amount *increases* the risk of heart disease substantially.[20] Wine is often credited with aiding heart health, but research indicates that even beer may reduce heart attack risk in some populations.[21]

While many studies support a beneficial effect of moderate alcohol intake on heart health, the matter is not yet settled. Researchers followed the alcohol intakes and health histories of almost 6,000 men for over 20 years.[22] The results showed no beneficial relationship between mortality from cardiovascular disease and *any* level of alcohol consumption. The study did show an increased risk of death from all causes with more than 22 drinks per week and that men drinking more than 35 drinks a week had double the mortality from stroke compared with nondrinkers. Strokes are associated with elevated blood pressure, and ingestion of alcohol is known to alter blood pressure, first lowering it, then raising it for a time.[23]

The Health Effects of Wine

Red wine has also been credited with special health-supporting properties. The following two statements concerning wine and health have been approved to appear on U.S. wine labels:

- "The proud people who made this wine encourage you to consult your family doctor about the health effects of wine consumption."
- "To learn the health effects of wine consumption, send for the Federal Government's Dietary Guidelines for Americans, Center for Nutrition Policy and Promotion, USDA, 1120 20th Street, NW, Washington, DC 20036 or visit its website."

These statements seem to promise that good news about wine and health may await the information seeker, but the science on wine and health is mixed.* For example, the high potassium content of grape juice may lower high blood pressure, and potassium persists when the grape juice is made into wine. Since alcohol in large amounts raises blood pressure, however, the grape juice may be more suitable than the wine for people with hypertension. Dealcoholized wine also facilitates the absorption of potassium, calcium, phosphorus, magnesium, and zinc. So does wine, but the alcohol in it promotes the quick *excretion* of these minerals, so the dealcoholized version is preferred.

In addition to alcohol, wine contains flavonoids that are under study as antioxidants that protect the cardiovascular system against damage from oxidation that leads to heart disease. Details about oxidation and flavonoids and other phytochemicals are offered in Controversies 2 and 7. The antioxidant flavonoids in wine are offered as an explanation for why the wine-drinking French and other Mediterranean peoples have a lower incidence of heart disease despite having many risk factors.[24] However, compared with other food sources such as onions and other vegetables, wine may deliver only small amounts of flavonoids to the body.[25] (Controversy 5 comes back to the issue of Mediterranean diets and disease.) Alcohol itself may increase oxidation in ways that are damaging to the tissues of the liver and pancreas.[26] Dealcoholized wine, purple grape juice, and the grapes themselves contain similar flavonoids to those of wine, but do not increase oxidation.[27]

*For more information on wine labels, visit the Internet website of the Bureau of Alcohol, Tobacco, and Firearms: www.atf.treas.gov/press/label_ab.htm.

Alcohol and Appetite

Alcoholic beverages affect the appetite. Usually, they reduce it, making people unaware that they are hungry. But in people who are tense and unable to eat or in the elderly who have lost interest in food, small doses of wine taken 20 minutes before meals improve appetite. The congeners of wine are credited with this effect. For undernourished people and for people with severely depressed appetites, wine may facilitate eating even when psychotherapy fails to do so.

Another example of the beneficial use of alcohol comes from research showing that moderate use of wine in later life improves morale, stimulates social interaction, and promotes restful sleep. In nursing homes, improved patient and staff relations have been attributed to greater self-esteem among elderly patients who drink moderate amounts of wine. Researchers hypothesize that chronic fatigue may be responsible for some behaviors associated with old age. The positive effects of wine on sleep may alleviate the fatigue, easing social interactions.

The Final Word

This discussion has explored some of the ways alcohol affects health and nutrition. In contrast to some possible benefits of moderate alcohol consumption, excessive alcohol consumption presents a great potential for harm. Alcohol is guilty of contributing not only to deaths from health problems, but also to most of the other deaths of young people, including car crashes, falls, suicides, homicides, drownings, and other accidents. The surest way to escape the harmful effects of alcohol is, of course, to refuse alcohol altogether. If you choose to drink, do so with care and strictly in moderation.

Shiva Dayal Lal, *Women Selling Grains and Vegetables*, mid-19th century

The Carbohydrates: Sugar, Starch, Glycogen, and Fiber

Frequently Asked Questions

If I want to lose weight, should I avoid carbohydrates? p. 104

How does fiber in food affect my health, and how much do I need to stay healthy? p. 104

Can my diet have too much fiber? p. 107

Why do some people have trouble digesting milk? p. 113

What is diabetes? p. 117

If I feel dizzy between meals, do I have hypoglycemia? p. 119

Contents

carbohydrates compounds composed of single or multiple sugars. The name means "carbon and water," and a chemical shorthand for carbohydrate is CHO, signifying carbon (C), hydrogen (H), and oxygen (O).

complex carbohydrates long chains of sugar units arranged to form starch or fiber; also called *polysaccharides*.

simple carbohydrates sugars, including both single sugar units and linked pairs of sugar units. The basic sugar unit is a molecule containing six carbon atoms, together with oxygen and hydrogen atoms.

photosynthesis the process by which green plants make carbohydrates from carbon dioxide and water using the green pigment chlorophyll to capture the sun's energy (*photo* means "light"; *synthesis* means "making").

chlorophyll the green pigment of plants that captures energy from sunlight for use in photosynthesis.

Figure 4•1

CARBOHYDRATE—MAINLY GLUCOSE—IS MADE BY PHOTOSYNTHESIS

The sun's energy becomes part of the glucose molecule—its calories, in a sense. In the molecule of glucose on the leaf here, dots represent the carbon atoms; bars represent the chemical bonds that contain energy.

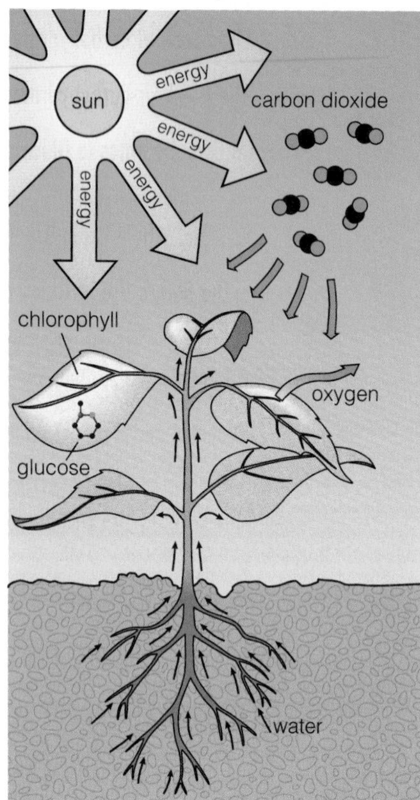

CARBOHYDRATES ARE THE ideal nutrients to meet your body's energy needs, keep your digestive system fit, feed your brain and nervous system, and, within calorie limits, help keep your body lean. False propaganda about carbohydrate's supposed "fattening power" misleads millions of weight-conscious people to avoid carbohydrate-rich foods, a counterproductive tactic. In truth, people who wish to lose fat and to maintain lean tissue can do no better than to control calories and design their diets around foods that supply carbohydrates in balance with other energy nutrients. Digestible carbohydrates, together with fats and protein, add bulk to foods and provide energy and other benefits for the body.[1] Indigestible carbohydrates, which include most of the fibers in foods, yield little or no energy but provide other important benefits.

All carbohydrates are not equal as far as nutrition is concerned. This chapter invites you to learn to distinguish between foods containing the **complex carbohydrates** (starch and fiber), which are put to good use in the body, and those made of the **simple carbohydrates** (sugars), which can be less valuable to people's health. The Controversy asks whether the sugar added to foods harms health and whether the alternative sweeteners designed to replace sugar are preferable.

This chapter on the carbohydrates is the first of three on the energy-yielding nutrients. Chapter 5 deals with the fats and Chapter 6 with protein. The Controversy of Chapter 3 addressed one other contributor of energy, alcohol.

A Close Look at Carbohydrates

Carbohydrates contain the sun's radiant energy, captured in a form that living things can use to drive the processes of life. Green plants make carbohydrate through **photosynthesis** in the presence of **chlorophyll** and sunlight. In this process, water (H_2O), absorbed by the plant's roots, donates hydrogen and oxygen, and carbon dioxide gas (CO_2), absorbed into its leaves, donates carbon and oxygen. Water and carbon dioxide combine to yield the most common of the **sugars,** the single sugar **glucose.** Scientists know the reaction in the minutest detail but have never been able to reproduce it from scratch; green plants are required to make it happen (see Figure 4-1).

Light energy from the sun drives the photosynthesis reaction. The light energy becomes the chemical energy of the bonds that hold six atoms of carbon together in the sugar glucose. Glucose provides energy for the work of all cells of the stem, roots, flowers, and fruits of the plant. For example, in the roots, far from the energy-giving rays of the sun, each cell draws upon some of the glucose made in the leaves, breaks it down (to carbon dioxide and water), and uses the energy thus released to fuel its own growth and water-gathering activities.

Plants do not use all of the energy stored in their sugars, so it remains available for use by the animal or human being that consumes the plant. Thus, carbohydrates form the first link in the food chain that supports all life on earth. Carbohydrate-rich foods come almost exclusively from plants; milk is the only animal-derived food that contains significant amounts of carbohydrate. The next few sections describe the forms assumed by carbohydrates: sugars, starch, glycogen, and fiber.

key point *Through photosynthesis, plants combine carbon dioxide, water, and the sun's energy to form glucose. Carbohydrates are made of carbon, hydrogen, and oxygen held together by energy-containing bonds:* carbo *means "carbon";* hydrate *means "water."*

Sugars

Six sugar molecules are important in nutrition. Three are single sugars, or **monosaccharides.** The other three are double sugars, or **disaccharides.** All of their chemical names end in *ose*, which means "sugar." Although they all sound alike at first, they exhibit distinct characteristics to the nutrition enthusiast who gets to know each individually. Figure 4-2 shows the relationships among the sugars.

The three monosaccharides are glucose, **fructose,** and **galactose.** Fructose or fruit sugar, the intensely sweet sugar of fruit, is made by rearranging the atoms in glucose molecules. Fructose occurs mostly in fruits, in honey, and as part of table sugar. Glucose and fructose are the most common monosaccharides in nature.

The other monosaccharide, galactose, has the same number and kind of atoms as glucose and fructose but in yet another arrangement. Galactose is one of two single sugars that are bound together to make up the sugar of milk. It rarely occurs free in nature but is tied up in milk sugar until it is freed during digestion.

The three other sugars important in nutrition are disaccharides, which are linked pairs of single sugars. All three contain glucose. In **lactose,** the sugar of milk just mentioned, glucose is linked to galactose.

Malt sugar, or **maltose,** has two glucose units. Maltose appears wherever starch is being broken down. It occurs in germinating seeds and arises during the digestion of starch in the human body.

The last of the six sugars, **sucrose,** is familiar table sugar, the product most people think of when they refer to *sugar*. In sucrose, fructose and glucose are bonded together. Table sugar is obtained by refining the juice from sugar beets or sugarcane, but sucrose also occurs naturally in many vegetables and fruits. It tastes sweet, as does fruit sugar, because it contains the sweet monosaccharide fructose. Sucrose is of major importance in human nutrition, and research about its effects on the human body is a topic of Controversy 4.

When you eat a food containing single sugars, you can absorb them directly into your blood. When you eat disaccharides, though, you must digest them first. Enzymes in your intestine must split the disaccharides into separate monosaccharides so they can enter the bloodstream. The blood delivers all products of digestion first to the liver, which possesses enzymes to modify nutrients, making them useful to the body. Glucose is the most-used monosaccharide inside the body, so the liver quickly converts fructose or galactose to glucose or to smaller pieces that can serve as building blocks for either glucose or fat.

Just because the energy of fruits and many vegetables comes from sugars doesn't mean that eating them is the same as eating concentrated sweets such as candy or cola beverages. Vegetables and fruits have higher nutrient density—their sugars arrive in the body diluted in large volumes of water, packaged with fiber, and mixed with many needed vitamins and minerals. In contrast, all types of refined sugars, even

sugars simple carbohydrates, that is, molecules of either single sugar units or pairs of those sugar units bonded together.

glucose (GLOO-cose) a single sugar used in both plant and animal tissues for energy; sometimes known as blood sugar or *dextrose*.

monosaccharides (mon-oh-SACK-ah-rides) single sugar units (*mono* means "one"; *saccharide* means "sugar unit").

disaccharides pairs of single sugars linked together (*di* means "two").

fructose (FROOK-tose) a monosaccharide; sometimes known as fruit sugar (*fruct* means "fruit"; *ose* means "sugar").

galactose (ga-LACK-tose) a monosaccharide; part of the disaccharide lactose (milk sugar).

lactose a disaccharide composed of glucose and galactose; sometimes known as milk sugar (*lact* means "milk"; *ose* means "sugar").

maltose a disaccharide composed of two glucose units; sometimes known as malt sugar.

sucrose (SOO-crose) a disaccharide composed of glucose and fructose; sometimes known as table, beet, or cane sugar.

Single sugars are monosaccharides.

Pairs of sugars are disaccharides.

Figure 4•2

HOW MONOSACCHARIDES JOIN TO FORM DISACCHARIDES

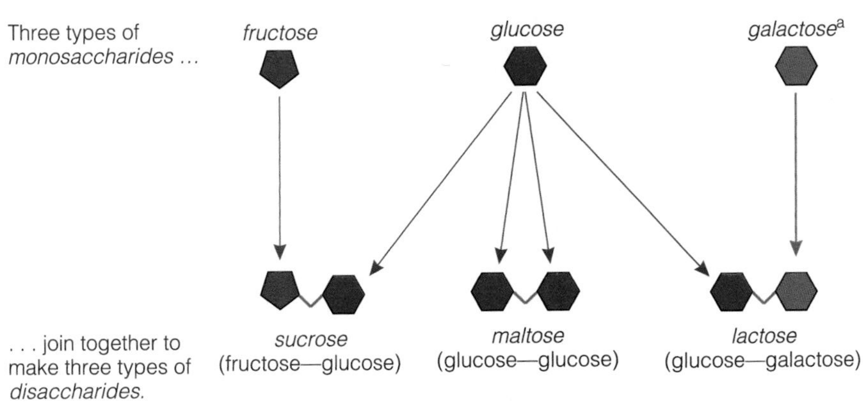

[a]Galactose does not occur in foods singly but only as part of lactose.

A note on the glucose symbol:
The glucose molecule is really a ring of 5 carbons and 1 oxygen plus a carbon "flag."

carbons — oxygen

For convenience, glucose is symbolized as [ring] or [hexagon].

polysaccharides another term for complex carbohydrates; compounds composed of long strands of glucose units linked together (*poly* means "many"). Also called *complex carbohydrates*.

starch a plant polysaccharide composed of glucose. After cooking, starch is highly digestible by human beings; raw starch often resists digestion.

granules small grains. Starch granules are packages of starch molecules. Various plant species make starch granules of varying shapes.

Strands of many sugar units are polysaccharides.

honey, arrive in the body in concentrated form, practically devoid of nutrients. From the body's point of view, fruits are vastly different from purified sugars, except that both provide glucose in abundance.

Glucose is the most important monosaccharide in the human body. Most other monosaccharides and disaccharides become glucose in the body.

Starch

In addition to occurring in sugars, the glucose in foods also occurs in long strands of thousands of glucose units. These are the **polysaccharides** (see Figure 4-3). Starch is a polysaccharide as are glycogen and most of the fibers.

Starch is a plant's storage form of glucose. As a plant matures, it not only provides energy for its own needs but also stores energy in its seeds for the next generation. For example, after a corn plant reaches its full growth and has many leaves manufacturing glucose, it stores packed clusters of **starch** molecules in **granules** and packs the granules into its seeds (the kernels). Glucose is soluble in water and would be washed away by rains while the seed lay in the soil. Starch is an insoluble substance that will stay with the seed and nourish it until it forms shoots with leaves that can catch the sun's rays. Most of the starch of corn and other plant foods is nutritive for human beings, too, because they can digest the starch to glucose and extract the sun's energy stored in its chemical bonds. A later section describes starch digestion in greater detail.

Figure 4•3

HOW GLUCOSE MOLECULES JOIN TO FORM POLYSACCHARIDES

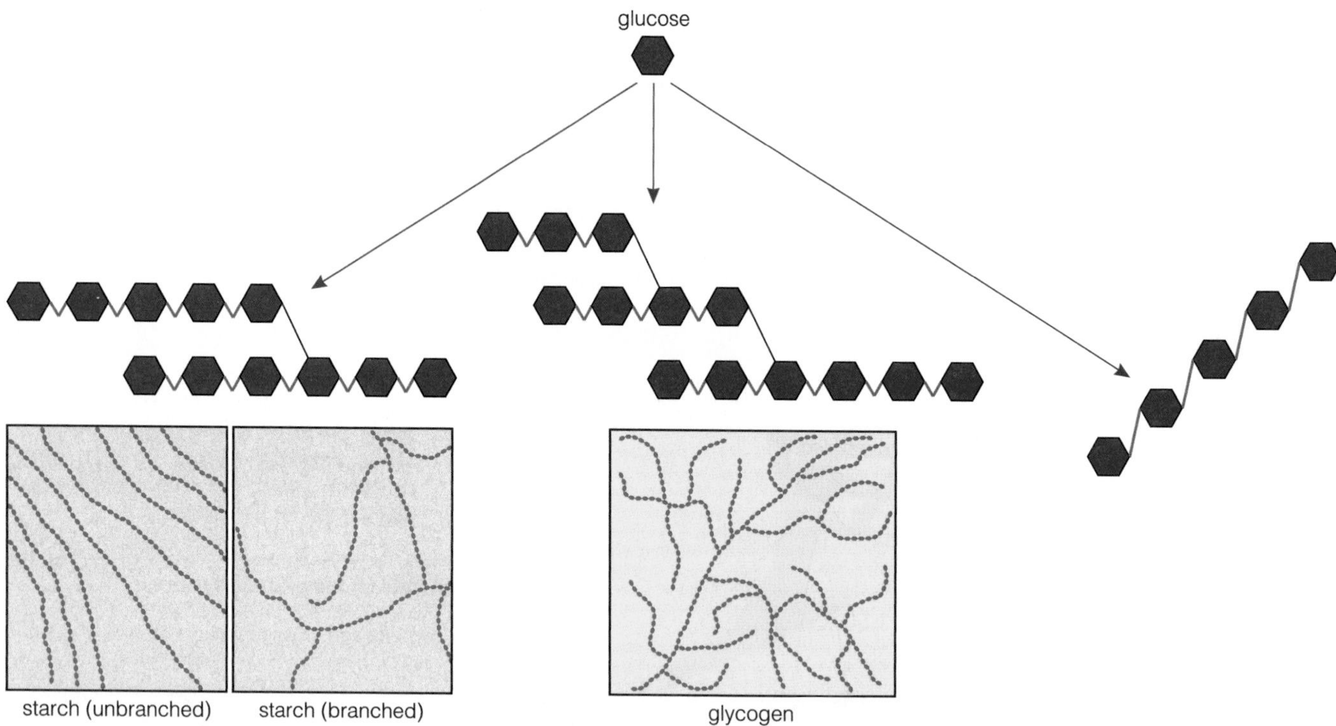

Starch Glucose units are linked in long, occasionally branched chains to make starch. Human digestive enzymes can digest these bonds, retrieving glucose. Real glucose units are so tiny that you can't see them, even with the highest-power light microscope.

Glycogen Glycogen resembles starch in that the bonds between its glucose units can be broken by human enzymes, but the chains of glycogen are more highly branched.

Cellulose (fiber) The bonds that link glucose units together in cellulose are different from the bonds in starch or glycogen. Human enzymes cannot digest them.

Starch is the storage form of glucose in plants and is also nutritive for human beings.

Glycogen

Just as plants store glucose in long chains of starch, animal bodies store glucose in long chains of **glycogen.** Glycogen resembles starch in that it consists of glucose molecules linked together to form chains, but its chains are longer and more highly branched (see Figure 4-3). Unlike starch, which is abundant in grains, potatoes, and other foods from plants, glycogen is nearly undetectable in meats because glycogen breaks down rapidly upon the slaughter of the animal. A later section describes how the human body handles its own packages of stored glucose.

Glycogen is the storage form of glucose in animals and human beings.

Fiber

The **fibers** of a plant form the supporting structures of its leaves, stems, and seeds. Most fibers are polysaccharides—chains of sugars—but the sugar units are held together by bonds that human digestive enzymes cannot break. Most fibers therefore pass through the human body without providing energy for its use. The best-known fibers are *cellulose* (shown in Figure 4-3), *hemicellulose*, and *pectin*. (Other fibers are *gums*, *mucilages*, and *lignins*.) Cellulose and hemicellulose are found in the familiar strings of celery, the skins of corn kernels, and the membranes surrounding kernels of wheat. In the body, these fibers provide **roughage**—fiber that aids in digestion and elimination. Pectin, isolated from plants such as apples or citrus fruits, is used as a food additive to thicken jelly, keep salad dressing from separating, and otherwise alter the texture and consistency of processed foods.

The term *dietary fiber* refers to substances that cannot be broken down by human digestive enzymes but are somewhat vulnerable to breakdown by the enzymes of bacteria that reside in our digestive tracts. Intestinal bacteria change some fibers and some other undigested substances into products that can be absorbed, contribute a few calories' worth of energy, and may provide health benefits. The amount of fiber that is broken down depends on the nature of both the fiber and the bacteria in the tract.

Some animals, such as cattle, depend heavily on their intestinal bacteria to make the energy of glucose available from the abundant cellulose in their fodder. When we eat beef, we indirectly receive some of the sun's energy that was originally stored in the fiber of the plants the cattle ate and converted into the energy contained in meat. Beef contains no fiber itself; no meats or dairy products contain fiber.

One way to classify fibers is according to how readily they dissolve in water.* Some fibers are **insoluble fibers;** others are **soluble fibers.** Each type of fiber exerts important effects on people's health, which will be described later.

Little fiber is digested by the enzymes in the human digestive tract. Much of the fiber passes through the digestive tract unchanged.

In summary, carbohydrate plays a prominent role in the global carbon cycle. Carbon dioxide, water, and energy are combined in plants to form glucose; the plants may store the glucose in the polysaccharide starch. Then animals or people eat the plants and retrieve the glucose. In the body, the liver and muscles may store the glucose as the polysaccharide glycogen, but ultimately it becomes glucose again. The glucose delivers the sun's energy to fuel the body's activities. In the process, glucose breaks down to waste products, carbon dioxide and water, which are excreted. Later, these compounds are used again by plants as raw materials to make carbohydrate.

*Another way to classify fibers is by chemical name without reference to solubility. See joint FAO/WHO Expert Consultation, *Carbohydrates in Human Nutrition* (Geneva: Food and Agriculture Organization, World Health Organization, 1998).

glycogen (GLY-co-gen) a polysaccharide composed of glucose that is made and stored by liver and muscle tissues of human beings and animals as a storage form of glucose. Glycogen is not a significant food source of carbohydrate and is not counted as one of the complex carbohydrates in foods.

fibers the indigestible polysaccharides in food, consisting mostly of cellulose, hemicellulose, and pectin; also called *nonstarch polysaccharides*.

roughage (RUFF-idge) the rough parts of food; an imprecise term that has largely been replaced by the term *fiber*.

insoluble fibers the tough, fibrous structures of fruits, vegetables, and grains; indigestible food components that do not dissolve in water.

soluble fibers food components that readily dissolve in water and often impart gummy or gel-like characteristics to foods. An example is pectin from fruit, which is used to thicken jellies. Soluble fibers are indigestible by human enzymes but may be broken down to absorbable products by bacteria in the digestive tract.

Chapter 15 revisits humankind's relationship with the earth's food chain.

The sugars in these carrots are diluted with water and packaged with vitamins, minerals, and fiber.

The Need for Carbohydrates

Glucose from carbohydrate is the preferred fuel for most body functions. Only two other nutrients provide energy to the body: protein and fats. Protein-rich foods are usually expensive, and when used to make fuel for the body, they provide no advantage over carbohydrates. Their overuse has disadvantages, as explained in Chapter 6. Fats normally are not used as fuel by the brain and central nervous system, and diets high in fats are associated with many disease states. Thus, glucose is the preferred energy source. Nerve cells, including those of the brain, depend almost exclusively on glucose for their energy. And starchy foods, or complex carbohydrates, are the preferred source of glucose in the diet.

If I Want to Lose Weight, Should I Avoid Carbohydrates?

Complex carbohydrates are often wrongly accused of being the "fattening" ingredients of foods. Most people need to consume *more* starchy foods rather than less. Gram for gram, carbohydrates donate fewer calories than do dietary fats, so a moderate, balanced diet based on high-carbohydrate foods is likely to be lower in calories than a diet of high-fat foods. Also, to convert glucose to fat in the body requires chemical conversions that cost many of the glucose's original calories, which makes glucose even less fattening. Government agencies in many countries urge their citizens to consume foods that contain abundant complex carbohydrates. Table 4-1 reviews the U.S. recommendations and goals first presented in Chapters 1 and 2, as well as the World Health Organization's recommended upper and lower limits for carbohydrate intakes.

Details about controlling body fatness are presented in Chapter 9.

Unlike complex carbohydrates, pure sugars displace nutrient-dense foods from the diet. Purified, refined sugar (sucrose) contains no other nutrients—protein, vitamins, minerals, or fiber—and is termed an empty-calorie food. If you choose 400 calories of sugar in place of 400 calories of starchy food such as whole-grain bread, you lose the starch, vitamins, minerals, and fiber of the bread. You can afford to do this only if you have already met your nutrient needs for the day and still have calories to spend. This chapter's Food Feature offers more about the sugars in foods.

key point *Complex carbohydrates are the preferred energy source for the body.*

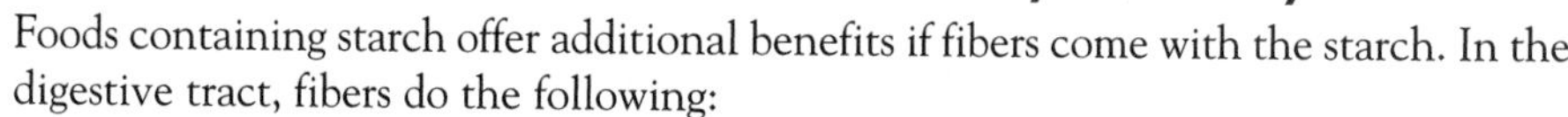

How Does Fiber in Food Affect My Health, and How Much Do I Need to Stay Healthy?

Foods containing starch offer additional benefits if fibers come with the starch. In the digestive tract, fibers do the following:

1. *Slow the absorption of nutrients* and other molecules by entrapping molecules and preventing their contact with absorptive surfaces.
2. *Delay cholesterol absorption*, probably by the same mechanism.
3. *Bind bile for excretion*.
4. *Stimulate bacterial fermentation* in the colon (described below).
5. *Increase stool weight* by holding water within the feces.

These five actions underlie the many health benefits attributed to dietary fibers:

- Possibly reduce the risk of heart and artery disease and stroke:
  - Soluble fibers recommended for lowering blood cholesterol may do so by delaying absorption in the digestive tract and because bile binds to the fiber and is excreted (see Figure 4-4 on page 106).[2]

Table 4•1

RECOMMENDATIONS CONCERNING INTAKES OF CARBOHYDRATES

1. Recommendations for complex carbohydrates

   Dietary Guidelines
   - Every day eat 5 to 9 servings[a] of a combination of vegetables (including legumes) and fruits. Also, choose a variety of grains daily, especially whole grains (6 to 11 daily servings of a combination of breads, crackers, popcorn, cereals, and cooked grains).

   Daily Values[b]
   - 300 grams of complex carbohydrate, or 60% of total calories.

   Healthy People 2010
   - Increase intakes of fruits and vegetables, including legumes, to at least 5 servings a day.
   - Increase intakes of grain products to at least 6 servings a day.

   World Health Organization
   - Lower limit: 50% of total calories from complex carbohydrates.
   - Upper limit: 75% of total calories from complex carbohydrates.

2. Recommendations for refined sugars

   Dietary Guidelines
   - Use sugars only in moderation.

   World Health Organization
   - Lower limit: 0% of total calories from refined sugars.
   - Upper limit: 10% of total calories from refined sugars.

3. Recommendations for dietary fiber

   Dietary Guidelines
   - Increase your fiber intake by eating more of a variety of foods that contain fiber naturally.

   Daily Values[b]
   - 25 grams of fiber per day.

   World Health Organization
   - Lower limit: 27 grams of dietary fiber a day.
   - Upper limit: 40 grams of dietary fiber a day.

[a]Serving sizes were presented in Figure 2-4 of Chapter 2.
[b]Daily Values are for a 2,000-calorie diet.

diverticulosis (dye-ver-tic-you-LOH-sis) outpocketing or ballooning out of areas of the intestinal wall, caused by weakening of the muscle layers that encase the intestine.

constipation hardness and dryness of bowel movements, associated with discomfort in passing them from the body.

hemorrhoids (HEM-or-oids) swollen, hardened (varicose) veins in the rectum, usually caused by the pressure resulting from constipation.

appendicitis inflammation and/or infection of the appendix, a sac protruding from the intestine.

  - Some soluble fibers are digested by intestinal bacteria to yield small, fatlike products that, when absorbed, may lower LDL cholesterol (the harmful kind).[3]
  - Fibers displace fatty, cholesterol-raising foods from the diet.
- Improve the body's handling of glucose and the hormone insulin.[4] A habitual lack of fiber in the diet along with an abundance of highly refined carbohydrate foods is likely to increase the risk of developing diabetes.[5]
- Help in maintaining a healthy body weight:
  - Fibers reduce energy consumption by displacing calorie-dense concentrated fats and sweets from the diet while donating little energy.
  - Because fibers absorb water and swell, they promote feelings of fullness. By slowing the movement of food through the upper digestive tract, fiber also delays hunger's return.[6]
- Improve the health of the digestive tract:
  - Insoluble fiber stimulates the muscles of the digestive tract so that they retain their health and tone, preventing **diverticulosis,** in which the intestinal walls become weak and bulge out in places (see Figure 4-5 in the margin on the next page).
  - Insoluble fiber helps prevent **constipation** and **hemorrhoids.**
  - Insoluble fiber helps prevent bacterial infection of the appendix **(appendicitis).**

See Figure 4-14 on page 126 or Appendix A for lists of the fiber contents of foods.

On the strength of findings from laboratory experiments and population studies, it was surmised that a high-fiber diet might help prevent colon cancer. Some types of

Figure 4•4

ONE WAY FIBER IN FOOD MAY LOWER CHOLESTEROL IN THE BLOOD

In some ways, the liver is like a vacuum cleaner, sucking up cholesterol from the blood, converting the cholesterol to bile, and discharging the bile into its storage bag, the gallbladder. The gallbladder empties its bile into the intestine, where bile performs necessary digestive tasks. In the intestine, some of the bile links up with fiber and is carried out of the body in feces.

A. *When the diet is rich in fiber, much of the cholesterol (as bile) is carried out of the body.*
B. *When the diet is low in fiber, most of the cholesterol is reabsorbed and returned to the bloodstream.*

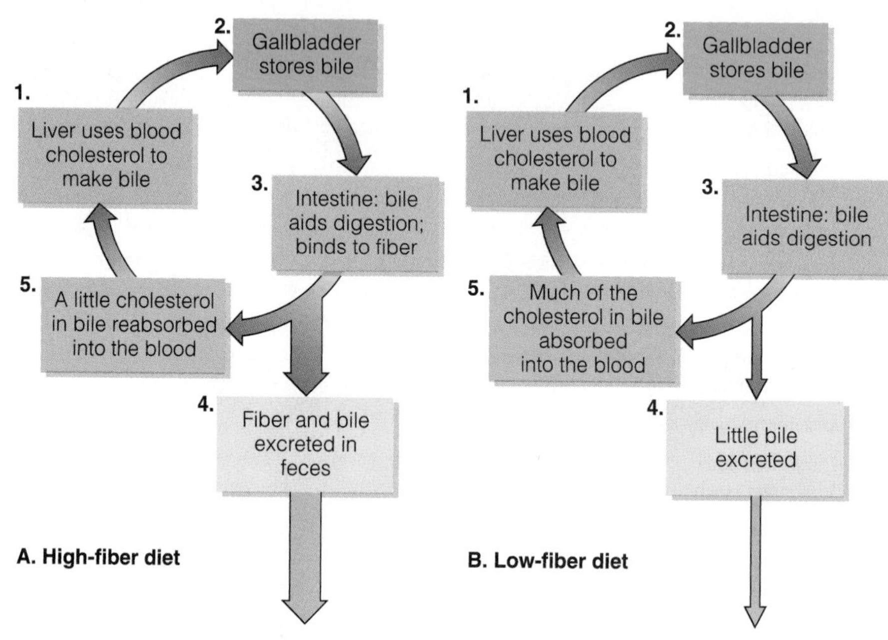
1. Liver uses blood cholesterol to make bile
2. Gallbladder stores bile
3. Intestine: bile aids digestion; binds to fiber
4. Fiber and bile excreted in feces
5. A little cholesterol in bile reabsorbed into the blood
A. High-fiber diet
1. Liver uses blood cholesterol to make bile
2. Gallbladder stores bile
3. Intestine: bile aids digestion
4. Little bile excreted
5. Much of the cholesterol in bile absorbed into the blood
B. Low-fiber diet

Chapter 11 and Controversy 2 have more about foods and cancer.

fiber, fiber extracts, and products of fiber from bacterial digestion reduce indicators of colon cancer in animal and tissue studies.[7] Studies of people who are free to eat as they choose, however, have revealed little correlation between fiber-rich diets and development or recurrence of colon cancer.[8] Until research clarifies this issue, a high-fiber diet still has much to recommend it.[9] Fiber-rich whole grains, fruits, vegetables, and legumes are concentrated sources of antioxidants, phytochemicals, vitamins, and minerals that also may be important in cancer prevention.[10]

People choosing high-fiber foods are wise to seek out a variety of fiber sources and to drink extra fluids to help the fiber do its job. Wheat bran, which is composed mostly of insoluble fibers, is one of the most effective stool-softening fibers; oat bran and other more soluble fibers have a greater cholesterol-lowering effect. The fibers of legumes, apples, and carrots may also lower blood cholesterol. Table 4-2 shows the diverse effects of different fibers. Note that most unrefined plant foods contain a mix of fiber types.

In the United States, people report an average intake of about 14 to 15 grams of fiber per day—too low to meet recommendations.[11] The Daily Value for fiber is 25 or 30 grams per day for people eating a 2,000-calorie or 2,500-calorie diet, respectively. Most experts agree that healthy people should meet their fiber need by eating unprocessed, fiber-containing foods such as whole grains, legumes, fruits, and vegetables, and not by eating refined fiber sources such as fiber supplements, because many of the benefits attributed to fiber may come from other constituents of fiber-containing foods, and not from fiber alone.[12] This chapter's Consumer Corner provides detailed information about choosing wisely among grain foods, and the Do It! feature guides you in analyzing the fiber in your own diet. You can get a quick approximation of the fiber in your diet right now by following the instructions in Table 4-3. Use your food record from the Do It! section of Chapter 2.

Figure 4•5

DIVERTICULOSIS

Diverticula are abnormal bulging pockets formed in the colon wall. These pockets can entrap feces and become painfully infected and inflamed, requiring hospitalization, antibiotic therapy, or surgery.

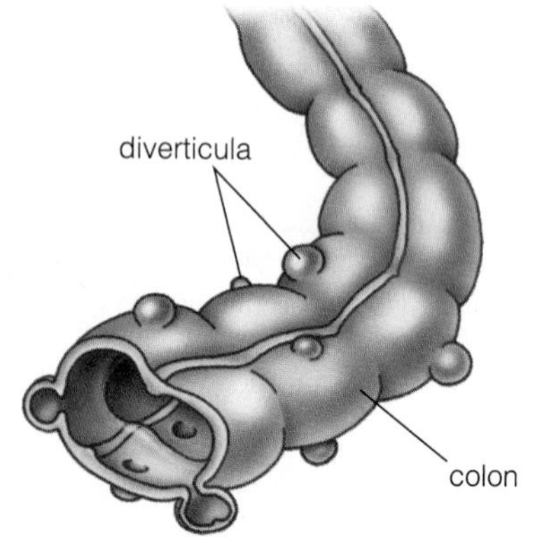
diverticula
colon

key point *Fibers aid in maintaining the health of the digestive tract and help to prevent or control certain diseases. Most people need between 20 and 40 grams of fiber each day.*

Table 4•2

WATER SOLUBILITIES, SOURCES, AND HEALTH EFFECTS OF FIBER

Fiber Type	Major Food Sources	Possible Health Effects
Soluble Gums, mucilages, pectins, psyllium,[a] some hemicellulose	Barley, fruits, legumes, oats, oat bran, rye, seeds, vegetables	Lower blood cholesterol Slow glucose absorption Slow transit of food through upper digestive tract Hold moisture in stools, softening them Are partly fermentable into fragments the body can use
Insoluble Cellulose, lignin, some hemicellulose	Brown rice, fruits, legumes, seeds, vegetables, wheat bran, whole grains	Soften stools Regulate bowel movements Speed transit of material through small intestine Increase fecal weight and speed fecal passage through colon Reduce risks of diverticulosis, hemorrhoids, and appendicitis

[a]Psyllium, a fiber laxative and a cereal additive, has both soluble and insoluble properties.

chelating (KEE-late-ing) **agents** molecules that surround other molecules and are therefore useful in either preventing or promoting movement of substances from place to place.

Chelating agents are often sold by supplement vendors to "remove poisons" from the body. Some valid medical uses such as treatment of lead poisoning exist, but most of the chelating agents sold over-the-counter are based on unproven claims.

Q Can My Diet Have Too Much Fiber?

Adding purified fibers, such as oat or wheat bran, to foods can be taken to extremes. One enthusiastic eater of oat bran muffins required emergency surgery for a blocked intestine; he had eaten so much bran that his digestive system was unable to function. This doesn't mean that you should avoid bran-containing foods, but that you should use bran with moderation. Less extreme concerns are that purified fiber might displace nutrients from the diet or cause them to be lost from the digestive tract. Purified fibers are like refined sugars in one way: the nutrients that originally accompanied the fibers have been lost. Also, purified fiber may not affect the body the same way as the fiber in its original food product.

Binders in some fibers act as **chelating agents** and link chemically with nutrient minerals (iron, zinc, calcium, and others) and then carry them out of the body. The mineral iron is mostly absorbed at the beginning of the intestinal tract, and excess insoluble fibers may limit absorption by speeding foods through the upper part of the digestive tract. Too much bulk in the diet can also limit the total amount of food consumed and cause deficiencies of both nutrients and energy. The malnourished, the elderly, and children who consume no animal products are particularly vulnerable to this chain of events. Fibers also carry water out of the body and can cause dehydration. Add an extra glass or two of water to go along with the fiber added to your diet.

Fiber needs are best met with whole foods. Purified fiber in large doses can have undesirable effects.

Table 4•3

A QUICK METHOD FOR ESTIMATING FIBER INTAKE

To quickly estimate fiber in a day's meals:

1. Multiply servings[a] of fruits and vegetables by 1.5 g.[b]
   Example: 5 servings of fruits and vegetables × 1.5 = 7.5 g fiber
2. Multiply servings of refined grains by 1.0 g.
   Example: 4 servings of refined grains × 1.0 = 4.0 g fiber
3. Multiply servings of whole grains by 2.5 g.
   Example: 3 servings of whole grains × 2.5 = 7.5 g fiber
4. Add fiber values for servings of legumes, nuts, seeds, and high-fiber cereals and breads; look these up in Appendix A.
   Example: ½ c black beans = 8.0 g fiber
5. Add up the grams of fiber from the previous lines.
   Example: 7.5 + 4.0 + 7.5 + 8.0 = 27 g fiber

Day's total fiber = 27 g fiber

[a]Use standard serving sizes presented in Chapter 2.
[b]Juices do not count toward this total.
SOURCE: Adapted from J. A. Marlett and T.-F. Cheung, Database and quick methods of assessing typical dietary fiber intakes using data for 228 commonly consumed foods, *Journal of the American Dietetic Association* 97 (1997): 1139–1148, 1151.

From Carbohydrates to Glucose

The body's cells cannot use foods such as bread or even whole molecules of lactose, sucrose, or starch for energy, but they need the glucose in those molecules, and they need it continuously. The various body systems must make glucose available to the cells, not all at once when it is eaten, but at a steady rate all day.

Consumer Corner

Refined, Enriched, and Whole-Grain Bread

FOR MANY PEOPLE, bread supplies much of the carbohydrate, or at least most of the starch, in a day's meals. Any food used in such abundance in the diet should be scrutinized closely, and if it doesn't measure up to high nutrition standards, it should be replaced with a food that does. The meanings of the words **refined, enriched, fortified,** and **whole grain** hold the key to understanding the nutritional levels of wheat bread (see Table 4-4).

The part of the wheat plant that is made into flour and then into bread and other baked goods is the seed or kernel. The wheat kernel (a whole grain) has four main parts: the **germ,** the **endosperm,** the **bran,** and the **husk,** as shown in Figure 4-6. The germ is the part that grows into a wheat plant and therefore contains concentrated food to support the new life—it is especially rich in vitamins and minerals. The endosperm is the soft, white, inside portion of the kernel, containing starch and proteins that help nourish the seed as it sprouts. The kernel is encased in the bran, a protective coating that is similar in function to the shell of a nut; the bran is also rich in nutrients and fiber. The husk, commonly called chaff, is the dry outermost layer and is inedible for human beings but can be used in animal feed.

In earlier times, people milled wheat by grinding it between two stones, blowing or sifting out the chaff, and retaining the nutrient-rich bran and germ as well as the endosperm. Then milling machinery was "improved," and it became possible to remove the dark, heavy germ and bran as well, leaving a whiter, smoother-textured flour. People

Table 4•4

TERMS THAT DESCRIBE GRAIN FOODS

- **bran** the protective fibrous coating around a grain; the chief fiber donator of a grain.
- **brown bread** bread containing ingredients such as molasses that lend a brown color; may be made with any kind of flour, including white flour.
- **endosperm** the bulk of the edible part of a grain, the starchy part.
- **enriched, fortified** refers to the addition of nutrients to a refined food product. As defined by U.S. law, these terms mean that specified levels of thiamin, riboflavin, niacin, folate, and iron have been added to refined grains and grain products. The terms *enriched* and *fortified* can refer to the addition of more nutrients than just these five; read the label.[a]
- **germ** the nutrient-rich inner part of a grain.
- **husk** the outer, inedible part of a grain.
- **refined** refers to the process by which the coarse parts of food products are removed. For example, the refining of wheat into flour involves removing three of the four parts of the kernel—the chaff, the bran, and the germ—leaving only the endosperm, composed mainly of starch and a little protein.
- **stone ground** refers to a milling process using limestone to grind any grain, including refined grains, into flour.
- **unbleached flour** a beige-colored endosperm flour with texture and nutritive qualities that approximate those of regular white flour.
- **wheat flour** any flour made from wheat, including white flour.
- **white flour** an endosperm flour that has been refined and bleached for maximum softness and whiteness.
- **whole grain** refers to a grain milled in its entirety (all but the husk), not refined.
- **whole-wheat flour** flour made from whole-wheat kernels; a whole-grain flour.

[a]Formerly, *enriched* and *fortified* carried distinct meanings with regard to the nutrient amounts added to foods, but a change in the law has made these terms virtually synonymous.

Figure 4•6

A WHEAT PLANT AND A SINGLE KERNEL OF WHEAT

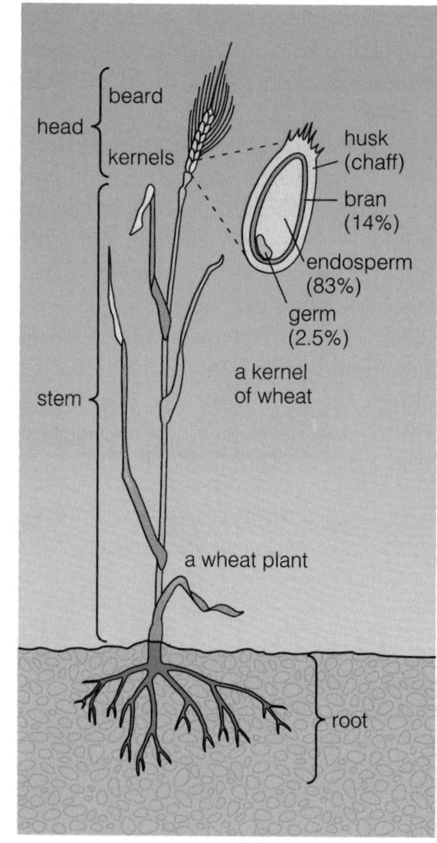

In Western societies, bread is the staff of life.

products still contain less magnesium, zinc, vitamin B_6, vitamin E, and chromium than whole-grain products do. When a grain is refined, fiber is lost, too (see Table 4-5). Bread sold for weight-reduction

Table 4•5

GRAMS OF FIBER IN ONE CUP OF FLOUR

Dark rye, 18 g
Whole wheat, 15 g
Light rye, 14 g
Buckwheat, 12 g
Whole-grain cornmeal, 9 g
Enriched white, 3 g

looked on this flour as more desirable than the crunchy, dark brown, "old-fashioned" flour.

In turning to white bread, bread eaters suffered a tragic loss of needed nutrients. Many people developed deficiencies of iron, thiamin, riboflavin, and niacin—nutrients formerly received from whole-grain bread. Finally, the problem was recognized, and Congress passed the Enrichment Act requiring that iron, niacin, thiamin, and riboflavin be added to refined grain products before they were sold. The Enrichment Act of 1942 is still in effect in the United States today but was amended in 1996 to include the vitamin folate (sometimes called folic acid on food labels). A single slice of refined bread is not "rich" in these nutrients, but people who eat several slices of bread a day obtain significantly more of the nutrients than they would from unenriched white bread, as Figure 4-7 on this page shows. Today, breads, grain products such as rice, macaroni, and spaghetti, and all types of cereals have been enriched with at least these nutrients.

To a great extent, the enrichment of grain products eliminated known deficiency problems, but other deficiencies went undetected for many more years. The trouble with enriched flour is that it is comparable to whole grain only with respect to the added nutrients and not with respect to others. Enriched

Figure 4•7

NUTRIENTS IN WHOLE-GRAIN, ENRICHED WHITE, AND UNENRICHED WHITE BREAD

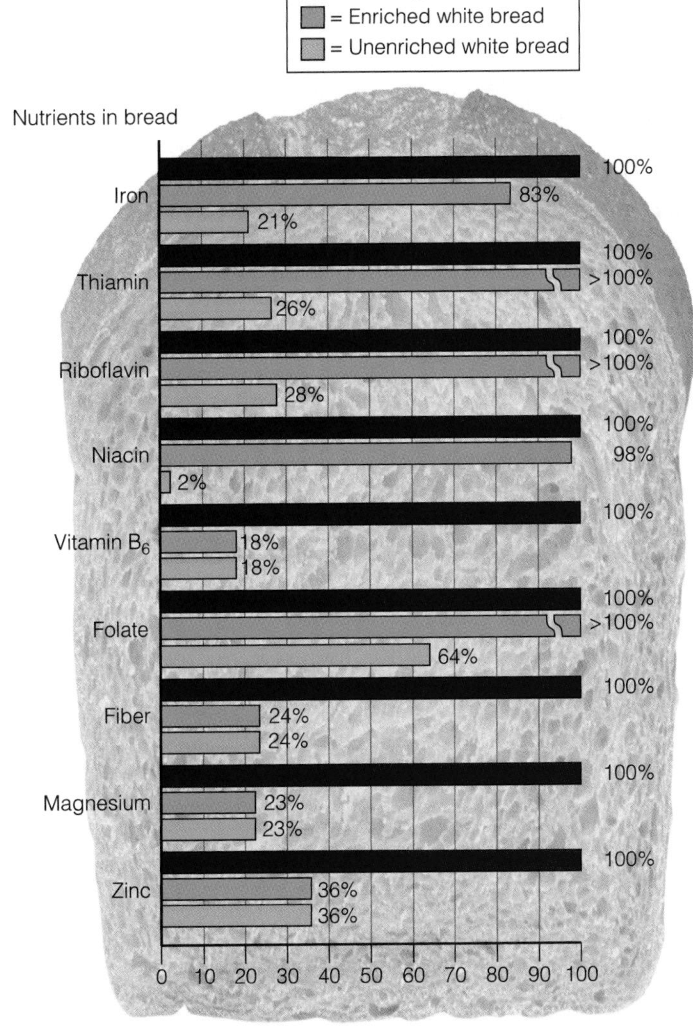

dieting may be fortified with pure cellulose, but adding cellulose alone is not enough; the bread still lacks other fibers.

Only *whole-grain* flour contains all the nutritive portions of the grain. Notice the distinctions between **wheat flour** and **whole-wheat flour** and **white flour** and **unbleached flour** among the terms that describe grain foods; also notice that the terms wheat bread, **brown bread,** and **stone ground** on a label do not guarantee that the bread has been made with whole-grain flour (see Figure 4-8). If bread is a staple food in your diet—that is, if you eat it every day—you are well advised to learn to like the hearty flavor of whole-grain bread.

Figure 4•8

BREAD LABELS COMPARED

The packages appear similar, but the labels reveal that bread made from whole wheat flour provides almost three times the fiber as the one made mostly from refined wheat flour. When the words "whole wheat" or "whole grain" appear on the label, the bread inside contains all of the nutrients that bread can provide.

WHOLE WHEAT

Nutrition Facts

Serving size 1 slice (30g)
Servings Per Container

Amount per serving	
Calories 90	Calories from Fat 14
	% Daily Value*
Total Fat 1.5g	**2%**
Sodium 135mg	**6%**
Total Carbohydrate 15g	**5%**
Dietary fiber 2g	**8%**
Sugars 2g	
Protein 4g	

MADE FROM: UNBROMATED STONE GROUND 100% WHOLE WHEAT FLOUR, WATER, CRUSHED WHEAT, HIGH FRUCTOSE CORN SYRUP, PARTIALLY HYDROGENATED VEGETABLE SHORTENING (SOYBEAN AND COTTONSEED OILS), RAISIN JUICE CONCENTRATE, WHEAT GLUTEN, YEAST, WHOLE WHEAT FLAKES, UNSULPHURED MOLASSES, SALT, HONEY, VINEGAR, ENZYME MODIFIED SOY LECITHIN, CULTURED WHEY, UNBLEACHED WHEAT FLOUR AND SOY LECITHIN.

Wheat Bread

Nutrition Facts

Serving size 1 slice (30g)
Servings Per Container 15

Amount per serving	
Calories 90	Calories from Fat 14
	% Daily Value*
Total Fat 1.5g	**2%**
Sodium 220mg	**9%**
Total Carbohydrate 15g	**5%**
Dietary fiber less than 1g	**2%**
Sugars 2g	
Protein 4g	

INGREDIENTS: UNBLEACHED ENRICHED WHEAT FLOUR [MALTED BARLEY FLOUR, NIACIN, REDUCED IRON, THIAMIN MONONITRATE (VITAMIN B1), RIBOFLAVIN (VITAMIN B2), FOLIC ACID], WATER, HIGH FRUCTOSE CORN SYRUP, MOLASSES, PARTIALLY HYDROGENATED SOYBEAN OIL, YEAST, CORN FLOUR, SALT, GROUND CARAWAY, WHEAT GLUTEN, CALCIUM PROPIONATE (PRESERVATIVE), MONOGLYCERIDES, SOY LECITHIN.

Digestion and Absorption of Carbohydrate

To obtain glucose from newly eaten food, the digestive system must first render the starch and disaccharides from the food into monosaccharides that can be absorbed through the cells that line the small intestine. The largest of the digestible carbohydrate molecules, starch, requires the most extensive breakdown. Disaccharides, in contrast, must be split only once before they can be absorbed.

resistant starch the fraction of starch in a food that is digested slowly, or not at all, by human enzymes.

Starch Digestion of most starch begins in the mouth, where an enzyme in saliva mixes with food and begins to split starch into maltose. The salivary enzyme continues to act on the starch, such as in a bite of bread, while it remains tucked in the stomach's storage area together with other bites. Slowly, each chewed lump is pushed downward, to be thoroughly mixed with the stomach's acid and other juices. Enzyme molecules are made of protein, and most are eventually deactivated by the stomach's protein-digesting acid. Starch digestion then ceases in the stomach, but it resumes at full speed in the small intestine, where another starch-splitting enzyme is delivered by the pancreas. This enzyme breaks starch down entirely into disaccharides and small polysaccharides.

While chewing a bite of bread, you may notice that a slightly sweet taste develops—maltose is being liberated from starch by the enzyme.

One protein-digesting enzyme produced by the stomach is designed to work in highly acidic conditions—its structure protects it from the stomach's acid.

Some forms of starch are easily digested. The starch in bread made of refined white flour, for example, breaks down rapidly to glucose that is absorbed high up in the small intestine.[13] Some starch, such as that of cooked beans, digests more slowly and releases its glucose later in the digestion process. Other starch, called **resistant starch,** is found in raw starches such as cornstarch, inside the impenetrable hulls of seeds, or in foods subjected to intense heating that renders the starch indigestible. Some resistant starch may be digested, but slowly, or it may remain intact until the bacteria of the colon eventually break it down.[14]

The rate of starch digestion may affect the body's handling of its glucose, as a later section explains.

Sugars Sucrose and lactose from food, along with maltose and small polysaccharides freed from starch, undergo one more split to yield free monosaccharides before they are absorbed. This split is accomplished by enzymes attached to the cells of the lining of the small intestine. The conversion of a bite of bread to nutrients for the body is completed when monosaccharides cross these cells and are washed away in a rush of circulating blood that carries them to the waiting liver. Figure 4–9 on the next page presents a quick review of carbohydrate digestion.

Once in the bloodstream, the absorbed carbohydrates (glucose, galactose, and fructose) travel to the liver, which converts fructose and galactose to glucose or products of glucose metabolism (such as fats). The circulatory system transports the glucose and fats to the cells. Liver and muscle cells may store circulating glucose as glycogen; all cells may split glucose for energy.

Fiber Molecules of most fiber and of some resistant starch are not changed by human digestive enzymes, but can be digested by the billions of living inhabitants of the human digestive tract, the resident bacteria. So active are these inhabitants in breaking down substances from food that one expert calls them "an organ of intense metabolic activity that is involved in nutrient salvage." Digestion of soluble fibers and resistant starch by resident bacteria yields waste products, mainly small fat fragments that the body absorbs and can use to provide a tiny bit of energy.

A by-product of metabolism by the bacteria in the colon can be any of several odorous gases. Don't give up on high-fiber foods if they cause gas. Instead, start with small servings and gradually increase the serving size over several weeks; chew foods thoroughly to break up hard-to-digest lumps that can ferment in the intestine; and try a variety of fiber-rich foods until you find some that do not cause the problem. Persistent painful gas may indicate that the digestive tract has undergone a change in its ability to digest the sugar in milk, a condition known as lactose intolerance.

Some people also find relief from excessive gas by using commercial enzyme preparations sold for use with beans. Such products contain enzymes that help to break down some of the indigestible fibers in foods before they reach the colon.

Figure 4•9

HOW CARBOHYDRATE IN FOOD BECOMES GLUCOSE IN THE BODY

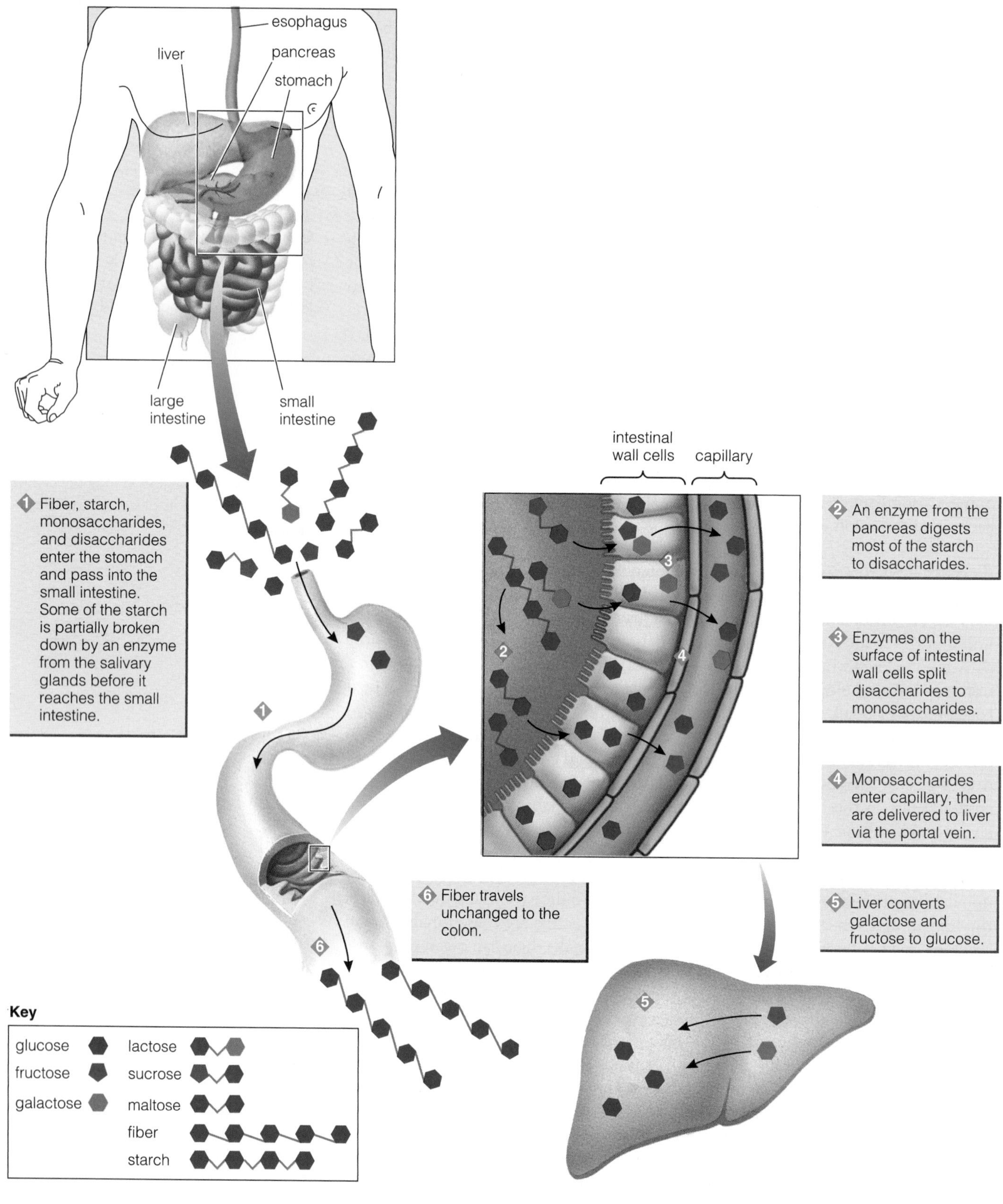

esophagus
liver
pancreas
stomach
large intestine
small intestine
intestinal wall cells
capillary
1 Fiber, starch, monosaccharides, and disaccharides enter the stomach and pass into the small intestine. Some of the starch is partially broken down by an enzyme from the salivary glands before it reaches the small intestine.
2 An enzyme from the pancreas digests most of the starch to disaccharides.
3 Enzymes on the surface of intestinal wall cells split disaccharides to monosaccharides.
4 Monosaccharides enter capillary, then are delivered to liver via the portal vein.
5 Liver converts galactose and fructose to glucose.
6 Fiber travels unchanged to the colon.
Key
glucose
fructose
galactose
lactose
sucrose
maltose
fiber
starch

With respect to starch and sugars, the main task of the various body systems is to convert them to glucose to fuel the cells' work. Fibers help regulate digestion and contribute a little energy.

lactase the intestinal enzyme that splits the disaccharide lactose to monosaccharides during digestion.

lactose intolerance inability to digest lactose due to a lack of the enzyme lactase.

Why Do Some People Have Trouble Digesting Milk?

About 75 percent of the world's people, as they age, lose most of their ability to produce enough of the enzyme **lactase** to digest the milk sugar lactose.[15] Lactase, which is made by the small intestine, splits the disaccharide lactose into its component monosaccharides glucose and galactose, which are then absorbed.

People with **lactose intolerance** experience nausea, pain, diarrhea, and excessive gas on drinking milk or eating lactose-containing products. The undigested lactose remaining in the intestine demands dilution with fluid from surrounding tissue and the bloodstream. Intestinal bacteria use the undigested lactose for their own energy, a process that produces gas and intestinal irritants.

Approximate percentages of people with lactose intolerance:

90% Asian Americans.
80% Native Americans.
80% African Americans.
70% Mediterranean peoples.
60% Inuits (Alaskan/Canadian natives).
50% Hispanics.
25% U.S. population.
<15% Northern Europeans.

The failure to digest lactose affects people to differing degrees. Many can tolerate as much as a cup or two of milk a day; some can tolerate lactose-reduced milk; only a rare few cannot tolerate lactose in any amount.[16] Often people overestimate the severity of their lactose intolerance, blaming it for symptoms most probably caused by something else.[17] Disadvantaged young children of the developing world sustain the most severe consequences of lactose intolerance when it combines with disease, malnutrition, or parasites to produce a loss of nutrients that greatly reduces the children's chances of survival.

Infants produce abundant lactase, which helps them absorb the sugar of breast milk and milk-based formulas; a few suffer inborn lactose intolerance and must be fed solely on lactose-free formulas.[18] Because milk is an almost indispensable source of the calcium every child needs for growth, a milk substitute must be found for any child who becomes lactose intolerant. Women who fail to consume enough calcium during youth may later develop weak bones, so young women must find substitutes if they become unable to tolerate milk. Yogurt or aged cheese may be acceptable—the bacteria or molds that help create these products digest lactose as they convert milk to a fermented product.[19] "Probiotic" bacterial cultures in some yogurts may take up residence in the intestinal tract where they seem to reduce symptoms of lactose intolerance.[20] Yogurts that contain added milk solids also contain added lactose; milk solids and live cultures are listed among the ingredients on the label.

Controversy 2 discusses research on other potential effects of probiotic bacteria.

People with lactose intolerance can also choose milk products that have undergone treatment with lactose-digesting enzymes, or they can treat the products themselves with over-the-counter enzyme pills and drops. The pills are taken with milk-containing meals, and the drops are added to milk-based foods; both products help to digest lactose by replacing the missing natural enzymes. In all cases, the trick is to find ways of splitting lactose to glucose and galactose so that the body can absorb the products, rather than leaving the lactose undigested to feed the bacteria of the colon.

Food allergies are a topic of Chapter 13.

Sometimes sensitivity to milk is due not to lactose intolerance but to an allergic reaction to the protein in milk. Milk allergy arises the same way other allergies do—from sensitization of the immune system to a substance. In this case, the immune system overreacts when it encounters the protein of milk. Children and adults with milk allergy often cannot tolerate cheese or yogurt either, and they have to find nondairy calcium sources. Good choices are calcium-fortified orange juice, calcium- and vitamin-fortified soy drink, and canned sardines or salmon with the bones. Controversy 8 examines the topic of milk in adult diets in relation to the adult bone disease osteoporosis.

Lactose intolerance is a common condition in which the body fails to produce sufficient amounts of the enzyme needed to digest the sugar of milk. Uncomfortable symptoms result and can lead to milk avoidance. Lactose-intolerant people and those allergic to milk need milk alternatives that contain the calcium and the vitamins of milk.

protein-sparing action the action of carbohydrate and fat in providing energy that allows protein to be used for purposes it alone can serve.

ketosis (kee-TOE-sis) an undesirable high concentration of ketone bodies, such as acetone, in the blood or urine.

ketone (kee-tone) **bodies** acidic, fat-related compounds that can arise from the incomplete breakdown of fat when carbohydrate is not available.

insulin a hormone secreted by the pancreas in response to a high blood glucose concentration. It assists cells in drawing glucose from the blood.

The Body's Use of Glucose

Carbohydrates serve structural roles in the body, such as forming part of the mucus that provides a protective coating for the internal organs, but their main role is providing energy. Glucose is not only the main original unit from which carbohydrate-rich foods are made, but it is also the basic carbohydrate unit that each cell of the body uses for energy. The body handles its glucose judiciously. It maintains an internal supply for use in case of need, and it tightly controls its blood glucose concentration to ensure that glucose remains available for ongoing use.

Splitting Glucose for Energy

Glucose fuels the work of most of the body's cells. When a cell splits glucose for energy, it performs an intricate sequence of maneuvers that are of great interest to the biochemist—and of no interest whatever to most people who eat bread and potatoes. What everybody needs to understand, though, is that there is no good substitute for carbohydrate. Carbohydrate is *essential,* as the following details illustrate.

The Point of No Return At a certain point, glucose is forever lost to the body. Inside a cell, glucose is broken in half, releasing some energy. Two pathways are then open to these glucose halves. They can be put back together to make glucose again, or they can be broken into smaller fragments. If they are broken further, they cannot be reassembled to form glucose. The smaller fragments can yield still more energy and in the process break down completely to carbon dioxide and water; or they can be hitched together into units of body fat. Figure 4-10 shows how glucose is broken down to yield energy and carbon dioxide.

Figure 4•10

THE BREAKDOWN OF GLUCOSE YIELDS ENERGY AND CARBON DIOXIDE

Cell enzymes split the bonds between the carbon atoms in glucose, liberating the energy stored there for the cell's use. The first split yields two 3-carbon fragments. The two-way arrows mean that these fragments can also be rejoined to make glucose again. Once they are broken down further into 2-carbon fragments, however, they cannot rejoin to make glucose. The carbon atoms liberated when the bonds split are combined with oxygen and released into the air, via the lungs, as carbon dioxide. Although not shown here, water is also produced at each split.

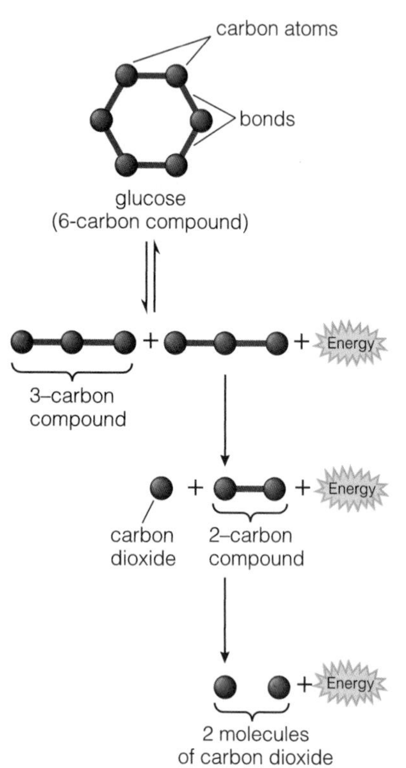

Below a Healthy Minimum Although glucose can be converted into body fat, body fat can never be converted into glucose to feed the brain adequately. This is one reason why fasting and low-carbohydrate diets are dangerous. When the body faces a severe carbohydrate deficit, it has two problems. Having no glucose, it must turn to protein to make some (the body has this ability), diverting protein from critical functions of its own such as maintaining the body's immune defenses. Protein's functions in the body are so indispensable that carbohydrate should be kept available precisely to prevent the use of protein for energy. This is called the **protein-sparing action** of carbohydrate.

Fat fragments have to combine with carbohydrate before they can be used for energy. Using fat without the help of carbohydrate causes the body to go into **ketosis,** a condition in which unusual products of fat breakdown **(ketone bodies)** accumulate in the blood, disturbing the normal acid-base balance. Ketosis during pregnancy can cause brain damage to the fetus, resulting in irreversible mental retardation after birth.

The minimum amount of carbohydrate needed to ensure complete sparing of body protein and avoidance of ketosis is around 100 grams of digestible carbohydrate a day for an average-sized person. Three or four times this minimum is recommended. The 5 to 9 servings of vegetables, fruits, and grains recommended in Table 4-1 (page 105) would deliver 125 grams at a minimum and 200 to 400 grams on average.

key point *Without glucose, the body is forced to alter its uses of protein and fats. The body breaks down its own muscles and other protein tissues to make glucose and converts its fats into ketone bodies, incurring ketosis.*

Storing Glucose as Glycogen

After a meal, as blood glucose rises, the pancreas is the first organ to respond. It releases the hormone **insulin,** which signals the body's tissues to take up surplus glucose. Muscle and liver cells use some of this excess glucose to build the polysaccharide glycogen. The muscles hoard two-thirds of the body's total glycogen and use it just for themselves. The liver stores the other one-third and is more generous with its glycogen, making it available as blood glucose for the brain or other organs when the supply runs low.

Glycogen is well designed for its task of releasing glucose on demand. Unlike starch, which has long chains with occasional branches that are cleaved linearly during digestion, glycogen is many branched with hundreds of ends extending from each molecule's surface (refer back to Figure 4-3 on page 102). When the blood glucose concentration drops and cells need energy, a pancreatic hormone, **glucagon,** floods the bloodstream. Thousands of enzymes within the liver cells respond by attacking a multitude of glycogen ends simultaneously to release a surge of glucose into the blood for use by all the other body cells.

glucagon a hormone secreted by the pancreas that stimulates the liver to release glucose into the blood when blood glucose concentration dips.

glycemic (gligh-SEEM-ic) **effect** the extent to which a food raises the blood glucose concentration and elicits an insulin response as compared with pure glucose.

key point *Glycogen is the body's storage form of glucose. The liver stores glycogen for use by the whole body. Muscles have their own private glycogen stock for their exclusive use. The hormone glucagon acts to liberate stored glucose from liver glycogen.*

Returning Glucose to the Blood

Should your glucose supplies ever fall too low, you would feel dizzy and weak. Should your blood glucose ever climb abnormally high, you might become confused or have difficulty breathing. The healthy body guards against both conditions.

Regulation of Blood Glucose Maintaining normal blood glucose concentration depends on two safeguards: replenishment from liver glycogen stores and siphoning off of excess glucose into the liver (to be converted to glycogen or fat) and into the muscles (to be converted to glycogen).

When blood glucose starts to fall too low, the hormone glucagon triggers the breakdown of liver glycogen to free glucose. Hormones that promote the conversion of protein to glucose are also released, but only a little protein can be spared. When body protein is used, it is taken from blood, organ, or muscle proteins; no surplus of protein is stored specifically for emergencies. As for fat, it cannot regenerate enough glucose to make a difference.

Another hormone, epinephrine, also breaks down liver glycogen as part of the body's defense mechanism in times of danger. To a person living in the Stone Age, this internal source of quick energy was indispensable. Life was fraught with physical peril. The person who stopped and ate before running from a saber-toothed tiger did not survive to produce our ancestors. The quick-energy response in a stress situation works to our advantage today as well. For example, it accounts for the energy you suddenly have to clean up your room when you learn that a special person is coming to visit. To meet such emergencies, we are well advised to eat and to store carbohydrate every four to six waking hours because the liver's glycogen stores can be depleted within half a waking day.

You may rightly ask, "What kind of carbohydrate?" Candy bars and sugary beverages supply sugar energy quickly, but are not the best choices. Balanced meals, eaten on a regular schedule, help the body to maintain its blood glucose. Meals containing starch and fiber along with some protein and a little fat slow digestion so that glucose enters the blood gradually in an ongoing steady supply. Such meals also provide an assortment of other nutrients, not found in candy and soft drinks, that help cells to use their glucose.

The Glycemic Effect of Foods Some carbohydrate-rich foods elevate blood glucose and insulin concentrations higher than others do. This **glycemic effect** of foods is currently under study for its potential influences on health. So far, slow absorption of carbohydrate that produces a modest rise in blood glucose and a smooth return to normal is considered desirable (a low glycemic effect). Fast absorption, a surge in blood glucose, and an overreaction that plunges glucose below normal (a high glycemic effect) are less desirable.

Many factors work together to determine a food's glycemic effect, and the result is not always what we might expect. Ice cream, for example, is a high-sugar food, but it

glycemic index (GI) a ranked measure of the glycemic effect of foods.

Low gycemic index foods:

- Basmati rice, long grain (brown or white).
- Bran cereals, muesli cereal (toasted), whole oats.
- Pasta.
- Pumpernickel or whole-grain bread (heavy textured).
- Baked beans, lentils, soybeans.
- Apple, apple juice, carrot, orange, peach.
- Milk, yogurt.
- "Snickers" candy bar, "M&Ms" candy, and "Dove" chocolate bar.

Moderate glycemic index foods:

- Barley flour bread, sourdough bread, rye bread.
- "Bran Chex" cereal, muesli cereal (not toasted), shredded wheat.
- Banana, orange juice, pawpaw, pineapple.
- Ice cream (full fat).
- "Kudos" and "Mars" candy bars.

High glycemic index foods:

- Cornflakes, "Cheerios," "Rice Krispies" cereals.
- French, white, and whole-meal breads (soft textured).
- Sticky rice, medium grain (brown or white).
- Waffle.
- Mashed potatoes, watermelon.
- Honey, jelly beans, "Life Savers" and "Skittles" candies, sugar-sweetened soft drinks.

SOURCE: Adapted from J. Brand-Miller and K. Foster-Powell, Diets with a low glycemic index: From theory to practice, *Nutrition Today* 34 (1999): 64–72.

produces less of a response than baked potatoes, a high-starch food. Mashed potatoes produce a large glycemic effect, but pure sugar (sucrose) produces only a moderate rise in blood glucose, probably by virtue of its fructose content, which has little effect on blood glucose. The margin lists selected foods according to their ranking on the **glycemic index (GI),** a measure of glycemic effect.

Evidence indicates that lowering the GI of the entire *diet* reduces insulin secretion and improves glucose control, which could help to control diabetes.[21] Diminishing the insulin response with a low-GI diet may also help to short-circuit the cycle of obesity and diabetes (described later), whereas the rapid absorption of glucose from a high-GI diet seems to promote overeating in some overweight people.[22] A low-GI diet may also improve blood lipids with regard to indicators of heart disease, and it may help to prevent obesity because foods high in fiber and other slowly digested carbohydrates stay in the digestive tract longer and provide greater satiety.[23]

Lowering the diet's glycemic effect is problematic, however. A food's GI ranking is unpredictable, so each food must be tested individually; currently, few foods have been tested. Most relevant to real life, a food's glycemic effect depends on whether it is eaten alone or as part of a mixed meal. Most people eat a variety of foods in a meal and so need not worry much about the GI of the foods they choose.[24] Current guidelines already suggest many low-GI choices—whole-grain breads, legumes, fruits, vegetables, and milk products—so following the advice of the Food Guide Pyramid and the *Dietary Guidelines for Americans* is a reasonable way to ensure a diet with a moderate glycemic effect.[25] This chapter's Controversy discusses research concerning possible relationships between a high-GI diet and disease development.[26]

key point *Blood glucose regulation depends mainly on the hormones insulin and glucagon. Certain carbohydrate foods produce a greater rise and fall in blood glucose than others do. Most people have no problem regulating their blood glucose when they consume regular mixed meals.*

Converting Glucose to Fat

Suppose you have eaten dinner and are now sitting on the couch, munching pretzels and drinking cola as you watch a ball game on television. Your digestive tract is delivering molecules of glucose to your bloodstream, and your blood is carrying these molecules to your liver and other body cells. The body cells use all the glucose they require for their energy needs of the moment. Excess glucose is linked together and stored as glycogen until the muscles and liver are full to capacity with glycogen. Still, the glucose keeps coming, and the liver has no choice but to handle the excess. The liver breaks the extra glucose into small fragments and puts them together into more permanent energy-storage compounds—fats. (This would happen with excess protein or fat, too.) The fats are then released into the blood, carried to the fatty tissues of the body, and deposited. Unlike the liver cells, which can store only about four to six hours' worth of glycogen, the fat cells can store practically unlimited quantities of fats. Moral: you had better play the game if you are going to eat the food. (The Think Fitness feature offers tips to help you play.)

What Can I Eat to Make Workouts Easier?

A working body needs carbohydrate fuel to replenish glycogen, and when it runs low, physical activity can seem more difficult. If your workouts seem to drag and never get easier, take a look at your diet. Are your meals regularly timed? Do they provide abundant carbohydrate to fill up glycogen stores to last through a workout?

Here's a trick: about two hours before your workout, eat a small snack of about 300 calories of foods rich in complex carbohydrates, and drink some extra fluid (see Chapter 10 for ideas). The snack provides glucose at a steady rate to spare glycogen, and the fluid helps to maintain hydration.

THINK FITNESS

Even though excess carbohydrate is converted to fat and stored, a balanced diet that is high in complex carbohydrates helps control body weight and maintain lean tissue. Chapter 5 presents a few more details, but the main point is that, calorie for calorie, carbohydrate-rich foods contribute less to body fatness than do fat-rich foods. Had you chosen fatty potato chips instead of low-fat pretzels for your ball game snack, your body would have stored even greater amounts of fat for the calories taken in. Thus, if you want to eat until full, stay within your calorie limits, never skip a meal, and remain lean, you should make every effort to choose foods that provide a diet with 55 percent or more of its calories from mostly unrefined sources of complex carbohydrates and 30 percent or less from fats. This chapter's Food Feature provides the first set of tools required for the job of designing such a diet. Once you have learned to identify the carbohydrates in foods, you must then learn where the fats come in (Chapter 5's Food Feature) and how to obtain adequate protein without overdoing it (Chapter 6).

diabetes (dye-uh-BEET-eez) a disease (technically termed *diabetes mellitus*) characterized by elevated blood glucose and inadequate or ineffective insulin, which renders a person unable to regulate blood glucose normally.

type 1 diabetes the type of diabetes in which the pancreas produces no or very little insulin; often diagnosed in childhood, although some cases arise in adulthood. Formerly called *juvenile-onset* or *insulin-dependent diabetes*.

key point *The liver converts extra energy compounds into fat, a more permanent energy-storage compound than glycogen and one that can be stored in almost unlimited quantities.*

Diabetes and Hypoglycemia

What happens if the body cannot handle carbohydrates normally? One result is **diabetes,** which is common in developed nations and can be detected by a timed blood test. Another is hypoglycemia, which is rare as a true disease condition, but many people believe they experience its symptoms at times.

What Is Diabetes?

Diabetes is a chronic disease characterized by elevated blood glucose concentrations that can lead to or contribute to a number of other serious diseases. Diabetes is one of the top ten killers of adults, and its incidence is rising both here and worldwide.[27] Diabetes causes more new cases of blindness in the United States than any other cause, and possible complications include amputations, heart disease, kidney disease, and premature death.[28] Half of those suffering from diabetes are unaware of their condition and so fail to take action to prevent its damaging effects on the body. The early stage of the most common form of diabetes often presents few or no warning signs typically associated with diabetes (see Table 4-6). Therefore, the American Diabetes Association is calling for everyone over 45 years old and younger people with risk factors such as overweight to be tested regularly for diabetes.[29]

Diabetes mellitus, the most common type of diabetes, occurs in two main forms: type 1 and type 2. Both are disorders of blood glucose regulation. Their characteristics are summarized in Table 4-7 on the next page.

Table 4•6

WARNING SIGNS OF DIABETES[a]

- Excessive urination and thirst
- Glucose in the urine
- Weight loss with nausea, easy tiring, weakness, or irritability
- Cravings for food, especially for sweets
- Frequent infections of the skin, gums, vagina, or urinary tract
- Vision disturbances; blurred vision
- Pain in the legs, feet, or fingers
- Slow healing of cuts and bruises
- Itching
- Drowsiness
- Abnormally high glucose in the blood

[a]These signs appear reliably in type 1 diabetes and, often, in the later stages of type 2 diabetes.

Type 1 Diabetes **Type 1 diabetes** is less common overall (about 10 to 20 percent of cases) but is the leading chronic disease among children and young adults.[30] In this type of diabetes, the person's own immune system attacks the cells of the pancreas that synthesize the hormone insulin. Soon the pancreas can no longer produce insulin, and after each meal, blood glucose remains elevated, even though body tissues are starving for glucose. The person must receive insulin from an external source to assist the cells in taking up the needed glucose from the blood; therefore, an older name for this type of diabetes is *insulin-dependent diabetes mellitus (IDDM)*.

Researchers concur that genetics, viral infection, other diseases, toxins, allergens, and a disordered immune system are all probable culprits in provoking an immune system attack on the pancreas.[31] A controversial idea is

You had better play the game if you are going to eat the food.

type 2 diabetes the type of diabetes in which the person makes plenty of insulin, but the body cells resist insulin's action; often diagnosed in adulthood. Formerly called *adult-onset* or *noninsulin-dependent diabetes*.

insulin resistance a condition in which a normal amount of insulin produces a less than normal response by the tissues; thought to be a metabolic consequence of obesity.

resistin a hormone made and released by fat cells under conditions of obesity and thought to increase tissue resistance to the glucose uptake effect of insulin.

that feeding formula based on cow's milk to infants younger than six months of age may sometimes trigger an immune response associated with the development of type 1 diabetes.[32] Controversy 12 revisits this and other possible causes of diabetes.

Insulin is a protein, and if it were taken orally, the digestive system would digest it. Insulin must therefore be injected, either by daily shots or by an insulin pump about the size of a personal pager worn next to the abdomen that delivers insulin through an implanted needle or tube. New fast-acting and long-lasting forms of insulin and other drugs allow more flexibility in managing meals and treatments, but users must still plan ahead to balance blood insulin and glucose concentrations. Medical advances may eliminate the need for insulin shots—an insulin nasal spray and an inhaler that delivers insulin to the lungs are proving useful in clinical studies, and transplantation of insulin-producing cells is under development.[33] Medical researchers are also coming closer to their goal of developing a vaccine or other therapy to prevent type 1 diabetes.[34]

Type 2 Diabetes The predominant type of diabetes mellitus, **type 2 diabetes** (about 90 percent of cases), is associated with **insulin resistance** of the body's cells, including fat cells. Insulin may be present, often in abnormally large amounts, and it may stimulate cells to take up glucose, but they respond less sensitively than normal. Blood glucose rises too high, as in type 1, but in this case blood insulin also rises. Eventually, the pancreas becomes less able to make insulin. At some point, some people with type 2 diabetes must take insulin to supplement their own supply. If drugs are necessary, a preferred therapy is to take a drug that stimulates the person's own pancreas to secrete insulin or one that improves the uptake of glucose by the tissues.[35]

Type 2 diabetes tends to occur late in life, but overweight children and adolescents are now being diagnosed with the condition, and no one really knows why (see Controversy 12).[36] Type 2 diabetes tends to run in families, and people with the disease often tend to be obese. Figure 4-11 depicts one theory on how obesity and insulin resistance may become a cycle—the larger the fat cells become, the more insulin resistant the tissues become, and the more obese the person becomes. A hormone, **resistin,** secreted by fat cells in obesity, may perpetuate this cycle.[37] In mice, resistin increases body tissues' resistance to the actions of insulin but whether the same is true in people is unknown. Weight loss alone in overweight people with diabetes often helps control the disease. Even moderate weight gain in adults has been observed to predict diabetes.

Incidence of type 2 diabetes also increases with age because the pancreatic cells that produce insulin progressively lose their function with time. Nevertheless, a great majority of cases are probably preventable through personal choice. Recent findings indicate that middle-aged men and women who maintain a healthy body weight, eat a diet high in vegetables, fruit, fish, poultry, and whole grains, exercise regularly, restrict alcohol, and abstain from smoking have a greatly reduced incidence of type 2 diabetes compared with similar people who choose other, less healthy behaviors.[38]

Figure 4•11

THE OBESITY-DIABETES CYCLE

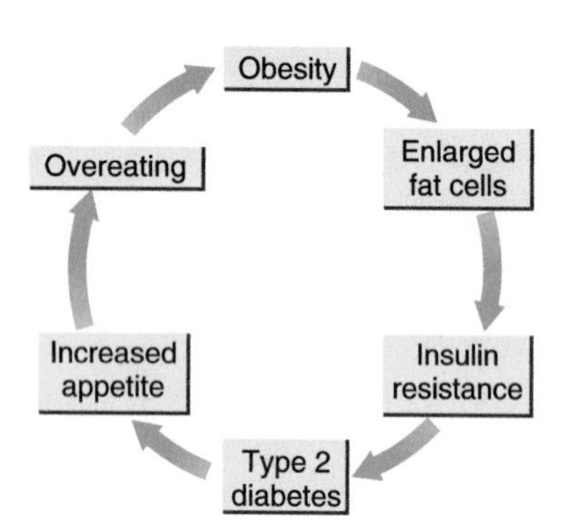

Table 4•7

DIABETES TYPES 1 AND 2 COMPARED

	Type 1 Diabetes	Type 2 Diabetes
Age of onset	Childhood or mid-life	Adulthood or, increasingly, childhood
Body cells	Responsive to insulin action	Resistant to insulin action
Body fatness	Generally low to average	Generally high
Insulin shots required	Yes	Possibly[a]
Insulin-stimulating drugs or other drugs may be effective	No	Yes
Natural insulin	Pancreas makes too little or none.	Pancreas makes enough or too much.
Pancreatic function	Insulin-producing cells impaired or nonfunctional	Insulin-producing cells normal
Severity of symptoms	Relatively severe; many are apparent on diagnosis.	Relatively mild; few or none may be present on diagnosis.

[a]People past age 40 who suffer from type 2 diabetes may lose pancreatic function and become dependent on insulin.

Diagnosing and Controlling Diabetes Diabetes can be diagnosed with more than one positive result from a fasting blood glucose test. Blood is drawn after a night of fasting so that a clinician can measure an indicator of blood glucose to determine whether it falls within the normal range for fasting.

Although the symptoms of diabetes are controllable for the most part, its effects can be severe and may progress even when drugs control blood glucose. Diabetes causes blockage or destruction of capillaries that feed the body organs, and tissues die from lack of nourishment.[39] Problems include impaired circulation leading to disease of the feet and legs, often necessitating amputation; kidney disease, sometimes requiring hospital care or kidney transplant; impaired vision or blindness due to cataracts and damaged retinas; nerve damage; skin damage; and strokes and heart attacks. A new line of thinking holds that some damage may result from oxidation that accompanies elevated blood glucose.[40] The hope is that a diet high in antioxidant-containing vegetables and fruits may be protective, but much more research is needed to confirm the oxidation theory. Meanwhile the person with diabetes is advised to control not only weight but also all possible risk factors that might contribute to heart and blood vessel disease (atherosclerosis and hypertension, discussed in Chapter 11).

Constructed of a balanced pattern of foods, the diet best for controlling diabetes and also for controlling weight and supporting physical activity:

- Is adequate (deficiencies in trace minerals, especially chromium, may hasten diabetes onset).
- Provides the recommended amount of fiber (fiber helps maintain the health of the body).
- Is moderate in concentrated sugar (the amount allowed varies with an individual's blood glucose response).
- Provides a controlled amount of carbohydrate (to regulate blood glucose concentration).[41]
- Is low in saturated fat (thought to worsen insulin response and cardiovascular disease) and may provide some monounsaturated fat (thought to be harmless to the heart).[42]
- Is not too high in protein (to protect the kidneys).

Such a diet also has all the characteristics important to prevention of chronic diseases and meets most of the recommendations of the *Dietary Guidelines for Americans*. The diet can vary, depending on personal tastes and on how much restriction is required to control an individual's blood glucose and lipid values. A person at risk for diabetes can do no better than to adopt such a diet long before any symptoms appear.

The roles of regular physical activity in preventing and controlling diabetes cannot be overstated.[43] Not only does exercise help to maintain a desirable body weight, but it also heightens tissue sensitivity to insulin and may help prevent or forestall type 2 diabetes.[44] Like a juggler who keeps three balls in motion, the person with diabetes must constantly balance three lifestyle factors—diet, exercise, and medication—to control the blood glucose level.

key point *Diabetes is an example of the body's abnormal handling of glucose. Inadequate or ineffective insulin leaves blood glucose high and cells undersupplied with glucose energy. This causes blood vessel and tissue damage. Weight control and exercise may be effective in preventing the predominant form of diabetes (type 2) and the illnesses that accompany it. A person diagnosed with diabetes must establish patterns of eating, exercise, and medication to control blood glucose.*

Q If I Feel Dizzy between Meals, Do I Have Hypoglycemia?

The term **hypoglycemia** refers to abnormally low blood glucose. People with the condition **postprandial hypoglycemia,** literally, "low blood glucose after a meal," experience fatigue, weakness, dizziness, irritability, a rapid heartbeat, anxiety, sweating, trembling,

hypoglycemia (HIGH-poh-gly-SEE-mee-ah) a blood glucose concentration below normal, a symptom that may indicate any of several diseases, including impending diabetes.

postprandial hypoglycemia a drop in blood glucose that follows a meal and is accompanied by symptoms of the stress response; also called *reactive hypoglycemia*.

Chapter 12 discusses a form of diabetes seen only in pregnancy—gestational diabetes.

The exchange system introduced in Chapter 2 and presented in full in Appendix D was developed to help people with diabetes control calorie, carbohydrate, sugar, and fat intakes.

Physical activity: a key player in controlling diabetes.

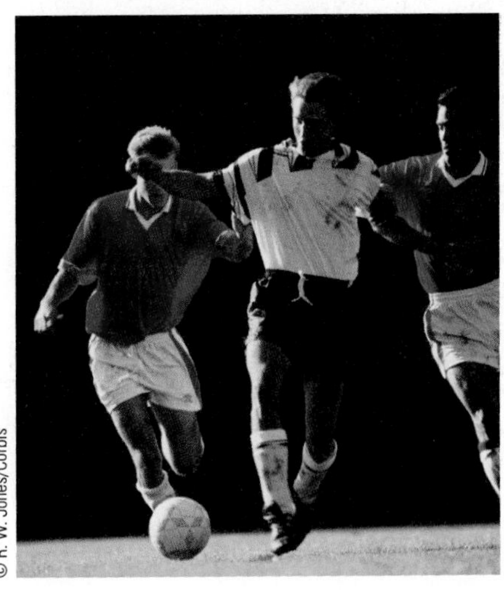

fasting hypoglycemia hypoglycemia that occurs after 8 to 14 hours of fasting.

hunger, and headaches. Mental symptoms, such as confusion and impairment of intellectual tasks, are also evident.[45] These symptoms are so general and common to so many conditions that people can easily misdiagnose themselves as having postprandial hypoglycemia.[46] The condition is rare, however, except among very lean people or those who have recently lost a great deal of weight, often due to an exaggerated sensitivity to insulin.[47] A true diagnosis requires a test to detect low blood glucose while the symptoms are present to confirm that both occur simultaneously.

A person who has symptoms while well advanced into the fasting state (for example, overnight) has a different kind of hypoglycemia—**fasting hypoglycemia.** Its symptoms are headache, mental dullness, fatigue, confusion, amnesia, and even seizures and unconsciousness. Serious diseases and conditions, such as cancer, pancreatic damage, unbalanced treatment of diabetes, infection of the liver with accompanying damage (hepatitis), and advanced alcohol-induced liver disease, can all produce hypoglycemia.

To produce even mild hypoglycemia and its symptoms in normal, healthy people requires extreme measures—administering drugs that overwhelm the body's glucose-controlling hormones, insulin and glucagon. Without such intervention, those hormones hardly ever fail to keep blood glucose within normal limits. Symptoms that people ascribe to hypoglycemia rarely correlate to low blood glucose in blood tests, and such conditions may best be classified as *non*hypoglycemia. Still, people who feel such symptoms may benefit from eating regularly timed, balanced, protein-containing meals; they especially should avoid oscillating between low-carbohydrate dieting and sudden large refined carbohydrate doses.[48] It may also help to avoid alcoholic beverages because alcohol impairs normal regulation of blood glucose in otherwise healthy people.[49]

key point *Postprandial hypoglycemia is a rare medical condition in which blood glucose falls too low. It can be a warning of organ damage or disease. Many people believe they experience symptoms of hypoglycemia, but their symptoms do not accompany below-normal blood glucose.*

Part of eating right is choosing wisely among the many foods available. This chapter has explored the body's responses to carbohydrate, processes that occur largely without your awareness. Now you take the controls by learning which foods supply the carbohydrates your body needs. The Food Feature explains how to integrate foods into a diet that meets the body's needs for carbohydrates. The Do It! section then asks you to critique your diet with regard to the fiber it contains.

Regularly timed, balanced meals help to hold blood glucose steady.

Finding the Carbohydrates in Foods

TO SUPPORT health, a diet must supply enough carbohydrate-rich foods to meet the body's needs. The Daily Value suggests a goal of 300 grams of mostly complex carbohydrate each day for a person who eats a 2,000-calorie diet. Using the Food Guide Pyramid, this Food Feature illustrates how you can obtain the carbohydrate-rich foods you need. Breads, cereals, vegetables, fruits, and milk are all good contributors of starches and dilute sugars, both valuable energy-yielding carbohydrates.

Bread, Cereal, Rice, and Pasta

A serving of most foods in this group—a slice of whole-wheat bread, half an English muffin or bagel, a 6-inch tortilla, or a half-cup of rice, pasta, or cooked cereal—provides about 15 grams of carbohydrate, mostly as starch.* People who like breads and other starchy foods are happy to learn that nutrition authorities encourage people to use them in abundance. If calories are a problem, people should first limit foods with added sugar or fat and then limit total calories. Some foods in this group, especially baked goods such as biscuits, croissants, muffins, and snack crackers, do contain added sugar and/or fat.

Vegetables

Some vegetables are major contributors of starch in the diet. Just one small white or sweet potato or a half-cup of cooked dry beans, corn, peas, plantain, or winter squash provides 15 grams of carbohydrate, as much as in a slice of bread, though as a mixture of sugars and starch. A half-cup of carrots, okra, onions, tomatoes, cooked greens, or most other nonstarchy vegetables or a cup of salad greens provides about 5 grams as a mixture of starch and sugars.

*Gram values in this section are adapted from the 1995 exchange system.

Fruits

Different forms of fruit are assigned different serving sizes. A typical fruit serving—three-quarters of a cup of juice; a small banana, apple, or orange; a half-cup of most canned or fresh fruit; or a quarter-cup of dried fruit—contains an average of about 15 grams of carbohydrate, mostly as sugars, including the fruit sugar fructose. Fruits vary greatly in their water and fiber contents, and therefore in their sugar concentrations. With the exception of avocado, which is high in fat, fruits contain insignificant amounts of fat and protein.

Milk, Cheese, and Yogurt

A serving (a cup) of milk or yogurt is a generous contributor of carbohydrate, donating about 12 grams. Cottage cheese provides about 6 grams of carbohydrate per cup, but most other cheeses contain little if any carbohydrate. These foods also contribute high-quality protein (a point in their favor), as well as several important vitamins and minerals. Milk products vary in fat content, an important consideration in choosing among them; Chapter 5 provides the details.

Cream and butter, although dairy products, are not equivalent with milk because they contain little or no carbohydrate and insignificant amounts of the other nutrients important in milk. They are appropriately placed with the fats at the top of the Pyramid.

Meat, Poultry, Fish, Dry Beans, Eggs, and Nuts

With two exceptions, foods of this group provide almost no carbohydrate to the diet. The exceptions are nuts, which provide a little starch and fiber along with their abundant fat, and dry beans, revered by diet watchers as low-fat sources of both

added sugars sugars added to a food for any purpose, such as to add sweetness or bulk or to aid in browning (baked goods).

naturally occurring sugars sugars that are not added to a food but are present as its original constituents, such as the sugars of fruit or milk.

The U.S. Food Exchange System (Appendix D) and the Canadian Choice System (Appendix B) list carbohydrate values for a variety of foods.

starch and fiber. Just a half-cup serving of beans provides 15 grams of carbohydrate, an amount equaling the richest sources in the Food Guide Pyramid. Among providers of fiber, beans and other legumes are peerless, totaling 8 grams in a half-cup.

Fats, Oils, and Sweets

Fats are devoid of carbohydrate, of course, but sweets supply carbohydrate, so it is useful to account for them in the diet. Most people enjoy sweets, so it is useful to learn something of their nature.

In the last decade, scientists have been arguing about what constitutes "sugar" and how to measure it in the diet. (Table 4-8 defines sugar terms.) Some experts wish to abandon the idea of measuring the sugars in foods. They accurately point out that a sugar molecule arising in an orange by way of photosynthesis is indistinguishable in laboratory tests from one added at the jam factory to sweeten orange marmalade. How can we measure the **added sugars,** they ask, when we cannot separate them from the **naturally occurring sugars** in foods? Besides, they say, the body handles all the sugars in the same ways, whatever the source.

Nutritionists argue back that chemical structures of sugars are not at issue; the addition of a concentrated energy source reduces the nutrient density of

Table 4•8

TERMS THAT DESCRIBE SUGAR

Note: The term *sugars* here refers to all of the monosaccharides and disaccharides. On a label's ingredient list, the term *sugar* means sucrose. See Controversy 4 for terms related to *artificial sweeteners* and *sugar alcohols*.

- **brown sugar** white sugar with molasses added, 95% pure sucrose.
- **concentrated fruit juice sweetener** a concentrated sugar syrup made from dehydrated, deflavored fruit juice, commonly grape juice; used to sweeten products that can then claim to be "all fruit."
- **confectioner's sugar** finely powdered sucrose, 99.9% pure.
- **corn sweeteners** corn syrup and sugar solutions derived from corn.
- **corn syrup** a syrup, mostly glucose, partly maltose, produced by the action of enzymes on cornstarch. *High-fructose corn syrup (HFCS)* is mostly fructose; glucose (dextrose) and maltose make up the balance.
- **dextrose** an older name for glucose.
- **fructose, galactose, glucose** the monosaccharides.
- **granulated sugar** common table sugar, crystalline sucrose, 99.9% pure.
- **honey** a concentrated solution primarily composed of glucose and fructose produced by enzymatic digestion of the sucrose in nectar by bees.
- **invert sugar** a mixture of glucose and fructose formed by the splitting of sucrose in an industrial process. Sold only in liquid form and sweeter than sucrose, invert sugar forms during certain cooking procedures and works to prevent crystallization of sucrose in soft candies and sweets.
- **lactose, maltose, sucrose** the disaccharides.
- **levulose** an older name for fructose.
- **maple sugar** a concentrated solution of sucrose derived from the sap of the sugar maple tree, mostly sucrose. This sugar was once common but is now usually replaced by sucrose and artificial maple flavoring.
- **molasses** a syrup left over from the refining of sucrose from sugarcane; a thick, brown syrup. The major nutrient in molasses is iron, a contaminant from the machinery used in processing it.
- **raw sugar** the first crop of crystals harvested during sugar processing. Raw sugar cannot be sold in the United States because it contains too much filth (dirt, insect fragments, and the like). Sugar sold as "raw sugar" domestically is not actually raw but has gone through more than half of the refining steps.
- **turbinado** (ter-bih-NOD-oh) **sugar** raw sugar from which the filth has been washed; legal to sell in the United States.
- **white sugar** pure sucrose, produced by dissolving, concentrating, and recrystallizing raw sugar.

the foods, and this can affect the body adversely. Manufacturers measure sugar when they add it to food products, so laboratory tests are not needed to determine the amounts in foods.

Both sides of the argument have validity, and as the spat continues, some useful distinctions between sugars are emerging. The term **carbohydrate sweeteners** shifts the focus from separating sugars by their chemical structures to including sweet sugars from all sources. The carbohydrate sweeteners from beets, corn, grapes, honey, and sugarcane are alike. All arise naturally and, through processing, are purified of most or all of the original plant material—bees process honey, and machines process the other types. Nutrition authorities discourage the liberal use of added sugars, but Americans consume dramatically more added sugar than a century ago when most forms of purified sugar were unknown. Today, our intakes of added sugars exceed recommendations, and nutrition authorities recommend that we use them more sparingly (see Controversy 4).

The Nature of Sugar

Each teaspoonful of any sweet can be assumed to supply about 20 calories and 4 grams of carbohydrate. You may not think of candy or molasses in terms of *teaspoons*, but this helps to emphasize that all sugary items are like white sugar—in spite of many people's belief that some are different or "better." If you use ketchup liberally, remember that a tablespoon of it contains a teaspoon of sugar. And for the soft drink user, a 12-ounce can of sugar-sweetened cola contains about 8 or more teaspoons of sugar. Figure 4-12 shows that processed foods contain surprisingly large amounts of sugar. Figure 4-13 shows that strawberry jam claiming to be "100% fruit" can contain even more sugars per serving than regular sucrose-sweetened jam.

What about the nutritional value of a product such as molasses, honey, or concentrated fruit juice sweetener compared to white sugar? Molasses contains 1 milligram of iron per tablespoon so, if used frequently, it can contribute some of this important nutrient. Molasses is

carbohydrate sweeteners ingredients containing sugars from any source that are used for sweetening food products. They include glucose, fructose, corn syrup, concentrated fruit juice, and other sweet carbohydrates.

Figure 4•12

SUGAR IN PROCESSED FOODS

½ c canned corn = 3 tsp sugar[a]
12 oz cola = 8 tsp sugar
1 tbs ketchup = 1 tsp sugar
1 tbs creamer = 2 tsp sugar
8 oz sweetened yogurt = 7 tsp sugar
2 oz chocolate = 8 tsp sugar

[a]Values based on 1 tsp = 4 g.

Figure 4•13

JAM LABELS COMPARED

Notice that a product claiming to contain "100% fruit" can contain concentrated fruit juice sweeteners that contribute sweet flavor from sugars, just as ordinary sugar does.

Strawberry Jam

Nutrition Facts	
Serving size 1 Tbsp (20g)	
Servings Per Container About 14	
Amount per serving	
Calories 40	Calories from Fat 0
	% Daily Value*
Total Fat 0g	0%
Sodium 1mg	1%
Total Carbohydrate 10g	4%
Sugars 7g	
Protein 0g	

*Percent Daily Values are based on a 2,000 calorie diet.

INGREDIENTS: Strawberries, Corn Syrup, Sugar, High Fructose Corn Syrup, Citric Acid, Fruit Pectin.

Strawberry 100% Fruit Spread

Nutrition Facts	
Serving size 1 Tbsp (18g)	
Servings Per Container About 16	
Amount per serving	
Calories 40	Calories from Fat 0
	% Daily Value*
Total Fat 0g	0%
Sodium 0mg	0%
Total Carbohydrate 10g	3%
Sugars 8g	
Protein 0g	

*Percent Daily Values are based on a 2,000 calorie diet.

INGREDIENTS: Clarified Grape Juice Concentrate, Strawberries, Clarified Pear Juice Concentrate, Pectin, Natural Flavor, Citric Acid.

Sugars on the Nutrition Facts panel of a food label reflect both added and naturally occurring sugars in foods. Sugars listed among the ingredients are all added. Products listing sugars among the first few ingredients contain substantial amounts per serving.

Sugar alcohols, discussed in this chapter's Controversy, help protect against tooth decay.

less sweet than the other sweeteners, however, so more molasses is needed to provide the same sweetness as sugar. Also, the iron comes from the iron machinery in which the molasses is made and is in the form of an iron salt that is not easily absorbed by the body.

Honey is no better for health than other sugars by virtue of being "natural"—honey is chemically almost indistinguishable from sucrose. Honey contains the two monosaccharides glucose and fructose in approximately equal amounts. Sucrose contains the same monosaccharides but joined together in the disaccharide form. Spoon for spoon, however, sugar contains fewer calories than honey because the dry crystals of sugar take up more space than the sugars of honey dissolved in its water.

As for concentrated juice sweeteners, these are highly refined and have lost virtually all of the beneficial nutrients and nonnutrients of the original fruit. No form of sugar is "more healthy" than white sugar, as Table 4-9 shows.

It would be absurd to rely on any sugar for nutrient contributions. A tablespoon of honey (64 calories) does offer 0.1 milligram of iron, but it would take 180 tablespoons of honey—11,500 calories—to provide 100 percent of the Daily Value of 18 milligrams of iron. The nutrients of honey just don't add up as fast as its calories. Thus, if you choose molasses, brown sugar, or honey, choose them not for their nutrient contributions but for the pleasure they give. These tricks can help magnify the sweetness of foods without boosting their calories:

- Serve sweet food warm (heat enhances sweet tastes).
- Add sweet spices such as cinnamon, nutmeg, allspice, or clove.
- Add a tiny pinch of salt; it will make food taste sweeter.
- Try reducing the sugar added to recipes by one-third.
- Select fresh fruits or fruit juice, or those prepared without added sugar.
- Use small amounts of sugar substitutes in place of sucrose.
- Read food labels for clues on sugar content.

Finally, enjoy whatever sugar you do eat. Sweetness is one of life's great sensations, and you need not forgo it completely.

Table 4•9

THE EMPTY CALORIES OF SUGAR

At first glance, honey, jelly, and brown sugar look more nutritious than plain sugar, but when compared with a person's nutrient needs, none contributes anything to speak of. The cola beverage is clearly an empty-calorie item, too.

Food	Energy (cal)	Protein (g)	Fiber (g)	Calcium (mg)	Iron (mg)	Magnesium (mg)	Potassium (mg)	Zinc (mg)	Vitamin A (µg)	Thiamin (mg)	Riboflavin (mg)	Niacin (mg)	Vitamin B_6 (mg)	Folate (µg)	Vitamin C (mg)
Sugar (1 tbs)	46	0	0	0	0.0	0	0	0.0	0	0	0	0.0	0	0	0
Honey (1 tbs)	64	0	0	1	0.1	0	11	0.0	0	0	0	0.0	0	<1	0
Molasses (1 tbs)	55	0	0	42	1.0	50	300	0.1	0	0	0	0.2	0.1	0	0
Concentrated grape or fruit juice sweetener (1 tbs)	30	0	0	0	0	0	0	0	0	0	0	0	0	0	
Jelly (1 tbs)	49	0	0	1	0.0	1	12	0.0	0	0	0	0.0	0	0	<1
Brown sugar (1 tbs)	34	0	0	8	0.2	3	31	0.0	0	0	0	0.0	0	0	0
Cola beverage (12 fl oz)	153	0	0	11	0.1	4	4	0	0	0	0	0.0	0	0	0
Daily Values	2,000	56	25	1,000	18.0	400	3,500	15.0	1,000	1.5	1.7	20.0	2.0	400	60

INVESTIGATE YOUR FIBER INTAKE

This activity guides you in estimating the grams of fiber in your diet and helps you to recognize the fiber in the foods of the Food Guide Pyramid.

Preparing Your Food and Fiber Record

Step 1. On a copy of Form 4-1, list all the foods you ate and beverages you drank in one day. Take care to record portion sizes accurately. It makes a big difference whether you ate a half-cup or a quarter-cup of beans, berries, or any other foods. If you need guidance in doing this, reread "Preparing Your Food Report" in the Do It! section of Chapter 2 (page 53).

Form 4•1

FOOD AND FIBER RECORD

Instructions: In the left-hand column, list the foods and amounts that you ate. Then list the grams of fiber in the food in the appropriate column to the right. For example, for a piece of wheat toast eaten at breakfast, list 2 grams of fiber in the Breads, Cereals, Rice, and Pasta column. For a glass of milk, list 0 grams in the Milk, Yogurt, and Cheese column.

Food/Amount	**Breads, Cereals, Rice, and Pasta**	**Vegetables**	**Fruit**	**Milk, Yogurt, and Cheese**	**Meat, Poultry, Fish, Dry Beans, Eggs, and Nuts**
Breakfast:					
Snack:					
Lunch:					
Supper:					
Snack:					
Your subtotals:					

Daily Value (2,000-calorie diet): 25 g

Your grand total: ______ g Your percent of Daily Value ______ % Calculation: (Your intake ÷ Daily Value) × 100 = % Daily Value

Step 2. List the fiber grams in each of the foods. Fiber values for many common foods are listed in Figure 4-14 below; look there first for convenience. The fiber values of other foods can be found in Appendix A.

Step 3. Add the numbers in the vertical columns to obtain subtotals of fiber; then add the subtotals across to obtain your grand total fiber intake for the day. Find the percentage of the Daily Value for fiber that your diet provided. Keep in mind that a person eating more than 2,000 calories a day needs proportionally more fiber.

Figure 4•14

FINDING THE FIBER IN FOODS

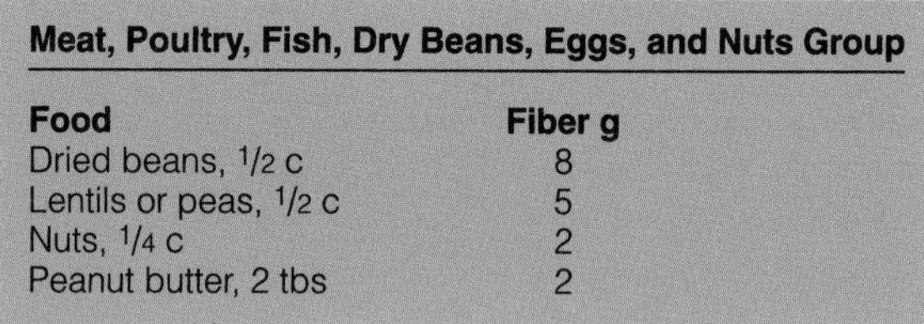

Meat, Poultry, Fish, Dry Beans, Eggs, and Nuts Group

Food	Fiber g
Dried beans, 1/2 c	8
Lentils or peas, 1/2 c	5
Nuts, 1/4 c	2
Peanut butter, 2 tbs	2

Vegetable Group

Food	Fiber g
Baked potato with skin, 1	5
Brussels sprouts, 1/2 c	3
Carrot juice, 3/4 c	2
Broccoli, 1/2 c	2
Asparagus, 1/2 c	2
Corn, 1/2 c	2
Celery, 1/2 c	2
Green beans, 1/2 c	2
Spinach, 1/2 c	2
Baked potato, no skin, 1	2
Cauliflower, 1/2 c	2
Carrots, 1/2 c	2
Cabbage, 1/2 c	2
Onions, 1/2 c	1
Tomato, raw, 1 medium	1
Eggplant, 1/2 c	1
Lettuce, raw, 1 c	1
Bell peppers, 1/2 c	1
Dill pickle, 1 whole	1
Tomato juice, canned, 3/4 c	1

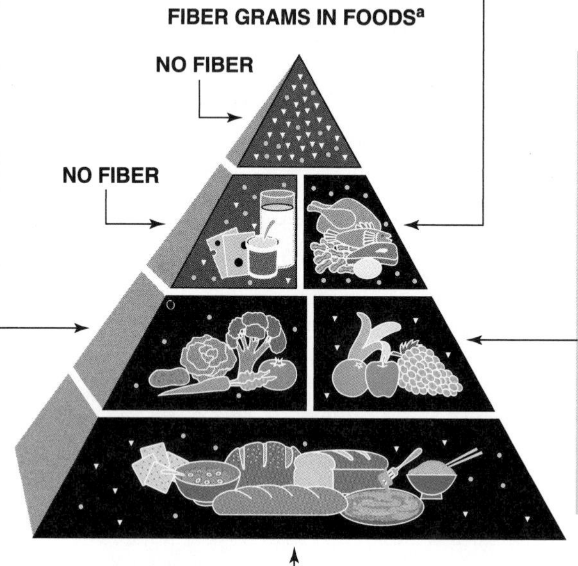

Fruit Group

Food	Fiber g
Prunes, cooked, 1/4 c	4
Pear, raw, 1 medium	4
Blackberries/raspberries, raw, 1/2 c	4
Apple/orange, raw, 1 medium	3
Apricots, raw, 3 each	3
Banana, raw, 1	2
Other berries, raw, 1/2 c	2
Peach, raw, 1 medium	2
Fruit cocktail, canned, 1/2 c	1
Raisins, dry, 1/4 c	1
Cantaloupe, raw, 1/2 c	1
Cherries, raw, 1/2 c	1
Apple juice, 3/4 c	<1
Orange juice, 3/4 c	<1

Breads, Cereals, Rice, and Pasta Group

Food	Fiber g
100% bran cereal, 1 oz	8
Barley, whole grain, 1/2 c	7
Muffin, bran, 1	4
Wheat flakes, 1 oz	3
Shredded wheat, 1 large biscuit	2
Oatmeal, 1/2 c	2
Puffed wheat, 1 1/2 c	2
Whole-wheat bread, 1 slice	2
Light rye bread, 1 slice	2
Pumpernickel bread, 1 slice	2
Popcorn, 2 c	2
Brown rice, 1/2 c	2
Cheerios, 1 oz	2
Corn flakes, 1 oz	1
Pasta,[b] 1/2 c	1
Muffin, blueberry, 1	1
White rice, 1/2 c	<1
White bread, 1 slice	<1

[a]All values are for ready-to-eat or cooked foods, unless otherwise noted. Fruit values include edible skins. All values are rounded values.

[b]Pasta includes spaghetti noodles, lasagna noodles, and other noodles made from enriched white flour. Whole-wheat pastas have significantly more fiber.

Analysis

Answer the following questions:

1. Did your fiber total meet 100 percent of the Daily Value for a 2,000-calorie diet (25 grams)? Did it approach or exceed the maximum value of 40 grams?
2. Do you think your intake was too low, just right, or too high? Why do you think so?
3. Did your diet meet the minimum number of servings of foods from each fiber-containing group? One of the reasons the Food Guide Pyramid recommends a minimum of 3 servings of vegetables, 2 of fruit, and 6 of grains is to meet fiber needs.
4. Which fiber-containing groups fell short of the recommended intake?
5. Analyze the fiber in the foods you listed on Form 4-1 using the quick method described in Table 4-3 on page 107. How do the results of the two methods of estimating fiber compare? Which method would you prefer to use, and why?
6. Which specific foods provided the most fiber to the day's meals? Which provided the least? Identify trends in your food choices that would affect your fiber intakes. For example, if you consistently choose whole grains, your fiber intake benefits.
7. What alterations might you make among your vegetable, fruit, meat and alternates, or grain choices to increase the fiber in your meals? (Remember to increase your fluid intake when you increase fiber.)
8. Are any of the foods listed in Figure 4-14 relatively rich in fiber? List them.
9. Are any of the juices listed in Figure 4-14 relatively rich in fiber? What type of fiber do you think juices might contain? (Hint: The main constituent of juice is water.)
10. What contributions do meats and milk products make to the day's fiber total? What advice about fiber would you give to someone who emphasizes meat and milk products at each meal? (Hint: A quick way to direct someone to the fiber in foods is to guide them to the *bottom half* of the Food Guide Pyramid, which accounts for 11 of the 15 food servings recommended for a day.)
11. Did your meals include any fiber-rich legume dishes, such as chili, bean burritos, beans in a salad, or split pea soup? Anyone interested in obtaining fiber could do no better than to find ways to eat some legumes each day.

Constructed mostly of refined foods and meats and lacking in plant-derived foods, the average U.S. diet does a disservice to the health of the eater. Does this mean that you should never consume low-fiber foods such as white rolls, white rice, meats, potato chips, or apple juice? No, but such foods should be included in moderation among abundant fruits, vegetables, legumes, and whole grains. Constructed from a variety of whole foods, the diet can easily meet the recommended 25 or more grams of dietary fiber daily.

Answers to these Self Check questions are in Appendix G.

1. The dietary disaccharides include:
   a. sucrose, fructose, and glucose
   b. maltose, lactose, and sucrose
   c. lactose, maltose, and glucose
   d. glycogen, starch, and fiber
2. The polysaccharide that helps form the supporting structures of plants is:
   a. cellulose
   b. maltose
   c. glycogen
   d. sucrose
3. Digestible carbohydrates are absorbed as ______ through the small intestinal wall and are delivered to the liver where they are converted to ________ .
   a. disaccharides; sucrose
   b. glucose; glycogen
   c. monosaccharides; glucose
   d. galactose; cellulose
4. When blood glucose concentration rises, the pancreas secretes ______ , and when blood glucose levels fall, the pancreas secretes _____ .
   a. glycogen; insulin
   b. insulin; glucagon
   c. glucagon; glycogen
   d. insulin; fructose
5. When the body uses fat for fuel without the help of carbohydrate, this results in the production of:
   a. ketone bodies
   b. glucose
   c. starch
   d. galactose
6. Type 2 diabetes is characterized by insulin resistance of the body's cells. T F
7. Type 1 diabetes is most often controlled by successful weight-loss management. T F
8. By law, enriched white bread must equal whole-grain bread in nutrient content. T F
9. The fiber-rich portion of the wheat kernel is the bran layer. T F
10. Around the world, most people are lactose intolerant. T F

nutrition on the net

For further study of the topics of this chapter, access these websites and search for the phrases or words in quotation marks:

1. Find updates and quick links to these and other nutrition-related sites at our website: www.wadsworth.com/nutrition
2. Search for "lactose intolerance" at the U.S. Government health information site: www.healthfinder.gov
3. Search for "artificial sweeteners" at the U.S. Government health information site: www.healthfinder.gov
4. Search for "sugars" and "fiber" at the International Food Information Council site: http//:ificinfo.health.org
5. Learn more about dental caries from the American Dental Association and the National Institute of Dental Research: www.ada.org and www.nidr.nih.gov
6. Learn more about diabetes from the American Diabetes Association, the Canadian Diabetes Association, and the National Diabetes Information Clearinghouse: www.diabetes.org, www.diabetes.ca, and www.niddk.nih.gov
7. Visit the National Institute of Diabetes & Digestive & Kidney Diseases site: www.niddk.nih.gov

INTERNET ACTIVITY

The incidence of diabetes, especially type 2 diabetes, has been increasing even among the young. Determine your own risk of developing diabetes by completing the Diabetes Risk Test.

1. Go to the site of the American Diabetes Association: http://www.diabetes.org
2. On the left side of the page, click on "Basic Diabetes Information."
3. Scroll down the left side of page, click on "Risk Factors."
4. Take the Diabetes Risk Test.

What were your results? Which risk factors can be modified? Which cannot be altered? What actions can be taken to reduce your risk?

CONTROVERSY 4

SUGAR AND ALTERNATIVE SWEETENERS: ARE THEY BAD FOR YOU?

ALMOST EVERYONE finds sweet tastes pleasing—after all, the preference for sweets is inborn. To a child's taste, the sweeter the food, the better.[1] In adults, the preference for sweets is somewhat diminished, although adult consumption of sugars and sweeteners is increasing in the United States.

Imagine pouring almost three-quarters of a measuring cup (32 teaspoons) of sugar onto your foods and into your beverages before consuming them each day.*[2] This represents the amount of added sugars in the U.S. food supply, enough on average to provide every man, woman, and child with more than 100 pounds per year—an all-time high by some ways of reckoning.[3] This number may be somewhat higher than actual intakes because it does not account for waste, such as the syrup drained from sweet pickles or jam that molds and is tossed out.[4] Still, such numbers are useful for approximating the average U.S. sugars consumption and for tracking changes from year to year (see Figure C4-1). The steady upward trend in U.S. sugars consumption parallels a dramatic increase in the purchase of commercially prepared foods and beverages to which sugars have been added.[5] In contrast, people are adding less sugar to foods from the sugar bowl at home.

In addition to consuming more sugars, people are also consuming more artificial sweeteners, such as aspartame and saccharine.[6] These substances are intended to reduce sugar intakes, but instead of substituting artificial sweeteners for sugar, people seem to be choosing more sweet foods and beverages in all their forms.

Does all this sugar harm people's health? And if it does, are sugar substitutes a better choice? This Controversy addresses these questions and, in the process, demonstrates how nutrition researchers pursue their answers, step by step, via scientific inquiry.

Evidence Concerning Sugar

Sugar has been accused of causing these nutrition problems: (1) promoting and maintaining obesity, (2) causing and aggravating diabetes, (3) increasing the risk of heart disease, (4) disrupting behavior in children and adults, and (5) causing dental decay and gum disease. Is sugar guilty or innocent of these charges?

Does Sugar Cause Obesity?

Over the past decade, obesity has risen rapidly in the United States; during the same time frame, sugar consumption reached an all-time high. Does this mean that increasing sugar intake increases obesity? The answer from research is probably not, unless the calories of sugar are taken in addition to an already calorie-replete diet. For example, rats fed a sucrose-rich diet do not become fatter than rats eating the same number of calories of other foods, but their fat distribution changes. The sugar-fed rats seem to develop more belly fat than do rats fed a diet of regular rat chow. This change might be significant if it held true for people because central obesity is associated with human heart disease. So far, evidence for the effect in people is lacking.

Some researchers are studying whether a diet's glycemic index (GI) ranking affects body fatness. Overweight children who were given a low-GI diet lost three times as much weight as children fed a low-fat diet.[7] Many studies also suggest that low-GI foods, such as beans, sustain feelings of fullness for much longer than mashed potatoes, a food with a high GI ranking.[8] No conclusions can be drawn from such studies

Figure C4•1

ADDED SUGARS: AVERAGE SUPPLY PER PERSON IN THE UNITED STATES, 1890–1996

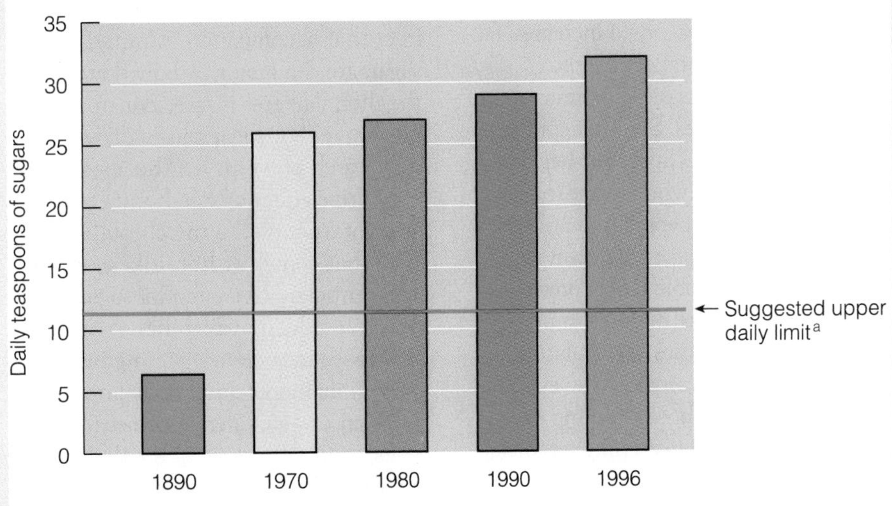

[a]Recommended upper limit for a 2,200-calorie diet.
SOURCE: U.S. Department of Agriculture, Economic Research Service, *A Dietary Assessment of the U.S. Food Supply: Comparing Per Capita Food Consumption with the Food Guide Pyramid Serving Recommendations*, AER no. 772 (Washington, D.C.: Government Printing Office 1998), p. 25.

*This estimate from the USDA Economic Research Service includes all caloric sweeteners in the U.S. human food supply, including cane and beet sugars, corn sweeteners, honey, and syrups. Other estimates derived from self-reporting of sugars eaten or other databases may differ.

The average U.S. citizen is supplied with 100 pounds of added sugars each year.

to implicate sucrose as a cause of obesity, however, because sucrose itself was not studied, and sucrose evokes only a moderate glycemic response. Another common sugar, the high-fructose corn syrup used to sweeten soft drinks, ranks even lower on the glycemic index because the monosaccharide fructose elicits a low glycemic response. Thus, if a high-GI diet is one day implicated as a cause of obesity, sugars are unlikely to play a major role other than through the calories they provide.

More direct evidence on obesity in people comes from population studies. In many developing countries, incidence of obesity increases as sugar consumption rises, but this evidence is not sufficient to name sugar as the cause. In general, when sugar intake increases, fat and total calorie intakes also rise because sweet treats often contain abundant fat and calories. Simultaneously, physical activity declines. Furthermore, obesity also occurs where sugar intakes are low, and obese people in many instances eat less sugar than thin people do. Also, obese people report preferring sweet foods no more often than normal-weight people. Studies of populations by themselves cannot separate the effects of eating sugar from those of eating too many calories or of exercising too little.

Concentrated sweets and soft drinks do make it easy to consume large amounts of calories quickly, however, which is why most diet plans recommend avoiding such items. Some people believe that eating even small amounts of sugar triggers eating binges; for them, conscientious sugar avoidance is an important part of weight management. For others, including small amounts of sugar in a weight-loss plan makes the plan easier to follow. In short, the effects of sugar on a person's eating style and body weight depend on the user.

Does Sugar Cause Type 2 Diabetes?

Recall from the chapter that in diabetes, insulin secretion or tissue responsiveness to it becomes abnormal, affecting the body's ability to manage sugar. At one time, people thought that eating sugar caused diabetes by "overstraining the pancreas," but now we know that this is not the case. Body fatness is more closely related to diabetes than diet is; high rates of diabetes have not been reported in any society where obesity is rare.

In populations around the world, a profound increase, by as much as tenfold, in the incidence of diabetes has occurred simultaneously with an increase in sugar consumption. This has been true for the Japanese, Israelis, Africans, Native Americans, Eskimos, Polynesians, and Micronesians. Yet, in some other populations, no relationship has been found between sugar intake and diabetes. Wherever starch and the fiber that accompanies it, rather than sugar, are the major carbohydrates in the diet, diabetes is rare. But this does not prove that sugar causes diabetes or that starch prevents it. The apparent protective effect of starch might be due, for example, to the chromium or fiber that comes with it. Likewise, the apparent causative effect of sugar may reflect a sudden availability of calorie-rich foods in societies gaining in both disposable income and body fatness.

Supporting evidence comes from researchers who tracked the dietary habits of over 65,000 women and almost 43,000 men for six years.[9] The researchers were trying to discover whether diabetes development is related to a diet providing many servings of foods with a high glycemic effect. The researchers discovered that women who most often chose foods with a high glycemic effect, such as mashed potatoes, white rice, highly refined cold breakfast cereals, and white bread, developed diabetes more often than those who consistently ate foods with a low glycemic effect, such as fiber-rich whole-grain breads and cereals, legumes, and fruits and vegetables. Sucrose itself elicits only a moderate glycemic effect, so the fairest conclusion is that sugar alone is not culpable in type 2 diabetes causation. Consumed in excess of energy need, however, sugar may contribute both to obesity, a major cause of diabetes, and to a total diet profile that may increase diabetes risk.

Can a person who already has diabetes use moderate amounts of sugar? Most authorities agree that an amount of sucrose equaling 5 to 10 percent of total calories consumed is acceptable as part of the carbohydrate in a controlled diet. *Other* body responses must also be considered, however, among them raised blood lipids, which suggest a high risk of heart disease.

Does Eating Sugar Increase the Risk of Heart Disease?

Research using rats provides some clues to the relationship between sugar and heart disease. When researchers fed rats a diet with sucrose as the only carbohydrate source, the rats sustained microscopic damage to their arteries, and their blood tested high for both saturated fat and cholesterol. Rats fed starch instead of sugar did not develop the damage or the elevated blood lipids. When researchers tried a similar study on human beings, they confirmed that people, too, respond to diets with extremely high sugar contents by releasing saturated fat, probably made from the sugar, into the blood. When fed diets high in starch instead of sugar, the same subjects dramatically reduced their synthesis and release of saturated fats.[10]

Keep in mind that both the rats and the people in the studies just described consumed sucrose amounting to 75 percent or more of their daily calories; even people with a highly unusual craving for sweets wouldn't choose to live on such a diet. Other studies among human beings also show that blood lipids associated with heart disease rise in response to diets high in sucrose and fructose and that those believed to be protective fall.[11] The diets evoking this response often contain almost twice the nation's average intake of added sugars, though; current intake levels are thought to be below the amount that may adversely affect blood lipids. Also, some people may inherit a genetic susceptibility that makes a greater lipid response to dietary sugars likely.[12] For most people, though, moderate sugar intakes do not elevate blood lipids.[13]

Saturated fat is clearly the major *dietary* culprit in heart disease susceptibility, but there is a *hereditary* culprit, too: some people may have inherited the tendency to develop raised blood lipid levels in response to dietary sugars, carbohydrate, or alcohol. If their heart disease risk is assessed as high, they are told to restrict their intakes of carbohydrate and alcohol. Throughout many years of research, no one has shown conclusively that moderate amounts of sugar (10 percent of total calories) affect the disease process in healthy human beings.

What about Sugar and Behavior?

Twenty years ago, claims appeared that eating sugary foods caused children to become unruly and adolescents and adults to exhibit antisocial and even criminal behavior. Research since then has yielded only mixed or negative results. Most experts agree that the "sugar-behavior" theory has been put to rest, but many teachers, parents, grandparents, and others still believe that the children they know react behaviorally to sugar. Sugar might influence behavior in many ways: by altering the levels of chemicals in the brain that affect mood, by inducing nutrient deficiencies, by stimulating the release of the series of hormones the body secretes after consuming sugar, or by providing energy (the Halloween effect).

We know that blood glucose is regulated not by diet but by hormones, and one of those is the stress hormone, norepinephrine. To test for any relationship between sugar intake and norepinephrine, researchers fed a syrupy beverage to 9 adults and 14 children and then tested their blood norepinephrine levels three hours later, after insulin had had time to store the sugar. The blood *sugar* levels of both groups had dropped only slightly, but the blood *norepinephrine* was elevated. In the children, it had shot up to double the level seen in the adults. The children also complained of symptoms such as weakness and nervousness during the test period. Although it is tempting to declare that sugar elevates blood norepinephrine levels in children and that this effect leads to behavior changes, many more studies are needed before the theory can be confirmed as fact.

Studies do not suggest that sugar has a negative effect on behavior. Indeed, several well-controlled studies have shown that sugar calms normal children, a finding consistent with convincing biochemical evidence. One such study showed no differences in activity, social interactions, learning performance, or mood in children given artificial sweeteners, but found that sugar made the children less active. In other studies, sugar has calmed juvenile delinquents with pronounced behavioral problems. Conclusion: occasional behavioral reactions to sugar may be possible, but studies have overwhelmingly failed to demonstrate any consistent effects of sucrose on behavior in either normal or hyperactive children.[14]

Does Sugar Cause Dental Caries?

Dental caries are a serious public health problem afflicting the majority of people in the country, half by the age of two (see Figure C4-2). A very lucky few *never* get dental caries because they have an inherited resistance; others have had sealants applied to teeth that stop caries before they can begin. One of the most successful measures taken to reduce the incidence of dental decay is fluoridation of community water. But sugar has something to do with dental caries, too.

Caries develop as acids produced by bacterial growth in the mouth eat into tooth enamel. Bacteria form colonies known as **plaque** whenever they can get established on tooth surfaces. Once established, they multiply and affix themselves more and more firmly unless they are brushed, flossed, or scraped away. Eventually, the acid of plaque creates pits that deepen into cavities. Below the gum line, plaque works its way down until the acid erodes the roots of teeth and the jawbone in which they are embedded, loosening the teeth and leading to infections of the gums. Gum disease severe enough to threaten tooth loss afflicts the majority of our population by their later years. Table C4-1 on the next page defines some terms related to caries.

Bacteria thrive on carbohydrate. Carbohydrate as sugar has been named as the main causative factor in the formation of caries. However, starch also supports bacterial growth if the bacteria are allowed sufficient time to work on it. Of prime importance is the length of time the food stays in your mouth, and this depends on the food's composition, how sticky it is, how often you eat it,

Figure C4•2

DENTAL CARIES

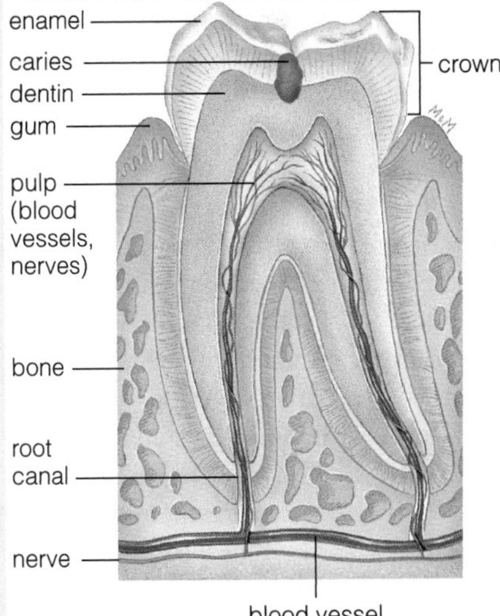

Caries begin when acid dissolves the enamel that covers the tooth. If not repaired, the decay may penetrate the dentin and spread into the pulp of the tooth, causing inflammation and an abscess.

Table C4•1

DENTAL TERMS

- **dental caries** decay of the teeth (*caries* means "rottenness").
- **plaque** (PLACK) a mass of microorganisms and their deposits on the crowns and roots of the teeth, a forerunner of dental caries and gum disease. (The term *plaque* is also used in another connection—arterial plaque in atherosclerosis. See Chapter 11.)

and especially on whether you brush your teeth afterward.[15] For most people, good oral hygiene will prevent dental caries. Regular brushing (twice a day, with a fluoride toothpaste) and flossing (once a day) may be more effective in preventing dental caries and gum disease than restricting sugary foods.

Bacteria produce acid for 20 to 30 minutes after exposure to sugar. Thus, when you eat three pieces of candy, one right after the other, your teeth are exposed to approximately 30 minutes of acid demineralization. When you eat the candy pieces at half-hour intervals, the acid exposure time is 90 minutes. Likewise, slowly sipping a sugary soft drink may be more harmful than drinking quickly and emptying the mouth of sugar.

Some forms of candy, such as milk chocolate and caramels, may be less harmful than once believed because the sugar dissolves completely and is washed away in saliva. Breads, granola bars, sugary cereals, oatmeal cookies, raisins, salted crackers, and chips, however, may be worse than once thought because particles get stuck in the teeth and do not dissolve. These particles may remain in contact with tooth surfaces for hours, providing a feast for bacteria and greatly increasing the likelihood of caries. A table in Chapter 13 lists foods of both high and low caries potential. The punchline seems to be: Brush your teeth after eating.

Total sugar intakes play a major role in caries incidence, though, and populations whose diets provide more than 10 percent of calories from sugar have an unacceptably high incidence of dental caries. Worldwide, many governing agencies urge their citizens to consume no more than 10 percent of calories from sugar because of sugar's link with dental caries. Sugar is an energy source for the bacteria that cause tooth decay, and when exposure is sufficient in susceptible people, sugar is guilty as charged.

Personal Strategy for Using Sugar

Research indicates that sugar is not guilty of the first four accusations and guilty of the fifth—sugar does cause dental caries. While waiting for scientific clarification on other health effects of sugar, should you avoid sugar or reduce your intakes? The *Dietary Guidelines* suggest only to "moderate your intake of sugars," not avoid it altogether. The World Health Organization recommends that added sugars contribute no more than 10 percent of a person's total calorie intake. A person who eats 2,000 calories of energy a day is allowed 200 calories from sugar. Those 200 calories of sugar, about 13 teaspoons, sound like quite a lot. But when you add up all the teaspoons of sugars added to common foods, 200 calories may seem restrictive.

To meet sugar intake recommendations, many people should reduce their intakes. One option may be to replace sugar's sweetness by choosing from two sets of alternative sweeteners: the sugar alcohols, which are energy-yielding sweeteners sometimes referred to as nutritive sweeteners, and the artificial sweeteners, which provide virtually no energy and are thus sometimes referred to as nonnutritive sweeteners. These options do provide sweetness without sucrose, but what about their safety?

Evidence Concerning Sugar Alcohols

The sugar alcohols are familiar to people who use special dietary products. Many new low-sugar food products depend on sugar alcohols for their bulking and sweetening powers: **mannitol, sorbitol, isomalt,** and **xylitol.**[16] These and other sugar alcohols can be metabolized by human beings, and most provide about as much energy as sucrose or a little less. An approximation of calories from sugar alcohols can be obtained by dividing grams of sugar alcohols by 2.[17]

A proven benefit of sugar alcohols is that ordinary mouth bacteria metabolize them less rapidly than other carbohydrates. As a result, sugar alcohols do not contribute to dental caries, and foods that contain them can say so on their labels.

Mannitol is the least satisfactory of the sugar alcohols. Because it is less sweet than sucrose, large amounts have to be used to obtain the same sweetness (see Table C4-2). It lingers unabsorbed in the intestine for a long time, available as energy for intestinal bacteria. As they consume the mannitol, the bacteria multiply, attract water, produce irritating waste, and cause diarrhea.

More commonly used than mannitol, sorbitol sweetens sugar-free gums and candies, but it, too, has drawbacks. At least two teaspoons as much sorbitol (with twice the calories) must be used to deliver the sweetness of one teaspoon of sucrose. Like mannitol, it causes diarrhea when consumed in large amounts. One woman, after seven years of unsuccessful treatments for her severe chronic diarrhea, finally received a correct diagnosis: her daily chewing gum, sweetened with sorbitol, was causing her problem.[18] When she cut back on the gum, her diarrhea vanished.

A relative newcomer to the sugar alcohol list is isomalt. Chemically, isomalt combines molecules of the sugar alcohols mannitol and sorbitol with molecules of glucose, yielding a product that provides about half the sweetness and half the calories of sucrose. Isomalt's weak sweetening power necessitates adding other sweeteners to products to make them taste acceptably sweet to consumers. Because it is heat stable, isomalt can be used in cooked or baked products. Like other sugar alcohols, isomalt reduces the risks of tooth decay. Isomalt may be less likely than the others to cause diarrhea because its

Table C4•2

SWEETNESS OF SUGAR SUBSTITUTES

Sugar Substitute	Relative Sweetness[a]
Sugars	
Sucrose	1.0
Fructose	1.7
Sugar alcohols	
Sorbitol	0.5
Isomalt	0.6
Mannitol	0.7
Xylitol	1.0
Noncaloric sweeteners	
Cyclamate	45.0
Acesulfame-K	200.0
Aspartame	200.0
Saccharin	300.0
Sucralose	600.0
Alitame	2,000.0

[a]The relative sweetness depends on the temperature, acidity, and other flavors of the foods in which the substance occurs. The sweetness of pure sucrose is the standard with which the approximate sweetness of sugar substitutes is compared.

larger chemical structure attracts less water into the colon.[19]

Xylitol is popular, especially in chewing gums. It does not support caries-producing bacteria and may inhibit their production of acid and prevent them from adhering to the teeth. Xylitol occurs naturally in many fruits and also arises in the body during normal metabolic processes. Most people can tolerate small amounts, and it may be a useful sugar replacer in diets for treating diabetes.[20] In large amounts, xylitol slows down the emptying of the stomach, but also stimulates release of a hormone (motilin) that speeds up intestinal activity and so causes diarrhea.

If you want to reduce energy intake, remember that the sugar alcohols *do* provide energy. The body handles them differently from sugar, but unlike artificial sweeteners, they are not calorie-free.

Evidence Concerning Artificial Sweeteners

Like the sugar alcohols, artificial sweeteners make foods taste sweet without promoting dental decay. Unlike sugar alcohols, they are calorie-free, and the human taste buds perceive them as supersweet. But are they safe?

All substances are toxic if high enough doses are consumed. Artificial sweeteners, their components, and their metabolic by-products are not exceptions. The questions to ask are whether artificial sweeteners are harmful to human beings at the levels normally used, and how much is too much. The Food and Drug Administration (FDA) has proposed answers by setting **acceptable daily intake (ADI)** levels for some of the artificial sweeteners used in the United States. Table C4-3 defines some sugar substitute terms. The major synthetic sweeteners today are **saccharin, aspartame, acesulfame-K,** and **sucralose.**

Saccharin

Saccharin has had a rocky history of acceptance, although it is now consumed by millions of Americans, primarily in prepared foods and beverages and secondarily as a tabletop sweetener. Questions about its safety surfaced in the late 1970s, when experiments suggested that it caused bladder tumors in rats. As a result, the FDA proposed banning it. The public outcry in favor of retaining it was so loud, however, that Congress imposed a moratorium on any action, and the ban proposal was eventually withdrawn. For 25 years, saccharin was widely used but remained on the government's roster of "anticipated carcinogens." Products containing it carried a warning label about saccharin as a cancer hazard. In the year 2000, government officials reviewed the research and reversed their opinion, removing saccharin from the carcinogen list and freeing it from the labeling requirement.[21] Opponents to these changes, however, maintain that saccharin causes cancer in mice and rats and so should be avoided.

Does saccharin cause cancer? The evidence in animals is as follows. Rats

Table C4•3

SUGAR SUBSTITUTE TERMS

- **acceptable daily intake (ADI)** the estimated amount of sweetener that can be consumed daily over a person's lifetime without any adverse effects.
- **acesulfame** (AY-sul-fame) **potassium,** also called **acesulfame-K** a zero-calorie sweetener approved by the FDA and Health Canada.
- **alitame** a noncaloric sweetener formed from the amino acids L-aspartic acid and L-alanine. In the United States, the FDA is considering its approval.
- **aspartame** a compound of phenylalanine and aspartic acid that tastes like the sugar sucrose but is much sweeter. It is used in both the United States and Canada.
- **cyclamate** a zero-calorie sweetener under consideration for use in the United States and used with restrictions in Canada.
- **isomalt, mannitol, sorbitol, xylitol** sugar alcohols that can be derived from fruits or commercially produced from dextrose; absorbed more slowly and metabolized differently than other sugars in the human body and not readily used by ordinary mouth bacteria.
- **saccharin** a zero-calorie sweetener used freely in the United States but restricted in Canada.
- **sucralose** a noncaloric sweetener derived from a chlorinated form of sugar that travels through the digestive tract unabsorbed. Approved by the FDA for use in the United States.

that had been fed diets containing saccharin from the time of weaning to adulthood were mated. The offspring of those rats were then fed saccharin throughout their lives and were found to have a higher incidence of bladder tumors than comparable animals not fed saccharin. In Canada, on the basis of these findings, saccharin was banned except for use as a tabletop sweetener to be sold in pharmacies with a warning label.

In human beings, a large-scale population study involving 9,000 people seemed to show a slightly elevated risk of cancers in women who drank two or more saccharin-sweetened diet sodas a day and in men and women who both smoked heavily and used artificial sweeteners. Other studies involving more than 5,000 people showed no excess risk of bladder cancers.[22]

A solid clue from the laboratory is based on some physiological differences between the urinary systems of rats and human beings. Proportionally, rats excrete far less water in their urine than people do. As a result, rats can make highly concentrated solutions of substances in just small amounts of water in their urine. Dissolved substances in such high concentrations are likely to crystallize.[23] In safety tests, saccharin overdoses caused crystals to form in the rats' bladders, and the crystals probably caused the tumors. Human beings cannot concentrate urinary substances to such a degree, so they would never form saccharin crystals, even if they consumed larger-than-normal doses of saccharin. They would, however, lose large amounts of water as the kidneys struggled to free the blood of the overload.

Overloading on huge saccharin doses is probably not safe, but consuming moderate amounts almost certainly does not cause bladder cancer in human beings. An ADI has been set for saccharin in the amount of 5 milligrams per kilogram of body weight (see Table C4-4). The amount of saccharin that can be commercially added to foods or drinks is limited to about 30 milligrams per serving.

Aspartame

Aspartame is one of the most thoroughly studied substances ever to be approved for use in foods.[24] Manufacturers use aspartame under the name *NutraSweet* to sweeten foods that require sweetness but are not exposed to cooking temperatures, as aspartame is not heat stable.[25] A gram of aspartame provides 4 calories, as does a gram of protein, but because so little is needed, calories are negligible. Under various brand names, such as *Equal* or *NutraSweet*, aspartame is also available as a powder to use at home in place of sugar. In powdered form, it is mixed with lactose, so a 1-gram packet contains 4 calories.

The amazing popularity of aspartame is mostly due to its flavor, which is almost identical to that of sugar. Furthermore, aspartame is touted as safe for children, so families wishing to limit their children's sugar intakes are offering them NutraSweet products instead.

Aspartame is a simple chemical compound: two protein fragments (the amino acids phenylalanine and aspartic acid) joined together. In the digestive tract, the two fragments are split apart, absorbed, and metabolized just as they would be if they had come from protein in food. The flavors of the components give no clue to the combined effect; one of them tastes bitter, and the other is tasteless. Yet aspartame is 200 times sweeter than sucrose.

With its phenylalanine base, aspartame poses problems for people with an inherited metabolic disease known as phenylketonuria (PKU). People with PKU have the hereditary inability to dispose of phenylalanine eaten in excess of need. Unusual products made from phenylalanine build up and damage the tissues. PKU causes irreversible,

Table C4•4

U.S. APPROVED ARTIFICIAL SWEETENERS

Artificial Sweeteners	Energy (cal/g)	Acceptable Daily Intake (ADI)	Average Amount to Replace 1 tsp Sugar[a]	Approved Uses
Saccharin (SugarTwin, others)	0	5 mg/kg body weight (341 mg for a 150 lb person)	12 mg	Tabletop sweeteners, wide range of foods, beverages, cosmetics, and pharmaceutical products
Aspartame (Nutrasweet, Equal, others)	4	50 mg/kg body weight (3,409 mg for a 150 lb person)	18 mg	General-purpose sweetener in all foods and beverages. Warning to population with PKU
Acesulfame-K (Sunette, Sweet One)	0	15 mg/kg body weight (1,023 mg for a 150 lb person)	25 mg	Tabletop sweeteners, puddings, gelatins, chewing gum, candies, baked goods, desserts, alcoholic beverages
Sucralose (Splenda)	0	5 mg/kg body weight (341 mg for a 150 lb person)	6 mg	Carbonated beverages, dairy products, baked goods, coffee and tea, fruit spreads, syrups, tabletop sweeteners, chewing gum, frozen desserts, salad dressing

[a]Rounded values.

progressive brain damage if left untreated in early life. Newborns are tested for PKU; if they have it, the treatment is to limit dietary intake of phenylalanine.

For a compelling reason, children with PKU should not get their phenylalanine from aspartame. Phenylalanine occurs in such protein-rich and nutrient-rich foods as milk and meat, and the PKU child is allowed only a limited amount of these foods. The child has difficulty obtaining the many essential nutrients, such as calcium, iron, and the B vitamins, found along with phenylalanine in these foods. To suggest that such a child squander any of the limited phenylalanine allowance on the purified phenylalanine of aspartame, with none of the associated nutrients to support normal growth, would be to invite nutritional disaster. Product labels carry special warnings for people with PKU.

Other concerns about aspartame's safety have had to do with compounds that arise briefly during its metabolism. These compounds (methyl alcohol, formaldehyde, and diketopiperazine) are not toxic at the levels generated from the ADI amount of aspartame, and concerns about them have been laid to rest.

A recent uproar about aspartame's safety began when a scientist wrote an article in a scientific journal noting a parallel between an increasing rate of brain tumors beginning in the 1980s and the approval of aspartame for public use in 1981.† The article put forth an observation, but presented no data to support a scientifically based relationship between aspartame and brain tumors; in fact, no documentation exists to show whether any of the brain cancer patients cited in the article had ever consumed aspartame.[26] Still, newspapers, magazines, and other media reported the story with gusto and "taught" the whole nation, wrongly, that a scientist had proved that aspartame causes brain cancer. Soon, stories circled the globe on the Internet accusing aspartame of causing everything from Alzheimer's disease and brain cancer to nerve disorders and skin warts.

†The article appeared in the *Journal of Neuropathology and Experimental Neurology*.

Meanwhile, a year after the release of the original published opinion, other researchers had finished their orderly scientific investigations into the theory and found no relationship between aspartame intake and brain tumors, behavior, mood, or brain chemistry.[27] They did, in fact, find a disturbing unexplained nationwide acceleration in the incidence of brain tumors, but the greatest increase was recorded several years before aspartame's approval. Of course, the media had lost interest by this time—the finding that aspartame is safe will not sell newspapers.

Another concern about aspartame use is headaches. No experimental evidence has shown a connection, but the Centers for Disease Control (CDC) has received many thousands of individual complaints. Every day, millions of people use aspartame. Every day, millions of people have headaches. Anyone who claims, on this basis, that aspartame causes headaches is using personal experience to jump to conclusions. Some of the headache sufferers might indeed be reacting to the artificial sweetener, but they might also be reacting to another substance such as caffeine or to factors in their lives unrelated to foods. People who believe aspartame gives them headaches should use a different sweetener.

On approving aspartame for U.S. consumers, the FDA set the ADI at 50 milligrams per kilogram of body weight in a day. In Canada, the acceptable level is set at 40 milligrams per kilogram. These are reasonable numbers, but the ADI amount can be exceeded. For a 132-pound person, 50 milligrams equals 80 packets of aspartame sweetener or 15 soft drinks sweetened only with aspartame. A child who drinks a quart of Kool-Aid on a hot day and who also has pudding, chewing gum, cereal, and other products sweetened with aspartame can pack in more than the daily ADI limit. Infants or toddlers under two years old should not be fed artificially sweetened foods and drinks.

Acesulfame-K

During 15 years of testing and use, the artificial sweetener acesulfame potassium (or acesulfame-K) has been used without reported health problems. An ADI of 15 milligrams per kilogram of body weight was set for acesulfame-K on its approval. Marketed under the trade names *Sunette* and *Sweet One*, this sweetener is about as sweet as aspartame and is used in chewing gum, beverages, instant coffee and tea, gelatins, and puddings, as well as for table use. Acesulfame-K holds up well during cooking.

Acesulfame-K is 200 times as sweet as sucrose but, to some, it leaves a slight aftertaste. Blending it with other sweeteners solves the problem. Acesulfame-K is not recognized by the body's metabolic equipment and therefore is excreted unchanged by the kidneys.

Sucralose

Approved in the 1990s for use as a sweetener in the United States, sucralose (trade name *Splenda*) is the only artificial sweetener made from sucrose. Three chlorine atoms substitute for three hydrogen and oxygen groups on the structure of sucrose, making a product that provides 600 times the sweetness of sugar. Many years of testing have deemed sucralose safe to use and, specifically, not a cause of cancer. Sucralose is not recognized by the body as sugar and therefore passes through unchanged. The ADI for sucralose has been set at 5 milligrams per kilogram of body weight per day for all ages, including pregnant and lactating women. Sucralose is heat stable and so is useful for cooking and baking; it is used in commercially prepared products and as a tabletop sweetener. Its sugarlike taste and versatility are earning sucralose a degree of popularity among consumers.

Other Artificial Sweeteners

Two other artificial sweeteners are awaiting FDA approval—**cyclamate** and **alitame.** Cyclamate was once approved in the United States but later was banned when it was suspected, but never proved, to cause cancer in rats. In Canada, cyclamate is restricted to use as a tabletop sweetener on the advice of a physician and as a sweetening additive in medicines. Alitame resembles aspartame in being composed of two amino acids, but unlike aspartame, it remains stable when heated.

Do Artificial Sweeteners Help with Weight Control?

Many people eat and drink products sweetened with artificial sweeteners in the belief that the products help control weight. Do they work? Ironically, studies of rats report that intense sweeteners, such as saccharin, stimulate appetite and lead to weight *gain* instead of loss. Many studies on *people*, however, find either no change or a decline in feelings of hunger. Researchers conclude that most people's food intakes do not change much when they use artificial sweeteners.

In studying the effects of artificial sweeteners on food intake and body weight, different questions are asked and different approaches taken. It matters, for example, whether the subjects of a study are of a healthy weight or obese and whether they are on weight-loss diets. Motivations for using sweeteners also influence a person's actions. For example, a person might drink a low-calorie beverage now so as to be able to eat a high-calorie food later, so energy intake might stay the same or increase. In contrast, a person trying to control food energy intake might use the artificial sweetener and then choose low-calorie foods consistently, so energy intake might be lower than normal.

Researchers must also distinguish between the effects of the experience of tasting something sweet and the physiological effects of a particular substance on the body. If a person experiences hunger or feels full shortly after eating an artificially sweetened snack, is that because tasting something sweet stimulates or depresses the appetite? Or is it because the artificial sweetener itself somehow affects the appetite through nervous, hormonal, or other means? Furthermore, if appetite is stimulated, does that actually lead to increased food intake?

In one recent study on appetite and artificial sweeteners, researchers fed normal-weight people one of four breakfasts and then measured their food intakes at later meals throughout the day. Two of the breakfasts provided 700 calories: one contained sucrose, and the other aspartame with enough starch to equalize the calories. The other two breakfasts provided 300 calories: one was plain, and the other contained aspartame. Subjects who ate either lower-calorie breakfast, regardless of sweetness, were hungrier later. Those who ate either higher-calorie breakfast stayed full longer. Sweet taste and the presence of aspartame seemed to have no effect on hunger or subsequent food energy intakes. At the end of the day, both groups who had eaten the 700-calorie breakfasts had higher total energy intakes because, although they were less hungry at lunch, they still consumed ample food energy at lunch and supper.

Overall, artificial sweeteners alone do not seem to stimulate or depress appetite. Another sweetener, however, has the ability to cut the appetite and to reduce later food intake. That sweetener is sugar. The common belief that sugar "spoils the appetite" has proved true.

Can artificial sweeteners help people lose weight?

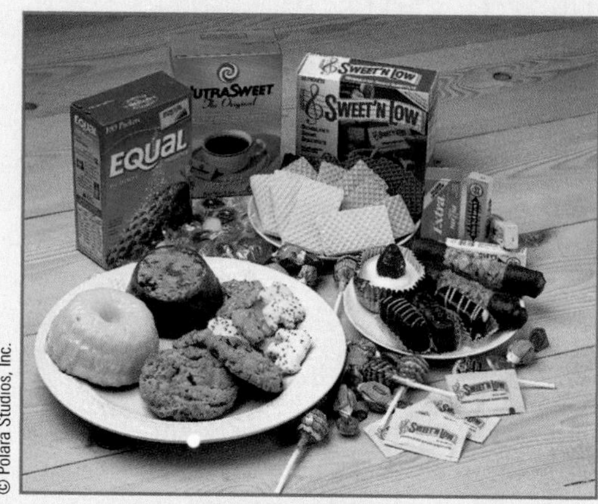

© Polara Studios, Inc.

Personal Strategies for Using Artificial Sweeteners

Current evidence indicates that moderate intakes of artificial sweeteners pose no health risks.[28] For those who choose to include artificial sweeteners in their diets, moderation is the key. Using artificial sweeteners does not automatically lower energy intake; to control energy intake successfully, a person needs to make informed diet and activity decisions throughout the day (as Chapter 9 explains). Although not magic bullets in fighting overweight, artificial sweeteners probably do not hinder weight-loss efforts either, and they are safer for the teeth than carbohydrate sweeteners.

David Tindle, R. A. (b. 1932) *Plate with Avocado Pears*

The Lipids: Fats, Oils, Phospholipids, and Sterols

Frequently Asked Questions

If fat floats, how does it travel around the body in the watery blood? p. 145

How can I use my stored fat for energy? p. 147

What, exactly, do fat and cholesterol have to do with health? p. 148

What is the significance of LDL and HDL in blood? p. 151

What is "hydrogenated vegetable oil," and what's it doing in my chocolate chip cookies? p. 155

What are *trans*-fatty acids, and are they harmful? p. 156

Contents

lipid (LIP-id) a family of organic (carbon-containing) compounds soluble in organic solvents but not in water. Lipids include triglycerides (fats and oils), phospholipids, and sterols.

cholesterol (koh-LESS-ter-all) a member of the group of lipids known as sterols; a soft, waxy substance made in the body for a variety of purposes and also found in animal-derived foods.

fats lipids that are solid at room temperature (70°F or 25°C).

oils lipids that are liquid at room temperature (70°F or 25°C).

cardiovascular disease (CVD) disease of the heart and blood vessels; also called *coronary heart disease (CHD)*. The two most common forms of CVD are atherosclerosis and hypertension.

triglycerides (try-GLISS-er-ides) one of the three main classes of dietary lipids and the chief form of fat in foods. A triglyceride is made up of three units of fatty acids and one unit of glycerol (fatty acids and glycerol are defined later). Triglycerides are also called *triacylglycerols*.

phospholipids (FOSS-foh-LIP-ids) one of the three main classes of dietary lipids. These lipids are similar to triglycerides, but each has a phosphorus-containing acid in place of one of the fatty acids. Phospholipids are present in all cell membranes.

lecithin (LESS-ih-thin) a phospholipid manufactured by the liver and also found in many foods; a major constituent of cell membranes.

sterols (STEER-alls) one of the three main classes of dietary lipids. Sterols have a structure similar to that of cholesterol.

essential fatty acids fatty acids that the body needs but cannot make in amounts sufficient to meet physiological needs.

Fat tissue is also called *adipose tissue;* Chapter 9 has more about the activities of adipose tissue.

A reminder from Chapter 1:

- 1 g carbohydrate = 4 calories.
- 1 g fat = 9 calories.
- 1 g protein = 4 calories.

YOUR BILL FROM a medical laboratory reads, "Blood **lipid** profile—$165." A health-care provider reports, "Your blood **cholesterol** is high." Your physician advises, "You must cut down on the saturated **fats** in your diet and replace them with **oils** to lower your risk of **cardiovascular disease (CVD).**" Blood lipids, cholesterol, saturated fats, and oils—what do they all mean and how do they relate to health?

Introducing the Lipids

The lipids in foods and in the human body fall into three classes. About 95 percent are **triglycerides.*** The other classes of the lipid family are the **phospholipids** (of which **lecithin** is one) and the **sterols** (cholesterol is the best known of these).

No doubt you are expecting to hear that these fat-related compounds have the potential to harm your health, but lipids are also valuable. In fact, lipids are absolutely necessary, and some lipids must be present in your foods if you are to maintain good health. The low-fat diet recommended for health doesn't mean a "no-fat" diet.[1] Luckily, traces of fats and oils are present in almost all foods, so you needn't make an effort to eat any extra.

Usefulness of Fats in the Body

When people speak of fat, they are usually talking about triglycerides. The term *fat* is more familiar, though, and we will use it in this discussion. Fat is the body's chief storage form for the energy from food eaten in excess of need. The storage of fat is a valuable survival mechanism for people who live a feast-or-famine existence: stored during times of plenty, fat enables them to remain alive during times of famine. In addition, fats provide most of the energy needed to perform much of the body's work, especially muscular work. Most body cells can store only limited fat, but some cells are specialized for storing fat. These fat cells seem able to expand almost indefinitely—the more fat they store, the larger they grow. An obese person's fat cells may be many times the size of a thin person's. Far from being an inert collection of sacks of fat, however, adipose (fat) tissue secretes hormones and produces enzymes that influence the body's intake of food and affect its use of energy nutrients.[2] A fat cell is shown in Figure 5-1.

You may be wondering why the carbohydrate glucose is not the body's major form of stored energy. As mentioned in Chapter 4, glucose is stored in the form of glycogen. Because glycogen holds a great deal of water, it is quite bulky and heavy, and the body cannot store enough to provide energy for very long. Fats, however, pack tightly together without water and can store much more energy in a small space. The body fat found on a normal-weight, healthy person contains sufficient energy to fuel a marathon run to the finish or to give a sick person who cannot eat the energy to battle disease.

Fat serves many other purposes in the body, too. Pads of fat surrounding the vital organs serve as shock absorbers. Thanks to these fat pads, you can ride a horse or a motorcycle for many hours with no serious internal injuries. The fat blanket under the skin also insulates the body from extremes of temperature, thus assisting with internal climate control.

Some essential nutrients are soluble in fat and therefore are found mainly in foods that contain fat and are absorbed most efficiently from them. These nutrients are the fat-soluble vitamins: A, D, E, and K. Other essential nutrients, the **essential fatty acids,** constitute parts of the fats themselves. As a later section explains, the essential fatty acids serve as raw materials from which the body makes molecules it needs. Lipids are also important to all the body's cells as part of their surrounding envelopes, the cell membranes.

*Another name for triglyceride is *triacylglycerol*.

satiety (sat-EYE-uh-tee) the feeling of fullness or satisfaction that people experience after meals.

Figure 5•1

A FAT CELL

Within the fat cell, lipid is stored in a droplet. This droplet can greatly enlarge, and the fat cell membrane will grow to accommodate its swollen contents. More about fat cells and obesity in Chapter 9.

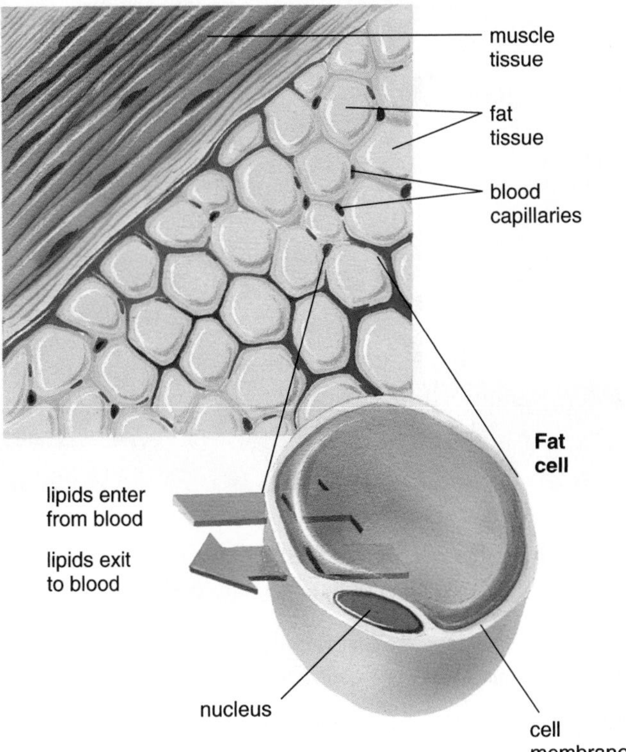

key point · · *Lipids not only serve as energy reserves but also cushion the vital organs, protect the body from temperature extremes, carry the fat-soluble nutrients, serve as raw materials, and provide the major component of cell membranes.*

Usefulness of Fats in Food

The energy density of fats makes foods rich in fat valuable in many situations. A gram of fat or oil delivers more than twice as many calories as a gram of carbohydrate. A hunter or hiker needs to consume a large amount of food energy to travel long distances or to survive in intensely cold weather. As Figure 5-2 shows, such a person can carry more energy in fat-rich foods than in carbohydrate-rich foods. But for a person who is not expending much energy in physical work, those same high-fat foods may deliver many unneeded calories in only a few bites.

People naturally like high-fat foods.[3] Around the world, as fat becomes less expensive and more available in a given food supply, people seem to choose diets providing greatly increased amounts of fat.[4] Fat carries with it many dissolved compounds that give foods enticing aromas and flavors, such as the aroma of frying bacon or french fries. In fact, when a person refuses food, foods flavored with some fat may tempt that person to eat again. Fat also lends tenderness to foods such as meats and baked goods.

Fat also contributes to **satiety,** the satisfaction of feeling full after a meal. The fat of swallowed food triggers a series of physiological events that slow down the emptying of the stomach and promote satiety.[5] Even so, before the sensation of fullness stops them, people can easily overeat on fat-rich foods because the delicious taste of fat stimulates eating and each bite can deliver many calories of fat. Glucose, fiber, fluid, and protein also contribute to satiety, and researchers are defining the

Figure 5•2

Two Lunches

Both lunches contain the same number of calories, but the fat-rich lunch takes up less space and weighs less.

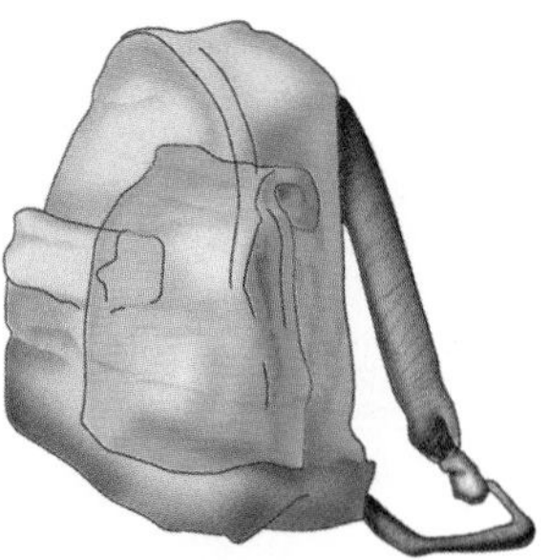

carbohydrate-rich lunch
1 low-fat muffin
1 banana
2 oz carrot sticks
8 oz fruit yogurt

calories = 550
weight (g) = 500

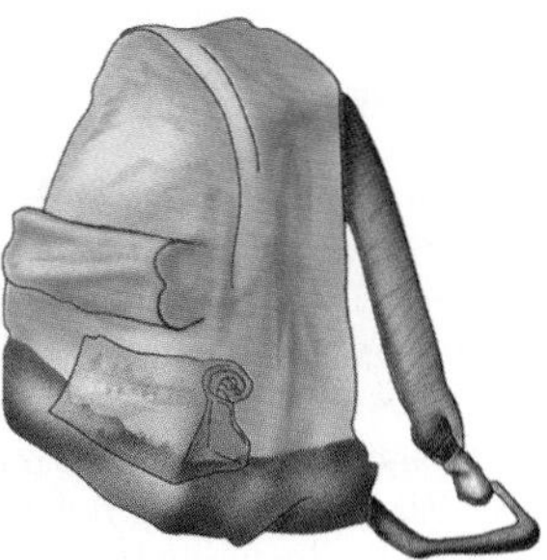

fat-rich lunch
6 butter-style crackers
1 1/2 oz American cheese
2 oz trail mix with candy

calories = 550
weight (g) = 115

fatty acids organic acids composed of carbon chains of various lengths. Each fatty acid has an acid end and hydrogens attached to all of the carbon atoms of the chain.

glycerol (GLISS-er-all) an organic compound, three carbons long, of interest here because it serves as the backbone for triglycerides.

Small amounts of fat offer eaters both pleasure and needed nutrients.

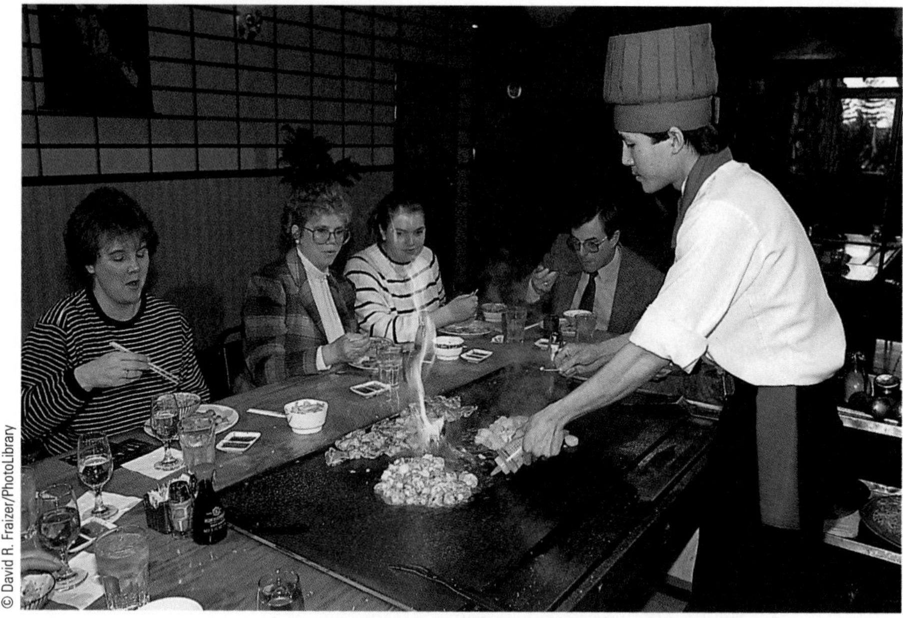

© David R. Fraizer/PhotoLibrary

relationships among their effects. Chapter 9 revisits the topic of appetite and its control. Table 5-1 sums up the usefulness of fats, both in foods and in the body.

key point *Lipids provide more energy per gram than carbohydrate and protein, enhance food's aroma and flavor, and contribute to satiety, or a feeling of fullness, after a meal.*

A Close Look at Fats

As mentioned, the term *fat* refers to triglycerides, the major form of lipid found in foods. Triglycerides, in turn, are made of fatty acids and glycerol.

Triglycerides: Fatty Acids and Glycerol

Very few **fatty acids** are found free in the body or in foods; most are incorporated into large, complex compounds: triglycerides. The name almost explains itself: three fatty acids (*tri*) are attached to a molecule of **glycerol** to form a triglyceride molecule

Table 5•1

THE USEFULNESS OF FATS

Fats in the Body	Fats in Food
• *energy stores* Fats are the body's chief form of stored energy. • *muscle fuel* Fats provide most of the energy to fuel muscular work. • *emergency reserve* Fats serve as an emergency fuel supply in times of illness and diminished food intake. • *padding* Fats protect the internal organs from shock through fat pads inside the body cavity. • *insulation* Fats insulate against temperature extremes through a fat layer under the skin. • *cell membranes* Fats form the major material of cell membranes. • *raw materials* Fats are converted to other compounds, such as hormones, bile, and vitamin D, as needed.	• *nutrient* Fats provide essential fatty acids. • *energy* Fats provide a concentrated energy source in foods. • *transport* Fats carry fat-soluble vitamins A, D, E, and K, and assist in their absorption. • *raw materials* Fats provide raw material for making needed products. • *sensory appeal* Fats contribute to taste and smell of foods. • *appetite* Fats stimulate the appetite. • *satiety* Fats contribute to feelings of fullness. • *texture* Fats help make foods tender.

Figure 5•3

TRIGLYCERIDE FORMATION

Glycerol, a small, water-soluble carbohydrate derivative, plus three fatty acids, equals a triglyceride.

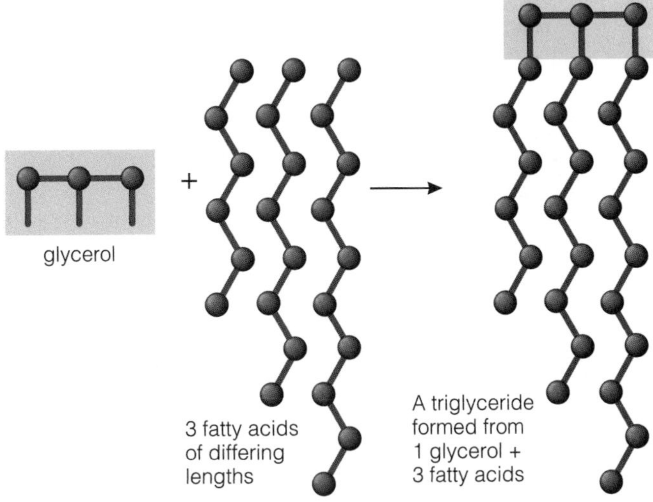

saturated fatty acid a fatty acid carrying the maximum possible number of hydrogen atoms (having no points of unsaturation). A saturated fat is a triglyceride that contains three saturated fatty acids.

point of unsaturation a site in a molecule where the bonding is such that additional hydrogen atoms can easily be attached.

unsaturated fatty acid a fatty acid that lacks some hydrogen atoms and has one or more points of unsaturation. An unsaturated fat is a triglyceride that contains one or more unsaturated fatty acids.

monounsaturated fatty acid a fatty acid containing one point of unsaturation.

polyunsaturated fatty acid (PUFA) a fatty acid with two or more points of unsaturation.

(Figure 5-3). Tissues all over the body can easily assemble triglycerides or disassemble them as needed.

Fatty acids can differ from one another in two ways: in chain length and in degree of saturation (explained next). Triglycerides usually include mixtures of various fatty acids. Depending on which fatty acids are incorporated into a triglyceride, the resulting fat will be soft or hard. Triglycerides containing mostly the shorter-chain fatty acids or the more unsaturated ones are softer and melt more readily. Each species of animal (including people) makes its own characteristic kinds of triglycerides, a function governed by genetics. Fats in the diet, though, can affect the types of triglycerides made because dietary fatty acids can be incorporated into triglycerides in the body. For example, many animals raised for food can be fed diets containing softer or harder triglycerides to give the animals softer or harder fat, whichever consumers demand.

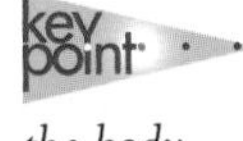

The body combines three fatty acids with one glycerol to make a triglyceride, its storage form of fat. Fatty acids in food influence the composition of fats in the body.

Three types of fatty acids: saturated, monounsaturated, and polyunsaturated.

Saturated versus Unsaturated Fatty Acids

Saturation refers to the number of hydrogens a fatty acid chain is holding. If every available bond from the carbons is holding a hydrogen, the chain forms a **saturated fatty acid;** it is filled to capacity with hydrogen. The zigzag structure on the left in Figure 5-4 on the next page represents a saturated fatty acid.

Sometimes, especially in the fatty acids of plants and fish, the chain has a place where hydrogens are missing, an "empty spot," or **point of unsaturation.** A fatty acid carbon chain that possesses one or more points of unsaturation is an **unsaturated fatty acid.** With one point of unsaturation, the fatty acid is a **monounsaturated fatty acid** (see the second structure in Figure 5-4). With two or more points of unsaturation, it is a **polyunsaturated fatty acid,** sometimes abbreviated **PUFA** (see the third structure in Figure 5-4; other examples are given later in the chapter).

The degree of saturation of fatty acids in a fat affects the temperature at which the fat melts. Generally, the more unsaturated the fatty acids, the more liquid the fat is at room temperature. In contrast, the more saturated the fatty acids, the firmer the fat. Thus, looking at three fats—lard (which comes from pork), chicken fat, and

Fats melt at different temperatures. The more unsaturated a fat, the more liquid it is at room temperature. The more saturated a fat, the higher the temperature at which it melts.

saturated fats triglycerides in which most of the fatty acids are saturated.

monounsaturated fats triglycerides in which most of the fatty acids have one point of unsaturation (are monounsaturated).

polyunsaturated fats triglycerides in which most of the fatty acids have two or more points of unsaturation (are polyunsaturated).

Figure 5•4

THREE TYPES OF FATTY ACIDS

The more carbon atoms in a fatty acid, the longer it is. The more hydrogen atoms attached to those carbons, the more saturated the fatty acid is.

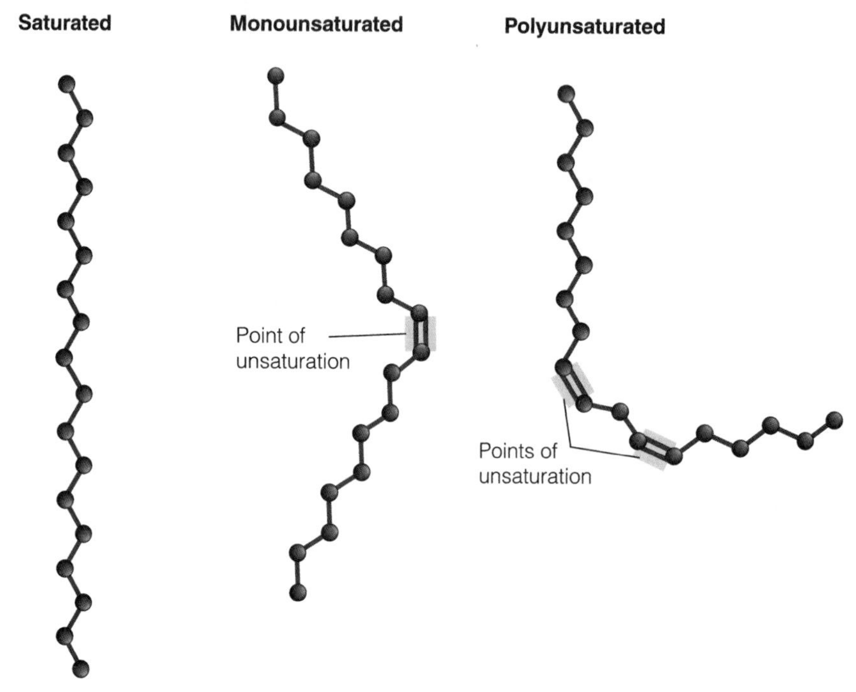

safflower oil—lard is the most saturated and the hardest; chicken fat is less saturated and somewhat soft; and safflower oil, which is the most unsaturated, is a liquid at room temperature. If a health-care provider recommends limiting **saturated fats** and using **monounsaturated fats** or **polyunsaturated fats** instead, you can generally judge by the hardness of the fats which ones to choose. To determine whether an oil you use contains saturated fats, place the oil in a clear container in the refrigerator and watch for cloudiness. The least saturated oils remain the clearest.

In general, vegetable and fish oils are rich in polyunsaturates; some vegetable oils, olive oil in particular, are also rich in monounsaturates; and animal fats are the most saturated. But you have to know your oils—it is not enough to choose foods with labels claiming plant oils over those containing animal fats. Some nondairy whipped dessert toppings use coconut oil in place of cream (butterfat). Coconut oil does come from a plant, but it disobeys the rule that plant oils are less saturated than animal fats; the fatty acids of coconut oil are actually more saturated than those of cream and are of a type that seems to add to heart disease risk. Palm oil, a vegetable oil used frequently in food processing, is also highly saturated and has been added to the list of fats that elevate blood cholesterol.[6]

A benefit to health is seen when monounsaturated fat is used in place of saturated fat in the diet.[7] Olive oil does not harm the health of the heart, and when it replaces other fats in the diet, it may offer a degree of protection against heart disease and breast cancer, as evidence from people of Mediterranean regions indicates (see the Controversy section).[8] Canola oil is another rich source of monounsaturated fatty acids. Figure 5-5 compares the percentages of saturated, monounsaturated, and polyunsaturated fatty acids in various fats and oils, and Figure 5-6 identifies the sources of saturated fats in the U.S. diet. Note that the bread, cereal, rice, and pasta group is not represented in the figure, but as a later section points out, manufacturers often add fats to these foods.

Fatty acids are energy-rich carbon chains that can be saturated (filled with hydrogens) or monounsaturated (with one point of unsaturation) or polyun-

Figure 5•5

FATTY ACID COMPOSITION OF COMMON FOOD FATS

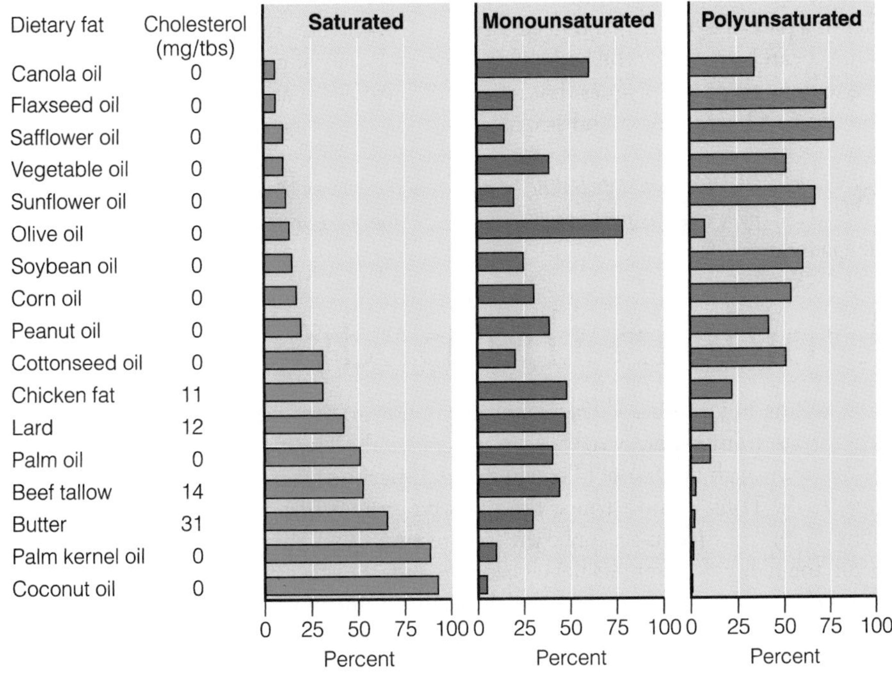

emulsifier a substance that mixes with both fat and water and permanently disperses the fat in the water, forming an emulsion.

emulsification the process of mixing lipid with water by adding an emulsifier.

bile an emulsifier made by the liver from cholesterol and stored in the gallbladder. Bile does not digest fat as enzymes do but emulsifies it so that enzymes in the watery fluids may contact it and split the fatty acids from their glycerol for absorption.

saturated (with more than one point of unsaturation). The degree of saturation of the fatty acids in a fat determines the fat's softness or hardness.

Other Members of the Lipid Family

Thus far we have dealt with the largest of the three classes of lipids—the triglycerides and their component fatty acids, which represent 95 percent of all the lipids in the diet and in the body. The other two classes—phospholipids and sterols—play important roles in the body.

A phospholipid, like a triglyceride, consists of a molecule of glycerol with fatty acids attached, but it contains two, rather than three, fatty acids. In place of the third is a molecule containing phosphorus, which makes the phospholipid soluble in water, while its fatty acids make it soluble in fat. This versatility permits any phospholipid to play a role in keeping fats dispersed in water; it can serve as an **emulsifier.**

Food processors blend fat with watery ingredients by way of **emulsification.** Some salad dressings separate to form two layers—vinegar on the bottom, oil on the top. Other dressings, such as mayonnaise, are also made from vinegar and oil but never separate. The difference lies in a special ingredient of mayonnaise, the emulsifier lecithin in egg yolks. Lecithin, a phospholipid, blends the vinegar and oil in a stable emulsion.

Lecithin and other phospholipids also play key roles in the structure of cell membranes. Because phospholipids are emulsifiers, they have both water-loving and fat-loving characteristics, which enable them to help fats travel back and forth across the lipid-containing membranes of cells into the watery fluids on both sides. Health-promoting properties, such as the ability to lower blood cholesterol, are sometimes attributed to lecithin, but the people making the claims stand to gain from selling supplements. Although an important lipid to the body, lecithin has no special ability to promote health.

Sterols such as cholesterol are large, complicated molecules consisting of interconnected *rings* of carbon atoms with side chains of carbon, hydrogen, and oxygen attached. Cholesterol serves as the raw material for making another emulsifier, **bile,**

Figure 5•6

SATURATED FATS IN THE U.S. DIET

Note that fruits, grains, and vegetables are insignificant sources, unless saturated fats are intentionally added to them during preparation.

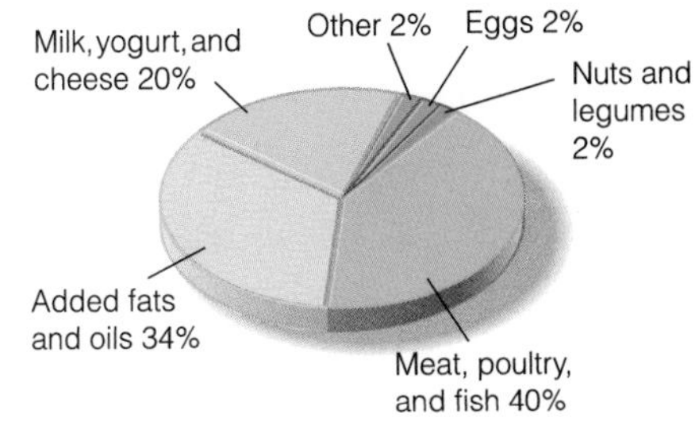

monoglycerides (mon-oh-GLISS-er-ides) products of the digestion of lipids; consist of glycerol molecules with one fatty acid attached (*mono* means "one"; *glyceride* means "a compound of glycerol").

which is important to digestion (see the next section for details). Other sterols are vitamin D, which is made from cholesterol, and several important hormones, the so-called steroid hormones, including the sex hormones.

Cholesterol is an important sterol in the structure of brain and nerve cells and, in fact, is a part of every cell and necessary to the body's functioning. Like lecithin, cholesterol can be made by the body, so it is not an essential nutrient. Cholesterol is also the major part of the plaques that narrow the arteries in atherosclerosis, the underlying cause of heart attacks and strokes.

Phospholipids, including lecithin, play key roles in cell membranes; sterols play roles as part of bile, vitamin D, the sex hormones, and other important compounds.

Lipids in the Body

In handling lipids, the body faces a problem: how to thoroughly mix fats, which tend to separate from water, with its own watery fluids. The digestive system solves this problem through the use of bile, which emulsifies the fat in food in the watery digestive fluids. Thus, to digest fats, the digestive system first mixes them with its bile-containing digestive juices; once the fats are emulsified, the fat-digesting enzymes can break them down. After fats have been digested, they face another watery barrier, the watery layer of mucus that coats the absorptive lining of the digestive tract. Fats must traverse this layer to enter the cells of the digestive tract lining. The final challenge is to package lipids so that they can travel in the watery blood and lymph of the circulatory system. The next two sections describe the body's superb adaptations to meet these needs for lipid digestion and transport.

Digestion of Fats

A bite of food in the mouth first encounters the enzymes of saliva. One enzyme, produced by the tongue, acts on triglycerides to liberate fatty acids, especially those of milk.[9] The enzyme plays a major role in milk fat digestion in infants, but is thought to be of little importance to fat digestion in adults. After being chewed and swallowed, the food travels to the stomach, where the fat separates from other components and floats as a layer on the top. Since fat does not mix with the stomach fluids, little fat digestion takes place in the stomach.

Chapter 3 first described the action of bile and gave details of the digestive system.

By the time fat enters the small intestine, the gallbladder, which stores the liver's output of bile, has contracted and squirted its bile into the intestine. Bile mixes fat particles with watery fluid by emulsifying them (see Figure 5-7), suspending them in the fluid until the fat-digesting enzymes contributed by the pancreas can split them into smaller particles for absorption. A bile molecule, made from cholesterol, works because one of its ends attracts and holds fat, while the other end is attracted to and held by water.

People sometimes wonder how a person without a gallbladder can digest food. The gallbladder is just a storage organ. Without it, the liver still produces bile, but delivers it continuously into the small intestine. People who have had their gallbladders removed must reduce their fat intakes because they can no longer store bile and release it at mealtimes. As a result, their systems can handle only a little fat at a time.

Once the intestine's contents are emulsified, fat-splitting enzymes act on triglycerides to split fatty acids from their glycerol backbones. Free fatty acids, glycerol, and **monoglycerides** cling together in balls surrounded by bile and are shuttled across the watery layer of mucus to the waiting absorptive cells of the intestinal villi. The cells then extract the lipids. The bile may be absorbed and reused by the body, or it may exit with the feces as shown in Figure 4-4 of Chapter 4.

The small products of lipid digestion, glycerol and shorter-chain fatty acids, can pass directly through the cells of the digestive tract lining into the bloodstream. From there they can travel without help to the liver and to the tissues that need them.

Figure 5•7

THE ACTION OF BILE IN FAT DIGESTION

Detergents are emulsifiers and work the same way, which is why they are effective in removing grease spots from clothes. Molecule by molecule, the grease is dissolved out of the spot and suspended in the water, where it can be rinsed away.

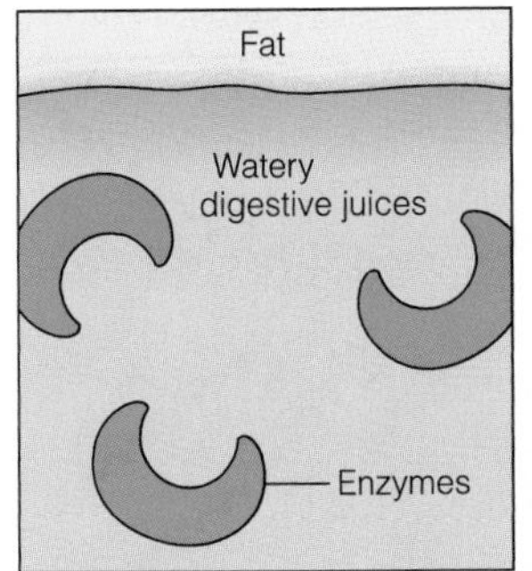

In the stomach, the fat and watery digestive juices tend to separate. Enzymes are in the water and can't get at the fat.

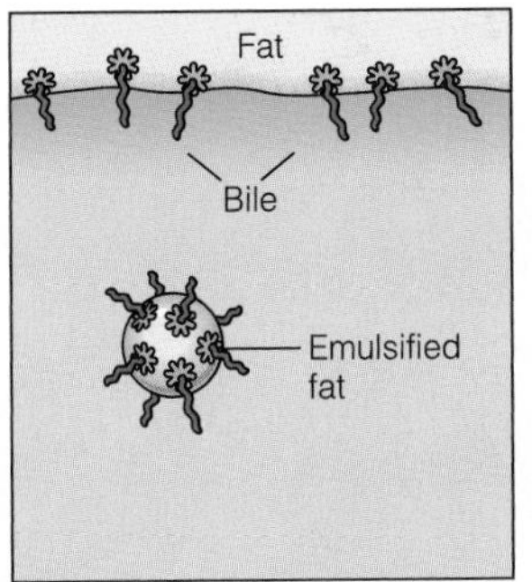

When fat enters the small intestine, the gallbladder secretes bile. Bile has an affinity for both fat and water, so it can bring the fat into the water.

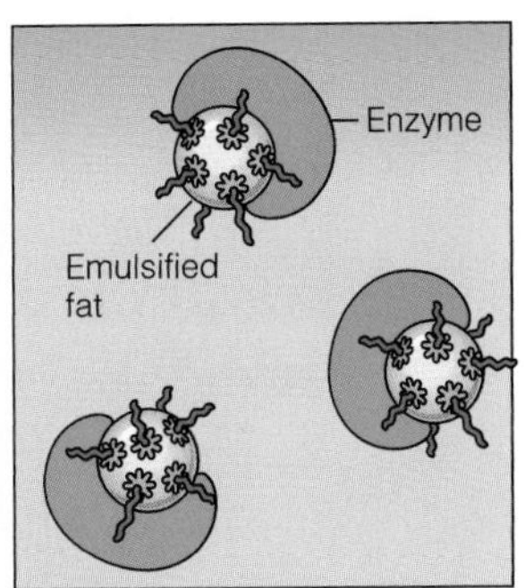

After emulsification, the enzymes have easy access to the fat droplets.

The larger products of lipid digestion, monoglycerides and long-chain fatty acids, need some help to get to their destinations via the bloodstream. Once inside the intestinal cells, they are re-formed into triglycerides and incorporated into **chylomicrons,** clusters of proteins and lipids. Chylomicrons are a class of **lipoproteins,** described in the next section.

The digestive tract absorbs triglycerides from a meal with up to 98 percent efficiency. In other words, little fat is excreted by a healthy system. The process of fat digestion takes time, though, so the more fat taken in at a meal, the slower the digestive system action becomes. The efficient series of events just described is depicted in Figure 5-8 on the next page.

key point *In the stomach, fats separate from other food components. In the small intestine, bile emulsifies the fats, enzymes digest them, and the intestinal cells absorb them.*

If Fat Floats, How Does It Travel around the Body in the Watery Blood?

Within the body, many fats travel from place to place in blood as passengers in lipoproteins. For example, the monoglycerides and long-chain fatty acids liberated from digested food fat are too large to be released into the bloodstream. Without some mechanism to keep them dispersed, these lipids would separate out and float in globules, disrupting the blood's normal functions. Therefore, before releasing the lipids, the intestinal cells allow them to cluster together, rejoin them as triglycerides, and combine them with protein to form chylomicrons. The protein and phospholipid in the clusters act as emulsifiers, attracting both water and fat. Their association with both substances enables chylomicrons to transport lipids in the watery lymph and blood. The tissues of the body can extract whatever fat they need from these clusters. The remnants are then picked up by the liver, which dismantles them and reuses their parts.

In addition to the chylomicrons, the body uses three other types of lipoproteins to carry fats: the **very-low-density lipoproteins (VLDL),** which carry triglycerides and other lipids made in the liver to the body cells for their use; the **low-density lipoproteins (LDL),** which are made from VLDL after they have donated much of their fat to body cells and picked up cholesterol; and the **high-density lipoproteins (HDL),** which are

chylomicrons (KYE-low-MY-krons) clusters formed when lipids from a meal are combined with carrier proteins in the intestinal lining. Chylomicrons transport food fats through the watery body fluids to the liver and other tissues.

lipoproteins (LYE-poh-PRO-teens, LIH-poh-PRO-teens) clusters of lipids associated with protein, which serve as transport vehicles for lipids in blood and lymph. Major lipoprotein classes are the chylomicrons, the VLDL, the LDL, and the HDL.

very-low-density lipoproteins (VLDL) lipoproteins that transport triglycerides and other lipids from the liver to various tissues in the body.

low-density lipoproteins (LDL) lipoproteins that transport lipids from the liver to other tissues such as muscle and fat; contain a large proportion of cholesterol.

high-density lipoproteins (HDL) lipoproteins that are critical in the process of carrying cholesterol from body cells to the liver for dismantling and disposal; contain a large proportion of protein.

Figure 5•8

THE PROCESS OF LIPID DIGESTION AND ABSORPTION

Inside the digestive tract:

Digestive enzymes accomplish most fat digestion in the small intestine where bile emulsifies fat, making it available for enzyme action. The enzymes cleave triglycerides into free fatty acids, glycerol, and monoglycerides.

At the intestinal lining:

The parts are absorbed by intestinal villi. Large lipid fragments, such as monoglycerides and long-chain fatty acids, are converted back into triglycerides and combined with protein, forming chylomicrons that travel in the lymph vessels to the bloodstream. Small lipid particles such as glycerol and short-chain fatty acids are small enough to enter directly into the bloodstream without further processing.

In this diagram, molecules of fatty acids are shown as large objects, but, in reality, molecules of fatty acids are too small to see even with a powerful microscope, while villi are visible to the naked eye.

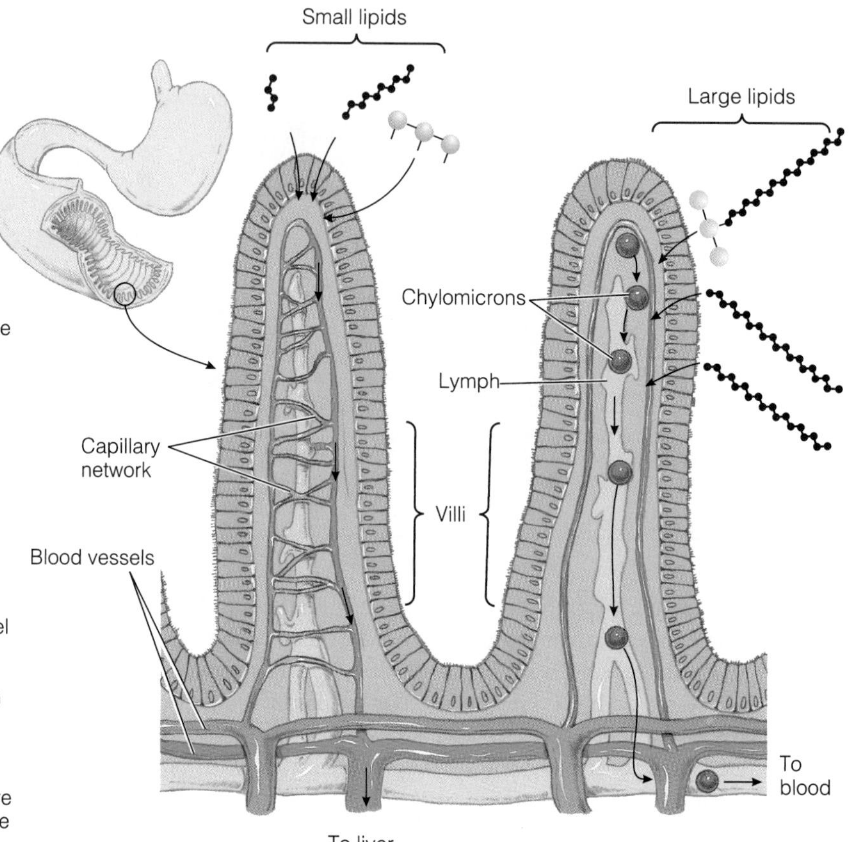

critical in the process of carrying cholesterol from body cells to the liver for disposal.[10] The proteins for both LDL and HDL are made in the liver. Figure 5-9 depicts a typical lipoprotein and demonstrates how a lipoprotein's density changes with its lipid and protein contents.

Your health-care provider may order a medical test, the blood lipid profile, to reveal not only the triglycerides and cholesterol in your blood but also the lipoproteins that carry them. LDL and HDL have separate functions, which have important implications for the health of your heart and blood vessels. Both LDL and HDL carry lipids in the blood, but LDL are larger, lighter, and more lipid filled; HDL are smaller, denser, and packaged with more protein. LDL deliver triglycerides and cholesterol from the liver to the tissues; HDL scavenge excess cholesterol and phospholipids from the tissues for disposal. Both LDL and HDL carry cholesterol, but elevated LDL concentrations in the blood are a sign of high risk of heart attack, whereas elevated HDL concentrations are associated with a low risk. Thus, some people refer to LDL as "bad" cholesterol and HDL as "good" cholesterol—even though they carry the same kind of cholesterol. The difference to the heart between LDL and HDL lies in the proportions of lipids they contain and the tasks they perform, not in the *type* of cholesterol they carry. A later section examines the significance of lipoproteins to the health of the heart.

key point *Blood and other body fluids are watery, so fats need special transport vehicles—the lipoproteins—to carry them around the body in these fluids. The chief lipoproteins are chylomicrons, VLDL, LDL, and HDL.*

Figure 5•9

LIPOPROTEINS

As the graph shows, the density of a lipoprotein is determined by its lipid-to-protein ratio. All lipoproteins contain protein, cholesterol, phospholipids, and triglycerides in varying amounts. An LDL has a high ratio of lipid to protein (about 80 percent lipid to 20 percent protein) and is especially high in cholesterol. An HDL has more protein relative to its lipid content (about equal parts lipid and protein).

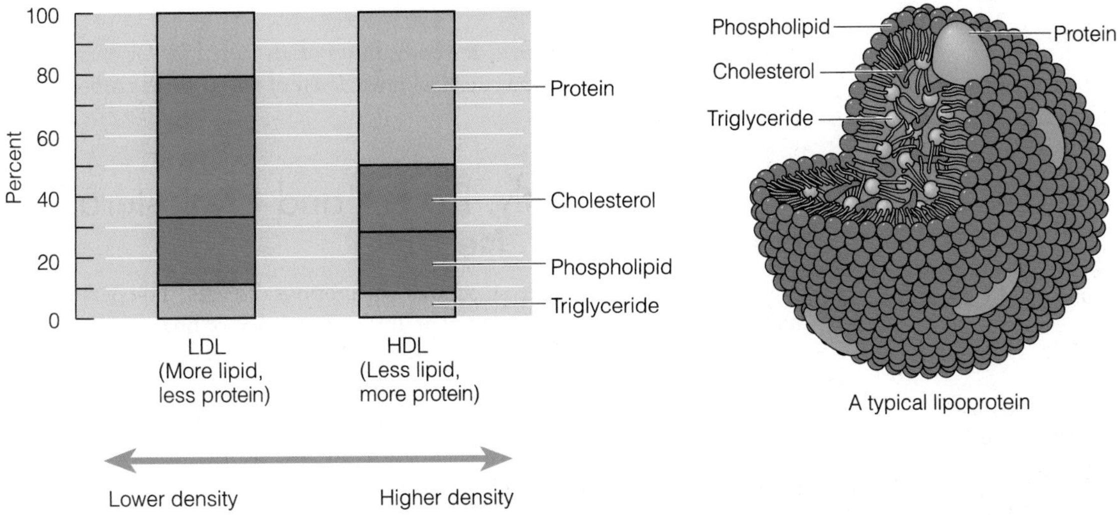

How Can I Use My Stored Fat for Energy?

Many triglycerides eaten in foods are transported by the LDL to the fat depots—muscles, breasts, the insulating fat layer under the skin, and others—where they are stored by the body's fat cells for later use. When a person's body starts to run out of fuel available from food, it begins to retrieve this stored fat to use for energy. (It also draws on its stored glycogen, as the last chapter described.) Fat cells respond to the call for energy by dismantling stored fat molecules and releasing fat components into the blood. Upon receiving these components, the energy-hungry cells break them down further into small fragments. Finally, each fat fragment is combined with a fragment derived from glucose, and the energy-releasing process continues, liberating energy, carbon dioxide, and water. The way to use stored fat for energy, then, is to create a demand for it in the tissues by decreasing intake of food energy, by increasing the body's expenditure of energy, or both.

Whenever body fat is broken down to provide energy, carbohydrate must be available as well. Without carbohydrate, ketosis will occur, as described in the last chapter, and products of incomplete fat breakdown (ketones) will appear in the blood and urine. Because this process and its consequences are so important in weight control, Chapter 9 describes them in greater detail.

The body can also store excess glucose as fat, but this conversion is not energy efficient. Figure 5-10 illustrates a simplified series of steps from carbohydrate to fat.

Figure 5•10

GLUCOSE TO FAT

Glucose can be used for energy, or it can be changed into fat and stored.

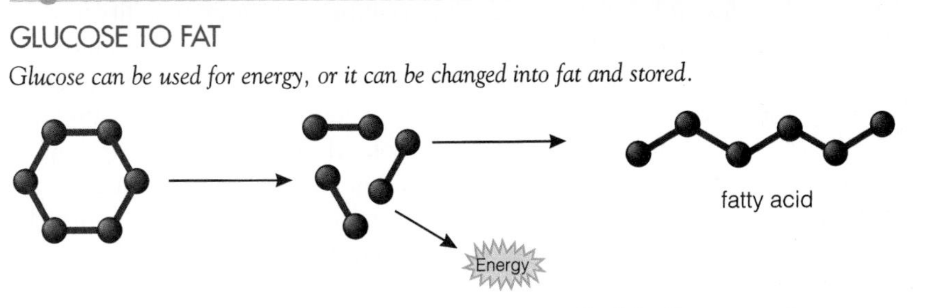

fatty acid
Energy
Glucose is broken down into fragments.
The fragments can provide immediate energy for the tissues.
Or, if the tissues need no more energy, the fragments can be reassembled, not back to glucose but into fatty acid chains.

Body fat supplies much of the fuel that muscles need to do their work.

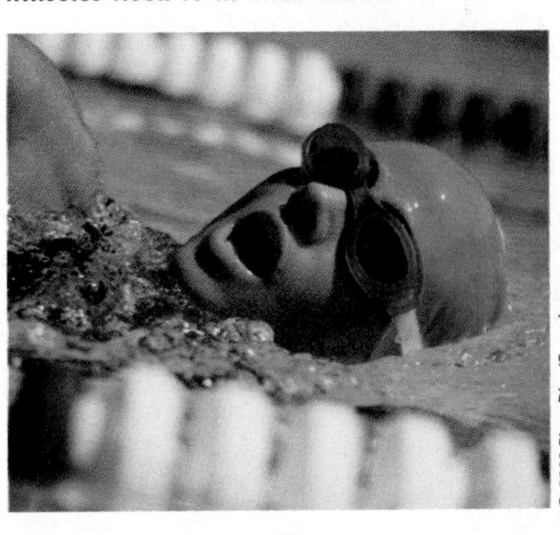

Before excess glucose can be stored as fat, it must first be broken into tiny fragments and then reassembled into fatty acids, steps that require energy to perform. Fat goes through fewer chemical steps before storage. Thus, given the same number of calories from excess dietary fat or from carbohydrate, the body stores more calories from the fat than from the carbohydrate. In short, you may get fatter on fat calories than on the same number of carbohydrate calories.

When low on fuel, the body draws on its stored fat for energy. Glucose is necessary for the complete breakdown of fat; without carbohydrate, ketosis occurs.

To read about dietary fats and cancer, see Chapter 11.

The 2000 American Heart Association Dietary Guidelines for Healthy American Adults are found in Table 11-10 of Chapter 11.

Six major CVD risk factors

1. Diagnosis of CVD or diabetes, or a family history of premature CVD.
2. Age (men older than 45 and women older than 55 years).
3. Cigarette smoking.
4. Elevated LDL cholesterol (explained later).
5. Low HDL cholesterol (explained later).
6. High blood pressure (hypertension).

Three lifestyle CVD risk factors:

1. Obesity.
2. Physical inactivity.
3. A diet high in saturated fats and low in vegetables, fruits, and whole grains.

SOURCE: Executive Summary of the Third Report of the National Cholesterol Education Program (NCEP) Expert Panel on Detection, Evaluation, and Treatment of High Blood Cholesterol in Adults (Adult Treatment Panel III), *Journal of the American Medical Association* 285 (2001): 2486–2497.

What, Exactly, Do Fat and Cholesterol Have to Do with Health?

High dietary fat intakes are associated with serious diseases. The person who chooses a diet too high in certain fats may be inviting the risk of heart and artery disease, or CVD.[11] Heart disease is the number-one killer of adults in the United States. The person who eats a high-fat diet also incurs a greater-than-average risk of developing some forms of cancer, another leading killer disease. Obesity carries serious risks to health, and fat's energy density makes it likely that people who eat high-fat diets will exceed their energy needs and so gain weight. Much research has focused on the links between diet and disease, and Chapter 11 is devoted to these connections. A few points about fats and heart health are presented here because they underlie dietary recommendations concerning fats (see Table 5-2).

The blood lipid profile, a medical test mentioned earlier, tells much about a person's risk category for CVD.[12] Most important in regard to CVD are total blood cholesterol and the lipoproteins that carry it.* Blood cholesterol concentration is a predictor of that person's likelihood of suffering a fatal heart attack or stroke, and the higher the cholesterol, the earlier the episode is expected to occur. Blood lipoproteins account for two of the six major risk factors for CVD (listed in the margin). The importance of blood cholesterol cannot be overemphasized. The more of these factors present in a person's life, the more urgent the need for lifestyle changes and other therapies to reduce the risk of CVD.

Now, what does *food* cholesterol have to do with *blood* cholesterol? The answer may be, "Not as much as most people think." Most saturated food *fats* (triglycerides) raise blood cholesterol more than food *cholesterol* does. When told that cholesterol doesn't matter as much as fat, people may then jump to the wrong conclusion—that blood cholesterol doesn't matter. It does matter. High *blood* cholesterol is an indicator of risk for CVD. The main dietary factor associated with elevated blood cholesterol is a high *saturated fat* intake. In comparison, *dietary cholesterol* alone makes a smaller contribution.

Genetics modifies everyone's ability to handle food cholesterol somewhat.[13] About 10 percent of people exhibit little increase in their blood cholesterol even with a high dietary intake. Another 10 percent respond to the same diet with greatly increased blood cholesterol. A few individuals have inherited a total inability to clear from their blood the cholesterol they have eaten and absorbed. This condition is rare, but studying it led to the discovery of how cholesterol is transported in the body. People with a genetic tendency toward high blood cholesterol must strictly limit fats and refrain from eating foods rich in cholesterol. Most people can eat limited amounts of eggs, liver, and other cholesterol-containing foods without fear of incurring high blood cholesterol because the body slows its cholesterol synthesis when the diet provides greater amounts. For most, moderation, not elimination, is key where cholesterol-containing foods are concerned.

***Blood, plasma,* and *serum* all refer to about the same thing; this book uses the term *blood* cholesterol. Plasma is blood with the cells removed; in serum, the clotting factors are also removed. The concentration is not much altered by these treatments.

Table 5•2
RECOMMENDATIONS CONCERNING INTAKES OF FATS FOR HEALTHY PEOPLE

1. Total Fat[a]

   Dietary Guidelines
   - Choose a diet moderate in total fat.

   Daily Values[b]
   - 65 grams fat per day.

   Healthy People 2010
   - Increase the proportion of people aged 2 and older who meet the average daily goal of no more than 30% of calories from fat.

   American Heart Association[c]
   - Limit fat to 30% or less of total energy.

   World Health Organization
   - Lower limit for total fat intake: 15% of total calories from fat.[d]
   - Upper limit for total fat intake: 30% of total calories from fat.[e]

2. Saturated Fat

   Dietary Guidelines
   - Choose a diet low in saturated fat.

   Daily Values[b]
   - 20 grams of saturated fat per day.

   Healthy People 2010
   - Increase the proportion of people aged 2 and older who meet the average daily goal of no more than 10% of calories from saturated fat.

   American Heart Association[c]
   - Limit saturated fat to less than 10% of total energy.

   World Health Organization
   - Lower limit for saturated fat intake: 0% of total calories from saturated fat.
   - Upper limit for saturated fat intake: 10% of total calories from saturated fat.

3. Polyunsaturated Fatty Acids

   World Health Organization
   - Lower limit for polyunsaturated fat intake: 3% of total calories from polyunsaturated fatty acids.
   - Upper limit for polyunsaturated fat intake: 7% of total calories from polyunsaturated fatty acids.

4. Cholesterol

   Dietary Guidelines
   - Choose a diet low in cholesterol.

   Daily Values[b]
   - 300 milligrams cholesterol per day.

   American Heart Association[c]
   - Limit cholesterol to less than 300 milligrams per day.

   World Health Organization
   - Lower limit for cholesterol intake: 0 milligrams cholesterol per day.
   - Upper limit for cholesterol intake: 300 milligrams cholesterol per day.

[a]Includes monounsaturated fatty acids.
[b]The Daily Values are for a 2,000-calorie diet.
[c]American Heart Association guidelines from R. M. Krauss and coauthors, AHA Dietary Guidelines: Revision 2000: A statement for healthcare professionals from the nutrition committee of the American Heart Association, *Circulation* 102 (2000): 2284–2299.
[d]Except for women of reproductive age, who should consume at least 20% of energy from fat.
[e]Sedentary individuals. Active individuals may consume up to 35% of total energy from fat if saturated fat does not exceed 10% of calories and the diet is otherwise adequate.

A dietary tactic often effective against high blood cholesterol is to trim the fat, and especially the saturated fat, from foods. This tactic is also affected by genetics; some people respond better than others to attempts to control blood lipids through diet.[14] The photos of Figure 5-11 show that food trimmed of fat is also trimmed of much of its energy. A pork chop trimmed of its border of fat loses 220 calories. A plain baked potato has about 40 percent of the calories of one with butter and sour cream. Choosing fat-free milk over whole milk provides large savings of fat, saturated fat, and calories. The single most effective step you can take to reduce a food's potential for elevating blood cholesterol is to eat it without the fat.

Figure 5•11

FOOD FAT AND CALORIES

Fat hides calories in food. When you trim fat, you trim calories

Nutrition Facts

Amount Per Serving

Pork chop (5 ounces) with 1/2 inch of fat	
Calories 450	Calories from Fat 315
	% Daily Value*
Total Fat 35g	**54%**
Saturated Fat 13g	**65%**

Pork chop (4 ounces) with fat trimmed off	
Calories 230	Calories from Fat 100
	% Daily Value*
Total Fat 11g	**17%**
Saturated Fat 4g	**20%**

Potato (5 ounces) with 1 tablespoon butter and 1 tablespoon sour cream	
Calories 400	Calories from Fat 250
	% Daily Value*
Total Fat 28g	**43%**
Saturated Fat 18g	**90%**

Plain potato (5 ounces)	
Calories 150	Calories from Fat 0
	% Daily Value*
Total Fat 0g	**0%**
Saturated Fat 0g	**0%**

Whole milk (1 cup)	
Calories 150	Calories from Fat 70
	% Daily Value*
Total Fat 8g	**12%**
Saturated Fat 5g	**25%**

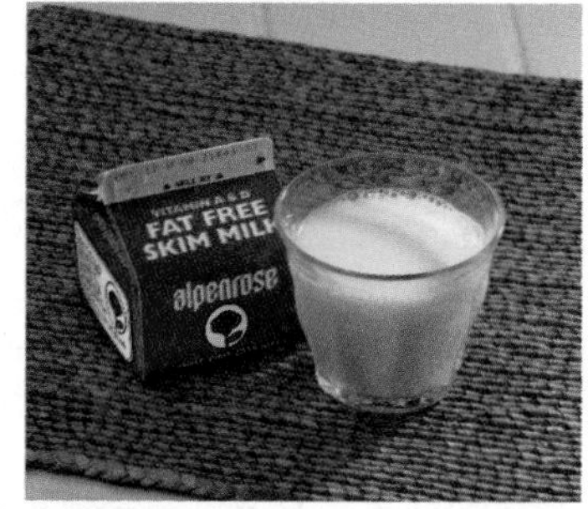

Fat-free milk (1 cup)	
Calories 90	Calories from Fat 0
	% Daily Value*
Total Fat 0g	**0%**
Saturated Fat 0g	**0%**

key point *The major dietary factor that raises blood cholesterol is the saturated fat intake, not the intake of cholesterol in foods. Elevated blood cholesterol is a risk factor for cardiovascular disease. Trimming fat from food trims calories and, often, saturated fat as well.*

What Is the Significance of LDL and HDL in Blood?

To repeat, cholesterol in foods contributes somewhat to cholesterol in the blood, and excesses of food cholesterol should be avoided. Dietary cholesterol is not as influential in raising blood cholesterol, however, as is dietary fat, especially saturated fat, which triggers a rise in LDL cholesterol.[15] Now the link to LDL can be explained. When high blood cholesterol signifies a risk of heart disease, it is because the LDL are carrying cholesterol to the body tissues to be deposited there. When high blood cholesterol is in HDL, that is cause for celebration because HDL carry cholesterol *away* from body tissues.[16] The vehicle matters.

Elevated LDL forecast heart and artery disease; elevated HDL signify a low disease risk. As a general rule, a minimum of 40 milligrams HDL per deciliter of blood or plasma is associated with a low risk of heart attack. A 2001 guideline recognizes a too-low HDL cholesterol reading as a major independent risk factor for CVD and suggests that raising HDL may be as important as lowering LDL in reducing risk.[17] Physical activity is a major weapon against heart disease and is an effective way to raise HDL, as the Think Fitness feature on the next page points out. So powerful against heart disease is HDL cholesterol that a reading of 60 or above counts as a *negative* risk factor mathematically; that is, when adding up risk factors, HDL over 60 removes one other risk factor from the total count.

The dietary changes that reduce blood cholesterol concentration do so mostly by reducing LDL and do not affect HDL concentration for the most part. Because the most influential dietary factor thought to raise LDL cholesterol is a diet high in saturated fat, reducing only the total fat in the diet may not lower LDL cholesterol.[18] Better to reduce total fat *and* replace cholesterol-raising saturated fat with monounsaturated or polyunsaturated fat. No beneficial change in blood lipids is seen when monounsaturated or polyunsaturated fat is *added* to a diet rich in saturated fat. That is, an eater of a bacon double cheeseburger cannot expect health protection from adding olive oil dressing to a side salad.

An important detail about LDL concerns its susceptibility to damage by **oxidation.** Oxidation of the lipid part of LDL is thought to play a role in injury to the arteries of the heart. **Dietary antioxidants,** such as vitamin C, vitamin E, the mineral selenium, and antioxidant phytochemicals, may slow LDL oxidation.

Some health authorities say all adults should take steps to reduce their blood cholesterol; others say that only those medically identified as at risk for heart disease should do so. In any case, most people are wise to limit saturated fat intake. A diet that provides less than 30 percent of its calories from fat and less than 10 percent from saturated fat and is rich in vegetables offers many health advantages by supplying abundant nutrients and antioxidants along with beneficial fiber.

What about cholesterol intake? The best course is to proceed with caution and moderation. Eggs, shellfish, liver, and other cholesterol-containing foods are nutritious. Cholesterol differs from salt and added fats and sugar in this respect: it cannot be omitted from the diet without omitting nutritious foods.

key point *Dietary measures to lower LDL in the blood involve reducing saturated fat and substituting monounsaturated and polyunsaturated fats for saturated fat. A few people must also reduce cholesterol intake. Cholesterol-containing foods are nutritious and are best used in moderation by most people.*

oxidation interaction of a compound with oxygen; in this case, a damaging effect by a chemically reactive form of oxygen (Controversy 7 provides details).

dietary antioxidants (anti-OX-ih-dants) substances in food that significantly decrease the damaging effects of reactive compounds, such as reactive forms of oxygen and nitrogen on tissue functioning (*anti* means "against"; *oxy* means "oxygen").

Desirable blood lipid values (milligrams per deciliter):

Total cholesterol <200 mg/dL.
LDL <100 mg/dL.
HDL ≥60 mg/dL.
Triglycerides <200 mg/dL.

Other standards for blood lipids are found in Table 11-3 of Chapter 11.

Here's a trick:
Remember **H**DL is **H**ealthy.
LDL is **L**ess healthy.

Antioxidant nutrients are topics of Chapter 7 and its Controversy.

linoleic (lin-oh-LAY-ic) **acid** and **linolenic** (lin-oh-LEN-ic) **acid** polyunsaturated fatty acids that are essential nutrients for human beings.

omega-6 fatty acid a polyunsaturated fatty acid with its endmost double bond six carbons from the end of the carbon chain; also called *n-6 fatty acid*. Linoleic acid is an example.

Why Exercise the Body for the Health of the Heart?

Every leading authority recommends physical activity to promote and maintain the health of the heart. The blood, arteries, heart, and other body tissues respond to exercise in these ways:

- The blood develops higher HDL concentrations, shifting blood lipids in a healthy direction.
- The circulation improves, easing the delivery of blood to the heart.
- A larger volume of blood is pumped with each heartbeat, reducing the heart's workload.
- The body grows leaner, reducing overall risk of cardiovascular disease.

THINK FITNESS

To estimate the approximate number of fat grams allowed in a day's intake that limits fat calories to 30% of the total, use this simple method:

- Estimate the total energy need (in calories, see inside front cover).
- Drop the last digit, and divide by 3.

For example, for a 2,300-calorie energy need:

$$\frac{230}{3} = 77 \text{ g fat/day.}$$

Essential Polyunsaturated Fatty Acids

The human body needs fatty acids, and it can use carbohydrate, fat, or protein to synthesize nearly all of them. Two are well-known exceptions: **linoleic acid** and **linolenic acid.** These two polyunsaturated fatty acids, which the body needs for its basic functions, cannot be made from scratch by body cells; nor can the cells convert one to the other. Linoleic and linolenic acid must be supplied by the diet and are therefore essential nutrients. These essential fatty acids are found in the oils of plants while derivatives are found in the oil of coldwater fish. The compounds are stored in the adult body. They serve as raw materials from which the body makes substances that act somewhat like hormones, regulating a wide range of body functions: blood pressure, blood clot formation, blood lipids, the immune response, the inflammation response to injury and infection, and others.[19] Essential fatty acids also serve as structural parts of cell membranes, constitute a major part of the lipids of the brain and nerves, and are essential to normal growth and vision in infants and children.[20]

To summarize then, essential polyunsaturated fatty acids: provide raw materials for regulatory substances, serve as structural parts of cell membranes, contribute lipids to the brain and nerves, are essential to normal growth.

To fail to obtain enough of the essential fatty acids is to invite ills of many kinds.

Deficiencies of Essential Fatty Acids A deficiency of an essential fatty acid in the diet leads to observable changes in cells, some more subtle than others. When the diet is deficient in *all* of the polyunsaturated fatty acids, symptoms of reproductive failure, skin abnormalities, and kidney and liver disorders appear. In infants and children, growth is retarded. These extreme deficiency disorders are seldom seen except when intentionally induced in research or on rare occasions when inadequate diets have been provided by mistake. Historically, one such mistake was to feed hospital clients a formula lacking essential fatty acids through a vein for long periods. Another was to feed infants exclusively on a formula that lacked essential polyunsaturated fatty acids. These formulas are now enriched with essential fatty acids. For healthy children and adults, a normal balanced diet that includes grains, seeds, nuts, leafy vegetables, and fish supplies all the needed forms of fatty acids in abundance and prevents deficiencies.

The Omega-6 and Omega-3 Difference Linoleic acid is the primary member of a group of fatty acids named the **omega-6 fatty acid** family after their chemical structure. The body can convert linoleic acid to the other members of its omega-6 family, and from one of these it makes diverse hormonelike substances known as **eicosanoids** that serve many, sometimes opposing, functions.* One eicosanoid causes muscles to relax and blood vessels to open; another causes muscles to contract and blood vessels to constrict. Others participate in the immune response to injury and infection, producing fever, inflammation, and pain. Aspirin relieves fever, inflammation, and pain by slowing the synthesis of these eicosanoids.

Linolenic acid is the primary member of the **omega-3 fatty acid** family. Like linoleic acid, it cannot be made in the body and must be supplied by foods. Given dietary linolenic acid, however, the body can make other members of the omega-3 series.

The omega-3 fatty acid family has come to be appreciated for its role in health. Someone thought to ask why the native people of Greenland and Alaska, who eat a diet very high in fat, have such a low death rate from heart disease.[21] The trail led to the abundance of fish and other marine life that they eat, then to the oils in those fish, and finally to **EPA** and **DHA** in the fish oils. Since then, research has revealed that omega-3 fatty acids:

- Make up a large portion of the brain's cerebral cortex (the brain's conscious, thinking part) and are thought to be required for its development.
- Help to form the eye's retina and are likely required for development of normal vision.[22]
- May benefit the health of the heart (by reducing inflammation, reducing the tendency of the blood to clot, and other effects) and the functioning of the immune system.[23]

eicosanoids (eye-COSS-ah-noyds) biologically active compounds that regulate body functions.

omega-3 fatty acid a polyunsaturated fatty acid with its endmost double bond three carbons from the end of the carbon chain; also called *n-3 fatty acid*. Linolenic acid is an example.

EPA, DHA eicosapentaenoic (EYE-cossa-PENTA-ee-NO-ick) acid, docosahexaenoic (DOE-cossa-HEXA-ee-NO-ick) acid; omega-3 fatty acids made from linolenic acid in the tissues of fish.

Recommendations and Intakes

No intake recommendations for omega-6 and omega-3 fatty acids exist yet, but they soon will. Experts proclaim the U.S. diet deficient in omega-3 fatty acids, but excessive in omega-6, and they recommend more balanced intakes.[24] Their dietary advice is to eat meals of fish two or three times a week, totaling about 10 ounces of fish, as well as small amounts of vegetable oils, to obtain the right balance between omega-3 and omega-6 intakes.[25] Even one fish meal per week has been associated with a reduced risk of heart attack, and people who make it a point to eat more than this amount suffer from strokes only half as often as those who eat no fish.[26] The ratio of omega-3 to omega-6 fatty acids should be about 1 to 5. This ratio may even turn out to be the key to human requirements; more omega-3 acids may not necessarily be better. To obtain a healthy balance, most people in the United States should choose a few more servings of fish each week and a few less of meat. Table 5-3 (next page) lists sources of both omega-6 and omega-3 fatty acids.

Learn more about vitamin E's antioxidant effects in Controversy 7.

Fish Oil Supplements

Taking fish oil supplements is not recommended, although many claims are made for their power to cure diseases. The Food and Drug Administration (FDA) does not allow labels to claim that fish oil supplements can prevent or cure diseases, but allows the claim that research is suggestive but inconclusive regarding heart disease. In Canada, fish oil supplements require a physician's prescription.

Supplements of fish oil may raise LDL cholesterol and may not be safe for other reasons.[27] In animal studies, an excess of omega-3 fatty acids during fetal development lowered brain weight and altered neurological development.[28] Omega-3 fatty acids are among the most vulnerable of the lipids to damage by oxidation, and researchers are investigating whether people taking fish oil capsules may experience an increase in potentially harmful oxidative reactions.[29] Vitamin E in high doses can prevent the damage, but scientists are still investigating how much vitamin E is needed to do so.[30] There have been reports of excessive bruising, bleeding, and strokes occurring with high doses of fish oil. Preliminary evidence on cancer and supplemental fish oil suggests both decreased and increased cancer in laboratory animals

Fish is a good source of omega-3 fatty acids.

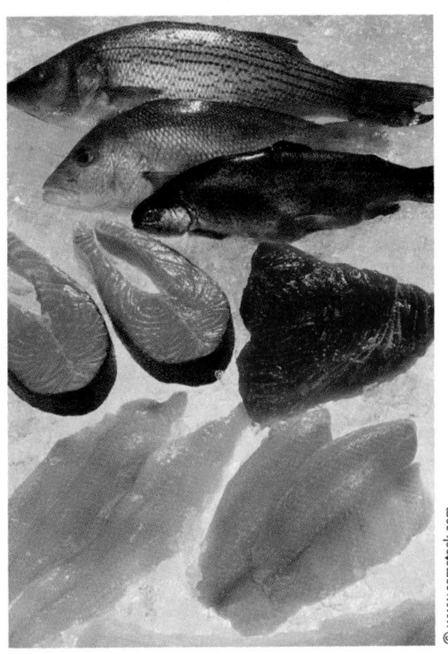

*Eicosanoid synthesis from linolenic acid is limited. Arachidonic acid provides the majority of substrate for eicosanoids; other fatty acids, including the omega-3 fatty acid EPA (see text), can also contribute to the total.

Table 5•3

SOURCES OF OMEGA-6 AND OMEGA-3 FATTY ACIDS

Omega-6	
Linoleic acid	Leafy vegetables, seeds, nuts, grains, vegetable oils (corn, cottonseed, safflower, sesame, soybean, sunflower), poultry fat
Omega-3	
Linolenic acid	Oils (canola, flaxseed, soybean, walnut, wheat germ; liquid or soft margarine made from canola or soybean oil) Nuts and seeds (butternuts, walnuts, soybean kernels) Vegetables (soybeans)
EPA and DHA	Human milk Coldwater fish[a] (mackerel, salmon, bluefish, mullet, sablefish, menhaden, anchovy, herring, lake trout, sardines, tuna) (or can be made from linolenic acid)

[a]All of these fish except tuna provide at least 1 gram of omega-3 fatty acids in 100 grams of fish (3.5 ounces); the fish oil content of each species varies with the season and site of harvest. Tuna provides fewer omega-3 fatty acids, but because it is commonly consumed, its contribution can be significant.

fed chow enriched with fish oil.[31] No one knows what such findings mean for people, but with regard to supplements of fish oil, the old axiom "let the buyer beware" seems to apply.

Fish most likely to be contaminated with mercury are shark, swordfish, king mackerel, and tilefish. Smaller ocean fish are generally safer.

A proven drawback is that fish oil supplements are made from fish skins and livers, which may have accumulated toxic concentrations of pesticides, heavy metals such as mercury, and other ocean contaminants that may be further concentrated in the pills. This caution must also now be applied to the flesh of some fish living in mercury-polluted waters (see Chapter 14 for advice on which fish are generally safe).[32] Even without contamination, fish oil naturally contains high levels of the two most potentially toxic vitamins, A and D. Lastly, supplements of fish oil are expensive. So little is known about the long-term effects of fish oil supplements that taking them is chancy. Better to go to the source for fish oils: eat safe varieties of fish.[33]

key point *Two polyunsaturated fatty acids, linoleic acid (an omega-6 acid) and linolenic acid (an omega-3 acid), are essential nutrients used to make substances that perform many functions. The omega-6 family includes linoleic acid. The omega-3 family includes linolenic acid, EPA, and DHA.*

The Effects of Processing on Unsaturated Fats

Vegetable oils make up most of the added fat in the U.S. diet because fast-food chains use them for frying, food manufacturers add them to processed foods, and consumers tend to choose margarine over butter.[34] Consumers of vegetable oils may feel safe in choosing them because they are generally less saturated than animal fats. If consumers choose a liquid oil, they may be justified in feeling secure. If the choice is a processed food, however, their security may be questionable, especially if the word *hydrogenated* appears on the label's ingredient list.

What Is "Hydrogenated Vegetable Oil," and What's It Doing in My Chocolate Chip Cookies?

When manufacturers process foods, they often alter the fatty acids in the fat (triglycerides) the foods contain through a process called **hydrogenation.** Hydrogenation of fats makes them stay fresher longer and also changes their physical properties.

Points of unsaturation in fatty acids are weak spots that are vulnerable to attack by oxygen. Oxidative damage is not confined to fats within body tissues but occurs anywhere oxygen mixes with fats. When the unsaturated points in the oils of food are oxidized, the oils become rancid and the food tastes "off."[35] This is why cooking oils should be stored in tightly covered containers that exclude air. If stored for long periods, they need refrigeration to retard oxidation.

One way to prevent spoilage of unsaturated fats and also to make them harder and more stable when heated to high temperatures is to change their fatty acids chemically by hydrogenation, as shown in Figure 5-12. When food producers want to use a polyunsaturated oil such as corn oil to make a spreadable margarine, for example, they hydrogenate it by forcing hydrogen into the oil. Some of the unsaturated fatty acids become more saturated as they accept the hydrogen, and the oil hardens. The resulting product is more saturated and more spreadable than the original oil. It is also more resistant to damage from oxidation or breakdown from high cooking temperatures.

Hydrogenated oils are thus easy to handle, easy to spread, and store well. Makers of cookies and other baked goods often use hydrogenated fats to protect their products against rancidity and ensure a long shelf life. Hydrogenated oils also have a high **smoking point,** so they are suitable for purposes such as frying.

Once hydrogenated, oils lose their unsaturated character and the health benefits that go with it. If you, the consumer, are looking for polyunsaturated oils to include in your diet, hydrogenated oils such as those in shortening or stick margarine will not

hydrogenation (high-dro-gen-AY-shun) the process of adding hydrogen to unsaturated fatty acids to make fat more solid and resistant to the chemical change of oxidation.

smoking point the temperature at which fat gives off an acrid blue gas.

Baked goods often contain hydrogenated fats.

Figure 5•12

HOW HYDROGENATION MAKES FATS MORE SATURATED

Points of unsaturation are places on fatty acid chains where hydrogen is missing. The bonds that would normally be occupied by hydrogen in a saturated fatty acid are shared, reluctantly, as a double bond between two carbons that both carry a slightly negative charge.

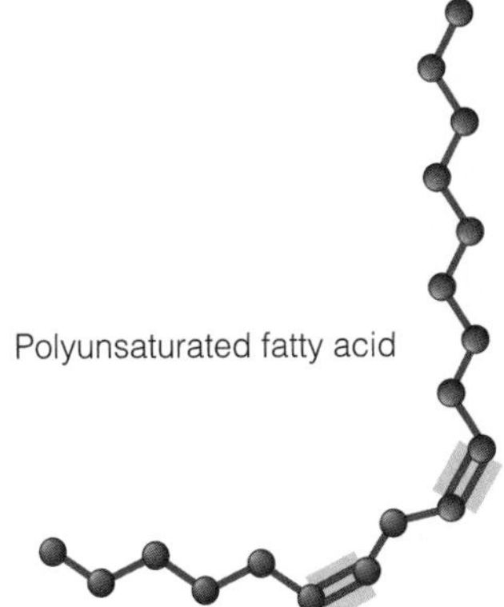

Polyunsaturated fatty acid

When positively charged hydrogen is made available to one of those bonds, it readily accepts the hydrogen molecules and, in the process, becomes saturated. It no longer has a point of unsaturation.

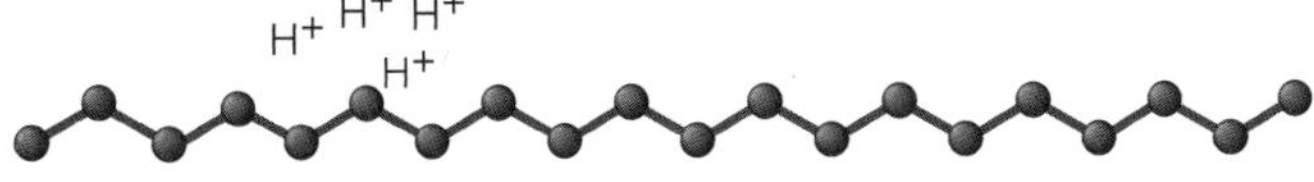

Hydrogenated fatty acid (now saturated)

***trans*-fatty acids** fatty acids with unusual shapes that can arise when polyunsaturated oils are hydrogenated.

stanol esters plant-derived compounds belonging to the sterol family of lipids that have been shown experimentally to reduce blood cholesterol when consumed in place of other fats in a low-fat diet.

meet your need. An alternative to hydrogenation is to add a chemical preservative that will compete for oxygen and thus protect the oil. The additives are antioxidants, and they work just as vitamin E does, by reacting with oxygen before it can do damage. Examples are the additives BHA and BHT* listed on snack food labels. Another alternative, already mentioned, is to keep the product refrigerated.

key point *Vegetable oils become more saturated when they are hydrogenated. Hydrogenated fats resist rancidity better, are firmer textured, and have a higher smoking point than unsaturated oils, but they also lose the health benefits of unsaturated oils.*

What Are *Trans*-Fatty Acids, and Are They Harmful?

When polyunsaturated oils are hardened by hydrogenation, some of the unsaturated fatty acids end up changing their shapes instead of becoming saturated (see Figure 5-13). This change in chemical structure creates unusual products that are not made by the body and that occur naturally in tiny amounts, mainly in dairy foods and beef. These changed fatty acids, or ***trans*-fatty acids**, may affect the body's health. Many researchers suggest that *trans*-fatty acids carry a risk to the health of the heart and arteries by raising LDL and lowering HDL cholesterol.[36] Epidemiological studies also suggest an association between dietary *trans*-fatty acids and heart disease risk.[37]

Finally, a high *total* fat consumption is also associated with cancer susceptibility, and *trans*-fatty acids contribute to the total fat intake. No strong evidence suggests that *trans*-fatty acids by *themselves* play a specific role in promoting or causing cancer, but research is continuing.[38] When processing changes essential fatty acids into their *trans* counterparts, the eater derives none of their associated benefits.

When news of *trans*-fatty acids' possible effects on heart health first emerged, some people switched from using margarine back to butter, believing oversimplified reports that margarine provided no heart health advantage over butter. Hardened margarines and virtually all shortenings *are* made largely from hydrogenated fats and therefore are saturated and contain substantial *trans*-fatty acids—up to 40 percent. Some margarines, however, especially the soft or liquid varieties, are made from unhydrogenated oils, which have long proved to be less saturated and so less likely to elevate blood cholesterol than the saturated fats of butter. When oils (but not hydrogenated oils) are the first ingredient listed on a margarine label, that margarine is, in all probability, low in *trans*-fatty acids as well as in saturated fat.

In addition to soft and liquid margarine choices, some margarines contain few or no *trans*-fatty acids. Some of these also contain a functional food ingredient, **stanol esters,** that reduces blood cholesterol when consumed in addition to a low-fat diet.† Stanol esters are not recognized by the intestine and therefore are not absorbed, and they also block the absorption of cholesterol.[39] Simply adding the margarine to a high-fat diet is unlikely to bring benefits, however. Stanol esters work only when people are also willing to cut their fat intakes. Drawbacks include the price (three or four times higher than regular margarine), a high fat content (the full fat kind equals the fat in regular margarine), and an unproven record of safety for use by consumers of all ages.

Foods other than margarine also contribute *trans*-fatty acids. Fast foods, chips, baked goods, and other commercially prepared foods are high in fats containing up to 50 percent *trans*-fatty acids. Fast-food chains fry foods in hydrogenated vegetable oil that contains abundant *trans*-fatty acids. Overall, consumers are now eating more fats containing *trans*-fatty acids than ever before, amounting to about 2 to 4 percent of daily calories, and they are eating these *trans*-fatty acids in the form of processed foods (see the margin list on the facing page).[40]

Figure 5•13

A *TRANS*-FATTY ACID

Note that an unsaturated trans*-fatty acid is similar in shape to a saturated fatty acid (review Figure 5-12); it behaves similarly in the body, too.*

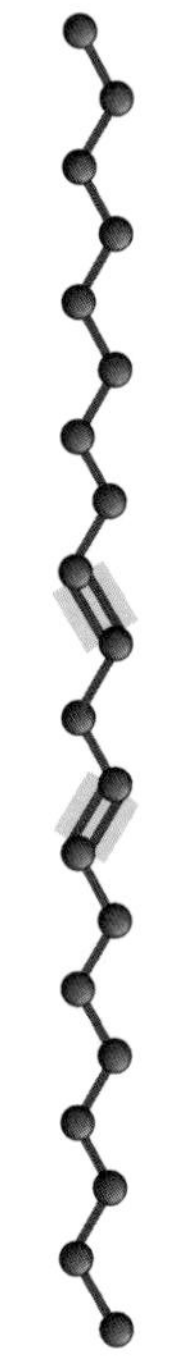

*BHA and BHT are butylated hydroxyanisole and butylated hydroxytoluene.

†Two brand names of margarines with stanol esters currently on the market are *Benecol* and *Take Control*.

Current food labels can be misleading in this regard, but changes are in the works. Now, grams of *trans*-fatty acids are counted with the polyunsaturated fats from which they arose, and not with the saturated fats whose health effects they mimic. Soon, though, a separate statement of *trans*-fatty acids on food labels may help consumers make informed choices. Reducing total fat and replacing both saturated and *trans* fats with monounsaturated and polyunsaturated fats may be the wisest strategies for preventing heart disease.[41]

The words *hydrogenated vegetable oil* or *shortening* in an ingredient list indicate trans-fatty acids in the product.

The process of hydrogenation creates trans-*fatty acids.* Trans-*fatty acids act somewhat like saturated fats in the body.*

Fat in the Diet

The remainder of this chapter shows you how to choose the right kinds of fat, and the right amounts, to provide optimal health and pleasure in eating. As you read, notice which foods offer unsaturated fat and which offer saturated fat. Your choices can make a difference in the unseen condition of your arteries.

Remember that some fat is necessary for health. People who take fat recommendations to an extreme and try to eliminate all traces of fat from food do so at their peril. Most adults need about 15 percent of their daily energy in the form of fat, an amount that is far, far smaller than the average intakes in this country. Fat occupies, on average, 34 percent of total daily calories, down from 45 percent in 1965. At first glance, this appears to be a healthy trend—until the actual grams of fat and carbohydrate are inspected. The total number of fat grams people take in has actually increased, and not decreased at all, but the carbohydrate grams increased even more during the same time period. The result is a misleading drop in the percentage of fat calories from total energy intakes but not a drop in actual fat consumed.[42] Bottom line: most people must work to learn to recognize the fats in foods and reduce their intakes.

Some common food sources of *trans*-fatty acids:

- Most hardened margarines and shortenings.
- Salad dressing, mayonnaise.
- Biscuits, rolls, cakes, cookies, crackers.
- Corn snacks and chips.
- Other fried snacks and chips.
- Cookies, doughnuts.
- French fries, fried chicken or fish.
- Fried fast foods, even those fried in commercial "vegetable oils."

In the Food Guide Pyramid, two groups always contain fat (the fats and the meats and nuts), and two sometimes contain fat (the milk and milk products and the breads). Most unprocessed vegetables and fruits are fat-free, but two exceptions are rich in monounsaturated fat—avocados and olives. In their natural states, grains are like fruits and vegetables in that they contain little or no fat. Keep in mind that fats may be visible on foods, such as the fat trimmed from a steak, or they may be invisible, such as the fats in the marbling of meat, the fat ground into lunch meats and hamburger, the fats blended into sauces of mixed dishes, and the fats in avocados, biscuits, cheese, coconuts, other nuts, olives, and fried foods.* Many such invisible fats are on the rise in U.S. diets.[43]

Added Fats

A dollop of dessert topping, a spread of butter on bread, oil or shortening in a recipe, dressing on a salad—all of these are examples of *added* fats. All sorts of fats can be added to foods during commercial or home preparation or at the table. The following amounts of these fats contain about 5 grams of pure fat, providing 45 calories and negligible protein and carbohydrate:

- 1 teaspoon oil or shortening.
- 1½ teaspoons mayonnaise, butter, or margarine.
- 1 tablespoon regular salad dressing, cream cheese, or heavy cream.
- 1½ tablespoons sour cream.

These foods provide the majority of added fats to the diet. They are the hidden fats of fried foods or baked goods, sauces and mixed dishes, and dips and spreads.

Ten small olives or a sixth of an avocado each provide about 5 grams of mostly monounsaturated fat.

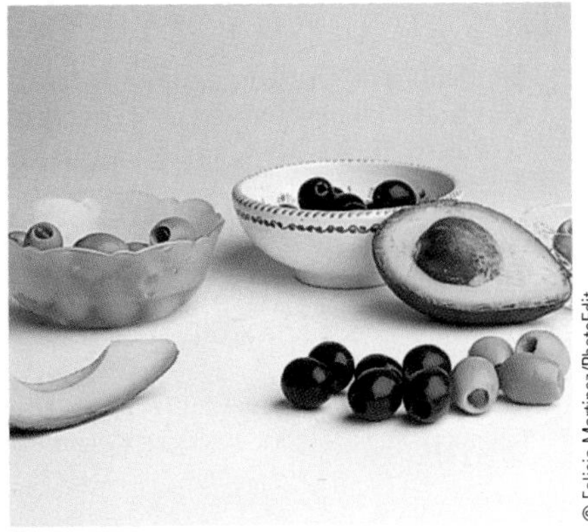

© Felicia Martinez/PhotoEdit

Fats added to foods during preparation or at the table are a major source of fat in the diet.

*Coconut is a nut, not a vegetable, although its oil is listed among vegetable oils.

Fat Replacers

TODAY, CONSUMERS can choose from thousands of fat-reduced products. Many bakery goods, cheeses, frozen desserts, and other products made with **fat replacers** (see Table 5-4) offer less than half a gram of fat in a serving. Some of these products contain **artificial fats,** and others use conventional ingredients in unconventional ways to reduce fat and calories. Among the latter, manufacturers can:

- Add water or whip air into foods.
- Add fat-free milk to creamy foods.
- Use lean meats and soy protein to replace high-fat meats.
- Bake foods instead of frying them.

Common food ingredients such as fibers, sugars, or proteins can also take the place of fats in some foods. These products still provide calories, but far fewer calories than fat.

Manufactured fat replacers consist of chemical derivatives of carbohydrate, protein, or fat, or modified versions of foods rich in those constituents (see Table 5-5). To gain the FDA's consent for the use of a new fat replacer, U.S. manufacturers must prove that their fat replacer contributes little food energy, is nontoxic, is not stored in body tissues, and does not rob the body of needed nutrients. Olestra serves as an example of an artificial fat that has received FDA approval, although it has not yet won the approval of consumers.

An Artificial Fat: Olestra

Olestra, brand name Olean, is a member of the **sucrose polyester** chemical family. Chemically, olestra bears some resemblance to ordinary fat; it consists of a core molecule of the carbohydrate sucrose to which up to eight fatty acid molecules are bonded (see Figure 5-14). In comparison, ordinary triglycerides consist of a core of the carbohydrate glycerol to which three fatty acids are bonded. The human digestive enzymes that break down triglycerides in the digestive tract do not recognize the shape of the olestra molecule and so cannot split the fatty acids from their sucrose. All of the olestra eaten in a food remains undigested and passes through the digestive system intact.

Table 5•4

TERMS RELATED TO FAT REPLACERS

- **artificial fats** zero-energy fat replacers that are chemically synthesized to mimic the sensory and cooking qualities of naturally occurring fats, but are totally or partially resistant to digestion. Also called *fat analogues*.
- **fat replacers** ingredients that replace some or all of the functions of fat and may or may not provide energy. Often used interchangeably with *fat substitutes*, but the latter technically applies only to ingredients that replace all of the functions of fat and provide no energy.
- **olestra** a noncaloric artificial fat made from sucrose and fatty acids; formerly called *sucrose polyester*.
- **sucrose polyester** any of a family of compounds in which fatty acids are bonded with sugars or sugar alcohols. Olestra is an example.

Figure 5•14

OLESTRA AND TRIGLYCERIDE COMPARED

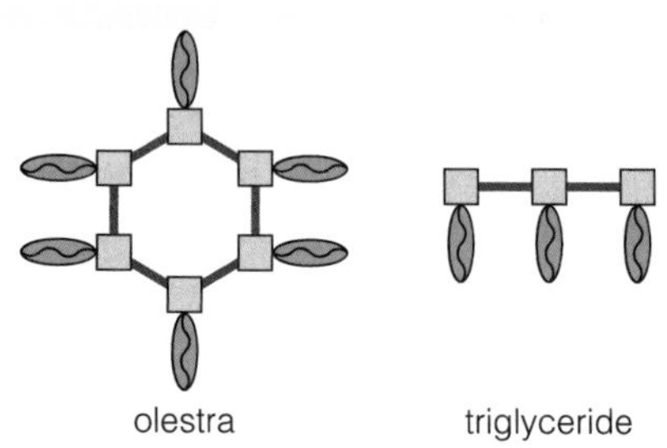

Olestra's Advantages

From some points of view, olestra is the most successful of the artificial fats because its properties are identical to those of fats and oils when used in frying, cooking, and baking. It can be heated to frying temperatures without breaking down; it performs all of the functions of fat in cakes, pie crusts, and other baked goods; and most remarkably—aside from a slight aftertaste—it tastes like fat. Nevertheless, olestra has not proved popular among consumers.

Concerns about Olestra

More than two decades of research have revealed that olestra is safe in most regards, but when consumed in large quantities, it can cause digestive distress, nutrient losses, and losses of phytochemicals. The presence of indigestible olestra in the large intestine can theoretically cause diarrhea, gas, cramping, and an urgent need for defecation. Oily olestra can creep through the feces and leak uncontrollably from the anus, producing smelly dark yellow stains on underwear. No significant increase in digestive distress was detected in an experiment with some 3,000 volunteers who ate olestra-

Table 5•5

A SAMPLING OF FAT REPLACERS

For comparison, remember that fat has 9 calories per gram.

Fat Replacers	Energy (cal/g)	Properties	Uses in Foods
Carbohydrate-Based Fat Replacers			
• *Fruits* purees and pastes of apples, bananas, cherries, plums, or prunes; add bulk and tenderness to baked goods.	1–4	Replace bulk of fat; add moisture and tenderness.	Baked goods, candy, dairy products.
• *Gels* derived from cellulose or starch to mimic the texture of fats in regular margarine and other products.	0–4[a]	Replace bulk; lend thickness.	Fat-free margarines, salad dressing, frozen desserts.
• *Gums* extracted from beans, sea vegetables, or other sources.	0–4	Add bulk; thicken salad dressings.	Salad dressing, processed meats, desserts.
• *Maltodextrins* made from corn; powdered and flavored to resemble the taste of butter.	1–4	Add "buttery" flavor.	Butter-flavored "sprinkles" for melting on hot foods.
• *Oatrim* derived from oat fiber; has the added advantage of providing satiety.	4	Creamy, replaces bulk of fat; can be used in baking but not frying.	Dips, dressings, baked goods.
• *Z-trim* a modified form of insoluble fiber; is powdered and feels like fat in the mouth.	0	Creamy, replaces bulk of fat; can be used in baking but not frying.	Cheese, ground beef, chocolates, baked goods.
Fat-Based Fat Replacers			
• *Olestra*[b] a noncaloric artificial fat made from sucrose and fatty acids; formerly called ***sucrose polyester.***	0	Same properties as fats; heat stable in frying, cooking, and baking.	Potato chips, tortilla chips, crackers.
• *Salatrim*[c] derived from fat and contains short- and long-chain fatty acids.	5	Same properties as fats; can be used in baking but not frying.	Chocolate coatings, dairy products, spreads.
Protein-Based Fat Replacers			
• *Microparticulated protein*[d] processed from the proteins of milk or egg white into mistlike particles that roll over the tongue, making it feel and taste like fat.	4	Creamy; heat stable in some cooking and baking but not frying.	Ice cream, dairy products, mayonnaise, salad dressing, baked goods, spreads.

[a]Energy made available by action of colonic bacteria.
[b]Trade name: Olean.
[c]Trade name: Benefat.
[d]Trade names: Simplesse and K-Blazer.

© Polara Studios, Inc.

containing snacks.[1] In free-living subjects, however, about 10 percent of olestra users calling a toll-free number reported digestive disturbances after consuming olestra-containing foods.[2] Researchers have dismissed these effects as not serious or dangerous. The FDA concurs, but requires olestra-containing foods to bear this warning:

> **"This Product Contains Olestra.** Olestra may cause abdominal cramping and loose stools. Olestra inhibits the absorption of some vitamins and other nutrients. Vitamins A, D, E, and K have been added."

Olestra acts as a potent solvent for some of the fat-soluble substances in foods, such as the vitamins that dissolve in fat (vitamins A, D, E, and K). The absorption of these vitamins from a meal that includes olestra is reduced because olestra dissolves them and carries them out of the digestive tract unabsorbed. To compensate for this effect, olestra is fortified with vitamins A, D, E, and K. The FDA ruled that fortification removes the threat of harm from malnutrition that olestra could otherwise cause.

Figure 5•15

OLESTRA'S PROS AND CONS

Pros of Olestra

- Zero calories
- Zero fat and saturated fat
- Zero cholesterol
- Withstands frying
- Withstands baking
- Tastes like fat

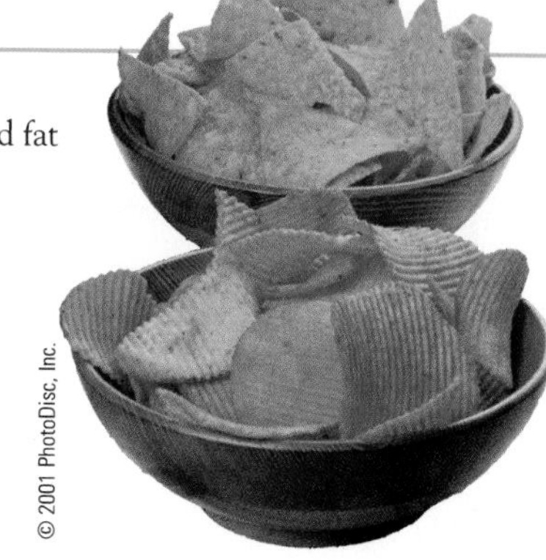

Cons of Olestra

- Vitamin losses
- Phytochemical losses
- Possible digestive upset
- Possible anal leakage
- Slight aftertaste
- Expensive
- No long-term studies in children

Olestra also causes the loss of health-promoting phytochemicals from foods. One study showed that just 3 grams of olestra a day strongly reduced (by about 40 percent) the blood concentrations of lycopene, a phytochemical believed to defend the body against some forms of cancer (more on phytochemicals in foods in the Controversy of Chapter 2).[3] To date, no studies exist to predict the effects, if any, of lifelong olestra exposure or the effects of olestra on growing children, although children often favor the foods in which olestra is allowed. Olestra's pros and cons are summed up in Figure 5-15.

Fat Replacers and Weight Control

People hope, of course, that fat replacers will help them fight both obesity and heart disease by lowering fat intakes. People who choose foods in which fat content is reduced, generally consume fewer calories, less fat, and more nutrients than nonusers of such foods.[4] Whether eating fat replacers will assist in weight control, however, is an open question because people may compensate for the energy reduction by eating more food later on. As for blood lipids, people who were "heavy users" of olestra, consuming more than 2 grams each day for a year, were found to have significantly lowered their blood cholesterol compared with their values before consuming olestra, but their weights did not change significantly.[5]

In truth, many reduced-fat foods deliver appreciable fat and about as many calories as a comparable regular product (see Table 5-6). A food made with fat replacers may not be low enough in fat to be called a "low-fat" food according to labeling requirements. If the U.S. experience with artificial sweeteners, described in Controversy 4, is any guide, consumers are likely to eat fat-replacer products *in addition* to other high-fat foods they prefer, negating the potential benefits of the fat replacers. Used wisely, though, fat replacers can help consumers achieve some of their dietary goals, especially when the user learns to cut fat from the diet in other ways as well.[6]

Table 5•6

FAT AND ENERGY IN REGULAR AND REDUCED-FAT FOODS

Fat and calories can be reduced by using fat replacers or by using reduced-fat recipes with ordinary ingredients.

Food	Fat (g)	Energy (cal)
Corn tortilla chips (1 oz)		
Made with vegetable oil	6	140
Baked	1	110
Made with olestra	1	90
Crackers, woven whole wheat (7 crackers)		
Regular	5	140
Low-fat recipe	2.6	114
Ice cream dessert (½ cup)		
Super premium	19	274
Regular	7	135
Ice milk	3	92
Frozen dessert made with artificial fat[a]	<1	120
Butter and substitutes (1 tsp equivalent)		
Butter	4	36
Margarine	4	34
Maltodextrin sprinkles	<1	3
French fried potatoes (about 15, fast-food type)		
Fried in vegetable oil	12	237
Fried in 75% olestra blend	3	144
Fried chicken (thigh)		
Fried in vegetable oil	15	252
Fried in 75% olestra blend	8	198

[a]Simplesse or Oatrim.

Meat, Poultry, Fish, Dry Beans, Eggs, and Nuts

Meats conceal much of the fat—mostly saturated fat—that people consume. To help you "see" the fat in meats, the exchange lists in Appendix D present the meats in four categories according to their fat contents: very lean, lean, medium-fat, and high-fat meats. Meats in all four categories contain about equal amounts of protein, but because their fat contents differ, their calorie amounts vary significantly. Figure 5-16 shows fat and calorie data for some ground meats.

According to the Food Guide Pyramid, a serving of meat amounts to just 2 or 3 ounces—very small by average consumption standards. A small, fast-food hamburger, for example, weighs about 3 ounces. A steak served in a restaurant averages 12 to 16 ounces, more than a whole day's meat allowance. You may have to weigh a serving or two of meat to see how much you are eating.

People think of meat as protein food, but calculation of its nutrient content reveals a surprising fact. A big (4-ounce), fast-food hamburger sandwich contains 23 grams of protein and 20 grams of fat. Because protein offers 4 calories per gram and fat offers 9, the sandwich provides 92 calories from protein and 180 calories from fat. The calorie total, counting carbohydrates from the bun and condiments, is over 400 calories, with more than 50 percent of them from fat. Hot dogs, fried chicken sandwiches, and fried fish sandwiches are also high-fat choices. Because so much of the energy in a meat eater's diet is hidden from view, people can easily overeat on high-fat food, making weight control difficult.

Figure 5•16

FAT IN GROUND MEATS

Only the ground round, at 7 percent fat by weight, qualifies to bear the "lean" label. To be called "lean," products must contain fewer than 10 grams fat, 4 grams saturated fat, and 95 milligrams cholesterol per 100 grams of food. The numbers that qualify products to be called "extra lean" are, respectively, 5, 2, and 95. The red labels on these packages list rules for safe meat handling, explained in Chapter 14.

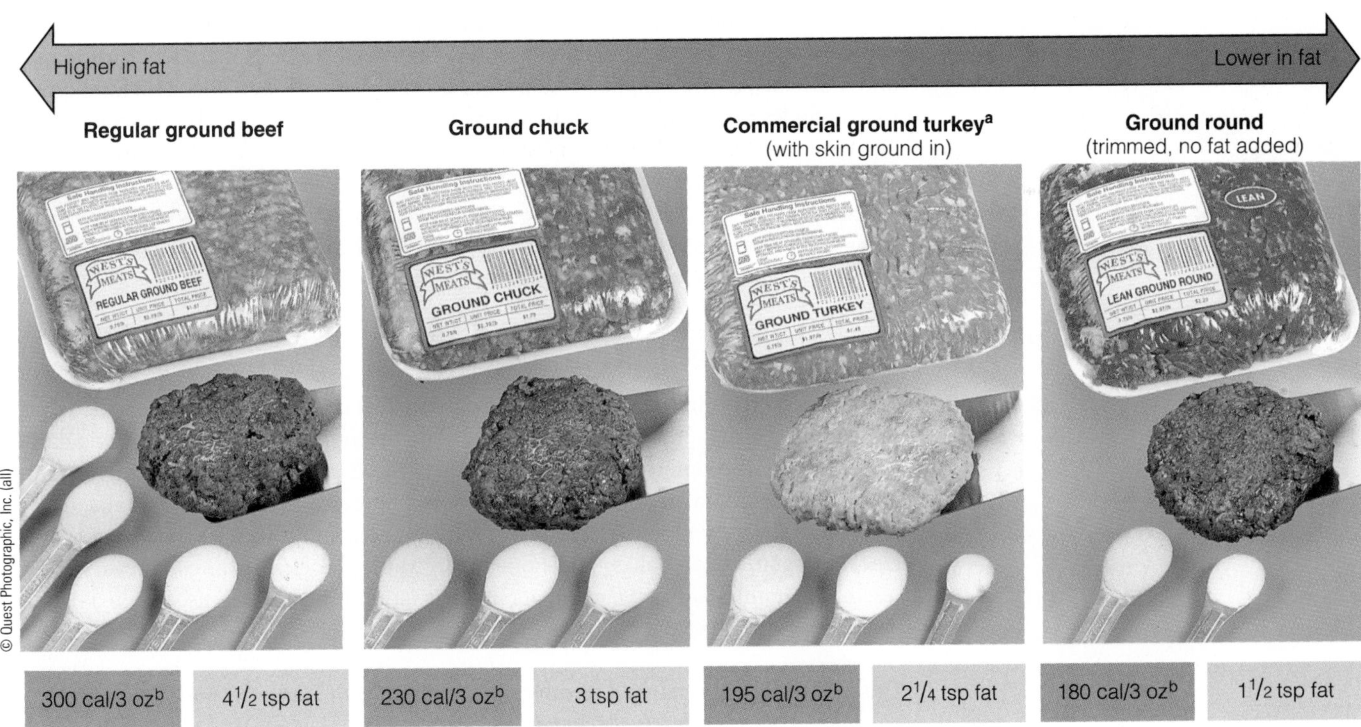

[a]Values for 3 ounces of cooked turkey breast ground without skin are 108 calories and 1/2 teaspoon fat (25 percent calories from fat). This type is not typically offered in many areas, but can be specially ordered from the butcher.

[b]The 3-ounce cooked serving used here may seem small to some, but it is the largest allowable meat serving according to the Daily Food Guide.Larger servings will, of course, provide more fat and calories than the values listed here.

Animal breeders have been striving to produce beef and pork that are lower in fat.[44] Thus, shoppers who choose lean cuts receive less fat in the same quantity of meat. When choosing beef or pork, look for lean cuts named *loin* or *round* from which the fat can be trimmed. Eat small portions, too.

Chicken and turkey meat is naturally lean, but processing and frying add fats, especially in "patties," "nuggets," "fingers," or "wings." Chicken wings are mostly skin, and a chicken stores most of its fat just under its skin. The tastiest wing snacks have also been fried in cooking fat (often a saturated type), smothered with a buttery, spicy sauce, and then dipped in blue cheese dressing, making wings an extraordinarily high-fat snack. If you snack on wings, plan on eating low-fat foods at other meals to balance them out.

Watch out for ground turkey or chicken products. The skin is often ground in to add moistness when cooked, and these products can be much higher in fat than even lean beef—see Figure 5-16 (previous page). Table 2-6 on pages 48 and 49 in Chapter 2 provided some definitions concerning the fat contents of meats.

Meats account for a large proportion of the hidden fat in many people's diets. Most people consume meat in larger servings than those recommended.

Milk, Yogurt, and Cheese

Some milk products contain fat. In homogenizing whole milk, milk processors blend in the cream, which otherwise would float and could be removed by skimming. A cup of whole milk contains the protein and carbohydrate of fat-free milk, but in addition it contains about 60 extra calories from fat. A cup of reduced-fat (2 percent fat) milk falls between whole and fat-free, with 45 calories of fat. The fat of whole milk occupies only a teaspoon or two of the volume but nearly doubles the calories in the milk. Depending upon its fat content, milk bears one of the names listed in the margin (older names are listed in parentheses).

Milk's names:

- Milk, whole milk (whole milk).
- Reduced-fat, less-fat milk (2% milk).
- Low-fat milk (1% milk).
- Fat-free, zero-fat, or no-fat (skim or nonfat milk).

Yogurt and cheese appear in the milk group, but cream and butter do not. Yogurt and cheese are rich in calcium and protein, but cream and butter are not. Cream and butter are fats, as are whipped cream, sour cream, and cream cheese. The food group that includes milk is thus carefully called the "milk, yogurt, and cheese group," and not the "dairy group." Figure 5-17 shows where the lipids are found in various kinds of milk, yogurt, and cheese.

The choice between whole and fat-free milk products can make a large difference to the fat content of a diet.

Bread, Cereal, Rice, and Pasta

Breads and cereals in their natural state are very low in fat, but fat may be added during manufacturing, processing, or cooking. The fat in these foods can be particularly hard to detect, so diners must remember which foods stand out as being high in fat. Notable are granola and certain other ready-to-eat cereals, croissants, biscuits, cornbread, fried rice, pasta with creamy or oily sauces, quick breads, snack and party crackers, muffins, pancakes, and homemade waffles. Packaged breakfast bars often resemble vitamin-fortified candy bars in their fat and sugar contents. Figure 5-18 on page 164 shows the lipid contents of many grain products.

Fat in breads and cereals can be well hidden. Consumers must learn which foods of this group contain fats.

Now that you know where the fats in foods are found, how can you reduce or eliminate them from your diet? The Food Feature provides some pointers.

Figure 5•17

LIPIDS IN MILK, YOGURT, AND CHEESE

Red boxes below indicate foods with higher lipid contents that warrant moderation in their use. Green indicates lower-fat choices.

Nutrition Facts

Amount Per Serving

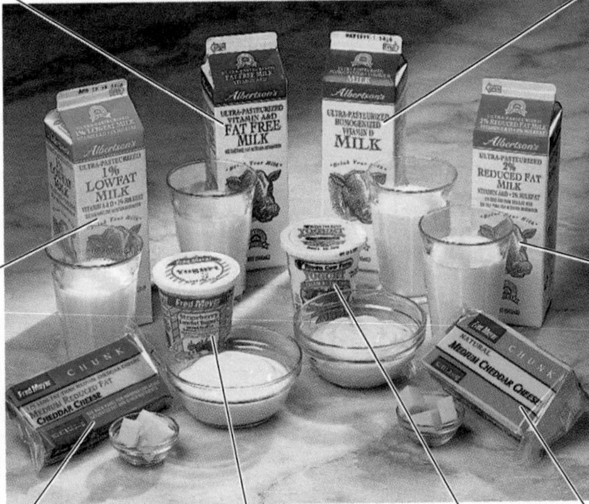

© Polara Studios, Inc.

Fat-free, skim, zero-fat, no-fat, or nonfat milk, 8 oz (<0.5% fat by weight)

Calories 90	Calories from Fat 0
	% Daily Value*
Total Fat 0g	**0%**
Saturated Fat 0g	**0%**
Cholesterol 5mg	**2%**

Low-fat milk, 8 oz (1% fat by weight)

Calories 110	Calories from Fat 25
	% Daily Value*
Total Fat 2.5g	**4%**
Saturated Fat 1.5g	**8%**
Cholesterol 15mg	**5%**

Low-fat cheddar cheese, 1.5 oz

Calories 70	Calories from Fat 30
	% Daily Value*
Total Fat 3g	**5%**
Saturated Fat 2g	**10%**
Cholesterol 10mg	**3%**

Strawberry yogurt, 8 oz

Calories 250	Calories from Fat 40
	% Daily Value*
Total Fat 4g	**6%**
Saturated Fat 2.5g	**13%**
Cholesterol 15mg	**5%**

Whole milk, 8 oz (3.3% fat by weight)

Calories 150	Calories from Fat 70
	% Daily Value*
Total Fat 8g	**12%**
Saturated Fat 5g	**25%**
Cholesterol 35mg	**12%**

Reduced fat, less-fat milk, 8 oz (2% fat by weight)

Calories 120	Calories from Fat 45
	% Daily Value*
Total Fat 5g	**8%**
Saturated Fat 3g	**15%**
Cholesterol 20mg	**7%**

Cheddar cheese, 1.5 oz

Calories 170	Calories from Fat 130
	% Daily Value*
Total Fat 14g	**22%**
Saturated Fat 9g	**45%**
Cholesterol 45mg	**15%**

Low-fat strawberry yogurt, 8 oz

Calories 200	Calories from Fat 20
	% Daily Value*
Total Fat 2.5g	**4%**
Saturated Fat 1.5g	**8%**
Cholesterol 15mg	**5%**

Figure 5•18

LIPIDS IN BREAD, CEREAL, RICE, AND PASTA

Red boxes below indicate foods with higher lipid contents that warrant moderation in their use. Green indicates lower-fat choices.

Nutrition Facts

Amount Per Serving

Low-fat granola, ½ c	
Calories 190	Calories from Fat 25
	% Daily Value*
Total Fat 3g	**5%**
Saturated Fat 0.5g	**3%**
Cholesterol 0mg	**0%**

A home-made waffle	
Calories 220	Calories from Fat 100
	% Daily Value*
Total Fat 11g	**17%**
Saturated Fat 2g	**10%**
Cholesterol 50mg	**17%**

A dinner roll	
Calories 80	Calories from Fat 20
	% Daily Value*
Total Fat 2g	**3%**
Saturated Fat 0g	**0%**
Cholesterol 0mg	**0%**

Crispy oat bran, ½ c	
Calories 130	Calories from Fat 35
	% Daily Value*
Total Fat 4g	**6%**
Saturated Fat 1.5g	**8%**
Cholesterol 0mg	**0%**

Fettuccine alfredo, ½ c	
Calories 250	Calories from Fat 130
	% Daily Value*
Total Fat 14g	**22%**
Saturated Fat 8g	**40%**
Cholesterol 60mg	**20%**

Buttery crackers, 4 crackers	
Calories 80	Calories from Fat 35
	% Daily Value*
Total Fat 4g	**6%**
Saturated Fat 1g	**5%**
Cholesterol 0mg	**0%**

A breakfast bar	
Calories 150	Calories from Fat 55
	% Daily Value*
Total Fat 6g	**9%**
Saturated Fat 2.5g	**13%**
Cholesterol 0mg	**0%**

© Polara Studios, Inc.

Fried rice, ½ c[a]	
Calories 140	Calories from Fat 65
	% Daily Value*
Total Fat 7g	**11%**
Saturated Fat 1g	**5%**
Cholesterol 20mg	**7%**

A muffin	
Calories 170	Calories from Fat 65
	% Daily Value*
Total Fat 7g	**11%**
Saturated Fat 1.5g	**8%**
Cholesterol 25mg	**8%**

A biscuit	
Calories 190	Calories from Fat 80
	% Daily Value*
Total Fat 9g	**14%**
Saturated Fat 2.5g	**13%**
Cholesterol 0mg	**0%**

A croissant	
Calories 260	Calories from Fat 140
	% Daily Value*
Total Fat 16g	**25%**
Saturated Fat 10g	**50%**
Cholesterol 50mg	**17%**

[a]The fat content of fried rice varies by preparation method.

Defensive Dining

To meet the most important recommendation of almost every nutrition authority—to reduce dietary fats—most people would have to make changes according to the following five principles. The changes would also lower intakes of saturated fat.

1. Eliminate much of the fat used as a seasoning and in cooking.
2. Cut down on intake of red meat.
3. Remove the fat from high-fat foods.
4. Replace high-fat foods with specially manufactured lower-fat versions of those foods.
5. Replace high-fat foods with naturally occurring low-fat alternatives.

With these principles in mind, you can make choices about foods in your diet.

The first arena of choice is the grocery store. The right choices here can save you many grams of fat at the dinner table. Food labels can reveal much about a processed food's fat content, and then the choice of whether to consume it depends on how you intend to use it in your diet: as a staple item, or as an occasional treat.

Once at home, one of the most effective steps for reducing fats is principle #1: eliminate fats used as seasonings. This means eating cooked vegetables without butter, bacon, or margarine; omitting high-fat gravies and sauces; and leaving off other last-minute fat additions. Butter and regular margarine contain the same number of calories (about 35 per teaspoon); diet margarine contains fewer calories because water, air, or fillers have been added. Imitation butter flavoring contains no fat and few calories.

For snacks, use an air popper for popcorn, and then add butter flavoring to the popcorn, if you like it. Keep that flavoring on hand together with other low-fat cooking substitutes such as diet margarine, low-fat salad dressings, fat-free sauce mixes or recipes, and nonstick spray for frying. Check Table 5-7 for substitutes for high-fat ingredients in recipes. These replacements will not change the taste or appearance of the finished product very much, but will dramatically lower its contents of fat and saturated fat.

If you must add fats, be sure that they are detectable in the food and that you enjoy them. For example, if you use strongly flavored fat, a little goes a long way. Sesame oil, peanut butter, and the fats of strong cheeses

Table 5•7

SUBSTITUTES FOR HIGH-FAT INGREDIENTS

Use	Instead of
Fat-free milk products	Whole-milk products
Evaporated fat-free ("skim") milk (canned)	Cream
Yogurt[a] or fat-free sour cream replacer	Sour cream
Reduced-calorie margarine; butter replacers	Butter
Wine, lemon juice, or broth	Butter
Fruit butters	Butter
Part-skim or fat-free ricotta; low-fat or fat-free cottage cheese[a]	Whole-milk ricotta
Part-skim, low-fat, or fat-free cheeses	Regular cheeses
1 tbs cornstarch (for thickening sauces)	1 egg yolk
Low-fat or fat-free mayonnaise	Regular mayonnaise
Low-fat or fat-free salad dressing (for salads and marinades)	Regular salad dressing
Water-packed canned fish and meats	Oil-packed fish and meats
Lean ground meat and grain mixture	Ground beef
Low-fat frozen yogurt or sherbet	Ice cream
Herbs, lemons, spices, fruits, liquid smoke flavoring, or ham-flavored bouillon cubes	Butter, bacon, bacon fat

[a]If the recipe calls for the food to be boiled, the yogurt or cottage cheese must be stabilized with a small amount of cornstarch or flour.

USDA facts:

- 70% of teenage males eat meals away from home each day.
- 57% of all Americans, and 40% of those over 60 years old, do so.
- The foods chosen away from home are higher in fat, saturated fat, and cholesterol and lower in vitamins and minerals than meals eaten at home.

are equal in calories to others, but they are so strongly flavored that you can use much less. Try small amounts of grated sapsago, romano, or other hard cheeses to replace larger amounts of less flavorful cheeses.

If you use oils, trade off among types to obtain the benefits different oils offer. Peanut and safflower oils are especially rich in vitamin E. Olive and canola oil offer the heart health benefits associated with monounsaturates, and canola oil also contains omega-3 fatty acids. High temperatures, such as those used in frying, destroy omega-3 acids, however. Take care to substitute oils for saturated fats in the diet, not to add oils to an already fat-rich diet.

Here are some other tips to update old, high-fat recipes:

- Grill, roast, broil, boil, bake, stir-fry, microwave, or poach foods. Don't fry in fat.
- Add a little water or fat-free yogurt to thick, bottled salad dressings, and then apply them sparingly. They'll go farther this way, and you'll use less oil.
- Cut recipe amounts of meat in half; use only lean meats. Fill in the lost bulk with shredded vegetables, legumes, pasta, grains, or other low-fat items.
- Trim all visible fat and skin from meat and poultry.
- Refrigerate meat pan drippings and broth, and lift off the fat when it solidifies. Then add the defatted broth to a recipe.
- Make prepared mixes, such as rice or potato mixtures, without the fats called for on the label.

All of these suggestions work well when a person plans, selects, purchases, and prepares each meal at home. But in the real world, people fall behind schedule and don't have time to cook, so they eat fast food. Figure 5-19 compares some fast-food choices, offering tips to reduce the fat in the high-fat choices and showing some lower-fat choices.

Keep these facts about fast food in mind:

- Salads are a good choice. Avoid mixed salad bar items, such as macaroni salad. Use only about a quarter of the dressing provided or use low-fat dressing.
- If you are really hungry, order a small hamburger on the side. Hold the mayonnaise: use mustard or ketchup instead. A small bowl of chili or a plain baked potato can also satisfy a bigger appetite.
- Fried fish or chicken sandwiches are at least as high in fat as hamburgers. Broiled sandwiches are far less fatty—if you order them made without spreads, dressings, cheese, bacon, or mayonnaise.

Because fast foods are short on variety, let them be part of a lifestyle in which they complement the other parts. Eat differently elsewhere, often.

By this time you may be wondering if you can realistically make all the changes recommended for your diet and keep high-fat foods under control. Be assured that most of the needed changes can become habits after a few repetitions. You do not have to give up all high-fat foods; you need only learn to exercise moderation. The famous French chef Julia Child makes this point about moderation:

> An imaginary shelf labeled INDULGENCES is a good idea. It contains the best butter, jumbo-size eggs, heavy cream, marbled steaks, sausages and pâtés, hollandaise and butter sauces, French butter-cream fillings, gooey chocolate cakes, and all those lovely items that demand disciplined rationing. Thus, with these items high up and almost out of reach, we are ever conscious that they are not everyday foods. They are for special occasions, and when that occasion comes we can enjoy every mouthful.
>
> —Julia Child, *The Way to Cook* (1989)

You decide what the treats should be and then choose them judiciously, just for pure pleasure. Meanwhile, make sure that your everyday, ordinary choices are those whole, nutrient-dense foods suggested throughout this book. That way you'll meet all your body's needs for nutrients and never feel deprived.

Figure 5•19

FAST-FOOD CHOICES

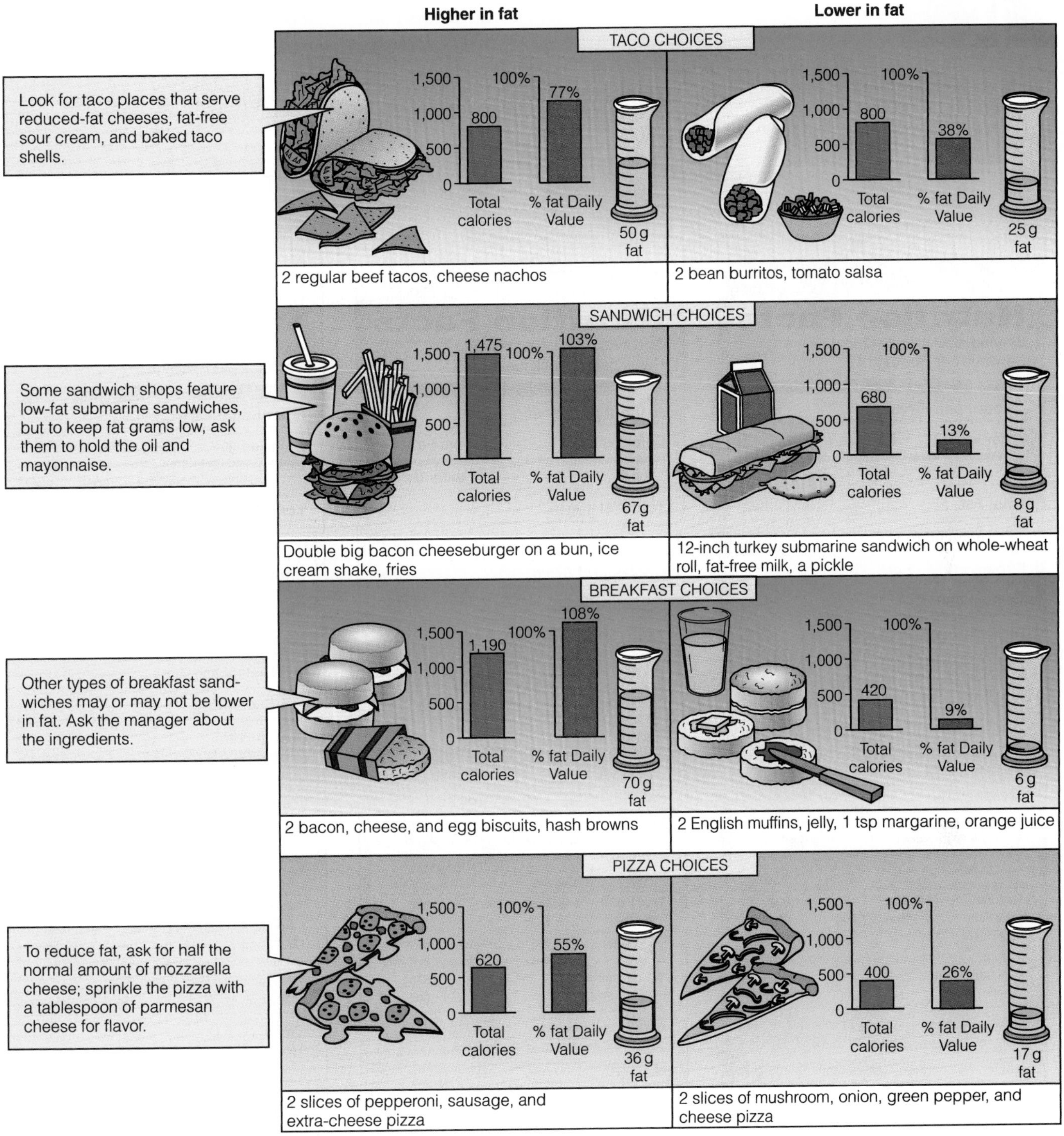

Higher in fat | **Lower in fat**

TACO CHOICES

Look for taco places that serve reduced-fat cheeses, fat-free sour cream, and baked taco shells.

	Higher in fat	Lower in fat
Total calories	800	800
% fat Daily Value	77%	38%
Fat	50 g fat	25 g fat
	2 regular beef tacos, cheese nachos	2 bean burritos, tomato salsa

SANDWICH CHOICES

Some sandwich shops feature low-fat submarine sandwiches, but to keep fat grams low, ask them to hold the oil and mayonnaise.

	Higher in fat	Lower in fat
Total calories	1,475	680
% fat Daily Value	103%	13%
Fat	67g fat	8 g fat
	Double big bacon cheeseburger on a bun, ice cream shake, fries	12-inch turkey submarine sandwich on whole-wheat roll, fat-free milk, a pickle

BREAKFAST CHOICES

Other types of breakfast sandwiches may or may not be lower in fat. Ask the manager about the ingredients.

	Higher in fat	Lower in fat
Total calories	1,190	420
% fat Daily Value	108%	9%
Fat	70 g fat	6 g fat
	2 bacon, cheese, and egg biscuits, hash browns	2 English muffins, jelly, 1 tsp margarine, orange juice

PIZZA CHOICES

To reduce fat, ask for half the normal amount of mozzarella cheese; sprinkle the pizza with a tablespoon of parmesan cheese for flavor.

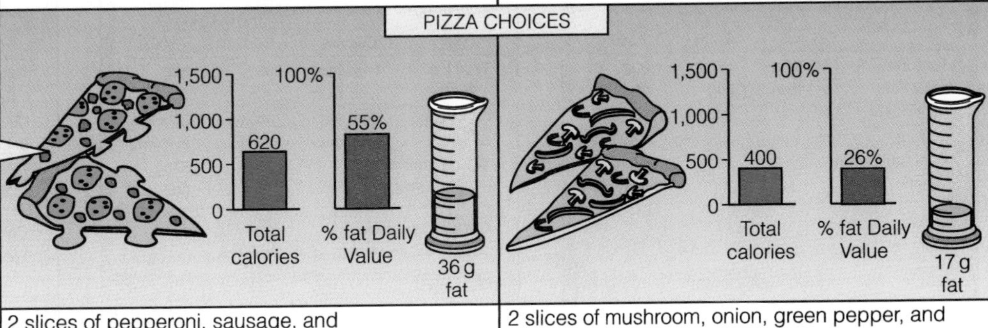

	Higher in fat	Lower in fat
Total calories	620	400
% fat Daily Value	55%	26%
Fat	36 g fat	17 g fat
	2 slices of pepperoni, sausage, and extra-cheese pizza	2 slices of mushroom, onion, green pepper, and cheese pizza

Note: Fat Daily Value based on a 2,000-calorie diet.

READ ABOUT FATS ON FOOD LABELS

Figure 5-20 presents three labels from packages of lasagna and asks you to compare the fat in one serving of each with your day's allowances for fat and saturated fat. Food labels make this comparison easy for people whose energy needs are about 2,000 calories a day. For those people, the Daily Values for fat

Figure 5•20

COMPARISON OF THREE DIFFERENT LASAGNAS

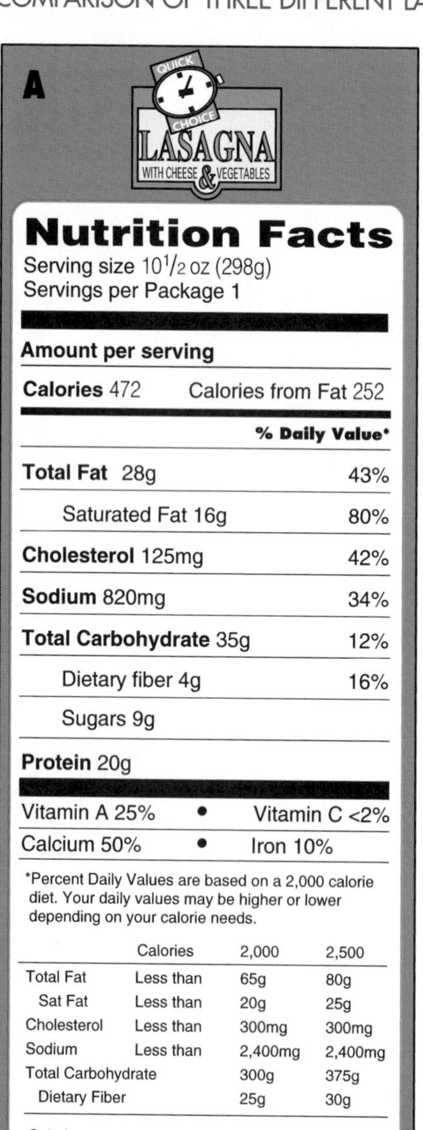

A

QUICK CHOICE LASAGNA WITH CHEESE & VEGETABLES

Nutrition Facts
Serving size 10 1/2 oz (298g)
Servings per Package 1

Amount per serving	
Calories 472	Calories from Fat 252
	% Daily Value*
Total Fat 28g	43%
Saturated Fat 16g	80%
Cholesterol 125mg	42%
Sodium 820mg	34%
Total Carbohydrate 35g	12%
Dietary fiber 4g	16%
Sugars 9g	
Protein 20g	
Vitamin A 25% •	Vitamin C <2%
Calcium 50% •	Iron 10%

*Percent Daily Values are based on a 2,000 calorie diet. Your daily values may be higher or lower depending on your calorie needs.

	Calories	2,000	2,500
Total Fat	Less than	65g	80g
Sat Fat	Less than	20g	25g
Cholesterol	Less than	300mg	300mg
Sodium	Less than	2,400mg	2,400mg
Total Carbohydrate		300g	375g
Dietary Fiber		25g	30g

Calories per gram
Fat 9 • Carbohydrate 4 • Protein 4

INGREDIENTS, Skim Milk, Ricotta Cheese (Whole Milk, Cream, Skim Milk, Vinegar and Salt), Cooked Macaroni, Spinach, Parmesan Cheese, Carrots, Onions, Butter, Soybean Oil, Modified Cornstarch, Bread Crumbs (Enriched Bleached Wheat Flour, Sugar, Corn Syrup, Partially Hydrogenated Soybean Oil, Salt, Yeast, Calcium Propionate, Spice Extractives and BHT), Corn Syrup, Long Grain Rice Meal, Potato Flakes, Malt, Yeast, Vegetable Shortening (Partially Hydrogenated Soybean Oil), Salt, Calcium Propionate, Fat-Free Dry Milk Solids, Salt, Romano Cheese (made from Cow's Milk), Mushrooms, Sugar, Salt, Mono- and Diglycerides, Xanthan Gum, Spices, Garlic Salt.

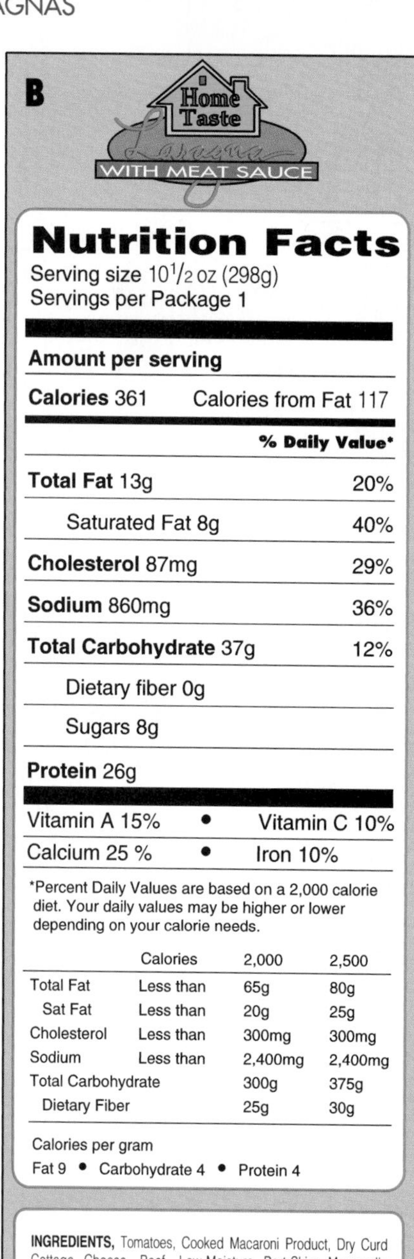

B

Home Taste Lasagna WITH MEAT SAUCE

Nutrition Facts
Serving size 10 1/2 oz (298g)
Servings per Package 1

Amount per serving	
Calories 361	Calories from Fat 117
	% Daily Value*
Total Fat 13g	20%
Saturated Fat 8g	40%
Cholesterol 87mg	29%
Sodium 860mg	36%
Total Carbohydrate 37g	12%
Dietary fiber 0g	
Sugars 8g	
Protein 26g	
Vitamin A 15% •	Vitamin C 10%
Calcium 25 % •	Iron 10%

*Percent Daily Values are based on a 2,000 calorie diet. Your daily values may be higher or lower depending on your calorie needs.

	Calories	2,000	2,500
Total Fat	Less than	65g	80g
Sat Fat	Less than	20g	25g
Cholesterol	Less than	300mg	300mg
Sodium	Less than	2,400mg	2,400mg
Total Carbohydrate		300g	375g
Dietary Fiber		25g	30g

Calories per gram
Fat 9 • Carbohydrate 4 • Protein 4

INGREDIENTS, Tomatoes, Cooked Macaroni Product, Dry Curd Cottage Cheese, Beef, Low-Moisture Part-Skim Mozzarella Cheese, Dehydrated Onions, Modified Cornstarch, Salt, Parmesan Cheese, Enriched Wheat Flour, Sugar, Spices, Tomato Flavor (Salt, Tomato Paste and Flavorings), Dehydrated Garlic.

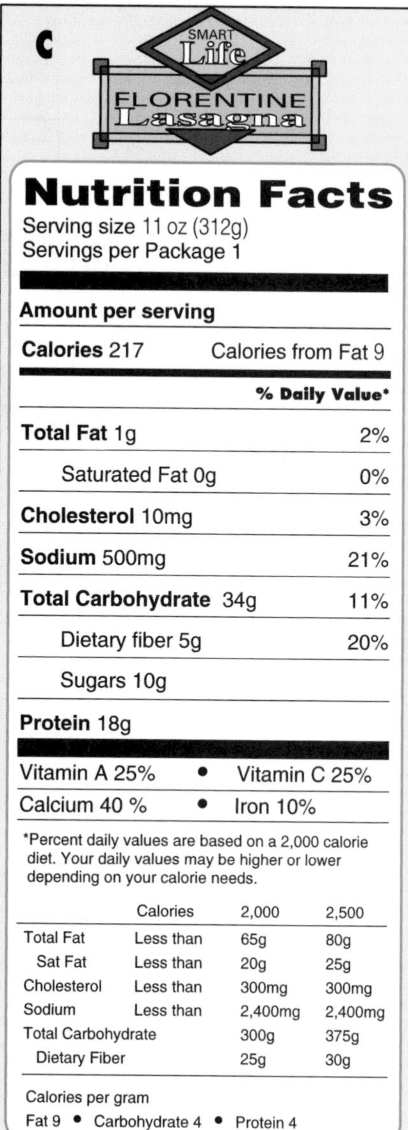

C

SMART Life FLORENTINE Lasagna

Nutrition Facts
Serving size 11 oz (312g)
Servings per Package 1

Amount per serving	
Calories 217	Calories from Fat 9
	% Daily Value*
Total Fat 1g	2%
Saturated Fat 0g	0%
Cholesterol 10mg	3%
Sodium 500mg	21%
Total Carbohydrate 34g	11%
Dietary fiber 5g	20%
Sugars 10g	
Protein 18g	
Vitamin A 25% •	Vitamin C 25%
Calcium 40 % •	Iron 10%

*Percent daily values are based on a 2,000 calorie diet. Your daily values may be higher or lower depending on your calorie needs.

	Calories	2,000	2,500
Total Fat	Less than	65g	80g
Sat Fat	Less than	20g	25g
Cholesterol	Less than	300mg	300mg
Sodium	Less than	2,400mg	2,400mg
Total Carbohydrate		300g	375g
Dietary Fiber		25g	30g

Calories per gram
Fat 9 • Carbohydrate 4 • Protein 4

INGREDIENTS, Tomato Puree, Cooked Enriched Macaroni Product (Durum Semolina, [Niacin, Ferrous Sulfate, Thiamin Mononitrate, Riboflavin], Water, Egg White Solids, Disodium Phosphate, Powdered Cellulose, Soy Protein Isolate, Soy Protein, Vital Wheat Gluten, Guar Gum), Ricotta Cheese (Pasteurized Whey, Pasteurized Milk, Vinegar, Xanthum Gum) Tomatoes, Zucchini, Cheese (Pasteurized Skim-Milk, Water, Natural Flavors, Enzyme, Calcium Chloride, Salt and Vitamin A & D), Carrots, Spinach, Onions, Water, Mushrooms, Concentrated Dealcoholized Burgundy Wine, Sugar, Modified Food Starch, Salt, Spices, Microcrystalline Cellulose, Methylcellulose, Maltodextrin, Hydrolyzed Corn Protein, Xanthum Gum, Guar Gum, Autolyzed Yeast, Calcium Chloride, Citric Acid, Garlic Extractives, Dextrin.

and saturated fat are listed right on the food labels. Other people must perform a few calculations to arrive at meaningful numbers for themselves. This activity asks that you do three things:

- Calculate your own personal Daily Values for *total fat* and *saturated fat*.
- Calculate the percentage of your personal Daily Value for *total fat* contributed by each of the lasagnas.
- Calculate the percentage of your personal Daily Value for *saturated fat* contributed by each of the lasagnas.

Then it asks you some questions.

In calculating your Daily Values for fats, you will use three numbers. First is your recommended intake for energy (from the inside back cover, page Y). Second is the recommendation to consume no more than 30 percent of calories from fat, or 10 percent of calories from saturated fat. The third number arises from the calorie value of fats: 9 calories for every 1 gram of fat.

Find Your Personal Daily Values for Total Fat and Saturated Fat

Step 1. Calculate your personal Daily Value for total fat. On Form 5-1, section 1, part A, copy your energy intake recommendation from the inside back cover, page Y. Transfer your energy recommendation to part B and calculate 30 percent of it as shown. In part C, copy the answer of part B and divide by 9 calories per gram to determine the number of grams of fat as shown. The answer is your personal Daily Value for total fat. Copy it into part D for later use.

Step 2. Calculate your personal Daily Value for saturated fat. On Form 5-1, section 2, part A, write in your personal energy recommendation (from part A of section 1.) Transfer your energy recommendation to part B and calculate 10 percent of it as shown. Transfer this number from part B to part C and divide by 9 calories per gram to find grams of saturated fat. The answer is your personal Daily Value for saturated fat. Copy it into part D for later use.

It's wise to memorize your Daily Values for fat and saturated fat and make food choices each day that do not exceed them. These are two of the most valuable numbers you can learn.

Compare the Fat in a Serving of Each of Three Lasagnas with Your Daily Values

Step 3. Using Form 5-2, part A (next page), copy the grams of total fat and saturated fat per serving of lasagna from the three package labels of Figure 5-20. Then, in part B, copy your personal Daily Values for both total fat and saturated fat from Form 5-1. Enter these values on part C of Form 5-2 and

Form 5•1

CALCULATE YOUR DAILY VALUE FOR FAT AND SATURATED FAT

Section 1—Total Fat

A. Copy your energy recommendation in calories from inside back cover, page Y:
__________________ calories.
(energy recommendation)

B. Calculate the number of calories you can consume as fat in a day:
__________________ cal × .3 = ____________
(energy recommendation) (fat calories)

C. How many grams is this?
______________ cal ÷ 9 cal per gram = ____________ grams fat
(fat calories, from B)

D. Your personal Daily Value for total fat = ____________ g
(from C)

Section 2—Saturated Fat

A. Copy your energy recommendation from A in section 1: __________________ calories.
(energy recommendation)

B. Calculate the number of calories you can consume as saturated fat each day:
__________________ cal × .1 = ______________
(energy recommendation) (saturated fat calories)

C. How many grams is this?
__________________ cal ÷ 9 cal per gram = ____________ grams saturated fat
(saturated fat calories from B)

D. Your personal Daily Value for saturated fat = ____________ g
(from C)

calculate the percentage of your Daily Value for total fat presented by each lasagna. Repeat the process for saturated fat in part D.

Analysis

Now use the information you have generated to respond to the following questions:

1. How many grams of fat can you consume in a day and not exceed 30 percent of calories from fat?
2. How many grams of saturated fat can you consume in a day and not exceed 10 percent of calories from saturated fat?
3. Which lasagna is highest in fat per serving? What percentage of your personal Daily Value for fat would this lasagna contribute to your day's intake?
4. What percentage of your personal Daily Value for fat do the other two lasagnas contribute?
5. If you ate a serving of the highest-fat lasagna, how could you avoid exceeding the recommended fat intake for the day?
6. If you substituted a serving of the lowest-fat lasagna for the highest-fat choice, what effects would this have on your other food choices and on your calorie and nutrient intakes that day?
7. How does the saturated fat in each of the three lasagnas compare with your personal Daily Value for saturated fat?

Consider the Ingredients

Read the ingredient list for each lasagna. Keep in mind that manufacturers list ingredients in descending order of predominance. Ingredients listed first are present in the largest quantities; those listed last are present in the smallest quantities. Now respond to the remaining questions:

8. Which ingredients contributed most to the total fat, saturated fat, and cholesterol in the highest-fat lasagna?
9. In the lowest-fat lasagna, which ingredients replaced or substituted for high-fat ingredients present in the other lasagnas? How did this help to lower the total fat content?
10. Do you agree or disagree with the following statement: "No food is good or bad based on its fat content alone." Justify your stance on this issue.

Form 5•2

COMPARE YOUR PERSONAL DAILY VALUES WITH FATS IN THREE LASAGNAS

A. Grams total fat per serving:

__________ (lasagna A) __________ (lasagna B) __________ (lasagna C)

Grams saturated fat per serving:

__________ (lasagna A) __________ (lasagna B) __________ (lasagna C)

B. Your personal Daily Value for total fat (from Form 5-1, section 1, part D) __________ g
Your Daily Value for saturated fat (from Form 5-1, section 2, part D) __________ g

C. What percentage of your Daily Value for total fat does a serving of each lasagna present?

Lasagna A __________ g (total fat) ÷ __________ g (total fat Daily Value) × 100 = __________ % of personal Daily Value for total fat

Lasagna B __________ g (total fat) ÷ __________ g (total fat Daily Value) × 100 = __________ % of personal Daily Value for total fat

Lasagna C __________ g (total fat) ÷ __________ g (total fat Daily Value) × 100 = __________ % of personal Daily Value for total fat

D. What percentage of your Daily Value for saturated fat does a serving of each lasagna present?

Lasagna A __________ g (saturated fat) ÷ __________ g (saturated fat Daily Value) × 100 = __________ % of personal Daily Value for saturated fat

Lasagna B __________ g (saturated fat) ÷ __________ g (saturated fat Daily Value) × 100 = __________ % of personal Daily Value for saturated fat

Lasagna C __________ g (saturated fat) ÷ __________ g (saturated fat Daily Value) × 100 = __________ % of personal Daily Value for saturated fat

Answers to these Self Check questions are in Appendix G.

1. Which of the following is *not* one of the ways fats are useful in foods?
   a. Fats contribute to the taste and smell of foods.
   b. Fats carry fat-soluble vitamins.
   c. Fats provide a low-calorie source of energy compared to carbohydrates.
   d. Fats provide essential fatty acids.
2. Generally speaking, vegetable and fish oils are rich in:
   a. polyunsaturated fat.
   b. saturated fat.
   c. cholesterol.
   d. *trans*-fatty acids.
3. A benefit to health is seen when _____ is used in place of _____ in the diet.
   a. saturated fat/monosaturated fat
   b. saturated fat/polyunsaturated fat
   c. monosaturated fat/saturated fat
   d. polyunsaturated fat/cholesterol
4. Chylomicrons, a class of lipoproteins, are produced in the:
   a. gall bladder
   b. small intestinal cells
   c. large intestinal cells
   d. liver
5. Which food(s) from the bread, cereals, rice, and pasta group generally contain fat?
   a. biscuits
   b. muffins
   c. pasta
   d. (a) and (b)
6. LDL deliver triglycerides and cholesterol from the liver to the body's tissues. T F
7. Coconut oil is one of the healthier types of oils when considering a heart healthy diet. T F
8. Consuming large amounts of *trans*-fatty acids lowers LDL cholesterol and thus lowers the risk of heart disease and heart attack. T F
9. When olestra is present in the digestive tract it enhances the absorption of vitamin E. T F
10. Any diet that contains vegetable oils, seeds, nuts, and whole grain products supplies enough linolelic acid to meet the body's needs. T F

nutrition on the net

For further study of the topics of this chapter, access these websites and search for the phrases or words in quotation marks:

1. Find updates and quick links to these and other nutrition-related sites at our website:
   www.wadsworth.com/nutrition
2. Search for "cholesterol" and "dietary fat" at the U.S. Government health information site:
   www.healthfinder.gov
3. Review the American Dietetic Association's *ABC's of Fats, Oils, and Cholesterol:*
   www.eatright.org/nfs2.html
4. Search for "fat" and "fat replacers" at the International Food Information Council site:
   ificinfo.health.org
5. Search for "fat substitutes" and "olestra" in the foods section of the Food and Drug Administration's site:
   www.fda.gov

INTERNET ACTIVITY

While we may be aware of the health connections between our dietary fat intake and health, it is often hard to follow the recommended guidelines to consume fewer calories from dietary fat and 300 mg or less from dietary cholesterol. It is particularly hard if we eat out a lot in restaurants—which most of us do these days. Take the Restaurant Quiz from the Center For Science In the Public Interest to learn more about your food selections.

1. Go to:
   http://www.cspinet.org/nah/quiz/index.html
2. Complete the Restaurant Quiz.

Discuss your results. List three ways your score can be improved to reduce your dietary fat intake.

CONTROVERSY 5

THE MEDITERRANEAN DIET: DOES IT HOLD THE SECRET FOR A HEALTHY HEART?

PEOPLE WHO EAT traditional Mediterranean diets that have evolved over thousands of years die much less frequently from heart disease and certain cancers than do people who eat diets typical of northern Europe and North America.[1] They may even live longer.[2] Does this ancient diet of ample grains, vegetables, fruits, olive oil, and cheese, accompanied by wine, hold special benefits? Some experts think so, but the issues have sparked some lively debates.[3] This Controversy explores this idea from the scientific point of view.

Anyone beginning to explore these ideas runs into some paradoxes. For example, the diet of one Mediterranean country, Greece, provides up to 42 percent of its calories as fat, mostly from olive oil and olives. Many Greeks carry more body fat than is considered prudent in the United States. According to U.S. guidelines, then, the Greek population would be urged to eat less fat and to lose weight to protect their hearts. Yet Greeks living in Greece enjoy one of the longest life expectancies worldwide and die from cardiovascular disease far less often than do Americans.[4]

Should we be eating the Mediterranean way?

Should Americans Eat like Greeks?

Some have suggested that a diet similar to that of the Greeks might achieve heart disease rates here as low as those in Greece. Such a diet might also be easier to follow than the very-low-fat diet traditionally prescribed for heart health in the United States.[5] This does not mean that Greek ethnic restaurants in the United States provide the healthiest dining choices—their menu items often have been adapted to Western tastes by adding red meat and saturated fats.[6] The diet of Greece in the mid-twentieth century serves as the model for the Mediterranean diet.

Greece is not alone in having heart disease risks lower than those of the United States; many nations of the Mediterranean region do, and nobody knows why. The idea that genetics might bestow resistance to heart disease upon Mediterranean people has been abandoned because the risks seem to follow regions, not individual people. Mediterranean immigrants to the United States who adopt an American diet and lifestyle suffer heart disease and cancer at the same rates as native-born U.S. citizens. Further, as "American-style" fast foods advance across the Mediterranean region, disease rates shift toward those more typical of northern Europe and North America.[7] This evidence supports the idea that diet, not genetics, is the primary variable that confers disease risk or protection.

Defining the Mediterranean Diet

Scientists who try to define a single "Mediterranean" diet run into a problem.[8] The vast Mediterranean region includes Spain, Portugal, France, Syria, Israel, and many other nations, all with different diets. As an example, Greeks eat abundant fat as olive oil, but southern Italians keep total fat intakes low.

About 5,000 years ago, the diet of the region was based on grain foods such as crusty breads, honey-sweetened cakes, rice, and seeds; legumes including beans, peas, and lentils; fish and other seafood; goat cheese; olives and olive oil; vegetables; fruits (especially grapes and figs); and wine mixed half-and-half with water. Over time, oranges and lemons arrived from East Asia, and tomatoes, eggplant, and potatoes from other lands also became central to Mediterranean meals. Meats were for special occasions only; the favored cooking fat was oil pressed from olives; butter was shunned. For most of the twentieth century, Mediterranean people followed a diet with the following eight attributes, and many continue to do so today:

- Low in saturated fats (such as butter or meat fat), but high in monounsaturated fats (such as olive oil).
- Low in meat and meat products.
- High in legumes.
- High in grains, including bread.
- High in fruits.
- High in vegetables.
- Moderate in milk and milk products.
- Moderate in alcohol.[9]

Based on these observations, a Mediterranean Pyramid emerged, similar to the U.S. Food Guide Pyramid. Figure C5-1 shows the pyramids side by side, and the rest of this Controversy compares them. Later chapters revisit many of the issues discussed here because they are among the top concerns of nutrition scientists today.

The Weighted Base—Carbohydrates and Fiber

Some similarities between the pyramids of Figure C5-1 are apparent—both have wide bases of breads, cereals, pasta, rice, and other grains, signifying that the two diets are built on the same foundation. This wise choice supplies abundant carbohydrate-rich foods—foods that in their unrefined state provide needed energy, vitamins, minerals, and fiber with almost no fat. Such

Figure C5•1

TWO PYRAMID PLANS

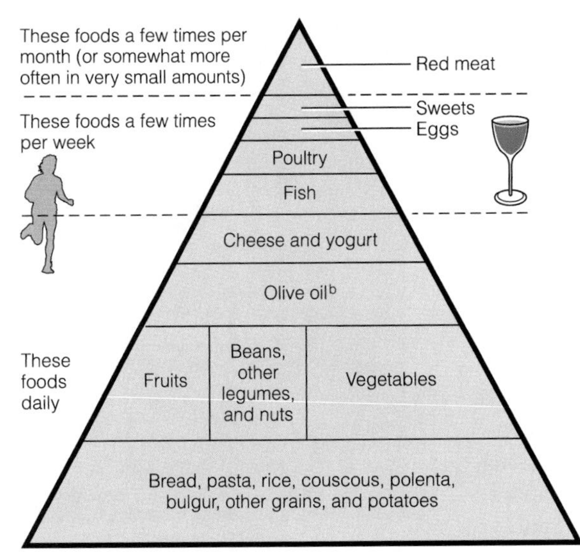

[a]The authors of this pyramid also recommend regular physical exercise and moderate consumption of wine.

[b]Other oils rich in monounsaturated fats, such as canola or peanut oil, can be substituted for olive oil. People who are watching their weight should limit their oil consumption.

SOURCE: 1994 Oldways Preservation & Exchange Trust, by permission.

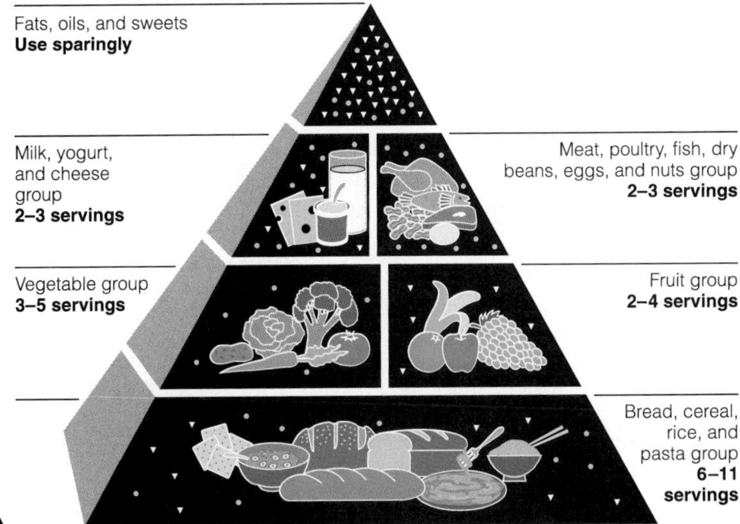

SOURCE: U.S. Department of Agriculture/U.S. Department of Health and Human Services

KEY TO SYMBOLS
Fat (naturally occurring and added) **Sugars** (added)
These symbols show that fat and added sugars come mostly from fats, oils, and sweets, but can be part of or added to foods from the other food groups as well.

foods meet the body's needs for these nutrients without elevating the person's risk of disease. So far, both plans agree.

Vegetables and Fruit

In both pyramids, vegetables and fruits appear on the next level but with one important difference: the Mediterranean plan calls for servings of legumes (dried beans) or nuts every day (more on this in the next section). Peoples of the Mediterranean region stand out as regularly consuming more than the recommended five servings of fruit and vegetables daily.[10] In Spain, for example, a plateful of raw or cooked vegetables commonly precedes the main meal; melon or figs serve as dessert.

Later chapters cover the virtues of vegetables and fruits in the diet; let it suffice here to say that the case for their role in disease prevention is virtually ironclad. Vegetables provide fiber, phytochemicals, antioxidants, and other nutrients (see Controversies 2 and 7), all of which correlate with reduced incidence of chronic diseases.[11] Further, with few exceptions, these foods are extraordinarily low in fat. Without a doubt, vegetables and fruits are allies to those who seek a healthful diet.

Problem: To Which Group Do Legumes Belong?

Legumes are difficult to classify in any food grouping plan because they have desirable qualities in common with so many other foods. For example, the U.S. Pyramid classes legumes with meats because, like meats, legumes are high in protein and also with the vegetables by virtue of their fiber, carbohydrate, and vitamins. Although legume protein is not quite identical to that of meat, a serving or two of either meat or legumes in a balanced diet amply meets the body's need for protein.

Legumes also supply iron and other minerals typically associated with meats. The similarities end there, however. Many meats are notoriously high in fat, much of it the heart-clogging saturated kind, and they provide no fiber or carbohydrate. Legumes, on the other hand, supply abundant fiber and carbohydrate with little or no fat, a combination the heart prefers. In some ways, certain legumes even resemble dairy products: they combine calcium and ample protein in one food.

In light of legumes' high fiber and mineral contents, the Mediterranean system lists legumes separately from both the meats and the vegetables. Mediterranean main dishes often include some form of legumes, sometimes as part of the dish, sometimes as garnish. In Spain, for example, cooked legumes are deep-fried in olive oil and served as snack foods, similar to potato chips in this country. As the next section describes, the evidence for the health effects of legumes leaves no doubt about their importance.

Arguments for Requiring Legumes

"Why should I eat a bowl of beans every day when I can afford to eat meat?" asks a student who observes that legumes and meats are interchangeable in the U.S. Pyramid. The answer is found in evidence on legumes' fiber and their health effects.

The fiber of legumes is a constituent often lacking in U.S. diets. Dry beans have more dietary fiber per serving than almost any other unprocessed food, and they provide a balance of the types of fibers that exert various beneficial effects on health. Fiber alone cannot explain all of the findings about legumes and health, however. Consider the evidence:

- Endurance athletes who eat legumes in the hours before competition may extend their glucose fuel availability during competition.
- Legumes contain a form of starch believed to help control blood glucose in people with diabetes.
- Legumes are rich in the type of fiber associated with low blood cholesterol and low rates of heart disease.
- Replacing meat *protein* in the diet with soybean protein may lower blood lipids, thereby lowering heart disease risk, even when dietary *fats* are held constant.
- Some legumes contain phytochemicals that may, under certain conditions, inhibit the growth of some cancers (see Controversy 2).
- A meal that includes legumes satisfies hunger longer than other meals.

This last attribute may have importance for weight control. A person who achieves a feeling of fullness sooner during a meal and stays full longer has less trouble resisting unneeded snacks.

As wonderful as legumes are, however, they are not "perfect foods," nor do they comprise an adequate diet when eaten alone. They lack the vitamins A and C of vegetables and fruits and the vitamin B_{12} of meats and dairy products. To overconsume legumes to the exclusion of other foods would be as poor a choice as never to include a bean in the diet.

Legumes: remarkable foods

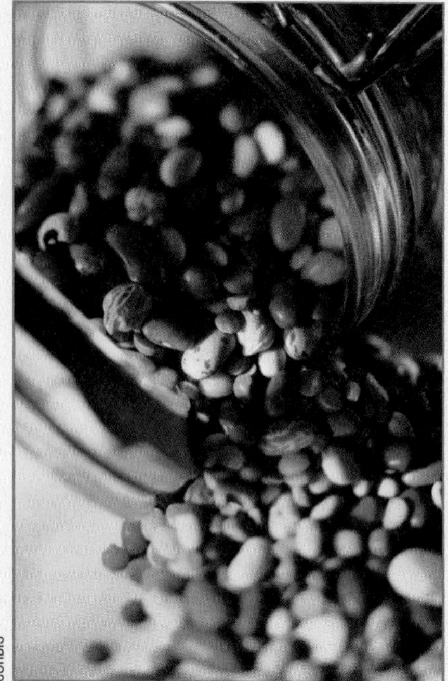

CORBIS

Arguments Concerning Olive Oil

One of the most startling differences between the two pyramids, and the one that draws the most controversy, concerns fat. The Mediterranean plan suggests that the next most predominant daily dietary constituent after fruits, vegetables, and legumes should be olive oil—a source of pure fat. In the U.S. Pyramid, olive oil joins other fats at the tip-top, flagged with a warning to use all fats sparingly. People following the Mediterranean plan receive some 40 percent of their calories from fat, mostly from olive oil and not from butter, margarine, cooking oils, or other fats. Advocates of this high fat intake believe that a diet high in olive oil but low in other fats may reduce the risk of cancer and heart disease and help control diabetes, while providing more variety and greater satisfaction to the eaters.[12]

Some scientific evidence supports this view. For example, researchers have found that olive oil is metabolized differently in the body than the oil of soybeans. They theorize that olive oil may therefore conserve a beneficial blood lipid (HDL) associated with low risk of heart disease, while lowering levels of the riskier types of blood lipids. Olives and the dark oils made from them contain potent antioxidants that may be partly responsible for low disease risks in those who consume them regularly.[13] Furthermore, olive oil stands out among unsaturated fats in being resistant to oxidation (see Chapter 5), which is a possible cause of heart disease.

Not everyone agrees that research to date has made a sufficient case for a protective effect from olive oil. Controlled clinical trials of the Mediterranean diet and heart disease are lacking, causing some to say that it is too early to make recommendations for its use.[14] Critics of the Mediterranean Pyramid note that avocados, nuts, and canola oil have oils similar to the oil of olives, but these foods are ignored by the plan. Also, critics point out that olive oil is fattening, providing 9 calories per gram, as do all fats. Though some very active people of the Mediterranean region may burn off the excess calories from their high-fat diets, many people from that region are also overweight. Obesity and its many associated ills remain major threats to health in the United States. Critics therefore conclude that the recommendation to consume any sort of high-fat diet is unwise, even if the fat itself is easy on the heart.

Olives and their oil contain phytochemicals that may benefit health.

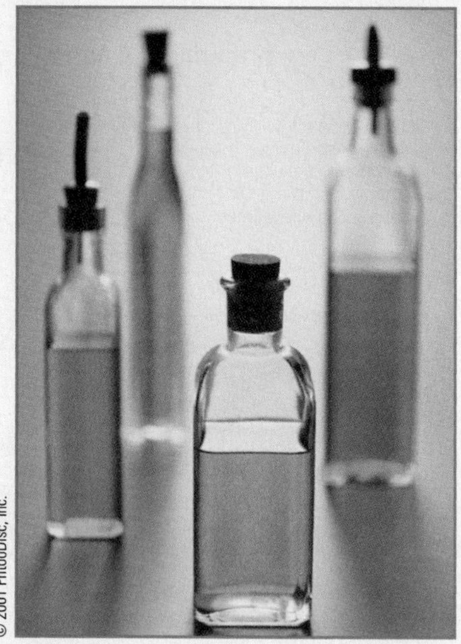

© 2001 PhotoDisc, Inc.

A Mediterranean Approach to Meats and Fish

Returning to the differences between the two pyramids, the Mediterranean plan does not restrict total fat, but does limit *animal* fat (and this is mostly saturated fat). The U.S. Pyramid makes no such distinction. The U.S. Pyramid allows high-fat red meats at every meal, yet the diner who eats this much meat cannot help exceeding the recommended limit on fat and especially saturated fat intake.

People following the Mediterranean plan typically consume less than 10 percent of their calories from saturated fat—a goal of both plans. People following the U.S. plan find meeting this goal difficult, however. Notice that fish dominates poultry in the Mediterranean plan and that eggs and sweets dominate red meats. Notice, too, that recommendations for consuming poultry and red meats are in terms of servings *per week or month,* not two to three servings *per day* as suggested by the U.S. plan. The average person in the United States consumes more than half a pound of meat per day; the average person in the Mediterranean region consumes about half a pound per *week.* This difference in meat intake, and therefore in saturated fat intake, is significant.

The choice of fish for some servings of animal protein may protect the heart. Consistently, research shows that people who consume fish regularly are less likely to suffer heart disease.[15] Chapter 5 discusses the topic of fish and the essential fats they contain, as well as their profound influence on human health.

Milk Products

The Mediterranean plan suggests daily cheese or yogurt but fails to mention fat-free milk, the staple milk source in the U.S. plan. Mediterranean people eat yogurt or cheese each day, and they may use milk to lighten their coffee, but adults do not use milk by itself as a beverage. Cheese contains the nutrients of milk but is typically very high in saturated fat.

The Mediterranean diet emphasizes yogurt over milk, possibly because adults of the region may have problems digesting the milk sugar lactose and therefore experience digestive distress when they consume milk. Yogurt contains somewhat less lactose than does fresh milk (Chapter 4 provides details), and it may provide health benefits from its probiotic organisms (see Controversy 2). People of northern European descent, however, retain milk-digesting abilities throughout adult life. In whatever form, milk and its products provide calcium in abundance, and most adults need more calcium in their diets.

Evidence about Wine

The most controversial suggestion of the Mediterranean Pyramid is the advice to partake moderately of wine. The major concern is that encouraging wine drinking may invite development of alcoholism or an increase in alcohol-related traffic accidents. These consequences would be too severe to risk even if wine's action against heart disease was a certainty—and it is not.

Does red wine protect the French and other Mediterranean people from heart disease? Studies reveal that populations using moderate amounts of wine (red or white) as a beverage with meals have low rates of heart disease. Controversy 2 mentioned the phytochemicals of grapes that survive the wine-making process, but compared with other food sources such as onions, a serving of wine turns out to be a relatively poor source of these compounds.[16] Indeed, wine-drinking peoples have also been observed to consume a *diet* rich in the phytochemicals of fiber-rich, low-fat fruits, vegetables, garlic, and herbs, and such a diet is known to be protective of the health of the heart.[17]

Alcohol itself (one or two drinks a day) may reduce the risk of heart disease by raising levels of beneficial blood lipids (HDL) and preventing blood clot formation. These benefits are most apparent in people over age 50 and in those most likely to develop heart disease. Studies also report, however, that heavy alcohol consumption (three or more drinks a day) *increases* the risk of death from other causes. Regular alcohol consumption also increases the risk of breast cancer in women and heavy drinking has many negative effects on body systems; the Controversy section of Chapter 3 describes these effects.

Should Americans Eat the Mediterranean Way?

Many questions surrounding the Mediterranean Pyramid remain unanswered. First, though heart disease rates are low in Mediterranean countries, no one can explain why the incidence of stroke is almost double that of the United States. If the diet can receive the credit for heart health, should it also be blamed for the increased incidence of stroke?

Second, although the link between diet and chronic diseases overall is strong, no one knows for sure what parts of the diet form that link.[18] Perhaps the olive oil is a key factor, but both nutrients and phytochemicals found in vegetables, legumes, seafood, and seasonings such as garlic or herbs offer protective effects, too. The overall pattern of the Mediterranean diet, rather than individual foods or constituents, may be the key to health and is under study.[19] Active lifestyles may play a role, as may differences in tobacco use; exercise reduces heart disease risk, and smoking greatly increases it.

Should everyone abandon the U.S. Pyramid, then, and take up eating Mediterranean style? Critics of the Mediterranean plan have expressed concerns that it may be inadequate in calcium and iron—two problem nutrients for many people, especially women and children. Because these nutrients are typically lacking in many people's diets, it seems unwise to restrict selections of calcium-rich cheeses and yogurt and iron-rich meats to a few times a month.

Certainly, though, benefits are likely to follow if a few suggestions from the Greeks and other Mediterranean peoples are adopted. Including legumes as part of a balanced daily diet seems like a good idea, as does *replacing* saturated fats such as butter and meat fat with unsaturated phytochemical-rich fats like olive oil. The authors of this book would not stop there, however. They would urge you to reduce fats from all sources; choose small portions of the leanest meats, fish, and poultry; and include fresh foods from all the groups each day. Also, exercise daily, as the Mediterraneans do. As for wine, this is a personal choice, but one to approach with extreme caution, for no other dietary choice has such destructive potential.

Conclusion

Recommendations will continue to evolve as nutrition science unfolds. Meanwhile, you must choose foods every day. The next chapter on protein completes a set of three on the energy-yielding nutrients and provides answers to meeting protein needs in the real world.

chapter 6

Sénèque Obin (1893–1997) *Marché Poissons (Fish Market)*, before 1997

The Proteins and Amino Acids

Frequently Asked Questions

How do amino acids build proteins? p. 179

After protein is digested, what happens to amino acids? p. 184

Which kinds of protein-rich foods are easiest to digest and use? p. 192

How much protein do people really need? p. 194

What happens when people consume too little protein? p. 196

Is it possible to consume too much protein? p. 198

Contents

proteins compounds composed of carbon, hydrogen, oxygen, and nitrogen and arranged as strands of amino acids. Some amino acids also contain the element sulfur.

amino (a-MEEN-o) **acids** the building blocks of protein. Each has an amine group at one end, an acid group at the other, and a distinctive side chain.

amine (a-MEEN) **group** the nitrogen-containing portion of an amino acid.

side chain the unique chemical structure attached to the backbone of each amino acid that differentiates one amino acid from another.

essential amino acids amino acids that either cannot be synthesized at all by the body or cannot be synthesized in amounts sufficient to meet physiological needs. Also called *indispensable amino acids*.

conditionally essential amino acid an amino acid that is normally nonessential, but must be supplied by the diet in special circumstances when the need for it exceeds the body's ability to produce it.

THE PROTEINS are versatile and vital cellular working molecules. Without them, life would not exist. First named 150 years ago after the Greek word *proteios* ("of prime importance"), **proteins** have revealed countless secrets of the life processes, and they account for many nutrition concerns. How do we grow? How do our bodies replace the materials they lose? How does blood clot? What gives us immunity? Understanding the nature of the proteins gives us the answers to these questions.

Some proteins are working proteins; others form structures. Working proteins include the body's enzymes, antibodies, transport vehicles, hormones, cellular "pumps," and oxygen carriers. Structural proteins include tendons and ligaments, scars, the cores of bones and teeth, the filaments of hair, the materials of nails, and more.

The Structure of Proteins

The structure of proteins enables them to perform many vital functions. One key difference from carbohydrates and fats, which likewise contain carbon, hydrogen, and oxygen atoms, is that proteins also contain nitrogen atoms. These nitrogen atoms give the name *amino* ("nitrogen containing") to the **amino acids,** the building blocks of protein. Another key difference is that in contrast to the carbohydrates, whose repeating units, glucose molecules, are identical, the amino acids in a strand of protein are different from one another. A strand of amino acids that makes up a protein may contain 20 *different* kinds of amino acids.

Amino Acids

All amino acids have the same simple chemical backbone consisting of a single carbon atom with both an **amine group** (the nitrogen-containing part) and an acid group attached to it. Each amino acid also has a distinctive chemical **side chain** attached to the center carbon of the backbone (see Figure 6-1). It is this side chain that gives each amino acid its identity and chemical nature. About 20 amino acids, each with its different side chain, make up most of the proteins of living tissue. Other rare amino acids appear in a few proteins.

The side chains make the amino acids differ in size, shape, and electrical charge. Some are negative, some are positive, and some have no charge (are neutral). The first part of Figure 6-2 is a diagram of three amino acids, each with a different side chain attached to its backbone. The rest of the figure shows how amino acids link to form protein strands. Long strands of amino acids form large protein molecules, and the side chains of the amino acids ultimately help to determine the molecules' shapes and behaviors.[1]

More about phenylketonuria, or PKU, in Controversy 4

Essential Amino Acids The body can make about half of the 20 amino acids for itself, given the needed parts: fragments derived from carbohydrate or fat to form the backbones and nitrogen from other sources to form the amine groups. The healthy adult body makes some other amino acids too slowly to meet its needs, however, or cannot make them at all. These are the **essential amino acids** (see the margin, facing page). Without these essential nutrients, the body cannot make the proteins it needs to do its work. Because the essential amino acids can be obtained only from foods, a person must often eat the foods that provide them.

Sometimes a nonessential amino acid becomes essential under special circumstances. For example, the body normally makes tyrosine (a nonessential amino acid) from the essential amino acid phenylalanine. If the diet fails to supply enough phenylalanine, or if the body cannot make the conversion for some reason (as happens in the inherited disease phenylketonuria), then tyrosine becomes a **conditionally essential amino acid.**

Figure 6•1

AN AMINO ACID

The "backbone" is the same for all amino acids. The side chain differs from one amino acid to the next. The nitrogen is in the amine group.

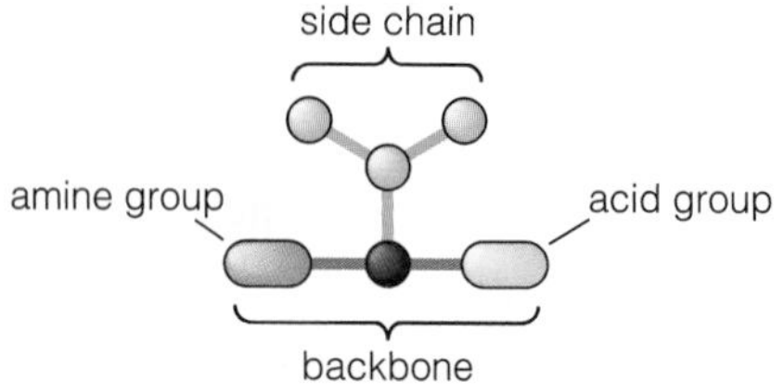

Recycling Amino Acids The body not only makes some amino acids but also breaks protein molecules apart and reuses those amino acids. Both food proteins, after digestion, and body proteins, when they have finished their cellular work, are

Hair, skin, eyesight, and the health of the whole body depend on protein from food.

© Gary Conners/PhotoEdit

dismantled to liberate their component amino acids. Pools of such amino acids provide the cells with raw materials from which they can build the protein molecules they need. Cells can also use the amino acids for energy and discard the nitrogen atoms as wastes. By reusing amino acids to build proteins, however, the body recycles and conserves a valuable commodity while easing its nitrogen disposal burden.

This recycling system also ensures that the body has access to an emergency fund of amino acids in times of fuel or protein deprivation. At such times, tissues can break down their own proteins, sacrificing working molecules before the ends of their normal lifetimes, to supply energy and protein to the body's cells. The body employs a priority system in selecting the tissue proteins to dismantle—it uses the most dispensable ones first, such as the small proteins of the blood. It guards the protein structures of the heart and other organs until forced, by dire need, to relinquish them.

key point *Proteins are unique among the energy nutrients in that they possess nitrogen-containing amine groups and are composed of as many as 20 different amino acid units. Of the 20 amino acids, some are essential and some are essential only in special circumstances.*

How Do Amino Acids Build Proteins?

In the first step of making a protein, each amino acid is hooked to the next (as shown in Figure 6-2). A chemical bond, called a **peptide bond,** is formed between the amine group end of one amino acid and the acid group end of the next. The side chains bristle out from the backbone of the structure, giving the protein molecule its unique character.

The strand of protein does not remain a straight chain. Figure 6-2 shows only the first step in making proteins—the linking of from several dozen to as many as 300 amino acid units with peptide bonds. In the next step, amino acids at different places along the strand are attracted to each other, and this attraction causes some segments of the strand to coil, somewhat like a metal spring. Also, each spot along the coiled strand is attracted to, or repelled from, other spots along its length. These interactions cause the entire coil to fold this way and that, forming either a globular structure, as shown in Figure 6-3 on the next page, or a fibrous structure (not shown).

The amino acids whose side chains are electrically charged are attracted to water. Therefore, in the body's watery fluids, they orient themselves on the outside of the protein structure. The amino acids whose side chains are neutral are repelled by water and are attracted to one another; these tuck themselves into the center, away from the body fluids. All these interactions among the amino acids and the surrounding fluids give each protein a unique architecture.

peptide bond a bond that connects one amino acid with another, forming a link in a protein chain.

The essential amino acids:

- Histidine
- Isoleucine
- Leucine
- Lysine
- Methionine
- Phenylalanine
- Threonine
- Tryptophan
- Valine

Other amino acids important in nutrition:

- Alanine
- Arginine
- Asparagine
- Aspartic acid
- Cysteine
- Glutamic acid
- Glutamine
- Glycine
- Proline
- Serine
- Tyrosine

NOTE: In special cases, some nonessential amino acids may become conditionally essential (see the text).

Figure 6•2

DIFFERENT AMINO ACIDS JOIN TOGETHER

This is the basic process by which proteins are assembled.

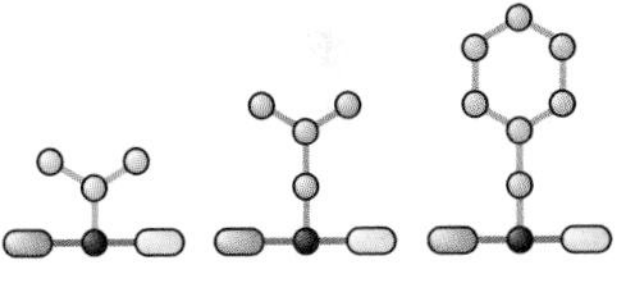

Single amino acids with different side chains…

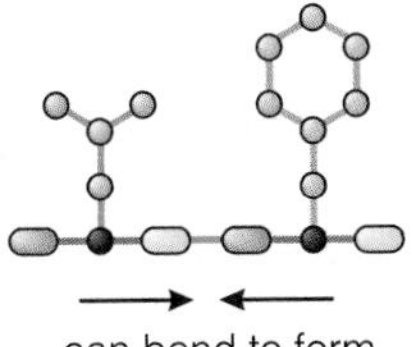

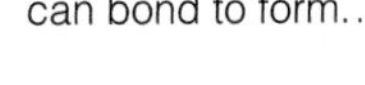

can bond to form…

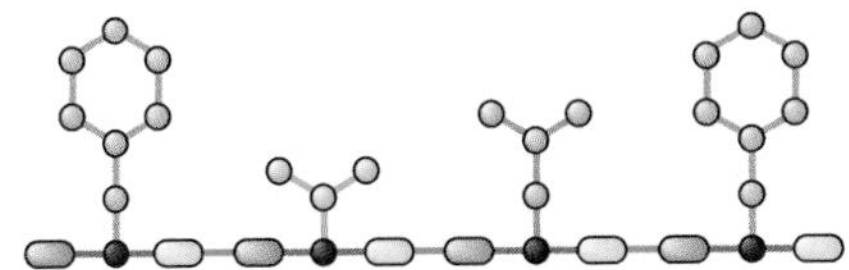

a strand of amino acids, part of a protein.

enzymes (EN-zimes) protein catalysts. A catalyst is a compound that facilitates a chemical reaction without itself being altered in the process.

hemoglobin the globular protein of red blood cells, whose iron atoms carry oxygen around the body. (More about hemoglobin in Chapter 8.)

Figure 6•3

THE COILING AND FOLDING OF A PROTEIN MOLECULE

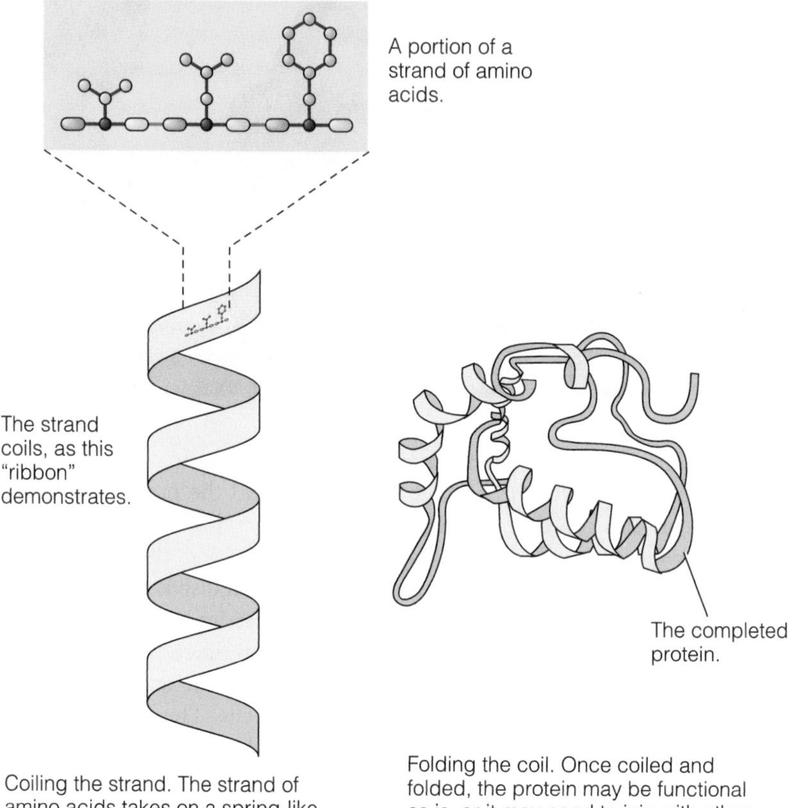
A portion of a strand of amino acids.
The strand coils, as this "ribbon" demonstrates.
The completed protein.

Coiling the strand. The strand of amino acids takes on a spring-like shape as their side chains variously attract and repel each other.

Folding the coil. Once coiled and folded, the protein may be functional as is, or it may need to join with other proteins or add a vitamin or mineral to become active.

Figure 6•4

THE PROTEIN HEMOGLOBIN

The coiled and looped red structures are the globular proteins: the flat, jagged-edged objects are heme structures; the balls in the center of the heme structures are iron atoms. This model represents a molecule of hemoglobin magnified 27 million times.

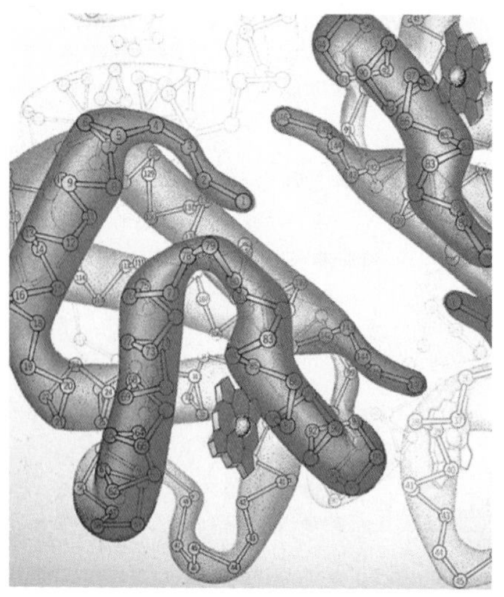

One final detail may be needed for the protein to become functional. Several strands may cluster together into a functioning unit, or a metal ion (mineral) or a vitamin may join to the unit and activate it.

key point

Amino acids link into long strands that coil and fold to make a wide variety of different proteins.

The Variety of Proteins

The particular shapes of proteins enable them to perform different tasks in the body. Those of globular shape, such as some proteins of blood, are water soluble. Some are hollow balls, which can carry and store materials in their interiors. In some proteins, several coils of amino acids wind around each other and form ropelike fibers that can give strength and elasticity to body parts. Some, such as those that form tendons, are more than ten times as long as they are wide, forming stiff, rodlike structures that are somewhat insoluble in water and very strong. Still others act like glue. Among the most fascinating proteins are the **enzymes,** which act on other substances to change them chemically. A model of a single large globular protein molecule, the **hemoglobin** that carries oxygen in the red blood cells, is shown in Figure 6-4.

The great variety of proteins in the world is possible because an infinite number of sequences of amino acids can be formed. To understand how so many different proteins can be designed from only 20 or so amino acids, think of how many words are in an unabridged dictionary—all of them constructed from just 26 letters. The letters in a word must alternate between consonant and vowel sounds, but the amino acids in a protein need follow no such rules. Nor is there any restriction on the length of the chain of amino acids. Thus, the number of possible proteins is much greater than the number of possible English words. A single human cell may contain as many as 10,000 different proteins, each one present in thousands of copies.

denaturation the change in a protein's shape brought about by heat, acids, bases, alcohol, salts of heavy metals, or other agents.

Inherited Amino Acid Sequences For each protein, there is only one proper amino acid sequence, and that sequence is specified by heredity. If a wrong amino acid is inserted, the result can be disastrous to health.

Sickle-cell disease, in which hemoglobin, the oxygen-carrying protein of the red blood cells, is abnormal, is an example of an inherited mistake in the amino acid sequence. Normal hemoglobin contains two kinds of chains. In sickle-cell hemoglobin, one of the chains is an exact copy of that in normal hemoglobin, but in the other chain, the sixth amino acid is valine, rather than glutamine as it should be. This replacement of one amino acid so alters the protein that it is unable to carry and release oxygen. The red blood cells collapse from the normal disk shape into crescent shapes (see Figure 6-5). If too many crescent-shaped cells appear in the blood, the result is abnormal blood clotting, strokes, bouts of severe pain, susceptibility to infection, and early death.[2]

You are unique among human beings because of minute differences in your body proteins. These differences are determined by the amino acid sequences of your proteins, which are written into the genetic code you inherited from your parents and they from theirs. Your unique combination of genes directs the making of all your body's proteins, as shown in Figure 6-6 on the next page. Notice that what the genes determine is the sequence of the amino acids in the finished protein.

Nutrients and Gene Expression When a cell makes a protein as shown in Figure 6-6, scientists say that the gene for that protein has been "expressed." Every cell nucleus contains the DNA for making every human protein, but each cell does not make all types of protein. Instead, cells specialize in making only the proteins typical of their cell types.* For example, only cells of the pancreas express the gene for the protein hormone insulin; in other cells, that gene is idle. Cells regulate the expression of genes, and so the synthesis of various proteins, in response to changing conditions within the body. Often, nutrients signal the need for more or less of a protein. For example, when the presence of glucose in the blood indicates that more of the hormone insulin is needed, pancreas cells speed up production. Likewise, when the body's iron stores run low, bone marrow cells slow down their synthesis of hemoglobin. Nutrients often act to modulate gene expression, and later chapters provide more examples. The Think Fitness feature on page 183 addresses the question of whether increasing dietary protein consumption can signal the genes in muscle cells to begin building larger muscles.

Each type of protein has a distinctive sequence of amino acids and so has great specificity. Cells specialize in synthesizing only particular types of proteins.

Figure 6•5

NORMAL RED BLOOD CELLS AND SICKLE CELLS

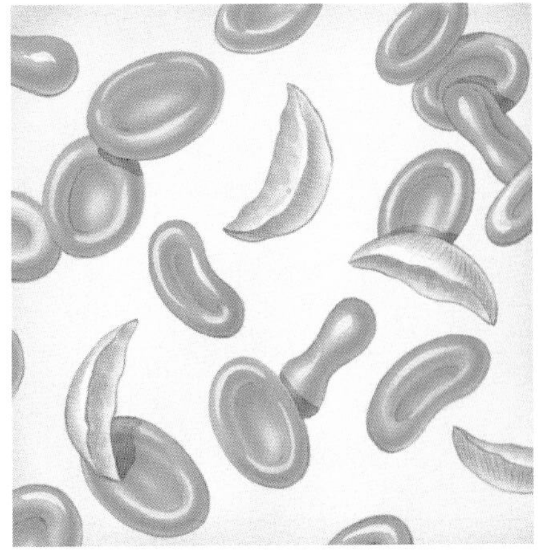

Denaturation of Proteins

Proteins can be denatured (distorted in shape) by heat, alcohol, acids, bases, or the salts of heavy metals. The **denaturation** of a protein is the first step in its destruction; thus, these agents are dangerous because they damage the body's proteins. In digestion, however, denaturation is useful to the body. During the digestion of a food protein, the stomach acid opens up the protein's structure, permitting digestive enzymes to make contact with the peptide bonds and cleave them. Denaturation also occurs during the cooking of foods. Cooking an egg denatures the proteins of the egg and makes it firm. More important for nutrition is that heat denatures two proteins in raw eggs: one binds the B vitamin biotin and the mineral iron, and the other slows protein digestion. Thus, cooking eggs liberates biotin and iron and aids digestion.

*Red blood cells lack nuclei and so also lack the DNA needed for making proteins.

Figure 6•6

PROTEIN SYNTHESIS

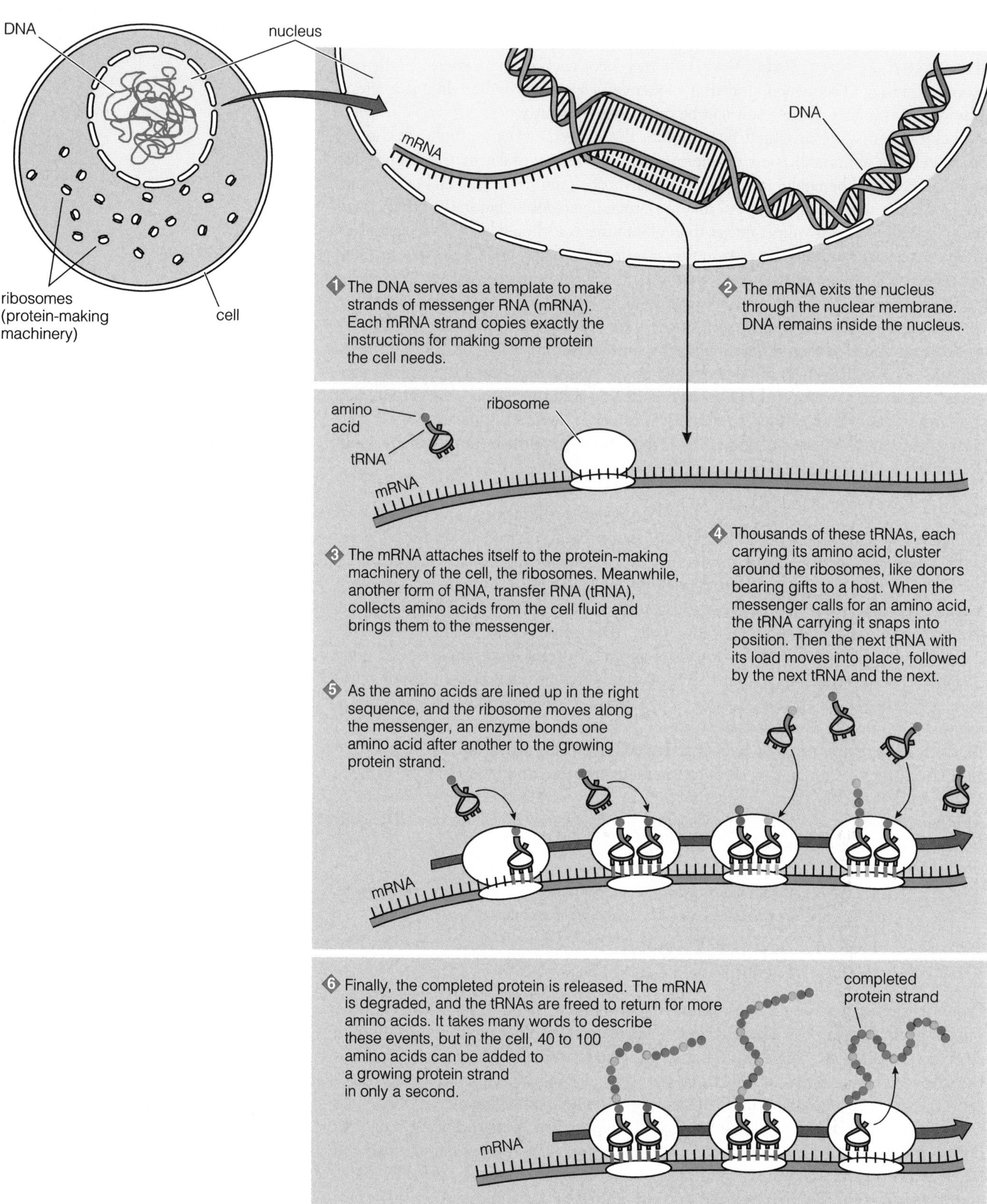

DNA
nucleus
ribosomes
(protein-making
machinery)
cell
mRNA
DNA
1 The DNA serves as a template to make strands of messenger RNA (mRNA). Each mRNA strand copies exactly the instructions for making some protein the cell needs.
2 The mRNA exits the nucleus through the nuclear membrane. DNA remains inside the nucleus.
amino
acid
ribosome
tRNA
mRNA
3 The mRNA attaches itself to the protein-making machinery of the cell, the ribosomes. Meanwhile, another form of RNA, transfer RNA (tRNA), collects amino acids from the cell fluid and brings them to the messenger.
4 Thousands of these tRNAs, each carrying its amino acid, cluster around the ribosomes, like donors bearing gifts to a host. When the messenger calls for an amino acid, the tRNA carrying it snaps into position. Then the next tRNA with its load moves into place, followed by the next tRNA and the next.
5 As the amino acids are lined up in the right sequence, and the ribosome moves along the messenger, an enzyme bonds one amino acid after another to the growing protein strand.
mRNA
6 Finally, the completed protein is released. The mRNA is degraded, and the tRNAs are freed to return for more amino acids. It takes many words to describe these events, but in the cell, 40 to 100 amino acids can be added to a growing protein strand in only a second.
completed
protein strand
mRNA

Many well-known poisons are salts of heavy metals like mercury and silver; these denature proteins wherever they touch them. The common first-aid antidote for swallowing a heavy-metal poison is to drink milk. The poison then acts on the protein of the milk rather than on the protein tissues of the mouth, esophagus, and stomach. Later, vomiting can be induced to expel the poison that has combined with the milk.

key point *Proteins can be denatured by heat, acids, bases, alcohol, or the salts of heavy metals. Denaturation begins the process of digesting food protein and can also destroy body proteins.*

dipeptides (dye-PEP-tides) protein fragments that are two amino acids long. A peptide is a strand of amino acids (*di* means "two").

tripeptides (try-PEP-tides) protein fragments that are three amino acids long (*tri* means "three").

polypeptides protein fragments of many (more than ten) amino acids bonded together (*poly* means "many"). A chain of between four and ten amino acids is called an *oligopeptide*.

Can Eating Extra Protein Make Muscles Grow Larger?

Can athletes and fitness seekers stimulate their muscles to grow larger by consuming more protein or amino acids? No. Only hard work, not excess dietary protein, triggers the genes to build more muscle tissue. Exercise generates cellular messages that stimulate DNA to begin the process of building up muscle fibers (muscle fibers are made of protein). An excess of amino acids or other nutrients does not generate these messages.

THINK FITNESS

Digestion and Absorption of Protein

Each protein is designed for a special purpose in a particular tissue of a specific kind of animal or plant. When a person eats food proteins, whether from cereals, vegetables, beef, fish, or cheese, the body must first alter them by breaking them down into amino acids; only then can it rearrange them into specific human body proteins.

Chapter 10 discusses the nutrient needs of athletes in detail.

Other than being crushed and moistened with saliva in the mouth, nothing happens to protein until it reaches the very strong acid of the stomach. There the acid helps to uncoil the protein's tangled strands so that molecules of the stomach's protein-digesting enzyme can attack the peptide bonds. You might expect that the stomach enzyme itself, being a protein, would be denatured by the stomach's acid. Unlike most enzymes, though, the stomach enzyme functions best in an acid environment. Its job is to break other protein strands into smaller pieces. The stomach lining, which is also made partly of protein, is protected against attack by acid and enzymes by a coat of mucus, secreted by its cells.

Protein Digestion

The whole process of digestion is an ingenious solution to a complex problem. Proteins (enzymes), activated by acid, digest proteins from food, denatured by acid. The coating of mucus secreted by the stomach wall protects its proteins from attack by either acid or enzymes. The normal acid in the stomach is so strong (pH 1.5) that no food is acid enough to make it stronger; for comparison, the pH of pure vinegar is about 3.

Chapter 3 discussed the use of medicines to control the stomach's acidity and also defined pH as a measure of acidity. See pages 80 and 81.

By the time most proteins slip from the stomach into the small intestine, they are already broken into smaller pieces. Some are single amino acids; others are strands of two or three amino acids (**dipeptides** and **tripeptides**—see Figure 6-7, next page). The majority are longer chains (**polypeptides**), and a few are whole proteins. In the small intestine, alkaline juice from the pancreas neutralizes the acid delivered by the stomach. The pH rises to about 7 (neutral), enabling the next enzyme team to accomplish the final breakdown of the strands. Protein-digesting enzymes from the pancreas and intestine continue working until almost all pieces of protein are broken into small fragments or single amino acids. Figure 6-8 summarizes the whole process.

Figure 6•7

A DIPEPTIDE AND TRIPEPTIDE

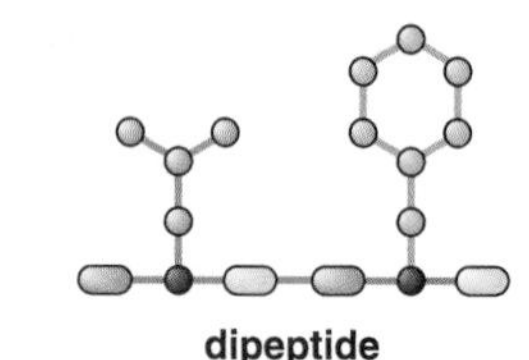

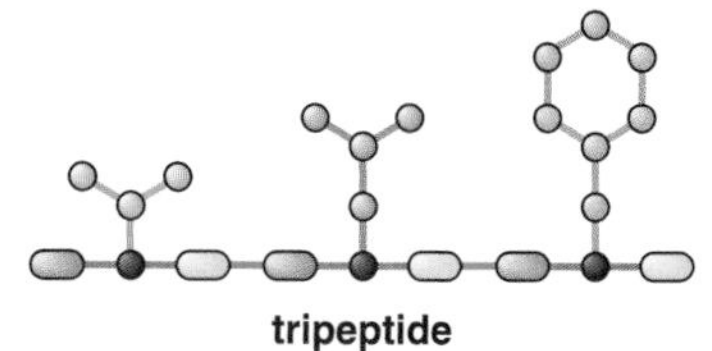

Digestion of protein involves denaturation by stomach acid, then enzymatic digestion in the stomach and small intestine to amino acids, dipeptides, and tripeptides.

After Protein Is Digested, What Happens to Amino Acids?

The cells all along the lining of the small intestine absorb single amino acids. The cells also have enzymes on their surfaces that split most of the dipeptides and tripeptides into single amino acids, and the cells absorb them, too. Some dipeptides and tripeptides are also absorbed into the cells where they are split into amino acids and released with others into the bloodstream. A few larger peptide molecules escape the digestive process altogether and enter the bloodstream intact.[3] Scientists think that these larger particles may act as hormones to regulate body functions and provide the body with information about the environment. The larger molecules may also play a role in food allergy via the immune response.

Figure 6•8

HOW PROTEIN IN FOOD BECOMES AMINO ACIDS IN THE BODY

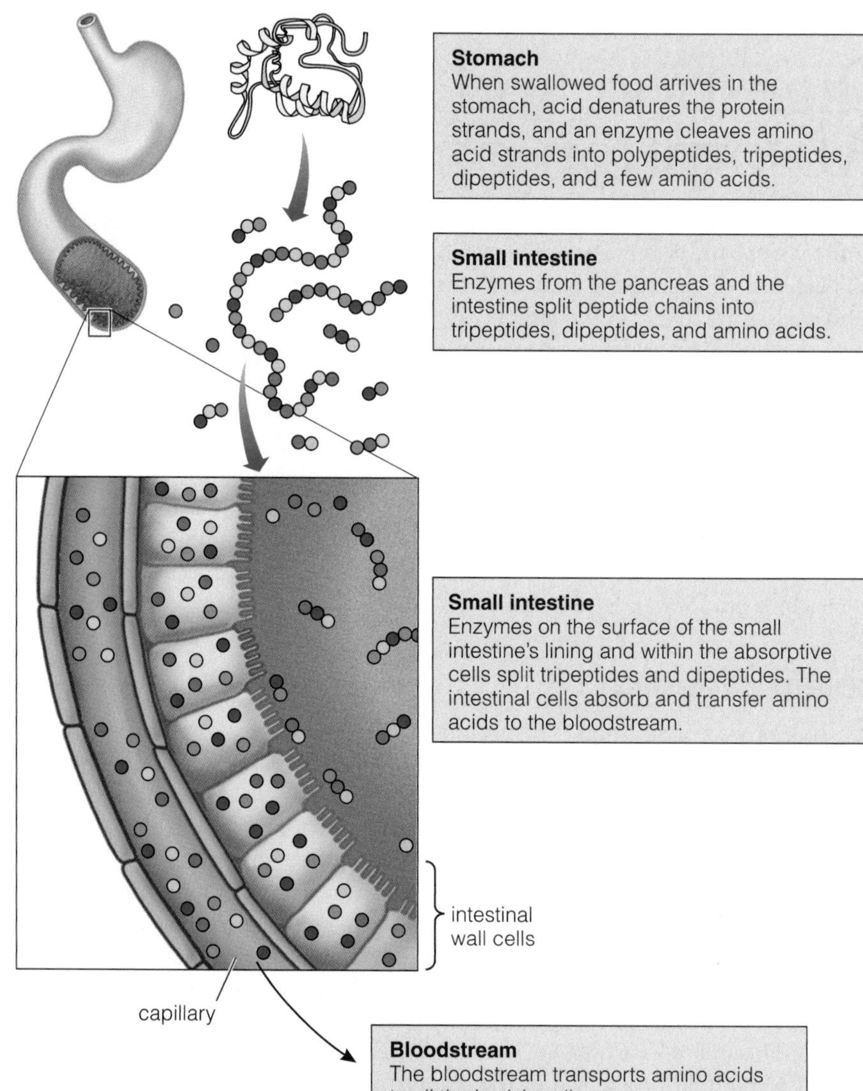

Stomach
When swallowed food arrives in the stomach, acid denatures the protein strands, and an enzyme cleaves amino acid strands into polypeptides, tripeptides, dipeptides, and a few amino acids.
Small intestine
Enzymes from the pancreas and the intestine split peptide chains into tripeptides, dipeptides, and amino acids.
Small intestine
Enzymes on the surface of the small intestine's lining and within the absorptive cells split tripeptides and dipeptides. The intestinal cells absorb and transfer amino acids to the bloodstream.
intestinal wall cells
capillary
Bloodstream
The bloodstream transports amino acids to all the body's cells.

The cells of the small intestine possess separate sites for absorbing different types of amino acids. Amino acids of the same type compete for the same absorption sites. Consequently, when a person ingests a large dose of any single amino acid, that amino acid may limit absorption of others of its general type. The Consumer Corner later in this chapter cautions against taking single amino acids as supplements, partly for this reason.

Once they are circulating in the bloodstream, amino acids are carried to the liver where they may be used or released into the blood to be taken up by other cells of the body. The body cells then use the amino acids to make proteins, either for their own use or for secretion into lymph or blood, or for other uses. When necessary, the body cells can also use amino acids for energy.

The cells of the small intestine complete digestion, absorb amino acids and some larger peptides, and release them into the bloodstream.

hormones chemical messengers secreted by a number of body organs in response to conditions that require regulation. Each hormone affects a specific organ or tissue and elicits a specific response. Also defined in Chapter 3.

The Roles of Proteins in the Body

Only a sampling of the many roles proteins play can be described here, but these illustrate the versatility, uniqueness, and importance of proteins in the body. No wonder their discoverers called proteins the primary material of life.

Supporting Growth and Maintenance

Amino acids must be continuously available to build the proteins of new tissue. The new tissue may be in an embryo; in a growing child; in new blood needed to replace blood lost in burns, hemorrhage, or surgery; in the scar tissue that heals wounds; or in new hair and nails.

Less obvious is the protein that helps to replace worn-out cells and internal cell structures. Each of your millions of red blood cells lives for only three or four months. Then it must be replaced by a new cell produced by the bone marrow. The millions of cells lining your intestinal tract live for only three days; they are constantly being shed and replaced. The cells of your skin die and rub off, and new ones grow from underneath. Nearly all cells arise, live, and die this way, and while they are living, they constantly make and break down their proteins. In addition, cells must continuously replace their own internal working proteins as old ones wear out. Amino acids from food support all the new growth and maintenance of cells and the making of the working parts within them.

The body needs amino acids to grow new cells and to replace worn-out ones.

Building Enzymes, Hormones, and Other Compounds

Enzymes are among the most important of the proteins formed in living cells. Thousands of enzymes reside inside a single cell, each one a catalyst that facilitates a specific chemical reaction. Figure 6-9 (next page) shows how a hypothetical enzyme works.

The body's many **hormones** are messenger molecules, and some are made from amino acids. (Recall from Chapter 5 that some hormones are made from lipids.) Various body glands release hormones in response to changes in the internal environment; the hormones then elicit the responses necessary to restore normal conditions. Among the hormones made of amino acids is the thyroid hormone, which regulates the body's metabolism. An opposing pair of hormones, insulin and glucagon, maintain blood glucose levels, as described in Chapter 4. For interest, Figure 6-10 shows how

antibodies (AN-te-bod-ees) large proteins of the blood, produced by the immune system in response to an invasion of the body by foreign substances (antigens). Antibodies combine with and inactivate the antigens. Also defined in Chapter 3.

immunity specific disease resistance, derived from the immune system's memory of prior exposure to specific disease agents and its ability to mount a swift defense against them.

Figure 6•9

ENZYME ACTION

Enzymes are catalysts: they speed up reactions that would happen anyway, but much more slowly. This enzyme works by positioning two compounds, A and B, so that the reaction between them will be especially likely to take place.

Compounds A and B are attracted to the enzyme's active site and park there for a moment in the exact position that makes the reaction between them most likely to occur. They react by bonding together and leave the enzyme as the new compound, AB.

A single enzyme can facilitate several hundred such synthetic reactions in a second. Other enzymes break compounds apart into two or more products or rearrange the atoms in one compound to make another one.

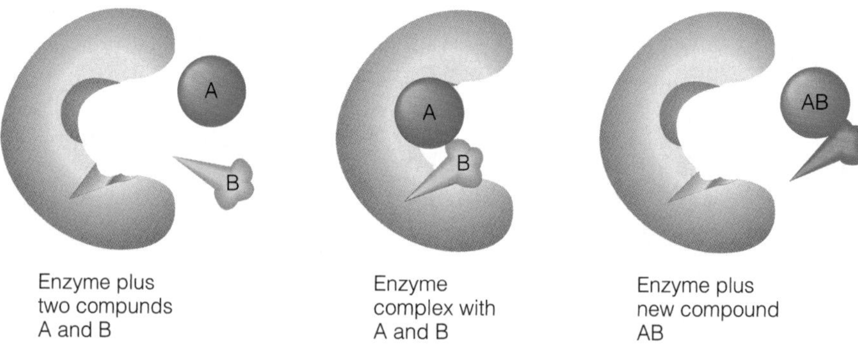

Controversy 13 presents some details about tryptophan's relationship to serotonin; the vitamin niacin is discussed in Chapter 7.

many amino acids are linked in sequence to form human insulin. It also shows how certain side groups attract one another to complete the insulin molecule and make it functional.

In addition to serving as building blocks for proteins, amino acids also perform tasks in the body as amino acids. For example, the amino acid tyrosine forms parts of the chemical messengers epinephrine and norepinephrine, which relay nervous system messages throughout the body. The body also uses tyrosine to make the brown pigment melanin, which is responsible for skin, hair, and eye color. It also becomes the hormone thyroxine, which helps to regulate the body's metabolic rate. The amino acid tryptophan serves as starting material for the neurotransmitter serotonin and the vitamin niacin.

The body makes enzymes, hormones, and chemical messengers of the nervous system from its amino acids.

Figure 6•10

AMINO ACID SEQUENCE OF HUMAN INSULIN

This picture shows a refinement of protein structure not mentioned in the text. The amino acid cysteine (cys) has a sulfur-containing side group. The sulfur groups on two cysteine molecules can bond together, creating a bridge between two protein strands or two parts of the same strand. Insulin contains three such bridges.

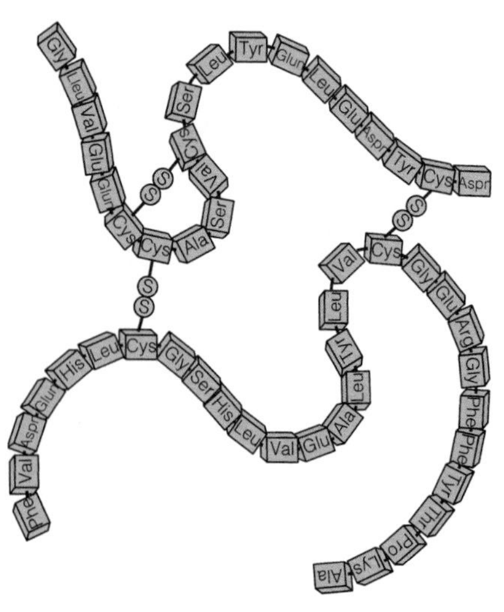

Building Antibodies

Of all the proteins in living organisms, the **antibodies** best demonstrate that proteins are specific to one organism. Antibodies recognize every protein that belongs in "their" body and leave those proteins alone, but they attack foreign particles (usually proteins) that invade that body. The foreign protein may be part of a bacterium, a virus, or a toxin, or it may be present in a food that causes allergy. The body, upon recognizing that it has been invaded, manufactures antibodies specially designed to inactivate the foreign protein.

Each antibody is designed to destroy one specific invader. An antibody active against one strain of influenza is of no help to a person ill with another strain. Once the body has learned to make a particular antibody, it remembers. The next time the body encounters that same invader, it destroys the invader even more rapidly. In other words, the body develops **immunity** to the invader. This molecular memory underlies the principle of immunizations, injections of drugs made from destroyed and inactivated microbes or their products that activate the body's immune defenses. Some immunities are lifelong; others, such as that to tetanus, must be "boosted" at intervals.

Antibodies are formed from amino acids to defend against foreign proteins and other foreign substances within the body.

Maintaining Fluid and Electrolyte Balance

Proteins help to maintain the **fluid and electrolyte balance** by regulating the quantity of fluids in the compartments of the body. To remain alive, cells must contain a constant amount of fluid. Too much can cause them to rupture; too little makes them unable to function. Although water can diffuse freely into and out of cells, proteins cannot; and proteins attract water. By maintaining stores of internal proteins and also of some minerals, cells retain the fluid they need. The cells also keep the fluid volume constant in the spaces between them by secreting proteins (and minerals) into those spaces. Should this system begin to fail, too much fluid would collect outside the cells, causing **edema.**

Not only is the quantity of the body fluids vital to life, but so is their composition. Transport proteins in the membranes of cells maintain this composition by continuously transferring substances into and out of cells (see Figure 6-11). For example, sodium is concentrated outside the cells, and potassium is concentrated inside. A disturbance of this balance can impair the action of the heart, lungs, and brain, triggering a major medical emergency. Cell proteins avert such a disaster by holding fluids and electrolytes in their proper chambers.

Proteins help to regulate the body's electrolytes and fluids.

Maintaining Acid-Base Balance

Normal processes of the body continually produce **acids** and their opposite, **bases,** that must be carried by the blood to the organs of excretion. The blood must do this without allowing its own **acid-base balance** to be affected. This feat is another trick of the blood proteins, which act as **buffers** to maintain the blood's normal pH. The buffers pick up hydrogens (acid) when there are too many and release them again when there are too few. The secret is that negatively charged side chains of amino acids can accommodate additional hydrogens, which are positively charged.

Blood pH is one of the most rigidly controlled conditions in the body. If blood pH changes too much, **acidosis** or the opposite basic condition, **alkalosis,** can cause coma or death. These conditions are hazardous because of their effect on proteins. When the proteins' buffering capacity is filled—that is, when they have taken on all the acid hydrogens they can accommodate—additional acid pulls them out of shape, denaturing them and disrupting many body processes.

fluid and electrolyte balance the distribution of fluid and dissolved particles among body compartments (see also Chapter 8).

edema (eh-DEEM-uh) swelling of body tissue caused by leakage of fluid from the blood vessels; seen in protein deficiency (among other conditions).

acids compounds that release hydrogens in a watery solution.

bases compounds that accept hydrogens from solutions.

acid-base balance equilibrium between acid and base concentrations in the body fluids.

buffers compounds that help keep a solution's acidity or alkalinity constant.

acidosis (acid-DOH-sis) blood acidity above normal, indicating excess acid (*osis* means "too much in the blood").

alkalosis (al-kah-LOH-sis) blood alkalinity above normal (*alka* means "base"; *osis* means "too much in the blood").

Figure 6•11

PROTEINS TRANSPORT SUBSTANCES INTO AND OUT OF CELLS

A transport protein within a cell membrane acts as a sort of two-door passageway—substances enter on one side and are released on the other, but the protein never leaves the membrane. The protein differs from a simple passageway in that it actively escorts the substances in and out of cells; therefore, this form of transport is often called active transport.

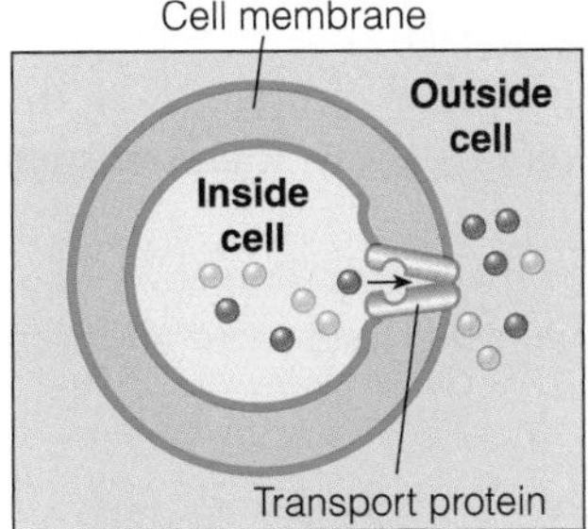

Molecule enters protein from inside cell.

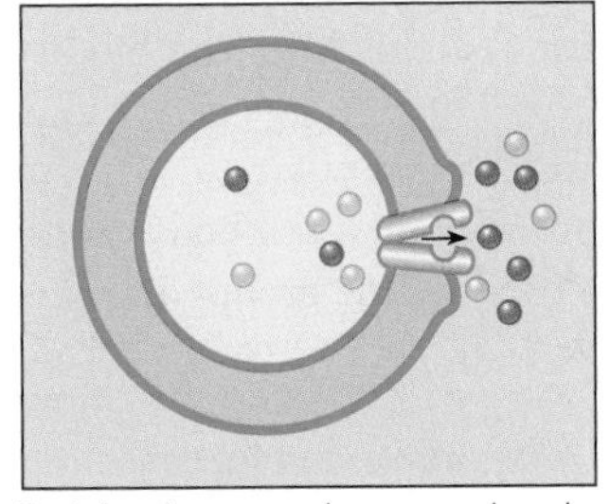
Protein changes shape; molecule exits protein outside the cell.

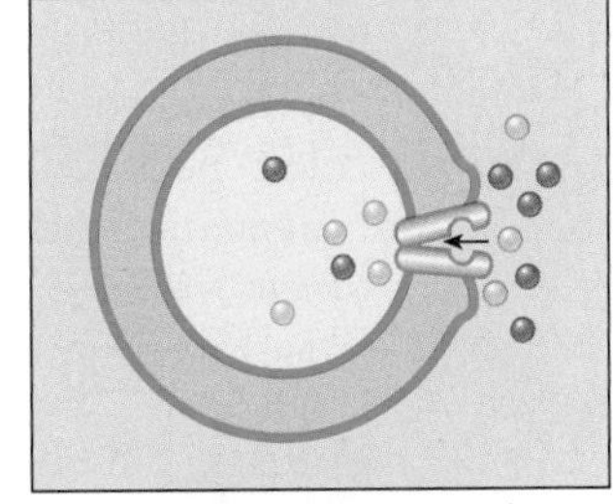
Molecule enters protein from outside cell.

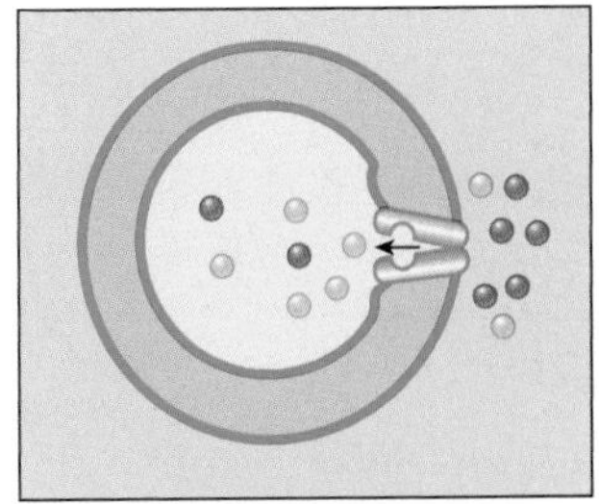
Molecule exits protein; proper balance restored.

urea (yoo-REE-uh) the principal nitrogen-excretion product of metabolism; generated mostly by removal of amine groups from unneeded amino acids or from amino acids being sacrificed to a need for energy.

Table 6•1

SUMMARY OF FUNCTIONS OF PROTEINS

- **growth and maintenance** Proteins serve as building materials for growth and repair of body tissues.
- **enzymes** Proteins facilitate needed chemical reactions.
- **hormones** Proteins regulate body processes. Some hormones are proteins or are made from amino acids.
- **antibodies** Proteins form the immune system molecules that fight diseases.
- **fluid and electrolyte balance** Proteins help to maintain the fluid and mineral composition of various body fluids.
- **acid-base balance** Proteins help maintain the acid-base balance of various body fluids by acting as buffers.
- **energy** Proteins provide some fuel for the body's energy needs.
- **transportation** Proteins help transport needed substances, such as lipids, minerals, and oxygen, around the body.
- **blood clotting** Proteins provide the netting on which blood clots are built.
- **structural components** Proteins form integral parts of most body structures such as skin, tendons, ligaments, membranes, muscles, organs, and bones.

Table 6-1 sums up the functions of protein we have discussed and adds several others.

Proteins buffer the blood against excess acidity or alkalinity.

Providing Energy

Only protein can perform all the functions just described, but protein will be surrendered to provide energy if need be. Under normal conditions, protein provides just a little of the body's energy; under conditions of energy or carbohydrate deficiency, protein use speeds up. The body must have energy to live from moment to moment, so obtaining that energy is a top priority.

When amino acids are degraded for energy, their amine groups are stripped off and used elsewhere or are incorporated by the liver into **urea** and sent to the kidneys for excretion in the urine. The fragments that remain are composed of carbon, hydrogen, and oxygen, as are carbohydrate and fat, and can be used to build those substances or can be metabolized like them.

Not only can amino acids supply energy, but many of them can be converted to glucose, as fatty acids can never be. Thus, if the need arises, protein can help to maintain a steady blood glucose level and serve the glucose need of the brain.

The similarities and differences of the three energy-yielding nutrients should now be clear. Carbohydrate offers energy; fat offers concentrated energy; and protein, if needed, can offer energy plus nitrogen (see Figure 6-12).

Glucose is stored as glycogen and fat as triglycerides, but there is no specialized protein energy-storage compound. Body protein is available only as the active working molecular and structural components of the tissues. When protein-sparing energy from carbohydrate and fat is lacking and the need becomes urgent, the body must dismantle its tissue proteins to obtain amino acids for energy. Each protein is taken in its own time: first, small proteins from the blood and liver; then, proteins from the muscles and other organs. Thus, energy deficiency (starvation) always incurs wasting of lean body tissue as well as loss of fat.

If amino acids are oversupplied, the body cannot store them. It has no choice but to remove and excrete their amine groups and then convert the residues to glycogen or fat for energy storage.

key point

When the carbohydrate and fat consumed are insufficient to meet the body's energy need, food protein and body protein are sacrificed to supply energy. The nitrogen part is removed from each amino acid, and the resulting fragment is oxidized for energy.

Figure 6•12

THREE DIFFERENT ENERGY SOURCES

Carbohydrate offers energy; fat offers concentrated energy; and protein, if necessary, can offer energy plus nitrogen. The compounds at the left yield the 2-carbon fragments shown at the right. These fragments oxidize quickly in the presence of oxygen to yield carbon dioxide, water, and energy.

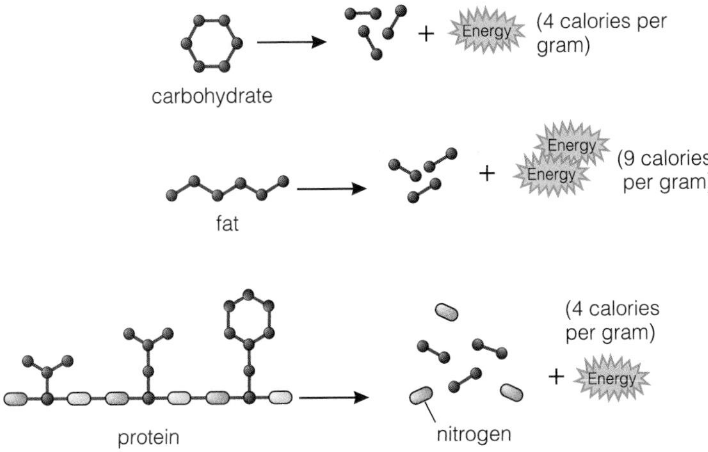

The Fate of an Amino Acid

To review the body's handling of amino acids, let us follow the fate of an amino acid that was originally part of a protein-containing food. When the amino acid arrives in a cell, it can be used in one of several ways, depending on the cell's needs at the time:

- The amino acid can be used as is to build part of a growing protein.
- The amino acid can be altered somewhat to make another needed compound, such as the vitamin niacin.
- The cell can dismantle the amino acid in order to use its amine group to build a different amino acid. The remainder can be used for fuel or, if not needed, converted to glucose or fat.
- In a cell that is starved for energy but has no glucose or fatty acids, the cell strips the amino acid of its amine group (the nitrogen part) and uses the remainder of its structure for energy. The amine group is excreted from the body in the urine.
- When the body has a surplus of amino acids and energy-yielding nutrients, the cell takes the amino acid apart, excretes the amine group, converts the rest to fat, and then stores the fat in the fat cells.

Amino acids in a cell can be:
- Used to build protein.
- Converted to other amino acids or small nitrogen-containing compounds.

Stripped of their nitrogen, amino acids can be:
- Converted to glucose.
- Burned as fuel.
- Stored as fat.

In the last two cases, when not used to build protein or make other nitrogen-containing compounds, amino acids are "wasted" in a sense. This wasting occurs under any of four conditions:

1. When the body does not have enough energy from other sources.
2. When the body has more protein than it needs.
3. When the body has too much of any single amino acid, such as from a supplement.
4. When the diet supplies protein of low quality, with too few essential amino acids, as described in the next section.

More about the effects of too little energy from food (fasting) in Chapter 9.

Amino acids are wasted when:
- Energy is lacking.
- Protein is overabundant.
- An amino acid is oversupplied in supplement form.
- The quality of the diet's protein is poor (too few essential amino acids).

To prevent the wasting of dietary protein and permit the synthesis of needed body protein, three conditions must be met. First, the dietary protein must be adequate in quantity. Second, it must supply all essential amino acids in the proper amounts. Third, enough energy-yielding carbohydrate and fat must be present to permit the dietary protein to be used as such.

key point *Amino acids can be metabolized to protein, nitrogen plus energy, glucose, or fat. They will be metabolized to protein only if sufficient energy is present from other sources. The diet should supply all essential amino acids and an adequate quantity of protein.*

Protein and Amino Acid Supplements

WHY DO PEOPLE take protein or amino acid supplements? Athletes take them to build muscle. Dieters take them to speed up the process of losing weight. Some consumers believe the products will cure herpes virus infections, induce restful sleep, or relieve pain or depression. Do protein and amino acid supplements really do any of these things? Probably not. Are they safe? Not always.

Enthusiastic popular reports have led to widespread use of certain amino acids. One is lysine, touted to prevent or relieve the infections that cause herpes sores on the mouth or genital organs. Lysine does not cure herpes infections, however. Whether it reduces outbreaks or even whether it is safe is unknown because scientific studies are lacking.[1] The branched-chain amino acids leucine, isoleucine, and valine have become popular with athletes hoping to feed their muscles extra energy, but the value of these potions also remains unclear (see Controversy 10). Tryptophan, advertised to relieve pain, depression, and insomnia, has some interesting effects with respect to pain and sleep in responsive individuals, as Controversy 13 explains later.

One problem with taking supplements of single amino acids, though, is that the body is designed to handle whole proteins. It breaks them into manageable pieces (dipeptides and tripeptides), then splits these a few at a time, simultaneously releasing them into the blood. This slow bit-by-bit assimilation is ideal because groups of chemically similar amino acids compete for the carriers that absorb them into the blood. An excess of one amino acid can tie up a carrier and prevent the absorption of another similar amino acid; as a result, needed amino acids may pass through the body unabsorbed. The result is a deficiency. The human body evolved without encountering the unbalanced arrays of highly concentrated amino acids found in supplements and therefore lacks the equipment to handle them.

When used as a replacement for foods, especially in weight-loss diets, protein supplements can be downright dangerous. The "liquid protein" diet, advocated some years ago for weight loss, caused the deaths of many users. Even the physician-supervised "protein-sparing" fast, also based on liquid protein, has caused abnormal heart rhythms (Chapter 9 has more about the effects of fasting).

Other problems can arise as well. In the not-too-distant past, people who elected to take manufactured tryptophan supplements developed a blood disorder (EMS, short for *eosinophilia-myalgia syndrome*), and at least 15 of the supplement takers died. Contaminants in the supplement were determined to be the cause of the disease, and the Food and Drug Administration (FDA) recalled tryptophan supplements and formulas to which it was added. Impurities of the same sort are still detected in some tryptophan-containing products currently on the market.[2] The lesson is that supplements of all sorts go to market in a largely untested state, and consumers become unsuspecting experimental subjects, sometimes with dire results.

Currently, the DRI committee is working toward establishing DRI values for protein and amino acids. With next to no safety research in existence, however, the panel may be unable to set Tolerable Upper Intake Levels for supplemental doses. Until research becomes available, no level of amino acid supplementation can be assumed safe for all people. Those who are growing or have altered metabolism are especially likely to suffer harm from self-prescribed amino acid supplements. They include:

- All women of childbearing age.
- Pregnant or lactating women.
- Infants, children, and adolescents.
- Elderly people.
- People with inborn errors of metabolism that affect their bodies' handling of amino acids.
- Smokers.
- People on low-protein diets.
- People with chronic or acute mental or physical illnesses.

A recent review of the literature has concluded that not enough research exists to support recommending long-term consumption of amino acid supplements by healthy people.[3] The study's author calls for scientific investigation to determine whether high doses of supplemental amino acids may cause harm similar to that observed with high-dose amino acid administration or with inborn errors of metabolism. A few examples are offered in Table 6–2. Whether the average supplement taker is at risk of suffering these effects is unknown, but the warning is clear; we have much yet to learn about the safety of amino acid supplements.

Many chapters of this book present evidence on purified nutrients added to

Table 6•2

REPORTED ADVERSE EFFECTS OF AMINO ACID IMBALANCES

Mental retardation is known to result from genetic abnormalities, such as PKU, that cause high levels of amino acids to accumulate in the blood and tissues. Other effects listed were reported in association with amino acid administration.

Amino Acid	Effects Reported in Scientific Literature
Arginine	Low blood pressure, increased tumor growth, increased blood acidity, increased blood potassium, heart failure
Aspartic acid, glycine	Nerve toxicity
Cysteine	Increased blood cholesterol, fatty liver, nerve toxicity
Glutamate[a]	Nerve toxicity, glutamate sensitivity syndrome (also called *Chinese restaurant syndrome*)
Histidine	Increased blood lipids
Isoleucine,[b] valine[b]	Increased blood ammonia, mental retardation
Leucine[b]	Hypoglycemia, increased blood ammonia, mental retardation, reduced blood levels of other branched-chain amino acids
Methionine	Increased homocysteine in the blood (a compound related to heart disease)
Phenylalanine	Mental retardation, worsened symptoms of schizophrenia
Tryptophan	Altered brain serotonin, toxicity from impurities
Tyrosine	Eye and skin lesions, mental retardation

[a]As monosodium glutamate, a sodium salt of glutamic acid.
[b]Branched-chain amino acids.
SOURCE: Data from P. J. Garlick, Assessment of the safety of amino acids, *Journal of Nutrition* 131 (2001): S2556–S2561.

PKU was described in Controversy 4.

foods or taken singly. The Consumer Corner in Chapter 4 showed that a nutritionally inferior food (refined bread) enriched with a few added nutrients is still deficient in many others. The Consumer Corner in Chapter 5 showed that artificial fats can have side effects. The same is true of amino acids. Even with all that we know about science, it is hard to improve on nature.

legumes (leg-GOOMS, LEG-yooms) plants of the bean, pea, and lentil family that have roots with nodules containing special bacteria. These bacteria can trap nitrogen from the air in the soil and make it into compounds that become part of the plant's seeds. The seeds are rich in protein compared with those of most other plant foods. Also defined in Chapter 1.

amino acid pools amino acids dissolved in cellular fluid that provide cells with ready raw materials from which to build new proteins or other molecules.

limiting amino acid an essential amino acid that is present in dietary protein in an insufficient amount, thereby limiting the body's ability to build protein.

Food Proteins: Quality, Use, and Need

The body's response to proteins depends on many factors: the body's state of health, the food source of the protein, its digestibility, the other nutrients taken with it, and its amino acid assortment. To know whether, say, 30 grams of a particular protein is enough to meet a person's daily needs, one must consider the effects of these other factors on the body's use of the protein.

Regarding a person's state of health, malnutrition or infection may greatly increase the need for protein while making it hard to eat even normal amounts of food. In malnutrition, secretion of digestive enzymes slows as the tract's lining degenerates, impairing protein digestion and absorption. When infection is present, extra protein is needed for enhanced immune functions.

Which Kinds of Protein-Rich Foods Are Easiest to Digest and Use?

Some concern exists about the formation of carcinogens in foods when meats are exposed to open flame—see Chapter 11.

The digestibility of protein varies from food to food and affects protein quality profoundly. The protein of oats, for example, is less digestible than that of eggs. In general, amino acids from animal proteins are most easily digested and absorbed (over 90 percent). Those from **legumes** are next (about 80 percent). Those from grains and other plant foods vary (from 60 to 90 percent). Cooking with moist heat improves protein digestibility, whereas dry heat methods can impair it.[4]

As for taking the other nutrients with protein, the need for carbohydrate and fat has already been emphasized. To be used efficiently, protein must be accompanied by the full array of vitamins and minerals. In addition, as the next section makes clear, a plentiful supply of all the needed amino acids is required.

key point *The body's use of a protein depends in part on the user's health and on the protein's digestibility. To be used efficiently, protein should be accompanied by all the other nutrients.*

Protein Quality

The quality of a food protein depends largely on its amino acid content. In making their own proteins, the cells need a full array of amino acids from food, from their own **amino acid pools,** or from both. If a nonessential amino acid (that is, one the cell can make) is unavailable from food, the cell synthesizes it and continues attaching amino acids to the protein strands being manufactured. If an essential amino acid (one the cell cannot make) is missing from food, the cell begins to adjust its activities almost immediately. Within a single day of restricted essential amino acid intake, the cells begin to conserve it by limiting the breakdown of their working proteins and by reducing their use of amino acids for fuel.

Cooking with moist heat improves protein digestibility, whereas frying makes protein harder to digest.

© Japack Company/CORBIS

Limiting Amino Acids Can Limit Protein Synthesis The measures just described help the cells to channel the available **limiting amino acid** to its wisest use: making new proteins. Even so, the normally fast rate of protein synthesis slows to a crawl, as the cells make do with the proteins on hand. When the limiting amino acid once again becomes available in abundance, the cells resume their normal protein-related activities. If the shortage becomes chronic, however, the cells begin to break down their protein-making machinery. Consequently, when protein intakes become adequate, protein synthesis lags behind until the needed machinery can be rebuilt. Meanwhile, the cells function less and less effectively as their proteins wear out and are only partially replaced.

Thus, a diet that is short in any of the essential amino acids limits protein synthesis. An earlier analogy likened amino acids to letters of the alphabet. To be meaningful, words must contain all the right letters. For example, a print shop that has no letter "N" cannot make personalized stationery for Jana Johnson. No matter how many J's, A's, O's, H's, and S's are in the printer's possession, they cannot replace the missing N's. Likewise, in building a protein molecule, no amino acid can fill another's spot. If a cell that is building a protein cannot find a needed amino acid, synthesis stops, and the partial protein is released.

Partially completed proteins are not held for completion at a later time when the diet may improve. Rather, they are dismantled, and the component amino acids are returned to the circulation to be made available to other cells. If they are not soon inserted into protein, their amine groups are removed and excreted, and the residues are used for other purposes. The need that prompted the call for that particular protein will not be met. Since the other amino acids are wasted, the amine groups are excreted, and the body cannot resynthesize the amino acids later.

mutual supplementation the strategy of combining two incomplete protein sources so that the amino acids in one food make up for those lacking in the other food. Such protein combinations are sometimes called *complementary proteins*.

complementary proteins two or more proteins whose amino acid assortments complement each other in such a way that the essential amino acids missing from one are supplied by the other.

Mutual Supplementation and Complementary Proteins It follows that if all the essential amino acids are not consumed in proportion to the body's needs, the body's pools of essential amino acids will dwindle until body organs are compromised. Consuming the essential amino acids presents no problem for people who regularly eat proteins containing ample amounts of all of the essential amino acids such as those of meat, fish, poultry, cheese, eggs, milk, and many soybean products. An equally sound choice is to eat a combination of foods from plants so that amino acids that may be low in some foods are supplied by the others. In this strategy, called **mutual supplementation,** two protein-rich foods are combined to yield **complementary proteins** (see Table 6-3), or proteins containing all the essential amino acids in amounts sufficient to support health. This concept is illustrated in Figure 6-13 on the next page. The two proteins need not even be eaten together, so long as the day's meals supply them both, and the diet provides enough energy and total protein from a variety of sources.[5]

Concern about the quality of individual food proteins is of only theoretical interest in settings where food is abundant. Most people in the United States and Canada eat a variety of nutritious foods to meet their energy needs—not just, say, cookies, potato chips, or alcoholic beverages. They would find it next to impossible *not* to meet their protein requirements, even if they were to eat no meat, fish, poultry, eggs, cheese, or soy products.

Table 6•3

COMPLEMENTARY PROTEIN COMBINATIONS

Combine foods from two or more of these categories to obtain complete protein.

Grains	Seeds and Nuts	Legumes	Vegetables
Barley	Cashews	Dried beans	Broccoli
Bulgur	Nut butters	Dried lentils	Leafy greens
Cornmeal	Other nuts	Dried peas	Other vegetables
Oats	Sesame seeds	Peanuts	
Pasta	Sunflower seeds	Soy products	
Rice	Walnuts		
Whole-grain breads			

Just as each letter of the alphabet is important in forming whole words, each amino acid must be available to build finished proteins.

protein digestibility–corrected amino acid score (PDCAAS) a measuring tool used to determine protein quality. The PDCAAS reflects a protein's digestibility as well as the proportions of amino acids that it provides.

protein efficiency ratio (PER) a measure of protein quality assessed by determining how well a given protein supports weight gain in growing rats. The PER is used to judge the quality of protein in infant formulas and baby foods.

nitrogen balance the amount of nitrogen consumed compared with the amount excreted in a given time period.

Protein quality can make the difference between health and disease when food energy intake is limited (where malnutrition is widespread) or when the selection of foods available is severely limited (where a single food such as potatoes or rice provides 90 percent of the calories). In these cases, the primary food source of protein must be checked since its quality is crucial.

key point *A protein's amino acid assortment greatly influences its usefulness to the body. Proteins lacking needed amino acids can be used only if those amino acids are present from other sources.*

Measuring Protein Quality

Researchers have developed many methods of evaluating the quality of food protein. The most important one for consumers is the **protein digestibility–corrected amino acid score,** or **PDCAAS.** The protein values that U.S. consumers read on food labels are based on the PDCAAS. Another measure of protein quality, the **protein efficiency ratio (PER),** is used for measuring the protein quality of infant formulas and baby foods.

The PDCAAS for protein digestibility is important. Simple measures of the total protein in foods are not useful by themselves—even animal hair or hooves would receive a top score by those measures alone. On the PDCAAS scale of 100 to 0, with 100 representing protein sources that are most readily digested and most perfectly balanced for meeting human needs, egg white, ground beef, chicken products, fat-free milk, and tuna fish all score 100. Soybean protein isn't far behind at 94. Most legumes rank in the 60s and 50s. The wheat protein, gluten, formed during bread making, ranks 25. Something interesting happens when pea flour (67) and whole-wheat flour (40) are combined: the score for the resulting flour is 82. Why? Mutual supplementation is at work. In deciding between peanut butter and chili in the grocery store, you may have no use for the PDCAAS method, but scientists who must establish adequacy of protein sources for human health worldwide rely on it heavily.[6] More relevant to the average well-fed North American, however, is the protein RDA, discussed next.

The quality of a protein is measured by its amino acids, by its digestibility, or by how well it supports growth.

How Much Protein Do People Really Need?

The 1989 RDA for protein stands as the current recommendation for protein intake. It is designed to cover the need to replace protein-containing tissue that people lose and wear out every day. Therefore it depends on body size: larger people have a higher protein need. The recommendation is also adjusted to cover additional needs for building new tissue and so is higher for growing children and pregnant and lactating women. The Canadian recommendation for protein is similar and is based on similar assumptions. Table 6-4 reviews the recommendations concerning dietary protein. These ensure that the body is well supplied with the protein it needs.

Underlying the protein recommendation are **nitrogen balance** studies, which compare nitrogen lost by excretion with nitrogen eaten in food. In healthy adults, nitrogen-in (consumed) must equal nitrogen-out (excreted). Scientists measure the body's daily nitrogen losses in urine, feces, sweat, and skin under controlled conditions and then estimate the amount of protein needed to replace these losses.*

Under normal circumstances, healthy adults are in nitrogen equilibrium, or zero balance; that is, they have the same amount of total protein in their bodies at all times. When nitrogen-in exceeds nitrogen-out, people are said to be in positive nitrogen balance; somewhere in their bodies more proteins are being built than are being broken down and lost. When nitrogen-in is less than nitrogen-out, people are said to

Figure 6•13

AN EXAMPLE OF MUTUAL SUPPLEMENTATION

In general, legumes provide plenty of the amino acids isoleucine (Ile) and lysine (Lys), but fall short in methionine (Met) and tryptophan (Trp). Grains have the opposite strengths and weaknesses, making them a perfect match for legumes.

	Ile	Lys	Met	Trp
Legumes	■	■		
Grains			■	■
Together	■	■	■	■

*The average protein is 16 percent nitrogen by weight; that is, each 100 grams of protein contain 16 grams of nitrogen. For an estimate of the protein's weight, multiply the nitrogen's weight by 6.25.

Table 6•4

RECOMMENDATIONS CONCERNING INTAKES OF PROTEIN FOR ADULTS[a]

1989 Recommended Dietary Allowance (RDA)
- 0.8 gram protein per kilogram body weight per day.

Dietary Guidelines
- Every day eat 2 to 3 servings to total 4 to 9 ounces of cooked dry beans and peas, lean beef or other lean meats, poultry without the skin, fish and shellfish, and occasionally eggs and organ meats.
- Every day choose 2 to 3 servings of low-fat or fat-free milk, yogurt, or cheese.
- Eat a variety of foods to provide small amounts of protein from other sources.

Daily Values[b]
- 50 grams protein per day.

World Health Organization
- Lower limit: 10% of total calories from protein.
- Upper limit: 15% of total calories from protein.

[a]Protein recommendations for infants, children, and pregnant and lactating women are higher.
[b]The Daily Value is for a 2,000-calorie diet.

be in negative nitrogen balance; they are losing protein. Figure 6-14 illustrates these different states.

Growing children add new blood, bone, and muscle cells to their bodies every day, so children must have more protein, and therefore more nitrogen, in their bodies at the end of each day than they had at the beginning. A growing child is therefore in positive nitrogen balance. Similarly, when a woman is pregnant, she must be in positive nitrogen balance until after the birth when she once again reaches equilibrium.

Negative nitrogen balance occurs when muscle or other protein tissue is broken down and lost.[7] Illness or injury triggers the release of powerful messengers that signal the body to break down some of the less vital proteins, such as those of the skin.* This action floods the blood with amino acids needed for building antibodies to fight the illness and for energy to fuel the body's defenses. The result is negative nitrogen

*The messengers are cytokines.

Figure 6•14

NITROGEN BALANCE

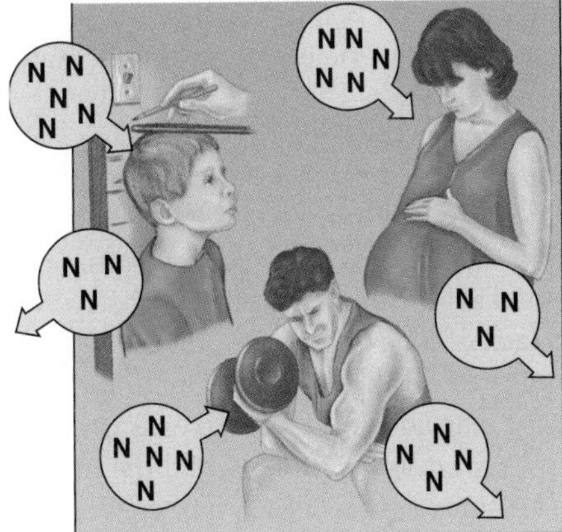

Positive Nitrogen Balance
These people, a growing child, a person building muscle, and a pregnant woman, are all retaining more nitrogen than they are excreting.

Nitrogen Equilibrium
These people, a healthy college student and a young retiree, are in nitrogen equilibrium.

Negative Nitrogen Balance
These people, an astronaut and a surgery patient, are losing more nitrogen than they are taking in.

protein-energy malnutrition (PEM) the world's most widespread malnutrition problem, including both marasmus and kwashiorkor and states in which they overlap; also called *protein-calorie malnutrition (PCM)*.

hunger the physiological craving for food; the progressive discomfort, illness, and pain resulting from the lack of food. (See also Chapters 9 and 15.)

marasmus (ma-RAZ-mus) the calorie-deficiency disease; starvation.

kwashiorkor (kwash-ee-OR-core, kwashee-or-CORE) a disease related to protein malnutrition, with a set of recognizable symptoms, such as edema.

balance. Astronauts, too, experience negative nitrogen balance.[8] In the stress of space flight and with no need to support the body's weight against gravity, the astronauts' muscles waste and weaken.[9] To minimize the inevitable loss of muscle tissue, the astronauts must do special exercises in space.

For healthy adults, the 1989 RDA for protein has been set at 0.8 gram for each kilogram (or 2.2 pounds) of body weight. Athletes need slightly more, but the increased need is well covered by a regular diet (see Chapter 10).[10] For infants and growing children, the protein recommendation, like all nutrient recommendations, is higher per unit of body weight. U.S. protein recommendations set an upper limit for protein intake of no more than twice the 1989 RDA amount, or 1.6 grams per kilogram of body weight per day.

Recommendations for protein intake assume a normal mixed diet, that is, a diet that includes a combination of animal and plant protein. Not all proteins are used with 100 percent efficiency. Accordingly, the recommendation is quite generous, and many healthy people can consume less than this amount and still meet their bodies' protein needs. What this means in terms of food selections is discussed in this chapter's Food Feature.

1989 RDA for protein (adult) = 0.8 g/kg.

To figure your protein need:

1. Find your body weight in pounds.[a]
2. Convert pounds to kilograms (pounds divided by 2.2 equal kilograms).
3. Multiply kilograms by 0.8 to find total grams of protein recommended.

For example:

1. Weight = 110 lb.
2. 110 lb ÷ 2.2 = 50 kg.
3. 50 kg × 0.8 = 40 g.

[a]If your weight falls outside the range marked "Healthy Weight" on the Body Mass Index table (see the inside back cover), use the midpoint of the "healthy weight" range for your height.

key point *Nitrogen balance compares nitrogen excreted from the body with nitrogen ingested in food. The amount of protein needed daily depends on size and stage of growth. The 1989 RDA for adults is 0.8 gram of protein per kilogram of body weight.*

Protein Deficiency and Excess

Protein deficiencies are well known because, together with energy deficiencies, they are the world's leading form of malnutrition. The health effects of too much protein are far less well known. Both deficiency and excess are of concern.

What Happens When People Consume Too Little Protein?

Protein deficiency and energy deficiency go hand in hand. This combination—**protein-energy malnutrition (PEM)**—is the most widespread form of malnutrition in the world today. Over 500 million children face imminent starvation and suffer the effects of severe malnutrition and **hunger.** Most of the 33,000 children who die each day are malnourished.[11] PEM is prevalent in Africa, Central America, South America, the Middle East, and East and Southeast Asia, but developed countries including the United States are not immune to it.

PEM strikes early in childhood, but it endangers many adults as well. Inadequate food intake leads to poor growth in children and to weight loss and wasting in adults. Stunted growth due to PEM is easy to overlook because a small child can look perfectly normal. The small stature of children in impoverished nations was once thought to be a normal adaptation to the limited availability of food; now it is known to be an avoidable failure of growth due to a lack of food during the growing years.[12]

PEM takes two different forms, with some cases exhibiting a combination of the two. In one form, the person is shriveled and lean all over—this disease is called **marasmus.** In the second, a swollen belly and skin rash are present, and the disease is named **kwashiorkor.**[*13] In the combination, some features of each type are present. Marasmus reflects a chronic inadequate food intake and therefore inadequate energy, vitamins, and minerals as well as too little protein. Kwashiorkor may result from severe acute malnutrition, with too little protein to support body functions.[14]

Scant supplies of donated food save some from starvation, but many others go hungry.

United Nations Food and Agriculture Organization/Photo Library

[*]A term gaining acceptance for use in place of kwashiorkor is *hypoalbuminemic-type PEM*.

Marasmus Marasmus occurs most commonly in children from 6 to 18 months of age in overpopulated city slums. Children in impoverished nations subsist on a weak cereal drink with scant energy and protein of low quality; such food can barely sustain life, much less support growth. A starving child often looks like a wizened little old person—just skin and bones.

dysentery (DISS-en-terry) an infection of the digestive tract that causes diarrhea.

Without adequate nutrition, muscles, including the heart muscle, waste and weaken.[15] Brain development is stunted and learning is impaired. Metabolism is so slow that body temperature is subnormal. There is little or no fat under the skin to insulate against cold, and hospital workers have found that children with marasmus need to be wrapped up and kept warm. They also need love because they have often been deprived of parental attention as well as food.

The starving child faces this threat to life by engaging in as little activity as possible—not even crying for food. The body collects all its forces to meet the crisis and so cuts down on any expenditure of protein not needed for the functioning of the heart, lungs, and brain. Growth ceases; the child is no larger at age four than at age two. The skin loses its elasticity and moisture, so it tends to crack; when sores develop, they fail to heal. Digestive enzymes are in short supply, the digestive tract lining deteriorates, and absorption fails. The child can't assimilate what little food is eaten.

Blood proteins, including hemoglobin, are no longer produced, so the child becomes anemic and weak. If a bone breaks, healing is delayed because the protein needed to heal it is lacking.[16] Antibodies to fight off invading bacteria are degraded to provide amino acids for other uses, leaving the child an easy target for infection.[17] Then **dysentery,** an infection of the digestive tract, causes diarrhea, further depleting the body of nutrients, especially minerals. Measles, which might make a healthy child sick for a week or two, kills a child with PEM within two or three days. Infections that occur with malnutrition are responsible for two-thirds of the deaths of young children in developing countries.

Ultimately, marasmus progresses to the point of no return, when the body's machinery for protein synthesis, itself made of protein, has been degraded. At this point, attempts to correct the situation by giving food or protein fail to prevent death. If caught before this time, however, the starvation of a child can be reversed by careful nutrition therapy.[18] The fluid balances are most critical. Diarrhea will have depleted the body's potassium and upset other electrolyte balances. The combination of electrolyte imbalances, anemia, fever, and infections often leads to heart failure and sudden death. Careful correction of fluid and electrolyte balances usually raises the blood pressure and strengthens the heartbeat within a few days. Later, fat-free milk, providing protein and carbohydrate, can safely be given; fat is introduced still later, when body protein is sufficient to provide carriers. Years after PEM is corrected, a child may experience deficits in thinking and achievement in school compared with well-nourished peers.[19]

The term *electrolyte balance* refers to the proper concentrations of salts within the body fluids (see Chapter 8 for details).

The extreme loss of muscle and fat characteristic of marasmus is apparent in this child's "matchstick" arms.

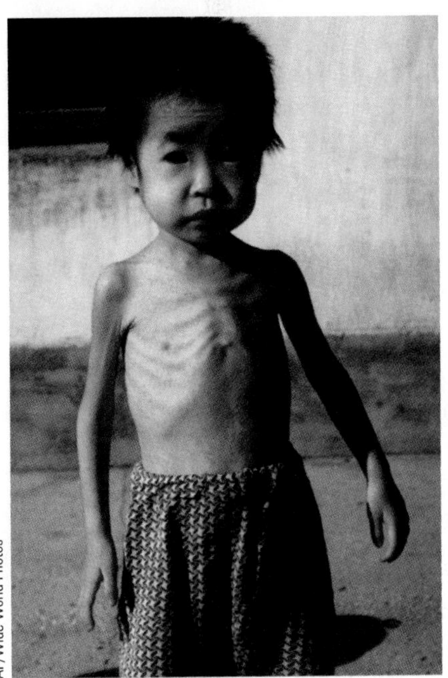

AP/Wide World Photos

Kwashiorkor Kwashiorkor is the Ghanaian name for "the evil spirit that infects the first child when the second child is born." In countries where kwashiorkor is prevalent, each baby is weaned from breast milk as soon as the next one comes along. The older baby no longer receives breast milk, which contains high-quality protein designed perfectly to support growth, but is given a watery cereal with scant protein of low quality. Small wonder the just-weaned child sickens when the new baby arrives. Though rare in the United States, kwashiorkor has recently been diagnosed in more than a dozen children fed ill-conceived vegetarian or "anti-allergy" diets or given a protein-poor "health-food" rice drink instead of cow's milk.[20]

Some kwashiorkor symptoms very much resemble those of marasmus (see Table 6-5 on the next page) but often without severe wasting of body fat. Proteins and hormones that previously maintained fluid balance are now diminished, so fluid leaks out of the blood and accumulates in the belly and legs, causing edema, a distinguishing feature of kwashiorkor. The kwashiorkor victim's belly often bulges with a fatty liver, caused by lack of the protein carriers that transport fat out of the liver. The fatty liver loses some of its ability to clear poisons from the body, prolonging their toxic effects.

Melanin, a brown pigment of hair, skin, and eyes, was mentioned earlier as a product made from tyrosine.

Without sufficient tyrosine to make melanin, the child's hair loses its color; inadequate protein synthesis leaves the skin patchy and scaly; sores fail to heal.

PEM at Home PEM occurs among some groups in the United States and Canada: the poor living on U.S. Indian reservations, in inner cities, and in rural areas; many elderly people; hungry and homeless children; and those suffering from the eating disorder anorexia nervosa.[21] Recently, some well-meaning but misinformed parents inflicted PEM and other deficiency diseases on their toddlers by replacing their milk with unenriched, protein-poor soy or rice drinks.[22] Also at risk for PEM are those with wasting diseases such as cancer or AIDS and those addicted to drugs and alcohol. In a downward spiral, PEM and serious illness worsen each other, so treating the PEM often reduces medical complications and suffering even when the underlying disease is untreatable.[23]

Today, in the United States, millions of people who work to support their children earn so little that they cannot afford nutritious food—well over two and a half million children go hungry at least some of the time.[24] Hunger, especially in children, threatens everyone's future. Hungry children do not learn as well as fed children, nor are they competitive. They are ill more often and more likely to be absent from school; when they attend, they cannot concentrate for long. The forces driving poverty and hunger will require many great minds working together to find solutions. In fighting hunger, programs that tailor interventions to the local people and involve them in the process of identifying problems and devising solutions report the most success (Chapter 15 comes back to topics relating to hunger at home and abroad).[25]

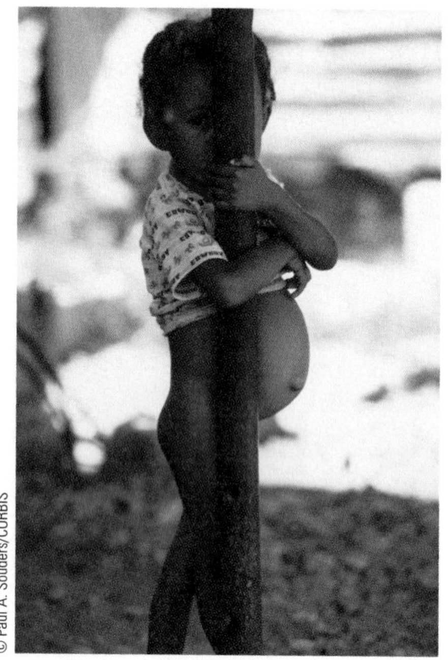

The edema and enlarged liver characteristic of kwashiorkor are apparent in this child's swollen belly. Malnourished children commonly have an enlarged abdomen from parasites as well.

key point *Protein-deficiency symptoms are always observed when either protein or energy is deficient. Extreme food-energy deficiency is marasmus; extreme protein deficiency is kwashiorkor. The two diseases overlap most of the time, and together are called PEM.*

Is It Possible to Consume Too Much Protein?

Overconsumption of protein offers no benefits and may pose health risks for the heart, kidneys, and bones. Protein-rich foods are often high-fat foods that contribute to obesity with its accompanying health risks; animal-protein sources in particular

Table 6•5

FEATURES OF MARASMUS AND KWASHIORKOR IN CHILDREN

Separating PEM into two classifications oversimplifies the condition, but at the extremes, marasmus and kwashiorkor exhibit marked differences. Marasmus-kwashiorkor mix presents symptoms common to both marasmus and kwashiorkor. In all cases, children are likely to develop diarrhea, infections, and multiple nutrient deficiencies.

Marasmus	Kwashiorkor
Infants and toddlers (less than 2 yr)	Older infants and young children (1 to 3 yr)
Severe deprivation or impaired absorption of protein, energy, vitamins, and minerals	Inadequate protein intake or, more commonly, infections
Develops slowly; chronic PEM	Rapid onset; acute PEM
Severe weight loss	Some weight loss
Severe muscle wasting with fat loss	Some muscle wasting, with retention of some body fat
Growth: <60% weight-for-age	Growth: 60 to 80% weight-for-age
No detectable edema	Edema
No fatty liver	Enlarged, fatty liver
Anxiety, apathy	Apathy, misery, irritability, sadness
Appetite may be normal or impaired	Loss of appetite
Hair is sparse, thin, and dry; easily pulled out	Hair is dry and brittle; easily pulled out; changes color; becomes straight
Skin is dry, thin, and wrinkled	Skin develops lesions

can be high in saturated fat, a known contributor to atherosclerosis and heart disease. Whether animal protein itself may affect heart health is still uncertain, as this chapter's Controversy explains.

Animals fed experimentally on high-protein diets often develop enlarged kidneys or livers. Pregnant rats fed a high-protein diet produce offspring that later develop energy imbalances and excess body fatness.[26] In human beings, a high-protein diet consumed over a lifetime worsens existing kidney problems such as kidney stones and may also significantly alter the functioning of healthy kidneys.[27] One of the most effective treatments for people with established kidney problems is to reduce protein intakes to prevent recurrence of kidney stones and slow down the progression of their disease.[28]

Evidence is mixed about whether high protein intakes from animal sources, especially when accompanied by low calcium intakes, can accelerate adult bone loss.[29] No doubt exists about the effect of feeding purified protein to human subjects—purified protein causes calcium to be spilled from the urine.[30] Also, eating diets high in animal protein, but not plant protein, correlates with a greater incidence of hip fractures in some populations.[31] In one population, however, the reverse is true—in malnourished elderly individuals, protein deficiency and hip fractures often occur together, and restoring protein can often improve bone status.[32]

Given appropriate nutrition and care, this child has successfully recovered from kwashiorkor.

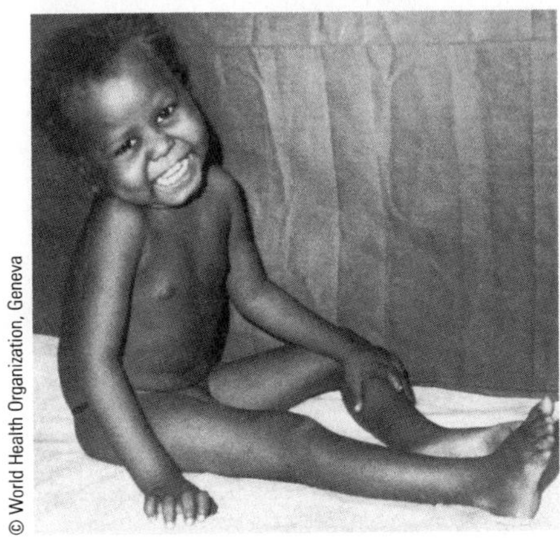
© World Health Organization, Geneva

Getting Enough, but Not Too Much, Protein

1989 RDA for protein for women aged 19 to 24 = 46 g.
Average protein intake = 65 g.

1989 RDA for protein for men aged 19 to 24 = 58 g.
Average protein intake = 105 g.

IT IS CLEAR by now that people in developed nations usually eat more than ample protein. The protein RDA is generous and more than adequately covers the estimated needs of most people, even those with unusually high requirements.

Protein-Rich Foods

Foods in the meat, poultry, fish, dry beans, eggs, and nuts group and in the milk, yogurt, and cheese group contribute an abundance of high-quality protein. Two other groups, the vegetables and the bread, cereals, rice, and pasta group, contribute smaller amounts of protein, but they can add up to significant quantities. What about the fruit group? Don't rely on fruit for protein—fruit contains only small amounts. Figure 6-15 shows the wide variety of foods that contribute most of the protein to the diet.

Protein is critical in nutrition, but too many protein-rich foods can displace other important foods from the diet. Foods richest in protein carry with them a characteristic array of vitamins and minerals, including vitamin B_{12} and iron, but they lack others—vitamin C and folate, for example. In addition, many protein-rich foods such as meat are high in calories, and to overconsume them is to invite obesity. Because American consumption of protein is ample, you can plan meatless or reduced-meat meals with pleasure. Of the many interesting, protein-rich meat alternates available, one has already been mentioned: the legumes.

Figure 6•15

FINDING THE PROTEIN IN FOODS[a]

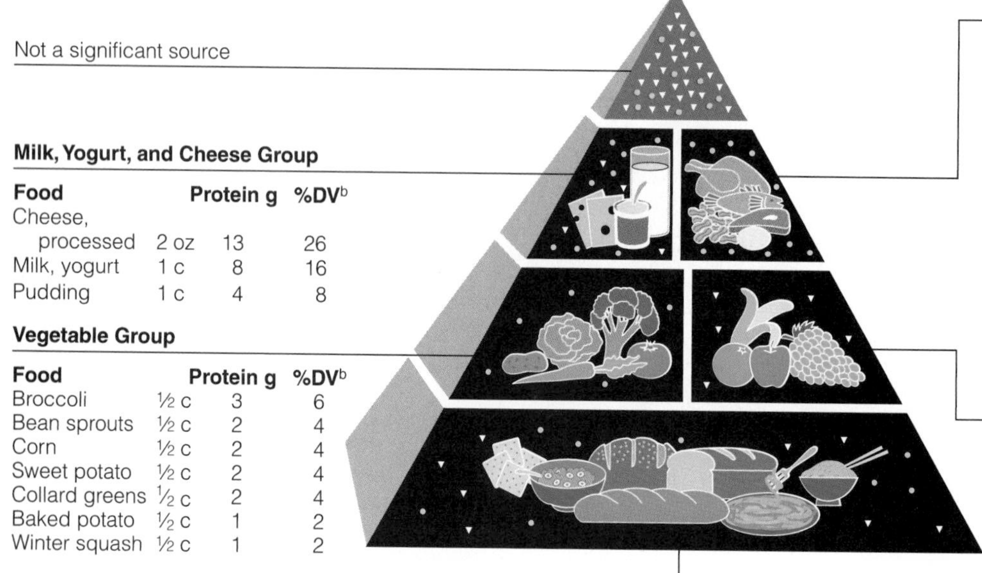

Not a significant source

Milk, Yogurt, and Cheese Group

Food		Protein g	%DV[b]
Cheese, processed	2 oz	13	26
Milk, yogurt	1 c	8	16
Pudding	1 c	4	8

Vegetable Group

Food		Protein g	%DV[b]
Broccoli	½ c	3	6
Bean sprouts	½ c	2	4
Corn	½ c	2	4
Sweet potato	½ c	2	4
Collard greens	½ c	2	4
Baked potato	½ c	1	2
Winter squash	½ c	1	2

Meat, Poultry, Fish, Dry Beans, Eggs, and Nuts Group

Food		Protein g	%DV[b]
Chicken breast	2 oz	18	36
Roast beef	2 oz	16	32
Pork meat	2 oz	16	32
Turkey leg	2 oz	16	32
Tuna	2 oz	14	28
Lentils, beans, peas	½ c	9	18
Peanut butter	2 tbs	8	16
Almonds	¼ c	7	14
Lunch meat	2 oz	7	14
Egg	1 lg	6	12
Hot dog	1 reg	6	12
Cashew nuts	¼ c	5	10

Fruit Group

Food		Protein g	%DV[b]
Avocado	½ c	2	4
Cantaloupe	½ c	1	2
Orange sections	½ c	1	2
Strawberries	½ c	0	0

Breads, Cereals, Rice, and Pasta Group

Food		Protein g	%DV[b]
Pancakes	2 sm	4	8
Bagel	½	4	8
Fried rice	½ c	3	6
Grain bread	1 sl	3	6
Noodles, pasta	½ c	3	6
Oatmeal	½ c	3	6
Barley	½ c	2	4
Cereal flakes	1 oz	2	4

[a] All foods are prepared and ready to eat.
[b] The Daily Value (DV) for protein is 50 g, and is based on an energy intake of 2,000 calories per day.

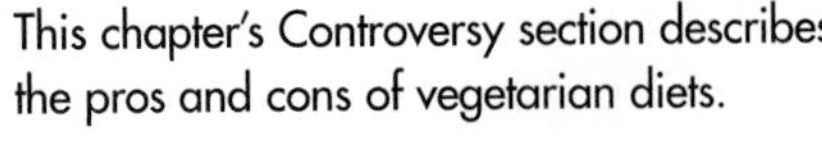

The Advantages of Legumes

The protein of some legumes is of a quality almost comparable to that of meat. For practical purposes, the quality of soy protein can be considered equivalent to that of meat. Figure 6-16 shows a legume plant's special root system that enables it to make abundant protein. Legumes are also excellent sources of fiber, many B vitamins, iron, calcium, and other minerals. On average, a cup of cooked legumes contains about 30 percent of the Daily Values for both protein and iron.* Like meats, though, legumes do not offer every nutrient, and they do not make a complete meal by themselves. They contain no vitamin A, vitamin C, or vitamin B_{12}, and their balance of amino acids can be much improved by using grains and other vegetables with them.

Soybeans are versatile legumes, and many nutritious products are made from them. The heavy use of soy products in place of meat, however, inhibits iron absorption. The effect can be alleviated by using small amounts of meat and/or foods rich in vitamin C in the same meal with soy products. Vegetarians sometimes use convenience foods made from **textured vegetable protein** (soy protein) formulated to look and taste like hamburgers or breakfast sausages. Many of these are intended to match the known nutrient contents of animal-protein foods, but often they fall short.† A wise vegetarian uses such foods sparingly and learns to use combinations of whole foods to supply the needed nutrients.

The nutrients of soybeans are also available as bean curd, or **tofu,** a staple used in many Asian dishes. Thanks to the use of calcium salts when some tofu is made, it can be high in calcium. Check the Nutrition Facts panel on the label.

The Food Features presented so far show that the recommendations for the three energy-yielding nutrients occur in balance with each other. If you reduce fat and increase carbohydrate, protein totals automatically come into line with the requirements. To help you accept that protein is abundant in most foods, the Do It! section asks you to complete a day's meals and watch the protein grams add up.

*Data from the *Food Processor Plus*, ESHA research, version 7.11.

†In Canada, regulations govern the nutrient contents of such products.

This chapter's Controversy section describes the pros and cons of vegetarian diets.

textured vegetable protein processed soybean protein used in products formulated to look and taste like meat, fish, or poultry.

tofu (TOE-foo) a curd made from soybeans that is rich in protein, often rich in calcium, and variable in fat content; used in many Asian and vegetarian dishes in place of meat.

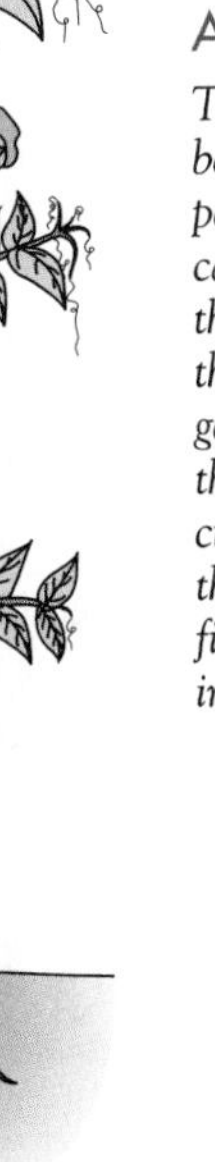

Figure 6•16

A LEGUME

The legumes include such plants as the kidney bean, soybean, garden pea, lentil, black-eyed pea, and lima bean. Bacteria in the root nodules can "fix" nitrogen from the air, contributing it to the beans. Ultimately, thanks to these bacteria, the plant accumulates more nitrogen than it can get from the soil and also leaves more nitrogen in the soil than it takes out. The legumes are so efficient at trapping nitrogen that farmers often grow them in rotation with other crops to fertilize fields. Legumes are shown with the meat group in Figure 2-4 of Chapter 2.

ADD UP THE PROTEIN IN A DAY'S MEALS

Consider the sources of protein in a day's meals. Look at the meals in Figure 6-17. Breakfast and lunch are given, but supper is yet to be planned. A simple breakfast of cereal, milk, and juice provides 14 grams of protein. Lunch is a bit heartier with a ham and cheese sandwich contributing most of its 18 protein grams. Now comes a puzzle—which supper to choose? After picking a supper from among the choices, check Table 6-6 to find out how much protein each supper contains (the totals are printed upside down to prevent you from peeking before you guess). Then compare the protein in the meals with your protein recommendation (use the 1989 RDA value from the inside front cover).

Two of the supper options are meatless (the spaghetti supper and the vegetable-rice supper), and one contains meat. To quickly assess the protein in such meals, remember that fruits provide only a little protein, but meats, milk, and cheeses are the richest sources, followed by legumes, grains, breads, and vegetables.

Settle on a supper choice and consider these questions:

1. Did the supper you chose add up to a protein total that meets, but does not exceed, your protein RDA? Which other suppers also qualify?
2. If your answer to the first part of question 1 was no, which foods might you substitute to achieve your goal without shorting yourself on nutrients? Hint: If your pro-

Figure 6•17

A PROTEIN PUZZLE

The protein values listed in this exercise are from the Food Processor Plus, 7.11, *a computerized diet analysis software program developed by* ESHA Research.

Breakfast
1 c orange juice
2 g protein
Cheerios cereal, 1 oz
4 g protein
1 c low-fat milk
8 g protein
Breakfast total = 14 g protein

© Polara Studios, Inc. (all)

Lunch
iced tea
0 g protein
ham and cheese sandwich (1 slice lunch meat; 1 slice cheese; ¾ c lettuce and tomato; 2 slices whole-wheat bread)
17 g protein
peaches ½ c
1 g protein
Lunch total = 18 g protein

So far, this day's meals have contributed 32 grams of protein. On this basis, what would you choose for supper? After choosing, turn Table 6-6 upside down to see how much protein each supper adds to the day's intake.

Supper A = ? protein

Supper B = ? protein

Supper C = ? protein

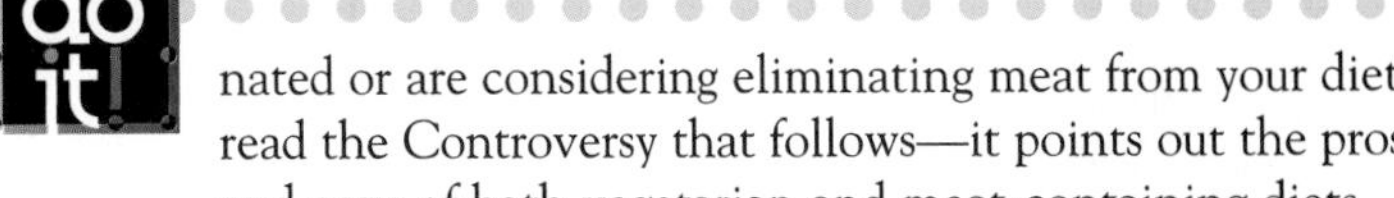

tein exceeded your goal, you may need to restrict something, and an obvious "something" to restrict is some of the meat.

3. Make some educated guesses concerning the other two energy-yielding nutrients, fat and carbohydrate. Which foods contribute abundant fat and saturated fat to this day's meals? Which contribute carbohydrates and fiber?
4. Note that breakfast, though it contains no meat, provides almost as much protein as the ham and cheese sandwich at lunch. Which foods in this breakfast provide protein? Describe how the amino acids in some foods might complement those of other foods in the breakfast.
5. Which plant food shown in Figure 6-17 is the richest in protein? Which is next richest? Hint: If you have eliminated or are considering eliminating meat from your diet, read the Controversy that follows—it points out the pros and cons of both vegetarian and meat-containing diets.
6. If you were to design a day's meals around the lamb supper, yet did not want to consume too much protein, how would you change breakfast and lunch? What foods would you substitute for some of the protein-rich foods listed in Figure 6-17? Another hint: If you are considering doing away with the milk or cheese, remember that you must then provide other sources of calcium (you may reconsider this decision when you discover in Chapter 8 that few foods other than milk supply an abundance of calcium).

Table 6•6

ANSWERS TO PROTEIN PUZZLE

Supper Choice A
iced water/lemon
0 g protein
garlic bread, 2 pieces
6 g protein
large salad with ¼ c each garbanzo beans, artichoke, and cucumber
6 g protein
spaghetti, 1 c; parmesan cheese, 1 tbs
13 g protein
sherbet, 1 c
2 g protein
Totals:

Supper A total = 27 g protein
Entire day's protein = 59 g

Supper Choice B
iced tea/lemon
0 g protein
tomato slices, ½ c
1 g protein
grated cheese, 1 tbs
3 g protein
mixed vegetables, 1 c
4 g protein
brown rice, 1 c
6 g protein
carrot cake, 1 pce
4 g protein
Totals:

Supper B total = 18 g protein
Entire day's protein = 50 g

© Polara Studios, Inc. (all)

Supper Choice C
coffee, black, 1 c
0 g protein
asparagus, ½ c
2 g protein
potatoes au gratin, ½ c
6 g protein
lamb chops, 2 (2 oz each)
35 g protein
sliced beets, ½ c
1 g protein
bread pudding, ½ c
7 g protein
Totals:

Supper C total = 51 g protein
Entire day's protein = 83 g

Answers to these Self Check questions are in Appendix G.

1. The basic building blocks for protein are
   a. glucose units
   b. amino acids
   c. side chains
   d. saturated bonds
2. Protein digestion begins in the
   a. mouth
   b. stomach
   c. small intestine
   d. large intestine
3. Which of the following can form enzymes?
   a. carbohydrates
   b. lipids
   c. proteins
   d. b and c
4. For healthy adults, the 1989 RDA for protein has been set at
   a. 0.8 gram per kilogram of body weight
   b. 2.2 pounds per kilogram of body weight
   c. 12 to 15 percent of total calories
   d. 100 grams per day
5. Which of the following statements is correct regarding protein and amino acid supplements?
   a. They help athletes build muscle.
   b. They help dieters lose weight quicker.
   c. They can assist in relieving depression.
   d. None of the above.
6. Under certain circumstances, protein can be converted to glucose and so serve the energy needs of the brain. T F
7. Too little protein in the diet can have severe consequences but excess protein has not been proven to have adverse effects. T F
8. Although protein-energy malnutrition (PEM) is prevalent in underdeveloped nations, it is not seen in the United States. T F
9. Partially completed proteins are not held for completion at a later time when the diet may improve. T F
10. An example of a person in positive nitrogen balance is an astronaut. T F

nutrition on the net

For further study of the topics of this chapter, access these websites and search for the phrases or words in quotation marks:

1. Find updates and quick links to these and other nutrition-related sites at our website:
   www.wadsworth.com/nutrition
2. To learn more about "protein" in foods, visit the American Dietetic Association (ADA) site:
   www.eatright.org
3. Search the World Health Organization (WHO) site for information on "protein-energy malnutrition":
   www.who.org
4. For more on "vegetarian" diets, search the Food and Drug Administration (FDA) site for foods:
   www.fda.gov
5. Search among thousands of current scientific and medical abstracts for any topic related to protein at:
   www.ncbi.nlm.nih.gov/PubMed/
6. Learn more about sickle-cell anemia from the National Heart, Lung, and Blood Institute or the Sickle Cell Disease Association of America:
   www.nhlbi.nih.gov or www.sicklecelldisease.org
7. Learn more about protein-energy malnutrition and world hunger from the World Health Organization Nutrition Programme:
   www.who.ch/nut/prot
8. Search for "amino acid supplements" at the National Council for Reliable Health Information site:
   www.ncrhi.org
9. For statistics on hunger and other concerns about U.S. children, visit the Federal Interagency Forum on child and family statistics:
   www.childstats.gov

INTERNET ACTIVITY

After reading Chapter 6 Controversy *Vegetarians versus Meat Eaters: Whose Diet Is Best?*, you will be familiar with the different categories of vegetarian dietary patterns.

1. Go to the website of the Vegetarian Resource Group at:
   http://www.vrg.org
2. Explore the website.

Suggest four ways to incorporate vegetarian dietary patterns into your eating style. Would you actually implement these vegetarian strategies? Explain why or why not.

CONTROVERSY 6

VEGETARIANS VERSUS MEAT EATERS: WHOSE DIET IS BEST?

ONE YOUNG person rejects all animal products, shuns grains, and seeks out vegetables, fruits, and herbs. Another young person relishes meat at every meal and usually orders "a steak and potato: hold the rabbit food." These two have a lot more in common than either would probably believe. Both are extremists in their choices of foods. Both may be jeopardizing their health by their rigid, unbalanced eating styles.[1] But both **vegetarian** diets and meat-containing diets have elements in their favor, provided that they are not taken to extremes.

Vegetarianism is often mistakenly associated with a particular culture, but individuals choose it for many different reasons. Some believe that we should not kill animals to eat their meat. Some believe that we should not even partake of animal products such as milk, cheese, eggs, or honey or use items made from leather, wool, or silk. Many people don't want to endorse the inhumane treatment of livestock animals in pens, feedlots, and slaughterhouses by eating meat. Some believe we should eat less meat for health reasons or for environmental reasons. Some fear contracting diseases, such as food poisoning or "mad cow disease" from meats. (The effects of food choices on the earth's resources are topics of Chapter 15 and its Controversy, and Chapter 14 provides the whole story on the threat from food-borne illnesses.)

People who eat meat also do so for a variety of reasons. Some find that a hamburger makes a convenient lunch while providing a concentrated source of energy and nutrients. Others enjoy the taste of roast chicken or beef stew. Others wouldn't know what to eat without meat; they are accustomed to seeing it on the plate. Whatever your reasons for choosing a particular diet, these daily choices have implications for your health.

This Controversy looks first at the positive health aspects of vegetarian diets, then at the positive aspects of meat eaters' diets. Both types of eaters can maximize the benefits and minimize the risks of their diets.

Positive Health Aspects of Vegetarian Diets

A statement of support for some vegetarian eating styles comes from the *Dietary Guidelines for Americans*, which states that a well-chosen vegetarian diet can meet nutrient needs while staying within the recommendations of the *Guidelines*.[2] This statement agrees with strong evidence linking vegetarian diets with reduced incidences of chronic diseases.[3]

Such evidence, though abundant, is not easily obtained. It would be easy if vegetarians differed from others only in not consuming meat, but they often have *increased* intakes of fruits and vegetables as well, and these foods are the primary contributors of phytochemicals believed to reduce disease risks. Vegetarian diets may also contain more fiber, another factor associated with reduced disease risks. Also, though there are exceptions, vegetarians typically use no tobacco, use alcohol in moderation if at all, and may be more physically active than other adults. Researchers must account for the effects of a total health-conscious lifestyle on disease development before they can see how diet correlates with health. Even then, *correlation* is not cause. Without more evidence, conclusions must be tentative.

Still, with all these qualifications, research findings are intriguing. They indicate that a vegetarian diet may offer some protection against six conditions: obesity, diabetes, high blood pressure, heart disease, digestive disorders, and some forms of cancer. It matters, however, what form the vegetarian diet takes. Vegetarians differ, as Table C6-1 on the next page demonstrates. The following sections outline what is known about the relationships between vegetarianism and protection against disease.

Less Obesity and Diabetes

Vegetarians tend to be leaner than nonvegetarians.[4] Perhaps they

A balanced meal need not include meat to be nutritious.

Table C6•1

TERMS USED TO DESCRIBE VEGETARIANS

- **fruitarian** includes only raw or dried fruits, seeds, and nuts in the diet.
- **lacto-ovo vegetarian** includes dairy products, eggs, vegetables, grains, legumes, fruits, and nuts; excludes flesh and seafood.
- **lacto-vegetarian** includes dairy products, vegetables, grains, legumes, fruits, and nuts; excludes flesh, seafood, and eggs.
- **macrobiotic diet** a vegan diet that progressively eliminates more and more foods. Ultimately, only brown rice and small amounts of water or herbal tea are consumed; taken to extremes, macrobiotic diets have resulted in malnutrition and even death.
- **ovo-vegetarian** includes eggs, vegetables, grains, legumes, fruits, and nuts; excludes flesh, seafood, and milk products.
- **partial vegetarian** includes seafood, poultry, eggs, dairy products, vegetables, grains, legumes, fruits, and nuts; excludes or strictly limits certain meats, such as red meats. Also called *semivegetarian*.
- **pesco-vegetarian** same as partial vegetarian, but eliminates poultry.
- **vegan** includes only food from plant sources: vegetables, grains, legumes, fruits, seeds, and nuts; also called *strict vegetarian*.
- **vegetarian** includes plant-based foods and eliminates some or all animal-derived foods.

consciously control their calorie intakes and make an effort to exercise regularly. Perhaps their diet, which tends to be high in fiber-rich bulky foods, is automatically lower in calories than the average diet based on meat.

Vegetarians tend to have low intakes of fat, which is high in calories and efficiently stored in the body, and high intakes of high-fiber carbohydrate, which is lower in calories, gram for gram, and costs more energy for the body to convert into fat for storage.[5] Leanness may also result from a higher metabolic rate (the rate at which the body burns fuel, see Chapter 9) in vegetarians compared with nonvegetarians. This higher rate may occur because, when supplied with a mixed diet, the body may preferentially burn off more of the extra carbohydrate calories for fuel (increase carbohydrate metabolism) and store excess fat calories.[6] In any case, a healthy body weight combined with high intakes of complex carbohydrates and fiber reduces the risks of diabetes and several other obesity-related diseases. A limited body of research suggests a connection between meat-containing diets and increased incidence of diabetes, even without obesity.

Lower Blood Pressure

Vegetarians are often found to have lower blood pressure than nonvegetarians. Various combinations of lifestyle factors and diet influence blood pressure. Among lifestyle factors, smoking and alcohol intake raise blood pressure, and exercise lowers it. Diet alone may be significant, however. A review of the literature in this area concluded that there is convincing evidence for a blood-pressure-lowering effect of some vegetarian diets. The authors say that the effect is associated with diets low in fat and saturated fat and high in fiber, fruits, and vegetables, and they say that meat itself need not be totally excluded.

Less Heart Disease

Fewer vegetarians than meat eaters suffer from diseases of the heart and arteries or exhibit indicators of those diseases, even when the people being compared are all nonsmokers.[7] The dietary factor most directly related to coronary artery disease is saturated fat intake, although other factors may play a role. When vegetarians are fed meat, which contains saturated fat, their lipid profiles change for the worse; when meat eaters are fed a low-fat vegetarian diet, their lipid profiles improve. One study compared two low-fat diets, one vegetarian and the other containing lean meats. Both diets lowered blood cholesterol, but the vegetarian diet's effects were greater. This result implies that the source of protein in the diet—animal or vegetable—affects blood cholesterol and therefore heart disease risk.

An analysis of the results from 38 experiments revealed that when soy protein *replaces* animal protein in the diet, total blood cholesterol concentration drops, blood LDL concentration is significantly reduced, blood triglycerides fall, and HDL cholesterol level rises.[8] The effect may be partly attributable to the amino acids characteristic of soy protein or to the heart-protecting effects of soy's accompanying phytochemicals. However, even the protein of wheat bread, gluten, when it replaces animal protein in the diet, reduces cholesterol oxidation and other indicators of heart disease risk.[9] Likewise, fat-free milk *lowers* blood cholesterol just as soy does, and meals of fish provide benefits to heart health, as Chapter 5 made clear.[10] These findings have led researchers to look beyond dietary protein for answers about heart disease risk.

Researchers found some surprising results from a 17-year study of dietary habits and causes of death among 11,000 health-conscious people.[11] About half of the study's subjects were vegetarians and half were meat eaters. The researchers reported a slight, but not statistically significant, correlation between lowered rates of heart disease and omission of meat from the diet. They reported a much stronger and more significant association between reduced risk of death from heart disease and the inclusion of fresh fruit and, to a lesser extent, green salads, regardless of meat consumption. Other researchers agree with this line of speculation: reduced risk of heart disease in vegetarians may result from what they *obtain* from the diet rather than entirely from what they *omit*. Many vegetarians replace the meat in their diets with foods made with fruits, vegetables, whole grains, tree nuts, and soybeans.[12] The case for health benefits from consuming such foods is growing more solid each day.[13] Current research is concentrating on the sources of protein in the diet and on the antioxidant nutrients and phytochemicals found in plants (details on this line of research can be found in Controversies 2 and 7).[14] Against heart disease, then, a vegetarian diet that includes fruit, vegetables, and other nutritious foods offers these advantages:

- Promotes leaner body composition and lower blood pressure.
- Provides less saturated fat.
- Provides more vegetable protein, fiber, antioxidant nutrients, and phytochemicals.

Fewer Digestive Disorders

Constipation and diverticular disease are less common in people who consume high-fiber vegetarian or **partial vegetarian** diets than in people who consume typical meat-based diets. Chapter 4 presented fiber's positive influence on the health of the digestive tract.

Lower Risk of Cancer

Seventh-Day Adventists, an often-studied health-conscious vegetarian group, enjoy a significantly lower cancer rate than the rest of the population, even when cancers linked to smoking and alcohol are taken out of the picture.[15] Their low cancer mortality may be due to their low meat intakes, to their high intakes of fruits, vegetables, and cereal grains, to both, or to other lifestyle factors.[16]

Some scientific findings support the idea that vegetarian diets may reduce the risks of colon cancer. People with colon cancer seem to eat more meat, less fiber, and more saturated fat than others without colon cancer. Something about high-fat, high-protein, low-fiber diets creates an environment in the human colon that may promote the development of cancer. Such a diet has also been associated with a form of cancer of the lymphatic system.[17] In general, vegetarians tend to consume less fat and protein, and they seem to produce fewer carcinogens in the body than do meat eaters.[18] Further, they take in more carbohydrate, fiber, and water, and these dietary constituents add bulk to the stools, diluting any carcinogens that may be present. Meat also forms carcinogenic substances when it is browned during cooking (more about these in Chapter 11), and vegetarians are spared any risks these substances may present.[19]

Overall, then, many vegetarians have lower risks of developing obesity, diabetes, high blood pressure, heart disease, digestive disorders, and cancer than do meat eaters. In the United States, some 12 million people report following vegetarian diets.[20] If they plan their diets wisely, they obtain all the nutrients they need to support good health.

Positive Health Aspects of the Meat Eater's Diet

The meat lover introduced at the start of this Controversy was exaggerated to make a point. Those who really eat like that place themselves in immediate peril of malnutrition. Few people shun all green and yellow vegetables, fruits, and grains. To be healthy, people must either eat foods from all groups of the Food Guide Pyramid, or if they omit foods from one group, they must make careful substitutions to compensate. No substitutes can take the place of fruits and vegetables (not even antioxidant supplement pills, as the next chapter points out). This section considers a balanced diet of which meat is a part.

Growth

Meats, eggs, milk, and other foods from animal sources support growth well. Without them, children's growth often lags behind the growth of peers.[21] Populations existing on monotonous grain diets, because of strict meat taboos or economic necessity, are often malnourished, as revealed by their short stature, low resistance to diseases, short life span, and high infant mortality.

Even in populations with somewhat more varied diets, the children who eat the most animal-derived products have been observed to grow the best. Even when the protein amounts are equal, children whose protein intakes are from plant sources may not grow as well as those eating animal products.[22] Protein may not be the only nutrient affecting growth, however; families who can afford to buy animal products are also likely to consume a larger variety of fruits and vegetables. These foods provide the vitamins and minerals also needed for growth. In developed nations, children whose parents are strict vegetarians may face two diet-related threats: lack of variety and reliance on nutritionally inadequate convenience foods.[23]

Foods of plant origin offer much less energy for their bulk than do foods of animal origin. A child's small stomach can hold only so much food, and a vegetarian child may feel full before eating enough food to supply nutrients and energy sufficient to support growth. For obesity-prone adults, a bulky diet can be advantageous, but a child fed without meat, milk, or eggs may experience stunted growth that lasts a lifetime.

Are animal and dairy products superior to plants as protein-rich foods? Meats, eggs, and dairy foods do contain complete, more digestible proteins. Children of milk- and meat-eating populations are generally larger, fatter, and more resistant to infections than are those of grain-eating populations. They are also protected from the vitamin D–deficiency disease rickets, which is especially likely to strike **vegan** children in cold climates who are rarely exposed to the sun.

Meat, however, is not necessary for children to achieve healthy growth.

This 5-ounce steak provides over half the daily maximum of meat recommended by the Food Guide Pyramid.

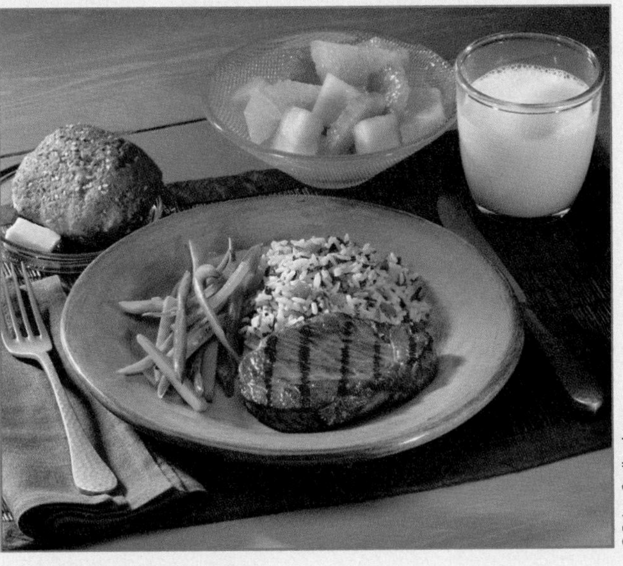

Healthy children grow normally when milk and eggs accompany a vegetarian diet and when knowledgeable adults plan and deliver the diet with care. An example is found among the children of a **lacto-ovo vegetarian** community in England: their growth is practically identical to the growth of children who eat meat.[24]

Support during Critical Times

Both meat eaters and lacto-ovo vegetarians can generally rely on their diets during critical times of life. In contrast, a vegan woman who doesn't meet her nutrient needs may enter pregnancy too thin; have inadequate stores of iron, zinc, and vitamin B_{12}; fail to supply the omega-3 fatty acids needed for normal fetal development; and fail to gain enough weight during pregnancy to support the normal growth and development of her fetus.

Obtaining enough vitamin B_{12} is often a challenge for vegans.[25] The importance of attention to this crucial vitamin is underscored by a severe, often irreversible, sometimes fatal disorder reported among breastfed infants of vegan mothers who fail to obtain sufficient vitamin B_{12}.[26] The infants exhibit a striking syndrome of body tremors and facial twitches involving the tongue and throat that combine to make nursing difficult. After some months, deprivation of vitamin B_{12} leads to severe psychomotor retardation of the infant, accompanied by shrinkage of the brain.[27] In some cases, the retardation caused by the illness lingers on even after treatment with the missing vitamin.[28] Urban legends may hold that vitamin B_{12} poses no concern for vegans, but they are far off the mark, especially where growing children are concerned.

Parents who take on the challenge of rearing a child on a vegan diet must educate themselves to feed the child adequately. As the chapter mentioned, protein malnutrition, though rare in developed nations, is cropping up among vegan babies fed protein-poor "rice drinks" instead of nutritious breast milk or formula that babies need for growth and development.[29] Kwashiorkor, formerly a rarity in the United States, is being diagnosed among children fed ill-conceived vegetarian diets.[30] With careful planning and appropriate nutrient supplements or fortified cereals, fortified soy beverages, or fortified meat analogs, children can be raised successfully on vegan diets.[31]

Some vegetarians worry, and rightly so, about the health of their bones. Children who were fed a **macrobiotic diet,** an extreme form of vegetarian diet, were observed to have lower bone mineral density in adolescence, a time of great importance to the bones.[32] Nevertheless, parents who make sure that their vegetarian children consume milk, cheese, and other milk products regularly can be assured that their children's bones will receive the calcium they need.

Vegetarian women of childbearing age may benefit from increasing their intakes of sources of omega-3 fatty acids and reducing their intakes of omega-6 fatty acids (vegetable oils) to nudge the fatty acid ratio toward a balance that best supports normal fetal and infant growth and development (see Chapter 5 for details).[33] Vegetable sources of omega-3 fatty acids include sea vegetables, ground flaxseed, flaxseed oil, canola oil, and soybean oil.[34] Eggs from chickens fed on omega-3–rich feed deliver these scarce nutrients to egg eaters as well. Labels of egg cartons specify which are rich in omega-3 fatty acids.

Unlike vegans, women who eat meat, eggs, and milk products can be sure of receiving enough vitamin B_{12}, vitamin D, calcium, iron, and zinc, as well as protein, to support pregnancy and breastfeeding. Women following lacto-ovo vegetarian diets can also receive all of these nutrients in abundance, and they have the added advantage of habitually consuming more folate and other nutrients associated with vegetables and fruits than typical meat eaters do. The importance of adequate folate, especially for women in the childbearing years, will be shown in the next chapter.

Obtaining Nutrients

To obtain calcium, U.S. consumers who use no milk can purchase calcium-fortified soy drink or calcium-fortified orange juice. Alternatively, several large servings of calcium-rich green vegetables such as broccoli or kale can make a sizable calcium donation to the daily diet.

The nutrients iron, zinc, vitamin D, and vitamin B_{12} also require special attention from strict vegetarians. Meat provides much of the iron and zinc in the meat eater's diet, and vitamin D and vitamin B_{12} are found reliably only in animal-derived foods. Vegetarian sources of iron and zinc, such as legumes, dark green, leafy vegetables, fortified cereals, and whole-grain breads and cereals, provide less of these minerals than meat and in a less absorbable form.[35] To obtain enough iron and zinc, an emphasis on whole grains and legumes in the diet is important. A vegetarian relying heavily on white rice, refined white flour, and sweets for carbohydrates can easily obtain too few minerals for health. A strict vegetarian diet cannot meet vitamin D needs without a supplemental source or adequate exposure to sunlight (the next chapter provides details). These substitutions take planning but yield results.

Eggs, for those who eat them, can meet vitamin B_{12} needs; vegans, especially women of childbearing age, must take particular care to choose vitamin B_{12}–fortified cereals and other sources or must rely on supplements. Fermented plant products such as tempeh, made from soybeans, may contain some vitamin B_{12} contributed by the bacteria that did the fermenting, but much of the vitamin B_{12} in these foods may be inactive. Vegans who worry about a lack of omega-3 fatty acids in their foods might try the sources that fish themselves use: sea vegetables, which can be rich in omega-3 fatty acids.[36]

Have you noticed the lack of concern about protein for adult vegetarians? Protein is not the problem it was once thought to be for adults eating a varied diet. Even in vegans, protein deficiency is rare in those who consume adequate calories in various nutritious foods.

Well-planned diets of both meat eaters and vegetarians can contribute to good health, but both diets can damage it if haphazardly chosen. Both diets can be high in saturated fat and so pose a threat to the health of the heart. A vegetarian who dines on cheddar cheese, butter sauces, sour cream, and deep-fried vegetables invites the same health hazards as the overeater of high-fat meats. And both diets, if not properly balanced, can lack nutrients.

For both diets, then, planning is the key to obtaining adequate nutrients. Those who eliminate meats can adapt the Food Guide Pyramid to their needs (see Figure C6-1).

Figure C6•1

VEGETARIAN FOOD GUIDE PYRAMID[a]

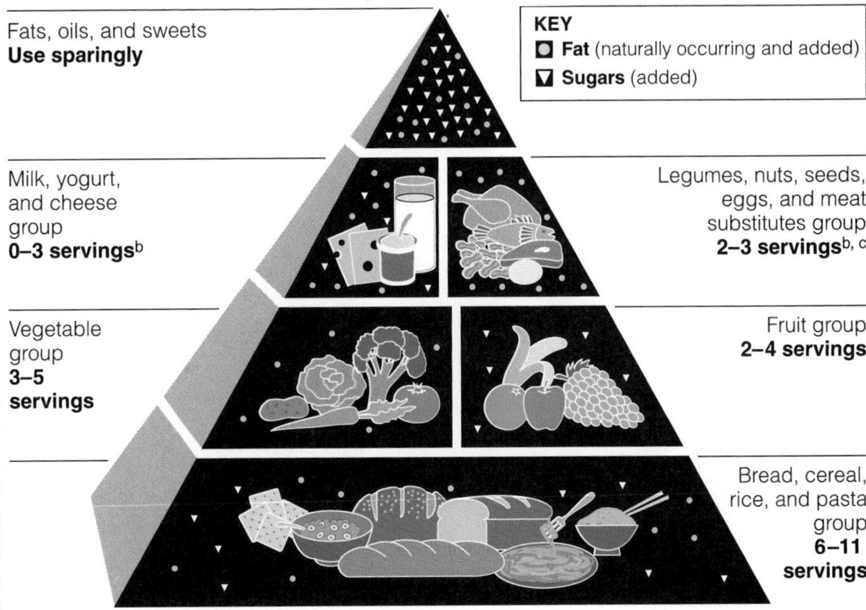

[a] Serving sizes are listed in Figure 2-4 of Chapter 2.

[b] Vegans and other vegetarians who include no eggs, milk, or milk products in the diet must obtain calcium, vitamin D, and vitamin B_{12} from other sources. See text for details.

[c] Meat substitutes include soybean products, such as tofu, tempeh, and soy milk, and commercial meat replacers.

SOURCE: Adapted from the USDA Food Guide Pyramid and Food Guide Pyramid for Vegetarian Meal Planning in Position of the American Dietetic Association: Vegetarian diets, *Journal of the American Dietetic Association* 97 (1997): 1317–1321.

Conclusion

This comparison has shown that both a meat eater's diet and a vegetarian's diet are best approached scientifically. Vegetarianism is not a religion like Buddhism or Hinduism; it is merely an eating plan that selects plant foods to deliver needed nutrients. Some people make much of the distinctions between types of vegetarians; although these distinctions are useful academically, they do not represent uncrossable lines. Some people use meat as a condiment or seasoning for vegetable or grain dishes. Some people eat meat only once a week and use plant-protein foods the rest of the time. Many people rely mostly on milk products to meet their protein needs, but eat fish occasionally, and so forth. To force people into the categories of "vegetarians" and "meat eaters" leaves out all these in-between styles of eating that have much to recommend them.

If you are just beginning to study nutrition, consider adopting the attitude that the choice to make is not whether to be a meat eater or a vegetarian, but where along the spectrum to locate yourself. Your preferences, whatever they are, should be honored, and the only caveats are that you make your diet adequate, balanced, and varied, and that you use moderation when choosing foods high in saturated fat or calories.

Mattie Lou O'Kelly (1908–1997), *Spring Vegetable Scene*, 1968

The Vitamins

Frequently Asked Questions

Does eating carrots really protect the health of the eyes? p. 218

How can people make a vitamin from sunlight? p. 219

How are B vitamins related to heart disease? p. 234

Is too much vitamin C hazardous to health? p. 237

Do I need to take a vitamin supplement? p. 240

Which supplement doses may be helpful, and which should I avoid? p. 240

Supplements must be safe, or the government would not allow their sales, right? p. 242

Contents

vitamins organic compounds that are vital to life and indispensable to body functions, but are needed only in minute amounts; noncaloric essential nutrients.

precursors, provitamins compounds that can be converted into active vitamins.

The only disease a vitamin can cure is the one caused by a deficiency of that vitamin.

AT THE BEGINNING of the twentieth century, the thrill of the discovery of the first **vitamins** captured the world's imagination as seemingly miraculous cures took place. In the usual scenario, a whole group of people were unable to walk (or were going blind or bleeding profusely) until an alert scientist stumbled onto the substance missing from their diets. The scientist confirmed the discovery by feeding vitamin-deficient feed to laboratory animals, which responded by becoming unable to walk (or going blind or bleeding profusely). When the missing ingredient was restored to their diet, they soon recovered. People, too, were quickly cured when they received the vitamins they lacked.

Over the next decades, the growing sophistication of chemistry and biology allowed scientists to isolate the vitamins and define their chemical structures. More scientific advances brought an understanding of the biological roles that vitamins play in maintaining health and preventing deficiency diseases. Today, research hints that two of the major scourges of humankind, cardiovascular disease (CVD) and cancer, may be linked with low intakes of vitamins. A respected thinker in genetics has suggested that deficiencies of vitamins and minerals may be a major cause of genetic damage that can lead to cancer in people who consistently choose diets lacking in these substances.[1] Can it be that foods rich in vitamins will protect us from life-threatening diseases? What about vitamin pills? For now, we can say only this with certainty: the only disease a vitamin will *cure* is the one caused by a deficiency of that vitamin. As for chronic disease *prevention*, the evidence is still emerging and is the topic of this chapter's Controversy.

Vitamins fall into two classes—fat soluble and water soluble.

Definition and Classification of Vitamins

A child once defined a vitamin as "what, if you don't eat, you get sick." Although the grammar left something to be desired, the definition was accurate. Less imaginatively, a *vitamin* is defined as an essential, noncaloric, organic nutrient needed in a tiny amount in the diet. Many vitamins play the role of facilitator: they help make possible the processes by which other nutrients are digested, absorbed, and metabolized or built into body structures. Although small in size and quantity, the vitamins accomplish mighty tasks.

As they were discovered, the vitamins were named, and many were also given letters and numbers. This led to the confusing variety of vitamin names that still exists today. This chapter uses the names in Table 7-1; alternative names are given in Tables 7-5 and 7-6, summaries that appear on pages 245 through 249 of this chapter.

Some vitamins occur in foods in a form known as **precursors,** or **provitamins.** Once inside the body, these are transformed chemically to one or more active vitamin forms. Thus, to measure the amount of a vitamin found in food, we often must count not only the amount of the true vitamin but also the vitamin activity potentially available from its precursors. Tables 7-5 and 7-6 specify which vitamins have precursors.

The vitamins fall naturally into two classes: fat soluble and water soluble. Solubility confers on vitamins many of their characteristics. It determines how they are absorbed into and transported by the bloodstream, whether they can be stored in the body, and how easily they are lost from the body. In general, like other lipids, fat-soluble vitamins are absorbed into the lymph, and they travel in the blood in association with protein carriers. Fat-soluble vitamins can be stored in the liver or with other lipids in fatty tissues, and some can build up to toxic concentrations. The water-soluble vitamins are absorbed directly into the bloodstream, where they travel freely. Most are not stored in tissues to any great extent; rather, excesses are excreted in the urine. Thus, the risks of immediate toxicities are not as great as for fat-soluble vitamins. This chapter examines the fat-soluble vitamins first and then the water-soluble ones. The tables at the end of the chapter sum up the basic facts about all of them.

Table 7•1

VITAMIN NAMES[a]

Fat-Soluble Vitamins
- Vitamin A
- Vitamin D
- Vitamin E
- Vitamin K

Water-Soluble Vitamins
- B vitamins
  - Thiamin (B_1)
  - Riboflavin (B_2)
  - Niacin (B_3)
  - Folate
  - Vitamin B_{12}
  - Vitamin B_6
  - Biotin
  - Pantothenic acid
- Vitamin C

[a]Vitamin names established by the International Union of Nutritional Sciences Committee on Nomenclature. Other names are listed in Tables 7-5 and 7-6.

key point *Vitamins are essential, noncaloric nutrients, needed in tiny amounts in the diet, that help to drive cell processes. The fat-soluble vitamins are vitamins A, D, E, and K; the water-soluble vitamins are the B vitamins and vitamin C.*

The Fat-Soluble Vitamins

The fat-soluble vitamins—A, D, E, and K—are found in the fats and oils of foods and require bile for absorption. Once absorbed, these vitamins are stored in the liver and fatty tissues until the body needs them. The body can survive weeks of consuming foods that lack these vitamins, as long as the *average* amounts provided by the diet over several months approximate the recommended intakes. This capacity to be stored also sets the stage for toxic buildup if you take in too much. Excesses of vitamins A and D from supplements can reach toxic levels especially easily.

Deficiencies of the fat-soluble vitamins are likely when the diet is consistently low in them. We also know that any disease that produces fat malabsorption (such as liver disease that prevents bile production) can cause the loss of vitamins dissolved in undigested fat and so bring about deficiencies. In the same way, a person who uses mineral oil (which the body can't absorb) as a laxative risks losing fat-soluble vitamins because they dissolve into the oil and are excreted. Deficiencies are also likely when people eat diets that are extraordinarily low in fat because such diets interfere with absorption of these vitamins.

Fat-soluble vitamins play diverse roles in the body. Vitamins A and D may act somewhat like hormones, directing cells to convert one substance to another, to store this, or to release that.[2] Many of their effects are exerted at the level of the genes, directing and regulating protein production.[3] Vitamin E flows throughout the body preventing oxidative destruction of tissues. Vitamin K is necessary for blood to clot. Each vitamin is worth a book in itself.

beta-carotene an orange pigment with antioxidant activity; a vitamin A precursor made by plants and stored in human fat tissue.

retinol one of the active forms of vitamin A made from beta-carotene in animal and human bodies; an antioxidant nutrient. Other active forms are *retinal* and *retinoic acid*.

retina (RET-in-uh) the layer of light-sensitive nerve cells lining the back of the inside of the eye.

cornea (KOR-nee-uh) the hard, transparent membrane covering the outside of the eye.

rhodopsin (roh-DOP-sin) the light-sensitive pigment of the cells in the retina, containing vitamin A (*rhod* refers to the rod-shaped cells; *opsin* means "visual protein").

night blindness slow recovery of vision after exposure to flashes of bright light at night; an early symptom of vitamin A deficiency.

Characteristics fat-soluble vitamins share:

- Dissolve in lipid.
- Require bile for absorption.
- Are stored in tissues.
- May be toxic in excess.

Look back at Figure 5-7 of Chapter 5 to see how bile acts in fat absorption.

Vitamin A

Vitamin A has the distinction of being the first fat-soluble vitamin to be recognized. Today, after almost a century of research, vitamin A and its plant-derived precursor, **beta-carotene,** are still very much a focus of research.

Three forms of vitamin A are active in the body; one of the active forms, **retinol,** is stored in the liver. The liver makes retinol available to the bloodstream and thereby to the body cells. The cells convert retinol to its other two active forms, retinal and retinoic acid, as needed.

An acne medication and a wrinkle cream contain retinoic acid—see Chapter 13.

A Jack of All Trades—Vitamin A Vitamin A is a versatile vitamin, with roles in vision, immune defenses, maintenance of body linings and skin, bone and body growth, normal cell development, and reproduction.[4] In short, vitamin A is needed everywhere (for a listing of vitamin A's chief functions in the body, consult Table 7-5 on page 245).

The most familiar function of vitamin A is in eyesight. Vitamin A plays indispensable roles in two areas: in the perception of light at the **retina** and in the maintenance of a healthy, crystal-clear outer window, the **cornea** (see the margin drawing).

When light falls on the eye, it passes through the clear cornea and strikes the cells of the retina, bleaching many molecules of the pigment **rhodopsin** that lie within those cells. Vitamin A is a part of the rhodopsin molecule. When bleaching occurs, the vitamin is broken off, initiating the signal that conveys the sensation of sight to the optic center in the brain. The vitamin then reunites with the pigment, but a little vitamin A is destroyed each time this reaction takes place, and fresh vitamin A is needed to replenish the supply. If the supply begins to run low, a lag occurs before the eye can see again after a flash of bright light at night (see Figure 7-1 on the next page). This lag in the recovery of night vision, termed **night blindness,** may indicate a vitamin A deficiency.[5] A bright flash of light can temporarily blind even normal,

An eye (sectioned).

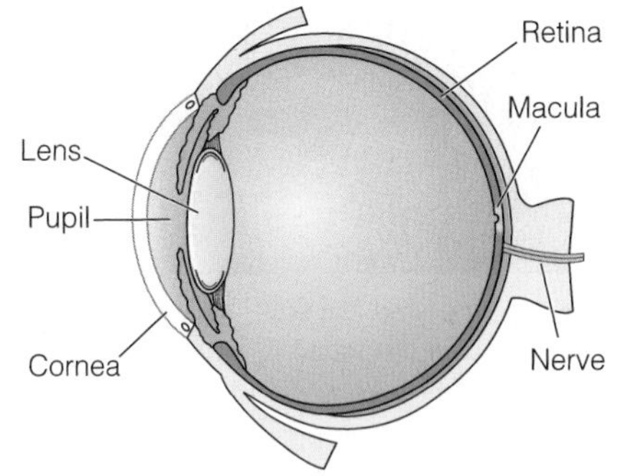

keratin (KERR-uh-tin) the normal protein of hair and nails.

keratinization accumulation of keratin in a tissue; a sign of vitamin A deficiency.

xerosis (zeer-OH-sis) drying of the cornea; a symptom of vitamin A deficiency.

xerophthalmia (ZEER-ahf-THALL-me-uh) hardening of the cornea of the eye in advanced vitamin A deficiency that can lead to blindness (*xero* means "dry"; *ophthalm* means "eye").

epithelial (ep-ith-THEE-lee-ull) **tissue** the layers of the body that serve as selective barriers to environmental factors. Examples are the cornea, the skin, the respiratory tract lining, and the lining of the digestive tract.

cell differentiation (dih-fer-en-she-AY-shun) the process by which immature cells are stimulated to mature and gain the ability to perform functions characteristic of their cell type.

well-nourished eyes, but if you experience a long recovery period before vision returns, your health-care provider may want to check your vitamin A intake.

A more profound deficiency of vitamin A is exhibited when the protein **keratin** accumulates and clouds the eye's outer vitamin A–dependent part, the cornea. The condition is known as **keratinization,** and if the deficiency of vitamin A is not corrected, it can worsen to **xerosis** (drying) and then to thickening and permanent blindness, **xerophthalmia.** Tragically, a half million of the world's vitamin A–deprived children become blind each year from this often preventable condition.[6] If the deficiency is discovered early, capsules containing 60,000 micrograms of vitamin A taken twice each year can reverse it. Better still, a child fed fruits and vegetables regularly is virtually assured protection.

Vitamin A is needed by all **epithelial tissue** (external skin and internal linings), not just by the cornea. The skin and all of the protective linings of the lungs, intestines, vagina, urinary tract, and bladder serve as barriers to infection by bacteria and to damage from other sources. Vitamin A works behind the scenes at the genetic level to promote the process of **cell differentiation,** in which each type of cell develops to perform a different specific function. For example, when goblet cells (cells found in linings of body organs) mature, they specialize in synthesizing and releasing mucus to protect those tissues from toxic particles or bacteria and other microbial invaders.

If vitamin A is deficient, the differentiation and maturing process is impaired. Goblet cells, among others, fail to mature, then fail to make protective mucus, and eventually die off. Some of the cells in these areas are displaced by cells that secrete keratin, the protein mentioned above. Keratin is the same protein that provides toughness in hair and fingernails, but in the wrong place, like our skin, keratin makes the tissue surfaces dry, hard, and cracked (see Figure 7-2). As dead cells accumulate on the surface, the tissue becomes vulnerable to infection. In the cornea, keratinization leads to xerophthalmia; in the lungs, the displacement of mucus-producing cells makes respiratory infections likely; in the vagina, the same process leads to vaginal infections.

Vitamin A has gained a reputation as an "anti-infective" vitamin because so many of the body's defenses against infection depend on an adequate supply.[7] An emerging area of research concerns the need for vitamin A in the regulation of the genes that produce proteins involved in immunity.[8] Without sufficient vitamin A, these genetic interactions produce an altered response to infection that weakens the body's defenses against disease.

Figure 7•1

NIGHT BLINDNESS

This is one of the earliest signs of vitamin A deficiency.

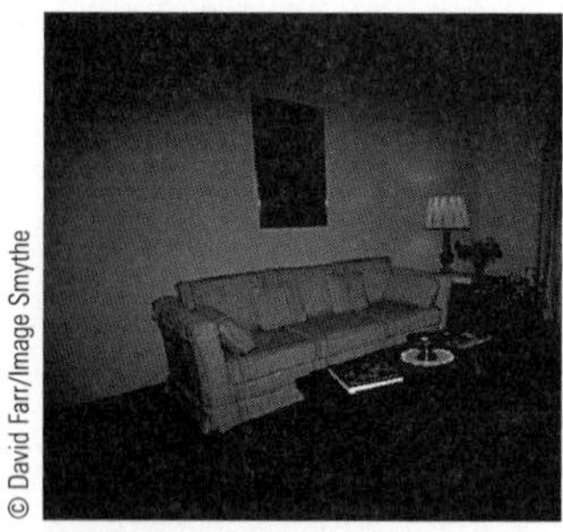

In dim light, you can make out the details in this room.

A flash of bright light momentarily blinds you as the pigment in the retina is bleached.

You quickly recover and can see the details again in a few seconds.

With inadequate vitamin A, you do not recover but remain blind for many seconds; this is night blindness.

Colorful foods often are rich in vitamins.

© David Ulmer/Stock, Boston/PictureQuest

When the defenses are weak, especially in vitamin A–deficient children, an illness such as measles can become severe. A downward spiral of malnutrition and infection can set in. The child's body must devote its scanty store of vitamin A to the immune system's fight against measles virus, but the infection causes vitamin A to be lost from the body.[9] As vitamin A dwindles, the infection worsens. Even if the child survives the measles infection, blindness is likely. The corneas, already damaged by the chronic vitamin A shortage, degenerate rapidly as their meager supply of vitamin A is diverted to the immune system.

Vitamin A also assists in bone growth. Normal children's bones grow longer, and the children grow taller, by remodeling each old bone into a new, bigger version. To do so, the body dismantles the old bone structures and replaces them with new, larger bone parts. Growth cannot take place just by adding on to the original small bone; vitamin A is needed in the critical dismantling steps. In children, failure to grow is one of the first signs of poor vitamin A status. Restoring vitamin A to such children is imperative, but correcting dietary deficiencies may be more effective than giving vitamin A supplements alone; other nutrients from nutritious food are also needed for children to gain weight and grow taller.[10]

Figure 7•2

THE SKIN IN VITAMIN A DEFICIENCY

The hard lumps on the skin of this person's arm reflect accumulations of keratin in the epithelial cells.

H. Sanstead, University of Texas

Vitamin A Deficiency around the World Although uncommon in developed countries, vitamin A deficiency remains a vast problem worldwide, placing a heavy burden on society.[11] Between three and ten million of the world's children suffer from signs of severe vitamin A deficiency—not only xerophthalmia and blindness but also diarrhea and reduced food intake that rapidly worsen their condition.[12] A staggering 275 million more children suffer from milder deficiency that impairs immunity and promotes infections. In some areas of the world, vitamin A and other deficiencies seem to be the rule, rather than the exception, among new mothers and infants.[13] In countries where supplements of vitamin A have been made available to children, childhood death rates have declined by half. The World Health Organization (WHO) and UNICEF (United Nations International Children's Emergency Fund) are working to eliminate vitamin A deficiency and thereby improve child survival throughout the developing world.

Through genetic technology, researchers have developed rice that is rich in beta-carotene to provide vitamin A to many of the world's malnourished people. See Controversy 14.

Vitamin A Toxicity For people who take excess vitamin A in supplements, Figure 7-3 (next page) shows that toxicity presents a danger equal to that of deficiency. Toxicity's many symptoms include abdominal pain, hair loss, joint pain, stunted growth,

carotenoid (CARE-oh-ten-oyd) any of a group of pigments in foods ranging from light yellow to reddish orange. Carotenoids are chemical relatives of beta-carotene; many have a degree of vitamin A activity in the body. Also defined in Controversy 2.

The effects of excessive vitamin A intakes during pregnancy are discussed in Chapter 12.

Figure 7•3

VITAMIN A DEFICIENCY AND TOXICITY

Danger lies both above and below a normal range of intakes of vitamin A.

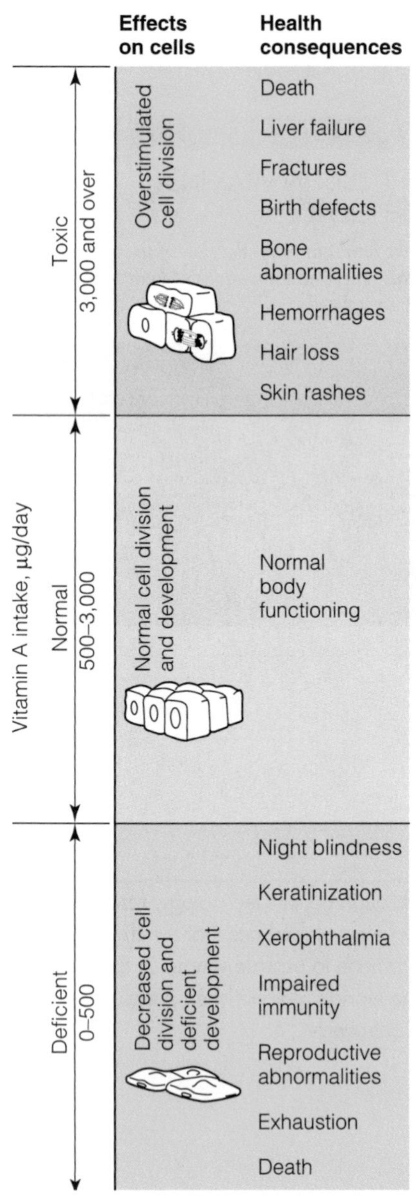

bone and muscle soreness, cessation of menstruation, nausea, diarrhea, rashes, and serious damage to the liver. Chronic high intake may weaken the bones of older women.[14] The earliest symptoms of overdoses are loss of appetite, blurred vision, growth failure in children, headache, fever, itching of the skin, and irritability. Pregnant women should be wary—chronic use of vitamin A supplements providing three to four times the amount recommended for pregnancy has caused malformations of the fetus. Even a single massive vitamin A dose (100 times the need) will do so. Children, who often mistake chewable vitamin pills for candy, are also likely to be hurt from vitamin A excesses because they need less and are more sensitive to overdoses. Adolescents may take massive vitamin A doses in the mistaken belief that vitamin A can correct acne. An effective acne medicine, Accutane, is *derived* from vitamin A, but it is chemically altered and given in carefully controlled dosages—vitamin A itself has no effect on acne.

Healthy people can eat vitamin A–rich foods in large amounts without risking toxicity, with the possible exception of liver. When laboratory pigs eat chow made from salmon parts, including the livers, the animals stop growing and fall ill from vitamin A toxicity.[15] Inuit people and Arctic explorers know that polar bear livers contain large enough amounts of the vitamin to be a dangerous food source because the bears eat fish whole with the livers.

Sources of Vitamin A Active vitamin A is provided in foods of animal origin. The richest sources are liver and fish oil, but milk and milk products and fortified cereals can also be good sources. Even butter and eggs provide some vitamin A to the diet. Plants contain no active vitamin A, but many vegetables and fruits contain the vitamin A precursor, beta-carotene. Snapshot 7-1 is the first of a series of figures that show a sampling of foods that provide more than 10 percent of the Daily Value for a vitamin in a standard-size serving and therefore qualify to be called "good" or "rich" sources.

The definitive fast-food meal—a hamburger, fries, and cola—lacks vitamin A. Many fast-food restaurants, however, now offer salads with cheese and carrots and other vitamin A–rich foods. These selections greatly improve the nutritional quality of a fast-food meal.

Vitamin A Recommendations The amount of vitamin A you need is proportional to your body weight. According to the DRI committee, a man needs a daily average of about 900 micrograms; a woman needs about 700 micrograms and more during lactation; children need less. A regular balanced diet that includes the recommended five or more servings of vegetables and fruits each day supplies more than adequate amounts. Although the vitamin A intake recommendation is given as a daily amount, it is not necessary to consume the vitamin every day. An average intake that meets the daily need over several months is sufficient.

As for vitamin A supplements, the National Research Council (NRC) and other nutrition agencies recommend that people avoid taking supplements that exceed DRI intake recommendations. The Tolerable Upper Intake Level for vitamin A has been set at 3,000 micrograms for adults over age 18. The best way to ensure a safe intake of vitamin A is to steer clear of supplements and obtain it instead from foods.

key point *Vitamin A is essential to vision, integrity of epithelial tissue, bone growth, reproduction, and more. Vitamin A deficiency causes blindness, sickness, and death and is a major problem worldwide. Overdoses are possible and cause many serious symptoms. Foods are preferable to supplements for supplying vitamin A.*

Beta-Carotene In plants, vitamin A exists only in its precursor forms. Beta-carotene, the most abundant of these **carotenoid** precursors, has the highest vitamin A activity. The conversion of beta-carotene to retinol in the body entails losses, however. It takes about 12 micrograms of beta-carotene from food to supply the equivalent of one microgram of retinol to the body. Scientists also recognize beta-carotene and its other carotene relatives for their antioxidant actions in the body.

7.1

VITAMIN A AND BETA-CAROTENE

These foods provide 10 percent or more of the vitamin A Daily Value (DV = 900 μg/day).[a]

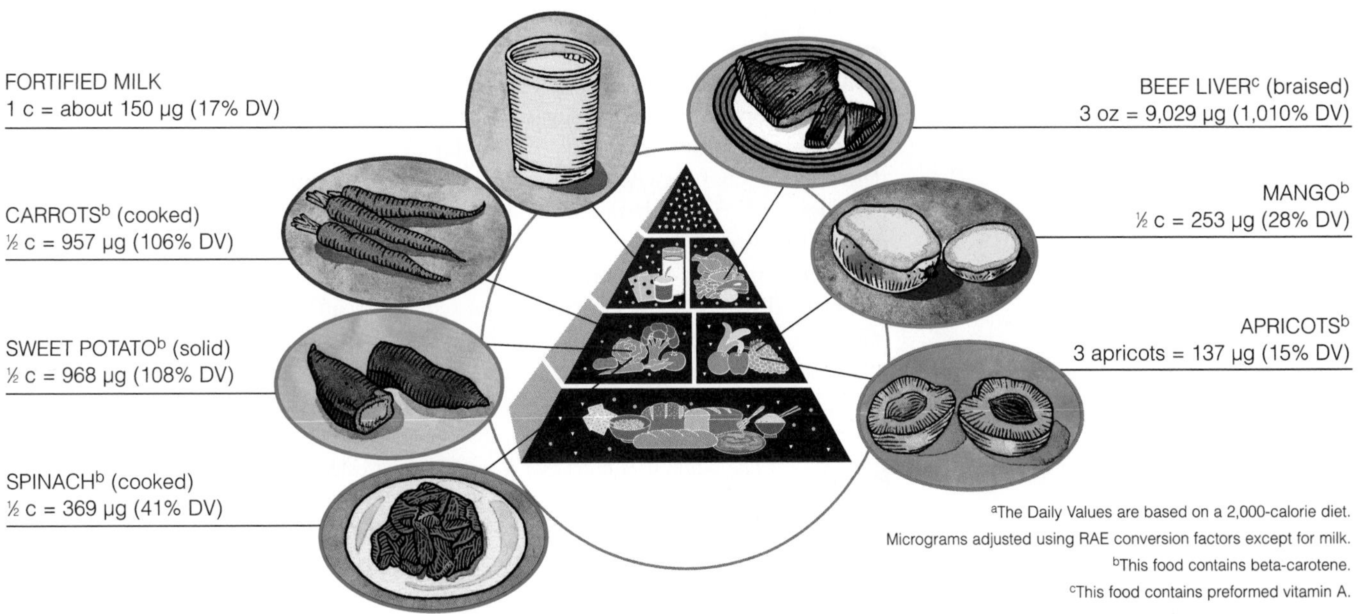

[a]The Daily Values are based on a 2,000-calorie diet. Micrograms adjusted using RAE conversion factors except for milk.
[b]This food contains beta-carotene.
[c]This food contains preformed vitamin A.

Retinol in excess is toxic, but beta-carotene is not—it is not converted to retinol efficiently enough to cause toxicity symptoms. Beta-carotene has, however, been known to turn people bright yellow if they eat too much because it builds up in the fat just beneath the skin and imparts a yellow cast.

Many foods from plants contain beta-carotene. Bright orange fruits and vegetables—so bright in color that they decorate the plate—are rich sources. They include carrots, sweet potatoes, pumpkins, mango, cantaloupe, and apricots. Another colorful group, *dark* green vegetables, such as spinach, other greens, and broccoli, owe their color to the green pigment chlorophyll and to beta-carotene. The green and orange pigments together give a deep dark green color to the vegetables.

Other colorful vegetables, such as red cabbage, beets, and sweet corn, can fool you into thinking they contain beta-carotene, but these foods derive their colors from other pigments and are poor sources of beta-carotene. As for "white" plant foods such as grains and potatoes, they have none. Some confusion exists concerning the term *yam*. The white-fleshed Mexican root vegetable called "yam" is devoid of beta-carotene, but the orange-fleshed sweet potato called "yam" in the United States is one of the richest beta-carotene sources known. Recommendations say that a person should eat *deep* orange or *dark* green vegetables and fruits regularly.

A Note about Vitamin A Values In making recommendations for vitamin A, the DRI committee ran into a problem—how to account for the difference in vitamin A activity that the body derives from a microgram of plant beta-carotene and a microgram of retinol from animal sources and supplements. The answer was **retinol activity equivalents (RAE).*** RAE units adjust for vitamin A losses incurred when the body converts micrograms of beta-carotene and other carotenoids from plant foods to microgram equivalents of active retinol. In case you are curious, the RAE conversion factors are listed in the margin. For convenience, the vitamin A micrograms used in this book, including most entries in the table of food composition

IU (international unit) a measure of fat-soluble vitamin activity sometimes used on supplement labels.

retinol activity equivalents (RAE) a measure of vitamin A activity of beta-carotene and other vitamin A precursors that reflects the amount of retinol that the body will derive from a food containing the precursors.

Standards for vitamin A intake are listed on the inside front cover, page B.

Controversy 2 discussed the carotene family of phytochemicals and explored their disease-fighting potential.

RAE conversion factors

1 retinol activity equivalent (RAE) equals:

- 1 μg retinol, or
- 12 μg beta-carotene, or
- 24 μg other carotenoids.

*An older unit of measure, retinol equivalents (RE), has been replaced by the DRI committee.

macular degeneration a common, progressive loss of function of the part of the retina that is most crucial to focused vision (the macula is shown on page 213). This degeneration often leads to blindness.

rickets the vitamin D–deficiency disease in children; characterized by abnormal growth of bone and manifested in bowed legs or knock-knees, outward-bowed chest, and knobs on the ribs.

of Appendix A, are expressed as RAE units to make them compatible with DRI recommendations.

Some food tables and supplement labels still express vitamin A contents using a different unit, the **IU (international unit).** When comparing vitamin A in foods or supplements, make sure that the amounts are all expressed in the same units. See the Aids to Calculations (Appendix C) for help in converting many kinds of units.

Does Eating Carrots Really Protect the Health of the Eyes? Population studies suggest that people whose diets lack foods rich in beta-carotene have a high incidence of **macular degeneration,** a common form of blindness in the aged. Likewise, certain types of cancer are linked with a lifelong diet that excludes foods rich in beta-carotene. Research has not supported a protective effect of supplements of beta-carotene, however.[16] The important link seems to exist between eating beta-carotene–rich *foods*, such as carrots, and low rates of eye diseases and other diseases. So yes, carrots do protect the health of the eyes. Even there, the effect is probably not attributable to beta-carotene itself but to a related carotenoid compound, also present in carrots and other beta-carotene-rich foods, as this chapter's Controversy concludes.

key point *The vitamin A precursor in plants, beta-carotene, is an effective antioxidant in the body. Brightly colored plant foods are richest in beta-carotene, and diets containing these foods are associated with eye health.*

Vitamin D

Vitamin D is different from all the other nutrients in the body in that the body can synthesize all it needs with the help of sunlight. Therefore, in a sense, vitamin D is not an essential nutrient. Given enough sun each day, most people need consume no vitamin D at all from foods.

The sunshine vitamin: vitamin D.

© Fotografia/CORBIS

Roles of Vitamin D The best-known role of vitamin D is as a member of a large cast of nutrients and hormones that interact to maintain blood calcium and phosphorus levels and thereby bone integrity, which is especially important during growth. Many of these interactions take place at the genetic level of cellular function in ways that are just now beginning to be understood.[17]

Calcium is indispensable to the proper functioning of all tissues of the body; cells of muscles, nerves, glands, and others all draw calcium from the blood as they need it. The skeleton serves as a vast warehouse of stored calcium that can be tapped when the supply in the bloodstream begins to fall even slightly. To raise the level of blood calcium, the body can draw from only two other places besides the bones: the digestive tract, where food brings calcium in, and the kidneys, which can recycle calcium into the body from blood filtrate destined to become urine. When calcium is needed, vitamin D acts at all three locations to raise the blood calcium level.

Vitamin D functions as a hormone, that is, a compound manufactured by one organ of the body that acts on other organs or tissues. In addition to its actions in the bones, intestines, and kidneys, vitamin D plays roles in the workings of the brain, heart, pancreas, skin, and reproductive organs. Like vitamin A, vitamin D stimulates maturation of cells, including cells of the immune system.

Too Little Vitamin D—A Danger to Bones The most obvious sign of vitamin D deficiency is abnormality of the bones in the disease **rickets.** Children with rickets develop bowed legs because their leg bones are too weak to support their body weight, and they have a protruding belly because of lax abdominal muscles.

Chapter 8 and Controversy 8 present more about bone minerals and their regulation, and about osteoporosis, the bone-weakening disease.

As early as the 1700s, rickets was known to be curable with cod-liver oil, which is rich in vitamin D. More than a hundred years later, a Polish physician linked sunlight exposure to the prevention and cure of rickets. Today, the bowed legs, knock-knees, beaded ribs, and protruding (pigeon) chests of children with rickets are no longer common sights in the United States, although several new cases of rickets have been reported among infants and toddlers fed unfortified soy and rice beverages instead of formula or milk.[18] Many children worldwide suffer the ravages of rickets because of inadequate food combined with a lack of sunlight.

Adult rickets, or **osteomalacia,** occurs most often in women with low calcium intakes and little exposure to the sun who have repeated pregnancies and then breast-feed their babies. Under these conditions, calcium is withdrawn from the bones, but is not picked up efficiently from the intestine or recycled by the kidneys. The bones of the legs and spine can soften to such an extent that a young woman who is tall and straight at the age of 20 years may, after several pregnancies, become bowlegged and bent by age 30.

osteomalacia (OS-tee-o-mal-AY-shuh) the vitamin D–deficiency disease in adults (*osteo* means "bone"; *mal* means "bad"). Symptoms include bending of the spine and bowing of the legs.

Too much sun is dangerous—it may trigger the start of skin cancer.

Too Much Vitamin D—A Danger to Soft Tissues Vitamin D is the most potentially toxic of all vitamins. Chronic ingestion of excesses may be directly toxic to the bones, kidneys, brain, nerves, and the heart and arteries. Symptoms of toxicity include appetite loss, nausea, vomiting, and increased urination and thirst, and a severe form of psychological depression can result from vitamin D's effects on the central nervous system.[19] If overdoses continue, vitamin D raises the blood mineral level to dangerous extremes, forcing calcium to be deposited in soft tissues such as the heart, blood vessels, lungs, and kidneys. Even the soft pulp of the teeth hardens, while tooth enamel thins. Calcium deposited in critical organs may cause them to malfunction, with serious consequences for health and life.[20]

The likeliest victims of vitamin D poisoning are infants whose well-intentioned but misguided parents think that if some is good, more is better. Also, older people who take vitamin D supplements to stem adult bone loss may also easily overdose, not realizing that their tissues are building up stockpiles of the vitamin. Intakes of only five times the recommended amount have been associated with signs of vitamin D toxicity in young children and adults. Two people died and others became ill after drinking milk from a dairy that had mistakenly overfortified the milk with up to 500 times the usual dosage of vitamin D. One infant survivor later developed dental problems because the vitamin D overdose disturbed the normal development of her permanent teeth.[21] Such instances are rare, but the incident renewed awareness of the potential for harm from vitamin D and the need for close monitoring of those who fortify the nation's foods with vitamins. The DRI committee has set a Tolerable Upper Intake Level for vitamin D at 50 micrograms per day (2,000 IU on supplement labels).

This child has the bowed legs of the vitamin D–deficiency disease rickets.

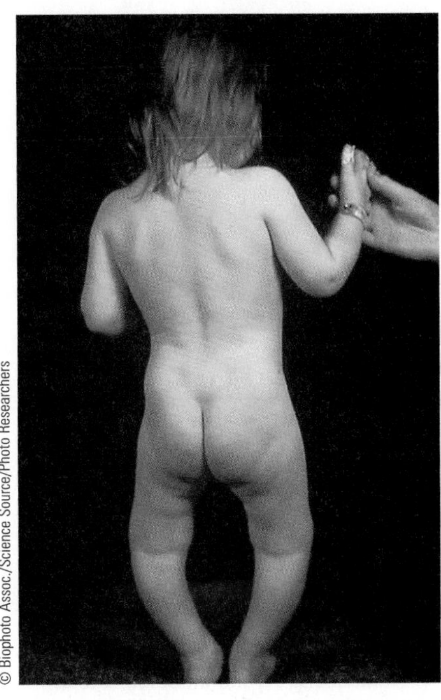

How Can People Make a Vitamin from Sunlight? Most of the world's population relies on natural exposure to sunlight to maintain adequate vitamin D nutrition. When ultraviolet light from the sun shines on a cholesterol compound in human skin, the compound is transformed into a vitamin D precursor and is absorbed directly into the blood. Slowly, over the next day and a half, the liver and kidneys finish converting the precursor to the active form of vitamin D. Diseases that affect either the liver or the kidneys can impair the conversion of the inactive precursor to the active vitamin and therefore lead to symptoms of vitamin D deficiency.

Unlike concentrated supplements, sunlight presents no risk of vitamin D toxicity; the sun itself begins breaking down excess vitamin D made in the skin. Sunbathers run *other* risks, of course, such as premature wrinkling of the skin and the increased risk of skin cancer. Sunscreens with sun protection factors (SPF) of 8 and above can reduce these risks, but they also prevent vitamin D synthesis. Making vitamin D doesn't require many idle hours sunbathing, however. In the warmer months in most locations, just being outdoors, even in lightweight clothing, is sufficient. The pigments of dark skin provide protection from ultraviolet radiation but also reduce vitamin D synthesis.[22] Dark-skinned people therefore require longer exposure to direct sun (up to three hours, depending on the climate) to obtain several days' worth of vitamin D, but light-skinned people need much less time (10 or 15 minutes). One successful strategy is to wait until just enough time has elapsed to make some vitamin D and then apply sunscreen. Tanning booths may or may not promote vitamin D

This child displays the beaded ribs common in rickets.

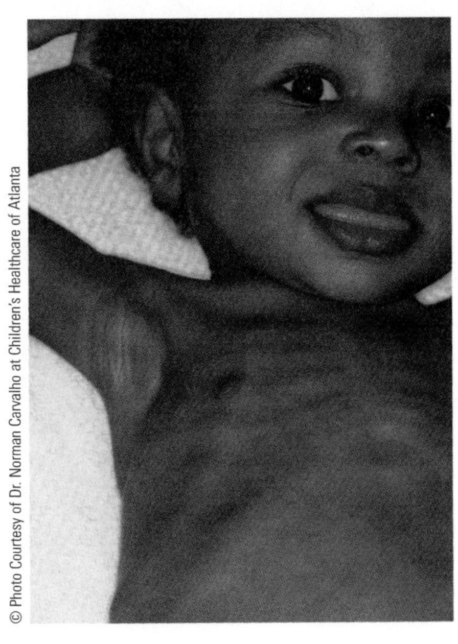

VITAMIN D

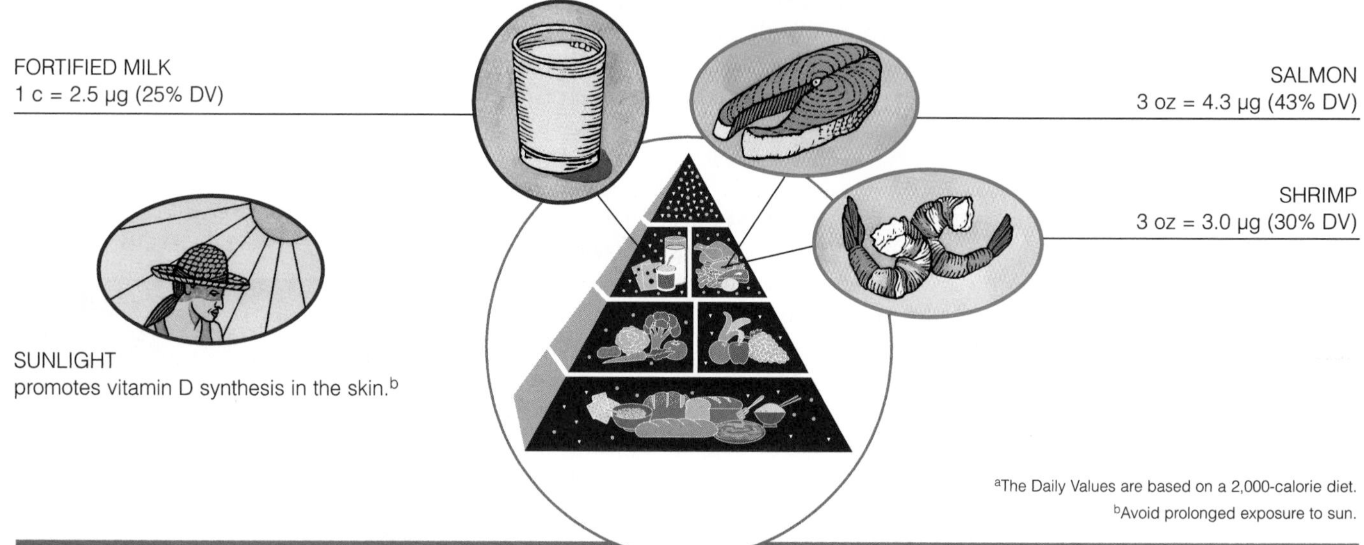

Factors affecting sun exposure and vitamin D synthesis:

- *Air pollution.* Particles in the air screen out the sun's rays.
- *City living.* Tall buildings block sunlight.
- *Clothing.* Heavy clothing blocks sunlight.
- *Geography.* Southern locations receive more direct sun exposure.
- *Homebound.* Living indoors prevents sun exposure.
- *Season.* Warmer seasons of the year bring more direct sun rays.
- *Sunscreen.* Use reduces or prevents skin exposure to sun's rays.
- *Time of day.* Midday hours provide maximum direct sun exposure.

synthesis, but the Food and Drug Administration (FDA) has declared them risky because their unfiltered rays may promote skin cancer and damage the blood vessels and eyes. Daily doses of vitamin D are not necessary because the body stores enough vitamin D in its fat tissue to last through the dark winter months.

The ultraviolet rays of the sun that promote vitamin D synthesis cannot penetrate clouds, smoke, smog, heavy clothing, window glass, or even window screens. In the United States and Canada, almost all cases of rickets show up in dark-skinned people who live in smoggy northern cities or who lack exposure to sunlight. People who are housebound, institutionalized, or work at night may incur (over years) a vitamin D deficiency, as may elderly adults who drink little milk, have limited exposure to sunlight, and become less efficient at activating vitamin D as they age. Because of these risks, the DRI committee set recommended intakes for vitamin D that increase with age: 5 micrograms per day for adults 19 to 50 years, 10 micrograms for those 51 to 70 years, and 15 micrograms for those over 70.

Snapshot 7-2 shows the few significant food sources of vitamin D. Butter, cream, and fortified margarine contribute small amounts. In the United States and Canada, milk, whether fluid, dried, or evaporated, is fortified with vitamin D. Young adults who drink the recommended 2 cups a day receive half their daily requirement; the other half comes from exposure to sunlight and other food sources. A daily quart (or liter) of milk will supply the entire recommended amount. Children who drink 2 cups or more of milk a day will have a head start toward meeting their vitamin D needs for growth. Yogurt and cheese products are often not fortified, so read the labels. Strict vegetarians and their children may have low vitamin D intakes because only two fortified plant sources exist: some margarines and, in the United States, certain fortified cereals.

key point *Vitamin D raises mineral levels in the blood, notably calcium and phosphorus, permitting bone formation and maintenance. A deficiency can cause rickets in childhood or osteomalacia in later life. Vitamin D is the most toxic of all the vitamins, and excesses are dangerous or deadly. People exposed to the sun make vitamin D from a cholesterol-like compound in their skin; fortified milk is an important food source.*

Vitamin E

More than 80 years ago, researchers discovered a compound in vegetable oils necessary for reproduction in rats. This compound was named **tocopherol** from *tokos*, a Greek word meaning "offspring." A few years later, the compound was named vitamin E. Four tocopherol compounds have been identified, and each is designated by one of the first four letters of the Greek alphabet: alpha, beta, gamma, and delta. Of these, alpha-tocopherol is the gold standard for vitamin E activity in the body; the DRI intake recommendations are based on alpha-tocopherol.*

tocopherol (tuh-KOFF-er-all) a kind of alcohol. The active form of vitamin E is alpha-tocopherol.

erythrocyte (eh-REETH-ro-sight) **hemolysis** (HE-moh-LIE-sis, he-MOLL-ih-sis) rupture of the red blood cells, caused by vitamin E deficiency (*erythro* means "red"; *cyte* means "cell"; *hemo* means "blood"; *lysis* means "breaking").

The Extraordinary Bodyguard Vitamin E is an antioxidant and thus serves as one of the body's main defenders against oxidative damage. By being oxidized itself, vitamin E protects the polyunsaturated fats and other vulnerable components of the cells and their membranes from destruction. Vitamin E protects all the cells' lipids and related compounds, such as vitamin A, from oxidation. Vitamin E's antioxidant effect is crucial in the lungs, where the cells are exposed to high oxygen concentrations that can destroy molecules in their membranes. As the red blood cells carry oxygen from the lungs to other tissues, vitamin E protects their cell membranes, too.

Normal nerve development depends on vitamin E. Vitamin E also protects the white blood cells that defend the body against disease, and it may play other roles in normal immunity. Supplements of the vitamin were found to improve the immune response in healthy elderly people.[23] Vitamin E may also help defend against heart disease; this chapter's Controversy provides details.

Vitamin E Deficiency A deficiency of vitamin E produces a wide variety of symptoms in laboratory animals. Most of these symptoms have not been reproduced in human beings, however, despite many attempts. Three reasons have been given for this. First, the vitamin is so widespread in food that it is almost impossible to create a vitamin E–deficient diet. Second, the body stores so much vitamin E in its fatty tissues that a person would find it difficult to eat a vitamin E–free diet for long enough to deplete these stores and produce a deficiency. Third, the cells recycle their working supply of vitamin E, using the same molecules over and over to ward off deficiency.

The classic vitamin E–deficiency symptom in human beings occurs in premature babies who are born before the transfer of the vitamin from the mother to the infant, which takes place in the last weeks of pregnancy. Without sufficient vitamin E, the infant's red blood cells rupture **(erythrocyte hemolysis),** and the infant becomes anemic. The few symptoms of vitamin E deficiency that have been observed in adults include loss of muscle coordination and reflexes with impaired movement, vision, and speech. All of these symptoms may be caused by oxidative damage; vitamin E treatment corrects them.[24]

In adults, vitamin E deficiency is usually associated with diseases that cause malabsorption of fat, including disease or injury of the liver (which makes bile, necessary for digestion of fat), the gallbladder (which delivers bile into the intestine), and the pancreas (which makes fat-digesting enzymes). In people without diseases, low intakes of vitamin E are most likely in those who for years eat diets extremely low in fat. People who rely solely on fat replacers, such as diet margarines and fat-free salad dressings, to the exclusion of real fat may have low vitamin E intakes. Those consuming diets of highly processed or "convenience" foods may lack sufficient vitamin E because vitamin E is destroyed by extensive heating in the processing of these foods.

Researchers are currently exploring links between vitamin E and diseases. One line of research suggests that when body stores of vitamin E and the mineral selenium are low, viruses respond by becoming more virulent—even normally harmless viruses appear to undergo changes that make them more likely to cause diseases.[25] No one yet knows the details of these effects, but they may be related to "oxidative stress," caused when oxidative activities outstrip the capacity of the tissues' antioxidant defenses.

More about oxidative stress in the Controversy section of this chapter.

*The DRI committee has replaced the unit *alpha-tocopherol equivalents* with *milligrams alpha-tocopherol*.

VITAMIN E

These foods provide 10 percent or more of the vitamin E Daily Value (DV = 30 IU or 20 mg/day).[a,b]

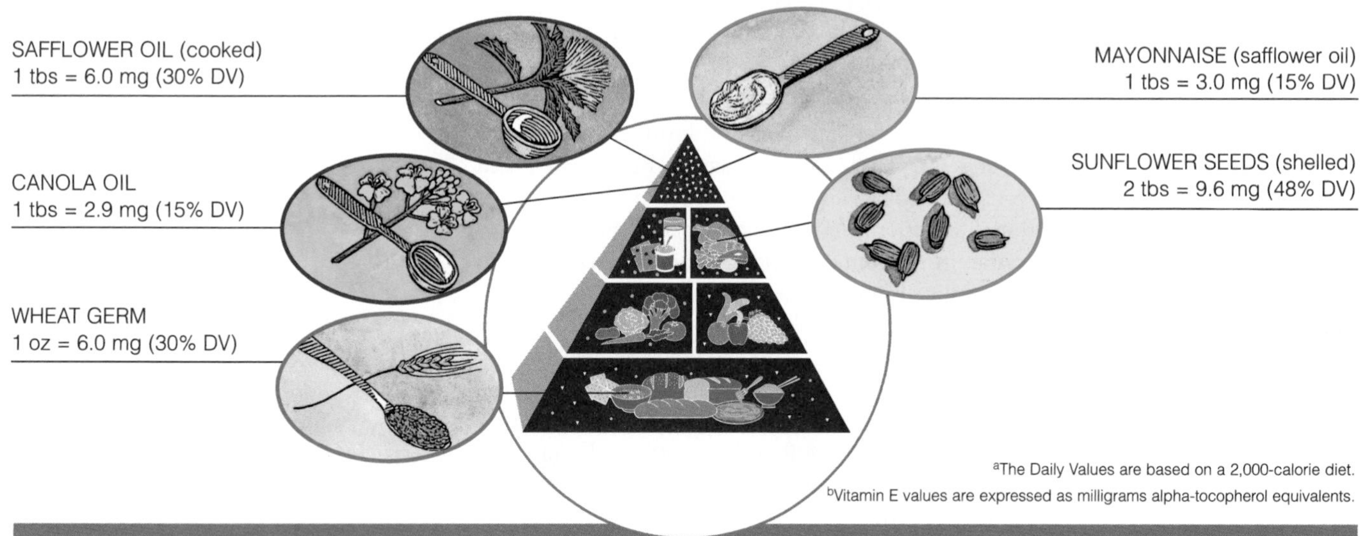

[a]The Daily Values are based on a 2,000-calorie diet.

[b]Vitamin E values are expressed as milligrams alpha-tocopherol equivalents.

Extravagant claims are often made for vitamin E because its deficiency affects animals' muscles and reproductive systems. Research in human beings has discredited all claims that vitamin E improves athletic endurance and skill, enhances sexual performance, or cures sexual dysfunction in males, although such claims are still being used to sell supplements containing vitamin E.

Vitamin E conversion factors:

1 IU natural vitamin E = .67 mg alpha-tocopherol.
1 IU synthetic vitamin E = .45 mg alpha-tocopherol.

Vitamin E Requirements, Toxicity, and Sources The DRI intake recommendation (inside front cover) for vitamin E is 15 milligrams a day for adults. This amount seems sufficient to maintain blood values for both vitamin E and indicators of oxidation reactions within healthy, normal limits. The need for vitamin E rises as people consume more polyunsaturated oil because the oil requires antioxidant protection by the vitamin. Luckily, most raw oils also contain vitamin E, so people who eat the oil also receive the vitamin. As mentioned, heat processing, such as frying, destroys vitamin E, as does oxidation, so most processed, fast, deep-fried, and convenience foods retain little intact vitamin E.

No adverse effects are known to occur from naturally occurring vitamin E in foods. Ordinary supplemental doses of vitamin E taken over a period of months seem to have no adverse effects on most people's health.[26] The medical literature contains isolated reports of adverse effects from very high doses in laboratory animals and occasional reports of nausea, intestinal distress, fatigue, blurring of vision, and other vague complaints in human beings. Large doses may increase the effects of anticoagulant medication used to oppose unwanted blood clotting; people taking such drugs risk uncontrollable bleeding when they also take large doses of vitamin E. Even low doses taken over time may have unwanted effects. One large study from the early 1990s reported a 50 percent increase in brain hemorrhages, a form of stroke, among smokers taking just 50 milligrams a day of vitamin E over six years. After weighing the bulk of the research, the DRI committee suggests that most individuals can safely take daily doses of alpha-tocopherol in amounts up to 800 milligrams (530 IU).

For perspectives on possible risks and benefits of vitamin E supplements, see the Controversy.

Vitamin E is widespread in foods. About 20 percent of the vitamin E people consume comes from vegetable oils and products made from them, salad dressings, and shortening (see Snapshot 7-3). Another 20 percent comes from fruits and vegetables although none of these is a good source by itself. Fortified cereals* and other grain products contribute about 15 percent of vitamin E in the diet, and meats, poultry, fish, eggs, milk products, nuts, and seeds contribute smaller percentages. Wheat germ is a good source of vitamin E; animal fats have almost none.

key point *Vitamin E acts as an antioxidant in cell membranes and is especially important for the integrity of cells that are constantly exposed to high oxygen concentrations, namely, the lungs and blood cells, both red and white. Vitamin E deficiency is rare in human beings, but it does occur in newborn premature infants. The vitamin is widely distributed in plant foods; it is destroyed by high heat; toxicity is rare.*

Vitamin K

Have you ever thought about how remarkable it is that blood can clot? The liquid turns solid in a life-saving series of reactions—if blood did not clot, wounds would just keep bleeding. The main function of vitamin K is to help synthesize proteins that help clot the blood. Hospitals measure the clotting time of a person's blood before surgery and sometimes administer vitamin K before operations to reduce bleeding in surgery. Vitamin K may be of value at this time, but only if a vitamin K deficiency exists. Vitamin K does not improve clotting in those with other bleeding disorders, such as the genetic disease hemophilia.

K stands for the Danish word *koagulation* (clotting).

Some people with heart problems need to *prevent* the formation of clots within their circulatory system—this is popularly referred to as "thinning" the blood. One of the best-known medicines for this purpose is warfarin, which interferes with the action of vitamin K in promoting clotting. Vitamin K therapy may be needed for people on warfarin if uncontrolled bleeding should occur.[27] People taking warfarin who self-prescribe vitamin K supplements risk interfering with the action of the drug.

Vitamin K is also necessary for the synthesis of a key protein needed in bone formation.[28] Together with the more famous bone vitamin, vitamin D, vitamin K ensures that the bones produce this protein, which enables them to properly bind the minerals they need. Vitamin K intake may also play a part in reducing the risk of hip fracture: in one large study, women who ate abundant green vegetables, known sources of vitamin K, suffered hip fractures less often than those with lower intakes.[29]

Like vitamin D, vitamin K can be obtained from a nonfood source—in this case, the intestinal bacteria. Billions of bacteria normally reside in the intestines, and some of them synthesize vitamin K. Scientists have not determined how much the body uses the vitamin K synthesized by these bacteria.

Also like vitamin D, just a few types of food supply significant amounts of vitamin K. As Snapshot 7-4 on the next page shows, vitamin K's richest food sources are dark green, leafy vegetables such as cooked spinach and collard greens, which provide an average of 300 micrograms per 3-ounce serving. Lettuce, broccoli, brussels sprouts, and other members of the cabbage family provide about 100 micrograms per 3-ounce serving. Canola and soybean oils also provide significant amounts, while fortified cereals can be rich sources of added vitamin K. Tables of food composition do not include the vitamin K contents of foods although they may do so in the future.

The DRI recommendation for daily intake of vitamin K is 120 micrograms for adult men and 90 micrograms for adult women.

Few U.S. adults are likely to experience vitamin K deficiency, even if they seldom eat vitamin K–rich foods. Exceptions are people who have taken antibiotics that have killed both the beneficial and harmful bacteria in their intestinal tracts, and newborn infants whose intestinal tracts are not yet inhabited by bacteria. Supplements of the vitamin are needed in these cases.

Nutrient needs of infants are discussed in Chapter 12.

Reports of vitamin K toxicity among healthy adults are rare, and the DRI committee has set no Tolerable Upper Intake Level. For infants and pregnant women, however, vitamin K toxicity can result when supplements of a synthetic version of

*Cereals fortified with vitamin E may not be available in Canada.

7.4

VITAMIN K

The Daily Value (DV) for vitamin K is 80 μg/day.[a]

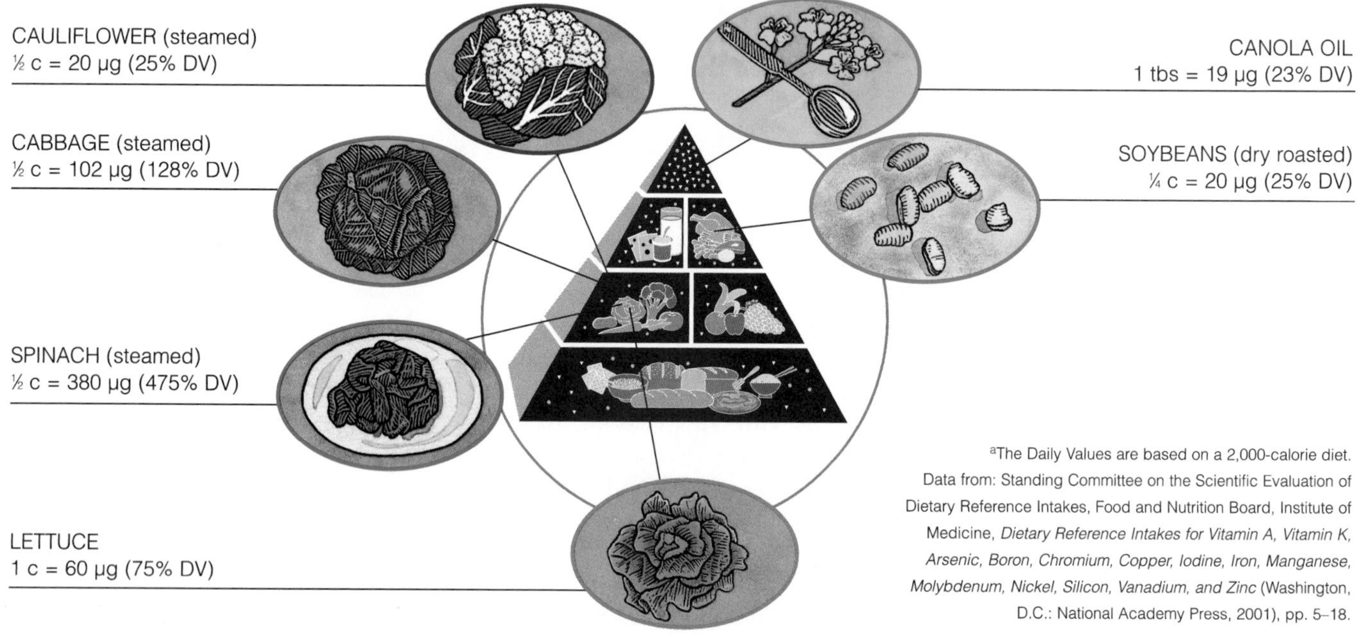

[a]The Daily Values are based on a 2,000-calorie diet.
Data from: Standing Committee on the Scientific Evaluation of Dietary Reference Intakes, Food and Nutrition Board, Institute of Medicine, *Dietary Reference Intakes for Vitamin A, Vitamin K, Arsenic, Boron, Chromium, Copper, Iodine, Iron, Manganese, Molybdenum, Nickel, Silicon, Vanadium, and Zinc* (Washington, D.C.: National Academy Press, 2001), pp. 5–18.

vitamin K are given too enthusiastically.* Toxicity induces breakage of the red blood cells and release of their pigment, which colors the skin yellow. A toxic dose of synthetic vitamin K causes the liver to release the blood cell pigment (bilirubin) into the blood (instead of excreting it into the bile) and leads to jaundice. When bilirubin invades the brain of an infant, the condition may lead to brain damage or death.

key point *Vitamin K is necessary for blood to clot; deficiency causes uncontrolled bleeding. The bacterial inhabitants of the digestive tract produce vitamin K, but the extent to which the body uses this intestinal vitamin K has not been determined.*

The Water-Soluble Vitamins

The water-soluble vitamins require special consideration in food preparation to avoid losing or destroying them. See the Food Feature of Chapter 14.

Characteristics water-soluble vitamins share:

- Dissolve in water.
- Are easily absorbed and excreted.
- Are not stored extensively in tissues.
- Seldom reach toxic levels.

The B vitamins and vitamin C are water soluble. Cooking and washing with water can leach them out of foods. The body absorbs these vitamins easily and just as easily excretes them in the urine. Some of the water-soluble vitamins can remain in the lean tissues for a month or more, but these tissues are actively exchanging materials with the body fluids at all times. At any time, the vitamins may be picked up by the extracellular fluids, washed away by the blood, and excreted in the urine. Advice for meeting the need for these nutrients is straightforward: choose foods that are rich in water-soluble vitamins to achieve an average of the recommended intakes over three days' time. The snapshots in this section can help to guide your choices.

Foods never deliver toxic doses of the water-soluble vitamins, but the large doses concentrated in some vitamin supplements can reach toxic levels. Normally, though, the most likely hazard to the supplement taker is to the wallet: "If you take supplements of the water-soluble vitamins, you may have the most expensive urine in town." The nearby Think Fitness features asks whether athletes may need supplements.

*The version of vitamin K responsible for this effect is menadione.

coenzyme (co-EN-zime) a small molecule that works with an enzyme to promote the enzyme's activity. Many coenzymes have B vitamins as part of their structure (*co* means "with").

Vitamins for Athletes

Do athletes who strive for top performance need more vitamins than foods can supply? Competitive athletes who choose their diets with reasonable care almost never need nutrient supplements. The reason is elegantly simple. The need for energy to fuel exercise requires that people eat extra calories of food, and if that extra food is of the kind shown in this chapter's snapshots—fruits, vegetables, milk, eggs, whole or enriched grains, lean meats, and even some oils—then the vitamins to support activity follow automatically. Chapter 10 explains the roles of vitamins in physical activity.

THINK FITNESS

The B Vitamins

The B vitamins act as part of coenzymes. A **coenzyme** is a small molecule that combines with an enzyme and activates it. (Recall from Chapter 6 that enzymes are large proteins that do the body's building, dismantling, and other work.) Figure 7-4 shows how a coenzyme enables an enzyme to do its job. Sometimes the vitamin part of the enzyme is the active site, where the chemical reaction takes place. The substance to be worked on is attracted to the active site and snaps into place; the reaction proceeds instantaneously. The shape of each enzyme predestines it to accomplish just one kind of job. Without its coenzyme, however, the enzyme is as useless as a car without wheels.

Each of the B vitamins has its own special nature, and the amount of detail known about each one is overwhelming. To simplify things, this introduction describes the teamwork of the B vitamins and emphasizes the consequences of deficiencies. The sections that follow present more details about the vitamins as individuals.

key point *As part of coenzymes, the B vitamins help enzymes do their jobs.*

B Vitamin Roles in Metabolism

Figure 7-5 shows some body organs and tissues in which the B vitamins help the body metabolize carbohydrates, lipids, and amino acids. The purpose of the figure is not to present a detailed account of metabolism but to give you an impression of where the B vitamins work together with enzymes in the metabolism of energy nutrients and in the making of new cells.

Many people mistakenly believe that B vitamins supply the body with energy. They do not, at least not directly. Remember that B vitamins are "helpers." The energy-yielding nutrients—carbohydrate, fat, and protein—give the body fuel for energy; the B vitamins *help* the body use that fuel. More specifically, active forms of five of the B vitamin—thiamin, riboflavin, niacin, pantothenic acid, and biotin—participate in the release of energy from carbohydrate, fat, and protein. Vitamin B_6 helps the body use amino acids to make protein; the body then puts the protein to work in many ways—to build new tissues, to make hormones, to fight infections, or to serve as fuel for energy, to name only a few.

Folate and vitamin B_{12} help cells to multiply, which is especially important for cells with short life spans that must replace themselves rapidly. Such cells include both the red blood cells (which live for about 120 days) and the cells that line the digestive tract (which replace themselves every three days). These cells deliver energy to all the others. In short, each and every B vitamin is involved, directly or indirectly, in energy metabolism.

key point *The B vitamins facilitate the work of every cell. Some help generate energy; others help make protein and new cells. B vitamins work everywhere in the body tissue to metabolize carbohydrate, fat, and protein.*

Figure 7•4

COENZYME ACTION

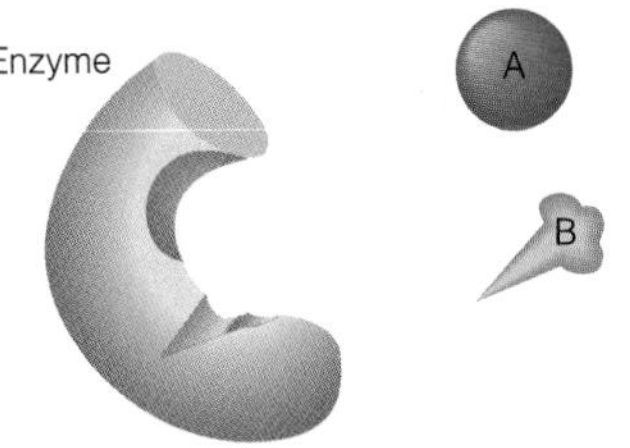

Without the coenzyme, compounds A and B don't respond to the enzyme.

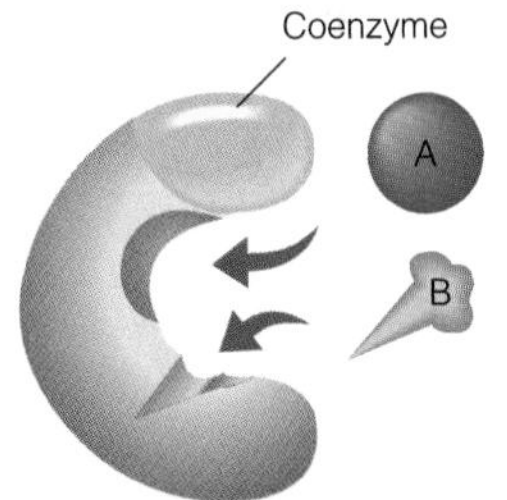

With the coenzyme in place, compounds A and B are attracted to the active site on the enzyme, and they react.

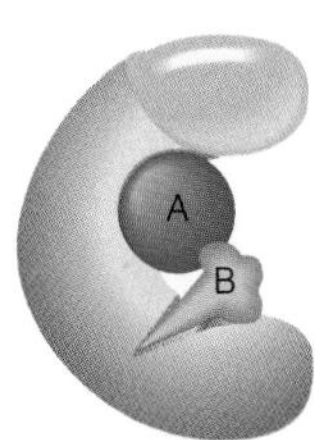

The reaction is completed with the formation of a new product. In this case the product is AB.

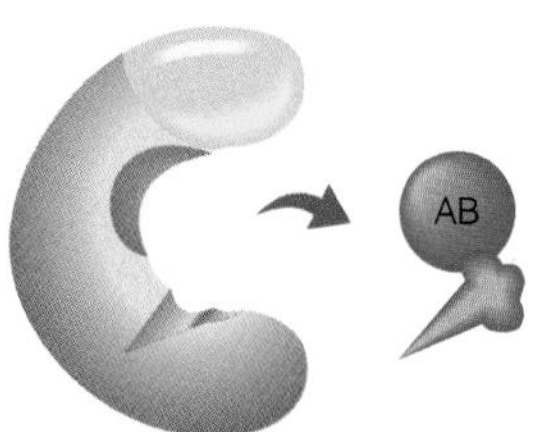

The product AB is released.

Figure 7•5

SOME ROLES OF THE B VITAMINS IN METABOLISM: EXAMPLES

This figure does not attempt to teach intricate biochemical pathways or names of B vitamin–containing enzymes. Its sole purpose is to show a few of the many tissue functions that depend on B vitamin–containing enzymes. The B vitamins work in every cell, and this figure displays less than a thousandth of what they actually do.

Every B vitamin is part of one or more coenzymes that make possible the body's chemical work. For example, the niacin, thiamin, and riboflavin coenzymes are important in the energy pathways. The folate and vitamin B_{12} coenzymes are necessary for making RNA and DNA and thus new cells. The vitamin B_6 coenzyme is necessary for processing amino acids and, therefore, protein. Many other relationships are also critical to metabolism.

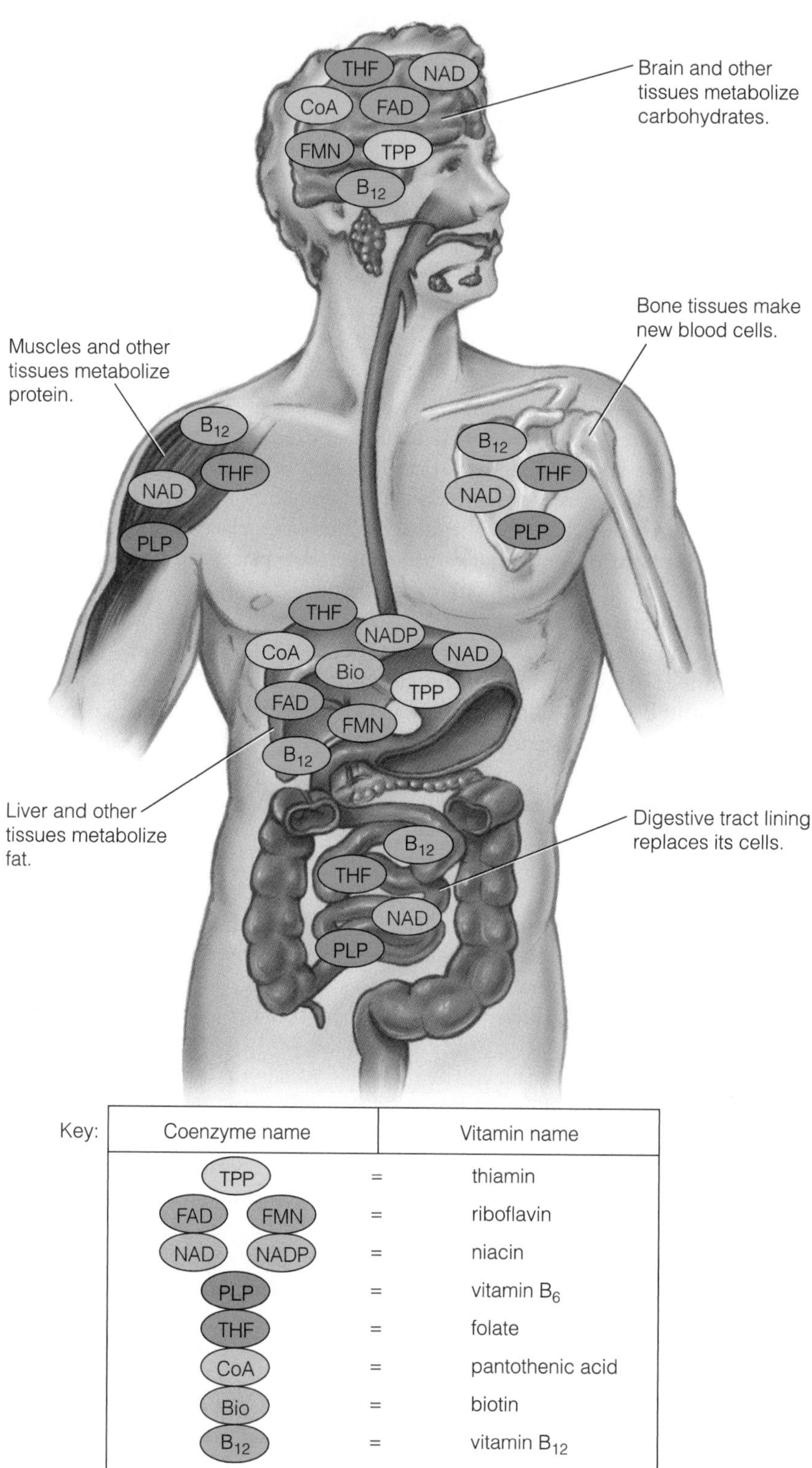

B Vitamin Deficiencies

As long as B vitamins are present, their presence is not felt. Only when they are missing does their absence manifest itself in a lack of energy and a multitude of other symptoms, as you can imagine after looking at Figure 7-5. The reactions by which B vitamins facilitate energy release take place in every cell, and no cell can do its work without energy. Thus, in a B vitamin deficiency, every cell is affected. Among the symptoms of B vitamin deficiencies are nausea, severe exhaustion, irritability, depression, forgetfulness, loss of appetite and weight, pain in muscles, impairment of the immune response, loss of control of the limbs, abnormal heart action, severe skin problems, swollen red tongue, and teary or bloodshot eyes. Because cell renewal depends on energy and protein, which in turn depend on the B vitamins, the digestive tract and the blood are invariably damaged. In children, full recovery may be impossible. In the case of a thiamin deficiency during growth, permanent brain damage can result.

In academic discussions of the vitamins, different sets of deficiency symptoms are given for each one. Such clear-cut sets of symptoms, however, are found only in laboratory animals that have been fed contrived diets that lack just one ingredient. In real life, a deficiency of any one B vitamin seldom shows up by itself because people don't eat nutrients singly; they eat foods that contain mixtures of nutrients. A deficiency of one B vitamin may appear to be responsible for a cluster of symptoms, but subtler, undetected deficiencies may accompany it. If treatment involves giving wholesome food rather than a single supplement, the subtler deficiencies will be corrected along with the major one. The symptoms of B vitamin deficiencies and toxicities are listed in Table 7-6 at the end of the chapter.

Every cell is affected by a B vitamin deficiency. Deficiencies of single B vitamins are rare; a person deficient in one is likely to be deficient in others.

thiamin (THIGH-uh-min) a B vitamin involved in the body's use of fuels.

beriberi the thiamin-deficiency disease; characterized by loss of sensation in the hands and feet, muscular weakness, advancing paralysis, and abnormal heart action.

To help remember the names of the eight B vitamins, try memorizing this sentence or make up one of your own:

Tender	(thiamin)
Romance	(riboflavin)
Never	(niacin)
Fails,	(folate)
with 6 or 12	(B_6 and B_{12})
Beautiful	(biotin)
Pearls.	(pantothenic acid)

The B Vitamins as Individuals

Although the B vitamins all work as part of coenzymes and share other characteristics, each also has its own special qualities. The next sections provide details about each B vitamin.

Thiamin and Riboflavin **Thiamin** plays a critical role in the energy metabolism of all cells. Thiamin also occupies a special site on nerve cell membranes. Consequently, nerve processes and their responding tissues, the muscles, depend heavily on thiamin.

The classic thiamin-deficiency disease **beriberi** was first observed in East Asia, where rice provided 80 to 90 percent of the total calories most people consumed and was therefore their principal source of thiamin. When the custom of polishing rice (removing its brown coat, which contained the thiamin) became widespread, beriberi swept through the population like an epidemic. Scientists wasted years of effort hunting for a microbial cause of beriberi before they realized that the cause was not something present in the environment but something absent from it. Figure 7-6 depicts beriberi and describes its two forms.

Just before 1900, an observant physician working in a prison in East Asia discovered that beriberi could be cured with proper diet. The physician noticed that the chickens at the prison had developed a stiffness and weakness similar to that of the prisoners who had beriberi. The chickens were being fed the rice left on prisoners' plates. When the rice bran, which had been discarded in the kitchen, was given to the chickens, their paralysis was cured. The physician met resistance when he tried to feed the rice bran, the "garbage," to the prisoners, but it worked—it produced a miracle cure like those described at the beginning of the chapter. Later, extracts of rice bran were used to prevent infantile beriberi; still later, thiamin was synthesized.

In developed countries today, alcohol abuse often leads to a severe form of thiamin deficiency, Wernicke-Korsakoff syndrome, defined in Controversy 3. Alcohol

Figure 7•6

BERIBERI

Beriberi takes two forms: wet beriberi, characterized by edema (fluid accumulation), and dry beriberi, without edema but with muscle wasting. This woman's leg retains the imprint of her physician's thumb, showing the edema of wet beriberi.

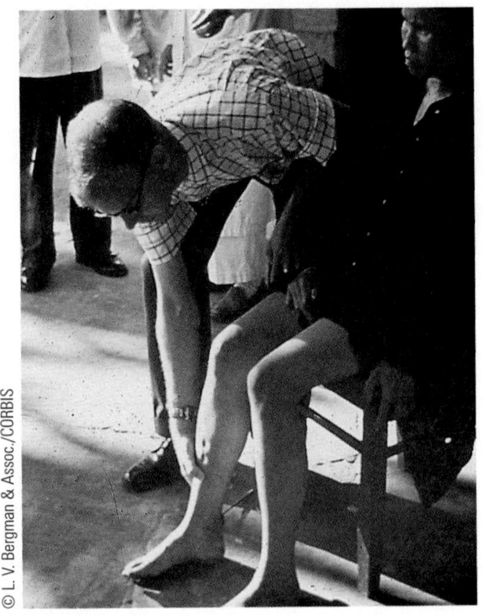

THIAMIN

These foods provide 10 percent or more of the thiamin Daily Value (DV = 1.5 mg/day).[a]

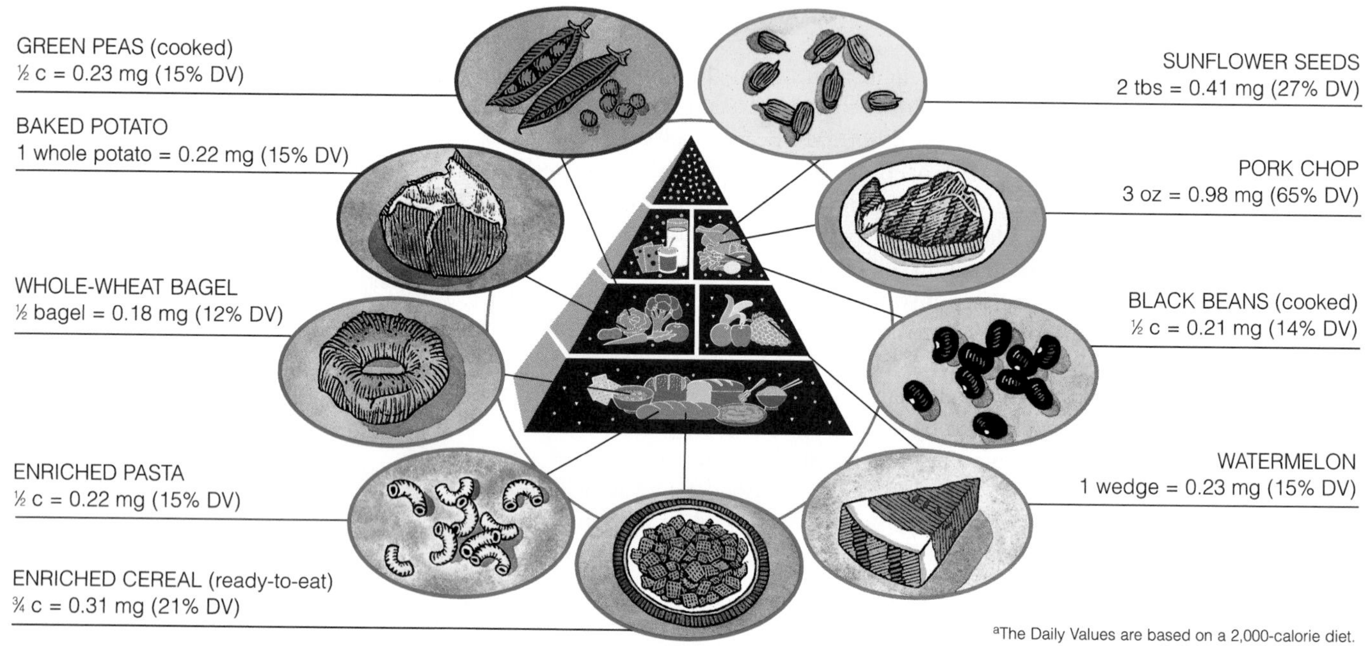

[a]The Daily Values are based on a 2,000-calorie diet.

riboflavin (RIBE-o-flay-vin) a B vitamin active in the body's energy-releasing mechanisms.

niacin (NYE-ah-sin) a B vitamin needed in energy metabolism. Niacin can be eaten preformed or can be made in the body from tryptophan, one of the amino acids. Other forms of niacin are *nicotinic acid, niacinamide*, and *nicotinamide*.

pellagra (pell-AY-gra) the niacin-deficiency disease (*pellis* means "skin"; *agra* means "rough"). Symptoms include the "4 Ds": diarrhea, dermatitis, dementia, and, ultimately, death.

The symptoms of riboflavin deficiency are listed on page 247 of Table 7-6.

contributes energy but carries almost no nutrients with it and often displaces food. In addition, alcohol impairs absorption of thiamin from the digestive tract and hastens its excretion in the urine, tripling the risk of deficiency. The syndrome is characterized by symptoms almost indistinguishable from alcohol abuse itself: apathy, irritability, mental confusion, disorientation, loss of memory, jerky eye movements, and a staggering gait. Unlike alcohol toxicity, the syndrome responds quickly to an injection of thiamin, and some experts recommend a precautionary dose for any patients suspected of having the syndrome.[30]

Thiamin occurs in small amounts in many nutritious foods. Ham and other pork products, leafy green vegetables, whole-grain cereals, and legumes are especially rich in thiamin (see Snapshot 7-5). If you keep empty-calorie foods to a minimum and include ten or more servings of nutritious foods each day, you will easily meet your thiamin needs. The DRI committee set the thiamin intake recommendation at 1.2 milligrams per day for men and at 1.1 milligrams per day for women. Pregnancy and lactation require somewhat more thiamin (see the DRI, inside front cover, page A).

Like thiamin, **riboflavin** plays a role in the energy metabolism of all cells. When thiamin is deficient, riboflavin may be lacking, too, but its deficiency symptoms may go undetected because those of thiamin deficiency are more severe. Foods that remedy the thiamin deficiency invariably also contain some riboflavin, so they clear up both deficiencies. People obtain as much as half of their riboflavin from milk and milk products. Leafy green vegetables, whole-grain breads and fortified cereals, and some meats contribute the rest of the riboflavin in people's diets (see Snapshot 7-6).

Niacin The vitamin **niacin,** like thiamin and riboflavin, participates in the energy metabolism of every cell of the body. The niacin-deficiency disease **pellagra** appeared in Europe in the 1700s when corn from the New World became a staple food. In the early 1900s in the United States, pellagra was devastating lives throughout the South and Midwest. Hundreds of thousands of pellagra victims were thought

RIBOFLAVIN

These foods provide 10 percent or more of the riboflavin Daily Value (DV = 1.7 mg/day).[a]

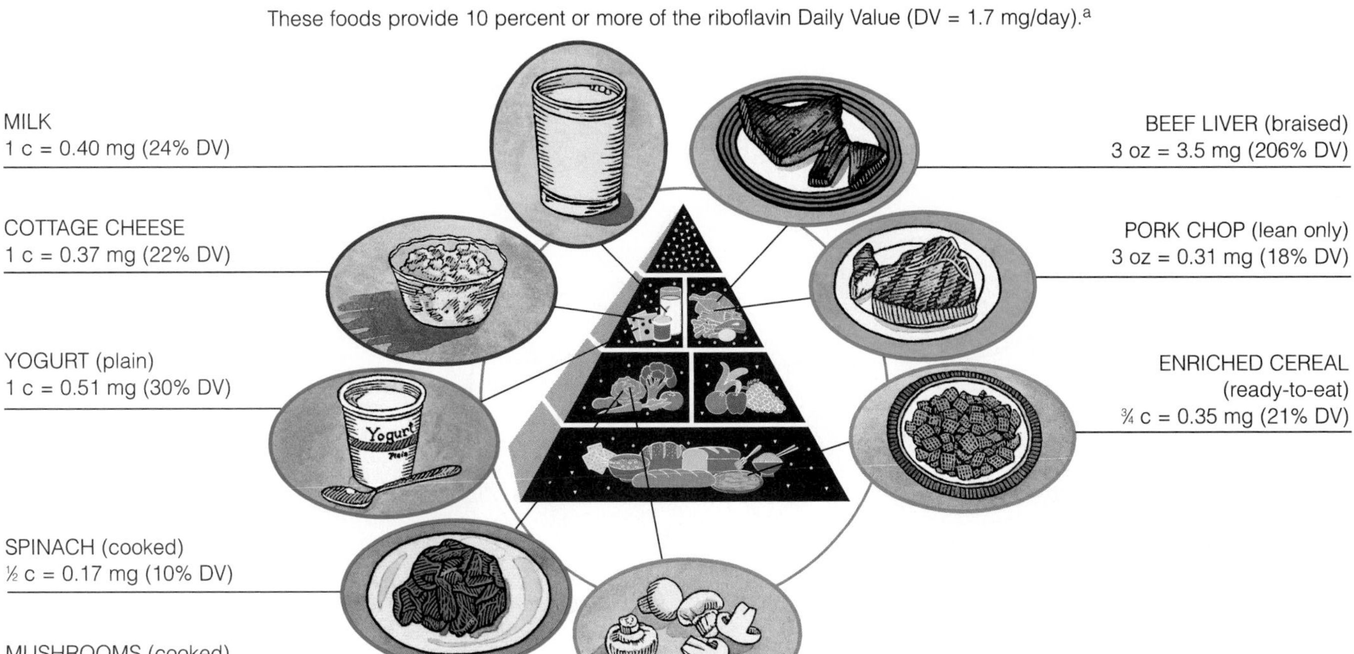

[a]The Daily Values are based on a 2,000-calorie diet.

to be suffering from a contagious disease until this dietary deficiency was identified. The disease still occurs among poorly nourished people living in urban slums and particularly in those with alcohol addiction. Pellagra is also still common in parts of Africa and Asia.

The key nutrient that prevents pellagra is niacin, but any protein containing sufficient amounts of the amino acid tryptophan will serve in its place. Tryptophan, which is abundant in almost all proteins (but is limited in the protein of corn), is converted to niacin in the body, and it is possible to cure pellagra by administering tryptophan alone. Thus, a person eating adequate protein (as most people in developed nations do) will not be deficient in niacin. The amount of niacin in a diet is stated in terms of **niacin equivalents,** a measure that takes available tryptophan into account.

niacin equivalents the amount of niacin present in food, including the niacin that can theoretically be made from its precursor tryptophan that is present in the food.

Early workers seeking the cause of pellagra observed that well-fed people never got it. From there the researchers defined a diet that reliably produced the disease—one of cornmeal, salted pork fat, and molasses. Corn not only is low in protein, but corn protein also lacks tryptophan. Salt pork is almost pure fat and contains too little protein to compensate; and molasses is virtually protein-free. Snapshot 7-7 on the next page shows some good food sources of niacin.

Figure 7-7 shows the skin disorder associated with pellagra. For comparison, Figure 7-9 (on page 234) and Figure 7-2 show skin disorders associated with vitamin B_6 and vitamin A deficiency, respectively, a reminder that any nutrient deficiency affects the skin and all other cells. The skin just happens to be the organ you can see. Table 7-6 at the end of the chapter lists many more symptoms of niacin deficiency.

Physicians often administer large doses of a form of niacin to help achieve an anticlotting effect of the blood and lower blood lipids associated with cardiovascular disease.*[31] When used this way, niacin leaves the realm of nutrition to become a

Figure 7•7

PELLAGRA

The typical dermatitis of pellagra develops on skin that is exposed to light.

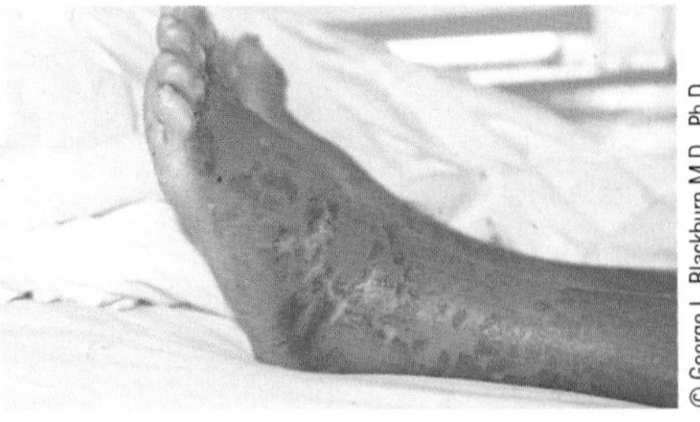

*The form of niacin is nicotinic acid.

NIACIN[a]

These foods provide 10 percent or more of the niacin Daily Value (DV = 20 mg/day).[b]

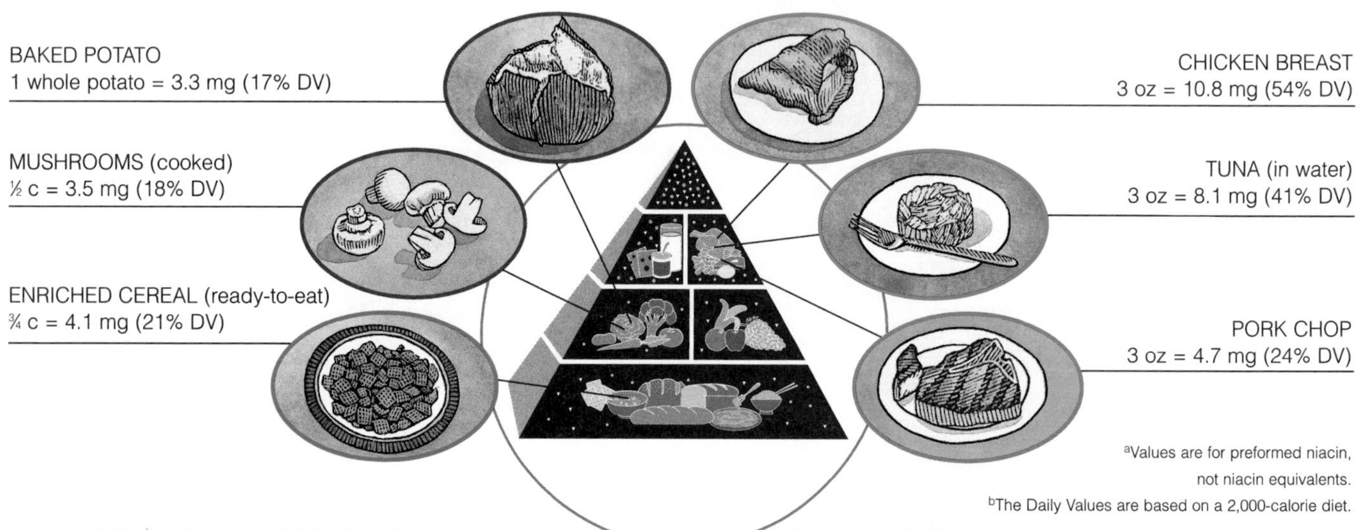

[a]Values are for preformed niacin, not niacin equivalents.

[b]The Daily Values are based on a 2,000-calorie diet.

folate (FOH-late) a B vitamin that acts as part of a coenzyme important in the manufacture of new cells. The form added to foods and supplements is *folic acid*.

neural tube defects abnormalities of the brain and spinal cord apparent at birth and believed to be related to a woman's folate intake before and during pregnancy. Also defined in Chapter 12.

pharmacological agent—a drug. As with any drug, self-dosing with niacin is ill-advised; large doses may injure the liver, cause peptic ulcers, or cause vision loss.[32] Certain forms of niacin supplements in amounts two to three times the DRI intake recommendation cause "niacin flush," a dilation of the capillaries of the skin with perceptible tingling that, if intense, can be painful.[33] For safety's sake, anyone taking large doses of niacin should do so only under the care of a physician.[34]

Folate To make new cells, tissues must have the vitamin **folate.** Each new cell must be equipped with new DNA copies, and folate helps to synthesize DNA. Because the red and white blood cells and the cells of the digestive tract divide most rapidly, they are most vulnerable to deficiency. As a result, deficiencies of folate cause anemia, diminished immunity, and abnormal digestive function. In the United States, a significant number of cases of folate-deficiency anemia occur yearly. This anemia is related to the anemia of vitamin B_{12} malabsorption because the two vitamins work as teammates in producing red blood cells—see Figure 7-8, later. Research suggests that folate deficiency may also elevate the risk of cardiovascular disease and cancer of the colon and increase a woman's risk for cervical cancer. Folate deficiencies may result from an inadequate intake or from illnesses that impair folate's absorption, increase its excretion, require medication that interacts with folate, or otherwise increase the body's need.

The DRI committee advises all women of childbearing age to consume 400 micrograms of synthetic folate, or *folic acid*, each day in addition to the folate that occurs naturally in their foods.[35] The reason is that folate deficiency is associated with a group of devastating birth defects known as **neural tube defects,** which affect approximately 3,000 U.S. births each year, making them the second most common form of birth defect after Down syndrome.[36] Neural tube defects range from slight problems in the spine to mental retardation, severely diminished brain size, and death shortly after birth.

Neural tube defects arise in the first days or weeks of pregnancy, long before most women suspect that they are pregnant, and most women eat too few fruits and vegetables from day to day to supply even half the folate needed to prevent neural tube defects. In the late 1990s, the FDA ordered fortification of all enriched grain prod-

FOLATE[a]

These foods provide 10 percent or more of the folate Daily Value (DV = 400 µg/day).[b]

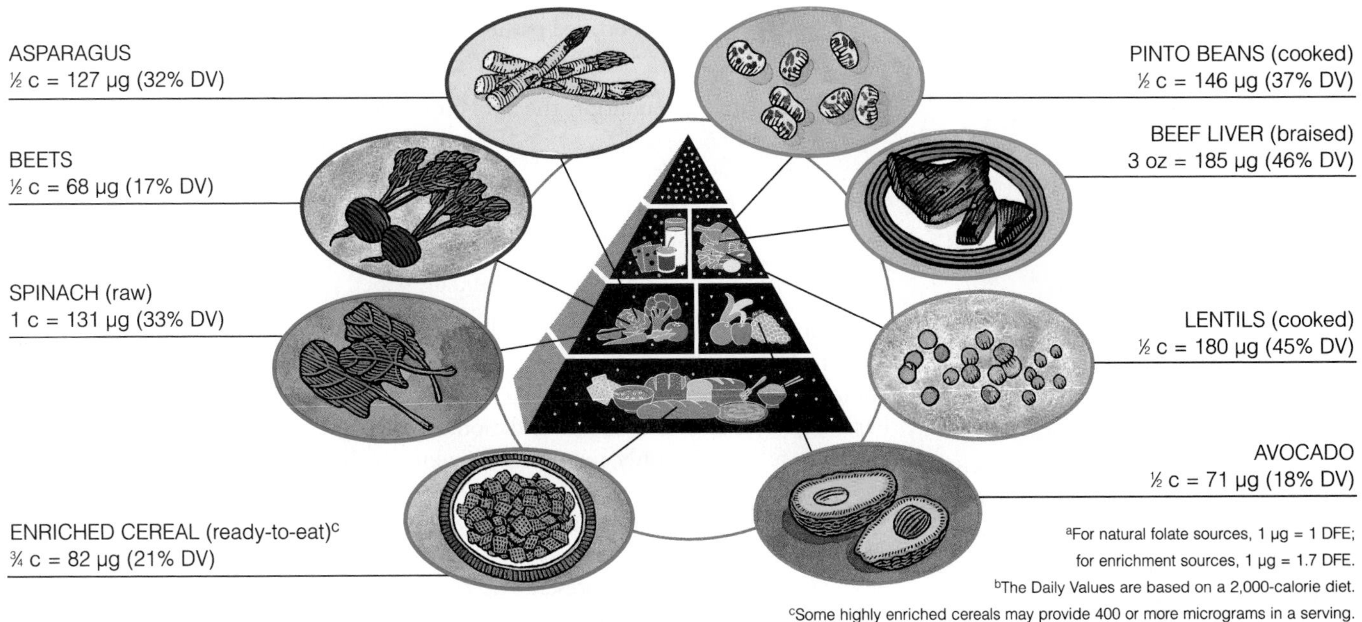

[a]For natural folate sources, 1 µg = 1 DFE; for enrichment sources, 1 µg = 1.7 DFE.
[b]The Daily Values are based on a 2,000-calorie diet.
[c]Some highly enriched cereals may provide 400 or more micrograms in a serving.

ucts such as breads, cereals, and pastas with an especially absorbable synthetic form of folate, folic acid.[37] Since this fortification began, women's folate intakes have been increasing, and observers report an almost 20 percent drop in the national incidence of neural tube defects (about half are from causes other than folate deficiency), even among women receiving late or no prenatal care.[38] Researchers expect to see signs of decline in rates of some other birth defects and miscarriages as well.[39]

Dietary Folate Equivalent (DFE) a unit of measure expressing the amount of folate available to the body from naturally occurring sources. The measure mathematically accounts for the greater absorption of synthetic folate added to enriched foods and supplements.

Folate's name is derived from the word *foliage*, and sure enough, folate is naturally abundant in leafy green vegetables such as spinach and turnip greens (see Snapshot 7-8). Fresh, uncooked vegetables and fruits are the best natural sources because the heat of cooking and the oxidation that occurs during storage destroy much of the folate in foods. Eggs also contain some folate. Orange juice and legumes contain folate, but they also contain factors that may interfere with folate absorption, limiting their usefulness as folate contributors. Milk may enhance the absorption of folate.

A Tolerable Upper Intake Level for synthetic folate from supplements and enriched foods is set at 1,000 micrograms a day for adults. Of major importance are concerns about folate's ability to mask deficiencies of vitamin B_{12} (more about this effect later). The possibility also exists that, once in the blood, excess folate may negate actions of some anticancer drugs that work by blocking the activities of folate in rapidly dividing cancer cells. Time will tell whether the apparent benefits of folate enrichment outweigh the risks.

The difference in absorption between naturally occurring food folate and the synthetic folate that enriches foods and is added to supplements necessitated a conversion factor for folate: the **Dietary Folate Equivalent,** or **DFE.**[40] The DFE converts all forms of folate into units that are equivalent to the folate in foods. Folate is expressed in terms of DFE (dietary folate equivalents) because synthetic folate from supplements and fortified foods is absorbed at almost twice (1.7 times) the rate of naturally occurring folate from other foods. Use the following equation to calculate:

$$\text{DFE} = \mu\text{g food folate} + (1.7 \times \mu\text{g synthetic folate}).$$

The B vitamins thiamin, riboflavin, niacin, and folate (as folic acid) are among the enrichment nutrients added to grain foods such as breads and cereals sold in the United States. Chapter 4 presented more details on enrichment of grain foods.

vitamin B_{12} a B vitamin that helps to convert folate to its active form and also helps maintain the sheaths around nerve cells. Vitamin B_{12}'s scientific name, not often used, is *cyanocobalamin*.

intrinsic factor a factor found inside a system. The intrinsic factor necessary to prevent pernicious anemia is now known to be a compound that helps in the absorption of vitamin B_{12}.

pernicious (per-NISH-us) **anemia** a vitamin B_{12}–deficiency disease, caused by lack of intrinsic factor and characterized by large, immature red blood cells and damage to the nervous system (*pernicious* means "highly injurious or destructive").

The text example shows that without converting folate to DFE units, folate intakes are underestimated. Measured in micrograms, the woman's intake appears to fall short of the 600 μg recommended for pregnancy. Converted to DFE untis, her intake proves ample.

Figure 7•8

ANEMIC AND NORMAL BLOOD CELLS

The anemia of folate deficiency is indistinguishable from that of vitamin B_{12} deficiency.

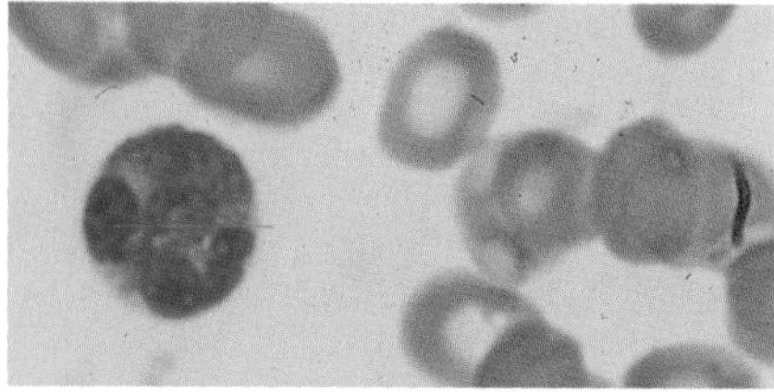

Blood cells of pernicious anemia. The cells are larger than normal and irregular in shape.

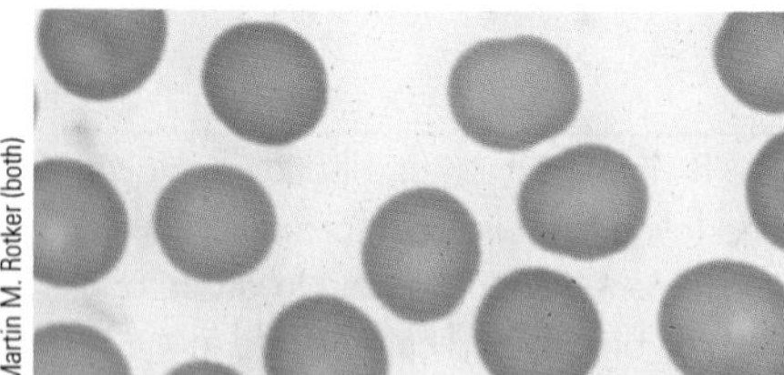

Martin M. Rotker (both)

Normal blood cells. The size, shape, and color of the red blood cells show that they are normal.

Consider, for example, a pregnant woman who takes a supplement and eats a bowl of fortified cornflakes, 2 slices of fortified bread, and a cup of fortified pasta. From the supplement and fortified foods, she obtains synthetic folate:

Supplement	100 μg folate
Fortified cornflakes	100 μg folate
Fortified bread	40 μg folate
Fortified pasta	60 μg folate
	300 μg folate

To calculate the DFE, multiply the amount of synthetic folate by 1.7:

$$300\ \mu g \times 1.7 = 510\ \mu g\ \text{DFE}.$$

Now add the naturally occurring folate from the other foods in her diet—in this example, another 90 μg of folate.

$$510\ \mu g\ \text{DFE} + 90\ \mu g = 600\ \mu g\ \text{DFE}.$$

Notice that if we had not converted synthetic folate from supplements and fortified foods to DFE, then this woman's intake would appear to fall short of the 600 μg recommendation for pregnancy (300 μg + 90 μg = 390 μg). But as our example shows, her intake does meet the recommendation. At this time, supplement and fortified food labels list folate in μg only, not μg DFE, making such calculations necessary.

Of all the vitamins, folate is most likely to interact with medications. Ten major groups of drugs, including antacids and aspirin and its relatives, have been shown to interfere with the body's use of folate. Occasional use of these drugs to relieve headache or upset stomach presents no concern, but frequent users may need to pay attention to their folate intakes. These include people with chronic pain or ulcers who rely heavily on aspirin or antacids as well as those who smoke or take oral contraceptives or anticonvulsant medications.

Vitamin B_{12} Vitamin B_{12} and folate are closely related: each depends on the other for activation. By itself vitamin B_{12} also helps to maintain the sheaths that surround and protect nerve fibers. Without sufficient vitamin B_{12}, nerves become damaged and folate fails to do its blood-building work, so vitamin B_{12} deficiency causes an anemia identical to that caused by folate deficiency. The blood symptoms of a deficiency of either folate or vitamin B_{12} include the presence of large, immature red blood cells. Administering extra folate often clears up this blood condition but allows the deficiency of vitamin B_{12} to continue undetected.[41] Vitamin B_{12}'s other functions then become compromised, and the results can be devastating: damaged nerve sheaths, creeping paralysis, and general malfunctioning of nerves and muscles.

Absorption of vitamin B_{12} requires an **intrinsic factor,** a compound made by the stomach with instructions from the genes. With the help of the stomach's acid to liberate vitamin B_{12} from the food proteins that bind it, intrinsic factor attaches to the vitamin; the complex is then absorbed from the small intestine into the bloodstream. A few people have an inherited defect in the gene for intrinsic factor, which makes vitamin B_{12} absorption abnormal, beginning in mid-adulthood. Without normal absorption of vitamin B_{12} from food, they develop deficiency symptoms. In this case or in the case of stomach injury that limits production of intrinsic factor, vitamin B_{12} must be supplied by injection to bypass the defective absorptive system. The anemia of the vitamin B_{12} deficiency caused by lack of intrinsic factor is known as **pernicious anemia** (see Figure 7-8).

Diagnosing a vitamin B_{12} problem is difficult, and often the damage will proceed unchecked. In an effort to prevent excessive folate intakes that could mask symptoms of a vitamin B_{12} deficiency, the FDA specifies exact amounts of folate that can be added to enriched foods.

As Snapshot 7-9 shows, vitamin B_{12} is present only in foods of animal origin, not in foods from plants. The uninformed, strict vegetarian is at special risk and may not show signs of deficiency right away because the body stores up to six years' worth of

VITAMIN B_{12}

These foods provide 10 percent or more of the vitamin B_{12} Daily Value (DV = 6 μg/day).[a]

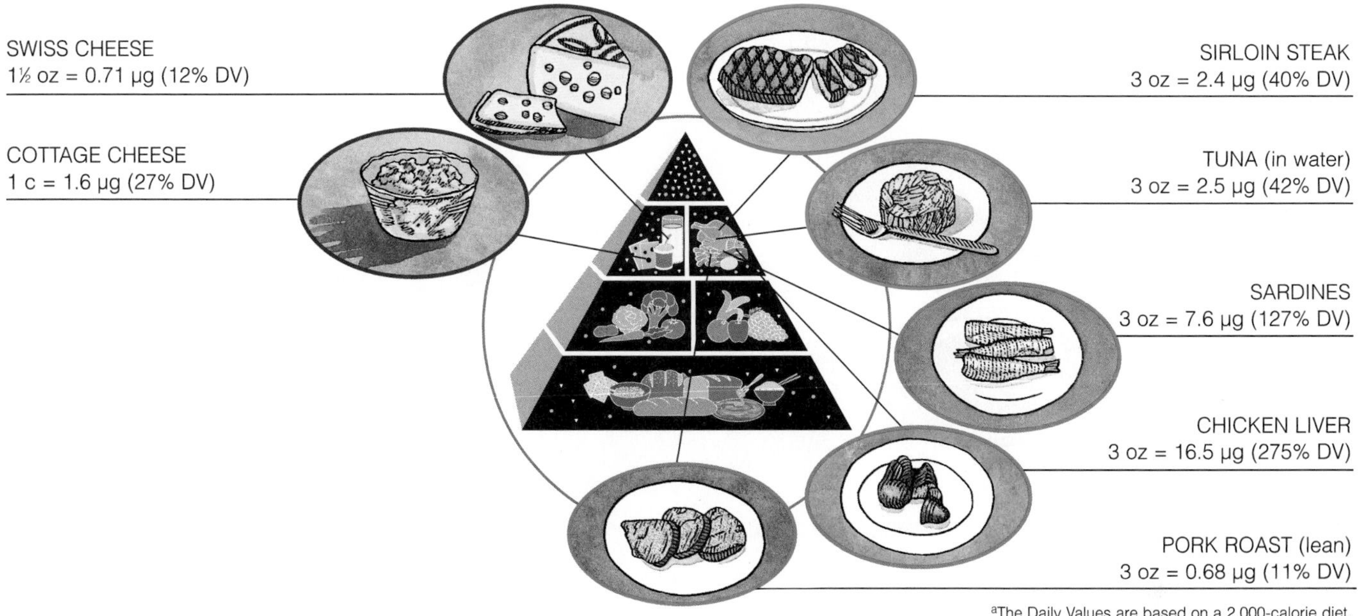

[a]The Daily Values are based on a 2,000-calorie diet.

vitamin B_{12}.[42] A pregnant or lactating woman who is not eating any foods of animal origin should be aware that her infant can develop a vitamin B_{12} deficiency, even if the mother appears healthy. A deficiency of this vitamin can cause irreversible nervous system damage in the developing fetus, which can only be diagnosed after birth when the infant displays nerve problems. All strict vegetarians, and especially pregnant women, must be sure to use vitamin B_{12}–fortified products, such as vitamin B_{12}–fortified soy "milk," or to take the appropriate supplements.

The way folate masks the anemia of vitamin B_{12} deficiency underscores a point worth repeating. It takes a skilled professional to correctly diagnose a nutrient deficiency or imbalance, and you take a serious risk when you diagnose yourself or listen to self-proclaimed experts. A second point: Since vitamin B_{12} deficiency in the body may be caused by either a lack of the vitamin in the diet or a lack of the intrinsic factor necessary to absorb the vitamin, a change in diet alone may not correct the deficiency, another reason for seeking professional diagnosis when you have physical symptoms.

vitamin B_6 a B vitamin needed in protein metabolism. Its three active forms are *pyridoxine*, *pyridoxal*, and *pyridoxamine*.

Vitamin B_6 Vitamin B_6 helps the cells to convert one kind of amino acid, which cells have in abundance, to other nonessential amino acids that the cells lack. In addition, vitamin B_6 functions in these ways:

- Aids in the conversion of tryptophan to niacin.
- Plays important roles in the synthesis of hemoglobin and neurotransmitters, the communication molecules of the brain.
- Assists in releasing stored glucose from glycogen and thus contributes to the regulation of blood glucose.
- Has roles in immune function and steroid hormone activity.[43]
- Is critical to the developing brain and nervous system of a fetus. Deficiency during this stage causes behavioral problems later.

VITAMIN B_6

These foods provide 10 percent or more of the vitamin B_6 Daily Value (DV = 2 mg/day).[a]

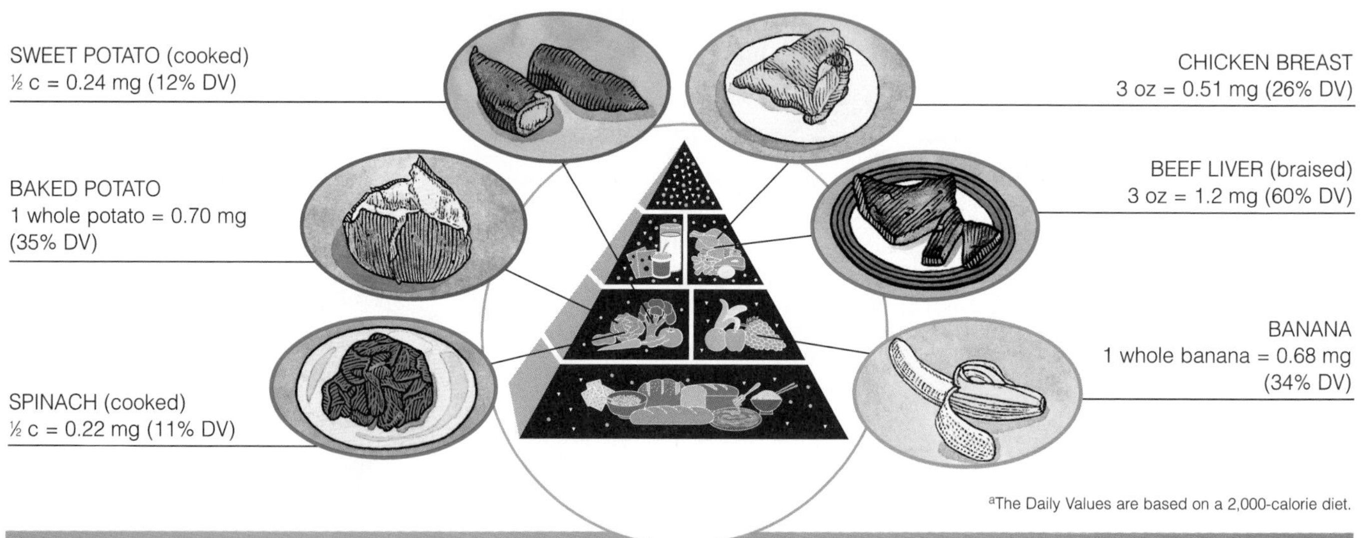

[a]The Daily Values are based on a 2,000-calorie diet.

homocysteine (home-oh-SIS-teen) an amino acid produced as an intermediate compound during amino acid metabolism. A buildup of homocysteine in the blood is associated with deficiencies of folate and other B vitamins and may increase the risk of diseases.

The links between PMS and vitamin B_6 are explored in Chapter 13.

Figure 7•9

VITAMIN B_6 DEFICIENCY

In this dermatitis, the skin is greasy and flaky, unlike the skin affected by the dermatitis of pellagra.

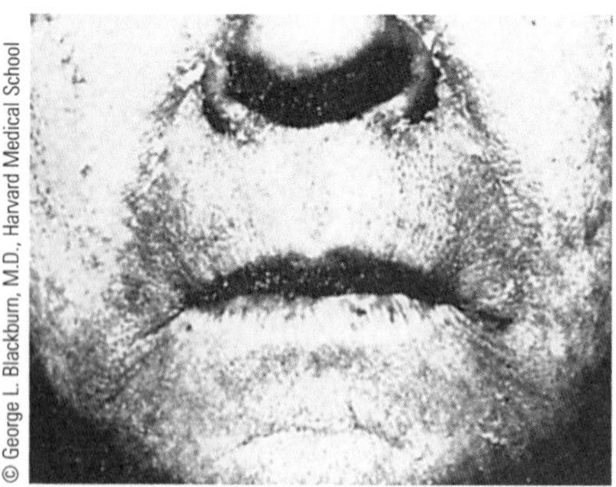

© George L. Blackburn, M.D., Harvard Medical School

Because of these diverse functions, vitamin B_6 deficiency is expressed in general symptoms, such as weakness, psychological depression, confusion, irritability, and insomnia. Other symptoms include anemia, the greasy dermatitis depicted in Figure 7-9, and, in advanced cases of deficiency, convulsions. A shortage of vitamin B_6 may also weaken the immune response.[44] Some evidence suggests that low vitamin B_6 intakes may also be related to increased incidence of heart disease.[45]

Large doses of vitamin B_6 can be dangerous. Years ago it was generally believed that, like most of the other water-soluble vitamins, vitamin B_6 could not reach toxic concentrations in the body. Then a report told of women who took more than 2 grams of vitamin B_6 daily (the Tolerable Upper Intake Level is set at 100 *milligrams*, or 0.1 gram) for two months or more in an attempt to cure the symptoms of premenstrual syndrome (PMS). The women developed numb feet, then lost sensation in their hands, and eventually became unable to walk or work. Since the first report of vitamin B_6 toxicity, researchers have seen toxicity symptoms in more than 100 women who took vitamin B_6 supplements for more than five years. The women recovered after they stopped taking the supplements. The potential toxicity of vitamin B_6 is yet another reason why people should not self-diagnose and self-prescribe high doses of vitamins for their illnesses. Table 7-6 on pages 247–249 lists common deficiency and toxicity symptoms of vitamin B_6 and other water-soluble vitamins.

Vitamin B_6 plays so many roles in protein metabolism that the body's requirement for vitamin B_6 is roughly proportional to protein intakes.[46] The DRI committee set the vitamin B_6 intake recommendation high enough to cover most people's needs, regardless of differences in protein intakes (see the inside front cover).[47] Meats, fish, and poultry (protein-rich foods), potatoes, leafy green vegetables, and some fruits are good sources of vitamins B_6 (that is, one serving provides at least 10 percent of the Daily Value; see Snapshot 7-10). Other foods, such as legumes and peanut butter provide smaller amounts.

Q How Are B Vitamins Related to Heart Disease? People who inherit a rare disorder that raises the level of a special amino acid, **homocysteine,** in the blood almost invariably suffer from a severe early form of cardiovascular disease

(CVD).* Also, some other CVD sufferers without the inherited disorder accumulate homocysteine in the blood. Researchers are now investigating the possibility that elevated homocysteine might be a risk factor for CVD and that a lack of some B vitamins in the diet might contribute to elevating it.[48]

So far, they have shown that a deficiency of folate, vitamin B_{12}, and possibly vitamin B_6 causes excess homocysteine to build up in the blood.[49] When healthy men are given supplements of B vitamins (folate, vitamin B_6, and vitamin B_{12}), their homocysteine values drop significantly.[50] The research to date is mixed, however, on whether such a drop in homocysteine reduces a person's risk of developing CVD.

Folate enrichment of food, in effect since the late 1990s, may soon shed some light on this issue. Emerging data indicate that U.S. blood values for folate are rising while those for homocysteine are dropping.[51] Not every study has shown a relationship between high folate intakes and healthier hearts, and none has produced evidence that giving B vitamins to a population reduces heart disease risks.[52] Also, some groups of people are known to clear blood homocysteine more readily than others, regardless of their vitamin status, leading researchers to postulate that genetics may be at work.[53] Until more firm evidence exists, caution is in order when considering taking B vitamin supplements to improve health.

biotin (BY-o-tin) a B vitamin; a coenzyme necessary for fat synthesis and other metabolic reactions.

pantothenic (PAN-to-THEN-ic) **acid** a B vitamin.

choline (KOH-leen) a nonessential nutrient used to make the phospholipid lecithin and other molecules.

carnitine a nonessential nutrient that functions in cellular activities.

inositol (in-OSS-ih-tall) a nonessential nutrient found in cell membranes.

lipoic (lip-OH-ic) **acid** a nonessential nutrient.

Biotin and Pantothenic Acid Two other B vitamins, **biotin** and **pantothenic acid,** are, like thiamin, riboflavin, and niacin, important in energy metabolism. Biotin is a cofactor for several enzymes in the metabolism of carbohydrate, fat, and protein. Pantothenic acid is a component of a key coenzyme that makes possible the release of energy from the energy nutrients. It also participates in more than 100 steps in the synthesis of lipids, neurotransmitters, steroid hormones, and hemoglobin.

Although rare diseases may precipitate deficiencies of biotin and pantothenic acid, both vitamins are widespread in foods. A steady diet of raw egg whites, which contain a protein that binds biotin, can produce biotin deficiency, but you would have to consume more than two dozen egg whites daily to produce the effect. Cooking eggs denatures the protein. Healthy people eating ordinary diets are not at risk for deficiencies.

The DRI recommended intakes for biotin and pantothenic acid are listed on the inside front cover, page A.

key point *Historically, famous B vitamin–deficiency diseases are beriberi (thiamin), pellagra (niacin), and pernicious anemia (vitamin B_{12}). Pellagra can be prevented by adequate protein because the amino acid tryptophan can be converted to niacin in the body. A high intake of folate can mask the blood symptom of vitamin B_{12} deficiency but will not prevent the associated nerve damage. Vitamin B_6 is important in amino acid metabolism and can be toxic in excess. Biotin and pantothenic acid are important to the body and are abundant in food.*

Non-B Vitamins

In addition to the B vitamins just discussed, a few compounds that are topics of debate among researchers deserve mention. **Choline** could be considered an essential nutrient because when the diet is devoid of choline, the body cannot make enough of the compound to meet its need. Choline is common in foods, though, and deficiencies are practically unheard of outside the laboratory. DRI intake recommendations have been set for choline (see inside front cover).

Links between choline and brain function are discussed in Chapter 13.

The compounds **carnitine, inositol,** and **lipoic acid** might appropriately be called *nonvitamins* because they are not essential nutrients for human beings. Carnitine, sometimes called "vitamin B_T," is an important piece of cell machinery, but it is not a vitamin. Although deficiencies can be induced in laboratory animals for experimental purposes, these substances are abundant in ordinary foods. Even if these compounds were essential in human nutrition, supplements would be unnecessary for

*Although chemically an amino acid, homocysteine is not incorporated into body proteins but is metabolized to other compounds.

scurvy the vitamin C–deficiency disease.

ascorbic acid one of the active forms of vitamin C (the other is *dehydroascorbic* acid); an antioxidant nutrient.

collagen (COLL-a-jen) the chief protein of most connective tissues, including scars, ligaments, and tendons, and the underlying matrix on which bones and teeth are built.

prooxidant (proh-OX-ih-dant) a compound that triggers reactions involving oxygen.

healthy people eating a balanced diet. Vitamin companies often include these substances to make their formulas appear more "complete," but there is no physiological reason to do so.

In addition to carnitine, inositol, and lipoic acid, other substances have been mistakenly thought essential in human nutrition because they are needed for growth by bacteria or other life-forms. These substances include PABA (para-aminobenzoic acid), bioflavonoids ("vitamin P" or hesperidin), and ubiquinone (coenzyme Q). Other names you may hear are "vitamin B_{15}" and pangamic acid (hoaxes) and "vitamin B_{17}" (laetrile or amygdalin, not a cancer cure and not a vitamin by any stretch of the imagination).

key point *Choline is needed in the diet, but it is not a vitamin and deficiencies are unheard of outside the laboratory. Many other substances that people claim are B vitamins are not. Among these substances are carnitine, inositol, and lipoic acid.*

Vitamin C

More than two hundred years ago, any sailors who joined the crew of a seagoing ship knew they had only half a chance of returning alive—not because they might be slain by pirates or die in a storm but because they might contract **scurvy,** a disease that might kill as many as two-thirds of a ship's crew on a long voyage. Only ships that sailed on short voyages, especially around the Mediterranean Sea, were safe from this disease. The special hazard of long ocean voyages was that the ship's cook used up the fresh fruits and vegetables early and relied for the duration of the voyage on cereals and live animals.

The first nutrition experiment to be conducted on human beings was devised nearly 250 years ago to find a cure for scurvy. A physician divided some British sailors with scurvy into groups. Each group received a different test substance: vinegar, sulfuric acid, seawater, oranges, or lemons. Those receiving the citrus fruits were cured within a short time. Sadly, it took 50 years for the British navy to make use of the information and require all its vessels to provide lime juice to every sailor daily. British sailors were mocked with the term *limey* because of this requirement. The name later given to the vitamin, **ascorbic acid,** literally means "no-scurvy acid."

The Work of Vitamin C Since vitamin C is also a water-soluble vitamin, you might expect its mode of action to resemble that of the B vitamins. Vitamin C does help specific enzymes perform their jobs—for example, the enzymes involved in formation and maintenance of the tissue protein **collagen** depend on vitamin C for their activity. Collagen forms the base for all of the connective tissues in the body: bones, teeth, skin, and tendons. Collagen also forms the scar tissue that heals wounds, the reinforcing structure that mends fractures, and the supporting material of capillaries that prevents bruises. Vitamin C also acts as a cofactor in the production of carnitine, important for transporting fatty acids within the cells.

Long voyages without fresh fruits and vegetables spelled death by scurvy for the crew.

In addition to its role as a vitamin assisting enzymes, vitamin C also acts in a more general way as an antioxidant. Vitamin C protects substances found in foods and in the body from oxidation by being oxidized itself. Much of the oxidized vitamin C is not lost, however; it is readily recycled back to the active form for reuse.[54]

In the intestines, vitamin C protects iron from oxidation and so promotes its absorption. In the blood, vitamin C protects sensitive blood constituents from oxidation and helps to protect and recycle vitamin E. The antioxidant roles of vitamin C are the focus of extensive study, especially in relation to disease prevention (see this chapter's Controversy). In test tubes, however, high concentrations of vitamin C have the opposite effect; that is, they act as a **prooxidant** by activating oxidizing elements, such as iron and copper. One study revealed an increase in markers of oxidation in men given 500 milligrams of vitamin C daily.[55] The question of what, if anything, such findings may mean to human health remains unanswered.

Vitamin C also supports immune system functions and so protects against infection. A long-claimed relationship between vitamin C and the common cold is the topic of this chapter's Consumer Corner.

The Need for Vitamin C The adult DRI intake recommendation for vitamin C is 90 milligrams for men and 75 milligrams for women. These amounts are far higher than the 10 or so milligrams per day needed to prevent the symptoms of scurvy. In fact, they are close to the amount at which the body's pool of vitamin C is full to overflowing: about 100 milligrams per day.

Cigarette smoking, among its many harmful effects, introduces oxidants that deplete the body's vitamin C. Smokers, and "passive smokers" who live and work with smokers, need more vitamin C than others. Intake recommendations for smokers are set higher, at 125 milligrams for men and 110 milligrams for women, in order to maintain blood levels comparable to those of nonsmokers. Sufficient intake of vitamin C can normalize blood levels, but it cannot protect against the damage caused by exposure to tobacco smoke.

Most of the symptoms of scurvy can be attributed to the breakdown of collagen in the absence of vitamin C: loss of appetite, growth cessation, tenderness to touch, weakness, bleeding gums (shown in Figure 7-10), loose teeth, swollen ankles and wrists, and tiny red spots in the skin where blood has leaked out of capillaries. One symptom, anemia, reflects an important role worth repeating—Vitamin C helps the body to absorb and use iron.

In the United States, scurvy is seldom seen today except in a few elderly people, people addicted to alcohol or other drugs, and a few infants who are fed only cow's milk. Breast milk and infant formula supply enough vitamin C, but infants who are fed cow's milk and receive no vitamin C in formula, fruit juice, or other outside sources are at risk. Low intakes of fruits and vegetables and a poor appetite overall lead to low vitamin C intakes and are not uncommon among people aged 65 and older.

Is Too Much Vitamin C Hazardous to Health? The easy availability of vitamin C in pill form and the publication of books recommending vitamin C as a "nutraceutical" treatment to prevent and cure colds and cancer have led thousands of people to take huge doses of vitamin C. These "volunteer" subjects enabled researchers to study potential adverse effects of large vitamin C doses. One effect observed with a 2-gram dose is alteration of the insulin response to carbohydrate in people with otherwise normal glucose tolerances. Other adverse effects include nausea, abdominal cramps, excessive gas, and diarrhea.

Several instances of interference with medical regimens are known. Large amounts of vitamin C excreted in the urine have obscured the results of tests used

4,000 — Nutraceutical recommendation
2,000 — Tolerable Upper Intake Level
1,000
800
600
400
200
125 — DRI recommended intake for smokers (men)
110 — DRI recommended intake for smokers (women)
100 — Maintains full body pool
90 — DRI recommended intake for
75 — DRI recommended intake for
60 — Daily Value on food and supplement labels
30 — Supports metabolism
10 — Prevents scurvy
0

Figure 7•10

SCURVY SYMPTOMS—GUMS AND SKIN

Vitamin C deficiency causes the breakdown of collagen, which supports the teeth.

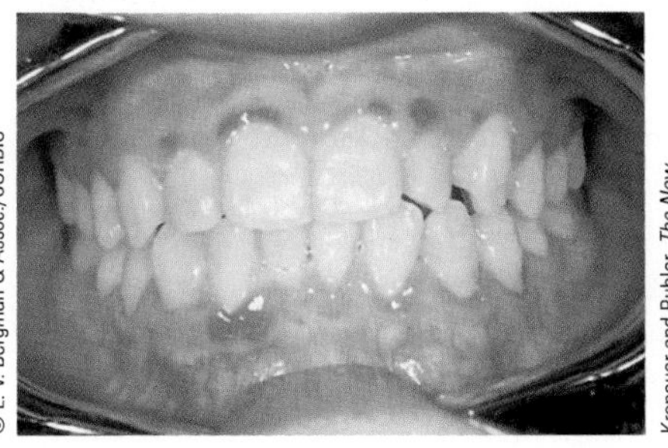

© L. V. Bergman & Assoc./CORBIS

Small pinpoint hemorrhages (red spots) appear in the skin indicating that invisible internal bleeding may also be occurring.

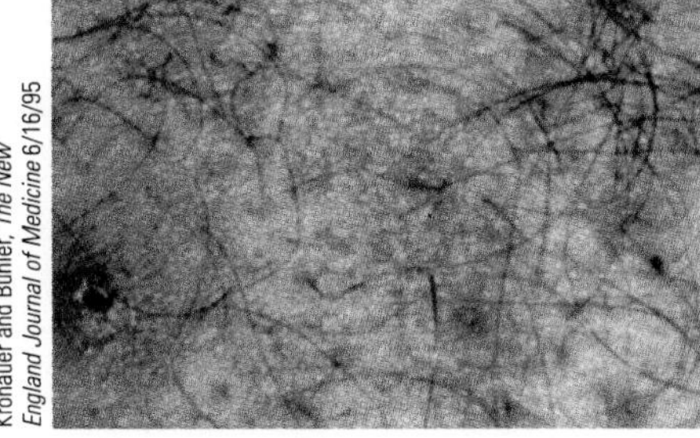

Kronauer and Buhler, *The New England Journal of Medicine* 6/16/95

The marketing term *nutraceutical* is often used to refer to nutrients having pharmacological effects.

7.11

VITAMIN C

These foods provide 10 percent or more of the vitamin C Daily Value (DV = 60 mg/day).[a]

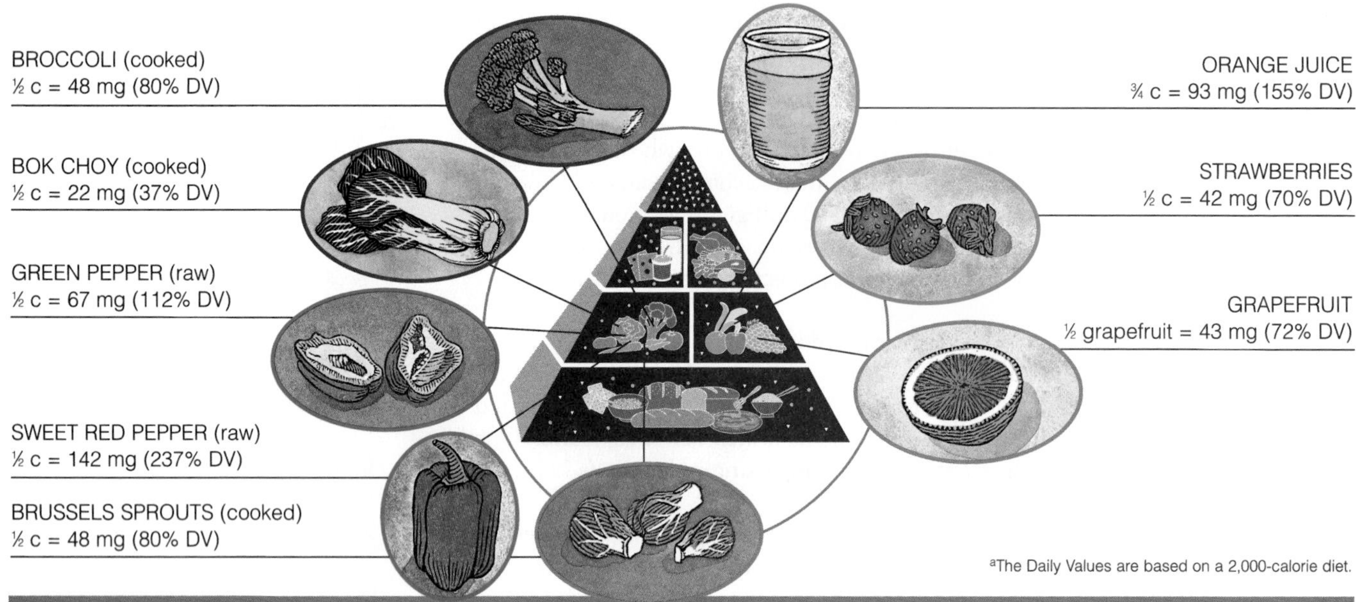

[a]The Daily Values are based on a 2,000-calorie diet.

to detect diabetes, giving false positive or false negative results. Vitamin C in amounts over 250 milligrams has produced false negative results on tests for blood in the digestive system, masking the presence of potentially dangerous medical conditions.[56] Massive doses of vitamin C interfere with medications to prevent blood clotting. Finally, vitamin C supplements in any dosage may be dangerous for people with an overload of iron in the body because vitamin C increases iron absorption from the intestine and releases iron from storage. Effects that are theoretically possible (but have not been seen with intakes as high as 3 grams a day) include formation of kidney stones, alteration of the acid-base balance, and interference with the action of vitamin E.

The published research on large doses of vitamin C reveals few instances in which consuming more than 100 to 300 milligrams a day is beneficial, although the range of safe vitamin C intakes seems to be broad. Between the absolute minimum of 10 milligrams a day and the DRI maximum of 2,000 milligrams (2 grams) should be a suitable intake for most people. People with kidney disorders and those with a condition of too much iron in their blood may be more susceptible to adverse effects and should avoid vitamin C supplements altogether. According to the committee on DRI, doses of 3 or more grams can be expected to be unsafe. Vitamin C from food sources, such as those shown in Snapshot 7-11, is always safe for healthy people.

key point *Vitamin C, an antioxidant, helps to maintain the connective tissue protein collagen, protects against infection, and helps in iron absorption. The theory that vitamin C prevents or cures colds or cancer is not well supported by research. Taking high vitamin C doses may be unwise. Ample vitamin C can be obtained from foods.*

Consumer Corner

Vitamin C and the Common Cold

FOR YEARS, people have claimed that taking supplements of vitamin C helps to cure their colds, but no study to date has shown conclusively that vitamin C can prevent colds or reduce their severity.[1] Why, then, do so many people continue taking vitamin C supplements to relieve colds?

More than 30 years ago, claims by Linus Pauling, a Nobel Prize winner and vocal supporter of vitamin C supplements, were all but discounted by the scientific community because research suggested that regular ingestion of vitamin C supplements is not effective in reducing the number of colds people suffer each year.[2] However, a pair of researchers recently revisited the cornerstone review of the literature on vitamin C and colds. By making some changes in the underlying assumptions and methodology of the original review, they concluded that vitamin C in amounts up to 1 gram per day may indeed shorten the duration of a cold by about one day and reduce the severity of its symptoms by about 23 percent.[3] This effect may be greater in children than in adults; in adults, doses teetering on the edge of the Tolerable Upper Intake Level (2 grams a day) may be required to produce an effect.[4]

It could turn out that vitamin C's antioxidant or other activities boost the body's immunity or somehow improve its defenses. In a test tube, antioxidant nutrients, including vitamin C, stimulate cells of the immune system to move and work more efficiently.[5] In people, regularly taking in more vitamin C from food and supplements does not seem to affect the chances of coming down with a cold.[6]

However, when taken in large doses for a short time (2 grams taken daily for two weeks) vitamin C has been observed to reduce blood histamine. Anyone who has ever had a cold knows the effects of histamine: sneezing, a runny or stuffy nose, and swollen sinuses. Antihistamines provide relief from just those symptoms. In druglike doses, vitamin C may work like a weak antihistamine by deactivating histamine.

One other effect is hard at work with supplements of all kinds: the placebo effect. One study vividly demonstrated its effects with regard to vitamin C and colds. Half of the experimental subjects received a placebo but thought they were receiving vitamin C. These subjects had fewer colds than the group who had received vitamin C but thought they were receiving the placebo. At work was the powerful healing effect of faith—the placebo effect.

Much more research is required on vitamin C and the common cold before any recommendations are possible. One thing is certain, though—no drug is risk-free, and vitamin C in large doses qualifies as a drug that may have side effects.[7]

Can vitamin C ease the suffering of a person with a cold?

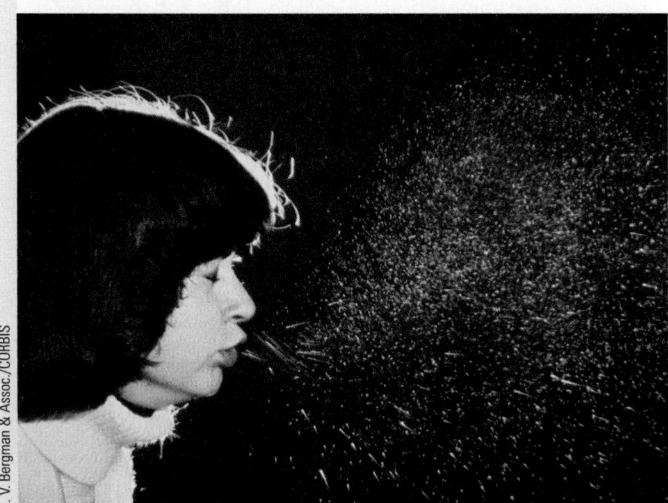

L. V. Bergman & Assoc./CORBIS

Table 7•2

SOME VALID AND INVALID REASONS FOR TAKING SUPPLEMENTS

These People May Need Supplements:

- People with nutrient deficiencies.
- Women in their childbearing years (supplemental folate is recommended to reduce risk of neural tube defects in infants).
- Pregnant or lactating women (they may need iron and folate).
- Newborns (they are routinely given a vitamin K dose).
- Infants (they may need various supplements, see Chapter 12).
- Those who are lactose intolerant (they need calcium to forestall osteoporosis).
- Habitual dieters (they may eat insufficient food).
- Elderly people (they may choose poorly, have trouble chewing, or absorb or metabolize less efficiently; see Chapter 13).
- Victims of AIDS or other wasting illnesses (they lose nutrients faster than foods can supply them).
- Those addicted to drugs or alcohol (they absorb fewer and excrete more nutrients; nutrients cannot undo damage from drugs or alcohol).
- Those recovering from surgery, burns, injury, or illness (they need extra nutrients to help regenerate tissues).
- Strict vegetarians (they may need vitamin B_{12}, vitamin D, iron, and zinc).
- People taking medications that interfere with the body's use of nutrients.

These People Don't:

- Those who feel insecure about the amounts of nutrients in the food supply.
- Those who feel tired and falsely believe that supplements can provide energy.
- Those who believe that supplements will help them cope with stress.
- Those who wish to build lean body tissue faster or without physical work.
- Those who want to prevent or cure self-diagnosed conditions, from the common cold to cancer.
- Those with kidney or liver diseases (they are susceptible to vitamin toxicity).
- Those who hope that excess nutrients will produce mysterious beneficial reactions in the body.
- Those who are taking certain medications (supplements may interfere with the action of the medications).
- Those who smoke and take beta-carotene supplements (such supplements are associated with an increase in the risk of lung cancer in smokers).

Vitamin Supplements

When people dose themselves with supplements, they leave the realm of nutrition and enter that of pharmacology. Like drugs, large doses of nutrients can have medicinal effects on the body and can present serious side effects. Nevertheless, almost 40 percent of the U.S. population does take vitamin supplements, collectively spending billions of dollars a year on them. But who really needs supplements? And which supplements should they take?

Do I Need to Take a Vitamin Supplement?

People who may need vitamin or mineral supplements are listed in the column at the top of Table 7-2. You may wonder if you are in the first group, "People with nutrient deficiencies," or perhaps you are one of the "Habitual dieters." Some people's diets put them at risk of developing a **subclinical,** or **marginal, deficiency:** a state of unwellness shy of a classical, full-blown nutrient deficiency. Subclinical deficiencies are subtle and easy to overlook; for example, a lack of vitamin C may bring an undetectable increase in oxidative stress to the tissues long before the symptoms of scurvy appear. However, the appropriate remedy in these instances is to improve the diet so that it supplies the needed nutrients. Nutrition experts say that if you are a generally healthy adult between the ages of 20 and 70 who eats a reasonably nutritious diet and is not in one of the categories in the top column of Table 7-2, you do not need to take vitamin supplements.

In some special cases, nutrient supplements may be appropriate. Consider a woman who loses a lot of blood and therefore a lot of iron and other blood-building nutrients in menstruation each month. This woman may be able to eat in such a way as to make up all nutrient losses except that of iron, and for iron, she may need a supplement prescribed by a health-care provider.

key point *People who routinely fail to obtain the recommended amounts of vitamins and minerals from the diet and people with special needs, such as those who are pregnant or elderly, may be at risk for deficiencies and may benefit from a multivitamin-mineral supplement.*

Which Supplement Doses May Be Helpful, and Which Should I Avoid?

When people self-prescribe supplements, they have to choose doses. The higher the dose, the greater the risk of toxicity. Tolerance for high doses of nutrients varies, and no one can know what level may be safe for an individual person. Toxic overdoses of vitamins and minerals may be more common than we realize. Table 7-3 compares the DRI Tolerable Upper Intake Levels and Daily Values with nutrient amounts in typical vitamin and mineral supplements.

Even more worrisome than the possibility of short-term, acute overdoses is the potential for chronic, low-level nutrient toxicity in which the subtle effects develop slowly and go unrecognized. For example, a woman took just 1,500 micrograms (5,000 IU) of vitamin A a day, an amount typically found in vitamin-mineral supplements—but she took this dose daily for ten years. Then she was diagnosed with liver disease. When she discontinued the supplement, the condition cleared up. Vitamin A reliably produces liver injury at doses greater than 10,000 micrograms, and even at 5,000 micrograms, abnormal levels of liver enzymes are detectable in the blood. Because of the potential hazards that supplements present, some authorities believe supplements should be required to bear warning labels, but such labels have not been seriously considered.

Because tolerances vary, people in the first column of Table 7-2 should err on the conservative side in taking supplements. The DRI committee sets upper intake val-

Table 7•3

VITAMIN AND MINERAL INTAKES FOR ADULTS

Nutrient	Tolerable Upper Intake Levels[a]	Daily Values	Typical Multivitamin-Mineral Supplement	Average Single-Nutrient Supplement
Vitamins				
Vitamin A	3,000 μg (10,000 IU)	5,000 IU	5,000 IU	8,000 to 10,000 IU
Vitamin D	50 μg (2,000 IU)	400 IU	400 IU	400 IU
Vitamin E	1,000 mg (1,500 to 2,200 IU)[b]	30 IU	30 IU	100 to 1,000 IU
Vitamin K	—[c]	80 μg	40 μg	—[e]
Thiamin	—[c]	1.5 mg	1.5 mg	50 mg
Riboflavin	—[c]	1.7 mg	1.7 mg	25 mg
Niacin (as niacinamide)	35 mg[b]	20 mg	20 mg	100 to 500 mg
Vitamin B_6	100 mg	2 mg	2 mg	100 to 200 mg
Folate	1,000 μg[b]	400 μg	400 μg	400 μg
Vitamin B_{12}	—[c]	6 μg	6 μg	100 to 1,000 μg
Pantothenic acid	—[c]	10 mg	10 mg	100 to 500 mg
Biotin	—[c]	300 μg	30 μg	300 to 600 μg
Vitamin C	2,000 mg	60 mg	10 mg	500 to 2,000 mg
Choline	3,500 mg	—	10 mg	250 mg
Minerals				
Calcium	2,500 mg	1,000 mg	160 mg	250 to 600 mg
Phosphorus	4,000 mg	1,000 mg	110 mg	—[e]
Magnesium	350 mg[d]	400 mg	100 mg	250 mg
Iron	45 mg	18 mg	18 mg	18 to 30 mg
Zinc	40 mg	15 mg	15 mg	10 to 100 mg
Iodine	1,100 μg	150 μg	150 μg	—[e]
Selenium	400 μg	70 μg	10 μg	50 to 200 μg
Fluoride	10 mg	—	—	—[e]
Copper	10 mg	2 mg	0.5 mg	—[e]
Manganese	11 mg	2 mg	5 mg	—[e]
Chromium	—[c]	120 μg	25 μg	200 to 400 μg
Molybdenum	2,000 μg	75 μg	25 μg	—[e]

[a]Unless otherwise noted, Tolerable Upper Intake Levels represent total intakes from food, water, and supplements combined.
[b]Tolerable Upper Intake Levels represent intakes from supplements, fortified foods, or both.
[c]These nutrients have been evaluated by the DRI committee for Tolerable Upper Intake Levels, but none were established because of insufficient data.
[d]Tolerable Upper Intake Level represents intake from supplements only.
[e]Available as a single supplement by prescription.

ues only when sufficient research makes it scientifically reasonable to do so. A lack of a Tolerable Upper Intake Level (inside front cover) for a nutrient indicates only that insufficient research exists, not that the nutrient is safe in any amount.

Another potential problem is that supplements may lull their takers into a false sense of security. A person may eat irresponsibly, thinking, "My supplement will cover my needs." More often, supplements supply precisely the nutrients people need least—those they consume in food—while failing to provide those missing from the diet.

Another problem is **bioavailability.** In most cases, nutrients are absorbed best from foods in which they are dispersed among other ingredients that facilitate their absorption. In contrast, nutrients taken in pure, concentrated form are likely to interfere with the absorption of other nutrients. Minerals provide examples: zinc hinders copper and calcium absorption, iron hinders zinc absorption, calcium hinders magnesium and iron absorption, and so on. Interactions between vitamins also exist.

In view of all the negatives and uncertainties associated with supplement taking, we repeat that several nutrition societies have stated that most healthy people should not use supplements. If a person's diet is inadequate, the remedy is to improve food choices and eating patterns. Failing this, supplements rank a distant second choice.

subclinical, or **marginal, deficiency** a nutrient deficiency that has no outward clinical symptoms. The term is often used to scare consumers into buying unneeded nutrient supplements.

bioavailability absorbability; the individual differences in the proportion of a nutrient that is available for absorption from various sources.

dietary supplement a product, other than tobacco, that is added to the diet and contains one of the following ingredients: a vitamin, mineral, herb, botanical (plant extract), amino acid, metabolite, constituent, or extract, or a combination of any of these ingredients.

Before you decide to take supplements, make sure to recognize the potential for toxicity.

Supplements Must Be Safe, or the Government Would Not Allow Their Sales, Right?

"Structure/function" claims were described in the Controversy section of Chapter 2.

The Dietary Supplement Health and Education Act (DSHEA) of 1994 was intended to enable consumers to make informed choices about nutrient supplements without much regulation from the government. The act subjects supplements to some of the same general labeling requirements that apply to foods:

- They must provide nutrient information (see Figure 7-11).
- They may not make unapproved health claims.
- They may not claim to diagnose, prevent, cure, or treat specific illnesses.
- They may, however, bear "structure/function" claims—claims that nutrients or other substances in the product affect the structure or function of the body—and indicate that consuming the compound is associated with general well-being.

Figure 7•11

A SUPPLEMENT LABEL

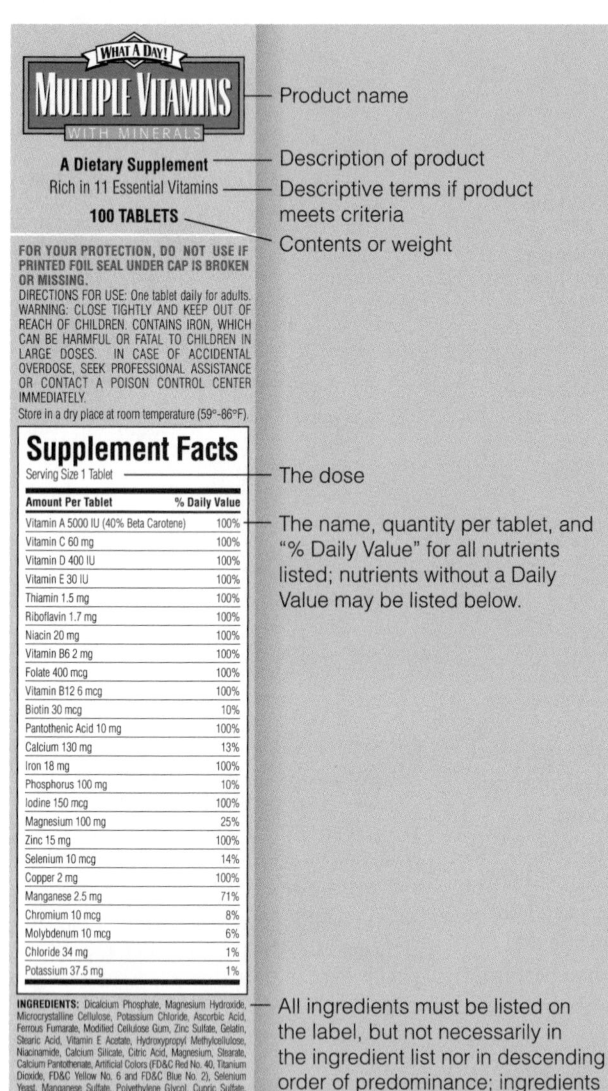

The DSHEA also defined the term **dietary supplement,** but the definition is so broad as to be almost meaningless.

In effect, the DSHEA resulted in the deregulation of the supplement industry, and the FDA does not ensure their safety.[57] Unlike foods, food additives, and drugs, supplements do not require government approval before entering the market, and manufacturers alone decide whether their products are safe and effective. The FDA does have the burden of proving that a supplement *ingredient* is unsafe and should be removed from the market, but the FDA lacks the resources to test all the ingredients and remove unsafe ones from the market before they cause harm.

Table 7-4 provides a sampling of the substances sold in the United States as "dietary supplements." Herbs, amino acids, dried organ tissues of animals, microorganisms, and concentrated hormones can be sold freely over the counter. Chapter 9 tells the story of ephedrine (Table 7-4), a dangerous ingredient still available in weight-loss products; Chapter 11 describes other herbs and dietary supplements.

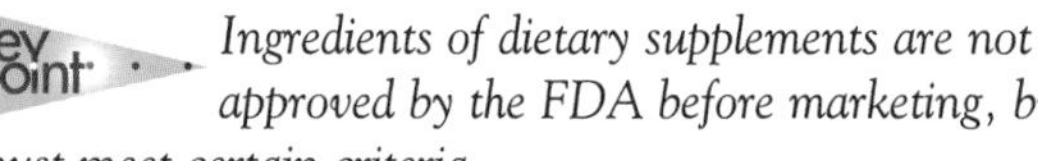

Ingredients of dietary supplements are not tested or approved by the FDA before marketing, but their labels must meet certain criteria.

Selection of a Multinutrient Supplement

As the preceding sections have indicated, if you choose to take a vitamin supplement, you do so at some risk. If you fall into one of the categories in the top column of Table 7-2, however, and if you cannot meet your nutrient needs from foods, a supplement containing nutrients only may be in order.

Which supplement to choose? The first step in escaping the clutches of the health hustlers is to use your imagination and delete the picture on the label of sexy people on the beach and the meaningless, glittering generalities like "new and improved." Now all you have left is the list of ingredients, the form they are in, and the price—the plain facts.

Table 7•4

A SAMPLE OF SUBSTANCES SOLD AS DIETARY SUPPLEMENTS

According to legal definitions, all of these substances qualify as dietary supplements, even though some appear to have the effects of drugs, not nutrients. Chapter 11 defines many more medicinal herbs sold as dietary supplements.

- **DHEA**[a] a hormone secretion of the adrenal gland whose level falls with advancing age. DHEA may protect antioxidant nutrients. Theories that DHEA might stimulate hormone-responsive cancers such as breast or prostate are unproved. Real DHEA is available only by prescription; herbal DHEA imitator for sale in health-food stores is not active in the body. No safety information exists.
- **desiccated liver** a powder sold in health-food stores and supposed to contain in concentrated form all the nutrients found in liver (*desiccated* means "totally dried").
- **ephedrine** One of a group of compounds with dangerous amphetamine-like stimulant effects; commonly added to herbal preparations such as Ma huang, to weight-loss products, and to products claiming to imitate the effects of illegal drugs of abuse. The most severe reported side effects of ephedrine include sudden death, heart attack, and stroke; other reported effects include chest pain, dizziness, fatigue, headache, insomnia, nausea, psychosis, seizure, tremor, and vomiting. The World Health Organization has called for worldwide controls on ephedrine-containing products.
- **garlic oil** an extract of garlic; may or may not contain the chemicals associated with garlic; claims for health benefits unproved.
- **green pills, fruit pills** pills containing dehydrated, crushed vegetable or fruit matter. An advertisement may claim that each pill equals a *pound* of fresh produce, but in reality a pill may equal one small forkful—minus nutrient losses incurred in processing.
- **kelp tablets** tablets made from dehydrated kelp, a kind of seaweed used by the Japanese as a foodstuff.
- **ma huang** an evergreen plant derivative that supposedly boosts energy and helps with weight control; Ma huang contains ephedrine (see above), especially dangerous in combination with kola nut or other caffeine-containing substances.
- **melatonin** a hormone of the pineal gland believed to help regulate the body's daily rhythms, to reverse the effects of jet lag, and to promote sleep. Claims for life extension or enhancement of sexual prowess are without merit.
- **nutritional yeast** a preparation of yeast cells, often praised for its high nutrient content. Yeast is a source of B vitamins, as are many other foods. Also called *brewer's yeast;* not the yeast used in baking.
- **SAM-e** an amino acid derivative that may have an antidepressant effect on the brain in some people, but is not recommended as a substitute for standard antidepressant therapy.
- 1,000 others

[a]Dehydroepiandrosterone.

See Table 11-9 on page 421 for medicinal herbs and their effects.

You have two basic questions to answer. The first question: What form do you want—chewable, liquid, or pills? If you'd rather drink your vitamins and minerals than chew them, fine. If you choose a fortified liquid or bar-type "energy" meal replacer, you must then proportionately reduce the calories you consume as food, or you may gain unwanted weight. If you choose chewable pills, be aware that vitamin C can erode tooth enamel. Swallow promptly and flush the teeth with a drink of water.

The second question: Who are you? What vitamins and minerals do you need? The DRI nutrient intake recommendations listed in the tables on the inside front cover are the standards appropriate for all reasonably healthy people.

An appropriate supplement provides all the vitamins and minerals in amounts smaller than, equal to, or very close to the intake recommendations. Avoid any preparation that, in a daily dose, provides more than the DRI recommended intake of vitamin A, vitamin D, or any mineral or more than the Tolerable Upper Intake Level for any nutrient. Warning: Expect to reject about 80 percent of available preparations when you choose according to these criteria; be choosy where your health is concerned. Avoid these:

- High doses of iron (more than 10 milligrams per day) except for menstruating women. People who menstruate need more iron, but people who don't, don't.

To get the most from a supplement of vitamins and minerals:

- Take the supplement with food. A full stomach retains the pill and dissolves it with churning action.
- If you take an iron supplement, choose foods that will assist in its absorption, such as meats, fish, poultry, or foods that contain vitamin C.

- "Organic" or "natural" preparations with added substances. They are no better than standard types, but they cost much more.
- "High-potency" or "therapeutic dose" supplements. More is not better.
- Items not needed in human nutrition, such as carnitine and inositol. These particular items won't harm you, but they reveal a marketing strategy that makes the whole mix suspect. The manufacturer wants you to believe that its pills contain the latest "new" nutrient that other brands omit, but, in fact, for every valid discovery of this kind, there are 999,999 frauds.
- "Time release." Medications such as some antibiotics or pain relievers often must be sustained at a steady concentration in the blood to be effective, but nutrients are incorporated into the tissues where they are needed whenever they arrive.
- "Stress formulas." Although the stress response depends on certain B vitamins and vitamin C, the recommended amount provides all that is needed of these nutrients. If you are under stress (and who isn't?), generous servings of fruits and vegetables will more than cover your need.
- Pills containing extracts of parsley, alfalfa, and other vegetable components.
- Geriatric "tonics." They are generally poor in vitamins and minerals and yet may be so high in alcohol as to threaten inebriation.
- Any supplement sold with claims that today's foods lack the nutrients they once contained. Truth: Plants make vitamins for their own needs, not ours. A plant lacking a mineral or failing to make a needed vitamin dies before it can bear food for our consumption.

People in developed nations are far more likely to suffer from *overnutrition* and poor lifestyle choices than from nutrient deficiencies. People wish that swallowing vitamin pills would boost their health. The truth—that they need to improve their eating and exercise habits—is harder to swallow.

If you decide to take a supplement, examine the ingredients on supplement labels and choose one that satisfies your needs.

This chapter has addressed all 13 of the vitamins. The basic facts about each one are summed up in Tables 7-5 and 7-6.

Table 7•5

THE FAT-SOLUBLE VITAMINS—FUNCTIONS, DEFICIENCIES, AND TOXICITIES

Vitamin A

Other Names
Retinol, retinal, retinoic acid; main precursor is beta-carotene

Chief Functions in the Body
Vision; health of cornea, epithelial cells, mucous membranes, skin; bone and tooth growth; reproduction; immunity
Beta-carotene: antioxidant

Deficiency Disease Name
Hypovitaminosis A

Significant Sources
Retinol: fortified milk, cheese, cream, butter, fortified margarine, eggs, liver
Beta-carotene: spinach and other dark, leafy greens; broccoli; deep orange fruits (apricots, cantaloupe) and vegetables (squash, carrots, sweet potatoes, pumpkin)

	Deficiency Symptoms	Toxicity Symptoms
Blood/Circulatory System	Anemia (small-cell type)[a]	Red blood cell breakage, cessation of menstruation, nosebleeds
Bones/Teeth	Cessation of bone growth, painful joints; impaired enamel formation, cracks in teeth, tendency toward tooth decay	Bone pain; growth retardation; increased pressure inside skull; headaches; possible bone mineral loss
Digestive System	Diarrhea, changes in intestinal and other body linings	Abdominal pain, nausea, vomiting, diarrhea, weight loss
Immune System	Depression; frequent respiratory, digestive, bladder, vaginal, and other infections	Overreactivity
Nervous/Muscular Systems	Night blindness (retinal)	Blurred vision, muscle weakness, fatigue, irritability, loss of appetite
Skin and Cornea	Keratinization, corneal degeneration leading to blindness,[b] rashes	Dry skin, rashes, loss of hair; cracking and bleeding lips, brittle nails; hair loss
Other	Kidney stones, impaired growth	Liver enlargement and liver damage; birth defects

Vitamin D

Other Names
Calciferol, cholecalciferol, dihydroxy vitamin D; precursor is cholesterol

Chief Functions in the Body
Mineralization of bones (raises blood calcium and phosphorus via absorption from digestive tract and by withdrawing calcium from bones and stimulating retention by kidneys)

Deficiency Disease Name
Rickets, osteomalacia

Significant Sources
Self-synthesis with sunlight; fortified milk or margarine, eggs, liver, sardines

	Deficiency Symptoms	Toxicity Symptoms
Blood/Circulatory System		Raised blood calcium
Bones/Teeth	Abnormal growth, misshapen bones (bowing of legs), soft bones, joint pain, malformed teeth	Calcification of tooth soft tissue; thinning of tooth enamel
Nervous System	Muscle spasms	Excessive thirst, headaches, irritability, loss of appetite, weakness, nausea
Other		Kidney stones, stones in arteries,

[a]Small-cell anemia is termed *microcytic anemia;* large-cell type is *macrocytic* or *megaloblastic anemia.*
[b]Corneal degeneration progresses from *keratinization* (hardening) to *xerosis* (drying) to *xerophthalmia* (thickening, opacity, and irreversible blindness).

(*Continued on next page*)

Table 7•5

THE FAT-SOLUBLE VITAMINS—FUNCTIONS, DEFICIENCIES, AND TOXICITIES—(Continued)

Vitamin E

Other Names	Deficiency Symptoms	Toxicity Symptoms
Alpha-tocopherol, tocopherol		mental and physical retardation
Chief Functions in the Body	**Blood/Circulatory System**	
Antioxidant (detoxification of strong oxidants), stabilization of cell membranes, regulation of oxidation reactions, protection of PUFA and vitamin A	Red blood cell breakage, anemia	Augments the effects of anticlotting medication
	Digestive System	
		General discomfort, nausea
	Eyes	
		Blurred vision
Deficiency Disease Name	**Nervous/Muscular Systems**	
(No name)	Degeneration, weakness, difficulty walking, leg cramps	Fatgue
Significant Sources		
Polyunsaturated plant oils (margarine, salad dressings, shortenings), green and leafy vegetables, wheat germ, whole-grain products, nuts, seeds		

Vitamin K

Other Names	Deficiency Symptoms	Toxicity Symptoms
Phylloquinone, naphthoquinone		
Chief Functions in the Body	**Blood/Circulatory System**	
Synthesis of blood-clotting proteins and proteins important in bone mineralization	Hemorrhage	Interference with anticlotting medication; vitamin K analogues may cause jaundice
	Bones	
Deficiency Disease Name	Poor skeletal mineralization	
(No name)		
Significant Sources		
Bacterial synthesis in the digestive tract; green leafy vegetables, cabbage-type vegetables, soybeans, vegetable oils		

Table 7•6

THE WATER-SOLUBLE VITAMINS—FUNCTIONS, DEFICIENCIES, AND TOXICITIES

Thiamin

Other Names	Deficiency Symptoms	Toxicity Symptoms
Vitamin B_1	**Blood/Circulatory System**	
Chief Functions in the Body	Edema, enlarged heart, abnormal heart rhythms, heart failure	(No symptoms reported)
Part of a coenzyme used in energy metabolism, supports normal appetite and nervous system function	**Nervous/Muscular Systems**	
Deficiency Disease Name	Degeneration, wasting, weakness, pain, apathy, irritability, low morale, difficulty walking, loss of reflexes, mental confusion, paralysis	(No symptoms reported)
Beriberi (wet and dry)	**Other**	
Significant Sources	Anorexia; weight loss	
Occurs in all nutritious foods in moderate amounts; pork, ham, bacon, liver, whole and enriched grains, legumes, nuts		

Riboflavin

Other Names	Deficiency Symptoms	Toxicity Symptoms
Vitamin B_2	**Mouth, Gums, Tongue**	
Chief Functions in the Body	Cracks at corners of mouth,[a] magenta tongue; sore throat	(No symptoms reported)
Part of a coenzyme used in energy metabolism, supports normal vision and skin health	**Nervous System and Eyes**	
	Hypersensitivity to light,[b] reddening of cornea	(No symptoms reported)
Deficiency Disease Name	**Skin**	
Ariboflavinosis	Skin rash	(No symptoms reported)
Significant Sources		
Milk, yogurt, cottage cheese, meat, liver, leafy green vegetables, whole-grain or enriched breads and cereals		

Niacin

Other Names	Deficiency Symptoms	Toxicity Symptoms
Nicotinic acid, nicotinamide, niacinamide, vitamin B_3; precursor is dietary tryptophan	**Digestive System**	
	Diarrhea; abdominal pain	Nausea, vomiting
Chief Functions in the Body	**Mouth, Gums, Tongue**	
Part of a coenzyme used in energy metabolism	Black or bright red swollen smooth tongue[c]	
Deficiency Disease Name	**Nervous System**	
Pellagra	Irritability, loss of appetite, weakness, headache, dizziness, mental confusion progressing to psychosis or delirium	
Significant Sources	**Skin**	
Synthesized from the amino acid tryptophan; milk, eggs, meat, poultry, fish, whole-grain and enriched breads and cereals, nuts, and all protein-containing foods	Flaky skin rash on areas exposed to sun	Painful flush and rash, sweating
	Other	
		Liver damage; impaired glucose tolerance

[a]Cracks at the corners of the mouth are termed *cheilosis* (kee-LOH-sis).
[b]Hypersensitivity to light is *photophobia*.
[c]Smoothness of the tongue is caused by loss of its surface structures and is termed *glossitis* (gloss-EYE-tis).

(Continued on next page)

Table 7•6

THE WATER-SOLUBLE VITAMINS—FUNCTIONS, DEFICIENCIES, AND TOXICITIES—(Continued)

Folate

Other Names / Chief Functions / Deficiency Disease Name / Significant Sources	Deficiency Symptoms	Toxicity Symptoms
Other Names: Folic acid, folacin, pteroyglutamic acid		
	Blood/Circulatory System	
Chief Functions in the Body: Part of a coenzyme needed for new cell synthesis	Anemia (large-cell type)[d], elevated homocysteine	Masks vitamin B_{12} deficiency
	Digestive System	
Deficiency Disease Name: (No name)	Heartburn, diarrhea, constipation	
	Immune System	
Significant Sources: Leafy green vegetables, legumes, seeds, liver, enriched breads, cereal, pasta, and grains	Suppression, frequent infections	
	Mouth, Gums, Tongue	
	Smooth red tongue[c]	
	Nervous System	
	Depression, mental confusion, fatigue, irritability, headache	

Vitamin B_{12}

Other Names / Chief Functions / Deficiency Disease Name / Significant Sources	Deficiency Symptoms	Toxicity Symptoms
Other Names: Cyanocobalamin		
	Blood/Circulatory System	
Chief Functions in the Body: Part of a coenzyme used in new cell synthesis, helps maintain nerve cells	Anemia (large-cell type)[d]	(No toxicity symptoms known)
	Mouth, Gums, Tongue	
Deficiency Disease Name: (No name)[e]	Smooth tongue[c]	
	Nervous System	
Significant Sources: Animal products (meat, fish, poultry, milk, cheese, eggs)	Fatigue, degeneration progressing to paralysis	
	Skin	
	Hypersensitivity	

Vitamin B_6

Other Names / Chief Functions / Deficiency Disease Name / Significant Sources	Deficiency Symptoms	Toxicity Symptoms
Other Names: Pyridoxine, pyridoxal, pyridoxamine		
	Blood/Circulatory System	
Chief Functions in the Body: Part of a coenzyme used in amino acid and fatty acid metabolism, helps convert tryptophan to niacin, helps make red blood cells	Anemia (small-cell type)[d]	Bloating
	Nervous/Muscular Systems	
Deficiency Disease Name: (No name)	Depression, confusion, abnormal brain wave pattern, convulsions	Depression, fatigue, impaired memory, irritability, headaches, numbness, damage to nerves, difficulty walking, loss of reflexes, restlessness, convulsions
	Skin	
Significant Sources: Meats, fish, poultry, liver, legumes, fruits, potatoes, whole grains, soy products	Rashes, greasy, scaly dermatitis	Lesions

Table 7•6

THE WATER-SOLUBLE VITAMINS—FUNCTIONS, DEFICIENCIES, AND TOXICITIES—(Continued)

Pantothenic Acid

Other Names
(None)

Chief Functions in the Body
Part of a coenzyme used in energy metabolism

Deficiency Disease Name
(No name)

Significant Sources
Widespread in foods

	Deficiency Symptoms	Toxicity Symptoms
Digestive System	Vomiting, intestinal distress	
Nervous/Muscular Systems	Insomnia, fatigue	
Other	Hypoglycemia, increased sensitivity to insulin	Water retention (infrequent)

Biotin

Other Names
(None)

Chief Functions in the Body
A cofactor for several enzymes used in energy metabolism, fat synthesis, amino acid metabolism, and glycogen synthesis

Deficiency Disease Name
(No name)

Significant Sources
Widespread in foods

	Deficiency Symptoms	Toxicity Symptoms
Blood/Circulatory System	Abnormal heart action	(No toxicity symptoms reported)
Digestive System	Loss of appetite, nausea	
Nervous/Muscular Systems	Depression, muscle pain, weakness, fatigue, numbness of extremities	
Skin	Dry around eyes, nose, and mouth	

Vitamin C

Other Names
Ascorbic acid

Chief Functions in the Body
Collagen synthesis (strengthens blood vessel walls, forms scar tissue, matrix for bone growth), antioxidant, thyroxine synthesis, amino acid metabolism, strengthens resistance to infection, helps in absorption of iron

Deficiency Disease Name
Scurvy

Significant Sources
Citrus fruits, cabbage-type vegetables, dark green vegetables, cantaloupe, strawberries, peppers, lettuce, tomatoes, potatoes, papayas, mangoes

	Deficiency Symptoms	Toxicity Symptoms
Blood/Circulatory System	Anemia (small-cell type),[d] pinpoint hemorrhages	
Digestive System		Nausea, abdominal cramps, diarrhea, excessive urination
Immune System	Suppression, frequent infections	
Mouth, Gums, Tongue	Bleeding gums, loosened teeth	
Nervous/Muscular Systems	Muscle degeneration and pain, hysteria, depression	Headache, fatigue, insomnia
Skeletal System	Bone fragility, joint pain	
Skin	Rough skin, blotchy bruises	Rashes
Other	Failure of wounds to heal	Interference with medical tests; aggravation of gout symptoms; deficiency symptoms may appear at first on withdrawal of high doses

[a]Cracks at the corners of the mouth are termed *cheilosis* (kee-LOH-sis).
[b]Hypersensitivity to light is *photophobia*.
[c]Smoothness of the tongue is caused by loss of its surface structures and is termed *glossitis* (gloss-EYE-tis).
[d]Small-cell anemia is termed *microcytic anemia;* large-cell type is *macrocytic* or *megaloblastic anemia*.
[e]The name *pernicious anemia* refers to the vitamin B_{12} deficiency caused by lack of intrinsic factor, but not to that caused by inadequate dietary intake.

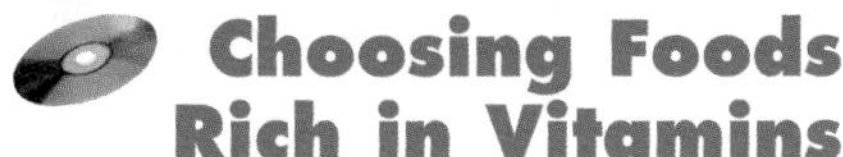

Choosing Foods Rich in Vitamins

On learning how important the vitamins are to their health, most people want to choose foods that are vitamin-rich. Look down the columns of vitamins and calories in Table 7-7 to identify some of these foods. If you are interested in folate, for instance, you can see that cornflakes are an especially good source (folic acid is added to cornflakes), as are many vegetables (folate occurs naturally in them). This table is used again in the Do It! section.

Another way of looking at such data appears in Figure 7-12 (pages 252–253)—the long bars show some foods that are rich sources of a particular vitamin, and the short or nonexistent bars indicate poor sources. The serving sizes in Figure 7-12 are those recommended by the Food Guide Pyramid. For example, 3 ounces is used for most cooked meats. The serving size for most cooked or chopped vegetables is ½ cup. The colors of the bars represent the various food groups.

Which Foods Should I Choose?

After looking at Figure 7-12, don't think that you must memorize the richest sources of each vitamin and include those foods daily. That false notion would lead you to limit your variety of

Table 7•7

VITAMIN CONTENTS OF RESTAURANT MEALS

	Energy (cal)	Vitamin A (µg)	Thiamin (mg)	Niacin (mg)	Vitamin C (mg)	Folate (µg)	Vitamin B_{12} (µg)
Breakfast Foods							
Hotcakes with syrup and butter, scrambled egg and sausage patty	1,140	150	0.41	4	0	32	1.7
Egg, ham, and cheese muffin	290	100	0.49	3	0	30	0.7
Oatmeal, brown sugar	107	2	0.13	0	0	5	0
Cornflakes	110	225	0.87	5	15	100	0
Hash browned potatoes	130	0	0.08	1	3	8	0
2 small cinnamon sweet rolls	300	50	0.25	2	2	20	0.1
Large blueberry muffin	400	10	0.2	2	2	20	0.1
English muffin	130	0	0.25	2	0	20	0.1
Orange juice	80	40	0.17	1	90	60	0
Milk (low-fat)	120	140	0.1	0	2	10	0.8
Lunch Foods							
Homemade chili/crackers	350	150	0.26	5	25	40	0.5
Cold cut hoagie sandwich/chips	460	80	1.0	6	12	55	1.1
Peanut butter and jelly sandwich on whole wheat; fruit cocktail	450	30	0.3	7	3	62	0
Tuna sandwich on white; banana	470	40	0.32	6	11	49	0.6
Chef's salad with cheese, ham, turkey, and dressing/crackers	580	140	0.43	7	16	100	0.8
Fat-free milk	85	150	0.1	0	0	10	0.9
Apple juice	120	0	0.05	0	2	0	0
Supper Foods							
New England boiled dinner (corned beef, potatoes, brussels sprouts)	440	150	0.27	5	80	70	1.4
Vegetable plate	400	680	0.26	2	53	120	0.1
Spaghetti and meatballs; small salad	600	220	0.38	6	25	50	1.3
Fried fish, tartar sauce, corn, and macaroni salad	510	70	0.26	3	6	83	1
Rolls	100	0	0.15	1	0	0	0
Garlic bread	190	4	0.4	3	0	0	0
Corn muffins	300	0	0.12	2	0	0	0
Lemon pie	350	60	0.16	1	4	11	0.2
Chocolate cake	250	20	0.02	0	0	5	0.1

foods while overemphasizing the components of a few foods. Although it is reassuring to know that your carrot-raisin salad at lunch amply provided your entire day's need for vitamin A, it is a mistake to think that you must then select equally rich sources of all the other vitamins. Such rich sources do not exist for many vitamins—rather, foods work in harmony to provide most nutrients. For example, a baked potato, not a star performer among vitamin C providers, contributes substantially to a day's need for this nutrient and contributes some thiamin, too. By the end of the day, assuming that your food choices were made with reasonable care, the bits of thiamin, vitamin B_6, and vitamin C from each serving of food have accumulated to make a more-than-adequate total diet.

A Variety of Foods Works Best

With a few exceptions, nutritious foods provide small quantities of thiamin, as shown in Figure 7-12. Members of the pork family are an exceptionally good thiamin source, with one small pork chop (275 calories) providing over half of the Daily Value for thiamin—but again, this does not suggest that you must eat pork every day. Legumes and grains are also good, low-fat sources, and they provide beneficial fiber and nutrients lacking from meats. Beans lack the vitamin B_{12} provided by meats, however. Peanut butter is a good source of thiamin, as it is of most B vitamins, but its high fat and calorie contents call for moderation in its use.

The vitamin B_6 data provide another insight to support the argument for variety. From just the few foods listed here, you can see that no one source can provide the whole day's requirement, but that a variety of meats, fish, and poultry along with potatoes and a few other vegetables and fruits can work together to supply it.

The last two graphs of Figure 7-12 show sources of folate and vitamin C. These nutrients are both richly supplied by fruits and vegetables. The richest source of either may be only a moderate source of the other, but the recommended servings of fruits and vegetables in food group plans cover both needs amply. As for vitamin E, vegetable oils are the richest sources, and some vegetables, nuts, and fruits contribute some, too.

Are you ready to try out what you've learned about the food sources of vitamins? The Do It! section that follows gives you a chance to test your skills.

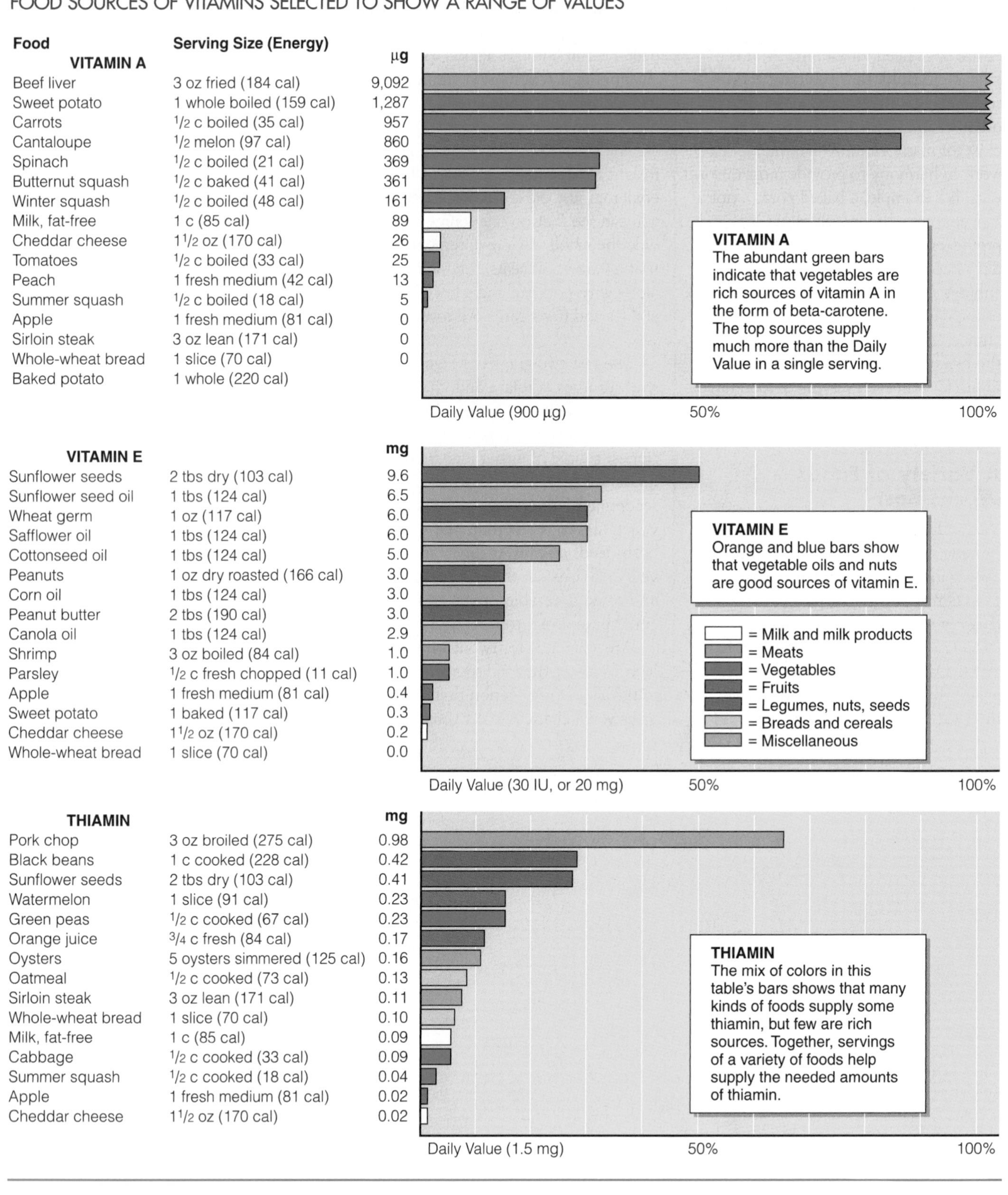
Figure 7•12
FOOD SOURCES OF VITAMINS SELECTED TO SHOW A RANGE OF VALUES
Daily Value (900 µg)
50%
100%
VITAMIN A
The abundant green bars indicate that vegetables are rich sources of vitamin A in the form of beta-carotene. The top sources supply much more than the Daily Value in a single serving.
Daily Value (30 IU, or 20 mg)
50%
100%
VITAMIN E
Orange and blue bars show that vegetable oils and nuts are good sources of vitamin E.
= Milk and milk products
= Meats
= Vegetables
= Fruits
= Legumes, nuts, seeds
= Breads and cereals
= Miscellaneous
Daily Value (1.5 mg)
50%
100%
THIAMIN
The mix of colors in this table's bars shows that many kinds of foods supply some thiamin, but few are rich sources. Together, servings of a variety of foods help supply the needed amounts of thiamin.

Food	Serving Size (Energy)	µg
VITAMIN A		
Beef liver	3 oz fried (184 cal)	9,092
Sweet potato	1 whole boiled (159 cal)	1,287
Carrots	1/2 c boiled (35 cal)	957
Cantaloupe	1/2 melon (97 cal)	860
Spinach	1/2 c boiled (21 cal)	369
Butternut squash	1/2 c baked (41 cal)	361
Winter squash	1/2 c boiled (48 cal)	161
Milk, fat-free	1 c (85 cal)	89
Cheddar cheese	1 1/2 oz (170 cal)	26
Tomatoes	1/2 c boiled (33 cal)	25
Peach	1 fresh medium (42 cal)	13
Summer squash	1/2 c boiled (18 cal)	5
Apple	1 fresh medium (81 cal)	0
Sirloin steak	3 oz lean (171 cal)	0
Whole-wheat bread	1 slice (70 cal)	0
Baked potato	1 whole (220 cal)	

Food	Serving Size (Energy)	mg
VITAMIN E		
Sunflower seeds	2 tbs dry (103 cal)	9.6
Sunflower seed oil	1 tbs (124 cal)	6.5
Wheat germ	1 oz (117 cal)	6.0
Safflower oil	1 tbs (124 cal)	6.0
Cottonseed oil	1 tbs (124 cal)	5.0
Peanuts	1 oz dry roasted (166 cal)	3.0
Corn oil	1 tbs (124 cal)	3.0
Peanut butter	2 tbs (190 cal)	3.0
Canola oil	1 tbs (124 cal)	2.9
Shrimp	3 oz boiled (84 cal)	1.0
Parsley	1/2 c fresh chopped (11 cal)	1.0
Apple	1 fresh medium (81 cal)	0.4
Sweet potato	1 baked (117 cal)	0.3
Cheddar cheese	1 1/2 oz (170 cal)	0.2
Whole-wheat bread	1 slice (70 cal)	0.0

Food	Serving Size (Energy)	mg
THIAMIN		
Pork chop	3 oz broiled (275 cal)	0.98
Black beans	1 c cooked (228 cal)	0.42
Sunflower seeds	2 tbs dry (103 cal)	0.41
Watermelon	1 slice (91 cal)	0.23
Green peas	1/2 c cooked (67 cal)	0.23
Orange juice	3/4 c fresh (84 cal)	0.17
Oysters	5 oysters simmered (125 cal)	0.16
Oatmeal	1/2 c cooked (73 cal)	0.13
Sirloin steak	3 oz lean (171 cal)	0.11
Whole-wheat bread	1 slice (70 cal)	0.10
Milk, fat-free	1 c (85 cal)	0.09
Cabbage	1/2 c cooked (33 cal)	0.09
Summer squash	1/2 c cooked (18 cal)	0.04
Apple	1 fresh medium (81 cal)	0.02
Cheddar cheese	1 1/2 oz (170 cal)	0.02

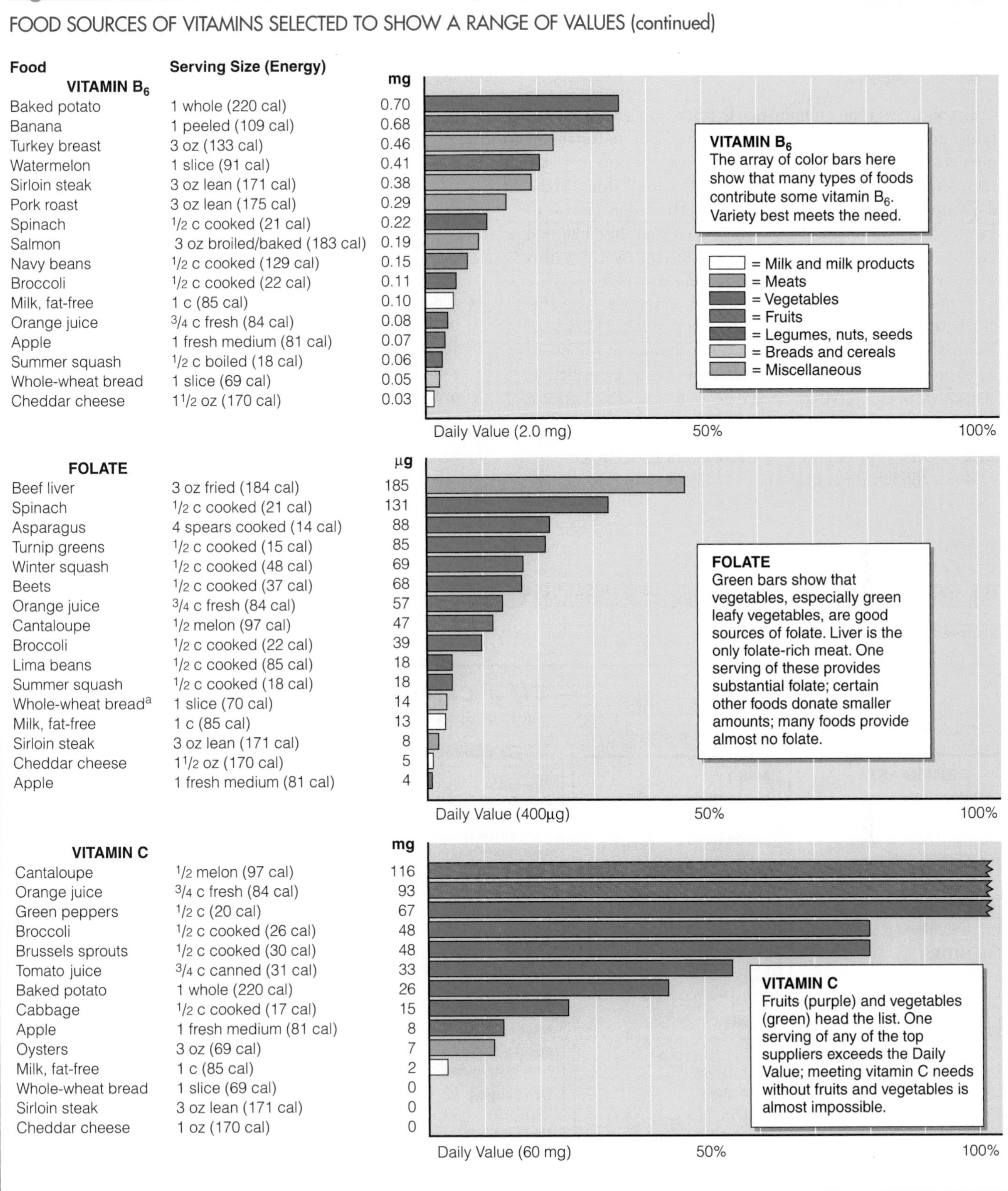

Figure 7•12

FOOD SOURCES OF VITAMINS SELECTED TO SHOW A RANGE OF VALUES (continued)

Food	Serving Size (Energy)	mg
VITAMIN B_6		
Baked potato	1 whole (220 cal)	0.70
Banana	1 peeled (109 cal)	0.68
Turkey breast	3 oz (133 cal)	0.46
Watermelon	1 slice (91 cal)	0.41
Sirloin steak	3 oz lean (171 cal)	0.38
Pork roast	3 oz lean (175 cal)	0.29
Spinach	1/2 c cooked (21 cal)	0.22
Salmon	3 oz broiled/baked (183 cal)	0.19
Navy beans	1/2 c cooked (129 cal)	0.15
Broccoli	1/2 c cooked (22 cal)	0.11
Milk, fat-free	1 c (85 cal)	0.10
Orange juice	3/4 c fresh (84 cal)	0.08
Apple	1 fresh medium (81 cal)	0.07
Summer squash	1/2 c boiled (18 cal)	0.06
Whole-wheat bread	1 slice (69 cal)	0.05
Cheddar cheese	1 1/2 oz (170 cal)	0.03

Food	Serving Size (Energy)	µg
FOLATE		
Beef liver	3 oz fried (184 cal)	185
Spinach	1/2 c cooked (21 cal)	131
Asparagus	4 spears cooked (14 cal)	88
Turnip greens	1/2 c cooked (15 cal)	85
Winter squash	1/2 c cooked (48 cal)	69
Beets	1/2 c cooked (37 cal)	68
Orange juice	3/4 c fresh (84 cal)	57
Cantaloupe	1/2 melon (97 cal)	47
Broccoli	1/2 c cooked (22 cal)	39
Lima beans	1/2 c cooked (85 cal)	18
Summer squash	1/2 c cooked (18 cal)	18
Whole-wheat bread[a]	1 slice (70 cal)	14
Milk, fat-free	1 c (85 cal)	13
Sirloin steak	3 oz lean (171 cal)	8
Cheddar cheese	1 1/2 oz (170 cal)	5
Apple	1 fresh medium (81 cal)	4

Food	Serving Size (Energy)	mg
VITAMIN C		
Cantaloupe	1/2 melon (97 cal)	116
Orange juice	3/4 c fresh (84 cal)	93
Green peppers	1/2 c (20 cal)	67
Broccoli	1/2 c cooked (26 cal)	48
Brussels sprouts	1/2 c cooked (30 cal)	48
Tomato juice	3/4 c canned (31 cal)	33
Baked potato	1 whole (220 cal)	26
Cabbage	1/2 c cooked (17 cal)	15
Apple	1 fresh medium (81 cal)	8
Oysters	3 oz (69 cal)	7
Milk, fat-free	1 c (85 cal)	2
Whole-wheat bread	1 slice (69 cal)	0
Sirloin steak	3 oz lean (171 cal)	0
Cheddar cheese	1 oz (170 cal)	0

Daily Value (60 mg)
50%
100%
VITAMIN B6
The array of color bars here show that many types of foods contribute some vitamin B6. Variety best meets the need.
= Milk and milk products
= Meats
= Vegetables
= Fruits
= Legumes, nuts, seeds
= Breads and cereals
= Miscellaneous
FOLATE
Green bars show that vegetables, especially green leafy vegetables, are good sources of folate. Liver is the only folate-rich meat. One serving of these provides substantial folate; certain other foods donate smaller amounts; many foods provide almost no folate.
VITAMIN C
Fruits (purple) and vegetables (green) head the list. One serving of any of the top suppliers exceeds the Daily Value; meeting vitamin C needs without fruits and vegetables is almost impossible.

do it! FIND THE VITAMINS ON A MENU

Can you meet your vitamin needs when you eat in restaurants? You can if you learn to identify the foods on restaurant menus that are rich sources of vitamins. Assume you are spending a day on the road and have to eat all three of the day's meals in restaurants. Read over the three menus in Figure 7-13 and create a meal from each one. Just like real menus, these menus lack serving size information. For this exercise, take for granted that the serving sizes of foods agree with those in the Food Guide Pyramid. (Never assume this about real menus, however—commercial serving sizes vary enormously.) Most vegetable and grain servings in the meals listed in Figure 7-13 are ½-cup servings. Those identified as "large" are 1-cup servings; meats are 2- to 3-ounce portions; milk is an 8-ounce serving.

Step 1. On a copy of Form 7-1, record your food choices down the left-hand column.

Step 2. Consult Table 7-7 on page 250 to determine the values for calories and nutrients in the day's meals. Fill in the values for the foods you listed on Form 7-1. Coffee, tea, and water contribute negligible energy and vitamins; use zeros for their values on Form 7-1.

Step 3. Enter the DRI intake recommendations that apply to your age and gender (see the inside front cover for nutrients, inside back cover for energy) in the spaces provided on Form 7-1.

Step 4. Divide the day's total intakes by the recommendation amounts and multiply by 100 to determine the percentages of your nutrient and energy needs contributed by the foods chosen this day.

- Example: If total vitamin C intake equals 40 milligrams and your vitamin C recommendation equals 90 milligrams, then (40 ÷ 90) × 100 = 44%. The meals provided less than one-half of the daily recommended amount of vitamin C.

Repeat this process for all five vitamins and for energy. For folate, identify the foods that have been fortified, such as breads and cereals, and apply the DFE conversion factor demonstrated on page 231.

Figure 7•13

RESTAURANT MENUS

Welcome to
BURGER DOODLE
★ ★ ★

★ **BREAKFAST**
Hotcakes, eggs, and sausage ....
Egg, ham, and cheese muffin ....
Oatmeal with brown sugar ....
Cornflakes ....

★ **SIDES**
Hash browned potatoes ....
A pair of cinnamon sweet rolls ....
Large blueberry muffin ....
Plain English muffin ....

★ **BEVERAGES**
Orange juice ....
Low-fat milk ....
Coffee or tea ....

The Box Lunch
EXPRESS
Soups
Homemade chili with crackers
Sandwiches
Cold cut hoagie sandwich with chips
Peanut butter and jelly sandwich on whole-wheat bread, served with fruit cocktail
Tuna sandwich on white bread, served with a fresh banana
Salads
Large chef's salad with cheese, ham, and turkey with crackers
Beverages
Fat-free milk
Apple juice
Sparkling water

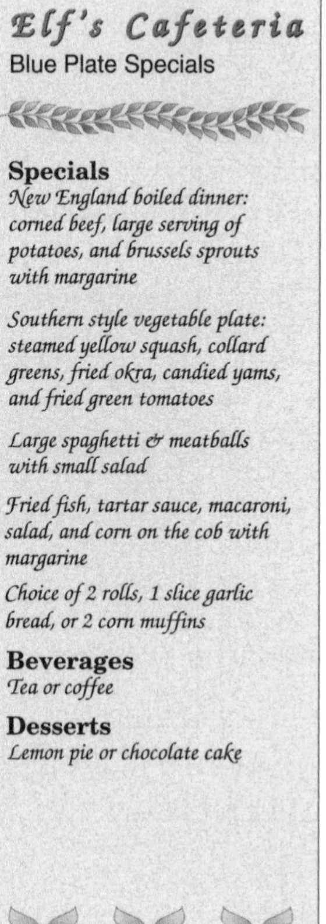

Elf's Cafeteria
Blue Plate Specials
Specials
New England boiled dinner: corned beef, large serving of potatoes, and brussels sprouts with margarine
Southern style vegetable plate: steamed yellow squash, collard greens, fried okra, candied yams, and fried green tomatoes
Large spaghetti & meatballs with small salad
Fried fish, tartar sauce, macaroni, salad, and corn on the cob with margarine
Choice of 2 rolls, 1 slice garlic bread, or 2 corn muffins
Beverages
Tea or coffee
Desserts
Lemon pie or chocolate cake

Analysis

Answer the following questions:

1. How did the day's totals for calories and vitamins compare with your recommended amounts? Did the day's meals meet or exceed your need for energy or any of these vitamins? Which ones?
2. Did the meals present too little energy or any of the nutrients listed? Which ones? What changes in your choices among those foods would have improved the vitamin totals for the day?
3. Which foods listed on the menus contributed significant amounts of vitamin A? Remembering that brightly colored fruits and vegetables offer the vitamin A precursor beta-carotene, name any foods in your meal choices that provide beta-carotene (turn back to Snapshot 7-1 on page 217 for hints). Which foods contributed little or no vitamin A in any form?
4. Did your choices supply enough folate to meet your requirement? Remember to use the DFE conversion factor (p. 232) for folate from fortified foods. What percentage of your folate need did the day's meals provide? Which individual foods

were richest in folate? Turn to the Table of Food Composition, Appendix A at the back of the book, and look down the folate column of the fruits and vegetables sections. Identify some fruits or vegetables that, in a single serving, provide 10 percent or more of the recommended 400 micrograms of folate.

5. What are the sources of niacin in the day's meals? Rich sources are shown in Snapshot 7-7 on page 230.
6. What about vitamin C? What percentage of your daily need of vitamin C did these meals provide? Which individual foods were the main contributors? To what food groups do they belong?
7. How did your total energy intake compare with your energy need? Express your answer as a percentage, and use it to answer the next two questions.
8. Find the meals that are vitamin "bargains." Compare the vitamin contents (as a percentage of your need) with the energy contents of the chosen meals (also as a percentage of your need). Which are the most vitamin-dense meals, providing the most vitamins for the fewest calories?
9. Which of the breakfast choices is highest in vitamin A? Which is highest in folate and niacin? What characteristics of this food make its vitamin content so high?
10. Which foods on the menus, though low in vitamins, possess other valuable constituents that make them desirable as part of a health-promoting diet?

If you are wondering about the vitamins in other foods you choose, you can find their vitamin and other nutrient contents in several references in this book:

1. The vitamin snapshots appearing throughout this chapter depict the richest vitamin sources.
2. The Daily Food Guide, on pages 36 and 37, notes which vitamins characterize foods of each group.
3. The Table of Food Composition, Appendix A at the back of the book, provides actual values for vitamins in about 2,000 foods.

With these references, you can judge the vitamin values of foods on virtually any menu.

Form 7•1

VITAMIN TALLY

	Energy (cal)	Vitamin A (μg)	Thiamin (mg)	Niacin (mg)	Vitamin C (mg)	Folate (μg)	Vitamin B_{12} (μg)
Breakfast:							
Lunch:							
Supper:							
Day's totals							
Energy and vitamin recommendations							
% of recommendations							

Answers to these Self Check questions are in Appendix G.

1. Which of the following vitamins are classified as water-soluble?
   a. vitamins B, D,
   b. vitamins B, E and C
   c. vitamins A, C, E, and K
   d. vitamins B and C
2. Night blindness and xerophthalmia are the result of a deficiency of which vitamin?
   a. niacin
   b. vitamin C
   c. vitamin A
   d. vitamin K
3. Which of the following foods is (are) rich in beta-carotene?
   a. sweet potatoes
   b. pumpkin
   c. cantaloupe
   d. all of the above
4. A deficiency of vitamin D may result in which disease?
   a. rickets
   b. beriberi
   c. scurvy
   d. pellagra
5. Which of the following describes the fat-soluble vitamins?
   a. vitamins A, D, E, and K
   b. easily absorbed and excreted
   c. stored extensively in tissues
   d. (a) and (c)
6. Which vitamin(s) is (are) present only in foods of animal origin?
   a. the active form of vitamin A
   b. vitamin B_{12}
   c. riboflavin
   d. (a) and (b)
7. The theory that vitamin C prevents or cures colds is well supported by research. T F
8. Almost any substance, including herbs, amino acids, dried animal organ tissues, microorganisms, and hormones can be sold freely over the counter in the United States. T F
9. In general, nutrients are absorbed equally well from foods and supplements. T F
10. People in developed nations are more likely to suffer from overnutrition than from nutrient deficiencies. T F

nutrition on the net

For further study of the topics of this chapter, access these websites and search for the phrases or words in quotation marks:

1. Find updates and quick links to these and other nutrition-related sites at our website: www.wadsworth.com/nutrition
2. Search for "vitamins" at the American Dietetic Association: www.eatright.org
3. Review the Dietary Reference Intakes for vitamins: www.nap.edu/readingroom
4. Explore dietary supplement information from the National Institute of Health, Office of Dietary Supplements: http://dietary-supplements.info.nih.gov
5. Visit the World Health Organization to learn about "vitamin deficiencies" around the world: www.who.int
6. Search for "vitamins" at the U.S. Government health information site: www.healthfinder.gov
7. Learn more about neural tube defects from the Spina Bifida Association of America: www.sbaa.org
8. Read about Dr. Joseph Goldberger and his ground-breaking discovery linking pellagra to diet by searching for his name at: www.nih.gov or www.pbs.org
9. Review the Dietary Reference Intakes for vitamin E and the carotenoids by searching for "DRI": www.nap.edu

INTERNET ACTIVITY

The *5 A Day* initiative of the Produce for Health Foundation and the National Cancer Institute aims to increase the consumption of fruits and vegetables by Americans to improve health and reduce the risk of developing certain cancers. Only 9 percent of Americans consume at least 5 servings of fruits and vegetables each day. Students often say that they would buy more fruits and vegetables if only they knew how. When is a peach ripe? How do you know when a melon is ready? Try this exercise to find out:

1. Go to the *5 A Day* site at: http://www.5aday.org/
2. Click on "Fruits & Vegetables: By Popular Demand."
3. Scroll down the page and click on "At the Supermarket."
4. Click on the first line "Click here for selection tips for a variety of fruits and vegetables."

Pick two fruits and two vegetables and, in your own words, describe how to choose these foods for maximum nutrients and flavor.

CONTROVERSY 7

DIETARY ANTIOXIDANTS: BEST FROM FOOD OR FROM PILLS?

DIETARY ANTIOXIDANTS have become household words. Consumers want to know which of these metabolic busybodies might be beneficial and how much they need to obtain the benefits. Knowledgeable people claim that eating foods rich in vitamin C, vitamin E, beta-carotene and other carotenoids, and the mineral selenium is the best path to disease prevention. Others claim that supplements providing large doses are more reliable allies. To tell who is right, consumers must weigh the evidence on both sides. This Controversy offers a way to score Foods versus Supplements as sources of these beneficial compounds.

Most of the results presented here are preliminary. The roles some of these compounds play in promoting health have not been fully defined, and research has provided only mixed results for others. Much of the evidence is from epidemiological or observational studies, which suggest interesting possibilities but have no specific implications for individuals. So proceed with caution and demand rigorous, repeated testing before you consider a finding confirmed—but do proceed. The evidence that lies before you holds secrets that are quickly evolving into tomorrow's nutrition concepts.

Free Radicals, Oxidative Stress, and Disease

The body's cells use oxygen to produce energy. In the process, oxygen sometimes reacts with body compounds to produce highly unstable molecules known as **free radicals** (see Table C7-1 for the definition of this and related terms). In addition to normal body processes, environmental factors such as radiation, pollution, tobacco smoke, and others can act as **oxidants** and cause free-radical formation (see Table C7-2 on the next page).[1] The trouble begins when free radicals in the body exceed its defenses against them, a condition known as **oxidative stress.**

A free radical is a molecule with one or more unpaired **electrons.*** An electron without a partner is unstable and highly reactive. To regain stability, the free radical finds a stable but vulnerable compound from which to steal an electron. With the loss of an electron, the formerly stable molecule becomes a free radical itself and steals an electron from some other nearby molecule, setting off an electron-snatching chain reaction.

*Oxygen-derived free radicals are common in the human body. Examples are superoxide radical (O_2•2), hydroxyl radical (OH •), and nitric oxide (NO •). The dots in the symbols represent the unpaired electrons. Scientists sometimes use the term *reactive oxygen species* (ROS) to describe all of these compounds.

Free radicals are like sparks, starting wildfires that lead to widespread damage by oxidative stress. Free-radical damage commonly disrupts unsaturated fatty acids in cell membranes, damaging the membranes' ability to transport substances into and out of cells. Free radicals also cause damage to cell proteins and to DNA, disrupting all cells that inherit the damaged DNA. Quantifying the body's level of oxidative stress involves testing the blood for increases of metabolic products of oxidized fatty acid or protein molecules.[2]

Researchers have identified tentative links between oxidative stress and the development of more than 200 diseases. Among them are age-related blindness, arthritis, cancers, cardiovascular disease, cataracts, and the kidney disease and other complications of diabetes.[3] Physical aging itself is thought by some to be the result of unrepaired free-radical damage that accumulates over the years, although no evidence to date clearly supports this idea.[4] Still, antioxidants seem to be part of the answer, whatever the question about nutrition.

While research focuses on damage caused by free radicals, they are not all bad. Their destructive properties are put to good use by some cells of the immune system. These cells stockpile

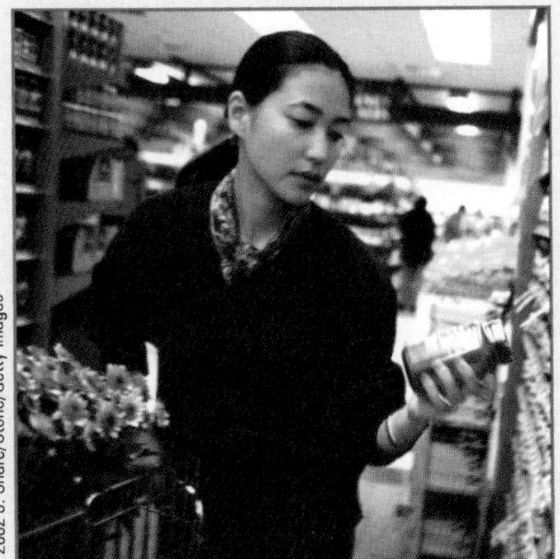

Table C7•1

ANTIOXIDANT TERMS

- **dietary antioxidants** compounds typically found in foods that significantly decrease the adverse effects of oxidants on human physical functions. The antioxidant vitamins are vitamin E, vitamin C, and beta-carotene, the plant precursor of vitamin A.
- **electrons** parts of an atom; negatively charged particles. Stable atoms (and molecules, which are made of atoms) have even numbers of electrons in pairs. An atom or molecule with an unpaired electron is an unstable *free radical.*
- **free radicals** atoms or molecules with one or more unpaired electrons that make the atom or molecule unstable and highly reactive.
- **oxidants** compounds (such as oxygen itself) that oxidize other compounds. Compounds that prevent oxidation are called *anti*oxidants, whereas those that promote it are called *pro*oxidants (*anti* means "against"; *pro* means "for").
- **oxidative stress** damage inflicted on living systems by free radicals.

Table C7•2

FACTORS THAT INCREASE FREE-RADICAL FORMATION

Body Factors	Environmental Factors
Energy metabolism	Air pollution
Diabetes	Asbestos
Exercise (immediate effect)	High levels of vitamin C
Acute illness	High levels of oxygen
Immune response	Radioactive emissions (for example, from radon gas)
Injury	Some herbicides
Obesity	Tobacco smoke
Other diseases	Trace minerals (iron, copper)
Other metabolic reactions	Ultraviolet light rays

SOURCE: Data from B. N. Ames and coauthors, Oxidants, antioxidants, and the degenerative disease of aging, *Proceedings of the National Academy of Sciences* 90 (1993): 7915–7920.

free radicals to use as ammunition in an "oxidative burst" against the viruses and bacteria that might otherwise cause diseases. Thus, infections cause a detectable increase in free-radical activity all over the body.

The Body's Defenses against Free Radicals

The body's two main systems of defense against damage from free radicals are its reserves of antioxidants and its enzyme systems that oppose oxidation.* These defense systems try to handle all free radicals, but they are not 100 percent effective. If insufficient radical-fighting agents are present in the body, if free radicals become excessive, or if the body's repair systems cannot undo all of the damage, health problems can develop.

*Internal enzyme systems include an enzyme-selenium complex and the superoxide dismutase (SOD) system. Chapter 8 offers more on minerals and oxidation.

Antioxidant Nutrients

Vitamin E and vitamin C actively scavenge and quench free radicals in the body, becoming oxidized themselves in the process. Once oxidized, vitamin C and vitamin E can to some extent be regenerated to become active antioxidants again, but some are dismantled and discarded. Free radicals attack the body continuously, so to maintain defenses, a person's supplies of dietary antioxidants must be replenished as rapidly as they are used up.

Vitamin E's special antioxidant role includes protecting body lipids by breaking the free-radical chain reaction at a rate 200 times faster than BHT,† a commercial antioxidant added to baked goods to prevent rancidity from fat oxidation (see Figure C7-1). The fat-soluble vitamin E defends lipids in cell membranes to maintain optimal functioning. Vitamin C is adept at neutralizing free radicals from polluted air and cigarette smoke and restoring oxidized vitamin E to its active state. Though a water-soluble compound itself, vitamin C may also protect blood lipids against

†BHT is butylated hydroxytoluene.

Figure C7•1

THE THEORY OF FREE RADICALS AND DISEASE

Free-radical formation occurs during metabolic processes, and it accelerates when diseases or other stresses strike.

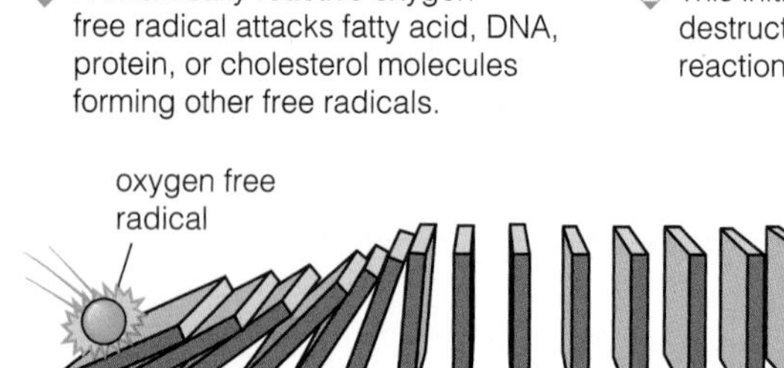
1 A chemically reactive oxygen free radical attacks fatty acid, DNA, protein, or cholesterol molecules forming other free radicals.
oxygen free radical
fatty acids, DNA, or cholesterol

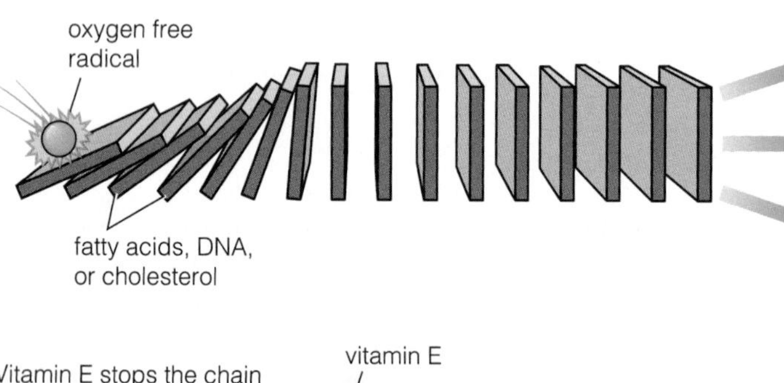
2 This initiates a rapid, destructive chain reaction.
3 The result is injury to tissues and the formation of more free radicals:
damage to cell membrane lipids and proteins, disabling them
precancerous changes in DNA
oxidation of blood cholesterol initiating steps leading to heart disease
4 And ultimately, diseases and tissue aging:
cancer
heart disease
macular degeneration
other diseases
aging

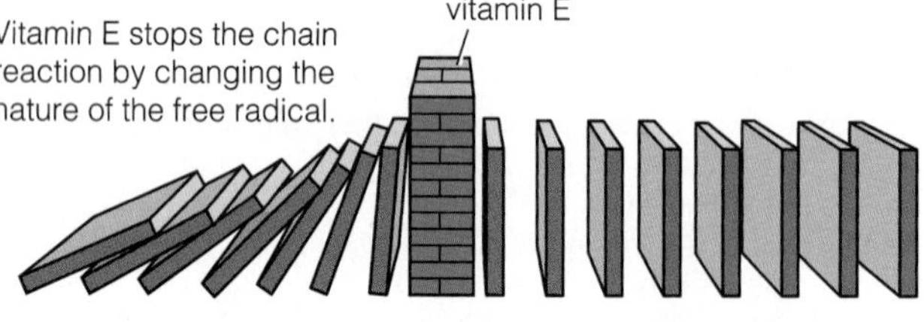
Vitamin E stops the chain reaction by changing the nature of the free radical.
vitamin E

oxidation by maximizing the total antioxidant capacity of the tissues.

A logical argument states that if body tissues could be drenched in extra amounts of antioxidants, then more free radicals could be quenched, and less damage to cells would bring better health (see Figure C7-2). This line of thinking is used to sell millions of dollars worth of antioxidant supplements each year. Remember, though, that logic is not science, and current evidence does not support such conclusions. Most positive findings come from antioxidants as they occur naturally in foods, not in supplements, and so the protection may arise from other features of a diet rich in antioxidant-containing foods or from some other unidentified lifestyle factor.[5] Recognizing this uncertainty, we begin our tally of points for Foods versus Supplements at zero points for both.

Internal Defense Systems

In addition to using antioxidant compounds from foods, the body defends itself against oxidative stress by making a powerful set of cellular enzymes that specialize in neutralizing free radicals.* These enzymes are proteins whose concentrations are controlled both by inherited genes and by influences affecting those genes. One type of enzyme that contains the mineral selenium breaks down oxidizing free radicals.[6] Another radical-quenching enzyme is superoxide dismutase (SOD), which has been purified for sale as an anti-aging supplement. Because these enzymes are proteins made by the body, however, they are useless as dietary supplements. Enzymes taken by mouth are digested in the stomach and small intestine long before they reach the bloodstream. In our scoring of Foods versus Supplements, then, neither gains a point here.

Phytochemicals

Some phytochemicals have antioxidant activity although they work in many other ways, too, as Controversy 2 made clear. One phytochemical filters incoming light, another slows blood clotting, some act as hormones, others inhibit harmful chemical reactions, still others stimulate immunity, and many act by mechanisms as yet unknown. Based on the evidence in Controversy 2, which is the better choice—foods or supplements? The complex combinations of phytochemicals in foods score a major point in favor of foods: Foods,1; Supplements, 0.

*Proteins that bind the minerals iron and copper also help in controlling oxidation.

Do Dietary Antioxidants Protect against Cancer?

Cancers arise when cellular DNA is damaged—sometimes by free-radical attacks. If dietary antioxidants protect DNA from this damage, then they probably reduce cancer risks. The strongest evidence that they do is found in studies showing that populations with high intakes of vegetables and fruits, foods rich in antioxidants, most often have low rates of cancer.[7] Laboratory studies with animals and with cell cultures seem to support such findings.

Beta-Carotene and the Carotenoids

Populations with high cancer rates have been found to consume few vegetables and fruits, especially those containing beta-carotene and other carotenoids. This evidence is strengthened by the finding that people with the highest concentrations of beta-carotene in their blood suffer less often from cancers of the mouth, throat, cervix, ovaries, and lung than people with lower beta-carotene values.[8] For a while, this evidence was touted as conclusive proof of beta-carotene's protective anticancer effect, and consumers across the nation bought and took beta-carotene supplements in hopes of preventing cancer.

Support for beta-carotene supplements soon crumbled, however, as studies showed no lessening of cancer incidence with the use of supplements.[9] For example, a long-term study reported no differences in disease rates between physicians who took beta-carotene for 14 years and a matched group who took placebos.[10] A similar study of almost 20,000 women detected no benefit or harm from taking

Figure C7•2

THE ANTIOXIDANT THEORY OF DISEASE PREVENTION

A. *Normally, the body's antioxidant enzymes, vitamins, and other molecules are sufficient to neutralize free-radical molecules before they do much damage.*
B. *An increased free-radical load can overwhelm the body's antioxidant systems. Free radicals then damage the tissues.*
C. *The body stocked with extra antioxidants is best equipped to handle an increase in free radicals. The two ways to obtain more antioxidants are to build them into cells by exercising regularly, which over time stimulates production of more antioxidant enzymes, and to eat them in foods or supplements, which supply antioxidant vitamins and phytochemicals.*

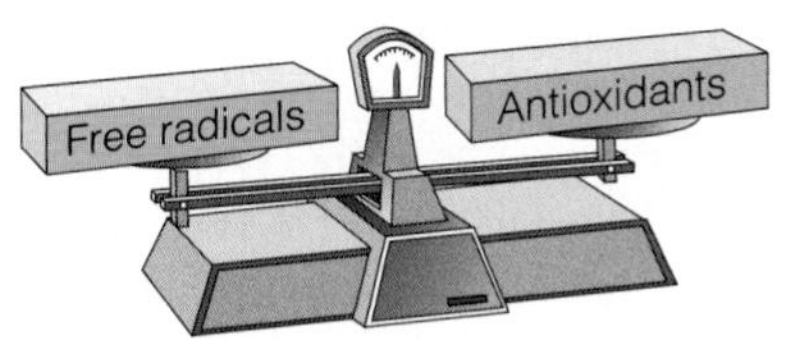

(a) Balanced system

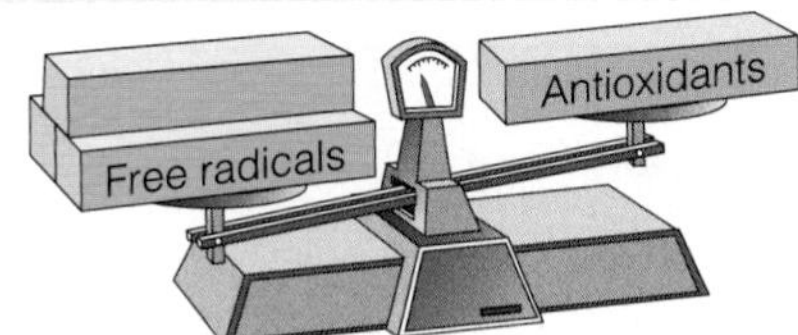

(b) Additional free-radical load—damage

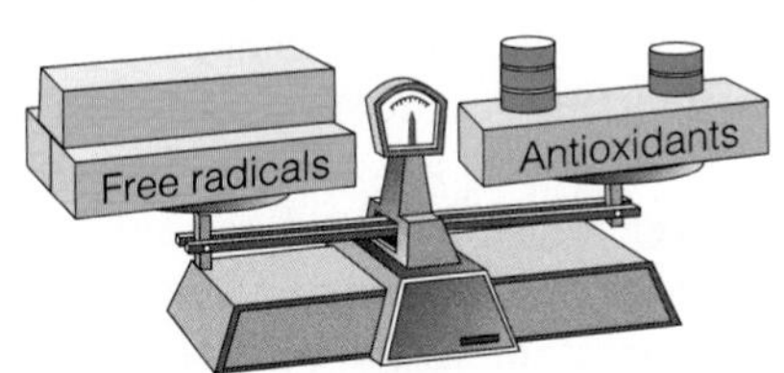
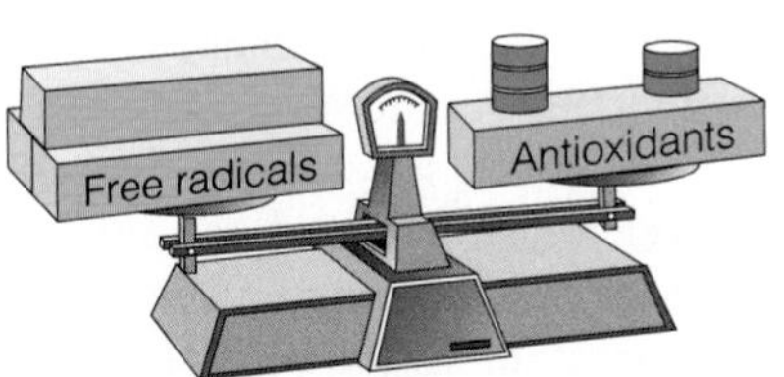

(c) Additional antioxidants added and balance restored

beta-carotene supplements over two years' time.[11] And a disquieting result came from a study of beta-carotene supplements and the incidence of lung cancer among smokers: major clinical trials of beta-carotene supplements were immediately ceased upon finding a 28 percent jump in lung cancer among the participants taking beta-carotene, but not among the controls taking placebos.[12]

Subsequent research has been mixed in this regard, with some confirming an excess of lung cancer among smokers who take beta-carotene and some finding no effect. At the same time, several studies have provided evidence that beta-carotene from foods correlates with a lower lung cancer risk, but there's no telling whether the beta-carotene itself or some other constituent of the food deserves the credit.[13] On the basis of such findings, the DRI committee concluded that supplements of beta-carotene provide no benefits and may cause harm to certain people.

Thus, we have an apparent contradiction: abundant beta-carotene from food and elevated beta-carotene in the blood are associated with lower cancer incidence, but the taking of beta-carotene supplements is not. These results lead to the conclusion that beta-carotene itself is not responsible for an anticancer effect, but simply tags along as a marker for another unknown factor that occurs along with it. Beta-carotene is just one of the dietary antioxidants present in fruits and vegetables, and such foods also contain other disease-fighting nutrients. These include vitamin A itself (which can arise from beta-carotene), vitamin B_6, folate, pantothenic acid, vitamin B_{12}, zinc, iron, copper, selenium, and more.

In addition to all of these nutrients, hundreds of phytochemicals are present in fruits and vegetables, too. Health effects attributed to beta-carotene may, in reality, be the work of one or a number of phytochemicals or the fiber provided by these foods. In truth, it could be an entire *diet* chosen by eaters of fruits and vegetables that makes the difference, or even an entire lifestyle. Still, evidence leans in favor of consuming increased fruits and vegetables for lowering disease risks. As Figure C7-3 demonstrates, population data link high intakes of fruits and vegetables with low cancer incidence. Foods now have 2 points in their favor; Supplements still score 0.

Vitamins C and E

Research on vitamin C and cancer is mixed: many studies indicate that when people's diets include foods rich in vitamin C, they seem to develop fewer cancers, but other studies detect no effect from dietary vitamin C. Like beta-carotene, vitamin C occurs in foods together with other cancer-fighting constituents. For example, broccoli and its sprouted seeds, leafy greens, and citrus fruits, which are all vitamin C–rich foods, also contain beta-carotene, other carotenoids, and a host of other powerful phytochemicals and nutrients thought to be active against cancer. These foods are also fiber-rich and low in fat—two other dietary characteristics believed to reduce cancer risk. Like beta-carotene, vitamin C may simply be a marker for a diet rich in fruits and vegetables. If so, the taking of vitamin C pills alone will do nothing to prevent cancer. Foods, 3; Supplements, still 0.

What about vitamin E? Because cancer may result from DNA damage, and vitamin E protects DNA, researchers have searched for an inhibitory effect of vitamin E on the development of cancer. The great majority of studies demonstrate no effect, with the exception of a single study showing a reduction of prostate cancer in heavy smokers taking vitamin E supplements.[14] More research is required to verify this finding before conclusions can be drawn concerning vitamin E, smoking, and prostate cancer risk.

The Food and Drug Administration (FDA) agrees that foods, not supplements, provide anticancer benefits. After reviewing the evidence, the agency concluded that diets high in fruits and vegetables, which are particularly good sources of beta-carotene and vitamin C, are strongly associated with reduced risks of several types of cancer. Because the reductions in risk could not be attributed solely to the named vitamins, the FDA rejected a request by the supplement industry to allow health claims on the labels of antioxidant supplements and ruled that

Figure C7•3

VEGETABLE AND FRUIT INTAKES AND CANCER IN POPULATION STUDIES

Groups of people with high fruit and vegetable intakes often have low rates of cancer. Well-controlled laboratory and clinical studies are needed to verify these findings.

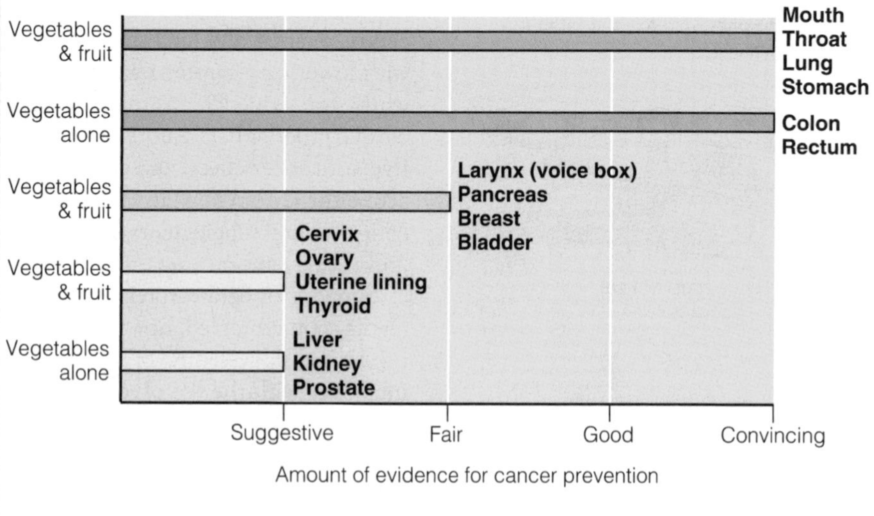

SOURCE: Data from World Cancer Research Fund/American Institute for Cancer Research, *Food, Nutrition and the Prevention of Cancer: A Global Perspective*, American Institute for Cancer Research, 1997, p. 437.

only food labels can make health claims in terms of fruits and vegetables and cancer. At this point, the score still stands: Foods, 3; Supplements, 0.

Selenium

The selenium content of food depends partly on the selenium content of the soil, which varies from region to region. In low-selenium areas of the United States, people suffer from higher rates of some cancers.[15] Studies of animals confirm these results and suggest that selenium may play roles in cancer prevention. In an experiment with more than 1,300 people with a history of skin cancer, half were given 200 micrograms of selenium over several years while the others received a placebo during the same period.[16] The researchers found that the incidence of recurrence of skin cancer was about the same between the two groups. However, the selenium-treated group had fewer cancers of the prostate, colon, and lung. In light of these seeming benefits, the researchers stopped their study to allow all study participants access to the selenium treatment.

Selenium supporters hail this study as proof that selenium supplements prevent cancer, but a problem remains. Though the selenium takers did indeed suffer fewer cancers of some types, their rate of death from all causes was virtually identical to that of the placebo group. Dr. Victor Herbert, a researcher famous for his vigorous antisupplement stance, concluded that the selenium given in the study must have *increased* the rate of deadly diseases other than cancer, thus equalizing the overall death rates between the two groups.[17] Evidence for this conclusion remains to be established.

An adequate intake of selenium is not to be neglected, however, especially by men. In another study, men who tested low on selenium were reported as being more likely to develop prostate cancer than men whose selenium stores were full.[18] Future research may reveal whether selenium affects cancer and how it may do so—through enzymatic antioxidant activity, through improved immunity, or through some mechanism of direct inhibition of cancer cell development.[19]

The DRI committee has set a Tolerable Upper Intake Level for selenium at 400 micrograms per day; doses somewhat above this amount may prove toxic to some individuals. In animals grazing on selenium-rich pastures, toxicity has caused hoof loss and nerve and muscle damage known as "blind staggers." In people, toxicities have caused nausea, loss of hair and fingernails, and nerve damage. Deaths from extremely large doses have occurred. However, no harm can come from including in your diet nutritious foods such as fish, vegetables, and whole grains, which provide forms of selenium that may be especially effective in the body.[20] From the evidence so far, then, give supplements another zero, but score another point for foods. Foods, 4; Supplements, 0.

Evidence concerning Age-Related Blindness

The leading cause of age-related blindness in the United States is macular degeneration. The macula, a yellow spot located at the focal center of the retina, loses integrity, which causes the loss of the most important field of vision, the area of central focus (peripheral vision remains unimpaired).

This blindness has been untreatable and unpreventable, but a discovery about the yellow color of the macula offers some promise—the macular pigments derive from dietary carotenoids other than beta-carotene, including lutein, first mentioned in Controversy 2.* These carotenoids are believed to filter out damaging light rays before they can harm the macula and may improve visual abilities.[21] This protection may account for a doubled rate of macular degeneration among people who consume diets low in carotenoids compared with people whose diets are carotenoid-rich.[22] Table C7-3 on the next page lists some foods in which carotenoids are plentiful. Many of the same foods also provide vitamin C, another protector of eye health.[23] Evidence for the taking of supplements to support eye health is lacking. Foods now have 5 points; Supplements, still 0.

*The other carotenoid is zeaxanthin.

Do Dietary Antioxidants Protect against Heart Disease?

Antioxidant nutrients, especially vitamin E, may help protect against cardiovascular disease.[24] One theory suggests that vitamin E prevents oxidation of low-density lipoproteins (LDL). Cholesterol carried in LDL in the blood correlates directly with cardiovascular disease, and most of the cholesterol

For the latest cancer fighters, visit your local produce center.

collected from damaged arteries has turned out to be oxidized cholesterol. The theory suggests that once inside the artery wall, LDL undergo oxidation by free radicals, thereby promoting the formation of artery-clogging plaques.

Other theories pit vitamin E against heart disease in other ways. Vitamin E may act upon cells to reduce the inflammation associated with arterial damage or reduce proliferation of smooth muscle cells (see Chapter 11 for details about the development of clogged arteries). It may also reduce the likelihood of heart attacks by interfering with blood clotting or by dilating the arteries and allowing the blood to flow through them.[25] A high level of vitamin E in the blood correlates with a lower concentration of an enzyme released when heart attack damages heart muscle tissue. Researchers have taken this to mean that a person with a high blood concentration of vitamin E may escape some of the usual heart damage from a heart attack.[26] The following evidence represents a large body of work that seems to indicate that vitamin E in amounts greater than the DRI intake recommendation of 15 milligrams per day may offer some measure of protection against heart disease.[27]

Vitamin E and Heart Disease

In one approach to studying vitamin E and heart disease, scientists selected groups of men in 16 European regions where rates of death from heart disease varied sixfold. The researchers compared the plasma vitamin E, cholesterol, and blood pressure among the men from each region. The men with the lowest vitamin E values died more often from heart disease. The correlation of heart disease mortality was stronger with low vitamin E than with high cholesterol or high blood pressure, supporting the "antioxidant hypothesis" of heart disease.

In another approach, researchers inspected the arteries and blood lipids of young male heart attack victims and found the arteries more severely narrowed when the men's LDL measured low for vitamin E.[28] In women, a diet of foods rich in vitamin E has been associated with fewer heart disease deaths.[29] Such evidence is suggestive, but indirect: Was vitamin E actually protecting the heart? Or could it have been some other factor, such as another dietary constituent that follows along with vitamin E into the diet?

To eliminate some of these factors, researchers focus on supplements and not on vitamin E in the diet. Two classic large-scale studies reported a significant reduction in heart disease in middle-aged men and women who took supplements of vitamin E for two or more years.[30] In a smaller study of elderly people, vitamin E supplements also seemed to offer some protection against death from heart disease.[31] An experiment called CHAOS (Cambridge Heart Antioxidant Study) strengthened these findings. Researchers were so impressed with the effect of taking 400 to 800 milligrams of supplemental vitamin E daily on the risk of *nonfatal* heart attacks that they terminated CHAOS earlier than expected to allow the placebo group to begin taking vitamin E if they wished. Interestingly, the rate of *fatal* heart attacks and total deaths was about the same among those taking vitamin E and the placebo.[32]

Recent clinical studies on vitamin E supplements and cardiovascular outcomes have been disappointing. A large study of patients at high risk for heart attack, stroke, or cardiovascular-related death revealed no fewer cardiovascular events among subjects taking vitamin E supplements than among those taking a placebo.[33] Other clinical studies have also revealed no benefit from vitamin E supplementation on outcomes of heart disease.[34]

Because supplements of vitamin E may turn out to have side effects when taken over many years and because despite early promise, research on vitamin E is mixed with regard to efficacy, the rationale for taking supplemental vitamin E is crumbling. We give zero points to vitamin E supplements but because vitamin E–rich foods are safe and may be protective, they receive a point: Foods, 6; Supplements, 0.

Vitamin C and Heart Disease

Research results are mixed on vitamin C's effect on susceptibility to heart disease.[35] Vitamin C and vitamin E work in tandem in defending LDL against oxidation: both vitamins defend against free radicals in cells.[36] Vitamin C regenerates vitamin E from its oxidized form, making it available to act again as an antioxidant.[37] Some studies also suggest that vitamin C works with vitamin E to reduce the damage from artery-clogging plaques.[38] After measuring the vitamin C status of 1,600 men and tracking them for an average of five years, researchers observed that a mild vitamin C deficiency increased a man's risk of fatal heart attack by two

Table C7•3

FOODS RICH IN CAROTENOIDS[a]

Strive to consume several servings of a variety of these foods each day.

Apricots, fresh and dried	Kale
Arugula (roquette)	Leek
Asparagus	Lettuce, all except iceberg type
Avocado	Mango
Broccoli	Parsley
Brussels sprouts	Peaches, dried
Cantaloupe	Peas, green
Carrots	Pepper, green, yellow, or red bell
Chanterelle mushrooms	Plums
Chicory leaves	Pumpkin
Cooked leafy greens, all varieties	Spinach
Endive	Squash, all varieties
Fennel leaves	Sweet potato
Grapefruit, pink	Swiss chard
Green beans	Tomato, fresh
Guavas, guava juice	Tomato products, ketchup, sauce, paste, juice
Herbs, fresh leaves	Watermelon

[a]Include beta- and alpha-carotenes, lutein, and lycopene.

and a half times.[39] However, two other factors measured in this study also correlated with both low blood vitamin C and heart attack—a low intake of fruits and vegetables and low blood carotene levels. In addition, an inverse relationship between consuming carotene-rich vegetables and incidence of cardiovascular disease was noted among over 15,000 U.S. physicians—the more of these vegetables they consumed, the lower their heart disease risk.[40] From these results, it is impossible to isolate vitamin C as the only factor at play. Something else lacking from a diet low in fruits and vegetables could have elevated the men's risks. Or perhaps a healthy lifestyle that includes eating fruits and vegetables, as well as exercising regularly and sleeping adequately, is the key.

Since fruits and vegetables supply adequate vitamin C, and these foods supply a host of other nutrients and phytochemicals, a few servings a day of vitamin C–rich foods make high-dose vitamin C supplements, along with any associated risks, unnecessary. Prudence dictates that only foods receive a point. Foods, now 7; Supplements, 0.

Should We Take Supplements Anyway, Just to Be Safe?

Although supplement manufacturers have proclaimed antioxidant pills to be magic bullets against aging, disease, and even death itself, our scientific reckoning has foods beating supplements by a score of 7 to 0. Dr. Victor Herbert, the scientist introduced earlier, energetically opposes the taking of supplements. He points to a host of side effects that might endanger supplement takers' health:

- Vitamin E supplements, taken over a period of time, may increase the risk of brain hemorrhage (a form of stroke).
- Vitamin E supplements delay blood clotting.
- Vitamin E supplements may worsen autoimmune diseases, such as asthma or rheumatoid arthritis.
- Vitamin C supplements enhance iron absorption, making iron overload likely in some people.
- Vitamin C supplements may increase markers of oxidation in the blood.
- Daily supplements of vitamin E, beta-carotene, or both do not reduce the incidence of lung cancer among smokers, and beta-carotene may increase it.
- Selenium supplements can be toxic (see the earlier section).

Besides, while orange juice and pills may both contain vitamin C, the orange juice presents a balanced array of chemicals that modulate vitamin C's effects. The pill provides only vitamin C, a lone chemical. And although fruits and vegetables rich in antioxidant nutrients have been associated with a diminished risk of many cancers, supplements of beta-carotene and vitamins C and E have not always proved beneficial. Most scientists agree that it is too early to recommend that people start taking antioxidant supplements now, even those of vitamin E. The risks are real, and clinical studies to quantify them and clarify the benefits have not yet been completed.

Should We Try to Eat More Antioxidant-Rich Foods?

You probably know our answer to this question. Every credible agency making recommendations for chronic disease prevention includes advice to increase consumption of fruits and vegetables, for the antioxidant nutrients, other nutrients, and phytochemicals they supply. We conclude by recommending this personal strategy: Don't try to single out a few magic nutrients to take as supplements. Instead, invest energy in eating a wide variety of fruits and vegetables in generous quantities every day. This is one of the most important favors you can do yourself, and the benefits are well backed by research.

Teresa Fasolino, *Rice Paddy*, ca. 1985

Water and Minerals

Frequently Asked Questions

Why is water the most indispensable nutrient? p. 267

How much water do I need to drink in a day? p. 269

Are some kinds of water better for my health than others? p. 270

How much calcium do I need? p. 277

How are salt and "water weight" related? p. 281

What happens to a person who lacks iron? p. 287

Can a person take in too much iron? p. 288

Contents

minerals naturally occurring, inorganic, homogeneous substances; chemical elements.

major minerals essential mineral nutrients found in the human body in amounts larger than 5 grams.

trace minerals essential mineral nutrients found in the human body in amounts less than 5 grams.

"ASHES TO ASHES and dust to dust"—it is true that when the life force leaves the body, what is left behind becomes nothing but a small pile of ashes. Carbohydrates, proteins, fats, vitamins, and water are present at first, but they soon disappear.

The carbon atoms in all the carbohydrates, fats, proteins, and vitamins combine with oxygen to produce carbon dioxide, which vanishes into the air; the hydrogens and oxygens of those compounds unite to form water; and this water, along with the water that was a large part of the body weight, evaporates. The ashes left behind are the **minerals,** a small pile that weighs only about 5 pounds. The pile is not impressive in size, but those minerals are critical to the functioning of living tissue.

Consider calcium and phosphorus. If you could separate these two minerals from the rest of the pile, you would take away about three-fourths of the total. Crystals made of these two minerals, plus a few others, form the structure of the bones and so provide the architecture of the skeleton.

Run a magnet through the pile that remains and you pick up the iron. It doesn't fill a teaspoon, but it consists of billions and billions of iron atoms. As part of hemoglobin, these iron atoms are able to attach to oxygen and make it available at the sites inside the cells where metabolic work is taking place.

If you then extract all the other minerals from the pile of ashes, leaving only copper and iodine, close the windows first. A slight breeze would blow these remaining bits of dust away. Yet the amount of copper in the dust is necessary for iron to hold and to release oxygen, and iodine is the critical mineral in the thyroid hormones. Figure 8-1 shows the amounts of the seven **major minerals** and a few of the **trace minerals** in the human body. Other minerals such as gold and aluminum are present in the body but are not known to be nutrients.

The distinction between major and trace minerals doesn't mean that one group is more important in the body than the other. A daily deficiency of a few micrograms of iodine is just as serious as a deficiency of several hundred milligrams of calcium. The major minerals are present in larger total quantities, however, and so they influence the body fluids, which in turn affect the whole body.

This chapter begins with a discussion of water. Water is unique among the nutrients—standing alone as the most indispensable of all. The body needs more water

Figure 8•1

MINERALS IN A 60-KILOGRAM (132-POUND) PERSON (GRAMS)

The major minerals are those present in amounts larger than 5 grams (a teaspoon). The essential trace minerals number a dozen or more: only four are shown. A pound is about 454 grams; thus only calcium and phosphorus appear in amounts larger than a pound.

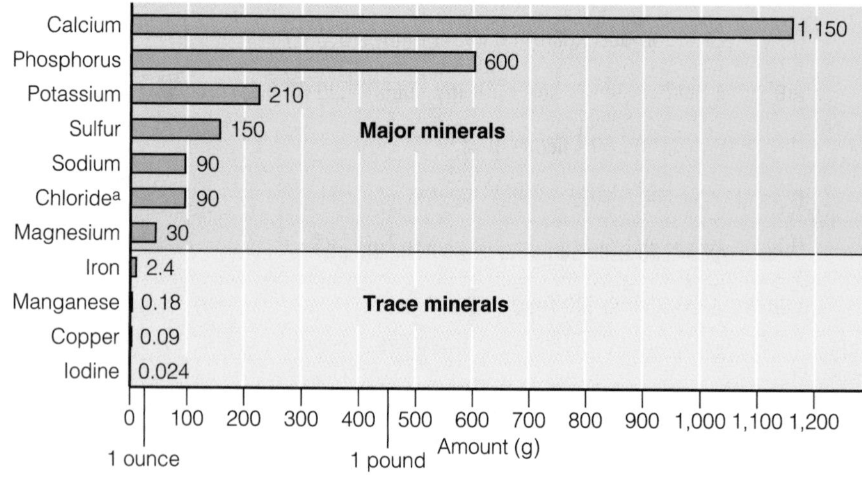

[a]Chlorine appears in the body as the chloride ion.

each day than any other nutrient—50 times more water than protein and 5,000 times more water than vitamin C. You can survive a deficiency of any of the other nutrients for a long time, in some cases for months or years, but you can survive only a few days without water. In less than a day, a lack of water alters the body's chemistry and metabolism.

Our discussion begins with water's many functions. Next we examine how water and the major minerals mingle to form the body's fluids and how cells regulate the distribution of those fluids. Then we take up the specialized roles of each of the minerals.

Water is the most indispensable nutrient.

© Lawrence Migdale/Stock, Boston

Water

You began as a single cell bathed in a nourishing fluid. As you became a beautifully organized, air-breathing body of trillions of cells, each of your cells had to remain next to water to stay alive.

Water makes up about 60 percent of a person's weight—that's almost 80 pounds of water in a 130-pound person. All this water in the body is not simply a river coursing through the arteries, capillaries, and veins. Some of the water is incorporated into the chemical structures of compounds that form the cells, tissues, and organs of the body. For example, proteins hold water molecules within them, water that is locked in and not readily available for any other use. Water also participates actively in many chemical reactions.

Why Is Water the Most Indispensable Nutrient?

Water brings to each cell the exact ingredients the cell requires and carries away the end products of the cell's life-sustaining reactions. Without water, cells quickly die. The water of the body fluids is thus the transport vehicle for all the nutrients.

Boasting scientist: "I'm working on discovering the universal solvent."

Skeptic: "Is that so? Well, when you've got it, what are you going to keep it in?"

Water is nearly a universal solvent: it dissolves amino acids, glucose, minerals, and many other substances needed by the cells. Fatty substances are specially packaged with water-soluble proteins so that they too can travel freely in the blood and lymph.

Water is also the body's cleansing agent. Small molecules, such as the nitrogen wastes generated during protein metabolism, dissolve in the watery blood and must be removed before they build up to toxic concentrations. The kidneys filter these wastes from the blood and excrete them, mixed with water, as urine. When the kidneys become diseased, as can happen in diabetes and other disorders, toxins can build to life-threatening levels. A machine must then take over the task of cleansing the blood by filtering wastes into water contained in the machine.*

Water molecules resist being crowded together. Thanks to this incompressibility, water can act as a lubricant and a cushion for the joints, and it can protect sensitive tissue such as the spinal cord from shock. The fluid that fills the eye serves in a similar way to keep optimal pressure on the retina and lens. From the start of human life, a fetus is cushioned against shock by the bag of amniotic fluid in the mother's uterus. Water also lubricates the digestive tract and all tissues that are moistened with mucus.

Yet another of water's special features is its ability to help maintain body temperature. The water of sweat is the body's coolant. Heat is produced as a by-product of energy metabolism and can build up dangerously in the body. To rid itself of this excess heat, the body routes its blood supply through the capillaries just under the skin. At the same time, the skin secretes sweat and its water evaporates. Converting water to vapor takes energy; therefore, as sweat evaporates, heat energy dissipates, cooling the skin and the underlying blood. The cooled blood then flows back to cool the body's core. Sweat evaporates continuously from the skin, usually in slight amounts that go unnoticed; thus, the skin is a major organ through which water is lost from the body.

Human life begins in water.

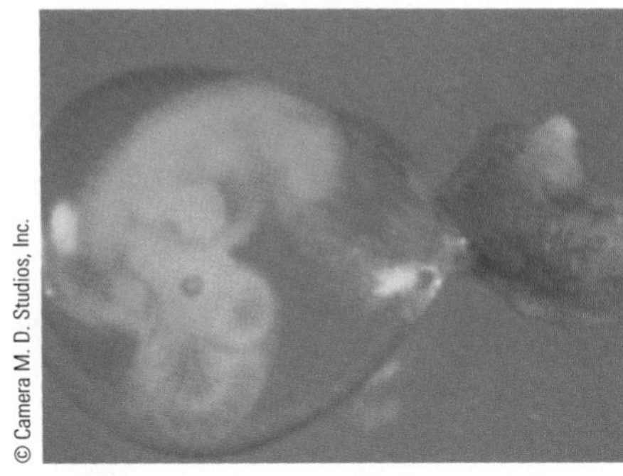

© Camera M. D. Studios, Inc.

*The machine that cleanses the blood is a kidney dialysis machine.

water balance the balance between water intake and water excretion, which keeps the body's water content constant.

dehydration loss of water. The symptoms progress rapidly, from thirst to weakness to exhaustion and delirium, and end in death.

water intoxication the rare condition in which body water content is too high. Symptoms are headache, muscular weakness, lack of concentration, poor memory, and loss of appetite.

To sum up, water:

- Carries nutrients throughout the body.
- Cleanses the tissues and blood of wastes.
- Serves as the solvent for minerals, vitamins, amino acids, glucose, and other small molecules.
- Actively participates in many chemical reactions.
- Acts as a lubricant around joints.
- Serves as a shock absorber inside the eyes, spinal cord, joints, and amniotic sac surrounding a fetus in the womb.
- Aids in maintaining the body's temperature.

key point *Water acts as a solvent, provides the medium for transportation, participates in chemical reactions, provides lubrication and shock protection, and aids in temperature regulation in the human body.*

The Body's Water Balance

Water is such an integral part of us that people seldom are conscious of water's importance, unless they are deprived of it. Since the body must excrete some water every day to cleanse its fluids, a person must consume at least the same amount to avoid life-threatening losses, that is, to maintain **water balance.**

The total amount of fluid in the body is kept balanced by delicate mechanisms. Imbalances such as **dehydration** and **water intoxication** can occur, but the balance is restored as promptly as the body can manage it. The body controls both intake and excretion to maintain water equilibrium.

The amount of the body's water varies by pounds at a time, especially in women who retain water during menstruation. Eating a meal high in salt can temporarily increase the body's water content; the body sheds the excess over the next day or so as the sodium is excreted. These temporary fluctuations in body water show up on the scale, but gaining or losing water weight does not reflect a change in body fat. Fat weight takes days or weeks to change noticeably, whereas water weight can change overnight.

Water makes up about 60 percent of the body's weight. A change in the body's water content can bring a change in body weight.

An extra drink of water benefits both young and old.

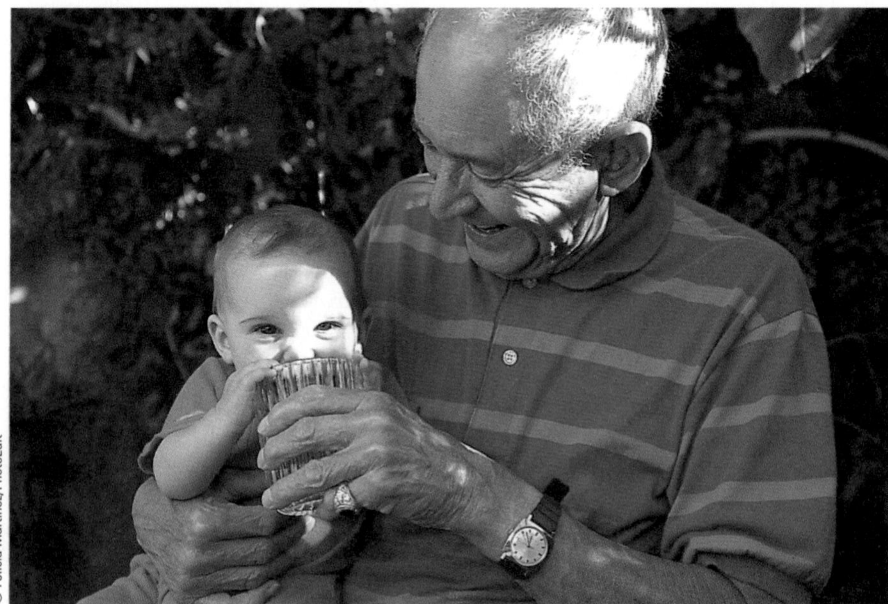

© Felicia Martinez/PhotoEdit

Quenching Thirst and Balancing Losses

Thirst and satiety govern water intake. When the blood is too concentrated (having lost water but not salt and other dissolved substances), the molecules and particles in the blood attract water out of the salivary glands, and the mouth becomes dry. The brain center known as the hypothalamus (described in Chapter 3) monitors the concentration of the blood. When the blood is too concentrated, or when the blood volume or pressure is too low, the hypothalamus initiates nerve impulses to the brain that stimulate drinking behavior. The hypothalamus also signals a hormone from the pituitary gland to direct the kidneys to shift water back into the bloodstream from the pool destined for excretion. The kidneys themselves respond to the sodium concentration in the blood passing through them and secrete regulatory substances of their own. The net result is that the more water the body needs, the less it excretes. Figure 8-2 shows how intake and excretion naturally balance out.

Thirst lags behind a lack of water. When too much water is lost from the body and is not replaced, dehydration can threaten survival. A first sign of dehydration is thirst, the signal that the body has already lost up to 2 cups of its total fluid. But suppose a person is unable to obtain fluid or, as in many elderly people, fails to perceive the thirst message. With a loss of just 5 percent of body fluid, perceptible symptoms appear: headache, fatigue, confusion or forgetfulness, and an elevated heart rate. Instead of "wasting" precious water in sweat, the dehydrated body diverts most of its water into the blood vessels to maintain the life-supporting blood pressure. Meanwhile, body heat builds up because sweating has ceased, creating the possibility of serious consequences (see Table 8-1). A water deficiency that develops slowly can switch on drinking behavior in time to prevent serious dehydration, but one that develops quickly may not. Rather than waiting until thirst sets in, people should drink regularly throughout the day.

key point *Water losses from the body necessitate intake equal to output to maintain balance. The brain regulates water intake; the brain and kidneys regulate water excretion. Dehydration can have serious consequences.*

Figure 8•2

WATER BALANCE

Water enters the body in liquids and foods, and some water is created in the body as a by-product of metabolic processes. Water leaves the body through the evaporation of sweat, in the moisture of exhaled breath, in the urine, and in the feces.

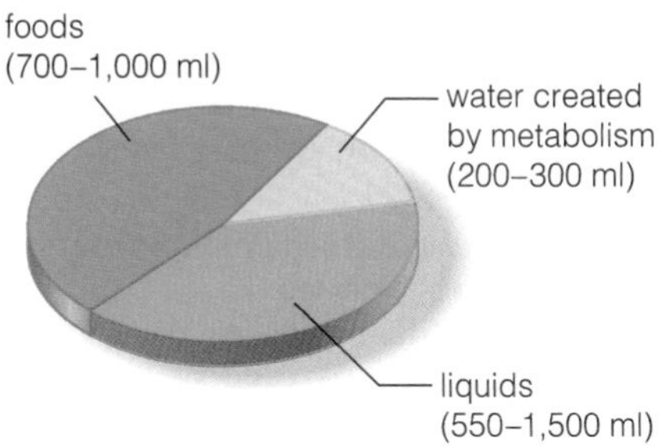

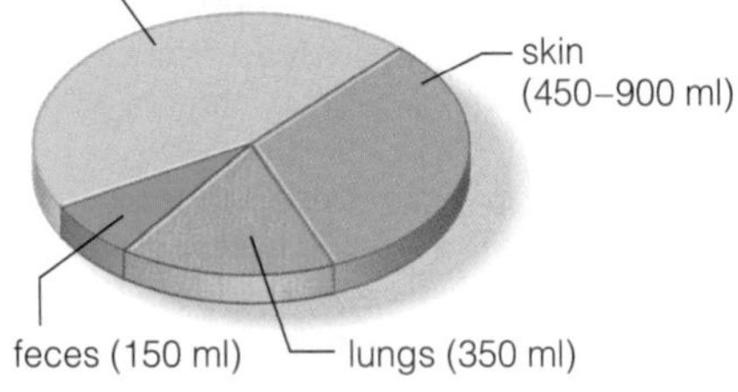

How Much Water Do I Need to Drink in a Day?

Water needs vary greatly depending on the foods a person eats, the environmental temperature and humidity, the person's activity level, and other factors. Under normal dietary and environmental conditions, adults need between 1 and 1.5 milliliters of water from all sources for each calorie spent in the day. A person who expends about 2,000 calories a day needs a fluid intake of about 2 to 3 liters (about 8 to 12 cups). Sweating increases water needs.

Factors that increase water needs:

- Alcohol consumption.
- Diseases that disturb water balance, such as diabetes.
- Physical activity (see Chapter 10).
- Forced air environments, such as airplanes or sealed buildings.
- Heated environments.
- Hot weather.
- Increased dietary fiber, protein, salt, or sugar.
- Medications (diuretics).
- Pregnancy and breastfeeding (see Chapter 12).
- Prolonged diarrhea, vomiting, or fever.
- Surgery, blood loss, or burns.
- Very young or old age.

Table 8•1

SIGNS OF MILD AND SEVERE DEHYDRATION

Mild (Loss of <5% Body Weight)	Severe (Loss of >5% Body Weight)
Thirst	Pale skin
Sudden weight loss	Bluish lips and fingertips
Rough dry skin	Confusion; disorientation
Dry mouth, throat, body linings	Rapid, shallow breathing
Rapid pulse	Weak, rapid, irregular pulse
Low blood pressure	Thickening of blood
Lack of energy; weakness	Shock; seizures
Impaired kidney function	Coma; death
Reduced quantity of urine; concentrated urine	

diuretics (dye-you-RET-ics) compounds, usually medications, causing increased urinary water excretion; "water pills."

hard water water with high calcium and magnesium concentrations.

soft water water with a high sodium concentration.

bottled water drinking water sold in bottles.

In addition to water and beverages made of water, nearly all foods contain water. Most fruits and vegetables contain large quantities of water, up to 95 percent of their volume; many meats and cheeses contain at least 50 percent. The energy-yielding nutrients in foods release additional water as the body breaks them down. Beverages containing alcohol or caffeine have a negative effect on the body's water balance—they are **diuretics,** compounds that cause water excretion.[1] A person drinking beer can end up with a net fluid loss rather than a gain. Caffeinated soft drinks, coffee, or tea may have a smaller effect on fluid balance.[2] Better choices are foods and beverages that contain abundant water without unwanted alcohol or caffeine (see the list in the margin). The Table of Food Composition, Appendix A, lists the water contents of most other foods and beverages.

Water content of various foods and beverages:

- 100%: water, diet soft drinks, seltzer (unflavored), plain tea.
- 95–99%: sugar-free gelatin dessert, clear broth, Chinese cabbage, celery, cucumber, lettuce, summer squash, decaffeinated black coffee.
- 90–94%: Gatorade, grapefruit, fresh strawberries, broccoli, tomato.
- 80–89%: sugar-sweetened soft drinks, milk, yogurt, egg white, fruit juices, low-fat cottage cheese, fresh apple, carrot.
- 60–79%: low-calorie mayonnaise, instant pudding, banana, shrimp, lean steak, pork chop, baked potato.
- 40–59%: diet margarine, sausage, chicken, macaroni and cheese.
- 20–39%: bread, cake, cheddar cheese, bagel, cooked oatmeal.
- 10–19%: butter, margarine, regular mayonnaise, cooked rice.
- 5–9%: peanut butter, popcorn.
- 1–4%: ready-to-eat cereals, pretzels.
- 0%: cooking oils, meat fats, shortening, white sugar.

key point *Many factors influence a person's need for water. The water of beverages and foods helps meet water needs, as does the water formed during cellular breakdown of energy nutrients.*

Are Some Kinds of Water Better for My Health than Others?

Water occurs as **hard water** or **soft water,** a distinction that affects your health with regard to three minerals. Hard water has high concentrations of calcium and magnesium. Soft water's principal mineral is sodium. In practical terms, soft water makes more bubbles with less soap; hard water leaves a ring on the tub, a jumble of rocklike crystals in the teakettle, and a gray residue in the wash.

Soft water may seem more desirable, and some homeowners purchase water softeners that remove magnesium and calcium and replace them with sodium. However, some evidence suggests that soft water may aggravate hypertension and heart disease.[3] Mineral-rich hard water may oppose these conditions by virtue of its calcium content.[4]

Soft water also more easily dissolves certain contaminant metals, such as cadmium and lead, from pipes. Cadmium can harm the body, affecting enzymes by displacing zinc from its normal sites of action and disturbing iron and copper transfer during pregnancy.[5] Cadmium is also suspected of promoting hypertension. Lead is another toxic metal, and the body seems to absorb it more readily from soft water than from hard water, possibly because the calcium in hard water protects against its absorption. Old plumbing may contain cadmium or lead. People who live in old buildings should run the cold-water tap a minute to flush out harmful minerals before drawing water for the first use in the morning and whenever no water has been drawn during the previous six hours.

Lead poisoning is especially harmful to children (see Chapter 13).

Many people turn to **bottled water** as an alternative to tap water. Read the Consumer Corner to find out if bottled water is a better option.

Hard water is high in calcium and magnesium. Soft water is high in sodium, and it dissolves cadmium and lead from pipes.

Consumer Corner

Which Type of Water Is Safest?

REMEMBER THAT water is practically a universal solvent: it dissolves almost anything it encounters to some degree. Hundreds of contaminants—including disease-causing bacteria and viruses from human wastes, toxic pollutants from highway fuel runoff, spills and heavy metals from industry, organic chemicals such as pesticides from agriculture, and manure bacteria from farm animals—have been detected in public drinking water.

Public water systems remove some hazards; treatment includes the addition of a disinfectant (usually chlorine) to kill most microorganisms. Private well water is usually not chlorinated, so the 40 million Americans who drink water from private wells are likely to encounter microorganisms, mostly harmless, in their water.

All public drinking water must be tested regularly for contamination, and the Environmental Protection Agency (EPA) is responsible for ensuring that public water systems meet minimum standards for protection of public health. Public utilities are required to provide their customers with a yearly statement, written in plain language, that names the chemicals and bacteria found in local water. This document makes fascinating reading for those interested in the purity of their tap water.

The law also requires a utility to notify the public within 24 hours of discovering any dangerous contaminants in drinking water. The intent is to reduce the threat from such harmful contaminants as *Cryptosporidium*, a chlorine-resistant parasite common in lakes and rivers. Several years ago, *Cryptosporidium* invaded the public water supply of Milwaukee, Wisconsin, and caused 400,000 people to fall ill.[1] Some even died. Since that time, awareness of the threat from *Cryptosporidium* has prevented all but a few outbreaks of illness from this organism, and no more lives have been lost.[2] Nevertheless, the incident stands as testimony that even our sophisticated water systems cannot always guarantee 100 percent safety, especially during floods and other natural disasters, excessive runoff, chemical spills, or intentional tampering.* The EPA's suggestions for purifying water in an emergency are listed to the right.

Some people fear that chlorine itself presents a danger to health. Large doses of by-products of water chlorination have been found to cause cancer-related changes in human cells and cancer in laboratory animals.[3] People consuming large amounts of chlorinated tap water have been reported to be somewhat more likely to develop colon, brain, and other cancers.[4] Conversely, men who take in 8 cups of water a day from any source, chlorinated or not, have been found to be half as likely to develop bladder cancer as men who restrict water intake to less than a cupful.[5] Although most investigators acknowledge the possibility of a connection between consumption of chlorinated drinking water and cancer incidence, they also passionately defend chlorination as a benefit to public health. In areas of the world without chlorination, an estimated 25,000 people die *each day* from diseases caused by organisms carried by water and easily killed by chlorine. Substitutes for chlorine exist, but they are currently too expensive or too

*Concerned consumers can call the Safe Drinking Water Hotline toll-free at (800) 426-4791, ask experts water safety questions by e-mail at hotline-sdwa@epamail.epa.gov, or visit the EPA's Drinking Water Homepage at www.epa.gov.

In an extreme emergency, when other safe water is unavailable, EPA advises disinfecting available water for drinking, cooking, and brushing teeth. Well water is safest but lake or stream water may be used. Clear water is most easily treated; filter cloudy or discolored water through several layers of clean cloth or a coffee filter before use.

1. Preferred method: Boil water vigorously for at least one minute to kill *all* disease-causing organisms.
2. If boiling is not possible, add disinfectants to kill *most* disease-causing microorganisms.
   - Use chlorine or iodine disinfecting tablets available from drugstores and sporting goods stores. Follow label directions.
   - Laundry bleach contains a chlorine compound that disinfects water. Follow directions on the bottle. If no directions, mix five drops of regular (not concentrated, scented, or color-safe) bleach with each quart of clear water. If water is cloudy or colored, double the amount. Let stand for at least 30 minutes before use. Properly treated water smells slightly of chlorine; if no chlorine odor is present, repeat dosage. To remove odor, pour water back and forth between clean containers to aerate it.
   - Less effective than chlorine, iodine tincture, a common first aid antiseptic for wounds, kills *some* disease-causing organisms. Add five drops of 2 percent iodine tincture to each quart of water (add ten drops if water is cloudy). Let stand for at least 30 minutes before use.

slow to be practical for treating a city's water, and some may create their own by-products.

Meanwhile what is a consumer to drink? One option is to drink tap water because municipal water is held to minimum standards for purity. Another option is to further purify tap water with home purifying equipment, which ranges in price from about $20 to $5,000. Some home systems do an adequate job of removing lead, chlorine, and other contaminants, but others only improve the water's taste. Many are not designed to remove microorganisms that are not affected by chlorine. Each system has advantages and drawbacks, and all require periodic maintenance or filter replacements that vary in price. Not all companies or representatives are legitimate—some perform water tests that yield dramatic-appearing but meaningless results to sell unneeded systems. Verify all claims of contamination by checking reports from local municipal water agencies or by testing well water before buying any purifying system.

A third option is to use bottled water. About 1 in 15 households uses bottled water as its main drinking water source, believing it to be safer than tap water and therefore worth its substantial price—typically 250 to 10,000 times the cost of tap water.[6] A consumer group tested bottled water, however, and disproved the notion of superior safety. Of 1,000 bottles and 103 brands tested, about a third were contaminated with bacteria, arsenic, or synthetic organic chemicals.*[7] At least a quarter of bottled water is drawn directly from the tap. In reality, whether it comes from the tap or is poured from a bottle, all water comes from the same sources—**surface water** and **ground water** (see Table 8-2).

Surface water flowing from lakes, rivers, and reservoirs fills about half of the nation's need for drinking water, mostly in major cities. Surface water is exposed to contamination by acid rain, petroleum products, pesticides, fertilizer, human and animal wastes, and industrial wastes that run directly from pavements, septic tanks, farmlands, and industrial areas into streams that feed surface water bodies. Surface water generally moves faster than ground water and stays above ground where aeration and exposure to sunlight can cleanse it. The plants and microorganisms that live in surface water also filter it. These processes can remove some contaminants, but others stay in the water.

Table 8•2

WATER SOURCES

- **aquifers** underground rock formations containing water that can be drawn to the surface for use.
- **ground water** water that comes from underground aquifers.
- **surface water** water that comes from lakes, rivers, and reservoirs.

Ground water comes from protected **aquifers,** deep underground rock formations saturated with water. People in rural areas rely mostly on ground water pumped from private wells, and some cities tap this resource, too. Ground water can become contaminated from hazardous waste sites, dumps, oil and gas pipelines, and landfills, as well as downward seepage from surface water bodies. Ground water moves slowly and is not aerated or exposed to sunlight, so contaminants break down more slowly than in surface water. To mingle with water in the aquifer, surface water must first "percolate," or seep, through soil, sand, or rock, which filters out some contaminants.

Bottled water sold in interstate commerce is regulated by the Food and Drug Administration (FDA).[8] The FDA requires yearly tests of bottled water for purity and sanitation standards, but the standards are substantially less rigorous than those applied to U.S. tap water. For example, bottled water does not have to be filtered to remove disease-causing organisms, such as *Cryptosporidium*, or tested for the presence of asbestos contamination as tap water sources must be. Still, the great majority of people who buy bottled water say that it tastes better than the water from their taps. Most water-bottling plants disinfect their products with ozone, which, unlike chlorine, leaves no flavor or odor in the water.

As a consumer, what should you look for? Look for the trademark of the International Bottled Water Association (IBWA), a trade organization supporting the FDA's regulations and enforcement efforts. Also look for the water's place of origin. Water bottled in another state might be the safest choice because only water sold across state lines must meet the FDA's sanitation and safety requirements. Then try to determine the water's source. If the water you buy is from a spring or a stream in your state, is the area agricultural, residential, industrial, or undeveloped? Agricultural, indus-

Surface water is easily contaminated by acid rain, pesticides, and other pollutants that fall or wash into streams, rivers, and lakes.

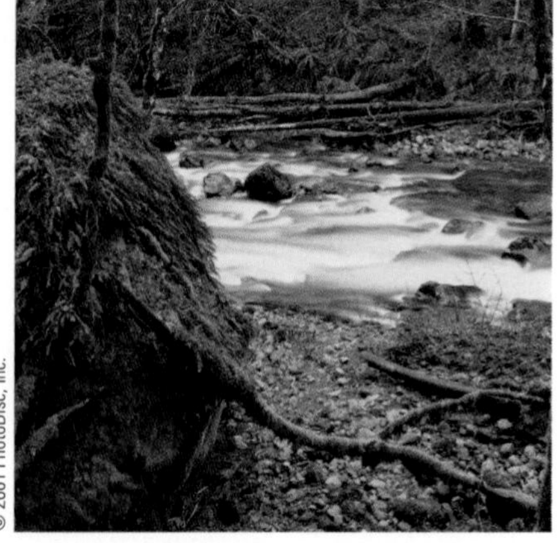

The label on a water bottle may imply purity, but what counts is the purity of the product inside the bottle.

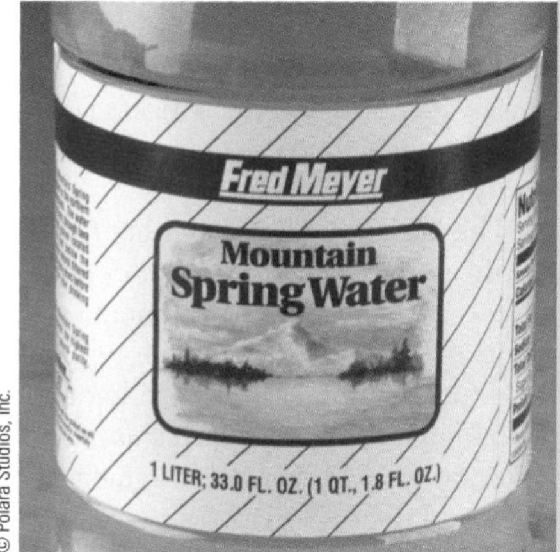

*The group was the Natural Resources Defense Council, and their report, *Bottled Water: Pure Drink or Pure Hype?* can be found at www.nrdc.org/water/drinking/nbw.asp.

Table 8•3

WATER TERMS THAT MAY APPEAR ON LABELS

- **artesian water** water drawn from a well that taps a confined aquifer in which the water is under pressure.
- **carbonated water** water that contains carbon dioxide gas, either naturally occurring or added, that causes bubbles to form in it; also called *bubbling* or *sparkling water*. Seltzer, soda, and tonic waters are legally soft drinks and are not regulated as water.
- **distilled water** water that has been vaporized and recondensed, leaving it free of dissolved minerals.
- **filtered water** water treated by filtration, usually through *activated carbon filters* that reduce the lead in tap water, or by *reverse osmosis* units that force pressurized water across a membrane removing lead, arsenic, and some microorganisms from tap water.
- **mineral water** water from a spring or well that typically contains 250 to 500 parts per million (ppm) of minerals. Minerals give water a distinctive flavor. Many mineral waters are high in sodium.
- **natural water** water obtained from a spring or well that is certified to be safe and sanitary. The mineral content may not be changed, but the water may be treated in other ways such as with ozone or by filtration.
- **public water** water from a municipal or county water system that has been treated and disinfected.
- **purified water** water that has been treated by distillation or other physical or chemical processes that remove dissolved solids. Because purified water contains no minerals or contaminants, it is useful for medical and research purposes.
- **spring water** water originating from an underground spring or well. It may be bubbly (carbonated), or "flat" or "still," meaning not carbonated. Brand names such as "Spring Pure" do not necessarily mean that the water comes from a spring.
- **well water** water drawn from ground water by tapping into an aquifer.

trial, and residential activities can expose water sources to contamination. Finally, ask whether your state strictly enforces standards for purity and sanitation of bottled water. Many states have unenforced rules on the books.

Table 8-3 defines some terms that appear on labels. What you are unlikely to find on the label, however, is the water's mineral content. For nutrition's sake, the best choice of any water is one rich in magnesium and calcium, but low in sodium.[9] Bottled waters vary in their content of fluoride, a mineral important to the health of teeth and bones. Some bottling companies will provide mineral information if a consumer requests it.

If your water is dispensed from a water cooler, cleanse the cooler once a month by running half a gallon of white vinegar through it. Remove the vinegar residue by rinsing the cooler with 4 or 5 gallons of tap water. The microbial content of water coolers has been found to be considerably higher than that recommended by the government. Regular cleaning reduces bacterial and mold growths that can cause serious infection and disease in those who ingest water contaminated with them.

salts compounds composed of charged particles (ions). An example is potassium chloride (K^+Cl^-).

ions (EYE-ons) electrically charged particles, such as sodium (positively charged) or chloride (negatively charged).

electrolytes compounds that partly dissociate in water to form ions, such as the potassium ion (K^+) and the chloride ion (Cl^-).

fluid and electrolyte balance maintenance of the proper amounts and kinds of fluids and minerals in each compartment of the body.

Figure 8•3

HOW ELECTROLYTES GOVERN WATER FLOW

Water flows in the direction of the more highly concentrated solution.

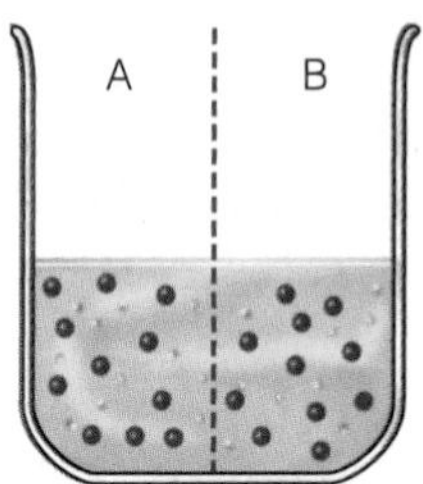

1. With equal numbers of dissolved particles on both sides of a water-permeable divider, water levels remain equal.

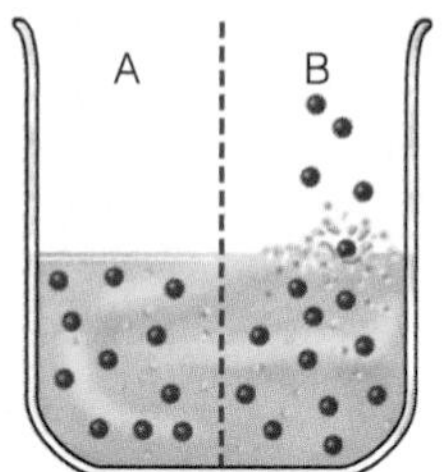

2. Now additional particles are added to increase the concentration on side B. Particles cannot flow across the divider (in the case of fluid inside and outside a cell, the divider is a cell membrane).

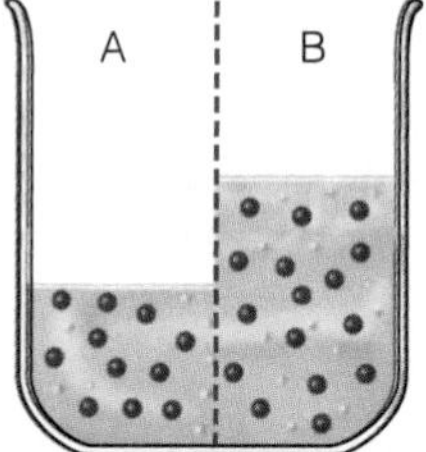

3. Water can flow both ways across the divider, but tends to move from side A to side B, where there is a greater concentration of dissolved particles. The *volume* of water increases on side B, and the *concentrations* on sides A and B become equal.

Body Fluids and Minerals

Most of the body's water weight is contained inside the cells, and some water bathes the outsides of the cells. The remainder fills the blood vessels. How do cells keep themselves from collapsing when water leaves them and from swelling up when too much water enters them? The cells cannot regulate the amount of water directly by pumping it in and out because water slips across membranes freely. The cells can, however, pump minerals across their membranes. The major minerals form **salts** that dissolve in the body fluids; the cells direct where the salts go; and this determines where the fluids flow because water follows salt.

When mineral (or other) salts dissolve in water, they separate into single, electrically charged particles known as **ions.** Unlike pure water, which conducts electricity poorly, ions dissolved in water carry electrical current; for this reason, these electrically charged ions are called **electrolytes.** As Figure 8-3 shows, when dissolved particles, such as electrolytes, are present in unequal concentrations on either side of a water-permeable membrane, water flows toward the more concentrated side to equalize the concentrations. Cells and their surrounding fluids work in the same way. Think of a cell as a sack made of a water-permeable membrane. The sack is filled with watery fluid and suspended in a dilute solution of salts and other dissolved particles. Water flows freely between the fluids inside and outside the cell, but generally moves from the more dilute solution toward the more concentrated one (the photo of salted eggplant slices on the next page shows this effect). To control the flow of this water, the body must spend energy moving its electrolytes from one compartment to another (see Figure 8-4). Figure 6-11 of Chapter 6 (page 187) introduced the proteins that form the pumps that move mineral ions across cell membranes. The result is **fluid and electrolyte balance,** the proper amount and kind of fluid in every body compartment.

If the fluid balance is disturbed, severe illness can develop quickly because fluid can shift rapidly from one compartment to another. For example, in vomiting or diarrhea, the loss of water from the intestinal tract pulls fluid from between the cells in every part of the body. Fluid then leaves the cell interiors to restore balance. Meanwhile the kidneys detect the water loss and attempt to retrieve water from the pool destined for excretion. To do this, they raise the sodium concentration outside

Figure 8•4

ELECTROLYTE BALANCE

Transport proteins in cell membranes maintain the proper balance of sodium (mostly outside the cells) and potassium (mostly inside the cells).

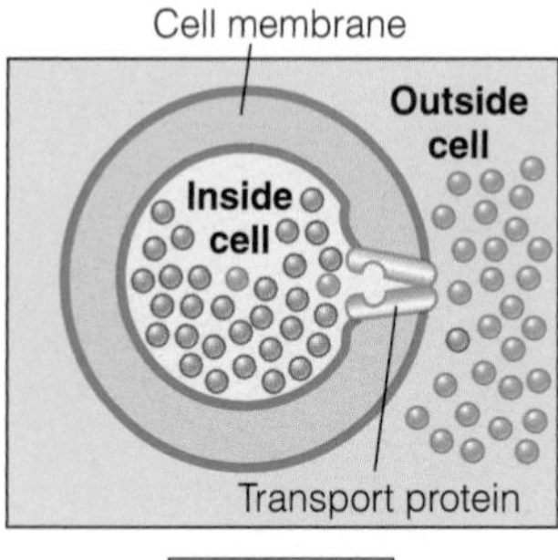

the cells, and this pulls still more water out of them. The result is **fluid and electrolyte imbalance,** a medical emergency. Water and minerals lost in vomiting or diarrhea ultimately come from every body cell. This loss disrupts the heartbeat and threatens life. It is a cause of death among those with eating disorders.

The minerals help manage still another balancing act, the **acid-base balance,** or pH, mentioned in Chapters 3 and 6. In pure water, a small percentage of water molecules (H_2O) exist as positive (H) and negative (OH) ions, but they exist in equilibrium—the positive charges exactly equal the negatives. When dissolved in watery body fluids, some of the major minerals give rise to acids (H, or hydrogen, ions), and others to bases (OH). Excess H ions in a solution make it an acid; they lower the pH. Excess OH ions in a solution make it a base; they raise the pH.

Maintenance of body fluids at a nearly constant pH is critical to life. Even slight changes in pH drastically change the structure and chemical functions of most biologically important molecules. The body's proteins and some of its mineral salts help prevent changes in the acid-base balance of its fluids by serving as **buffers**—molecules that gather up or release H ions as needed to maintain the correct pH. The kidneys help to control the pH balance by excreting more or less acid (H ions). The lungs also help by excreting more or less carbon dioxide. (Dissolved in the blood, carbon dioxide forms an acid, carbonic acid.) This tight control of the acid-base balance permits all other life processes to continue.

fluid and electrolyte imbalance failure to maintain the proper amount and kind of fluid in every body compartment; a medical emergency.

acid-base balance maintenance of the proper degree of acidity in each of the body's fluids.

buffers molecules that can help to keep the pH of a solution from changing by gathering or releasing H ions.

hydroxyapatite (hi-DROX-ee-APP-uh-tight) the chief crystal of bone, formed from calcium and phosphorus.

fluorapatite (floor-APP-uh-tight) a crystal of bones and teeth, formed when fluoride displaces the hydroxy portion of hydroxyapatite. Fluorapatite resists being dissolved back into body fluid.

Electrolytes help keep fluids in their proper compartments and buffer these fluids, permitting all life processes to take place.

Figure 3-10 of Chapter 3 showed the pH of common substances; Figure 3-3 depicted fluid movement in and around cells.

The Major Minerals

Though all the major minerals help to maintain the fluid balance, each one also has some special duties of its own. Table 8-10 on pages 296–297 summarizes the roles of the minerals discussed below.

Major minerals:

- Calcium
- Chloride
- Magnesium
- Phosphorus
- Potassium
- Sodium
- Sulfur

Calcium

As Figure 8-1 showed, calcium is by far the most abundant mineral in the body. Nearly all (99 percent) of the body's calcium is stored in the bones and teeth, where it plays two important roles. First, it is an integral part of bone structure. Second, bone calcium serves as a bank that can release calcium to the body fluids if even the slightest drop in blood calcium concentration occurs. Many people have the idea that, once deposited in bone, calcium (together with the other minerals of bone) stays there forever—that once a bone is built, it is inert, like a rock. Not so. The minerals of bones are in constant flux, with formation and dissolution taking place every minute of the day and night, (see Figure 8-5 on the next page).

Calcium and phosphorus are both essential to bone formation: calcium phosphate salts crystallize on a foundation material composed of the protein collagen. The resulting **hydroxyapatite** crystals invade the collagen and gradually lend more and more rigidity to a younster's maturing bones until they are able to support the weight they will have to carry. During and after the bone-strengthening processes, fluoride may displace the "hydroxy" parts of these crystals, making **fluorapatite.** Fluorapatite resists bone-dismantling forces to help maintain bone integrity.

Teeth are formed in a similar way: hydroxyapatite crystals form on a collagen matrix to create the dentin that gives strength to the teeth (see Figure 8-6). The turnover of minerals in teeth is not as rapid as in bone, but some withdrawal and redepositing do take place throughout life. As in bone, fluoride hardens and stabilizes the crystals of teeth and makes the enamel resistant to decay.

The slices of eggplant on the right were sprinkled with salt. Notice their beads of "sweat," formed as cellular water moves across each cell's membrane (water-permeable divider) toward the higher concentration of salt (dissolved particles) on the surface.

osteoporosis (OSS-tee-oh-pore-OH-sis) a reduction of the bone mass of older persons in which the bones become porous and fragile (*osteo* means "bones"; *poros* means "porous"); also known as **adult bone loss.**

Figure 8•5

A BONE

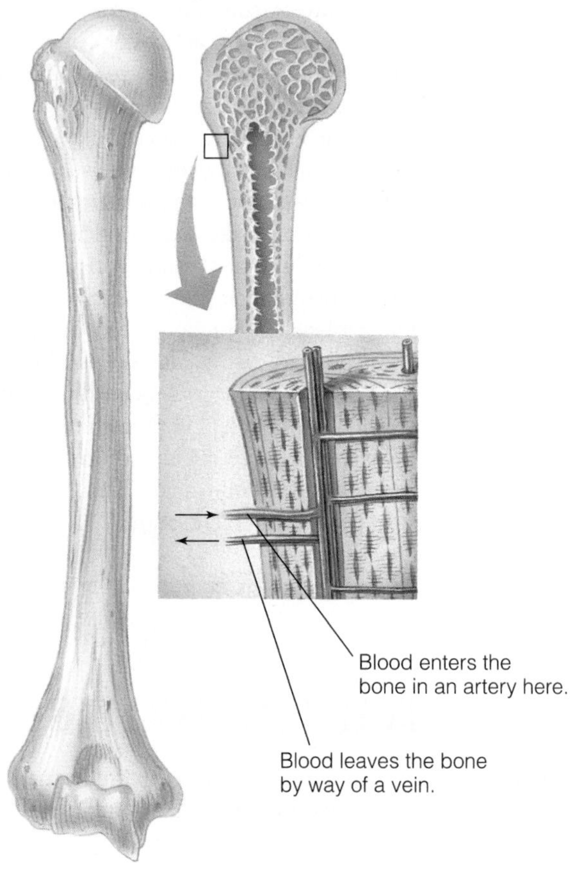

Bone is active, living tissue. Blood travels in capillaries throughout the bone, bringing nutrients to the cells that maintain the bone's structure and carrying away waste materials from those cells. It picks up and deposits minerals as instructed by hormones.
Bone derives its structural strength from the lacy network of crystals that lie along its lines of stress. If minerals are withdrawn to cover deficits elsewhere in the body, the bone will grow weak and ultimately will bend or crumble.
Blood enters the bone in an artery here.
Blood leaves the bone by way of a vein.

In osteoporosis:

- Bones of older adults become brittle and fragile.

Figure 8•6

A TOOTH

The inner layer of dentin is bonelike material that forms on a protein (collagen) matrix. The outer layer of enamel is harder than bone. Both dentin and enamel contain hydroxyapatite crystals (made of calcium and phosphorus). The crystals of enamel may become even harder when exposed to the trace mineral fluoride.

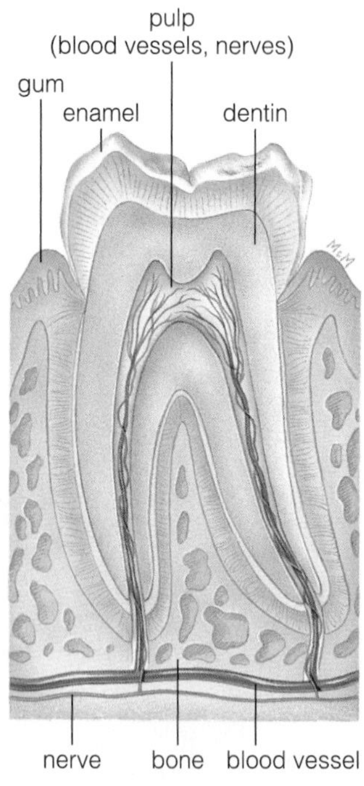

Calcium in Body Fluids Only about 1 percent of the body's calcium is in the fluids that bathe and fill the cells, but this tiny amount plays these major roles:

- Regulates the transport of ions across cell membranes and is particularly important in nerve transmission.
- Helps maintain normal blood pressure (see Chapter 11).
- Plays an essential role in the clotting of blood.
- Is essential for muscle contraction and therefore for the heartbeat.
- Allows secretion of hormones, digestive enzymes, and neurotransmitters.
- Activates cellular enzymes that regulate many processes.

Because of its importance, blood calcium is tightly controlled.

Calcium and the Bones The key to bone health lies in the body's calcium balance. Cells need continuous access to calcium, so the body maintains a constant calcium concentration in the blood. The skeleton serves as a bank from which the blood can borrow and return calcium as needed. Blood calcium is regulated, not by a person's daily calcium intake or bone density, but by hormones sensitive to blood calcium.* One of the consequences of aging is bone loss. If your calcium savings account is not sufficient, you will develop the fragile bones of **osteoporosis,** or **adult bone loss.** Osteoporosis constitutes a major health problem for many older people—its possible causes and prevention are the topics of this chapter's Controversy.

*Calcitonin, made in the thyroid gland, is secreted whenever the calcium concentration in the blood rises too high. It acts to stop withdrawal from bone and to slow absorption somewhat from the intestine. Parathormone, from the parathyroid glands, has the opposite effect.

To protect against bone loss, high calcium intakes early in life are recommended. A calcium-poor diet during the growing years may prevent the person from achieving maximum **peak bone mass.**[6] Too little calcium packed into the skeleton during childhood and young adulthood strongly predicts susceptibility to osteoporosis in adulthood.

peak bone mass the highest attainable bone density for an individual; developed during the first three decades of life.

The body is sensitive to an increased need for calcium, although it sends no signals to the conscious brain indicating calcium need. Instead, the body quietly increases the absorption of calcium from the intestine and prevents its loss from the kidneys. For example, more calcium is needed for growth, so infants and children absorb about 60 percent of ingested calcium and pregnant women absorb about 50 percent. The body of an adolescent hungers for calcium, absorbing and retaining more calcium from the calcium in each meal than does the body of an adult.[7] Although adults absorb only about 25 percent, the body absorbs a higher percentage of calcium when less total calcium is provided in the diet.[8] Deprived of calcium for months or years, an adult may double the calcium absorbed; when supplied for years with abundant calcium, the same person may absorb only about one-third the normal amount. These adjustments take time, though, and increased absorption cannot fully compensate for a reduced calcium intake. A person who suddenly cuts back on calcium is likely to lose calcium from the bones.

How Much Calcium Do I Need? Setting recommended intakes for calcium is difficult because absorption varies not only with age, but also with a person's vitamin D status and the calcium content of the diet. The DRI committee took such variations into account and set recommendations for calcium at levels that produce maximum calcium retention. At lower intakes, the body does not store calcium to capacity; at greater intakes, the excess calcium is excreted and thus is wasted.

The importance of vitamin D in calcium absorption was described in Chapter 7.

Recommended intakes are high for children and adolescents because people develop their peak bone mass during their growing years. Obtaining enough calcium at that time helps to ensure that the skeleton starts adulthood with a high bone density. By the late twenties, or 10 years after adult height is achieved, the skeleton no longer adds significantly to bone density.[9] After about 40 years of age, regardless of calcium intake, bones begin to lose density, but the loss can be slowed somewhat by a diet high in calcium along with sufficient physical activity. Table 8-4 offers the DRI recommendations and other calcium goals. Snapshot 8-1 (next page) provides a look at some foods that are good or excellent sources of calcium, and the Food Feature at the end of this chapter focuses on using foods to meet calcium needs.

key point *Calcium makes up bone and tooth structure and plays roles in nerve transmission, muscle contraction, and blood clotting. Calcium absorption rises when there is a dietary deficiency or an increased need such as during growth.*

Phosphorus

Phosphorus is the second most abundant mineral in the body, but its concentration in the blood is less than half that of calcium. About 85 percent of the body's

Table 8•4

CALCIUM INTAKE RECOMMENDATIONS

Healthy People 2010

- Increase to at least 90 percent the proportion of people aged 2 and older who meet the DRI dietary recommendations for calcium.

DRI Recommended Intakes[a]

- Adolescents: 1,300 milligrams per day.
- Women and men (19–50 years): 1,000 milligrams per day.
- Women and men (51 years and older): 1,200 milligrams per day.

[a]For values for other groups, see the inside front cover.

8.1

CALCIUM

All except broccoli provide 10 percent or more of the calcium Daily Value (DV = 1,000 mg/day).[a]

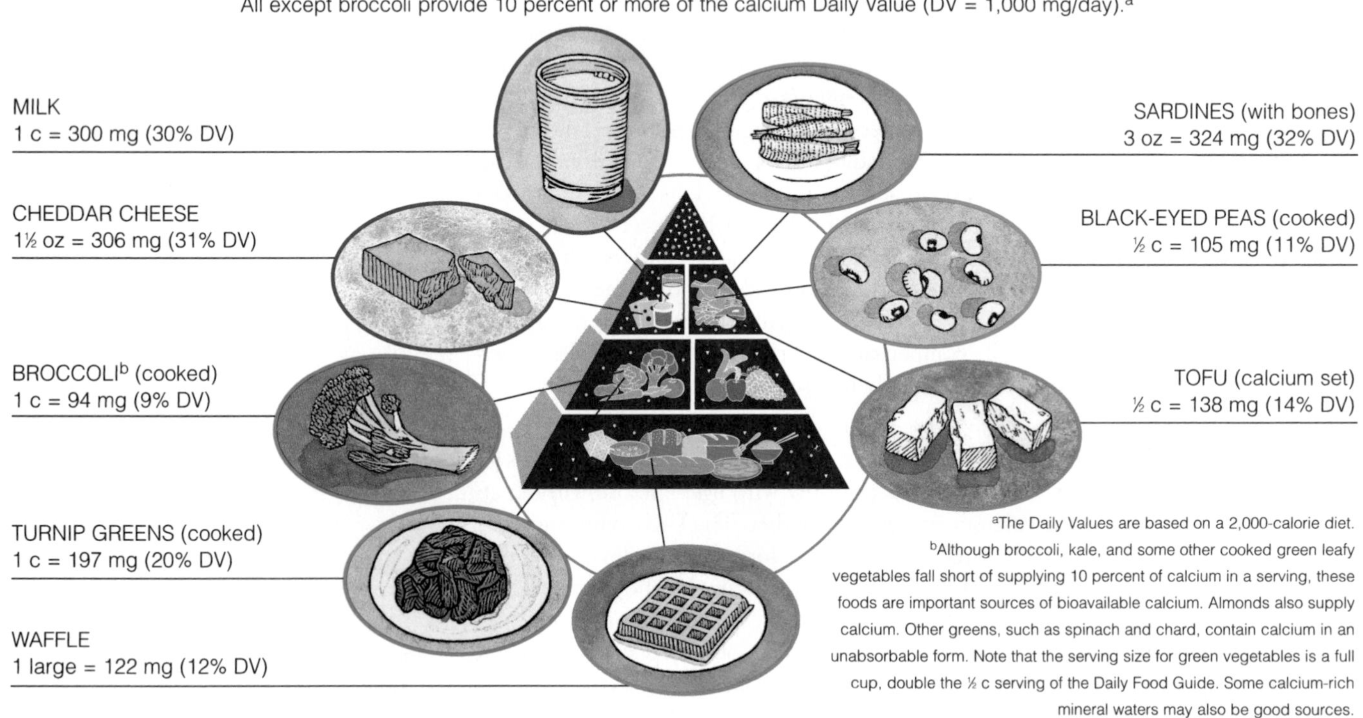

[a]The Daily Values are based on a 2,000-calorie diet.

[b]Although broccoli, kale, and some other cooked green leafy vegetables fall short of supplying 10 percent of calcium in a serving, these foods are important sources of bioavailable calcium. Almonds also supply calcium. Other greens, such as spinach and chard, contain calcium in an unabsorbable form. Note that the serving size for green vegetables is a full cup, double the ½ c serving of the Daily Food Guide. Some calcium-rich mineral waters may also be good sources.

The mineral is *phosphorus*. The adjective form is spelled with an *-ous* (as in *phosphorous salts*).

phosphorus is found combined with calcium in the crystals of the bones and teeth. The rest is everywhere else:

- Phosphorous salts are critical buffers, helping to maintain the acid-base balance of cellular fluids.
- Phosphorus is part of the DNA and RNA of every cell and thus is essential for growth and renewal of tissues.
- Phosphorous compounds carry, store, and release energy in the metabolism of energy nutrients.
- Phosphorous compounds assist many enzymes and vitamins in extracting the energy from nutrients.
- Phosphorus forms part of the molecules of the phospholipids that are principal components of cell membranes (discussed in Chapter 5).
- Phosphorus is present in some proteins.

Despite all of these critical roles, the body's needs for phosphorus are easily met by almost any diet, and deficiencies are unknown. As Snapshot 8-2 shows, animal protein is the best source of phosphorus (because phosphorus is abundant in the cells of animals).

key point *Most of the phosphorus in the body is in the bones and teeth. Phosphorus helps maintain acid-base balance, is part of the genetic material in cells, assists in energy metabolism, and forms part of cell membranes. Under normal circumstances, deficiencies of phosphorus are unknown.*

Magnesium

Magnesium barely qualifies as a major mineral: only about 1 ounce is present in the body of a 130-pound person, over half of it in the bones. Most of the rest is in the muscles, heart, liver, and other soft tissues, with only 1 percent in the body fluids.

PHOSPHORUS

These foods provide 10 percent or more of the phosphorus Daily Value (DV = 1,000 mg/day).[a]

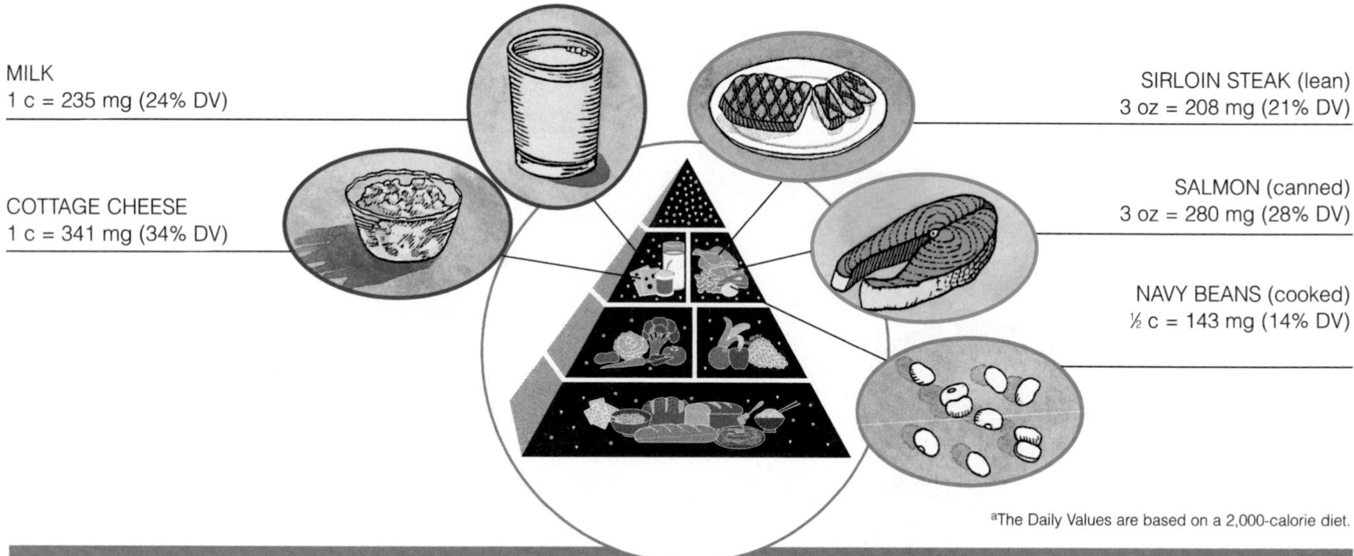

[a]The Daily Values are based on a 2,000-calorie diet.

The supply of magnesium in the bones can be tapped to maintain a constant blood level whenever dietary intake falls too low. The kidneys can also act to conserve magnesium.

Like phosphorus, magnesium is critical to many cell functions. It assists in the operation of more than 300 enzymes, is needed for the release and use of energy from the energy-yielding nutrients, and directly affects the metabolism of potassium, calcium, and vitamin D. Magnesium acts in the cells of all the soft tissues, where it forms part of the protein-making machinery and is necessary for the release of energy. Magnesium and calcium work together for proper functioning of the muscles: calcium promotes contraction, and magnesium helps the muscles relax afterward. In the teeth, magnesium promotes resistance to tooth decay by holding calcium in tooth enamel.

A magnesium deficiency may occur as a result of inadequate intake, vomiting, diarrhea, alcoholism, or protein malnutrition. It may also occur in hospital clients who have been fed magnesium-poor fluids through a vein for too long or in people who are using diuretics. People whose drinking water has a high magnesium content experience a lower incidence of sudden death from heart failure than other people. It seems likely that magnesium deficiency makes the heart unable to stop itself from spasms once it starts. Magnesium deficiency may also be related to cardiovascular disease, heart attack, and high blood pressure. A deficiency also causes hallucinations that can be mistaken for mental illness or drunkenness. Although intakes are often below those recommended, overt deficiency symptoms are rare in normal, healthy people.

Most Americans receive only about three-quarters of the recommended magnesium from their diets. Snapshot 8-3 on the next page shows magnesium-rich foods. Magnesium is easily washed and peeled away from foods during processing, so slightly processed or unprocessed foods are the best sources. In some parts of the country, water contributes significantly to magnesium intakes, so people living in those regions need less from food.

Magnesium toxicities are most often reported in older people who abuse magnesium-containing laxatives, antacids, and other medications. The consequences can be severe

MAGNESIUM

These foods provide 10 percent or more of the magnesium Daily Value (DV = 400 mg/day).[a]

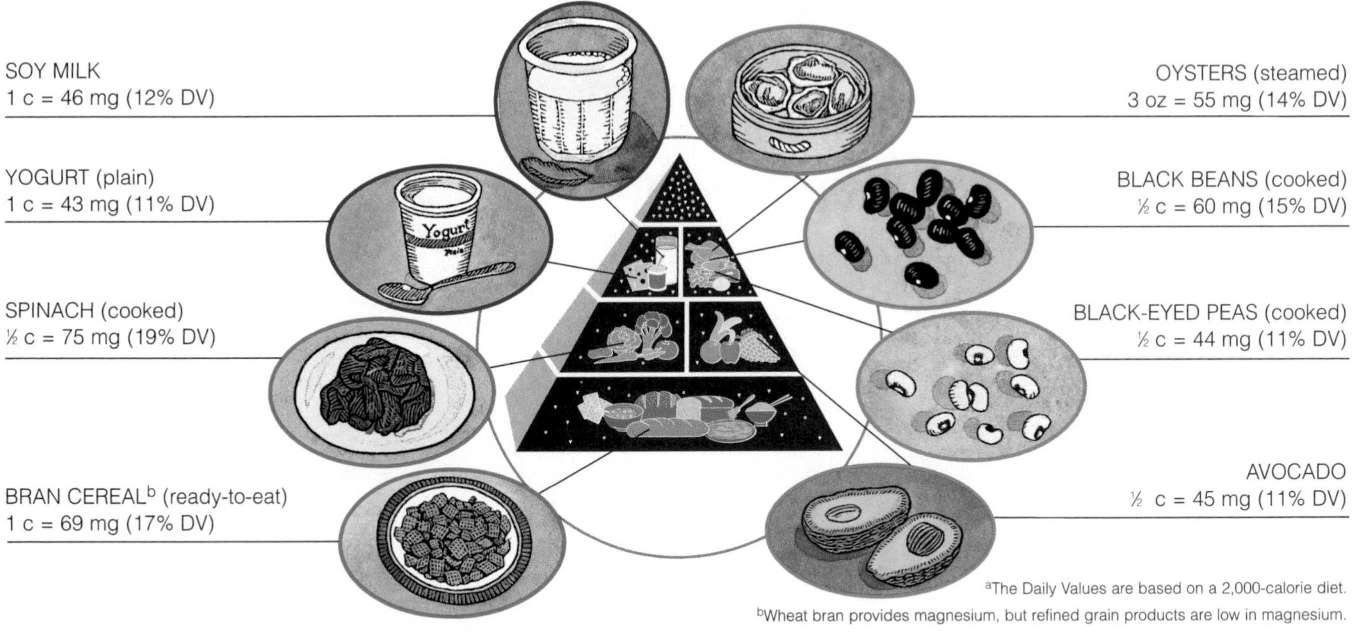

[a]The Daily Values are based on a 2,000-calorie diet.

[b]Wheat bran provides magnesium, but refined grain products are low in magnesium.

diarrhea, acid-base imbalance, and dehydration. For safety, use magnesium-containing laxatives with discretion.

key point *Most of the body's magnesium is in the bones and can be drawn out for all the cells to use in building protein and using energy. Most people in the United States fail to obtain enough magnesium from their food.*

To the chemist, a salt results from the neutralization of an acid and a base. Sodium chloride, table salt, results from the reaction between hydrochloric acid and the base sodium hydroxide. The positive sodium ion unites with the negative chloride ion to form the salt. The positive hydrogen ion unites with the negative hydroxide ion to form water.

Base + acid = salt + water.
Sodium hydroxide + hydrochloric acid = sodium chloride + water.

Sodium

Salt has been known and valued throughout recorded history. "You are the salt of the earth" means that you are valuable. If "you are not worth your salt," you are worthless. Even our word *salary* comes from the Latin word for *salt*. Chemically, sodium is the positive ion in the compound sodium chloride (table salt) and makes up 40 percent of its weight: a gram of salt contains 400 milligrams of sodium.

Sodium is a major part of the body's fluid and electrolyte balance system because it is the chief ion used to maintain the volume of fluid outside cells. Sodium also helps maintain acid-base balance and is essential to muscle contraction and nerve transmission. Scientists think that 30 to 40 percent of the body's sodium is stored on the surface of the bone crystals, where the body can easily draw on it to replenish the blood concentration.

A deficiency of sodium would be harmful, but few diets lack sodium. Most foods include more salt than is needed, and the body absorbs it freely. The kidneys filter the surplus out of the blood into the urine. They can also sensitively conserve sodium. In the rare event of a deficiency, they can return to the bloodstream the exact amount needed. Small sodium losses occur in sweat, but the amount of sodium you excrete in a day equals the amount you have ingested that day. But, if sodium is so well controlled by the body, why do authorities urge people to limit their intakes? To understand why, you must first understand how sodium interacts with body fluids.

hypertension high blood pressure.

How Are Salt and "Water Weight" Related?

If blood sodium rises, as it will after a person eats salted foods, thirst ensures that the person will drink water until the sodium-to-water ratio is restored. Then the kidneys excrete the extra water along with the extra sodium.

Dieters sometimes think that eating too much salt or drinking too much water will make them gain weight, but they do not gain fat, of course. They gain water, but a healthy body excretes this excess water immediately. Excess salt is excreted as soon as enough water is drunk to carry the salt out of the body. From this perspective, then, the way to keep body salt (and "water weight") under control is to control salt intake and drink more, not less, water.

For a brief summary of the kidneys' actions, see Chapter 3.

If blood sodium drops, body water is lost, and both water and sodium must be replenished to avert an emergency. Overly strict use of low-sodium diets in the treatment of hypertension, kidney disease, or heart disease can deplete the body of needed sodium; so can vomiting, diarrhea, or extremely heavy sweating. The sodium lost through normal sweating due to exercise is easily replaced later in the day with ordinary foods (see Chapter 10).

See Chapter 10 for more on sodium, sweating, and exercise.

Sodium Intakes

No known human diet lacks sodium. For this reason, no intake recommendation has been set. Instead, the *minimum* sodium requirement for U.S. adults is estimated to be 500 milligrams, 115 milligrams in Canada. Both these amounts are provided by a diet of plain foods with no salt added. The *Dietary Guidelines for Americans* urge people to consume less salt (see Table 8-5).

Adults in the United States consume an average of 3,300 milligrams of sodium, or more than 8 grams of salt, a day (see Figure 8-7). Asian people, whose staple sauces and flavorings are based on soy sauce and monosodium glutamate (MSG or Accent), may consume the equivalent of 30 to 40 grams of salt per day.

Sodium and Blood Pressure

Around the world, communities with high intakes of salt experience high rates of **hypertension,** cardiovascular disease, and cerebral hemorrhage, a hypertension-related stroke.[10] As blood pressure rises, the risk of death from cardiovascular disease climbs steadily.[11] Over 30 years of observational evidence point to a relationship between elevated blood pressure and sodium intakes of over 2,400 milligrams. As mentioned, the average sodium intake in the United States exceeds the maximum recommended intake of 2,400 milligrams by about a third.[12]

Some people with hypertension respond to reduced sodium intakes with lowered blood pressure (they are salt-sensitive), but some other individuals do not (they are not salt-sensitive). The connection between salt and blood pressure in salt-sensitive people is direct: the more salt they eat, the higher their blood pressure goes. People

Figure 8•7

SODIUM INTAKES OF U.S. ADULTS

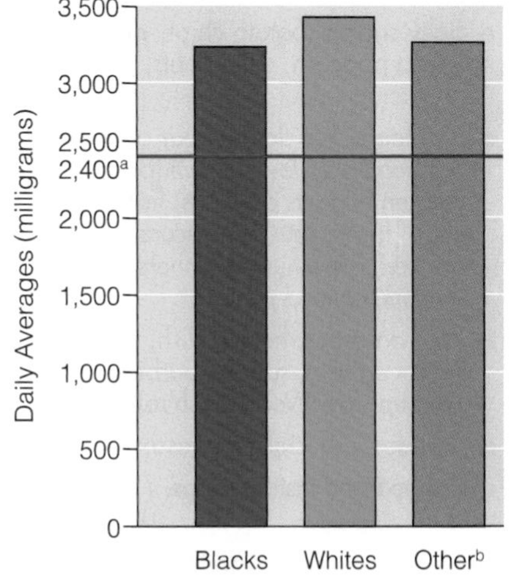

[a]Recommended maximum intake.
[b]Except Alaskan natives who average over 4,500 milligrams per day.
SOURCE: Adapted from USDA, Nutrition Insights, Issue 3, May 1997.

Table 8•5

SALT AND SODIUM INTAKE GUIDELINES

Estimated Safe and Adequate Daily Intakes
- Adolescents and adults: 500 milligrams per day.

Healthy People 2010
- Increase to at least 65 percent the proportion of people aged 2 and older who consume 2,400 milligrams or less of sodium daily.[a]

World Health Organization
- Upper limit: 6 grams salt from mixed food sources per day. Lower limit not defined.

Dietary Guidelines
- Choose and prepare foods with less salt.

[a]Fruit and vegetable intake also affects blood pressure.

DASH Diet
Daily Food Servings[a]

- Grains: 7 to 8.
- Vegetables: 4 to 5.
- Fruits: 4 to 5.
- Low-fat or fat-free milk products: 2 to 3.
- Lean meats, poultry, and fish: 2 or fewer.
- Nuts, seeds, and dry beans: 4 to 5 *per week*.
- Fats and oils: 2 to 3 teaspoons (5 grams fat per serving).
- Sweets: 5 *per week* (15 grams sugar per serving).

[a]Unless otherwise noted, serving sizes equal those of the Daily Food Guide, pages 36 and 37.

More about hypertension and the DASH diet in Chapter 11.

Cut down on salt by minimizing intakes of:

- Foods prepared in brine (pickles, olives, sauerkraut).
- Salted or smoked meats (bologna, corned or chipped beef, franks, ham, lunch meats, salt pork, sausages, bacon).
- Salted or smoked fish (anchovies, caviar, salted dried cod, pickled herring, canned sardines, smoked salmon).
- Salty snacks (potato chips, pretzels, salted popcorn, salted nuts, most crackers).
- Fast foods that do not bear a label stating *healthy* or *low-sodium* (pizza, chicken nuggets or wings, fish and chips, tacos, sausage biscuits, fried chicken, convenience dinners, frozen TV dinners, canned pastas).
- Bouillon cubes, horseradish, mustard, seasoned salts, sauces (barbeque, ketchup, soy, Worcestershire).
- Cheeses, especially processed types.
- Canned and instant soups.

Here's a short cut: look for the words *healthy* or *low-sodium* on the labels of packaged foods.

tending toward salt sensitivity usually include those with kidney disease, those of African descent, those whose parents had high blood pressure, and anyone over 50 because salt sensitivity becomes more pronounced in older age. It might seem, then, that only salt-sensitive people should be advised to cut down on salt. Unfortunately, there is no feasible way to tell to which group people belong before they become seriously ill with hypertension—hence, the recommendation that people in general should consume less salt.

Critics of current guidelines point out that *non*-salt-sensitive individuals with hypertension will likely not benefit as much from restricting dietary sodium and salt as they might from taking other steps to bring their blood pressure down. For example, weight loss in overweight people reliably reduces blood pressure. The addition of fruits, vegetables, and milk and milk products to the diet may also cause blood pressure to fall.[13]

One dietary approach may help salt-sensitive and *non*-salt-sensitive people alike. The DASH (Dietary Approaches to Stop Hypertension) diet often achieves a lower blood pressure than restriction of sodium intake alone.[14] The DASH approach calls for increasing intakes of fruit, vegetables, nuts, fish, whole grains, and low-fat dairy products while reducing intakes of red meat, butter, and other high-fat foods as well as lowering salt and sodium in the diet (see the margin).

In a study of people who ate three diets with progressively lower sodium, the subjects' blood pressures fell in response to the declining sodium. The diets were also rich in magnesium, potassium, and calcium, as well as adequate in protein and fiber. When researchers added the remaining DASH components to all three diets, the average blood pressure dropped even lower at each level of sodium intake. For controlling hypertension, then, a move to reduce sodium and implement the other dietary changes of the DASH plan constitutes a sound policy.

Many authorities support the idea that limiting salt intake to the recommended level can reasonably be expected to reduce the rate of death from cardiovascular disease. Both the *Dietary Guidelines for Americans* and the *Healthy People 2010 Objectives for the Nation* have restated their goals to recommend that people choose a diet moderate in salt and sodium.[15] Many Americans may have much to gain in terms of cardiovascular health and nothing to lose from cutting back on salt as part of an overall lifestyle strategy to reduce blood pressure.[16] Physical activity should also be part of that lifestyle, for regular moderate exercise reliably lowers the blood pressure.

There are also other valid reasons for most people to hold their salt intakes at or below the recommended maximum. For example, older people without clinical hypertension often die of stroke, and reducing dietary sodium may lower their blood pressure enough to reduce their stroke risk.[17] Excess sodium in the diet causes increased calcium excretion—just the wrong effect for preserving the integrity of the bones. Excessive salt may also directly stress a weakened heart or aggravate kidney problems. Asians' high salt intakes have been suggested as a possible cause for their greatly elevated rate of stomach cancer.

Controlling Salt Intake For all the reasons just presented, cutting down on salt and sodium in the diet may be wise, and the margin offers tips for doing so. Foods eaten without salt may seem less tasty at first, but with repetition, tastes adjust and the natural flavor becomes the preferred taste. Also, remember that the recommendation is to reduce, not eliminate, salt intake.

An obvious step is to control the saltshaker, but this source may contribute as little as 15 percent of the total salt consumed. A more productive step is to cut down on processed and fast foods, the source of almost 75 percent of salt in the U.S. diet—see the list in the margin. In Table 8-6, notice that the least processed foods in each food group are not only lowest in sodium but also highest in potassium. Low potassium intakes are thought to play an important role in the development of hypertension. Many people are unaware that foods high in sodium do not always taste salty.[18] Who could guess by taste alone that a serving of instant chocolate pudding provides a full third of the daily allowable sodium? Moral: Read the labels. Figure 8-8 and this chapter's Do It! section identify some other sodium sources in the U.S. diet.

Table 8•6

PROCESSING REDUCES POTASSIUM, INCREASES SODIUM IN FOODS

Food	Potassium (mg)	Sodium (mg)	Ratio
Milk Products			
Milk (whole), 1 c	371	120	3:1
Chocolate pudding (home cooked), 1 c	506	274	2:1
Chocolate pudding (instant), 1 c	488	834	1:2
Meats			
Beef roast (cooked), 3 oz	250	53	5:1
Corned beef (canned), 3 oz	115	855	1:7
Frankfurter, 1 large	95	638	1:7
Chipped beef, 3 oz	377	2,953	1:8
Vegetables			
Corn (cooked), 1 c	242	8	30:1
Creamed corn (canned), 1 c	390	572	1:2
Cornflakes, 1 c	25	300	1:12
Fruit			
Peaches (fresh), 1	193	<1	193:1
Peaches (canned), 1 c	241	16	15:1
Peach pie, 1 piece	131	253	1:2
Grains			
Whole-wheat flour, 1 c	486	6	81:1
Shredded wheat cereal, 1 c	155	4	39:1
Whole-wheat bread, 1 slice	71	148	1:2
Wheat crackers, 4	16	70	1:4

Figure 8•8

SOURCES OF SODIUM IN THE U.S. DIET

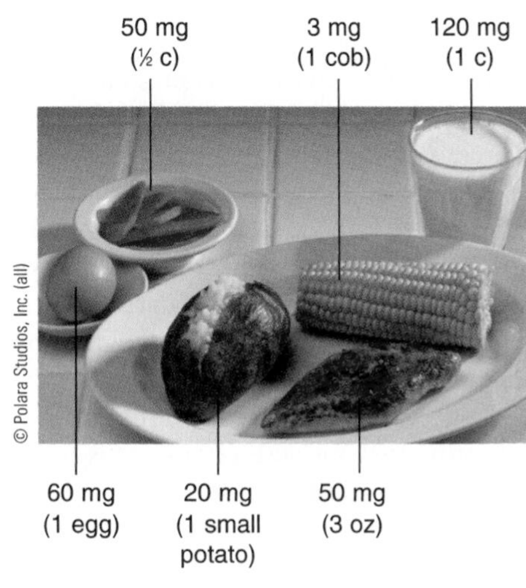

Unprocessed foods that are low in sodium contribute less than 10 percent of the total sodium in the U.S. diet.

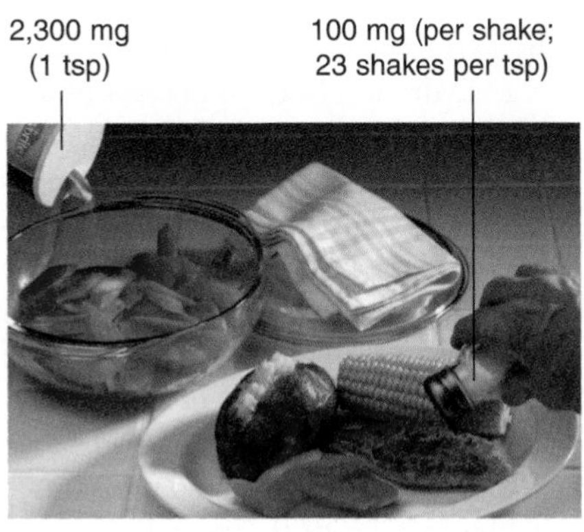

Salt added at home, in cooking or at the table, contributes 15 percent of the total sodium in the U.S. diet.

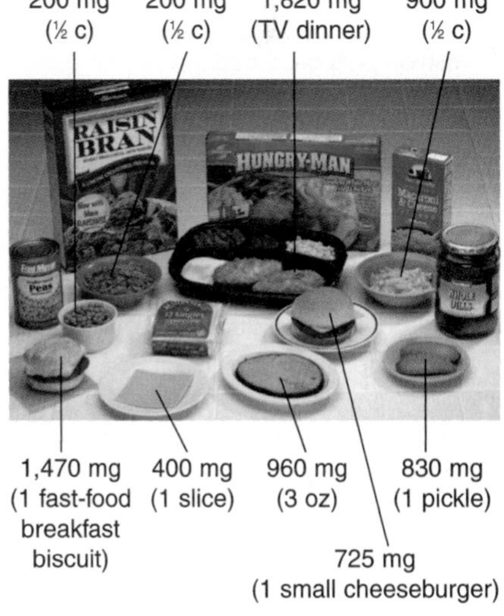

Processed foods such as these contribute 75 percent of the sodium in the U.S. diet.

8.4

POTASSIUM

These foods provide 10 percent or more of the potassium Daily Value (DV = 3,500 mg/day).[a]

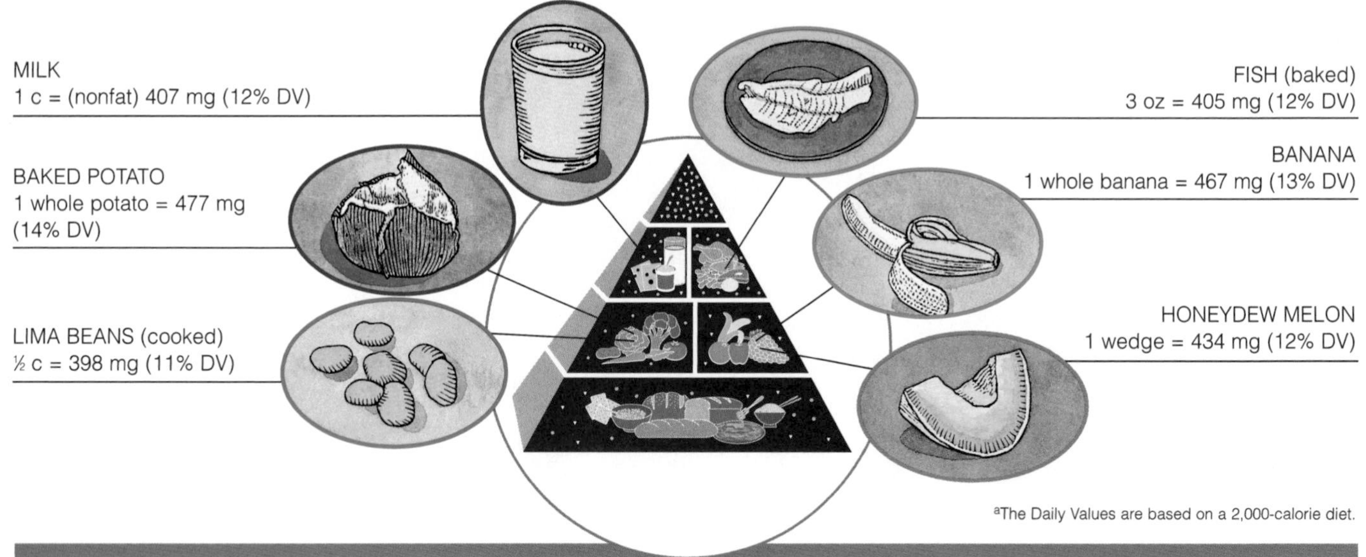

[a]The Daily Values are based on a 2,000-calorie diet.

Sodium is the main positively charged ion outside the body's cells. Sodium attracts water. Thus, too much sodium (or salt) may aggravate hypertension. Diets rarely lack sodium.

Potassium

Potassium is the principal positively charged ion *inside* body cells. It plays a major role in maintaining fluid and electrolyte balance and cell integrity, and it is critical to maintaining the heartbeat. The sudden deaths that occur during fasting or severe diarrhea and in children with kwashiorkor or people with eating disorders are thought to be due to heart failure caused by potassium loss.

Kwashiorkor is described in Chapter 6.

Dehydration leads to potassium loss from inside cells. This condition is dangerous because when brain cells lose potassium, the victim loses the ability to notice the need for water. Adults are warned not to take diuretics (water pills) that cause potassium loss or to give them to children, except under a physician's supervision. Physicians prescribing diuretics will tell clients to eat potassium-rich foods to compensate for the losses. Depending on the diuretic, physicians may also advise a lower sodium intake. When taking diuretics, a person should alert all other health-care providers.

Unlike sodium, potassium may exert a positive effect against hypertension and related ills. See Chapter 11 for details.

A dietary deficiency of potassium is unlikely in healthy people, although a low potassium intake is possible with a steady diet of highly processed foods. Some of the credit for the healthy blood pressure–lowering effect of high intakes of fruits and vegetables may go to the potassium these foods provide.[19] Because potassium is found inside all living cells and because cells remain intact unless foods are processed, the richest sources of potassium are *fresh* foods of all kinds (see Snapshot 8-4). Most whole vegetables and fruits are outstanding. Bananas, despite their fame as the richest potassium source, are just one among many rich sources. However, they are readily available, are easy to chew, and have a sweet taste that almost everyone likes, so health-care professionals often recommend them.

Potassium from foods is safe, but potassium injected into a vein can stop the heart. Potassium chloride pills are available over the counter and are sold in health-food stores without a warning label, but they should *not* be used except on a physician's advice. Potassium overdoses normally are not life-threatening as long as they are taken by mouth because the presence of excess potassium in the stomach triggers a vomiting reflex that expels the unwanted substance. A person with a weak heart, however, should not go through this trauma, and a baby may not be able to withstand it. Several infants have died when well-meaning parents overdosed them with potassium supplements.

key point *Potassium, the major positive ion inside cells, is important in many metabolic functions. Fresh foods are the best sources of potassium. Diuretics can deplete the body's potassium and so can be dangerous; potassium excess can also be dangerous.*

Chloride

In its elemental form, chlorine forms a deadly green gas. In the body, the chloride ion plays important roles as the major negative ion. In the fluids outside the cells, it accompanies sodium; inside the cells, it occurs primarily in association with potassium. Thus, it helps to maintain the crucial fluid balances (acid-base and electrolyte balances). The chloride ion also plays a special role as part of hydrochloric acid, which maintains the strong acidity of the stomach necessary to digest protein. The principal food source of chloride is salt, both added and naturally occurring in foods.

Chloride is the body's major negative ion; it is responsible for stomach acidity and assists in maintaining proper body chemistry.

Sulfur

The body does not use sulfur by itself as a nutrient, but it is present in essential nutrients that the body does use, such as thiamin and all proteins. Sulfur plays its most important role in helping strands of protein to assume a functional shape. Skin, hair, and nails contain some of the body's more rigid proteins, which have high sulfur contents.

There is no recommended intake for sulfur, and deficiencies are unknown. The summary table at the end of this chapter presents the main facts about the major minerals.

Sulfur plays important roles in body proteins.

The Trace Minerals

An obstacle to determining the precise roles of the trace elements in humans has been the difficulty of providing an experimental diet lacking in the one element under study. Thus, research in this area is limited mostly to the study of laboratory animals, which can be fed highly refined, purified diets in environments free of all contamination. New laboratory techniques have enabled scientists to detect minerals in smaller and smaller quantities in living cells, and research is now rapidly expanding our knowledge about them. Intake recommendations for human beings have been established for nine trace minerals—see Table 8-7. Others are recognized as essential nutrients for some animals, but have not been proved to be required for human beings.

Table 8•7

TRACE MINERALS

Human Intake Recommendations Established
Iodine
Iron
Zinc
Selenium
Fluoride
Chromium
Copper
Manganese
Molybdenum
Known Essential for Animals; Human Requirements under Study
Arsenic
Boron
Nickel
Silicon
Vanadium
Known Essential for Some Animals; No Evidence That Intake by Humans Is Ever Limiting
Cobalt

The evidence for requirements and essentiality is weak for the trace minerals cadmium, lead, lithium, and tin.

Iodine

The body needs only an infinitesimally small quantity of iodine, but obtaining this amount is critical. Iodine is a part of thyroxine, the hormone made by the thyroid gland responsible for regulating the basal metabolic rate. Iodine must be available for thyroxine to be synthesized.

goiter (GOY-ter) enlargement of the thyroid gland due to iodine deficiency is *simple goiter;* enlargement due to an iodine excess is *toxic goiter.*

cretinism (CREE-tin-ism) severe mental and physical retardation of an infant caused by the mother's iodine deficiency during pregnancy.

potassium iodide a medication approved by the Food and Drug Administration as safe and effective for the prevention of thyroid cancer caused by radioactive iodine known to be released during radiation emergencies.

hemoglobin (HEEM-oh-globe-in) the oxygen-carrying protein of the blood; found in the red blood cells (*hemo* means "blood"; *globin* means "spherical protein").

myoglobin (MYE-oh-globe-in) the oxygen-holding protein of the muscles (*myo* means "muscle").

The Federal Trade Commission has warned of bogus Internet marketing of dietary supplements as "defenses" against biological threats, such as anthrax. Read more on the Internet:
www.ftc.gov/bcp/conline/pubs/alerts/bioalrt.htm

In iodine deficiency, the thyroid gland enlarges—a condition known as simple goiter.

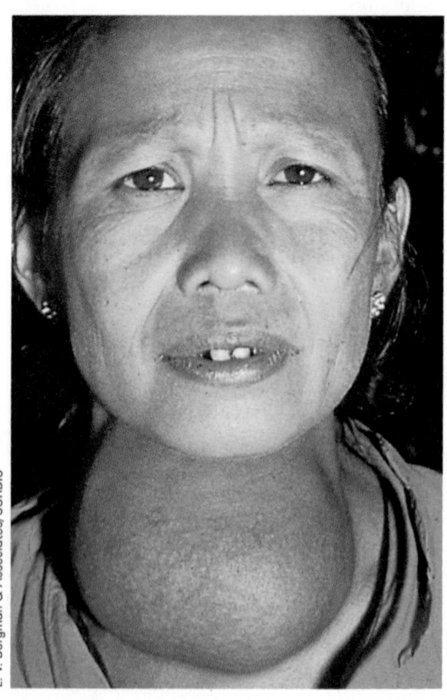

L. V. Bergman & Associates/CORBIS

In iodine deficiency, the cells of the thyroid gland enlarge in an attempt to trap as many particles of iodine as possible. Sometimes the gland enlarges until it makes a visible lump in the neck, a **goiter.** People with iodine deficiency this severe become sluggish and gain weight. Severe iodine deficiency during pregnancy causes extreme and irreversible mental and physical retardation in the infant, known as **cretinism.** Much of the mental retardation can be averted if the woman's deficiency is detected and treated within the first six months of pregnancy, but if treatment comes too late or not at all, the child may have an IQ as low as 20 (100 is average).[20] Iodine deficiency is one of the world's most common and most preventable causes of mental retardation.[21] Researchers hope to reverse the high rates of cretinism and goiter reported in developing nations by adding iodine to community food or water supplies.

The iodine in food varies because it reflects the soil in which plants are grown or on which animals graze. Iodine is plentiful in the ocean, so seafood is a dependable source. In the central parts of the United States that were never under the ocean, the soil is poor in iodine. In those areas, the use of iodized salt and the consumption of foods shipped in from iodine-rich areas have wiped out the iodine deficiency that once was widespread. Surprisingly, sea salt delivers little iodine because iodine becomes a gas and flies off into the air during the salt-drying process. In the United States, salt labels state whether the salt is iodized; in Canada, all table salt is iodized.

Excessive intakes of iodine can enlarge the thyroid gland just as a deficiency can. U.S. intakes are above the recommended intake of 150 micrograms, but still below the Tolerable Upper Intake Level of 1,100 micrograms per day for an adult.[22] Like chlorine and fluorine, iodine is a deadly poison in large amounts.

Much of the iodine in U.S. diets today comes from fast-food establishments, which use iodized salt with a liberal hand, and from bakery products and milk. The baking industry uses iodine-containing dough conditioners, and most dairies use iodine to disinfect milking equipment. One cup of milk supplies nearly half of one day's recommended intake of iodine, and less than a half-teaspoon of iodized salt meets the entire recommendation.

An iodine-containing medication, **potassium iodide,** effectively blocks damage to the thyroid gland caused during radiation emergencies, such as hostile attacks or malfunctions of nuclear power plants. When given in the correct dosage within a certain timeframe relative to radiation exposure, potassium iodide can greatly reduce the likelihood of thyroid cancer development.[23] When given in the wrong dosage or with faulty timing, potassium iodide is useless or toxic. For this reason, concerned people who live near power plants are urged to rely on health professionals for guidance.

key point *Iodine is part of the hormone thyroxine, which influences energy metabolism. The deficiency diseases are goiter and cretinism. Iodine occurs naturally in seafood and in foods grown on land that was once covered by oceans; it is an additive in milk and bakery products. Large amounts are poisonous.*

Iron

Every living cell, whether plant or animal, contains iron. Most of the iron in the body is a component of two proteins: **hemoglobin** in red blood cells and **myoglobin** in muscle cells. Hemoglobin in the red blood cells carries oxygen from the lungs to tissues throughout the body. Myoglobin carries and stores oxygen for the muscles. Iron helps these proteins to hold and carry oxygen and then release it.

All the body's cells need oxygen to combine with the carbon and hydrogen atoms the cells release as they break down energy nutrients. The oxygen combines with these atoms to form the waste products carbon dioxide and water; thus, the body constantly needs fresh oxygen to keep the cells going. As cells of tissues use up and excrete their oxygen (as carbon dioxide and water), red blood cells shuttle in fresh oxygen supplies from the lungs. In addition to this major task, iron helps many enzymes to use oxygen, and iron is needed to make new cells, amino acids, hormones, and neurotransmitters.

Iron is clearly the body's gold, a precious mineral to be hoarded. The liver packs iron sent from the bone marrow into new red blood cells, also from the bone marrow, and ships them out to the blood. Red blood cells live for about three to four months. When they die, the spleen and liver break them down, salvage their iron for recycling, and send it back to the bone marrow to be kept until it is reused. The body does lose iron in nail clippings, hair cuttings, and shed skin cells, but only in tiny amounts. Bleeding can cause significant iron loss from the body, however.

The body has special provisions for obtaining iron. Only about 10 to 15 percent of dietary iron is absorbed; but if the body's supply of iron is diminished or if the need increases (say, during pregnancy), absorption can increase several-fold.[24] Once inside the body, iron is difficult to excrete, so absorption of iron is carefully controlled.[25]

iron deficiency the condition of having depleted iron stores, which, at the extreme, causes iron-deficiency anemia.

iron-deficiency anemia a form of anemia caused by a lack of iron and characterized by red blood cell shrinkage and color loss. Accompanying symptoms are weakness, apathy, headaches, pallor, intolerance to cold, and inability to pay attention. (For other anemias, see the index.)

anemia the condition of inadequate or impaired red blood cells; a reduced number or volume of red blood cells along with too little hemoglobin in the blood. The red blood cells may be immature and therefore too large or too small to function properly. Anemia can result from blood loss, excessive red blood cell destruction, defective red blood cell formation, and many nutrient deficiencies. Anemia is not a disease, but a symptom of another problem; its name literally means "too little blood."

pica (PIE-ka) a craving for nonfood substances. Also known as *geophagia* (gee-oh-FAY-gee-uh) when referring to clay eating, and *pagophagia* (pag-oh-FAY-gee-uh) when referring to ice craving (*geo* means "earth"; *pago* means "frost"; *phagia* means "to eat").

What Happens to a Person Who Lacks Iron? If absorption cannot compensate for losses or low dietary intakes, then iron stores are used up and iron deficiency sets in. **Iron deficiency** and **iron-deficiency anemia** are not one and the same, though they often occur together. The distinction between iron deficiency and its anemia is a matter of degree. People may be iron deficient, meaning that they have depleted iron stores, without being anemic, or they may be iron deficient *and* anemic. With regard to iron, the term **anemia** refers to severe depletion of iron stores resulting in low blood hemoglobin.

A body severely deprived of iron becomes unable to make enough hemoglobin to fill new blood cells, and anemia results. A sample of iron-deficient blood examined under the microscope shows cells that are smaller and lighter red than normal (see Figure 8-9 on the next page). The undersized cells contain too little hemoglobin and thus deliver too little oxygen to the tissues. The diminished supply of oxygen limits the cells' energy metabolism, and causes tiredness, apathy, and a tendency to feel cold.

Even slightly lowered iron levels impair physical work capacity and productivity.[26] Many of the symptoms associated with iron deficiency are easily mistaken for behavioral or motivational problems. With reduced energy, people work less, play less, and think or learn less eagerly. (Lack of energy does not always mean an iron deficiency—see the Think Fitness feature below.) Children deprived of iron become restless, irritable, unwilling to work or play, and unable to pay attention, and they may fall behind their peers academically. Some symptoms disappear when iron intake improves, but others may linger after iron repletion.[27]

Iron deficiency makes children more susceptible to lead poisoning. See Chapter 13.

Think Fitness: Exercise Deficiency Fatigue

On hearing about symptoms of iron deficiency, tired people may jump to the conclusion that they need to take iron supplements to restore their pep. More likely, they can obtain help by simply getting to bed on time and getting enough exercise. Few realize that too little exercise over weeks and months is as exhausting as too much—the less you do, the less you're able to do, and the more fatigued you feel. The condition even has a name: "sedentary inertia."

THINK FITNESS

A curious symptom seen in some people with iron deficiency is an appetite for ice, clay, paste, or other nonnutritious substances. Such people have been known to eat as many as eight trays of ice in a day. This consumption of nonfood substances, most often observed in poverty-stricken women and children and in people whose blood is cleansed by dialysis machines, has been given the name **pica.**[28] In some cases, pica clears up within days after iron is given, even before the red blood cells have had a chance to respond. Other times, pica is unresponsive to iron.

Feeling fatigued, weak, and apathetic is a sign that something is wrong. It is not a sign that you necessarily need iron or other supplements. Three actions are called for: first, get your diet in order; second, get some exercise; third, if symptoms persist for more than a week or two, consult a physician for a diagnosis.

Causes of Iron Deficiency and Anemia Iron deficiency is usually caused by malnutrition, that is, inadequate iron intake, either from sheer lack of food or from high

iron overload the state of having more iron in the body than it needs or can handle, usually arising from a hereditary defect. Also called *hemochromatosis*.

consumption of the wrong foods. In the developed countries, overconsuming foods rich in sugar and fat and poor in nutrients is often responsible for low iron intakes. Snapshot 8-5 shows iron amounts in some foods that are good or excellent sources of iron.

Among nonnutritional causes of anemia, blood loss is number one. Because 80 percent of the iron in the body is in the blood, losing blood means losing iron. Because of menstrual losses, women need one and a half times as much iron as men do. Women are especially vulnerable to iron deficiency because they not only need more iron than men but they also, on average, eat less food. Infants over age six months, young children, adolescents, menstruating women, and pregnant women all have increased need for iron to support the growth of new body tissues or replace losses, or in the case of adolescent girls, both.

Worldwide and in the United States, iron deficiency is the most common nutrient deficiency.[29] Iron-deficiency anemia affects an estimated 40 percent of the world's population, with the highest prevalence in developing countries. In those countries, parasitic infections of the digestive tract cause people to lose blood daily. For their entire lives, they may feel fatigued and listless but never know why. Digestive tract problems such as ulcers, sores, and even inflammation can also cause blood loss severe enough to cause anemia.[30]

Among young U.S. women, about 10 percent are iron deficient and 3 to 5 percent have anemia. Happily, the iron status of U.S. infants and young children has improved over the last decade, thanks to more widespread breastfeeding, which promotes iron absorption, and greater use of iron-fortified infant formula and cereals. For low-income families, the Special Supplemental Food Program for Women, Infants, and Children (WIC) provides coupons redeemable for foods high in iron, giving another boost to the iron status of many U.S. children.

Figure 8•9

NORMAL AND ANEMIC BLOOD CELLS

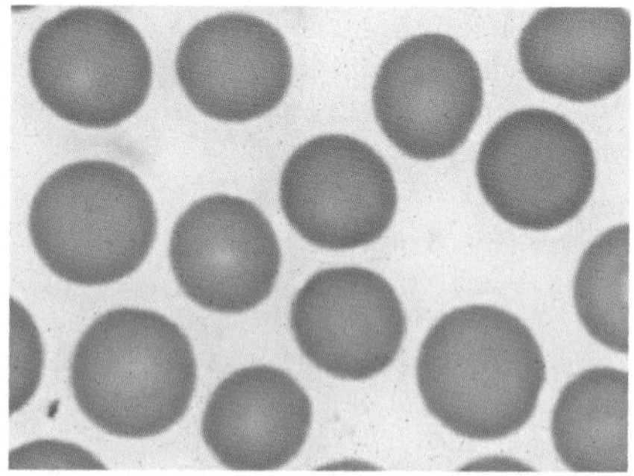

Normal red blood cells. Both size and color are normal.

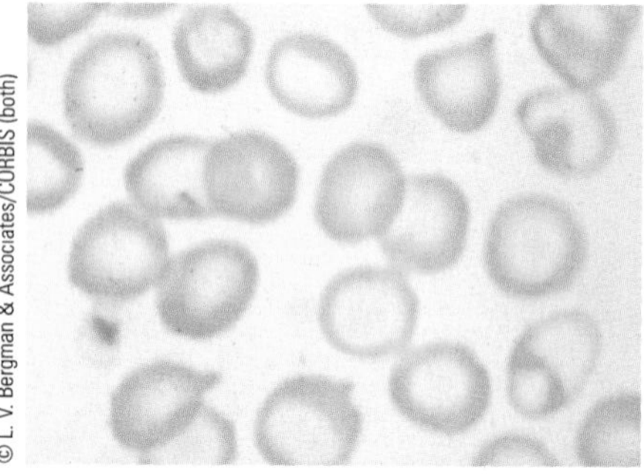

Blood cells in iron-deficiency anemia. These cells are small and pale because they contain less hemoglobin.

Q&A Can a Person Take in Too Much Iron?

Iron is toxic in large amounts, and once absorbed inside the body, it is difficult to excrete. The body defends against iron poisoning by controlling its entry: the intestinal cells trap some of the iron and hold it within their boundaries. When they are shed, these cells carry out of the intestinal tract the excess iron that they collected during their brief lives.[31] In healthy people, when iron stores fill up, less iron is absorbed.[32]

Once considered rare, **iron overload** has increased in frequency over the last few decades. Iron overload is often caused by a hereditary defect that causes the intestine to absorb excess iron.[33] Tissue damage occurs, especially in iron-storing organs such as the liver. Infections are also likely because bacteria thrive on iron-rich blood. The effects are most severe in alcohol abusers because alcohol damages the intestine, impairing its defense against absorbing too much iron.

The body does guard against iron's renegade nature. Left free, iron is a powerful oxidant that can start free-radical reactions that damage cellular structures.[34] Protein carriers guard the body's iron molecules and keep them away from vulnerable body compounds, thereby preventing damage. Iron's actions are thus tightly controlled.

Elevated iron stores might be related to the development of cardiovascular disease.[35] In rats fed a highly absorbable form of iron, excess iron causes an increase in oxidation of low-density lipoproteins (LDL) in the blood, a process believed important in heart disease development.[36] In a classic study of Finnish men, elevated serum ferritin (the iron-carrying protein in the blood) doubled the risk of heart attack. However, other investigations designed to replicate this finding have produced mixed results.[37]

The danger of iron overload is an argument against high-level iron fortification of foods. Susceptible people would find it difficult to follow a low-iron diet. Worsening the picture is the U.S. population's love of vitamin C supplements because vitamin C greatly enhances iron absorption.[38]

Iron supplements are a leading cause of fatal accidental poisonings among U.S. children under six years old.[39] High-dose iron pills may soon come packaged in individually sealed units to help prevent such poisonings. Keep iron supplements out of children's reach.

snapshot

8.5

IRON

These foods provide 10 percent or more of the iron Daily Value (DV = 18 mg/day).[a, b]

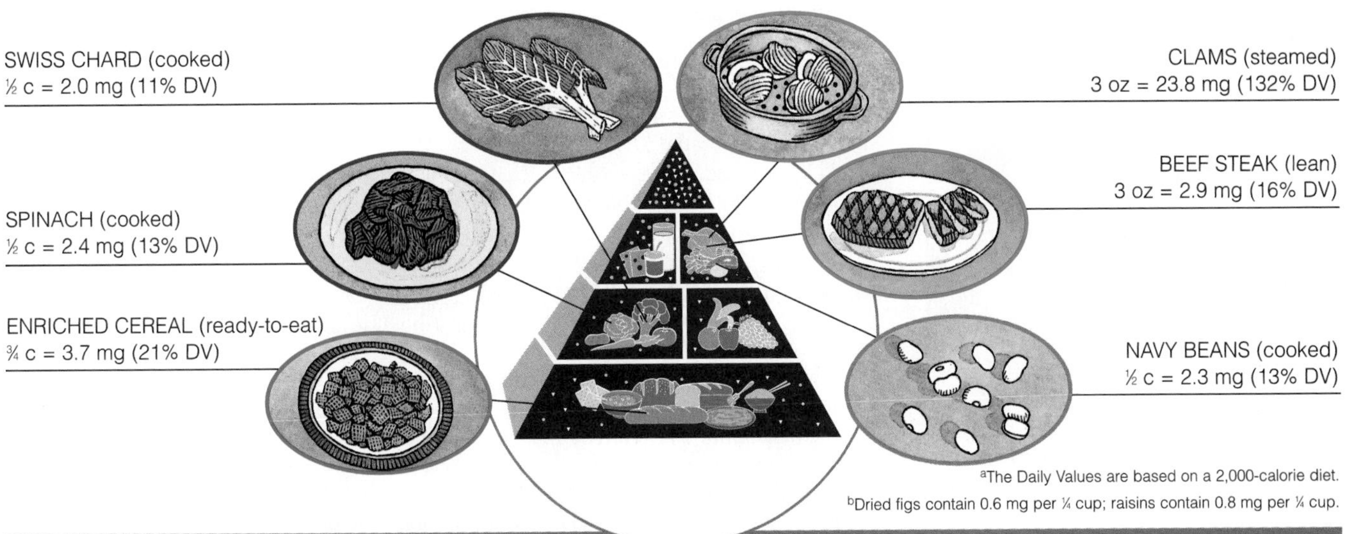

[a]The Daily Values are based on a 2,000-calorie diet.

[b]Dried figs contain 0.6 mg per ¼ cup; raisins contain 0.8 mg per ¼ cup.

Iron Recommendations and Sources Men need 8 milligrams of iron each day, and so do women past age 51. For women of childbearing age, the recommendation is higher—18 milligrams—to replace menstrual losses. During pregnancy, a woman needs significantly more—27 milligrams. Adult men rarely experience iron-deficiency anemia. If a man has a low hemoglobin concentration, his health-care provider should examine him for a blood-loss site. Table 8-8 sums up iron recommendations.

To meet your iron needs, it is best to rely on foods because the iron from supplements is much less well absorbed than that from food. The usual Western mixed diet provides only about 5 to 6 milligrams of iron in each 1,000 calories. An adult male who eats 2,500 calories or more a day has no trouble obtaining his needed 8 milligrams or more, but a woman who eats fewer calories and needs more iron won't obtain her needed 18 milligrams unless she selects high-iron, low-calorie foods from each food group. And pregnant women need an iron supplement.[40] No one should take iron supplements without a physician's recommendation, however.

Absorbing Iron Iron occurs in two forms in foods. Some is bound into **heme,** the iron-containing part of hemoglobin and myoglobin in meat, poultry, and fish (look back at Figure 6-4 in Chapter 6). Some is nonheme iron, found in foods from plants and in the nonheme iron in meats. The form affects absorption. Heme iron is much more reliably absorbed than nonheme iron. Healthy people with adequate iron stores absorb heme iron at a rate of about 23 percent over a wide range of meat intakes. People absorb nonheme iron at rates of 2 to 20 percent, depending on dietary factors and iron stores.

Meat, fish, and poultry contain a factor **(MFP factor)** that promotes the absorption of nonheme iron from other foods eaten at the same time. Vitamin C can triple nonheme iron absorption from foods eaten in the same meal.[41] A system of calculating the amount of iron absorbed from a meal, based on these factors, is presented in Table 8-9 on the next page.

Some substances impair iron absorption. They include the **tannins** of tea and coffee, the calcium and phosphorus in milk, and the **phytates** that accompany fiber in whole-grain cereals. Ordinary black tea is exceptional in its efficiency at reducing

heme (HEEM) the iron-containing portion of the hemoglobin and myoglobin molecules.

MFP factor a factor (identity unknown) present in meat, fish, and poultry that enhances the absorption of nonheme iron present in the same foods or in other foods eaten at the same time.

tannins compounds in tea (especially black tea) and coffee that bind iron. Tannins also denature proteins.

phytates (FYE-tates) compounds present in plant foods (particularly whole grains) that bind iron and may prevent its absorption.

Table 8•8

IRON DRI VALUES (milligrams per day)

Intake Recommendations
- Men, 8 mg
- Women
  childbearing years, 18 mg
  51 years and older, 8 mg
  pregnancy, 27 mg

Tolerable Upper Intake Levels
- Infants and children, 40 mg
- Adolescents and adults, 45 mg

Table 8-10, pages 296–297 summarizes the effects of iron toxicity.

This chili dinner provides iron and MFP factor from meat, iron from legumes, and vitamin C from tomatoes. The combination of heme iron, nonheme iron, MFP factor, and vitamin C helps to achieve maximum iron absorption.

Dietary factors that increase iron absorption:

- Vitamin C.
- MFP factor.

Factors that hinder iron absorption:

- Tea.
- Coffee.
- Calcium and phosphorus.
- Phytates, tannins, and fiber.

The old-fashioned iron skillet adds supplemental iron to foods.

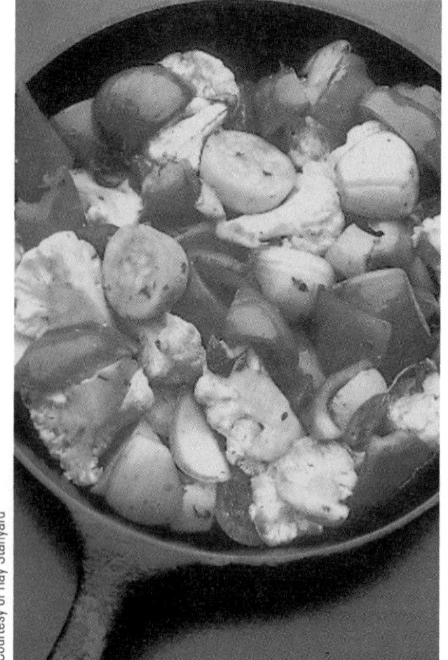

Table 8•9

CALCULATION OF IRON ABSORBED FROM MEALS

You need to know three factors to calculate the amount of iron absorbed from a meal:

1. How much of the iron in the meal was heme iron, and how much was nonheme iron?
2. How much vitamin C was in the meal?
3. How much total meat, fish, and poultry (MFP factor) was consumed?

(It is assumed that your iron stores are moderate; otherwise, you'd have to take this into consideration, too.) Write down the foods you eat at a typical meal, look up their iron content in the Table of Food Composition, Appendix A, and then answer these questions:

1. How much iron was from animal tissues (MFP)? ______ mg
2. 40% of (1), on the average, is heme iron:
   (1) ________ mg × 0.40 = ________ mg heme iron
3. How much iron was from other sources? ________ mg
4. Add (3) and 60% of (1) to get nonheme iron:
   (3) ________ mg + 0.60 × (1) ________ mg = ________ mg nonheme iron
5. How much vitamin C was in the meal? ________ mg Less than 25 mg is low; 25 to 75 mg is medium; more than 75 mg is high: ________
6. How much MFP factor was in the meal? ________ oz Less than 1 oz lean MFP is low; 1 to 3 oz is medium; more than 3 oz is high: ________
7. Now you can calculate the amount of each type of iron that was absorbed. Start with the heme iron. You absorbed 25% of the heme iron:
   (2) ________ mg × 0.25 = ________ mg heme iron absorbed
8. Consider your scores from (5) and (6). If either vitamin C or MFP factor was high or if both were medium, the availability of your nonheme iron was **high.** If neither was high, but one was medium, the availability of your nonheme iron was **medium.** If both were low, your nonheme iron had **poor** availability. You absorbed:
   - High availability: 10% of the nonheme iron
   - Medium availabilty: 8% of the nonheme iron
   - Poor availability: 5% of the nonheme iron
9. Now you can calculate the nonheme iron absorbed. You absorbed ________ % of the nonheme iron, or:
   (4) ________ mg × (8) ________ % = ________ mg nonheme iron absorbed
10. Add the heme and nonheme iron from (7) and (9) together:
    (7) ________ mg heme iron absorbed
    + (9) ________ mg nonheme iron absorbed
    Total = ________ mg iron absorbed.

According to the DRI committee, if you are a man age 19 or older or a woman age 51 or older, you need to absorb about 1 mg per day. If you are a woman 11 to 50 years old, you need to absorb 1.5 mg per day on average. If you have higher menstrual losses than the average woman, you may need still more.

absorption of iron—clinical dietitians advise people with iron overload to drink it with their meals. For those who need more iron, the opposite advice applies—drink tea between meals, not with food.[42]

Thus, the amount of iron *absorbed* from a meal depends on the interaction between promoters of iron absorption and inhibitors. When you eat meat with legumes (for example, ham and beans or chili with beans and meat), the iron from the meat is well absorbed, and MFP factor enhances iron absorption from the beans. The vitamin C from a slice of tomato and a leaf of lettuce in a sandwich will enhance iron absorption from the bread. The meat and tomato in spaghetti sauce help the digestive tract to absorb the iron from the spaghetti.

Cooking the sauce in an iron pan also adds more iron. Foods cooked in iron pans contain iron salts somewhat like those in supplements. The iron content of 100 grams of spaghetti sauce simmered in a glass dish is 3 milligrams, but it increases to 87 milligrams when the sauce is cooked in a black iron skillet. In the short time it takes to scramble eggs, a cook can triple the eggs' iron content by scrambling them in an iron pan. Dried peaches and raisins contain more iron than the fresh fruit because they are dried in iron pans. This iron salt is not as well absorbed as iron from meat, but some does get into the body, especially if the meal also contains MFP factor or vitamin C.

key point *Most iron in the body is contained in hemoglobin and myoglobin or occurs as part of enzymes in the energy-yielding pathways. Iron-deficiency anemia is a problem worldwide; too much iron is toxic. Iron is lost through menstruation and other bleeding; reduced absorption and the shedding of intestinal cells protect against overload. For maximum iron absorption, use meat, other iron sources, and vitamin C together.*

Zinc

Zinc occurs in a very small quantity in the human body, but works with proteins in every organ, helping nearly 100 enzymes to:

- Make parts of the cells' genetic material.
- Make heme in hemoglobin.
- Help the pancreas with its digestive functions.
- Help metabolize carbohydrate, protein, and fat.
- Liberate vitamin A from storage in the liver.
- Dispose of damaging free radicals.

Besides helping enzymes function, zinc helps to regulate gene expression in protein synthesis. Zinc also affects behavior and learning; assists in immune function; and is essential to wound healing, sperm production, taste perception, fetal development, and growth and development in children. Zinc is needed to produce the active form of vitamin A in visual pigments. When zinc deficiency occurs, it packs a wallop to the body, impairing all these functions.[43] Even a mild zinc deficiency can result in impaired immunity, abnormal taste, and abnormal vision in the dark.

Problem: Too Little Zinc Zinc deficiency in human beings was first reported in the 1960s from studies with growth-delayed children and adolescent boys in the Middle East. Their native diets were typically low in animal protein and high in whole grains and beans; consequently, the diets were high in fiber and phytates, which bind zinc as well as iron. Furthermore, the bread of the region was not **leavened;** in leavened bread, yeast breaks down phytates as the bread rises.

Since the first reports, zinc deficiency has been recognized elsewhere, and it affects much more than growth. It alters digestive function profoundly and causes diarrhea, which worsens the malnutrition already present, with respect not only to zinc but to all nutrients. It drastically impairs the immune response, making infections likely.[44] Infections of the intestinal tract worsen malnutrition, including zinc malnutrition. Even mild zinc deficiency, brought on after one month of consuming a low-zinc diet, causes imbalances in the body's immune system that can increase susceptibility to infections.[45] In developing countries, zinc treatment for children with infectious diseases reduces diarrhea and death.[46]

Normal vitamin metabolism depends on zinc, so zinc-deficiency symptoms often include vitamin-deficiency symptoms. Zinc deficiency also disturbs thyroid function and slows the body's energy metabolism, causing loss of appetite and slowing wound healing. In laboratory animals, a mild deficiency may reduce physical activity, memory, and attention span.[47] The symptoms are so pervasive that when faced with zinc deficiency, physicians are more likely to diagnose it as general malnutrition and sickness than as zinc deficiency.

Although severe zinc deficiencies are not widespread in developed countries, they occur among some groups, including pregnant women, young children, the elderly, and the poor. When pediatricians or other health workers evaluating children's health note poor growth accompanied by poor appetite, they should think zinc.

leavened (LEV-end) literally, "lightened" by yeast cells, which digest some carbohydrate components of the dough and leave behind bubbles of gas that make the bread rise.

ZINC

These foods provide 10 percent or more of the zinc Daily Value (DV = 15 mg/day).[a]

[a]The Daily Values are based on a 2,000-calorie diet.

Problem: Too Much Zinc Zinc is toxic in large quantities, and zinc supplements can cause serious illness or even death in high enough doses. Regular doses of zinc only a few milligrams above the recommended intake, taken over time, block copper absorption and lower the body's copper content. In animals, this effect leads to degeneration of the heart muscle. In high doses, zinc may reduce the concentration of beneficial high-density lipoproteins (HDL) in the blood.

High doses of zinc can also inhibit iron absorption from the digestive tract. A protein in the blood that carries iron from the digestive tract to tissues also carries some zinc. If this protein is burdened with excess zinc, little or no room is left for iron to be picked up from the intestine. The opposite is also true: too much iron leaves little room for zinc to be picked up, thus impairing zinc absorption. Zinc and iron are often found together in foods, but food sources are safe and never cause imbalances in the body. Supplements, in contrast, can easily do so.

Unlike excess iron, excess zinc has a normal escape route from the body. The pancreas secretes zinc-rich juices into the digestive tract, and some of these are excreted. Still, large doses from zinc supplements can overwhelm the escape route and cause toxicity. Small doses of zinc may be relatively nontoxic compared to other more hazardous trace minerals, but supplements should be approached with caution.[48] Zinc lozenges sold for relief of sore throats and colds may be safe if used sparingly. Mixed results are achieved with zinc lozenges so don't take higher doses when the lozenges fail to bring relief.[49] The Tolerable Upper Intake Level for adults is 40 milligrams per day.

How old does the boy in the picture appear to be? He is 17 years old but is only 4 feet tall, the height of a seven-year-old in the United States. His genitalia are like those of a six-year-old. The retardation is rightly ascribed to zinc deficiency because it is partially reversible when zinc is restored to the diet. The photo was taken in Egypt.

H. Sanstead, University of Texas-Galveston

Food Sources of Zinc Meats, shellfish, and poultry are among the top providers of zinc (see Snapshot 8-6). Among plant sources, some legumes and whole grains are rich in zinc, but the zinc is not as well absorbed as from meat. Most people meet the recommended 11 milligrams per day for men and 8 milligrams per day for women.[50] Vegetarians are advised to eat varied diets that include whole-grain breads well leavened with yeast, which helps make zinc available for absorption.

key point *Zinc assists enzymes in all cells. Deficiencies in children cause growth retardation with sexual immaturity. Zinc supplements can reach toxic doses, but zinc in foods is nontoxic. Animal foods are the best sources.*

Selenium

Selenium has earned the attention of the world's scientists for its role in protecting vulnerable body chemicals against oxidative destruction. Selenium assists a group of enzymes that, in concert with vitamin E, works to prevent the formation of free radicals and prevent oxidative harm to cells and tissues.[51] For example, cells of the immune system generate oxidizing compounds when they destroy foreign invaders, and the selenium-dependent enzymes reduce these compounds to harmless byproducts that can be safely metabolized by body tissues. Selenium also plays roles in activating thyroid hormone, the hormone that regulates the body's rate of metabolism.

Whether selenium protects against the development of some forms of cancers is under investigation, and some, but not all, preliminary results seem encouraging.[52] See the Controversy in Chapter 7 for news about selenium's possible role in cancer prevention.

A deficiency of selenium can open the way for a specific type of heart disease (unrelated to the heart disease discussed in Chapters 5 and 11). The condition, first identified in China among people from areas with selenium-deficient soils, prompted researchers to place this mineral among the essential nutrients (see the inside front cover for selenium's intake recommendation). Foods grown on U.S. and Canadian soils supply plenty of selenium.

If you eat a normal diet composed of mostly unprocessed foods, you do not need to worry about selenium. It is widely distributed in foods such as meats and shellfish and in vegetables and grains grown on selenium-rich soil. Most people in the United States receive plenty of selenium because they eat supermarket foods transported from many regions and because they eat meat, a food rich in selenium.

Toxicity is possible when people take selenium supplements over a long period. Selenium toxicity brings on symptoms such as hair loss, diarrhea, and nerve abnormalities. The Tolerable Upper Intake Level for selenium is set at 400 micrograms per day.

key point *Selenium works with an enzyme system to protect body compounds from oxidation. A deficiency induces a disease of the heart. Deficiencies are rare in developed countries, but toxicities occur from overuse of supplements.*

fluorosis (floor-OH-sis) discoloration of the teeth due to ingestion of too much fluoride during tooth development.

Fluoride

Fluoride is not essential to life, but it is beneficial in the diet because of its ability to inhibit the development of dental caries in both children and adults. Only a trace of fluoride occurs in the human body, but the crystalline deposits in bones and teeth are larger and more perfectly formed because this fluoride replaces the hydroxy portion of hydroxyapatite, forming the more decay-resistant fluorapatite in developing teeth. Once teeth have erupted through the gums, fluoride helps prevent dental caries by promoting the remineralization of early lesions of the enamel that might otherwise progress to form caries. Fluoride also acts directly on the bacteria of plaque, suppressing their metabolism and reducing the amount of acid they produce.[53]

Drinking water is the usual source of fluoride. In communities where the water contains too much—2 to 8 parts per million—discoloration of the teeth, or **fluorosis,** may occur. Fluorosis occurs only during tooth development, never after the teeth have formed—and it is irreversible. Widespread availability of fluoridated toothpaste and mouthwash, foods made with fluoridated water, and fluoride-containing supplements has led to an increase in the mildest form of fluorosis.[54] In this condition, characteristic white spots form in the tooth enamel; a more severe form is shown in Figure 8-10. To prevent fluorosis, people in areas with fluoridated water should limit other sources, such as fluoride supplements for infants or children, unless prescribed by a physician. Children younger than six years should use only a pea-sized squeeze of toothpaste and should be taught not to swallow their toothpaste when brushing their teeth.

Fluoride helps prevent caries in three ways:

In developing teeth:
- Forms decay-resistant crystals.

In erupted teeth:
- Promotes remineralization.
- Reduces acidity of plaque.

Figure 8•10

FLUOROSIS

The brown mottled stains on these teeth indicate exposure to high concentrations of fluoride during development.

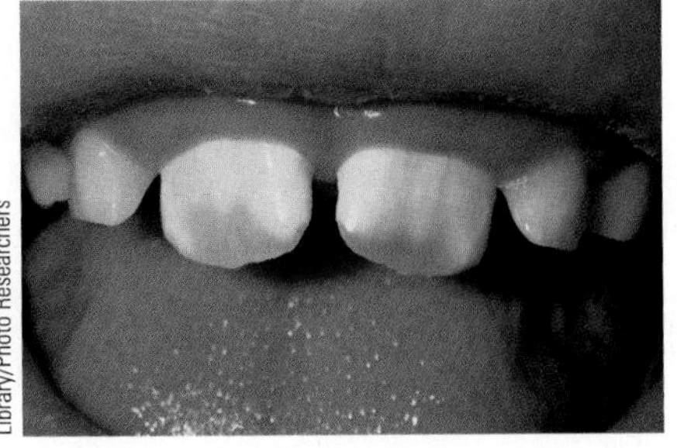

The Tolerable Upper Intake Level for fluoride for all people older than 8 years is 10 milligrams per day.

To prevent fluorosis, young children should not swallow toothpaste.

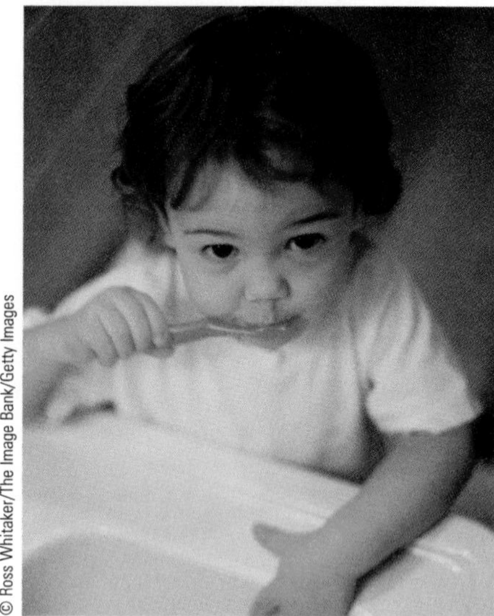

© Ross Whitaker/The Image Bank/Getty Images

Where fluoride is lacking, the incidence of dental decay is very high. Fluoridation of water to raise fluoride concentration to 1 part per million is recommended as an important public health measure.[55] Fluoridation is a practical, safe, and cost-effective way to help prevent dental caries in the young, and its widespread use has been a major factor in reducing U.S. dental caries.[56] Sufficient fluoride during the tooth-forming years of infancy and childhood gives lifetime protection against tooth decay. Some uninformed fluoride opponents claim that communities using fluoridated water have an increased cancer rate, but studies show no connection. Based on the accumulated evidence of its beneficial effects, fluoridation has been endorsed by the National Institute of Dental Health, the American Dietetic Association, the American Medical Association, the National Cancer Institute, and the Centers for Disease Control and Prevention. Figure 8-11 shows the percentage of the population in each state with access to fluoridated water.

Fluoride stabilizes bones and makes teeth resistant to decay. Excess fluoride discolors teeth; large doses are toxic.

Chromium

Chromium works closely with the hormone insulin to regulate and release energy from glucose. When chromium is lacking, insulin action is impaired, resulting in a diabetes-like condition of high blood glucose that resolves with chromium supplementation. Supplements of chromium cannot cure the common forms of diabetes, but researchers are currently investigating whether such supplements may help in the management of type 2 diabetes.[57] Diets high in simple sugars and low in whole, nutrient-dense foods deplete the body's supply of chromium.

Unfortunately, chromium-containing supplements will *not* build extra muscle tissue or melt off body fat or ward off its regain as popular magazines may profess.[58] Chromium supplements have been reported to slightly increase lean body mass in laboratory animals and sometimes in human beings tested under laboratory conditions. These results led to exaggerated claims for chromium's ability to bring about weight loss and muscle gain, but follow-up studies have shown no effect of chromium on body fat and lean tissue. Likewise, chromium supplementation does not lower blood cholesterol in people, an effect that has been implied in the "information" provided by supplement marketers.

Chromium compounds used in various industrial processes are known carcinogens and are responsible for many cases of cancer in exposed workers.[59] The form of chromium in foods and supplements is nontoxic by comparison, and amounts of 200 micrograms per day seem to be safe.[60] Supplements may cause skin eruptions, however, and taking large doses is ill-advised.[61]

Chromium is widely distributed in the food supply, especially in unrefined foods and whole grains. It exists in foods in complexes with other compounds that make it easily controlled and used by the body. Researchers use the terms *biologically active chromium* or *glucose tolerance factor* to describe these chromium-containing compounds.[62]

Although chromium is present in a variety of foods, it is estimated that 90 percent of U.S. adults consume less than the recommended minimum intake of 50 micrograms a day. Chromium is lost during food processing, and chromium deficiencies become more likely as people depend more heavily on refined foods. The best chromium food sources are liver, whole grains, nuts, and cheeses.

Chromium works with the hormone insulin to control blood glucose concentrations. Chromium is present in a variety of unrefined foods.

Figure 8•11

PERCENTAGE OF POPULATIONS WITH ACCESS TO FLUORIDATED WATER THROUGH PUBLIC WATER SYSTEMS

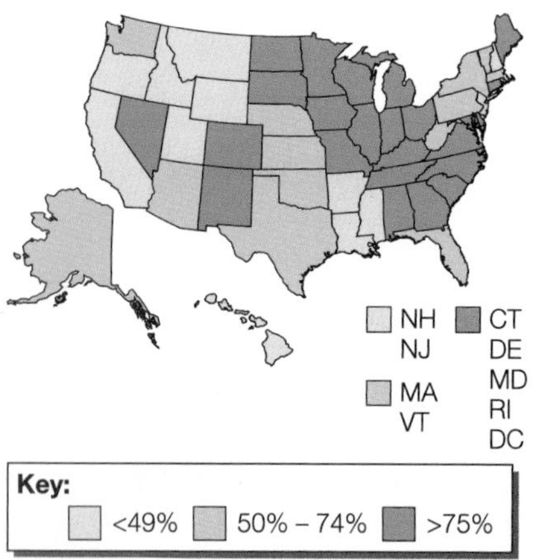

SOURCE: Centers for Disease Control and Prevention, Recommendations for using fluoride to prevent and control dental caries in the United States, *Morbidity and Mortality Weekly Report* (supplement), 17 August 2001, p. 10.

Copper

One of copper's most vital roles is to help form hemoglobin and collagen. In addition, many enzymes depend on copper for its oxygen-handling ability. Copper, like iron, assists in reactions leading to the release of energy. It also works with proteins to regulate the activity of certain genes.[63] One copper-dependent enzyme helps to control damage from free-radical activity in the tissues.* Researchers are investigating the possibility that a low-copper diet may contribute to heart disease by suppressing the activity of this enzyme.

Copper deficiency is rare but not unknown: it has been seen in severely malnourished infants fed a copper-poor formula.[64] Deficiency can severely disturb growth and metabolism, and in adults, it can impair immunity and blood flow through the arteries. Excess zinc interferes with copper absorption and can cause deficiency.

Copper toxicity from foods is unlikely, but supplements can cause it—the Tolerable Upper Intake Level for adults is set at 10,000 micrograms (10 milligrams) per day. The best food sources of copper include organ meats, seafood, nuts, and seeds. Water may also supply copper, especially where copper plumbing pipes are used.[65] In the United States, copper intakes are thought to be adequate.[66]

Copper is needed to form hemoglobin and collagen and assists in many other body processes. Copper deficiency is rare.

Other Trace Minerals and Some Candidates

DRI intake recommendations have been established for two other trace minerals, molybdenum and manganese. Molybdenum functions as part of several metal-containing enzymes, some of which are giant proteins. Manganese works with dozens of different enzymes that facilitate body processes.

Several other trace minerals are now recognized as important to health. Research suggests that a low intake of boron may enhance susceptibility to osteoporosis by way of its effects on calcium metabolism. The richest food sources of boron are noncitrus fruits, leafy vegetables, nuts, and legumes. Cobalt is the mineral in the large vitamin B_{12} molecule; the alternative name for vitamin B_{12}, *cobalamin*, reflects cobalt's presence. Nickel is important for the health of many body tissues; deficiencies harm the liver and other organs. Silicon is known to be involved in bone calcification in animals. Future research may reveal key roles played by other trace minerals including barium, cadmium, lead, lithium, mercury, silver, tin, and vanadium. Even arsenic, a known poison and carcinogen, may turn out to be essential in tiny quantities.

All trace minerals are toxic in excess, and Tolerable Upper Intake Levels exist for boron, nickel, and vanadium (see the inside front cover). Overdoses are most likely to occur in people who take multiple nutrient supplements. The way to obtain the trace minerals is from food, which is not hard to do—just eat a variety of whole foods in the amounts recommended in the Food Guide Pyramid. Some claim that organically grown foods contain more trace minerals than those grown with chemical fertilizers. Organic fertilizers do contain more trace minerals than do refined chemical fertilizers, and plants do take up some of the minerals they are given; Controversy 14 considers the merits and demerits of foods grown organically.

Research on the trace minerals is uncovering many interactions among them: an excess of one may cause a deficiency of another. A slight manganese overload, for example, may aggravate an iron deficiency. A deficiency of one mineral may open the way for another to cause a toxic reaction. Iron deficiency, for example, makes the body much more susceptible to lead poisoning. Good food sources of one are poor food sources of another, and factors that cooperate with some trace elements oppose others. Vitamin C, for example, enhances the absorption of iron and depresses that of copper. The continuous outpouring of new information about the trace minerals is a sign that we have much more to learn. Table 8-10 on pages 296–297 sums up what this chapter has said about the minerals and fills in some additional information.

Many different trace elements play important roles in the body. All of the trace minerals are toxic in excess.

*The enzyme is superoxide dismutase, first mentioned in Controversy 7.

Table 8•10

THE MINERALS—A SUMMARY

Mineral and Chief Functions in the Body	Deficiency Symptoms	Toxicity Symptoms	Significant Sources
Major Minerals			
Calcium			
The principal mineral of bones and teeth. Also acts in normal muscle contraction and relaxation, nerve functioning, regulation of cell activities, blood clotting, blood pressure, and immune defenses.	Stunted growth in children; adult bone loss (osteoporosis).	Constipation; urinary tract stone formation; kidney dysfunction; interference with absorption of other minerals.	Milk and milk products, oysters, small fish (with bones), tofu (bean curd), greens, legumes.
Phosphorus			
Phosphorus is important in cells' genetic material, in cell membranes as phospholipids, in energy transfer, and in buffering systems.	Appetite loss, bone pain, muscle weakness, impaired growth, and rickets in infants.[a]	Excess phosphorus may cause calcium excretion.	All animal tissues.
Magnesium			
A factor involved in bone mineralization, the building of protein, enzyme action, normal muscular contraction, transmission of nerve impulses, and maintenance of teeth.	Weakness; muscle twitches; appetite loss; confusion; depressed pancreatic hormone secretion; if extreme, convulsions, bizarre movements (especially of eyes and face), hallucinations, and difficulty in swallowing. In children, growth failure.[b]	Excess magnesium from abuse of laxatives (Epsom salts) causes diarrhea with fluid and electrolyte imbalances.	Nuts, legumes, whole grains, dark green vegetables, seafoods, chocolate, cocoa.
Sodium			
Sodium, chloride, and potassium (electrolytes) maintain cells' normal fluid balance and acid-base balance in the body. Sodium is critical to nerve impulse transmission.	Muscle cramps, mental apathy, loss of appetite.	Hypertension.	Salt, soy sauce, processed foods.
Potassium			
Potassium facilitates reactions, including the making of protein; the maintenance of fluid and electrolyte balance; the support of cell integrity; the transmission of nerve impulses; and the contraction of muscles, including the heart.	Deficiency accompanies dehydration; causes muscular weakness, paralysis, and confusion; can cause death.	Causes muscular weakness; triggers vomiting; if given into a vein, can stop the heart.	All whole foods: meats, milk, fruits, vegetables, grains, legumes.
Chloride			
Chloride is also part of the hydrochloric acid found in the stomach, necessary for proper digestion.	Growth failure in children; muscle cramps, mental apathy, loss of appetite; can cause death (uncommon).	Normally harmless (the gas chlorine is a poison but evaporates from water); can cause vomiting.	Salt, soy sauce; moderate quantities in whole, unprocessed foods, large amounts in processed foods.

[a]Seen only rarely in infants fed phosphorus-free formula or in adults taking medications that interact with phosphorus.

[b]A still more severe deficiency causes tetany, an extreme, prolonged contraction of the muscles similar to that caused by low blood calcium.

Table 8•10

THE MINERALS—A SUMMARY—(Continued)

Mineral and Chief Functions in the Body	Deficiency Symptoms	Toxicity Symptoms	Significant Sources
Major Minerals			
Sulfur			
A component of certain amino acids; part of the vitamins biotin and thiamin and the hormone insulin; combines with toxic substances to form harmless compounds; stabilizes protein shape by forming sulfur-sulfur bridges (see Figure 6-10 in Chapter 6).	None known; protein deficiency would occur first.	Would occur only if sulfur amino acids were eaten in excess; this (in animals) depresses growth.	All protein-containing foods.
Trace Minerals			
Iodine			
A component of the thyroid hormone thyroxine, which helps to regulate growth, development, and metabolic rate.	Goiter, cretinism.	Depressed thyroid activity; goiter-like thyroid enlargement.	Iodized salt; seafood; bread; plants grown in most parts of the country and animals fed those plants.
Iron			
Part of the protein hemoglobin, which carries oxygen in the blood; part of the protein myoglobin in muscles, which makes oxygen available for muscle contraction; necessary for the use of energy.	Anemia: weakness, pallor, headaches, inability to concentrate, impaired cognitive function (children), lowered cold tolerance.	Iron overload: fatigue, infections, liver injury, possible increased risk of colon cancer, growth retardation in children, acidosis, bloody stools, shock.	Red meats, fish, poultry, shellfish, eggs, legumes, dried fruits.
Zinc			
Part of insulin and many enzymes; involved in making genetic material and proteins, immune reactions, transport of vitamin A, taste perception, wound healing, the making of sperm, and normal fetal development.	Growth failure in children, dermatitis, sexual retardation, loss of taste, poor wound healing.	Fever, nausea, vomiting, diarrhea, muscle incoordination, dizziness, anemia, accelerated atherosclerosis, kidney failure.	Protein-containing foods: meats, fish, shellfish, poultry, grains, vegetables.
Selenium			
Assists a group of enzymes that defend against oxidation.	Predisposition to a form of heart disease characterized by fibrous cardiac tissue (uncommon).	Nausea; abdominal pain; nail and hair changes; nerve, liver, and muscle damage.	Seafoods, organ meats; other meats, whole grains, and vegetables depending on soil content.
Fluoride			
Helps form bones and teeth; confers decay resistance on teeth.	Susceptibility to tooth decay.	Fluorosis (discoloration) of teeth, nausea, vomiting, diarrhea, chest pain, itching.	Drinking water if fluoride containing or fluoridated; tea; seafood.
Chromium			
Associated with insulin; needed for energy release from glucose.	Abnormal glucose metabolism.	Possibly skin eruptions.	Meat, unrefined grains, vegetable oils.
Copper			
Helps form hemoglobin; part of several enzymes.	Anemia; bone abnormalities.	Vomiting, diarrhea; liver damage.	Organ meats, seafood, nuts, seeds, whole grains, drinking water.

Meeting the Need for Calcium

kefir a yogurt-based beverage.

nori a type of seaweed popular in Asian, particularly Japanese, cooking.

Figure 8•12

FOOD SOURCES OF CALCIUM IN THE U.S. DIET

Milk and milk products contribute almost three-quarters of the calcium in a typical U.S. diet.

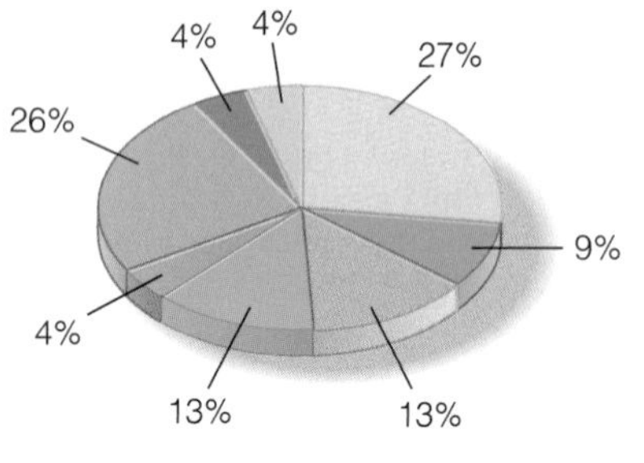

SOURCE: Data from National Dairy Council, 2001.
[a]Breads and most cereals provide only small amounts.

THE AVERAGE WOMAN consumes just a third of her recommended amount of calcium; men do somewhat better, with calcium intakes close to three-fourths of the recommendation. Low calcium intakes are associated with all sorts of major illnesses, including adult bone loss (see the following Controversy), high blood pressure and colon cancer (see Chapter 11), kidney stones, and even lead poisoning.

Consumption of one of the best sources of calcium—milk—has decreased in recent years while consumption of other beverages, such as soft drinks, has increased dramatically. One national dairy group calls the situation a "national calcium crisis," and 250 representatives of many national health and nutrition organizations agree.[67] This Food Feature focuses on sources of calcium in the diet and provides guidance about how to meet the need for calcium.

Milk, Yogurt, and Cheese Group

Milk and milk products are traditional sources of calcium for people who can tolerate them (see Figure 8-12). Table 8-11 shows the current milk recommendations that help to meet the calcium needs of various age groups. People who do not use milk because of lactose intolerance, dislike, or allergy must obtain calcium from other sources. Care is needed, though; *wise* substitutions must be made. Most of milk's many relatives are recommended choices: yogurt, **kefir,** buttermilk, cheese (especially the low-fat or fat-free varieties), and, for people who can afford the calories, ice milk. Cottage cheese and frozen yogurt desserts contain about half the calcium of milk, with 2 cups being equivalent in calcium to 1 cup of milk. Butter, cream, and cream cheese are almost pure fat and contain negligible calcium.

Tinker with milk products to make them more appealing. Add cocoa to milk and fruit to yogurt, make your own fruit smoothies from milk or yogurt, or add fat-free milk powder to any dish. The cocoa powder added to make chocolate milk does contain a small amount of oxalic acid, which binds with some of milk's calcium and inhibits its absorption, but the effect is insignificant. Sugar lends both sweetness and calories to chocolate milk, so mix your chocolate milk at home where you control the amount of sugary chocolate added to the milk.

Table 8•11

SUGGESTED MINIMUM DAILY FLUID MILK INTAKES

Young children	2 cups
Teenagers	3 cups
Adults	2 cups
Pregnant or lactating women	3 cups
Pregnant or lactating teens	4 cups
Women past menopause	3 cups

Vegetables

Among vegetables, rutabaga, broccoli, beet greens, turnip greens, mustard greens, bok choy (a Chinese cabbage), and kale are good sources of available calcium. So are collard greens, green cabbage, kohlrabi, watercress, parsley, and probably some seaweeds, such as the **nori** popular in Japanese cookery. Certain other foods, including spinach, Swiss chard, and rhubarb, appear equal to milk in calcium content but provide very little or no calcium to the body because they contain binders that prevent calcium's absorption (see Figure 8-13). The presence of calcium binders does not make spinach an inferior food.

Chocolate milk is an excellent source of calcium for those who can afford the extra calories.

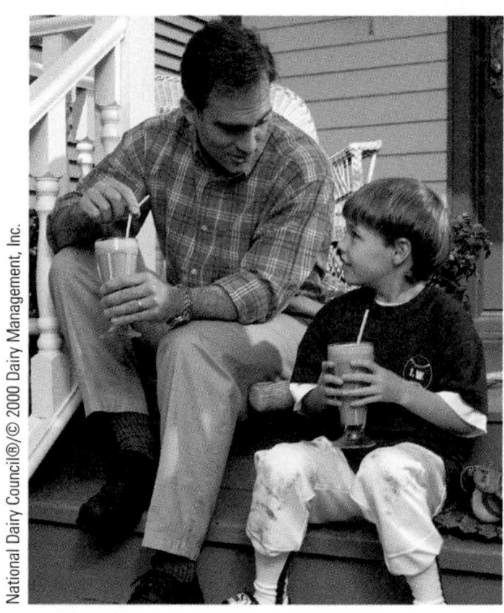

National Dairy Council®/© 2000 Dairy Management, Inc.

Spinach is also rich in iron, beta-carotene, and dozens of other essential nutrients and potentially helpful phytochemicals. Just don't rely on it for calcium. Dark greens of all kinds are superb sources of riboflavin and indispensable for the vegan or anyone else who does not drink milk.

Calcium in Other Foods

For the many people who cannot use milk and milk products, small fish such as canned sardines and other canned fishes prepared with their bones are rich sources of calcium. Stocks or extracts made from bones are another rich source. The Vietnamese tradition of making fish stock helps account for their adequate calcium intake without the use of milk. Almonds are a high-energy calcium source—one-third cup supplies about 100 milligrams of calcium along with almost 300 calories. Calcium-rich mineral water may also be a useful calcium source. Recent evidence seems to indicate that the calcium from mineral water, including hard tap water, may be as absorbable as the calcium from milk.[68] Many other foods contribute smaller, but still significant, amounts of calcium to the diet.

Figure 8•13

CALCIUM ABSORPTION FROM FOOD SOURCES

≥ 50% absorbed	cauliflower, watercress, Chinese cabbage, head cabbage, brussels sprouts, rutabaga, kolhrabi, kale, mustard greens, bok choy, broccoli, turnip greens
≃ 30% absorbed	milk, calcium-fortified soy milk, calcium-set tofu, calcium-fortified juices and drinks
≃ 20% absorbed	almonds, sesame seeds, beans (pinto, red, and white)
≤ 5% absorbed	spinach, rhubarb, Swiss chard

Calcium-Fortified Foods

Next in order of preference among non-milk sources of calcium are foods that contain large amounts of calcium salts by an accident of processing or by intentional fortification. In the processed category are soybean curd, or tofu (calcium salt is often used to coagulate it, so check the label); canned tomatoes (firming agents donate 63 milligrams per cup of tomatoes); **stone-ground flour** and self-rising flour; stone-ground whole and self-rising cornmeal; and blackstrap molasses.

Some food products available to U.S. consumers are fortified to add calcium to people's diets. The richest in calcium is high-calcium milk, that is, milk with extra calcium added; it provides more calcium per cup than any natural milk, 500 milligrams per 8 ounces. Then comes calcium-fortified orange juice, with 300 milligrams per 8 ounces, a good choice because the bioavailability of its calcium is comparable to that of milk. Calcium-fortified soy milk can also be prepared so that it contains more calcium than whole cow's milk. Soy-based infant formula is fortified with calcium, and no law prevents adults from using it in cooking for themselves.

Finally, calcium supplements are available, sold mostly to people hoping to ward off osteoporosis. Controversy 8 points out, however, that, while often useful, supplements are not magic bullets against bone loss.

Making Meals Rich in Calcium

Many cooks slip extra calcium into meals by sprinkling a tablespoon or two of fat-free dry milk into almost everything. The added calorie value is small, changes to taste and texture of the dish are practically nil, but each 2 tablespoons adds about 100 extra milligrams of calcium and moves people closer to meeting the recommendation to obtain about 2 servings of milk each day (see Figure 8-14). Here are some more tips for including calcium-rich foods in your meals:

At Breakfast

- Choose calcium-fortified orange or vegetable juice.
- Serve tea or coffee, hot or iced, with milk.

Figure 8•14

MILK, YOGURT, AND CHEESE GROUP: FOOD SUPPLY SERVINGS, 1970–1996[a]

On average, people in the United States fall far short of meeting the recommendation to obtain 2 or 3 servings of milk, yogurt, or cheese each day. The picture is worse for the dark green vegetables that supply calcium—only 3 percent of the vegetables consumed each day meet this description.

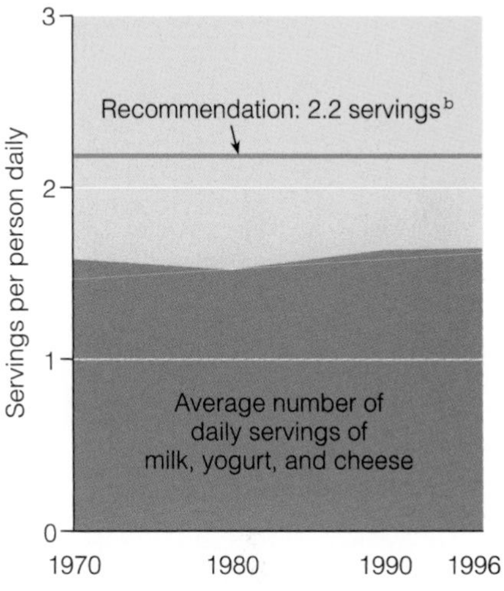

[a]Includes all forms of milk, yogurt, cheese, and frozen dairy desserts.

[b]Recommended servings based on weighted average of recommended servings for different age groups of the U.S. population, excluding the higher needs of pregnant and lactating women.

SOURCE: U.S. Department of Agriculture, Economic Research Service, 1998.

stone-ground flour flour made by grinding kernels of grain between heavy wheels made of limestone, a kind of rock derived from the shells and bones of marine animals. As the stones scrape together, bits of the limestone mix with the flour, enriching it with calcium.

- Choose cereals, hot or cold, with milk.
- Cook hot cereals with milk instead of water; then mix in 2 tablespoons of fat-free dry milk.
- Make muffins or quick breads with milk and extra powdered milk.
- Add milk to scrambled eggs.
- Moisten cereals with flavored yogurt.

At Lunch

- Add low-fat cheeses to sandwiches, burgers, or salads.
- Use a variety of green vegetables, such as watercress or kale, in salads and on sandwiches.
- Drink fat-free milk or calcium-fortified soy milk as a beverage or in a smoothie.
- Drink calcium-rich mineral water as a beverage (studies suggest significant calcium absorption).
- Marinate cabbage shreds or broccoli spears in low-fat Italian dressing for an interesting salad.
- Choose coleslaw over potato and macaroni salads.
- Mix the mashed bones of canned salmon into salmon salad or patties.
- Eat sardines with their bones.
- Stuff potatoes with broccoli and low-fat cheese.
- Try pasta such as ravioli stuffed with low-fat ricotta cheese instead of meat.
- Sprinkle parmesan cheese on pasta salads.

At Supper

- Toss a handful of thinly sliced green vegetables, such as kale or young turnip greens, with hot pasta; the greens wilt pleasingly in the steam of the freshly cooked pasta.
- Serve a green vegetable every night and try new ones—how about kohlrabi? It tastes delicious when cooked like broccoli.
- Learn to stir-fry Chinese cabbage and other Asian foods.
- Try tofu (the calcium-set kind); this versatile food has inspired whole cookbooks devoted to creative uses.
- Add fat-free powdered milk to almost anything—meat loaf, sauces, gravies, soups, stuffings, casseroles, blended beverages, puddings, quick breads, cookies, brownies. Be creative.
- Choose frozen yogurt, ice milk, or custards for dessert.

Here is a shortcut for tracking the amount of calcium in a day's meals. To start, memorize these two facts.

1. A cup of milk provides about 300 milligrams of calcium.
2. Adults need 1,000 to 1,200 milligrams each day. Broken down in terms of "cups of milk," the need is 3½ to 4 cups each day.

To estimate calcium from an entire day's foods, not just milk, assign "cups of milk" points to various calcium sources. The goal is to achieve 3½ to 4 points per day:

- 1 point = 1 cup milk, yogurt, calcium-fortified beverage, or 1½ ounces of cheese.
- 1 point = 4 ounces of canned fish with bones.
- ½ point = 1 cup ice cream, cottage cheese, or calcium-rich vegetables (see the discussion under Vegetables above).

Also, because bits of calcium are present in many foods (a bagel has about 50 milligrams, for example):

- 1 point = a well-balanced, adequate, and varied diet.

Example: Say a day's calcium-rich foods include cereal and a cup of milk, a ham and cheese sandwich, and broccoli and pasta salad:

1 point (cup of milk)
+ 1 point (cheese)
+ ½ point (broccoli)
= 2½ points.

Add 1 point for the other foods eaten that day:

1 point + 2½ points = 3½ points.

This day's foods provided a calcium intake approximately equal to the lower of the DRI committee's recommendations. The tips in this section have suggested many ways to aim higher. The Do It! section provides practice at finding calcium and other minerals in foods when you are in a hurry—from a convenience store.

FIND THE MINERALS IN SNACK FOODS

In an ideal world, you would set aside time every day for planning, shopping, and cooking nutritious meals. In the real world, you've had nothing to eat, your research papers are due, your room is topsy turvy, and your car needs fuel as you're rushing to class. A convenience store seems like a good idea; you can fill up your car and grab a bite to eat in one stop. What you grab makes a difference to your day's calorie and mineral intakes, however—to the benefit or detriment of your nutritional health.

The labels on the munchies list their contents of energy, calcium, iron, and sodium. For the sake of learning, we've added two other minerals to this exercise: magnesium and zinc.

List Your Nutrient Intake Goals and Choose Some Foods

Step 1. List your intake goals for energy and calcium, iron, zinc, and magnesium (see the DRI, inside front cover) as indicated on Form 8-1. For sodium, use the recommended upper limit of 2,400 milligrams as your target for the day.

Step 2. Scan the variety of items on the convenience store shelves represented in Figure 8-15. Choose a snack and enter your choices on Form 8-1. The portion sizes are those commonly used for such foods, and not those recommended in the Food Guide Pyramid. For example, a can of vienna sausages contains 5 ounces of sausages, not the recommended 3 ounces.

Record Energy and Mineral Values

Step 3. Obtain energy and mineral values for your snack choices from Table 8-12 (next page) and enter these values on Form 8-1. Total the columns for energy and minerals.

Step 4. Calculate the percentages of energy and mineral requirements contributed by this snack. Divide each nutrient total by the recommendation for that nutrient, and multiply by 100. Example: For a snack providing 84

Figure 8•15

SNACKS AT THE FUEL 'N' FEED

Snacks at the *Fuel 'n' Feed*

Sandwiches/ Canned foods	Snacks
Cheese pizza slice	Apple pie, packaged
Ham biscuit	Banana
Pork and beans	Cheese slice
Roast beef sandwich	Chocolate candy bar
Sardines	Dill pickle
Vienna sausage	Dried fruit mix
	Fig bar cookies
	Ice cream and sherbet bar
	Potato chips, small bag
	Pretzels, regular
Beverages	Pretzels, unsalted
Milk, 1% fat	Roasted almonds
Chocolate milk, 2% fat	Sunflower seeds, shelled
Cola	Wheat crackers
Pineapple orange juice	Yogurt, low fat, with fruit

Form 8•1

ENERGY AND MINERAL TALLY

Your Snack Foods	Energy (cal)	Calcium (mg)	Iron (mg)	Magnesium (mg)	Sodium (mg)	Zinc (mg)
Snack totals						
Personal intake goals[a]					2,400	
% of recommendation						

[a]See inside front cover.

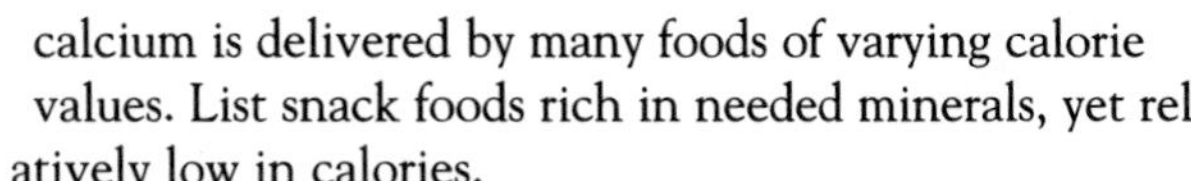

milligrams of calcium with an intake recommendation of 1,000 milligrams:

$$(84 \div 1{,}000) \times 100 = 8.4\%.$$

The snack would meet more than 8 percent of your DRI recommended intake.

Analysis

Now answer these questions:

1. What percentage of your energy need did the snack contribute? Did the snack also contribute a proportional amount of calcium, iron, magnesium, or zinc?
2. What about sodium? Did the sodium in this snack exceed 10 percent of the maximum of 2,400 milligrams? Did it exceed 30 percent? If so, look for low-sodium supper choices to round out the day. Consult Figure 8-8 to review the principles of high- and low-sodium foods.
3. Which of the foods listed in Table 8-12 are the best "bargains" in terms of providing less sodium and more of the other minerals? A food that supplies much of the daily sodium allowance and few other nutrients may support nutrition goals less well than a similar food with less sodium. List some high- and low-sodium choices from among the foods.
4. Did you find any "bargains" among the foods in terms of providing fewer calories and more minerals? For example, calcium is delivered by many foods of varying calorie values. List snack foods rich in needed minerals, yet relatively low in calories.
5. Where is the calcium in your snack foods? List the snack foods on your list that supply substantial calcium (10 percent of the recommendation). If your choices lack calcium, look back and find some calcium sources on the snack list.
6. Can a person looking for iron do well in a convenience store? Which foods or ingredients contribute iron? See Snapshot 8-5 for good iron sources.
7. Did any of your choices provide 10 percent or more of your magnesium need? Magnesium is easily lost in processing foods and is difficult to obtain from highly processed snacks.
8. Consider the zinc in your snack. Did any foods provide 10 percent or more of the daily recommendation? If so, which ones? If not, consult Snapshot 8-6 on page 292 for ideas about the kinds of foods that are good sources of zinc, and list some snack foods likely to contain significant amounts.
9. What might a whole day's mineral intake look like if a person ate convenience foods at every meal?

Americans are snacking more as the pace of life quickens. The next chapter makes clear that your choices of snack foods can make a difference of several hundred calories a day to energy intake.

Table 8•12

ENERGY AND MINERAL CONTENTS OF SELECTED CONVENIENCE STORE FOODS

	Energy (cal)	Calcium (mg)	Iron (mg)	Magnesium (mg)	Sodium (mg)	Zinc (mg)
Cheese pizza, slice	243	184	0.7	16	467	0.8
Ham biscuit	386	160	2.7	23	1,432	1.7
Pork and beans, 5 oz	139	80	4.7	50	624	8.4
Roast beef sandwich	318	47	3.1	22	1,252	3.0
Sardines, 3¾ oz	221	405	3.1	41	536	1.4
Vienna sausages, 5 oz	395	14	1.3	10	1,347	2.3
Milk, 1% fat, 1 c	102	300	0.1	34	123	1.0
Low-fat chocolate milk, 2% fat, 1 c	179	285	0.6	33	151	1.0
Cola, 12 oz	186	14	0.1	5	18	0.0
Pineapple orange drink, 12 oz	170	17	0.9	20	10	0.2
Apple pie, fast-food type	225	6	1.0	6	179	0.2
Banana	105	7	0.4	33	1	0.2
Cheese slice, 1 oz	69	121	0.2	6	250	0.6
Chocolate candy bar, 1½ oz	226	84	0.6	26	36	0.6
Dill pickle	12	6	0.3	7	833	0.1
Dried fruit mix, 3¾ oz	258	40	2.9	41	19	0.5
Fig bar cookie, 4	195	36	1.3	15	214	0.2
Ice cream and sherbet bar	92	62	0.1	7	43	0.4
Potato chips, salt and vinegar, 1 oz	150	7	0.5	19	380	0.3
Pretzels, salted, 1 oz	108	10	1.2	10	486	0.2
Pretzels, unsalted, 1 oz	110	10	1.2	10	60	0.2
Roasted almonds, salted, 3¾ oz	660	326	4.0	273	828	3.3
Sunflower seeds, shelled, 3¾ oz	654	123	7.2	375	3	5.4
Wheat crackers, 1 oz	134	14	1.3	18	225	0.5
Low-fat yogurt, with fruit, 8 oz	232	345	0.2	33	133	1.7

Answers to these Self Check questions are in Appendix G.

1. Water excretion is governed by the:
   a. liver
   b. kidneys
   c. brain
   d. (b) and (c)
2. Which agency is responsible for ensuring that public water systems are tested regularly for contamination?
   a. RDA
   b. FDA
   c. EPA
   d. USDA
3. Which two minerals are the major constituents of bone?
   a. calcium and zinc
   b. phosphorus and calcium
   c. sodium and magnesium
   d. magnesium and calcium
4. A deficiency of _______ is one of the world's most common preventable causes of mental retardation.
   a. zinc
   b. magnesium
   c. selenium
   d. iodine
5. Which mineral in excess is the number one cause of fatal accidental poisonings of U.S. children under six years old?
   a. iron
   b. sodium
   c. chloride
   d. potassium
6. You can survive being deprived of water for about a week. T F
7. The best way to control salt intake is to cut down on processed and fast foods. T F
8. The most abundant mineral in the body is iron. T F
9. Dairy foods such as butter, cream, and cream cheese are good sources of calcium whereas vegetables such as broccoli are poor sources. T F
10. Bottled water must meet higher standards for purity and sanitation than U.S. tap water. T F

nutrition on the net

For further study of the topics of this chapter, access these websites and search for the phrases or words in quotation marks:

1. Find updates and quick links to these and other nutrition-related sites at our website:
   www.wadsworth.com/nutrition
2. Find information about mineral supplements at:
   http://dietary-supplements.info.nih.gov
3. Search for "minerals" at the American Dietetic Association site:
   www.eatright.org
4. To read about the quality of drinking water in your area, visit this site:
   www.epa.gov/ebtpages/
5. Learn about sodium in foods and on food labels from the Food and Drug Administration:
   www.fda.gov/fdac/foodlabel/sodium.html
6. Find tips and recipes for including more milk in the diet:
   www.whymilk.com
7. Learn about the benefits of calcium from the National Dairy Council:
   www.nationaldairycouncil.org
8. Find information about U.S. intakes of minerals and other nutrients at:
   www.cdc.gov/nchs/fastats/diet.htm
9. Search for the individual minerals by name at the U.S. Government Health Information site:
   www.healthfinder.org
10. For guidelines concerning use of "potassium iodide," search:
    www.fda.gov
11. Learn more about iron overload from the Iron Overload Diseases Association:
    www.ironoverload.org
12. Learn more about iodine deficiency and thyroid disease from the American Thyroid Association:
    www.thyroid.org

INTERNET ACTIVITY

Sales of vitamin/mineral supplements have skyrocketed. While supplementation may be appropriate for some individuals who are at risk for deficiencies or for disease prevention, many do not know why they are taking supplements but do so anyway. Do you presently take any vitamin/mineral supplements? If so, what nutrients are they and in what quantities? Why do you take them?

1. Go to the website of the American Dietetic Association at:
   http://www.eatright.org/nfs/nfs81.html
2. Take the Vitamin / Mineral Supplement Quiz.

What was your score? What was your opinion of the results? Were you surprised?

CONTROVERSY 8

OSTEOPOROSIS: CAN LIFESTYLE CHOICES REDUCE THE RISKS?

MORE THAN 28 million people in the United States—many of them women—are suffering from or are developing osteoporosis.[1] A dangerous misconception is that men are immune to this devastation of the later years.[2] Each year, a million and a half people—a quarter of them men—suffer broken hips, pelvis, legs, arms, hands, and ankles attributable to osteoporosis.[3] Of these, hip fractures prove most serious. The break is rarely clean; the bone explodes into fragments that cannot be reassembled. Just removing them is a struggle, and replacing them with an artificial joint requires major surgery. Many elderly people with hip fracture never walk or live independently again. About a fifth die from related complications within a year. Both men and women are urged to do whatever they can to prevent fractures related to osteoporosis.

Fractures from osteoporosis occur during the later years, but osteoporosis itself develops silently much earlier.[4] Few young adults are aware that osteoporosis is sapping the strength of their bones; then suddenly, 40 years later, the hip gives way. People say, "She fell and broke her hip," but in fact the hip may have been so fragile that it broke *before* she fell.

Photograph of sectioned bone.

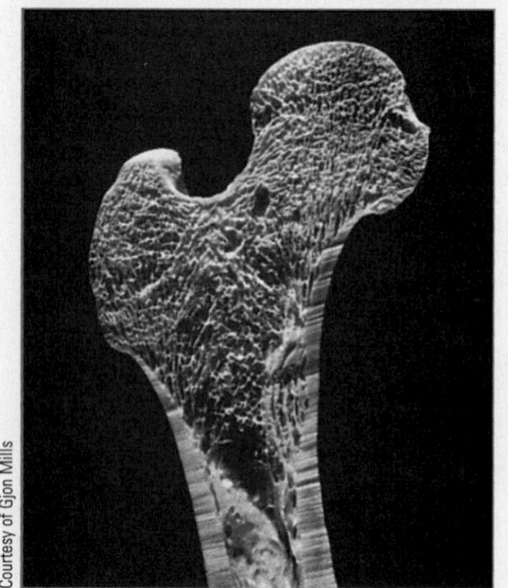

Courtesy of Gjon Mills

The causes of osteoporosis are tangled. Insufficient dietary calcium certainly plays a role, but physical activity, genetics, and other factors are also major potential players. This Controversy addresses several questions about osteoporosis: What is it? Who gets it? What can people do to reduce their risks? And where do dietary calcium and calcium supplements fit into the picture?

The Problem of Osteoporosis

To understand how the skeleton loses minerals in later years, you must first know a few things about bones. Table C8-1 offers definitions of relevant terms. The photograph on this page shows a human leg bone sliced lengthwise, exposing the lattice of calcium-containing crystals (the **trabecular bone**) inside that are part of the body's calcium bank. Invested as savings during the milk-drinking years of youth, these deposits provide a nearly inexhaustible fund of calcium. **Cortical bone** is the dense, ivorylike bone that forms the exterior shell of a bone and the shaft of a long bone (look closely at the photograph). Both types of bone are crucial to overall bone strength. Cortical bone forms a sturdy outer wall, and trabecular bone provides strength along the lines of stress.

The two types of bone handle calcium in different ways. The lacy crystals of the trabecular bone are tapped to raise blood calcium when the supply from the day's diet runs short; the calcium crystals are redeposited in bone when dietary calcium is plentiful. Trabecular bone is (1) generously supplied with blood vessels and is more metabolically active than is cortical bone; (2) is more sensitive to hormones that govern calcium deposits and withdrawals from day to day; and (3) readily gives up its minerals at the necessary rate whenever blood calcium needs replenishing. Losses of trabecular bone begin to be significant for men and women in their mid-20s. Cortical bone's calcium can also be withdrawn, but slowly. Cortical bone loss begins at about age 40, and bone tissue dwindles steadily thereafter.

As bone loss continues and osteoporosis progresses (Figure C8-1), **bone density** declines, and bones become so fragile that the body's weight can overburden the spine; vertebrae may suddenly disintegrate and crush down, painfully pinching major nerves. Or they may compress into wedges, forming what is insensitively called "dowager's hump," the posture of many men and women as they "grow shorter" (see Figure C8-2). Wrists may break as trabecula-rich bone ends weaken, and teeth may loosen or fall out as the trabecular bone of the jaw recedes.[5] As the cortical bone shell weakens as well, breaks often occur in the hip.

Table C8•1

OSTEOPOROSIS TERMS

- **bone density** a measure of bone strength; the degree of mineralization of the bone matrix.
- **cortical bone** the ivorylike outer bone layer that forms a shell surrounding trabecular bone and that comprises the shaft of a long bone.
- **trabecular** (tra-BECK-you-lar) **bone** the weblike structure composed of calcium-containing crystals inside a bone's solid outer shell. It provides strength and acts as a calcium storage bank.

Figure C8•1

LOSSES OF TRABECULAR BONE

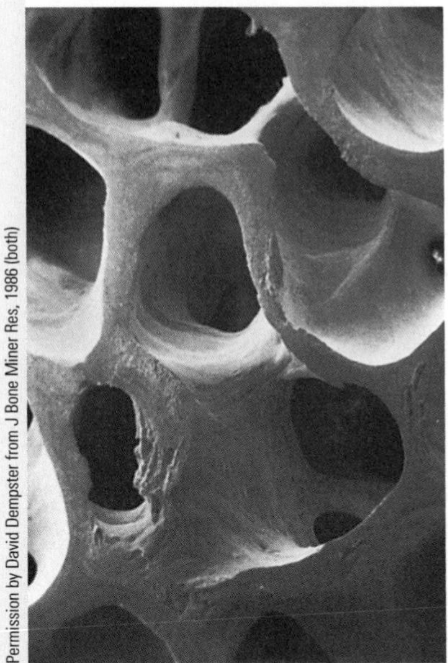

Electron micrograph of healthy trabeculae.

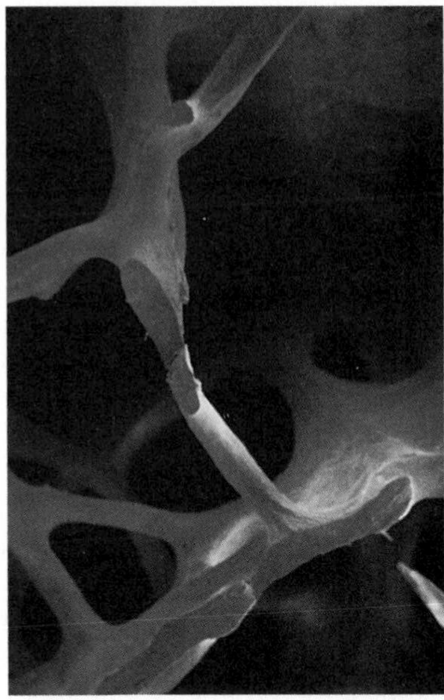

Electron micrograph of trabecular bone affected by osteoporosis.

Toward Prevention—Understanding the Causes of Osteoporosis

Scientists are searching for ways to prevent osteoporosis, but they must first discover its causes. In addition to the obvious factors of gender and advanced age, it seems that a person's chances of developing osteoporosis also depend on genetics and on the environment. Environmental factors under study for their roles in lowering bone density include:

- Poor nutrition involving calcium and vitamin D.
- Estrogen deficiency in women.
- Lack of physical activity.
- Being underweight.
- Use of tobacco and alcohol.
- Possibly, excess protein, sodium, caffeine, vitamin A, and soft drinks and inadequate vitamin K.

Genetics and Bone Density

Studies of mothers and daughters and of twins confirm that genetics plays a major role in bone density.[6] Most likely, genetic inheritance influences both the maximum bone mass possible during growth and the extent of a woman's bone loss during menopause, the time when women's estrogen production declines and menstruation ceases. The extent to which that genetic potential is realized, however, depends on individual life experiences. For example, those who attend to nutrition and physical activity attain their maximum bone density during growth, whereas those who overuse alcohol or use tobacco accelerate their bone losses.

Risks of osteoporosis also differ by race and ethnicity.[7] People of African descent have denser bones than do those of northern European extraction, and these differences are evident even before birth in X-ray images of fetuses.[8] Mexican Americans' bone density falls somewhere in between. These differences in bone density hold true for both sexes of all ages: hip fractures are three times more likely in 80-year-old white women than in black women of the same age.

Other ethnic groups have lower bone densities than do northern Europeans. Asians from China and Japan, Hispanic people from Central and South America, and Inuit people from St. Lawrence Island all have lower bone density than do northern Europeans. Do lower bone densities forecast a higher rate of fractures in these groups? Not always. Chinese people living in Singapore have low bone density, but their hip fracture

Figure C8•2

LOSS OF HEIGHT IN A WOMAN CAUSED BY OSTEOPOROSIS

The woman on the left is about 50 years old. On the right, she is 80 years old. Her legs have not grown shorter; only her back has lost length, due to collapse of her spinal bones (vertebrae). When collapsed vertebrae cannot protect the spinal nerves, the pressure of bones pinching the nerves causes excruciating pain.

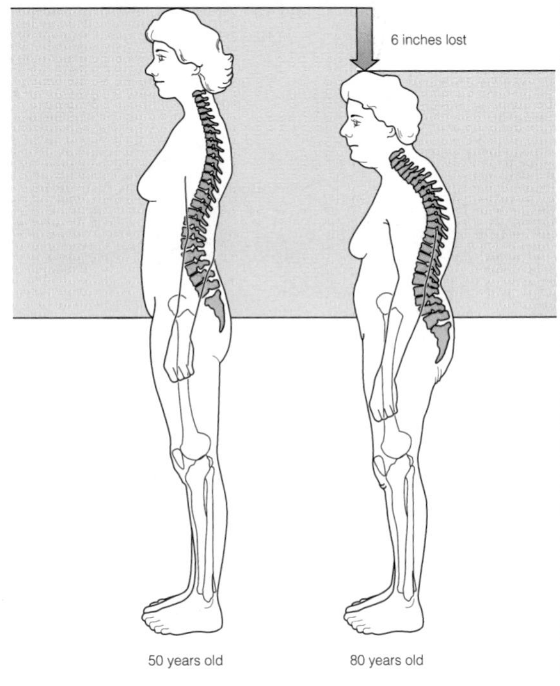

rates are among the lowest in the world. Understanding the reason why would be a breakthrough in current understanding.

Chinese women absorb calcium at a rate two to three times greater than that of people of northern European or even African descent.[9] Further, Chinese women's rate of calcium absorption seems to hold steady throughout menopause and into old age. Understanding why would be another breakthrough.

Calcium and Vitamin D

An environmental factor that affects bone withdrawal and deposition is calcium and vitamin D nutrition during childhood, adolescence, and early adult life. Preteen children who consume extra calcium together with adequate vitamin D lay more calcium into the structure of their bones than children with less adequate intakes.

When people reach the bone-losing years of middle age, those who formed dense bones during youth have more bone tissue to lose before suffering ill effects—see Figure C8-3. Therefore, whatever factors help build strong bones in youth, including calcium nutrition, also protect against osteoporosis much later.

Older people take in less calcium and vitamin D than others. After about age 65, they absorb less calcium, too, probably because aging skin is less efficient at making vitamin D. Also, many older people fail to go outdoors and so are deprived of the sunlight necessary to form vitamin D. Some of the hormones that regulate bone maintenance and calcium metabolism also change with age and accelerate bone mineral withdrawal.

Gender and Hormones

After age, gender is the next strongest predictor of osteoporosis: men have greater bone density than women at maturity, and women have great losses during menopause. Women thus account for four out of five cases of osteoporosis. Bone dwindles rapidly when the hormone estrogen diminishes as menstruation ceases. Accelerated losses continue for six to eight years following menopause, then taper off, so that women again lose bone at the same rate as their male counterparts. Losses of bone minerals continue throughout the remainder of a woman's lifetime, but not at the free-fall pace of the menopause years (see Figure C8-3).

When *young* women fail to produce enough estrogen and cease menstruating (amenorrhea), they lose bone rapidly. In some cases, diseased ovaries are to blame and must be removed; in others, the ovaries fail to produce sufficient estrogen because the women suffer from anorexia nervosa and have unreasonably restricted their body weight (see Controversy 9). Amenorrhea and low body weights may explain much of the bone loss seen in these women, and that loss remains after recovery from the eating disorder.[10]

Figure C8•3

TWO WOMEN'S BONE MASS HISTORY COMPARED

Woman A entered adulthood with enough calcium in her bones to last a lifetime. Woman B had less bone mass starting out and so suffered ill effects from bone loss later on.

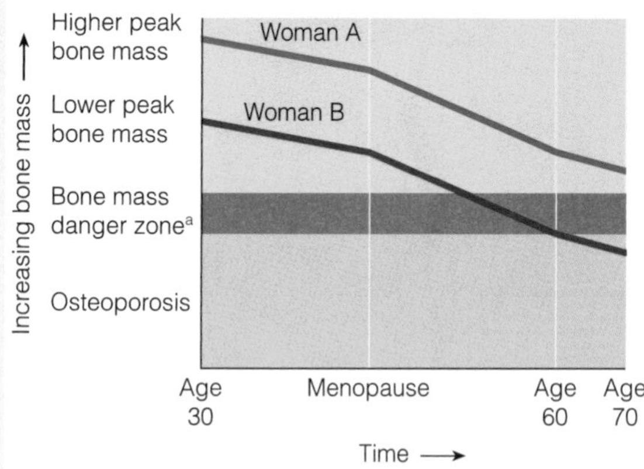

[a]People with a moderate degree of bone mass reduction are said to have osteopenia and are at increased risk of fractures.

SOURCE: Data from Standing Committee on the Scientific Evaluation of Dietary Reference Intakes, Food and Nutrition Board, Institute of Medicine, *Dietary Reference Intakes for Calcium, Phosphorus, Magnesium, Vitamin D, and Fluoride* (Washington, D.C.: National Academy Press, 1997), pp. 4–10.

If estrogen deficiency is a major cause of osteoporosis in women, what is the cause of bone loss in men? Men produce only a little estrogen, yet they resist osteoporosis better than women. Does the male sex hormone testosterone play a role? Perhaps so, because men suffer more fractures after removal of diseased testes or when their testes lose function with aging. Thus, both male and female sex hormones appear to play roles in the development of osteoporosis.

Physical Activity

When people lie idle—for example, when they are confined to bed—the bones lose strength just as the muscles do.[11] Astronauts who live without gravity for days or weeks at a time experience rapid and extensive bone losses.

Muscle strength and bone strength go together, and muscle use seems to promote bone strength.[12] When cross sections of bones of sedentary and active people are compared, the active bones are denser by far. The hormones that promote synthesis of new muscle tissue also favor the building of bone, and flexibility and muscle strength improve balance and help to prevent falls from occurring.[13]

To keep the bones healthy and to prevent falls, include weight-bearing exercises such as calisthenics, dancing, jogging, kick-boxing, vigorous walking, or weight training every day.[14] Even gardening, performed with vigor, can build bone strength and improve balance.

Body Weight

After age and gender, the next risk factor for osteoporosis is being underweight or losing weight.[15] Women who are thin throughout life, and especially those who lose 10 percent or more of their body weight after the age of menopause, face a hip fracture rate twice as high as that of most other women. Heavier body weights and higher body fatness stress the bones and promote their maintenance.[16] Also, fat tissue serves as a storage depot for hormones, and abundant body fat may mean greater hormone stores. An appetite-controlling hormone, leptin, is

These young people are putting bone in the bank.

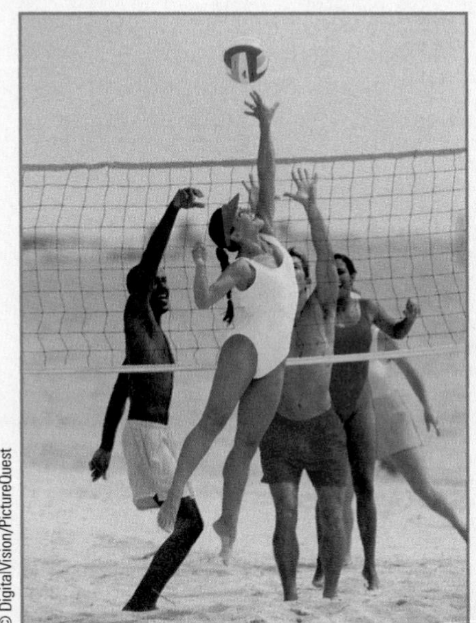

© DigitalVision/PictureQuest

produced by fat cells and may also have an effect on bone biology (more about leptin in Chapter 9).[17]

Tobacco Smoke and Alcohol

Smoking seems to be hard on the bones. Women who smoke lose 5 to 10 percent more of their bone density than nonsmokers by the time they reach menopause. Compared with nonsmokers, users of tobacco suffer more bone degeneration and spinal injuries, and surgeries to repair their backs are less often successful.[18]

People who are addicted to alcohol also experience more frequent fractures, and drinking can contribute to accidents and falls. Because alcohol (a diuretic) causes fluid excretion, it may induce excessive calcium losses through the urine. Heavy drinking may also upset the hormonal balance required for healthy bones or may be directly toxic to the bone-building cells, preventing their reproduction and diminishing their numbers.[19]

Table C8-2 summarizes the risk factors covered so far and includes some others. The more risk factors that apply to you, the greater your chances of developing osteoporosis in the future and the more seriously you should take the advice offered in a later section.

Protein, Sodium, Caffeine, Soft Drinks, Vitamin K, and Others

Researchers have discovered that excess dietary protein causes the body to excrete calcium in the urine. This finding suggests that a lifetime of consuming excess dietary protein could accelerate bone loss, especially when calcium intake is low.

Does a high-protein diet cause bone loss? It may, especially when the diet is rich in animal protein and low in vegetables.[20] In experiments, people fed increasing amounts of purified animal protein developed an increasingly negative calcium balance (they lost more calcium in the urine than they consumed from food). Other studies have found that protein from animal, but not vegetable, sources increases calcium in the urine.[21] Figure C8-4 on the next page shows that in women around the world, hip fracture incidence drops off dramatically when the ratio of vegetable to animal protein in the diet increases.[22]

Vegetables themselves, and not their protein, may be the protective factor, however, because of their role in providing alkaline substances to the body. Before the body can use or store energy from the carbon backbones of amino acids from any source, it must first cleave off the sulfur, which then forms sulfuric acid. The body must quickly neutralize any excess acidity and does so by withdrawing alkaline calcium salts from the bones and using the noncalcium part to buffer the acid. The calcium part is excreted in the urine. Vegetables are rich sources of alkaline constituents that help to neutralize acid and thereby protect the bones.

Of course, obtaining enough protein is essential to bone health because protein deprivation also stimulates calcium losses and weakens bones. In *malnourished* elderly people, restoring protein can often improve bone status and reduce the incidence of hip fractures.[23] In addition, increased intake of protein-rich foods from animals does

Table C8•2

RISK AND PROTECTIVE FACTORS THAT CORRELATE WITH OSTEOPOROSIS

Risk Factors	Protective Factors
High Correlation	
Advanced age	Black race
Alcoholism, heavy drinking	Estrogens, long-term use
Chronic steroid use	
Female gender	
Rheumatoid arthritis	
Surgical removal of ovaries or testes	
Thinness or weight loss	
White race	
Moderate Correlation	
Chronic thyroid hormone use	Having given birth
Cigarette smoking	High body weight
Diabetes (insulin-dependent, type 1)	High-calcium diet
Early menopause	Regular physical activity
Excessive antacid use	
Family history of osteoporosis	
Low-calcium diet	
Sedentary lifestyle	
Vitamin D deficiency	
Probably Important but Not Yet Proved	
Alcohol taken in moderation	Adequate vitamin K intake
Caffeine intake	Low-sodium diet (later years)
High-fiber diet	
High-protein diet	
Lactose intolerance	

not always produce negative calcium balance. Milk is a good example. A milk-rich diet provides both protein *and* calcium, and may help to *oppose* withdrawal of calcium from the skeleton.[24] With protein, as with other nutrients, then, it seems wise to follow a familiar nutrition principle—a diet based on adequacy, moderation, and variety is best.[25]

Sodium intake may also be a concern because some studies have reported that increased urinary calcium loss parallels increasing dietary sodium.[26] A recent review of the literature reports that there is currently insufficient evidence to recommend reducing sodium intakes to prevent osteoporosis.[27] Still, a prudent diet plan would include sufficient calcium throughout life and moderate sodium intakes in the later years for many reasons.[28]

Heavy users of caffeinated beverages, such as coffee, tea, and colas, should be aware that some evidence suggests a link between caffeine use and osteoporosis, although other findings do not point to caffeine use as a risk factor.[29] Caffeine may exert an effect only when calcium intakes are low.

Cola beverages and other soft drinks may also have adverse effects on calcium, although the reasons why are unclear. In rats, feeding colas, but not other soft drinks, results in significant urinary losses of calcium.[30] Rats given cola instead of water drank three times the normal amount of fluid while eating half the normal amount of solid food, and despite gaining more weight than water-drinking controls, their bones were significantly less dense.[31] In children, an unexplained tendency to develop bone problems such as stress fractures has been observed to accompany high intakes of fruit juices or cola beverages, but not other soft drinks.[32] Soft drinks also displace milk from the diet.

Vitamin K plays important roles in the production of at least one bone protein that participates in bone maintenance.*[33] People with hip fractures often have low intakes of vitamin K.[34] In a study of female athletes in whom strenuous exercise had lowered estrogen production, an increased vitamin K intake reduced markers of bone loss and increased markers of bone formation.[35]

Magnesium and potassium also help to maintain bone mineral density.[36] Vitamin A is needed in the bone-remodeling process, but too much may be associated with osteoporosis and hip fractures.[37] Clearly, a well-balanced diet that supplies abundant fruits and vegetables and a full array of nutrients without toxicity from nutrient supplements is central to bone health.[38]

Figure C8•4

RATIO OF VEGETABLE TO ANIMAL PROTEIN IN THE DIET AND HIP FRACTURE INCIDENCE WORLDWIDE

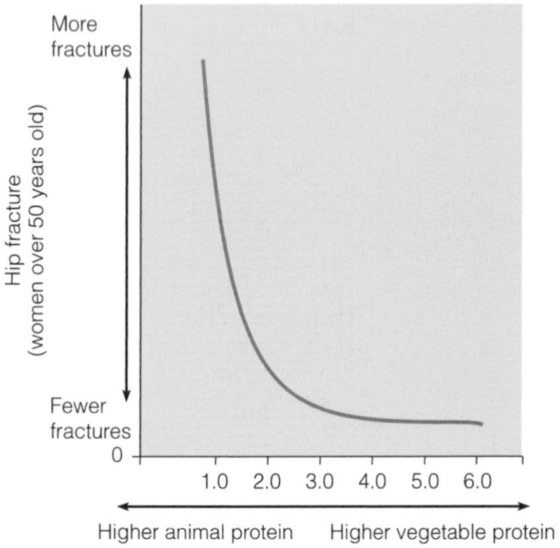

SOURCE: Adapted from L. A. Frassetto and coauthors, Worldwide incidence of hip fractures in elderly women. Relation to consumption of animal to vegetable foods, *Journals of Gerontology Series A: Biological Sciences and Medical Sciences* 55 (2000): M585–M592.

Diagnosis and Medical Treatment

Diagnosis of osteoporosis includes measuring bone density using an advanced form of X-ray (DEXA, described in Chapter 9) or ultrasound.† A thorough examination also includes evaluating risk factors for fractures, such as race, family history, and physical inactivity.[39] Men with a family history of osteoporosis and all women should have a bone density test after they reach age 50.

Hormone-replacement therapy may help nonmenstruating women prevent further bone loss and reduce the incidence of fractures.[40] Evidence also suggests that a phytochemical in soybeans may mimic the actions of estrogen in the body. If true, soy products may offer some benefits to women with reduced estrogen, but not without risks as Controversy 2 made clear.[41] Treatment for men may include testosterone-replacement therapy.[42]

Several drugs are proving powerful in the struggle to reverse bone loss. Such drugs inhibit the activities of the bone-dismantling cells, thus allowing the bone-building cells to slowly shore up bone tissue with new calcium deposits. The drugs work well for some, but not all, people with osteoporosis.

Calcium Recommendations

Bone strength later in life depends most on how well the bones were developed and maintained during youth. Adequate calcium nutrition during the growing years is essential to achieving optimal peak bone mass.[43] Only 10 percent of girls and 25 percent of boys meet the recommendation for calcium during their bone-forming years.[44] The DRI committee recommends 1,300 milligrams of calcium, the amount in about 4 cups of milk, each day for everyone 9 through 18 years of age; 1,000 milligrams through age 50; and 1,200 thereafter.

How should you obtain this calcium? We strongly recommend foods and beverages as your source of calcium and that you take supplements only when advised to do so by a physician. People can best support their bones' health by following the lifetime recommendations for healthy bones in Table C8-3. Calcium supplements cannot equal any of the actions listed in the table.

*The vitamin K–dependent bone protein is osteocalcin.

†DEXA stands for *dual X-ray absorptiometry*.

Table C8•3

A LIFETIME PLAN FOR HEALTHY BONES

0–18 Years	• Use milk as the primary beverage to meet the need for calcium within a balanced diet that provides all nutrients. • Play actively in sports or other activities. • Limit television. • Do not start smoking or drinking alcohol. • Drink fluoridated water.
19–25 Years	• Choose milk as the primary beverage, or if milk causes distress, include other calcium sources. • Commit to a lifelong program of physical activity. • Do not smoke or drink alcohol—if you have started, quit. • Drink fluoridated water.
26–50 Years	• Continue as for 19- to 25-year-olds. • At menopause, women should be evaluated for possible estrogen replacement therapy. • Obtain the recommended amount of calcium from food. • Take calcium and fluoride supplements only if prescribed by a physician.
51 Years and Above	• Continue as for 19- to 25-year-olds. • Follow a physician's advice concerning estrogen and supplements. • Continue striving to meet the calcium need from diet. • Continue bone-strengthening exercises.

Bone loss is not a calcium-deficiency disease comparable to iron-deficiency anemia. In iron-deficiency anemia, high iron intakes reliably reverse the condition. With respect to calcium balance, though, calcium intakes alone do little or nothing to reverse bone loss. Calcium and vitamin D supplements after the age of 50, however, do produce small but still beneficial effects on the bone mass and fracture rates.[45] During the menopausal years, calcium supplements of 1 gram may slow, but cannot fully prevent, the inevitable bone loss.

Taking self-prescribed calcium supplements entails possible risks (see Table C8-4) and cannot take the place of sound food choices and other healthy habits.[46] Whether you choose to take a calcium supplement is for you to decide. The next section provides some details about the variety of calcium supplements and the benefits and risks of taking them.

Table C8•4

CALCIUM SUPPLEMENT RISKS

People who take calcium supplements risk:

- *Impaired iron status*. Calcium inhibits iron absorption.
- *Accelerated calcium loss*. Calcium-containing antacids that also contain aluminum and magnesium hydroxide cause a net calcium loss.
- *Urinary tract stones or kidney damage in susceptible individuals*. People who have a history of kidney stones should be monitored by a physician and choose calcium citrate if they must take supplements.
- *Exposure to contaminants*. Some preparations of bone meal and dolomites are contaminated with hazardous amounts of arsenic, cadmium, mercury, and lead.
- *Vitamin D toxicity*. Vitamin D, which is present in many calcium supplements, can be toxic. Users must eliminate other concentrated vitamin D sources.
- *Excess blood calcium*. This complication is seen only with doses of calcium fourfold or more greater than customarily prescribed.
- *Milk alkali syndrome*. This condition is rare, but not absent. It is characterized by high blood calcium, metabolic alkalosis, and renal failure. Early symptoms include irritability, headaches, and apathy.
- *Other nutrient interactions*. Calcium inhibits absorption of magnesium, phosphorus, and zinc.
- *Drug interactions*. Calcium and tetracycline form an insoluble complex that impairs both mineral and drug absorption.
- *GI distress*. Constipation, intestinal bloating, and excess gas are common.

Calcium Supplements

Calcium supplements are available in three chemical forms. Simplest are the purified **calcium compounds,** such as calcium carbonate, citrate, gluconate, lactate, malate, or phosphate, and compounds of calcium with amino acids (called **amino acid chelates**). Second are mixtures of calcium with other compounds, such as calcium carbonate with magnesium carbonate, with aluminum salts (as in some **antacids**), or with vitamin D. Third are powdered, calcium-rich materials such as **bone meal, powdered bone, oyster shell,** or **dolomite** (limestone). See Table C8-5 on the next page for supplement terms.

Question 1: How well does the body absorb and use the calcium from various supplements? Based on research to date, many people seem to absorb calcium reasonably well—and about as well as from milk—from amino acid chelates and from calcium compounds such as calcium phosphate dibasic, calcium acetate, calcium carbonate, calcium citrate, calcium gluconate, and calcium lactate. People absorb calcium less well from a mixture of calcium and magnesium carbonates, from oyster shell calcium fortified with inorganic magnesium, from a chelated calcium-magnesium combination, and from calcium carbonate fortified with vitamins and iron. Some people absorb calcium better from milk and milk products than from even the most absorbable supplements. Early research indicates that calcium from mineral water may also be highly absorbable.[47]

Question 2: How much calcium does the supplement provide? The Tolerable Upper Level for calcium has been set at 2,500 milligrams. To be safe, supplements should provide less than this, since foods also provide calcium. Read the label to find out how much a dose supplies. Calcium carbonate is 40 percent elemental calcium,

Table C8•5

CALCIUM SUPPLEMENT TERMS

- **amino acid chelates** (KEY-lates) compounds of minerals (such as calcium) combined with amino acids in a form that favors their absorption. A *chelating agent* is a molecule that surrounds another molecule and can then either promote or prevent its movement from place to place (*chele* means "claw").
- **antacids** acid-buffering agents used to counter excess acidity in the stomach. Calcium-containing preparations (such as Tums) contain available calcium. Antacids with aluminum or magnesium hydroxides (such as Rolaids) can accelerate calcium losses.
- **bone meal** or **powdered bone** crushed or ground bone preparations intended to supply calcium to the diet. Calcium from bone is not well absorbed and is often contaminated with toxic materials such as arsenic, mercury, lead, and cadmium.
- **calcium compounds** the simplest forms of purified calcium. They include calcium carbonate, citrate, gluconate, lactate, malate, and phosphate. These supplements vary in the amount of calcium they contain, so read the labels carefully. A 500-milligram tablet of calcium gluconate may provide only 45 milligrams of calcium, for example.
- **dolomite** a compound of minerals (calcium magnesium carbonate) found in limestone and marble. Dolomite is powdered and is sold as a calcium-magnesium supplement, but may be contaminated with toxic minerals, is not well absorbed, and interacts adversely with absorption of other essential minerals.
- **oyster shell** a product made from the powdered shells of oysters that is sold as a calcium supplement, but is not well absorbed by the digestive system.

whereas calcium gluconate is only 9 percent. The user should select a low-dose supplement and take it several times a day rather than taking a large-dose supplement all at once: divided doses can improve a day's total absorption up to 20 percent.

Question 3: Will the supplement be digested and the calcium available for absorption? M310anufacturers compress large quantities of calcium into small pills, and stomach acid often has difficulty penetrating the pill. To test whether a supplement will dissolve, drop it into a 6-ounce cup of vinegar and stir occasionally. A high-quality formulation will dissolve within half an hour. The chewable kind, because they are broken into bits before swallowing, and calcium-fortified foods and beverages are not prone to this problem.

One last pitch: Think one more time before you decide to take supplements for calcium. The Consensus Conference on Osteoporosis recommends milk. The American Society for Bone and Mineral Research recommends foods as the source of calcium in preference to supplements. The National Institutes of Health concludes that foods are best and recommends supplements only when intake from food is insufficient. The authors of this book are so impressed with the importance of using abundant, calcium-rich foods that they have worked out ways to do so at every meal. Seldom do nutritionists agree so unanimously.

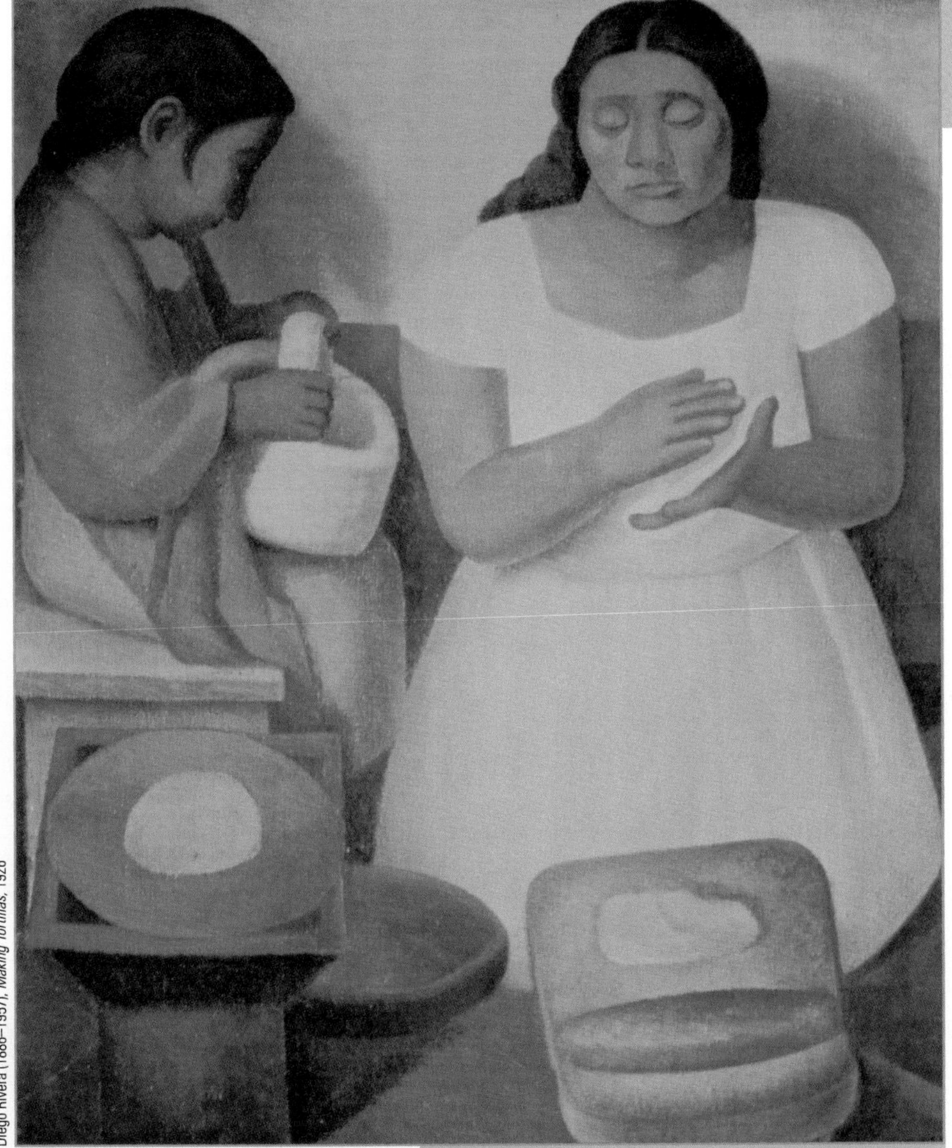

Diego Rivera (1886–1957), *Making Tortillas*, 1926

Energy Balance and Healthy Body Weight

Frequently Asked Questions

Contents

Figure 9•1

THE INCREASING PREVALENCE OF OBESITY AMONG U.S. ADULTS

In 1991, only four states had an obesity rate (BMI ≥30) of greater than 15 percent. By 2000, all states except Colorado reported an obesity rate of at least 15 percent, many reported a rate of greater than 20 percent, and the trend is steadily upward.

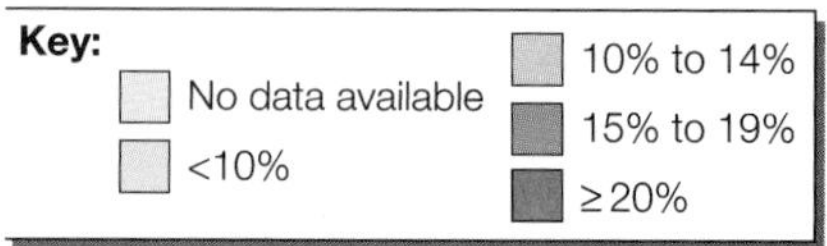

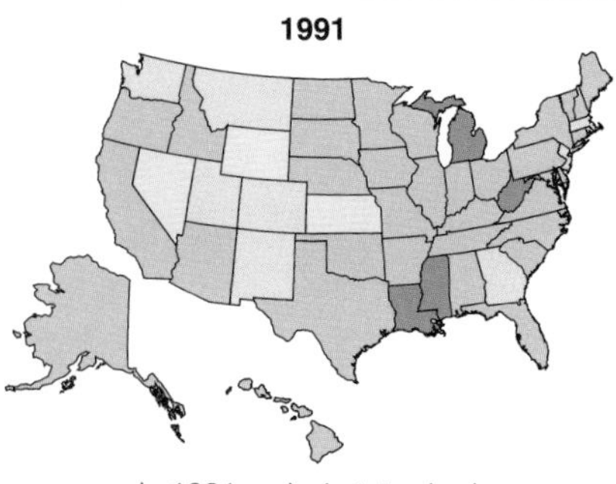

In 1991, only 4 states had obesity rates >15 percent.

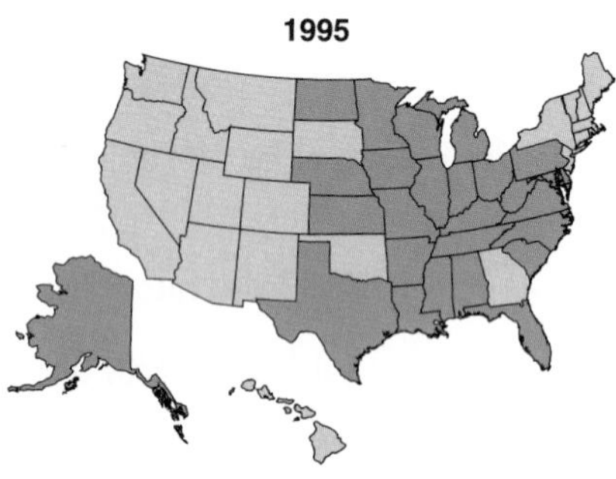

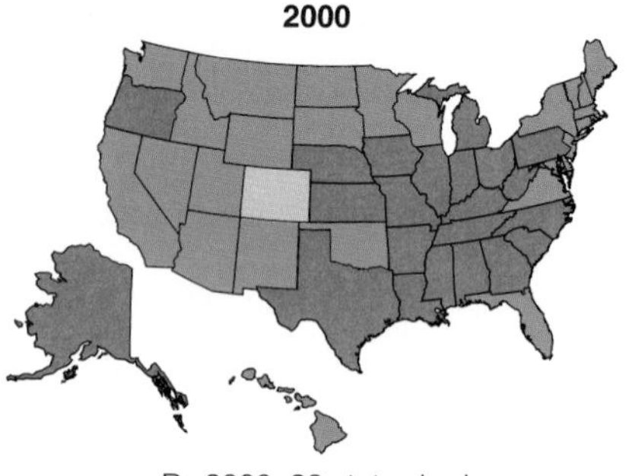

By 2000, 22 states had obesity rates >20 percent.

SOURCE: U.S. Obesity Trends 1985 to 2000, Centers for Disease Control and Prevention, available at www.cdc.gov/nccdphp/dnpa/obesity/trend/maps/slide/001.htm

ARE YOU PLEASED with your body weight? If you answered yes, you are a rare individual. Nearly all people in our society think they should weigh more or less (mostly less) than they do. Their primary concern is usually appearance, but they often correctly perceive that physical health is somehow related to weight. At the extremes, both **overweight** and **underweight** present definite health risks.

People also think of their weight as something they should control, once and for all. Three misconceptions in that sentence frustrate their efforts, however—the focus on weight, the focus on *controlling* weight, and the focus on a short-term endeavor. Simply put, it isn't your weight you need to control; it's the fat in your body in proportion to the lean—your **body composition.** And controlling body composition directly isn't possible—you can only control your *behavior*. Sporadic bursts of activity, such as "dieting," are not effective; the behaviors that achieve and maintain a healthy body weight take a lifetime of commitment.

This chapter starts out by presenting the problems associated with deficient and excessive body fatness and then examines how the body manages its energy budget. The following sections show how to judge body weight on the sound basis of health and describe theories about the causes of obesity. Finally, the chapter reveals how the body gains and loses weight and suggests lifestyle strategies for achieving and maintaining a healthy body weight.

The Problems of Too Little or Too Much Body Fat

Both deficient and excessive body fat present risks to health. In the United States, too little body fat is not a widespread problem. **Obesity,** in contrast, is an escalating epidemic—see Figure 9-1.[1] An estimated 35 percent of U.S. adults age 20 to 74 years are overweight, while 27 percent of adults are dangerously obese.[2] Additionally, one of every seven children and one of every nine teenagers in the United States are overweight.[3] The *Dietary Guidelines* and the *Healthy People 2010* objectives include mandates to reduce obesity and overweight (see Table 9-1).[4] To accomplish these goals requires first that Americans reverse their current trend toward *gaining* body weight.

What Are the Risks from Underweight?

Thin people die first during a siege or in a famine. Overly thin people are also at a disadvantage in the hospital, where they may have to refrain from eating food for days at a time while undergoing tests or surgery, and their nutrient status can easily deteriorate.[5] Underweight also increases the risk for any person fighting a **wasting** disease. People with cancer often die, not from the cancer itself, but from starvation. Thus, excessively underweight people are urged to gain body fat as an energy reserve and to acquire protective amounts of all the nutrients that can be stored. Risks to life increase for those at either extreme of body weight—see Figure 9-2 on page 314.

key point · *Deficient body fatness threatens survival during a famine or in wasting diseases.*

Q How Fat Is Too Fat?

People want to know exactly how fat is too fat for health, but no one cutoff point is appropriate for everyone. Obesity is definitely harmful and constitutes a major lifestyle risk factor for cardiovascular disease. Evidence suggests that being moderately overweight also increases the risk, assuming that excess weight is composed of fat, not muscle.[6] A few obese people, however, remain healthy and live long despite their excess body fatness. Physical fitness, despite body fatness, may lend some protection against early death from disease.[7] Genetics and other risk factors such as smoking can also help determine who among the overweight are most likely to stay well and who should make lifestyle changes for health's sake.

In the United States, more than 70 percent of obese people suffer from at least one other major health problem.[8] As one example among many, excess weight may cause up to half of all cases of hypertension, thereby contributing to an increased risk of stroke. Often a loss of just a few pounds can normalize the blood pressure of an overfat person; some people with hypertension can tell you at exactly what weight their blood pressure begins to rise. Modest weight loss can also bring other benefits: an older obese man, for example, might gain relief from the arthritis in his knees by losing just 25 pounds.[9]

The loss of life to obesity is enormous: an estimated 300,000 people die each year from obesity-related diseases.[10] Obesity is second only to tobacco in causing preventable illnesses and premature deaths, with increases in mortality proportionally following increases in excess weight.

overweight overfatness of a moderate degree; defined as a body mass index (BMI) of 25.0 through 29.9. BMI is defined later.

underweight too little body fat for health; defined as having a body mass index of less than 18.5.

body composition the proportions of muscle, bone, fat, and other tissue that make up a person's total body weight.

obesity overfatness with adverse health effects, as determined by reliable measures and interpreted with good medical judgment. Obesity is officially defined as a body mass index of 30 or higher.

wasting the progressive, relentless loss of the body's tissues that accompanies certain diseases and shortens survival time.

visceral fat fat stored within the abdominal cavity in association with the internal abdominal organs; also called *intra-abdominal fat*.

central obesity excess fat in the abdomen (visceral fat) and around the trunk.

subcutaneous fat fat stored directly under the skin (*sub* means "beneath"; *cutaneous* refers to the skin).

Obesity as a Chronic Disease The health risks of overfatness are so many that obesity has been declared a chronic disease. In addition to cardiovascular disease and hypertension, being overfat triples a person's risk of developing diabetes and all of its associated ills. Even a modest weight gain during adulthood may increase the risk of diabetes among women. If hypertension, cardiovascular disease, or diabetes runs in your family, you urgently need to attend to controlling body fatness. The risks from obesity appear to be greater for people of European descent than for African Americans.[11]

Obese adults are also threatened by other risks. Among them are abdominal hernias, arthritis, complications in pregnancy and surgery, flat feet, gallbladder disease, gout, high blood lipids, liver malfunction, respiratory problems (including Pickwickian syndrome, a breathing blockage linked to sudden death), sleep apnea (abnormal pauses in breathing during sleep), some cancers, varicose veins, and even a high accident rate. Moreover, after the effects of diagnosed diseases are taken into account, the risk of death from other causes remains almost twice as high for people with lifelong obesity as for others.

Risks from Central Obesity Even more than total fatness, fat that collects deep within the central abdominal area of the body, called **visceral fat,** may be especially dangerous with regard to risks of diabetes, stroke, hypertension, and coronary artery disease (see Figure 9-3 on the next page). The risk of death from *all* causes may be higher in those with **central obesity** than in those whose fat accumulates elsewhere in the body. The health risks of obesity seem to run on a continuum: normal weight brings no extra risk, central obesity carries severe risks, and other forms of obesity fall somewhere in between.

Factors affecting body fat distribution:

- Menopause in women.
- Smoking.
- Alcohol intake.
- Physical activity.

Why should fat in the abdomen bring extra risk to the heart? Some researchers suspect that differences in fat mobility are part of the explanation. Visceral fat, which is readily released into the bloodstream, may make a significant contribution to the blood's daily burden of cholesterol-carrying lipoproteins, LDL, thereby increasing heart disease risk.[12] Fat layers lying just beneath the skin **(subcutaneous fat)** of the abdomen, thighs, hips, and legs also release fat, but sluggishly and so, theoretically, may contribute less to blood lipids.[13]

Men of all ages and women who are past menopause are more prone to develop the "apple" profile of central obesity, whereas women in their reproductive years develop more of a "pear" profile (fat around the hips and thighs). Some women change profile at menopause, and lifelong "pears" may suddenly face increased risks of diseases that accompany excess visceral fat. Smokers, too, may carry more of their body fat centrally. Although a smoker may weigh less than the average nonsmoker, the smoker's waist measurement may be greater, leading to the theory that smoking directly affects body fat distribution. Two other factors may also affect body fat distribution. Moderate-to-high intakes of alcohol have a positive association with central obesity, and higher levels of physical activity have a negative association. A later section explains how to judge whether a person carries too much fat around the middle.

Table 9•1

RECOMMENDATIONS CONCERNING BODY WEIGHT

Dietary Guidelines

- Aim for a healthy body weight—balance the calories you eat with physical activity.

Healthy People 2010

- Increase to 60 percent the prevalence of healthy weight (BMI from 19 to 25) among all people age 20 or older.
- Reduce to less than 15 percent the prevalence of BMI of 30 or above in people age 20 and older.

NOTE: BMI values are on the inside back cover.

Figure 9•2

UNDERWEIGHT, OVERWEIGHT, AND MORTALITY

This J-shaped curve describes the relationship between an indicator of body weight for height, the body mass index (BMI, see text), and mortality. It shows that both underweight and overweight present risks of a premature death.

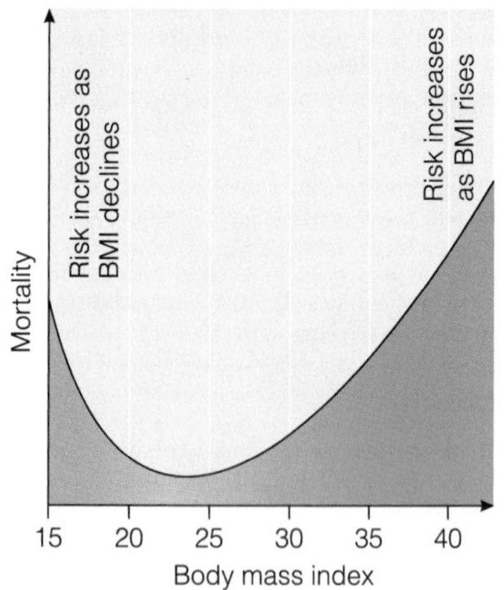

Figure 9•3

VISCERAL FAT AND SUBCUTANEOUS FAT

The fat deep within the body's abdominal cavity may pose an especially high risk to health.

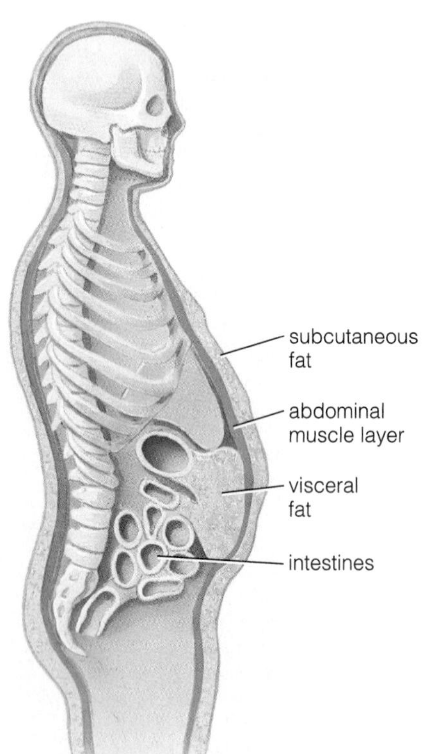

National Guidelines for Identifying Those at Risk from Obesity In 1998, obesity experts developed U.S. guidelines for evaluation of risks to health from obesity, as determined by three indicators.[14] The first indicator is a person's **BMI,** or **body mass index.** The BMI, which defines average relative weight for height in people older than 20 years, usually correlates with body fatness and degree of disease risks (see the inside back cover for the BMI table).[15] As a general guideline, overweight for adults is defined as BMI of 25.0 through 29.9 and obesity as BMI equal to or greater than 30.

The second indicator is waist circumference, reflecting the degree of visceral fatness in proportion to body fatness (see Table 9-2).* The third indicator is the person's disease risk profile, which takes into account whether the person has hypertension, type 2 diabetes, or elevated lipoproteins; whether the person is a smoker; and so forth—see the list in the margin on page 315. The more risk factors and the greater the obesity, the greater the urgency to control body fatness.

Social and Economic Costs of Obesity Although some overfat people escape health problems, no one who is fat in our society quite escapes the social and economic handicaps. Our society places enormous value on thinness, especially for women, and fat people are less sought after for romance, less often hired, and less often admitted to college. They pay higher insurance premiums, and they pay more for clothing.[16]

An estimated 30 to 40 percent of all U.S. women (and 20 to 25 percent of U.S. men) are trying to lose weight at any given time, spending up to $40 billion each year to do so. The assumption is that every overweight person can and should achieve slenderness. However, most overweight people cannot—for whatever reason—become slender. By one reckoning, only 5 percent of all people who successfully lose weight maintain their losses; other estimates are more optimistic but still predict that the majority of dieters' lost weight will creep back over time.[17]

Prejudice defines obese people by their appearance rather than by their character, often stereotyping them as lazy, stupid, and self-indulgent. Obese people suffer emotional pain when others treat them with hostility and contempt. Subtle blaming for an apparent lack of the discipline to resolve their weight problem often becomes internalized as guilt and self-deprecation. Even health care professionals, including

*The National Heart, Lung, and Blood Institute has replaced waist-hip ratio with waist circumference for assessment of obesity health risks.

Table 9•2

DISEASE RISKS ACCORDING TO BMI AND WASTE CIRCUMFERENCE[a]

The degree of risk is heightened by the presence of specific diseases, other risk factors (such as elevated blood LDL cholesterol, see Table 11-3 in Chapter 11), or smoking (see the margin of the next page).

BMI		Waist ≤40 in (Men) or ≤35 in (Women)	Waist ≥40 in (Men) or ≥35 in (Women)
18.5 or less	Underweight	Low	—
18.5–24.9	Normal	Low	—
25.0–29.9	Overweight	Increased	High
30.0–34.9	Obese, class I	High	Very high
35.0–39.9	Obese, class II	Very high	Very high
40 or greater	Extremely obese, class III	Extremely high	Extremely high

[a]Risk for type 2 diabetes, hypertension, and cardiovascular disease.

SOURCE: National Heart, Lung, and Blood Institute, National Institutes of Health, *The Practical Guide: Identification, Evaluation, and Treatment of Overweight and Obesity in Adults*, NIH publication no. 00-4084 (Washington, D.C.: Government Printing Office, 2000).

dietitians, can be among the chief offenders. To free our society of its obsession with body weight and prejudice against obesity, activists are speaking out for acceptance of body weight and respect for individuals.

key point *Obesity has been named a chronic disease. Central obesity may be more hazardous to health than other forms of obesity. National guidelines for evaluating risks to health from obesity define obesity as a BMI of 30 or above. Overfatness presents social and economic handicaps as well as physical ills. Judging people by their body weight is a form of prejudice in our society.*

body mass index (BMI) an indicator of obesity, calculated by dividing the weight of a person by the square of the person's height.

adipose tissue the body's fat tissue, consisting of masses of fat-storing cells and blood vessels to nourish them. Adipose tissue produces and releases hormones—among them, the hormone leptin involved in appetite regulation.

The Body's Energy Balance

What happens inside the body when you eat too much or too little food? The body ends up with an unbalanced energy budget—you have taken in more or less food energy than you spent. The mechanisms by which the body handles its energy underlie changes that occur in body composition.

When more food energy is consumed than is needed, excess fat enters the fat cells in the body's **adipose tissue** for storage. When energy supplies run low, stored fat is withdrawn. The daily energy balance can therefore be stated like this:

> Change in energy stores equals food energy taken in minus energy spent on metabolism and muscle activities.

More simply:

> Change in energy stores = energy in − energy out.

Too much or too little fat on the body today does not necessarily reflect today's energy budget. Small imbalances in the energy budget compound over time.[18]

The National Heart, Lung, and Blood Institute states that aggressive treatment may be needed for critically obese people who also fit any of the following:

- Established cardiovascular disease (CVD).
- Established type 2 diabetes, or impaired glucose tolerance.
- Sleep apnea, a disturbance of breathing in sleep, including temporary stopping of breathing.

The same urgency for treatment exists for an obese person with any *three* of the following.

- Hypertension.
- High LDL.
- Smoking.
- Low HDL cholesterol.
- Sedentary lifestyle.
- Age older than 45 years (men) or 55 years (women).
- Heart disease of an immediate family member before age 55 (male) or 65 (female).

SOURCE: National Heart, Lung, and Blood Institute, National Institutes of Health, *The Practical Guide: Identification, Evaluation, and Treatment of Overweight and Obesity in Adults*, NIH publication no. 00-4084 (Washington, D.C.: Government Printing Office, 2000).

Energy In

The energy in foods and beverages is the only contributor to the "energy in" side of the energy balance equation. Before you can decide how much food energy you need in a day, you must first become familiar with the amounts of energy in foods and beverages. One way to do so is to look up calorie amounts associated with foods and beverages in the Table of Food Composition (Appendix A). Or computer programs can provide this information in the blink of an eye for those with computer access. Such numbers are always fascinating to people concerned with managing body weight.

For example, an apple gives you 125 calories from carbohydrate; a regular-size candy bar gives you about 250 calories mostly from fat and carbohydrate. You may already know that for each 3,500 calories you eat in excess of expenditures, you store approximately 1 pound of body fat.

key point *The "energy in" side of the body's energy budget is measured in calories taken in each day in the form of foods and beverages. The number of calories in foods and beverages can be obtained from published tables or computer diet analysis programs.*

Camryn Manheim, award-winning actress on the television series* The Practice *and author of* Wake Up, I'm Fat, *envisions a society that accepts people of all shapes and sizes.

© George Sesota/Newsmakers/Getty Images

Energy Out

No easy method exists for determining the energy an individual spends and therefore needs. Recommended energy intakes for various age-gender groups are found on the inside front cover, page C, or in Appendix B. These recommendations are useful for population studies, but energy needs vary so widely among individuals in a group that it is impossible to guess any individual person's need without knowing something about the person's lifestyle and metabolism.

basal metabolism the sum total of all the involuntary activities that are necessary to sustain life, including circulation, respiration, temperature maintenance, hormone secretion, nerve activity, and new tissue synthesis, but excluding digestion and voluntary activities. Basal metabolism is the largest component of the average person's daily energy expenditure.

voluntary activities intentional activities (such as walking, sitting, or running) conducted by voluntary muscles.

thermic effect of food (TEF) the body's speeded-up metabolism in response to having eaten a meal; also called *diet-induced thermogenesis*.

basal metabolic rate (BMR) the rate at which the body uses energy to support its basal metabolism.

Energy Intake Recommendations The recommendations for energy intake are based on average people. For example, an intake of 2,200 calories per day is recommended for a woman who is assumed to be 20 years old, standing 5 feet 5 inches tall, weighing about 128 pounds, of average body fatness, and engaging in light activity. The man needing 2,900 calories per day is a healthy 20-year-old of average body fatness who stands 5 feet 10 inches tall, weighs 160 pounds, and engages in light activity. Taller people need proportionately more energy than shorter people to balance their energy budgets because their greater surface area allows more energy to escape as heat. Older people generally need less than younger people due to slowed metabolism and reduced muscle mass, which occur in part because of reduced physical activity.[19] As Chapter 13 points out, these losses may not be inevitable for people who stay active. On average, though, energy need diminishes by 5 percent per decade beyond the age of 30 years.

In reality, no one is average. In any group of 20 similar people with similar activity levels, one may expend twice as much energy per day as another. A 60-year-old person who bikes, swims, or walks briskly each day may need as many calories as a sedendary person of 30. Clearly, with such a wide range of variation, a necessary step in determining any person's energy need is to study that person.

1 lb body fat = 3,500 cal.

Pure fat is worth 9 calories per gram. A pound of it (450 grams), then, would store 4,050 calories. A pound of *body* fat is not pure fat, though; it contains water, protein, and other materials—hence the lower calorie value.

Methods of Estimating Energy Needs One way to estimate your energy needs is to monitor your food intake and body weight over a period of time in which your activities are typical of your lifestyle. If you keep an accurate record of all the foods and beverages you consume for a week or two and if your weight has not changed during the past few months, you can conclude that your energy budget is balanced. At least three days of honest record keeping are necessary because intakes fluctuate from day to day. A week or two is better still. (On about half the days, you eat less food energy than the average; on the other half, more.)

An alternative method of determining energy need is based on energy output. The two major ways in which the body spends energy are (1) to fuel its **basal metabolism** and (2) to fuel its **voluntary activities.** Basal metabolism generates energy to support the body's work that goes on all the time without our conscious awareness. To estimate output, you must compute the amount of these two major components and then add them together. This method leaves out a third energy component, the body's metabolic response to food. About 5 to 10 percent of a meal's energy value is used up in stepped-up metabolism in the five or so hours after the meal; this category of energy expenditure is called the **thermic effect of food.** Although this amount of energy could affect expenditures over the long run, most experts believe its effects are negligible. For our purposes, it can be ignored (see Figure 9-4).

Basal metabolism consumes a surprisingly large amount of fuel, and the **basal metabolic rate (BMR)** varies from person to person. A person whose total energy needs are 2,000 calories a day spends as many as 1,200 to 1,400 of them to support basal metabolism. The hormone thyroxine directly controls basal metabolism—the less secreted, the lower the energy requirements for basal functions. The rate is lowest during sleep.* Many other factors also affect the BMR (see Table 9-3).

Balancing food energy intake with physical activity can add to life's enjoyment.

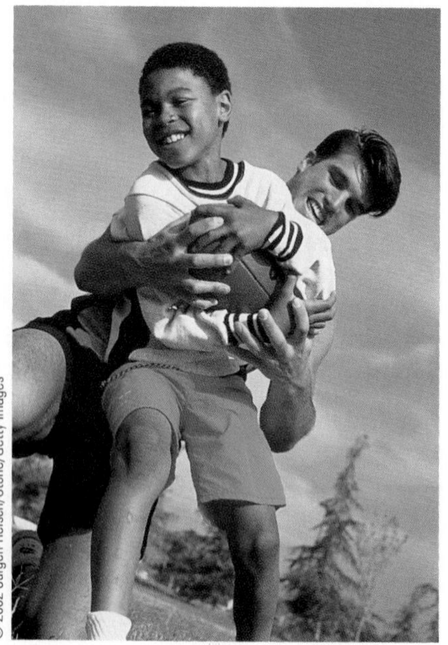

Q Can I Speed Up My Metabolism to Promote Fat Loss? You cannot speed up your BMR very much *today*. You can, however, amplify the second component of your energy expenditure—your voluntary activities. If you do, you will spend more calories today, and if you keep doing so day after day, your BMR will also increase.[20] Lean tissue is more metabolically active than fat tissue, so a way to speed up your BMR to the maximum possible rate is to make endurance and strength-building activities a daily habit so that your body composition becomes as lean as possible. A warning: some ads for weight-loss diets claim that eating certain foods, such as grapefruit or vinegar, can elevate the BMR and thus promote weight loss. This

*A measure of energy output taken during relaxation while awake yields a slightly higher number called the *resting metabolic rate*.

claim is false. Any meal promotes a temporary stepped-up energy expenditure in the form of the thermic effect of food; in the context of a mixed diet, the differences among foods are not large enough to be worth notice.

For voluntary activities, the amount of energy you spend depends somewhat on your personal style. In general, the heavier the weight of the body parts you move in your activity and the longer the time you invest, the more calories you expend. So important to energy balance is physical activity that much of Chapter 10 is devoted to presenting details of how the body spends its energy during activity. The next section shows how to calculate an approximation of your daily energy output.

key point *Two major components of the "energy out" side of the body's energy budget are basal metabolism and voluntary activities. A third minor component of energy expenditures is the thermic effect of food.*

How Many Calories Do I Need Each Day?

Simply put, you need enough to cover your energy expenditure. Here is one way to estimate your daily energy expenditure. First, estimate the two major components of energy expenditure; then, add them together. The first component is the energy in basal metabolism. Follow these steps. Use the BMR factor 1.0 calorie per kilogram of body weight per hour for men or 0.9 for women; the factors differ because men usually have more muscle (metabolically active tissue) than women do. Example (for a 150-pound man):

Step 1. Change pounds to kilograms:

150 pounds ÷ 2.2 pounds per kilogram = 68 kilograms.

Step 2. Multiply weight in kilograms by the BMR factor:

68 kilograms × 1 calorie per kilogram per hour = 68 calories per hour.

Step 3. Multiply the calories used in one hour by the hours in a day:

68 calories per hour × 24 hours per day = 1,632 calories per day.

Now calculate the second major component, physical activity, by multiplying the BMR calories by a percentage that varies by activity level. These percentages are estimates or approximations of energy expenditure based on the amount of muscular work a person typically performs in a day:

- Sedentary lifestyle: Men, 25 to 40 percent; women, 25 to 35 percent.
- Light activity: Men, 50 to 70 percent; women, 40 to 60 percent.

Table 9•3

FACTORS THAT AFFECT THE BMR

Factor	Effect on BMR
Age	The BMR is higher in youth; as lean body mass declines with age, the BMR slows. Continued physical activity may prevent some of this decline.
Height	Tall people have a larger surface area, so their BMRs are higher.
Growth	Children and pregnant women have higher BMRs.
Body composition	The more lean tissue, the higher the BMR.
Fever	Fever raises the BMR.
Stress	Stress hormones raise the BMR.
Environmental temperature	Adjusting to either heat or cold raises the BMR.
Fasting/starvation	Fasting/starvation hormones lower the BMR.
Malnutrition	Malnutrition lowers the BMR.
Thyroxine	The thyroid hormone thyroxine is a key BMR regulator; the more thyroxine produced, the higher the BMR.

Figure 9•4

COMPONENTS OF ENERGY EXPENDITURE

Generally, basal metabolism represents a person's largest expenditure of energy, followed by physical activity and the thermic effect of food. In estimating energy needs, physical activity and BMR are most important; the thermic effect of food can be ignored.

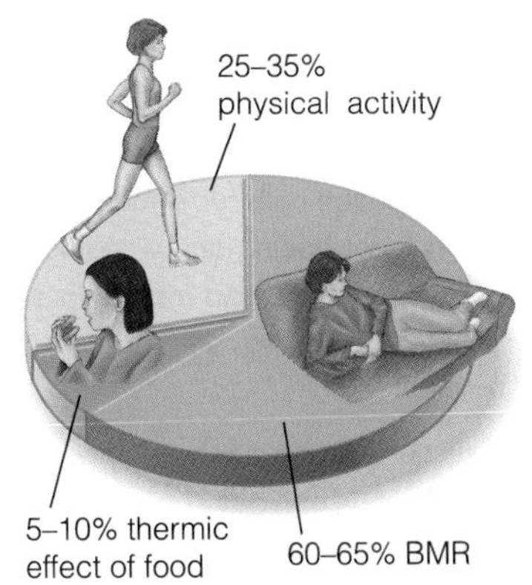

For both women and men, *light* activity means sleeping or lying down for eight hours a day, sitting for seven hours, standing for five, walking for two, and spending two hours in light physical activity.

An example for a woman weighing 128 pounds:

Step 1. 128 lb ÷ 2.2 = 58 kg.
Step 2. 58 kg × 0.9 = 52 cal/hour.
Step 3. 52 cal/hour × 24 hours = 1,248 cal/day spent in basal metabolism.

A quick and easy estimate of energy need:

Men: kg body weight × 24 = cal/day.
Women: kg body weight × 22 = cal/day.

- *Sedentary:* You sit down most of the day and drive or ride whenever possible; includes playing a musical instrument, standing still, ironing, painting.
- *Light activity:* You move around some of the time, as a teacher might during working hours; includes house cleaning, child care, golf (with a cart), sailing, table tennis.
- *Moderate activity:* You engage in some intentional exercise, such as an hour of fast walking four or five times a week, or your occupation calls for some physical work; includes hoeing and weeding a garden, carrying a load for distances, recreational bicycling, skiing, tennis, dancing.
- *Heavy activity:* Your job requires much physical labor, such as a roofer or a carpenter; includes fast jogging or running, hauling heavy loads uphill, heavy sustained manual digging, basketball, soccer.
- *Exceptional activity:* The exceptional category is reserved for those few who spend many hours a day in intense physical training, such as professional or college athletes during their seasons.

- Moderate activity: Men, 65 to 80 percent; women, 50 to 70 percent.
- Heavy activity: Men, 90 to 120 percent; women, 80 to 100 percent.
- Exceptional activity: Men, 130 to 145 percent; women, 110 to 130 percent.*

To select the activity level appropriate for you, consult the list in the margin. Think in terms of the amount of *muscular* work performed—don't confuse being *busy* with being *active*. Table 9-5 later in this chapter provides more exact energy costs per minute of activity based on body weight, and Chapter 10 comes back to activity for control of body fatness.

Calculate your energy expenditure using both the upper and lower ends of the range of percentages given for your gender and activity level. Suppose the 150-pound man used as an example earlier is a student who bikes about ten minutes a day and walks to classes but otherwise sits and studies. He falls into the light activity category, so we can estimate the range of energy he needs by multiplying his BMR calories per day by both 50 and 70 percent:

$$1{,}632 \text{ calories per day} \times 0.50 = 816 \text{ calories per day.}$$
$$1{,}632 \text{ calories per day} \times 0.70 = 1{,}142 \text{ calories per day.}$$

The man needs from 816 to 1,142 calories per day for his activities. Now add the metabolic component to each of these activity components. The total daily energy expenditure of the man in our example is between:

$$1{,}632 \text{ calories per day} + 816 \text{ calories per day} = 2{,}448 \text{ calories per day}$$

and

$$1{,}632 \text{ calories per day} + 1{,}142 \text{ calories per day} = 2{,}774 \text{ calories per day.}$$

Express the man's needs as a range of rounded values: 2,400 to 2,800 calories per day.

key point *To estimate the energy spent on basal metabolism, use the factor 1.0 calorie (for men) or 0.9 calorie (for women) per kilogram of body weight per hour for a 24-hour period. Then add a percentage of this amount to account for daily expenditures in muscular activity.*

Body Weight versus Body Fatness

In the past, to determine if a person had a healthy body weight, the person's height and weight were compared to a list of suggested weights for heights deemed by nutrition authorities to reflect health. But weight-for-height tables are not the most accurate method of evaluation, and the BMI has replaced these tables in clinical settings.

Finding your BMI:

$$\text{BMI} = \frac{\text{weight (kg)}}{\text{height (m)}^2}.$$

or

$$\text{BMI} = \frac{\text{weight (lb)}}{\text{height (in)}^2} \times 705.$$

Example: A person 5 feet 10 inches tall weighing 150 pounds has a BMI of 21.6.

$$\text{BMI} = \frac{150}{70^2} \times 705.$$

$$\text{BMI} = \frac{150}{4{,}900} \times 705.$$

$$\text{BMI} = .0306 \times 705.$$

$$\text{BMI} = 21.6 \text{ (rounded).}$$

Body Mass Index

BMI values correlate significantly with body fatness, and experts use them to help evaluate a person's health risks associated with underweight or overweight. The inside back cover of this book provides an easy way to find and evaluate BMI in adults and adolescents; the margin provides the calculation to demonstrate how BMI values are derived.

The BMI values are most accurate in assessing degrees of obesity and are less useful for evaluating nonobese people's body fatness. The BMI values have two major drawbacks: they fail to indicate how much of the weight is fat and where that fat is located. These drawbacks make the BMI unsuitable for use with:

- Athletes (because their highly developed musculature falsely increases their BMI values).

*Percentages are derived from the RDA (1989) formula for energy expenditure allowing a 15 to 30 percent range.

- Pregnant and lactating women (because their increased weight is normal during childbearing).
- Adults over 65 (because BMI values are based on data collected from younger people and because people "grow shorter" with age).

The bodybuilder in the margin proves this point: with a BMI over 30, he would be classified as obese by BMI standards alone. However, a clinician would find that his percentage of body fat is well below average and his waist circumference is within a healthy range. A diagnosis of obesity or overweight requires a BMI value *plus* some measure of body composition and fat distribution. There is no easy way to look inside a living person to measure bones and muscles, but indirect measures can reveal some clues.

key point *The body mass index mathematically correlates heights and weights with risks to health. It is especially useful for evaluating health risks of obesity but fails to measure body composition or fat distribution.*

fatfold test measurement of the thickness of a fold of skin on the back of the arm (over the triceps muscle), below the shoulder blade (subscapular), or in other places, using a caliper (depicted in Figure 9-6); also called *skinfold test*.

Estimating Body Fatness

A person who stands about 5 feet 10 inches tall and weighs 150 pounds carries about 30 of those pounds as fat. The rest is mostly water and lean tissues: muscles; organs such as the heart, brain, and liver; and the bones of the skeleton (see Figure 9-5). This lean tissue is vital to health. The person who seeks to lose weight wants to lose fat, not this precious lean tissue. And for someone who wants to gain weight, it is desirable to gain lean and fat in proportion, not just fat.

Laboratory techniques for estimating body fatness include these:

- *Anthropometry*. Direct body measurements include the **fatfold test** and waist circumference (see Figures 9-6 and 9-7 on pp. 320 and 321). Fatfold measurements

Figure 9•5

AVERAGE BODY COMPOSITION OF MEN AND WOMEN

The substantially greater fat tissue of women is normal and necessary for reproduction.

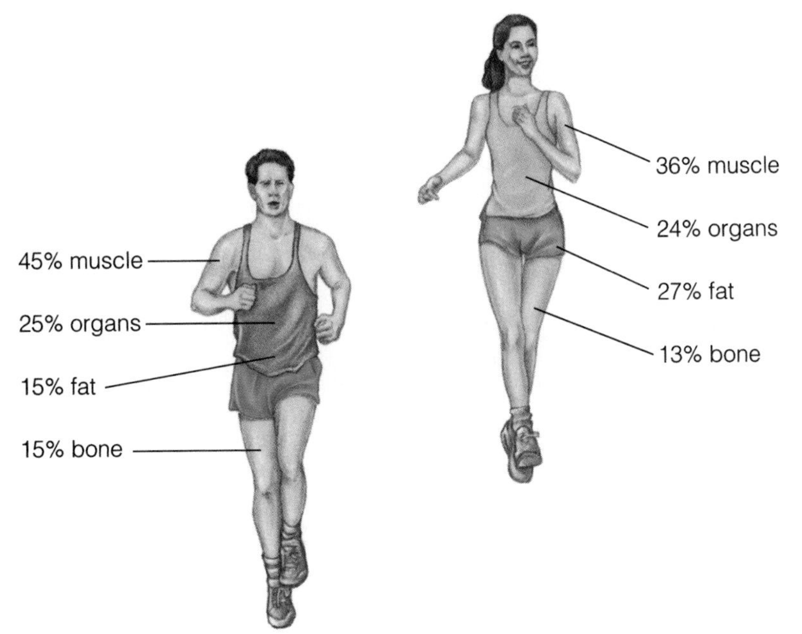

SOURCE: Data from R. E. C. Wildman and D. M. Medeiros, *Advanced Human Nutrition* (Boca Raton, Fla.: CRC Press, 2000), pp. 321–323.

The BMI standards are not accurate for athletes. At 6 feet 3 inches tall and 245 pounds, Mike O'Hearn would be judged to be obese by BMI standards alone. Further measures reveal that his body contains only 8 percent of its weight as fat, less than the average fat percentage for men, and that his waist circumference is within a healthy range.

underwater weighing a measure of density and volume used to determine body fat content.

bioelectrical impedance (im-PEE-dense) a technique for measuring body fatness by measuring the body's electrical conductivity.

dual energy X-ray absorptiometry (ab-sorp-tee-OM-eh-tree) **(DEXA)** a noninvasive method of determining total body fat, fat distribution, and bone density by passing two low-dose X-ray beams through the body. Also used in evaluation of osteoporosis.

For a fair indication of whether you develop fat centrally, measure your waist as shown in Figure 9-7; then compare your measurement with these cutoff points:

- Men: 102 centimeters (40 inches).
- Women: 88 centimeters (35 inches).

Anyone with a waist measurement larger than these standards may carry an increased risk of disease along with the extra girth.

taken by a trained technician with standard calipers provide an accurate estimate of total body fat and a fair assessment of the fat's location. Measures taken from central-body sites (around the abdomen) better reflect changes in fatness than those taken from upper sites (arm and back). Waist circumference indicates visceral fatness (Figure 9-7), and above a certain girth, disease risks rise—even when BMI values are normal.[21] An increasing number of people have too much body fat to be accurately measured with today's calipers.[22]

- *Density* (the measurement of body weight compared with volume). Lean tissue is denser than fat tissue, so the denser a person's body is, the more lean tissue it must contain. Density can be determined by **underwater weighing** or air displacement methods.
- *Conductivity.* Only lean tissue and water conduct electrical current; **bioelectrical impedance** measures how well a tiny harmless electrical charge is conducted through the lean tissue of the body and so reflects the body's contents of lean tissue, including water. A drawback is that temporary changes in body water content can produce changes in the measurement.
- *Radiographic techniques.* New technology yields images of body tissues and a more accurate assessment of body composition. For example, **dual energy X-ray absorptiometry (DEXA)** measures two beams of X-ray energy as they pass harmlessly through body tissues, giving high-quality assessments of total body fatness, fat distribution, and bone density. Drawbacks include the high cost of the equipment and a possible tendency to overestimate body fatness in people with a thick body shape.[23]

Each technique has strengths and weaknesses, but in all cases, the accuracy of the results depends on the skill of the clinician employing the technique and interpreting the results.

key point *A clinician can determine the percentage of fat in a person's body by measuring fatfolds, body density, or other parameters. Distribution of fat can be estimated by radiographic techniques and central adiposity by measuring waist circumference.*

Figure 9•6

THREE METHODS OF ASSESSING BODY FATNESS[a]

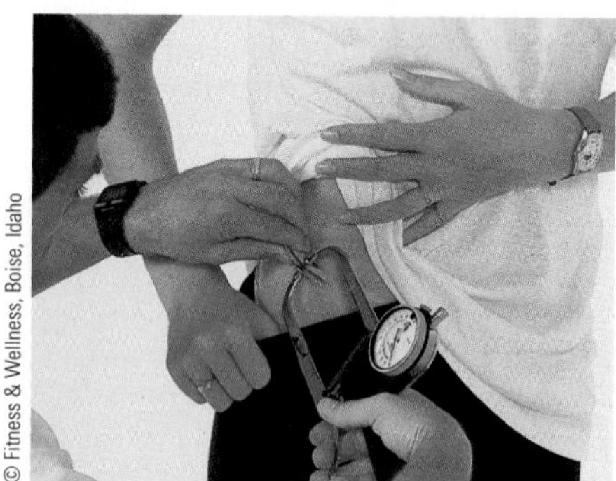

© Fitness & Wellness, Boise, Idaho

Fatfold measures can yield accurate results when a trained technician measures body fat by using a caliper to gauge the thickness of a fold of skin. Measurements are taken on the back of the arm (over the triceps), below the shoulder blade (subscapular), and in other places (including lower body sites) and are then compared with standards.

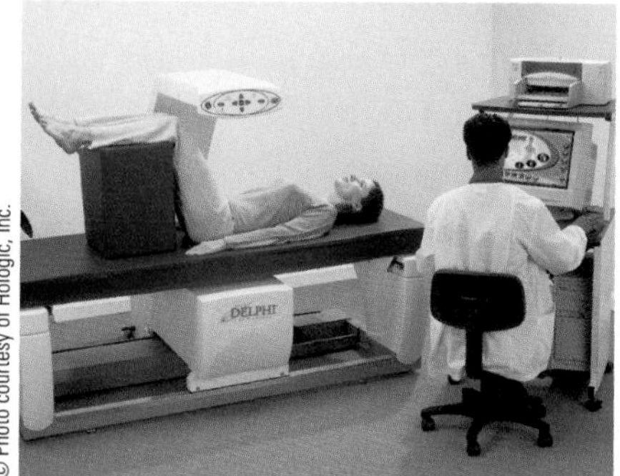

© Photo courtesy of Hologic, Inc.

Dual energy X-ray absorptiometry (DEXA) employs two low-dose X-rays that differentiate among fat-free soft tissue (lean body mass), fat tissue, and bone tissue, providing a precise measurement of total fat and its distribution in all but extremely obese subjects.

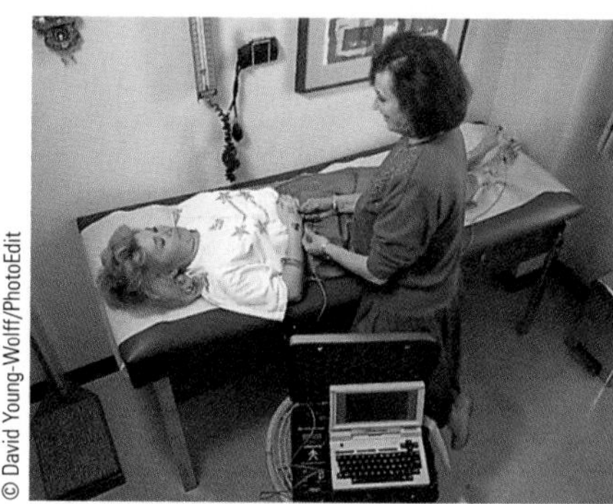

© David Young-Wolff/PhotoEdit

Bioelectrical impedance is simple, painless, and accurate when properly administered; the method determines body fatness by measuring conductivity. Lean tissue conducts a mild electric current; fat tissue does not.

[a]Other methods inlcude underwater weighing (hydrodensitometry), computed tomography, and magnetic resonance imaging.

How Much Body Fat Is Ideal?

After you have a body fatness estimate, questions arise. What is the "ideal" amount of fat for a body to have? This prompts another question: Ideal for what? If the answer is "society's approval," be aware that fashion is fickle. Body shapes valued for looks may have little to do with health, and many of the most popular body shapes are not achievable goals for most people.

If the answer is "health," then the ideal depends partly on who you are. A man of normal weight will have betweeen 12 and 20 percent of body weight as fat, and a woman will have between 20 and 30 percent. Researchers draw the line when body fat exceeds 22 percent in young men, 25 percent in older men, 32 percent in younger women, and 35 percent in older women; age 40 is the dividing line between younger and older.

Besides gender and age, standards differ because of lifestyle and stage of life. For example, competitive endurance athletes need just enough body fat to provide fuel, insulate the body, and permit normal fat-soluble hormone activity, but not so much fat as to weigh them down. An Alaskan fisherman, in contrast, needs a blanket of extra fat to insulate against the cold. For a woman starting pregnancy, the outcome is compromised if she begins with too much or too little body fat (see Chapter 12). Below a threshold for body fat content set by heredity, some individuals become infertile, develop depression or abnormal hunger regulation, or become unable to keep warm. These thresholds are not the same for each function or in all individuals, and much remains to be learned about them.

Seeking a single, authoritative answer to the question "How much should I weigh?" will bring disappointment. No one can tell you exactly how much you should weigh; but with health as a value, you have a starting framework in the BMI table. Your weight should fall within the range that best supports your health.

No single body composition suits everyone; needs vary by gender, lifestyle, and stage of life.

Figure 9•7

MEASURING WAIST CIRCUMFERENCE

Using a nonstretching tape measure, measure the body around the point near the belly button. (The skeleton shows the tape position relative to the hip bone.) Exhale normally while taking the measurement. A healthy waist circumference for men is no larger than 102 centimeters (40 inches); for women, no larger than 88 centimeters (35 inches).

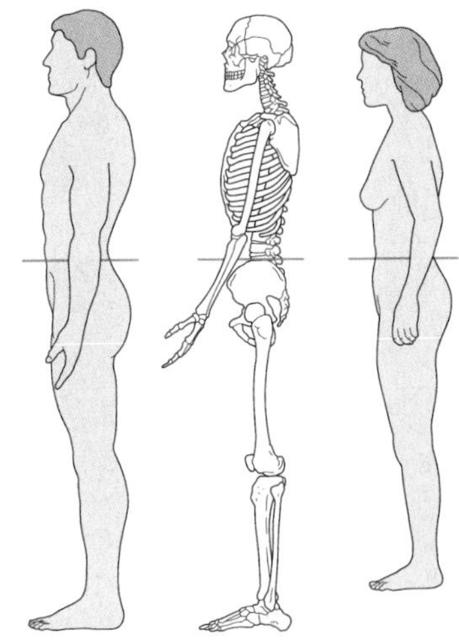

SOURCE: National Institutes of Health Obesity Education Initiative, *Clinical Guidelines on the Identification, Evaluation, and Treatment of Overweight and Obesity in Adults* (Washington, D.C.: U.S. Department of Health and Human Services, 1998), p. 59.

The Mystery of Obesity

Why do some people get fat? Why do some stay thin? Is weight controlled by hereditary metabolic factors or by environmental influences? Is it a matter of eating behaviors—and if so, what directs these behaviors, internal controls or a person's free will? Many factors, some of them conflicting, *correlate* with obesity (see the margin), but *cause* is elusive. This section sorts through the pieces of the obesity puzzle, but no law says that only one cause must prevail. In all likelihood, internal and external factors operate together and in different combinations. The following sections address obesity theories with substantial evidence to support them. We begin by examining the appetite and its controls.

Research has linked obesity with:

- Birth order, number of brothers.
- Divorced/single parents, nonprofessional or unemployed parents.
- Early menstruation.
- Ethnicity.
- Fat intake, protein intake, carbohydrate intake.
- Less leisure time, international travel, geographic location.
- Exposure to a variety of foods, fast-food consumption.
- Lower education level, lower social class.
- Increased wealth.
- Maternal famine or obesity during gestation.
- Meal skipping, meals eaten away from home.
- Napping habits.
- Reduced alcohol intake, increased alcohol intake.
- Sedentary behavior, television watching.
- Substandard housing.
- Many more.

SOURCE: Adapted from L. Grivetti, Psychology and cultural aspects of energy, *Nutrition Reviews* 59 (2001): S5–S12.

Why Did I Eat That?

Obesity researchers are interested in why people eat and what they eat, and especially why some people overeat. Overeating is thought to explain much of people's overweight, and scientists hope that by discovering how food intake is regulated in the body, they can devise effective strategies for obesity prevention and treatment.[24] Eating behavior seems to be regulated by a series of signals that fall into two broad functional categories: "go" mechanisms that stimulate eating and "stop" mechanisms that signal the body to cease or refrain from eating. One view of the process of food intake regulation is summarized in Figure 9-8 on the next page.

hunger the physiological need to eat, experienced as a drive for obtaining food; an unpleasant sensation that demands relief.

Figure 9•8

A CASCADE OF REGULATION: HUNGER, APPETITE, SATIATION, AND SATIETY

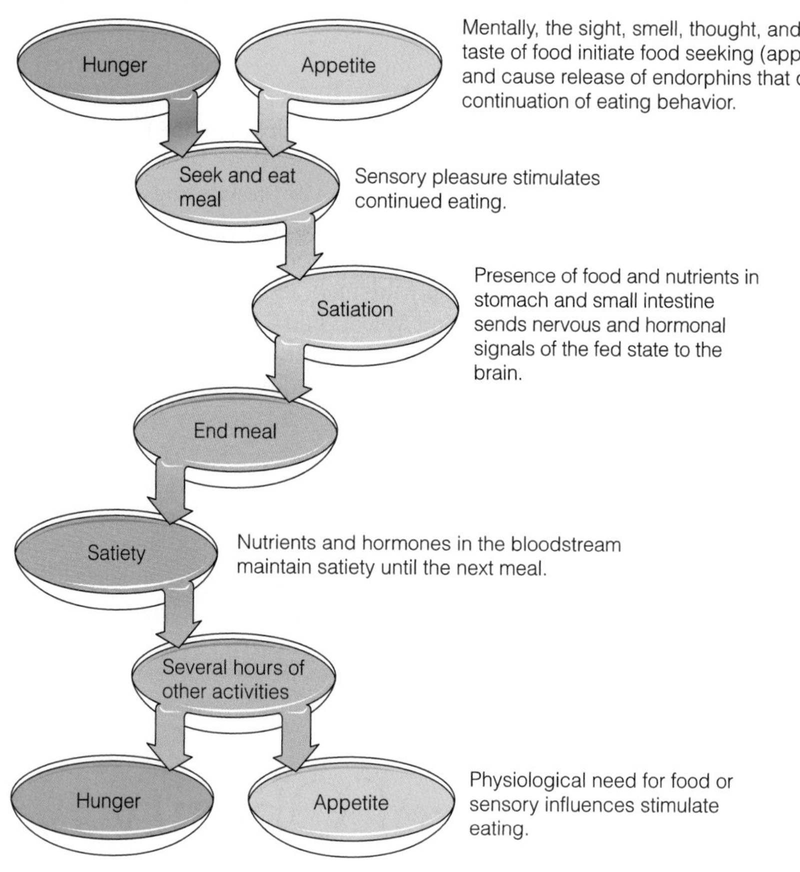
Physically, hunger is produced by hormones, absence of food from the digestive tract, and neurotransmitters in the brain.
Hunger
Appetite
Mentally, the sight, smell, thought, and taste of food initiate food seeking (appetite) and cause release of endorphins that drive continuation of eating behavior.
Seek and eat meal
Sensory pleasure stimulates continued eating.
Satiation
Presence of food and nutrients in stomach and small intestine sends nervous and hormonal signals of the fed state to the brain.
End meal
Satiety
Nutrients and hormones in the bloodstream maintain satiety until the next meal.
Several hours of other activities
Hunger
Appetite
Physiological need for food or sensory influences stimulate eating.

SOURCE: Adapted from ideas in J. E. Blundell and coauthors, Control of human appetite: Implications for the intake of dietary fat, *Annual Review of Nutrition* 16 (1996): 285–319.

Hunger is physiological and asks, "Is there anything to eat?" Appetite is psychological and asks, "What do I feel like eating?"

Chapter 3 described the brain's hypothalamus.

"Go" Signals—Hunger and Appetite Most people recognize **hunger** as an irritating feeling that prompts them to search out food and to start eating. Hunger is the physiological response to a need for food triggered by chemical messengers acting in the brain, especially in the brain's hypothalamus.[25] Other factors, including the nutrients present in the bloodstream, the size and composition of the preceding meal, customary eating patterns, the weather (heat reduces food intake; cold increases it), exercise, sex hormones, and physical and mental disease states, also influence hunger. Hunger makes itself known roughly four to six hours after eating, after the food has left the stomach and much of the nutrient mixture has been absorbed by the intestine.

The body's hunger response adapts quickly to changes in food intake. A person who restricts the amount of food consumed at each meal may feel extra hungry for a few days, but then hunger may diminish for a time. This dimming of hunger represents an adaptation to smaller quantities of food. During this period, a large meal may make the person feel uncomfortably full, partly because the stomach's capacity has adapted to a smaller quantity of food.[26] At this time a dieter may report "My stomach has shrunk," but the stomach organ itself doesn't shrink except in cases of chronic starvation.

Stomach adaptation may seem to be good news for dieters, but at some point in food deprivation, hunger returns with a vengeance and can lead to bouts of overeating that more than make up for the calories lost during the deprivation period. And just as the stomach's capacity can adapt to small meals, it can also adapt to larger and larger quantities of food, until a meal of normal size no longer feels satisfying. This observation may partly explain why obesity is on the rise: popular demand has led to larger and larger platefuls of pasta, super-sized fast-food meals, and huge candy bars and soft drinks, and stomachs have adapted to accommodate them.

Like hunger, **appetite** also initiates eating, but unlike hunger, appetite is learned. When the two coincide, appetite intensifies hunger, but a person can experience appetite without hunger. For example, seeing and smelling a freshly baked apple pie after finishing a big meal can stimulate the appetite for the pie—despite an already full stomach. In contrast, a person who is under stress or ill may feel hunger but have no appetite for food. Other factors affecting appetite include:

- Learned preferences, aversions, and timings (cravings for favorite foods, fear of new foods, eating according to the clock).
- Environmental conditions (people prefer hot foods in cold weather and vice versa).
- Brain compounds.
  - **Endorphins,** the brain's pleasure chemicals that enhance desire for the taste of delicious foods and may be triggered by the smell, sight, or taste of foods, as mentioned in Figure 9-8.
  - **Serotonin,** a compound chemically related to the amino acid tryptophan that confers feelings of relaxation and satiation in response to meals (Controversy 13 provides details).
- Inborn appetites (inborn preferences for fatty, salty, and sweet tastes).
- Social interactions (cultural or religious acceptability of foods, companionship).
- Some disease states (obesity may be associated with increased taste sensitivity, whereas colds, flu, and zinc deficiency reduce taste sensitivity).
- Some drugs (appetite stimulants or depressants and some mood-altering drugs affect food intake).

appetite the psychological desire to eat; a learned motivation and a positive sensation that accompanies the sight, smell, or thought of appealing foods.

endorphins (en-DORE-fins), **endogenous opiates** compounds of the brain whose actions mimic those of opiate drugs (morphine, heroin) in reducing pain and producing pleasure. In appetite control, endorphins are released on seeing, smelling, or tasting delicious food and are believed to enhance the drive to eat or continue eating.

serotonin (SER-eh-TONE-in) a compound related in structure to (and made from) the amino acid tryptophan. It serves as one of the brain's principal neurotransmitters and reduces appetite. Some weight-loss drugs aim to elevate the serotonin levels in the brain.

satiation (SAY-she-AY-shun) the perception of fullness that builds throughout a meal, eventually reaching the degree of fullness and satisfaction that halts eating. Satiation generally determines how much food is consumed at one sitting.

satiety (sah-TIE-eh-tee) the perception of fullness that lingers in the hours after a meal and inhibits eating until the next mealtime. Satiety generally determines the length of time between meals.

"Stop" Signals—Satiation and Satiety At some point during a meal, the brain receives messages from several sources that enough food has been eaten. Called **satiation,** this condition originates from the presence of food in the upper GI tract (consult Figure 9-8). When the stomach stretches to accommodate a meal, nerve receptors in the stomach fire, sending a signal to the brain that the stomach is full and causing the person to stop eating.[27] As nutrients from the meal enter the small intestine, they stimulate receptor nerves and trigger the release of hormones that provide the brain with information about the meal just eaten. The brain also detects absorbed nutrients passing by in the blood and releases neurotransmitters that suppress food intake in response.[28] Together, stomach distention, nutrients in the small intestine, and hormonal and neural signals inform the brain's hypothalamus about the size and nature of the meal. The response: satiation occurs; the eater feels full and stops eating.

Figure 2-7 in Chapter 2 demonstrated how portion sizes have increased over recent decades.

After a meal, the feeling of **satiety** continues to suppress hunger and allows for a period of some hours in which the person is free to dance, study, converse, wonder, fall in love, and concentrate on endeavors other than eating. Whereas satiation informs the body when to stop eating, satiety allows the body to stay stopped for a while.

A lack of satiety between meals can cause problems when people attempt to reduce their food intakes. After hours of annoying hunger pains, people may overeat regrettably when mealtime arrives, setting up a destructive cycle of starving and binge eating, with no weight loss to show for the effort. The choice of foods may affect satiety—some foods seem to sustain feelings of satiety for longer periods than others.

One attempt to rank individual foods by their satiety value produced the data of Figure 9-9. Here, foods high in fiber or protein seemed to sustain satiety for longer than those high in fat, white flour, or sugar.[29] The test portions were equal in calories,

set-point theory the theory that the body tends to maintain a certain weight by means of its own internal controls.

thermogenesis the generation and release of body heat associated with the breakdown of body fuels. *Adaptive thermogenesis* describes adjustments in energy expenditure related to changes in environment such as cold and to physiological events such as underfeeding or trauma.

and fat-rich foods pack many calories into a small bulk, so the greater volume of bulky foods in the digestive tract may have kept subjects feeling full longer. Fat in food is known to trigger the release of a hormone that slows digestion and produces satiety.

Another possible connection to satiety lies in the glycemic index, first explained in Chapter 4. Foods that elicit minimal insulin release (they rank low on the glycemic index scale, see the margin on page 116 of Chapter 4) may sustain satiety longer than foods ranking higher on the glycemic index scale.[30] Preliminary studies suggest that a diet of low-glycemic foods may even help to control obesity, but long-term experiments are needed to confirm this idea.[31]

A food's water content may also be a factor. Water adds volume to food, and foods with high water content, such as soups, seem to lend satiety to a meal, at least in some studies.[32] Even puffing up the volume of a food with air may add to satiety, but the resulting air gurgling through the digestive tract may make this option less than attractive.[33]

key point *Food intake is regulated by hunger, appetite, satiation, and satiety. Discoveries of neural and hormonal regulators of eating behaviors may lead to new effective pharmacological treatments of obesity. Some foods may confer greater satiety than others.*

Inside-the-Body Causes of Obesity

Although interesting and important, findings about appetite regulation do not fully explain why some people gain too much body fatness while others stay lean. Research on the metabolic pathways that affect weight gain has resulted in many theories; several are listed in Table 9-4. At the bottom of all of them lies a person's genetic inheritance.

Genetics and Weight Gain Genetic makeup influences the body's tendency to consume or store too much energy or burn too little, and a child's body fatness is often closely related to that of the parents.[34] For a person who has one obese parent, the chance of becoming obese is 60 percent; if both parents are obese, the probability may rise to as high as 90 percent. Adopted children tend to be similar in weight to their biological parents, not to their adoptive parents. Studies of twins bear this out: identical twins are twice as likely as fraternal twins to weigh the same when reared apart in similar households. When identical twins are reared apart in households that differ in terms of smoking, eating behaviors, nutrient intakes, and physical activity, however, their weights more often differ according to their adopted lifestyle habits.[35] Such findings indicate that although genetics strongly influences a person's susceptibility to obesity, lifestyle choices help to determine whether the inborn tendency to gain weight is expressed.[36]

Genetics does affect a person's tendency to gain or lose weight when eating more or less food energy than needed.[37] When given an extra 1,000 calories of food a day for 100 days, some people gain 30 pounds, but others gain less than 10 pounds. Similarly, some people lose more weight faster than others on comparable exercise regimens. This phenomenon is part of the **set-point theory** (see Table 9-4). Researchers are focusing on a genetic link to obesity that may produce a "thrifty" metabolism and so predispose people to conserve energy and store fat when lifestyle conditions are right.[38]

Figure 9•9

SATIETY SCORES OF FOODS

Higher satiety

Food	Score
Boiled white potatoes	323
Baked fish	225
Oatmeal with milk[a]	209
Orange, apple	200
Whole-grain pasta	188
Beefsteak, baked beans	170
Popcorn, eggs, bran cereal with milk[a]	150
Brown rice or white rice	135
White bread	100
Snack chips, ice cream	94
Candy bar	70
Cake, doughnuts	67
Croissant	47

Lower satiety

[a]Cereals were served with 1.5% fat milk.

SOURCE: Data from S. H. A. Holt and coauthors, A satiety index of common foods, *European Journal of Clinical Nutrition* 49 (1995): 675–690.

Energy-Wasting Proteins All proteins made in the body are products of genetic inheritance, and some of these proteins, enzymes involved in energy use, influence the tissues to store or spend energy with different efficiencies. Table 9-4 describes some obesity theories centering on enzymes that expend energy in heat production, or **thermogenesis,** but produce no other useful work in the process. Radiating energy away as heat enables the body to spend, rather than store, excess energy.

Genes code for the cellular protein products involved in energy metabolism. Researchers have discovered one of these genes, active in many body tissues.[39] Its actions

Table 9•4

SELECTED THEORIES OF METABOLIC CAUSES OF OBESITY

Theory	Mechanism of Action
Enzyme theory	Excess fat storage may stem from elevated concentrations of an enzyme, **lipoprotein lipase (LPL),** that enables fat cells to store triglycerides. The more LPL, the more easily fat cells store lipid, and the more likely the body will remain obese. The fat cells of obese people contain more LPL than the fat cells of lean people, and therefore reach a large size quickly.
Fat cell number theory	Body fatness is determined by both the number and the size of fat cells. Fat cells increase in number during the growing years, tapering off in adulthood. Fat cell number may increase more rapidly in obese children than in lean children, leading to a lifelong tendency toward obesity.
Set-point theory	The body may "choose" a weight it wants to be and defend that weight by regulating behaviors and hormonal actions. Just as a thermostat setting triggers a heater to run when air temperature falls and turn off when warmth is restored, whenever weight is lost or gained, the set-point mechanism changes metabolic energy expenditure to restore the "chosen" body weight. The theories of thermogenesis, below, explain possible mechanisms by which the body defends its set point.
Thermogenesis I: Energy-wasting proteins and brown fat theory	Proteins control the body's heat production, or thermogenesis. A type of adipose tissue, brown fat, has abundant energy-wasting proteins that specialize in converting energy to heat. Whereas regular white fat cells have a sluggish metabolism and conserve and store fat energy, brown fat actively metabolizes fat, releasing its stored energy as heat. Brown fat is more abundant in lean animals than in fat ones, and this theory states that a person with more brown fat and therefore more energy-wasting proteins may stay leaner. Human infants have abundant brown fat, but the amount dwindles with age.
Thermogenesis II: Adaptive thermogenesis theory	Many tissues, such as muscle, spleen, and bone marrow, convert stored energy into heat in response to cold temperature, physical conditioning, overfeeding, starvation, trauma, and other stress. Heat is also produced to "waste" fuel without useful work when energy supplies are too high; conversely, with low energy supplies, energy is conserved. Genetic inheritance is thought to determine the efficiency of this system. Dieters' efforts are often thwarted when, on reducing food intake, metabolism slows and heat production diminishes.
Thermogenesis III: Diet-induced thermogenesis theory	The thermic effect of food varies between obese and nonobese people. In lean people who have just eaten a meal, energy use speeds up for a while, but in many obese people, no change in energy use occurs after eating. In theory, this small difference in energy expenditure may account for an accumulation of body fat, but no studies have shown this conclusively. Overweight people often spend more energy each day than lean people do because their heavier bodies require more energy to move and maintain.

seem to influence the basal metabolic rate (BMR) and oppose the development of obesity.[40] In animals, an abundance of this protein predicts lean body weight. Children who possess a variant of the gene resulting in a slightly altered protein are reliably overweight.[41] Researchers attempting to step up the body's energy-wasting systems by manipulating this gene or its product must proceed with caution, however. At a level of activity not far beyond that of normal functioning, energy-wasting activity kills the cells.[42]

Leptin **Leptin** is a hormone produced by the adipose tissue that is directly linked to both appetite and body fatness. Leptin's functions include suppression of appetite in harmony with the brain's appetite-regulating chemistry and stimulation of the muscles to increase energy expenditure.[43] Further, the adipose tissue that is ultimately controlled by leptin also produces it, affording a precise feedback mechanism.[44] A gain in body fatness stimulates the production of leptin, which reduces food consumption and produces fat loss. Fat losses bring the opposite effect—a suppression of leptin and an increase in appetite.

Scientists are still exploring leptin in hopes that it may one day help in treating human obesity, but trials of leptin injections have so far been disappointing.[45] One reason may be that an increased amount of leptin in the blood does not necessarily produce a concurrent increase of leptin in the brain, except in a tiny percentage of obese people who fail to produce leptin at all.[46] (The mice in the photo on the next page demonstrate this abnormality.) This would explain why most overweight people, with their large amounts of fat tissue and thus greater output of leptin, still feel hungry and continue to grow fatter.[47] Another reason may be that leptin signals when enough, rather than too much, energy has been stored.[48]

lipoprotein lipase (LPL) an enzyme mounted on the surfaces of fat cells that splits triglycerides in the blood into fatty acids and glycerol to be absorbed into the cells for reassembly and storage.

leptin an appetite-suppressing hormone produced in the fat cells that conveys information about body fatness to the brain; believed to be involved in the maintenance of body composition (*leptos* means "slender").

Controversy 13 discusses the brain's blood-brain barrier.

Even if leptin never proves useful as an antiobesity drug, its discovery has contributed much to our understanding of the workings of the human body. For example, scientists once viewed adipose tissue solely as a metabolically sluggish storage depot for lipids. Now, adipose tissue is recognized as a hormonally active regulatory tissue with widespread effects on the body.[49] In addition to its appetite function, leptin signals the female reproductive system as to whether body fat reserves are adequate for reproduction; stimulates growth of new blood vessels in the cornea of the eye and elsewhere; acts on bone marrow cells to enhance their maturation into specialized cells; promotes formation of red blood cells; and helps support a normal immune response.[50]

If obesity results from a person's genetic makeup, does this mean that obesity is inevitable and unpreventable? This fatalistic view is put to rest by a simple observation: obesity rates have increased dramatically in recent years wherever in the world prosperity and abundance have emerged. During the same time, the human gene pool has remained the same.[51] An individual's genes may make obesity likely, but the disease of obesity cannot develop unless the environment provides the means of doing so.

key point *Inside-the-body causes of obesity include a person's genetic inheritance and its expression in terms of the handling of energy by the tissues. Energy-wasting proteins help to maintain normal body weight when food energy intake exceeds need. Leptin, a peptide hormone secreted by the adipose tissue, suppresses the appetite.*

Outside-the-Body Causes of Obesity

Food is a pleasure and people like to indulge. From traditional restaurants to fast-food chains and the grocery store, we have more food choices than ever before. And researchers suspect a link between increased food availability and the nation's increasing obesity. Another factor may be our increasing dependence on labor-saving inventions, such as automobiles and elevators.

External Cues to Overeating Being creatures of free will, people can override signals of satiety and hunger and eat whenever they wish, especially when presented with delicious options. Almost everyone has had the experience of walking into a store, not feeling particularly hungry, and walking out snacking on a favorite treat. A classic experiment showed that animals will eat when presented with delectable foods, even if they are not hungry. Rats, known to maintain body weight with precision when fed standard rat chow, rapidly became obese when fed "cafeteria style" on a variety of rich, palatable foods. Many people, too, are prone to gain weight when their diets provide a wide variety of rich, palatable foods, such as sweets, snacks, condiments, and main dishes.[52] Another truth pairs with this one: consumption of a wide variety of vegetables, but not many treats, correlates with lower body fatness.

One food constituent stands out as being palatable to most people—fat. Not only does fat entice people to eat, but it also delivers more than twice the calories, gram for gram, as protein and carbohydrate, and the body stores fat preferentially and with great efficiency in many people.[53] Of the three energy nutrients, fat also stimulates the least energy expenditure in diet-induced thermogenesis and may be the least satiating during a meal, thus leading to overconsumption of calories at meals. A person whose diet contains much fat is often a person who battles against overweight.[54]

The mouse on the right is genetically obese—it lacks the gene for producing leptin. The mouse on the left is also genetically obese but remains lean because it receives leptin.

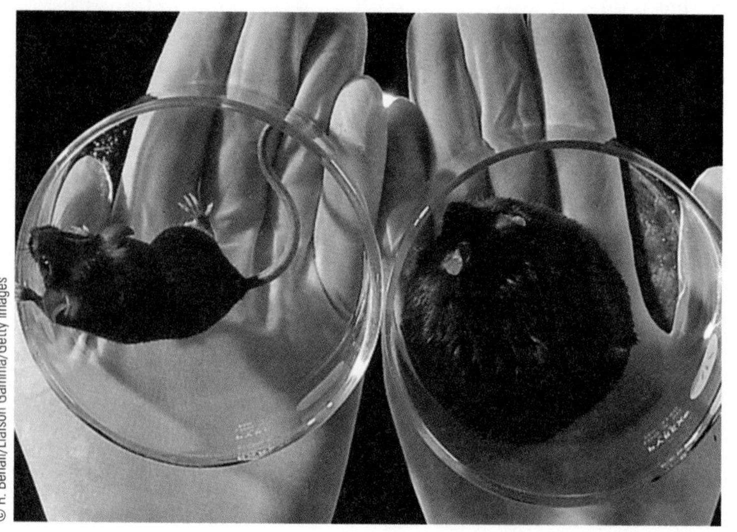

Eating behavior also occurs in response to complex human sensations such as loneliness, yearning, craving, addiction, or compulsion. Some people experience food cravings when feeling down or depressed. Food picks them up for a while. Some respond to other external stimuli such as the time of day ("I'm not hungry, but it's time for lunch.")

Any kind of stress can cause overeating, perhaps because **arousal** feelings are mistaken for hunger. ("What do I do when I'm grieving? Eat. What do I do when I'm celebrating? Eat!") The opposite can also be true, however. Some people undereat or cannot eat at all when under stress. Although all of these behaviors can lead to the overconsumption of food energy, they cannot fully explain obesity development because even thin people are susceptible to them.

arousal heightened activity of certain brain centers associated with attention, excitement, and anxiety.

Food Availability Most people living in the United States find high-calorie, high-fat fast foods readily available, relatively inexpensive, heavily advertised, and wonderfully delicious—not just in stand-alone restaurants, but in malls, grocery stores, filling stations, airports, and even the nation's school lunch rooms.[55] The Consumer Corner questions the wisdom of such attractive foods being available everywhere, at every hour of the day and night.

A steady diet of high-calorie, high-fat fast food probably encourages obesity.[56] Such foods lack the bulk of fiber, and their low water content may encourage overeating at a meal.[57] Does this mean that everyone who eats fast food is destined to become obese or that people should never partake of a fast-food meal? Of course not. Moderation remains a bedrock of nutrition common sense and holds true with regard to fast food.

Physical Inactivity One hundred years ago, 30 percent of the energy used in farm and factory work came from human muscle power; today, only 1 percent does. The same trend follows at home, at work, and in transportation. Inactivity contributes to weight gain and poor health.[58]

Some people may be obese, not because they eat too much, but because they move too little. Diet histories from obese people often report energy intakes that are similar to, or even less than, those of others. (Diet histories may not be accurate records of actual intakes, though; both normal-weight and obese people commonly underreport their dietary intakes, especially of high-fat, high-sugar foods.)[59] Some obese people are so inactive that even when they eat less than lean people, they still have an energy surplus.

Physical activity allows people to eat enough food to obtain the nutrients they need without weight gain. In fact, some threshold of physical activity seems to protect against weight gain.[60] This activity must be the active kind, not passive motion such as being jiggled by a machine at a health spa or being massaged. The threshold for weight maintenance appears near the moderate-to-heavy level of activity intensity (as defined in the margin list on p. 318).

Despite the benefits of body composition and reduced disease risk from physical activity, the United States seems locked in to what has been called an epidemic of inactivity.[61] For many people, television watching has all but replaced outdoor work and play as the major leisure time activity. One study showed that in children, obesity increases by 2 percent per hour of television watching per day. Another revealed that watching television costs *less* energy than simply doing nothing. The Think Fitness feature on page 330 underscores the importance of physical activity in weight management.

Choices in an Obesity-Prone Society

WHAT'S A CONSUMER to do? The easiest, most accessible, most inviting, and oftentimes least expensive consumer options encourage unhealthy behaviors.[1] Automobile-dependent communities offer efficient roadways but no sidewalks, thereby inviting a sedentary lifestyle; urban sprawl locates shopping malls and workplaces within commuting distances, but not within walking or biking distances, from residences; buildings with conveniently located elevators often hide stairwells from view; fast-food fried chicken, hamburger, and pizza places are convenient, speedy, and affordable; more healthful choices may require more effort and time to locate and are often more expensive. Consumers who would prefer to walk to work, use the stairs, or choose foods to support health often find roadblocks in their way.

Expanding U.S. food portions are probably also contributing to expanding U.S. waistlines.[2] If you doubt that small daily decisions such as the choice of lunch foods can influence body weight and health, try this: turn back to p. 44 in Chapter 2, and look at the foods depicted in Figure 2-7. Add up the calories in a 1970s hamburger, cola, and french fries (similar to today's "small" sizes). Do the same for the calories in the "colossal" size hamburger, cola, and french fries of present-day meals.

Now find the calorie difference between the two meals (subtract the smaller sum from the larger) and multiply the difference by 52 (for weeks in a year).

Today's calorie total =	2,020
− 1970s calorie total =	925
Calorie difference =	1,095

Multiplied by 52 weeks in a year, this adds up to an extra 56,940 calories. If a pound of body fat can be gained with each excess 3,500 calories, then a person who "super-sizes" a fast-food meal *just once per week* stands to gain over 16 pounds in a year's time. Many little daily decisions such as this add up over time.

One public health group suggests increasing incentives for healthy behaviors by placing a monetary tax on high-calorie, high-fat, high-salt fast-food choices, sugary sodas, candies, and fried snack foods.[3] These tax revenues could then be used to subsidize the prices of health-promoting salads, low-fat chili, whole-grain breads and buns, broiled chicken and fish, low-fat milk, and fruit and vegetables to make the healthiest choices also the most affordable and available. And why stop there, they ask. We could tax automobiles and other labor-saving devices that rob people of the exercise they need, while giving price incentives for purchasing and using bicycles, exercise equipment, and walking shoes. They propose restructuring our whole society (see Figure 9-10) to promote physical activity and reduce stress by developing small, centralized communities where people walk to work, restaurants, and shops in pleasant environments and where building designs invite the use of stairways by most people, while elevators are made available for the handicapped.

Such a plan would meet resistance in our free-market society, but consumers can choose to *live* as though such incentives were already in place:

- We can turn off the television and take up a physically active recreation.
- We can choose to live close enough to shopping, work, and school to allow walking as an option.
- We can park farther away than usual and walk through the parking lot.
- We can find stairways and use them.
- We can choose to consistently follow the Food Guide Pyramid and the *Dietary Guidelines*, even if it means paying more for our foods.
- We can save enticing fast foods and snacks for special occasions.

People who consistently make such choices often find that body weights fall into line with recommendations and disease risks drop dramatically. A good place to start is by reading the suggestions in the final sections of this chapter and then following them.

Figure 9•10

OBESITY-PROMOTING AND FITNESS-PROMOTING ENVIRONMENTS COMPARED

Obesity-Promoting Environment *This environment promotes a sedentary lifestyle and overconsumption of high-energy foods.*

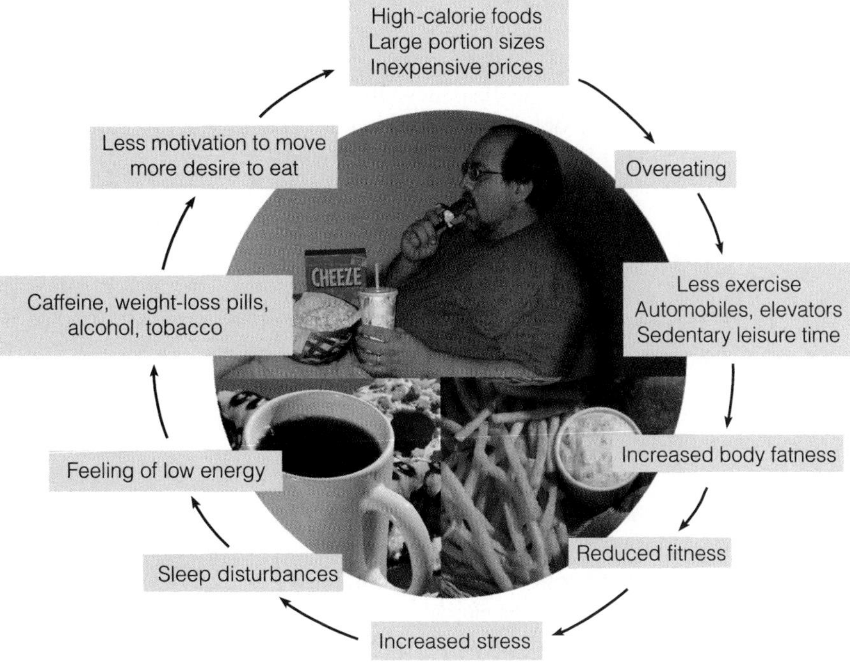

Fitness-Promoting Environment *This environment promotes sound nutrition and physical activity choices that lead to fitness.*

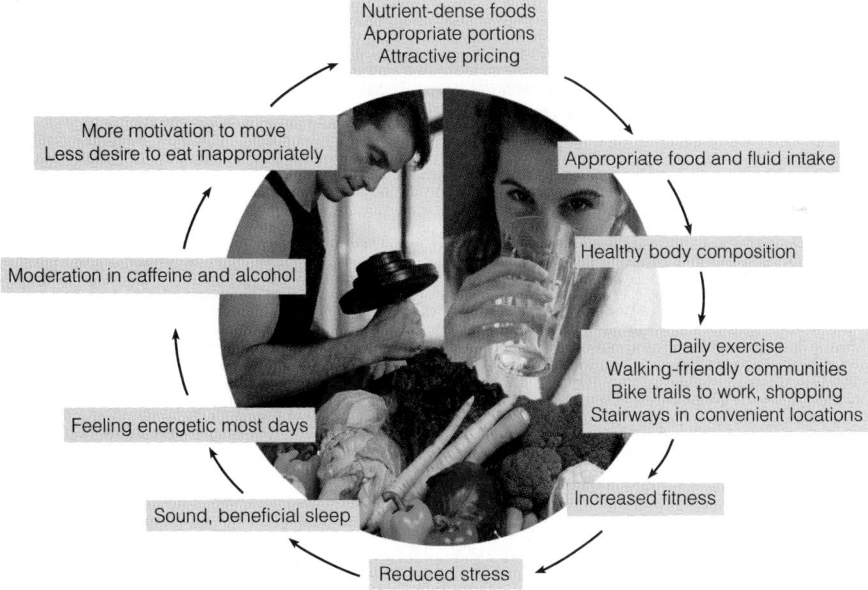

SOURCE: Obesity-promoting environment adapted from ideas in J. P. Foreyt and G. K. Goodrick, Dieting and weight loss: The energy perspective, *Nutrition Review* 59 (2001): S25–S26.

Activity for a Healthy Body Weight

According to the National Institutes of Health, exercise using about 150 calories of energy per day, or about 1,000 calories per week, constitutes moderate activity that can help build and maintain lean tissue while reducing body fatness and helping to protect against cardiovascular disease. Table 9-5 shows the approximate energy costs of various activities.

Some people believe that physical activity must be long and arduous to achieve fat loss. Not so. A brisk, 30-minute walk four or five days a week can help. Another strategy is to incorporate bits of physical activity into your daily schedule in many simple, small-scale ways. Work in the garden; work your abdominal muscles while you stand in line; stand up straight; walk up stairs; fidget while sitting down; tighten your buttocks each time you get up from your chair. Small energy expenditures add up to significant contributions. The margin provides some tips for staying active, and many more details are found in Chapter 10.

THINK FITNESS

Physical activity for weight loss or maintenance:

1. Choose moderate activities.
2. Move large muscle groups.
3. Invest longer times in physical activity.
4. Adopt informal strategies to be more active.

Physical activity for building body mass:

1. Choose strength-building exercises.
2. Use a balanced exercise routine.
3. Perform exercises with increasing intensity.
4. Adopt informal strategies to be more active.

Physical activity can help to regulate the appetite, help overweight people lose fat, and help underweight people gain muscle.

End of Story? Anyone involved in a good mystery wants to know how it ends. In the case of the causes of obesity, no one yet knows which of the suspects is the real culprit, and until evidence proves otherwise, any or all may be guilty as charged. In real life, the best way for most people to attain a healthy body weight boils down to control in three areas: diet, physical activity, behavior modification.[62] Later sections focus on these three areas. The next section delves into the details of how, exactly, the body loses and gains weight.

Reminder: The three lifestyle components leading to healthy body weight are diet, physical activity, and behavior modification.

key point *Studies of human behavior identify stimuli that lead to eating and exercise habits. Physical inactivity is clearly linked with overfatness. Obesity treatment involves diet, exercise, and behavior modification.*

Table 9•5

ENERGY DEMANDS OF ACTIVITIES

Activity	Energy per Pound of Body Weight per Minute	110	125	Body Weight (lb) 150	175	200
	cal/lb/min[a]			cal/min[b]		
Aerobic dance (vigorous)	.062	6.8	7.8	9.3	10.9	12.4
Basketball (vigorous, full court)	.097	10.7	12.1	14.6	17.0	19.4
Bicycling						
13 miles per hour	.045	5.0	5.6	6.8	7.9	9.0
15 miles per hour	.049	5.4	6.1	7.4	8.6	9.8
17 miles per hour	.057	6.3	7.1	8.6	10.0	11.4
19 miles per hour	.076	8.4	9.5	11.4	13.3	15.2
21 miles per hour	.090	9.9	11.3	13.5	15.8	18.0
23 miles per hour	.109	12.0	13.6	16.4	19.0	21.8
25 miles per hour	.139	15.3	17.4	20.9	24.3	27.8
Canoeing (flat water, moderate pace)	.045	5.0	5.6	6.8	7.9	9.0
Cross-country skiing (8 miles per hour)	.104	11.4	13.0	15.6	18.2	20.8
Golf (carrying clubs)	.045	5.0	5.6	6.8	7.9	9.0
Handball	.078	8.6	9.8	11.7	13.7	15.6
Horseback riding (trot)	.052	5.7	6.5	7.8	9.1	10.4
Rowing (vigorous)	.097	10.7	12.1	14.6	17.0	19.4
Running						
5 miles per hour	.061	6.7	7.6	9.2	10.7	12.2
6 miles per hour	.074	8.1	9.2	11.1	13.0	14.8
7.5 miles per hour	.094	10.3	11.8	14.1	16.4	18.8
9 miles per hour	.103	11.3	12.9	15.5	18.0	20.6
10 miles per hour	.114	12.5	14.3	17.1	20.0	22.9
11 miles per hour	.131	14.4	16.4	19.7	22.9	26.2
Soccer (vigorous)	.097	10.7	12.1	14.6	17.0	19.4
Studying	.011	1.2	1.4	1.7	1.9	2.2
Swimming						
20 yards per minute	.032	3.5	4.0	4.8	5.6	6.4
45 yards per minute	.058	6.4	7.3	8.7	10.2	11.6
50 yards per minute	.070	7.7	8.8	10.5	12.3	14.0
Table tennis (skilled)	.045	5.0	5.6	6.8	7.9	9.0
Tennis (beginner)	.032	3.5	4.0	4.8	5.6	6.4
Walking (brisk pace)						
3.5 miles per hour	.035	3.9	4.4	5.2	6.1	7.0
4.5 miles per hour	.048	5.3	6.0	7.2	8.4	9.6
Wheeling self in wheelchair[c]	.030	3.3	3.75	4.5	5.25	6
Wheelchair basketball[c]	.084	9.2	10.5	12.6	14.7	16.8

[a]Use this column if you want to calculate calories spent for your own exact body weight. Multiply cal/lb/min by your exact weight and then multiply that number by the number of minutes spent in the activity. For example, if you weigh 142 pounds, and you want to know how many calories you spent doing 30 minutes of vigorous aerobic dance: .062 × 142 = 8.8 calories per minute. 8.8 × 30 (minutes) = 264 total calories spent.

[b]Use this column if you weigh 110, 125, 150, 175, or 200 pounds. This eliminates the need to calculate from column 1.

[c]Wheelchair values estimated from data in Table IV-4 of the National Heart, Lung, and Blood Institute Expert Panel, National Institutes of Health, *Clinical Guidelines on the Identification, Evaluation, and Treatment of Overweight and Obesity in Adults* (Washington, D.C: Government Printing Office, 1998).

How the Body Loses and Gains Weight

The causes of obesity may be complex, but the body's energy balance is straightforward. The balance between the energy you take in and the energy you spend determines whether you will gain, lose, or maintain body *fat*. A change in body *weight* of

Chapter 8 gave details about the body's water balance.

a pound or two may not indicate a change in body fat—it can reflect shifts in body fluid content, in bone minerals, in lean tissues such as muscles, or in the contents of the bladder or digestive tract. A change often correlates with the time of day: people generally weigh the least before breakfast. One of the most important things for people concerned with weight control to realize is that quick, large changes in weight are usually not changes in fat alone, or even at all.

The type of tissue lost or gained depends on how you go about losing or gaining it. To lose fluid, for example, you can take a "water pill" (diuretic), causing the kidneys to siphon extra water from the blood into the urine. Or you can engage in intense exercise while wearing heavy clothing in hot weather and lose abundant fluid in sweat. (Both practices are dangerous and are not being recommended here.) To gain water weight, you can overconsume salt and water; for a few hours, your body will retain water until it manages to excrete the salt. (This, too, is not recommended.) Most quick weight-change schemes promote large changes in body fluids that register dramatic, but temporary, changes on the scale and accomplish little weight change in the long run.

Smoking may keep some people's weight down, but at what cost?

- Heart disease.
- Cancer.
- Osteoporosis.
- Chronic lung diseases.
- Shortened life span.
- Low-birthweight babies.
- Miscarriage.
- Sudden infant death.
- Many others.

A further word about practices not recommended: smoking. Each year, many adolescents, especially girls, take up smoking to control weight. Nicotine blunts feelings of hunger, so when hunger strikes, a smoker can reach for a cigarette instead of food. Fear of weight gain sometimes deters people from quitting smoking. Smokers do tend to weigh less than nonsmokers, and many gain weight when they stop smoking. The best advice to smokers wanting to quit seems to be to adjust diet and exercise habits to maintain weight during and after cessation. The best advice to a person flirting with the idea of taking up smoking for weight control is don't do it—many thousands of people who became addicted as teenagers die from tobacco-related illnesses each year.

Moderate Weight Loss versus Rapid Weight Loss

Being able to eat periodically, store fuel, and then use up that fuel between meals is a great advantage. The between-meal interval is normally about 4 to 6 waking hours—about the length of time the body takes to use up most of the readily available fuel—or 12 to 18 hours at night, when body systems slow down and the need is less.

When you eat less food energy than you need, your body draws on its stored fuel to keep going. If a person exercises appropriately, moderately restricts calories, and consumes an otherwise balanced diet that meets protein and carbohydrate needs, the body is forced to use up its stored fat for energy. Gradual weight loss will occur. This is preferred to rapid weight loss because lean body mass is spared and fat is lost.

In early food deprivation:

- The nervous system cannot use fat as fuel; it can only use glucose.
- Body fat cannot be converted to glucose.
- Body protein can be converted to glucose.

In later food deprivation:

- Ketone bodies help feed the nervous system and so help spare tissue protein.

The Body's Response to Fasting If a person doesn't eat for, say, three whole days, then the body makes one adjustment after another. In less than a day into the fast, the liver's glycogen becomes essentially exhausted.[63] Where, then, can the body obtain *glucose* to keep its nervous system going? Not from the muscles' glycogen because that is reserved for the muscles' own use. Not from the abundant fat stores most people carry because these are of no use to the nervous system. Fat cannot be converted to glucose—the body lacks enzymes for this conversion.* The muscles, heart, and other organs use fat as fuel, but at this stage the nervous system needs glucose. The body does, however, possess enzymes that can convert protein to glucose. Therefore, the underfed body sacrifices the proteins in its lean tissue to supply raw materials from which to make glucose.

If the body were to continue to consume its lean tissue unchecked, death would ensue within about ten days. After all, in addition to skeletal muscle, the blood proteins, liver, digestive tract linings, heart muscle, and lung tissue—all vital tissues—are being burned as fuel. (Fasting or starving people remain alive only until their stores of fat are gone or until half their lean tissue is gone, whichever comes first.) To

*Glycerol, which makes up 5 percent of fat, can yield glucose but is a negligible source.

prevent this, the body plays its last ace: it begins converting fat into compounds that the nervous system can adapt for use and so forestall the end. This process is ketosis, first mentioned in Chapter 4 as an adaptation to prolonged fasting or carbohydrate deprivation.

ketone bodies acidic compounds derived from fat and certain amino acids. Normally rare in the blood, they help to feed the brain during times when too little carbohydrate is available. Also defined in Chapter 4.

In ketosis, instead of breaking down fat molecules to carbon dioxide and water as it normally does, the body takes partially broken-down fat fragments and combines them to make **ketone bodies,** compounds that are normally rare in the blood. It converts some amino acids—those that cannot be used to make glucose—to ketone bodies, too. These ketone bodies circulate in the bloodstream and help to feed the brain, since about half of the brain's cells can make the enzymes needed to use ketone bodies for energy. After about ten days of fasting, the brain and nervous system can meet most of their energy needs using ketone bodies.

Thus, indirectly, the nervous system begins to feed on the body's fat stores. Ketosis reduces the nervous system's need for glucose, spares the muscle and other lean tissue from being devoured quickly, and prolongs the starving person's life. Thanks to ketosis, a healthy person starting with average body fat content can live totally deprived of food for as long as six to eight weeks. Figure 9-11 on p. 334 reviews how energy is used during both feasting and fasting.

Respected, wise people in many cultures have practiced fasting as a periodic discipline. The body tolerates short-term fasting, although there is no evidence that the body becomes internally "cleansed," as some believe. Ketosis may harm the body by upsetting the acid-base balance of the blood and by promoting mineral losses in the urine. In as little as 24 hours of fasting, the intestinal lining begins to lose its integrity.[64] Food deprivation also leads to a tendency to overeat or even binge when food becomes available.[65] The effect seems to last beyond the point when weight is restored to normal; people with eating disorders often report that a fast or a severely restricted diet heralded the beginning of their loss of control over eating. This indictment applies to extreme dieting and fasting, but not to the moderate weight-management strategies described later in this chapter.[66]

Names of some low-carbohydrate diets: Atkins New Diet Revolution, Calories Don't Count Diet, Drinking Man's Diet, Mayo Diet, Protein-Sparing Fast, Scarsdale Diet, Ski Team Diet, Stillman Diet, and the Zone Diet. New ones keep coming out, but they are essentially the same diet.

If you want to lose weight, fasting is not the best way. The body's lean tissues continue to be degraded. The body is deprived of nutrients it needs to assemble new enzymes, red and white blood cells, and other vital components. The body also slows its metabolism to conserve energy. A diet only moderately restricted in calories promotes a greater rate of *weight* loss, a faster rate of *fat* loss, and the retention of more lean tissue than a severely restricted fast.

The Body's Response to a Low-Carbohydrate Diet Any diet too low in carbohydrate brings about responses that are similar to fasting. Low-carbohydrate diets have been promoted in many different guises, and each diet has enjoyed a surge of popularity, thanks largely to a sizable initial weight loss. These diets are designed to throw a person into ketosis. The sales pitch is that "you'll never feel hungry" and that "you'll lose weight fast—faster than you would on any ordinary diet." Both claims are true but also misleading. Loss of appetite accompanies any low-calorie diet. Severe calorie restriction means loss of water and lean tissue, and the water is rapidly regained when people begin eating normally again. Nationally, people who successfully lost an average of 30 pounds and kept them off for five years did so by eating more, not less, carbohydrate.[67] They also consumed substantially less fat. Other studies agree that a diet high in complex carbohydrates, low in fat and sugar, supports weight loss efforts.[68] Recently, the USDA released results from an ambitious and continuing evaluation of weight-loss diets.[69] U.S. dieters who lost substantial weight consumed diets high in carbohydrates and low in fat, and their diets scored highest in nutrition quality.*

People who lose weight and keep it off report eating more, not less, carbohydrate-rich foods.

Even if a person can stick to a low-carbohydrate diet long enough to lose some body fat, the loss results from reduced intakes of food and calories, and not from any

*Based on the Healthy Eating Index, a measure of how well the diet meets the recommendations of the Food Guide Pyramid and the *Dietary Guidelines for Americans*.

Figure 9•11

FEASTING AND FASTING

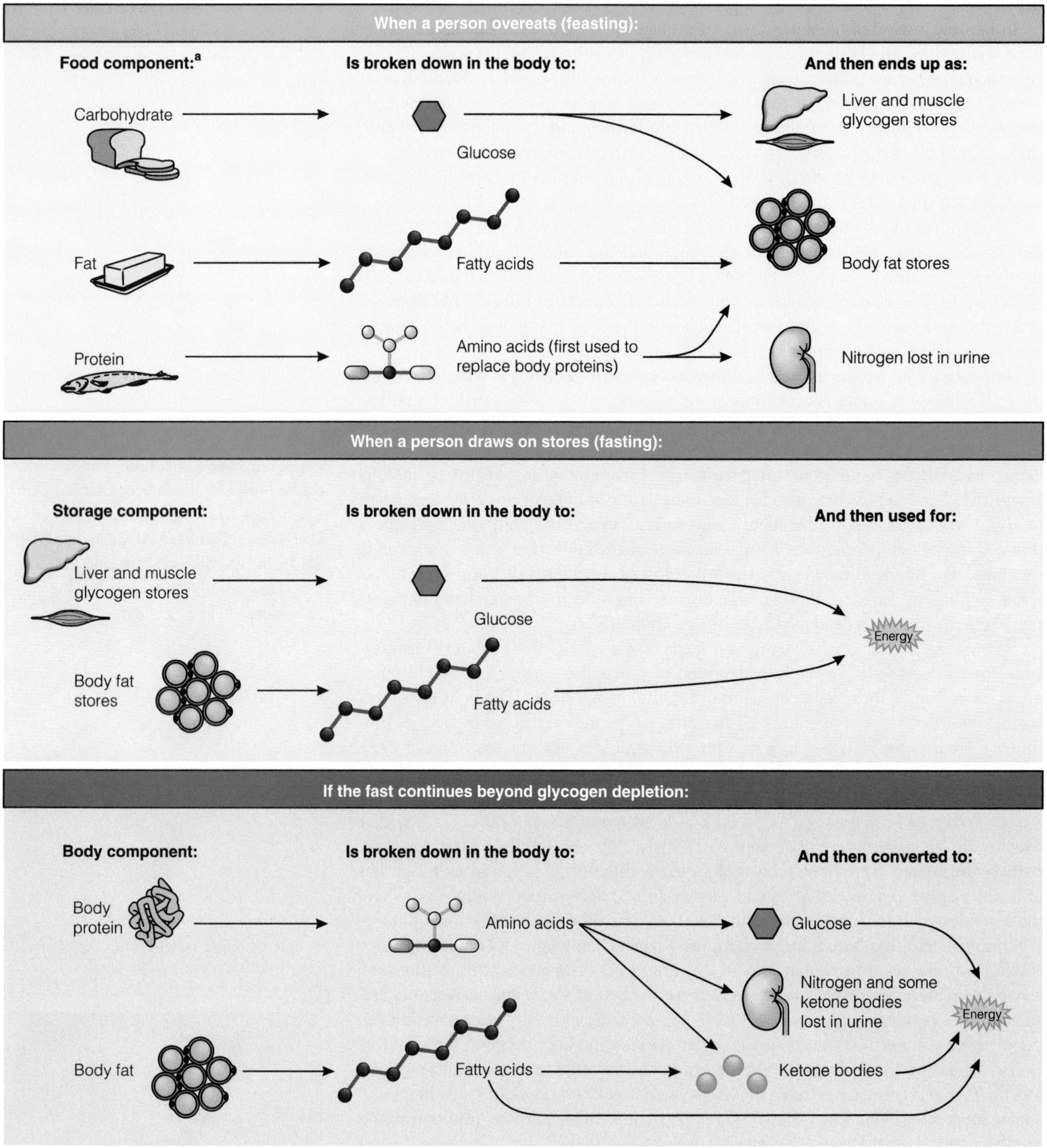

When a person overeats (feasting):
Food component:[a]
Is broken down in the body to:
And then ends up as:
Carbohydrate
Glucose
Liver and muscle glycogen stores
Fat
Fatty acids
Body fat stores
Protein
Amino acids (first used to replace body proteins)
Nitrogen lost in urine
When a person draws on stores (fasting):
Storage component:
Is broken down in the body to:
And then used for:
Liver and muscle glycogen stores
Glucose
Energy
Body fat stores
Fatty acids
If the fast continues beyond glycogen depletion:
Body component:
Is broken down in the body to:
And then converted to:
Body protein
Amino acids
Glucose
Nitrogen and some ketone bodies lost in urine
Energy
Body fat
Fatty acids
Ketone bodies

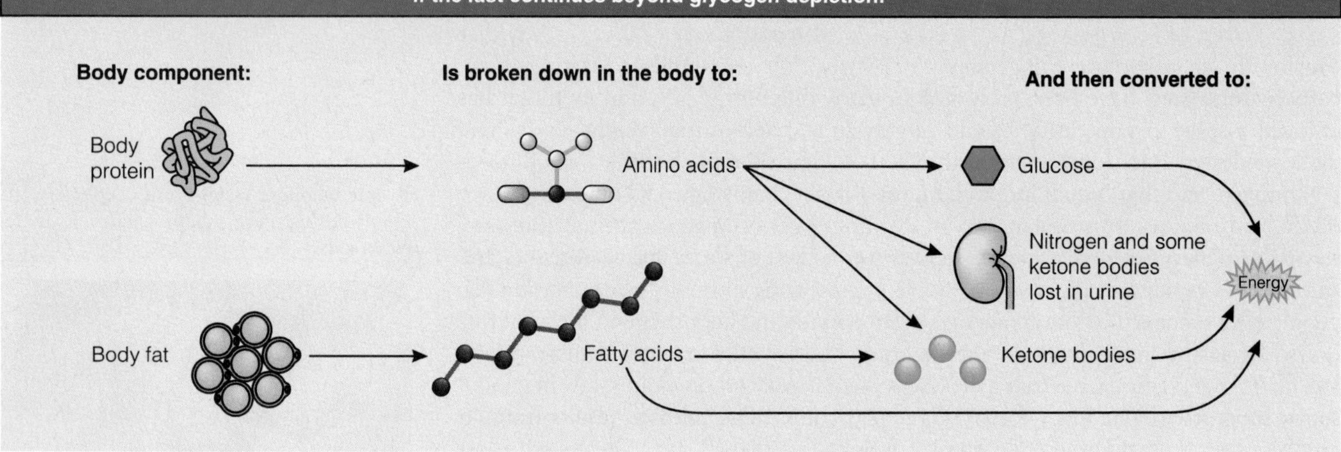

[a]Alcohol is not included because it is a toxin and not a nutrient, but it does contribute energy to the body. After detoxifying the alcohol, the body uses the remaining two-carbon fragments to build fatty acids and stores them as fat.

metabolic hocus-pocus as claimed in popular diet books.[70] It's simple arithmetic: stripped of its carbohydrate-rich beans, tortilla wrapper, and chopped vegetables, a burrito is reduced to a tiny pile of ground beef. Likewise, a steak dinner emerges as a lone piece of meat and a green salad without the calories (and nutrients) of a whole-grain roll, fresh fruit, milk, or even a baked potato.

Aside from the obvious threats to nutrition, many hazards accompany low-carbohydrate diets: extraordinarily high intakes of cholesterol, fat, and saturated fat set the stage for high blood cholesterol and gallbladder trouble; too little carbohydrate can bring on hypoglycemia; no milk or fruits and minimal vegetables rob the body of needed nutrients and phytochemicals; a lack of fiber may invite digestive tract ailments ranging from constipation to colon diseases. Some low-carbohydrate diets, particularly those called protein-sparing fasts, have caused heart failure. These diets are never recommended by knowledgeable practitioners.

Side effects of low-carbohydrate diets may include:

- Bad breath.
- Constipation or diarrhea.
- Elevated uric acid (contributor to gout, a painful condition of the joints).
- Fatigue.
- Foul taste in the mouth.
- Gastric pain.
- Low blood pressure.
- Nausea.

key point *When energy balance is negative, glycogen returns glucose to the body. When glycogen runs out, body protein is called upon for glucose. Fat also supplies fuel as fatty acids. If glucose runs out, fat supplies fuel as ketone bodies, but ketosis can be dangerous. Both fasts and low-carbohydrate diets are ill-advised. People who successfully lose weight and keep it off eat abundant carbohydrates.*

Read about the phytochemicals in Controversy 2.

Weight Gain

What happens inside the body when a person does not use up all of the food energy taken in? Previous chapters have already provided the answer—the energy-yielding nutrients contribute the excess to body stores as follows:

- Carbohydrate (other than fiber) is broken down to sugars for absorption. In the body tissues, excesses of these may be built up to *glycogen* and stored, burned off as heat, or converted to *fat* and stored.
- Fat is broken down to glycerol and fatty acids for absorption. Inside the body, storage as body *fat* is very efficient.
- Protein is broken down to amino acids for absorption. Inside the body, these may be used to replace lost body *protein* and, in a person who is exercising, to build new muscle and other lean tissue. This protein must be functioning protein; excess protein is not passively stored. Excess amino acids have their nitrogen removed and are used for energy or are converted to *glucose* or *fat,* mostly fat.
- Alcohol is easily absorbed intact and is converted into body fat for storage.

Chapter 10 further discusses muscle gains in response to exercise.

Although three kinds of energy-yielding nutrients and alcohol may enter the body, they become only two kinds of energy stores: glycogen and fat. Glycogen stores amount to about three-fourths of a pound; fat stores can amount to many pounds. Note that when excess protein is converted to fat, it cannot be recovered later as protein because the nitrogen is stripped from the amino acids and excreted in the urine. Thus, if you eat enough of any food, whether it's steak, brownies, or baked beans, any excess will be turned to fat within hours. Weight gain comes from spending less food energy than is taken in. Weight can be gained as body fat or as lean tissue, depending mostly upon whether the eater is also exercising.

Ethanol, the alcohol of alcoholic beverages, has been shown to slow down the body's use of fat for fuel by as much as a third, causing more fat to be stored. The storage is primarily in the visceral fat tissue of the "beer drinker's belly" and also on the thighs, legs, or anywhere the person tends to store surplus fat. Alcohol therefore is fattening, both through the calories it provides and through its effects on fat metabolism.* The obvious conclusion is that weight control and abundant alcohol intake cannot easily coexist.[71]

Each gram of alcohol presents 7 calories of energy to the body—energy that is easily stored as body fat.

CORBIS

*People addicted to alcohol are often overly thin because of diseased organs, depressed appetite, and subsequent malnutrition.

These points are worth repeating:

- Any food can make you fat if you eat enough of it. A net excess of energy is almost all stored in the body as fat in fat tissue.
- Fat, as opposed to carbohydrate or protein, from food is particularly easy for the body to store as fat tissue.
- Protein is not stored in the body except in response to exercise; it is present only as working tissue. Excess protein is converted to fat and stored as such.
- Alcohol both delivers calories and encourages storage of body fat.
- Too little physical activity encourages body fat accumulation.

key point *When energy balance is positive, carbohydrate is converted to glycogen or fat, protein is converted to fat, and food fat is stored as fat. Alcohol delivers calories and encourages fat storage.*

Achieving and Maintaining a Healthy Body Weight

Before setting out to change your body weight, think about your motivation for doing so. Many people in our society are dissatisfied with their body weight, not because of potential health risks but because their weight fails to meet society's idea of attractiveness. Yet people arrive in this world with varying weight tendencies; just as some tend to be tall and others short, some tend to be lean and others stout. No one expects tall people to grow shorter or short people to grow taller to become "normal."

Adopting health as the ideal, rather than some ill-conceived image of beauty, can avert much misery, and Table 9-6 offers some tips to that end. The human body is not infinitely malleable—few overweight people will ever become rail-thin, even with the right diet, exercise habits, and behaviors. Likewise, most underweight people will remain on the slim side even after putting on some heft. The rest of this chapter stresses health as a goal and provides practical strategies to help aim for a healthy body weight.

What Diet Strategies Are Best for a Healthy Body Weight?

This section reveals diet-related changes that most often lead to successful weight change and maintenance. Written as advice to "you," the idea is not to pressure you to take the advice personally but to give you the illusion of listening in on a conversation in which an overweight person is benefiting from competent dietary counseling.

Table 9•6

TIPS FOR ACCEPTING A HEALTHY BODY WEIGHT

- Adopt a new value system. Value yourself and others for human attributes other than body weight. Realize that prejudging people by weight is as harmful as prejudging them by race, religion, or gender.
- Use supportive, nonjudgmental descriptions of your body; never use degrading negative descriptions.
- Take compliments seriously. Positive comments from others probably reflect an objective viewpoint.
- Avoid frequent checking of your weight or appearance; focus on your whole self including your intelligence, social grace, and professional and scholastic accomplishments.
- Accept that no magic diet exists.
- Stop dieting to lose weight. Adopt a healthy eating and exercise lifestyle permanently.
- Memorize and employ the Food Guide Pyramid. Never restrict food intake beyond the minimum levels that meet nutrient needs.
- Become physically active, not because it will help you get thin but because it will enhance your health.
- Seek support from loved ones. Tell them of your plan for a healthy life in the body you have been given.
- Seek professional counseling, *not* from a weight-loss counselor, but from someone who can help you make gains in self-esteem without weight as a factor.
- Join with others to fight weight discrimination and fashion stereotypes. (Search your local paper, or see the Nutrition Resources Appendix for names of groups.)

Setting Goals The best diet is one tailored to your individual needs. Setting goals helps to achieve the desired result in controlling weight. For example, three broad goals are important for obese people who must reduce their weight to reduce their disease risks:

1. Reduce body weight by about 10 percent over half a year's time.
2. Maintain a lower body weight over the long term.
3. At a minimum, prevent further weight gain.

Whether your goals involve weight gain, loss, or maintenance, set broad goals that reflect the end results you wish to achieve, and then set smaller step goals for the dietary, physical activity, and behavioral changes necessary to achieve the desired result. These changes do not produce a dramatic weight loss overnight, but if you faithfully employ them, you can lose a pound or two of body fat each week, safely and effectively. Losses greater or faster than this are not recommended because they are almost invariably followed by just as rapid regain. Also, rapid weight loss and excessive fat restriction can cause gallbladder stones or dangerous electrolyte imbalances.[72] It's better to take your time and achieve a lasting change.

Keeping Records Keeping records can help you to meet your goal. Recording your food intake and exercise habits can help you to spot trends and identify areas needing improvement. To monitor changes in body composition, body weight can serve as a quick indicator of changes in body fatness. In addition to weight, measure your waist circumference to track changes in central adiposity.

Planning a Diet for Weight Loss and Maintenance No particular food plan is magical, and no particular food must be either included or excluded. You are the one who will have to live with the plan, so you had better be the one to design it. Remember, you are not "going on a diet" because you will not be "going off" your eating plan. Instead, you are adopting a healthy eating plan for life, so it must consist of foods that you like, that are readily available, and that you can afford. Table 9-7 on the next page provides a way to judge weight-loss plans according to standard nutrition principles.

Guidelines for a weight-loss diet are outlined in Table 9-8 (next page). For those with a BMI greater than 35, a deficit of 500 to 1,000 calories per day will produce the desired loss. This amounts to an intake of about 1,000 to 1,200 calories per day for most women and 1,200 to 1,600 calories per day for most men.[73] Figure 9-12 on page 339 suggests daily food servings from which to build a balanced 1,200-calorie diet. Diets providing energy intakes lower than about 800 calories, the so-called very-low-calorie diets (VLCD), are notoriously unsuccessful at achieving lasting weight loss and can be dangerous, and so are not recommended.[74] For those with a BMI ranging from 27 to 35, a deficit of 300 to 500 calories per day will result in losing ½ to 1 pound per week.[75]

If you plan resolutely to include the number of servings of food from each food group that you need each day, you will find that you will have little appetite left for high-fat or empty-calorie foods. Foods such as fruits, vegetables, and whole grains are high in carbohydrates and fiber and low in fat, and they take a lot of chewing, too. Crunchy, wholesome, unprocessed or lightly processed foods offer bulk and satiety for far fewer calories than smooth, refined foods. Limit, but don't eliminate, lean meats or other low-fat protein sources: an ounce of lean ham contains about the same number of calories as an ounce of bread, but the ham produces greater satiety.

Remember to pay careful attention to portion sizes—the monstrous helpings served by restaurants are the enemy of the person striving to control weight. And don't lose track of the fat in foods—fat calories add up quickly and probably contribute more to body fat stores than do carbohydrate calories.[76] The Food Feature of Chapter 5 is an excellent resource for the person trying to control fat. Beware, though, of overdoing foods that are manufactured to be low in fat but make up the calories with sugar. Many people have found out the hard way that reducing fat without restricting calories does not produce weight loss. They may eat less fat, but calorie intakes remain high or even increase. (Compare the calories of yogurts in Figure

Table 9•7

RATING SOUND AND UNSOUND WEIGHT-LOSS SCHEMES

Each diet or program starts with 160 points and is rated on 12 factors. Whenever a plan falls short of ideals, subtract points in the third column as instructed. A plan that loses more than 20 points might still be of value, but deserves careful scrutiny.

Factor	Does the Diet or Program:	Start:
		160 points
Calories	Provide a reasonable number of calories (not fewer than 1,200 calories for an average-size person)? If not, give it a minus 10.	________
Protein	Provide enough, but not too much, protein (at least the recommended intake, but not more than twice that much)? If no, minus 10.	________
Fat	Provide enough fat for balance but not so much fat as to go against current recommendations (about 30% of calories from fat)? If no, minus 10.	________
Carbohydrate	Provide enough carbohydrate to spare protein and prevent ketosis (100 grams of carbohydrate for the average-size person)? Is it mostly complex carbohydrate (not more than 10% of the calories as concentrated sugar)? If no to either or both, minus 10.	________
Vitamins and minerals	Offer a balanced assortment of vitamins and minerals by including foods from all food groups? If it omits a food group (for example, meats), does it provide a suitable food (not supplement) substitute? Count five food groups in all: milk/milk products; meat/fish/poultry/eggs/legumes; fruits; vegetables; and breads/cereals/grains. For each food group omitted and not adequately substituted for, subtract 10 points.	________
Variety	Offer variety, in the sense that different foods can be selected each day? If you'd classify it as boring or monotonous, give it a minus 10.	________
Ordinary foods	Consist of ordinary foods that are available locally (for example, in the main grocery stores) at the prices people normally pay? Or does the dieter have to buy special, expensive, or unusual foods to adhere to the diet? If you would class it as "bizarre" or "requiring special foods," minus 10.	________
False promises	Promise dramatic, rapid weight loss (substantially more than 1% of total body weight per week)? If yes, minus 10.	________
Lifestyle changes	Encourage permanent, realistic lifestyle changes, including regular exercise and the behavioral changes needed for weight maintenance? If not, minus 10.	________
Reasonable costs	Misrepresent salespeople as "counselors" supposedly qualified to give guidance in nutrition and/or general health without a profit motive, or collect large sums of money at the start, or require that clients sign contracts for expensive, long-term programs? If so, minus 10.	
Warnings of risks	Fail to inform clients about the risks associated with weight loss in general or the specific program being promoted? If so, minus 10.	
No gimmicks or mandatory supplements	Promote unproven or spurious weight-loss aids such as human chorionic gonadotrophin hormone (hormones can be dangerous; reject such a plan immediately), starch blockers, diuretics, sauna belts, body wraps, passive exercise, ear stapling, any type of injections, acupuncture, electric muscle stimulating devices, spirulina, amino acid supplements (e.g., arginine, ornithine), glucomannan, appetite suppressants, "unique" ingredients, and so forth? If so, minus 10.	________
	Total points:	________

9-13 on page 340.) Read labels, compare the calories per serving among similar foods, and choose the lowest-calorie item of the bunch.

If you drink alcoholic beverages, cutting down or eliminating alcohol is an obvious way to eliminate many unneeded calories and to enhance the use of fat for energy. This step can also make room in the diet for more nutritious foods.

Three meals a day is standard in our society, but no law says you can't have four or five—be sure they are smaller, of course. People who eat small, frequent meals are reported to be successful at weight loss and maintenance.[77] Make sure that mild hunger, not appetite, is prompting you to eat. Eat regularly, and eat before you become extremely hungry. When you do decide to eat, eat the entire meal you have planned for yourself. Then don't eat again until the next meal or snack. Save calorie-free or favorite foods or beverages for a planned snack at the end of the day if you need insurance against late-evening hunger.

One meal you should strive to include is breakfast. Much evidence supports the health effects of breakfast, and people who eat breakfast seem to need fewer snacks and consume less fat all day long.

Table 9•8

RECOMMENDATIONS FOR A WEIGHT-LOSS DIET

Nutrient	Recommended Intake
Calories	
For people with BMI ≥35	Approximately 500 to 1,000 calories per day reduction from usual intake
For people with BMI between 27 and 35	Approximately 300 to 500 calories per day reduction from usual intake
Total fat	30% or less of total calories
Saturated fatty acids[a]	8 to 10% of total calories
Monounsaturated fatty acids	Up to 15% of total calories
Polyunsaturated fatty acids	Up to 10% of total calories
Cholesterol[a]	300 mg or less per day
Protein[b]	Approximately 15% of total calories
Carbohydrate[c]	55% or more of total calories
Sodium chloride	No more than 2,400 mg of sodium or approximately 6 g of sodium chloride (salt) per day
Calcium	1,000 to 1,500 mg per day
Fiber[c]	20 to 30 g per day

[a]People with high blood cholesterol should aim for less than 7 percent calories from saturated fat and 200 milligrams cholesterol per day.

[b]Protein should be derived from plant sources and lean sources of animal protein.

[c]Carbohydrates and fiber should be derived from vegetables, fruits, and whole grains.

SOURCE: National Heart, Lung, and Blood Institute Expert Panel, National Institutes of Health, *Clinincal Guidelines on the Identification, Evaluation, and Treatment of Overweight and Obesity in Adults* (Washington, D.C.: Government Printing Office, 1998), p. 74.

Figure 9•12

SUGGESTED DAILY SERVINGS FOR AN ADEQUATE 1,200-CALORIE DIET[a]

Choose the lowest-calorie, fat-free, or lowest-fat options from each group. Strictly limit serving sizes to those specified in Figure 2-4 of Chapter 2. To further reduce calories, reduce servings of fats and added sugars.

[a]Assumes no alcohol intake.

key point *To achieve and maintain a healthy body weight, set realistic goals, keep records, and expect to progress slowly. Make the diet adequate, limit fat and calories, reduce alcohol, and eat regularly.*

What Diet Strategies Can Help Me to Gain Weight?

Should an underweight person try to gain weight? Not necessarily. If you are healthy at your present weight, stay there. If your physician has advised you to gain, if you are excessively tired, if you are unable to keep warm, if you fall into the "underweight" category of the BMI table (see inside back cover), or if, for women, you have missed at least three consecutive menstrual periods, you may be in danger from a too-low body weight.

For the person who needs to gain, a healthful weight gain can be achieved through physical activity, particularly strength training (see Chapter 10 for details), combined with a high-calorie diet. Diet alone can bring about weight gain, but the gain will be mostly fat. For someone facing a wasting disease, the gain of fat tissue may be a welcome sign of improvement. For an athlete, however, such a gain can impair performance. For most people, physical activity is an essential component of a sound weight-gain plan. Many an underweight person has simply been too busy (for months) to eat or to exercise enough to gain or to maintain weight.

As important to weight gain as exercise are the calories to support that activity—otherwise you will lose weight (body fat). If you eat just enough to fuel the activity, you will build muscle, but at the expense of body fat; that is, fat will be burned to support the muscle building. If you eat more, you will gain both muscle and fat. To gain a pound of muscle and fat requires taking in about 3,000 extra calories.*

*Theoretically, it takes an excess of 2,000 to 2,500 calories to gain a pound of pure lean tissue and about 3,500 calories to gain a pound of fat.

Figure 9•13

COMPARING YOGURTS: FAT, CARBOHYDRATES, CALORIES

The calorie values of the whole-milk and fat-free yogurts are almost the same. When sugar is controlled as well as the fat, calories are significantly reduced, so the wise calorie watcher pays attention to more than just fat.

Whole-Milk Fruit Yogurt

Nutrition Facts
Serving Size 1 container (170g)

Amount Per Serving	
Calories 175	Calories from Fat 15
	% Daily Value*
Total Fat 1.5g	3%
Cholesterol 10mg	3%
Sodium 105mg	4%
Total Carbohydrate 33g	11%
Sugars 27g	
Protein 7g	16%

Fat-Free Fruit Yogurt

Nutrition Facts
Serving Size 1 container (170g)

Amount Per Serving	
Calories 160	Calories from Fat 0
	% Daily Value*
Total Fat 0g	0%
Cholesterol <5mg	1%
Sodium 105mg	4%
Total Carbohydrate 33g	11%
Sugars 27g	
Protein 7g	16%

Fat-Free, Reduced-Sugar Fruit Yogurt

Nutrition Facts
Serving Size 1 container (170g)

Amount Per Serving	
Calories 90	Calories from Fat 0
	% Daily Value*
Total Fat 0g	0%
Cholesterol <5mg	1%
Sodium 105mg	4%
Total Carbohydrate 16g	5%
Sugars 8g	
Protein 7g	16%

Conventional advice on diet to the person building muscle is to eat about 700 to 1,000 calories a day above normal energy needs; this range supports both the added activity and the formation of new muscle.

The weight gainer needs nutritious calorie-dense foods. No matter how many sticks of celery you consume, you won't gain weight because celery simply doesn't offer enough calories. Calorie-dense foods (the very ones the weight-loss dieter is trying to avoid) are high in fat, but if they are contributing energy that will be spent building new tissue and if the fat is mostly unsaturated, they will not contribute to heart disease. Choose peanut butter instead of lean meat, avocado instead of cucumber, olives instead of pickles, whole-wheat muffins instead of whole-wheat bread, and milkshakes instead of milk. When you do eat celery, stuff it with tuna salad (use oil-packed tuna); choose flavored coffee drinks over plain coffee; use olive oil or canola oil dressings on salads, whipped toppings on fruit, and soft or liquid margarine on potatoes. Because fat contains more than twice as many calories per teaspoon as sugar, it adds calories without adding much bulk, and its energy is in a form that is easy for the body to store.

Expect to feel full. Most underweight individuals are accustomed to small quantities of food. When they begin eating significantly more food, they complain of uncomfortable fullness. This feeling is normal, and it passes as the stomach gradually adapts to the extra food.

Eat frequently. Make three sandwiches in the morning and eat them between classes in addition to the day's three regular meals. Spend time making foods appealing—the more varied and palatable, the better. If you fill up fast during a meal, start with the main course or a meat- or cheese-filled appetizer, not carrot sticks. Drink between meals, not with them, to save space for higher-calorie foods. Make milkshakes of milk, frozen bananas, and flavorings to drink between meals. Always finish with dessert. Be aware that most "weight-gain" supplements designed to add body weight are useless without physical activity and confer no special benefits. Of body weight gained in a day, only a half ounce to an ounce is protein tissue, so no special protein supplements can help speed weight gain. Ordinary food in abundance along with exercise to work the nutrients into place supports efforts to gain weight.

Smoking tobacco depresses the appetite and makes taste buds and olfactory (smelling) organs less sensitive. A person who smokes should quit before trying to gain weight. Quitters find that appetite picks up, food tastes and smells better, and the body reaps numerous benefits.

Weight gain requires a diet of calorie-dense foods, eaten frequently throughout the day. Physical activity builds lean tissue.

Drugs and Surgery

In their struggle to achieve an altered body size, people fall for weight-change gimmicks that abound in the marketplace (see the margin of the facing page). Most are ineffective, and some are truly dangerous. To someone fatigued from years of battling overweight, the idea of taking pills or undergoing surgery might seem attractive. These extreme approaches carry serious risks to health, however, and are reserved for efforts at saving the lives of obese people at critical risk.

Each year, a million and a half U.S. citizens take prescription weight-loss medications. Of these, a quarter are not overweight.[78] Physicians and patients alike seem willing to risk substantial drug side effects in pursuit of a particular body form, a point made clear when thousands of people of healthy body weight suffered heart-valve injury from the now-banned weight-loss drug combination "fen-phen."* Table 9-9 on page 342 presents some of the known side effects and other details about weight-loss medications currently on the market.

*Fenfluramine and phentermine. In 1997, fenfluramine was removed from the market, and over 4,000 lawsuits have been filed against makers of fen-phen.

People with a BMI of 30 or above and those with other disease risk factors may benefit from prescription medication, along with diet, exercise, and behavior therapy, to bring their weight down. Obesity with a BMI of 40 or above (or 35 with coexisting diseases) indicates an urgent need to reduce body fatness, and surgery may be an option for those healthy enough to withstand it. Surgical procedures effectively limit food intake by reducing the size of the stomach and delaying the passage of food from the stomach into the intestine, leading to significant weight loss in most patients. Surgeries include **gastric bypass, gastroplasty,** and **gastric banding.**[79] The risks of surgery are often justified in people with **extreme obesity.**[80]

The long-term safety and effectiveness of gastric surgery depend, in large part, on compliance with dietary instructions. Complications immediately following surgery often include infections, nausea, vomiting, and dehydration; in the long term, vitamin and mineral deficiencies and psychological problems may develop. Lifelong medical supervision is necessary for those who choose the surgical route, but in suitable candidates, the benefits of weight loss may prove worth the substantial risks.

Another surgical procedure is used not to treat obesity but to remove external pads of body fat. Plastic surgeons can extract some fat deposits by lipectomy, or "liposuction." People consenting to this cosmetic procedure expect an improved body shape, but an unusually large number of lawsuits against surgeons performing it indicates that many people are disappointed with the outcome.[81] If the fat is gained back after liposuction, as often happens, it can form a lumpy, dimpled layer that is worse in appearance than the original fat. Lipectomy is popular in part because it seems safe, but there can be serious complications, even death, from the surgery.[82]

Herbal Products Wildly popular but unproven for effectiveness or safety are herbal weight-loss products. People may falsely believe that "natural" herbs are never harmful to the body, but, of course, there are many poisonous herbs and toxins, such as belladonna and hemlock. Furthermore, because weight-loss herbs are marketed as "dietary supplements," manufacturers need not present a shred of evidence of their safety or effectiveness to the Food and Drug Administration (FDA) before marketing them. Evidence about safety is gathered only through reports of actual consumers who sicken or die after purchasing and using herbal remedies.

A good example is ephedrine, which showed promise as a weight-loss drug in preliminary studies. The traditional Chinese medicine ma huang and the Western "herbal fen-phen" are examples of ephedrine-containing "dietary supplements" that often include caffeine from coffee extract or aspirin as willow extract in a dangerous mixture. More than 1,000 consumers of these products have reported ill effects, ranging from headaches and vomiting to heart attacks and brain hemorrhage, and at least 35 people have died.[83] Canada has banned ma huang, the FDA has issued a warning against weight-loss products containing ephedrine but has not banned them, and the World Health Organization has called for worldwide controls on its sale.[84] Another diet aid to avoid is known as TRIAC, a powerful hormone that interferes with normal thyroid functioning and has caused heart attack and stroke (see the Controversy of Chapter 10).* Another "herbal" additive, aristolochic acid, is a known kidney toxin and cancer-causing agent. Read labels and don't take these products. The risks of doing so are too high.

Herbal laxatives containing senna, aloe, rhubarb root, cascara, castor oil, or buckthorn are sold as "dieter's tea" because they can cause a temporary water loss of a pound or two. Users commonly report nausea, vomiting, diarrhea, cramping, and fainting. Such "teas" are suspected of contributing to the deaths of four women who used them and also drastically reduced their food intakes.[85] Some herbs may be useful for other purposes, but those sold for weight loss clearly fail the safety test.

*The product name was Triax Metabolic Accelerator.

gastric bypass surgery that reroutes food from the stomach to the lower part of the small intestine; creates a chronic, lifelong state of malabsorption by preventing normal digestion and absorption of nutrients.

gastroplasty surgery that partitions the stomach by stapling off a "pouch" or otherwise modifying the stomach, thereby reducing total food intake.

gastric banding a surgical means of producing weight loss by restricting stomach size with a contricting band or pouch; used in people whose severe obesity brings extreme health risks.

extreme obesity clinically severe overweight, presenting very high risks to health; the condition of having a BMI of 40 or above; also called *morbid obesity*.

Ineffective or dangerous weight-loss and weight-gain gimmicks:

- Amino acid pills.
- Diet pills.
- Energy increasers.
- Expanding pills.
- Glucomannan, bee pollen, spirulina.
- Herbs, such as herbal fen-phen, ma huang, ephedrine, TRIAC, and dieter's teas.
- Hormones (see Controversy 10).
- Laxatives.
- Liposuction, lipectomy.
- Massages, muscle stimulators.
- Protein supplements.
- Spa belts, rollers, saunas, whirlpools.
- Thigh-reducing cream.
- Many more.

To gain weight and keep it on, plan for frequent meals throughout the day.

Table 9•9

CURRENT PHARMACEUTICAL TREATMENTS OF OBESITY[a]

Names	Actions	Known Side Effects[b]	Comments
Prescription Drugs			
Sibutramine Trade name: Meridia	Suppresses appetite by inhibiting the uptake of the neurotransmitter serotonin in the brain	Dry mouth, headache, constipation, insomnia, and high blood pressure	Effective when used in combination with a reduced-calorie diet and increased physical activity. The FDA advises those with high blood pressure against its use; others should monitor their blood pressure.
Orlistat Trade name: Xenical	Inhibits pancreatic lipase activity, thus blocking dietary fat absorption by about 30% when taken with meals	Gas, frequent bowel movements, and reduced absorption of fat-soluble vitamins	Effective when accompanied by a nutritionally balanced, reduced-calorie, low-fat diet
Other prescription drugs currently under study	Some block appetite-stimulating brain chemicals, and others stimulate thermogenesis.	Unknown	
Over-the-Counter Drugs and Products			
Benzocaine Trade names: Diet Ayds (candy) or Slim Mint (gum)	Anesthetizes the tongue, reducing taste sensations	None known	Only over-the-counter weight-loss medication with FDA approval
Phenylpropanolamine (PPA) (also called norephedrine) Some trade names of weight-loss products with PPA:[c] • Acutrim • Dex-A-Diet • Dexatrim • Dieutrim • LipoKinetix • Permathene • Phenyldrine • Super Ordinex • Thinz • Unitrol	Appetite suppressant; nasal and sinus decongestant	Dry mouth, rapid pulse, nervousness, sleeplessness, hypertension, irregular heartbeat, kidney failure, liver damage, liver failure, seizures, and hemorrhagic strokes (bleeding in the brain)	The FDA has asked drug manufacturers to discontinue making over-the-counter products containing PPA and has issued warnings to consumers. Ephedrine (below) is a related compound.
Ephedrine, ephedra, or ma huang (often combined with caffeine) Trade names include: • Diet Fuel products • Metabolife • Nature's Nutrition Formula One • Many others	Enhancement of effects of the "stress hormone" norepinephrine, including reduced appetite.	Nervousness, headache, insomnia, dizziness, palpitations, skin flush, possibly serious symptoms associated with PPA (see above); over 1,000 reported adverse events are currently under FDA scrutiny.	Sold without FDA regulation

[a]For answers to drug-related questions, call FDA Information toll free: (888) 463-6332 or visit www.fda.gov.
[b]Debilitating heart valve injury was associated with the appetite suppressant fenfluramine which is no longer available.
[c]As of this writing, many manufacturers are voluntarily reformulating their products without PPA in keeping with a request from the Food and Drug Administration. Read the labels.

Other Gimmicks Gimmicks don't help with weight loss. Manufacturers of pills containing **chitin,** the substance forming the shells of insects and shellfish, say the pills bind fat and prevent its absorption but chitin did not produce weight loss in an unpublished study.* Hot baths do not speed up metabolism so that pounds can be lost in hours. Steam baths and saunas do not melt the fat off the body, although they may dehydrate you so that you lose water weight. Brushes, sponges, wraps, creams, and massages intended to move, burn, or break up **"cellulite"** are useless for fat loss.

Reject "dietary supplements" containing these ingredients: ephedrine, TRIAC (triiodothyroacetic acid), or aristolochic acid. Be suspicious of the rest.

*Trade name Chitosan.

Cellulite—the rumpled, dimpled fat tissue on the thighs and buttocks—is simply fat, awaiting the body's call for energy.

As for fad diets, only those that reduce calorie intake produce weight loss, according to the USDA study mentioned earlier.[86] The study will soon provide needed details about the efficacy of the many hundreds of new diet books and plans developed each year. For now, Table 9-10 presents some common lies put forth in popular diet books and follows with a truthful response.

Surgery and drugs to reduce body fatness may be risky, but severe obesity may be more risky still. Herbs and other gimmicks are useless, and some are dangerous.

chitin an indigestible polysaccharide forming the hard shells of insects, lobsters, and shrimp; used in medicine to slow the absorption of drugs and sold in pill form as an unproven weight-loss aid.

cellulite a term popularly used to describe dimpled fat tissue on the thighs and buttocks; not recognized in science.

In 2001, 1,214 diet books were available for sale on one Internet book seller's website. Of the 50 top-selling diet books, well over half were written since 1999.

Q Once I've Changed My Weight, How Can I Stay Changed?

One reason gimmicks fail at weight control is that they fail to produce lasting change. Millions have experienced the frustration of achieving a desired change in weight only to see their hard work visibly slipping away: "I have lost 200 pounds, but I was never more than 20 pounds overweight." Disappointment, frustration, and self-condemnation are common in dieters who find they have slipped back to their original weight or even higher. What makes the difference between a successful, long-term weight-control program and one that doesn't stick? How can you maintain a healthy body weight?

A key to weight maintenance is accepting it as a lifelong endeavor, and not a goal to be achieved and then forgotten. Acceptance helps prepare the mind for making permanent changes. People who maintain a loss continue to employ the behaviors that reduce calorie intakes and increase expenditures through exercise.[87] They cultivate the habits of people who maintain a healthy weight, such as eating diets higher in carbohydrate, lower in fat, and higher in fruits and vegetables than average.[88] Those who maintain healthy weight also:

- Are more physically active than the average person.[89]
- Monitor fat grams, calorie intake, and body weight.

Table 9•10

LIES AND TRUTHS OF FAD DIETS

Lie:	*You can lose weight with "exceptionally easy rules."*
Truth:	Most fad diet plans have complicated rules that require you to calculate protein requirements, count carbohydrate grams, combine certain foods, time meal intervals, purchase special products, and plan daily menus.
Lie:	*You can lose weight by eating a specific ratio of carbohydrates, protein, and fat.*
Truth:	Weight loss depends on spending more energy than you take in, not on the energy nutrient composition of the diet.
Lie:	*This "revolutionary diet" can "reset your genetic code."*
Truth:	You inherited your genes and no diet can alter your genetic code.
Lie:	*High-protein, low-carbohydrate diets are so popular because they work.*
Truth:	Weight-loss books are popular because people hope for quick fixes and simple solutions. The diet books have been selling widely since the 1970s. If they worked, we would be a lean nation; instead, obesity is rising rapidly.
Lie:	*People gain weight on low-fat diets.*
Truth:	People gain weight when they overindulge on calories; low-fat diets are not necessarily low-calorie diets.
Lie:	*High-protein diets energize the brain.*
Truth:	The brain depends on the carbohydrate glucose for its energy, not protein.
Lie:	*Thousands of people have been successful with this plan.*
Truth:	Success stories are usually anecdotal and not scientific; failures are not reported.
Lie:	*Sugar dramatically raises blood glucose and insulin levels, triggering fat storage.*
Truth:	Sugar elicits only a moderate rise in blood glucose and insulin, and fat storage occurs only when energy intake exceeds energy needs.
Lie:	*Dietitians know nothing about "modern" nutrition.*
Truth:	Dietitians are, by training and experience, nutrition experts who rely on scientific approaches and cannot be swayed by the claims of quacks.

self-efficacy a person's belief in his or her ability to succeed in an undertaking.

weight cycling repeated rounds of weight loss and subsequent regain, with reduced ability to lose weight with each attempt; also called *yo-yo dieting*.

- Believe they have the ability to control their weight, an attribute known as **self-efficacy,** even in the face of previous failures.
- Develop social support systems.
- Eat controlled portions at planned times, and eat them at a leisurely pace.[90]
- Eat high-fiber foods, and consume sufficient water each day.[91]
- Cultivate and honor realistic expectations regarding body size and shape.

The importance of exercise cannot be overstated. Those who endeavor to lose weight without exercise often become trapped in **weight cycling,** endless repeating rounds of weight loss and regain, sometimes called "yo-yo" dieting. For example, periodic weight gains incurred during food-centered winter holidays alternating with periods of dieting lead to the accumulation of pounds over the years.[92] A history of such weight cycling can predict a person's future success (or lack thereof) in maintaining weight. For reasons unknown, weight cycling also seems to incur a higher mortality rate than weight maintenance.[93] A sedentary lifestyle correlates with both weight cycling and high mortality, and so may be at the bottom of both.

Self-acceptance also predicts success, while self-hate predicts failure. A paradox of behavior change is that it takes self-acceptance (loving the overweight self) to lay the foundation for changing that self. Once body weight is changed, further improvements are seen in many areas of life.[94] Self-acceptance is the basis of a beneficial cycle. The Food Feature explores how a person can modify daily behaviors into healthy lifelong habits.

key point *People who succeed at maintaining lost weight keep to their eating routines, keep exercising, and keep track of calorie and fat intakes and body weight. The more traits related to positive self-image and self-efficacy a person possesses or cultivates, the more likely that person will succeed.*

Behavior Modification for Weight Control

behavior modification alteration of behavior using methods based on the theory that actions can be controlled by manipulating the environmental factors that cue, or trigger, the actions.

Activities and rewards to substitute for eating:

- Attending sporting events.
- Enjoying leisure activities.
- Exercising or playing sports.
- Gardening.
- Getting praise from others.
- Going to a movie or play.
- Listening to music.
- Napping.
- Praising yourself.
- Reading.
- Receiving token rewards (stickers, stars).
- Redecorating.
- Relaxing.
- Saving money for future treats.
- Shopping.
- Taking a hot bath.
- Telephoning.
- Tidying your room or house.
- Vacationing.
- Working on hobbies or crafts.

Supporting both diet and exercise is the technique of **behavior modification,** which cements into place all the behaviors that lead to and perpetuate the desired body composition. Behavior modification is based on the knowledge that habits drive behaviors.

How Does Behavior Modification Work?

Suppose a friend tells you about a shortcut to class. To take it, you must make a left-hand turn at a corner where you now turn right. You decide to try the shortcut the next day, but when you arrive at the familiar corner, you turn right as always. Not until you arrive at class do you realize that you failed to turn left, as you had planned. You can learn to turn left, of course, but at first you will have to make an effort to remember to do so. After a while, the new behavior will become as automatic as the old one was.

For those striving to lose weight, learning to say "No, thank you" might be among the first habits to establish. Learning not to "clean your plate" might be another. Once you identify the behaviors you need to change, do not attempt to modify all of them at once. No one who attempts too many changes at one time is successful. Set your priorities and begin with a behavior you can handle—then practice it until it becomes habitual and automatic. Then select another.

Applying Behavior Modification

Behavior researchers have identified six elements of behavior modification to use to replace old eating habits with new ones:

1. Eliminate inappropriate eating cues.
2. Suppress the cues you cannot eliminate.
3. Strengthen cues to appropriate eating and exercise.
4. Repeat the desired eating and exercise behaviors.
5. Arrange or emphasize negative consequences of inappropriate eating.
6. Arrange or emphasize positive consequences of appropriate eating and exercise behaviors.

Table 9-11 on the next page provides specific examples of putting these six elements into action. Before doing so, however, you must establish a baseline, a record of your present eating behaviors against which to measure future progress. Keep a diary so that you can learn what particular eating stimuli, or cues, affect you.

To begin, set about eliminating or suppressing the cues that prompt you to eat inappropriately. An overeater's life may include many such cues: watching television, talking on the telephone, entering a convenience store, studying late at night. Resolve that you will no longer respond to such cues by eating. Respond only to one set of cues designed by you, in one particular place in one particular room. If some cues to inappropriate eating behavior cannot be eliminated, suppress them, as described in Table 9-11; then strengthen the appropriate cues, and reward yourself for doing so. The list in the margin suggests some activities and rewards to substitute for eating.

As you progress in your new behaviors of physical activity and sensible eating, enjoy your new, emerging fit and healthy self.

Table 9•11

APPLYING BEHAVIOR MODIFICATION TO CONTROL BODY FATNESS

1. Eliminate inappropriate cues:
   - Don't buy problem foods.
   - Eat only in one room at the designated time.
   - Shop when not hungry.
   - Avoid vending machines, fast-food restaurants, and convenience stores.
   - Turn off the television, video games, and computer.
2. Suppress the cues you cannot eliminate:
   - Serve individual plates; don't serve "family style."
   - Make small portions look large by spreading them over the plate.
   - Create obstacles to consuming problem foods—wrap them and freeze them, making them less quickly accessible.
   - Control deprivation; plan and eat regular meals.
   - Likewise, plan to spend one hour in sedentary activities, such as watching television or using a computer.
3. Strengthen cues to appropriate behaviors:
   - Share appropriate foods with others.
   - Store appropriate foods in convenient spots in the refrigerator.
   - Learn appropriate portion sizes.
   - Plan appropriate snacks.
   - Keep sports and play equipment by the door.
4. Repeat desired behaviors:
   - Slow down eating—put down utensils between bites.
   - Always use utensils.
   - Leave some food on your plate.
   - Move more—shake a leg, pace, stretch often.
   - Join groups of active people and participate.
5. Arrange negative consequences for negative behavior:
   - Ask that others respond neutrally to your deviations (make no comments—even negative attention is a reward).
   - If you slip, don't punish yourself.
6. Reward yourself personally and immediately for positive behaviors:
   - Buy tickets to sports events, movies, concerts, or other nonfood amusement.
   - Indulge in a new small purchase.
   - Get a massage; buy some flowers.
   - Take a hot bath; read a good book.
   - Treat yourself to a lesson in a new active pursuit such as horseback riding, handball, or tennis.
   - Praise yourself; visit friends.
   - Nap; relax.

CONTROL THE CALORIES IN A DAY'S MEALS

This exercise speaks to those of you who want to control calorie intakes to control body fatness. To choose foods to meet nutrient needs while staying within a calorie limit, you can use any or all of these four lines of action:

- *Cut down food portions*, if they are significantly larger than those recommended in the Food Guide Pyramid.
- *Eliminate* high-calorie, low-nutrient foods from the everyday diet (save for special treats).
- *Remove the high-calorie constituents* from most foods (trim fat from meat, choose fat-free milk).
- *Replace* high-calorie foods with lower-calorie versions, either naturally occurring or manufactured.

Fat is a main target for cutting calories because of its high calorie density. When using manufactured low-fat and fat-free items, you must read labels carefully, however. These products may be lower in *fat*, but some contain as many *calories* as the originals.

Read Figure 9-14

The meals shown in Figure 9-14 demonstrate one person's attempt to control calories. Shown on the left are the meals with too many calories to permit weight maintenance for this person. On the right side of the figure, the meals have been modified to reduce their calories while still presenting the minimum number of servings from each food group recommended in the Food Guide Pyramid. Through these changes, the person has already reduced calories by over 800. To reduce calories further, however, takes more careful thought. The person must cut calories while still meeting the minimum number of servings from the food groups to maintain nutrient adequacy.

The three top contributors of both calories and fat in the high-calorie meals are:

french fries > large hamburger > brownie

Consider these foods when cutting calories. Also, remember from Chapter 5's Food Feature that breads, cereals, baked goods, rice, and pasta also vary in their fat and sugar contents and, therefore, in calories. For example, the breakfast waffles of Figure 9-14 contribute more calories than plain bread does because waffle mix contains fat, and the syrup provides more calories than the waffles do.

Figure 9•14

CALORIES IN TWO SETS OF MEALS

About 3,200 cal

2% milk, 1 c, 121 cal
Orange juice, 1 c 112 cal
Waffles, 2 each, 185 cal
Margarine, 2 tsp, 68 cal
Syrup, 4 tbs, 210 cal
Banana slices, ½ c, 69 cal
Breakfast total: 765

© Michael Newman/PhotoEdit (both)

About 2,300 cal

2% milk, 1 c, 121 cal
Orange juice, ¾ c, 84 cal
Waffle, 1 each, 93 cal
Margarine, 1 tsp, 34 cal
Syrup, 2 tbs, 105 cal
Banana slices, ½ c, 69 cal
Breakfast total: 506

2% milk, 1 c, 121 cal
Hamburger, quarter pound, 415 cal
French fries, large (about 50), 448 cal
Ketchup, 2 tbs, 32 cal
Apple pie, 1 each, 225 cal
Lunch total: 1,241

2% milk, 1 c, 121 cal
Hamburger, small, 266 cal
Green salad, 1 c, with light dressing, 1 tbs, 67 cal; croutons, ½ c, 50 cal
French fries, regular (about 30), 207 cal
Ketchup, 1 tbs, 16 cal
Gelatin dessert with fruit, 73 cal
Lunch total: 800

Italian bread, 2 slices, 163 cal
Margarine, 2 tsp, 68 cal
Stewed skinless chicken breast, 4 oz, 202 cal
Tomato sauce, ½ c, 37 cal
Rice, 1 c, 267 cal
Mixed vegetables, ½ c, 54 cal
Regular cheese sauce, ¼ c, 108 cal
Brownie, 1 each, 267 cal
Supper total: 1,166
Day's total: 3,172

© Polara Studios, Inc. (all)

Italian bread, 1 slice, 82 cal
Margarine, 1 tsp, 34 cal
Stewed skinless chicken breast, 4 oz, 202 cal
Tomato sauce, ½ c, 37 cal
Rice, 1 c, 267 cal
Mixed vegetables, ½ c, 54 cal
Low-fat cheese sauce, ¼ c, 85 cal
Brownie, 1 each, 267 cal
Super total: 1,028
Day's total: 2,334

Some 300 calories were trimmed from the 3,200-calorie meals by reducing sweets—reducing syrup served at breakfast and eliminating apple pie from lunch. (The diet planner kept the brownie, however—pleasure matters, too.) These two actions alone, repeated each day for one month, produce a calorie reduction more than sufficient to make a 3-pound difference in the person's body weight.

Reduce Calories Further

Try your hand at cutting calories further to 1,800 or even 1,600 calories for the day. The only "must" in cutting calories is to make the diet adequate. Replace some of the higher-calorie choices with better calorie bargains. Make substitutions with an eye for adequacy: for example, substituting diet cola for one of the two milk servings would compromise calcium adequacy and so is not allowable.

Step 1. Use Form 9-1, the column marked *Lower-Calorie Choices*, to record your changes in the 2,300-calorie day's meals (already listed for convenience at the left of the form).

Step 2. Record the calorie savings. As you make changes, turn to the Table of Food Composition, Appendix A, to find calorie values for foods you propose as substitutions for the originals. There is no need to look up every food on the menu—just the substitutes for those you choose to change. Subtract the calorie value of the new food from that of the original and write the calorie difference in the *Calories Saved* column.

Form 9•1

TRY YOUR HAND: REDUCE CALORIES FURTHER

2,300-Calorie Day	Lower-Calorie Choices	Calories Saved	Food Group Servings: Name of Group	Food Group Servings: Number of Servings
Breakfast				
2% milk, 1 c, 121 cal				
Orange juice, ¾ c, 84 cal				
Waffle, 1 each, 93 cal				
Margarine, 1 tsp, 34 cal				
Syrup, 2 tbs, 105 cal				
Banana slices, ½ c, 69 cal				
Lunch				
2% milk, 1 c, 121 cal				
Hamburger, small, 2 oz, 266 cal				
Green salad, 1 c,				
with 1 tbs light dressing, 67 cal				
croutons, ½ c, 50 cal				
French fries, regular (about 30), 207 cal				
Ketchup, 1 tbs, 16 cal				
Gelatin dessert with fruit, 73 cal				
Dinner				
Italian bread, 1 slice, 82 cal				
Margarine, 1 tsp, 34 cal				
Stewed skinless chicken breast, 4 oz, 202 cal				
Tomato sauce, ½ c, 37 cal				
Rice, 1 c, 267 cal				
Mixed vegetables, ½ c, 54 cal				
Low-fat cheese sauce, ¼ c, 85 cal				
Brownie, 1 each, 267 cal				
	Total calories saved:		2,334 − ______ (calories saved)	= ______ (new day's total calories)

New plan includes how many servings from each food group:
Milk, yogurt, cheese ________ Fruit ________ Vegetables ________ Meats and alternates ________
Bread, cereal, rice, pasta ________

do it!

Step 3. Add the *Calories Saved* column to obtain your total calorie savings for the day.

Step 4. Subtract the total savings from the original total of 2,334 calories to find the calorie value of the new day's meals.

Step 5. Assign each food from the new menu to its appropriate food group, and estimate the number of servings it represents. (Tips on how to estimate servings were given in Chapter 2's Do It! section.) Write the food group names on the left-hand side of the *Food Group Servings* column of Form 9-1. On the right-hand side of the same column, write your estimate of the number of servings each food item represents. Total the day's servings from each group, and write the totals in the spaces provided at the bottom of the form.

Analysis

Answer the following questions:

1. How many total calories did your changes save?
2. Assuming that a pound of body weight is worth 3,500 calories, how much weight would a dieter theoretically lose in a month by cutting every day's calories to this extent?
3. By what methods (see the four-item list in the first paragraph of this Do It!) did you reduce calories?
4. Did you remove high-calorie constituents from any foods? Which ones?
5. Which high-calorie foods did you replace with lower-calorie ones? Try to judge how these changes may have affected the saturated fat, vitamin, mineral, or fiber content of the meals; write your responses.
6. Which of the changes most significantly reduced calories in the meals? Which changed calories least?
7. Did you find it necessary to replace the fast-food lunch in the diet? Why or why not?
8. Were you able to cut calories significantly while still meeting the minimum number of servings from each food group? If not, which groups fell short? List ways of adjusting your choices to include the missing foods.
9. Are the reduced-calorie meals appealing? If not, how can you include more appealing foods without increasing the calorie values?
10. Did your meals include any sweets or other treats? If so, which ones? If not, why not?

When you develop skill in making these sorts of changes, they tend to come to mind whenever the opportunity arises. Choosing foods with an eye for their contributions of calories and nutrients becomes a natural part of living. In case you are curious about how the authors might reduce calories, Table 9-12 shows our ideas for changes.

Table 9•12

OUR ANSWERS TO FIGURE 9-14: NEW CALORIE LEVEL = ABOUT 1,900 CALORIES

Food	Changes	Calories Saved
Orange juice	Same	0
Milk	Replace with fat-free	35
Waffle	Same	0
Margarine/syrup	Replace with light margarine	12
Banana	Same	0
Milk	Replace with fat-free	35
Hamburger	Same	0
French fries	Omit	207
Ketchup	Same	0
Salad	Same	0
Croutons	Same	0
Gelatin	Omit	73
Italian bread with margarine	Same	0
Chicken	Same	0
Rice	Same	0
Tomato sauce	Same	0
Mixed vegetables	Same	0
Low-fat cheese sauce	Omit	85
Brownie	Same	0
Total calories saved		447

Food group totals
Milk, yogurt, cheese 2 Fruit 2
Vegetables 4 Meats and alternates 2 Bread, cereal, rice, pasta 7

Answers to these Self Check questions are in Appendix G.

1. All of the following are health risks associated with excessive body fat except:
   a. respiratory problems
   b. sleep apnea
   c. gallbladder disease
   d. low blood lipids
2. Which of the following statements about basal metabolic rate (BMR) is correct?
   a. The more fat tissue, the higher the BMR.
   b. The more thyroxine produced the higher the BMR.
   c. Fever lowers the BMR.
   d. Pregnant women have lower BMRs.
3. Body density (the measurement of body weight compared with volume) is determined by which technique?
   a. fatfold test
   b. bioelectrical impedance
   c. underwater weighing
   d. all of the above
4. The obesity theory that suggests that the body chooses to be at a specific desired weight is the:
   a. set-point theory
   b. enzyme theory
   c. fat cell number theory
   d. external cue theory
5. Which of the following is a possible physical consequence of fasting?
   a. loss of lean body tissues
   b. lasting weight loss
   c. body cleansing
   d. all of the above
6. Which of the following is a recommended weight-loss strategy?
   a. muscle stimulators
   b. stomach stapling
   c. herbs containing ephedrine
   d. none of the above
7. The thermic effect of food plays a major role in energy expenditure. T F
8. If you bike about ten minutes a day and walk to classes but otherwise sit and study, you are considered lightly active. T F
9. The BMI standard is an excellent tool for evaluating obesity in athletes and the elderly. T F
10. A diet too low in carbohydrate brings about responses that are similar to fasting. T F

nutrition on the net

For further study of the topics of this chapter, access these websites and search for the phrases or words in quotation marks:

1. Find updates and quick links to these and other nutrition-related sites at our website:
   www.wadsworth.com/nutrition
2. Type in the search word "obesity" at:
   www.ilsi.org/
3. Information on a variety of obesity topics is available at the American Obesity Association website:
   www.obesity.org/
4. Type in the search word "obesity" at:
   www.eatright.org/
5. Peruse the offerings of the Division of Nutrition and Physical Activity, National Center for Chronic Disease Prevention and Health Promotion, at their website:
   www.cdc.gov/nccdphp/dnpa
6. Many materials to help teach others about obesity are available at the Weight Control Information Network:
   www.niddk.nih.gov/health/nutrit/win.htm
7. Read guides on fitness and healthy weight, and access many other materials at Shape Up America:
   www.shapeup.org
8. Review a transcript of presentations and panel discussions of leading obesity experts and fad diet authors by searching for "Symposium on the Great Nutrition Debate" at the USDA's site:
   www.usda.gov
9. Read the latest materials on obesity diagnosis and treatment here:
   www.nhlbi.nih.gov/guidelines/obesity/practgde.htm
10. The Partnership for Healthy Weight Management offers practical advice on starting a weight-loss program at this site:
    www.consumer.gov/weightloss/

INTERNET ACTIVITY

Shape Up America is an organization started by a coalition of health and weight-loss organizations.

1. Go to the website of Shape Up America at:
   http://www.shapeup.org
2. Click "Body Fat Lab."
3. Click "Body Fat Basics."
4. Then click "Your Body Fat IQ Test."
5. Take the test.

Report your results. How does this new knowledge influence your understanding of body composition?

CONTROVERSY 9

THE PERILS OF EATING DISORDERS

An estimated 5 million people in the United States, primarily girls and women, suffer from the eating disorders of **anorexia nervosa** and **bulimia nervosa.** Many more suffer from **binge eating disorder** or other related conditions that imperil the sufferer's well-being. White women are most likely to have an **eating disorder,** although men and ethnic women are by no means immune.[1]

An estimated 85 percent of eating disorders start during adolescence.[2] Among U.S. adolescents, some markers of disordered eating such as restrained eating, binge eating, purging, fear of fatness, and distorted body image are extraordinarily common. In one national survey of over 6,700 adolescents in grades 5 through 12, almost half of girls and a fifth of the boys reported having dieted to lose weight.[3] Disordered eating among girls was at 13 percent, and among boys, 7 percent. A survey at a major university found that only 8 percent of students were overweight by objective measure (BMI), yet more than half of the students reported themselves to be overweight.[4] Half of the students found to be *underweight* according to the BMI charts considered themselves overweight. Such misconceptions about weight are much less prevalent in most other societies, but they often gain ground when body image becomes central to self-worth.[5]

Why do so many people in our society suffer from eating disorders? Excessive pressure to be thin is at least partly to blame. When low body weight becomes an important goal, people begin to view normal healthy body weight as too fat, and some take unhealthy actions to lose weight. Severe restriction of food intake can create intense stress and extreme hunger that lead to binges.[6] Painful emotions such as anger, jealousy, or disappointment may be turned inward by youngsters who express dissatisfaction with body weight or who say they "feel fat."[7] As weight loss and diet restraint become more and more a focus, psychological problems worsen, and the likelihood of developing full-blown eating disorders intensifies.[8] Table C9-1 defines eating disorder terms.

Eating Disorders in Athletes

Athletes and dancers are at special risk for eating disorders. In females, three associated medical problems form the **female athlete triad:** disordered eating, amenorrhea (cessation of menstruation), and osteoporosis.[9] In males, disordered eating brings on many of the physical problems of their female counterparts.

The Female Athlete Triad

At age 14, Suzanne was a top contender for a spot on the state gymnastics team. Each day her coach reminded team members that they must weigh no more than a few ounces above their assigned weights in order to qualify for competition. The coach chastised gymnasts who gained weight. Suzanne weighed herself several times a day to make sure that she had not exceeded her 80-pound limit. Suzanne dieted and exercised to an extreme, and unlike many of her friends, she never began to menstruate. A few months before her fifteenth birthday, Suzanne's coach dropped her back to the second-level team. Suzanne blamed

Anorexia nervosa.

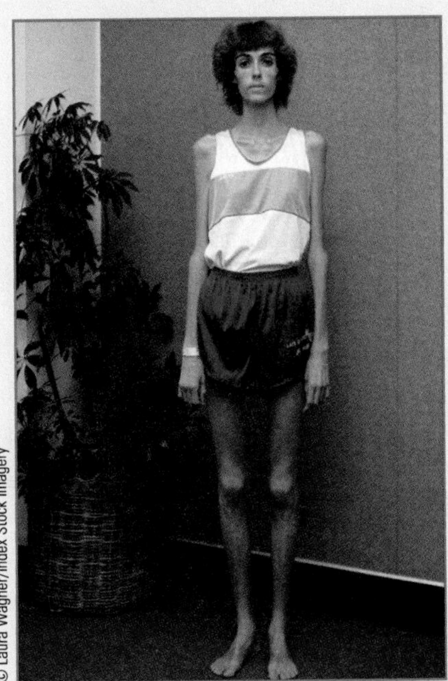

Table C9•1

EATING DISORDER TERMS

- **anorexia nervosa** an eating disorder characterized by a refusal to maintain a minimally normal body weight, self-starvation to the extreme, and a disturbed perception of body weight and shape; seen (usually) in teenage girls and young women (*anorexia* means "without appetite"; *nervos* means "of nervous origin").
- **binge eating disorder** an eating disorder whose criteria are similar to those of bulimia nervosa, excluding purging or other compensatory behaviors.
- **bulimia** (byoo-LEEM-ee-uh) **nervosa** recurring episodes of binge eating combined with a morbid fear of becoming fat; usually followed by self-induced vomiting or purging.
- **cathartic** a strong laxative.
- **cognitive therapy** psychological therapy aimed at changing undesirable behaviors by changing underlying thought processes contributing to these behaviors; in anorexia, a goal is to replace false beliefs about body weight, eating, and self-worth with health-promoting beliefs.
- **eating disorder** a disturbance in eating behavior that jeopardizes a person's physical or psychological health.
- **emetic** (em-ETT-ic) an agent that causes vomiting.
- **female athlete triad** a potentially fatal triad of medical problems seen in female athletes: disordered eating, amenorrhea, and osteoporosis.

her poor performance on a slow-healing stress fracture. Mentally stressed and physically exhausted, she quit gymnastics and began overeating between periods of self-starvation. Suzanne had developed the dangerous combination of problems that characterize the female athlete triad—disordered eating, amenorrhea, and weakening of the bones.

Female athletes, in keeping with their coaches' recommendations, often compare themselves to unsuitable weight standards. Extreme slimness has long been considered desirable in certain activities, such as dancing, gymnastics, and figure skating.[10] This puts the athlete in a bind because an athlete's body must be heavier than average for a given height, reflecting more muscle and bone tissue and less fat. Most weight standards appropriate for the general population fall far from the mark with regard to athletes' bodies. For athletes, body composition measures such as fatfold measures yield more useful information.

The prevalence of amenorrhea among premenopausal women in the United States is about 2 to 5 percent overall, but may be as high as 66 percent among female athletes. Amenorrhea is *not* a normal adaptation to strenuous physical training but a symptom of something going wrong, and it is particularly hazardous to the bones. For most people, weight-bearing exercise helps to strengthen bones. In anorexia nervosa, however, strenuous activity can imperil the bones.[11] Vigorous training and low food energy intakes may reduce estrogen levels and greatly increase the risks of stress fractures today and of osteoporosis in later life.[12] Bone tissue may not recover even after diagnosis and treatment. A recent report also links athletic amenorrhea with premature cardiovascular disease, possibly attributable to the same hormonal disturbances that weaken the bones.[14]

Male Athletes and Eating Disorders

Male athletes and dancers who face pressure to achieve a certain body weight often develop eating disorders, although they may deny having them in the mistaken belief that the disorders strike only women.[15] On average, male teenagers carry about 15 percent of body weight as fat, but some high school wrestlers, gymnasts, and figure skaters strive for only 5 percent body fatness.

Wrestlers, for example, are required to "make weight" to compete in the lowest possible weight class to face the smallest possible opponents. To that end, wrestlers starve themselves, don rubber suits, sweat in steam rooms, and take diuretics to shed water weight before weighing in for competition. These practices were responsible for the deaths of three college athletes in recent years and have caused untold misery and harm to many others.[16] Athletes engaging in these practices actually compromise their athletic abilities. The diminished anaerobic strength, reduced endurance, decreased oxygen capacity, and general weakness caused by food deprivation and dehydration can hobble performance, an effect lasting days after food and water are replenished.[17]

Male athletes are also susceptible to weight *gain* problems, in which athletes with well-muscled bodies see themselves as underweight and weak.[18] Such a distorted body image leads to frequent weighing, excessive exercise, overuse of special diets or protein supplements, or even the abuse of steroid drugs in the attempt to bulk up their muscles.

Young athletes and their coaches must be educated about links between inappropriate body-weight ideals, improper weight-loss techniques, eating disorder development, effective sports nutrition, and safe weight-control methods. For all young people's activities, idealistic artistic standards based on slim appearance or low body weight should be replaced with performance-based standards. Table C9-2 provides some suggestions to help athletes and dancers protect themselves against developing eating disorders. The next sections describe eating disorders that anyone, athlete or nonathlete, may experience.

Anorexia Nervosa

Julie is 18 years old and is a superachiever in school. She watches her diet with great care, and she exercises daily, maintaining a heroic schedule of self-discipline. She is thin, but she is determined to lose more weight. She is

Table C9•2

TIPS FOR COMBATING EATING DISORDERS

General Guidelines

- Never restrict food servings to below the numbers suggested for adequacy by the Daily Food Guide.
- Eat frequently. People often do not eat frequent meals because of time constraints, but eating can be incorporated into other activities, such as snacking while studying or commuting. The person who eats frequently never gets so hungry as to allow hunger to dictate food choices.
- If not at a healthy weight, establish a reasonable weight goal based on a healthy body composition. (Chapter 9 provides help in doing so.)
- Allow a reasonable time to achieve the goal. A reasonable rate for losing excess fat is about 1% of body weight per week.
- Establish a weight-maintenance support group with people who share interests.

Specific Guidelines for Athletes and Dancers

- Replace weight-based goals with performance-based goals.
- Remember that eating disorders impair physical performance. Seek confidential help in obtaining treatment if needed.
- Restrict weight-loss activities to the off-season.
- Focus on proper nutrition as an important facet of your training, as important as proper technique.

5 feet 6 inches tall and weighs 85 pounds. She has anorexia nervosa.

Characteristics of Anorexia Nervosa

Julie is unaware that she is undernourished, and she sees no need to obtain treatment. She stopped menstruating several months ago and is moody and chronically depressed. She insists that she is too fat, although her eyes are sunk in deep hollows in her face. Close to physical exhaustion, she no longer sleeps easily. Her family is concerned, and although reluctant to push her, they have finally insisted that she see a psychiatrist. Julie's psychiatrist has prescribed group therapy as a start, but warns that if Julie does not begin to gain weight soon, she will need to be hospitalized.

Most anorexia nervosa victims come from middle- or upper-class families. Men account for only 1 or 2 in 20 cases in the general population, although the incidence among male athletes and dancers may be much higher.

No one knows for certain what causes anorexia nervosa. Central to its diagnosis is a distorted body image that overestimates body fatness.[19] When Julie looks at herself in the mirror, she sees her 85-pound body as fat. The more Julie overestimates her body size, the more resistant she is to treatment, and the more unwilling to examine her faulty values and misconceptions. Malnutrition is known to affect brain functioning and judgment in this way. People with anorexia nervosa cannot recognize it in themselves; only professionals can diagnose it. Table C9-3 on the next page shows the criteria that experts use.

The Role of the Family

Certain family attitudes, and especially parental attitudes, stand accused of contributing to eating disorders. Families of persons with anorexia nervosa are likely to be critical and to overvalue outward appearances while undervaluing inner self-worth. Parents may oppose one another's authority and vacillate between defending the anorexic child's behavior and condemning it, confusing the child and disrupting normal parental control. In the extreme, parents may even be sexually abusive or abusive in other ways.[20]

Julie is a perfectionist, just as her parents are. She identifies so strongly with her parents' ideals and goals that she cannot get in touch with her own identity. She is respectful of authority but sometimes feels like a robot, and she may act that way, too: polite but controlled, rigid, and unspontaneous.[21] For Julie, rejecting food is a way of gaining control.

Women with anorexia nervosa see themselves as fat, even when they are dangerously underweight.

© David Kelly Crow/PhotoEdit

Self-Starvation

How can a person as thin as Julie continue to starve herself? Julie uses tremendous discipline to strictly limit her portions of low-calorie foods. She will deny her hunger, and having become accustomed to so little food, she feels full after eating only a half-dozen carrot sticks. She can recite the calorie contents of dozens of foods and the calorie costs of as many exercises. If she feels that she has gained an ounce of weight, she runs or jumps rope until she is sure she has exercised it off. She drinks water incessantly to fill her stomach, risking dangerous mineral imbalances.[22] If she fears that the food energy she has eaten exceeds the exercise she has done, she takes laxatives to hasten the passage of food from her system, not knowing that laxatives reduce water absorption, but not food energy absorption. Her other methods of staying thin are so effective that she is unaware that laxatives have no effect on body fat. She is desperately hungry. In fact, she is starving, but she doesn't eat because her need for self-control dominates other needs.

Physical Perils

Anorexia nervosa damages the body much as other forms of starvation do. In young people, growth ceases and normal development falters. They lose so much lean tissue that basal metabolic rate slows. In athletes, the loss of lean tissue handicaps physical performance.[23] The heart pumps inefficiently and irregularly, the heart muscle becomes weak and thin, the heart chambers diminish in size, and the blood pressure falls.[24] Electrolytes that help to regulate the heartbeat go out of balance. Many deaths in people with anorexia are due to heart failure.

Starvation brings neurological, digestive, and circulatory consequences as well. The brain loses significant amounts of tissue, nerves function abnormally, the electrical activity of the brain becomes abnormal, and insomnia is common. Digestive functioning becomes sluggish, the stomach empties slowly, and the lining of the intestinal tract shrinks. The ailing digestive tract fails to digest food adequately, even if the victim does eat. The pancreas slows its production of digestive enzymes. Diarrhea sets in, further worsening malnutrition.[25]

Changes in the blood include anemia, impaired immune response, altered blood lipids, high blood concentrations of vitamin A and vitamin E, and low blood proteins. Dry skin, low body temperature, and the development of fine body hair (the body's attempt to keep warm) also occur. In adulthood, both women and men lose their sex drives. Mothers with anorexia nervosa may severely underfeed their children who then fail to grow and suffer the other harms typical of starvation.[26]

Treatment of Anorexia Nervosa

Treatment of anorexia nervosa requires a multidisciplinary approach that addresses two issues and behaviors: those relating to food and weight and those involving relationships with oneself and others.[27] Teams of physicians, nurses, psychiatrists, family therapists, and dietitians work together to treat people with anorexia nervosa. Appropriate diet is crucial for normalizing body weight and must be crafted individually.[28] Clients are seldom willing to eat for themselves, but if they are, chances are they can recover without other interventions.

Professionals classify clients based on the risks posed by the degree of malnutrition present.* Clients with low risks may benefit from family counseling, **cognitive therapy,** behavior modification, and nutrition guidance; those with greater risks may also need other forms of psychotherapy and supplemental formulas to provide extra nutrients and energy. High-risk clients may require involuntary hospitalization and may need to be force-fed by tube at first to forestall death. This step causes psychological trauma.[29] Drugs are commonly prescribed, but to date, they play a limited role in treatment.

Stopping weight loss is a first goal of treatment; establishing regular eating patterns is next. At first, progress is slow partly owing to a speeded-up metabolic rate and an increased thermic response to food that occur upon refeeding.[30] As small gains of body fat occur, blood concentration of the appetite-suppressing hormone leptin begins creeping up, too, causing researchers to speculate that leptin may contribute to difficulties in weight restoration.[31]

Few people with anorexia nervosa seek treatment on their own, and denial makes treatment difficult. Almost half of the women who are treated can maintain their body weight within 15 percent of a healthy weight; at that weight, many of them begin menstruating again. The other half have poor or fair outcomes of treatment, and two-thirds of those treated continue a mental battle with recurring morbid thoughts about food and body weight. Weight gain often takes the form of fat, and fear of the growing pads of fat around their bodies forces many to relapse into abnormal eating behaviors.[32] About 5 percent die during treatment, 1 percent by suicide.[33]

Before drawing conclusions about someone who is extremely thin, remember that diagnosis of anorexia nervosa requires professional assessment. People seeking help with anorexia nervosa, either for themselves or for others, can call the National Anorexic Aid Society hotline.†

*Indicators of malnutrition include a low percentage of body fat, low blood proteins, and impaired immune response.

Bulimia Nervosa

Sophia is a 20-year-old flight attendant, and although her body weight is

†Phone numbers, addresses, and Internet addresses are in Appendix E.

Table C9•3

CRITERIA FOR DIAGNOSIS OF ANOREXIA NERVOSA

A person with anorexia nervosa demonstrates the following:

A. Refusal to maintain body weight at or above a minimal normal weight for age and height, e.g., weight loss leading to maintenance of body weight less than 85% of that expected; or failure to make expected weight gain during period of growth, leading to body weight less than 85% of that expected.
B. Intense fear of gaining weight or becoming fat, even though underweight.
C. Disturbance in the way in which one's body weight or shape is experienced; undue influence of body weight or shape on self-evaluation, or denial of the seriousness of the current low body weight.
D. In females past puberty, amenorrhea, i.e., the absence of at least three consecutive menstrual cycles. (A woman is considered to have amenorrhea if her periods occur only following hormone, e.g., estrogen, administration.)

Two types of anerexia nervosa include:

- Restricting type: during the episode of anorexia nervosa, the person does not regularly engage in binge eating or purging behavior (i.e., self-induced vomiting or the misuse of laxatives, diuretics, or enemas).
- Binge eating/purging type: during the episode of anorexia nervosa, the person regularly engages in binge eating or purging behavior (i.e., self-induced vomiting or the misuse of laxatives, diuretics, or enemas).

SOURCE: Reprinted with permission from the Diagnostic and Statistical Manual of Mental Disorders, Fourth Edition, Text Revision. Copyright 2000 American Psychiatric Association.

healthy, she thinks constantly about food. She alternately starves herself and then secretly binges; when she has eaten too much, she vomits. Few people would fail to recognize that these symptoms signify bulimia nervosa.

Characteristics of Bulimia Nervosa

Bulimia nervosa is distinct from anorexia nervosa and is much more prevalent, although the true incidence is difficult to establish. People with bulimia nervosa often suffer in secret and, when asked, may deny the existence of a problem. More men suffer from bulimia nervosa than from anorexia nervosa, but bulimia nervosa is still most common in women. Based on a questionnaire, one study estimates that 19 percent of female college students experience bulimic symptoms. A true diagnosis of bulimia nervosa is based on the criteria listed in Table C9-4.

Like the typical person with bulimia nervosa, Sophia is single, female, and white. She is well educated and close to her ideal body weight, although her weight fluctuates over a range of 10 pounds or so every few weeks.

Sophia seldom lets her bulimia nervosa interfere with her work or other activities. From early childhood she has been a high achiever but emotionally dependent on her parents. As a young teen, Sophia cycled on and off crash diets. She feels anxious at social events and cannot easily establish close relationships. She is usually depressed, is often impulsive, and has low self-esteem. When crisis hits, Sophia responds by replaying events, worrying excessively, seeking solace in alcohol and tobacco, and blaming herself but never asking for help—behaviors that are barriers to effective coping.[34]

A person may consume up to 10,000 calories during an eating binge.

© Felicia Martinez/PhotoEdit

The Role of the Family

Families of bulimic people are observed to be externally controlling but emotionally uninvolved with their children, resulting in a stifling negative self-image.[35] Dieting, arguments, criticism of body shape or weight, minimal affection and caring, and other weaknesses are common in the families of people with bulimia.[36] Typically, the family has "secrets " that are hidden from outsiders. Bulimic women who report having been abused sexually or physically by family members or friends may continually suffer a sense of being unable to gain control.[37]

Should the member with bulimia nervosa begin making the needed changes toward recovery, others in the family may feel threatened. Family cooperation is important, however, because making changes within a family requires effort from everyone. Such effort is well spent, for changing destructive family interactions can greatly benefit the person who has begun to fight against bulimia nervosa.

Table C9•4

CRITERIA FOR DIAGNOSIS OF BULIMIA NERVOSA

A person with bulimia nervosa demonstrates the following:

A. Recurrent episodes of binge eating. An episode of binge eating is characterized by both of the following:
   1. Eating, in a discrete period of time (e.g., within any two-hour period), an amount of food that is definitely larger than most people would eat during a similar period of time and under similar circumstances, and,
   2. A sense of lack of control over eating during the episode (e.g., a feeling that one cannot stop eating or control what or how much one is eating).

B. Recurrent inappropriate compensatory behavior in order to prevent weight gain, such as self-induced vomiting; misuse of laxatives, diuretics, enemas, or other medications; fasting; or excessive exercise.

C. Binge eating and inappropriate compensatory behaviors that both occur, on average, at least twice a week for three months.

D. Self-evaluation unduly influenced by body shape and weight.

E. The disturbance does not occur exclusively during episodes of anorexia nervosa.

Two types:

- Purging type: the person regularly engages in self-induced vomiting or the misuse of laxatives, diuretics, or enemas.
- Nonpurging type: the person uses other inappropriate compensatory behaviors, such as fasting or excessive exercise, but does not regularly engage in self-induced vomiting or the misuse of laxatives, diuretics, or enemas.

SOURCE: Reprinted with permission from the Diagnostic and Statistical Manual of Mental Disorders, Fourth Edition, Text Revision. Copyright 2000 American Psychiatric Association.

Binge Eating and Purging

A bulimic binge is unlike normal eating, and the food is not consumed for its nutritional value. During a binge, Sophia's eating is accelerated by her hunger from previous calorie restriction. She may take in anywhere from 1,000 to many thousands of calories of easy-to-eat, low-fiber, smooth-textured, high-fat, and high-carbohydrate foods. Typically, she chooses cookies, cakes, and ice cream; and she eats the entire bag of cookies, the whole cake, and every spoonful in a carton of ice cream.

By the end of the binge, she has vastly overcorrected for her attempts at calorie restriction at other times.[38]

The binge is a compulsion and usually occurs in several stages: "anticipation and planning, anxiety, urgency to begin, rapid and uncontrollable consumption of food, relief and relaxation, disappointment, and finally shame or disgust." Then, to purge the food from her body, she may use a **cathartic**—a strong laxative that can injure the lower intestinal tract. Or she may induce vomiting, using an **emetic**—a drug intended as first aid for poisoning. After the binge she pays the price with hands scraped raw against the teeth during induced vomiting, swollen neck glands and reddened eyes from straining to vomit, and the bloating, fatigue, headache, nausea, and pain that follow.

Physical and Psychological Perils

Purging may seem to offer a quick and easy solution to the problems of unwanted calories and body weight, but bingeing and purging have serious physical consequences. Fluid and electrolyte imbalances caused by vomiting or diarrhea can lead to abnormal heart rhythms and injury to the kidneys. Urinary tract infections can lead to kidney failure. Vomiting causes irritation and infection of the pharynx, esophagus, and salivary glands; erosion of the teeth; and dental caries. The esophagus or stomach may rupture or tear. Overuse of emetics can lead to death by heart failure.

Unlike Julie, Sophia is aware that her behavior is abnormal, and she is deeply ashamed of it. She wants to recover, and this makes recovery more likely for her than for Julie, who clings to denial.

Treatment of Bulimia Nervosa

To gain control over food and establish regular eating patterns requires adherence to a structured eating plan. Restrictive dieting is forbidden, for it almost always precedes and may even trigger binges. Steady maintenance of weight and prevention of relapse into cyclic gains and losses are the goals. Many a former bulimia nervosa sufferer has taken a major step toward recovery by learning to consistently eat enough food to satisfy hunger needs (at least 1,600 calories a day). Table C9-5 offers some ways to begin correcting the eating problems of bulimia nervosa. Daily physical activity may also be of key importance in recovery from bulimia.[39] About half of women receiving a diagnosis of bulimia may recover completely after five to ten years, with or without treatment, but treatment probably speeds the recovery process.[40] If Sophia's depression deepens, she may benefit from antidepressant medication.[41]

Table C9•5

SOME STRATEGIES FOR CONTROLLING BULIMIA

Planning principles:
- Plan meals and snacks; record plans in a food diary prior to eating.
- Plan meals and snacks that require eating at the table and using utensils.
- Refrain from finger foods.
- Refrain from "dieting" or skipping meals.

Nutrition principles:
- Eat a well-balanced diet and regularly timed meals consisting of a variety of foods.
- Include raw vegetables, salad, or raw fruit at meals to prolong eating times.
- Choose whole-grain, high-fiber breads, pasta, rice, and cereals to increase bulk.
- Consume adequate fluid, particularly water.

Other tips:
- Choose meals that provide protein and fat for satiety, and bulky, fiber-rich carbohydrates for immediate feelings of fullness.
- Try including soups and other water-rich foods (at the top of the list on page 270) for satiety.
- Choose portions that meet the definition of "a serving" according to the Daily Food Guide (pages 36–37).
- For convenience (and to reduce temptation) select foods that naturally divide into portions. Select one potato, rather than rice or pasta that can be overloaded onto the plate; purchase yogurt and cottage cheese in individual containers; look for small packages of precut steak or chicken; choose frozen dinners with metered portions.
- Include 30 minutes of physical activity every day—exercise may be an important tool in controlling bulimia.

Binge Eating Disorder

Anorexia nervosa and bulimia nervosa are distinct eating disorders, yet they sometimes overlap. People with both conditions share an overconcern with body weight and the tendency to drastically undereat. Both may purge. The two disorders can also appear in the same person, or one can lead to the other. Other people have eating disorders that fall short of anorexia nervosa or bulimia nervosa, but share some of their features, such as fear of body fatness. One such condition is binge eating disorder (defined earlier in Table C9-1).

Up to half of all people who restrict eating to lose weight periodically binge without purging, including about one-third of obese people who regularly engage in binge eating. Obesity itself, however, does not constitute an eating disorder. Table C9-6 lists the official diagnostic criteria for binge eating disorder.

Clinicians note the differences between people with bulimia nervosa and those with binge eating disorder. Binge eaters rarely purge, they consume less during a binge, and they are less restrained during nonbinge eating. Similarities also exist, including feeling out of control, or feeling disgusted, depressed, embarrassed, or guilty after bingeing.

Binge eating behavior responds more readily to treatment than other eating disorders, and resolving such behaviors can be a first step to authentic weight control. Successful treatment also improves physical health, mental health, and the chances of breaking the cycle of rapid weight losses and gains.

Eating Disorders in Society

Most experts agree that eating disorders have many causes: sociocultural, psy-

Table C9•6

CRITERIA FOR DIAGNOSIS OF BINGE EATING DISORDER

A person with a binge eating disorder demonstrates the following:

A. Recurrent episodes of binge eating. An episode of binge eating is characterized by both of the following:
  1. Eating, in a discrete period of time (e.g., within any two-hour period) an amount of food that is definitely larger than most people would eat in a similar period of time under similar circumstances.
  2. A sense of lack of control over eating during the episode (e.g., a feeling that one cannot stop eating or control what or how much one is eating).

B. Binge eating episodes are associated with at least three of the following:
  1. Eating much more rapidly than normal.
  2. Eating until feeling uncomfortably full.
  3. Eating large amounts of food when not feeling physically hungry.
  4. Eating alone because of being embarrassed by how much one is eating.
  5. Feeling disgusted with oneself, depressed, or very guilty after overeating.

C. The binge eating causes marked distress.

D. The binge eating occurs, on average, at least twice a week for six months.

E. The binge eating is not associated with the regular use of inappropriate compensatory behaviors (e.g., purging, fasting, excessive exercise) and does not occur exclusively during the course of anorexia nervosa or bulimia nervosa.

SOURCE: Reprinted with permission from the Diagnostic and Statistical Manual of Mental Disorders, Fourth Edition, Text Revision. Copyright 2000 American Psychiatric Association.

chological, hereditary, and probably also neurochemical. Proof that society plays a role in eating disorders is their demographic distribution: they are known only in developed nations, and they become more prevalent as wealth increases and food becomes plentiful.

No doubt our society sets unrealistic ideals for body weight, especially for women, and devalues those who do not conform to them. The Miss America beauty pageant, for example, puts forth a role model of female desirability. Even though no winner has ever been overweight, women chosen to wear the Miss America crown have been progressively thinner over the years.[42] Magazines and other media convey the message that to be thin is to be happy; eating disorders are not a form of rebellion against these unrealistic ideals, but rather an exaggerated acceptance of them.

Even professionals, including physicians and dietitians, tend to praise people for losing weight and to suggest weight loss to people who do not need it. As a result, normal-weight girls as young as 11 or 12 fear that they are too fat and are "on diets," and many are poorly nourished.[43] Some eat too little food to support normal growth; thus, they miss out on their adolescent growth spurts and may never catch up.

New research has uncovered potential roles for the brain's neurotransmitters in the development of eating disorders. These findings raise the hope that effective pharmacological treatments may one day be available.[44]

Perhaps a young person's best defense against these disorders is to learn about normal, expected growth patterns, especially the characteristic weight gain of adolescence (see Chapter 13), and to learn respect for the inherent wisdom of the body. When people discover and honor the body's real needs, they become unwilling to sacrifice health for conformity.

chapter 10

Fernand Leger (1881–1955), *The Cyclists*, 1944

Nutrients, Physical Activity, and the Body's Responses

Frequently Asked Questions

How do my muscles become physically fit? p. 361

Can I expect any health benefits from weight training? p. 362

How does cardiorespiratory training benefit the heart? p. 363

To burn more fat during activity, should athletes eat more fat? p. 369

How much protein should an athlete consume? p. 372

Do nutrient supplements benefit athletic performance? p. 373

Contents

IN THE BODY, nutrition and physical activity are tied together. The working body demands all three energy-yielding nutrients—carbohydrate, lipids, and protein—to fuel activity. The body also needs protein and a host of supporting nutrients to build lean tissue. Physical activity, in turn, benefits the body's nutrition by helping to regulate the use of fuels, by pushing the body composition toward the lean, and by increasing the daily calorie allowance. With more calories come more nutrients and other beneficial constituents of foods.

You don't have to run marathons to reap the health rewards of physical activity.[1] People who regularly engage in just moderate physical activity live longer on average than those who are physically inactive.[2] For health's sake, the American College of Sports Medicine (ACSM) and the *Dietary Guidelines for Americans* specify that people need to spend an accumulated minimum of 30 minutes in some sort of physical activity on most days of the week (see Figure 10-1). A sedentary lifestyle ranks with smoking and obesity as a powerful risk factor for developing the major killer diseases of our time—cardiovascular disease, some forms of cancer, stroke, diabetes, and hypertension.[3]

For many people, the health benefits of regular, moderate physical activity are reward enough. Others, however, seek the kinds and amount of physical activity that will not only benefit health, but improve their physical fitness or their performance in sports. In 1998, the ACSM issued recommendations for the quantity and quality of physical activity needed to develop and maintain fitness in healthy adults (see Table 10-1).[4]

If you are already physically fit, the following description applies: You move with ease and balance. You are strong and meet physical challenges without strain. You have endurance, and your energy lasts for hours. You meet daily physical challenges and have plenty of energy in reserve. What's more, you are prepared to meet mental and emotional challenges, too, for physical fitness also supports mental and emotional energy and resilience. Sounds good, doesn't it?

For those just beginning to increase fitness, be assured that improvement is not only possible, but is an inevitable result of becoming more active. As you improve your physical fitness, you not only *feel* better and stronger, but you *look* better, too.

Figure 10•1

PHYSICAL ACTIVITY PYRAMID

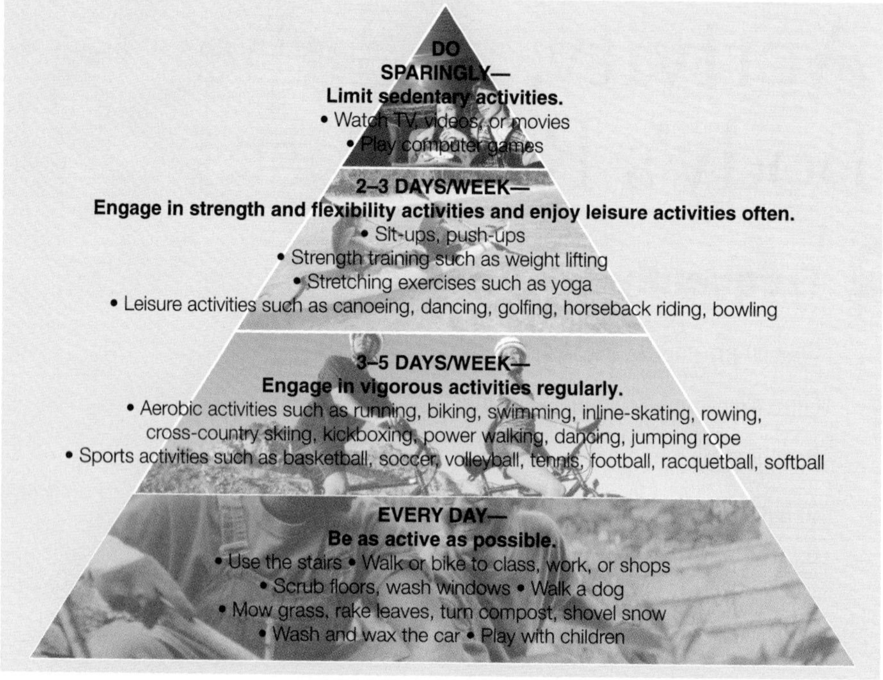

DO SPARINGLY—
Limit sedentary activities.
• Watch TV, videos, or movies
• Play computer games

2–3 DAYS/WEEK—
Engage in strength and flexibility activities and enjoy leisure activities often.
• Sit-ups, push-ups
• Strength training such as weight lifting
• Stretching exercises such as yoga
• Leisure activities such as canoeing, dancing, golfing, horseback riding, bowling

3–5 DAYS/WEEK—
Engage in vigorous activities regularly.
• Aerobic activities such as running, biking, swimming, inline-skating, rowing, cross-country skiing, kickboxing, power walking, dancing, jumping rope
• Sports activities such as basketball, soccer, volleyball, tennis, football, racquetball, softball

EVERY DAY—
Be as active as possible.
• Use the stairs • Walk or bike to class, work, or shops
• Scrub floors, wash windows • Walk a dog
• Mow grass, rake leaves, turn compost, shovel snow
• Wash and wax the car • Play with children

Table 10•1

GUIDELINES FOR PHYSICAL FITNESS

	Cardiorespiratory	Strength	Flexibility
	© Photo Disc, Inc.	© David Hanover Photography	© David Hanover Photography
Type of Activity	Aerobic activity that uses large-muscle groups and can be maintained continuously	Resistance activity that is performed at a controlled speed and through a full range of motion	Stretching activity that uses the major muscle groups
Frequency	3 to 5 days per week	2 to 3 days per week	2 to 3 days per week
Intensity	55 to 90% of maximum heart rate	Enough to enhance muscle strength and improve body composition	Enough to develop and maintain a full range of motion
Duration	20 to 60 minutes	8 to 12 repetitions of 8 to 10 different exercises (minimum)	4 repetitions of 10 to 30 seconds per muscle group (minimum)

SOURCE: Adapted from American College of Sports Medicine, Position stand: The recommended quantity and quality of exercise for developing and maintaining cardiorespiratory and muscular fitness, and flexibility in healthy adults, *Medicine and Science in Sports and Exercise* 30 (1998): 975–991.

Physically fit people walk with confidence and purpose because posture and self-image improve along with physical fitness. The kinds and amounts of physical activity that improve physical fitness also provide still greater health benefits (further reduction of cardiovascular disease risk and improved body composition, for example).[5]

This chapter is written for athletes and for active people who train like athletes. Casual athletes (those who compete only with their own goals) and competitive athletes (those who compete with others) are cut from the same cloth as far as their food and fluid needs are concerned. The chapter refers to "you" to make the connection between academic thinking and personal choices. To understand the interactions between physical activity and nutrition, you must first know a few things about physical fitness and **training.**

key point *Physical activity and fitness benefit people's physical and psychological well-being and improve their resistance to disease. Physical activity to improve physical fitness eases the tasks of daily living and offers additional personal benefits.*

training regular practice of an activity, which leads to physical adaptations of the body with improvement in flexibility, strength, or endurance.

flexibility the capacity of the joints to move through a full range of motion; the ability to bend and recover without injury.

muscle strength the ability of muscles to work against resistance.

muscle endurance the ability of a muscle to contract repeatedly within a given time without becoming exhausted.

cardiorespiratory endurance the ability to perform large-muscle dynamic exercise of moderate-to-high intensity for prolonged periods.

overload an extra physical demand placed on the body; an increase in the frequency, duration, or intensity of an activity. A principle of training is that for a body system to improve, it must be worked at frequencies, durations, or intensities that increase by increments.

hypertrophy (high-PURR-tro-fee) an increase in size (for example, of a muscle) in response to use.

The Essentials of Fitness and Training

To be physically fit, you need to achieve enough of the four components of fitness—**flexibility, muscle strength, muscle endurance,** and **cardiorespiratory endurance**—to allow you to meet the everyday demands of life with some to spare, and you need to achieve a reasonable body composition.

How Do My Muscles Become Physically Fit?

People shape their bodies by what they choose to do and not do. Muscle cells and tissues respond to an **overload** of physical activity by gaining strength and size, a response called **hypertrophy.** The opposite is also true: if not called on to perform,

atrophy (AT-tro-fee) a decrease in size (for example, of a muscle) because of disuse.

aerobic (air-ROE-bic) requiring oxygen. Aerobic activity strengthens the heart and lungs by requiring them to work harder than normal to deliver oxygen to the tissues.

myoglobin the muscles' iron-containing protein that stores and releases oxygen in response to the muscles' energy needs.

weight training the use of free weights or weight machines to provide resistance for developing muscle strength and endurance. A person's own body weight may also be used to provide resistance as when a person does push-ups, pull-ups, or sit-ups. Also called *resistance training*.

muscles dwindle and weaken, a response called **atrophy.** Thus, cyclists often have well-developed legs but less arm or chest strength; a tennis player may have one superbly strong arm, while the other is just average. A variety of physical activities produces the best overall fitness, and to this end, people need to work different muscle groups from day to day. For balanced fitness, stretching enhances flexibility, weight training develops muscle strength and endurance, and **aerobic** activity improves cardiorespiratory endurance. It makes sense to give muscles a rest, too, because it takes a day or two to replenish muscle fuel supplies and to repair wear and tear incurred through physical activity.

Periodic rest also gives muscles time to adapt to an activity. During rest, muscles build more of the equipment required to perform the activity that preceded the rest. The muscle cells of a superbly trained weight lifter, for example, store extra granules of glycogen, build up strong connective tissues, and add bulk to the special proteins that contract the muscles, thereby increasing the muscles' ability to perform.* In the same way, the muscle cells of a distance swimmer develop huge stocks of **myoglobin,** the muscles' oxygen-handling protein, and other equipment needed to burn fat and to sustain prolonged exertion. Therefore, if you wish to become a better jogger, swimmer, or biker, you should train mostly by jogging, swimming, or biking. Your performance will improve as your muscles develop the specific equipment they need to do the activity.

key point *The components of fitness are flexibility, muscle strength, muscle endurance, and cardiorespiratory endurance. To build fitness, a person must engage in physical activity. Muscles adapt to activities they are called upon to perform.*

Q Can I Expect Any Health Benefits from Weight Training?

Weight training, long recognized as a method to build and maintain muscle strength and endurance, was once considered stressful to the heart. Its benefits to health have emerged only recently. The American Heart Association's science advisory now endorses a program of progressive weight training to increase muscle strength and endurance; prevent and manage several chronic diseases, including cardiovascular disease; and enhance psychological well-being.[6]

By promoting strong muscles in the back and abdomen, weight training can improve posture and reduce the risk of back injury. Weight training can also help prevent the decline in physical mobility that often accompanies aging.[7] Older adults, even those in their eighties, who participate in weight training programs not only gain muscle strength, but also improve their muscle endurance, which enables them to walk significantly longer before exhaustion. Leg strength and walking endurance are powerful indicators of an older adult's physical abilities.

As an added benefit, weight training to improve muscle strength and endurance can also help to maximize and maintain bone mass.[8] Research shows that even in women past menopause (when most women are losing bone), a one-year program of weight training improves bone density.[9]

Weight training can emphasize either muscle strength or muscle endurance. To emphasize muscle strength, combine high resistance (heavy weight) with a low number of repetitions. To emphasize muscle endurance, combine less resistance (lighter weight) with more repetitions. Weight training enhances sports performance. Swimmers can develop a more efficient stroke, and tennis players a more powerful serve, when they train with weights.

Bodies are shaped by the activities they perform.

*All muscles contain a variety of muscle fibers, but there are two main types—slow-twitch (also called *red fibers*) and fast-twitch (also called *white fibers*). Slow-twitch fibers contain extra metabolic equipment to perform fat-burning aerobic work; the fast-twitch type store extra glycogen for anaerobic work. Muscle fibers of one type take on some of the characteristics of the other as an adaptation to exercise.

key point *Weight training offers health and fitness benefits to adults. Weight training reduces the risk of cardiovascular disease, improves older adults' physical mobility, and helps maximize and maintain bone mass.*

cardiac output the volume of blood discharged by the heart each minute.

stroke volume the amount of oxygenated blood ejected from the heart toward body tissues at each beat.

How Does Cardiorespiratory Training Benefit the Heart?

Although weight training provides some cardiovascular benefits, the kind of exercise most famous for improving the health of the heart is cardiorespiratory endurance training. Everyone has felt the heartbeat pick up its pace during physical activity. Cardiorespiratory endurance determines how long a person can remain active with an elevated heart rate—it is the ability of the heart and lungs to sustain a given physical demand. Working muscles need abundant oxygen to produce energy, and the heart and lungs work together to provide that oxygen. Cardiorespiratory endurance training, therefore, is aerobic.

The body's adaptation to the demands of aerobic activity involves a complex sequence of heart-healthy events. Cardiorespiratory endurance improves—the body delivers oxygen more efficiently. With cardiorespiratory endurance, the total blood volume and the number of red blood cells increase, so the blood can carry more oxygen. The heart muscle becomes stronger and larger, and its **cardiac output** increases. Each beat empties the heart's chambers more completely, so the heart pumps more blood per beat—its **stroke volume** increases. This makes fewer beats necessary, so the pulse rate falls. The muscles that inflate and deflate the lungs gain strength and endurance, so breathing becomes more efficient. Blood moves easily through the blood vessels because the muscles of the heart contract powerfully, and contraction of the skeletal muscles pushes the blood through the veins. Such improvements keep resting blood pressure normal. Figure 10-2 on the next page shows the major relationships among the heart, lungs, and muscles. The improvements that come with cardiorespiratory endurance also raise blood HDL, the lipoprotein associated with lower heart disease risk.[10]

Cardiorespiratory endurance is characterized by:

- Increased cardiac output and oxygen delivery.
- Increased heart strength and stroke volume.
- Slowed resting pulse.
- Increased breathing efficiency.
- Improved circulation.
- Reduced blood pressure.

The importance of HDL to heart health is a topic of the next chapter.

Which activities produce these beneficial changes? Effective activities elevate the heart rate, are sustained for longer than 20 minutes, and use most of the large-muscle groups of the body (legs, buttocks, and abdomen). Examples are swimming, cross-country skiing, rowing, fast walking, jogging, fast bicycling, soccer, hockey, basketball, inline-skating, lacrosse, and rugby.

An informal pulse check can give you some indication of how conditioned your heart is. The average resting pulse rate for adults is around 70 beats per minute. Active people can have resting pulse rates of 50 or even lower. To take your pulse, follow the directions in the margin.

To take your resting pulse:

Using a watch or clock with a second hand, place your hand over your heart or your finger firmly over an artery at the underside of the wrist or side of the throat under the jawbone. Start counting your pulse at a convenient second, and continue counting for ten seconds. If a heartbeat occurs exactly on the tenth second, count it as one-half beat. Multiply by 6 to obtain the beats per minute. To ensure a true count:

- Use only fingers, not your thumb, on the pulse point (the thumb has a pulse of its own).
- Press just firmly enough to feel the pulse. Too much pressure can interfere with the pulse rhythm.

The rest of this chapter describes the interactions between nutrients and physical activity. Nutrition alone cannot endow you with fitness or athletic ability, but along with the right mental attitude, it complements your effort to obtain them. Conversely, unwise food selections can stand in your way.

key point *Cardiorespiratory endurance training enhances the ability of the heart and lungs to deliver oxygen to the muscles. With cardiorespiratory endurance training, the heart becomes stronger, breathing becomes more efficient, and the health of the entire body improves.*

The Active Body's Use of Fuels

The fuels that support physical activity are glucose (from carbohydrate), fatty acids (from fat), and, to a small extent, amino acids (from protein). The body uses different mixtures of fuels depending on the intensity and duration of its activities and depending on its own prior training.

Epinephrine, discussed and defined in Chapter 3, is the major hormone that elicits the body's stress response, mobilizing fuels and readying the body for action.

$VO_{2\ max}$ the maximum rate of oxygen consumption by an individual (measured at sea level).

Figure 10•2

DELIVERY OF OXYGEN BY THE HEART AND LUNGS TO THE MUSCLES

The more fit a muscle is, the more oxygen it draws from the blood. This oxygen comes from the lungs, so the person with more fit muscles extracts oxygen from inhaled air more efficiently than a person with less fit muscles. The cardiovascular system responds to increased demand for oxygen by building up its capacity to deliver oxygen. Researchers can measure cardiovascular fitness by measuring the amount of oxygen a person consumes per minute while working out. This measure of fitness, which indicates the person's maximum rate of oxygen consumption, is called $VO_{2\ max}$.

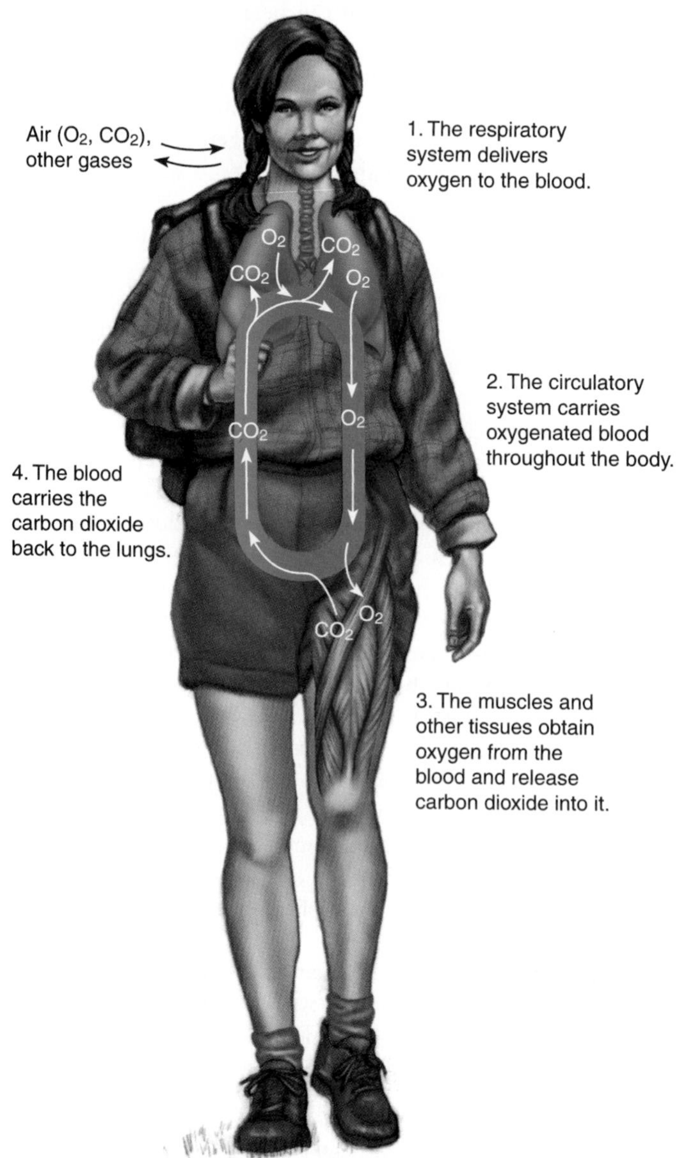

Air (O_2, CO_2), other gases
1. The respiratory system delivers oxygen to the blood.
O_2
CO_2
CO_2
O_2
2. The circulatory system carries oxygenated blood throughout the body.
CO_2
O_2
4. The blood carries the carbon dioxide back to the lungs.
CO_2
O_2
3. The muscles and other tissues obtain oxygen from the blood and release carbon dioxide into it.

During rest, the body derives a little more than half of its energy from fatty acids, most of the rest from glucose, and a little from amino acids. During physical activity, the body adjusts its fuel mix to use the stored glucose of muscle glycogen. In the early minutes of an activity, muscle glycogen provides the majority of energy the muscles use to go into action. As activity continues, messenger molecules, including the hormone epinephrine, flow into the bloodstream to signal the liver and fat cells to liberate their stored energy nutrients, primarily glucose and fatty acids. Thus, hormones set the table for the muscles' energy feast, and the muscles help themselves to the fuels passing by in the blood.

Figure 10•3

THE EFFECT OF DIET ON PHYSICAL ENDURANCE

A high-carbohydrate diet can increase an athlete's endurance. In this study, the high-fat diet provided 94 percent of calories from fat and 6 percent from protein; the normal mixed diet provided 55 percent of calories from carbohydrate; and the high-carbohydrate diet provided 83 percent of calories from carbohydrate.

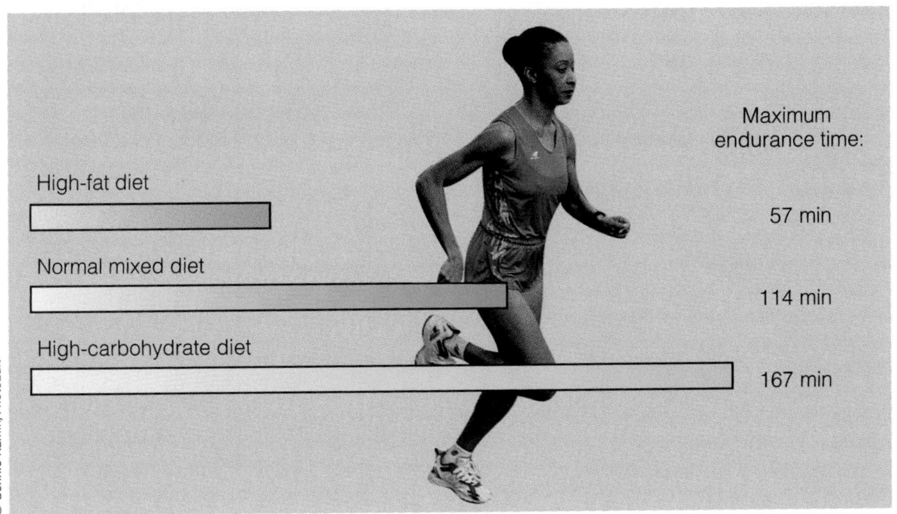

Glucose Use and Storage

Both the liver and muscles store glucose as glycogen; the liver can also make glucose from fragments of other nutrients. Muscles hoard their glycogen stores—they do not release their glucose into the bloodstream to share with other body tissues, as the liver does. This hoarding is fortunate because a muscle that conserves its glycogen is prepared to act in emergencies, say, when running from danger, because muscle glucose fuels quick action. As activity continues, glucose from the liver's stored glycogen and dietary glucose absorbed from the digestive tract also become important sources of fuel for muscle activity.

The body constantly uses and replenishes its glycogen. The more carbohydrate a person eats, the more glycogen muscles store (up to a limit), and the longer the stores will last to support physical activity.

A classic report compared fuel use during physical activity by three groups of runners, each on a different diet. For several days before testing, one of the groups ate a normal mixed diet (55 percent of calories from carbohydrate); a second group ate a high-carbohydrate diet (83 percent of calories from carbohydrate); and the third group ate a high-fat diet (94 percent of calories from fat). As Figure 10-3 shows, the high-carbohydrate diet enabled the athletes to work longer before exhaustion. This study and many others established that a high-carbohydrate diet enhances an athlete's endurance by ensuring ample glycogen stores.

key point *Glucose is supplied by dietary carbohydrate or made by the liver. It is stored in both liver and muscle tissue as glycogen. Total glycogen stores affect an athlete's endurance.*

Activity Intensity, Glucose Use, and Glycogen Stores

The body's glycogen stores are much more limited than its fat. A person with 30 pounds of body fat to spare may have only a pound or so of muscle and liver glycogen to draw on. How long a person's glycogen will last while exercising depends not only

Figure 10•4

GLUCOSE AND FATTY ACIDS IN THEIR ENERGY-RELEASING PATHWAYS

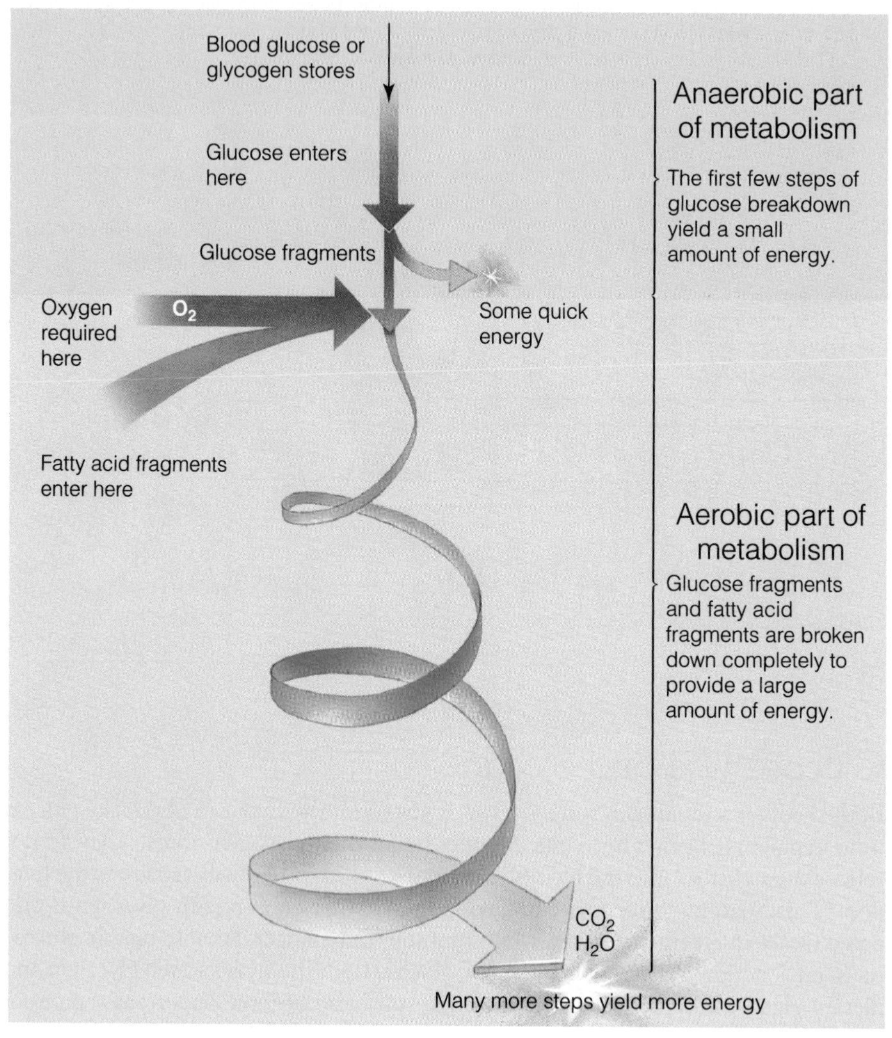
Blood glucose or glycogen stores
Glucose enters here
Glucose fragments
Oxygen required here
O_2
Some quick energy
Fatty acid fragments enter here
Anaerobic part of metabolism
The first few steps of glucose breakdown yield a small amount of energy.
Aerobic part of metabolism
Glucose fragments and fatty acid fragments are broken down completely to provide a large amount of energy.
CO_2
H_2O
Many more steps yield more energy

on diet but also on the intensity of the activity. The most intense activities, such as a quarter-mile run, use glycogen quickly. Less intense activities, such as jogging, during which breathing is steady and easy, use glycogen less rapidly. Thus, competitive athletes demand much more from their glycogen stores than do casual joggers. Joggers still use glycogen, however, and eventually they can run out of it. Glycogen depletion occurs after about two hours of vigorous activity.*

Aerobic Use of Glucose During *moderate* physical activity, the lungs and circulatory system have no trouble keeping up with the muscles' need for oxygen. The individual breathes easily, and the heart beats at a faster pace than at rest but steadily—the activity is aerobic. As the bottom half of Figure 10-4 shows, during aerobic activity muscles extract their energy from both glucose and fatty acids when both are present together with oxygen. In this way, a little glucose helps to metabolize a lot of fat. Fat yields a lot of energy, so moderate aerobic activity conserves glycogen stores.

*Here "vigorous exercise" means exercise at 75 percent of $VO_{2\ max}$.

Anaerobic Use of Glucose Intense activity presents a different picture. The heart and lungs can provide only so much oxygen only so fast. When muscle exertion is so great that the demand for energy outstrips the oxygen supply, fat cannot be used because oxygen is required for its breakdown. Instead, muscles must begin to rely more heavily on glucose, which can be partially broken down by **anaerobic** metabolism. Thus, the muscles begin drawing more heavily on their limited glycogen supply.

As the upper portion of Figure 10-4 shows, glucose can yield some energy in anaerobic metabolism, but not as much as in aerobic metabolism. Anaerobic breakdown of glycogen yields energy to muscle tissue when energy demands outstrip the body's ability to provide energy aerobically, but it does so by lavishly spending the muscles' glycogen reserves.

anaerobic (AN-air-ROE-bic) not requiring oxygen. Anaerobic activity may require strength but does not work the heart and lungs very hard for a sustained period.

lactic acid a product of the incomplete breakdown of glucose during anaerobic metabolism. When oxygen becomes available, lactic acid can be completely broken down for energy or converted back to glucose.

Lactic Acid Anaerobic breakdown of glucose produces **lactic acid,** fragments of glucose molecules that accumulate in the tissues and blood. When the nervous and hormonal systems detect these fragments in the blood, they respond by speeding up the heart and lungs to draw in more oxygen and break down the fragments. If the heart and lungs cannot keep up, lactic acid accumulates. Lactic acid causes burning muscle pain, followed within seconds by a type of muscle fatigue.

If you exercise intensely, you may have to slow down or even stop to "catch your breath" (replenish your oxygen supply). A strategy for dealing with lactic acid is to relax the muscles at every opportunity during activity so that the circulating blood can carry away the lactic acid and bring in more oxygen to sustain aerobic metabolism. At that point, lactic acid is burned for fuel or used by the liver to regenerate glucose.

The more intense an activity, the more glucose it demands. During anaerobic metabolism, the body spends glucose rapidly and accumulates lactic acid.

Activity Duration Affects Glucose Use

Glucose use during physical activity depends on the *duration* of the activity as well as its *intensity*. In the first 10 minutes or so of an activity, the active muscles rely almost completely on their own stores of glycogen. Within the first 20 minutes or so of moderate activity, a person uses up about one-fifth of the available glycogen. As the muscles devour their own glycogen, they become ravenous for more glucose and increase their uptake of blood glucose dramatically.[11] During moderate activity, your blood glucose declines slightly, reflecting its use by the muscles.

A person who exercises moderately for longer than 20 minutes begins to use less glucose and more fat for fuel. Still, glucose use continues, and if the activity goes on long enough and at a high enough intensity, muscle and liver glycogen stores will run out almost completely (see Figure 10-5, next page). Physical activity can continue for a short time thereafter only because the liver scrambles to produce some glucose from available lactic acid and certain amino acids. This minimum amount of glucose may briefly forestall exhaustion, but when hypoglycemia accompanies glycogen depletion, it brings nervous system function almost to a halt, making activity impossible. Marathon runners call this "hitting the wall."

Maintaining Blood Glucose for Activity To postpone exhaustion, endurance athletes must maintain their blood glucose concentrations for as long as they can. Three dietary strategies and one training strategy can help maintain glucose concentrations. One diet strategy is to eat a high-carbohydrate diet on a daily basis (see this chapter's Food Feature). Another is to take in some glucose during the activity, usually in fluid (see the next section). The third is to eat carbohydrate-rich foods after activity to boost the storage of glycogen. The training strategy involves training the muscles to store as much glycogen as they can, while supplying enough dietary glucose to enable them to do so (called *carbohydrate loading*, described in a later section).

Four strategies can help to maintain blood glucose to support sports performance (for endurance athletes only):

1. Eat a high-carbohydrate diet regularly.
2. Take glucose (usually in sports drinks, diluted fruit juice, or other sweet beverages) during endurance activity.
3. Eat carbohydrate-rich foods after performance.
4. Train the muscles to maximize glycogen stores.

carbohydrate loading a regimen of moderate exercise, followed by eating a high-carbohydrate diet, that enables muscles to temporarily store glycogen beyond their normal capacity; also called *glycogen loading* or *glycogen supercompensation*.

Figure 10•5

GLYCOGEN DEPLETION IN CYCLISTS

After three and a half hours of constant cycling, muscle glycogen is used up, but the demand for glucose fuel declines only slightly (dotted line).

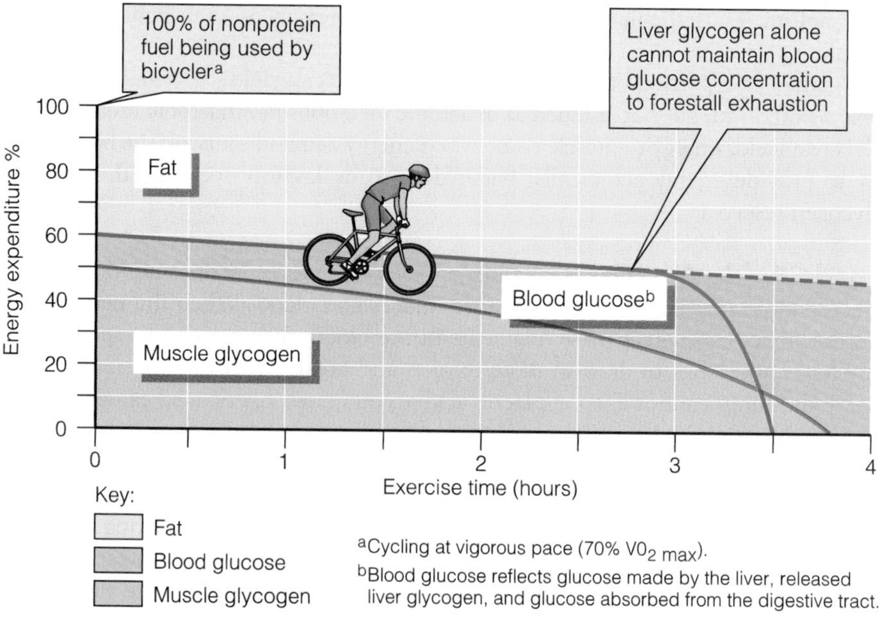

[a]Cycling at vigorous pace (70% VO_2 max).
[b]Blood glucose reflects glucose made by the liver, released liver glycogen, and glucose absorbed from the digestive tract.

To make glycogen, muscles need carbohydrate, but they also need rest. Vary daily activity routines to work different muscles on different days.

Glucose during Activity Glucose ingested before or during a long-duration competition makes its way from the digestive tract to the working muscles, augmenting dwindling internal glucose supplies from the muscle and liver glycogen stores.[12] Especially during games such as soccer or hockey, which last for hours and demand repeated bursts of intense activity, athletes benefit from carbohydrate-containing drinks taken during the activity.

Before concluding that sugar might be good for your own performance, consider first whether you engage in *endurance* activity. Do you run, swim, bike, or ski nonstop at a rapid pace for more than an hour at a time, or do you compete in games lasting for hours? If not, the sugar picture changes. For an everyday jog or swim lasting less than 60 minutes, sugar probably won't help (or harm) performance. Even in athletes, extra carbohydrate does not benefit those who engage in sports in which fatigue is unrelated to blood glucose, such as 100-meter sprinting, baseball, casual basketball, and weight lifting.

Those who compete in endurance activities require fluid and carbohydrate fuel.

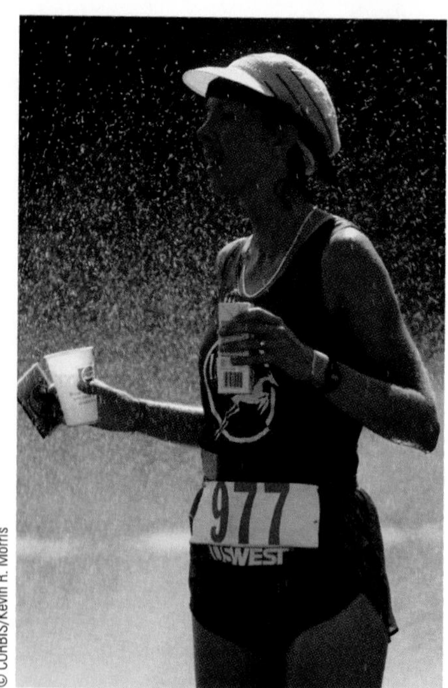

© CORBIS/Kevin R. Morris

Carbohydrate Loading Athletes whose sports routinely exhaust their glycogen stores sometimes use a technique called **carbohydrate loading** to trick their muscles into storing extra glycogen before competition. In general, the athlete tapers training during the week before the competition and then eats a high-carbohydrate diet during the three days just prior to the event.[13] See the carbohydrate-loading plan in Table 10-2.

Extra glycogen gained this way can benefit an athlete who must keep going for 90 minutes or longer. Those who exercise for shorter times simply need a regular high-carbohydrate diet. In a hot climate, extra glycogen confers an additional advantage: as glycogen breaks down, it releases water, which helps to meet the athlete's fluid needs.

A simpler method for gaining some extra glycogen involves eating carbohydrate after exercise. Train normally; then, within two hours after physical activity, consume a high-carbohydrate meal, such as a glass of orange juice and some graham crackers, toast, or cereal. This method accelerates the rate of glycogen storage by 300 percent

Table 10•2

CARBOHYDRATE LOADING

Before the Event	Training Intensity	Training Duration	Dietary Carbohydrate
6 days	Moderate[a]	90 min	Normal (5 g/kg body weight)
5 days 4 days	Moderate[a]	40 min	Normal (5 g/kg body weight)
3 days 2 days	Moderate[a]	20 min	High-carbohydrate (10 g/kg body weight)
1 day	Rest	—	High-carbohydrate (10 g/kg body weight)

[a]Moderate intensity equals 70 percent $VO_{2\ max}$.

for a while.[14] Timing is important—eating the meal after two hours have passed reduces the glycogen synthesis rate by almost half.

Chapter 4 introduced the glycemic effect and discussed some possible health benefits of eating a diet ranking low on the glycemic index. For athletes wishing to maximize muscle glycogen synthesis after strenuous training, however, eating foods with a high glycemic index (see the margin) may restore glycogen most rapidly.[15]

Foods with a high glycemic index:

- Cornflakes
- Mashed potatoes
- Short-grain rice
- Waffles
- Watermelon
- White bread

key point *Physical activity of long duration places demands on the body's glycogen stores. Carbohydrate ingested before and during long-duration activity may help to forestall hypoglycemia and fatigue. Carbohydrate loading is a regimen of physical activity and diet that enables an athlete's muscles to store larger-than-normal amounts of glycogen to extend endurance. After strenuous training, eating foods with a high glycemic index may help restore glycogen most rapidly.*

Degree of Training Affects Glycogen Use

Training affects glycogen use during activity in at least two ways. First, muscles that deplete their glycogen stores through work adapt to store greater amounts of glycogen to support that work. Second, trained muscles burn more fat, and at higher intensities, than untrained muscles, so they require less glucose to perform the same amount of work.[16] A person attempting an activity for the first time uses up much more glucose per minute than an athlete trained to perform it. A trained person can work at high intensities for longer periods than an untrained person while using the same amount of glycogen.

Factors that affect glucose use during physical activity:

- Carbohydrate intake.
- Intensity and duration of the activity.
- Degree of training.

People with diabetes should know how the moderating effect of physical training can influence their glucose metabolism. Those who must take insulin or insulin-eliciting drugs sometimes find that as their muscles adapt to physical activity, they can reduce their daily drug doses. Physical activity may also improve type 2 diabetes by helping the body lose excess fat. For those with type 1 diabetes, physical activity has been shown to lower the risk of cardiovascular disease, increase insulin sensitivity, lower blood pressure, and improve blood lipids.[17]

Chapter 4 described the action of insulin on blood sugar.

Highly trained muscles use less glucose and more fat than do untrained muscles to perform the same work, so their glycogen lasts longer.

To Burn More Fat during Activity, Should Athletes Eat More Fat?

An athlete who eats a fat-rich diet with little carbohydrate will burn more fat during activity but will sacrifice endurance, as Figure 10-3 showed. Although the importance of a high-carbohydrate diet for endurance has long been recognized, some research suggests that medium- and high-fat diets may benefit ultra-endurance performance and that severe dietary fat restriction may be detrimental for some athletes.[18] High-fat diets, however, carry risks of heart disease. Physical activity offers some protection against cardiovascular disease, but even athletes can suffer heart attacks and strokes. Most nutrition experts agree that the potential for adverse health effects from prolonged high-fat diets makes them an unwise choice for athletes.

Athletes who restrict fat below 20 percent of total energy intake may fail to consume adequate energy and nutrients. Sports nutrition experts recommend that endurance athletes consume 20 to 30 percent of their energy from fat.[19] One expert says the message is "not that high-fat diets improve performance, but rather that very low-fat diets inhibit performance."*

As fuel for activity, body fat stores are more important than fat in the diet. Unlike the body's glycogen stores, which are limited, fat stores can fuel hours of activity without running out; body fat is (theoretically) an unlimited source of energy. Even the lean bodies of elite runners carry enough fat to fuel several marathon runs.

Early in activity, muscles begin to draw on fatty acids from two sources—fats stored within the working muscles and fats from fat deposits such as the fat under the skin. Areas with the most fat to spare donate the greatest amounts of fatty acids to the blood (although they may not be the areas that one might choose to lose fat from). This is why "spot reducing" doesn't work: muscles do not own the fat that surrounds them. Fat cells release fatty acids into the blood for all the muscles to share. Proof is found in a tennis player's arms: the fatfolds measure the same in both arms, even though one arm has better-developed muscles than the other.

Factors that affect fat use during physical activity:

- Fat intake.
- Intensity and duration of the activity.
- Degree of training.

Intensity and Duration Affect Fat Use The *intensity* of physical activity also affects the percentage of energy contributed by fat because fat can be broken down for energy only by aerobic metabolism. When the intensity of activity becomes so great that energy demands surpass the ability to provide energy aerobically, the body cannot burn more fat. Instead, it burns more glucose.

The *duration* of activity also matters to fat use. At the start of activity, the blood fatty acid concentration falls, but a few minutes into an activity, the neurotransmitter norepinephrine signals the fat cells to break apart their stored triglycerides and to liberate fatty acids into the blood. After about 20 minutes of activity, the blood fatty acid concentration rises above the normal resting concentration. Only after the first 20 minutes, during this phase of sustained, submaximal activity, do the fat cells begin to shrink in size as they empty out their fat stores.

Degree of Training Affects Fat Use Training—repeated aerobic activity—stimulates the muscles to develop more fat-burning enzymes. Aerobically trained muscles burn fat more readily than untrained muscles. With aerobic training, the heart and lungs also become stronger and better able to deliver oxygen to the muscles during high-intensity activities. This improved oxygen supply, in turn, enables the muscles to burn more fat. Intense, prolonged activity may also increase your basal metabolic rate (BMR), as the Think Fitness feature explains.

key point *Athletes who eat high-fat diets may burn more fat during endurance activity, but the risks to health outweigh any possible performance benefits. The intensity and duration of activity, as well as the degree of training, affect fat use.*

Using Protein and Amino Acids to Build Muscles and Fuel Activity

Athletes use protein to build and maintain muscle and other lean tissue structures and, to a small extent, to fuel activity. The body handles protein differently during activity than during rest.

*The quotation is attributed to David R. Pendergast, in Cutting fat may crimp performance in endurance athletes, *Nutrition and the M.D.*, December 2000, pp. 3–4.

branched-chain amino acids amino acids that, unlike the others, can provide energy directly to muscle tissue: leucine, isoleucine, and valine.

Can Physical Training Speed Up an Athlete's Metabolism?

Athletes in training, whether endurance athletes or power athletes, expend huge amounts of energy each day while practicing their chosen activity. Common sense tells us that the harder an athlete works, the more energy the athlete spends. But what about *after* the work is done? Does the athlete continue to spend more energy at rest than a sedentary person or a casual exerciser? Research suggests the answer may be yes, for a limited time after intense, prolonged activity. For example, intense endurance activity (at greater than 70 percent of $VO_{2\ max}$) seems to increase the basal metabolic rate for anywhere from minutes to hours depending on the intensity and duration of the activity. The greater the intensity and the longer the duration of the activity, the longer the BMR remains elevated.

THINK FITNESS

Protein for Building Muscle Tissue In the hours of rest that follow physical activity, muscles speed up their rate of protein synthesis—they build more of the proteins they need to perform the activity. And whenever the body rebuilds a part of itself, it must tear down the old structures to make way for the new ones. Physical activity, with just a slight overload, calls into action both the protein-dismantling and the protein-synthesizing equipment of individual muscle cells that work together to remodel muscles.

Dietary protein provides the needed amino acids for synthesis of new muscle proteins. As Chapter 6 pointed out, however, the true director of synthesis of muscle protein is physical activity itself. Repeated activity signals the muscle cells' genetic material to begin producing more of the proteins needed to perform the work at hand.

The genetic protein-making equipment inside the nuclei of muscle cells seems to "know" when proteins are needed. Furthermore, it knows *which* proteins are needed to support each type of physical activity. Apparently, the intensity and pattern of muscle contractions initiate signals that direct the muscles' genetic material to make particular proteins. For example, a weight lifter's workout sends the information that muscle fibers need added bulk for strength and more enzymes for making and using glycogen. A jogger's workout stimulates production of proteins needed for aerobic oxidation of fat and glucose. Muscle cells are exquisitely responsive to the need for proteins, and they build them conservatively only as needed.

Finally, after muscle cells have made all the decisions about which proteins to build and when, protein nutrition comes into play. During active muscle-building phases of training, a weight lifter might add to existing muscle mass between ¼ ounce and 1 ounce (between 7 and 28 grams) of protein each day. This extra protein comes from ordinary food.

Factors that affect protein use during physical activity:

- Carbohydrate intake.
- Intensity and duration of the activity.
- Degree of training.

Protein for Fuel Not only do athletes retain more protein, but they also use a little more protein as fuel.[20] Studies of nitrogen balance show that the body speeds up its use of amino acids for energy during physical activity, just as it speeds up its use of glucose and fatty acids. Protein contributes about 10 percent of the total fuel used, both during activity and during rest.

Physical activity itself triggers the building of muscle proteins.

Diet Affects Protein Use during Activity The factors that regulate how much protein is used during activity seem to be the same ones that regulate the use of glucose and fat. One factor is diet—a carbohydrate-rich diet spares protein from being used as fuel. Some amino acids can be converted into glucose when needed. Others, the **branched-chain amino acids,** can stand in for glucose in energy pathways. If your diet is low in carbohydrate, much more protein will be used in place of glucose.

Intensity and Duration Affect Protein Use The intensity and duration of the activity also affect protein use.[21] Endurance athletes who train for over an hour a day, engaging in aerobic activity of moderate intensity and long duration, may deplete their glycogen stores by the end of their training and become more dependent on body protein for energy. The protein needs of bodybuilders and weight lifters are higher than those of sedentary people, but not as high as the protein intakes many bodybuilders consume.

Degree of Training Affects Protein Use Finally, the degree of training also affects the use of protein. The better trained a person is, the less protein used during activity at a given intensity.

How Much Protein Should an Athlete Consume? Although most athletes need somewhat more protein than do sedentary people, average protein intakes in the United States are high enough to cover those needs. Therefore, athletes in training should attend to protein needs, but should back up the protein with ample carbohydrate. Otherwise, they will burn off as fuel the very protein they wish to retain in muscle.

A joint position paper from the American Dietetic Association (ADA) and the Dietitians of Canada (DC) recommends protein intakes somewhat higher than the 0.8 gram of protein per kilogram of body weight recommended for sedentary people.[22] Table 10-3 lists some recommendations and translates them into daily intakes for athletes.

After considering these recommendations, you may wonder whether your diet provides the protein you need. This chapter's Food Feature answers questions about choosing a performance diet. Meanwhile, relax. Athletes who eat a balanced, high-carbohydrate diet that provides enough total energy also consume enough protein—they do not need special foods, protein shakes, or supplements.

key point *Physical activity stimulates muscle cells to break down and synthesize protein, resulting in muscle adaptation to activity. Athletes use protein both for building muscle tissue and for energy. Diet, intensity and duration of activity, and training affect protein use during activity.*

Vitamins and Minerals—Keys to Performance

Many vitamins and minerals assist in releasing energy from fuels and transporting oxygen. In addition, vitamin C is needed for the formation of the protein collagen, the foundation material of bones and the cartilage that forms the linings of the joints

Table 10•3

RECOMMENDED PROTEIN INTAKES FOR ATHLETES

	Recommendations (g/kg/day)	Protein Intakes (g/day)	
		Males	Females
RDA for adults	0.8	56	44
Recommended intake for power (strength or speed) athletes	1.6–1.7	112–119	88–94
Recommended intake for endurance athletes	1.2–1.6	84–112	66–88
U.S. average intake		95	65

NOTE: Daily protein intakes are based on a 70-kilogram (154-pound) man and a 55-kilogram (121-pound) woman.
SOURCE: Position of The American Dietetic Association, Dietitians of Canada, and the American College of Sports Medicine: Nutrition and athletic performance, *Journal of the American Dietetic Association* 100 (2000): 1543–1556.

and other connective tissues. Folate and vitamin B_{12} help build the red blood cells that carry oxygen to working muscles. Calcium and magnesium help make muscles contract, and so on. Do active people need extra nutrients to support their work? Do they need supplements?

Do Nutrient Supplements Benefit Athletic Performance?

An estimated 84 percent of world-class athletes take nutrient supplements. Many other athletes also take supplements in the hope of improving their performance.

Thiamin, Riboflavin, and Niacin This vitamin trio plays key roles in energy release. Scientists have concluded, however, that extra amounts of these vitamins from supplements do not benefit performance.[23] An adequate diet supplies all the thiamin, riboflavin, and niacin needed by an active person, even an athlete. Athletes, with their greater energy needs, eat more food, so most athletes' diets are adequate in these vitamins. Extra amounts provide no competitive advantage.

The summary tables listing functions of vitamins and minerals begin on pp. 245 and 296.

Niacin in excess of the recommended amount can affect performance adversely. Excess niacin suppresses the release of fatty acids and thus forces muscles to use extra glycogen during physical activity. Thus, glycogen may be depleted more rapidly, making the work seem more difficult.

Vitamin B_6 and Vitamin B_{12} Vitamin B_6 plays key roles in the release of energy from nutrients, in the liberation of glucose from glycogen, and in the formation of hemoglobin. To ensure that the diet is adequate in vitamin B_6, a person need only include some leafy green vegetables, meats, fish, legumes, fruits, and whole grains. Sellers of supplements claim that vitamin B_6 pills promote athletic performance, but scientific research proves otherwise.[24] Megadoses of vitamin B_6 provide no additional benefit, and large doses can be toxic.

Nutrients necessary to ward off anemias include vitamins A, B_6, B_{12}, and folate and the minerals iron, zinc, copper, and magnesium along with protein—in short, the perfect mix of nutrients that occurs naturally in whole, nutrient-dense foods.

The belief that vitamin B_{12} supplementation will enhance performance stems from its role in the production of red blood cells. Anemias of all kinds rob the blood of its oxygen-carrying capacity by reducing the number and impairing the function of circulating red blood cells. Vitamin B_{12} deficiency causes anemia, but so do iron and folate deficiencies (and others, see the margin). A diet low enough in vitamin B_{12} to bring on anemia will likely be low in other nutrients as well. A person with so poor a diet does not need to take pills; the person needs to eat right. For a well-nourished athlete, any perceived benefits from vitamin B_{12} supplements or shots taken before competition are based on psychology, not physiology.

Vitamins C and E, the Antioxidants During prolonged, high-intensity physical activity, the muscles' consumption of oxygen increases tenfold or more, enhancing the production of damaging free radicals in the body.[25] Vitamin E is a potent fat-soluble antioxidant that vigorously defends cell membranes against oxidative damage. Vitamin C, a water-soluble antioxidant, helps regenerate vitamin E. Vitamin E seems to be the most important antioxidant related to physical activity. Many athletes are taking antioxidant supplements, particularly vitamin E, and research suggests that such supplementation, either with vitamins C and E together or with vitamin E alone, does offer protection against exercise-induced oxidative stress.[26] Supplement doses in these studies varied considerably, however, and no one yet knows the precise dose that will offer the greatest benefits with the least risk of toxicity. Furthermore, there is little evidence that these supplements can improve performance.[27] More research is needed before drawing conclusions about antioxidant supplements for athletes.

Chapter 7 specified the risks of toxicity from vitamin pills and supplements.

Thus, with the possible exception of vitamin E, the working body gains no benefit from vitamin supplements.[28] To meet recommendations for vitamins, athletes need only consume sufficient nutrient-dense food. Athletes who must lose weight to meet

stress fractures bone injuries or breaks caused by the stress of exercise on the bone surface.

amenorrhea the absence or cessation of menstruation.

low body-weight requirements, however, may consume so little food that they fail to obtain all the nutrients they need. The practice of "making weight" is opposed by many health-minded groups, but for athletes who choose this course, a single daily multivitamin-mineral tablet that provides no more than the DRI recommendations for nutrients may be beneficial.

key point *Vitamins are essential for releasing the energy trapped in energy-yielding nutrients and for other functions that support physical activity. Active people can meet their vitamin needs if they eat enough nutrient-dense foods to meet their energy needs.*

Stringent weight requirements pose a risk of developing eating disorders. See Controversy 9.

Excessive Physical Activity and Bone Loss

Osteoporosis, the condition of reduced bone mass, increases susceptibility to bone damage, including **stress fractures.** Moderate physical activity and adequate calcium intakes protect against bone loss (see Controversy 8), but extremes in physical activity may be detrimental to bone health in some young women and adolescent girls. Many young female athletes and dancers restrict energy intakes to meet the weight guidelines of their sport and thus have calcium intakes below the recommendations. Such young women risk developing the potentially fatal "female athlete triad"—abnormal eating behaviors, **amenorrhea,** and osteoporosis—discussed in Controversy 9.[29]

Female athlete triad:

- Disordered eating
- Amenorrhea
- Osteoporosis

key point *Moderate physical activity strengthens the bones, but young female athletes who train strenuously, become amenorrheic, and practice abnormal eating behaviors are susceptible to stress fractures and osteoporosis.*

Iron and Performance

Endurance athletes, and female athletes in particular, are prone to iron deficiency.[30] Physical activity can impair iron status in any of several ways. For one, iron may be excreted in sweat.[31] For another, iron may be lost through red blood cell destruction; blood cells are squashed when body tissues (such as the soles of the feet) make high-impact contact with an unyielding surface (such as the ground). Third, physical activity may cause small blood losses through the digestive tract, at least in some athletes. Fourth, and perhaps more significant than the losses, are the high iron demands by muscles to make the iron-containing molecules of aerobic metabolism. Habitually low intakes of iron-rich foods as well as increased losses and extra demands can contribute to iron deficiency in young female athletes.

Female athletes may be at special risk of iron deficiency.

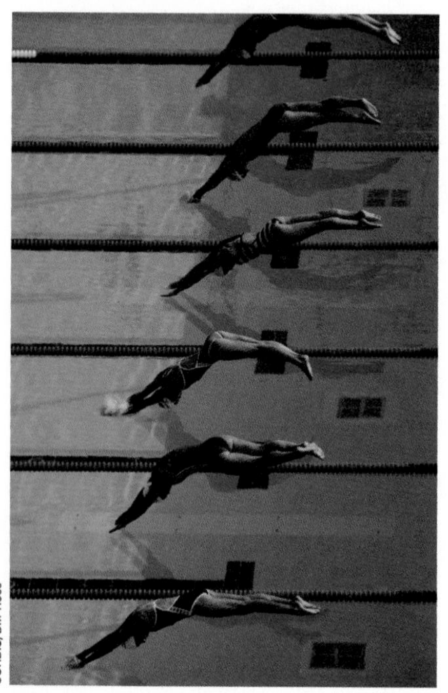

CORBIS/Bill Ross

Vegetarian female athletes are particularly vulnerable to iron insufficiency.[32] The bioavailability of iron is often poor in plant-based diets because such diets are high in fiber and phytic acid and because the nonheme iron in plant foods is not absorbed as well as the heme iron in animal-derived foods. Vegetarian diets are usually rich in vitamin C, however, which enhances iron absorption. To protect against iron deficiency, vegetarian athletes need to pay close attention to their intake of good dietary sources of iron (fortified cereals, legumes, nuts, and seeds) and include vitamin C–rich foods with each meal.[33] As long as vegetarian athletes, like all athletes, consume enough nutrient-dense foods, they can perform as well as anyone.

Iron deficiency impairs performance because iron helps deliver the muscles' oxygen. Insufficient oxygen delivery reduces aerobic work capacity, so the person tires easily. Whether marginal deficiency without clinical signs of anemia hinders physical performance is a point of debate among researchers.[34]

Early in training, athletes may develop low blood hemoglobin. This condition, sometimes called "sports anemia," is not a true iron-deficiency condition. Strenuous training promotes destruction of the more fragile, older red blood cells, and the resulting cleanup work reduces the blood's iron content temporarily. Strenuous activity also promotes increases in the fluid of the blood; with more fluid, the red blood cell count in a unit of blood drops. Most researchers view sports anemia as an *adaptive*, temporary response to endurance training. True iron-deficiency anemia requires treatment with prescribed iron supplements, but sports anemia goes away by itself, even with continued training.

The best strategy concerning iron is to determine individual needs. Many menstruating women border on iron deficiency even without the additional iron demand and losses incurred by physical activity. Teens of both genders, because they are growing, have high iron needs, too. For women and teens, then, prescribed supplements may be needed to correct a deficiency of iron that is confirmed by tests. (Medical testing is needed to eliminate nondietary causes of anemia, such as internal bleeding or cancer.)

key point *Iron-deficiency anemia impairs physical performance because iron is the blood's oxygen handler. Sports anemia is a harmless temporary adaptation to physical activity.*

Other Minerals

Three trace minerals—chromium, zinc, and magnesium—have specific roles in physical activity. Sweat and urine losses of all three minerals accelerate during physical training, but the increased losses are short term and diminish in the days following physical activity.[35]

Chromium has received a lot of media attention as a purported muscle builder or fat burner for athletes. This chapter's Controversy discusses chromium picolinate.

When athletes consume diets adequate in chromium, zinc, and magnesium (and most athletes do), supplementation with these minerals does not enhance performance.[36] If you think your diet lacks one of these nutrients, the best food sources of chromium are whole grains, nuts, and cheeses; zinc-rich foods are meats, shellfish, and poultry; and magnesium is abundant in leafy vegetables, legumes, and whole-wheat products. Convenience and highly processed snack foods lack magnesium.

During physical activity, the body loses electrolytes—the minerals sodium, potassium, and chloride—in sweat. Beginners lose these electrolytes to a much greater extent than do trained athletes. The body's adaptation to physical activity includes better conservation of these electrolytes.

Normally, potassium remains safely inside the cells where it does its work. In prolonged dehydration from profuse sweating, it may migrate outside the cells and be lost by excretion in the urine. Even so, potassium is easily replaced with just a few servings of fresh fruits and vegetables. Avoid potassium supplements unless prescribed by a physician; although they improve some conditions, they worsen others. Most times, a regular diet supplies all the electrolytes athletes need.

key point *The body adapts to compensate for sweat losses of electrolytes, but urinary potassium losses may persist. Athletes are advised to use foods, not supplements, to make up for these losses.*

Fluids and Temperature Regulation in Physical Activity

The body's need for water far surpasses its need for any other nutrient. If the body loses too much water, its life-supporting chemistry is compromised.

The exercising body loses water primarily via sweat; second to that, breathing costs water, exhaled as vapor. During physical activity, both routes can be significant, and dehydration is a real threat. The first symptom of dehydration is fatigue. A water loss of even 1 to 2 percent of body weight can reduce a person's capacity to do muscular work.[37] A person with a water loss of about 7 percent is likely to collapse. The athlete who arrives at an event even slightly dehydrated starts out at a competitive disadvantage.

Foods like these are packed with the nutrients that active people need.

heat stroke an acute and life-threatening reaction to heat buildup in the body.

hypothermia a below-normal body temperature.

Temperature Regulation

As Chapter 8 pointed out, sweat cools the body. The conversion of water to vapor uses up a great deal of heat, so as sweat evaporates, it cools the skin's surface and the blood flowing beneath it.

In hot, humid weather, sweat may fail to evaporate because the surrounding air is already laden with water. Little cooling takes place and body heat builds up. In such conditions, athletes must take precautions to avoid **heat stroke.** Heat stroke is an especially dangerous accumulation of body heat with accompanying loss of body fluid. Three measures to prevent heat stroke are to drink enough fluid before and during the activity, rest in the shade when tired, and wear lightweight clothing that encourages evaporation.[38] The rubber or heavy suits sold with promises of weight loss during physical activity are dangerous because they promote profuse sweating, prevent sweat evaporation, and invite heat stroke. If you experience any of the symptoms of heat stroke listed in the margin, stop your activity, sip cold fluids, seek shade, and ask for help. The condition demands medical attention—it can kill.

Symptoms of heat stroke:

- Headache.
- Nausea.
- Dizziness.
- Clumsiness.
- Stumbling.
- Sudden cessation of sweating (hot, dry skin).
- Internal (rectal) temperature above 104° Fahrenheit.
- Confusion or loss of consciousness.

In cold weather, **hypothermia,** or loss of body heat, can pose as serious a threat as heat stroke does in hot weather. Inexperienced runners participating in long races on cold or wet, chilly days are especially vulnerable to hypothermia. Slow runners can produce too little heat to keep warm, especially if their clothing is inadequate. Early symptoms of hypothermia include shivering and euphoria. As body temperature continues to fall, shivering stops, and weakness, disorientation, and apathy set in. People with these symptoms soon become helpless to protect themselves from further body heat losses. Even in cold weather, the body still sweats and needs fluids, but the fluids should be warm or at room temperature to help prevent hypothermia.

key point *Evaporation of sweat cools the body. Heat stroke can be a threat to physically active people in hot, humid weather. Hypothermia threatens those who exercise in the cold.*

Fluid Needs during Physical Activity

Endurance athletes can lose 2 or more quarts of fluid in every hour of activity, but the digestive system can absorb only about a quart or so an hour. Hence the athlete must hydrate before and rehydrate during and after activity to replace it all. In hot weather, the digestive tract may not be able to absorb enough water fast enough to keep up with an athlete's sweat losses, and some degree of dehydration becomes inevitable. Wise athletes preparing for competition drink extra fluids in the last few days of training before the event. The extra fluid is not stored in the body, but drinking extra ensures maximum tissue hydration at the start of the event. Any coach or athlete who withholds fluids during practice for any reason takes a great risk and is subject to sanctions by the American College of Sports Medicine.

Athletes who rely on thirst to govern fluid intake can easily become dehydrated. During activity thirst becomes detectable only *after* fluid stores are depleted. Don't wait to feel thirsty before drinking. Table 10-4 presents one schedule of hydration for physical activity. To find out how much water you need to replenish losses, weigh yourself before and after the activity. The difference is all water. Two cups (16 ounces) of fluid weigh about a pound.

Physically active people lose fluids and must replace them to avoid dehydration. Thirst indicates that water loss has already occurred.

Water

What is the best fluid to support physical activity? The best drink for most active bodies is just plain cool water, for two reasons: (1) water rapidly leaves the digestive tract to enter the tissues, and (2) it cools the body from the inside out. Endurance athletes

Table 10•4

HYDRATION SCHEDULE FOR PHYSICAL ACTIVITY

When to Drink	Approximate Amount of Fluid
2 hr before activity	2 to 3 c
15 min before activity	1 to 2 c
Every 15 min during activity	½ to 1 c
After activity	At least 2 c for each pound of body weight lost

SOURCE: R. Murray, Fluid and electrolytes, in C. A. Rosenbloom, ed., *Sports Nutrition: A Guide for the Professional Working with Active People*, 3rd ed. (Chicago: The American Dietetic Association, 2000), pp. 95–106.

are an exception: they need more from their fluids than water alone. The first priority for endurance athletes should always be replacement of fluids to prevent life-threatening heat stroke. But endurance athletes also need carbohydrate to supplement their limited glycogen stores, so glucose is important, too. This chapter's Consumer Corner compares water and sports drinks as fluid sources for endurance athletes.

Water is the best drink for most physically active people, but endurance athletes need drinks that supply glucose as well as fluids.

Active people need extra fluid, even in cold weather.

Fluid replacement tips:

- To ensure adequate fluid intake without being distracted during an event, try this technique. Before the event, fill a 32-ounce (4-cup) water bottle and place two colored rubber bands to mark the bottle into thirds. Finish off the first segment of the bottle in the first 30 minutes of activity; finish the next segment in the next 30 minutes, and the remainder in the next. Have someone refill the bottle if activity lasts longer than 90 minutes.
- The urine of a person who is adequately hydrated is the color of pale lemonade. Urine the color of apple juice indicates slight dehydration.

SOURCE: Ideas from J. Berning, nutrition professor and sports nutrition consultant, personal communication, 1999.

What Do Sports Drinks Have to Offer?

MORE THAN 20 sports drinks compete for their share of the $1 billion market. What do sports drinks offer you? First, and most important, sports drinks offer fluids to help you offset the loss of fluids during physical activity. However, plain water can do this, too.

Second, sports drinks supply glucose. A beverage that supplies glucose in some form can be useful during endurance activity lasting 60 minutes or more or during prolonged competitive games that demand repeated intermittent activity.[1] Not just any sweet beverage can meet this need, however, because a carbohydrate concentration greater than 8 percent can delay fluid emptying from the stomach and thereby slow down the delivery of water to the tissues. Most sports drinks contain an appropriate amount to ensure water absorption—about 7 percent glucose (about half the sugar of ordinary soft drinks, or about 5 teaspoons in each 12 ounces).

Third, sports drinks offer sodium and other electrolytes to help replace those lost during physical activity. Sodium in sports drinks also helps to improve palatability and fluid retention and maintains the osmotic drive for drinking fluid. This makes sense, physiologically, because the sensation of thirst is a function of changes in blood sodium concentration.[2]

Most athletes do not need to replace the minerals lost in sweat immediately; a meal eaten within hours of competition replaces these minerals soon enough. Most sports drinks are relatively low in sodium, however, so healthy people who choose to use these beverages run little risk of excessive intake.

In strenuous world-class competitions lasting for four hours or more, heavy sweating coupled with drinking large amounts of plain water has been reported to dangerously dilute blood sodium. Ultra-endurance athletes, therefore, may need to replace sodium, and sports drinks are the best method.[3] Electrolyte or salt tablets can increase potassium losses, irritate the stomach, cause vomiting, and always pull water out of the tissues into the digestive tract at first. Athletes should avoid them.

In addition, most sports drinks also taste good. Manufacturers reason that if a drink tastes good, people will drink more, thereby ensuring adequate hydration. Research backs up such reasoning—fluids that are flavored, sweetened, and cool stimulate fluid intake.[4] Finally, sports drinks can also provide a psychological edge to people who associate them with success in sports. Thus, for athletes who exercise for an hour or more, sports drinks offer some advantages over water.

Other Beverages

Some drinks, such as iced tea, deliver caffeine along with fluid. Moderate doses of caffeine (2 milligrams per pound of body weight or about the amount in 2 cups of coffee) one hour prior to activity sometimes seem to assist athletic performance and other times have no effect. (More about caffeine's drug effects can be found in this chapter's Controversy, and the amounts of caffeine in foods and beverages are listed in Controversy 11.)

In a hot environment, caffeine's diuretic effect is potentially hazardous. Physical activity slows the excretion of caffeine, prolonging its effects, so beverages containing caffeine should be used in moderation and in addition to other fluids, not as substitutes for them. In college, national, and international athletic competitions, the use of caffeine is forbidden in amounts greater than about 800 milligrams, the equivalent of drinking 5 or 6 cups of strong, brewed coffee in a two-hour period before the event.

Carbonated beverages are not a good choice for meeting an athlete's fluid needs. Although they are composed largely of water, the air bubbles from the carbonation take up room in the stomach that might otherwise be filled with fluid that the athlete can absorb.

Athletes, like others, sometimes drink beverages that contain alcohol, but these beverages are inappropriate as fluid replacements. Like caffeine, alcohol is a diuretic. Both substances promote the excretion of water; of vitamins such as thiamin, riboflavin, and folate; and of minerals such as calcium, magnesium, and potassium—exactly the wrong effects for fluid balance and nutrition. It is hard to overstate alcohol's detrimental effects on physical activity. It impairs temperature regulation, making hypothermia or heat stroke much more likely. It alters perceptions and slows reaction time. It depletes strength and endurance and deprives people of their judgment, thereby compromising their safety in sports. Many sports-related fatalities and injuries each year involve alcohol or other drugs.

Beer facts:

- Beer is not carbohydrate-rich. Beer is calorie-rich, but only one-third of its calories are from carbohydrates. The other two-thirds are from alcohol.
- Beer is mineral-poor. Beer contains a few minerals, but to replace those lost in sweat, athletes need good sources such as fruit juices.
- Beer is vitamin-poor. Beer contains tiny traces of some B vitamins, but it cannot compete with rich food sources.
- Beer causes fluid losses. Beer is a fluid, but alcohol is a diuretic and causes the body to lose more fluid in urine than is provided by the beer.

Read about alcohol's effects on the brain in Controversy 3.

key point *Caffeine-containing drinks within limits may not impair performance, but water and fruit juice are preferred. Alcohol use can impair performance in many ways and is not recommended.*

Choosing a Performance Diet

glucose polymers compounds that supply glucose, not as single molecules, but linked in chains somewhat like starch. The objective is to attract less water from the body into the digestive tract.

Small daily choices, when made consistently, enhance an athlete's nutritional health.

Compare and decide which best meets your needs:

- 1 sandwich of 2 slices bologna, 2 slices white bread, 2 tbs mayonnaise (525 calories, 9% protein, 23% carbohydrate, 68% fat).

or

- 2 sandwiches of 2 slices lean ham, 4 slices whole-wheat bread, 2 tsp mayonnaise (503 calories, 20% protein, 51% carbohydrate, 29% fat).

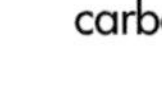

MANY DIFFERENT DIETS can support an athlete's performance. However, food choices must obey the rules for diet planning.

Nutrient Density

First, athletes need a diet composed mostly of nutrient-dense foods, the kind that supply a maximum of vitamins and minerals for the energy they provide. When athletes eat mostly refined, processed foods that have suffered nutrient losses and contain added sugar and fat, their nutrition status suffers. Even if foods are fortified or enriched, manufacturers cannot replace the whole range of nutrients and nonnutrients lost in refining. For example, manufacturers mill out much of a food's original magnesium and chromium but do not replace them. This doesn't mean that athletes can never choose a white bread, bologna, and mayonnaise sandwich, but only that later they should eat a large salad or big portions of vegetables and whole grains and drink a glass of milk to compensate. The nutrient-dense foods will provide the magnesium and chromium; the bologna sandwich provides extra energy, mostly from fat.

Balance

Athletes must eat for energy, and their energy needs can be immense. Athletes need full glycogen stores, and they need to strive to prevent heart disease and cancer by limiting fat and saturated fat. A diet that is high in carbohydrate (60 to 70 percent of total calories), low in fat (20 to 30 percent), and adequate in protein (12 to 15 percent) is best for all these purposes. Even if you do not compete in glycogen-depleting events, such a diet provides adequate fiber while supplying abundant nutrients and energy.

With these principles in mind, compare the two 500-calorie sandwich meals in the margin. The trick to getting enough carbohydrate energy is easy, at least in theory: just reduce the amount of fat and meat in a meal, and let carbohydrate-rich foods fill in for them.

Adding carbohydrate-rich foods is a sound and reasonable option for increasing energy intake, up to a point. It becomes unreasonable when the person cannot eat enough food to meet energy needs. At that point, the person can add more food energy into the diet only by eating refined sugars and fats or liquid meals. Still, these energy-rich additions must be superimposed on nutrient-rich choices; energy alone is not enough.

Some athletes use commercial high-carbohydrate liquid supplements to obtain the carbohydrate and energy needed for heavy training and top performance. Most of these products contain **glucose polymers** and about 18 to 24 percent carbohydrate. These supplements do not *replace* regular food; they are meant to be used in *addition* to it. Unlike the sports beverages discussed in the Consumer Corner, these high-carbohydrate supplements are too concentrated in carbohydrate to be used for fluid replacement.

Protein

In addition to carbohydrate, athletes need protein. Meats and milk products head the list of protein-rich foods, but suggesting that athletes eat more than the recommended servings of meat would be shortsighted advice. Athletes must protect themselves from heart disease, and even lean meats contain fat, much of it saturated fat. Besides, the extra servings of carbohydrate-rich foods such as legumes, grains, and vegetables that an athlete needs to meet energy requirements also boost protein intakes.

Earlier in this chapter, Table 10-3 showed recommended protein intakes for a 55-kilogram female athlete or a 70-kilogram male athlete. An athlete weighing 70 kilograms who engages in vigorous physical activity on a daily basis could require 3,000 to 5,000 calories per day. As a general rule, endurance athletes should aim for an average intake of 50 calories per kilogram (2.2 pounds) of body weight (23 calories per pound of body weight). Others may need more. To meet such an energy requirement, an athlete should select from a variety of nutrient-dense foods. Figure 10-6 provides an example of how foods that provide the extra nutrients athletes need can be added to regular meal selections to attain a 3,300-calorie diet. These meals supply over 130 grams of protein, more than the highest recommended intake for an athlete. For those with reasonable diets, protein is rarely a problem.

The meals in Figure 10-6 provide 63 percent of their calories from carbohydrate. Athletes who train exhaustively for endurance events may want to aim for somewhat higher carbohydrate levels—from 65 to 75 percent. Notice that breakfast, though light in fat, is filling and hearty. Current thinking supports the idea that athletes benefit from such a morning start. If you train early in the

Figure 10•6

AN ATHLETE'S MEALS

Regular Meals | **Modifications** | **Athlete's Meals**

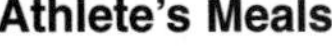

Breakfast:
1 c shredded wheat.
1 c 1% low-fat milk.
1 small banana.
1 c orange juice

The regular breakfast *plus*:
2 pieces whole-wheat toast.
1/2 c orange juice.
4 tsp jelly.

Lunch:
1 turkey sandwich.
1 c 1% low-fat milk.

The regular lunch *plus*:
1 turkey sandwich.
1/2 c 1% low-fat milk.
Large bunch of grapes.

Snack:
2 c plain popcorn.
A smoothie made from:
1 1/2 c apple juice.
1 1/2 frozen banana.

The regular snack *plus*:
1 c popcorn.

Dinner:
Salad:
1 c spinach, carrots, and mushrooms.
1/2 c garbanzo beans.
1 tbs sunflower seeds.
1 tbs ranch dressing.
1 c spaghetti with meat sauce.
1 c green beans.
1 slice Italian bread.
2 tsp butter.
1 1/4 c strawberries.
1 c 1% low-fat milk.

© Polara Studios, Inc. (all)

The regular dinner *plus*:
1 corn on the cob.
1 slice Italian bread.
2 tsp butter.
1 piece angel food cake.
1 tbs whipping cream.

Total cal: 2,600
62% cal from carbohydrate
23% cal from fat
15% cal from protein

Total cal: 3,300
63% cal from carbohydrate
22% cal from fat
15% cal from protein

All vitamin and mineral intakes exceed the recommendations for both men and women.

pregame meal a meal eaten three to four hours before athletic competition.

Looking for an amino acid supplement that rates a perfect score of 100 for protein quality? Try 1 ounce of chicken breast—it provides almost 10,000 milligrams of amino acids in perfect complement for use by the human body.

Good choices for pregame meals:

- Apricot nectar, pineapple juice, grape juice, banana, toast with jam or jelly, pancakes with syrup, baked white or sweet potatoes, pasta with steamed vegetables, lentils or other peas, raisins, figs, dates, frozen yogurt, graham crackers, sponge cake, angel food cake.

Not recommended:

- Stuffing, muffins, biscuits, croissants, french fries, onion rings, potato chips, meats, cheese, pies, ice cream, eggnog, creams, nuts, butter, gravy, mayonnaise, salad dressing, frosted cakes.

morning, try splitting breakfast into two parts. An hour or so before training, eat some toast, juice, and fruit. Later, after your workout, come back for the cereal and milk.

Planning an Athlete's Meals

Table 10-5 shows some sample food patterns for athletes at various high-energy and high-carbohydrate intakes. These plans are effective only if the user chooses foods to provide nutrients as well as energy: extra milk for calcium and riboflavin; many servings of fruit for folate and vitamin C; energy-rich vegetables such as sweet potatoes, peas, and legumes; modest portions of lean meat for iron and other vitamins and minerals; and whole grains for B vitamins, magnesium, zinc, and chromium. In addition, these foods provide plenty of electrolytes.

A trick used by professional sports nutritionists to maximize athletes' intakes of energy and carbohydrates is to make sure that vegetable and fruit choices are as dense as possible in both nutrients and energy. A whole cupful of iceberg lettuce supplies few calories or nutrients, but a half-cup portion of cooked sweet potatoes is a powerhouse of vitamins, minerals, and carbohydrate energy. Similarly, it takes a whole cup of cubed melon to equal the calories and carbohydrate in a half-cup of canned fruit. Small choices like these, made consistently, can contribute significantly to nutrient, energy, and carbohydrate intakes.

Before competition, athletes may eat particular foods or practice rituals that convey psychological advantages. One eats steak the night before; another spoons up honey at the start of the event. As long as these foods or rituals remain harmless, they should be respected. Still, science has recommendations for the **pregame meal.** The foods should be carbohydrate-rich and the meal light (300 to 800 calories). It should be easy to digest and should contain fluids. Breads, potatoes, pasta, and fruit juices—carbohydrate-rich foods low in fat, protein, and fiber—form the basis of the pregame meal. Bulky, fiber-rich foods such as raw vegetables or high-fiber cereals, although usually desirable, are best avoided just before competition. Such foods can cause stomach discomfort during performance. The competitor should finish eating three to four hours before competition to allow time for the stomach to empty before exertion.

What about drinks or candylike sport bars claiming to provide "complete" nutrition? These mixtures of carbohydrate, protein (usually amino acids), fat, some fiber, and certain vitamins and minerals usually taste good and provide additional food energy before a game or for those needing to gain weight. They fall short of providing "complete" nutrition, however, since they lack many of real food's nutrients and the nonnutri-

Table 10•5

HIGH-CARBOHYDRATE FOOD PATTERNS FOR ATHLETES

	Number of Servings for a Daily Energy Intake of:					
Food Group	1,500 cal	2,000 cal	2,500 cal	3,000 cal	3,500 cal	4,000[a] cal
Milk	3	3	4	4	4	4
Fruit	5	6	7	9	10	12
Vegetable	3	3	3	5	6	7
Grain	7	11	16	18	20	24
Fat[b]	2	3	5	6	8	10
Meat (ounces)	5	5	5	5	6	6
Percent carbohydrate:	58%	58%	63%	64%	60%	62%

[a]A way to add more energy to the diet without adding much bulk is to snack on milkshakes or "complete meal" liquid supplements (see the text).
[b]A fat serving is 1 teaspoon of butter, margarine, oil, or the equivalent.

ents that benefit health. These products provide no special advantage for active people except one—they are easy to eat in the hours before competition. They are expensive, however.

As for "complete" drinks, Table 10-6 demonstrates that there is no point in paying high prices for fancy brand-name drinks. Homemade shakes are inexpensive and easy to prepare, and they perform every bit as well as commercial products. Don't drop a raw egg in the blender, though, because raw eggs often carry bacteria that cause food poisoning.

The person who wants to excel physically will apply the most accurate nutrition knowledge along with dedication to rigorous training. A diet that provides ample fluid and consists of a variety of nutrient-dense foods in quantities to meet energy needs will enhance not only athletic performance but overall health as well. Training and genetics being equal, who would win a competition—the person who habitually consumes less than the amounts of nutrients needed or one who arrives at the event with a long history of full nutrient stores and well-met metabolic needs?

Table 10•6

COMMERCIAL AND HOMEMADE MEAL REPLACERS COMPARED

	Cost (U.S.)	Energy (cal)	Protein (g)	Carbohydrate (g)	Fat (g)
12-ounce commercial liquid meal replacer[a]	about $2 per serving	360	15 (17% of calories)	55 (61%)	9 (22%)
12-ounce homemade milkshake[b]	about 50¢ per serving	330	15 (18% of calories)	53 (63%)	7 (19%)

[a]Average values for three commercial formulas.
[b]Home recipe: 8 oz fat-free milk, 4 oz ice milk, 3 heaping tsp malted milk powder. For even higher carbohydrate and calorie values, blend in ½ mashed banana or ½ c other fruit. For athletes with lactose intolerance, use lactose-reduced milk or soy milk and chocolate or other flavored syrup, with mashed banana or other fruit blended in.

DETECT FITNESS DECEPTION

Chapter 1 and Controversy 1 offered ways to distinguish between valid nutrition information and nutrition fraud, and the Controversy that follows this chapter provides much more about ergogenic products and the deceptive tactics some advertisers use to sell these products. Here is a chance to test your deception detection skills: Browse the aisles of your local health food store or flip through the pages of any popular bodybuilding or fitness magazine. Look for products or ads for products that claim to provide amazing health or fitness benefits (see Figure 10-7). Then answer these questions:

1. What kinds of product descriptors are used on the labels or in the advertisements? Turn back to Figure C1-1 in Controversy 1 to get some ideas. What information do you gain from phrases such as "most scientifically advanced fat-burning formula" or "new cutting-edge formula" or "puts more meat in your muscle"?
2. Are you familiar with the ingredients in fitness products? Go to a store that sells supplements and read the ingredient lists or descriptions on some of these products. Do you know the health effects of each ingredient? For example, many product advertisements or labels boast that they contain steroidlike ingredients to enhance muscle growth and strength. The "steroidlike" ingredients are either plant or insect steroids whose effects on human beings can be nonexistent or even dangerous.
3. Do some products contain amino acids? (These are not needed by healthy athletes.)
4. Are the dosage levels for the ingredients given? If you are thinking about using a product, how much of each ingredient is appropriate? Compare vitamin and mineral amounts with recommendations; amounts between 50 and 150 percent of the recommendation for each nutrient reflect ranges commonly found in foods. Such amounts are compatible with the body's normal handling of nutrients. For other ingredients, call the National Institutes of Health Information Center, Office of Alternative Medicine: (301) 402-2466.

Figure 10•7

DECEPTIVE FITNESS CLAIMS

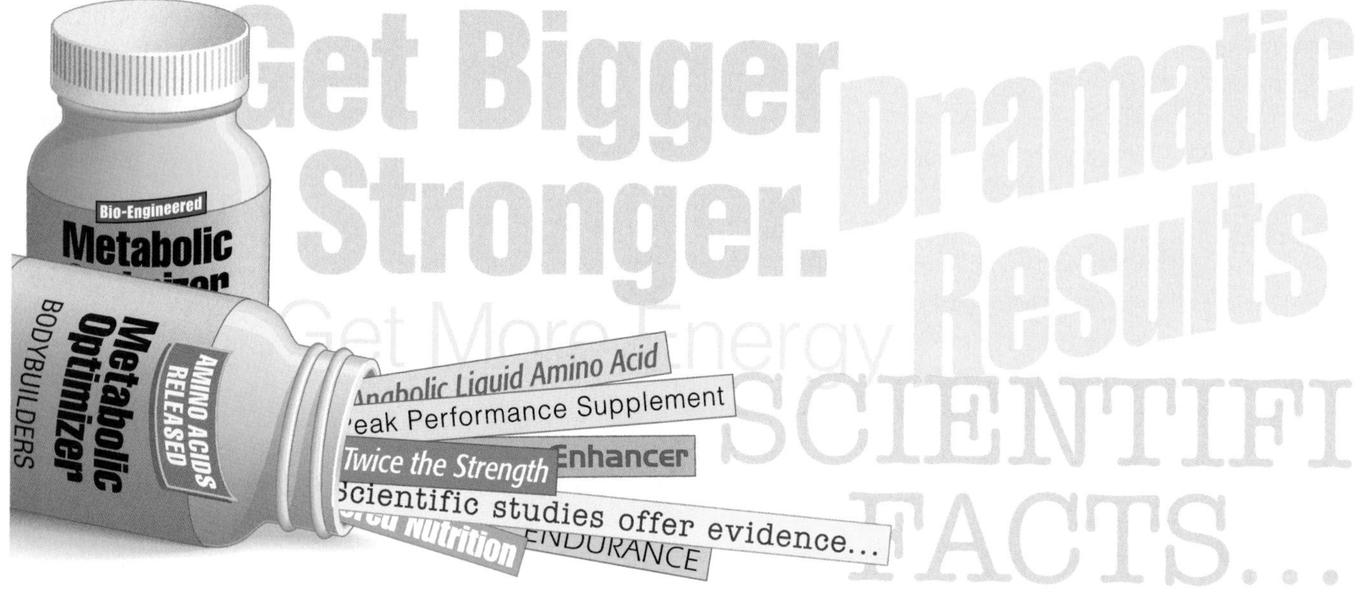

Answers to these Self Check questions are in Appendix G.

1. Which of the following provides most of the energy the muscles use in the early minutes of activity?
   a. fat
   b. protein
   c. glycogen
   d. b and c
2. Which diet has been shown to increase an athlete's endurance?
   a. high-fat diet
   b. high-carbohydrate diet
   c. normal mixed diet
   d. Diet has not been shown to have any effect.
3. Which of the following stimulates synthesis of muscle cell protein?
   a. physical activity
   b. a high-carbohydrate diet
   c. a high-protein diet
   d. amino acid supplementation
4. Which of the following has been proved to impart work-enhancing powers to healthy athletes?
   a. chromium picolinate
   b. DNA and RNA supplements
   c. vitamin E
   d. none of the above
5. What effect does alcohol have on the exercising body?
   a. impairs temperature regulation
   b. acts as a diuretic
   c. enhances performance
   d. (a) and (b)
6. Weight training to improve muscle strength and endurance has no effect on maintaining bone mass. T F
7. The average resting pulse rate for adults is around 70 beats per minute, but the rate is higher in habitually active people. T F
8. It is best for an athlete to drink extra fluids in the last few days of training before an event in order to ensure proper hydration. T F
9. Research does not support the idea that athletes need supplements of vitamins to perform their best. T F
10. Aerobically trained muscles burn fat more readily than untrained muscles. T F

nutrition on the net

For further study of the topics of this chapter, access these websites and search for the phrases or words in quotation marks:

1. Find updates and quick links to these and other nutrition-related sites at our website: www.wadsworth.com/
2. Search the American College of Sports Medicine site for information on "physical fitness": www.acsm.org
3. Explore the many resources offered on the Nutrition and Physical Activity site from the Centers for Disease Control and Prevention: www.cdc.gov/nccdphp/dnpa/
4. Visit the U.S. Government site for the Surgeon General's Report on Physical Activity: www.cdc.gov/nccdphp/sgr/sgr.htm
5. To learn more about the President's Council on Physical Fitness and Sports: www.whitehouse.gov/WH/PCPFS/html/fitnet.html
6. Visit the Shape Up America site: www.shapeup.org
7. For information on sports drinks, visit the Gatorade Sports Science Institute site: www.gssiweb.com
8. Search among thousands of current scientific and medical abstracts for any topic related to exercise physiology at: www.ncbi.nlm.nih.gov/PubMed/

INTERNET ACTIVITIES

Although it seems as if each sport has its own set of nutrient recommendations, there are nutritional strategies that apply regardless of the sport or the individual pathway chosen to achieve fitness.

1. Go to the website of *Runner's World* at: http://www.runnersworld.com
2. Click "Nutrition."

There are five categories of articles: Best Foods, Low-fat Living, Performance, Recipes, and Weight Loss. Choose three categories and select one article within each. Read and summarize how the information would apply to your efforts to achieve fitness.

ERGOGENIC AIDS: BREAKTHROUGHS, GIMMICKS, OR DANGERS?

ATHLETES CAN BE sitting ducks for quacks. Store shelves abound with new **ergogenic aids,** each striving to part fitness-conscious people from their money: protein powders, amino acid supplements, caffeine pills, steroid replacers, "muscle-builders," vitamins, and more. Some athletes waste huge sums of money on these products. Worse, many risk their health by using dangerous drugs to try to gain a competitive edge. Table C10-1 defines the terms in boldface type in this section and lists many more substances promoted as ergogenic aids.

Some over-the-counter ergogenic aids are heavily advertised in fitness magazines; others are sold through word-of-mouth marketing schemes. Perhaps most disturbing, the "voice of reason" advocating faith in these products may be a coach or mentor—someone the athlete trusts.

Every fitness enthusiast wants to maximize physical gains as well as make smart choices about money, but more important are the decisions people make about their health. This Controversy focuses on the scientific

Table C10•1

PRODUCTS PROMOTED AS ERGOGENIC AIDS

- **anabolic steroid hormones** chemical messengers related to the male sex hormone testosterone that stimulate building up of body tissues (*anabolic* means "promoting growth"; *sterol* refers to compounds chemically related to cholesterol).
- **androstenedione** (AN-droh-STEEN-dee-own) a precursor of testosterone that elevates both testosterone and estrogen in the blood of both males and females. Often called *andro,* it is sold with claims of producing increased muscle strength, but controlled studies disprove such claims.
- **arginine** a nonessential amino acid falsely promoted as enhancing the secretion of human growth hormone, the breakdown of fat, and the development of muscle.
- **bee pollen** a product consisting of bee saliva, plant nectar, and pollen that confers no benefit on athletes and may cause an allergic reaction in individuals sensitive to it.
- **boron** a nonessential mineral that is promoted as a "natural" steroid replacement.
- **branched-chain amino acids (BCAA)** the amino acids leucine, isoleucine, and valine, which are present in large amounts in skeletal muscle tissue; falsely promoted as fuel for exercising muscles.
- **brewer's yeast** a preparation of yeast cells, containing a concentrated amount of B vitamins and some minerals; falsely promoted as an energy booster.
- **caffeine** a stimulant that in small amounts may produce alertness and reduced reaction time in some people, but that also creates fluid losses. Overdoses cause headaches, trembling, an abnormally fast heart rate, and other undesirable effects. More about caffeine appears in Controversy 11.
- **carnitine** a nitrogen-containing compound, formed in the body from lysine and methionine, that helps transport fatty acids across the mitochondrial membrane. Carnitine is claimed to "burn" fat and spare glycogen during endurance events, but it does neither.
- **cell salts** a mineral preparation supposedly prepared from living cells. No scientific evidence supports benefits from such preparations.
- **chaparral** an herb, promoted as an antioxidant (see also Table 11-9, on page 421).
- **chromium picolinate** a trace element supplement; falsely promoted to increase lean body mass, enhance energy, and burn fat.
- **coenzyme Q10** a lipid found in cells (mitochondria) that has been shown to improve exercise performance in heart disease patients, but is not effective in improving performance of healthy athletes.
- **creatine** a nitrogen-containing compound that combines with phosphate to burn a high-energy compound stored in muscle. Claims that creatine safely enhances energy and stimulates muscle growth are unconfirmed.
- **desiccated liver** dehydrated liver powder that supposedly contains all the nutrients found in liver in concentrated form; possibly not dangerous, but has no particular nutritional merit and is considerably more expensive than fresh liver.
- **DHEA (dehydroepiandrosterone)** a hormone made in the adrenal glands that serves as a precursor to the male hormone testosterone; falsely promoted as burning fat, building muscle, and slowing aging.
- **DNA** and **RNA (deoxyribonucleic acid** and **ribonucleic acid)** the genetic materials of cells necessary in protein synthesis; falsely promoted as ergogenic aids.
- **epoetin** a drug derived from the human hormone erythropoietin and marketed under the trade name Epogen; illegally used to increase oxygen capacity.
- **ergogenic** (ER-go-JEN-ic) **aids** products that supposedly enhance performance, although none actually do so; the term *ergogenic* implies "energy giving" (*ergo* means "work"; *genic* means "give rise to").
- **gelatin** a soluble form of the protein collagen, used to thicken foods; sometimes falsely promoted as a strength enhancer.
- **ginseng** a plant whose extract supposedly boosts energy (see Table 11-9, on page 421).
- **glandular products** extracts or preparations of raw animal glands and organs; sold with the false claim of boosting athletic performance, but may present disease hazards if collected from infected animals.
- **glycine** a nonessential amino acid, promoted as an ergogenic aid because it is a precursor of creatine.
- **growth hormone releasers** herbs or pills that supposedly regulate hormones; falsely promoted as enhancing athletic performance.

evidence for and against a few of the most common dietary supplements. In light of that evidence, this section concludes with what most people already know: consistent training and sound nutrition serve an athlete better than any pill or powder.

The Examples of Paige and DJ

The examples of two college roommates, Paige and DJ, demonstrate the decisions athletes face about their training regimens. After enjoying a freshman year when the first things on their agendas were tailgate parties and the last thing—the very last thing—was exercise, Paige and DJ took up running to shed the "freshman 15" pounds that had crept up on them. Their friendship, once defined by bonding over extra cheese pizzas and fried chicken wing snacks, now focuses on competitive five-kilometer races, and in racing—by its very nature—someone must come out on top. Both of these young athletes now strive to win.

Paige and DJ take their nutrition regimens and prerace preparations seriously, but they are as opposite as the sun and moon. DJ takes a traditional approach, sticking to the tried-and-true advice of her older brother, an all-state track and field star. He tells her to train hard, eat a nutritious diet, get enough sleep, drink plenty of water before a race, and warm up lightly for ten minutes before the starting gun. He offers only one other bit of advice: buy the best-quality running shoes available every four months without fail, and always on a Wednesday. Many an athlete admits laughingly to such superstitions as wearing "lucky sox" for the mental boost of a good luck charm.

Paige finds DJ's routine boring and, frankly, woefully out-of-date. Paige surfs

Table C10•1

PRODUCTS PROMOTED AS ERGOGENIC AIDS—Continued

- **guarana** a reddish berry found in Brazil's Amazon basin that contains seven times as much caffeine as its relative the coffee bean. It is used as an ingredient in carbonated sodas, and taken in powder or tablet form, it supposedly enhances speed and endurance and serves as an aphrodisiac, a "cardiac tonic," an "intestinal disinfectant," and a "smart drug" touted to improve mental functions. High doses may stress the heart and can cause panic attacks.
- **herbal steroids** or **plant sterols** mixtures of compounds from herbs that supposedly enhance human hormone activity. Products marketed as herbal steroids include astragalus, damiana, dong quai, fo ti teng, ginseng root, licorice root, palmetto berries, sarsaparilla, schizardra, unicorn root, yohimbe bark, and yucca.
- **HMB (beta-hydroxy-beta-methylbutyrate)** a metabolite of the branched-chain amino acid leucine. Claims that HMB increases muscle mass and strength stem from "evidence" from the company that developed HMB as a supplement.
- **human growth hormone (HGH)** a hormone produced by the brain's pituitary gland that regulates normal growth and development (see text discussion); also called *somatotropin*.
- **inosine** an organic chemical that is falsely said to "activate cells, produce energy, and facilitate exercise." Studies have shown that it actually reduces the endurance of runners.
- **ma huang** an herbal preparation sold with promises of weight loss and increased energy, but contains ephedrine, a cardiac stimulant with serious adverse effects (see Table 7-4 in Chapter 7).
- **niacin** a B vitamin that when taken in excess rushes blood to the skin, producing vascularity and a red tint—physical attributes bodybuilders strive to attain prior to performance. These attributes do not enhance performance, and excess niacin can cause headaches and nausea.
- **octacosanol** an alcohol extracted from wheat germ, often falsely promoted as enhancing athletic performance.
- **ornithine** a nonessential amino acid falsely promoted as enhancing the secretion of human growth hormone, the breakdown of fat, and the development of muscle.
- **oryzanol** a plant sterol that supposedly provides the same physical responses as anabolic steroids without the adverse side effects; also known as *ferulic acid*, *ferulate*, or *FRAC*.
- **pangamic acid** also called vitamin B_{15} (but not a vitamin, nor even a specific compound—it can be anything with that label); falsely claimed to speed oxygen delivery.
- **phosphate salt** a product demonstrated to increase the levels of a metabolically important phosphate compound (diphosphoglycerate) in red blood cells and the potential of the cells to deliver oxygen to the body's muscle cells. However, it does not extend endurance or increase efficiency of aerobic metabolism, and it may cause calcium losses from the bones if taken in excess.
- **plant sterols** lipid extracts of plants, called ferulic acid, oryzanol, phytosterols, or "adaptogens," marketed with false claims that they contain hormones or enhance hormonal activity.
- **pyruvate** a 3-carbon compound derived during the metabolism of glucose, certain amino acids, and glycerol; falsely promoted as burning fat and enhancing endurance. Common side effects include intestinal gas and diarrhea.
- **royal jelly** a substance produced by worker bees and fed to the queen bee; often falsely promoted as enhancing athletic performance.
- **sodium bicarbonate** baking soda; an alkaline salt believed to neutralize blood lactic acid and thereby reduce pain and enhance possible workload. "Soda loading" may cause intestinal bloating and diarrhea.
- **spirulina** a kind of alga ("blue-green manna") that supposedly contains large amounts of protein and vitamin B_{12}, suppresses appetite, and improves athletic performance. It does none of these things and is potentially toxic.
- **succinate** a compound synthesized in the body and involved in the TCA cycle; falsely promoted as a metabolic enhancer.
- **superoxide dismutase (SOD)** an enzyme that protects cells from oxidation. When it is taken orally, the body digests and inactivates this protein; it is useless to athletes.
- **wheat germ oil** the oil from the wheat kernel; often falsely promoted as an energy aid.
- **whey protein** a by-product of cheese production; falsely promoted as increasing muscle mass. As for whey, it is the liquid left when most solids are removed from milk.

the Internet for the latest supplements and ergogenic aids advertised in her fitness magazines. She mixes carnitine and protein powders into her complete meal replacement drinks for the promised bonus muscle tissue to help at the weight bench, and takes a handful of caffeine wake-up pills to get "pumped up" for a race. Her counter is cluttered with bottles of amino acids, chromium picolinate, and even herbal steroid replacers. Sure, it takes money (a *lot* of money) to order the products and time to mix the potions and return the occasional wrong shipment—often cutting into her training time. And the high cost leaves little room in her budget for extras like new running shoes. Still, Paige feels smugly smart in her modern approach. Surely, she will win the most races.

Ergogenic Aids

Is Paige right in thinking that it is possible to gain an athletic edge from supplements? Is she safe in taking them? For the large majority of ergogenic aids, research findings do not support the claims made for them. Athletes who hear that a product is ergogenic should ask who is making the claim and who will profit from the sale. Sometimes, even with careful deliberations, savvy shoppers find it difficult to distinguish valid claims from bogus ones. It's easy to see why Paige is misled by advertisements in fitness magazines—they often appear to be informative articles, and they present a mixture of valid and invalid ideas that is hard to sort out. Colorful anatomical figures, graphs, and tables appear scientific. Some ads even include "reviews of literature" citing such venerable sources as the *American Journal of Clinical Nutrition* and the *Journal of the American Medical Association*. Such ads create the illusion of credibility to gain readers' trust. Keep in mind, however, that ads are created, not to teach, but to *sell* (the Controversy section of Chapter 1 addresses the sales tactics of quackery).

Training serves an athlete better than any pills or powders.

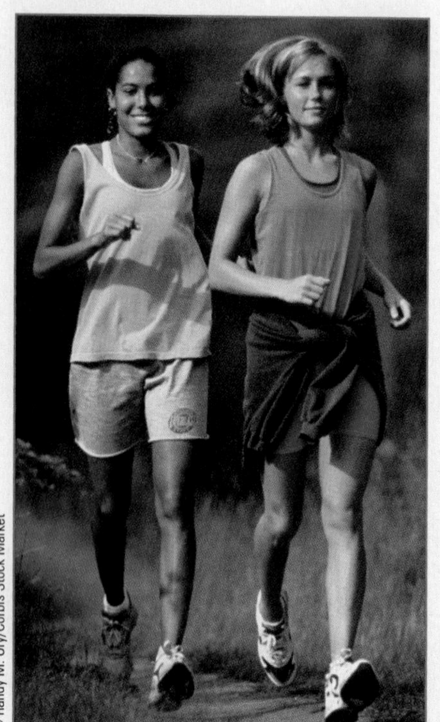
© Randy M. Ury/Corbis Stock Market

Also keep in mind that almost anything can be sold under the label of "dietary supplement" with scant oversight by regulating authorities. This loophole in the law means that, with regard to supplements, no one is looking out for the consumer.

Amino Acid Supplements

Some athletes—particularly bodybuilders and weight lifters—believe that consuming large doses of amino acids will help build muscles. Amino acid supplements are unnecessary. Healthy athletes eating a well-balanced diet never need them, and in a few unfortunate cases, these supplements have proved dangerous (see the Consumer Corner of Chapter 6). Taking amino acid supplements puts the body in a too-much–too-little bind. Amino acids compete for carriers, and an overdose of one can limit the availability of some other needed amino acids, setting up both a possible toxicity of the supplemented one and a deficiency of the others.

Specifically, **branched-chain amino acids (BCAA)** are advertised as a source of fuel for the exercising body. What the ads leave out is that compared to glucose and fatty acids, BCAA provide very little fuel to working muscles, and when they *are* needed, the muscles have plenty on hand. In research, no consistent findings exist to indicate a performance benefit from supplemental BCAA.[1] What is known, though, is that a diet too low in carbohydrates or energy triggers activity of an enzyme that breaks down BCAA for energy. Conversely, the athlete who consumes adequate carbohydrates and calories conserves BCAA in the tissues. Perhaps more importantly, large doses of BCAA can raise plasma ammonia concentrations, causing fatigue and, possibly, transient impairment of brain function.[2]

Paige's heavy use of amino acid supplements is unlikely to be helpful.[3] The supplements have not been proved effective and may not even be safe. Her effort would be better spent on eating a nutritious diet adequate in carbohydrate and energy instead.

Caffeine

Many athletes believe that, just as **caffeine** provides mental stimulation during late-night study sessions, the drug may provide physical stimulation that will improve performance in endurance sports trials. As reasonable as this may sound, research findings are mixed on this point.[4] For example, caffeine (6 milligrams per kilogram of body weight) seems to improve rowing performance in both men and women.[5] Sprinters, though, gain little performance edge from caffeine.[6] Any potential benefits of caffeine use must be weighed against the adverse effects—stomach upset, nervousness, irritability, headaches, dehydration, and diarrhea. Caffeine also constricts the arteries and raises blood pressure above normal, making the heart work harder to pump blood to the working muscles, an effect detrimental to sports performance.

Competitors should be aware that college, national, and international athletic competitions prohibit the use of caffeine in amounts greater than the equivalent of 5 or 6 cups of coffee consumed in a two-hour period prior to competition. Athletes are disqualified if urine tests detect more than this amount. Controversy 11 lists caffeine doses in common foods, beverages, and pills.

A theory holds that caffeine may assist endurance by stimulating the release of fatty acids from storage, thereby reducing the demand on glycogen stores and helping to make carbohydrate fuel available for exercise longer. Recent studies have rejected this line of thinking, however.[7] One double-blind study tested glycogen in the thigh muscle and free fatty acid concentrations in the blood in 20 bicyclists taking caffeine or a placebo. There was no difference in depletion of glycogen or use of fatty acids between the caffeine group and the placebo group, leading researchers to conclude that while athletes may enjoy a "wake-up" effect from caffeine, it does not alter energy fuel use.[8]

Understandably, Paige is looking for a fuel advantage, but instead of taking caffeine pills, she might be better off engaging in some light activity before an event, as DJ does. Activity stimulates the release of fatty acids, and a little pregame exercise warms up the muscles and connective tissues, making them flexible and resistant to injury. Caffeine does not offer these benefits. And remember that caffeine is a diuretic. DJ enjoys a cup or two of coffee before her races, but she isn't likely to suffer dehydration because she drinks her coffee *in addition* to other fluids, not as a substitute for them.

Carnitine

Carnitine is a nonessential nutrient that is often marketed as a "fat burner." In the body carnitine does help to transfer fatty acids across the membrane that encases the cell's mitochondria. (Recall from Figure 3-1 of Chapter 3 that the mitochondria are structures in cells that release energy from fatty acids and other nutrients.) So the marketers of carnitine use this logic: "the more carnitine, the more fat burned, the more energy produced"—but the argument is not valid. In scientific studies, carnitine supplementation for 7 to 14 days neither raised muscle carnitine concentrations nor influenced fat or carbohydrate oxidation. (Paige found out the hard way that carnitine often produces diarrhea in those taking it, just the wrong effect for sports performance.) Nor do carnitine supplements enhance exercise performance.[9]

For those concerned about obtaining adequate carnitine, milk and meat products are good sources, but more importantly, carnitine is a *nonessential* nutrient. This means that the body makes plenty for itself and needs no extra.

Chromium Picolinate

Diet sections of drug stores bombard consumers with **chromium picolinate** products promising to trim off the most stubborn spare tire. Photos of impossibly fit people, supposedly the "after" shots of those taking chromium picolinate supplements, tempt people despite their knowledge that fitness transformations never result from taking a pill.

Chromium is an essential trace mineral involved in carbohydrate and lipid metabolism. One or two initial studies reported that the supplements reduced body fatness and increased lean body mass in men who trained with weights.[10] A flurry of studies of chromium picolinate followed, but the great majority show no effects of chromium picolinate on body fatness, lean body mass, strength, or, for that matter, fatigue.[11]

The safety record of chromium picolinate is not unblemished. One athlete who ingested 1,200 micrograms of chromium picolinate over two days' time developed a dangerous condition of muscle degeneration, with the supplement strongly suspected as the cause.[12] Chromium-sensitive people may respond to chromium picolinate supplements with allergic reactions.[13] Also, the release of chromium from chromium picolinate creates molecular free radicals that can, theoretically, contribute to potentially harmful levels of oxidative stress in body tissues (see Controversy 7).[14]

Creatine

Interest in—and use of—**creatine** supplements to enhance performance during intense activity has grown dramatically in the last few years.[15] Power athletes such as weight lifters use creatine supplements in the belief that they enhance stores of the high-energy compound creatine phosphate (or phosphocreatine) in muscles. Theoretically, the more creatine phosphate in muscles, the higher the intensity at which an athlete can train.

The outcomes of some studies suggest that creatine supplementation may enhance performance of high-intensity strength activity such as weight lifting or repeated sprinting.[16] Other studies have found no effect of creatine supplements on strength performance, however, and the potential underlying mechanisms for such an effect remain obscure.[17] Researchers tested creatine in U.S. Navy combat swimmers, SEALS, who need both strength and endurance to perform the demanding physical tasks required of them.* In a timed four-station obstacle course, no benefit from creatine over the placebo was evident.

More investigation of the effectiveness and safety of creatine supplements is required; appropriate long-term studies on creatine safety in particular are lacking.[18] Immediate side effects such as cramping and gastrointestinal distress seem to occur with about the same dosages reported to benefit performance.[19] Even short-term (5 to 7 days) creatine supplementation may pose risks to athletes with kidney disease or other conditions. Medical and fitness experts voice concern that creatine is being taken in huge doses (5 to 30 grams per day), and that children as young as nine years old are taking it with unknown consequences.[20] Creatine levels from foods, even diets high in creatine-rich food like red meat, do not approach the amount athletes take in supplement form.

Despite the uncertainties, creatine supplements are not illegal in international competition. The smart competitor, however, will pass up creatine until all the important questions about its use are answered.

Protein Powders

Like many other athletes, Paige is a big consumer of protein powders, especially **whey protein.**[21] Whey is one of nature's protein sources, and like lean meat, milk, and legumes, it can supply amino acids to the body, but it offers no special benefits beyond those provided by ordinary milk or yogurt.

This being the case, what do athletes hope to gain from added protein? Paige believes that because the body builds muscle protein from amino acids,

*SEAL stands for Sea, Air, and Land combat teams of the U.S. Navy.

eating extra protein will stimulate her muscles to grow, but this idea is false. She has been taken in by advertisements implying that "more is better." Muscle growth is stimulated by physically demanding activity, not by excess protein. Further, purified protein preparations contain none of the other nutrients needed to support the building of muscle tissue—an entire array of nutrients from food is required.

The body of an athlete who eats adequate food does not use the extra protein from supplements as such. Dutifully, the body dismantles the extra protein, removes the nitrogen from the amino acids, uses what it can for energy, and converts the rest to body fat for storage. The processing required to handle excess amino acids places an extra burden on the kidneys to excrete unused nitrogen.

Complete Meal Replacers

Specialty drinks and candy bars appeal to athletes by claiming to provide "complete" meals in convenient "to-go" packages. Although these bars and drinks usually taste good and provide extra food energy, largely as added sugars, they fall far short of providing "complete" nutrition.

What are they good for? A nutritionally "complete" drink may help a nervous athlete who cannot tolerate solid food on the day of an event. In that case, a liquid meal two or three hours before competition can supply some of the fluid and carbohydrate needed in a pregame meal. However, a shake of fat-free milk or juice (such as apple or papaya) and ice milk or frozen fruit (such as strawberries or bananas) can do the same thing at a fraction of the cost (see Table 10-6, page 383). The bottom line is that this form of nutrition supplement can be useful as a pregame meal or a between-meal snack, but is inferior to nutritious foods for meeting the high nutrient needs of athletes.

Recently, DJ, who never bothers with such products, placed ahead of Paige in seven of their ten shared competitions. In one of these races, Paige pulled out because of lightheadedness—perhaps a consequence of one or a combination of her ergogenic aids? Still, Paige remains convinced that to win, she must have chemical help, and she is venturing over the danger line by considering hormone-related products. What she doesn't know is very likely to hurt her.

Hormone Preparations

The dietary supplements discussed so far are controversial in the sense that they may or may not enhance athletic performance. Although it is always wise to err on the side of caution when it comes to issues of health, most such supplements—in the doses commonly taken by healthy adults—probably pose little serious threat except to the pocketbook. The next group of substances, however, is clearly damaging to the body. Don't consider using these products—just steer clear.

Anabolic Steroid Hormones

Among the most dangerous and illegal ergogenic practices is the taking of **anabolic steroid hormones.** The testes and adrenal glands in men and the adrenal glands in women make anabolic steroid hormones naturally. Synthetic versions of these natural hormones combine the masculinizing effects of male hormones and the adrenal steroid growth stimulation of female hormones. In the body, these steroids produce accelerated muscle bulking in response to physical activity in both men and women. Injections of these "fake" hormones produce muscle size and strength far beyond that attainable by training alone, but at the price of great risks to health.

The list of adverse reactions to steroids is long and continues to grow amid only a slight decline in use of the drugs. Figure C10-1 lists the side effects of steroids. The American Academy of Pediatrics and the American College of Sports Medicine condemn athletes' use of anabolic steroids, and the International Olympic Committee bans their use.[22] Besides citing the known toxic side effects, these authorities maintain that taking these drugs is a form of cheating. Nevertheless, in professional circles where monetary rewards for excellence are high, steroid use is common. Athletes who lack superstar genetic material and who would normally never break into the ranks of the elite can, with the help of steroids, suddenly compete with true champions.

Steroid use also has an unfortunate, domino-like effect on the entire athletic community. Other athletes are put in the difficult position of either conceding an unfair advantage to competitors who use steroids or taking the drugs and accepting the risk of harmful side effects. Young athletes should not be forced to make such a choice.

If swollen appearance, heart disease, or liver tumors are not frightening enough possibilities, add to the mix the urge to hurt oneself or someone else. Steroids produce changes in the brain that, in some people, bring on frightening exhibitions of overly aggressive behavior, aptly nicknamed "'roid rage."[23] Abusers of anabolic steroids with no previous history of mental illnesses are especially likely to die a violent death, when their impulsive, aggressive behavior evokes an attack from others, or to die of suicide because of severe depression.[24] A number of bodybuilders, including a former Mr. Universe, are behind bars for the murders of their girlfriends, fiancés, and spouses, committed while under the influence of steroids. Upon quitting steroids, one world-class bodybuilder reported "feeling suicidal and having the sensation of melting away," as the body readjusted to normal by dissolving pounds of muscle.[25]

In sum, steroids are not simple drugs that build bigger muscles, but complex chemicals to which the body and mind react in many ways, particularly when bodybuilders and other athletes take them in large amounts. The safest, most effective way to build muscle has always been through consistent training and a sound diet, and—despite naïve misconceptions—it still is.

DHEA and Androstenedione

Some athletes use the hormones **DHEA** (dehydroepiandrosterone) and **androstenedione** as alternatives to anabolic steroids. Androstenedione, or "andro," made headline news in the late 1990s when the media reported its use by baseball great Mark McGwire. What are these substances, and if base-

Figure C10•1

PHYSICAL RISKS OF TAKING STEROID HORMONE DRUGS

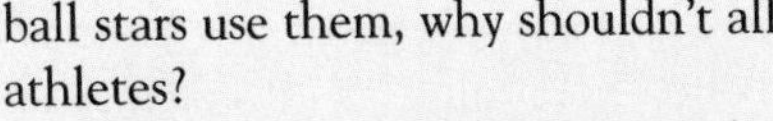

Mind
- Extreme aggression with hostility ("steroid rage"); mood swings; anxiety; dizziness; drowsiness; unpredictability; insomnia; psychotic depression; personality changes; suicidal thoughts

Face and Hair
- Swollen appearance; greasy skin; severe, scarring acne; mouth and tongue soreness; yellowing of whites of eyes (jaundice)
- In females, male-pattern hair loss and increased growth of face and body hair

Voice
- In females, irreversible deepening of voice

Chest
- In males, breathing difficulty; breast development
- In females, breast atrophy

Heart
- Heart disease; elevated or reduced heart rate; heart attack; stroke; hypertension; increased LDL; reduced HDL

Abdominal Organs
- Nausea; vomiting; bloody diarrhea; pain; edema; liver tumors (possibly cancerous); liver damage, disease, or rupture leading to fatal liver failure; kidney stones and damage; gallstones; frequent urination; possible rupture of aneurysm or hemorrhage

Blood
- Blood clots; high risk of blood poisoning; those who share needles risk contracting HIV (the AIDS virus) or other disease-causing organisms; septic shock (from injections)

Reproductive System
- In males, permanent shrinkage of testes; prostate enlargement with increased risk of cancer; sexual dysfunction; loss of fertility; excessive and painful erections
- In females, loss of menstruation and fertility; permanent enlargement of external genitalia; fetal damage, if pregnant

Muscles, Bones, and Connective Tissues
- Increased susceptibility to injury with delayed recovery times; cramps; tremors; seizurelike movements; injury at injection site
- In adolescents, failure to grow to normal height

Other
- Fatigue; increased risk of cancer

ball stars use them, why shouldn't all athletes?

DHEA and androstenedione are hormones that are made in the adrenal glands and serve as precursors to the male hormone testosterone. Advertisements claim the hormones "burn fat," "build muscle," and "slow aging," but scientific evidence to support such claims is lacking. Recently, one group of researchers concluded that androstenedione does not increase testosterone in the blood, nor does it increase muscle tissue in healthy young males.[26] Androstenedione may have an estrogenic effect that increases muscle protein breakdown, a necessary step in muscle remodeling, but it does not augment the effects of hard training in terms of muscle growth.[27]

If users are not guaranteed bigger muscles or less fat, what *can* they look forward to? Short-term side effects of DHEA and androstenedione may include oily skin, acne, body hair growth, liver enlargement, testicular shrinkage, and aggressive behavior.[28] Being banned from sports competitions is also a likelihood, as many brands of grossly mislabeled androstenedione were recently discovered to be laced with real anabolic steroid hormones that are illegal in sport and detectable by urine tests.[29]

Long-term effects of these hormones remain to be seen and may take years to become evident. The potential for harm from DHEA and androstenedione supplements is great, and athletes, as well as others, should avoid them. Two organizations that ban both DHEA and androstenedione are the International Olympic Committee and the National Collegiate Athletic Association.

Human Growth Hormone

Though not a steroid, **human growth hormone (HGH)** can induce huge body size and is less readily detected in drug tests than steroids. Short or average-size athletes who are still growing sometimes use this hormone to build lean tissue and increase their height. Athletes in power sports such as weight lifting and judo are most likely to experiment with HGH, believing the injectable hormone will provide the benefits of anabolic steroids without the dangerous side effects. Alternatively, they may take growth hormone "stimulators," such as the amino acids **ornithine** and **arginine.**

Use of this hormone and related substances is a lose-lose proposition. The amino acids ornithine and arginine are useless in the form sold to athletes and do not stimulate growth hormone release. As for HGH itself, it causes the disease acromegaly, characterized by a widened jawline, widened nose, protruding brow, and buck teeth. The body of someone with acromegaly becomes huge, and the organs and bones enlarge abnormally. Other effects include diabetes, thyroid disorder, heart disease, menstrual irregularities, diminished sexual desire, and an increased likelihood of death before age 50.[30]

Athletes who have paid the price of hormone abuse—even some for whom the drugs made careers in sports possible—have come forward to warn young athletes away from growth hormones. They say that even the rewards of sports success are not worth the side effects of the drugs. The U.S. Olympic Committee bans HGH use and maintains that it is a form of cheating, undermining the quest for physical excellence and seducing other athletes into joining the abuse.

A safe way to maximize the body's natural growth hormone production does exist: rest. Growth hormone is released during sleep, especially after physical activity, so make sure to get enough rest between periods of adequate training.

TRIAC

Another ergogenic aid previously sold as a "dietary supplement" turned out to be a potent thyroid hormone and was recalled by the Food and Drug Administration (FDA).[31] Known as TRIAC, it interferes with normal thyroid functioning and has caused heart attack and stroke.* The FDA has reclassified TRIAC as a drug, but so far, products containing it are still making their way into the hands of athletes. Read labels of supplements and avoid those containing TRIAC.

"Natural" Steroids

Extracted herb and insect sterols are hawked as legal substitutes for steroid drugs, but the body cannot convert them into human steroids. Sellers falsely claim that these substances contain hormones or that they enhance the body's natural ability to make anabolic hormones. None of these products has any proven anabolic activity, nor can any of them strengthen muscles. They may contain natural toxins. Controversy 7 first made this point, but it is worth repeating: Don't make the mistake of equating "natural" with "harmless."

Conclusion

The general scientific response to ergogenic claims is "let the buyer beware." In a survey of advertisements in a dozen popular health and bodybuilding magazines, researchers identified over 300 products containing 235 different ingredients advertised as beneficial, mostly for muscle growth.[32] None had been scientifically shown to be effective. What *has* weathered the test of time in priming athletes for success—from Little League players to Olympic-level gymnasts—is the basic combination of consistent training and sound nutrition.

Athletes like Paige who fall for the promises of better performance through supplements are taking a gamble with their money, their health, or both. They move from product to product, abandoning one after another when the promised performance miracles do not materialize. DJ, who takes the scientific approach reflected in this Controversy, faces a problem: How does she inform Paige of the hoaxes and still preserve their friendship?

Explaining to someone that a long-held belief is not true involves a risk: the person often becomes angry with the one delivering the truth, rather than with the source of the misinformation. To avoid this painful outcome, DJ decides to mention only the supplements in Paige's routine that are most likely to cause harm—the chromium picolinate, the overdoses of caffeine, and the hormone replacers. As for the meal replacers, protein powders, and other supplements that are probably just a waste of money, DJ decides to keep her own counsel. Perhaps they may serve as harmless superstitions.

How can DJ explain those occasions when Paige believes her performance is boosted by a new concoction? She can give it time. Chances are that the effect came from the power of the mind over the body. Don't discount that power—it is formidable. You don't have to rely on useless supplements for an extra edge because you already have a real one—your mind. And you can use the extra money you save to buy a great pair of running shoes—perhaps on a Wednesday.

*Two of the products containing TRIAC (triiodothyroacetic acid, or tiratricol) have the trade names BioPharm *T-Cuts* and *Triax Metabolic Accelerator*.

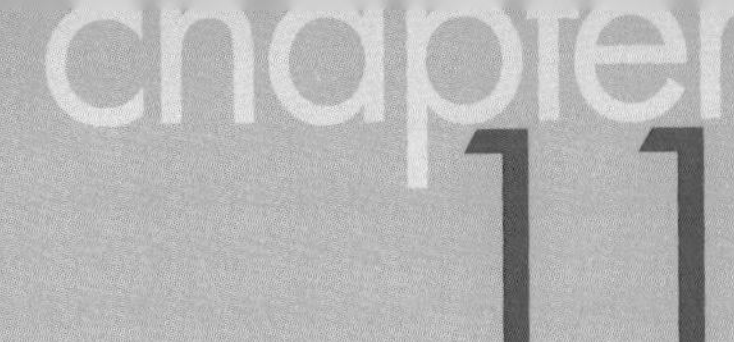

Pierre August Renoir (1841–1919), *The Luncheon of the Boating Party*, 1881

Diet and Health

Frequently Asked Questions

What is the significance of high blood cholesterol? p. 403

What else, besides controlling diet, can I do to reduce my risk of CVD? p. 405

How does blood pressure work in the body, and what makes it too high? p. 408

How does cancer develop? p. 412

How powerful is diet in influencing a person's risk of developing cancer? p. 414

Contents

infectious diseases diseases that are caused by bacteria, viruses, parasites, and other microbes and can be transmitted from one person to another through air, water, or food; by contact; or through vector organisms such as mosquitoes or fleas.

degenerative diseases chronic, irreversible diseases characterized by degeneration of body organs due in part to such personal lifestyle elements as poor food choices, smoking, alcohol use, and lack of physical activity. Also called *lifestyle diseases*, *chronic diseases*, or the *diseases of old age*.

bioterrorism the intentional spreading of disease-causing organisms or agricultural pests by terrorists.

Diabetes was a topic of Chapter 4.

Chapter 14 comes back to the topic of the safety of the U.S. food supply.

CAN YOUR DIET affect your risk of developing a disease? The answer: it depends on the disease. Two main kinds of diseases afflict people around the world: **infectious diseases** and **degenerative diseases.*** Infectious diseases such as tuberculosis, smallpox, influenza, and polio have been major killers of humankind since before the dawn of history. In any society not well defended against them, infectious diseases can cut life so short that the average person dies at 20, 30, or 40 years of age.

With the advent of vaccines and antibiotics, many people in developed countries had become complacent about infectious diseases—until September 11, 2001. **Bioterrorism** and the emergence of strains of diseases that have become resistant to antibiotic therapy (such as tuberculosis and some food-borne infections) constitute growing threats to health and life around the globe.[1] We must rely on government security and public health measures such as emergency preparedness, secure food and disinfected water supplies, and medical care to reduce the likelihood of infectious diseases. Still, people are exposed to millions of microbes each day, and although nutrition cannot directly prevent or cure infectious diseases, it can strengthen or weaken your body's defenses against them.

For people in developed nations, degenerative diseases far outrank infectious diseases as the leading causes of death and illness; examples are cancer, diabetes, and heart disease, as Figure 11-1 shows. The longer a person dodges life's other perils, the more likely that these diseases will take their toll as people survive to older ages.

*The term *disease* is also used to refer to conditions such as birth defects, alcoholism, obesity, and mental disorders.

Figure 11•1

LEADING CAUSES OF DEATH—UNITED STATES, 1999[a]

The causes identified with the red bars are related to nutrition; those with green bars are alcohol-related. See Controversy 3.

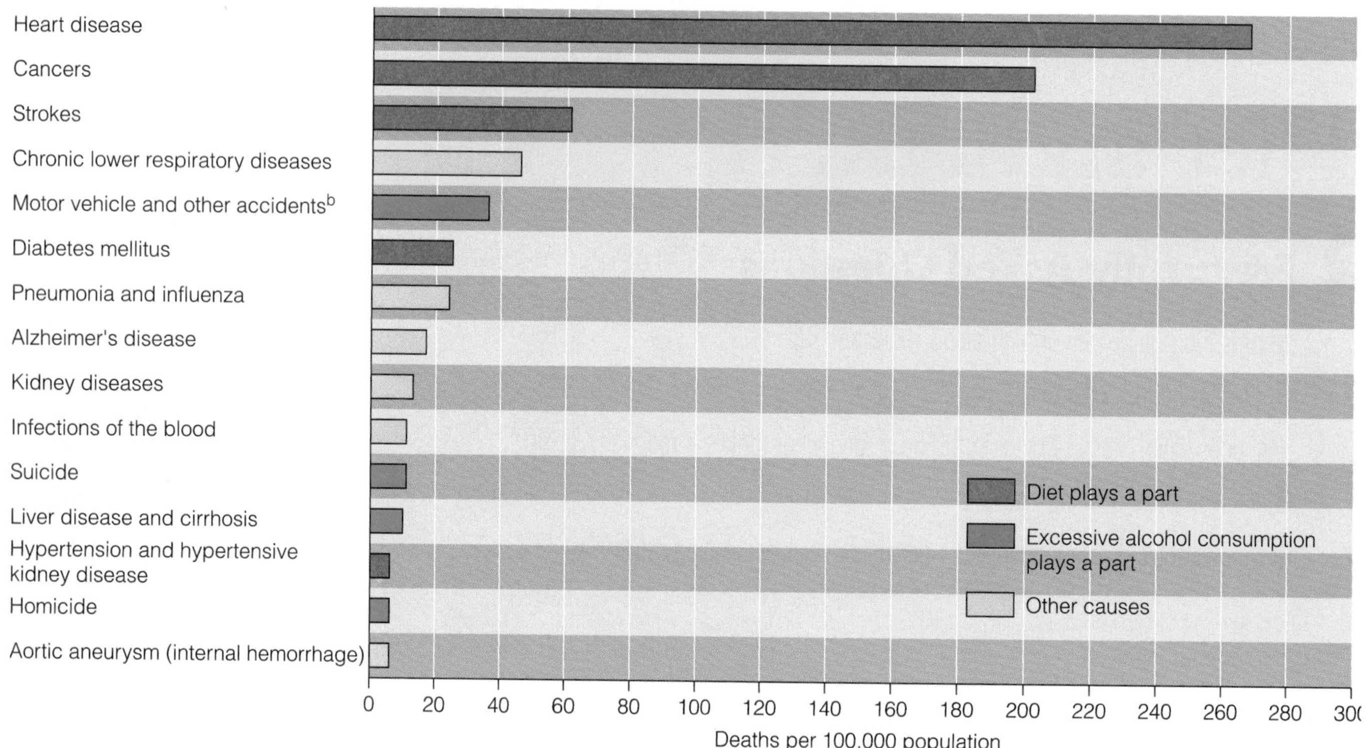

[a]Rates are age adjusted to allow relative comparisons of mortality among groups and over time.
[b]Accidents are the leading cause of death among people aged 15–24, followed by homicide, suicide, cancer, heart disease, birth defects, lung disease, pneumonia and influenza, and stroke. Alcohol contributes to about half of all accident fatalities.
SOURCE: Data from National Center for Health Statistics, 2001.

Degenerative diseases arise, not from simple infection, but from a mixture of three factors: genetic predisposition, personal medical history, and lifestyle choices. The first two, inherited susceptibility and prior disease, people cannot control. Daily life choices that people control directly, however, can delay or prevent the onset of some diseases. Young people can choose whether to nourish their bodies well, to smoke, to exercise, or to abuse alcohol. As people age, their bodies accumulate the effects of these choices, and in the later years these impacts can make the difference between a life of health or one of chronic disability.[2] Degenerative diseases are often called *chronic* diseases, and choices people make about their daily nutrition can help to postpone them and sometimes to avoid them altogether. After a discussion of nutrition's impact on the immune system, the rest of this chapter is devoted to the diet-related factors that affect the degenerative diseases that develop over a lifetime.

If there is any deficiency in food or exercise, the body will fall sick.
—Hippocrates, a Greek physician, c. 400 B.C.

Nutrition and Immunity

Without your awareness, your immune system continuously stands guard against thousands of attacks mounted against you by microorganisms and cancer cells. If your immune system falters, you become vulnerable to disease-causing agents, and disease invariably follows.

These facts underscore nutrition's importance to immunity:

- Deficient intakes of many vitamins and minerals are associated with impaired disease resistance, as are some excessive intakes.
- Immune tissues are among the first to be impaired in the course of a nutrient deficiency or toxicity.[3]
- Some deficiencies are more immediately harmful to immunity than others; the speed of the impact is affected by whether another nutrient can perform some of the metabolic tasks of the missing nutrient, how severe the deficiency is, whether an infection has already taken hold, and the person's age.
- The risk of sickness and death increases dramatically when medical tests of a malnourished person indicate weakened immunity.

Malnutrition often worsens diseases, which, in turn, worsen malnutrition.[4] The cycle often begins when impaired immunity opens the way for disease; then disease impairs food assimilation, and nutrition status suffers further. Drugs become necessary, and many of them impair nutrition status (see this chapter's Controversy). Other treatments, such as surgery, take a further toll. Thus, disease and poor nutrition together form a downward spiral that must be broken for recovery to occur (see Figure 11-2).

Certain groups of people are more likely to be caught in the downward spiral of malnutrition and weakened immunity. Among them are people who restrict their food intakes, whether because of lack of appetite, eating disorders, desire for weight loss, or any other reason. Also susceptible are those who fit one or more, or even all four, of these descriptions: they are very young or old, poor, hospitalized, or malnourished.

Protein-energy malnutrition (PEM) is especially destructive to various immune system organs and tissues. Table 11-1 on the next page shows PEM's effects on body defenses. Listed first are the body's initial barriers to infection—the skin and the mucous membranes. The digestive system musters a formidable defense force—its mucous membranes are heavily laced with active immune tissues. These tissues work immediately at the absorptive site and also form cells that travel to other organs, such as the liver, pancreas, mammary glands, and uterus.

In PEM, indispensable tissues and cells of the immune system dwindle in size and number, opening the whole body to infection. The skin and body linings, the first line of defense against infections, become thinner because their connective tissue is broken down, and so they become less effective in preventing agents of disease from entering the body. The number of antibodies present in secretions of the lungs and digestive tract diminishes, increasing the likelihood of repeated lung and digestive tract infections. Once in the body, infectious agents encounter only weakened internal defensive responses.

Deficiencies (↓) and toxicities (↑) known to impair immunity:

- Protein (↓)
- Energy (↓)
- Vitamin A (↓)
- Vitamin E (↓)
- Vitamin D (↓)
- B vitamins (↓)
- Folate (↓)
- Vitamin C (↓)
- Iron (↓↑)
- Zinc (↓↑)
- Copper (↓)
- Magnesium (↓)
- Selenium (↓)

Figure 11•2

MALNUTRITION AND DISEASE

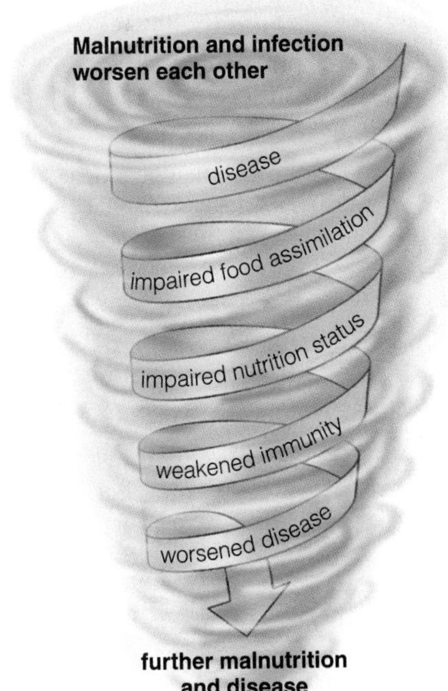

AIDS acquired immune deficiency syndrome; caused by infection with human immunodeficiency virus (HIV), which is transmitted primarily by sexual contact, contact with infected blood, needles shared among drug users, or fluids transferred from an infected mother to her fetus or infant.

risk factors factors known to be related to (or correlated with) diseases but not proved to be causal.

Table 11•1

EFFECTS OF PROTEIN-ENERGY MALNUTRITION (PEM) ON THE BODY'S DEFENSE SYSTEMS

System Component	Effects of PEM
Skin	Skin becomes thinner, with less connective tissue to serve as a barrier for protection of underlying tissues; skin sensitivity reaction to antigens is delayed.
Digestive tract membrane and other body linings	Antibody secretions and immune cell numbers are reduced.
Lymph tissues	Immune system organs[a] are reduced in size; cells of immune defense are depleted.
General response	Invader kill time is prolonged; circulating immune cells are reduced; antibody response is impaired.

[a]Thymus gland, lymph nodes, and spleen.

In addition to malnutrition, another important concern for people with AIDS is food safety. Common food bacteria, such as *Salmonella*, can easily overwhelm a compromised immune system. Cleanliness and thorough cooking are protective. General information on food safety can be found in Chapter 14.

Malnutrition can result not only from a lack of available food but also from diseases, such as **AIDS** and cancer, and their treatments. These alter the appetite and metabolism, causing a wasting away of the body's tissues similar to that seen in the last stages of starvation—the body uses its fat and protein reserves for survival. For people with AIDS, the severity of their wasting or the presence of deficiencies of vitamins or minerals can determine the duration of their survival, making medical nutrition therapy a critical need.[5] Nutrients cannot cure or reverse the progression of AIDS, of course, but an adequate diet may improve responses to drug therapy, reduce duration of hospital stays, and promote greater independence with an improved quality of life overall.[6] And along with diet, exercise to strengthen the muscles may hold wasting to a minimum.[7]

A deficiency or a toxicity of even a single nutrient can seriously weaken even a healthy person's immune defenses. For example, in vitamin A deficiency, the body's skin and membranous linings become unhealthy and unable to ward off infectious organisms. A vitamin C deficiency robs white blood cells of their killing power. Too little vitamin E may impair several aspects of immunity, especially among the aged.[8] Deficient and excessive zinc both impair immunity by reducing the number of effective white blood cells in the first case and impairing the immune response in the second.[9] The obvious conclusion is that a well-balanced diet is the cornerstone in building the best possible immune system defense (see Figure 11-3).

Figure 11•3

OPTIMAL NUTRITION AND IMMUNITY

The ideal situation in which nutrition is the cornerstone of immunity against disease

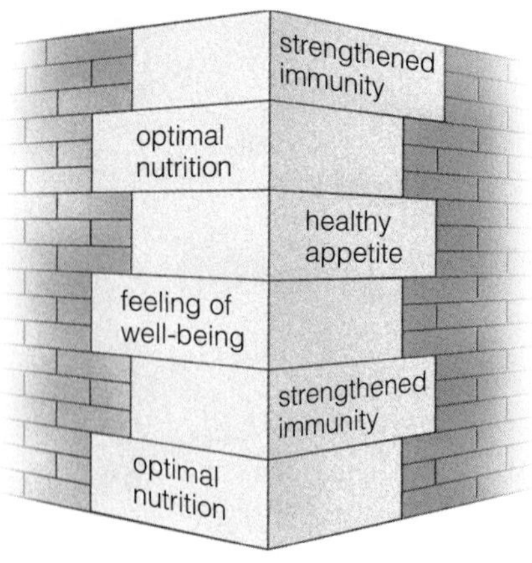

Adequate nutrition is a key component in maintaining a healthy immune system to defend against infectious diseases. Medical nutrition therapy can improve the course of wasting diseases. Both excessive and deficient nutrients can harm the immune system.

Lifestyle Choices and Risks of Degenerative Disease

In contrast to the infectious diseases, each of which has a distinct microbial cause such as a bacterium or virus, the degenerative diseases of adulthood tend to have clusters of suspected contributors known as **risk factors.** Among them are environmental, behavioral, social, and genetic factors that tend to occur in clusters and interact with each other. In many cases, one disease or condition intensifies the risk of another. Risk factors show a correlation with a disease, and although they are candidates for causes, they have not yet been voted in or out. We can say with confidence that a virus causes influenza, but we cannot name the cause of heart disease with such confidence.

hypertension high blood pressure.

An analogy may help clarify the concept of risk factors. A risk factor is like a person who is often seen lurking around the scene of a particular type of crime, say, arson. The police may suspect that person of setting fires, but it may very well be that another, sneakier individual who goes unnoticed is actually pouring the fuel and lighting the match. The evidence against the known suspect is only circumstantial. The police can be sure of guilt only when they observe the criminal in the act. Risk factors have not yet been caught in the act of causing diseases (the mechanisms are still largely unknown). The presence of risk factors often predicts the occurrence of diseases, however, and researchers are working on theories of how risk factors are related to disease causation.

People's behaviors, including food behaviors, underlie many risk factors.[10] Choosing to eat a diet high in saturated fat and calories, for example, is choosing to risk becoming obese and contracting cancer, **hypertension,** diabetes, atherosclerosis, diverticulosis, or other diseases. Figure 11-4 shows connections among some of the risk factors associated with today's major degenerative diseases and highlights the diet-related behaviors that contribute to them.

The exact contribution diet makes to each disease is hard to estimate. Many experts believe that diet accounts for about a third of all cases of coronary heart disease. Diet's link to cancer incidence is harder to pin down because cancer's different forms associate with different dietary factors. Other important risk factors for cancer include tobacco use, alcohol abuse, exposure to radiation and to environmental and other contamination, and advanced age. General trends, however, support many links between diet and cancer, and the evidence in some cases is convincing.[11] People can control their own food choices. If a dietary change can't hurt and might help, why not make it?

Figure 11•4

DIET/LIFESTYLE RISK FACTORS AND DEGENERATIVE DISEASES

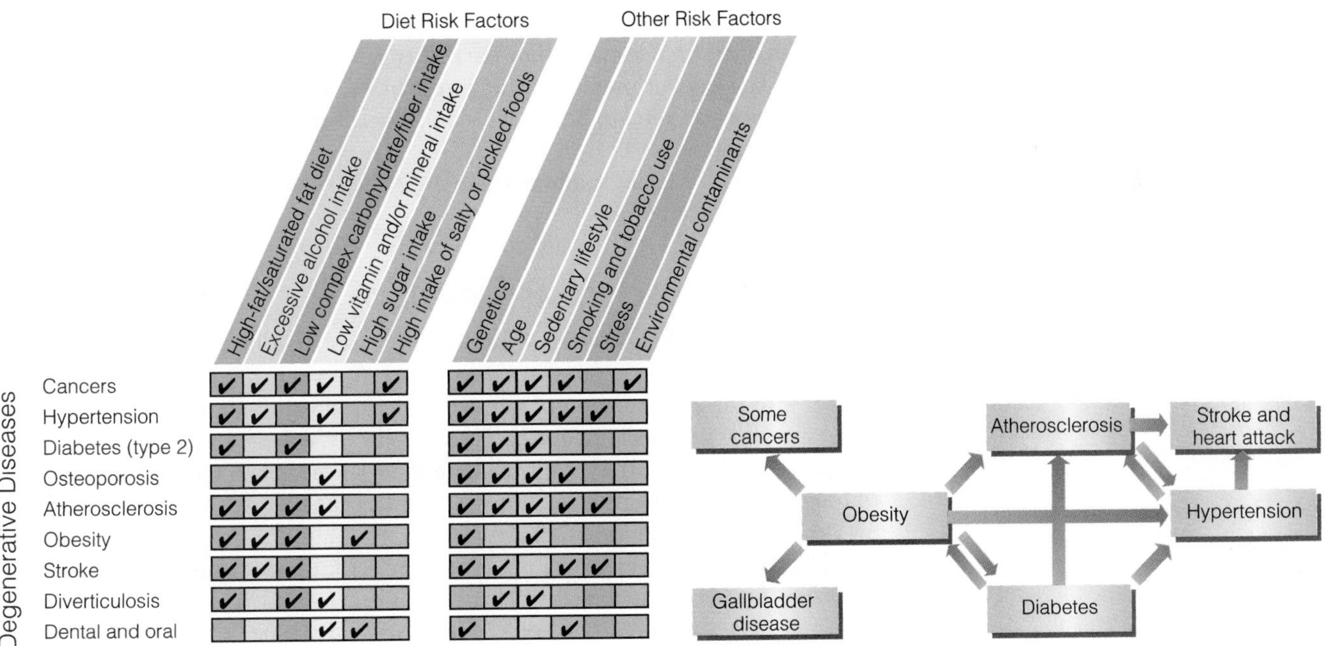

This chart shows that the same risk factor can affect many degenerative diseases. Notice, for example, how many diseases have been linked to a high-fat diet. The chart also shows that a particular disease, such as atherosclerosis, can have several risk factors.

This flowchart shows that many of these conditions are themselves risk factors for other degenerative diseases. For example, a person with diabetes is likely to develop atherosclerosis and hypertension. These two conditions, in turn, worsen each other. Notice how all of these degenerative diseases are linked to obesity.

atherosclerosis (ath-er-oh-scler-OH-sis) the most common form of cardiovascular disease; characterized by plaques along the inner walls of the arteries (*scleros* means "hard"; *osis* means "too much"). The term *arteriosclerosis* refers to all forms of hardening of the arteries and includes some rare diseases.

plaques (PLACKS) mounds of lipid material mixed with smooth muscle cells and calcium that develop in the artery walls in atherosclerosis (*placken* means "patch"). The same word is also used to describe the accumulation of a different kind of deposits on teeth, which promote dental caries.

A family history of these conditions in parents, grandparents, or siblings, especially when they occur early in life, may raise a warning flag for you:

- Alcoholism.
- Atherosclerosis.
- Cancer.
- Cardiovascular disease.
- Diabetes.
- Hypertension.
- Liver disease (cirrhosis).
- Osteoporosis.

Learn to recognize the signs of a heart attack. Should they occur, call for emergency medical help; in most areas, dial 911.

- *Chest discomfort:* Discomfort in the center of the chest that lasts more than a few minutes, or that goes away and comes back. An uncomfortable pressure, squeezing, fullness, or pain in the chest.
- *Discomfort in other areas of the upper body:* Pain or discomfort in one or both arms, the back, neck, jaw, or stomach.
- *Shortness of breath:* Accompanying chest discomfort or preceding it.
- *Other signs:* Cold sweat, nausea, or lightheadedness.

Call 911 also for these signs of stroke:

- *Sudden numbness or weakness* of the face, arm, or leg, especially on one side of the body.
- *Sudden confusion, trouble speaking,* or trouble understanding.
- *Sudden trouble seeing* in one or both eyes.
- *Sudden trouble walking,* dizziness, or loss of balance or coordination.
- *Sudden severe headache.*

Some choices, such as not smoking, are important to almost everyone's health. Other choices, such as those relating to diet, are more important for people who are genetically predisposed to certain diseases.[12] To pinpoint your own areas of concern, search your family's medical history for diseases common to your forebears.[13] Any condition that shows up in several close blood relatives may be a special concern for you. Also, after your next physical examination, find out which test results are out of line. Family history and lab test results together are powerful predictors of disease.

Accepting that everyone has certain unchangeable "givens," an effective strategy is to look to the things that can be changed and choose the most influential among them. For example, a person whose parents, grandparents, or other close blood relatives suffered from diabetes and heart disease is urgently advised to avoid becoming obese and not to smoke. The guidelines presented in later sections of this chapter can benefit most people.

The same diet and lifestyle risk factors may contribute to several degenerative diseases. A person's family history can reveal strategies for disease prevention.

Nutrition and Atherosclerosis

In developed countries, the major cause of death in men and women over 50 is disease of the heart and blood vessels (cardiovascular disease, abbreviated CVD). This remains true despite a steady and substantial downward trend in its occurrence since 1960.[14] CVD accounts for more of the world's deaths each year than any other single cause, mostly by way of heart attacks and strokes. The margin lists symptoms of these killers. Learning to recognize them can be lifesaving, for prompt medical attention is most effective.

How can you minimize your risks of CVD? Or, more positively, what actions can you take to help hold on to your cardiovascular health and vigor throughout life? Many people have changed their lifestyle to lower their risk—they have quit smoking or refrained from starting; they have changed their diets, consuming less fat, less saturated fat, less salt, more fruits and vegetables, and more fiber. Many people are still reluctant to exercise, however. Score yourself in Figure 11-5 to rate your own risks of developing CVD.

At the root of CVD is **atherosclerosis.** Atherosclerosis is the common form of hardening of the arteries.

How Atherosclerosis Develops

No one is free of atherosclerosis. The question is not whether you have it but how far advanced it is and what you can do to retard or reverse it. Atherosclerosis usually begins with the accumulation of soft, fatty streaks along the inner walls of the arteries, especially at branch points.[15] These gradually enlarge and become hardened **plaques** that damage artery walls, making them inelastic and narrowing the passage through them (see Figure 11-6 on p. 400). Most people have well-developed plaques by the time they reach age 30.[16]

What causes the plaques to form is a subject of intense scientific investigation. A suspected first step is an oxidative change that occurs in LDL particles. A destructive sequence may begin when the lipids of LDL become oxidized to form dangerous free-radical compounds—a condition known as *oxidative stress* (details in Controversy 7).[17] Inside the artery wall, immune cells (phagocytes, see Chapter 3) are attracted to the oxidized LDL and engulf them. Once engorged with oxidized LDL, the immune cells become known as foam cells, which themselves become sources of oxidation that attract more and more fresh immune scavengers to the scene. Smooth muscle cells of the arterial wall proliferate, mixing with the foam cells to form hardened areas of

Figure 11•5

ASSESS YOUR HEART DISEASE RISK

Do you know your heart disease risk score? Respond to the statements below, and score yourself as directed. Be aware that a high risk score does not mean you will develop heart disease, but it should warn you of the possibility. Consult your physician if you have questions about your score results.

In each category, circle the number next to the statement that's most true for you.

Cigarette Smoking

I never smoked or stopped smoking three or more years ago.	1
I don't smoke but live and/or work with smokers.	2
I stopped smoking within the past three years.	3
I smoke regularly.	4
I smoke regularly and live and/or work with other smokers.	5

Total Blood Cholesterol

Use the number from your most recent blood cholesterol measurement:

Less than 160	1
160–199	2
Don't know	3
200–239	4
240 or higher	5

HDL Cholesterol

Use the number from your most recent HDL cholesterol measurement:

Over 60	−1
50–59	0
Don't know	1
40–49	1
Less than 40	2

Systolic Blood Pressure

Use the first (highest) number from your most recent blood pressure measurement:

Less than 120	0
120–129	0
Don't know	3
140–159	4
160 or higher	5

Excess Body Weight[a]

I am within 10 pounds above my desirable weight.	1
I am 10–20 pounds above my desirable weight.	2
I am 21–30 pounds above my desirable weight.	3
I am 31–50 pounds above my desirable weight.	4
I am more than 50 pounds above my desirable weight.	5

Physical Activity

Determine which statements best describe your usual level of physical activity:

A: Highly Active
My job requires very hard physical labor (such as digging or loading heavy objects) at least four hours a day
or
I do vigorous activities (jogging, cycling, swimming, etc.) at least three times a week for 30 minutes or more
or
I do at least one hour of moderate activity such as brisk walking at least four days a week.

B: Moderately Active
My job requires that I walk, lift, carry, or do other moderately hard work for several hours a day (day-care worker, stock clerk, or busboy/waitress)
or
I spend much of my leisure time doing moderate activities (dancing, gardening, walking, or housework).

C: Inactive
My job requires that I sit at a desk most of the day
and
Much of my leisure time is spent in sedentary activities (watching TV, reading, etc.)
and
I seldom work up a sweat, and I cannot walk fast without having to stop to catch my breath.

Now circle the number that best describes your level of physical activity:

A: Highly Active	1
In between A and B	2
B: Moderately Active	3
In between B and C	4
C: Inactive	5

Scoring Your Heart Attack Risk

To learn your estimated risk, add the six numbers you've circled.

If Your Total Score Is:	*Your Heart Attack Risk Is:*
9–12	*Low*
13–20	*Moderate*
21–30	*High*

SOURCE: Adapted from the American Heart Association.

[a]At or below desirable weight: score equals zero.

plaque. The process is repeated until many inner artery walls become virtually covered with disfiguring plaques.[18]

Normally, the arteries expand with each heartbeat to accommodate the pulses of blood that flow through them. Arteries hardened and narrowed by plaques cannot expand, however, so the blood pressure rises. The increased pressure damages the artery walls further and strains the heart. Because plaques are more likely to form at damage sites, the development of atherosclerosis becomes a self-accelerating process.

As pressure builds up in an artery, the arterial wall may become weakened and balloon out, forming an **aneurysm.** An aneurysm can burst, and in a major artery such as the **aorta,** this leads to massive bleeding and death.

Abnormal blood clotting can also threaten life. Clots form and dissolve in the blood all the time, and the balance between these processes ensures that clots do no

aneurysm (AN-you-rism) the ballooning out of an artery wall at a point that is weakened by deterioration.

aorta (ay-OR-tuh) the large, primary artery that conducts blood from the heart to the body's smaller arteries.

platelets tiny cell-like fragments in the blood, important in blood clot formation (*platelet* means "little plate").

thrombus a stationary blood clot.

thrombosis a thrombus that has grown enough to close off a blood vessel. A *coronary thrombosis* is the closing off of a vessel that feeds the heart muscle. A *cerebral thrombosis* is the closing off of a vessel that feeds the brain (*coronary* means "crowning" [the heart]; *thrombo* means "clot"; the cerebrum is part of the brain).

embolus (EM-boh-luss) a thrombus that breaks loose (*embol* means "to insert").

embolism an embolus that causes sudden closure of a blood vessel.

heart attack the event in which the vessels that feed the heart muscle become closed off by an embolism, thrombus, or other cause with resulting sudden tissue death. A heart attack is also called a *myocardial infarction* (*myo* means "muscle"; *cardial* means "of the heart"; *infarct* means "tissue death").

stroke the sudden shutting off of the blood flow to the brain by a thrombus, embolism, or the bursting of a vessel (hemorrhage).

Chapter 5 described the effects of omega-6 and omega-3 fatty acids on heart health and identified some food sources of each.

Figure 11•6

THE FORMATION OF PLAQUES IN ATHEROSCLEROSIS

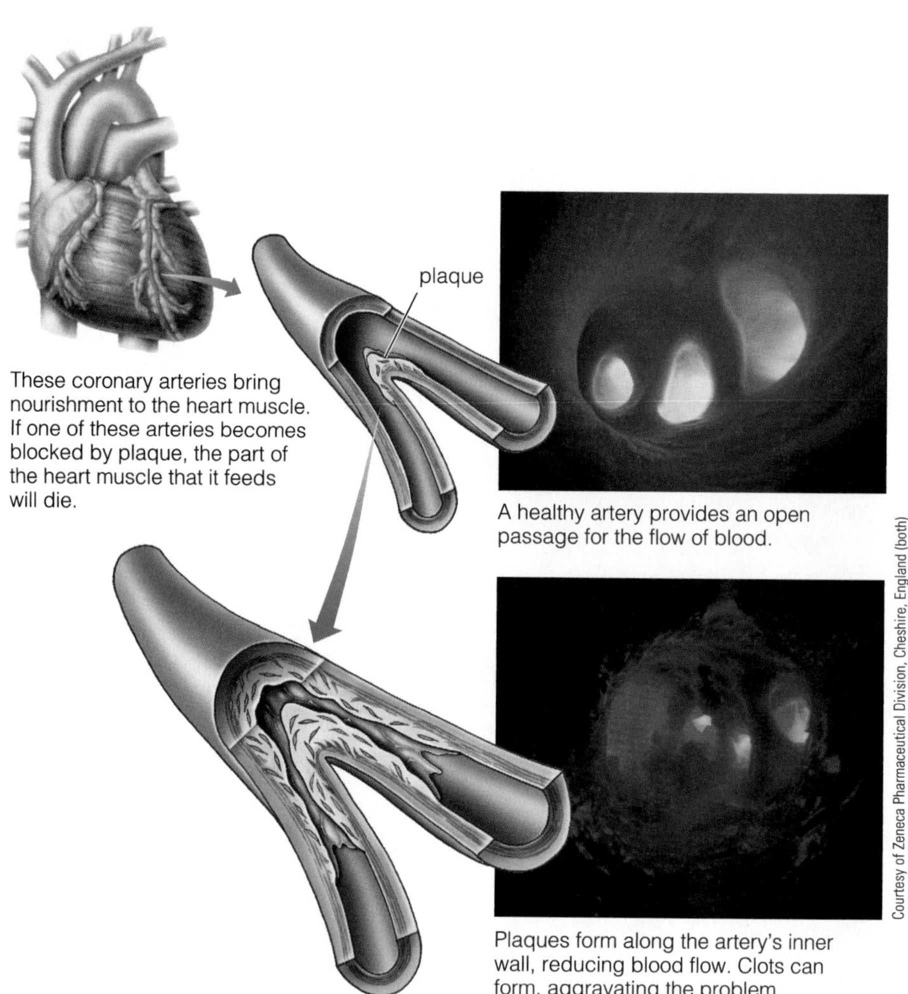

plaque
These coronary arteries bring nourishment to the heart muscle. If one of these arteries becomes blocked by plaque, the part of the heart muscle that it feeds will die.
A healthy artery provides an open passage for the flow of blood.
Plaques form along the artery's inner wall, reducing blood flow. Clots can form, aggravating the problem.

A blood clot in an artery, like the fatal heart embolism shown, blocks the blood flow to tissues fed by that artery.

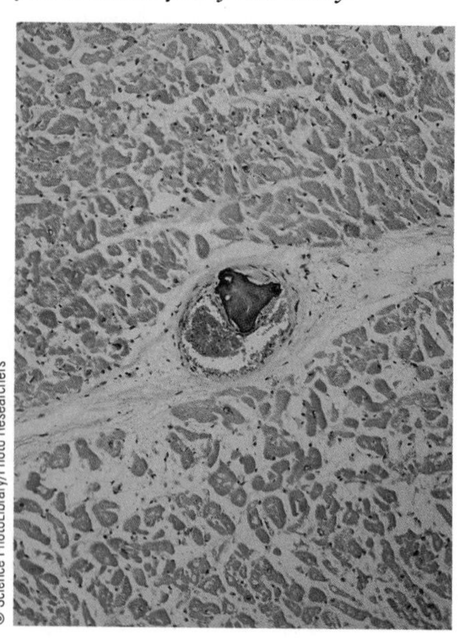

harm. That balance is disturbed in atherosclerosis. Small, cell-like bodies in the blood, known as **platelets,** cause clots to form when they encounter injuries in blood vessels. In atherosclerosis, the platelets respond to plaques as they do to injuries and form unneeded clots. Platelets also release substances that enlarge plaques. Opposing platelet action are the active products of omega-3 fatty acids. A diet lacking the seafoods that contain these essential fatty acids may contribute to clot formation.

A clot, once formed, may remain attached to a plaque in an artery and grow until it shuts off the blood supply to the surrounding tissue. That tissue may die slowly and be replaced by nonfunctional scar tissue. The stationary clot is called a **thrombus.** When it has grown large enough to close off a blood vessel, it is a **thrombosis.** A clot can also break loose, becoming an **embolus,** and travel along the system until it reaches an artery too small to allow its passage. There the clot becomes stuck and is referred to as an **embolism.** The tissues fed by this artery will be robbed of oxygen and nutrients and will die suddenly. Such a clot can lodge in an artery of the heart, causing sudden death of part of the heart muscle, a **heart attack.** A clot may also lodge in an artery of the brain, killing a portion of brain tissue, a **stroke.**

On many occasions, heart attacks and strokes occur with no apparent blockage. An artery may go into spasms, restricting or cutting off the blood supply to a portion of the heart muscle or brain. Much research today is devoted to finding out what causes plaques to form, what causes arteries to go into spasms, what governs the activ-

Table 11•2

RISK FACTORS FOR CVD

Six major medical risk factors for CVD:

1. Established CVD or diabetes, or family history of premature heart disease
2. Age (greater than 45 years for men, 55 years for women)
3. Cigarette smoking
4. High blood LDL cholesterol
5. Low blood HDL cholesterol
6. High blood pressure (hypertension)

Three lifestyle risk factors for CVD:

1. Obesity
2. Physical inactivity
3. An "atherogenic" diet that is high in saturated fats and low in vegetables, fruits, and whole grains

SOURCE: Executive summary of the third report of the National Cholesterol Education Program (NCEP) expert panel on detection, evaluation, and treatment of high blood cholesterol in adults (Adult Treatment Panel III), *Journal of the American Medical Association* 285 (2001): 2486–2497.

ities of platelets, and why the body allows clots to form unopposed by clot-dissolving cleanup activity.

Hypertension and atherosclerosis are twin demons that worsen CVD, and each worsens the other. Hypertension worsens atherosclerosis because a stiffened artery, already strained by each pulse of blood surging through it, is stressed further by high internal pressure. Injuries multiply, more plaques grow, and more weakened vessels become likely to burst and bleed.

Atherosclerosis also worsens hypertension. Since hardened arteries cannot expand, the heart's beats raise the blood pressure. Hardened arteries also fail to let blood flow freely through the kidneys, which control blood pressure. The kidneys sense the reduced flow of blood and respond as if the blood pressure were too low; they take steps to raise it further (see the discussion of hypertension later in the chapter).

key point *Plaques of atherosclerosis induce hypertension and trigger abnormal blood clotting, leading to heart attacks or strokes. Abnormal vessel spasms can also cause heart attacks and strokes.*

Risk Factors for CVD

Efforts to fight atherosclerosis and the resulting CVD have led to discoveries about their prevention. In 2001, an expert panel of the National Cholesterol Education Program defined major risk factors for CVD; already listed briefly in Chapter 5, they are presented in full in Table 11-2.[19] All people reaching middle age exhibit at least one of these factors (middle age is a risk factor), and many people have several factors, silently increasing their risks of CVD. Figure 11-7 on the next page gives one example of how rates of heart attacks for both men and women rise as risk factors mount. It befits a nutrition book to focus on dietary strategies to reduce these risks, but as Table 11-2 shows, diet is not the only, and perhaps not even the most important, factor in the development of CVD. Age, cigarette smoking, heart disease or family history of premature heart disease, diabetes, obesity, and physical inactivity predict CVD development as well.

The risk factors for CVD in Table 11-2 fall into two categories, "medical" and "lifestyle." In the medical group, the big *diet-related* risk factors are high blood LDL and low blood HDL (discussed in Chapter 5 and below), hypertension (discussed later in this chapter), and diabetes (discussed in Chapter 4 and in the following section). Among the lifestyle factors, those related to nutrition are an "atherogenic diet" (the opposite of the diet described in this chapter's Food Feature) and obesity (discussed in Chapter 9).

metabolic syndrome a combination of four risk factors—diabetes, obesity (especially central obesity), hypertension, and high blood cholesterol—that greatly increase a person's risk of developing CVD. Also called *insulin resistance syndrome* or *syndrome X*.

Figure 11•7

HOW RISK FACTORS COMPOUND A PERSON'S RISK OF HEART ATTACK

This graph shows how risk of heart attack rises in people with more than one risk factor. In the graph, "high cholesterol" is 260 or above, and "high blood pressure" is 150 or above (systolic pressure—the first figure in a blood pressure reading).

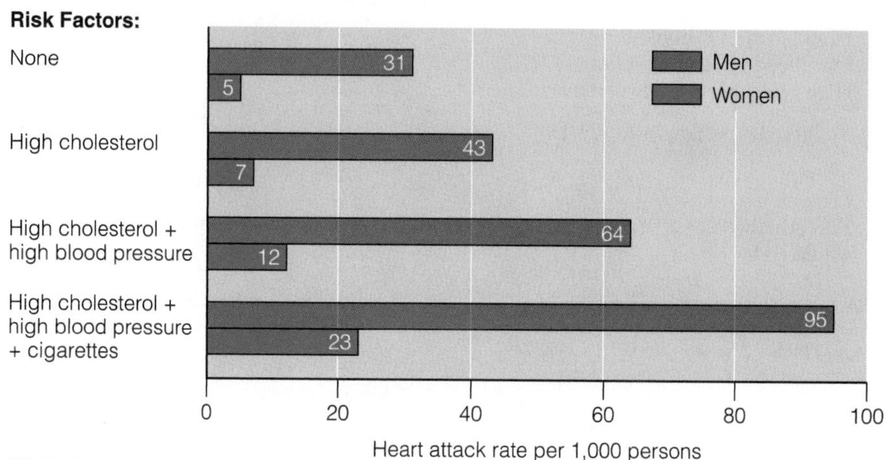

SOURCE: Framingham Heart Study. Personal communication, Thomas Thom, National Heart, Lung, and Blood Institute.

Chapter 7 discussed the scientific interest in the emerging links among the B vitamins, homocysteine, and atherosclerosis. Researchers are not yet ready to declare homocysteine an independent risk factor for CVD, however.[20]

Diabetes and Insulin Resistance Notice that the expert panel publishing the risk factors of Table 11-2 ranks having diabetes as equal to a previous diagnosis of CVD for predicting future CVD risk. This is because people with diabetes develop heart disease two to four times as often as others in the population, and when it occurs, it can be severe. The insulin resistance that characterizes type 2 diabetes is associated with an increased ratio of blood LDL to HDL and increased levels of triglycerides, which worsen atherosclerosis. Diabetes also damages the nerves of the heart, increasing the likelihood of a painless heart attack that is difficult to detect in time to save the person's life. Stroke is also more frequent among people with diabetes.

Insulin resistance was defined in Chapter 4 as a condition in which a normal amount of insulin produces a less-than-normal response in the tissues.

According to a 2001 survey by the American Heart Association, 63 percent of people with diabetes reported having been diagnosed with CVD, yet few recognize this threat to their health.[21] Even without diabetes, people with insulin resistance may face an elevated risk of heart disease.[22] About a third of middle-aged men may have a symptomless type of insulin resistance that greatly increases the likelihood of a heart attack—*without* elevated cholesterol values that might otherwise warn them.

Metabolic syndrome is also called *insulin resistance syndrome* or *syndrome X*.

Metabolic Syndrome People suffering from a constellation of four risk factors—central obesity, diabetes or altered insulin response, atherogenic blood lipids, and hypertension—have an especially high risk of developing cardiovascular disease. Each factor elevates CVD risk independently, and when they occur together, they synergistically elevate the risk.[23] This "deadly quartet" is often called **metabolic syndrome** by researchers and is a target for intensive therapy second only in importance to LDL cholesterol in modifying the risk of CVD.[24]

Researchers suspect that many people with metabolic syndrome may possess an underlying genetic predisposition that is expressed when diet and exercise habits result in obesity.[25] Experimentally, when people with the syndrome switched to a diet low in fats and rich in unrefined grains and vegetables, their body composition improved along with indicators of the deadly quartet, reducing their risk of CVD.[26] As appealing as this may sound, diet alone may not reduce CVD risk substantially enough in most people to avert a heart attack or stroke. Other actions, such as quitting smoking and taking up physical activity, are also needed to lower the risk.

Table 11•3

ADULT STANDARDS FOR BLOOD LIPIDS, BODY MASS INDEX (BMI), AND BLOOD PRESSURE

	Low CVD Risk	Borderline	Elevated CVD Risk
Total blood cholesterol (mg/dL)	<200	200–239	≥240
LDL cholesterol (mg/dL)	<100[a]	130–159	160–189[b]
HDL cholesterol (mg/dL)	≥60	59–40	<40
Triglycerides, fasting (mg/dL)	<150	150–199	200–499[c]
Body mass index (BMI)[d]	18.5–24.9	25–29.9	≥30
Blood pressure (systolic and/or diastolic pressure)	<120–129/<80–84	130–159/85–99	≥160/≥100

[a]100–129 mg/dL LDL indicates a near or above optimal level.
[b]≥190 mg/dL LDL indicates a very high risk.
[c]≥500 mg/dL triglycerides indicates very high risk.
[d]Body Mass Index (BMI) was defined in Chapter 9; BMI standards are found on the inside back cover.

Table 11-3 shows the standards by which blood lipids, blood pressure, and obesity are evaluated. Almost half of all deaths from CVD occur among men with blood cholesterol in the borderline range, so only the lowest values, if any, are "safe." The moral of the story: everyone, even those with normal cholesterol values, should follow the diet and exercise advice for reducing CVD risk.

A Word about Blood Triglycerides Though not named among the independent risk factors for heart disease in Table 11-2 (page 401), triglycerides are often elevated in the blood of people with CVD. Elevated blood triglycerides often accompany insulin resistance and reduced HDL levels, especially among the overweight.[27] In people with diabetes, central obesity, artery disease, hypertension, or kidney disease, elevated triglycerides may worsen atherosclerosis and accelerate clotting activity while slowing clot destruction in the blood. Experts suggest that even though some people with elevated triglycerides suffer no apparent harm, weight reduction in the overweight and increased physical activity should be recommended to lower even borderline triglyceride values (see Table 11-3).[28] Higher values may require drug therapy to bring them down.

What Is the Significance of High Blood Cholesterol? High blood cholesterol generally reflects elevated LDL cholesterol and predicts CVD. A population whose average blood cholesterol is 10 percent lower than another population's will suffer one-third less CVD; a 30 percent difference in blood cholesterol predicts a CVD rate that is four times lower. Table 11-3 showed that cholesterol carried in LDL correlates *directly* with risk of heart disease, whereas that carried in HDL correlates *inversely* with risk. The higher the ratio of LDL to HDL, the greater the risk (see Figure 11-8).

The exact reasons why high LDL increase the risk of CVD remain unclear. As mentioned earlier, however, mounting evidence suggests that atherosclerosis is accelerated by the oxidation of LDL by free radicals.[29] (Controversy 7 described how antioxidant nutrients and phytochemicals protect against free-radical formation and its damaging consequences.)

Diet and Blood Cholesterol Now, how does *diet* relate to high blood cholesterol? Diet's effects are felt in two opposing ways: first, a diet high in saturated fat and *trans*-fatty acids contributes to high blood LDL cholesterol, and second, reducing those fats in the diet lowers blood LDL cholesterol and may reduce the rate of CVD.[30]

Generally, wherever in the world diets are high in saturated fat and low in fish, fruits, and vegetables, blood cholesterol is high, and heart disease takes a great toll on health and life. Conversely, wherever dietary fat consists mostly of monounsaturated fats with abundant fish, fruits, and vegetables, blood cholesterol and the rate of death from heart disease are low.

Figure 11•8

LDL TO HDL RATIO AND RISK OF HEART DISEASE

Low HDL relative to LDL increases risk

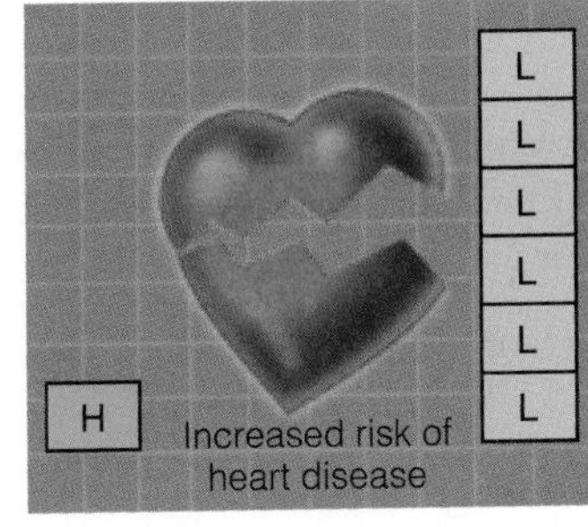

High HDL relative to LDL decreases risk

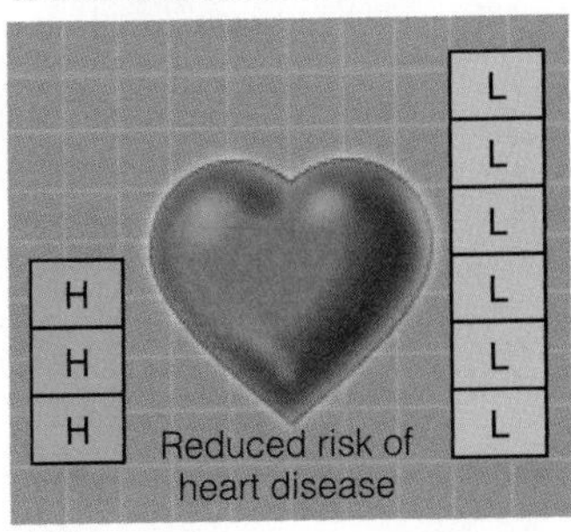

Table 11•4

AMERICAN HEART ASSOCIATION'S GOALS TO MINIMIZE RISKS OF CVD IN HEALTHY PEOPLE[a]

A Healthy Eating Pattern Including Foods from All Major Food Groups

- Consume a variety of fruits and vegetables and grain products, including whole grains.
- Include fat-free and low-fat dairy products, fish, legumes, poultry, and lean meats.

A Healthy Body Weight

- Match intake of energy (calories) to overall energy needs; limit consumption of foods with a high caloric density and/or low nutritional quality, including those with a high content of sugars.
- Maintain a level of physical activity that achieves fitness and balances energy expenditure with energy intake; for weight reduction, expenditure should exceed intake.

A Desirable Blood Cholesterol and Lipoprotein Profile

- Limit the intake of foods with a high content of saturated fatty acids and *trans*-fatty acids (<10 percent of total calories) and cholesterol (<300 milligrams per day).[b]
- Substitute unsaturated fats (both long-chain omega-3 polyunsaturated and monounsaturated fatty acids) from vegetables, fish, and nuts.

A Desirable Blood Pressure

- Limit the intake of salt (sodium chloride) to <6 grams per day.
- Limit alcohol consumption (no more than 1 drink per day for women and 2 drinks per day for men).
- Maintain a healthy body weight and a dietary pattern that emphasizes vegetables, fruits, and low-fat or fat-free milk products.

[a]People over 2 years of age.
[b]For individuals with elevated LDL cholesterol, cardiovascular disease, diabetes, or combinations of risk factors, saturated fat intakes should be less than 7 percent of total calories and cholesterol should be less than 200 milligrams per day.
SOURCES: American Heart Association Nutrition Committee, Dietary guidelines for healthy American adults, *Circulation* 102 (2000): 2284–2299; Third Report of the National Cholesterol Education Program Expert Panel on Detection, Evaluation, and Treatment of High Blood Cholesterol in Adults (Adult Treatment Panel III), *Journal of the American Medical Association* 285 (2001): 2486–2497.

More about the Mediterranean diet in Controversy 5; food sources of saturated fat and *trans*-fatty acids were listed in Chapter 5.

When diets are rich in vegetables and fruits, life expectancies are long.

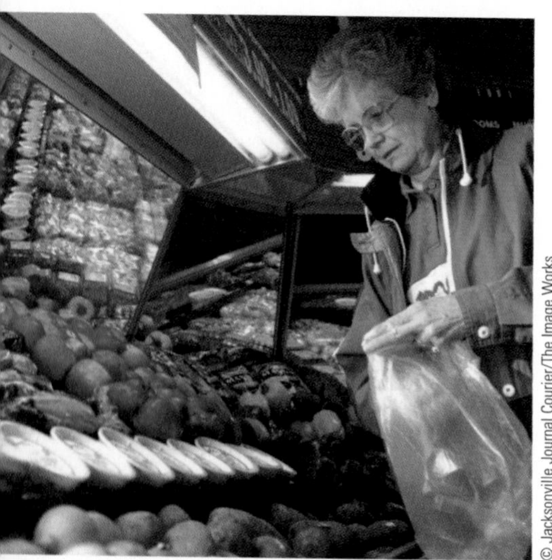

The bulk of research supports the idea that lowering saturated fat intakes will lead to lower blood cholesterol and reduced heart disease risks.[31] In addition, restricting intakes of *trans*-fatty acids along with saturated fats seems important for reducing heart disease risk.[32] Most authorities agree that for people living in the United States and Canada, saturated fat, including *trans*-fatty acids, should account for no more than 10 percent of calories. The American Heart Association recommends limiting the combined daily intake of saturated fat and *trans*-fatty acids to a level likely to be safe for most healthy people (see Table 11-4).[33]

Other recommendations urge that people limit total intakes of fats and oils to about 5 to 8 teaspoons (or 30 percent of total calories) and limit cholesterol from foods to 300 milligrams a day. These measures may be more important for some people than others. The links between intakes of total dietary fat, dietary cholesterol, and high blood cholesterol are not as firm as the links between saturated fat, *trans*-fatty acids, high blood cholesterol, and CVD. Data on the people of Mediterranean countries illustrate that diets high in total fat can coexist with low rates of heart disease so long as the diet is rich in fish, fruits, and vegetables and the fat is of the monounsaturated type.[34] Most people in the United States, however, eat diets rich in meats and hydrogenated fats, so if they reduce the *total* fat in their diets, no doubt their *saturated* fat and their *trans*-fatty acids will follow suit.

As previous chapters have already mentioned, other dietary factors seem to lower blood cholesterol or improve heart disease risks in other ways. Table 11-5 recounts them. Meanwhile food manufacturers are working to develop foods that have beneficial effects. The Controversy of Chapter 2 discussed the effects of margarines with sterol esters added. These compounds block absorption of cholesterol from the intestine and so may lower blood cholesterol by about 7 to 10 percent. Margarines made with stanol or sterol esters may be of use to those with elevated blood cholesterol or other risks of heart disease but only in the context of a low-fat diet and other cholesterol-controlling measures. Modified functional foods that contain novel ingredients should not be fed to children until the effects of doing so are determined by research.

To return to the main points: (1) high blood cholesterol indicates a risk of heart disease, and (2) it is possible to lower blood cholesterol, in part, by controlling dietary saturated fat. If people lower their blood cholesterol, they will reduce their risk of heart disease.

key point *Diet-related risk factors for CVD include high LDL and low HDL, hypertension, diabetes, an atherogenic diet, and obesity. The combination of risk factors called metabolic syndrome carries an especially high risk of CVD. Dietary measures to reduce fat, saturated fat, and cholesterol intakes are part of the first line of treatment for high blood cholesterol.*

What Else, Besides Controlling Diet, Can I Do to Reduce My Risk of CVD?

The *Healthy People 2010* goals for the nation include a reduction in deaths from CVD. Beyond the major role of a heart-healthy diet, other lifestyle factors, such as stopping smoking (or not starting), increased physical activity, and perhaps for some the moderate use of alcohol, may also be protective.[35] For those with established heart and artery disease, diabetes, or other high-risk conditions, medical therapies are essential as well.

Physical Activity Physical activity merits special attention. Endurance activities, such as brisk walking or jogging, are effective in lowering LDL and raising HDL concentrations. Resistance exercise, such as lifting weights, can contribute muscle

Table 11•5

DIETARY FACTORS PROTECTING AGAINST CVD

In addition to the dietary interventions mentioned in Table 11-4, these dietary factors may also protect against CVD.

Dietary Factor	Protection against CVD
Soluble fiber (apples and other fruits, oats, soy, barley, legumes)	• Lowers blood cholesterol, especially in those with high cholesterol • Lowers risk of heart attack • Improves LDL-to-HDL ratio
Omega-3 fatty acids (fish oils)	• Limit clot formation • Prevent irregular heartbeats • Lower risk of heart attack
Alcohol (in moderation)	• Raises HDL • Prevents clot formation
Folate, vitamin B_6, vitamin B_{12}	• Reduce homocysteine
Vitamin E (vegetable oils and margarines, some nuts, wheat germ)	• Slows progression of plaque formation • Lowers risk of heart attack in people with CVD • Limits LDL oxidation
Soy (protein and isoflavones)	• Lowers blood cholesterol • Raises HDL cholesterol • Improves LDL-to-HDL ratio

strength and improve the heart's response to everyday workloads.[36] Particularly when combined with a low-fat diet, physical activity may even help to reverse atherosclerosis. Men with moderate-to-high levels of cardiovascular fitness (determined by a treadmill test) were found to have a lower risk of dying from heart attack than men who were less fit, even when the fit men had other risk factors such as high blood cholesterol.[37] If pursued daily, even 30 minutes of light exercise, performed at intervals throughout the day, can improve the odds against heart disease considerably. The Think Fitness feature offers suggestions for incorporating physical activity into your daily routine.

Ways to Include Physical Activity in a Day

By now you know about the benefits of physical activity, so why not tie up your athletic shoes, head out the door, and get going? Here are some ideas to get you started:

- Coach a sport.
- Garden.
- Hike, bike, or walk to nearby stores or to classes.
- Mow, trim, and rake by hand.
- Park a block from your destination and walk.
- Play a sport.
- Play with children.
- Take classes for credit in dancing, sports, conditioning, or swimming.
- Take the stairs, not the elevator.
- Walk a dog.
- Walk 10,000 steps per day. This amounts to about 5 miles, enough to meet the daily activity that defines an active person. An inexpensive pedometer can record your steps.
- Wash your car with extra vigor, or bend and stretch to wash your toes in the bath.
- Work out at a fitness club.
- Work out with friends who help one another stay fit.

Also, try these:

- Give two labor-saving devices to charity.
- Lift small hand weights while talking on the phone or watching TV.
- Stretch often during the day.

THINK FITNESS

In addition to helping normalize blood lipids, regular exercise offers many other benefits. It can strengthen the heart and blood vessels, alter body composition in favor of lean over fat tissue, lower blood pressure, improve insulin response, and expand the volume of blood the heart can pump to the tissues at each beat and so reduce the heart's workload. Physical activity also stimulates development of new arteries to nourish the heart muscle, which may be a factor in the excellent recovery seen in some heart attack victims who exercise. These changes are so beneficial that some experts believe that physical activity should be the primary focus of CVD prevention efforts.[38]

Obesity worsens many disease risks, as discussed in Chapter 9. Chapter 10 specified exercise guidelines for health.

Both physical activity and the weight loss it induces raise HDL concentrations, and the effects of these two factors are additive. If exercise also reduces central obesity, the result is exceptionally beneficial because central obesity is an important determinant of CVD risk.[39]

Diet helps a little, physical activity helps more, and the combination is better still. People in a clinical setting have been able to reduce plaque buildup in their arteries by following a strict plan combining an extremely low-fat diet (less than 10 percent of calories from fat), no smoking, stress management, and exercise.[40] Without this program, atherosclerosis would likely have progressed; instead, it regressed and allowed the participants to avoid or postpone heart surgery.

Alcohol Consumption People want to know whether moderate consumption of alcohol will reduce their CVD risk. The answer is, "maybe, maybe not." Research on middle-aged or older people who drink one or two drinks a day, with no binge drinking supports the idea.[41] However, the effect is small and confined to certain population groups (Controversy 3 gave details). Also, the mechanism of action is unknown, although anticlotting effects, elevation of HDL, antioxidant effects of wine, and others are under study. Because of alcohol's destructive potential authorities do not suggest increasing U.S. alcohol intakes to reduce cardiovascular disease.

Heavy alcohol use (more than three drinks a day) is known to elevate blood pressure, to damage the heart muscle, and to have many other deleterious effects on the body's organs. Heart attacks among apparently healthy young people have been associated with alcohol intoxication from heavy weekend drinking.[42] Even moderate drinking (one drink a day for women) on a regular basis increases the risk of death from breast cancer, and heavy drinking is associated with a significant increase in deaths from cancers and other causes. A later section examines alcohol's link with cancer in more detail. The bottom line for young people is that the risks of consuming alcohol greatly outweigh any benefit to the heart, and they do their health no favor by drinking alcohol.

More details about alcohol's effects on the body are in Chapter 3's Controversy.

More Strategies Drug therapy can lower blood cholesterol, but it also presents risks and side effects that accumulate during years of therapy. The most common of these drugs, the "statin" drugs, efficiently lower the blood LDL values of users. Cholesterol-lowering drugs seem to work best in association with other efforts, such as proper diet and exercise.[43]

Periodically, the media repopularize the idea that the vitamin niacin can lower blood cholesterol. Experimentally, pharmaceutical doses of a specific form of niacin act like a drug in lowering blood cholesterol and prolonging life, but other drugs effective for this purpose probably have fewer side effects. Ordinary niacin supplements are useless in lowering blood cholesterol.

Although diet and exercise are not the easy route to heart health that everyone hopes for, they form a powerful and safe combination for improving health. Weight control may reduce blood pressure. So will eating a diet low in fat, restricted in cholesterol, and high in complex carbohydrates, whole grains, fruits, and vegetables. And even if such a diet does not lower cholesterol or blood pressure, it will help by normalizing blood glucose (diabetes). Remember, diabetes is a major risk factor for CVD. A meal of fish each week can help by favoring the right fatty acid balance so that clot formation is unlikely. The pattern of protection from the recommended diet and exercise regimen becomes clear—the effects of each small choice add to the beneficial whole. While you are at it, don't smoke. Relax. Meditate or pray. Control stress.[44] Play. Happy people have lower blood cholesterol levels.

Every day on average:

- 4 college students die from alcohol-related accidents.
- 192 college students are raped by their dates or sexually assaulted after drinking alcohol.
- 1,370 college students suffer drinking-related injuries.

Over 80 percent of college students say they drink.

SOURCE: National Institute of Alcohol and Alcoholism, *A Call to Action: Changing the Culture of Drinking at U.S. Colleges*, 2002 available from www.collegedrinkingprevention.gov/Reports/

Physical activity can reduce CVD risk. Moderate alcohol intake may also be associated with reduced risk, but its use can be problematic.

Nutrition and Hypertension

People with healthy low blood pressure generally enjoy a long life and suffer less often from heart disease. Chronic high blood pressure, or hypertension, remains one of the most prevalent forms of CVD, affecting about a quarter of the entire U.S. adult population, with the lifetime risk of developing it approaching 90 percent among people age 65 or older.[45] It contributes to half a million strokes and to over a million heart attacks each year, and its rate has been rising steadily over the past four decades.[46] The higher above normal the blood pressure, the greater the risk of heart disease. Paired with atherosclerosis, as it often is, hypertension is especially threatening.

systolic (sis-TOL-ik) **pressure** the first figure in a blood pressure reading (the "dub" of the heartbeat), which reflects arterial pressure caused by the contraction of the heart's left ventricle.

diastolic (dye-as-TOL-ik) **pressure** the second figure in a blood pressure reading (the "lub" of the heartbeat), which reflects the arterial pressure when the heart is between beats.

You cannot tell if you have high blood pressure—it presents no symptoms you can feel. The most effective single step you can take to protect yourself from hypertension is to find out whether you have it. At checkup time, a health-care professional can take an accurate resting blood pressure reading. Self-test machines in drugstores and other places are often inaccurate. If your resting blood pressure is above normal, the reading should be repeated before confirming the diagnosis of hypertension. Thereafter blood pressure should be checked at regular intervals.

When blood pressure is measured, two numbers are important: the pressure during contraction of the heart's ventricles (large pumping chambers) and the pressure during their relaxation. The numbers are given as a fraction, with the first number representing the **systolic pressure** (ventricular contraction) and the second number the **diastolic pressure** (relaxation). Return to Table 11-3 to see how to interpret your resting blood pressure.

Ideal resting blood pressure is 120 over 80 or lower, but a reading of 130 over 85 can be considered borderline normal, especially in the absence of other risk factors for CVD, such as diabetes.[47] Above this level, though, the risks of heart attacks and strokes increase in direct proportion to increasing blood pressure.

Hypertension is silent, progressively worsens atherosclerosis, and makes heart attacks and strokes likely. All adults should know their blood pressure.

How Does Blood Pressure Work in the Body, and What Makes It Too High?

Blood pressure is vital to life. It pushes the blood through the major arteries into smaller arteries and finally into tiny capillaries whose thin walls permit exchange of fluids between the blood and the tissues (see Figure 11-9). When the pressure is right, the cells receive a constant supply of nutrients and oxygen and can release their wastes.

The most effective single step you can take against hypertension is to learn your own blood pressure.

© Michael Keller/Index Stock Imagery

The Role of the Kidneys For the kidneys to filter waste materials out of the blood into the urine, blood pressure has to be high enough to force the blood's fluid out of the capillaries into the kidneys' filtering networks. If the blood pressure is too low, the kidneys act to increase it—they send hormones to constrict the peripheral blood vessels and to retain water and salt in the body. Dehydration sets these actions in motion, and in this case they are beneficial because when the blood volume is low, higher blood pressure is needed to deliver substances to the tissues. By constricting the blood vessels and conserving water and sodium, the kidneys ensure that normal blood pressure is maintained until the dehydrated person can drink water.

Atherosclerosis also sets this process in motion, however, and this is not beneficial. By obstructing blood vessels, atherosclerosis fools the kidneys, which react as if there were a water deficiency. The kidneys raise the blood pressure high enough to get the blood they need, but in the process they may make the pressure too high for the arteries and heart to withstand. Hypertension also aggravates atherosclerosis by mechanically injuring the artery linings, making plaques likely to form; plaques restrict blood flow to the kidneys, which may then act to raise the blood pressure still further; and the problem snowballs.

The Roles of Risk Factors Primary among the risk factors that precipitate or aggravate hypertension are atherosclerosis, obesity (particularly central obesity), and insulin resistance (which leads to type 2 diabetes).[48] Excess fat means miles of extra capillaries through which the blood must be pumped.[49]

Epidemiological studies have identified several other risk factors that predict hypertension. One is age: most people who develop hypertension do so in their 50s and 60s. Another is inherited genes, and researchers are working to ascertain the genetic determinants of hypertension.[50] A family history of hypertension and heart disease raises the risk of developing hypertension two to five times, and people of

Figure 11•9

THE BLOOD PRESSURE

Three major factors contribute to the pressure inside an artery. First, the heart pushes blood into the artery. Second, the small-diameter arteries and capillaries at the other end resist the blood's flow (peripheral resistance). Third, the volume of fluid in the circulatory system, which depends on the number of dissolved particles in that fluid, adds pressure.

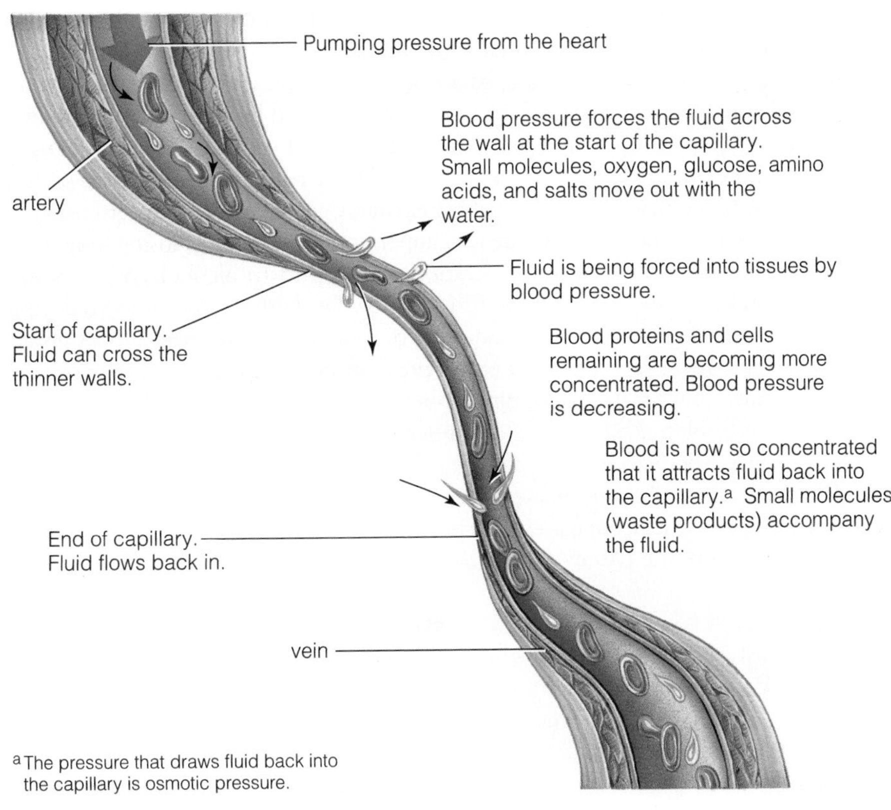

Pumping pressure from the heart
artery
Blood pressure forces the fluid across the wall at the start of the capillary. Small molecules, oxygen, glucose, amino acids, and salts move out with the water.
Fluid is being forced into tissues by blood pressure.
Start of capillary. Fluid can cross the thinner walls.
Blood proteins and cells remaining are becoming more concentrated. Blood pressure is decreasing.
Blood is now so concentrated that it attracts fluid back into the capillary.[a] Small molecules (waste products) accompany the fluid.
End of capillary. Fluid flows back in.
vein

[a]The pressure that draws fluid back into the capillary is osmotic pressure.

African American descent are likely to develop more severe hypertension earlier in life than people of European or Asian descent. Hypertension also bears some relation to insulin resistance, probably through a common genetic link, and measures to prevent diabetes may also protect against hypertension.

Environmental factors in the United States may also favor the development of hypertension. Africans living in Africa have a much lower rate of hypertension than do African Americans living in the United States.

Atherosclerosis, obesity, insulin resistance, age, family background, and race contribute to hypertension risks.

How Does Nutrition Affect Hypertension?

Even mild hypertension can be dangerous, but individuals who adhere to treatment are less likely to suffer illness or early death. Some people need medications to bring their blood pressure down, but diet and exercise alone can bring improvements for many and prevent hypertension for many others.[51] This section focuses on diet (salt intake, alcohol consumption, calcium, and magnesium) and on controlling weight and physical activity.

Salt (Sodium) and Prevention A controversy surrounds the relationship between excessive salt and sodium in the diet and hypertension. The benefit from reducing salt intake in *treatment* of hypertension is not questioned. For about half of people with hypertension, a lower salt (or sodium) intake leads to a reduction in blood pressure. Such people are said to be salt-sensitive, and they often fit one or more of these descriptions: have a family history of hypertension, are African American, have kidney problems or diabetes, are older, or have experienced sustained psychological stress.[52]

As for the role of salt in *prevention* of hypertension, the evidence is mixed. One major worldwide epidemiological study found that blood pressure rises whenever sodium or salt intakes increase.[53] Other studies, though, seem to suggest that the degree of sodium restriction needed to produce a meaningful reduction in blood pressure is neither achievable nor sustainable in our population.[54] The recommendation of many professionals and agencies, based on available research, is that everyone should moderately restrict salt intake to the level suggested by the *Dietary Guidelines for Americans*—that is, no more than 2,400 milligrams of sodium (6 grams salt) per day—and follow other standard dietary advice.[55] They reason that, at worst, such a diet cannot be harmful, and it will probably help many people to avoid hypertension.[56]

Details concerning sodium, salt-sensitivity, and hypertension were presented in Chapter 8.

Nutrition factors other than salt are also important, most notably, losing weight for those who are overweight; using moderation with regard to alcohol consumption; increasing intakes of fruit, vegetables, fish, and low-fat dairy products; and reducing intakes of fat. Calcium, magnesium, and other nutrients also play roles, as does physical activity. A blanket recommendation for prevention of hypertension, then, would center on controlling weight, consuming a nutritious diet, exercising, reducing intakes of alcohol, and holding salt intakes to prescribed levels. One such diet, known as DASH (Dietary Approaches to Stop Hypertension, mentioned in Chapter 8), recommends more servings of fruits and vegetables than the Food Guide Pyramid, provides 30 percent of its calories from fat, emphasizes legumes over red meats, and meets other recommendations of the *Dietary Guidelines* besides (see Table 11-6 in the margin).

Weight Control and Physical Activity For people who are overweight and hypertensive, a weight loss of as little as 10 pounds can significantly lower blood pressure.[57] Those who are using drugs to control their blood pressure can often cut down their doses if they lose weight.[58]

Moderate physical activity helps in weight loss and also helps to reduce hypertension directly. The right kind of regular physical activity can lower blood pressure in almost everyone, even in those without hypertension. Even a single session of exercise reduces the blood pressure, an effect that lasts 12 hours or more and intensifies as training improves physical condition.[59] The "right kind" of activity is the prolonged kind, such as walking, jogging, or cycling, also observed to increase blood HDL and lower LDL.[60] The exercise need not be strenuous—even substantial amounts of walking (about 5 miles per day) at any pace seem effective for bringing down high blood pressure.[61] Aerobic activity for cardiovascular fitness was described in Chapter 10.

Physical activity also changes the hormonal climate in which the body does its work. By reducing stress, physical activity reduces the secretion of stress hormones, and this lowers blood pressure. Physical activity also redistributes body water and eases transit of the blood through the peripheral arteries.

Table 11•6

THE DASH EATING PLAN AND THE FOOD GUIDE PYRAMID COMPARED

Food Group	Recommended Number of Daily Servings: DASH	Pyramid
Grains	7–8	6–11
Vegetables	4–5	3–5
Fruits	4–5	2–4
Milk (nonfat/low-fat)	2–3	2–3
Meat (lean)[a]	2 or less	2–3
Calories	2,000	1,600–2,800

NOTE: The DASH eating plan, like the Food Guide Pyramid, recommends that fats, oils, and sweets be used sparingly.

[a]The DASH eating plan also includes recommended servings for nuts, seeds, and dry beans (4 to 5 per week), whereas the Food Guide Pyramid includes these foods with the meat group.

Alcohol In moderate doses, alcohol initially relaxes the peripheral arteries and so reduces blood pressure, but high doses definitely raise blood pressure. Hypertension is common among people with alcoholism and is apparently caused directly by the alcohol. Hypertension caused by alcohol leads to CVD, the same as hypertension caused by any other factor. Furthermore, alcohol may cause strokes—even *without* hypertension.[62] The *Dietary Guidelines* urge moderation for those who drink alcohol. *Moderation* means no more than one drink a day for women or two drinks a day for men, an amount that seems safe relative to blood pressure.

Calcium, Magnesium, Potassium, and Vitamin C Other dietary factors may help to regulate blood pressure.[63] A diet providing enough calcium is certainly one such factor—calcium reduces blood pressure in both healthy people and those with hypertension.[64] Adding calcium-rich foods may lower blood pressure even in salt-sensitive people whose sodium intakes remain high.[65] If you are concerned about your blood pressure, include more calcium-rich foods in your diet.

cancer a disease in which cells multiply out of control and disrupt normal functioning of one or more organs.

Adequate potassium and magnesium also appear to help prevent and treat hypertension in certain populations. Diets low in potassium are often associated with hypertension, whereas high-potassium diets appear to both prevent and correct hypertension.[66] Magnesium deficiency causes the walls of the arteries and capillaries to constrict and so may raise the blood pressure. Similarly, vitamin C adequacy seems to help normalize blood pressure, while vitamin C deficiency may tend to raise it.[67]

How can people be sure of getting all of the nutrients needed to keep blood pressure low? The best answer is to consume a low-fat diet with abundant fruits, vegetables, and low-fat dairy products that provide the needed magnesium, potassium, vitamin C, and calcium.[68] In addition to reducing blood pressure, a diet such as the DASH diet mentioned earlier may also lower blood cholesterol values, providing a two-way benefit to the heart.[69]

Other dietary factors may also affect blood pressure; the roles of cadmium, selenium, lead, caffeine, protein, and fat are being studied. The Food Feature provides more details on dietary measures that help support normal blood pressure. Should diet and exercise fail to reduce blood pressure, drugs such as diuretics and other antihypertensive agents may be prescribed. Some of these work by increasing fluid loss in the urine, and they may also cause potassium losses. Chapter 8 specified foods that are rich in potassium, and people taking these drugs should make it a point to consume such foods daily.

key point *For most people, a healthy body weight, regular physical activity, moderation for those who use alcohol, and a diet high in fruits, vegetables, fish, and low-fat dairy products and low in fat work together to keep blood pressure normal. For some, salt restriction is also required.*

Nutrition and Cancer

Cancer ranks second only to heart disease as a leading cause of death and disability in the United States (see Figure 11-10, next page). Recently, advances in treatment have brought new hope that science may one day prevail over cancer. Produced through genetic engineering, new drugs can target and interrupt molecular processes in specific types of cancers. Unlike traditional chemotherapy and radiation treatments that injure healthy surrounding tissues, the new targeted therapies kill only cancerous cells, producing far fewer side effects. The results from early trials are promising. In one experiment, people with treatment-resistant leukemia benefited from a new targeted drug combining cancer-seeking antibodies from the immune system with the killing power of a bacterial toxin. The antibodies locate cancer cells among billions of healthy cells, and the toxin snuffs them out.[70]

The basics of genetic engineering are described in Controversy 14.

Though the potential for cure is exciting, prevention of cancer remains far and away preferable. Can an individual's chosen behaviors affect the risk of contracting cancer? Probably. Inherited tendencies exert only a modest effect on most people's risk of cancer development.[71] A very few rare cancers are known to be caused by genetics alone and will appear in members of an affected family regardless of lifestyle choices, but far more often behavioral and environmental factors play a role. For example, if everyone in the United States quit smoking right now and stayed quit, future cancer rates would probably drop by a third. Environmental tobacco smoke (secondhand smoke), overexposure to sun, infections, and exposure to water and air pollution or other toxic chemicals (possibly including pesticides that mimic estrogen in the body) are also responsible for a percentage of cancers. Lack of physical activity also probably plays a role in the development of some types of cancer.[72] Viral infections are associated with a few forms of cancer.

Some 20 to 50 percent of total cancers are influenced by diet, and these relationships are the focus of this section.[73] Dietary fat, meat, alcohol, excess calories, and low

Figure 11•10

TEN LEADING SITES OF NEW CASES OF CANCER FOR MEN AND WOMEN, UNITED STATES

The total of new cancer cases in 2001 was 1,268,000. The probability of developing any kind of invasive cancer during a lifetime, based on 1995 through 1997 rates, was 1 in 2 for men, 1 in 3 for women.

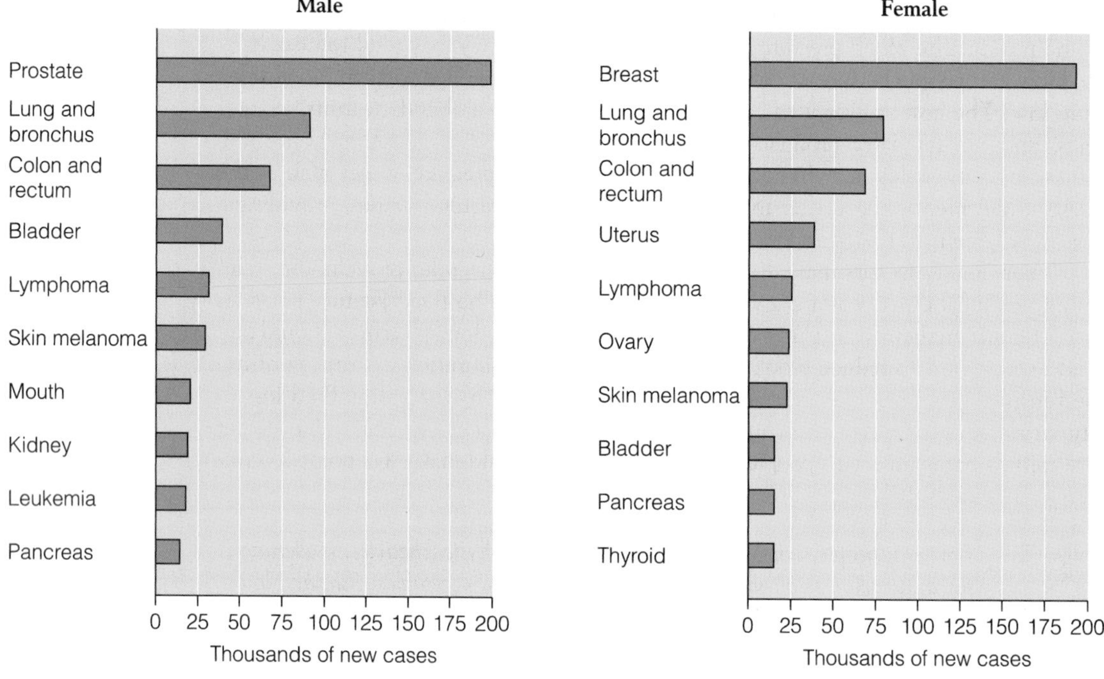

SOURCE: Data from American Cancer Society, *Cancer Facts and Figures*—2001, available from the American Cancer Society at (800) ACS-2345 or on the Internet at www.cancer.org.

carcinogen (car-SIN-oh-jen) a cancer-causing substance (*carcin* means "cancer"; *gen* means "gives rise to").

initiation an event, probably occurring in a cell's genetic material, caused by radiation or by a chemical carcinogen that can give rise to cancer.

carcinogenesis the origination or beginning of cancer.

promoters factors that do not initiate cancer but speed up its development once initiation has taken place.

metastasis (meh-TASS-ta-sis) movement of cancer cells from one body part to another, usually by way of the body fluids.

intakes of fruits and vegetables have been the targets of much research with regard to the occurrence of cancer. Such constituents of the diet relate to cancer in several ways:

- Foods or their components may cause cancer.
- Foods or their components may promote cancer.
- Foods or their components may protect against cancer.

Also, for the person who has cancer, diet can make a crucial difference in recovery. Table 11-7 lists some dietary and environmental factors believed to be important in cancer causation.

How Does Cancer Develop?

Cancer is thought to develop through the following steps (see Figure 11-11 on page 414):

1. Exposure to a **carcinogen.**
2. Entry of the carcinogen into a cell.
3. **Initiation** of cancer as the carcinogen alters the cell's genetic material in some way **(carcinogenesis).**
4. Acceleration by other carcinogens, called **promoters,** so that the cell begins to multiply out of control.
5. Spreading of cancer cells via blood and lymph **(metastasis);** disruption of normal body functions.

Researchers think that the first three steps, which culminate with initiation, are key to cancer prevention. On hearing this, many people mistakenly believe that they

Table 11•7

SOME FACTORS ASSOCIATED WITH CANCER AT SPECIFIC SITES

Cancer Sites	Incidence Associated with:	Protective Effect Associated with:
Bladder cancer	Weak associations with coffee, artificial sweeteners, and alcohol; stronger associations with cigarette smoking, chlorinated drinking water	Fruits and vegetables, especially green and yellow ones; adequate fluid intake
Breast cancer	High intakes of food energy and alcohol; sedentary lifestyle; probably not associated with dietary fat	Possibly soybeans and soy products, and fruits and vegetables;[a] physical activity
Cervical cancer	Folate deficiency; viral infection	Adequate folate intake
Colorectal cancer	High intakes of fat (particularly saturated fat), meat, alcohol (especially beer), and supplemental iron; low intakes of fiber, folate, and vegetables; inactivity	Vegetables; calcium, vitamin D, and dairy intake; whole wheat, wheat bran, and other fiber-rich foods; physical activity
Esophageal and mouth cancers	High alcohol, tobacco, and especially combined use; heavy use of preserved foods (such as pickles); low intakes of vitamins and minerals; high intakes of vitamin A supplements	Fruits and vegetables
Liver cancer	Infection with hepatitis virus; high intakes of alcohol; iron overload; toxins of a mold (aflatoxin) or other toxicity	
Lung cancer	Smoking; supplements of beta-carotene (in smokers)	Fruits and vegetables
Ovarian cancer	No dietary risk factors established; inversely correlated with oral contraceptive use	Fruits and vegetables, especially green ones
Pancreatic and lung cancer	No dietary risk factors established; correlated with cigarette smoking and air pollution	Fruits and vegetables, especially green and yellow ones
Prostate cancer	High intakes of fats, especially saturated fats from meats and possibly milk products	Fruits and vegetables, especially cooked tomatoes and green and yellow vegetables; soybeans and soy products; flaxseed; adequate selenium intake
Stomach cancer	High intakes of smoke- or salt-preserved foods (such as dried, salted fish); low intakes of fresh fruits and vegetables; infection with ulcer-causing bacteria	Fresh fruits and vegetables, especially tomatoes

NOTE: Findings based on epidemiological studies.

[a]Evidence on fruits and vegetables and prevention of breast cancer is mixed; S. A. Smith-Warner and coauthors, Intake of fruits and vegetables and risk of breast cancer, *Journal of the American Medical Association* 285 (2001): 769–776.

SOURCES: R. L. Nelson, Iron and colorectal risk: Human studies, *Nutrition Reviews* 59 (2001): 140–148; M. C. Jansen and coauthors, Dietary fiber and plant foods in relation to colorectal cancer mortality: The Seven Countries Study, *International Journal of Cancer* 81 (1999): 174–179; B. S. Reddy, Role of dietary fiber in colon cancer; An overview, *American Journal of Medicine* 106 (1999): S50–S51; G. J. Handelman, High-dose, vitamin supplements for cigarette smokers: Caution is indicated, *Nutrition Reviews* 55 (1997): 369–370; D. J. Hunter and coauthors, Cohort studies of fat intake and the risk of breast cancer—pooled analysis, *New England Journal of Medicine* 334 (1996): 356–361.

should avoid eating all foods that contain carcinogens. Doing so would be impossible, however, because most carcinogens occur naturally among thousands of other chemicals and nutrients the body needs. The body is well equipped to deal with the minute amounts of carcinogens occurring naturally in foods, such as those listed in the margin.

For those who suspect food additives of being carcinogenic, be assured that additives are held to strict standards, and no additive scientifically shown to cause cancer is approved for use in the United States. (Details concerning saccharin are found in Controversy 4.) Contaminants that enter foods by accident or toxins that arise naturally, for example, when food becomes moldy, may indeed be powerful carcinogens, or they may be converted to carcinogens by the body's attempts to metabolize them. Most such constituents are monitored in the U.S. food supply and are generally present, if at all, in amounts well below those that may pose significant cancer risks to consumers.[74]

key point *Cancer develops in steps including initiation and promotion, which are thought to be influenced by diet. The body is equipped to handle tiny doses of carcinogens that occur naturally in foods.*

Here are some chemicals and carcinogens occurring naturally in breakfast foods:

- Coffee: acetaldehyde, acetic acid, acetone, atractylosides, butanol, cafestol palmitate, chlorogenic acid, dimethyl sulfide, ethanol, furan, furfural, guaiacol, hydrogen sulfide, isoprene, methanol, methyl butanol, methyl formate, methyl glyoxal, propionaldehyde, pyridine, 1,3,7,-trimethylxanthine.
- Toast and coffee cake: acetic acid, acetone, butyric acid, caprionic acid, ethyl acetate, ethyl ketone, ethyl lactate, methyl ethyl ketone, propionic acid, valeric acid.

NOTE: Consuming coffee, toast, and coffee cake does not elevate a person's risk of developing cancer.

Figure 11•11

CANCER DEVELOPMENT

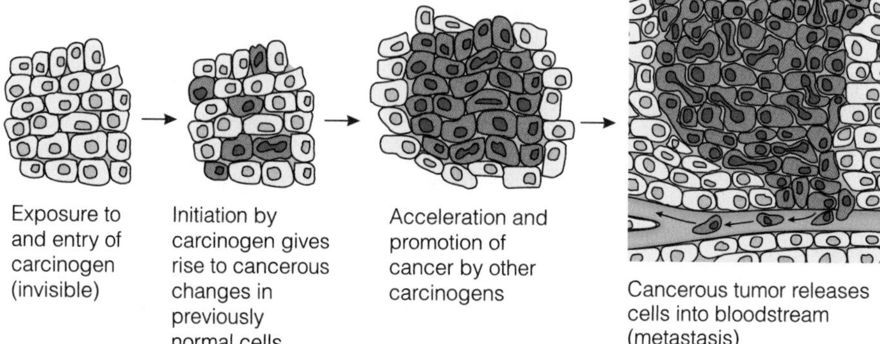
Exposure to and entry of carcinogen (invisible)
Initiation by carcinogen gives rise to cancerous changes in previously normal cells
Acceleration and promotion of cancer by other carcinogens
Cancerous tumor releases cells into bloodstream (metastasis)

How Powerful Is Diet in Influencing a Person's Risk of Developing Cancer?

From the evidence so far, it is almost certain that diet affects cancer rates—both for the better and for the worse.[75] Studies of populations suggest that low rates of many, but not all, kinds of cancer correlate with intakes of fiber-rich fruit, vegetables, and whole grains, particularly whole wheat.[76] Under study for possible effects regarding cancer are diets high in certain fats, daily use of alcohol, vitamins and minerals in foods and supplements, and diets in which meat plays a dominant role. The following sections explore current scientific thought on the effects of dietary constituents on cancer development.

Fat and Fatty Acids Laboratory studies using animals support the idea that high dietary fat intakes correlate with development of cancer. Simply feeding fat to experimental animals is not enough to get tumors started, however; an experimenter must also expose the animals to a known carcinogen. After that exposure, animals fed the high-fat diet develop more cancers faster than animals fed low-fat diets. Thus, fat appears to be a cancer promoter in animals.

In studies of human beings, however, evidence is mixed as to whether a diet high in fat promotes cancer.[77] Among world populations, comparisons reveal that high-fat diets often, but not always, correlate with high cancer rates. Within a single population, however, cancer rates do not reliably reflect fat intakes.[78] This finding does not exclude the possibility that dietary fat contributes to cancer, but more research is required before any conclusion can be reached. Today, scientific opinion seems to be leaning toward a statement of "no effect" on the question of whether dietary fat promotes breast cancer in particular.[79] For the risk of prostate cancer, evidence seems to implicate meat fats but not vegetable fats while consuming fatty fish may be protective.[80] An attribute of dietary fat is energy density—fat is extremely calorie dense. Diets high in *calories* do seem to promote cancer, especially in laboratory settings, so researchers still must untangle the effects of fat alone from those of the energy content of the diet.[81]

In addition to energy, dietary fat may be related to certain cancers because of its tendency to oxidize when exposed to high cooking temperatures. When these oxidized fat compounds enter the body, they may set up a condition of oxidative stress that may trigger cancerous changes in the tissues of the colon and rectum.[82] Finally, the type of fat in the diet may be important. Some laboratory evidence implicates omega-6 fatty acids in cancer promotion, while suggesting that omega-3 fatty acids from fish may protect against some cancers and may support recovery during treatment for cancer.[83] Countering this is a recent study reporting an increase in liver tumors in laboratory rats treated with fish oil.[84] In any case, moderation in fat intake remains a sound principle, if not for cancer protection, then for the prevention of obesity and the protection of the heart.

caloric effect the drop in cancer incidence seen whenever intake of food energy (calories) is restricted.

Food Energy When calorie intakes are reduced, cancer rates fall. In animal experiments, this **caloric effect** proves to be one of the most effective dietary interventions for cancer prevention. When researchers establish a cancer-causing condition and then restrict the energy in laboratory animals' feed, the onset of cancer in the restricted animals is delayed beyond the time when animals on normal feed have died. At the moment, no experimental evidence exists showing this effect in human subjects, but some population observations seem to imply that the effect seen in animals may hold true for human beings as well.[85] This effect occurs only in cancer prevention; once started, cancer continues advancing even in a person who is starving.

It is also true that when calorie intakes rise, cancer rates rise: excess calories from carbohydrate, fat, and protein all raise cancer rates.[86] The processes by which excess calories may stimulate cancer development remain obscure, but some researchers suspect that the hormones produced by the kidneys' adrenal gland are involved. High calorie intakes stimulate the release of these hormones, which cause inflammation, and inflammation stimulates the growth of tumors. Restricting energy intakes inhibits adrenal hormone release. Another idea is that obesity, and the insulin resistance it fosters, may promote the development of cancer.[87] Also, a high-calorie diet can augment the damaging actions of carcinogens that may be present in the tissues. Physical activity to balance energy intake may lower the risk of developing some cancers.[88]

Controversy 3 addressed the topic of alcoholic beverages and cancer risks.

To minimize risks from carcinogens formed during cooking:

- When grilling, line the grill with foil, or wrap the food in foil to minimize the formation of carcinogenic compounds.
- Take care not to burn foods while cooking by any method.
- Marinating meats beforehand may help to reduce the carcinogens formed during grilling.
- Limit intakes of fried, browned, and broiled foods.
- Limit intakes of smoked foods.

Alcohol, Smoked Foods, and Meats Cancers of the head and neck correlate strongly with the combination of alcohol and tobacco use and with low intakes of green and yellow fruits and vegetables. Alcohol intake alone is associated with cancers of the mouth, throat, and breast, and alcoholism often damages the liver and precedes the development of liver cancer.

Smoke generated from burning wood or charcoal, like smoke from burning tobacco, contains a multitude of chemical substances, some of which initiate cancer.* Some carcinogens from smoke settle on food during cooking; others form when meat fats or added oils land on the coals and then vaporize, creating carcinogens that rise and stick to the food. Just the process of browning foods triggers chemical changes among the sugars and amino acids of foods such as steak that are typically seared in cooking. These changes often improve the flavor, aroma, and appearance of foods, but they may also create carcinogens.[89] Eating smoked, grilled, charbroiled, or browned foods introduces the carcinogens into the digestive system, where they may affect the tissues of the intestinal lining. Once the compounds are absorbed and enter the body's tissues, however, they are quickly captured and detoxified by the body's competent detoxifying system.

Evidence from population studies spanning the globe over a period of over 20 years supports the theory that diets high in meat, and particularly red meat, are related to a greater risk of developing colon cancer.[90] In particular, processed meats and meats cooked to the crispy well-done stage may be at fault. Remember, however, that even strong correlation is not causation—although certain foods may appear at the scene of the colon cancer crime, no one knows whether eating such foods actually causes cancer or whether some other feature of a meat-containing diet is at fault. Still, a health-savvy diner replaces most servings of red meats with poultry, fish, or legumes, and chooses grilled, fried, highly browned, and smoked foods only occasionally, perhaps once or twice a month.

Smoked, grilled, charbroiled, or browned foods introduce carcinogens into the digestive system.

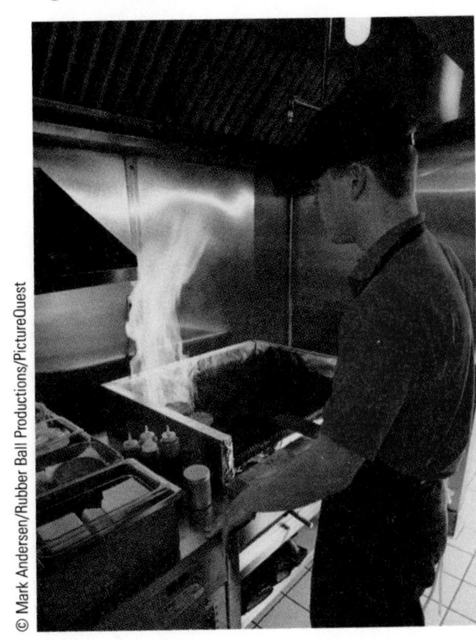

© Mark Andersen/Rubber Ball Productions/PictureQuest

Fiber and Fluid Epidemiological studies often report links between eating plenty of fruits and vegetables and a low incidence of cancers, but the reason for this is not clear. One prominent theory is that the fiber in fruits and vegetables helps to protect against some cancers by speeding up the transit time of all materials through

*The carcinogens of greatest concern are some of those called *polycyclic aromatic hydrocarbons*.

anticarcinogens compounds in foods that act in any of several ways to oppose the formation of cancer.

the colon so that the colon walls are not exposed to cancer-causing substances for long. Uncertainty developed when several well-designed studies refuted the idea that high-fiber diets protect against colon cancer.[91] Upon reviewing all the evidence, however, one respected medical group states that the majority of the evidence supports a role for fiber-rich diets as protective against colorectal cancer.[92]

Evidence weighs in favor of eating high-fiber, low-fat foods such as whole grains for many reasons. Foods rich in fiber and low in fat may not only reduce the risk of some cancers, but such foods are thought to protect health in these ways:

- The breakdown of fibers of whole grains by digestive tract bacteria yields short-chain fatty acids thought to protect against cancer and heart disease.
- High-fiber diets help to regulate blood glucose to control diabetes, and blood glucose may also play roles in cancer prevention.
- Whole grains are rich in antioxidants and other phytochemicals that reduce oxidative stress and promote overall health.[93]

If a meat-rich, calorie-dense diet is implicated in causation of certain cancers and if a vegetable-rich, whole-grain-rich diet is associated with prevention, then shouldn't vegetarians have a lower incidence of those cancers? They do, as the many studies cited in Controversy 6 have shown.

One type of cancer, bladder cancer, may be related to intake of fluids. Men who drink about 10 cups of fluid a day have been reported to develop substantially less bladder cancer than those drinking only about half this amount.[94] The most probable explanation involves carcinogens that form naturally in urine. A greater fluid intake dilutes these carcinogens and causes more frequent urination, thus reducing the likelihood that carcinogens will interact with the tissues of the bladder. Plain water seems most beneficial in this regard, but almost any kind of fluid, save one kind, will do. The exception is alcoholic beverages, which at best do not lower bladder cancer risks.

Folate and Other Vitamins Folate deficiency seems to make cancers of the cervix and colon more likely, and ample folate may ward off development of other cancers in certain populations.[95] One expert in the field of cancer causation estimates that 10 percent of the U.S. population, and a much larger percentage of people with low incomes, consume a diet low enough in folate to cause breaks in DNA that make cancer likely to develop.[*96] This reason alone is enough to warrant everyone attending to folate intake; many previous chapters gave other compelling reasons.

Some **anticarcinogens,** especially some phytochemicals, may protect against cancer by acting as mild toxins that force the body to build up its arsenal of carcinogen-destroying enzymes. Then, when a potent carcinogen arrives, the prepared body deals with it swiftly. Vitamin A regulates aspects of cell division and communication that go awry in cancer, and it helps to maintain the immune system. Immune cells can often identify cancerous cells and destroy them before cancer can develop. Folate, vitamin B_6, vitamin B_{12}, and pantothenic acid may oppose cancer in other ways.

Vitamin E, vitamin C, and beta-carotene received attention in Controversy 7, which included a discussion of their antioxidant roles and cancer-fighting effects. Suffice it to say here that taking supplements has not been proved to prevent or cure cancer. In fact, once cancer is established, antioxidants may do more harm than good according to a current line of investigation. As Controversy 7 explained, immune system cells release oxidative free radicals in an oxidative burst to kill off potentially dangerous microorganisms and cancer cells. Research suggests that some cancer cells may selectively stockpile antioxidants to defend against such assaults from the immune system.[97] Researchers found that when mice with brain tumors were fed antioxidant-depleted chow, the tumor growth slowed and cancer cells died off at a rapid rate.[98] Whether this intriguing laboratory finding translates into useful information for fighting cancer in people remains to be determined.

Often, it is foods like these, not individual chemicals, that lower people's cancer rates.

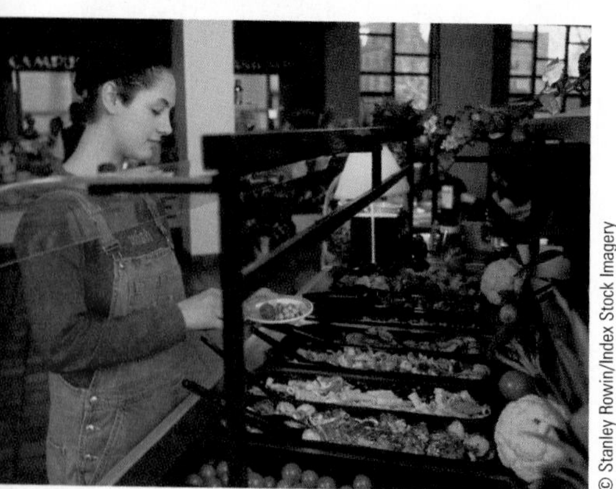

*The expert is Dr. Bruce Ames.

Calcium and Other Minerals Laboratory evidence suggests that a high-calcium diet may help to prevent colon cancer. Supplemental calcium seems to suppress changes in the lining of the colon associated with the onset of cancerous changes. In laboratory experiments, human colon cancer cells replicate rapidly when deprived of calcium but slow their replication when calcium is restored.[99] The findings from epidemiological studies are mixed, but a recent review of the literature indicated that, overall, people who develop colon cancer consume less calcium than people who do not develop the cancer.[100] Together, these studies have not yet proved that one can avoid colon cancer by increasing calcium intake, but with all the other points in calcium's favor, prudence dictates that everyone should arrange to meet calcium needs every day.

cruciferous vegetables vegetables with cross-shaped blossoms—the cabbage family. Their intake is associated with low cancer rates in human populations. Examples include broccoli, brussels sprouts, cabbage, cauliflower, rutabagas, and turnips.

For years, iron has been the subject of research with regard to colon cancer. A recent review revealed that the majority of studies since 1990 confirm an association between colon cancer and both increased dietary iron intake and high body iron stores.[101] Whether iron increases a person's risk for colon cancer may depend on whether the person has inherited a tendency to store too much iron (a relatively common genetic trait), the person's gender (being female may be protective), and degree of iron supplementation (higher intake poses greater risk).

Controversy 2 described the national "5 a Day" campaign designed to encourage consumers to eat at least five servings of fruits and vegetables every day to maintain good health.

How iron may facilitate cancer remains unanswered. Iron is a powerful oxidizing substance and may facilitate mutation of DNA in ways that initiate cancer. Alternatively, iron supplements are constipating, and constipation also raises a person's risk of colon cancer. Also, meat is a generous supplier of iron in the diet, and high-meat diets often correlate with colon cancer.[102] These findings have scientists questioning the wisdom of widespread enrichment of breads, cereals, and other grains with iron from a public health point of view. If the iron and cancer link holds up to further scrutiny, the benefits of enrichment to iron-deficient populations must be weighed against the harm to those prone to colorectal cancer.[103] Other minerals in foods, including zinc, copper, and selenium, are thought to play roles in cancer prevention, perhaps by helping antioxidant enzymes defend against its initiation.

Cruciferous vegetables belong to the cabbage family: bok choy, broccoli, broccoli sprouts, brussels sprouts, cabbages (all sorts), cauliflower, greens (collard, mustard, turnip), kale, kohlrabi, rutabaga, and turnip root.

Foods and Phytochemicals In the end, whole foods, not single nutrients, may be most influential on cancer development. For example, the phytochemicals of some fruits and vegetables are thought to be anticarcinogens. Almost without exception, population studies find that infrequent use of green and yellow fruits and vegetables and citrus fruits correlates with cancers of many types. Further evidence indicates that a low intake of fruits and vegetables increases DNA changes believed to be a first step in the development of bladder and other cancers.[104] Specifically, infrequent use of **cruciferous vegetables**—broccoli, brussels sprouts, cabbage, cauliflower, turnips, and the like—is common in colon cancer victims (see the margin).

One review of the literature found an almost unheard-of perfect association between reduced incidence of lung cancer and diets high in fruits and vegetables—every study included in the review reported a protective effect.[105] Incidence of stomach cancer, too, correlates with too few vegetables in the diet: in one study, with vegetables in general; in another, with fresh vegetables; and in others, with lettuce and other fresh greens or vegetables containing vitamin C. Unfortunately, adults in the United States are slow to get the message (see Figure 11-12). Herbs and spices often contain beneficial compounds, too, and their medicinal use is growing rapidly among those who search for natural remedies to ailments through alternative medicine (see the Consumer Corner).

Figure 11•12

PERCENTAGE OF U.S. ADULTS CONSUMING 5 OR MORE DAILY SERVINGS OF FRUITS AND VEGETABLES

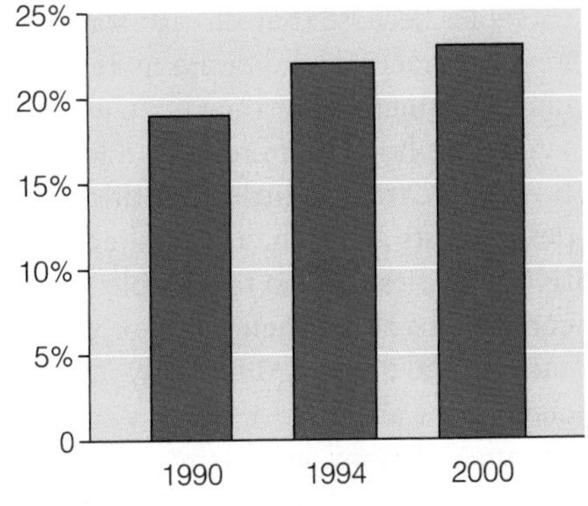

SOURCE: R. Li and coauthors, Trends in fruit and vegetable consumption among adults in 16 U.S. states: Behavioral risk factor surveillance system, 1990–1996, *American Journal of Public Health* 90 (2000): 777–781; Centers for Disease Control and Prevention—2000 Nutrition, available at http://apps.nccd.cdc.gov.

Foods that contain phytochemicals are believed to promote health and fight diseases. Such foods, often called *functional foods*, have been recognized as potentially beneficial by the National Academy of Sciences. The academy defines functional foods as "potentially healthful products [that include] any modified food or food ingredient that may provide a health benefit beyond the traditional nutrients it contains."[106] Controversy 2 explored the state of the science concerning functional foods and their phytochemical constituents. The Food Feature lists dietary guidelines for cancer prevention.

Complementary, Alternative, and Herbal Medicine

Where do you turn for help when illness strikes? Do you see a physician who offers treatment methods sanctioned by the established medical community? Or do you seek out an herbalist, an acupuncturist, or another practitioner of alternative therapy? People are turning to **alternative medical systems** more each year—in 1997, U.S. consumers spent upward of $21 billion on such treatments.[1]

Unlike conventional therapies, alternative therapies:

- Generally have not been well established by scientific experimentation to be safe and effective—they are used but unproven.
- Are not taught by most medical schools in the United States.
- Are not reimbursable by most health insurance providers in the United States.

Many people seek out alternative therapies because they distrust standard medical practices and desire more "natural" treatments that they hope will be safer. This hope is often ill-founded, however. Often, so little scientific evidence exists about the use of alternative therapies that no reasonable conclusions about their safety or efficacy can be drawn. Almost anyone can claim to be an expert in a "new" or "natural" therapy, and many practitioners act knowledgeable, when in reality they are untrained (see Controversy 1).

Stories abound that credit alternative therapies with miraculous cures of diseases. The listener may think that unless the speaker is lying, the therapies really do cure the diseases. But a third option also exists: remember from Chapter 1 that giving a placebo medication very often brings about a physical healing when the patient believes in the treatment.[2] No doubt many an alternative therapy has helped people to recover by engaging the ability of the body to heal itself.

This is not to say that all alternative therapies lack efficacy. Indeed, some of today's proven medical practices started out as alternative therapies. For example, cancer radiation therapy was once an unconventional therapy, but it quickly proved its clinical value and became part of mainstream cancer treatment. Efforts are now under way to investigate the efficacy of alternative therapies. In 1992, the prestigious National Institutes of Health established its National Center for Complementary and Alternative Medicine (NCCAM) for that purpose. Table 11-8 shows the major areas of complementary and alternative medicine as categorized by NCCAM. NCCAM has deemed **acupuncture** useful for quelling nausea associated with surgery, cancer chemotherapy, and pregnancy and for relieving pain during dental procedures. Acupuncture, when performed by skilled practitioners under sterile conditions, presents very little risk to the user, a claim that cannot be made for many standard drugs administered for relief of nausea and pain.[3]

Another alternative therapy that has attracted many followers is **herbal medicine.** Since the dawn of humankind, herbs have been harvested for use as medicines. Dozens of herbs, when investigated scientifically, are found to contain effective natural drugs. For example, ancient people greatly valued the resin they called myrrh, which contains an analgesic (pain-killing) compound.[4] Willow bark contains aspirin; the herb valerian contains a tranquilizing oil; senna leaves produce a powerful laxative. Medicinal herbs are showing up on grocery shelves in "functional foods," making consumer education about their effects a pressing need.

Herbal medicine also has some serious drawbacks. For one thing, few herbalists prescribing herbs have the understanding of botany, pharmacology, or human physiology necessary to use these drugs effectively and safely. Instead they rely on hearsay and folklore. Dangerous mistakes with herbs are extraordinarily likely. For example, most mint is safe when brewed as tea, but some may contain highly toxic pennyroyal oil. Folk medicine urges parents to soothe a colicky baby with mint tea, but one concoction laden with pennyroyal was blamed for liver and neurological injuries to at least two infants, one of whom died.[5]

Another example is *Aristolochia fangchi*, an herb recently discovered to cause severe kidney damage and cancer in those who take it. Its name in Chinese is similar to another, safer herb, and so it was mistakenly included in herbal "weight reduction pills," turning them into herbal poisons.[6] Some preparations may even include *Aristolochia* intentionally because no safety tests are required before herbs can be marketed. The Food and Drug Administration (FDA) has issued a national recall of products containing this substance, but the law dictates that the FDA may act only after substantial harm to unsuspecting consumers has already occurred.

Table 11•8

NATIONAL INSTITUTES OF HEALTH: ALTERNATIVE THERAPIES

I. **alternative medical systems** complete systems of theory and practice that have evolved independently and often prior to conventional biomedical approaches. Examples:
- **acupuncture** (AK-you-PUNK-cher) a technique that involves piercing the skin with long thin needles at specific anatomical points to relieve pain or illness. Acupuncture sometimes uses heat, pressure, friction, suction, or electromagnetic energy to stimulate the points.
- **ayurveda** (EYE-your-VAY-dah) traditional medical system of India that treats the body, mind, and spirit with breathing control, diet, exercise, herbs, massage, meditation, and sunlight.
- **homeopathic** (HOME-ee-oh-PATH-ick) **medicine** a practice based on the theory that "like cures like," that is, that substances that cause symptoms in healthy people can cure those symptoms when given in very dilute amounts (*homeo* means "like"; *pathos* means "suffering").
- **irridology** the study of changes in the iris of the eye and their purported relationships to disease.
- **naturopathic** (NAY-cher-oh-PATH-ick) **medicine** a system that integrates traditional medicine with botanical medicine, clinical nutrition, homeopathy, acupuncture, East Asian medicine, and manipulative therapy.

II. **mind-body interventions** techniques to facilitate the mind's capacity to affect bodily functions and symptoms. Examples:
- **aroma therapy** a technique that uses oil extracts from plants and flowers (usually applied by massage or baths) to try to enhance physical, psychological, and spiritual health.
- **biofeedback** the use of sensors to convey information about heart rate, blood pressure, skin temperature, muscle relaxation, and the like to enable a person to learn how to consciously control internal body functions.
- **faith healing** the practice of invoking divine intervention without the use of medical, surgical, or other traditional therapy.
- **hypnotherapy** a technique that uses hypnosis and the power of suggestion to improve health behaviors, relieve pain, and heal.
- **imagery** a technique to achieve a desired physical, emotional, or spiritual state by visualizing the state.

III. **biological-based therapies** use of untested dietary regimes, herbal preparations, and other substance-based therapies for treating physical illness or symptoms. Examples:
- **chelation** (kee-LAY-shun) **therapy** the use of ethylene diamine tetraacetic acid (EDTA), supposedly to heal the body by binding with metallic ions and removing toxic metals.
- **herbal medicine** the use of herbs and other natural substances with the intention of preventing or curing diseases.
- **orthomolecular medicine** the administration of large doses of vitamins to attempt treatment of chronic disease.
- **supplement therapy** administration of dietary supplements, such as laetrile or shark cartilage, to attempt to treat diseases such as cancer.

IV. **manipulative and body-based methods** methods of treatment involving manipulation and movement of the body. Examples:
- **chiropractic** (KYE-row-PRAK-tick) a manual healing method of manipulating vertebrae to relieve musculoskeletal pain.
- **massage therapy** manipulation of the soft tissues of the body to relax the tissues, reduce tension, and restore health.

V. **energy therapies** therapies focusing on energy fields assumed to be emanating from the body or from other sources. Examples:
- **bioelectromagnetic medical applications** the use of electrical energy, magnetic energy, or both in an attempt to stimulate bone repair, wound healing, and tissue regeneration.
- **biofield therapeutics** a manual healing method that supposedly directs a healing force from an outside source (commonly God or another supernatural being) through the practitioner and into the client's body; also called *Reike* (RAY-kee), or "laying on of hands."

In addition to cases of mistaken identity, purity of herbal products can be a problem. Quantities of mercury and arsenic detected in traditional Chinese herb balls for treating fever, rheumatism, and cataracts have exceeded the Environmental Protection Agency's maximum allowable levels by 20,000 and 1,000 times, respectively.

Another problem is lack of information about which herbs to use and *not* to use and when. For example, no one really knows if St. John's wort, said to have a calming effect, is safe to take during pregnancy. Another herb claimed to calm the nerves, kava kava, recently was reported as causing women to develop severe, irreversible Parkinson's disease.[7] Foxglove leaves contain digoxin, a compound that modifies the heart's action; a person taking cardiac medication who also decides to take foxglove may be headed for disaster from the combined effect on the heart.

Herbal medicines are sold as "dietary supplements" instead of drugs, which means their labels may legally include unproven claims as long as the label also includes the words "Has not been evaluated by the FDA." Not surprisingly, when a label claims that an herbal product *may* strengthen immunity, support eyesight, or maintain heart health, consumers believe that it *will*. Beware. Table 11-9 on the next page lists some additional herbs and their potential actions.*

*A reliable source of information about herbs is V. Tyler, *The Honest Herbal* (New York: Pharmaceutical Products Press). Look for the latest edition.

Another huge source of misinformation about herbs is the Internet. People tend to believe what they find there. All of the Internet cautions of Controversy 1 apply.

A growing number of health-care professionals are trying alternative therapies and incorporating the helpful ones into their practices. This open-minded approach, termed complementary medicine, takes advantage of the best of both kinds of medicine. As more becomes known, no doubt more beneficial therapies will be ushered into mainstream medicine. Until then, consumers are left in the dark about the potential risks and benefits of choosing alternative therapies.

Table 11•9

SELECTED HERBS AND THEIR EFFECTS[a]

- **aloe** a tropical plant with widely claimed value as a topical treatment for minor skin injury. Some scientific evidence supports this claim; evidence against its use in severe wounds also exists.
- **belladonna** any part of the deadly nightshade plant; a fatal poison.
- **cat's claw** an herb from the rain forests of Brazil and Peru; claimed, but not proved, to be an "all-purpose" remedy.
- **chamomile** flowers that may provide some limited medical value in soothing menstrual, intestinal, and stomach discomforts.
- **chaparral** an herbal product made from ground leaves of the creosote bush and sold in tea or capsule form; supposedly, this herb has antioxidant effects, delays aging, "cleanses" the bloodstream, and treats skin conditions—all unproven claims. Chaparral has been found to cause acute toxic hepatitis, a severe liver illness.
- **comfrey** leaves and roots of the comfrey plant; believed, but not proved, to have drug effects. Comfrey contains cancer-causing chemicals.
- **echinacea** an herb popular before the advent of antibiotics for its "anti-infectious" properties and as an all-purpose remedy, especially for colds and allergy and for healing of wounds. A small body of research seems to lend preliminary support for some of the claims, but also points to an insecticidal property, leading to questions about safety. Also called *cone-flower.*
- **feverfew** an herb sold as a migraine headache preventive. Some evidence exists to support this claim.
- **foxglove** a plant that contains a substance used in the heart medicine digoxin.
- **ginkgo biloba** an extract of a tree of the same name, claimed to enhance mental alertness, but not proved to be effective or safe.
- **ginseng** (JIN-seng) a plant root containing chemicals that have stimulant drug effects. *Ginseng abuse syndrome* is a group of symptoms associated with the overuse of ginseng, including high blood pressure, insomnia, nervousness, confusion, and depression.
- **hemlock** any part of the hemlock plant, which causes severe pain, convulsions, and death within 15 minutes.
- **kava-kava** the root of a tropical pepper plant, often brewed as a a tea consumed for its calming effects. Limited scientific research supports the effectiveness of kava-kava for treating anxiety. Adverse effects include skin rash, metabolic abnormalities, elevated blood cholesterol, lethargy, mental disorientation and possibly life-threatening Parkinson's disease.
- **kombucha** a product of fermentation of sugar-sweetened tea by various yeasts and bacteria. Proclaimed as a treatment for everything from AIDS to cancer but lacking scientific evidence. Microorganisms in home-brewed teas have caused serious illnesses in people with weakened immunity. Also known as *Manchurian tea, mushroom tea,* or *Kargasok tea.*
- **kudzu** a weedy vine, whose roots are harvested and used by Chinese herbalists as a treatment for alcoholism. Kudzu reportedly reduces alcohol absorption by up to 50 percent in rats.
- **medicinal herbs** nonwoody plants, plant parts, or extracts valued by some people for their medicinal qualities, both proved and unproved.
- **sassafras** root bark from the sassafras tree; once used in beverages but now banned as an ingredient in foods or beverages because it contains cancer-causing chemicals.
- **saw palmetto** the ripe fruit or extracts of the saw palmetto plant claimed to relieve symptoms associated with enlarged prostate; may act as a diuretic.
- **St. John's wort** an herb containing psychoactive substances that has been used for centuries to treat depression, insomnia, bedwetting, and "nervous conditions." Most scientific reports find St. John's wort equal in effectiveness to standard antidepressant medication for relief of depression. Long-term safety, however, has not been established.
- **valerian** a preparation of the root of an herb used as a sedative and sleep agent. Safety and effectiveness of valerian have not been scientifically established.
- **witch hazel** leaves or bark of a witch hazel tree; not proved to have healing powers.

[a]See also Table 7-4 of Chapter 7.

Can ginkgo biloba improve memory? Is it safe? (See Controversy 13.)

Is saw palmetto good for the prostate gland?

Can St. John's wort safely chase away the blues?

key point *Diets high in certain fats and red meats are associated with cancer development. Foods containing fiber, folate, calcium, many other vitamins and minerals, and phytochemicals, along with an ample intake of fluid, are thought to be protective.*

Conclusion

Nutrition is often associated with promoting health, and medicine with fighting disease, but no clear line separates nutrition and medicine. Every major agency involved with health recommends a healthful diet as part of a lifestyle that provides the best possible chance for a long and healthy life.

This chapter has summarized the major forms of disease and their links with nutrition. You may have noticed a philosophical shift from previous chapters. There, we could say "a deficiency of nutrient X causes disease Y." Here we could only cite theories and discuss research that illuminates current thinking. We can say with certainty, for example, that "a diet lacking vitamin C causes scurvy," but to say that a low-fiber diet that lacks vegetables causes cancer would be inaccurate. We can, however, recommend behaviors that are prudent and reduce the likelihood of illness. The Food Feature presents these recommendations.

Simple advice can be powerful. Think Food Guide Pyramid.

Diet as Preventive Medicine

A REMARK BY a former surgeon general is worth repeating: If you do not smoke or drink excessively, "your choice of diet can influence your long-term health prospects more than any other action you might take."[107] Indeed, healthy young adults today are privileged to be the first generation in history who can know enough now to lay the foundation for healthy later years through a lifetime of proper nutrition. Figure 11-13 illustrates this point.

Dietary Guidelines for Disease Prevention

An early chapter of this book presented dietary guidelines for the prevention of diseases. Chapters that followed focused on the "whys" and "hows" of those guidelines. This Food Feature comes full circle to revisit the guidelines with a broader and deeper understanding of their significance. Recent research has unveiled a simple truth: those who consume the low-fat meats and dairy products, fruits and vegetables, and whole grains recommended by the *Dietary Guidelines for Americans* (see Table 11-10) enjoy a longer, healthier life than those who do not.[108]

The American Heart Association and the American Cancer Society offer suggestions specifically for disease prevention. Table 11-10 presents guidelines from many sources and shows how similar they are, clinching the argument that it's time to get busy putting the recommendations into practice.
The following paragraphs review the specifics.

Reduce Saturated Fat Intake

Primary among the recommendations is to reduce saturated fat intake. The Food Feature of Chapter 5 showed how to keep saturated fat down by selecting low-fat foods. To meet recommendations to keep fat and saturated fat low, limit pure fat foods such as sour cream, butter, and margarine; high-fat foods

Figure 11•13

PROPER NUTRITION SHIELDS AGAINST DISEASES

A well-chosen diet can protect your health.

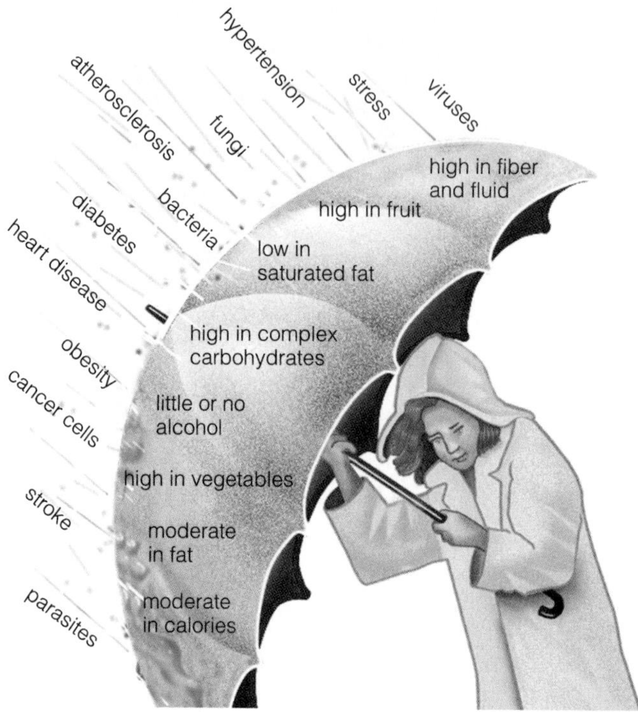

SOURCE: Adapted from an idea in R. K. Chandra, Nutrition and the immune system: An introduction, *American Journal of Clinical Nutrition* 66 (1997): S460–S463.

such as mayonnaise, cheese, and cream cheese; and foods high in hidden fat such as convenience foods, foods with sauces, fried foods, fat-marbled meat cuts, sausages, ground beef, whole milk, and others. Shop for foods whose labels indicate no more than 2 grams of saturated fat per 100 calories.

When you must add fat, use olive oil or canola oil, since these are high in monounsaturated fatty acids, but because they are high in calories use them, like all fats, sparingly. Eat meals of fish regularly, especially fatty fish such as salmon, to balance your intakes of omega-6 and omega-3 fatty acids. Consult the Table of Food Composition, Appendix A, for further details about the fatty acid contents of your favorite foods.

Include Fruits and Vegetables

Every legitimate source of dietary advice urges people to include a variety of fruits and vegetables in the diet, not just for nutrients but also for the phytochemicals that promote health. Vow to try a new fruit or vegetable each week. Who knows? Some of the foods still waiting for you on the produce shelves may become your favorites. An adventurous spirit is a plus in this regard. For example, most people are not familiar with soybeans and soy products, but calcium- and vitamin-fortified soy drinks can be added to casseroles or hot beverages; textured soy protein products can replace part of the hamburger in any recipe; tofu makes delicious stir-fried dishes; and soybeans themselves are good in many recipes calling for beans. Read some cookbooks for ideas on incorporating other new foods into the diet.

Go for Variety

Eat foods high in potassium (whole foods), high in calcium and magnesium (milk products and appropriate substitutes), low in fat, high in fiber (whole grains, legumes, vegetables, and fruits), and ample in fluids. If you are prone to hypertension, experts advise that you eat less salt.

Table 11•10

DIETARY GUIDELINES FOR DISEASE PREVENTION

Dietary Guidelines for Americans, 2000	American Heart Association Dietary Guidelines for Healthy American Adults, 2000	American Cancer Society Guidelines on Diet, Nutrition and Cancer Prevention, 1999
Aim for fitness: • Aim for a healthy weight. • Be physically active each day. Build a healthy base: • Let the Pyramid guide your food choices. • Choose a variety of grains, especially whole grains, and fruits and vegetables, daily. • Keep foods safe to eat. Choose sensibly: • Choose a diet low in saturated fat and cholesterol, and moderate in total fat. • Moderate your intake of sugars. • Prepare foods with less salt. • If you drink alcohol, do so in moderation.	• Overall healthy eating pattern: Include a variety of fruits, vegetables, grains, low-fat or nonfat dairy products, fish, legumes, poultry, and lean meats. • Appropriate body weight: Match energy intake to energy needs, with appropriate changes to achieve weight loss when indicated. • Desirable cholesterol profile: Limit foods high in saturated fat and cholesterol; and substitute unsaturated fat from vegetables, fish, legumes, and nuts. • Desirable blood pressure: Limit salt and alcohol; maintain a healthy body weight and a diet with emphasis on vegetables, fruits, and low-fat or nonfat dairy products.	Choose most of the foods you eat from plant sources. • Eat five or more servings of fruits and vegetables each day. • Eat other foods from plant sources, such as breads, cereals, grain products, rice, pasta, or beans, several times each day. Limit your intake of high-fat foods, particularly from animal sources. • Choose foods low in fat. • Limit consumption of meats, especially high-fat meats. Be physically active: achieve and maintain a healthy weight. • Be at least moderately active for 30 minutes or more each day. • Stay within your healthy weight range. Limit consumption of alcoholic beverages, if you drink at all. • For men, limit alcohol to two drinks a day; for women, one drink a day.

SOURCES: Adapted from American Heart Association Nutrition Committee, Dietary guidelines for healthy American adults, *Circulation* 102 (2000): 2284–2299, and from American Cancer Society, Nutrition and diet, *Prevention and Risk Factors*, 1999, available from www.cancer.org or upon request from the American Cancer Society at (800) ACS-2345.

Exercise regularly, all your life.

© Tom McCarthy/PhotoEdit

Advice on varying your diet is based on an important concept in the prevention of cancer initiation—dilution. Whenever you switch from food to food, you are diluting whatever is in one food with what is in the others. It is safe to eat *some* salt-cured foods or smoked or grilled meats, but don't eat them all the time. One study found that omission of several food groups from the diet brought extra risks of both cancer and CVD.

Be Physically Active

In addition to making wise food choices, maintain a proven program of weight control. Expend energy, so as to earn the right to eat more nutrient-dense foods; that is, be physically active. If the threat of CVD doesn't motivate you, then exercise to improve your self-image, to improve your morale, or to make friends—but do exercise.

In the end, people's choices are based on their own likes and dislikes within the limits that their own lives impose on them. Whoever you are, we encourage you to take the time to work out ways of making your diet meet the guidelines known to support health, at least on most days. If you include fruits, vegetables, and high-fiber grains and control your fat intake, you can feel confident that you are supporting your health. Take time to enjoy your meals, too: the sights, smells, and tastes of good foods are among life's greatest pleasures. Joy, even the simple joy of eating, contributes to a healthy life.

PRACTICE EYEBALLING A MEAL

Selecting nutritious foods from among pictures in a textbook is easy when the nutrient values of the foods appear on the page. Selecting foods is more difficult in the real world with nothing to go on but appearance. This Do It! section offers the chance to hone your "eyeballing" skills for judging meals according to nutrient ideals such as the *Dietary Guidelines*.

Step 1. Study the suppers shown in Figure 11-14.

Step 2. Consider each dietary goal of Form 11-1 on the next page individually, and judge each of the suppers according to that goal.

Step 3. Write into the blanks on Form 11-1 the letters that identify the three meals that you think are highest or lowest in the characteristics named. (For example, for "Lowest in calories," you might write in "A B C".) Instructions and hints follow.

- *Calories*. List the three meals *lowest* in calories (or if you need to gain weight, list those highest in calories). Without nutrient data, this question might stump even the most skilled eyeballer. The person who can identify both the visible and hidden fats in foods (see Chapter 5's Food Feature) can "see" excess calories in foods right away. Added sugar adds calories, too (Chapter 4's Food Feature can help). Don't forget about portion sizes—too much of almost anything pushes the calories of a meal into the high ranges.

Figure 11•14

SIX SUPPERS

Meal A
roasted chicken breast with skin, 1 average
creamed corn, ½ c
mashed potatoes, 1 c, with ¼ c gravy
bread, 1 slice with 1 tsp margarine
fruit punch, 10 oz

Meal B
red beans and rice, homemade, 1 c
zucchini and yellow squash mix, ½ c
whole-wheat roll with 1 tsp margarine
sweetened iced tea, 10 oz

Meal C
extra cheese, sausage, and pepperoni pizza, 3 slices
lettuce and tomato salad, 1½ c with 2 tbs dressing
cola, 10 oz

Meal D
macaroni and cheese, commercially prepared, 1½ c
green peas, canned, ½ c
iced water

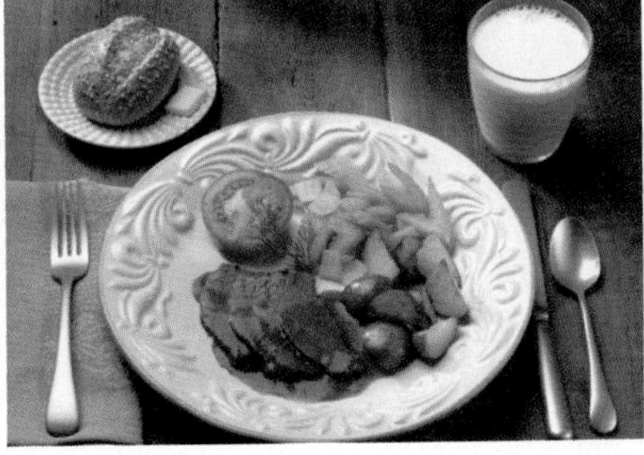

Meal E
pot roast, 2½ oz, with ½ c gravy
potatoes, 1 c
onions and celery mix, 1 c
whole-wheat roll with 1 tsp margarine
tomato, ½ c
fat-free milk, 1 c

Meal F
taco salad, fast food
lemon-lime soda pop, 10 oz

do it!

- *Fat, saturated fat, and cholesterol.* Select the three meals likely to be *lowest* in fat, those *lowest* in saturated fat, and those within limits for cholesterol. Fats and saturated fats can be as obvious on the plate as pats of margarine and added salad dressings, or they can be hidden in dishes with fat-laden butter sauces, cheeses, high-fat meats, creamy sauces, baked goods, toppings, and crusts. Cholesterol is harder to spot because it is part of lean meat, eggs, shellfish, and other nutritious foods. Chapter 5's Food Feature identifies the lipids in common foods.
- *Sodium.* Select the meals you think are *lowest* in sodium. As mentioned in Chapter 8, foods that contain the most salt, and therefore are highest in sodium, are often the most processed foods such as lunch meats, canned soups, chips, instant mixes, boxed foods, and condiments. It isn't always possible to guess which foods contain salt and sodium, but it's a safe bet that most farm-fresh, unprocessed foods are low in salt and sodium, unless salt was added during preparation.
- *Fiber.* Select the three meals *highest* in fiber. Fiber follows fruits, vegetables, and whole grains.
- *Vitamin A and calcium.* Pick out the three meals *highest* in vitamin A and the three *highest* in calcium. Foods rich in these nutrients are shown in the Snapshots on pp. 217 and 278.
- *Variety.* Choose the meals that provide the greatest variety of foods. Of all the nutrition recommendations, the one concerning variety of foods may be hardest to measure. The U.S. government's *Healthy Eating Index* suggests a minimum of eight different types of food in a day. Other experts believe that more variety best supports health.

Step 4. Score yourself. Compare your answers with the actual ranking of the meals as shown in Table 11-11. Score as instructed on Form 11-1, and then add the column to obtain a total score. Interpret your score as suggested at the bottom of Form 11-1.

Analysis

Answer the following questions:

1. Did you guess which meals were highest in calories, fats, salt, and the rest? For example, logic might predict that a taco salad, because of the word *salad*, would be low in fat and calories, but high-fat ingredients such as sour cream, cheddar cheese, and commercial ground beef change the nutrient picture dramatically.
2. Which areas did you find to be the hardest to judge? How can you hone your eyeballing skills in these problem areas? What information are you lacking?
3. What do you notice about meals that are lowest in fat? Do they often meet more than one of the nutrition goals of this exercise?
4. In Table 11-11, which meal fell into the top three most often? What were the flaws in that meal?
5. Look at the meals that fell at the very bottom of the various lists. Give some reasons why the meals fell so far from the nutrition goals in terms of the foods they included or lacked.
6. Does any meal rank at the top of the list for one goal and at the bottom of the list for another? Which one? What does this mean to the eater? Should you avoid such a meal, or might it bring benefits when used in moderation?

Form 11•1

DIETARY RECOMMENDATIONS SCOREBOARD

Write into the blanks the letters that identify the three meals that you judge to be highest or lowest in each characteristic.

The Three Meals	Your Score
Lowest in calories _____ _____ _____	__________
Lowest in fat _____ _____ _____	__________
Lowest in saturated fat _____ _____ _____	__________
Lowest in cholesterol _____ _____ _____	__________
Lowest in salt (sodium) _____ _____ _____	__________
Highest in fiber _____ _____ _____	__________
Highest in vitamin A _____ _____ _____	__________
Highest in calcium _____ _____ _____	__________
Highest in variety _____ _____ _____	__________
Total:	__________

Compare each of your responses with the ranking of meals in the charts of Table 11-11. Scoring: If you correctly identified the top three meals (in any order) for the category, give yourself three points. If you correctly identified two of the top-ranking meals, give yourself two points. For one meal, you get one point. If your total score adds up to be:

- 22 or more points—you have excellent eyeballing skills.
- 12–21 points—you are well on your way to developing your skills.
- 11 or fewer—you should identify your weakest areas and reread the corresponding material, especially the Food Features, in previous chapters. Then try this exercise again.

Keep in mind that not all the *Dietary Guidelines* focus on food; some also mention that being physically active and limiting alcohol are crucial to overall health. As you learn to identify meals that best meet your needs, choosing them will become second nature.

Table 11•11

RANKING MEALS ACCORDING TO DIETARY RECOMMENDATIONS

Important: The meal that comes closest to meeting each dietary ideal ranks highest on that chart; in other words, the meal *highest* in fiber is at the top of its list, as is the meal *lowest* in sodium, because both meals come closest to their ideals.

Ideal: Moderate in Calories[a]

Meal	
B	Lowest in calories
E	↑
D	
A	
F	↓
C	Highest in calories

Ideal: 30% or Fewer Calories from Fat

Meal	
B	Lowest in fat
A	↑
E	
D	
C	↓
F	Highest in fat

Ideal: 10% or Fewer Calories from Saturated Fat

Meal	
B	Lowest in saturated fat
A	↑
E	
F	
C	↓
D	Highest in saturated fat

Ideal: Moderate in Cholesterol[b]

Meal	
B	Lowest in cholesterol
E	↑
A	
D	
F	↓
C	Highest in cholesterol

Ideal: Moderate in Sodium[c]

Meal	
B	Lowest in sodium
F	↑
A	
D	
E	↓
C	Highest in sodium

Ideal: 25 Grams or More of Fiber

Meal	
B	Highest in fiber
F	↑
E	
C	
A	↓
D	Lowest in fiber

Ideal: Provides Significant Vitamin A[d]

Meal	
F	Highest in vitamin A
C	↑
D	
E	
A	↓
B	Lowest in vitamin A

Ideal: Provides Significant Calcium[e]

Meal	
C	Highest in calcium
D	↑
E	
F	
A	↓
B	Lowest in calcium

Ideal: More Variety

Meal	
E	Highest in variety
F	↑
C	
A	
B	↓
D	Lowest in variety

[a]This ranking reflects a need to reduce calorie intakes. A person needing to increase calorie intakes might want to change the order of this list to give higher ranks to higher-calorie meals that also meet other guidelines.
[b]Ideal for cholesterol: a day's meals should contain 300 milligrams or fewer of cholesterol.
[c]Ideal for sodium: a day's meals should contain 2,400 milligrams or fewer of sodium.
[d]Ideal for vitamin A: a day's meals should contain 800 to 1,000 micrograms vitamin A.
[e]Ideal for calcium: a day's meals should contain 800 to 1,200 milligrams calcium.

nutrition on the net

For further study of the topics of this chapter, access these websites and search for the phrases or words in quotation marks:

1. Find updates and quick links to these and other nutrition-related sites at our website: www.wadsworth.com/nutrition
2. Review resources offered by the National Center for Chronic Disease Prevention and Health Promotion: www.cdc.gov/nccdphp
3. Find information about health statistics from the National Center for Health Statistics site: www.cdc.gov/nchs
4. Search for "chronic diseases," "disease prevention," "men's health," "women's health, " "heart disease," "stroke," "high blood pressure," "cancer," and "diabetes" at the U.S. Government site: www.healthfinder.gov
5. Visit the American Heart Association: www.americanheart.org
6. Find information on the "DASH diet": dash.bwh.harvard.edu
7. Find resources to help you quit smoking: www.lungusa.org
8. Learn about complementary and alternative medicine from the National Center for Complementary and Alternative Medicine: http://nccam.nih.gov
9. Get dietary supplement information from the National Institutes of Health's Office of Dietary Supplements: http://dietary-supplements.info.nih.gov
10. Review the backgrounds and practices of many popular practitioners of alternative treatments: www.quackwatch.com

INTERNET ACTIVITY

This chapter discusses several disorders that may be diet-related but each also has other risk factors. Test your knowledge of cancer, heart disease, and hypertension risk factors by completing the following quizzes.

1. Go to the website of WebMD at http://www.webmd.com
2. Click on "Health Tools."
3. Click on "Quizzes."
4. Click on "Cancer Quiz," "Heart Disease Quiz," and "High Blood Pressure Quiz."
5. Complete the quizzes.

What were your scores? For which disorder did you score the highest? The lowest? For each disorder, were you least knowledgeable about risk factors, prevention, or treatment issues?

self check

Answers to these Self Check questions are in Appendix G.

1. Most people have well-developed plaques in their arteries by the time they reach the age of ______.
   a. 20 years
   b. 30 years
   c. 40 years
   d. 50 years
2. Which of the following is a risk factor for cardiovascular disease?
   a. high blood HDL cholesterol
   b. low blood pressure
   c. low LDL cholesterol
   d. diabetes
3. Which of the following dietary factors appears to influence heart disease risk the most?
   a. sodium
   b. cholesterol
   c. saturated fat
   d. total fat
4. Which of the following dietary factors may help to regulate blood pressure?
   a. calcium
   b. magnesium
   c. potassium
   d. all of the above
5. Which type of cancer is associated with overuse of alcohol?
   a. mouth
   b. breast
   c. throat
   d. all of the above
6. The best way to plan a diet to support the immune system is to exceed the recommended dietary allowances for each nutrient. T F
7. Resting blood pressure should ideally be 120 over 80 or lower. T F
8. Hypertension is more severe and occurs earlier in life among people of European or Asian descent than among African Americans. T F
9. Laboratory evidence suggests that a high-calcium diet may help to prevent colon cancer. T F
10. Alternative therapies, such as herbal medicine, have not been well established by scientific experimentation to be safe and effective. T F

CONTROVERSY 11

NUTRIENT-DRUG INTERACTIONS: WHO SHOULD BE CONCERNED?

A 45-YEAR-OLD CHICAGO business executive attempts to give up smoking with the help of nicotine gum. She replaces smoking breaks with beverage breaks, drinking frequent servings of tomato juice, coffee, and colas. She is discouraged when her stomach becomes upset and her craving for tobacco continues unabated despite the nicotine gum. Problem: nutrient-drug interaction.

A 14-year-old girl develops frequent and prolonged respiratory infections. Over the past six months, she has suffered constant fatigue despite adequate sleep, has had trouble completing school assignments, and has given up playing volleyball because she runs out of energy on the court. During the same six months, she has been taking huge doses of antacid pills each day because she heard this was a sure way to lose weight. Her pediatrician has diagnosed iron-deficiency anemia. Problem: nutrient-drug interaction.

A 30-year-old schoolteacher who benefits from antidepressant medication attends a faculty wine and cheese party. After sampling the cheese with a glass or two of red wine, his face becomes flushed. His behavior prompts others to drive him home. In the early morning hours, he awakens with severe dizziness, a migraine headache, vomiting, and trembling. An ambulance delivers him to an emergency room where a physician takes swift action to save his life. Problem: nutrient-drug interaction.

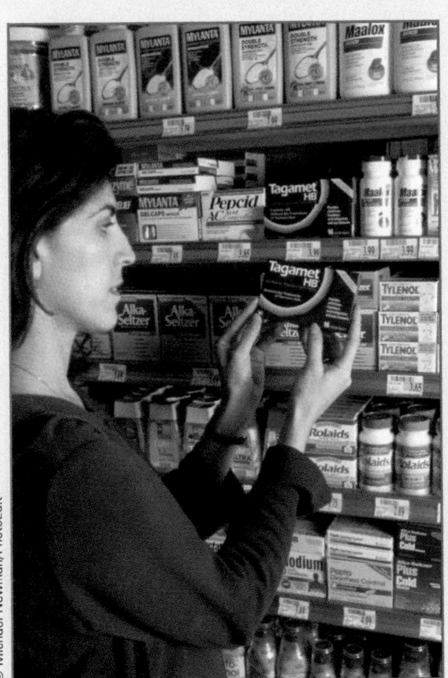

© Michael Newman/PhotoEdit

Medicines and Nutrition

People sometimes think that medical drugs do only good, not harm. As the opening stories illustrate, however, both prescription and over-the-counter (OTC) medicines can have unintended consequences, causing harm when they interact with the body's normal use of nutrients. As Figure C11-1 shows, drugs can interact with nutrients in the following ways:

- Foods can delay or prevent drug absorption.
- Drugs can delay or prevent nutrient absorption.
- Nutrients can interfere with drug action, metabolism, or excretion.
- Drugs can interfere with nutrient action or excretion.
- Drugs can modify taste, appetite, or food intake.

These interactions do not occur every time a person takes a drug. Some people are more vulnerable than others to nutrient-drug interactions. The potential for undesirable nutrient-drug interactions is greatest for those who:

- Take drugs (or medicines) for long times.
- Take two or more drugs at the same time.
- Are poorly nourished to begin with or are not eating well.

Herbal remedies also react with drugs, sometimes dangerously (see Table C11-1). Alcohol is infamous for its interactions with nutrients (see Controversy 3).

Figure C11•1

FOOD, DRUG, AND HERB INTERACTIONS

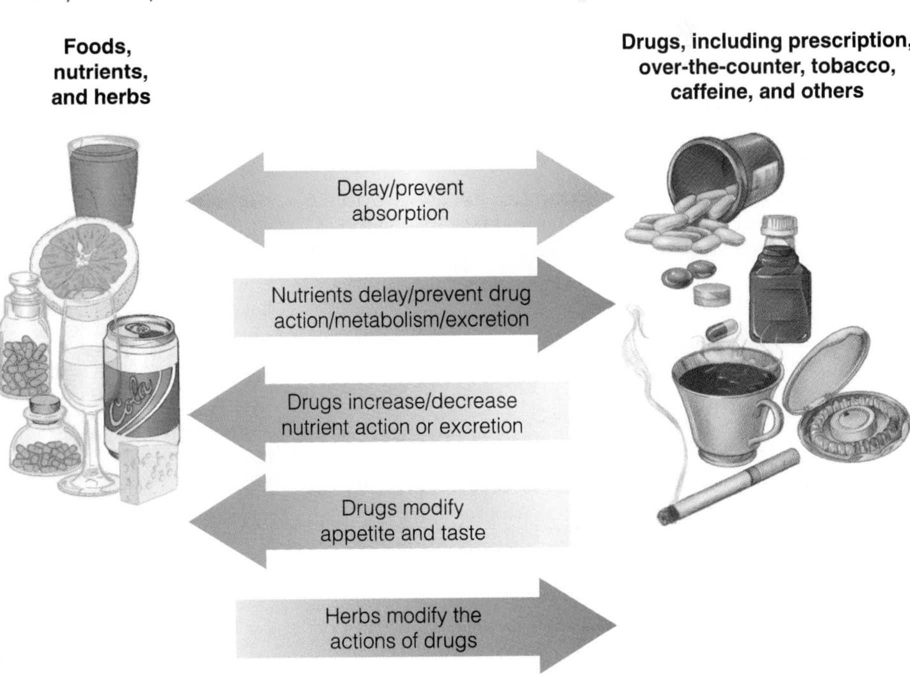

Absorption of Drugs and Nutrients

The business executive described earlier felt the effects of the first type of interaction in the list above. Acid from the tomato juice, coffee, and colas she drank before chewing the nicotine gum kept the nicotine from being absorbed through the lining of her mouth. With this route blocked, the nicotine traveled to her stomach, remained unabsorbed, and caused nausea. Other foods and beverages can have similar effects (see Table C11-2). Once identified, the problem is easy to prevent by waiting to eat or drink until after chewing the gum.

Drugs can also interfere with the small intestine's absorption of nutrients, particularly minerals. This interaction explains the experience of the tired 14-year-old. Her overuse of antacids eliminated the stomach's normal acidity, on which iron absorption depends. The medicine bound tightly to the iron molecules, forming an insoluble, unabsorbable complex. Her iron stores already bordered on deficiency, as iron stores for young girls typically do, so her misuse of antacids pushed her over the edge into outright deficiency.

Chronic laxative use can also lead to malnutrition. Laxatives can carry nutrients through the intestines so rapidly that many vitamins have no time to be absorbed. Mineral oil, a laxative the body cannot absorb, can rob a person of fat-soluble vitamins. Vitamin D deficiencies can occur this way; calcium can also be excreted with the oil, accelerating adult bone loss.

Metabolic Interactions and Nutrient Excretion

The teacher who landed in the emergency room was taking an antidepressant medicine, one of the monoamine oxidase inhibitors (MAOI). At the party, he suffered a dangerous chemical interaction between the medicine and the compound tyramine in his cheese and wine. Tyramine is produced during the fermenting process in cheese and wine manufacturing.

The MAOI medication works by depressing the activity of enzymes that destroy the brain neurotransmitter dopamine. With less enzyme activity, more dopamine is left, and depression lifts. At the same time, the drug also depresses enzymes in the liver that destroy tyramine. Ordinarily, the man's liver would have quickly destroyed the tyramine from the cheese and wine. But due to the MAOI medication, tyramine built up too high in the man's body and caused the potentially fatal reaction.

Other culprits that affect the metabolism of medication include grapefruit juice and one of the most popular herbal supplements in the United States, ginkgo biloba. Something in grapefruit juice suppresses an enzyme responsible for breaking down more than 20 kinds of medical drugs.[1] With less drug breakdown, doses build up in the blood to levels that can have undesirable effects on the body. For example, in a drinker of grapefruit juice, a normal dosage of the blood-thinning drug coumarin can lead to dangerously prolonged bleeding and delayed clotting of blood. As for ginkgo, people take it as a supplement in hopes of improving memory, but this effect is unproved. Takers may not know that it has been found to stimulate the activity of liver enzymes responsible for metabolizing many medications, and so may diminish their effects.[2]

Drugs often cause nutrient losses, too. Many people take large quantities of aspirin (10 to 12 tablets each day) to relieve the pain of arthritis, backaches, and headaches. This much aspirin can

Table C11•1

HERB AND DRUG INTERACTIONS

Herb	Drug	Interaction
Feverfew	Aspirin, ibuprofen, and other nonsteroidal anti-inflammatory drugs	Drugs negate the effect of the herb for headaches.
Feverfew, garlic, Ginkgo, ginger, and ginseng	Warfarin, coumarin (anticlotting drugs, "blood thinners")	Prolonged bleeding time; danger of hemorrhage
Ginseng	Estrogens, corticosteroids	Enhanced hormonal response
Kyushin, licorice, plantain, uzara root, hawthorn, ginseng	Digoxin (cardiac antiarrhythmic drug derived from the herb foxglove)	Herbs interfere with drug action and monitoring.
Ginseng, karela	Blood glucose regulators	Herbs affect blood glucose levels.
Kelp (iodine source)	Synthroid or other thyroid hormone replacers	Herb may interfere with drug action.
St. John's wort, saw palmetto, black tea	Iron	Tannins in herbs inhibit iron absorption.
Echinacea (possible immunostimulant)	Cyclosporine and corticosteroids (immunosuppressants)	May reduce drug effectiveness
Evening primrose oil, borage	Anticonvulsants	Seizures

SOURCE: L. G. Miller, Herbal medicinals: Selected clinical considerations focusing on known or potential herb-drug interactions, *Archives of Internal Medicine* 158 (1998): 2200–2211.

Table C11•2

FOODS AND BEVERAGES THAT LIMIT THE EFFECTIVENESS OF NICOTINE GUM

- Apple juice
- Beer
- Coffee
- Colas
- Grape juice
- Ketchup
- Lemon-lime soda
- Mustard
- Orange juice
- Pineapple juice
- Soy sauce
- Tomato juice

speed up blood loss from the stomach by as much as ten times, enough to cause iron-deficiency anemia in some people. People who take aspirin regularly should eat iron-rich foods regularly as well. Table C11-3 lists some examples of other possible nutrient-drug interactions, including both prescription and OTC medications. Details on some common interactions follow.

Oral Contraceptives and Estrogen

Millions of women use oral contraceptives, daily doses of hormones that prevent pregnancy. Oral contraceptive interactions with nutrients illustrate the complexity of nutrient-drug interactions.

Each nutrient responds differently to oral contraceptive use (see Table C11-3). The vitamin B_{12} status of oral contraceptive users may be slightly lower than in

Table C11•3

NUTRITION EFFECTS OF A FEW COMMONLY USED MEDICAL DRUGS

Medicines and Caffeine	Effects on Absorption	Effects on Excretion	Effects on Metabolism
Antacids (aluminum containing)	Reduce iron absorption	Increase calcium and phosphorus excretion	May accelerate destruction of thiamin
Antibiotics (long-term usage)	Reduce absorption of fats, amino acids, folate, fat-soluble vitamins, vitamin B_{12}, calcium, copper, iron, magnesium, potassium, phosphate, zinc	Increase excretion of folate, niacin, potassium, riboflavin, vitamin C	Destroy vitamin K–producing bacteria and reduce vitamin K production
Aspirin (large doses, long-term usage)	Lowers blood concentration of folate	Increases excretion of thiamin, vitamin C, vitamin K; causes iron and potassium losses through blood loss	
Caffeine		Increases secretion of small amounts of calcium and magnesium	Stimulates release of fatty acids into the blood
Diuretics		Raise blood calcium and zinc; lower blood folate, chloride, magnesium, phosphorus, potassium, vitamin B_{12}; increase excretion of calcium, sodium, thiamin, potassium, chloride, magnesium	Interfere with storage of zinc
Laxatives (effects vary with type)	Reduce absorption of glucose, fat, carotene, vitamin D, other fat-soluble vitamins, calcium, phosphate, potassium	Increase excretion of all unabsorbed nutrients	
Oral contraceptives	Reduce absorption of folate, may improve absorption of calcium	Cause sodium retention	Raise blood vitamin A, vitamin D, copper, iron; may lower blood beta-carotene, riboflavin, vitamin B_6, vitamin B_{12}, vitamin C; may elevate requirements for riboflavin and vitamin B_6; alter blood lipids elevating risk of heart disease in smokers and older women
Estrogen replacement therapy	May reduce absorption of folate	Causes sodium retention	May raise blood glucose, triglycerides, vitamin A, vitamin E, copper, and iron; may lower blood vitamin C, folate, vitamin B_6, riboflavin, calcium, magnesium, and zinc

SOURCES: Data from Z. M. Pronsky, *Food Medication Interactions*, 9th ed. (Pottstown, Pa.: Food-Medication Interactions, 1995); G. Berg, L. Kohlmeier, and H. Brenner, Use of oral contraceptives and serum beta-carotene, *European Journal of Clinical Nutrition* 51 (1997): 181–187; S. S. Harris and B. Dawson-Hughes, The association of oral contraceptive use with plasma 25-hydroxyvitamin D levels, *Journal of the American College of Nutrition* 17 (1998): 282–284; S. M. Vaziri and coauthors, The impact of female hormone usage on the lipid profile: The Framingham Offspring Study, *Archives of Internal Medicine* 153 (1993): 2200–2206.

others.[3] Beta-carotene values may also be reduced, leading researchers to wonder whether the lower levels of this antioxidant might influence some disease risks.[4] Vitamin D levels, on the other hand, may be higher in oral contraceptive users, with unknown effects.[5] At first glance these findings seem to indicate that women using oral contraceptives are on their way to suffering deficiencies of some nutrients and have somehow enlarged their body stores of others. Research has yielded conflicting results, however, so any such assumptions are premature.

Significantly, oral contraceptives alter blood lipids, possibly increasing the risk of cardiovascular disease for menstruating women.[6] Especially in women older than about 35 years, most oral contraceptives raise total cholesterol and triglyceride concentrations and lower HDL, amplifying the risk of stroke and heart disease. A few women using oral contraceptives also experience mild hypertension.

Some women lose weight when taking oral contraceptives, but others may gain as much as 20 pounds or more from fat deposited in the hips, thighs, and breasts or from retained fluid. Some lean tissue is also deposited in response to an androgenic (steroid) effect of the pills. Sometimes a switch to another form of pill can normalize body weight.

As with oral contraceptives, women's responses to estrogen replacement drugs must be assessed individually. Some women suffer edema because estrogen promotes sodium conservation by the kidneys; sodium restriction can correct this condition. Others develop abnormally low blood folate or vitamin B_6, indicating a need to include more vitamin-rich, nutrient-dense foods in the diet. All women taking estrogen should be aware that vitamin C doses of a gram or more may elevate serum estrogen and falsely suggest that a lower dose is needed.

If a woman taking any form of estrogen thinks she may have a nutrient deficiency, she should refrain from taking individual supplements and seek testing and a diagnosis from a health-care professional to rule out other causes of her symptoms. For most women, a nutritious diet is all that is needed. If a woman feels compelled to take a supplement, however, a standard multivitamin-mineral supplement is probably harmless, as long as it accompanies a well-balanced diet.

Caffeine

The well-known "wake-up" effect of caffeine is the primary reason people in every society use it in some form. Compared with the drugs discussed so far, though, caffeine's interactions with foods and nutrients are subtle. Yet caffeine's relationship to nutrition is important because caffeine is so widespread that people may be unaware that they are consuming it—see Table C11-4 for the caffeine contents of many beverages and foods. Many OTC cold and headache remedies contain caffeine because, in addition to being a mild pain reliever in its own right, caffeine remedies the headache caused by caffeine withdrawal that no other pain reliever can touch. Caffeine is present in chocolate bars, colas, and other foods children favor, and children are more sensitive to caffeine's effects because they are small and, at first, not adapted to its use.

Caffeine is the most popular and widely consumed drug in the United States. One in three U.S. citizens consumes about 200 milligrams of caffeine per day (as in 2 small cups of coffee), but many others consume much more. Many people's intake patterns fulfill

Table C11•4

CAFFEINE CONTENT OF BEVERAGES AND FOODS

Drinks and Foods	Average (mg)	Range (mg)
Coffee (5 oz cup)		
Brewed, drip method	130	110–150
Brewed, percolator	94	64–124
Instant	74	40–108
Instant "lite"	30	no data
Decaffeinated, brewed or instant	3	1–5
Tea (5 oz cup)		
Brewed, major U.S. brands	40	20–90
Brewed, imported brands	60	25–110
Instant	30	25–50
Iced (12 oz glass)	70	67–76
Herb teas (caffeine-free)	0	0
Soft drinks (12 oz can)		
Dr. Pepper		40
Colas and cherry colas:		
Regular		30–46
Diet		2–58
Clear and caffeine-free		0–trace
Extra caffeine (Jolt)		75–100
Mountain Dew, Mello Yello		52
Big Red		38
Fresca, 7-Up, Sprite, Squirt, Sunkist Orange, seltzers, root beers		0
Cocoa beverage (5 oz cup)	4	2–20
Chocolate milk beverage (8 oz)	5	2–7
Milk chocolate candy (1 oz)	6	1–15
Dark chocolate, semisweet (1 oz)	20	5–35
Baker's chocolate (1 oz)	26	26
Chocolate-flavored syrup (1 oz)	4	4
Carob	0	0

NOTE: Many over-the-counter medications such as pain relievers and cold medicines also contain caffeine. Their labels must list the milligram amounts of caffeine per dose of medicine. Read medicine labels carefully.

some of the accepted criteria for a diagnosis of drug dependence.[7] A single 200-milligram dose of caffeine significantly improves the ability to pay attention over a two-hour time period but more is probably not better.[8] A dose of 500 milligrams has been shown to worsen thinking abilities, and more than this may present some risk to health.[9]

Caffeine is a true stimulant drug. Like all stimulants, it increases the respiratory rate, heart rate, blood pressure, and secretion of stress and other hormones. A moderate dose of caffeine may speed up metabolic energy expenditures for several hours, and it stimulates the digestive tract, promoting efficient elimination. Because caffeine is a diuretic, it promotes water loss from the body as well.

Despite caffeine's tremendous popularity, many people today are consuming less because they fear possible harm to their health. Research in the last decade has yielded sporadic reports linking caffeine to health problems such as cancer, birth defects, and hypertension. Much other research, however, refutes any links between caffeine and cancer or birth defects and finds a weak link between caffeine and elevated blood pressure, an effect that may be significant for those with diagnosed hypertension.[10] Coffee and tea contain phytochemicals other than caffeine, some of which may reduce heart disease risk while others may worsen it.[11]

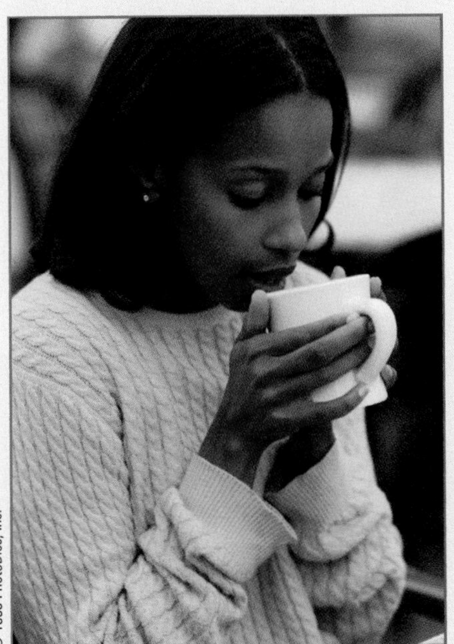

Caffeine seems relatively harmless when used in moderation (2 cups of coffee a day). One study reports that men consuming 2 to 3 cups of coffee a day run a 40 percent lower risk of developing active gallstone disease than men who avoid coffee.[12] However, *cola* consumption may increase the risk.[13] In higher doses, caffeine can cause symptoms associated with anxiety: sweating, tenseness, and inability to concentrate. High doses may also accelerate bone loss in women past mid-life, and caffeine may contribute to painful but benign fibrocystic breast disease. More controlled studies are needed to determine whether eliminating caffeine can help to reverse the condition.[14]

If you like caffeine-containing foods or beverages, the most reasonable approach is to limit your intake to the equivalent of 2 small cups of coffee per day. For most people, this is enough to reduce drowsiness and sharpen awareness without paying too high a price. Pregnant women should exercise moderation in using caffeine, and parents should monitor and control their children's intakes.

Tobacco

Cigarette and other tobacco use causes thousands of people to suffer from cancer and other diseases of the cardiovascular, digestive, and respiratory systems. These effects are beyond nutrition's scope, but smoking does depress hunger and body fatness and change nutrient status. Chapter 9 makes the point that any contribution smoking may make toward controlling body fatness is far exceeded by its destructive potential.

Nutrient intakes of smokers and nonsmokers differ. Smokers have lower intakes of dietary fiber, vitamins, and minerals, even when their energy intakes are quite similar to those of nonsmokers. The association between smoking and low vitamin intake may be important because studies have shown that smoking alters the metabolism of vitamin C. (Research has just begun on other nutrients.)

Research shows that the vitamin C requirement of smokers exceeds that of nonsmokers. Smokers break down vitamin C faster and so must take in more vitamin C–containing foods to achieve steady body pools comparable to those of nonsmokers. The effect is apparently related to an increase in oxidative stress produced by smoking, and unrelated to the tobacco drug nicotine.[15] The evidence is so strong that the vitamin C recommendation is set higher for smokers—smokers require an extra 35 milligrams of vitamin C per day.

Illicit Drugs

People know that illicit drugs are harmful, but many choose to abuse them anyway in spite of the risks. Like OTC and prescription drugs, illegal drugs modify body functions. Unlike medicines, however, no watchdog agency such as the Food and Drug Administration monitors them for safety, effectiveness, or even purity.

Smoking a marijuana cigarette affects several senses including the sense of taste. It produces an enhanced enjoyment of eating, especially of sweets, commonly known as "the munchies." Why or how this effect occurs is not known. Despite higher food intakes, marijuana abusers often consume fewer nutrients than do nonabusers because the extra foods they choose tend to be high-calorie, low-nutrient snack foods. Besides the nutrition effects, regular marijuana users face the same risk of lung cancer as people who smoke a pack of cigarettes a day.

Cocaine elicits effects such as intense euphoria, restlessness, heightened self-confidence, irritability, insomnia, and loss of appetite. Weight loss is a common side effect, and cocaine abusers often develop eating disorders. Repeated use can cause a rapid heart rate, irregular heartbeats, heart attacks, and death.

Unlike marijuana, cocaine causes serious malnutrition in those who use it. The stronger the craving for cocaine, the less a drug abuser wants nutritious food. Rats given unlimited access to cocaine will choose the drug over food until they die of starvation. The effects of the other addictive drugs vary in degree but are similar in kind to those of cocaine. A few are listed in

Table C11•5

NUTRITION EFFECTS OF FOUR NONMEDICAL DRUGS

Drug of Abuse	Possible Effects on Nutrition Status
Cocaine	Reduces intakes of nutritious foods; increases intakes of alcohol, coffee, and fat; may induce or aggravate eating disorders
Heroin	Heightens and delays insulin response to glucose; reduces intakes of nutritious foods
Marijuana	Increases intakes of foods, especially sweets; may cause weight gain
Nicotine[a]	Reduces intake of sweet foods and water; increases intakes of fat; reduces fetal weight; lowers blood concentration of beta-carotene.

[a]Other effects of smoking include increased vitamin C requirements.

SOURCES: Data from M. E. Mohs, R. R. Watson, and T. Leonard-Green, Nutritional effects of marijuana, heroin, cocaine, and nicotine, *Journal of the American Dietetic Association* 90 (1990): 1261–1267; G. van Poppel, S. Spanhaak, and T. Ockhuizen, Effects of beta carotene on immunological indexes in healthy male smokers, *American Journal of Clinical Nutrition* 57 (1993): 402–407.

Table C11-5. Drug abusers face multiple nutrition problems, and an important aspect of addiction recovery is the identification and correction of nutrition problems.

Personal Strategy

In conclusion, when you need to take a medicine, do so wisely. Ask your physician, pharmacist, or other health-care provider for specific instructions about the doses, times, and how to take the medication—for example, with meals or on an empty stomach. If you notice new symptoms or if a drug seems not to be working well, consult your physician. The only instruction people need about illicit drugs is to avoid them altogether for countless reasons. As for smoking and chewing tobacco, the same advice applies: don't take these habits up, or if you already have, take steps to quit. For drugs with lesser consequences to health, such as caffeine, use moderation.

Try to live life in a way that requires less chemical assistance. If you are sleepy, try a 15-minute nap or meditation instead of a 15-minute coffee break. The coffee will stimulate your nerves for an hour, but the alternatives will refresh your attitude for the rest of the day. If you suffer constipation, try getting enough exercise, fiber, and water for a few days. Chances are that a laxative will be unnecessary. The strategy being suggested here is to take control of your body, allowing your reliable, self-healing nature to make fine adjustments that you need not force with chemicals. Bodies have few requests: adequate nutrition, rest, exercise, and hygiene. Give your body what it asks for, and let it function naturally, day-to-day, without interference from drugs.

Michael Escoffery, *Mother and Son*, 1996

Life Cycle Nutrition: Mother and Infant

Frequently Asked Questions

Contents

ALL PEOPLE need the same nutrients, but the amounts we need change as we move through life. This chapter is the first of two on life's changing nutrient needs. It focuses on the two life stages that might be the most important to an individual's lifelong health—pregnancy and infancy.

Pregnancy: The Impact of Nutrition on the Future

We normally think of our nutrition as personal, affecting only our own lives. The woman who is pregnant, or who soon will be, must understand that her nutrition today is critical to the health of her child throughout life. The nutrition demands of pregnancy are extraordinary—as the saying goes, she's eating for two.

Preparing for Pregnancy

Before she becomes pregnant, a woman must establish eating habits that will optimally nourish both the growing fetus and herself. She must be well nourished at the outset because early in pregnancy the embryo undergoes rapid and significant developmental changes that depend on good nutrition.

Fathers-to-be are wise to examine their eating and drinking habits. Limited evidence suggests that men who consume too few fruits and vegetables containing vitamin C or who drink too much alcohol in the weeks before conception can sustain damage to their sperm's genetic material. This damage can cause birth defects in children.

low birthweight a birthweight of less than 5½ pounds (2,500 grams); used as a predictor of probable health problems in the newborn and as a probable indicator of poor nutrition status of the mother before and/or during pregnancy. Low-birthweight infants are of two different types. Some are premature; they are born early and are the right size for their gestational age. Others have suffered growth failure in the uterus; they may or may not be born early, but they are small for gestational age (small for date).

Underweight is defined as BMI <19.8. Obese is defined as BMI >29 (see Table 12-5 on page 444).

Both parents can prepare in advance for a healthy pregnancy.

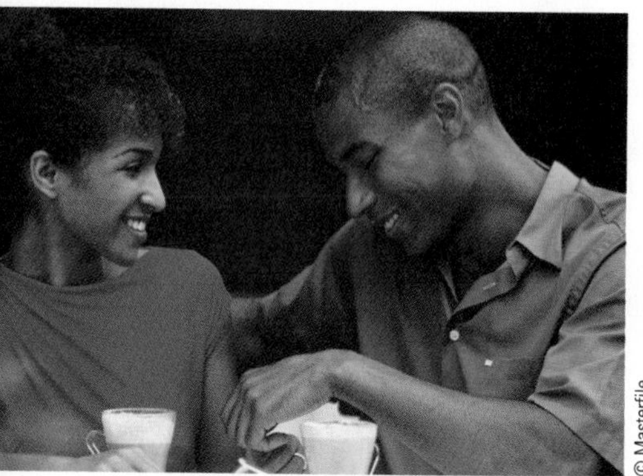

© Masterfile

Prepregnancy Weight Prior to pregnancy, all women, but underweight women in particular, should strive for appropriate body weights. An underweight woman who fails to gain enough weight during pregnancy is most likely to bear a baby with a dangerously low birthweight. Infant birthweight is the single most potent indicator of an infant's future health. A **low-birthweight** baby, defined as one who weighs less than 5½ pounds (2,500 grams), is nearly 40 times more likely to die in the first year of life than a normal-weight baby. Other hazards of low birthweight may include lower adult IQ and other brain and sensory impairments, short stature, and educational disadvantages.[1] Research suggests that when nutrient supplies during pregnancy fail to meet demands, permanent adaptations take place that may predispose the infant to chronic diseases in later life.[2] Controversy 12 addresses this topic in detail. Underweight women are therefore advised to gain weight before becoming pregnant and to strive to gain adequately during pregnancy.

Nutritional deficiency, coupled with low birthweight, is the underlying cause of more than half of all the deaths worldwide of children under five years of age. In 1998, the U.S. infant mortality rate was among the lowest the nation has ever recorded: 7.2 deaths per 1,000 live births.[3] This rate, though higher than some developed countries, is part of a significant steady decline over the last two decades and is a tribute to public health efforts aimed at reducing infant deaths (see Figure 12-1).

Figure 12•1

INFANT MORTALITY DECLINE IN THE TWENTIETH CENTURY

The graph shows infant deaths per 1,000 live briths.

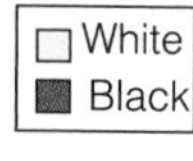

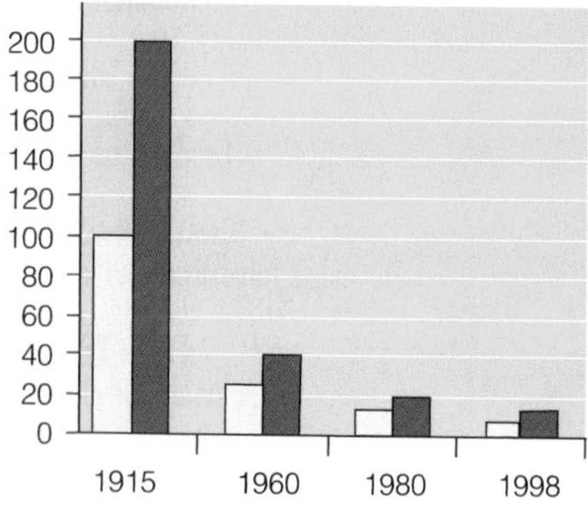

SOURCE: Data from B. Guyer and coauthors, Annual summary of vital statistics: Trends in the health of Americans during the 20th century, *Pediatrics*, 106 (2000): 1307–1317.

Not all cases of low birthweight reflect poor nutrition. Other factors are heredity, disease conditions, smoking, and drug (including alcohol) use during pregnancy. Even with optimal nutrition and health during pregnancy, some women give birth to small infants for reasons unknown. But poor nutrition is the major factor in low birthweight—and an avoidable one, as later sections make clear.

Obese women are also urged to attain healthy weights before pregnancy. The infant of an obese mother may be larger than normal and born late, or may be large even if born prematurely. The large early baby may not be recognized as premature and thus not receive the special medical care required. Maternal obesity may also double the risk for neural tube defects in the infant.[4] Obese women are more likely to require drugs to induce labor or require surgical intervention for the birth, and they suffer gestational diabetes, hypertension, and infections after the birth more often than do women of healthy weight.[5] An appropriate goal for the obese woman who wishes to become pregnant is to attain a prepregnancy body weight low enough to minimize her medical risks and those of her future child.

A Healthy Placenta and Other Organs A major reason the mother's nutrition before pregnancy is so crucial is that it determines whether her **uterus** will be able to support the growth of a healthy **placenta** during the first month of **gestation.** The placenta is both a supply depot and a waste-removal system for the fetus. If the placenta works perfectly, the fetus wants for nothing; if it doesn't, no alternative source of sustenance is available, and the fetus will fail to thrive. Figure 12-2 shows the placenta, a mass of tissue in which maternal and fetal blood vessels intertwine and exchange materials. The two bloods never mix, but the barrier between them is notably thin. To grasp how thin, picture your hands as fetal blood vessels, skintight surgical gloves as the tissue-thin placenta, and finally, your gloved hands immersed in water as the pool of maternal blood. Across this thin barrier, nutrients and oxygen move from the mother's blood into the fetus's blood, and wastes move out of the fetal blood, to be excreted by the mother. The umbilical cord is the pipeline from the placenta to the fetus. The **amniotic sac** surrounds and cradles the baby, cushioning it with fluids.

Far from being a passive transfer system, the placenta is an active metabolic organ with some 60 sets of enzymes of its own. It gathers up hormones, nutrients, and protein molecules such as antibodies and transfers them into the fetal bloodstream. The placenta also produces hormones that maintain pregnancy and prepare the mother's breasts for **lactation.**

If the mother's nutrient stores are inadequate during the period when her body is developing the placenta, then the placenta will never form and function properly. As a consequence, no matter how well the mother eats later, her fetus will not receive

uterus (YOO-ter-us) the womb, the muscular organ within which the infant develops before birth.

placenta (pla-SEN-tuh) the organ that develops inside the uterus in early pregnancy in which maternal and fetal blood circulate in close proximity and exchange materials. The fetus receives nutrients and oxygen across the placenta; the mother's blood picks up carbon dioxide and other waste materials to be excreted via her lungs and kidneys.

gestation the period of about 40 weeks (three trimesters) from conception to birth; the term of a pregnancy.

amniotic (am-nee-OTT-ic) **sac** the "bag of waters" in the uterus in which the fetus floats.

lactation production and secretion of breast milk for the purpose of nourishing an infant.

Neural tube defects and gestational diabetes are discussed in later sections.

Figure 12•2

THE PLACENTA

The placenta is composed of spongy tissue in which fetal blood and maternal blood flow side by side, each in its own vessels. The maternal blood transfers oxygen and nutrients to the fetus's blood and picks up fetal wastes to be excreted by the mother. Thus, the placenta performs the nutritive, respiratory, and excretory functions that the fetus's digestive system, lungs, and kidneys will provide after birth.

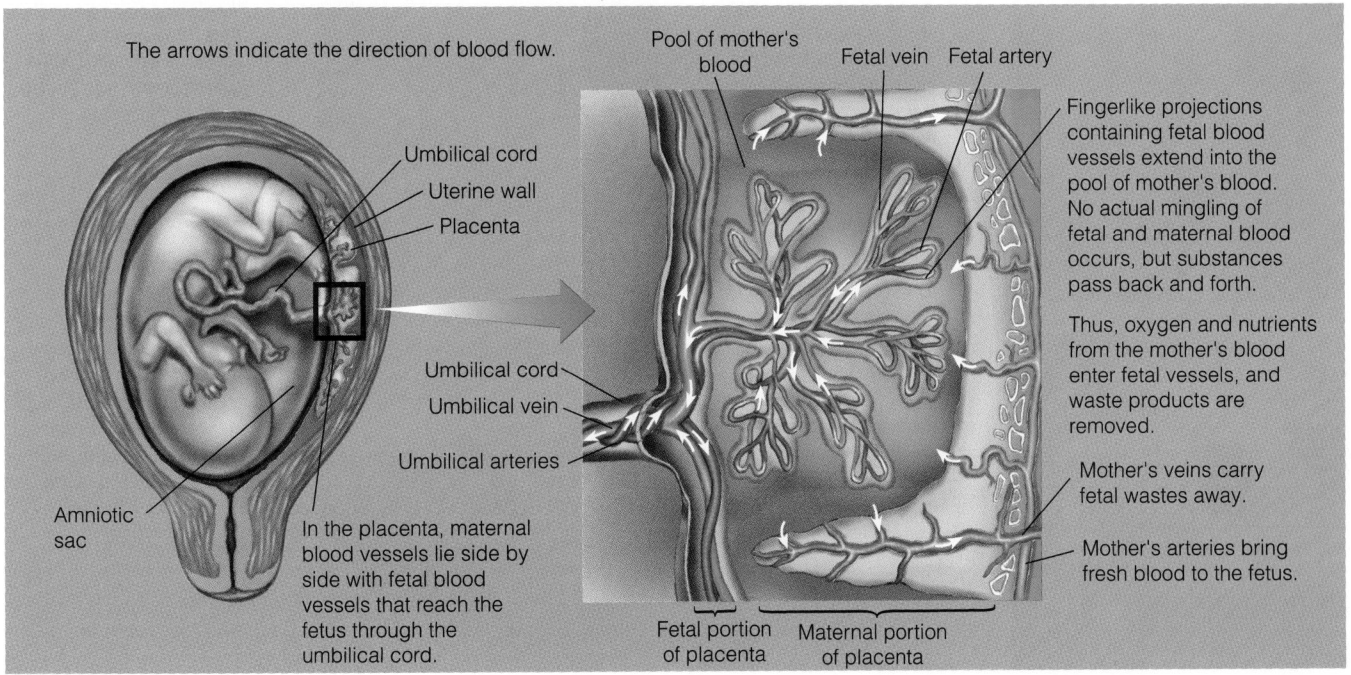

implantation the stage of development, during the first two weeks after conception, in which the fertilized egg (fertilized ovum or zygote) embeds itself in the wall of the uterus and begins to develop.

ovum the egg, produced by the mother, that unites with a sperm from the father to produce a new individual.

zygote (ZYE-goat) the term that describes the product of the union of ovum and sperm during the first two weeks after fertilization.

critical period a finite period during development in which certain events may occur that will have irreversible effects on later developmental stages. A critical period is usually a period of cell division in a body organ.

embryo (EM-bree-oh) the stage of human gestation from the third to the eighth week after conception.

fetus (FEET-us) the stage of human gestation from eight weeks after conception until the birth of an infant.

optimal nourishment, and a low-birthweight baby with all of the associated risks is likely. After getting such a poor start on life, children may be ill equipped, even as adults, to store sufficient nutrients, and a girl may later be unable to grow an adequate placenta. In turn, she may bear an infant who is unable to reach full potential.

key point *Adequate nutrition before pregnancy establishes physical readiness and nutrient stores to support fetal growth. Both underweight and overweight women should strive for appropriate body weights before pregnancy. Newborns who weigh less than 5½ pounds face greater health risks than normal-weight babies. The healthy development of the placenta depends on adequate nutrition before pregnancy.*

The Events of Pregnancy

On **implantation** of the newly fertilized **ovum** in the uterine wall, the placenta begins to grow inside the uterus. During the two weeks following fertilization, the **zygote** divides into many cells, and these cells sort themselves into three layers. Minimal growth in size takes place at this time, but it is a **critical period** in development. Adverse influences such as smoking, drug abuse, and malnutrition at this time lead to failure to implant or to abnormalities such as neural tube defects that can cause loss of the zygote, possibly before the woman knows she is pregnant.

The Embryo During the next six weeks of development, the **embryo** registers astonishing physical changes (see Figure 12-3). At eight weeks, the **fetus** has a complete central nervous system, a beating heart, a fully formed digestive system, well-defined fingers and toes, and the beginnings of facial features.

In the last seven months of pregnancy, the fetal period, the fetus grows 50 times heavier and 20 times longer. Critical periods of cell division and development occur in organ after organ. The amniotic sac fills with fluid and the mother's body changes. The uterus and its supporting muscles increase in size, the breasts may become ten-

Figure 12•3

STAGES OF EMBRYONIC AND FETAL DEVELOPMENT

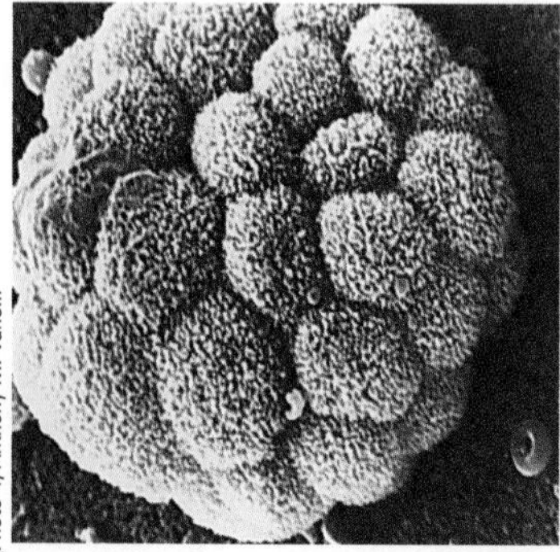

(1) A newly fertilized ovum is about the size of the period at the end of this sentence. This zygote at less than one week after fertilization is not much bigger and is ready for implantation.

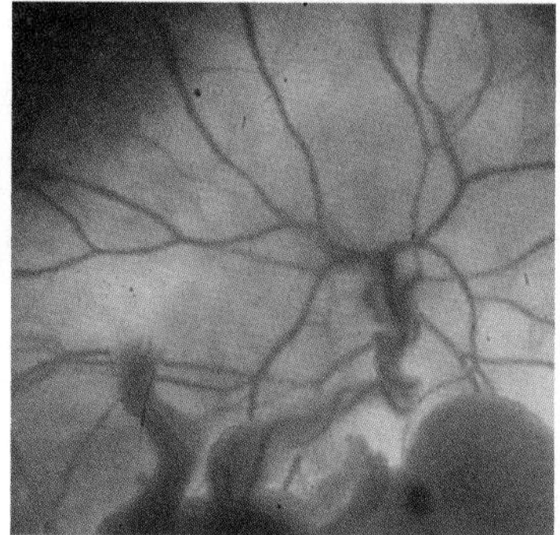

(3) A fetus after 11 weeks of development is just over an inch long. Notice the umbilical cord and blood vessels connecting the fetus with the placenta.

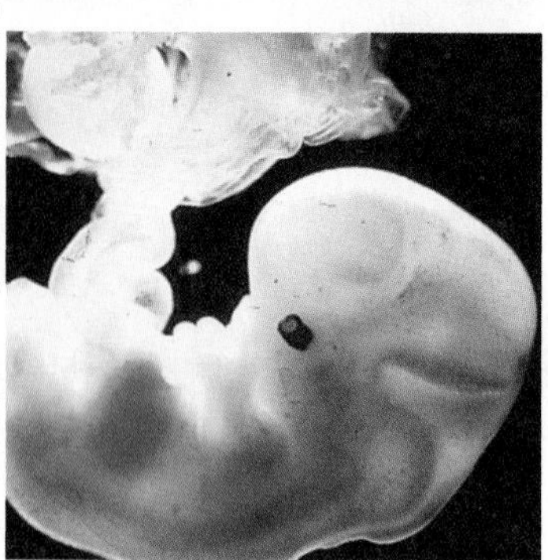

(2) After implantation, the placenta develops and begins to provide nourishment to the developing embryo. An embryo five weeks after fertilization is about ½ inch long.

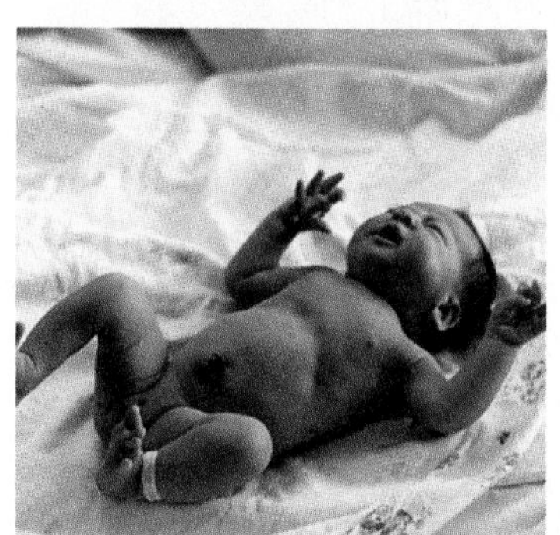

(4) A newborn infant after nine months of development measures close to 20 inches in length. From eight weeks to term, this infant grew 20 times longer and 50 times heavier.

der and full, the nipples may darken in preparation for lactation, and the mother's blood volume increases by half to accommodate the added load of materials it must carry. Gestation lasts approximately 40 weeks and ends with the birth of the infant.

trimesters periods representing one-third of gestation. A trimester is about 13 to 14 weeks.

Critical Periods Each organ and tissue type grows with its own characteristic pattern and timing. The development of each takes place only at a certain time—the critical period. Whatever nutrients and other environmental conditions are necessary during this period must be supplied on time if the organ is to reach its full potential. If the development of an organ is limited during a critical period, recovery is impossible.[6] For example, the fetus's heart and brain are well developed at 14 weeks; the lungs, 10 weeks later. Therefore, early malnutrition impairs the heart and brain; later malnutrition impairs the lungs.

The effects of malnutrition during critical periods of pregnancy are seen in neural tube defects of the nervous system of the embryo (explained later), in the child's poor dental health, and in the adult's vulnerability to infections and possibly higher risks of diabetes, hypertension, stroke, and heart disease.[7] The effects of malnutrition during critical periods are irreversible: abundant and nourishing food, fed after the critical time, cannot remedy harm already done.

DRI nutrient intake recommendations for pregnant women are listed on the inside front cover.

Table 12-1 provides a list of factors that make nutrient deficiencies likely during pregnancy. Notice that young age heads the list; a later section explains why pregnant adolescents are especially prone to malnutrition.

Placental development, implantation, and early critical periods depend on maternal nutrition before and during pregnancy.

Increased Need for Nutrition

During pregnancy, a woman's nutrient needs increase more for certain nutrients than for others. Figure 12-4 on the next page shows the percentage increase in nutrient intakes recommended for pregnant women compared to nonpregnant women.

Energy, Protein, and Fat A pregnant woman needs extra food energy, but only a little extra—300 calories above the allowance for nonpregnant women—and only during the second and third **trimesters.** Studies of energy balance in well-nourished pregnant women find they meet the energy demands of pregnancy in several ways: some women eat more food, some reduce their activity, and some store less fat than others.[8] A woman can get 300 calories with just one extra serving from each of the five food groups—a slice of bread, a serving of vegetables, an ounce of lean meat, a piece of fruit, and a cup of fat-free milk (see Table 12-2 on page 441 and the sample meal plan in the margin on page 442). Pregnant teenagers, underweight women, and physically active women may require more. This increment of extra energy is needed to spare protein for its all-important tissue-building work.

The increase recommended for protein is greater than for energy: from 45 to 50 grams of protein per day for a nonpregnant woman to 60 grams for a pregnant woman. Most women in the United States, however, need not add protein-rich foods to their diets because they already exceed the recommended protein intake for pregnancy. Pregnant women in the United States—even those with low incomes who are not participating in food assistance programs—consume between 75 and 110 grams of protein a day. Excess protein may also have adverse effects, as Chapter 6 explained.

Some vegetarian women limit or omit protein-rich meats, eggs, and dairy products from their diets. For them, meeting the recommendation for food energy each day and including several generous servings of plant-protein foods such as legumes, tofu, whole grains, nuts, and seeds are imperative. Use of high-protein supplements during pregnancy can be harmful and is discouraged. All pregnant women need generous amounts of carbohydrate-rich foods to spare their protein and to provide energy.

The high nutrient requirements of pregnancy leave little room in the diet for excess energy from added purified fats such as oil, margarine, and butter. The essential fatty

Table 12•1

FACTORS PLACING PREGNANT WOMEN AT NUTRITIONAL RISK

Women likely to develop nutrient deficiencies include those who:

- Are young (adolescents).
- Have had many previous pregnancies (3 or more to mothers under age 20; 4 or more to mothers age 20 or older).
- Have short intervals between pregnancies (<18 months).
- Lack nutrition knowledge, have too little money to purchase adequate food, or have too little family support.
- Consume an inadequate diet due to food faddism, preferences, weight-loss "dieting," uninformed vegetarianism, or eating disorders.
- Smoke cigarettes or use alcohol or illicit drugs.
- Are lactose intolerant or suffer chronic health conditions requiring special diets.
- Are underweight or overweight at conception.
- Are carrying twins or triplets.
- Gain insufficient or excessive weight during pregnancy.
- Have a low level of education.

Figure 12•4

COMPARISON OF NUTRIENT RECOMMENDATIONS FOR NONPREGNANT, PREGNANT, AND LACTATING WOMEN

For actual values, turn to the table on the inside front cover.

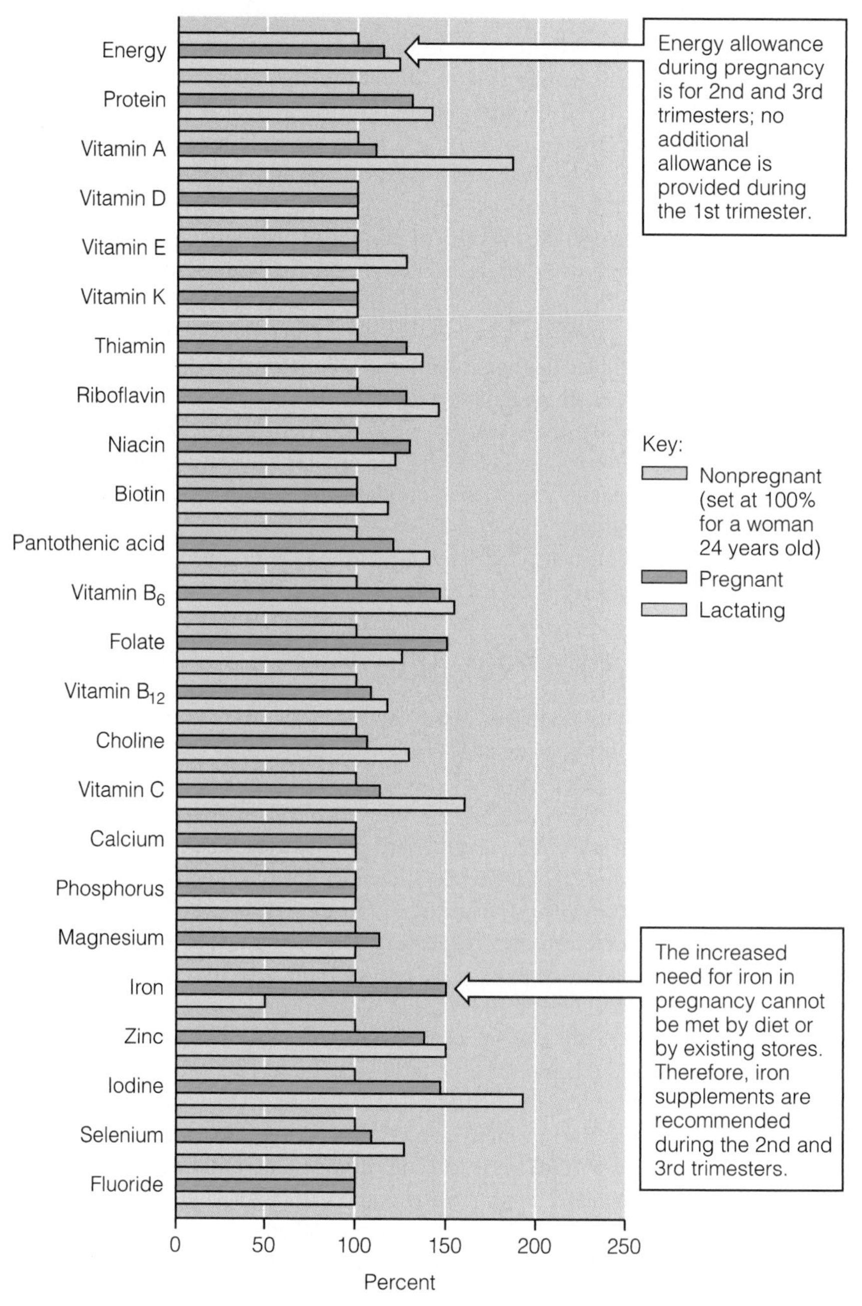

Recommended protein intake: 60 grams/day.

Recommended carbohydrate intake: about 50% of energy intake. In a 2,000-calorie per day intake, this represents 1,000 calories of carbohydrate, or about 250 grams.

acids, however, are important to the growth of the fetus and are regarded by some as essential nutrients in early human development.[9] The brain is composed mainly of lipid material and depends heavily on products of both omega-3 and omega-6 fatty acids for its growth, function, and structure. If a mother-to-be eats a diet that includes seafood, she receives a balance of the essential fatty acids and their derivatives. This benefits her pregnancy and later her infant by way of her milk. Supplements of fish oil are not recommended, however, both because they may carry concentrated toxins and because high fish oil intakes seem to alter the course of pregnancy and labor with unknown effects.

Table 12•2

DAILY FOOD GUIDE FOR PREGNANT AND LACTATING WOMEN

Food Group	Number of Servings[a]	
	Adults	Pregnant or Lactating Women
Breads/cereals/rice/pasta	6 to 11	7 to 11
Vegetables	3 to 5	4 to 5
Fruits	2 to 4	3 to 4
Meat/meat alternates	2 to 3	3
Milk/milk products	2	3 to 4

[a]Figure 2-4 in Chapter 2 provides examples of foods in each group and serving sizes. A sample menu appears in the margin of the next page.

neural tube the embryonic tissue that later forms the brain and spinal cord.

neural tube defects (NTD) any of a group of nervous system abnormalities caused by interruption of the normal early development of the neural tube.

anencephaly (an-en-SEFF-ah-lee) a severe neural tube defect in which the brain fails to form. Anencephaly leads to death soon after birth.

spina bifida (SPY-na BIFF-ih-duh) one of the most common types of neural tube defects. The infant is born with gaps in the bones of the spine, leaving the spinal cord protected only by a sheath of skin in those spots, or with no protection at all. The spinal cord may bulge and protrude through the gaps in the vertebral column.

Of Special Interest: Folate and Vitamin B_{12} The vitamins required for rapid cell reproduction—folate and vitamin B_{12}—are needed in large amounts during pregnancy. New cells are laid down at a tremendous pace as the fetus grows and develops. At the same time, the number of the mother's red blood cells must rise because her blood volume increases. So the recommendation for folate during pregnancy increases from 400 to 600 micrograms a day.

As described in Chapter 7, folate plays an important role in preventing neural tube defects. To review, the early weeks of pregnancy are a critical period for the formation and closure of the **neural tube** that will later develop to form the brain and spinal cord. By the time a woman suspects she is pregnant, usually around the sixth week, the embryo's neural tube should already have closed. A **neural tube defect (NDT)** occurs when the tube fails to close properly. In the United States, about 4,000 pregnancies each year are affected by neural tube defects.[10] When the upper end of the neural tube fails to close, a rare but lethal neural tube defect known as **anencephaly** occurs. All infants with anencephaly die shortly after birth. When the lower end of the neural tube fails to close and the spinal cord and backbone do not develop normally, a common NTD occurs—**spina bifida** (see Figure 12-5). The

Figure 12•5

SPINA BIFIDA—A NEURAL TUBE DEFECT

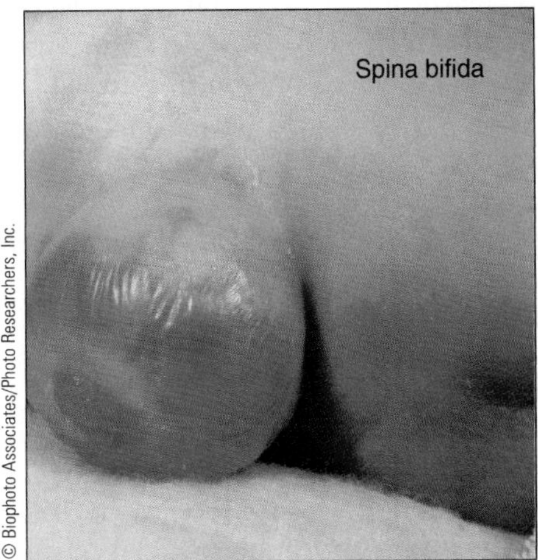

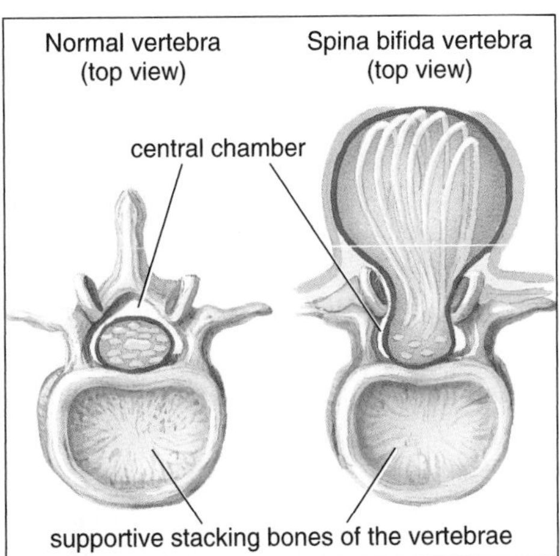

Normally, the bony central chamber closes fully to encase the spinal cord and its surrounding membranes and fluid. In spina bifida, the two halves of the slender bones that should complete the casement of the cord fail to join.

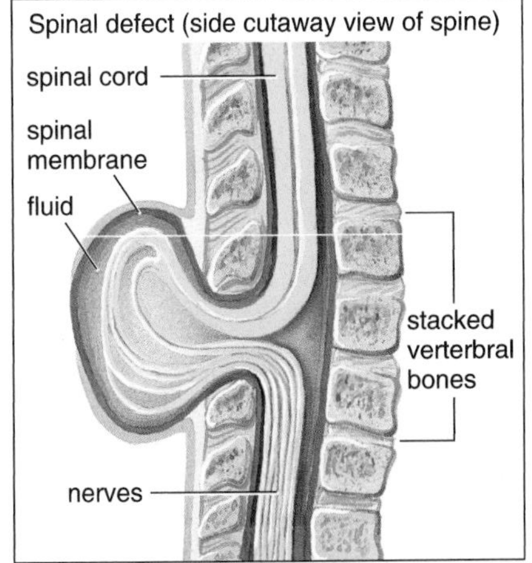

In the serious form shown here, membranes and fluid have bulged through the gap and nerves are exposed, invariably leading to some degree of paralysis and often to mental retardation.

Table 12•3

RICH FOLATE SOURCES[a]

Natural Folate Sources	Fortified Folate Sources
Liver (3 oz) 185 μg	Multi-Grain Cheerios Plus cereal (1 c) 400 μg[b]
Lentils (½ c) 180 μg	Product 19 cereal (1 c) 400 μg[b]
Chickpeas or pinto beans (½ c) 145 μg	Total cereal (1 c) 400 μg[b]
Asparagus (½ c) 125 μg	Pasta, cooked (1 c) 110 μg
Spinach (1 c raw) 115 μg	Rice, cooked (1 c) 80 μg
Avocado (½ c) 70 μg	Bagel (1 small whole) 50 μg
Orange juice (1 c) 60 μg	Waffles, frozen (2) 40 μg
Beets (½ c) 46 μg	Bread, white (1 slice) 20 μg

[a]Folate amounts for these and 2,000 other foods are listed in the Table of Food Composition in Appendix A.
[b]Folate in cereals varies; read the Nutrition Facts panel of the label.

membranes covering the spinal cord often protrude as a sac, and sometimes a portion of the spinal cord is contained in the sac. Spina bifida is accompanied by varying degrees of paralysis, depending on the extent of spinal cord damage. Mild cases may not be noticed, but severe cases lead to death. Common problems include clubfoot, dislocated hip, kidney disorders, curvature of the spine, muscle weakness, mental handicaps, and motor and sensory losses.

To obtain the folate that can reduce the risk of neural tube defects, women who are capable of becoming pregnant should:[11]

- Take a daily supplement containing 400 micrograms of folic acid.
- Eat folate-fortified foods (see Table 12-3).[12]
- Eat folate-rich foods.

Regular intake of supplements and fortified foods offers women a convenient way to ensure sufficient folate to benefit pregnancy.[13] Furthermore, *folic acid,* the synthetic form of folate in supplements and fortified food, is better absorbed than the naturally occurring folate in foods, thus improving folate status more.[14] Foods that naturally contain folate are still important, however, because they are rich sources of other vitamins, minerals, fiber, and phytochemicals.

As of 1999, all refined grain products (cereal, pasta, flour, bread, rolls, buns, farina, grits, cornmeal, and rice) sold commercially in the United States are fortified with folic acid. Folate fortification is expected to prevent half of all neural tube defects that occur each year. Already, folate fortification has improved folate status and reduced blood concentrations of a compound elevated in many women with NTD births, homocysteine.[15] Folate fortification, however, does raise at least one safety concern. The pregnant woman needs a greater amount of vitamin B_{12} to assist folate in the manufacture of new cells. High intakes of folate complicate the diagnosis of a vitamin B_{12} deficiency. For this reason, folate intakes should not exceed 1 milligram per day.[16]

Women who eat meat, eggs, or dairy products receive all the vitamin B_{12} they need, even for pregnancy. Those who exclude all animal products from the diet, however, need vitamin B_{12}–fortified foods or supplements.

Sample meal plan for pregnant and lactating women:

Breakfast
1 English muffin
2 tbs peanut butter
1 c low-fat vanilla yogurt
½ c fresh strawberries
1 c orange juice

Midmorning snack
1 c cranberry juice
1 oz pretzels

Lunch
Sandwich (tuna salad on whole-wheat bread)
½ carrot (sticks)
1 c low-fat milk

Dinner
Chicken cacciatore
4 oz chicken
¾ c stewed tomatoes
1 c rice
¾ c summer squash
1½ c salad (spinach, mushrooms, onions)
1 tbs salad dressing
2 slices Italian bread
2 tsp butter or margarine
1 c low-fat milk

Evening snack
1 c low-fat milk
3 oatmeal cookies

Calcium, Magnesium, Iron, and Zinc Among the minerals, calcium, phosphorus, and magnesium are in great demand during pregnancy because they are involved in building the skeleton. Insufficient intakes may result in abnormal development of fetal bones and teeth. Intestinal absorption of calcium doubles early in pregnancy, and the mineral is stored in the mother's bones. Later, when the fetal bones begin to calcify, the mother's bone stores are drawn upon, and there is a dramatic shift of calcium across the placenta. Whether calcium added to the mother's bones early in pregnancy is withdrawn to build the fetus's bones later is unclear.[17] In the final weeks of pregnancy, more than 300 milligrams of calcium a day are transferred to the fetus. Efforts to ensure an adequate calcium intake during pregnancy are aimed at conserving the mother's bone mass while supplying fetal needs.

Most women's prepregnancy calcium intakes are below recommendations, so increased calcium intakes are important.[18] Because bones are still actively depositing minerals until about age 25, pregnant women under age 25 who consume less than 600 milligrams of calcium a day need to increase their intakes of milk, cheese, yogurt, and other calcium-rich foods. Less preferred is a daily supplement of 600 milligrams of calcium. The DRI recommendation for calcium intake is the same for nonpregnant and pregnant women in the same age group. Magnesium for bone and tissue growth is needed during pregnancy in amounts slightly higher than recommendations for nonpregnant women.

The body conserves iron during pregnancy—menstruation ceases, and absorption of iron increases up to threefold. Despite these conservation measures, iron stores dwindle because the developing fetus draws on its mother's iron to create its own stores to carry it through the first three to six months of life. Maternal blood losses are also inevitable at birth, especially during a delivery by **cesarean section,** further draining the mother's iron supply. Few women enter pregnancy with adequate stores to meet pregnancy demands, so a daily iron supplement containing 30 milligrams is recommended during the second and third trimesters for all pregnant women.[19]

Zinc, required for protein synthesis and cell development, is vital during pregnancy. Severe zinc deficiency during pregnancy predicts low birthweight.[20] Zinc is most abundant in protein-rich foods such as shellfish, meat, and nuts, but the presence of other trace elements and fiber in foods may adversely affect zinc absorption. For example, iron interferes with the body's absorption and use of zinc, so women taking iron supplements in excess of the recommended 30 milligrams per day may also need zinc supplements to prevent zinc deficiency.[21]

Nutrient Supplements Women who make wise food choices during pregnancy can meet their nutrient needs except for iron. As discussed, iron supplements are recommended during the second and third trimesters for all pregnant women. Daily multivitamin-mineral supplements are also recommended for women who do not eat adequately and for those in high-risk groups: women carrying multiple fetuses, cigarette smokers, and alcohol and drug abusers. The use of prenatal supplements may help to reduce the risks of preterm delivery, low birthweights, and birth defects.[22] Table 12-4 lists recommended amounts of supplements for pregnant women at nutritional risk (listed in Table 12-1).

Food Assistance Programs Women of limited financial means may need help in obtaining the food and the nutrition counseling they need. At the federal level, the **Special Supplemental Food Program for Women, Infants, and Children (WIC)** provides nutrition education and vouchers redeemable for nutritious foods to low-income pregnant and breastfeeding women and their children. WIC provides food vouchers for milk and cheese, iron-fortified cereals, fruit or vegetable juices, eggs, dried beans, and peanut butter. These foods provide nutrients often lacking in diets of low-income women and children (calcium, iron, vitamins A and C, and protein). A recent study of about 100 pregnant women receiving WIC assistance found that 90 percent had consumed only two-thirds or less of their requirement for iron.[23] For infants given formula, WIC also provides iron-fortified infant formula.

Participation in the WIC program benefits both the iron status and the growth and development of infants and children. WIC participation during pregnancy has been shown to reduce the risks of delivering preterm or low-birthweight infants.[24]

Federal food stamps can also help to stretch the low-income pregnant woman's grocery dollars. Many communities provide educational services and materials, including nutrition, food budgeting, and shopping information, through the local agricultural extension service. Organizations such as the American Dietetic Association, the American Diabetes Association, and local hospitals also provide nutrition information.

cesarean (see-ZAIR-ee-un) **section** surgical childbirth, in which the infant is taken through an incision in the woman's abdomen.

Special Supplemental Food Program for Women, Infants, and Children (WIC) a USDA program to provide nutrition support to low-income women who are pregnant or have infants or preschool children. WIC offers coupons redeemable for specific foods to supply the nutrients deemed most needed for growth and development.

Table 12•4

NUTRIENT SUPPLEMENTS FOR PREGNANCY[a]

Nutrient	Amount
Folate	400 µg
Vitamin B_6	2 mg
Vitamin C	50 mg
Vitamin D	5 µg
Calcium	600 mg
Copper	2 mg
Iron	30 mg
Zinc	15 mg

[a]For pregnant women at nutritional risk (see Table 12-1).

SOURCE: Reprinted with permission from *Nutrition during Pregnancy* © 1990 by the National Academy of Sciences. Published by National Academy Press, Washington, D.C.

Pregnancy brings physiological adjustments that demand increased intakes of energy and nutrients. A daily iron supplement is recommended for all pregnant women during the second and third trimesters. Food assistance programs such as WIC can benefit pregnant women of limited financial means.

How Much Weight Should a Woman Gain during Pregnancy?

The pregnant woman must gain a certain amount of weight during pregnancy as a defense against bearing a low-birthweight baby. Ideally, she will have begun her pregnancy at the appropriate weight for her height, but even more importantly, she will gain enough weight based on her prepregnancy body mass index (BMI). Table 12-5 presents recommended weight gains for pregnancy. For the normal-weight woman, the ideal pattern is about 2 to 4 pounds during the first trimester and a pound per week thereafter.

Dieting during pregnancy is not recommended. Even an obese woman should gain about 15 pounds for the best chance of delivering a healthy infant. Weight gain for a teenager must be generous to meet her own needs and those of the fetus. Women who are carrying twins should strive for a weight gain of 35 to 45 pounds. Women have exceeded or undershot the recommended weight gains without ill effects, but meeting the recommendations offers the best chances of health. A sudden, large weight gain is a danger signal because it may indicate the onset of preeclampsia (see the section entitled "Troubleshooting").

The weight the pregnant woman puts on is mostly lean tissue: placenta, uterus, blood, milk-producing glands, and, the fetus itself (see Figure 12-6). The fat she gains is needed later for lactation. Physical activity can help a pregnant woman cope with the extra weight, as the next section explains. Some weight is lost at delivery, but many women retain a few pounds with each pregnancy.

Weight gain is essential for a healthy pregnancy. A woman's prepregnancy BMI, her own nutrient needs, and the number of fetuses she is carrying help to determine appropriate weight gain.

Table 12•5

RECOMMENDED WEIGHT GAINS FOR PREGNANCY[a]

- Underweight women (BMI <19.8):[b] 28 to 40 lb
- Normal-weight women (BMI 19.8 to 26): 25 to 35 lb
- Overweight women (BMI 26 to 29): 15 to 25 lb
- Obese women (BMI >29): 13 lb minimum

[a]The BMI cutoff points in this table were established by the Subcommittee on Nutritional Status and Weight Gain during Pregnancy. The cutoff points defining underweight, normal weight, overweight, and obesity differ slightly from those established by the National Heart, Lung, and Blood Institute Expert Panel on the Identification, Evaluation, and Treatment of Overweight and Obesity in Adults.

[b]BMI tables are on the inside back cover.

SOURCE: Committee on Nutritional Status during Pregnancy and Lactation, Food and Nutrition Board, *Nutrition during Pregnancy* (Washington, D.C.: National Academy Press, 1990), pp. 10, 12.

Figure 12•6

COMPONENTS OF WEIGHT GAIN DURING PREGNANCY

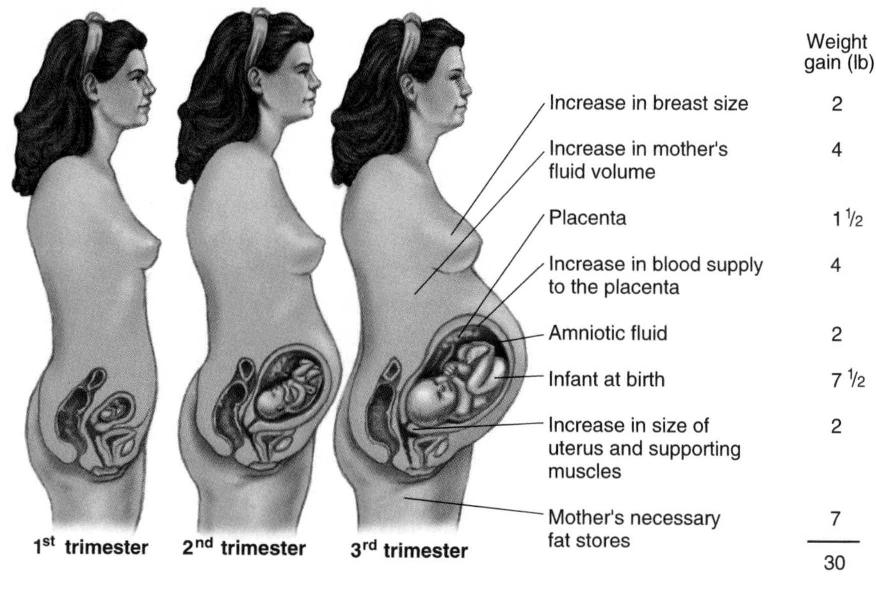

Should Pregnant Women Be Physically Active?

Staying active during the course of a normal, healthy pregnancy can improve the fitness of the mother-to-be, facilitate labor, and reduce psychological stress. Women who remain active during pregnancy report fewer discomforts throughout their pregnancies and gain less weight than those who are not physically active.[25] Pregnant women should take care in choosing their physical activities, however, participating in "low-impact" activities and avoiding sports in which they might fall or be hit by other people or objects (for some suggestions, see Think Fitness). A pregnant woman should consult her health-care provider before taking up additional activity. A few guidelines are offered in Table 12-6.

Physically fit women can continue to be physically active throughout pregnancy. Pregnant women should be cautious in their choice of activities.

Pregnant women can enjoy the benefits of physical activity.

© 2002 Tracy Frankel/Image Bank/Getty Images

Physical Activities for the Pregnant Woman

Is there an ideal physical activity for the pregnant woman? There might be. Swimming and water aerobics offer advantages over other activities during pregnancy. Water cools and supports the body, provides a natural resistance, and lessens the impact of the body's movement, especially in the later months. Water aerobics can often reduce the intensity of back pain during pregnancy. Other activities considered safe and comfortable for pregnant women include walking, light strength training, and the use of rowing machines, stair climbers, and treadmills.

THINK FITNESS

Teen Pregnancy

Each year in the United States, almost one million adolescent girls become pregnant.[26] Of these, about half choose to continue their pregnancies. A pregnant adolescent presents a special case of intense nutrient needs. Young teenage girls have not had time to store the nutrients needed to support their own rapid growth and development, much less nutrients to support pregnancy and a developing fetus. Many teens enter pregnancy deficient in vitamins A and D, folate, iron, calcium, and zinc—deficiences that place both mother and fetus at risk. Smoking also presents risks, and teens are more likely to smoke while pregnant than older women. Teenagers have more miscarriages, premature births, stillbirths, and low-birthweight

Table 12•6

GUIDELINES FOR PHYSICAL ACTIVITY DURING PREGNANCY

- Be physically active on a regular basis (at least three times a week), not intermittently.
- Warm up with 5 to 10 minutes of light activity.
- Stop exercising if you feel overheated.
- Drink plenty of fluids before, during, and after physical activity.
- Avoid exerting yourself in hot, humid weather; avoid overheating.
- Avoid jarring or jerky motions.
- Avoid any activity that has the potential to cause even mild abdominal trauma.
- Avoid prolonged periods of standing still.
- Discontinue any activity that causes discomfort.
- Do not exercise while lying on your back after the fourth month.
- Do not allow your heart rate to exceed 150 beats per minute.
- Cool down with 5 to 10 minutes of slow activity and gentle stretching.
- Eat enough to support the energy needs of pregnancy and physical activity.

infants than do adult women.[27] Their greatest risk, though, is death of the infant; mothers under age 16 bear more babies who die within the first year than do women in any other age group. These statistics show that teenage pregnancy is a major public health problem.

To support the needs of both mother and fetus, a pregnant teenager with a BMI in the normal range is encouraged to gain about 30 pounds. Teenagers who gain less have smaller infants with associated risks. Adequate nutrition is an indispensable component of prenatal care and can substantially improve the health of the mother and infant. Pregnant and lactating teenagers can follow the food guide presented earlier in Table 12-2 (page 441), making sure to choose at least 4 servings of milk or milk products daily.

Of all the population groups, pregnant teenage girls have the highest nutrient needs and an increased likelihood of having problem pregnancies.

Why Do Some Women Crave Pickles and Ice Cream While Others Can't Keep Anything Down?

Does pregnancy give a woman the right to demand pickles and ice cream at 2 A.M.? Not for nutrition's sake. Food cravings and aversions during pregnancy are common but do not seem to reflect real physiological needs. In other words, a woman who craves pickles is not in need of salt. Food cravings and aversions are due to changes in taste and smell sensitivities, and they quickly disappear after the baby's birth.

Sometimes cravings may occur in women with nutrient-poor diets. A pregnant woman who is deficient in iron, zinc, or other nutrients may crave and eat clay, ice, cornstarch, and other nonnutritious substances (pica, first mentioned in Chapter 8). Such cravings are not adaptive; the substances the woman craves do not deliver the nutrients she needs. In fact, clay and other substances can cling to the intestinal wall and form a barrier that interferes with normal nutrient absorption.

The nausea of "morning" (actually, anytime) sickness seems unavoidable because it arises from the hormonal changes of early pregnancy. Many women complain that smells, especially cooking smells, make them sick. Thus, minimizing odors can alleviate morning sickness. Sipping carbonated drinks and nibbling soda crackers or other salty snack foods before getting out of bed can sometimes prevent nausea. Some women do well by simply eating what they desire whenever they feel hungry. Table 12-7 offers some other suggestions, but morning sickness can be persistent. If morning sickness interferes with normal eating for more than a week or two, the woman should seek medical advice to prevent nutrient deficiencies.

As the hormones of pregnancy alter her muscle tone and the thriving fetus crowds her intestinal organs, an expectant mother may complain of heartburn or constipation. Raising the head of the bed with two or three pillows can help to relieve nighttime heartburn. A high-fiber diet, physical activity, and a plentiful water intake will help relieve constipation. The pregnant woman should use laxatives or heartburn medication only if her physician prescribes them.

Food cravings usually do not reflect physiological needs, and some may interfere with nutrition. Nausea arises from normal hormonal changes of pregnancy.

What Substances or Behaviors Should Pregnant Women Avoid?

Some substances in a woman's diet and environment can harm the fetus, and their potential impact is too great to ignore. Alcohol predominates, and we will devote the next section to it. Here we'll discuss smoking, medications, illegal drugs, environmental contaminants, vitamin-mineral megadoses, dieting, and caffeine.

environmental tobacco smoke (ETS) the combination of exhaled smoke (mainstream smoke) and smoke from lighted cigarettes, pipes, or cigars (sidestream smoke) that enters the air and may be inhaled by other people.

Table 12•7

TIPS FOR RELIEVING COMMON DISCOMFORTS OF PREGNANCY

To alleviate the nausea of pregnancy:

- On waking, arise slowly.
- Eat dry toast or crackers.
- Chew gum or suck hard candies.
- Eat small, frequent meals whenever hunger strikes.
- Avoid foods with offensive odors.
- When nauseated, do not drink citrus juice, water, milk, coffee, or tea.

To prevent or alleviate constipation:

- Eat foods high in fiber.
- Exercise daily.
- Drink at least 8 glasses of liquids a day.
- Respond promply to the urge to defecate.
- Use laxatives only as prescribed by a physician; avoid mineral oil—it carries needed fat-soluble vitamins out of the body.

To prevent or relieve heartburn:

- Relax and eat slowly.
- Eat small, frequent meals.
- Drink liquids between meals.
- Avoid spicy or greasy foods.
- Sit up while eating.
- Wait an hour after eating before lying down.
- Wait 2 hours after eating before exercising.

Cigarette Smoking The surgeon general has warned that parental smoking can kill an otherwise healthy fetus or newborn. Cigarette (and cigar) smoking adversely affects the pregnant woman's nutrition status, which, in turn, impairs fetal nutrition. Smokers tend to have lower intakes of dietary fiber, vitamin A, beta-carotene, folate, and vitamin C. Oxidants in cigarette smoke accelerate vitamin C metabolism, depleting the body's stores and further compromising smokers' vitamin C status.

Maternal smoking also affects the fetus directly. Constituents of cigarette smoking, such as nicotine and cyanide, are toxic to a fetus. Smoking restricts the blood supply to the growing fetus and so limits the delivery of oxygen and nutrients and the removal of wastes. It slows growth, thus retarding physical development of the fetus, and it may cause behavioral or intellectual problems later.[28]

A mother who smokes is more likely to have a complicated birth, and her infant is more likely to be of low birthweight.[29] The more a mother smokes, the smaller her baby will be. Of all preventable causes of low birthweight in the United States, smoking has the greatest impact. Sudden infant death syndrome (SIDS), the unexplained deaths that sometimes occur in otherwise healthy infants, has been linked to the mother's cigarette smoking during pregnancy.[30] Research suggests that even in women who do not smoke, exposure to **environmental tobacco smoke (ETS,** or secondhand smoke) during pregnancy increases the risk of low birthweight and the likelihood of SIDS.[31]

Medicinal Drugs Medicinal drugs taken during pregnancy can cause serious birth defects. Pregnant women should not take over-the-counter drugs, herbal preparations, or any medications not prescribed by a physician. Drug labels warn: "As with any drug, if you are pregnant or nursing a baby, seek the advice of a health professional before using this product." For aspirin and ibuprofen, there is an additional warning: "It is especially important not to use aspirin (or ibuprofen) during the last three months of pregnancy unless specifically directed to do so by a doctor because it

Fetal effects of illegal drugs:

- Amphetamines: Suspected nervous system damage; behavioral abnormalities.
- Barbiturates: Drug withdrawal symptoms in the newborn, lasting up to six months.
- Cocaine: Uncontrolled jerking motions; paralysis; permanent mental and physical damage.
- Marijuana: Short-term irritability at birth.
- Opiates (including heroin): Drug withdrawal symptoms in the newborn; permanent learning disability (attention deficit hyperactivity disorder).

may cause problems in the unborn child or excessive bleeding during delivery." Such warnings should be taken seriously.

Illegal Drugs Research shows that women who use illegal drugs such as marijuana and cocaine during pregnancy inflict serious health consequences, including nervous system disorders, on their fetuses.[32] Infants born to mothers who use crack and other forms of cocaine face low-birthweight complications, heartbeat abnormalities, the pain of withdrawal, or even death as they first experience life outside the womb. Some effects of other illegal drugs on the fetus are listed in the margin.

Environmental Contaminants Infants and young children of pregnant women exposed to environmental contaminants such as lead and mercury show signs of impaired cognitive development. During pregnancy, lead and mercury readily move across the placenta, inflicting severe damage on the developing fetal nervous system.

Unacceptably high concentrations of mercury in fish have prompted the Food and Drug Administration (FDA) to issue an advisory to all pregnant women, women who may become pregnant, lactating mothers, and young children against eating large ocean fish such as king mackerel, swordfish, shark, tuna, and tilefish.[33] Pregnant and lactating women are also advised to limit their consumption of canned tuna to one can per week, and young children are advised to eat less than a can per *month*. Furthermore, the Environmental Protection Agency (EPA) has warned the same groups of people to limit intakes of freshwater fish to one fish meal per week. Chapter 14 offers more details on contaminants in foods.

Vitamin-Mineral Megadoses Many vitamins are toxic when taken in excess and the minerals even more so. Among vitamins, a single massive dose of preformed vitamin A (100 times the recommended intake) has caused birth defects. Chronic use of lower doses of vitamin A supplements (three to four times the recommended intake) may also cause birth defects. Intakes before the seventh week of pregnancy appear to be the most damaging. For this reason, additional vitamin A is not recommended during pregnancy, and the vitamin is prescribed in the first trimester of pregnancy only upon evidence of deficiency, which is rare. Women taking supplements should take heed—experts urge pregnant women not to exceed three times the recommended daily intake of vitamin A.

Dieting Dieting, even for short periods, is hazardous during pregnancy. Low-carbohydrate diets or fasts that cause ketosis deprive the growing fetal brain of needed glucose and may impair its development. Such diets are also likely to be deficient in other nutrients vital to fetal growth. Energy restriction during pregnancy is dangerous, regardless of the woman's prepregnancy weight or the amount of weight gained in the previous month.

Caffeine Caffeine crosses the placenta, and the fetus has only a limited ability to metabolize it. No firm limit for caffeine intake is yet available. So far, research studies have not proved that caffeine causes birth defects in human beings (as it does in animals), but pregnant women who drink more than 3 cups of coffee a day may increase their risk of spontaneous abortion.[34] Some evidence suggests that moderate-to-heavy use of caffeine (more than 300 milligrams per day—the equivalent of about 2 to 3 cups of coffee a day) may lower infant birthweight.[35] Another study, however, found that daily caffeine intake is not a risk factor for fetal growth retardation.[36]

In light of this evidence, the most sensible course is to limit caffeine consumption to the equivalent of one cup of coffee or two 12-ounce cola beverages a day. Caffeine amounts in food and beverages are listed in Controversy 11, on page 432.

key point *Abstaining from smoking and other drugs, limiting intake of foods known to contain unsafe levels of contaminants such as mercury, avoiding large doses of nutrients, refraining from dieting, and limiting caffeine use are recommended during pregnancy.*

Drinking during Pregnancy

Alcohol is arguably the most hazardous drug to future generations because it is legally available, heavily promoted, and widely abused. Society often sends mixed messages concerning alcohol. Beverage companies promote an image of drinkers as wealthy, healthy, young, and active. Opposing this image, health authorities warn that alcohol may have adverse effects, especially during pregnancy (see Figure 12-7). Every container of beer, wine, or liquor for sale in the United States is required to warn pregnant women of the danger of drinking during pregnancy. It's that serious.

Apgar score a system of scoring an infant's physical condition right after birth. Heart rate, respiration, muscle tone, response to stimuli, and color are ranked 0, 1, or 2. A low score indicates that medical attention is required to facilitate survival.

fetal alcohol syndrome (FAS) the cluster of symptoms seen in an infant or child whose mother consumed excessive alcohol during her pregnancy. FAS includes, but is not limited to, brain damage, growth retardation, mental retardation, and facial abnormalities.

Alcohol's Effects

Women of childbearing age need to know about alcohol's harmful effects on a fetus. Alcohol crosses the placenta freely and is directly toxic:

- Oxygen is indispensable on a minute-to-minute basis to the development of the fetus's central nervous system. A sudden dose of alcohol can halt the delivery of oxygen through the umbilical cord.
- Alcohol slows cell division, reducing the number of cells produced and inflicting abnormalities on those that are produced.
- During the first month of pregnancy, the fetal brain is growing at the rate of 100,000 new brain cells a minute. Even a few minutes of alcohol exposure during this critical period can exert a major detrimental effect.
- Alcohol interferes with placental transport of nutrients to the fetus and can cause malnutrition in the mother; then, all of malnutrition's harmful effects compound the effects of the alcohol. Before fertilization, alcohol can damage the ovum or sperm in the mother- or father-to-be, which can lead to abnormalities in the child.

key point *Alcohol limits oxygen delivery to the fetus, slows cell division, and reduces the number of cells organs produce. Alcoholic beverages must bear warnings to pregnant women.*

Fetal Alcohol Syndrome

Drinking alcohol during pregnancy threatens the fetus with irreversible brain damage, growth retardation, mental retardation, facial abnormalities, vision abnormalities, a low **Apgar score,** and more than 40 identifiable health problems—a cluster of symptoms known as **fetal alcohol syndrome** or **FAS.**[37] The fetal brain is extremely vulnerable to a glucose or oxygen deficit, and alcohol causes both by disrupting placental functioning. In addition, alcohol itself crosses the placenta freely and is directly toxic to the defenseless fetal brain and nervous system.[38] The result is permanent brain damage and lifelong mental retardation. FAS can be prevented by abstaining from drinking alcohol during pregnancy, but children born with alcohol damage remain impaired.

Figure 12-8 on the next page shows the facial abnormalities of FAS, which are easy to depict. A visual picture of the internal harm is impossible, but that damage seals the fate of the child. An estimated 5 to 30 of every 10,000 children are victims of this preventable damage, making FAS the leading known cause of mental retardation in the world.

Between 1979 and 1993, incidence of FAS increased sixfold. About a fifth of women continue drinking alcohol after they learn that they are pregnant. One of every 29 pregnant women reports "frequent" drinking (seven or more drinks per week or five or more drinks on one occasion).[39] For women who want to drink during their pregnancies, then, the question is, how much alcohol is too much?

Birth defects have been observed in the children of some women who drank 2 ounces (4 drinks) of alcohol daily during pregnancy. Low birthweight has been

Figure 12•7

MIXED MESSAGES IN ALCOHOL ADVERTISEMENTS

Labels on alcoholic beverages often display "healthy" images, but their warnings tell the truth.

alcohol-related neurodevelopmental disorder (ARND) a condition caused by prenatal alcohol exposure. ARND is diagnosed when there is a confirmed history of substantial regular maternal alcohol intake or heavy episodic drinking, combined with behavioral, cognitive, or central nervous system abnormalities in the child that are known to be associated with alcohol exposure.

alcohol-related birth defects (ARBD) a condition caused by prenatal alcohol exposure. ARBD is diagnosed when there is a history of substantial regular maternal alcohol intake or heavy episodic drinking, combined with birth defects known to be associated with alcohol exposure.

Controversy 3 defined "a drink" as ½ ounce of pure ethanol, equivalent to:

- 3 to 4 ounces wine.
- 10 ounces wine cooler.
- 12 ounces beer.
- 1 ounce hard liquor.

observed in infants born to some women who drank 1 ounce (2 drinks) per day during pregnancy. At that level of alcohol intake, a sizable and significant increase in the rate of spontaneous abortions occurs; the reason is unclear, but perhaps the alcohol poisons the fetus or causes the placenta to detach. FAS is also known to occur with as few as 2 drinks a day.

The pattern of a woman's drinking may be as important as her average alcohol intake. For example, a woman whose average intake is only 1 ounce of alcohol a day might not drink at all during the week, but then have 14 drinks each weekend. Thus, the fetus might be intermittently exposed to high alcohol doses. No matter what the intake or pattern, the most severe impact is likely to occur in the first two months, before the woman is even aware that she is pregnant.

Research using animals shows that one-fifth of the amount of alcohol needed to produce major visible defects will produce learning impairment or other defects in the offspring. The term *fetal alcohol effects*, originally used to describe this damage, has been replaced with two more descriptive terms, **alcohol-related neurodevelopmental disorder (ARND)** and **alcohol-related birth defects (ARBD).**[40] The terms describe conditions in which there is a history of maternal alcohol intake, combined with evidence of abnormalities related to alcohol.

Some ARND and ARBD children show no outward sign of impairment, while others may be short in stature or display subtle facial abnormalities. Most perform poorly in school and in social interactions and suffer a subtle form of brain damage. Furthermore, anyone exposed to alcohol before birth may always respond differently to it, and also to certain drugs, than if no exposure had occurred.

Figure 12•8

TYPICAL FACIAL CHARACTERISTICS OF FAS

The severe facial abnormalities shown here are just outward signs of severe mental impairments and internal organ damage. These defects, though hidden, may create major health problems later.

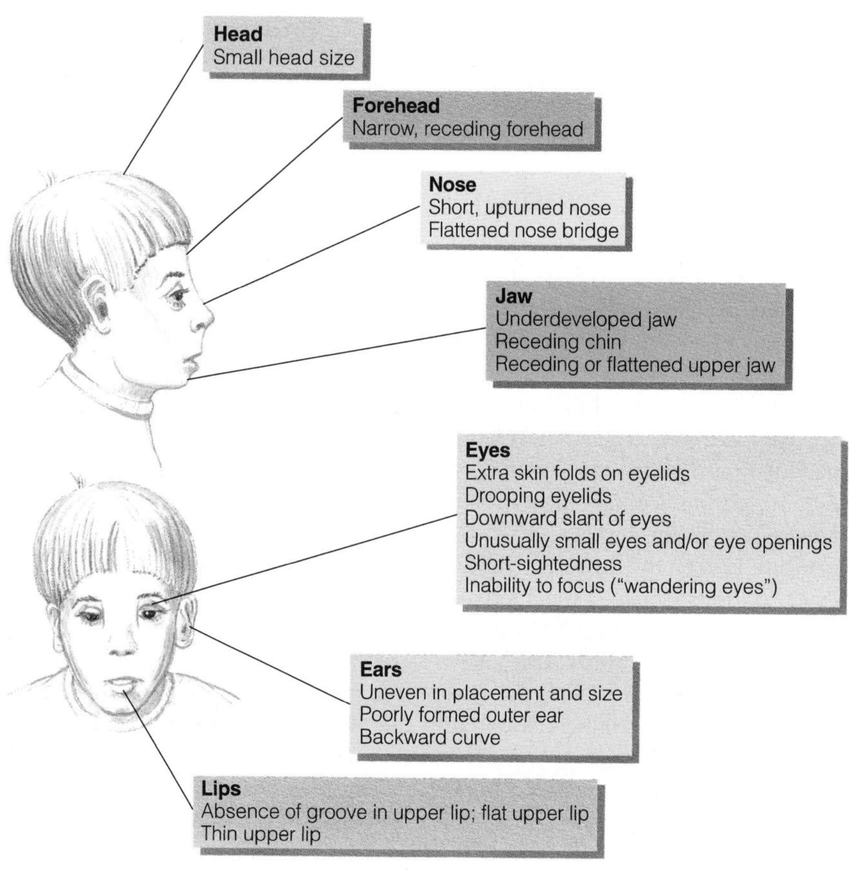

A child with FAS.

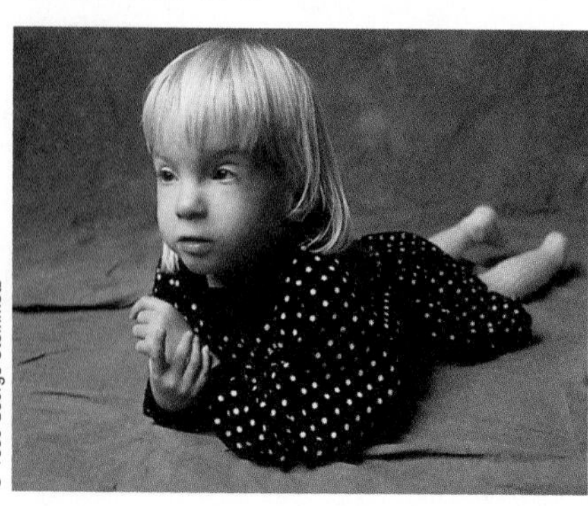

For every baby diagnosed with FAS, three or four with ARND or ARBD go undiagnosed until problems develop in the preschool years. Upon reaching adulthood, such children are ill equipped for employment, relationships, and the other facets of life most adults take for granted.

For years researchers have tried to uncover the mechanisms by which alcohol damages the fetus in hopes of finding ways to prevent such damage. Ironically, research on mice reveals that a different kind of alcohol may protect the fetus from damage caused by ethanol, the beverage alcohol.[41] It seems that the protective alcohol, known as 1-octonol, can block the abnormal cell development and cell death that ethanol causes. Such findings may someday lead to the development of drugs or other interventions to prevent FAS.

gestational diabetes abnormal glucose tolerance appearing during pregnancy, with subsequent return to normal after the end of pregnancy.

edema accumulation of fluid in the tissues (also defined in Chapter 6).

preeclampsia a potentially dangerous condition during pregnancy characterized by edema, hypertension, and protein in the urine.

key point *The birth defects of fetal alcohol syndrome arise from severe damage to the fetus caused by alcohol. Lesser conditions, ARND and ARBD, may be harder to diagnose, but also rob the child of a normal life.*

Experts' Advice

The American Academy of Pediatrics takes the position that women should stop drinking as soon as they *plan* to become pregnant.[42] This step is important for fathers-to-be as well. Researchers have looked for a "safe" alcohol intake limit during pregnancy and have found none. Their conclusion: Abstinence from alcohol is the best policy for pregnant women. We recommend this choice, too. For pregnant women who have already drunk alcohol, the advice is "Stop now." A woman who has drunk heavily during the first two-thirds of her pregnancy can still prevent some organ damage by stopping heavy drinking during the third trimester.

Pregnancy is a time of adjustment to major changes: physical, social, emotional, and financial. The couple who are expecting a baby will have to change their lifestyles as they take on the responsibility of caring for a child. The mother demonstrates her sense of responsibility by caring for herself and her developing fetus during pregnancy.

Abstinence from alcohol is critical to prevent irreversible damage to the fetus.

Troubleshooting

Maternal diseases detract from the health and growth of the mother and the fetus. If discovered early, many diseases can be controlled—another reason early prenatal care is recommended.

Gestational Diabetes Pregnancy precipitates the onset of **gestational diabetes** in some women.[43] With most cases of gestational diabetes, blood glucose becomes abnormal during pregnancy but returns to normal after the infant is born. Almost one-third of all women with gestational diabetes, however, develop type 2 diabetes later in life, especially if they are overweight. Without proper management, gestational diabetes can lead to fetal or infant sickness and death. Properly managed, it causes no harm except that surgical birth may be necessary.[44] The American Diabetes Association recommends that women be screened for diabetes between 24 and 28 weeks' gestation, with the exception of women who meet specific low-risk criteria such as no family history of diabetes.[45]

Preeclampsia A certain degree of **edema** is to be expected in late pregnancy, and some women also develop hypertension during that time. If a rise in blood pressure is mild, it may subside after childbirth and cause no harm. In some cases, however, hypertension may signal the onset of **preeclampsia,** a condition characterized not only by high blood pressure but by protein in the urine and fluid retention (edema). The

certified lactation consultant a health-care provider, often a registered nurse, with specialized training in breast and infant anatomy and physiology who teaches the mechanics of breastfeeding to new mothers. Certification is granted after passing a standardized post-training examination.

Warning signs of preeclampsia:

- Headaches.
- Swelling, especially facial swelling.
- Dizziness.
- Blurred vision.
- Sudden weight gain.

edema of preeclampsia is a severe, whole-body edema, distinct from the localized fluid retention women normally experience late in pregnancy. The normal edema of pregnancy is a response to gravity; fluid from blood pools in the ankles. The edema of preeclampsia causes swelling of the face and hands as well as of the feet and ankles.

Preeclampsia affects almost all of the mother's organs—the circulatory system, liver, kidneys, and brain. If the condition progresses and she experiences convulsions, the condition is called eclampsia. Maternal mortality during pregnancy is rare in developed countries, but eclampsia is the most common cause. Preeclampsia demands prompt medical attention. Treatment focuses on regulating blood pressure and preventing convulsions. Dietary factors have been studied over the years, but so far none have proved conclusive in preventing preeclampsia.[46]

Common medical problems associated with pregnancy are gestational diabetes and preeclampsia. These should be managed to minimize associated risks.

Lactation

As the time of childbirth nears, a woman must decide whether she will feed her baby breast milk, infant formula, or both. These options are the only recommended foods for an infant during the first four to six months of life. A woman who plans to breastfeed her baby should begin to prepare toward the end of her pregnancy. No elaborate or expensive preparations are needed, but the expectant mother can read one of the many handbooks available on breastfeeding or consult a **certified lactation consultant,** employed at many hospitals.* One way to prepare is to learn what dietary changes are needed because adequate nutrition is essential to successful lactation.

In rare cases, women produce too little milk to nourish their infants adequately. Severe consequences, including infant dehydration, malnutrition, and brain damage, can occur should the condition go undetected for long. Early warning signs of insufficient milk are dry diapers (a well-fed infant wets about six diapers a day) and infrequent bowel movements.

Nutrition during Lactation

A nursing mother produces about 25 ounces of milk a day, depending on the infant's demand for milk. Producing this milk costs a woman almost 650 calories per day above her regular need during the first six months of lactation. To meet this energy need, the woman is advised to eat an extra 500 calories of food each day. The other 150 calories can be drawn from the fat stores she accumulated during pregnancy. Energy needs of women who are breastfeeding exclusively range from 2,500 to 3,300 calories a day, depending on physical activity.[47] The food energy consumed by the nursing mother should carry with it abundant nutrients. Look back at Figure 12-4 (page 440) for a lactating woman's nutrient recommendations and Table 12-2 (page 441) for the suggested number of servings in each food group.

The volume of breast milk produced depends on how much milk the baby demands, not on how much fluid the mother drinks. The nursing mother is nevertheless advised to drink at least 2 quarts of liquids each day to protect herself from dehydration. To help themselves remember to drink enough liquid, many women make a habit of drinking a glass of milk, juice, or water each time the baby nurses as well as at mealtimes.

A common question is whether a mother's milk may lack a nutrient if she fails to get enough in her diet. The answer differs from one nutrient to the next, but in general, the effect of nutritional deprivation of the mother is to reduce the *quantity,* not the *quality,* of her milk. For protein, carbohydrate, fat, folate, and most minerals, the

*La Leche League is an international organization that helps women with breastfeeding concerns. See Appendix E for its address and website.

milk of a healthy mother has a fairly constant composition. Any excess water-soluble vitamins the mother takes in are excreted in the urine; the body does not release them into the milk. The amounts of fat-soluble vitamins in human milk, however, are affected by the mother's excessive or deficient intakes. For example, large doses of vitamin A raise the concentration of this vitamin in breast milk. Vitamin supplementation of undernourished women appears to help normalize the vitamin concentrations in their milk and may be beneficial.

Some infants may be sensitive to foods such as cow's milk, onions, or garlic in the mother's diet and become uncomfortable when she eats them. Nursing mothers are advised to eat whatever nutritious foods they choose. If a particular food seems to cause an infant discomfort, the mother can eliminate that food from her diet for a few days and see if the problem goes away.

Another common question is whether breastfeeding promotes a more rapid loss of the extra body fat accumulated during pregnancy. Results of studies on the relationship between feeding method and loss of body fat and body weight are inconsistent. In most studies where breastfeeding duration was three months or longer, researchers found that lactation did accelerate a woman's weight loss.[48] This does not mean that a breastfeeding woman can eat unlimited food and return to prepregnancy weight. Breastfeeding costs energy, true, but diet and physical activity are still the cornerstones of weight control. Physical activity in particular helps to reduce body fatness and improve fitness while having little effect on a woman's milk production or her infant's weight gain. A gradual weight loss (1 pound per week) is safe and does not reduce milk output.[49] Too large an energy deficit, however, especially soon after birth, will inhibit lactation.

key point *The lactating woman needs extra fluid and enough energy and nutrients to make sufficient milk each day. Malnutrition most often diminishes the quantity of the milk produced without altering quality. Lactation facilitates loss of the extra fat gained during pregnancy.*

Q When Should a Woman Not Breastfeed?

Some substances impair maternal milk production or enter breast milk and interfere with infant development, making breastfeeding an unwise choice. Some medical conditions also prohibit breastfeeding.

Alcohol, Nicotine, and Other Drugs Alcohol enters breast milk and can adversely affect production, volume, composition, and ejection of breast milk as well as overwhelm an infant's immature alcohol-degrading system.[50] Alcohol concentration peaks within one hour after ingestion of even moderate amounts (equivalent to a can of beer). This amount may alter the taste of the milk to the disapproval of the nursing infant, who may, in protest, drink less milk than normal. Drug addicts, including alcohol abusers, can take such high doses that their infants become addicts by way of breast milk. In these cases, breastfeeding is contraindicated.

As for cigarette smoking, research shows that lactating women who smoke produce less milk, and milk with a lower fat content, than mothers who do not smoke. Thus, their infants gain less weight than infants of nonsmokers. A lactating woman who smokes not only transfers nicotine and other chemicals to her infant via her breast milk, but also exposes the infant to secondhand smoke. Babies who are "smoked over" experience a wide array of health problems—poor growth, hearing impairment, vomiting, breathing difficulties, and even unexplained death.[51]

Excess caffeine can make a baby jittery and wakeful. Caffeine consumption should be moderate when breastfeeding.

If a nursing mother must take medication that is secreted in breast milk and is known to affect the infant, then breastfeeding must be put off for the duration of treatment. Meanwhile, the flow of milk can be sustained by pumping the breasts and discarding the milk. A nursing mother should consult with her physician before taking medicines.

Many women wonder about using oral contraceptives during lactation. One type that combines the hormones estrogen and progestin seems to suppress milk output, lower the nitrogen content of the milk, and shorten the duration of breastfeeding. In contrast, progestin-only pills have no effect on breast milk or breastfeeding and are considered appropriate for lactating women.

For more about contaminants and nutrition, turn to Chapter 14.

Environmental Contaminants A woman sometimes hesitates to breastfeed because she has heard that environmental contaminants may enter breast milk and harm her infant. Although some contaminants do enter breast milk, others may be filtered out. Because formula is made with water, formula-fed infants consume any contaminants that may be in the water supply. The decision of whether to breastfeed on this basis can be made after consultation with a physician or dietitian familiar with the local circumstances.

For safe breast milk storage:

- Wash hands thoroughly before pumping.
- Clean pumping equipment according to manufacturer's directions.
- Sterilize bottles, nipples, and rings before using.
- Refrigerate milk to be fed within 48 hours.
- Freeze milk to be stored longer than 48 hours.
- Thaw milk gently on defrost cycle of microwave or in refrigerator.
- Do not refreeze thawed milk.

Maternal Illness If a woman has an ordinary cold, she can continue nursing without worry. The infant will probably catch it from her anyway, and thanks to immunological protection, a breastfed baby may be less susceptible than a formula-fed baby. If a woman has a serious communicable disease such as tuberculosis or hepatitis, then mother and baby have to be separated. Breastfeeding can be continued by pumping the mother's breasts several times a day and letting the baby drink the milk from a bottle (see the margin for tips for safe handling).

The human immunodeficiency virus (HIV), responsible for causing AIDS, can be passed from an infected mother to her infant during pregnancy, at birth, or through breast milk, especially during the early months of breastfeeding.[52] Women in developed countries who have tested positive for HIV should not breastfeed if the infant is not infected. They should choose a safe alternative feeding method, such as breast milk from a milk bank.[53] Milk banks in the United States pasteurize donated human milk and make it available to infants who lack access to milk from their own mothers. Pasteurization destroys harmful organisms, such as HIV, but leaves intact most of the beneficial constituents of the milk.[54]

In developing countries, where feeding inappropriate or contaminated formulas causes 1.5 million infant deaths each year, breastfeeding can be critical to infant survival. This advantage, however, must be weighed against the following: in 1999, 200,000 to 300,000 infants became infected with HIV by way of breastfeeding.[55] Whether HIV-infected women in developing countries should breastfeed comes down to a delicate weighing of risks and benefits. For HIV-positive women in developing countries who are literate, have access to safe water, and have an uninterrupted supply of infant formula, replacement feeding may reduce the risk of infant illness and death by AIDS. For those mothers without safe water and with minimal education, the risk of replacement feeding may be substantial in terms of infant mortality. WHO and UNICEF, acknowledging the transmission of HIV by way of breast milk, recommend that babies of HIV-positive mothers in developing countries be fed formula if they can be ensured uninterrupted access to safe, nutritionally adequate breast milk substitutes.

key point *Breastfeeding is inadvisable if the mother's milk is contaminated with alcohol, drugs, or environmental pollutants. Most ordinary infections such as colds have no effect on breastfeeding. The decision to breastfeed or not in the case of an HIV-infected woman depends on availability of formula and clean water.*

Feeding the Infant

Early nutrition affects later development, and early feedings establish eating habits that influence nutrition throughout life. Trends change and experts may argue the fine points, but nourishing a baby is relatively simple. Common sense and a nurturing, relaxed environment go far to promote the infant's well-being.

Nutrient Needs

A baby grows faster during the first year of life than ever again, as Figure 12-9 shows. Pediatricians carefully monitor the growth of infants and children because growth directly reflects their nutrition status. An infant's birthweight doubles around four months of age and triples by the age of one year. (If a 150-pound adult were to grow like this, the person would weigh 450 pounds after a single year.) By the end of the first year, the growth rate slows considerably.

The rapid growth and metabolism of the infant demand an ample supply of all the nutrients. Of special importance during infancy are the energy nutrients and the vitamins and minerals critical to the growth process, such as vitamin A, vitamin D, and calcium.

Because they are small, babies need smaller *total* amounts of these nutrients than adults do, but as a percentage of body weight, babies need more than twice as much of most nutrients. Infants require about 100 calories per kilogram of body weight per day; most adults require fewer than 40. Figure 12-10 on the next page compares a five-month-old baby's needs (per unit of body weight) with those of an adult man. You can see that differences in vitamin D and iodine, for instance, are extraordinary. Around six months of age, energy needs begin to increase less rapidly as the growth rate begins to slow down, but some of the energy saved by slower growth is spent in increased activity. When their growth slows, infants spontaneously reduce their energy intakes. Parents should expect their babies to adjust their food intakes downward when appropriate and should not force or coax them to eat more.

Vitamin K nutrition for newborns presents a unique case. A newborn's digestive tract is sterile, and vitamin K–producing bacteria take weeks to establish themselves in the baby's intestines. To prevent uncontrolled bleeding in the newborn, the American Academy of Pediatrics (AAP) recommends that a single dose of vitamin K be given at birth.[56]

The most important nutrient of all is the one easiest to forget: water. The younger a child is, the more of its body weight is water and the faster the water is lost and replaced. Proportionately more of an infant's body water than an adult's is between the cells and in the vascular space, and this water is easy to lose. In early infancy, breast milk or infant formula provides enough water for a healthy infant to replace water losses from the skin, lungs, feces, and urine. Conditions that cause rapid fluid loss, such as hot weather, vomiting, diarrhea, or sweating, however, can propel an infant into life-threatening dehydration and justify offering water or, in severe cases, an electrolyte solution designed for infants. When the older infant starts eating solid foods, additional water is required for reasons described later.

Figure 12•9

WEIGHT GAIN OF HUMAN INFANTS AND CHILDREN IN THE FIRST FIVE YEARS OF LIFE

The colored vertical bars show how the yearly increase in weight gain slows its pace over the years.

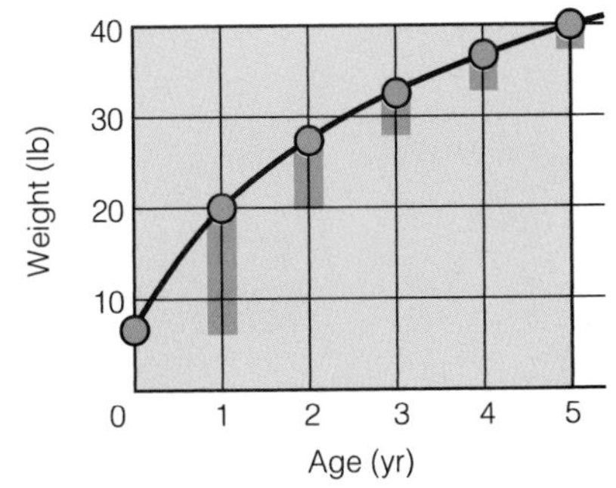

Infants' rapid growth and development depend on adequate nutrient supplies. Adequate water is also crucial.

Breastfeeding

Both the AAP and the Canadian Paediatric Society stand behind this statement: "Breastfeeding is strongly recommended for full term infants, except in the few instances where specific contraindications exist." The American Dietetic Association advocates breastfeeding for the nutritional health it confers on the infant as well as for the physiological, social, economic, and other benefits it gives to the mother.[57] The AAP recommends that infants receive breast milk for at least the first 12 months of life.[58] All legitimate nutrition authorities share this view, but some makers of baby formula try to convince women otherwise—see the Consumer Corner later in this chapter.

After six months of age, the energy saved by slower growth is spent on increased activity.

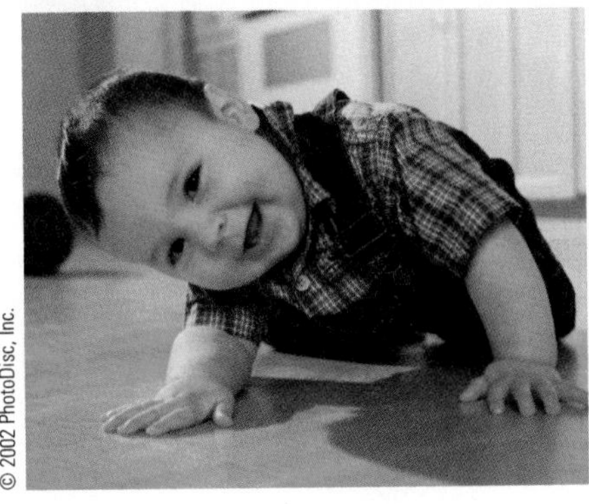

Why Is Breast Milk So Good for Babies?

Breast milk is tailor-made to meet the nutrient needs of the human infant. Its carbohydrate is lactose, and its fat provides a generous portion of the essential omega-6 fatty acid linoleic acid and its products. A mother who consumes food rich in omega-3 fatty

Figure 12•10

NUTRIENT RECOMMENDATIONS FOR A FIVE-MONTH-OLD INFANT AND AN ADULT MALE COMPARED ON THE BASIS OF BODY WEIGHT

Infants may be relatively small and inactive, but they use large amounts of energy and nutrients in proportion to their body size to keep all their metabolic processes going.

	Infants	Adults
Heart rate (beats/minute)	120 to 140	70 to 80
Respiration rate (breaths/minute)	20 to 40	15 to 20
Energy needs (cal/body weight)	45/lb (100/kg)	<18/lb (<40/kg)

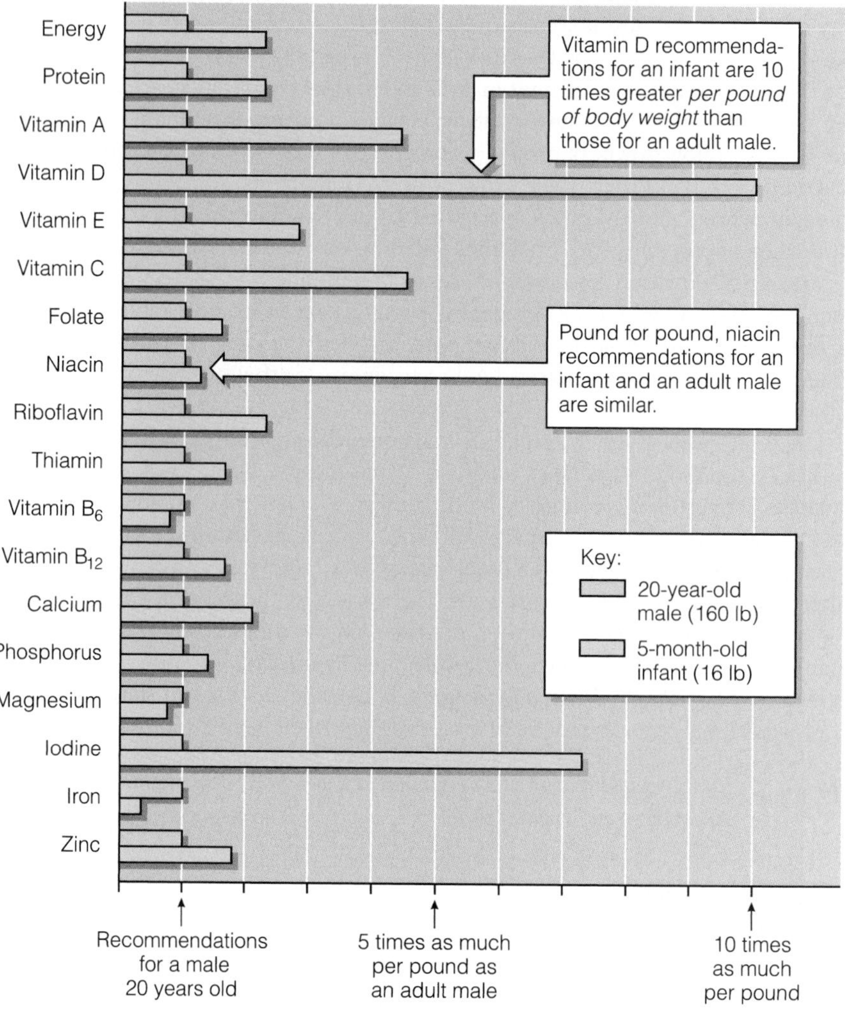

Breastfeeding is a natural extension of pregnancy—the mother's body continues to nourish the infant.

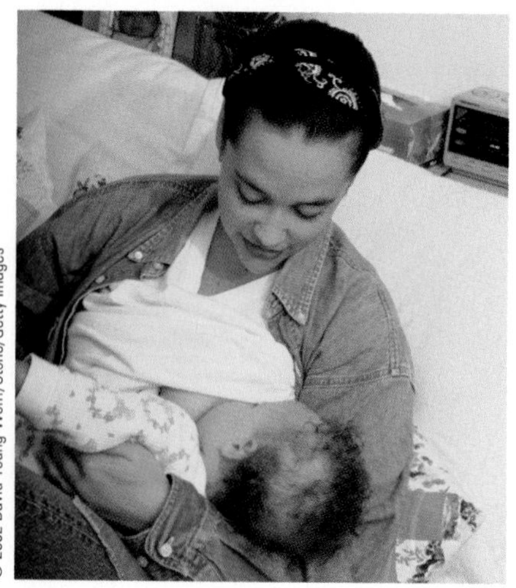

acids will also pass these beneficial nutrients on to her child through her milk. The protein of breast milk is particularly digestible and usable to support tissue growth. Breast milk contains fat-digesting enzymes that help ensure efficient fat absorption by the infant. Breast milk also conveys immune factors, which both protect an infant and inform its body about the outside environment.

Immune Factors in Breast Milk Breast milk offers an infant unsurpassed protection against infection, including antiviral agents, antibacterial agents, and infection inhibitors.[59] Some of these immune molecules are proteins that the infant absorbs whole, but the greatest protection may occur in the milk itself. For example, immune factors in breast milk interfere with the growth of bacteria that could otherwise attack the infant's vulnerable digestive tract linings. Breastfed babies are less prone to develop stomach and intestinal disorders during the first few months of life and so experience less vomiting and diarrhea than formula-fed babies.[60] Research shows that breast milk contains not only antibodies against the most common cause of diarrhea in infants and young children (rotavirus) but also another factor that binds to, and inhibits replication of, the infective agent.*[61] Breastfeeding reduces the severity and duration of symptoms associated with this infection. Breastfeeding also protects against other common illnesses of infancy such as middle ear infection and respiratory illness.[62]

During the first two or three days of lactation, the breasts produce **colostrum,** a premilk substance containing antibodies and white cells from the mother's blood. Because it contains immunity factors, colostrum helps protect the newborn infant from those infections against which the mother has developed immunity, precisely those in the environment likely to infect the infant.[63] Maternal antibodies from colostrum inactivate harmful bacteria within the infant's digestive tract. Later, breast milk also delivers antibodies, although not as many as colostrum.

Certain factors in colostrum and breast milk favor the growth of "friendly" probiotic bacteria in the infant's digestive tract, preventing other, harmful bacteria from thriving there. Hormones and other factors present in colostrum and breast milk stimulate the development of the infant's digestive tract. Clearly, breast milk is a very special substance.

Controversy 12 discusses research showing that children with diabetes (type 1) almost always have antibodies to cow's milk protein in their pancreatic tissues.[64] This finding leads some to suspect that early feeding of cow's milk formula may set the stage for abnormal immune functioning that causes diabetes later. Others say a virus, not milk protein, initiates the immune changes associated with diabetes. In any case, breastfeeding prevents early exposure to cow's milk protein and also confers extra immune cells that destroy the kind of virus suspected of involvement in diabetes.

colostrum (co-LAHS-trum) a milklike secretion from the breasts during the first day or so after delivery before milk appears; rich in protective factors.

alpha-lactalbumin (lact-AL-byoo-min) the chief protein in human breast milk. The chief protein in cow's milk is *casein* (CAY-seen).

lactoferrin (lack-toe-FERR-in) a factor in breast milk that binds iron and keeps it from supporting the growth of the infant's intestinal bacteria.

Probiotics are discussed in Controversy 2.

Nutrients in Breast Milk Breast milk composition changes throughout lactation to meet the infant's changing energy and nutrient needs. Milk from the mother of a premature infant meets the developmental needs of a preterm infant in ways that full-term mother's milk cannot match. For example, the milk for a premature infant provides more protein in less volume, just the right mix to support the rapid growth required to help a premature infant survive its first critical weeks.

The carbohydrate in breast milk (and infant formula) is lactose. Besides being easily digested, lactose enhances calcium absorption.

The lipids in breast milk—and infant formula—provide the main source of energy in the infant's diet. Breast milk contains a generous proportion of the essential fatty acids linoleic acid and linolenic acid, as well as their longer-chain derivatives arachidonic acid and docosahexaenoic acid (DHA), which are found abundantly in both the retina of the eye and the brain; infant formula contains only linoleic acid and linolenic acid. Research is under way to determine the physiological significance of this DHA differential.[65] One apparent benefit is that young children who were breastfed as infants have sharper vision than those who were fed formulas; this enhanced visual development is attributed to the DHA in breast milk.[66]

The protein in breast milk is largely **alpha-lactalbumin,** a protein the human infant can easily digest. Another breast milk protein, **lactoferrin,** is an iron-gathering compound that helps absorb iron into the infant's bloodstream, keeps intestinal bacteria from getting enough iron to grow out of control, and kills certain bacteria.

The vitamin content of the breast milk of a well-nourished mother is ample. Even vitamin C, for which cow's milk is a poor source, is supplied generously by this breast milk. The concentration of vitamin D in breast milk is low, but this is not a threat to

*More children are hospitalized for rotavirus infection than for any other single cause.

hypoallergenic formulas clinically tested infant formulas that do not provoke reactions in 90% of infants or children with confirmed cow's milk allergy. Like all infant formulas, hypoallergenic formulas must demonstrate nutritional suitability to support infant growth and development. Extensively hydrolyzed and free amino acid–based formulas are examples.

light-skinned infants who are taken out into the sunshine regularly. A dark-skinned infant, or one who has little exposure to sunlight, however, may not make enough vitamin D to prevent rickets. Because so many variables exist regarding vitamin D and sunlight exposure, the AAP recommends vitamin D supplementation beginning at birth for breastfed babies who do not receive sufficient exposure to sunlight.

As for minerals, the 2-to-1 calcium-to-phosphorus ratio of breast milk is ideal for calcium absorption, and both of these minerals, along with magnesium, support the rate of growth expected in a human infant. Breast milk is also low in sodium. The limited amount of iron in breast milk is highly absorbable, and its zinc, too, is absorbed better than from cow's milk, thanks to the presence of a zinc-binding protein.

An exclusively breastfed baby does not need supplements except possibly for vitamin D and, after four months, iron. Before four months, supplemental iron is unnecessary. As lactation progresses, the iron in breast milk dwindles, making iron a concern for the four- to six-month-old. Most babies are born with enough iron in their livers to last about half a year, and iron deficiency is rarely seen in very young infants. By six months, feeding the breastfed infant iron-fortified cereals is desirable. If the water supply is severely deficient in fluoride, both breastfed and formula-fed infants require fluoride supplementation after six months of age.

Breast milk is the ideal food for infants, with the needed nutrients in the right proportions and also protective factors. It is especially valuable for premature infants.

Formula Feeding

Formula options:

- Liquid concentrate (inexpensive, relatively easy)—mix with equal part water.
- Powdered formula (cheapest, lightest for travel)—follow label directions.
- Ready-to-feed (easiest, most expensive)—pour directly into clean bottles.
- Never an option—whole cow's milk before 12 months of age.

The substitution of formula feeding for breastfeeding involves striving to copy nature as closely as possible. Human milk and cow's milk differ; cow's milk is significantly higher in protein, calcium, and phosphorus, for example, to support the calf's faster growth rate. A formula can be prepared from cow's milk that does not differ much from human milk in these respects; the formula makers first dilute the milk and then add carbohydrate and nutrients to make the proportions comparable to those of human milk (see Table 12-8 for a comparison of human milk and standard formulas).

Standard formulas are inappropriate for some infants. For example, premature babies require special formulas, and infants allergic to milk protein can drink special **hypoallergenic formulas** or formulas based on soy protein.[67] Soy formulas are lactose-free and so can be used for infants with lactose intolerance; they are also useful as an alternative to milk-based formulas for vegetarian families. For infants with other special needs, many other variations are available.

Table 12•8

HUMAN MILK COMPARED WITH INFANT FORMULA FOR SELECTED NUTRIENTS

Content	Mature Human Milk	Fortified Infant Formula
Energy (cal/L)	680	680
Protein (% of cal)	6	8
Fat (% of cal)	50	50
Carbohydrate (% of cal)	42	43
Iron (mg/L)	0.5	1.5–12
Vitamin A (µg/L)	675	660
Niacin (mg/L)	1.5	7.1
Vitamin D (µg/L)	0.5	10
Inositol (mg/L)	149	32

SOURCE: Committee on Nutrition, American Academy of Pediatrics, *Pediatric Nutrition Handbook*, 4th ed., ed. R. E. Kleinman (Elk Grove, Ill.: American Academy of Pediatrics, 1998), Appendix E.

The AAP recommends iron-fortified formulas for all formula-fed infants.[68] Low-iron formulas have no role in infant feeding. Use of iron-fortified formulas has risen in recent decades and is credited with the decline of iron-deficiency anemia in U.S. infants.[69]

Formula feeding offers an acceptable alternative to breastfeeding. Nourishment for an infant from formula is adequate, and a mother can choose this course with confidence. One advantage is that parents can see how much milk the baby drinks during feedings. Another is that other family members can participate in feeding sessions, giving them a chance to develop the special closeness that feeding fosters. Mothers who resume employment soon after giving birth may choose formula for their infants, but they have another option. Breast milk can be pumped into bottles and given to the baby in day care. At home, mothers may breastfeed as usual. Many mothers use both methods—they breastfeed at first but wean to formula later on.

For as long as breast milk or formula is the baby's major food (until the first birthday), unmodified cow's milk is an inappropriate replacement because milk provides little iron and vitamin C. If an infant's digestive tract is sensitive to the protein content, it may bleed and worsen iron deficiency. Thus, plain cow's milk both causes iron loss and fails to replace iron. Also, the infant's immature kidneys are stressed by plain cow's milk. Once the baby is obtaining at least two-thirds of total daily food energy from a balanced mixture of cereals, vegetables, fruits, and other foods (usually after 12 months of age), whole cow's milk, fortified with vitamins A and D, is an acceptable accompanying beverage. Reduced-fat milk is not recommended before the age of two years. Table 12-9 defines some terms applied to types of milk.

key point *Infant formulas are designed to resemble breast milk and must meet an AAP standard for nutrient composition. Special formulas are available for premature babies, allergic babies, and others. Formula should be replaced with milk only after the baby's first birthday.*

Table 12•9
MILK TERMS

- **casein** or **sodium caseinate** the principal protein of cow's milk. Another milk protein found in human milk's whey is **lactalbumin.**
- **evaporated milk** milk concentrated to half volume by evaporation. Adding water reconstitutes the milk; the taste is altered by the processing, however.
- **homogenized milk** milk treated to mix the fat evenly with the watery part (fat ordinarily floats to the top as cream). Heated milk is forced under high pressure through small openings to emulsify the fat.
- **pasteurized milk** milk that is heat treated to eliminate disease-causing microbes and to reduce its total bacterial count to an acceptable level.
- **powdered milk** dehydrated milk solids. Some powdered milks rehydrate easily (instant milk); others require extensive blending. Both whole and fat-free milk can be powdered.
- **whey** the liquid that remains after milk has coagulated (see also *casein*).
- **whole milk** full-fat cow's milk.

The infant thrives on formula offered with affection.

© Myrleen Cate/Index Stock

Formula's Advertising Advantage

MOST WOMEN ARE free to choose whatever feeding method best suits their needs. For only a few is breastfeeding either prohibited for medical reasons or medically indicated for special needs of the infant. With the strong scientific consensus that breastfeeding is preferable for most infants, why do women who could breastfeed their infants choose formula? Some women find the time and logistics of breastfeeding burdensome. For many women, though, the decision to forgo breastfeeding is influenced by aggressive advertising of formulas.

Advertisers of infant formulas often strive to create the illusion that formula is identical to human milk. No formula can match the nutrients, agents of immunity, and environmental information conveyed to infants by human milk, but the ads are convincing: "Like mother's milk, our 'gentle' formula provides complete nutrition" or "Our brand is scientifically formulated to meet your baby's needs." These ads imply that breast milk is "unscientific," unknown, and therefore untrustworthy.

To increase market share, formula manufacturers give coupons and samples of free formula to pregnant women. After childbirth, women in the hospital receive "goody bags" with more coupons to tempt them to receive their "formula gifts." Drugstores dispense still more coupons whenever computerized cash registers ring up items related to breastfeeding, such as pads that protect clothing from milk. More coupons arrive by mail a couple of months later, at a time when many women give up breastfeeding, even though nutrition authorities urge continued breastfeeding for several more months. Aggressive marketing tactics can undermine a woman's confidence concerning her breastfeeding choice, and lack of confidence has a significant influence on early discontinuation of breastfeeding.[1]

National efforts to promote breastfeeding seem to be working to some extent.[2] Many hospitals employ certified lactation consultants who specialize in helping new mothers establish a healthy breastfeeding relationship with their newborns. Table 12-10 lists ten steps hospitals and birth centers can take to promote successful long-term breastfeeding. An encouraging trend of breastfeeding initiation is emerging, with 64 percent of women initiating breastfeeding today, up from 50 percent in 1990. Despite this trend toward increasing breastfeeding, the percentage of women breastfeeding their infants still falls short of the goal of *Healthy People 2010*. Few infants are breastfed beyond about two months of age.[3] The American Dietetic Association encourages and supports continued national efforts to increase breastfeeding initiation rates, but states that the new challenge is to communicate the importance of breastfeeding for six months or longer.[4]

Formula-fed infants in developed nations are healthy and grow normally, but they miss out on the breastfeeding advantages described in the text.[5] In developing nations, however, the consequence of choosing not to breastfeed can be tragic. Feeding formula is often fatal to the infant in nations where poverty limits access to formula mixes, clean water is unavailable for safe formula preparation, and medical help is limited. The World Health Organization (WHO) strongly supports breastfeeding for the world's infants in its "baby-friendly" initiative and opposes the marketing of infant formulas to new mothers.[6]

Women are free to choose between breast and bottle, but the decision should be made by weighing valid factual information and should not be influenced by sophisticated advertising ploys.

Table 12•10

TEN STEPS TO SUCCESSFUL BREASTFEEDING

To promote breastfeeding, every maternity facility should:

- Develop a written breastfeeding policy that is routinely communicated to all health-care staff.
- Train all health-care staff in the skills necessary to implement the breastfeeding policy.
- Inform all pregnant women about the benefits and management of breastfeeding.
- Help mothers initiate breastfeeding within ½ hour of birth.
- Show mothers how to breastfeed and how to maintain lactation, even if they need to be separated from their infants.
- Give newborn infants no food or drink other than breast milk, unless medically indicated.
- Practice rooming-in, allowing mothers and infants to remain together 24 hours a day.
- Encourage breastfeeding on demand.
- Give no artificial nipples or pacifiers to breastfeeding infants.[a]
- Foster the establishment of breastfeeding support groups and refer mothers to them at discharge from the facility.

[a]Compared with nonusers, infants who use pacifiers breastfeed less frequently and stop breastfeeding at a younger age. C. G. Victora and coauthors, Pacifier use and short breastfeeding duration: Cause, consequence, or coincidence? *Pediatrics* 99 (1997): 445–453.

SOURCE: United Nations Children's Fund and World Health Organization, *Barriers and Solutions to the Global Ten Steps to Successful Breast-feeding*, 1994.

An Infant's First Foods

Foods can be introduced into a baby's diet as the baby becomes physically ready to handle them. This readiness develops in stages. A newborn baby can swallow only liquids that are well back in the throat. Later (at four months or so), the baby's tongue can move against the palate to swallow semisolid food such as cooked cereal. The stomach and intestines are immature at first; they can digest milk sugar (lactose) but not starch. At about four months, most babies can begin to digest starchy foods. Still later, the first teeth erupt, but not until sometime during the second year can a baby begin to handle chewy food.

milk anemia iron-deficiency anemia caused by drinking so much milk that iron-rich foods are displaced from the diet.

The Need for Water Once solid foods are introduced, an infant's risk of dehydration increases. Infant kidneys are inefficient in concentrating waste, so a baby must excrete more water than an adult to carry off a comparable amount of waste. Foods high in protein or electrolytes such as meat and eggs can promote dehydration if offered without water. Water should be offered to infants regularly once they are eating solid food. If the weather is hot and the mother is thirsty, her infant probably is, too. Give infants plain water, and let them drink it until they quench their thirst.

When to Introduce Solid Food Babies who are ready for solid foods thrive on receiving them and develop new skills through handling the foods. Indications of readiness for solid foods include:

- The infant can sit with support and can control its head movements.
- The infant is about six months old.

Babies develop according to their own schedules, and although Table 12-11 presents a suggested sequence, individuality is important. Three considerations are relevant: the baby's nutrient needs, the baby's physical readiness to handle different forms of foods, and the need to detect and control allergic reactions. With respect to nutrient needs, the nutrient needed most is iron, then vitamin C.

Foods to Provide Iron and Vitamin C Iron deficiency is prevalent in children between the ages of six months and three years due to their rapid growth rate and the significant place that milk has in their diets. Excessive milk consumption (more than 3½ cups a day) displaces iron-rich foods and can lead to iron-deficiency anemia, popularly called **milk anemia.**

Iron ranks highest on the list of nutrients most needing attention in infant nutrition. A baby's stored iron supply from before birth runs out after the birthweight doubles, long before the end of the first year. Breast milk or iron-fortified formula, then iron-fortified cereals, and then meat or meat alternates such as legumes are recommended. Once babies are eating iron-fortified cereals, parents or caregivers should begin selecting vitamin C–rich foods to go with meals to enhance absorption. The best sources of vitamin C are fruits and vegetables.

Fruit juice is a source of vitamin C, but some research shows that babies and young children may fail to grow and thrive when they drink so much juice that more nutrient- and energy-dense foods are displaced from their diets.[70] Although other research shows no relationship between juice consumption and children's growth, the AAP has nevertheless issued recommendations setting limits on juice consumption for infants and children: 4 to 6 ounces per day.[71] Fruit juices should be served in a cup, not a bottle, and not before the infant is six months of age.

A first birthday party and the possibility of tasting whole, unmodified cow's milk for the first time.

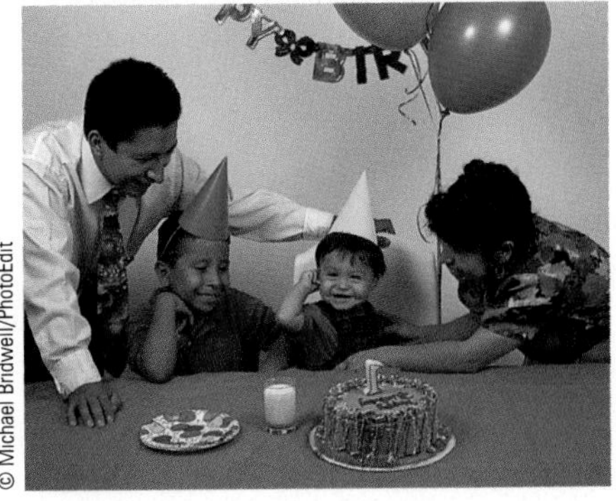

© Michael Bridwell/PhotoEdit

Physical Readiness for Solid Foods Foods introduced at the right times contribute to an infant's physical development. For example, experience with solid food at four to six months, when swallowing ability is developing, helps to desensitize the gag reflex. When the baby can sit up, can handle finger foods, and is teething, hard crackers and other hard finger foods may be introduced under the watchful eye of an adult. These foods promote the development of manual dexterity

Table 12•11

INFANT FEEDING SKILLS AND RECOMMENDED FOODS

Note: Because each stage of development builds on the previous stage, the foods from an earlier stage continue to be included in all later stages.

Age (mo)	Feeding Skill	Foods Introduced into the Diet
0–4	Turns head toward any object that brushes cheek. Initially swallows using back of tongue; gradually begins to swallow using front of tongue as well. Strong reflex (extrusion) to push food out during first 2 to 3 months.	Feed breast milk or infant formula.
4–6	Extrusion reflex diminishes, and the ability to swallow nonliquid foods develops. Indicates desire for food by opening mouth and leaning forward. Indicates satiety or disinterest by turning away and leaning back. Sits erect with support at 6 months. Begins chewing action. Brings hand to mouth. Grasps objects with palm of hand.	Begin iron-fortified cereal mixed with breast milk, formula, or water. Begin pureed vegetables and fruits.
6–8	Able to feed self with fingers. Develops pincher (finger to thumb) grasp. Begins to drink from cup.	Begin breads and other cereals and mashed vegetables and fruits. Begin plain, unsweetened fruit juices from cup.
8–10	Begins to hold own bottle. Reaches for and grabs food and spoon. Sits unsupported.	Begin yogurt. Begin pieces of soft, cooked vegetables and fruit from table. Gradually begin finely cut meats, fish, casseroles, cheese, eggs, and legumes.
10–12	Begins to master spoon, but still spills some.	Include at least 4 servings of breads and cereals from table, in addition to infant cereal; at least 2 servings of fruits and 3 servings of vegetables; and 2 servings of meat, fish, poultry, eggs, or legumes.[a]

[a]Serving sizes for infants and young children are smaller than those for an adult. For example, a serving might be ½ slice of bread instead of 1 slice, or ¼ cup rice instead of ½ cup.
SOURCE: Adapted in part from Committee on Nutrition, American Academy of Pediatrics, *Pediatric Nutrition Handbook*, 4th ed., ed. R. E. Kleinman (Elk Grove Village, Ill.: American Academy of Pediatrics, 1998), pp. 43–53.

Foods such as iron-fortified cereals and formulas, mashed legumes, and strained meats provide iron.

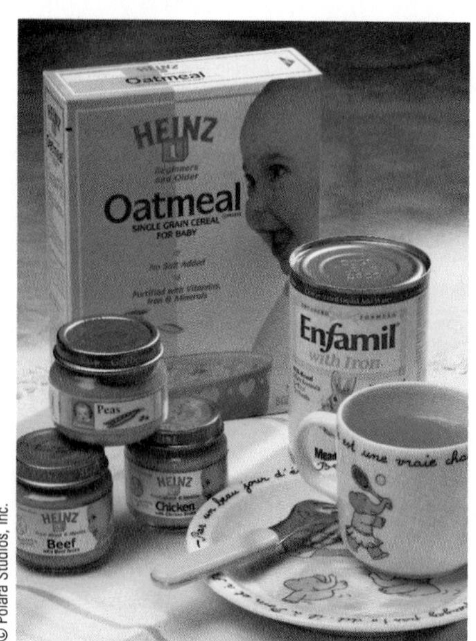

and control of the jaw muscles, but the caregiver must be careful that the infant does not choke on them. Hard crackers that melt slowly to a mush that is easy to swallow are best. Babies and young children can choke on popcorn, nuts, hot dogs, raw carrots, whole grapes, and hard candy; these foods are not worth the risk.

Some parents want to feed solids as early as possible on the theory that "stuffing the baby" at bedtime will promote sleeping through the night. There is no proof for this theory. Babies start to sleep through the night when they are ready, no matter when solid foods are introduced.

Food Allergies New foods should be introduced one at a time so that allergies or other sensitivities can be detected. For example, when fortified baby cereals are introduced, try rice cereal first for several days; it causes allergy least often. Try wheat-containing cereal last; it is a common offender. Introduce egg whites, soy products, peanut products, cow's milk, and citrus fruits still later for the same reason. If a food causes an allergic reaction (irritability due to skin rash, digestive upset, or respiratory discomfort), discontinue its use before going on to the next food. About nine times out of ten, the allergy won't be evident immediately but will manifest itself in vague symptoms occurring up to five days after the offending food is eaten. Wait a month or two to try the food again; many sensitivities disappear with maturity. If your family history indicates allergies, apply extra caution in introducing new foods. Parents or caregivers who detect allergies early in an infant's life can spare the whole family much grief.

Choice of Infant Foods Commercial baby foods in the United States and Canada are safe, and except for mixed dinners with added starch fillers and heavily sweetened desserts, they have high nutrient density. Brands vary in their use of starch and sugar—check the ingredient lists. Parents or caregivers should not feed directly from the jar—remove portions to a dish for feeding so as not to contaminate the unused food that will be stored in the jar.

Chapter 13 offers more information on allergies.

An alternative to commercial baby food is to process a small portion of the family's table food in a blender, food processor, or baby food grinder. This necessitates cooking without salt or sugar, though, as the best baby food manufacturers do. Adults can season their own food after taking out the baby's portion. Pureed food can be frozen in an ice cube tray to yield a dozen or so servings that can be quickly thawed, heated, and served on a busy day.

Appendix A includes the nutrient composition of many commercial baby foods.

Foods to Omit Sweets of any kind (including baby food "desserts") have no place in a baby's diet. The added food energy can promote obesity, and they convey few or no nutrients to support growth. Canned vegetables are inappropriate for babies because they often contain too much salt. Awareness of food-borne illness and precautions against it are imperative. Honey should never be fed to infants because of the risk of botulism, a form of food-borne illness (see Chapter 14).

Children love to eat what their families eat.

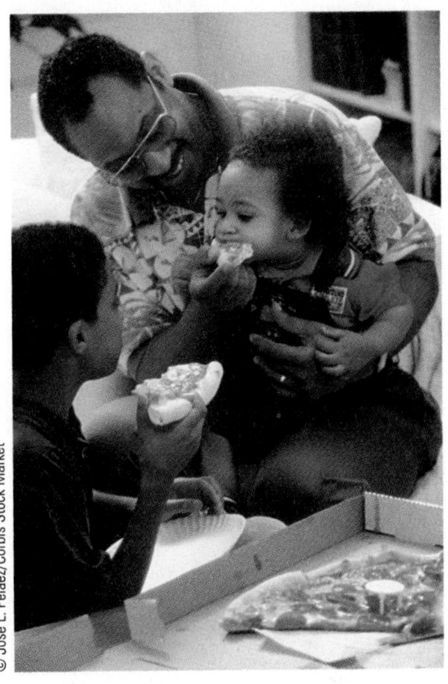

Foods at One Year For the baby weaned to whole milk after one year of age, whole milk can supply most of the nutrients the infant needs; 2 to 3½ cups a day meet those needs. Other foods—meat and meat alternates, iron-fortified cereal, enriched or whole-grain bread, fruits, and vegetables—should be supplied in variety and in amounts sufficient to round out total energy needs. Ideally, the one-year-old sits at the table, eats many of the same foods everyone else eats, and drinks liquids from a cup, not a bottle. A meal plan that meets the requirements for a one-year-old is shown in Table 12-12.

key point *Solid food additions to a baby's diet should begin at about six months and should be governed by the baby's nutrient needs and readiness to eat. By one year, the baby should be receiving foods from all food groups.*

Looking Ahead

The first year of a baby's life is the time to lay the foundation for future health. From the nutrition standpoint, the problems most common in later years are obesity and

Table 12•12

MEAL PLAN FOR A ONE-YEAR OLD

Breakfast
½ c whole milk
½ c iron-fortified cereal
½ c orange juice

Morning snack
½ c yogurt
¼ c fruit[a]

Lunch
½ c whole milk
½ c vegetables[b]
1 egg or ¼ c tofu
½ c noodles

Afternoon snack
½ c whole milk
½ slice toast
1 tbs peanut butter

Dinner
1 c whole milk
2 oz chopped meat or well-cooked mashed legumes
¼ c potato, rice, or pasta
½ c vegetables[b]
¼ c fruit[a]

[a]Include citrus fruits, melons, and berries.
[b]Include dark green, leafy, and deep yellow vegetables.

Nursing bottle syndrome in an early stage.

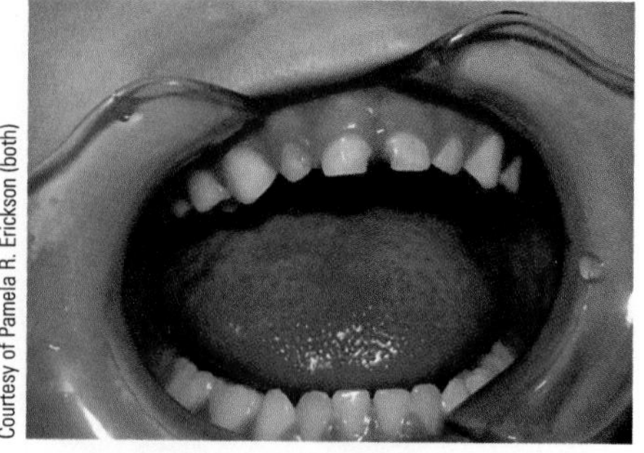

Courtesy of Pamela R. Erickson (both)

Nursing bottle syndrome, an extreme example. The lower teeth have decayed all the way to the gum line.

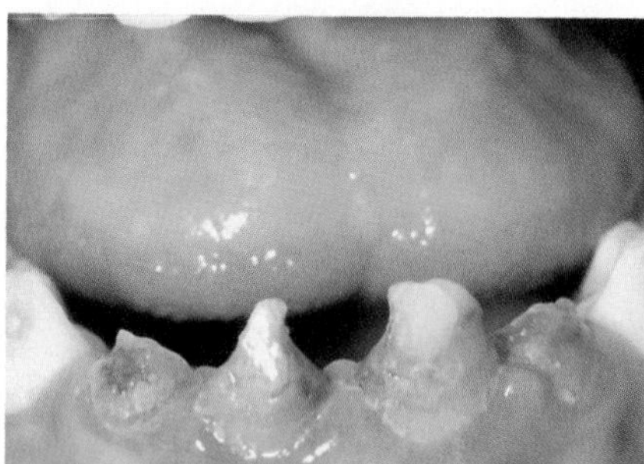

dental disease. Prevention of obesity may also help prevent the obesity-related diseases: atherosclerosis, diabetes, and cancer.

The most important single measure to undertake during the first year is to encourage eating habits that will support continued normal weight as the child grows. This means introducing a variety of nutritious foods in an inviting way, not forcing the baby to finish the bottle or baby food jar, avoiding concentrated sweets and empty-calorie foods, and encouraging physical activity. Parents should not teach babies to seek food as a reward, to expect food as comfort for unhappiness, or to associate food deprivation with punishment. If they cry for thirst, give them water, not milk or juice. If they cry for companionship, pick them up, don't feed them. If they are hungry, by all means, feed them appropriately. More pointers are offered in this chapter's Food Feature.

An irrational fear of obesity leads some parents to underfeed their infants, depriving them of the energy and nutrients they need to grow. Others wonder if they should feed their infants a low-fat diet to reduce heart disease risk, but the AAP recommends a fat intake of 40 to 50 percent of total calories for infants. A diet too low in fat hinders growth and development even when energy from carbohydrate and protein is ample. With rare exceptions, to be identified by physicians, babies from age one to two years need the food energy and fat of whole milk. They also need frequent servings of food containing the essential fatty acids.

The same strategies promote normal dental development: supply nutritious foods, avoid sweets, and discourage the association of food with reward or comfort. Dentists strongly discourage the practice of giving a baby a bottle as a pacifier. Sucking for long periods of time pushes the normal jawline out of shape and causes a bucktoothed profile: protruding upper and receding lower teeth. Prolonged sucking on a bottle of milk or juice also bathes the upper teeth in a carbohydrate-rich fluid that favors the growth of bacteria that produce acid that dissolves tooth material. Babies regularly put to bed with a bottle sometimes have teeth decayed all the way to the gum line, a condition known as nursing bottle syndrome, shown in the margin photos.

key point *The early feeding of the infant lays the foundation for lifelong eating habits. It is desirable to foster preferences that will support normal development and health throughout life.*

Mealtimes with Infants

THE WISE PARENT or caregiver of a one-year-old offers nutrition and affection together. "Feeding with love" produces better growth in both weight and height than feeding the same food in an emotionally negative climate.

Foster a Sense of Autonomy

The person feeding a one-year-old should be aware that the child's exploring and experimenting are normal and desirable behaviors. The child is developing a sense of autonomy that, if allowed to develop, will provide the foundation for later assertiveness in choosing when and how much to eat and when to stop eating. The child's self-direction, if consistently overridden, can later turn into shame and self-doubt.

Some Feeding Guidelines

In light of the developmental and nutrient needs of one-year-olds and in the face of their often contrary and willful behavior, a few feeding guidelines may be helpful:

- Discourage unacceptable behavior (such as standing at the table or throwing food) by removing the child from the table to wait until later to eat. Be consistent and firm, not punitive. The child will soon learn to sit and eat.
- Let the child explore and enjoy food. This may mean the child eats with fingers for a while. Use of the spoon will come in time.
- Don't force food on children. Provide children with nutritious foods, and let them choose which ones and how much they will eat. Gradually, they will acquire a taste for different foods. If children refuse milk, provide cheese, cream soups, and yogurt.
- Limit sweets strictly. Infants have little room in their 1,000-calorie daily energy allowance for empty-calorie sweets.

These recommendations reflect a spirit of tolerance that best serves the emotional and physical interests of the infant. This attitude, carried throughout childhood, helps the child to develop a healthy relationship with food. The next chapter finishes the story of growth and nutrition.

Let the child explore and enjoy food.

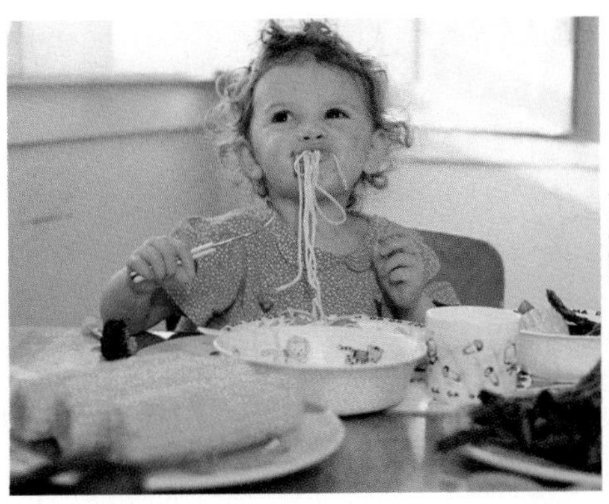

Answers to these Self Check questions are in Appendix G.

1. A pregnant woman needs an extra 300 calories above the allowance for nonpregnant women during whch trimester?
   a. First
   b. Second
   c. Third
   d. b and c
2. Which two vitamins, required for rapid cell reproduction, are needed in large amounts during pregnancy?
   a. folate and vitamin B_{12}
   b. vitamin C and riboflavin
   c. vitamin D and vitamin A
   d. biotin and vitamin A
3. A deficiency of which nutrient appears to be related to an increased risk of neural tube defects in the newborn?
   a. vitamin B_6
   b. folate
   c. calcium
   d. niacin
4. Which of the following may be hazardous during pregnancy?
   a. vitamin A supplementation
   b. dieting for weight loss
   c. drinking alcohol
   d. all of the above
5. Breastfeeding is contraindicated if a woman has:
   a. AIDS
   b. hepatitis
   c. tuberculosis
   d. all of the above
6. Breastfed infants may need supplements of:
   a. fluoride, iron, and vitamin D
   b. zinc, iron, and vitamin C
   c. vitamin E, calcium, and fluoride
   d. vitamin K, magnesium, and potassium
7. A major reason why a woman's nutrition before pregnancy is crucial is that it determines whether her uterus will support the growth of a normal placenta. T F
8. Fetal alcohol syndrome (FAS) is the leading known cause of mental retardation in the world. T F
9. In general, the effect of nutritional deprivation on a breastfeeding mother is to reduce the quality of her milk. T F
10. A sure way to get a baby to sleep through the night is to feed solid foods as soon as the baby can swallow them. T F

nutrition on the net

For further study of the topics of this chapter, access these websites and search for the phrases or words in quotation marks:

1. Find updates and quick links to these and other nutrition-related sites at our website: www.wadsworth.com/nutrition
2. Search for "birth defects," "pregnancy," "adolescent pregnancy," "maternal and infant health," and "breastfeeding" at the U.S. Government health information site: www.healthfinder.gov
3. Learn more about neural tube defects from the Spina Bifida Association of America: www.sbaa.org
4. For information about "gestational diabetes" and other topics, visit: www.eatright.org
5. Learn more about the WIC program: www.usda.gov/fns
6. Learn more about breastfeeding from La Leche League International: www.lalecheleague.org
7. Learn about nondietary approaches to weight loss from HUGS International: www.hugs.com
8. Visit the Nemours Foundation: www.kidshealth.org
9. Request information on drinking during pregnancy from the National Institute on Alcohol Abuse and Alcoholism: www.niaaa.nih.gov
10. Learn how to care for infants, children, and adolescents from the American Academy of Pediatrics and the Canadian Paediatric Society: www.aap.org and www.cps.ca
11. Download growth charts to monitor an infant's growth: www.cdc.gov/growthcharts

INTERNET ACTIVITY

The March of Dimes, an organization dedicated to reducing birth defects, provides up-to-date information regarding food-related concerns and pregnancy.

1. Go the the website of the March of Dimes at: http://www.modimes.org
2. Click on "Health Library."
3. Click on "Fact Sheets."
4. Under pregnancy, click on "Food-borne Illnesses in Pregnancy."

Read the Fact Sheet. Choose three food-related concerns and summarize the health issues of each during pregnancy. Were you previously aware of these health concerns?

CONTROVERSY 12

FETAL AND INFANT NUTRITION: EMERGING LINKS WITH CHRONIC DISEASES

WHEN PEOPLE think of health problems in children and adolescents, they typically think of measles and acne. They assign to adults diseases such as diabetes, heart disease, and high blood pressure. Lately, researchers have been surprised by an increasing number of children being diagnosed with diseases associated with adulthood, and they fear that such children may soon be headed for serious health problems.

A potential link between early life and later disease development may be rooted in conditions occurring during pregnancy and infancy. No doubt exists that conditions during gestation in the womb can affect the health, and even the life, of an infant. What researchers are now asking is whether conditions such as nutrition during pregnancy and infancy have far-reaching or even permanent effects that can alter a person's health much later, even during adulthood. The so-called fetal origins hypothesis proposes that fetal nutrition may change development in ways that may last a lifetime.[1]

In the past, such issues were framed as either-or questions: Are diseases caused by genetics or environment? Which is the culprit? Today, science is rapidly adopting a less mutually exclusive attitude as it finds an interactive rather than exclusive relationship between these two realms. No doubt exists that genetics lies at the base of a person's innate susceptibility to develop certain illnesses. However, environment, including nutrition, alters the expression of the genes, thereby altering the likelihood that disease will actually develop. Today, the important questions concern not whether, but how, when, and to what extent the environment modifies the expression of a person's genetic potential. Within this tangled net of influences lie the topics of this Controversy.

Disease Trends in Children

In a disturbing trend, the number of the world's children being diagnosed with type 2 diabetes has risen sharply in recent years.[2] The problem is strongly associated with obesity, hypertension, elevated blood lipids, and cardiovascular disease (CVD). Children with type 2 diabetes often show early signs of CVD, and the disease in full can be expected just a few years into adulthood.[3] Scientists fear that unless this trend is reversed, today's children will suffer much more CVD than the present adult generation.

Type 2 diabetes often makes itself known during the adolescent years, particularly among obese adolescents, with rates varying widely among U.S. ethnic groups. Eight percent of newly diagnosed cases appear among white children, 45 percent among Pima Indian children of Arizona. African American, Asian, and Hispanic children frequently develop the disease. Up to 85 percent of children with type 2 diabetes are obese, and more than 75 percent have close relatives with type 2 diabetes.[4]

Likewise, some 400,000 new cases of type 1 diabetes are diagnosed each year, with the great majority occurring in children.[5] Unlike type 2 diabetes, type 1 is typified by weight loss; still, it often leads to CVD in later life.

Untreated diabetes constitutes a severe threat to health. It foretells blindness, seizures, limb amputations, kidney failure, and CVD with heart attacks and strokes. The differences between the symptoms of type 2 and type 1 diabetes can be subtle, however, making accurate diagnosis and appropriate treatment difficult. In the child with type 2 diabetes, many of the telltale symptoms, such as glucose in the urine, ketones in the blood, and excessive thirst and urination, may be mild or absent. Also, physicians are trained to expect type 1 diabetes, not type 2, in children, but accurate diagnosis and appropriate treatment of both are essential to minimize risks of early CVD and other ills. Chapter 4 presented more details about diabetes.

If the causes of diabetes among children remain undiscovered and uncontrolled, the world may face major health disasters in the coming years. In searching for clues, researchers have theorized that an impaired nutrient supply to a fetus during gestation may trigger later development of type 2 diabetes. Another line of investigation concerns possible influences of early feeding practices on the development of type 1 diabetes. Later sections describe the evidence. The next section discusses what is known about CVD risks in children.

Children and Cardiovascular Disease

Obesity, high blood cholesterol, and hypertension stand with diabetes at the top of the list of factors leading to CVD. When these conditions appear in childhood, evidence indicates that CVD is perilously close on their heels.[6]

More and more children with obesity are being diagnosed with type 2 diabetes.

© Will Hart/PhotoEdit

The Development of Childhood Obesity

Children are heavier today than they were 20 or so years ago. Since the late 1970s, the prevalence of overweight has almost doubled for children and shows no sign of diminishing (see Figure C12-1).[7] Often, an obese child will also have high blood cholesterol, high blood pressure, and diabetes, which together may predict an elevated risk of heart disease in adult life.[8] Despite recent medical wonders in the prevention and cure of other childhood diseases, obesity remains a challenge to medical science.[9]

Parental obesity predicts excessive weight gain during childhood and more than doubles the chance that a young child will become an obese adult. Further, an infant born with a high birthweight is likely to remain heavy into adulthood.[10] The expression of inherited genes may be altered by specific conditions in the uterine environment, so conditions of pregnancy may bear some responsibility. For example, mothers who develop diabetes during their pregnancies (gestational diabetes) often give birth to infants who tend to be overweight. Researchers theorize that a fetus that develops in an environment of overnutrition will be born with more of the cellular and metabolic equipment that makes and stores body fat.

How parents choose to feed their infants may also influence later obesity. One recent well-controlled survey of over 15,000 adolescent children and their mothers indicates that infants who are mostly breastfed for the first six months of life are less likely to become overweight than infants fed mostly formula.[11] Another equally well-designed study of much younger children (three to five years of age) found no clear evidence that the duration of exclusive breastfeeding has an effect on body weight.[12] It may be that the effect becomes evident only in later childhood, or perhaps some other unknown factors may affect the study results.

Certainly learned behavior is a factor. Whole families may be eating too much, dieting inappropriately, and exercising too little, a pattern particularly common among mothers and daughters.[13]

Figure C12•1

PERCENTAGE OF YOUNG PEOPLE WHO ARE OVERWEIGHT

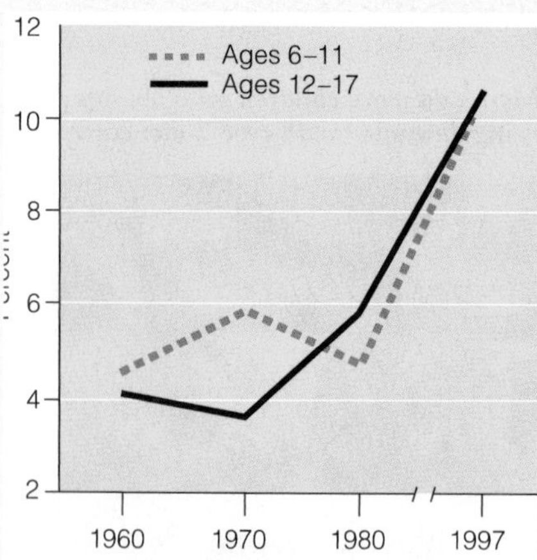

SOURCE: Data from National Center for Health Statistics, Centers for Disease Control and Prevention (CDC), and R. P. Troiano and K. M. Flegal, Overweight children and adolescents: Description, epidemiology, and demographics, *Pediatrics* 101 (1998): 497–504.

Physical Inactivity

When the question is asked, "Are children today consuming significantly more calories than in the past?" the answer is no. Children's energy intakes have remained relatively stable over the past 15 years, but the children have grown more sedentary.

A child who spends more than an hour or two a day in front of a television or computer monitor can become obese and develop unhealthy blood lipids even while eating fewer calories than a more active child (Figure C12-2).[14] Physically active children have higher HDL, lower LDL, and lower blood pressure than sedentary children, and these positive findings often persist in adulthood.[15] Likewise, active teenagers weigh less, smoke less, eat diets lower in saturated fats, and have healthier blood lipid profiles than sedentary peers.

Figure C12•2

PREVALENCE OF OBESITY BY HOURS OF TV PER DAY, CHILDREN AGES 10–15 YEARS

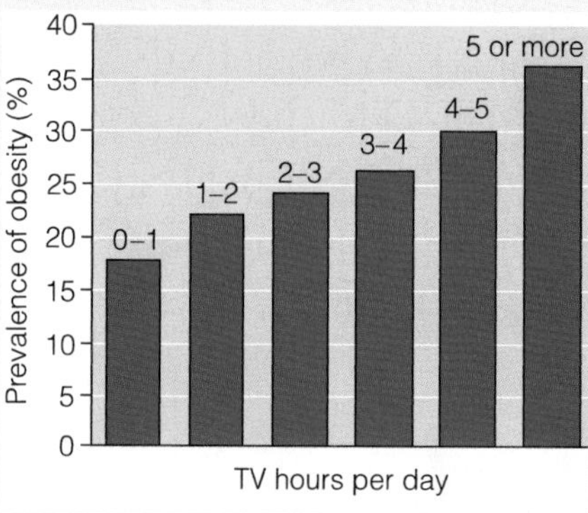

SOURCE: Centers for Disease Control and Prevention, Youth Risk Behavior Survey, 1999, available at www.cdc.gov

Elevated Blood Lipids

Like type 2 diabetes, the high blood cholesterol normally associated with heart disease in adults is showing up more and more frequently among the nation's children. High blood cholesterol correlates highly with childhood obesity and more highly still with central obesity (see Chapter 9).[16] In obese children, LDL cholesterol values are often too high, while HDL are too low for health. These relationships are apparent throughout childhood, and they increase with age. When LDL is high and HDL is low, fatty streaks and raised lesions in the arteries that foretell of heart disease often appear and grow significantly during the second decade of life.[17]

In deciding which children should be screened for elevated blood cholesterol, health professionals look to the child's family. When a parent or grandparent has high blood pressure, high blood lipids, and heart disease, the child is also likely to develop them.[18] For children with high blood lipids, a diet that meets the recommendations of the *Dietary Guidelines* for moderating fat, reducing saturated fat and cholesterol and increasing fiber while following the Food Guide Pyramid for adequacy may bring down the blood lipid values and still support normal growth.[19] Standard values for cholesterol screening in children and adolescents (ages 2 to 18 years) are listed in Table C12-1.[20]

High Blood Pressure

Hypertension can develop in a child's first decades of life, particularly in an obese child.[21] High blood pressure can signal underlying disease or early

Table C12•1

CHOLESTEROL VALUES FOR CHILDREN AND ADOLESCENTS

Disease Risk	Total Cholesterol (mg/dL)	LDL Cholesterol (mg/dL)
Acceptable	<170	<110
Borderline	170–199	110–129
High	≥200	≥130

NOTE: Adult values appear in Chapter 11.

CVD.[22] Table C12-2 specifies blood pressure standards for children up to age 18 years. Researchers are investigating the possibility of a link between breastfeeding and later development of hypertension, but the findings conflict. One study suggests that breastfeeding during infancy may help prevent hypertension during adolescence, but another finds that breastfeeding for longer than four months negatively affects the arteries in ways that make hypertension more likely.[23] More research is needed to make conclusions possible.

Type 2 Diabetes and Early Nutrition

In addition to the ongoing research into CVD, researchers are hastening to identify nutrition-related causes of type 2 diabetes in children. Currently under investigation are the following theories: childhood obesity, low birthweight, fetal overnutrition, and weight change after birth. Among adults, obesity is the number-one cause of type 2 diabetes, and the same may be true of children. As mentioned, the body weights of children are creeping up, while participation in physical activity is dropping off.[24] In addition to being overweight, children likely to develop type 2 diabetes typically:

- Are female.
- Have a family history of type 2 diabetes.
- Are of non-European descent.
- Were born to mothers who had diabetes while pregnant with them.
- Have insulin resistance.[25]

Another factor may also be in play—low birthweight.

Low Birthweight Theory

Low birthweight serves as an indicator of adverse conditions during fetal growth. Mothers who suffer malnutrition during pregnancy often give birth to babies of lower-than-average birthweights. Currently, scientists are debating whether research reflects a connection between low-birthweight babies and later development of diabetes, hypertension, and CVD.[26]

It seems ironic that a disease of abundance, type 2 diabetes, might be linked with a condition of deprivation, low birthweight. One explanation may lie in the basic functions of pregnancy. A malnourished woman who becomes pregnant may not support the development of a normal placenta and so may be unable to supply her fetus with needed nutrients. Under conditions of scarcity, the expression of fetal genes leading to conservation of both blood glucose and body energy stores temporarily best serves the needs of the growing fetus. The problem, however, the theory goes, is that upon emerging into a world of abundance, the child's body remains thrifty. The child grows quickly to catch up to and eventually surpass the size of peers, becomes fatter, develops insulin resistance and high blood glucose, and eventually suffers the effects of type 2 diabetes.

Observations of people are mixed, but intriguing.[27] In one study, men who were born small and who remained small at one year of age were more likely to die of CVD than men who were heftier as infants.[28] In a study of over 70,000 female nurses, the women who had been born with the lowest birthweights were later diagnosed with type 2 diabetes at about twice the rate of those with the highest birthweights, and the risk of CVD also climbed steadily as birthweights declined.[29] The association held true even for women whose mothers had never had diabetes. In other studies, being born to a mother who had diabetes during her pregnancy (gestational diabetes) predicts developing diabetes later on.[30] Other studies, however, indicate no association or even the reverse of the effect predicted by the originators of the low-birthweight theory.[31]

Like the United States, India is experiencing a rapidly escalating epidemic of diabetes and heart disease.[32] India provides a unique perspective on this issue because low birthweights have plagued its people for centuries. If low birthweight alone caused diabetes and later CVD, then India would have experienced high rates of the diseases throughout its history. This is not the case, however; diabetes and heart disease are newly emerging as problem diseases among the populations of India. Also, babies born in Indian cities are heavier than babies born in rural areas, but urban Indians have four to five times more diabetes than those living in the countryside.

These observations seem to point to life events after birth as the important

Table C12•2

BLOOD PRESSURE STANDARDS FOR CHILDREN AND ADOLESCENTS

	Systolic over Diastolic Pressure (mm Hg)			
	6 to 9 yr	10 to 12 yr	13 to 15 yr	16 to 18 yr
Mild hypertension	111–121 over 70–77	117–125 over 75–81	124–135 over 77–85	127–141 over 80–91
Moderate hypertension	122–129 over 70–85	126–133 over 82–89	136–143 over 86–91	142–149 over 92–97
Severe hypertension	>129 over >85	>133 over >89	>143 over >91	>149 over >97

factors driving the worldwide diabetes and CVD epidemic. One such factor may be a shift in eating patterns that reflects the global adoption of Western-style foods. As segments of the populations of developing countries gain wealth, they often abandon traditional eating patterns and active lifestyles for the fast foods and automobiles of the developed world. At the same time, countries with traditionally lean populations often experience a surge in obesity and all of its associated ills, including CVD.

Fetal Overnutrition Theory

Another theory blames *overnutrition* during fetal life for the expression of "thrifty" genes and the later development of obesity, type 2 diabetes, and hypertension. The idea is that an overabundance of energy nutrients encountered by the fetus may set in motion the development of efficient enzymatic systems for storing energy nutrients as body fat. A thrifty fetus would have an advantage if born into a famine environment, but in a food-rich world such a child faces a greatly increased risk for the illnesses of abundance.[33]

Weight Change Theory

Another suggestion is that perhaps the magnitude of weight gain after birth, and not just low birthweight, is important in the development of type 2 diabetes.[34] Thin infants who gain weight rapidly during childhood may be more at risk for diabetes and CVD, but thin infants who remain thin may not be.[35]

Whatever its origins, the problem of diabetes among children is severe and damaging. Urgently needed are studies to establish how many children have type 2 diabetes and whether the same treatments used for type 2 diabetes in adults are safe and effective for children.[36] The next section looks at a theory linking early nutrition to type 1 diabetes.

Type 1 Diabetes and Early Nutrition

Type 1 diabetes is thought to be an autoimmune disease in which the immune system somehow goes awry and makes antibodies that attack the protein insulin and destroy insulin-producing cells of the pancreas. It usually sets in during youth and necessitates insulin injections. Some evidence has linked some cases of type 1 diabetes to the practice of feeding cow's milk formula to young infants. A caution is in order: the research is preliminary and far from conclusive, and more study is needed before any inferences can be drawn concerning infant nutrition and development of type 1 diabetes.

Cow's Milk Formula Theory

Some researchers propose that feeding cow's milk formula may trigger the immune systems of susceptible infants to make antibodies against the insulin in cow's milk.[37] According to the theory, this may eventually lead to the development of antibodies to human insulin, thus initiating type 1 diabetes.[38] The researchers speculate that infants with a genetic predisposition to develop diabetes are most vulnerable.[39]

The American Academy of Pediatrics condemns the idea of feeding whole, unmodified cow's milk to infants before their first birthdays. It recommends infant formulas to avoid allergy and other adverse consequences (see Chapter 12). Yet, concerns have been raised about even standard baby formulas.

Formation of Antibodies

The presence of antibodies in the blood indicates to researchers that the immune system has reacted to a foreign substance. The proposed connection between formula and diabetes is based on the idea that antibodies formed in response to cow's milk proteins may trigger the autoimmune disease diabetes. How likely is this to occur? So far, no strong case can be made to support the idea. The evidence in support of the theory is as follows.

Some studies have found higher concentrations of bovine (cow) antibodies circulating in the blood of formula-fed infants than in the blood of exclusively breastfed infants.[40] Likewise, children with type 1 diabetes have also been reported to have higher blood concentrations of antibodies to cow's milk proteins compared with the blood of healthy children. In another line of evidence, laboratory animals exposed to conditions likely to produce diabetes were reported to be protected from development of indicators of the disease when they are fed a milk-free diet.[41]

Many other studies, however, reveal no link between formula feeding and diabetes, and instead of implicating milk in diabetes causation they conclude that antibodies to milk protein have no role in diabetes development.[42] For example, one Indian research team reported that the rate of diabetes in India is about the same for infants fed cow's milk formula and those fed exclusively their mother's breast milk.[43] Also, some animal studies disagree with positive findings mentioned earlier. They report no difference in diabetes development between animals fed milk proteins and those on milk-protein free diets.[44] To disentangle this mixture of findings promises to challenge researchers for some time to come.

Milk Protein Variations

A group of researchers has investigated cow's milk itself for clues to why some studies support the cow's milk theory while others do not. They began by searching infant-feeding histories of children with type 1 diabetes living in Iceland, an area of low incidence of type 1 diabetes, and in Scandinavia, an area with a higher incidence. They also analyzed the proteins in cow's milk of the two regions.[45] As expected, in Iceland they found no connection between feeding cow's milk formula and later development of type 1 diabetes. In Scandinavia, however, incidence of the disease was significantly greater with early use of cow's milk formula.

Analysis of the milk revealed that Scandinavian cow's milk contained one form of a milk protein, casein, whereas Icelandic cow's milk contained a different form. The researchers speculate that the form of casein in the milk from which formula is made may be a factor in determining whether feeding the formula triggers the processes leading to diabetes.

Currently, the collective studies are insufficient to conclude that feeding milk-based formulas to infants is related to the development of diabetes. In fact, the weight of evidence tips toward the negative. Despite this, some have stretched the truth about the evidence, declaring cow's milk to be a dangerous food that everyone, including children, should avoid. This is not the case. Other than lactose intolerance and occasional allergy, no adverse consequences are evident in most milk-drinking people over one year of age. For most people, cow's milk and milk products are likable, rich sources of some hard-to-get nutrients that may help *prevent* such chronic diseases as hypertension and osteoporosis.

Children and adolescents especially need the nutrients found in milk to ensure maximum bone strength (see the next chapter). If children do not drink milk, their caretakers must find adequate substitutes or the children's nutrition will surely suffer. Without a doubt, for practically all infants, human breast milk remains the best food in the first 12 months of life, including infants with a family medical history of type 1 diabetes.[46]

Conclusion

More and more parents are choosing unwise nutrition and other lifestyle behaviors for their families that injure the health of an alarming number of children around the globe.[47] This trend bodes ill for some 60 million children who, without immediate intervention, are destined to suffer the consequences of diabetes and subsequent CVD very early in adulthood.

To combat these trends in the United States, health education programs implemented in schools try to educate children directly. The trend in childhood diseases is not occurring in a vacuum, however. In the entire developed world, adult populations are also experiencing greatly increased incidences of obesity and related ills. To tackle the problem in children may require taking a long honest look at global trends, such as greater accessibility of inexpensive high-calorie fast foods, greater use of automobiles, and fewer incentives to be physically active (Chapter 9's Consumer Corner addressed these issues). Basic health-care and nutrition information must be made equally available to all citizens, and especially to women of childbearing age, regardless of income, and schools would do well to insist upon daily physical education for every able child in their care.[48]

It is tempting to look for simple answers to complex problems, but in truth, if everyone today would take up the advice of every legitimate health agency worldwide, many of these problems would disappear. Today, the best advice remains the easiest to give, and perhaps the most difficult to follow. Parents should set the example for their children. Don't smoke, choose a diet in accord with the *Dietary Guidelines for Americans*, follow the Food Guide Pyramid, and make it a habit to be physically active each day. Many times, as parents go, so go their children.

chapter 13

Kerry Damianakes, *Tall Sandwich*, 1996

Child, Teen, and Older Adult

Frequently Asked Questions

Can nutrient deficiencies impair a child's thinking or cause misbehavior? p. 477

Does diet affect hyperactivity? p. 482

Is breakfast really the most important meal of the day for children? p. 484

How nourishing are the lunches served at school? p. 484

Can nutrition help people to live longer? p. 495

Can foods or supplements affect the course of Alzheimer's disease? p. 497

Contents

TO GROW AND to function well in the adult world, children need a solid background of sound eating habits, which begin during babyhood with the introduction of solid foods. But at that point the person's nutrition story has just begun; the plot thickens. Nutrient needs change throughout life into old age, depending on the rate of growth, gender, activities, and many other factors. Nutrient needs also vary from individual to individual, but generalizations are possible and useful.

Early and Middle Childhood

Imagine growing 10 inches taller in just one year, as the average healthy baby does during the first dramatic year of life. At age one, infants have just learned to stand and toddle, and growth has slowed by half; by two years, they can take long strides with confidence and are learning to run, jump, and climb. These new accomplishments reflect the accumulation of a larger mass and greater density of bone and muscle tissue. These same growth trends—a lengthening of the long bones and an increase in musculature—continue until adolescence, but unevenly and more slowly.

A 174-pound adult male needs 900 milligrams of vitamin A, or 5.2 milligrams per pound. A 44-pound five-year-old needs 500 milligrams of vitamin A, or 11.4 milligrams per pound.

Growth and Nutrient Needs of Young Children

An infant's appetite decreases markedly near the first birthday, and the appetite fluctuates thereafter. At times children seem to be insatiable, and at other times they seem to live on air and water. Parents and other caregivers need not worry: internal appetite regulation in children of normal weight guarantees that their overall energy intakes will be right for each stage of growth. One caution: some children may disregard internal satiety signals and overeat, thereby inviting the onset of obesity (see Chapter 9 and Controversy 12).

A one-year-old child needs perhaps 1,000 calories a day; a three-year-old needs 300 calories more. The next seven years add 700 more calories for a total of about 2,000 calories a day. Although total energy needs have doubled by age ten, the child's energy need per pound of body weight has steadily declined as the rate of growth has slowed.

Growth enlarges the demand for all the nutrients per pound of body weight. On this basis, a five-year-old's need for, say, vitamin A is about double the need of an adult man (see the margin). Children accumulate stores of nutrients for the adolescent growth spurt when their nutrient intakes will not be able to meet the demands of rapid growth, and they will draw on those stored nutrients. Calcium storage is critical: the denser the bones are in childhood, the better prepared they are to support teen growth and to withstand the inevitable bone losses of later life.

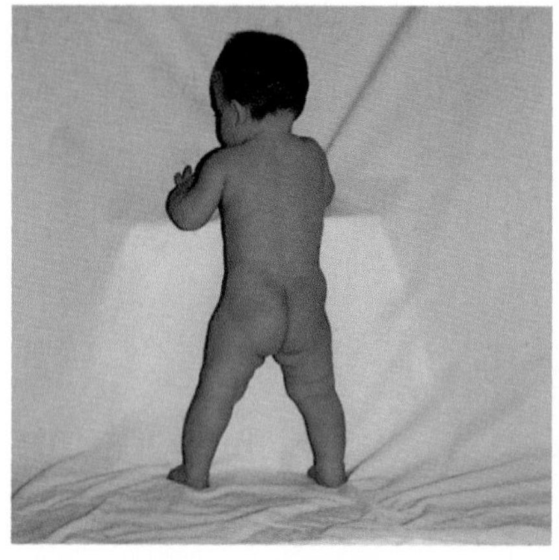

The body shape of a one-year-old (above) changes dramatically by age two (below). The two-year-old has lost much baby fat; the muscles (especially in the back, buttocks, and legs) have firmed and strengthened; and the leg bones have lengthened.

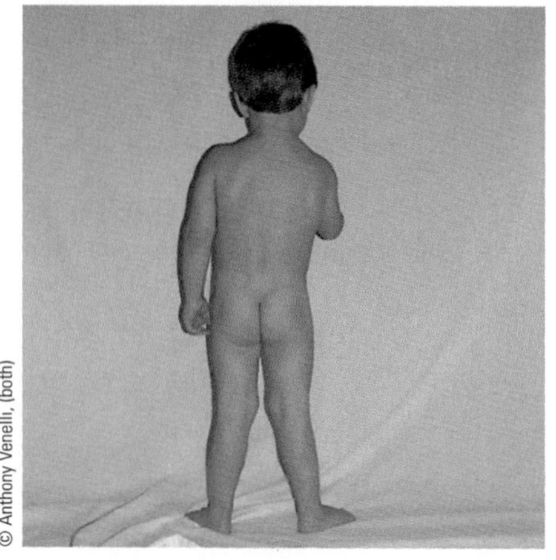

The Food Guide Pyramid for Young Children The Food Guide Pyramid for Young Children, shown in Figure 13-1, displays one means of providing the needed nutrients to children. This pyramid differs from the one for adults (see Chapter 2) by designating a set number of servings from each food group rather than a range. Children two to six years old need at least the specified number of servings, but the serving sizes should vary according to age. Children two to three years old should receive servings that are about two-thirds the size of an adult serving. Children in the four- to six-year-old group can generally eat the serving sizes recommended for adults. Older children and adolescents need additional servings. Few children's diets follow this nutritious pattern, however, and intakes of calcium, iron, and zinc fall far below recommendations.[1]

Careful food selection is essential to ensure that a child receives the right amounts of nutrients. When a child consistently skips breakfast or is allowed to consistently choose sugary foods (candy or marshmallows) in place of nourishing ones (whole-grain cereals), the child will fail to get enough of several nutrients. Nutrients missed from a skipped breakfast won't be "made up" at lunch and dinner but will be left out com-

Figure 13•1

FOOD GUIDE PYRAMID FOR YOUNG CHILDREN

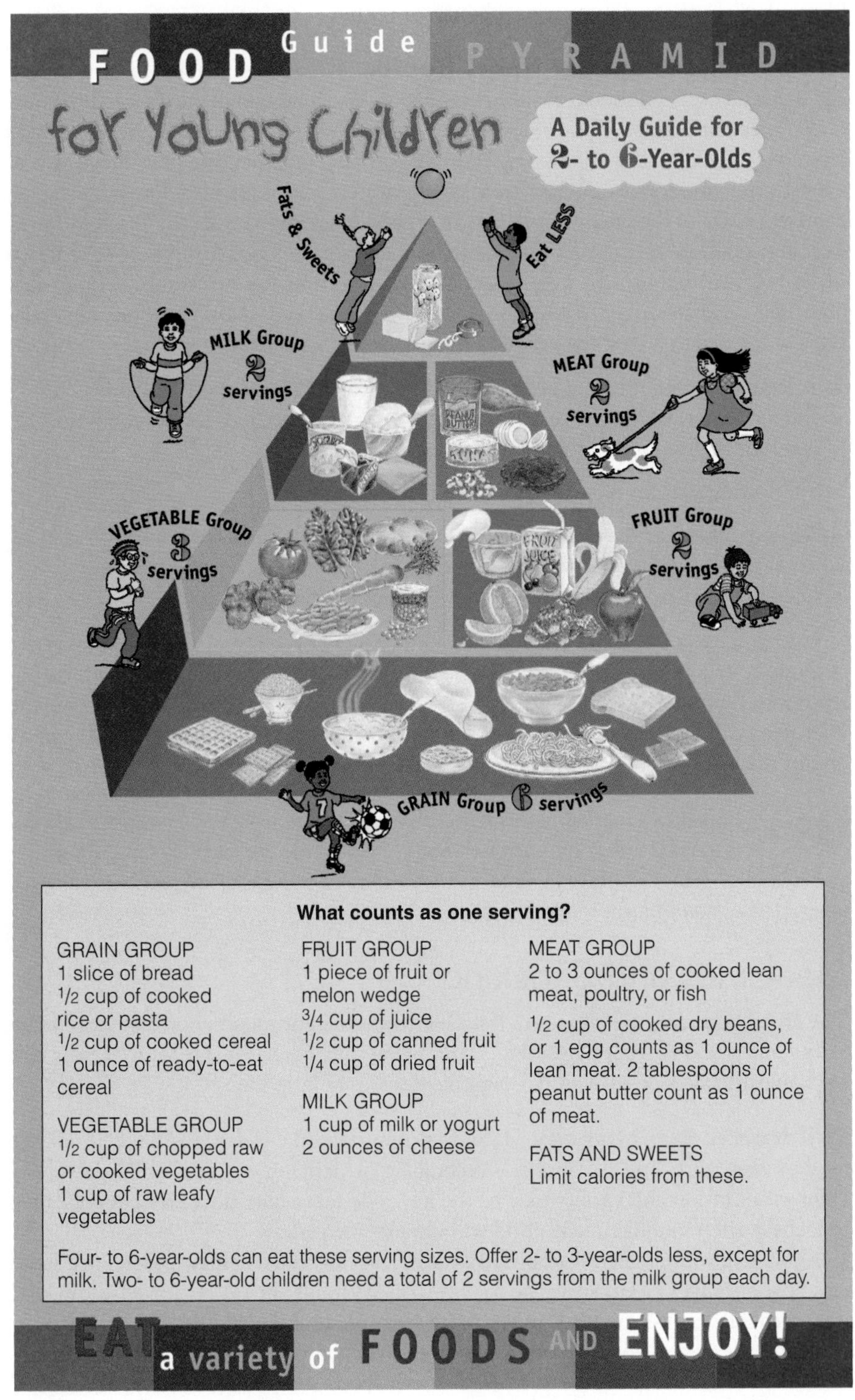

SOURCE: USDA Center for Nutrition and Policy Promotion, March 1999, Program AID 1649.

pletely that day.[2] Children naturally love sweets and may choose them to the exclusion of more nutritious foods if sweets are too often made available for choosing.

Active, normal-weight children may enjoy occasional treats of high-calorie but nutritious foods. From the milk group, ice cream or pudding is good now and then;

from the bread group, whole-grain or enriched cakes, cookies, or doughnuts are an acceptable addition to a balanced diet. These foods encourage a child to learn that pleasure in eating is important. Regular large quantities of these treats, however, can lead only to nutrient deficiencies, obesity, or both.

Recommendations for Heart Health Regardless of family history, experts agree that all children should eat a variety of foods and maintain desirable weight. There is less agreement, however, on restricting fat in the diets of children.[3] Limiting a child's fat intake does appear to improve blood lipids without compromising physical growth or neurological development.[4] However, a too-strict application of a low-fat diet can deprive a child of the energy and nutrients that are associated with the lipids in foods. Recommendations to limit fat and cholesterol are not intended for infants or children under two years old because infants and toddlers need a higher percentage of fat to support their rapid growth. The *Dietary Guidelines for Americans* make these recommendations for all children over the age of two years and for adolescents:

- Adequate nutrition should be achieved by eating a wide variety of foods.
- Energy (calories) should be adequate to support growth and development and to reach or maintain desirable body weight.

The following pattern of nutrient intake is recommended:

- Saturated fatty acids—less than 10 percent of total calories.
- Total fat—an *average* of no more than 30 percent of total calories.
- Dietary cholesterol—less than 300 milligrams per day.[5]

To determine the fiber need of a child, use the "age plus 5" method.

No harm can come from encouraging children over the age of two to eat a variety of foods, to reach or maintain a desirable weight, to obtain enough fiber, and, within reason, to limit saturated fat and cholesterol intakes. One easy way to determine the fiber needs of children is to use the "age plus 5" method: add 5 to the child's age to obtain the approximate number of fiber grams needed per day.

key point *Children's nutrient needs reflect their stage of growth. For a healthy child, use the Food Guide Pyramid for Young Children and the* Dietary Guidelines for Americans *as guides. Positive parental guidance can help establish food patterns that provide adequate nourishment for growth without obesity.*

Mealtimes and Snacking

The childhood years are a parent's last chance to influence food choices. Appropriate eating habits and attitudes toward food can help future adults emerge with healthy eating habits that reduce risks of degenerative diseases in later life.

Children's Preferences Children naturally like nutritious foods in all the food groups, with one exception—vegetables, which some young children refuse. Here presentation and variety may be the key. The more nutritious choices presented to a child, the more likely the child will choose adequately.

Many children prefer vegetables that are mild flavored, slightly undercooked and crunchy, bright in color, and easy to eat. Cooked foods should be served warm, not hot, because a child's mouth is much more sensitive than an adult's. The mild flavors of carrots, peas, and corn are often preferred over sharper tasting broccoli or turnips because a child has more taste buds. Smooth foods such as grits, oatmeal, mashed potatoes, and pea soup are often well received.

Fear of new foods is practically universal among children. Suggesting, rather than commanding, that a child try small amounts of new foods at the beginning of a meal when the child is hungry seems to work best. Offering the child samples of new foods that adults are enjoying can stimulate the child's natural curiosity and often produces the desired result: the child tastes the new food. Forcing or bribing a child to try new foods produces the opposite of the desired effect: the child will likely not try those foods again. Rewarding a child for eating a particular food, such as allowing extra television time for eating vegetables, erases any preference the child might have otherwise developed for that food.[6] Likewise, when children are banned from eating favorite foods, they yearn for them more.[7]

Little children prefer small portions of food served at little tables. If offered large portions, children may fill up on favorite foods, ignoring others. Toddlers often go on food jags, eating only one or two favored foods. For food jags lasting a week or so, make no response, since two-year-olds regard any form of attention as a reward. After two weeks of indulging the jag, try serving tiny portions of many foods, including the favored items. Distract the child with friends at meals, and make other foods as attractive as possible.

Just as parents are entitled to their likes and dislikes, a child who genuinely and consistently rejects a food should be allowed the same privilege. Children should be believed when they say they are full: the "clean-your-plate" dictum should be stamped out for all time. Children who are forced to override their own satiety signals are in training for obesity. Encourage children to listen to their bodies, and do not make an issue of food acceptance. The parent is responsible for *what* the child is offered to eat, but the child is responsible for *how much* and even *whether* to eat.

Little children like to eat small portions of food at little tables.

© Mary Kate Denny/PhotoEdit

Chapter 3 provides actions to take in case choking should occur.

Choking A child may make no sound when choking, so an adult should keep an eye on children when they are eating. Encourage the child to sit when eating—choking is more likely when children are running or reclining. Round foods such as grapes, nuts, hard candies, and pieces of hot dog can become lodged in a child's small windpipe. Other potentially dangerous foods include tough meat, popcorn, chips, and peanut butter eaten by the spoonful.

Snacking and Other Healthy Habits Parents may find that their children snack so much that they are not hungry at mealtimes. This is not a problem if children know how to snack—nutritious snacks are just as nutritious as small meals. Keep snack foods simple and available: milk, cheese, crackers, fruit, vegetable sticks, yogurt, peanut butter sandwiches, and whole-grain cereal.

A bright, unhurried atmosphere free of conflict is conducive to good appetite and provides a climate in which a child can learn to enjoy eating. Parents who beg, cajole, and demand that their children eat invite power struggles. A child may find mealtimes unbearable if they are accompanied by a barrage of accusations—"Susie, your hands are filthy . . . your report card . . . and clean your plate!" The child's stomach recoils as both body and mind react to stress of this kind.

Children love to be included in meal preparation, and they like to eat foods they helped to prepare (see Table 13-1 in the margin). A positive experience is most likely when tasks match developmental abilities and are undertaken in a spirit of enthusiasm and enjoyment, not criticism or drudgery. Praise for a job well done (or at least well attempted) expands a child's sense of pride and helps to develop skills and positive feelings toward healthy foods.

Many parents overlook perhaps the single most important influence on their child's food habits—their own habits. Parents who don't prepare, serve, and eat carrots shouldn't be surprised when their child refuses to eat carrots.[8]

Table 13•1

FOOD SKILLS OF PRESCHOOLERS[a]

Age 1–2 years, when large muscles develop, the child:

- uses short-shanked spoon.
- helps feed self.
- lifts and drinks from cup.
- helps scrub, tear, break, or dip foods.

Age 3 years, when medium hand muscles develop, the child:

- spears food with fork.
- feeds self independently.
- helps wrap, pour, mix, shake, or spread foods.
- helps crack nuts with supervision.

Age 4 years, when small finger muscles develop, the child:

- uses all utensils and napkin.
- helps roll, juice, mash, or peel foods.
- cracks egg shells.

Age 5 years, when fine coordination of fingers and hands develops, the child:

- helps measure, grind, grate, and cut (soft foods with dull knife).
- uses hand-cranked egg beater with supervision.

[a]These ages are approximate. Healthy, normal children develop at their own pace.

SOURCES: Adapted from M. Sigman-Grant, Feeding preschoolers: Balancing nutrition and developmental needs, *Nutrition Today*, July/August 1992, pp. 13–17; A. A. Hertzler, Preschoolers' food handling skills—Motor development, *Journal of Nutrition Education* 21 (1989): 100B–100C.

Healthy eating habits and positive relationships with food are learned in childhood. Parents teach children best by example. Choking hazards can often be avoided.

Can Nutrient Deficiencies Impair a Child's Thinking or Cause Misbehavior?

A child who suffers from nutrient deficiencies may exhibit physical and behavioral symptoms: the child will be sick and out of sorts. Sometimes, however, a child with a borderline deficiency will have few readily apparent symptoms. In either case, diet-behavior connections are of keen interest to caregivers who both feed children and live with them.

Deficiencies of protein, energy, vitamin A, iron, and zinc plague children the world over. In developing nations, such deficiencies cause or contribute to nearly half the deaths of children under age four and inflict blindness, stunted growth, and vulnerability to infections on millions more.

Table C13-2 of Controversy 13 provides details concerning the mental symptoms of anemia.

In developed countries such as the United States and Canada, most deficiencies have subtle, even unnoticeable, effects. A study of seemingly healthy British children revealed that about 40 percent of them had intakes of less than half the recommended amounts of folate, vitamin D, calcium, iron, magnesium, selenium, zinc, and other minerals. The researchers gave multinutrient supplements to some of the children and later administered intelligence tests to all of them. Those who had received the supplements scored significantly higher on the tests than the others did. Another study of school-aged children in a U.S. correctional institution noted that children receiving a low-dose supplement of vitamins and minerals required discipline for fighting, vandalism, disrespect, refusal to work, and other infractions about half as often as those receiving a placebo.[9] The authors of both studies interpreted the findings to mean that brain function may be sensitive to borderline deficiencies of some nutrients, even in children who are well nourished with protein and some vitamins. This conclusion has been supported by other findings.

Iron deficiency remains common in children and adolescents despite iron fortification of foods and other programs to combat this deficiency.[10] Besides carrying oxygen in the blood, iron works as part of large molecules to release energy within cells and plays key roles in many molecules of the brain and nervous system. A lack of iron not only causes an energy crisis but also affects behavior, mood, attention span, and learning ability.[11]

Iron deficiency is diagnosed by a deficit of iron in the *blood*, after anemia has developed. A child's *brain*, however, is sensitive to slightly lowered iron concentrations long before the blood effects appear. Distinguishing the effects of iron deficiency from those of other factors in children's lives is difficult, but studies have found connections between iron deficiency and behavior. Iron deficiency seems to manifest itself in a lowering of the motivation to persist in intellectually challenging tasks, a shortening of the attention span, and a reduction of overall intellectual performance. Furthermore, a child who had iron-deficiency anemia *as an infant* may continue to perform poorly as he or she grows older, even with improvement of iron status.[12] No one knows whether the poverty and poor health often associated with early iron deficiency or some lingering effect of the deficiency itself is to blame for the later cognition problems.[13] Iron-deficient children may be irritable, aggressive, and disagreeable or sad and withdrawn, and they may be labeled "hyperactive," "depressed," or "unlikable."

Inspection of a disruptive or apathetic child's diet by a qualified health-care professional can identify these reversible problems, and additions to the diet can correct them. Table 13-2 lists some iron-rich foods children often like to eat. Only a health-care provider should make the decision to give iron supplements, and supplements should be kept out of children's reach. Iron toxicity is a leading cause of poisoning each year in toddlers and other children who accidentally ingest iron pills.

Table 13•2

IRON-RICH FOODS KIDS LIKE[a]

Breads, Cereals, and Grains
Canned macaroni (½ c)
Canned spaghetti (½ c)
Cream of wheat (¼ c)
Fortified dry cereals (1 oz)[b]
Noodles, rice, or barley (½ c)
Tortillas (1 flour, 2 corn)
Whole-wheat, enriched, or fortified bread (1 slice)
Vegetables
Baked flavored potato skins (½ skin)
Cooked mushrooms (½ c)
Cooked mung bean sprouts or snow peas (½ c)
Green peas (½ c)
Mixed vegetable juice (1 c)
Fruits
Apple juice (1 c)
Canned plums (3 plums)
Cooked dried apricots (¼ c)
Dried peaches (4 halves)
Raisins (1 tbs)
Meats and Legumes
Bean dip (¼ c)
Canned pork and beans (⅓ c)
Mild chili or other bean/meat dishes (¼ c)
Liverwurst on crackers (½ oz)
Meat casseroles (½ c)
Peanut butter and jelly sandwich (½ sandwich)
Lean chopped roast beef or cooked ground beef (1 oz)
Sloppy joes (½ sandwich)

[a]Each serving provides at least 1 milligram iron, or one-tenth of a child's iron RDA. Vitamin C–rich foods included with these snacks increase iron absorption.
[b]Some fortified breakfast cereals contain more than 10 milligrams iron per half-cup serving (read the labels).

key point *The detrimental effects of nutrient deficiencies in children in developed nations can be subtle. Iron deficiency is the most widespread nutrition problem of children and causes abnormalities in both physical health and behavior. Iron toxicity is a major form of poisoning in children.*

The Problem of Lead

Another form of metal poisoning arises from ingestion of lead. Lead poisoning often occurs because babies like to explore and they put everything into their mouths, including things that may harm them, such as chips of old paint, pieces of metal, and other unlikely substances. These are normal baby activities, but they can silently cause lead to build up in the child's body. Not until much later, after lead toxicity has set in, do caretakers notice unusual symptoms. Tragically, once symptoms set in, even today's most effective medical treatment for lead toxicity may not reverse all of the functional damage. Impaired thinking, reasoning, perception, and other academic skills, as well as hearing impairments and decreased growth, are associated with even very low levels of lead toxicity.[14]

allergy an immune reaction to a foreign substance, such as a component of food. Also called *hypersensitivity* by researchers.

Joey's Story Joey was a normal-appearing baby who grew up in an industrial city where dust settled on his playthings, sprinkling lead from industrial emissions into his environment. He loved to taste everything—pets, toys, the spindles of old painted railings. And his mother often mixed his formula with the first water from the tap, water that had spent the night absorbing lead from the old building's lead pipes.

Joey became a cautious, quiet preschooler who clung to stair railings as he slowly climbed up and down. He was late in walking and talking, small for his age, seldom played vigorously, and was prone to diarrhea, fever, irritability, and lethargy. Finally, a pediatrician detected lead toxicity in Joey's blood and started treating him with lead-scavenging drugs. Except for persistent, minor learning disabilities, Joey is now growing and playing normally.

Lead poisoning often causes general illness of the child without easily diagnosable symptoms. As lead injures the kidneys, nerves, brain, and other organs, the child may slip into coma, have convulsions, and may even die if an accurate diagnosis is not made in time to prevent it.[15] Older children with high blood lead also suffer physical complaints and are often delinquent, aggressive, and distractible.[16]

Malnutrition makes lead poisoning more likely because children absorb more lead from empty stomachs or if they lack calcium, zinc, vitamin C, vitamin D, or iron.[17] A child with iron-deficiency anemia is three times as likely to have elevated blood lead as a child with normal iron status.

Figure 13•2

LEAD LEVELS IN CHILDREN

Percent children with >10 ug/dL blood lead, 1976–1999

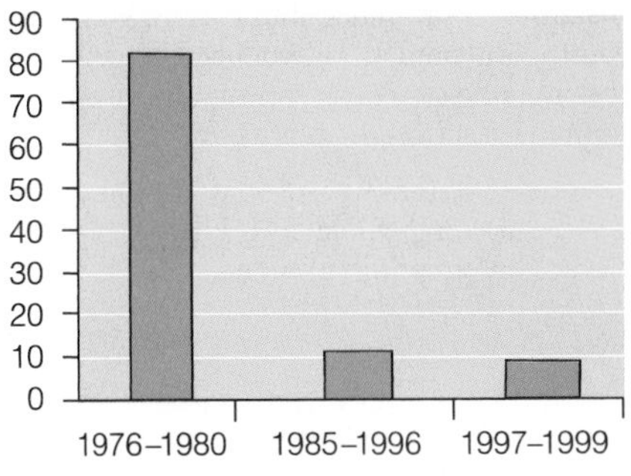

SOURCES: Data from the Centers for Disease Control and Prevention, National Center for Health Statistics, *National Health and Nutrition Examination Survey II and III*; and Blood lead levels in young children—United States and selected states, 1996–1999, *Morbidity and Mortality Weekly Report* 49 (2000): 1133–1137.

The Nature of Lead Lead is an indestructible metal element; the body cannot alter it. Its chemistry is similar to that of nutrient minerals like iron, calcium, and zinc, and lead displaces these minerals from their sites of action but cannot perform their biological functions. Thus, lead interferes with many of the body's systems, particularly the vulnerable tissues of the nervous system, kidneys, blood, and bone marrow. During pregnancy, lead crosses the placenta to inflict severe damage on the fetal nervous system. Infants and young children absorb five to ten times as much lead as do adults.

The Environmental Protection Agency (EPA) provides this toll-free hotline for lead information: 800-LEAD-FYI (800-532-3394).

A Public Health Success Bans on leaded gasoline, leaded house paint, and lead-soldered food cans have dramatically reduced the amount of lead in the U.S. environment and have resulted in a corresponding decline in children's average blood lead concentrations since the 1970s (see Figure 13-2).[18] A nationwide lead-monitoring system is now in place, and aggressive community programs are testing and treating children for lead poisoning.

Lead is still a problem, however, among children of low-income families who live in houses that contain old lead-based paint.[19] More than a million U.S. children today are estimated to have blood lead concentrations high enough to harm their health.[20] Some tips for avoiding lead toxicity are offered in Table 13-3.

key point *Lead poisoning has declined dramatically over the past two decades, but it can inflict severe, irreparable damage on growing children. Higher awareness of the remaining sources of lead poisoning can help to reduce its present rate of occurrence.*

Old paint is the main source of lead in most children's lives.

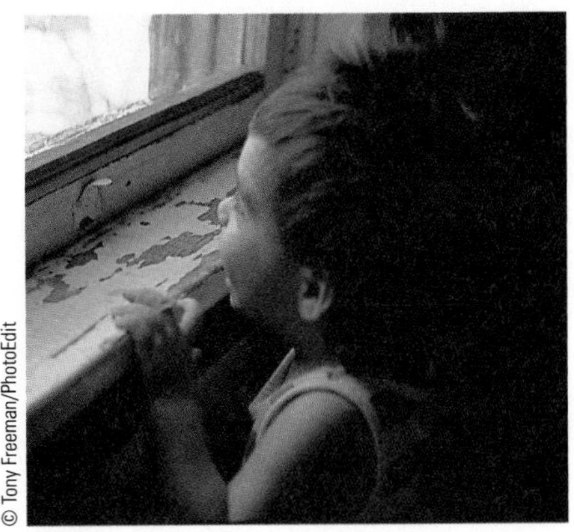

© Tony Freeman/PhotoEdit

Food Allergy, Intolerance, and Aversion

Food **allergy** is frequently blamed for physical and behavioral abnormalities in children, but just 3 or 4 percent of children are diagnosed with true food allergies.[21] Food

antigen a substance foreign to the body that elicits the formation of antibodies or an inflammation reaction from immune system cells. Food antigens are usually glycoproteins (large proteins with glucose molecules attached). Inflammation consists of local swelling and irritation and attracts white blood cells to the site.

antibodies as defined in Chapter 6, large protein molecules that are produced in response to the presence of antigens and then help to inactivate the antigens.

histamine a substance that participates in causing inflammation; produced by cells of the immune system as part of a local immune reaction to an antigen.

anaphylactic (an-AFF-ill-LAC-tic) **shock** a life-threatening whole-body allergic reaction to an offending substance.

A concern exists about allergic reactions to new genetically engineered foods. See Controversy 14.

These symptoms can occur in minutes or hours after ingesting an allergen:

- Tingling sensation in the mouth.
- Swelling of the tongue and throat.
- Irritated, reddened eyes.
- Difficulty breathing, asthma.
- Hives, swelling, rashes.
- Vomiting, abdominal cramps, diarrhea.
- Drop in blood pressure.
- Loss of consciousness.
- Death.

Table 13•3

STEPS TO PREVENT LEAD POISONING

To protect children:

- If your home was built before 1978, wash floors, windowsills, and other surfaces weekly with warm water and detergent to remove dust released by old lead paint; clean up flaking paint chips immediately.
- Feed children balanced, timely meals with ample iron and calcium.
- Prevent children from chewing on old painted surfaces.
- Wash children's hands, bottles, and toys often.
- Wipe soil off shoes before entering the home.
- Ask a pediatrician whether your child should be tested for lead.

To safeguard yourself:

- Avoid daily use of handmade, imported, or old ceramic mugs or pitchers for hot or acidic beverages, such as juices, coffee, or tea. Commercially made U.S. ceramic, porcelain, and glass dishes or cups are safe. If ceramic dishes or cups become chalky, use them for decorative purposes only.
- Do not use lead crystal decanters for storing alcoholic or other beverages.
- If your home is old and may have lead pipes, run the water for a minute before using, especially before the first use in the morning.
- Remove lead foil from wine bottles, and wipe the mouth of the bottle before pouring.

allergies diminish with age, until in adulthood they affect about 1 or 2 percent of the population.[22]

A true food allergy occurs when a whole food protein or other large molecule enters the body tissues. Recall that most large proteins from food are dismantled to smaller fragments in the digestive tract before absorption. Some, however, are not digested but enter the bloodstream whole. Once they are inside, the body's immune system reacts to undigested proteins as it does to any other **antigen:** it releases **antibodies, histamine,** or other defensive agents.

The life-threatening food allergy reaction of **anaphylactic shock** is most often caused by peanuts, tree nuts, milk, eggs, wheat, soybeans, fish, or shellfish.[23] Families and school personnel who tend to children with a life-threatening food allergy must guard them against any exposure to the allergen.[24] The child must learn to identify which foods pose a problem and then learn and use refusal skills for all foods that may

These normally wholesome foods may cause life-threatening symptoms in people with allergies.

Polara Studios, Inc.

contain the allergen. For example, a pork chop (an innocent food) may be breaded (wheat allergy) and dipped in egg (egg allergy) before being fried; a chocolate cookie tempting the child may contain peanut chunks (peanut allergy). Allergens often sneak into foods in unexpected ways.

Most parents of allergic children pack lunches and snacks of safe foods prepared at home and ask school officials to strictly enforce a "no swapping" policy in the lunchroom. The child must be able to recognize the symptoms of impending anaphylactic shock, such as a tingling of the tongue, throat, or skin, or difficulty breathing (see the margin).[25] Finally, the responsible child and the school staff should be prepared with injections of **epinephrine,** which prevents anaphylaxis after exposure to the allergen.[26] Too many preventable deaths occur each year among people with food allergies who accidentally ingest the allergen but have no access to epinephrine.[27]

To protect people with allergies, regulations require food manufacturers to declare common allergens, including food additives, on food labels.[28] Currently, the Food and Drug Administration (FDA) is working with industry to develop more clearly worded labels to alert consumers to the presence of food allergens.[29] Manufacturers must also prevent cross-contamination during production. For example, equipment used for making peanut butter must be disassembled and cleaned before being used to pulverize another kind of food, say, cashew nuts for cashew butter.[30] When cross-contamination is likely, food labels must state that the product may contain an allergy-producing food.

epinephrine (EP-ih-NEFF-rin) a hormone of the adrenal gland administered by injection to counteract anaphylactic shock by opening the airways and maintaining heartbeat and blood pressure. Also defined in Chapter 3.

food intolerance an adverse effect of a food or food additive not involving an immune response.

food aversion an intense dislike of a food, possibly biological in nature, resulting from an illness or other negative experience associated with that food.

Detecting Food Allergy Allergies have one or two components. They *always* involve antibodies; they *sometimes* involve symptoms. Symptoms without antibody production are not due to allergy, so allergies cannot be diagnosed from symptoms alone. Only testing for antibodies can provide a diagnosis.

Allergic reactions to food can occur with different timings; symptoms may appear within minutes or up to 24 hours later. Identifying a food that causes an immediate allergic reaction is easy because symptoms correlate with the time of eating the food. If the reaction is delayed, identifying the offending food is more difficult because other foods will have been eaten by the time the symptoms appear. Many people are allergic to just one food, but some are allergic to many.

A common food problem that is often confused with allergy is a **food intolerance.** Physicians must search for clues to discern food allergies from other digestive complaints such as lactose intolerance (see the margin).

Clues to food allergy as a cause of digestive distress:

- Young age.
- Personal or family history of food allergy, skin problems, or asthma.
- More than one occurrence.
- No identifiable cause, such as infection or ulcer.
- No improvement with treatments for other conditions, such as heartburn or lactose intolerance.
- Improvement with removal of suspected food from the diet.
- Positive allergy tests.

SOURCE: Adapted from the American Gastroenterological Association Medical Position Statement: Guidelines for the evaluation of food allergies, *Gastroenterology* 120 (2001): 1023–1025.

Allergy Testing A skin prick test and a double-blind food challenge under medical supervision can confirm a true food allergy.[31] These tests are time-consuming and often expensive, however, so people—and even physicians—often try to guess the cause of an adverse reaction and use the term *food allergy* loosely. A parent whose child has any kind of discomfort after eating, such as stomachache, headache, pain, rapid pulse rate, nausea, wheezing, hives, bronchial irritation, or cough, may decide that an allergy is responsible, when the actual cause is something else. Only skilled testing by a physician can distinguish the many possibilities, but such testing is seldom done.

With reliable tests being inconvenient and expensive, people are tempted by quacks offering quick and easy but sophisticated-sounding laboratory work. For example, "cytotoxic testing" involves mixing blood with foods to see what blood cells "react" to. As you might guess, this test cannot detect allergy because isolated blood cells are cut off from the body's immune system, which produces the allergic response. Other terms indicating allergy quackery are *brain allergy* and *metabolic rejectivity syndrome*.

Food Aversion A **food aversion,** an intense dislike of a food, may be a biological response to a food that once caused trouble. Parents are advised to watch for signs of food dislikes and to take them seriously. Such a dislike may turn out to be a whim or fancy, but it may turn out to be an allergy or other valid reason to avoid a certain food. Don't prejudge. Get the child tested. Then, if an important

hyperactivity (in children) a syndrome characterized by inattention, impulsiveness, and excess motor activity; usually diagnosed before age seven, lasts six months or more, and usually does not entail mental illness or mental retardation. Properly called *attention-deficit/hyperactivity disorder (ADHD)* and may be associated with minimal brain damage.

learning disability any of a group of conditions resulting in an altered ability to learn basic cognitive skills such as reading, writing, and mathematics.

A child with ADHD often:

- Has a short attention span, even while playing.
- Has trouble with tasks that require sustained mental effort.
- Has trouble learning and earns failing grades in school.
- Has poor impulse control and acts physically or verbally before thinking.
- Angers friends by not taking turns or playing by the rules.
- Runs instead of walks, climbs instead of sitting still, talks excessively.

These symptoms are clues to ADHD, but they are not enough to make a diagnosis. Tests by a knowledgeable clinician are needed for diagnosis and treatment.

Controversy 11 presented a table of the caffeine in some foods and beverages.

The placebo effect was defined earlier. It is the healing effect produced by faith in a treatment, rather than by the treatment itself.

staple food must be excluded from the diet, find other foods to provide the omitted nutrients.

Allergies are often blamed when behavior problems arise, but children who are sick from any cause are likely to be cranky. Evidence does not support the hypothesis that allergy can cause misbehavior without other symptoms. The next section singles out a type of misbehavior that is not caused by foods.

key point *Food allergies cause illness, but diagnosis is difficult. Tests are imperative to determine whether allergy exists. Food aversions can be related to food allergies or to adverse reactions to food.*

Does Diet Affect Hyperactivity?

Hyperactivity, or attention-deficit/hyperactivity disorder (ADHD), is a **learning disability** that occurs in 5 to 10 percent of young, school-aged children, that is, in 2 or 3 in a classroom of 30 children. ADHD is characterized by the chronic inability to pay attention, along with overly active behavior and poor impulse control (see the margin). It can delay growth, lead to academic failure, and cause major behavioral problems.[32] Although some children improve with age, many reach the college years or adulthood before receiving a diagnosis and with it the possibility of treatment.

Food allergies have been blamed for ADHD; research to date has shown no connection, but studies continue.[33] Research has all but dismissed the idea that sugar makes children hyperactive (see Controversy 4 for details). One study did find an association between doses of the food colorant tartrazine and increased irritability, restlessness, and sleep disturbances in a small percentage of hyperactive children. Parents who wish to avoid tartrazine can find it listed with the ingredients on food labels.

Parents often hope that a new diet or some other simple solution may improve children's behavior. Unfounded dietary "treatments" may seem to help for a while due to the placebo effect, but they fail to provide lasting cures. Common sense says that all children at times get unruly and "hyper." There are many normal, everyday causes of such behavior:

- Too much caffeine from colas or chocolate.
- Desire for attention.
- Lack of sleep.
- Overstimulation.
- Too much television.
- Lack of exercise.
- Chronic hunger.[34]

A child who often fills up on colas and chocolate, misses lunch, becomes too cranky to nap, misses out on outdoor play, and spends hours in front of a television suffers stresses that trigger chronic patterns of crankiness. This cycle of tension and fatigue resolves itself when the caregivers begin insisting on regular hours of sleep, regular mealtimes, and regular outdoor exercise. As for children living with chronic hunger, the issues are many and are topics of Chapter 15.

key point *Hyperactivity, properly named attention-deficit/hyperactivity disorder (ADHD), is not caused by food or poor nutrition; temporary "hyper" behavior may reflect excess caffeine consumption or inconsistent care. A wise parent will limit children's caffeine intakes and meet their needs for structure to prevent tension and fatigue.*

Television and Children's Nutrition Problems

More than 25 percent of children in the United States watch four or more hours of television every day, and 67 percent watch two or more hours every day.[35] Television exerts four major adverse impacts on children's nutrition. First, television viewing requires no energy. It seems to reduce the metabolic rate below the resting level, requiring even less energy than daydreaming. Second, it consumes time that could be spent in energetic play. Third, watching television correlates with between-meal snacking and with eating the calorically dense and fatty foods most heavily advertised on children's programs. Children who watch for more than four hours a day, or during meals, are least likely to eat fruits and vegetables and most likely to be obese.[36] Fourth, children who watch hours of television a day are prone to frequent snacking on foods that contribute to the formation of dental caries.

The Problem of Childhood Obesity Obesity is increasing among the world's children and poses hazards to the health of these children both now and in the future. Medical science, meanwhile, has little to offer in the way of prevention or cure.[37] The single most important problem for obese children is the likelihood of becoming obese adults with associated health effects. A few obese children may remain untouched by problems of overweight, but the majority must cope with problems arising from differences in their growth, physical health, and psychological development.

Obese children often grow and develop faster than their peers and begin puberty earlier. In the process, they develop the greater bone and muscle mass needed to carry their extra weight. Consequently, they tend to remain "stocky" even after losing their excess body fatness.

Such children often face an increased physical threat from cardiovascular disease (CVD) early in life. They may have high blood pressure, insulin resistance, and a blood lipid profile that warns of impending CVD. More often than other children, they develop diseases such as diabetes and asthma.[38]

Concern is growing for the future health of children worldwide with risk factors for CVD normally seen in adults. See Controversy 12.

Psychologically, obese children suffer from the thoughtless comments and prejudice of others. Adults may discriminate against them, while peers may reject them. An obese child may develop a poor self-image, a sense of failure, and a passive approach to life. Television shows, a major influence on children, often stereotype the fat person as a misfit. Children have few defenses against these unfair portrayals and may come to accept them as truth. Research shows that both overweight and normal-weight children respond unfavorably to bulky body sizes.

Prevention and Treatment of Childhood Obesity Help for the overweight child is possible, and an integrated approach is recommended involving diet, physical activity, psychological support, and behavioral changes. Parents play key roles in preventing or managing obesity, or making helpful changes, because the best solutions usually involve the whole family. Overweight in children must be addressed sensitively, however. Children are impressionable and can easily come to believe that their worth or lovability is somehow tied to their weight.

An initial goal for obese children is to slow their rate of gain—that is, to hold weight steady while they grow taller. Weight loss ordinarily is not recommended because diet restriction can easily interfere with normal growth. By feeding the whole family balanced meals, offering appropriate portion sizes, restricting treats, and boosting physical activity, the goal can often be accomplished without making the child feel singled out.[39] The margin lists some other strategies that often prove effective for producing change.

Family lifestyle choices to help the overweight child:

- Serve family meals that control both the calorie density of the foods and the serving sizes offered.
- Have fun and play vigorously outdoors every day, as a family or with friends.
- Learn and use appropriate food portions.
- Provide a wide variety of nutritious snacks that are low in fat and sugar.
- Limit high-sugar, high-fat foods, including sugar-sweetened soft drinks.
- Involve children in shopping for and preparing family meals.
- Slow down eating, and pause to enjoy table companions; stop eating when full.
- Do not use foods to reward or punish behaviors.
- Set a good example, and demonstrate positive behaviors for children to imitate.

Some researchers believe that sedentary lifestyles are almost entirely to blame for the upswing in obesity in children and adults. Fewer and fewer children are asked to participate in physical activities in school.[40] At home, limiting television watching and other sedentary activities is a good idea.[41] As mentioned earlier, the more television children watch, the less likely they are to eat a nutritious diet and the more likely to be overweight.[42] This is not surprising considering that the average U.S. child sees an estimated 30,000 TV commercials a year, many for high-sugar, high-fat foods such as sugar-coated cereals, candies, fried snack chips, fast foods, and sugar-sweetened soft drinks.

Physical activity of just 20 minutes' duration each day can help a child to achieve or maintain a healthy body weight.

Parents often need help in managing the obese child from a professional, such as a registered dietitian who specializes in childhood nutrition and weight management. In evaluating programs, those that involve parents most often prove helpful.[43]

Parents are among the most significant influences shaping self-concept, weight concerns, and dieting practices of children.[44] They can also be the most powerful forces in helping to instill healthy habits and a strong sense of self-worth that can benefit a child through a lifetime.

Dental Caries Sticky, high-carbohydrate snack foods cling to the teeth and provide an ideal environment for the growth of mouth bacteria that cause caries. Parents must combat the influence of television commercials by helping children to:

- Limit between-meal snacking.
- Brush and floss daily, and brush or rinse after eating meals and snacks.
- Choose foods that don't stick to the teeth and are swallowed quickly.
- Snack on crisp or fibrous foods to stimulate the release and rinsing action of saliva.

Table 13-4 in the margin lists foods that promote dental health and those that require speedy removal from the teeth.

Table 13•4

THE CARIES POTENTIAL OF FOODS

Low Caries Potential

These foods are less damaging to teeth:

- Eggs, legumes
- Fresh fruit, fruits packed in water
- Lean meats, fish, poultry
- Milk, cheese, plain yogurt
- Most cooked and raw vegetables
- Pizza
- Popcorn, pretzels
- Sugarless gum and candy,[a] diet soft drinks
- Toast, hard rolls, bagels

High Caries Potential

Brush teeth after eating these foods:

- Cakes, muffins, doughnuts, pies
- Candied sweet potatoes
- Chocolate milk
- Cookies, granola bars, crackers
- Dried fruits (raisins, figs, dates)
- Frozen or flavored yogurt
- Fruit juices or drinks
- Fruits in syrup
- Glazed carrots
- Ice cream or ice milk
- Jams, jellies, preserves
- Lunch meats with added sugar
- Meats with sugary glazes
- Oatmeal, oat cereals, oatmeal baked goods[b]
- Peanut butter with added sugar
- Potato and other snack chips
- Ready-to-eat sugared cereals
- Sugared gum, soft drinks, candies, honey, sugar, molasses, syrups
- Toaster pastries

[a]Cariogenic bacteria cannot efficiently metabolize the sugar alcohols in these products, so they do not contribute to dental caries.
[b]The soluble fiber in oats makes this grain particularly sticky and therefore cariogenic.

key point *The nation's children are becoming fatter and face growing risks of diseases. Childhood obesity demands careful family-centered management. Television viewing can contribute to obesity through lack of exercise and promoting overconsumption of calorie-dense snacks. Such snacks may also contribute to dental caries.*

Is Breakfast Really the Most Important Meal of the Day for Children?

Elders have long held that breakfast is the most important meal of the day, and for children, this bit of wisdom is now backed by science. A nutritious breakfast is a central feature of a child's diet that supports healthy growth and development.[45]

Children who eat no breakfast perform poorly in tasks requiring concentration, have shorter attention spans, achieve lower test scores, and are tardy or absent more often than their well-fed peers. Common sense tells us that it is unreasonable to expect anyone to study and learn when no fuel has been provided. Even children who have eaten breakfast suffer from distracting hunger by late morning. Chronically underfed children suffer all the more.[46]

The U.S. government funds several programs to provide nutritious, high-quality meals, including breakfast, to children at school. Schools that begin to participate in the federal school breakfast program observe higher achievement test scores and lower tardiness and absence rates.

Breakfast is critical to school performance. Not all children start the day with an adequate breakfast, but school breakfast programs help to fill the need.

How Nourishing Are the Lunches Served at School?

For the past 50 years, school lunches have been meeting the midday nutrient needs of the nation's children. Today, 66 percent of children from age 6 to 11 years partake of school lunches providing servings of milk, protein-rich foods (meat, poultry, fish, cheese, eggs, legumes, or peanut butter), vegetables, fruits, and breads or other grain foods each day.[47] The lunches are designed to provide at least a third of the recommended intake for each of the nutrients and may often provide more reliably for children's nutrition than do lunches brought from home.[48] Table 13-5 shows school lunch patterns for different ages.

Parents can rely on school lunches to meet a significant part of their children's nutrient needs on school days. Students who regularly eat school lunches have higher intakes of energy and nutrients than students who do not. Children don't always like what they are served, and school lunch programs must strike a balance between what

Table 13•5

SCHOOL LUNCH PATTERNS FOR DIFFERENT AGES

	Preschool (Age)		Grade School through High School (Grade)[a]		
Food Group	1 to 2	3 to 4	K to 3	4 to 6	7 to 12
Milk					
1 serving of fluid milk[b]	¾ c	¾ c	1 c	1 c	1 c
Meat or Meat Alternate					
1 serving:					
Lean meat, poultry, or fish	1 oz	1½ oz	1½ oz	2 oz	3 oz
Cheese	1 oz	1½ oz	1½ oz	2 oz	3 oz
Large egg(s)	½	¾	¾	1	1½
Cooked dry beans or peas	¼ c	⅜ c	⅜ c	½ c	¾ c
Peanut butter	2 tbs	3 tbs	3 tbs	4 tbs	6 tbs
Peanuts, soynuts, tree nuts, or seeds[c]	½ oz	¾ oz	¾ oz	1 oz	1½ oz
Vegetable and/or Fruit					
2 or more servings, both to total	½ c	½ c	½ c	¾ c (plus ½ c extra over a week)	¾ c
Bread or Bread Alternate Servings[d]	5 per week (minimum ½ per day)	8 per week (minimum 1 per day)	8 per week (minimum 1 per day)	8 per week (minimum 1 per day)	10 per week (minimum 1 per day)

[a]These patterns may be used so long as the meals served meet the *Dietary Guidelines for Americans* and provide one-third of the child's recommendations for nutrients.
[b]Whole milk and unflavored low-fat milk must be offered; flavored milks or fat-free milk may also be offered.
[c]These foods may meet no more than one-half a serving of meat and must be accompanied by other meat or alternate in the meal.
[d]A serving is 1 slice of whole-grain or enriched bread; a whole-grain or enriched biscuit, roll, muffin, or the like; or ½ cup cooked rice, pasta, or other grain.
SOURCE: U.S. Department of Agriculture, 1998.

children want to eat and what will nourish them and guard their health. Additionally, short lunch periods and long waiting lines prevent some students from eating a school lunch and leave others with too little time to complete their meal.[49]

To help reduce cardiovascular risk, the U.S. Department of Agriculture (USDA) has ruled that all government-funded meals served at schools must follow the *Dietary Guidelines for Americans*.[50] Often, however, private vendors offer unregulated meals, even fast foods, side-by-side with the nutritious school lunches.[51] U.S. children develop a taste for such foods early in life and may reject nutritious school meals when offered a choice of meals higher in fat, sugar, and salt.[52] Children receive a mixed message when they are left on their own to choose between the health-supporting school lunch and the less optimal foods that their taste buds may prefer.[53]

In some schools, children also have the additional option of choosing soft drinks, sweets, and other low-nutrient treats from school snack bars, vending machines, or school stores.[54] No federal laws exist to restrict sales of these items to schoolchildren. Administrators of the school lunch program have tried to outlaw such sales on school grounds, but have been defeated by the powerful lobbying efforts of industries that reap huge profits from children's pocket money. Given a choice, many children select nutritious snacks, such as yogurt or fruit juices, when available.

A word about juice: Some recommend limiting fruit juices in children's diets for fear that overconsumption may promote obesity and dental caries.[55] Research vindicates juice with regard to obesity—no correlation between obesity and juice consumption is evident.[56] In fact, juice often crowds calorie-dense, nutrient-void soft drinks out of the diet, and soft drinks *are* associated with childhood obesity.[57] Frequent exposure of the teeth to the natural sugars of juice does increase the likelihood of dental caries, however. Also, too much juice may crowd out *other* nutrient-dense choices, such as milk. Best advice: follow the recommendations of the Food Guide Pyramid on page 475.

Breakfast ideas for rushed mornings:

- Make ahead and freeze sandwiches to thaw and serve with juice. Fillings may include peanut butter, low-fat cream cheese, other cheeses, jams, fruit slices, or meats. Or use flour tortillas with cheese; roll up, wrap, and freeze for later heating in a toaster oven or microwave oven.
- Teach school-aged children to help themselves to dry cereals, milk, and juice. Keep unbreakable bowls and cups in low cupboards, and keep milk and juice in small unbreakable pitchers on a low refrigerator shelf.
- Keep a bowl of fresh fruit and small containers of shelled nuts, trail mix (the kind without candy), or roasted peanuts for grabbing. Granola or other grain cereal poured into an 8-ounce yogurt tub is easy to eat on the run. So are plain toasted whole-grain frozen waffles—no syrup needed.
- Untraditional choices are often acceptable. Purchase or make ahead enough carrot sticks to divide among several containers; serve with yogurt or bean dip. Leftover casseroles, stews, or pasta dishes are nutritious choices that children can eat hot or cold.

epiphyseal (eh-PIFF-ih-seal) **plate** a thick, cartilage-like layer at the ends of bones that forms new cells that are eventually calcified, lengthening the bone (*epiphysis* means "growing" in Greek).

key point *School lunches are designed to provide at least a third of the daily nutrients needed by growing children and to stay within limits set by the* Dietary Guidelines for Americans. *Soda and snack vending machines, fast-food and snack bars, and school stores tempt schoolchildren with foods high in fats and sugars. Fruit juice is a healthy food, but may cause dental problems if used to excess.*

The Teen Years

Teenagers are not fed; they eat. Nutrient needs are high during adolescence, and choices made during the teen years profoundly affect health, both now and in the future. They need reliable nutrition information to enable them to make healthy food choices.

The teen years bring a search for identity, which is acquired largely through trial and error. Teens face tremendous pressures from peers and the media regarding body image, and many adopt fads and scams offering promises of slenderness, good-looking muscles, freedom from acne, or control over symptoms that may accompany menstruation.

Growth and Nutrient Needs of Teenagers

Adolescent needs for all nutrients are greater than at any other time of life except pregnancy and lactation. The need for iron is particularly high to support menstruation in girls and to develop lean body mass in boys.

Food sources of iron and calcium are listed in Chapter 8, and peak bone mass was discussed in Controversy 8.

Adolescence and the Bones Adolescence is a crucial time for bone development. The bones are growing longer at a rapid rate (see Figure 13-3) thanks to a special bone organ, the **epiphyseal plate,** that disappears as a teenager reaches adult height. At the same time, the bones are gaining density, laying down the calcium needed later in life. Low calcium intakes are all too common and, paired with

Nutritious snacks play an important role in an active teen's diet.

© Henley and Savage/Tony Stone Worldwide

Figure 13•3

GROWTH OF LONG BONES

Bones grow longer as new cartilage cells accumulate at the top portion of the epiphyseal plate and older cartilage cells at the bottom of the plate are calcified.

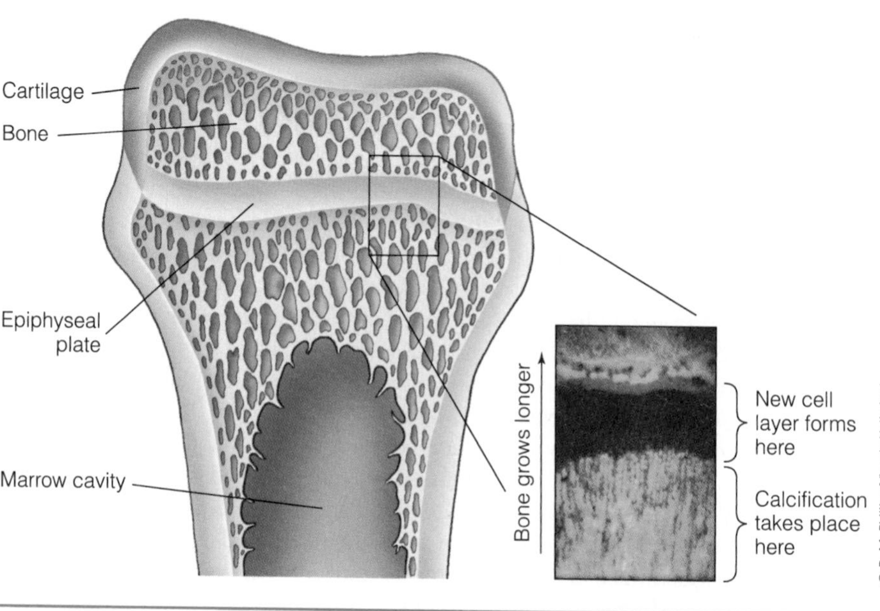

© D. M. Phillips/Visuals Unlimited

physical inactivity, can compromise the development of peak bone mass.[58] Milk as the primary beverage provides calcium and other bone-building nutrients, but teenagers often choose soft drinks over milk (see Table 13-6).[59] Soft drinks as the primary beverage can affect bone density because they displace milk from the diet and reduce calcium intakes. Bones grow stronger with physical activity, but few high schools require students to attend daily physical activity classes (see Figure 13-4), so most teenagers must make a point to be physically active during leisure hours. Attainment of maximal bone mass during youth and adolescence is the best protection against age-related bone loss and fractures in later life.

puberty the period in life when a person develops sexual maturity and the ability to reproduce.

premenstrual syndrome (PMS) a cluster of symptoms that some women experience prior to and during menstruation. They include, among others, abdominal cramps, back pain, swelling, headache, painful breasts, and mood changes.

The Body Changes of Adolescence The adolescent growth spurt brings rapid growth and hormonal changes that affect every organ of the body, including the brain. On average, girls' growth spurts begin at 10 or 11 years of age and peak at about 12 years. Boys' growth spurts begin at 12 or 13 years and peak at about 14 years, slowing down at about 19. Two boys of the same age may vary in height by a foot, but if growing steadily, then each is fulfilling his genetic destiny according to an inborn schedule of events. Weight standards meant for adults are useless for adolescents. Parents should watch only for smooth progress and guard against comparisons that can diminish the child's self-image. Health-care providers may compare the changes of **puberty** with standard rating scales.

The energy needs of adolescents vary tremendously. An active, growing boy of 15 may need 4,000 calories or more a day just to maintain his weight, but an inactive girl of the same age who is growing slowly may need fewer than 2,000 calories if she is to keep from becoming obese. Girls normally develop a somewhat higher percentage of body fat than boys do, a fact that causes much needless worry about becoming overweight. Teenagers of normal weight are often "on diets" and make all sorts of unhealthy weight-loss attempts. Even those without diagnosable eating disorders have been observed to stunt their own growth through "dieting." A low sense of self-worth may open the door to taking other risks such as using alcohol, tobacco, and drugs of abuse such as marijuana.[60] Parents, peers, and the media are the most influential in shaping self-concept, weight concerns, and dieting practices among teens.[61]

Girls face a major change with the onset of menstruation. The hormones that regulate the menstrual cycle affect not just the uterus and the ovaries but metabolic rate, glucose tolerance, appetite, food intake, mood, and behavior. Most women live easily with the cyclic rhythm of the menstrual cycle, but some are afflicted with physical and emotional pain prior to menstruation, a condition called **premenstrual syndrome,** or **PMS.** This chapter's Consumer Corner offers more on PMS.

Figure 13•4

U.S. HIGH SCHOOL STUDENTS ATTENDING DAILY PHYSICAL EDUCATION CLASSES, 1991–1999

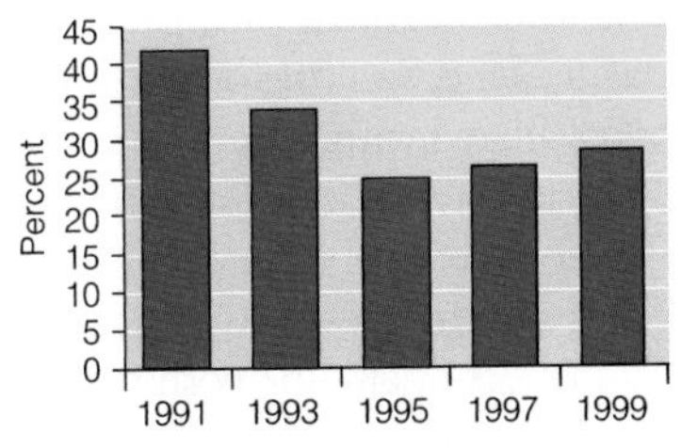

SOURCE: Centers for Disease Control, National Youth Risk Behavior Survey, 2000.

Growth patterns and nutrient and energy needs of teens vary widely with gender, body size, and activity level.

Table 13•6

SOFT DRINK CONSUMPTION OF U.S. ADOLESCENTS

In general, soft drink consumption is inversely associated with the consumption of the nutrients in milk and fruit juices.

This Percentage of Adolescents:	Consumes This Many Soft Drinks Each Day:
22%	More than 26 oz (more than 3¼ c or 2 cans)
28%	13 to 26 oz (1¾ c to 3¼ c or 2 cans)
32%	Up to 13 oz (about 1¾ c or 1 can)
18%	None

SOURCE: Data from L. Harnack, J. Stang, and M. Story, Soft drink consumption among U.S. children and adolescents: Nutritional consequences, *Journal of the American Dietetic Association* 99 (1999): 436–441.

People who abuse drugs or alcohol face multiple nutrition problems:

- Money is spent on drugs or alcohol instead of food.
- Substance-induced euphoria depresses the appetite.
- A lifestyle that focuses on drugs or alcohol does not include nutrition.
- Infections from sharing needles raise nutrient needs.
- Medical treatment for drug abuse may alter nutrition needs.

See Controversy 3 for details concerning alcohol and nutrition.

Consumer Corner

Nutrition and PMS

SYMPTOMS RELATED to PMS can include any or all of these: cramps and aches in the abdomen, back pain, headache, acne, swelling of the face and limbs associated with water retention, food cravings (especially for chocolate and other sweets), abnormal thirst, pain and lumps in the breasts, diarrhea, and mood changes, including both nervousness and depression. An official diagnosis of PMS requires that such symptoms impair some facet of a woman's life, true for about 5 to 10 percent of menstruating women, while as many as 85 percent complain of some degree of PMS.[1] The timing and pattern of symptoms are also important, and women who seek help must keep daily records.[2]

One of the candidates as a cause of PMS is altered response to the two major regulatory hormones of the menstrual cycle: estrogen and progesterone.[3] In particular, the hormone estrogen can affect the brain's neurotransmitters and, in turn, alter mood. Adequate serotonin in the brain buoys a person's mood, and deficient serotonin creates a depressed mood and is related to major depressive disorders. In PMS, the natural rise and fall of estrogen levels in the blood may affect the activities of serotonin in the brain during the last half of each menstrual cycle. Taking hormones in the form of oral contraceptives often improves mood by eliminating estrogen's peaks and valleys.[4]

The major connections between PMS and nutrition concern energy metabolism and vitamin and mineral status. Scientists believe that during the two weeks prior to menstruation, two events affect a woman's energy metabolism:

- The basal metabolic rate during sleep speeds up, although the daytime rate may not change.
- Appetite and calorie intakes increase.[5]

Most studies indicate that women take in an average of 300 calories a day more during the ten days prior to menstruation than during the ten days after it. As a consequence, a woman who wishes to control her weight may find it easier to restrict calories during the two weeks following menstruation. Limiting calories may be harder during the two weeks before the next menstruation, however, because she is fighting a natural, hormone-governed increase in appetite.

A common symptom of PMS is sodium retention, with the water retention that accompanies it. Some doctors prescribe diuretics to get rid of the excess sodium and water, with mixed results. Diuretic therapy causes loss of minerals such as potassium, possibly worsening PMS symptoms. Also, if women do retain sodium and water just before menstruation, that effect may be normal and desirable.

The role of vitamin B_6 in PMS has been heavily researched. In trials, a few subjects with PMS have responded favorably to treatment with vitamin B_6, but the improvement is not statistically meaningful. No need exists for megadoses of vitamin B_6, and the hazards associated with such doses are well documented (see Chapter 7).

Vitamin E deficiency is another possible contributor to PMS. One double-blind, placebo-controlled study of 75 women suggested that supplemental vitamin E brought relief from sore breasts associated with PMS, when the placebo did not. Some women without PMS also have sore breasts, however, and they, too, can sometimes be relieved by vitamin E. Possibly, then, vitamin E deficiency does not cause PMS, but can worsen symptoms associated with the menstrual period.

Magnesium supplements of 200 milligrams seem to reduce weight gain, swelling and water retention, and breast tenderness, at least in experiments.[6] Calcium supplements may help, too, especially with PMS-related moodiness, water retention, food cravings, and pain.[7] Few women in the United States consume enough magnesium or calcium from their diets, so an obvious first step for women with PMS is to make sure that the food they choose provides the nutrients they need.

Tea consumption has been strongly linked with PMS. Which component of tea—the caffeine, pigments, or other substances—is responsible is not known, but evidence indicates a role for caffeine. Data from questionnaires administered to more than 800 women correlated caffeine intakes with PMS in a linear fashion: the more caffeine-containing beverages the women reported drinking, up to 10 cups per day, the more symptoms of PMS they reported suffering. Thus, any woman who finds menstrual symptoms troublesome should try a caffeine-free lifestyle for a while and see if her symptoms improve.

The woman with PMS should examine her total lifestyle, of which diet is only a part. Adequate sleep and physical activity help, and controlling stress may be important.[8] She should moderate her intakes of caffeine, salt, alcohol, and any other abusable substances. Finally, she should watch out for bogus PMS "cures."

Eating Patterns and Food Choices

During adolescence food habits often change for the worse, and teenagers often miss out on nutrients they need. Teens may begin to skip breakfast; choose less milk, fruits, juices, and vegetables; and consume more soft drinks each day.[62] Ideally, the adult becomes a **gatekeeper,** controlling the type and availability of food in the teenager's environment. Teenage sons and daughters and their friends should find plenty of nutritious, easy-to-grab food in the refrigerator (meats for sandwiches, raw vegetables, milk, fruit and fruit juices) and more in the cupboards (breads, peanut butter, nuts, popcorn, cereals). In reality, in many households today, all the adults work outside the home, and teens perform some of the gatekeeper's roles, such as shopping for groceries or choosing fast foods or prepared foods.[63]

On average, about a fourth of a teenager's total daily energy intake comes from snacks, which also contribute some needed protein, thiamin, riboflavin, vitamin B_6, magnesium, and zinc. Often, however, teens choose foods that are too high in fat, sodium, and protein and too low in fiber to support the future health of their arteries.[64] Their calcium intakes often fall short unless they snack on dairy products, and they often fail to obtain enough iron and vitamin A. For iron and other nutrients, a teen could snack on iron-containing meat sandwiches, low-fat bran muffins, or tortillas with spicy bean spread along with a glass of orange juice to help maximize the iron's absorption.

Teenagers do a lot of eating away from home. They love fast food, and fortunately, some fast-food establishments offer nutritious choices. The gatekeeper can help by presenting the needed nutrition information in a way that is meaningful to the individual teen. Teens who are prone to gain weight will often open their ears to news about the fat and calorie contents of fast foods. Others attend best to information about the negative effects of an ill-chosen diet on sport performance.

The gatekeeper can set a good example, provide an environment with plenty of nutritious foods, keep lines of communication open, and stand by with reliable nutrition information and advice, but the rest is up to the teens themselves. Ultimately, they make the choices.

gatekeeper with respect to nutrition, a key person who controls other people's access to foods and thereby affects their nutrition profoundly. Examples are the spouse who buys and cooks the food, the parent who feeds the children, and the caretaker in a day-care center.

acne chronic inflammation of the skin's follicles and oil-producing glands, which leads to an accumulation of oils inside the ducts that surround hairs; usually associated with the maturation of young adults.

The nutritive values of selected fast foods are presented in the Table of Food Composition, Appendix A.

With planning, the gatekeeper can encourage teens to meet nutrient requirements by providing nutritious snacks.

Acne

No one knows why some people get **acne** while others do not, but heredity plays a role—acne runs in families. The hormones of adolescence also play a role by stimulating the glands in the skin. The skin's natural oil is made in deep glands and is supposed to flow out through tiny ducts to the skin's surface. In acne, the ducts become clogged, and oily secretions build up in the ducts.

One medical treatment for acne is the application of retinoic acid or Retin A (a vitamin A relative), directly to the skin. This loosens the plugs that form in the ducts, but the acid may burn the skin and cause pimples to form, making the acne look worse at first. Retin A is also available in topical wrinkle creams for older skin because retinoic acid can make fine lines and wrinkles less obvious in some people.

Prescribed antibiotic pills and ointments work for some. Some antibiotic creams contain zinc because it improves their staying power on the skin and may reduce inflammation.[65] Zinc supplements taken orally, however, are of no benefit against acne.

The oral prescription medicine Accutane, made from vitamin A, is effective against the deep lesions of cystic acne. Accutane is highly toxic and causes serious birth defects if taken during pregnancy. Although medicines made from vitamin A are successful in treating acne, vitamin A itself has no effect, and supplements of the vitamin can be toxic. Quacks, undaunted by these facts, market potentially toxic vitamin A supplements to young people who hope to cure acne.

Among foods charged with aggravating acne are chocolate, cola beverages, fatty or greasy foods, milk, nuts, sugar, and foods or salt containing iodine. None

life expectancy the average number of years lived by people in a given society.

life span the maximum number of years of life attainable by a member of a species.

longevity long duration of life.

of these factors has been proved to worsen acne, and two, chocolate and sugar, have been shown not to worsen it. Psychological stress, though, clearly worsens acne. Vacations from school often bring acne relief. Sun and swimming also help, perhaps because they are relaxing, the sun's rays kill bacteria, and water cleanses the skin.

One remedy always works: time. While waiting, attend to basic needs. Petal-smooth, healthy skin reflects a tended, cared-for body whose owner provides it with nutrients and fluids to sustain it, exercise to stimulate it, and rest to restore its cells.

Although no foods have been proved to aggravate acne, stress can worsen it. Supplements are useless against acne, but sunlight, proven medications, and relief from stress can help.

The Later Years

The title may imply a section about older people, but it is relevant even if you are only 20 years old. How you live and think at 20 years of age affects the quality of your life at 60 or 80 years. According to the old saying, "as the twig is bent, so grows the tree." Unlike a tree, however, you can bend your own twig.

Before you will adopt nutrition behaviors to enhance your health in old age, you must accept on a personal level that you yourself are aging. To learn what negative and positive views you hold about aging, try answering the questions in the margin. Your answers reveal not only what you think of older people now but also what will probably become of you. When older adults were asked to give tips to younger people on how to live life fully in the later years, they offered nine suggestions, also listed in the margin.[66]

How will you age?

- In what ways do you expect your appearance to change as you age?
- What physical activities do you see yourself enjoying at age 70?
- What will be your financial status? Will you be independent?
- What will your sex life be like? Will others see you as sexy?
- How many friends will you have? What will you do together?
- Will you be happy? Cheerful? Curious? Depressed? Uninterested in life or new things?

Tips for productive aging:

1. Simplify your life; identify priorities and set limits.
2. Pay attention to yourself—your body, your mind, and your spirit.
3. Continue to teach, continue to learn; take up leisure activities (painting, woodworking).
4. Let yourself laugh and cry; be flexible; learn to navigate change.
5. Be charitable; make it a practice to give (wisdom, experience, money, time, yourself).
6. Be financially astute; invest early for retirement.
7. Practice good nutrition and exercise.
8. Think about your past and future; but live in the moment. Accept your mortality.
9. Be involved; be positive; link with others.

SOURCE: Adapted with permission from H. Kerschner and J. M. Pegues, Productive aging: A quality of life agenda, *Journal of the American Dietetic Association* 98 (1998): 1445–1448.

The majority of the U.S. population is now middle-aged, and as that group ages, the ratio of old people to young people is growing larger, a trend called the "graying" of America. Since 1950, the number of people over age 65 has doubled, and people over 85 years old are the fastest-growing age group.

In the United States, the **life expectancy** for white women is 80 years and for black women, 75 years; for white men, it is 75 years and for black men, 68 years—all record highs, and much higher than the life expectancy of 47 years in 1900.[67] Once a person survives the perils of youth and middle age to reach age 80, women can expect to survive an additional nine years, on average; men, an additional seven.[68] Thanks largely to advances in medical science, including antibiotics and other treatments, the life expectancy almost doubled in the twentieth century. Still, the biological schedule that we call aging cuts off life at a genetically fixed point in time. The **life span** (the maximum length of life possible for a species) of human beings is now 130 years.[69] Even this limit may one day be challenged with advances of medical and genetic technologies.[70] One caution: to date, scientists who study the aging process have found no specific diet or nutrient supplement that will increase **longevity,** although there are hundreds of unproven claims to the contrary.

Life expectancy for U.S. adults increased in the twentieth century. No specific diet or supplement is proved to prolong human life.

Nutrition in the Later Years

Nutrient needs become more individual with age, depending on genetics and individual medical history. For example, one person's stomach acid secretion, which helps in iron absorption, may decline, so that person may need more iron. Another person may excrete more folate due to past liver disease and thus need a higher dose. Table 13-7 lists some changes that can affect nutrition.

Table 13•7

EXAMPLES OF PHYSICAL CHANGES OF AGING THAT AFFECT NUTRITION

Digestive Tract	Intestines lose muscle strength resulting in sluggish motility that leads to constipation. Stomach inflammation, abnormal bacterial growth, and greatly reduced acid output impair digestion and absorption. Pain and fear of choking may cause food avoidance or reduced intake.
Hormones	For example, the pancreas secretes less insulin and cells become less responsive, causing abnormal glucose metabolism.
Mouth	Tooth loss, gum disease, and reduced salivary output impede chewing and swallowing. Choking may become likely; pain may cause avoidance of hard-to-chew foods.
Sensory Organs	Diminished sight can make food shopping and preparation difficult; diminshed senses of smell and taste may reduce appetite, although research is needed to clarify this effect.
Body Composition	Weight loss and decline in lean body mass lead to lowered energy requirements. May be preventable or reversible through physical activity.

Energy and Activity

Energy needs often decrease with advancing age. One reason is that the number of active cells in each organ decreases, reducing the body's overall metabolic rate, although much of this loss may not be inevitable. Another reason is that older people often reduce their physical activity, and their lean tissue diminishes. For people older than 70 years, the best health and lowest risk of death have been observed in those who maintain a body mass index (BMI) between 25 and 32, which is higher than the optimal BMI for younger people (18.5 to 25).[71]

Body mass index was discussed in Chapter 9.

The DRI nutrient intake standards provide separate recommendations for those 51 to 70 years and for those 70 and older. See the inside front cover, page A.

After about the age of 50, the intake recommendation for energy assumes about a 5 percent reduction in energy output per decade (see the inside back cover). For those who must limit energy intake, there is little leeway in the diet for foods of low nutrient density such as sugars, fats, and, of course, alcohol.

Current thinking refutes the idea that steeply declining energy needs are unavoidable. Physical activity and a diet adequate in nutrients and rich in phytochemicals may not only maintain energy needs but also uphold other functions, such as a healthy immune response.[72] Physical activity and diet can often be effective against a destructive spiral of sedentary behavior and mental and physical losses in the elderly that one expert has called "the dwindles."[73] The "dwindles" refers to a complex of interacting failures in the elderly including:

- Weight loss.
- Diminished mental function.
- Decreased physical ability to function.
- Social withdrawal.
- Malnutrition.

The Think Fitness feature on the next page emphasizes the importance of physical activity to maintaining body tissue integrity throughout life. A nutrition expert phrases it this way:

> We now know that physically active elders can build and rebuild muscle mass. Even the frail elderly can improve function by a remarkable 200 percent on a short, focused exercise regimen. No single feature of aging can more dramatically affect basal metabolism, insulin sensitivity, calorie intake, appetite, breathing, ambulation, mobility, and independence than muscle mass.[74]

Institutionalized people in their *nineties* have gained muscle bulk, regained or improved their balance, and added pep to their walking steps after just eight weeks of weight training. People spending energy in physical activity can also eat more food, gaining nutrients.

Cross sections of two thighs. These two women's thighs may appear to be about the same size from the outside, but the 20-year-old woman's thigh (left) is dense with muscle tissue. The 64-year-old woman's thigh (right) has lost muscle and gained fat, changes that may be largely preventable with strength-building physical activities.

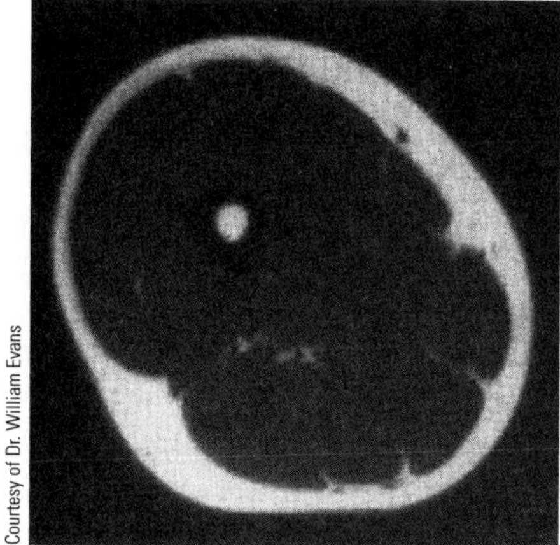

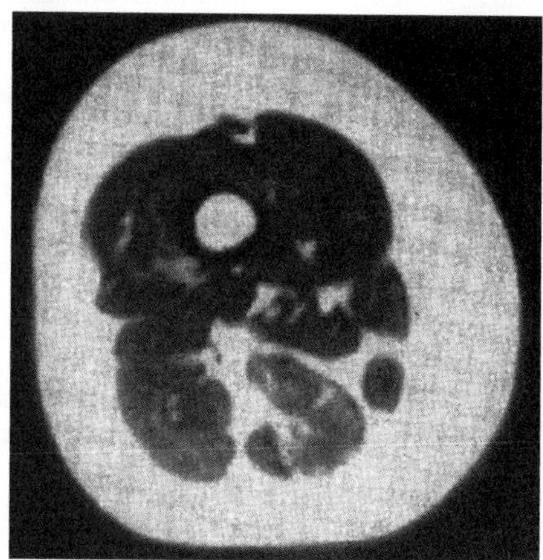

Courtesy of Dr. William Evans

The photos above emphasize this point: they compare cross sections of the thigh of a young woman and of an older woman to demonstrate the muscle loss of sedentary aging. Strength training helps to prevent at least some of this muscle loss.[75] Some unscrupulous practitioners try to sell elderly people a shortcut to muscle and bone tissue retention in the form of growth hormone (GH) "therapy." Although secretion of GH does decline with age, science does not yet support its usefulness or safety in reversing the tissue loss of aging.[76]

Energy needs decrease with age, but exercise burns off excess fuel, maintains lean tissue, and brings health benefits.

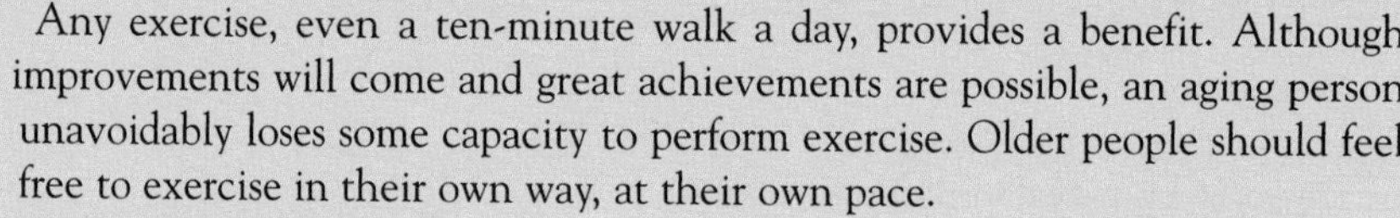

Benefits of Physical Activity in Aging

Physically active older adults have greater flexibility and endurance, greater lean body mass, a better sense of balance, greater blood flow to the brain, and stronger immune systems. They suffer fewer falls and broken bones, enjoy better overall health, and even live longer than their couch-loving peers.

Any exercise, even a ten-minute walk a day, provides a benefit. Although improvements will come and great achievements are possible, an aging person unavoidably loses some capacity to perform exercise. Older people should feel free to exercise in their own way, at their own pace.

THINK FITNESS

Carbohydrates and Fiber

The recommendation to obtain 6 to 11 servings of breads, grains, or pasta is appropriate for older people, and the majority of those servings should be from whole grains. With age, fiber takes on extra importance for its role against constipation, a common complaint among older adults and among nursing home residents in particular. Most older adults do not obtain the recommended 27 to 40 grams of fiber daily. When low fiber intakes are combined with low fluid intakes, inadequate exercise, and constipating medications, constipation becomes inevitable.

Generous carbohydrate intakes are recommended for older adults. Including fiber in the diet is important to avoid constipation.

Fats and Arthritis

arthritis a usually painful inflammation of the joints caused by many conditions, including infections, metabolic disturbances, or injury; usually results in altered joint structure and loss of function.

Older adults must limit overall fat in their diet for several reasons. Not only are the foods lowest in fat often richest in vitamins, minerals, and phytochemicals, but a diet high in certain fats is associated with many diseases (recall Chapter 11). A high-fat diet correlates closely with obesity, which, in turn, is a risk factor for developing **arthritis,** the painful deterioration and swelling of the joints that constitutes the leading cause of physical limitation in the United States.[77]

Two kinds of arthritis afflict the bones: osteoarthritis and rheumatoid arthritis. Osteoarthritis affects millions of older adults, setting in from unknown causes as people age. During movement, the ends of normal bones are protected by small sacs of fluid that act as lubricants. With arthritis, the sacs erode, cartilage and bone ends disintegrate, and joints become malformed and painful to move. Nutrition does not seem to play a role in the causation of osteoarthritis, but high dietary intakes of vitamins E and C may help to slow its progression and ease pain somewhat once it has started.[78] Low intake of vitamin D may speed its progression. Loss of body weight often brings relief, particularly in the knees.[79]

Rheumatoid arthritis can strike at any age. It probably arises from a malfunction of the immune system—the immune system mistakenly attacks the bone coverings as if they were foreign tissue. A positive nutrition link centers around dietary antioxidants and the omega-3 fatty acid, EPA, found in fish oil.[80] EPA may interfere with activities of the fatty acid–derived hormonelike chemicals involved in inflammation, or perhaps the antioxidants in vegetables and fruits interfere with inflammation by reducing oxidative stress in the joints. The same diet recommended for heart health—one low in fats and high in fruit, vegetables, and oils from fish—may help prevent or reduce the inflammation in the joints that makes arthritis so painful.[81] Supplements of vitamin E may help to reduce oxidative stress, but they do not improve active cases of rheumatoid arthritis.

No one universally effective diet for arthritis relief is known. Many *ineffective* or unproven "cures" are sold, however, as the margin list shows. Traditional medical intervention for arthritis includes medication and surgery. Two popular supplements—glucosamine and chondroitin—may indeed relieve pain and improve mobility as well as over-the-counter pain relievers, and scientists are awaiting information on the effectiveness of these two substances from a study in progress at the National Institutes of Health.[82]

Bogus or unproven arthritis treatments:

- Alfalfa tea.
- Aloe vera liquid.
- Any of the amino acids.
- Burdock root.
- Calcium.
- Celery juice.
- Copper or copper complexes.
- Dimethyl sulfoxide (DMSO).
- Fasting.
- Fresh fruit.
- Honey.
- Inositol.
- Kelp.
- Lecithin.
- Melatonin.
- Para-aminobenzoic acid (PABA).
- Raw liver.
- Selenium.
- Superoxide dismutase (SOD).
- Most vitamin or mineral supplements.
- Watercress.
- Yeast.
- Zinc.
- 100 other substances.

Medical drugs used to relieve arthritis can impose nutrition risks. Controversy 11 explains.

A diet high in fruits and vegetables and low in fats of meats and dairy products may improve some symptoms of arthritis. Omega-3 fatty acids may also have a positive effect.

Protein

Protein needs of older people remain about the same as for young adults. Too much protein can be hard on the kidneys of older adults because of the extra burden of excreting its nitrogen. Which protein-rich foods elders choose to eat takes on extra importance, too. For older people who have lost their teeth, chewing tough foods is next to impossible, and they need soft cooked beans or meats or chopped foods. Individuals with chronic constipation, heart disease, or diabetes may benefit from fiber-rich low-fat legumes and grains as sources of protein.

Protein needs remain about the same through adult life, but choosing low-fat fiber-rich protein foods may help control other health problems.

Vitamins

Vitamin A stands alone among the vitamins in that its absorption appears to increase with aging. For this reason some researchers have proposed lowering the vitamin A

cataracts (CAT-uh-racts) thickening of the lens of the eye that can lead to blindness. Cataracts can be caused by injury, viral infection, toxic substances, genetic disorders, and, possibly, some nutrient deficiencies or imbalances.

requirement for aged populations. Others resist this proposal because foods containing vitamin A and its precursor beta-carotene are under study for preventing oxidative damage to body tissues, an effect described in Chapter 7.

As people age, vitamin D synthesis declines fourfold, setting the stage for deficiency. Many older adults drink little or no vitamin D–fortified milk and get little or no exposure to sunlight. Thus, the recommendation for vitamin D intake for people 51 to 70 has been doubled to 10 micrograms daily and for people 71 and over, tripled to 15 micrograms. Every elderly person should obtain this amount of vitamin D and get outside more often (or sit by an open window for a while).[83]

The DRI committee has recommended that adults aged 51 years and older obtain 2.4 micrograms of vitamin B_{12} daily *and* that vitamin B_{12}–fortified foods (such as fortified cereals) or supplements be used to meet much of this intake.[84] The committee's recommendation reflects the finding that many people older than 50 years lose the ability to produce enough stomach acid to make the protein-bound form of vitamin B_{12} available for absorption and that 10 to 15 percent of those over 60 may suffer from an outright deficiency.[85] Synthetic vitamin B_{12} is reliably absorbed, however, and much misery can be averted by preventing deficiencies of vitamin B_{12} in elderly people.[86] In addition to its other functions, a sufficiency of vitamin B_{12} along with two other B vitamins, folate and vitamin B_6, may prevent some loss of mental ability that commonly occurs among older people.[87] Other nutrients, particularly the antioxidants such as vitamin E, may also play roles in conserving mental functions and eyesight in the aged.[88]

Several theories link nutrients and phytochemicals in foods to age-related changes in the eyes. One theory concerns the leading cause of permanent blindness in people over age 60—macular degeneration, described in Controversy 7. People with lifelong high intakes of vegetables, particularly dark green leafy vegetables such as spinach and collard greens, rarely suffer from macular degeneration. These vegetables are rich in certain carotenoid phytochemicals that may protect the eyes from this destructive disease.*[89] Diets high in fat may also pose a risk for macular degeneration, but the fatty acids of fish oils may be protective.[90]

The macula of the eye was described in Chapter 7.

Another theory concerns **cataracts.** A cataract is a thickening of the lens that impairs vision and leads to blindness. Cataracts can occur even in well-nourished individuals due to injury or other trauma, but most cataracts are vaguely called senile cataracts, meaning "caused by aging." Only 5 percent of people younger than 50 years have cataracts; by age 65, the percentage jumps to over 50 percent. The lens of the eye is easily oxidized. People who eat few fruits and green vegetables obtain too few antioxidants, and this causes oxidative stress in the eyes that puts them at risk of developing cataracts.[91] People taking supplements of vitamins C and E seem less likely to develop cataracts.[92] Sadly, many people needlessly endanger their vision by shunning the dark green leafy vegetables that may protect against both macular degeneration and cataracts.

Energy like this requires continued physical activity and all the nutrients to support it.

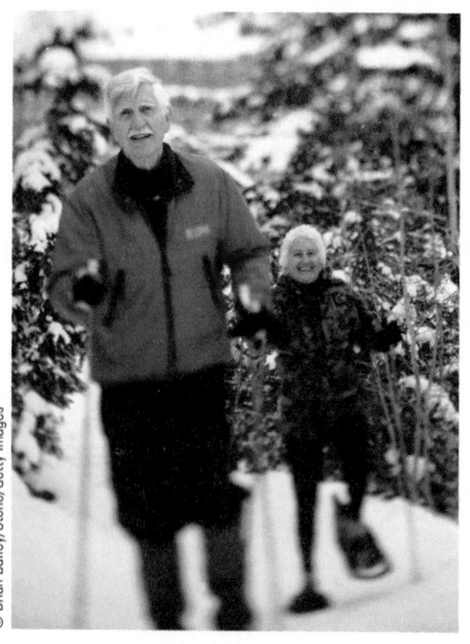

© Brian Bailey/Stone/Getty Images

key point *Vitamin A absorption increases with aging. Older people suffer more from deficiencies of vitamin D and vitamin B_{12} than young people do. Cataracts and macular degeneration often occur among those with low fruit and vegetable intakes.*

Water and the Minerals

Dehydration is a major risk for older adults: the thirst mechanism may become imprecise, and older people may go for long periods without drinking fluids. The kidneys also become less efficient in recapturing water before it is lost as urine. This water loss causes some problems and worsens others, such as constipation and bladder problems. In a person with asthma, dehydration thickens mucus in the lungs, which may then block airways. Dehydration can result in urinary tract infections, pneumonia, pressure ulcers (bed sores), and mental confusion.[93] Regardless of age, adults need to drink 6

*The carotenoids are lutein and zeaxanthin, which help to form pigments of the macula of the eye.

to 8 glasses of water each day.[94] A person we know uses this trick to ensure getting enough water: he keeps six inexpensive 8-ounce cups in the cupboard. Through the day he uses each to drink one cup of water, collecting them in the dish drain. In the afternoon he checks the cupboard and drinks from any remaining cups. For him, drinking water has become a habit, and seldom are any cups left in the cupboard after supper.

Adults of all ages need 6 to 8 glasses of water each day.

© Darama/Corbis Stock Market

Iron Iron status generally improves in later life, especially for women when menstruation ceases. When iron-deficiency anemia does occur, diminished appetite with low food intake is often the cause. Aside from diet, other factors make iron deficiency likely in older people:

- Chronic blood loss from ulcers, hemorrhoids, or the like.
- Poor iron absorption due to reduced stomach acid secretion.
- Antacid use, which interferes with iron absorption.
- Use of medicines that cause blood loss, including anticoagulants, aspirin, and arthritis medicines.

Older people take more medicines than others, and drug and nutrient interactions are common.

Zinc Zinc deficiencies are common in older people. Zinc deficiency can depress the appetite and blunt the sense of taste, thereby leading to low food intakes and worsening of zinc status. Many medications interfere with the body's absorption or use of zinc, and older adults' medicine load can worsen zinc deficiency.

Research on zinc supplements demonstrates that nutrient supplements taken by the elderly can bring unexpected results. Researchers studying the immune response of elderly people sometimes observe a reduced immune response, sometimes an enhanced response, and sometimes no effect in those given supplements of zinc.[95]

These foods provide iron and zinc together: meat, poultry, liver, oysters, whole grains, fortified breakfast cereals,* and legumes.

*Cereals fortified with iron and zinc may not be available in Canada.

Calcium Many sections of this book have emphasized the importance of abundant dietary calcium throughout life to protect against osteoporosis. The calcium intakes of many people, especially women, in the United States are well below the recommended amount. If fresh milk causes stomach discomfort, as the majority of older people report, then lactose-modified milk or other calcium-rich foods should take its place.

Controversy 8 discusses osteoporosis.

Overall, elderly people often benefit from a balanced low-dose vitamin and mineral supplement.[96] Older people taking such supplements suffer fewer sicknesses caused by infection.[97] Vitamin A has been seen to depress the immunity of elders, while vitamin E may enhance it.[98] A summary of the effects of aging on nutrient needs appears in Table 13-8 on page 496.

Aging alters vitamin and mineral needs. Some needs rise while others decline.

Can Nutrition Help People to Live Longer?

The evidence concerning nutrition and longevity is intriguing. One approach studies the combined effects of nutrition and other lifestyle habits on aging. In a classic study, researchers in California observed nearly 7,000 adults and noticed that some were young for their ages, while others were old for their ages.[99] To uncover what made the difference, the researchers focused on health habits and identified six factors that affect physiological age. Three of the six factors were related to nutrition:

- Abstinence from, or moderation in, alcohol use.
- Regular meals.
- Weight control.

The others were regular adequate sleep, abstinence from smoking, and regular physical activity. The physical health of those who reported all six positive health practices

Table 13•8

SUMMARY OF NUTRIENT CONCERNS IN AGING

Nutrient	Effects of Aging	Comments
Energy	Need decreases.	Physical activity moderates the decline.
Fiber	Low intakes make constipation likely.	Inadequate water intakes and physical activity, along with some medications, compound the problem.
Protein	Needs stay the same.	Low-fat, high-fiber legumes and grains meet both protein and other needs.
Vitamin A	Absorption increases.	Supplements normally not needed.
Vitamin D	Increased likelihood of inadequate intake; skin synthesis declines.	Daily moderate exposure to sunlight may be of benefit.
Vitamin B_{12}	Malabsorption of some forms.	Foods fortified with synthetic vitamin B_{12} or a low-dose supplement may be of benefit in addition to a balanced diet.
Water	Lack of thirst and increased urine output make dehydration likely.	Mild dehydration is a common cause of confusion.
Iron	In women, status improves after menopause; deficiencies linked to chronic blood losses and low stomach acid output.	Stomach acid required for absorption; antacid or other medicine use may aggravate iron deficiency; vitamin C and meat enhance absorption.
Zinc	Intakes are often inadequate and absorption may be poor, but needs may also increase.	Medications interfere with absorption; deficiency may depress appetite and sense of taste.
Calcium	Intakes may be low; osteoporosis becomes common.	Lactose intolerance commonly prevents milk intake; substitutes are needed.

was comparable to that of people 30 years younger who reported few or none. Numerous studies have confirmed the benefits of such lifestyle factors under personal control.[100] These findings suggest that even though people cannot alter the year of their birth, they can alter the probable length and quality of their lives. Tables 13-9 and 13-10 list some changes of aging that are unpreventable and some that may yield to lifestyle influences.

The first evidence that diet might extend life came more than half a century ago from experiments on rats. Researchers fed a group of young rats diets extremely limited in energy, while control rats ate normally. The starved rats stopped growing while the control rats grew normally. When the researchers increased the energy, growth resumed. Many of the starved group died young from malnutrition. A few survivors, though permanently deformed from their starvation ordeal, remained alive far beyond the normal life span for such animals and developed diseases of aging much later than normal.

Later studies repeated these findings using more moderate energy restriction with adequate nutrient intakes that did not inflict physical malformations. Restricted rats retained youthfulness longer and developed fewer of the factors associated with chronic diseases.[101] Energy-restricted animals are reported to have lower blood pressure, and news from an ongoing study of energy-restricted monkeys indicates that as the animals age, the restricted group has retained healthy blood glucose and insulin responses and has healthier blood lipid profiles than the freely fed controls.[102] Even genetically obese rats live longer when energy is restricted, even though their body fat remains similar to that of nonobese rats allowed to eat freely.[103] Evidence from other species (see the margin) suggests that this effect spans many biological systems.

The obvious question is whether findings from animal studies can be applied to human beings. Many of the physiological responses seen in rats during energy restriction do seem to occur in human beings who moderately restrict energy intakes. In experiments, when men of normal weight cut back on their usual energy intake by 20 percent, their body weight, body fat, and blood pressure dropped, and their HDL cholesterol rose—favorable changes for preventing obesity and chronic diseases. Much more evidence must be collected, however, before such findings can be applied to the general population.

Differences in maximum life span between animals eating normally and those that are energy restricted:

- *Rats:*
  Normal diet, 33 months.
  Restricted diet, 47 months.
- *Spiders:*
  Normal diet, 100 days.
  Restricted diet, 139 days.
- *Single-celled animals (protozoans):*
  Normal diet, 13 days.
  Restricted diet, 25 days.

SOURCE: R. Weindruch, Caloric restriction and aging, *Scientific American*, January 1996, pp. 46–52.

Investigators have proposed several mechanisms to explain how energy restriction prolongs life in rats, but none has been proved. A delay in the onset of age-related diseases has already been mentioned. Genetics may play a role: genes that modulate the aging process seem to become less active when food energy is scarce.[104] Current research is focused on the genetic response of both young and old rats to energy restriction in hopes of identifying genes that play key roles in aging. Some researchers predict that one day drugs and treatments may mimic the effects of energy restriction at the genetic level and so confer life-extending effects on experimental animals without true caloric restriction.[105]

The free-radical hypothesis of aging blames damage from oxidative stress for the physical deterioration associated with aging.[106] This theory finds support in research showing that the body's internal antioxidant enzymes diminish with age and that many "age-related" degenerative diseases are linked to free-radical damage.[107] This line of research has promoted a storm of worthless pills, supplements, and treatments with promises of life extension. Controversy 7 demonstrated that it's better to spend money on fresh fruit and green and yellow vegetables, which are known to provide the kinds of antioxidants the body can use.

key point *Lifestyle factors can make a difference in aging. In rats, food energy deprivation may lengthen the lives of individuals who survive the treatment. Claims for life extension through antioxidants or other supplements are common hoaxes.*

Can Foods or Supplements Affect the Course of Alzheimer's Disease?

Alzheimer's disease is now the third costliest health problem in the United States, following heart disease and cancer. In Alzheimer's disease, the most prevalent form of **senile dementia,** abnormal deterioration of the brain occurs in the areas that coordinate memory and cognition. The brain is littered with clumps of abnormal protein fragments that clog the brain and damage or kill certain nerve cells.* Dementia from conditions such as Alzheimer's may rob 6 to 10 percent of U.S. adults of productive life by age 65—and the rate doubles when milder cases are added to the count.[108] A cluster of symptoms justifies diagnosis: losses of memory and reasoning power, loss of the ability to communicate, and loss of physical capabilities. More research is needed, and quickly, to find a drug to block the destructive progression of the disease.[109]

Nutrition bears only weak links to Alzheimer's disease. Some research on Alzheimer's disease centers on a buildup of metals, including aluminum, in the brain tissue. Researchers often find elevated levels of copper, iron, and zinc in the brain tissues of those with Alzheimer's and theorize that these metals may accelerate the progression of the disease, possibly by increasing the formation of free radicals that produce oxidative stress.[110] Nerve cells in the brains of people with Alzheimer's disease show evidence of free-radical attack—damage to DNA, cell membranes, and proteins. Researchers are examining whether the antioxidant nutrients can limit free-radical damage and delay or prevent Alzheimer's disease.[111] Researchers even propose a link between mental deterioration in old age and twice weekly or greater tofu consumption, but this association requires much more research before conclusions can be drawn.[112]

There is conflicting evidence as to whether supplements of zinc or other trace minerals worsen Alzheimer's disease. To err on the safe side, food sources, not concentrated supplements, of trace minerals are advisable for people with the disease.[113]

A causal connection with the mineral aluminum seems unlikely. Brain aluminum in people with Alzheimer's exceeds normal brain aluminum by some 10 to 30 times, but blood and hair aluminum remains normal, indicating that the accumulation is caused by something in the brain itself, not by high aluminum in the diet. Other

*The protein fragments are called beta-amyloid.

senile dementia the loss of brain function beyond the normal loss of physical adeptness and memory that occurs with aging.

Table 13•9

CHANGES WITH AGE YOU PROBABLY CANNOT PREVENT

These changes are probably beyond your control:

- ✔ Graying of hair
- ✔ Balding
- ✔ Some drying and wrinkling of skin
- ✔ Impairment of near vision
- ✔ Some loss of hearing
- ✔ Reduced taste and smell sensitivity
- ✔ Reduced touch sensitivity
- ✔ Slowed reactions (reflexes)
- ✔ Slowed mental function
- ✔ Diminished visual memory
- ✔ Menopause (women)
- ✔ Loss of fertility (men)
- ✔ Loss of joint elasticity

Table 13•10

CHANGES WITH AGE YOU PROBABLY CAN SLOW OR PREVENT

By exercising, eating an adequate diet, reducing stress, and planning ahead, you may be able to slow or prevent:

- ✔ Wrinkling of skin due to sun damage
- ✔ Some forms of mental confusion
- ✔ Raised blood pressure
- ✔ Speeded-up resting heart rate
- ✔ Reduced breathing capacity and oxygen uptake
- ✔ Increased body fatness
- ✔ Raised blood cholesterol
- ✔ Slowed energy metabolism
- ✔ Decreased maximum work rate
- ✔ Loss of sexual functioning
- ✔ Loss of joint flexibility
- ✔ Oral health: loss of teeth, gum disease
- ✔ Bone loss
- ✔ Digestive problems, constipation

Some degree of memory loss is often simply a function of aging and is termed *benign* (meaning *harmless*) *senescent* (meaning *of aging*) *forgetfulness*. Occasional forgetful moments generally do not forecast the development of Alzheimer's disease in an older person.

"Smart" drugs, drinks, and supplements, sold with promises of brainpower enhancement, are discussed in this chapter's Controversy.

The Child and Adult Care Food Program (CACFP) is designed to help public and private nonresidential child and adult daycare programs provide nutritious meals to those younger than age 12 or older than 65, or people with disabilities.

Shared meals can be the high point of the day.

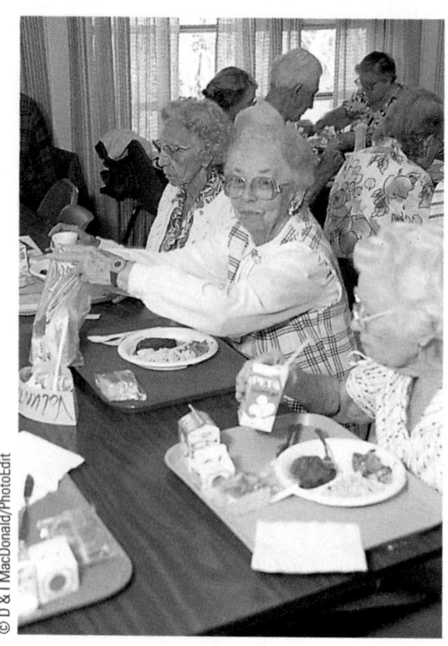

© D & I MacDonald/PhotoEdit

metals follow similar trends. Thus, the high levels of trace elements in the brain must be more a result, than a cause, of the disease.

The brain tissue of a person with Alzheimer's also contains an abnormally small amount of an enzyme that makes a compound from choline that is essential to memory (acetylcholine). To date, oral supplements of choline or lecithin (which contains choline, first mentioned in Chapter 5) have had no consistent effect on memory, mental functioning, or the progression of Alzheimer's. Some evidence suggests that vitamin E may help to slow its progression, but only in a dose (2,000 milligrams per day) high enough to adversely affect some people, especially those taking blood-thinning medications or those prone to strokes.[114] Deficiencies of folate or other B vitamins may worsen some effects of Alzheimer's, but research is just preliminary in these areas.[115]

The results of one small, yearlong study seem to indicate a modest benefit in cognitive and social functioning in Alzheimer's patients given a standardized extract of the medicinal herb ginkgo biloba.[116] To date, no proven benefits are available from herbs or other remedies, but claims from quacks are all too commonplace.

More promising are drugs that slow down the destruction of the choline compound just mentioned—they improve memory and other aspects of cognition.[117] Meanwhile, other drugs, such as the nicotine of tobacco and the hormone estrogen, seem to improve older people's ability to remember. Scientists also seem closer to developing a vaccine that may one day be able to prevent brain damage from Alzheimer's.[118]

Preventing weight loss may be the most important nutrition concern for the person with Alzheimer's disease.[119] Depression and forgetfulness can lead to skipped meals and poor food choices. Caregivers can help with food planning and mealtimes.[120] Well-liked and well-balanced meals and snacks served in a cheerful, peaceful atmosphere are welcome. As function diminishes, ready-to-eat foods in bite-size pieces may be most acceptable.

key point *Alzheimer's disease causes some degree of brain deterioration in as many as one-fifth of people past age 65. Current treatment helps only marginally; dietary aluminum is probably unrelated. Nutrition care gains importance as the disease progresses.*

Food Choices of Older Adults

Most older people are independent, socially sophisticated, mentally lucid, fully participating members of society who report themselves to be happy and healthy. The quality of life among the 85 and older group has improved, and their chronic disabilities have declined dramatically in recent years.[121]

Results of national surveys indicate that many older people have heard and heeded nutrition messages: they have cut down on saturated fats in dairy foods and meats and are eating slightly more vegetables and whole-grain breads. Store shelves prominently display good-tasting, low-fat, nutritious foods in easy-to-open, single-serving packages with labels that are easy to read. Many nutrient and other supplements are marketed for older adults. Whether to take a supplement is a personal choice, but evidence supports the idea that a single low-dose multivitamin-mineral tablet a day can improve resistance to disease in older people. The best choice would be a supplement low in vitamin A, with ample amounts of all of the other vitamins and minerals for which DRI values are set (see the inside front cover).

Obstacles to Adequacy Many factors affect the food choices and eating habits of older people, including whether they live alone or with others, at home or in an institution. Men living alone, for example, are likely to consume poorer-quality diets than those living with spouses. Older people who have difficulty chewing because of tooth loss or have lost taste sensitivity may no longer seek a wide variety of foods.[122] Medical conditions can also affect nutrition. Many older people experience unintentional weight loss, sometimes followed by illness or death. It may be that some of these outcomes could have been prevented or delayed if only the person had consumed an adequate diet.[123]

Two other factors seem to make older people vulnerable to malnutrition: use of multiple medications and abuse of alcohol. People over age 65 take about a fourth of all the medications, both prescription and over-the-counter, sold in the United States. Although these medications enable people with health problems to live longer and more comfortably, they also pose a threat to nutrition status because they may interact with nutrients, depress the appetite, or alter the perception of taste (see Controversy 11).

Estimates are that between 2 and 10 percent of the elderly in the United States suffer from alcoholism, alcohol abuse, or problem drinking.[124] Evidence is mounting that loneliness, isolation, and depression in the elderly accompany overuse of alcohol. It isn't possible to say whether the depression or the alcohol abuse comes first, for each worsens the other, and both detract from nutrient intakes. Table 13-11 and the margin list provide means of identifying those who might be at risk for malnutrition.

The DETERMINE predictors of malnutrition in the elderly:

- **D**isease.
- **E**ating poorly.
- **T**ooth loss or oral pain.
- **E**conomic hardship.
- **R**educed social contact.
- **M**ultiple medications.
- **I**nvoluntary weight loss or gain.
- **N**eed of assistance with self-care.
- **E**lderly person older than 80 years.

Programs That Help Federal programs can provide help for older people.[125] Social Security provides income to retired people over age 62 who paid into the system during their working years. The Food Stamp program assists the very poor by supplementing their monthly food budgets with a card similar to a credit card, encrypted with benefits and redeemable for food. The Elderly Nutrition Program effectively and efficiently provides nutritious meals in a social congregate setting, education and shopping assistance, counseling and referral to other needed services, and transportation to necessary appointments.[126] An estimated 25 percent of the nation's elderly poor benefit from meals provided by the program. For the homebound, Meals on Wheels volunteers deliver meals to the door, a benefit even though the recipients miss out on the social atmosphere of the congregate meals. Nutritionists are wise not to focus solely on nutrient and food intakes of the elderly because social interactions may be as important as food itself.

Many older people, even able-bodied ones with financial resources, find themselves unable to perform cooking, cleaning, and shopping tasks. For anyone living alone and particularly for those of advanced age, it is important to work through the problems that food preparation presents.[127] This chapter's Food Feature presents some ideas.

Federal sources of support for the elderly:

- Social Security.
- Food Stamps.
- Elderly Nutrition Program.
- Meals on Wheels.

key point *Food choices of the elderly are affected by aging, altered health status, and changed life circumstances. Assistance programs can help by providing nutritious meals, providing social interactions, and easing financial problems.*

Table 13•11

NUTRITION SCREENING INITIATIVE CHECKLIST FOR OLDER AMERICANS

Circle the number to the right if the statement applies to you.

Statement	Yes
I have an illness or condition that makes me eat different kinds and/or amounts of food.	2
I eat fewer than 2 meals per day.	3
I eat few fruits or vegetables and use few milk products.	2
I have 3 or more drinks of beer, liquor, or wine almost every day.	2
I have tooth or mouth problems that make it hard for me to eat.	2
I don't always have enough money to buy the food I need.	4
I eat alone most of the time.	1
I take 3 or more different prescribed or over-the-counter drugs a day.	1
Without wanting to, I have lost or gained 10 pounds in the last 6 months.	2
I am not always physically able to shop, cook, and/or feed myself.	2
Total	

Score:
0–2: Good. Recheck your score in 6 months.
3–5: Moderate nutritional risk. Visit your local office on aging, senior nutrition program, senior citizens center, or health department for tips on improving eating habits.
6 or more: High nutritional risk. See your doctor, dietitian, or other health-care professional for help in improving your nutrition status.

NOTE: The Nutrition Screening Initiative is part of a national effort to identify and treat nutrition problems in older Americans.

Single Survival and Nutrition on the Run

When it comes to feeding themselves wisely, singles of all ages face problems, ranging from the selection of fast foods to the purchase, storage, and preparation of food from the grocery store. Whether the single person is a busy student in a college dormitory, an elderly person in a retirement apartment, or a professional in an efficiency suite, the problems of preparing nourishing meals are often the same. People who live in places without kitchens and freezers find storing foods problematic. Following is a collection of ideas gathered from single people who have devised answers to some of these problems.

Is Eating in Restaurants the Answer?

For the single person as for others, restaurants mean convenience. On average, almost 40 percent of a U.S. household's food budget is spent on foods prepared and eaten away from home. And in any given month, up to 70 percent of households consume food from a carry-out restaurant.[128] Restaurant foods may be the quickest, easiest, and least taxing way to satisfy hunger at mealtime, but can they meet your body's nutrient needs or support health as well as homemade foods? The answer is "perhaps," if the diner makes the effort to meet nutritional needs.[129]

A few chefs and restaurant owners are concerned with the nutritional health of their patrons, but more often chefs strive to please the palate and leave nutrition-conscious diners on their own. Restaurant foods are often overly endowed with calories, fat, saturated fat, and salt, yet not overly generous with needed constituents such as fiber, iron, and calcium.[130] In addition, generous restaurant portions often equal three or more standard-size servings of the Food Guide Pyramid. If you are willing to restrict your portions to sizes that do not exceed your energy needs, ask that excess portions be placed in take-out containers, and judiciously choose foods that stay within intake guidelines for fat and salt, then restaurants can provide both convenience and nutrition. The Food Feature of Chapter 5 gave specific suggestions for ordering fast food and other foods with an eye to keeping fat intakes within bounds, and Chapter 8 gave a list of high-sodium foods.

Grocery Store Take-Out Choices

Take-out delicatessen-style foods offer convenience—they can be purchased while shopping for other items—and a modicum of control over their nutrient contents. They also often cost substantially less than similar foods from restaurants. Another bonus is control: you can specify the amount you need and portion it onto your plate at home.

Grocery store take-out can be an excellent bargain in terms of nutrient density, too. Choose from among roast chicken, smoked seafood, pasta with tomato sauce, steamed vegetables, precut salads and fruit without dressings, cooked beans, and plain baked potatoes for convenient nutrient bargains. Be aware that stuffing, macaroni and cheese, meat loaf and gravy, vegetables with creamy sauces, mayonnaise-dressed mixed salads, and fried chicken and fish can be as laden with saturated fat and calories as any traditional fast food.

More Grocery Store Know-How

Singles often face the quantity problem in the grocery store. Large packages of meat and vegetables, whether fresh or frozen, are suitable for a family of four or more, and even a head of lettuce can spoil before one person can use it all.

Buy only what you will use. Don't be timid about asking the grocer to break open a family-size package of wrapped meat or fresh vegetables. Look for bags of prepared salad greens to take the place of lettuce in both salads and sandwiches. Prepared salads and other small-size containers of food may be expensive, but it is also expensive to let the unused portion of a large container spoil. Buy only three pieces of each kind of fresh fruit: a ripe one, a medium-ripe one, and a green one. Eat the first right away and the second soon, and let the last one ripen to eat days later.

Think up ways to use a vegetable that you must buy in large quantity. For example, you can divide a head of cauliflower into thirds. Cook one-third and eat it as a hot vegetable. Toss another

Buy only what you will use.

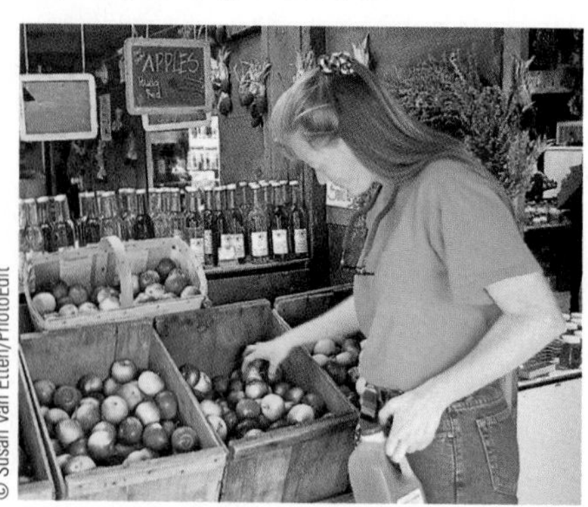

third into a salad dressing marinade for use as an appetizer. Blend up the rest, cooked, in a creamy soup.

Buy fresh milk in the size you can best use. If your grocer doesn't carry pints or quarts of milk, try a convenience store. If you eat lunch in a cafeteria, buy two pints of milk—one to drink and one to take home and store. Buy a loaf of bread and immediately store half, well wrapped, in the freezer (not the refrigerator, which will make it stale).

Food Preparation Hints

A wise person once said, "An hour spent organizing can save three hours later on."* This holds true in food preparation. For shelf-stable items, prepare a space for rows of glass jars (jars from spaghetti sauce, applesauce, or other foods work well). Use the jars to store pasta, rice, lentils, other dry beans, flour, cornbread or biscuit mix, dry fat-free milk, and cereal. Light destroys riboflavin, so use opaque jars for enriched pasta and dry milk. Cut the directions-for-use label from the package of each item and tape it to the jar. Place each jar, tightly sealed, in the freezer for a few days to kill any eggs or organisms before storing it on the shelf. Then the jars will keep bugs out of the foods indefinitely. The jars are also pretty to look at and will remind you of possibilities for variety in your menus.

Experiment with stir-fried foods. A large fry pan works well to stir-fry a variety of vegetables and meats. Inexpensive vegetables such as cabbage and celery are delicious when crisp cooked in a little oil with soy sauce or lemon juice. Interesting frozen vegetable mixtures are available, or cooked, leftover vegetables can be dropped into a stir-fry at the last minute. A bonus of a stir-fried meal is that you have only one pan to wash.

Make mixtures using what you have on hand. A thick stew prepared from any leftover vegetables and bits of meat, with some added frozen onions, peppers, celery, and potatoes, makes a complete and balanced meal, except for milk. If you like creamed gravy, add some fat-free dry milk to your stew.

If you can afford a microwave oven, buy one. Cooking times are quick, and you'll use fewer pots and pans. Be sure to use containers designed for microwaving, however. Margarine tubs, plastic bowls, and storage bags and containers can release potentially harmful chemicals into food when they are heated in the microwave oven. Use glass, or buy plastic containers that are labeled as safe for microwaving.

Depending on your freezer space, make a regular-size recipe of a dish that takes time to prepare: a casserole, vegetable pie, or meat loaf. Freeze individual portions in containers that can be heated later. Date these so you will use the oldest first.

Dealing with Loneliness

For nutrition's sake, it is important to attend to loneliness at mealtimes. The person who is living alone must learn to connect food with socializing. Invite guests, and make enough food so that you can enjoy the leftovers later on. If you know an older person who eats alone, you can bet that person would love to join you for a meal now and then.

*That wise person was Eva May Hamilton, one of the original authors of this textbook.

Invite guests to share a meal.

©2002 Larry Dale Gordon/Image Bank/Getty Images

Time-saving tips to turn convenience foods into nutritious meals:

- Add extra nutrients and a fresh flavor to canned stews and soups by tossing in some frozen ready-to-use mixed vegetables. Choose vegetables frozen without salty, fatty sauces—prepared foods generally contain enough salt to season the whole dish including added vegetables.
- Buy frozen vegetables in a bag, toss in a variety of herbs, and use as needed. Vary your choices to prevent boredom.
- When grilling burgers, wrap a mixture of frozen broccoli, onion, and carrots in a foil packet with a tablespoon of Italian dressing and grill alongside the meat for seasoned grilled vegetables.
- Use canned fruits in their own juices as desserts. Toss in some frozen berries or peach slices and top with flavored yogurt for an instant fruit salad.
- Prepared rice or noodle dishes are convenient, but those claiming to contain broccoli, spinach, or other vegetables really contain just a trifle—not nearly enough to qualify as a serving of vegetable. Pump up the nutrient value by adding a half-cup of your frozen vegetables per serving of pasta or rice just before cooking.
- Purchase frozen onion, mushroom, and pepper mixtures to embellish jarred spaghetti sauce or small frozen pizzas. Top with parmesan cheese.
- Use frozen shredded potatoes, sold for hash browns, in soups or stews, or mix with a handful of shredded reduced-fat cheese or a can of fat-free "cream of anything" soup and bake for a quick and hearty casserole.

Answers to these Self Check questions are in Appendix G.

1. Children naturally like nutritious foods in all the food groups, *except:*
   a. dairy
   b. meats
   c. vegetables
   d. fruits
2. Which of the following can contribute to choking in children?
   a. peanut butter eaten by itself
   b. reclining while eating
   c. popcorn and nuts
   d. all of the above
3. Which of the following is most commonly deficient in children and adolescents?
   a. folate
   b. zinc
   c. iron
   d. vitamin D
4. Which of the following may alleviate symptoms of PMS?
   a. adequate vitamin E and vitamin B_6
   b. exercise
   c. omitting caffeine from the diet
   d. all of the above
5. Which of the following have been shown to improve acne?
   a. avoiding chocolate and fatty foods
   b. retinoic acid or Retin A
   c. vitamin A supplements
   d. all of the above
6. Physical changes of aging that can affect nutrition include:
   a. reduced stomach acid
   b. increased saliva output
   c. reduced lean body mass
   d. a and c above
7. A food intolerance is the same problem as a food allergy. T F
8. Research to date supports the idea that food allergies or intolerances are common causes of hyperactivity in children. T F
9. The same diet recommended for heart health may help to alleviate the pain from arthritis. T F
10. Vitamin A absorption decreases with age. T F

nutrition on the net

For further study of the topics of this chapter, access these websites and search for the phrases or words in quotation marks:

1. Find updates and quick links to these and other nutrition-related sites at our website: www.wadsworth.com/nutrition
2. Search for "infants," "baby bottle tooth decay," "premature birth," "hyperactivity," "food allergies," and "adolescent health," at the U.S. Government health information site: www.healthfinder.gov
3. Learn how to care for infants, children, and adolescents from the American Academy of Pediatrics and the Canadian Paediatric Society: www.aap.org and www.cps.ca
4. Investigate many aspects of U.S. children's lives at: www.childstats.gov
5. Get information on the Food Guide Pyramid for Young Children from the USDA: www.usda.gov/cnpp
6. Get tips for feeding children from the American Dietetic Association, the I Am Your Child Program, and the Kids Food Cyber Club: www.eatright.org, www.iamyourchild.org, and www.kidsfood.org
7. Visit the National Center for Education in Maternal & Child Health and the National Institute of Child Health and Development: www.ncemch.org and www.nih.gov/nichd
8. Learn about the Child Nutrition Programs: www.fns.usda.gov/fns
9. Visit the Lead Program of the Centers for Disease Control: www.cdc.gov/nceh/programs/lead
10. Visit the Public Health Service's Office of Women's Health site for messages on positive self-images, good nutrition, and fitness for girls between the ages of 9 and 14: www.health.org/gpower/girlarea/bodywise

INTERNET ACTIVITY

Food Guide Pyramids have been developed to meet the needs of different life cycle stages. Two recent pyramids include one for young children and one for older adults.

1. Go to the website of the Food and Nutrition Information Center: www.nal.usda.gov/fnic/Fpyr/pyramid.html
2. Click "USDA's Food Guide Pyramid for Young Children." Study the Pyramid including the number of servings and serving sizes.
3. Scroll to the bottom of the page, click "Special Populations."
4. And then "Food Guide Pyramid for People over 70 Years Old." Study the Pyramid including the number of servings and serving sizes.

Compare the Pyramid for Young Children and the Pyramid for People Over 70 Years Old to the original Food Guide Pyramid. How are they the same? How are they different?

Consider a child and an older person that you know. If possible, discuss the appropriate pyramid with each. Would they be able to follow these new Pyramids? Why or why not?

CONTROVERSY 13

CAN FOOD OR SUPPLEMENTS AFFECT MIND AND MEMORY?

WHY DO YOU feel sleepy after lunch and not after dinner? Do some foods help you to think or to remember? Human behavior and the brain are still largely uncharted territory, but researchers no longer doubt that food and nutrients affect both mind and memory.[1]

The Brain and Its Neurotransmitters

Encased in the hard, bony, inelastic helmet of its protective skull, the brain cannot expand and contract as can, say, the liver or adipose tissue. It cannot store its own reserve supply of glycogen or fat because those fuels take up space. It cannot store oxygen to oxidize those fuels or nutrients to help it do so. The brain must depend on the passing blood supply for both its fuels and its oxygen.

The brain's needs for oxygen and fuel are extraordinary. Although it accounts for only 2 percent of an adult's body weight, at any given time the brain contains 15 percent of the body's blood, and it devours 20 to 30 percent of the fuels that support the basal metabolism. If the blood delivers too little oxygen or glucose, the brain's cells cease communicating with each other (see Figure C13-1 on the next page), and coma occurs within minutes. If the blood supply is interrupted altogether, coma ensues within 10 seconds.

Communication within the brain also requires other nutrients, for example:

- Amino acids to make its messenger molecules: some 30 to 40 **neurotransmitters** and related compounds.
- Electrically charged minerals to help transmit its electrical impulses.
- Vitamins and other minerals to facilitate these processes.
- Lipids to repair its cell membranes.
- Water to maintain the fluids in which many of the chemical processes take place.

The snack you choose may affect how you feel.

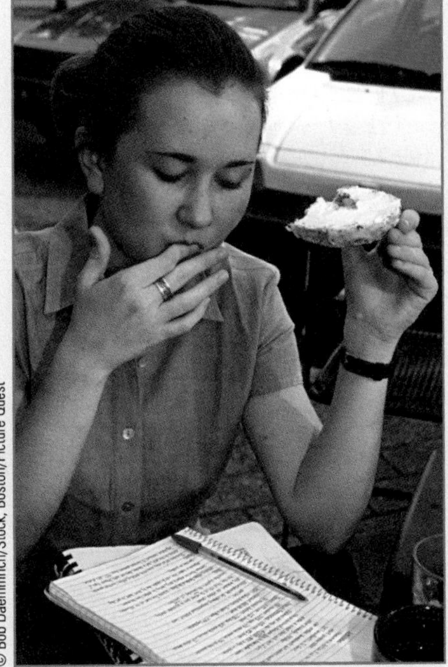

© Bob Daemmrich/Stock, Boston/Picture Quest

No matter how much the chemical composition of the blood fluctuates, the brain keeps its internal environment consistent. The **blood-brain barrier** acts as a molecular sieve to filter from the blood only the fluid and chemicals the brain needs and to keep everything else out (see Table C13-1, next page). The brain monitors the blood and sends messages to other organs when it needs fuels, oxygen, and nutrients. At one time the brain may need glucose; at another, amino acids.

Intake of lipids, vitamins, and minerals doesn't affect the brain's functioning immediately. Amino acids, in contrast, are used to form neurotransmitters the same day they are eaten, so their effects are seen within minutes or hours of ingestion. Most of the amino acids never exceed a given level in the brain no matter how much is consumed, but the brain's *regulatory* amino acids cross the blood-brain barrier freely. Once in the brain, these regulatory amino acids exert **precursor control;** that is, the brain responds to larger or smaller amounts of them by making larger or smaller amounts of neurotransmitters from them. Therein lies the biochemical chain that connects the food a person eats with brain chemistry by changing the rates at which the brain makes its neurotransmitters.[2] The next question: How does this biochemical chain of events influence a person's mood or memory?

One neurotransmitter whose brain concentration is especially sensitive to changes in precursor supply has been studied in depth: **serotonin,** whose precursor is the amino acid tryptophan. Similarly, a set of neurotransmitters, the **catecholamines,** depend on the availability of their precursor amino acid, tyrosine. The discussion that follows centers on serotonin.

Tryptophan to Serotonin

Ordinary meals of the kind people eat every day raise or lower the concentration of serotonin in the brain, depending on the meal's protein and carbohydrate content. Serotonin release, in turn, affects sensations and mood, so the ingredients of meals may have real effects on how people feel afterward. Many people report feeling relaxed, peaceful, or sleepy after eating certain foods, sensations that are sometimes attributed to serotonin's effects on the brain. Research shows that a lack of tryptophan flowing into the brain can manifest itself in wakefulness, depressed mood, a tendency to startle, and an enhanced sensitivity to pain. Tryptophan-deficient animals exhibit these symptoms, and when given tryptophan, they return to normal as their brain serotonin is restored.

The amount of tryptophan that enters the brain depends not only on the amount of tryptophan the person eats but also on the total protein and carbohydrates eaten with it. If tryptophan is taken as a single amino acid, then brain serotonin increases proportionately. Normally, however, whole proteins, not just tryptophan, are eaten.

Figure C13•1

COMMUNICATION WITHIN THE BRAIN

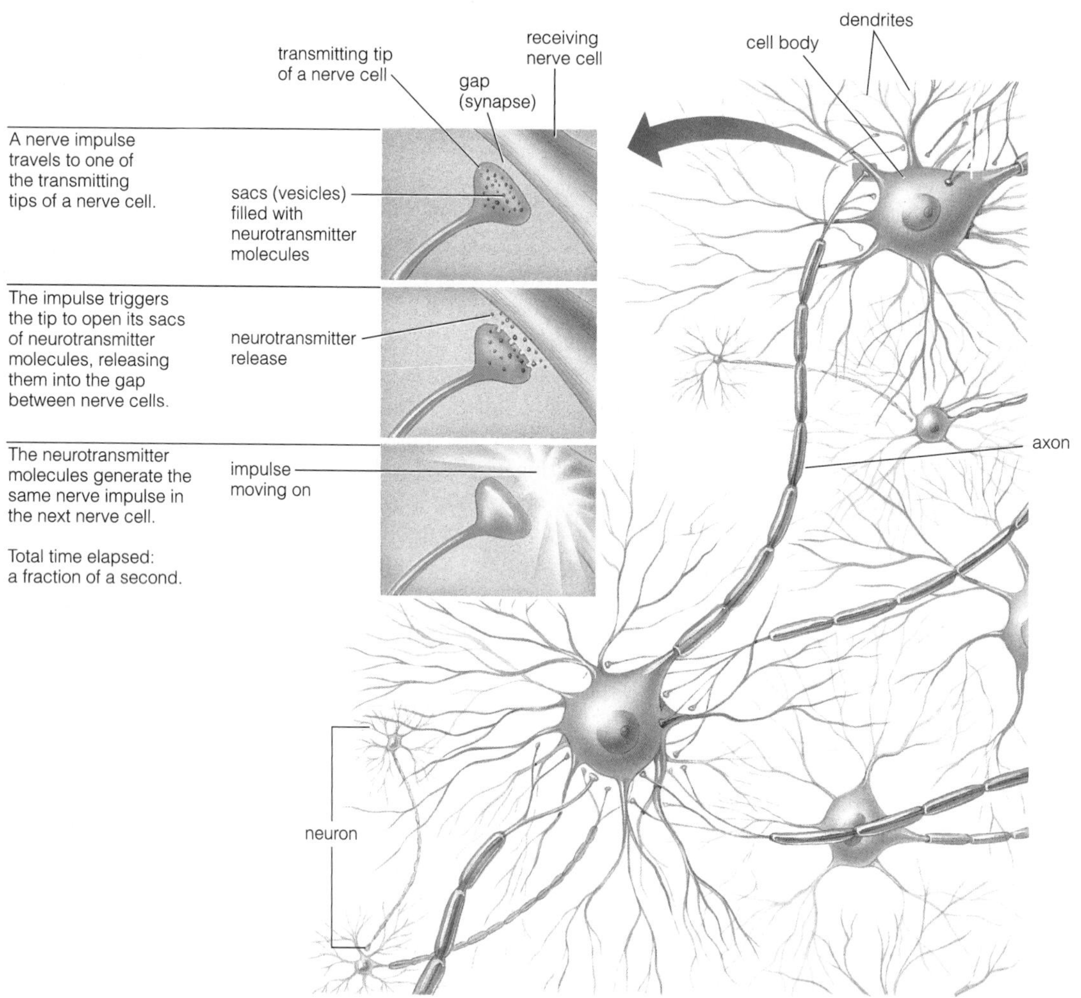
transmitting tip
of a nerve cell
gap
(synapse)
receiving
nerve cell
cell body
dendrites
A nerve impulse travels to one of the transmitting tips of a nerve cell.
sacs (vesicles) filled with neurotransmitter molecules
The impulse triggers the tip to open its sacs of neurotransmitter molecules, releasing them into the gap between nerve cells.
neurotransmitter release
The neurotransmitter molecules generate the same nerve impulse in the next nerve cell.
impulse moving on
Total time elapsed: a fraction of a second.
axon
neuron

Some of the other large amino acids in the proteins compete with tryptophan for entry into the brain because they use the same transport mechanism to get across the blood-brain barrier. Thus, tryptophan fails to enter the brain in increased quantities and so does not enhance brain serotonin synthesis.

If carbohydrate is eaten instead of protein, however, the carbohydrate can "help" to deliver tryptophan, already in the bloodstream, to the brain because it elicits secretion of the hormone insulin. Insulin drives the *other* amino acids, but not tryptophan, into *body* cells, leaving the tryptophan free to enter the brain without competition. Thus, paradoxically, food high in carbohydrate—*not* food high in protein—eases tryptophan's transport into the brain and so promotes serotonin synthesis. Should we eat high-carbohydrate meals to get that relaxed feeling produced by serotonin release? The brain is not that simple. Later the Applications section describes practical uses of this information. The amount of protein in most mixed meals, even those high in carbohydrate, is generally more than sufficient to block the tryptophan-delivering effect of carbohydrate.[3]

Serotonin and Appetite

Researchers speculate that people's food choices may be influenced by the brain's reactions to prior foods eaten. When a meal elicits top mental performance, the brain may be "trained" to prefer those foods and so to choose them often. This theory offers a partial explanation for the natural preference, in the great majority of societies in the world, for mixed meals over single foods. Meals with mixed protein and carbohydrate sources may provide just the right balance to support brain function.

Animals regulate their intakes of protein and carbohydrate, each proportional to the other, in response to promptings from the brain. People probably do this, too, unconsciously. Depending on what kinds of foods they have eaten last, animals and humans seem to know what foods to choose next time to ensure balance. According to one theory, by raising brain serotonin, a high-carbohydrate meal satisfies a need and so reduces the urge to eat more carbohydrate. Therefore an animal or person who has eaten plenty of carbohydrate will seek out more pro-

Table C13•1

BRAIN TERMS

- **blood-brain barrier** a barrier composed of the cells lining the blood vessels in the brain. These cells are so tightly glued to each other that blood-borne substances cannot get into the brain between the cells, but only by crossing the cell bodies themselves. Thus, the cells, using all their sophisticated equipment, can screen substances for entry.
- **catecholamines** (CAT-eh-COAL-ah-meenz) neurotransmitters made from the amino acid tyrosine: dopamine, epinephrine, and norepinephrine.
- **neurotransmitters** chemical messengers released by nerve cells when the cells are firing (conducting a nerve impulse). The neurotransmitter diffuses to the next nerve cell and alters the membrane of that cell, making it either less or more likely to fire. Exposed to enough neurotransmitter molecules, the next nerve cell will fire.
- **precursor control** control of a compound's synthesis by the availability of that compound's precursor. The more precursor there is, the more of the compound is made.
- **serotonin** (SARE-oh-TONE-in) a compound related in structure to (and made from) the amino acid tryptophan. It serves as one of the brain's principal neurotransmitters.

tein at the next meal. A high-protein meal creates a serotonin deficit, awakens the carbohydrate craving, and once again leads to the consumption of carbohydrate-containing foods.

A chief appeal of this theory is that it accounts for typical reactions to low-carbohydrate diets. The more dieters try to restrict carbohydrate, the more they seem to crave it. Serotonin is often below average in the brains of obese people and of normal-weight people who report craving carbohydrate-rich food. Though direct cause-and-effect conclusions are not yet possible, researchers suggest that high-carbohydrate weight-loss diets may be the most successful because they work *with* many people's brain chemistry rather than against it.

Dieting in general seems to disrupt mental functioning somewhat. During dieting, people are easily distracted from tasks requiring vigilance and have slower reaction times. They also score lower on memory tests than at times of normal eating. Whether these effects are related to neurotransmitter synthesis or are equally likely to result from any sort of change in regular eating habits remains a mystery.

Serotonin, Mood, and Sleep

Some people say they feel anxious, tense, and somewhat depressed before eating carbohydrate and feel the opposite afterward. The amino acid tryptophan given by itself often has similar effects, consistent with the notion that it is the indirect agent of carbohydrate's effect. Tryptophan (and perhaps carbohydrate) also induces fatigue or sleepiness. This effect from tryptophan is well known from at least 50 studies in people and animals. When tryptophan is restricted in the diet, some people report depressed feelings. When tryptophan is restored, depressed feelings lift, but in their place come drowsiness, clumsiness, and mental slowing.

Reports connect disturbed serotonin metabolism in people with major depression. Researchers fed tryptophan-deficient diets to two groups of young men: one with long family histories of depression and a matched (control) group with no reported depression in the family. As blood tryptophan concentrations dropped, the men with family histories of depression scored significantly lower on a mood scale, indicating depression. None of the controls scored lower on the test—their moods hadn't changed. Major depression is a serious condition that is not reversible through diet alone, but people who tend to be depressed would do well to eat balanced meals at regular intervals to supply the brain with the materials needed to make serotonin.

Single amino acids, when administered alone, act like drugs in the body. People should not dose themselves with amino acids seeking mental effects. The Consumer Corner of Chapter 6 warned that to do so is to imperil health.

Effects of Vitamins, Minerals, and Lipids on the Brain

Nutrients other than amino acids are also involved in the synthesis of neurotransmitters. Iron is needed in one of the first steps of neurotransmitter synthesis. Vitamin B_6 and riboflavin are needed in later steps. These nutrients are but three among many; among the first signs of deficiencies of them are fatigue, depressed mood, or impaired learning. Deficiencies of many nutrients also cause anemia, which produces mental symptoms of its own (see Table C13-2). Administration of the missing nutrient rapidly reverses these effects.

An interesting idea under study concerns the possibility that lipids may affect human emotions. In one study, fish oil was shown to reduce aggression in college students under the stress of final exams.[4] Also, a deficiency of omega-3 fatty acids (EPA and DHA) may be associated with depression.[5] It is unclear, however, which comes first—inadequate intake altering the brain's activity or depression that alters fatty acid metabolism. Studies examining the behavioral effects of lowered blood cholesterol levels are inconclusive: some find that people receiving cholesterol-lowering drugs or dietary treatments are twice as likely as controls to die from suicide or violence; others detect no correlation between blood cholesterol and mood.[6] The idea that cholesterol in the diet might

Table C13•2

THE MENTAL SYMPTOMS OF ANEMIA

Apathy, listlessness
Behavior disturbances
Clumsiness
Hyperactivity
Irritability
Lack of appetite
Learning disorders (vocabulary, perception)
Low scores on latency and associative reactions
Lowered IQ
Reduced physical work capacity
Repetitive hand and foot movements
Shortened attention span

NOTE: These symptoms are not caused by anemia itself but by iron deficiency in the brain. Children with much more severe anemias from other causes, such as sickle-cell anemia and thalassemia, show no reduction in IQ when compared with children without anemia.

control neurotransmitters in the brain seems unlikely for two reasons: dietary cholesterol plays only a minor role in determining blood cholesterol, and blood cholesterol has limited passage through the blood-brain barrier.

Can Certain Foods, Herbs, or "Smart" Supplements Enhance Brain Functioning?

If deficiencies of nutrients can cause mental disturbances, can extra amounts of some nutrients make brain function excel? The idea has appeal, especially to businesspeople, students, textbook authors, medical workers, and others who must remain alert despite long hours, little sleep, or international travel.

Purveyors of "smart" drugs, supplements, and drinks say their products can speed thinking and learning, jump-start a failing memory, and reverse aging processes.[7] Some drugs sold with these claims are being tested for reversal of the mental deterioration of Alzheimer's disease. The theory is that if a drug can reverse or slow deterioration in a diseased brain, then perhaps it can boost the thinking power of a normal brain. An expert in the neurobiology of learning and memory commented concisely on smart drugs: "I think they are silly."

Experts at the Food and Drug Administration (FDA) warn that little research exists to support taking these products, and many have well-known side effects such as gastrointestinal distress, headaches, ulcers of the nasal cavity, and insomnia. Some herbs are known to be toxic, and the FDA warns that long-term effects of many other herbs and unapproved drugs are not known.

Shopping Mall "Smart" Foods

Users tell convincing stories about the effects of herbal "smart" drinks, but little clinical proof of their effectiveness exists. Researchers who attempt to measure people's feelings of being smart, witty, energetic, or able to remember run into problems. The placebo effect and wishful thinking always cloud measurements of this sort. Also, mental functioning of a person with slight nutrient deficiencies may respond dramatically to a potion that provides missing nutrients.

Purveyors of herb-enriched foods and drinks not only claim they will make people smarter, but also often promote them to people too young to buy alcohol as a way to obtain a legal "high" (euphoria) or as legal stimulants or depressants. Touted as a "brain-power" enhancer, an extract of the evergreen herb Ginkgo biloba shows up in pills of all kinds, in "smart" drinks and herbal smoothies sold to children, in herbal candy bars sold in stores, and in innumerable other products. This widespread availability does not mean that this herb is beneficial and safe.

Ginkgo Biloba

Some studies lend support to a theory that Ginkgo biloba's antioxidant effects may bring slight improvement to the mental functioning of the impaired aging brain, but other studies refute the idea.[8] Researchers are exploring the herb's effects on memory and attention in healthy young people as well.[9] Reports of safety problems have been accumulating, however, and they seem to constitute a serious threat. For example, a healthy woman who took Ginkgo biloba suffered convulsions, although she had no previous history of such occurrences. A formerly healthy man suffered a brain hemorrhage (a form of stroke) after taking the herb. Excessive bleeding following surgery has also been attributed to Ginkgo biloba.[10] People taking blood thinning medications should avoid Ginkgo biloba. Until studies rule out such dangers and prove benefits, consumers should read labels and avoid supplements and foods that contain Ginkgo biloba.

Ginseng

Despite ginseng's long history of use and worldwide popularity, little research exists to back up claims that it can provide "extra energy" or "restored vigor." No scientific tools for measuring "mental energy" exist, but when measured with standardized, double-blind, placebo-controlled studies of mood, memory, and critical thinking, ginseng's reported enhancement of physical and mental states eludes detection.[11] In one recent study, after eight weeks of chronic ginseng ingestion, no psychological benefits were found as compared with nonginseng controls.[12] When taken in limited doses, ginseng probably isn't harmful for most healthy people, except perhaps to their wallets.

Ephedrine

Ephedrine is freely available as a "dietary supplement," mostly sold for weight loss, but it is recognized as a potent heart and nerve stimulant with a clouded safety history. The FDA has received more than 800 reports of significant ephedrine-related injury, including heart attack, stroke, and even death. Ephedrine is similar to the prescription stimulant drug amphetamine in some ways.[13] It produces the stress response, with increased heartbeat, raised blood pressure, dilated bronchial passages in the lungs, interference with the action of the bladder, and other effects on organ systems. Does the stimulant effect of ephedrine provide feelings of extra "get up and go"? Probably. Is it safe? No. Read labels carefully and avoid any products with ephedrine, ephedra, or any other derivative of the word.

Other Nutrition-Mind Connections

Other inquiries on the mental effects of nutrition have suggested a role for ingestion of carbohydrate on the formation of memory and performance of mental tasks.[14] An increase in blood glucose may somehow enhance brain functioning and spark the formation of memory or improve recall. The very act of eating also enhances memory, although how it does so is not known. People given a snack during a learning task exhibit better recall later than when no snack is given. Hungry mice, fed immediately after learning a task, later remember how to perform the task better than do mice fed before the learning session. This effect may be a survival adaptation; in the wild, individuals who learn something new and obtain food as a result benefit from remembering the new behavior. Table C13-3 offers a memory quiz, just for fun—take it while snacking, and see if you do better than your age group.

What can we deduce from the almost universal love for the taste of chocolate? Most chocolate lovers say that eating chocolate lifts their spirits.[15] Chocolate contains phytochemicals, such as caffeine (a central nervous system stimulant), theobromine (another stimulant), and phenylethylamine (a biologically active amine), all of which could affect mood. The caffeine in chocolate provides a familiar "wake up" effect that may improve the ability to pay attention when taken in moderate doses; in large doses, caffeine produces the stress response that interferes with clear thinking.[16] One group of researchers suggested that an amine of chocolate may activate some of the same brain areas as marijuana.[17] Chocolate has none of the mind-altering effects of marijuana, but if it stimulates the brain in ways that enhance the sensory qualities of food, as marijuana is known to do, this effect might make chocolate seem extra delicious. The vast majority of studies, however, have not found that chocolate affects the brain's functioning or lifts mood. As for caffeine, its administration has produced both positive and negative effects on learning and memory.[18]

Table C13•3

MATCHING MEMORY TEST

Here's a typical short-term memory test. Carefully read through this list of 15 foods just once. Concentrate on each word. Then turn the page and write down as many of the items as you can remember.

onions	shrimp	mangoes
plums	tonic water	pasta
eggs	mayonnaise	ham
blackberries	basil	brownies
hazelnuts	zucchini	oatmeal

How did you do? The average 18- to 39-year-old can remember 10 of the items. It's nine for the average 40- to 59-year-old, eight for the average 60- to 69-year-old, and seven if you're 70 or older.

SOURCE: The Memory Assessment Clinic, Bethesda, Maryland.

Applications

Can a person choose supplements or foods to maximize the brain's performance? With regard to supplements and "smart drugs," the best advice is to hold on to your money. They don't work. With some qualifications, though, food choices may minimize diet-induced sluggishness, clumsiness, or poorer-than-normal memory capacity. The opposite food choices may also bring on those effects in some people who want to go to sleep.

If you tend to nod off during exams, then avoid foods extremely high in carbohydrate and low in protein in the hours before the test.[19] Otherwise, the carbohydrate might speed up the brain's production of serotonin, and serotonin's calming, sleep-inducing effect is exactly the wrong effect for the test taker who needs to be alert. However, if you are nervous and may suffer from poor performance due to stress, such foods may be just the calming influence you need for clear thinking.[20] Foods that provide protein along with carbohydrate do not induce serotonin synthesis because the protein is more than sufficient to block the tryptophan-delivering effects of the carbohydrate. So regular, mixed meals will have no effect on your serotonin levels.

Tryptophan itself may also work to reduce stress. A group of researchers recently tested the effects of two chocolate drinks on the stress responses of about 60 men and women.[21] One of the drinks was laced with a tryptophan-rich protein from milk, alpha-lactalbumin, and the other drink with a tryptophan-poor milk protein, casein. After drinking one of the drinks, subjects were exposed to stress (loud noise) and asked to perform challenging mathematical computations. Some people were stress-sensitive and experienced a rapid pulse, deteriorating mood, a rise in stress hormones, and a drop in brain serotonin. These changes occurred only when they consumed the tryptophan-poor drink. The researchers concluded that some stress-vulnerable people might improve their ability to cope by enhancing their intake of tryptophan-containing foods.

Research, as well as common sense, tells us that breakfast is of prime importance to a person taking an exam in the morning hours, while lunch is linked to mental performance in the afternoon. Although any effects from food on the neurotransmitters of the brain take at least an hour to become manifest, food in the digestive tract seems to have an immediate, unexplained effect on performance. No one knows why, but a common finding is the "post-lunch dip"—an early afternoon period of less-than-optimal mental performance. A prudent action for those taking tests in the early afternoon is to eat lightly at the lunch hour and save heavier eating for after the test.

If you are accustomed to taking caffeine in beverages, then do so on exam day. People missing their normal caffeine intakes experience headaches, drowsiness, and fatigue that interfere with mental tasks. For people not accustomed to caffeine, its addition will not improve performance. Alcohol worsens mental performance.

For those seeking the calming effect of serotonin to induce sleep, fueling the brain with tryptophan might be in order. A meal of foods that provide less than 1 gram of protein per 100 calories might serve this purpose, but the high-carbohydrate foods that induce serotonin carry a lot of calories. A person who eats an extra meal of such foods before bedtime must reduce intakes at other meals or risk gaining body fat. A better choice for inducing sleep is daily physical activity, which burns off calories while promoting healthy sleep.

Finally, the diet should be adequate to supply all the precursor and supporting nutrients needed for mental functioning.[22] No manipulation of energy nutrients will correct less-than-optimal brain functioning if needed nutrients are missing. Fluid is also important. Even slight dehydration can cause confusion.

With all of its unsolved mysteries, the human brain remains fascinating to researchers. The ideas presented here, and others like them, offer plenty of grist for the mills of researchers for many years to come.

Richard H. Pettibone, *Warhol's Campbell Soup Cans*

Food Safety and Food Technology

Frequently Asked Questions

How do germs in food cause illness in the body? p. 511

Which foods are most likely to make people sick? p. 520

How can I avoid illness when traveling? p. 523

If irradiation is so great, why aren't more foods irradiated? p. 526

Do pesticides on foods pose a hazard to consumers? p. 531

Do canned foods contain any nutrients? p. 536

Are artificial flavors and the flavor enhancer MSG safe to eat? p. 542

Contents

safety the practical certainty that injury will not result from the use of a substance.

hazard a state of danger; used to refer to any circumstance in which harm is possible under normal conditions of use.

food-borne illness illness transmitted to human beings through food and water; caused by a poisonous substance (*food intoxication*) or an infectious agent (*food-borne infection*). Also called *food poisoning*.

The intentional spreading of illness-causing organisms or their toxins through the food supply or by other means is a form of bioterrorism, defined in Chapter 11.

CONSUMERS IN the United States and Canada enjoy the safest, most pleasing, and most abundant food supply in the world. With this benefit, though, comes the consumer's responsibility to distinguish between paths leading to food **safety** and choices that pose a **hazard.**

The Food and Drug Administration (FDA) is the major agency charged with monitoring the U.S. food supply, but other agencies are involved as well (see Table 14-1). The following list indicates the FDA's areas of concern regarding our food supply:

1. *Microbial food-borne illness* affects the most people every year.
2. *Natural toxins in foods* constitute a hazard whenever people consume single foods either by choice (fad diets) or by necessity (poverty).
3. *Residues in food* include three types:
   a. *Environmental contaminants* (other than pesticides) such as household and industrial chemicals are increasing yearly in number and concentration, and their impacts are hard to foresee and to forestall.
   b. *Pesticides* are a subclass of environmental contaminants but are listed separately because they are applied intentionally to foods and, in theory, can be controlled.
   c. *Animal drugs* include metabolically active proteins that increase growth or milk production in food animals and dairy cows.
4. *Nutrients in foods* require close attention as more and more artificially constituted foods appear on the market.
5. *Intentional food additives* are listed last because so much is known about them that they pose virtually no hazard to consumers and because their use is well regulated.

Microbial **food-borne illness,** commonly called *food poisoning,* is first on the list because episodes of food poisoning far outnumber any other kind of food contamination. Food additives, last on the list, are of least concern. The others fall somewhere in between.

After the events of September 11, 2001, the need to address the threat of deliberate microbial contamination of the U.S. food supply is pressing.[1] To tighten security around the nation's food supply, the FDA has established guidelines for firms that produce, process, transport, or otherwise handle food.* The USDA has also created a Food Biosecurity Action Team to protect agriculture and other aspects of the food supply. Other agencies are also taking action, but details of the war against domestic bioterrorism are beyond the scope of this discussion.

*FDA guidance documents on biosecurity are available on the Internet at www.cfsan.fda.gov//dms/guidance.html.

Table 14•1

AGENCIES THAT MONITOR THE U.S. FOOD SUPPLY

- **CDC (Centers for Disease Control and Prevention)** a branch of the Department of Health and Human Services that is responsible for monitoring food-borne diseases.
- **EPA (Environmental Protection Agency)** the federal agency that is responsible for regulating pesticides and establishing water quality standards.
- **FDA (Food and Drug Administration)** the part of the Department of Health and Human Services' Public Health Service that is responsible for ensuring the safety and wholesomeness of all foods sold in interstate commerce except meat, poultry, and eggs (which are under the jurisdiction of the USDA); inspecting food plants and imported foods; and setting standards for food consumption.
- **USDA (U.S. Department of Agriculture)** the federal agency that is responsible for enforcing standards for the wholesomeness and quality of meat, poultry, and eggs produced in the United States; conducting nutrition research; and educating the public about nutrition.
- **WHO (World Health Organization)** an international agency that develops standards to regulate pesticide use. A related organization is the FAO (Food and Agricultural Organization).

With the privilege of abundance comes the responsibility to choose wisely.

enterotoxins poisons that act upon mucous membranes, such as those of the digestive tract.

neurotoxins poisons that act upon the cells of the nervous system.

botulism an often-fatal food poisoning caused by botulinum toxin, a toxin produced by the *Clostridium botulinum* bacterium that grows without oxygen in nonacidic canned foods.

This chapter focuses on actions of the individual to promote food safety and addresses FDA's areas of concern listed on page 510. It begins with FDA's highest priority—the serious and prevalent threat of food-borne illness.

On average, each day:

- Over 200,000 people in the United States fall ill with a food-borne illness.
- Of those, 14 die from the illness.

Microbes and Food Safety

Some people brush off the threat from food-borne illnesses as less likely and less serious than the threat of flu, but they are misinformed—food-borne illnesses can be life-threatening and are increasingly unresponsive to standard antibiotics.[2] Each year in the United States, an estimated 76 million people become ill from food-borne diseases, and about 5,000 of them die.[3] Within these numbers lies some good news—because of improved safety procedures by food producers, incidences of disease from two problem organisms are on the decline.[4] By taking a few preventive steps, you can minimize your chances of contracting food-borne illnesses.

Get medical help when these symptoms occur:

- Bloody stools.
- Diarrhea of more than 3 days' duration.
- Difficulty breathing.
- Difficulty swallowing.
- Double vision.
- Fever of longer than 24 hours' duration.
- Headache accompanied by muscle stiffness and fever.
- Numbness, muscle weakness, and tingling sensations in the skin.
- Rapid heart rate, fainting, and dizziness.

How Do Germs in Food Cause Illness in the Body?

Microorganisms can cause food-borne illness either by infection or by intoxication. Infectious agents such as *Salmonella* multiply and infect the tissues of the human body. Other microorganisms in foods produce **enterotoxins** or **neurotoxins,** poisonous substances that are absorbed into the body causing food intoxication. These microorganisms can multiply or create toxins in food during improper preparation or storage or within the digestive tract after a person eats contaminated food. If the digestive tract disturbances listed in Table 14-2 (pages 512–514) are the major or only symptoms of your next bout of "stomach flu," chances are excellent that what you really have is a food-borne illness. For people who are ill or malnourished, have a compromised immune system, or are very old or very young, even mild disturbances can be fatal.

The symptoms of one neurotoxin poisoning stand out as severe and commonly fatal—those of **botulism.** Botulism is caused by the toxin of the *Clostridium botulinum* bacterium, which grows in improperly canned (and especially home-canned) foods, improperly prepared or stored vacuum-packed foods, or oils flavored with herbs, garlic, vegetables, or other flavoring agents and stored at room temperature. The

Table 14•2

FOOD-BORNE ILLNESSES

Disease and Organism That Causes It	Yearly Occurrence and Deaths[a]	Most Frequent Food Source	Onset and General Symptoms	Prevention Methods
Food-Borne Infections				
Campylobacteriosis *Campylobacter jejuni* bacterium	2 million; 100 deaths	Raw poultry, beef, lamb, unpasteurized milk (foods of animal origin eaten raw or undercooked or recontaminated after cooking).	Onset: 2 to 5 days. Diarrhea, nausea, vomiting, abdominal cramps, fever; sometimes bloody stools; lasts 7 to 10 days; rarely, nervous system paralysis (Guillain-Barré syndrome).	Cook foods (especially poultry) thoroughly; use pasteurized milk; use sanitary food-handling methods.
Cryptosporidiosis *Cryptosporidium parvum* microscopic parasite	30,000; 7 deaths. (In 1993, a Wisconsin water-borne outbreak affected more than 400,000 people.)	Commonly, swimming or drinking contaminated water, even from treated sources. Highly chlorine-resistant. Contaminated raw produce and unpasteurized juices and ciders.	Onset: 2 to 10 days. Diarrhea, loose or watery stools, stomach cramps, upset stomach, slight fever. Symptomless sufferers can pass the infection to others.	Wash all raw vegetables and fruits with uncontaminated water before peeling. Do not swallow drops of water while using pools, hot tubs, ponds, lakes, rivers, or streams for recreation.
Cyclosporiasis *Cyclospora cayetanensis* single-cell parasite	15,000; 8 deaths	Contaminated water; contaminated fresh produce.	Onset: average, 7 days. Watery diarrhea, loss of appetite, weight loss, stomach cramps, nausea, vomiting, muscle aches, low-grade fever, fatigue. Symptomless sufferers can spread the infection.	In areas of uncertain sanitation, drink only treated or boiled water, and eat only cooked hot foods or fruits you peel yourself.
Hemolytic-uremic syndrome *Escherichia coli (E. coli)* 0157:H7 bacterium	62,500; 50 deaths	Undercooked ground beef, unpasteurized milk and milk products, contaminated water, unpasteurized juices or cider, contaminated produce (especially alfalfa sprouts), and person-to-person contact.	Onset: 12 to 72 hr. Severe bloody diarrhea, abdominal cramps, acute kidney failure; death. Survivors may face kidney problems, hypertension, blindness, paralysis, and colon problems.	Cook ground beef thoroughly; avoid unpasteurized milk and juice products; use sanitary food-handling methods; use treated, boiled, or bottled water. Susceptible people should avoid alfalfa sprouts.
Hepatitis Hepatitis A virus	4,200; 4 deaths	Undercooked or raw shellfish; baked goods or other foods contaminated by infected food handlers.	Onset: 15 to 50 days (28 to 30 days average). Inflammation of the liver; fatigue; nausea, vomiting, or indigestion; jaundice (yellowed skin and eyes from buildup of wastes); muscle pain.	Cook foods thoroughly.
Listeriosis *Listeria monocytogenes* bacterium	2,500; 500 deaths	Raw meat and seafood, raw milk, and soft cheeses.	Onset: 7 to 30 days. Mimics flu; blood poisoning; meningitis (stiff neck, severe headache, and fever); miscarriage of pregnancy; severe illness or death of newborn.	Use sanitary food-handling methods; cook foods thoroughly; use pasteurized milk.

[a]Estimated data from Diseases and pathogens under surveillance, Centers for Disease Control and Prevention available at www.cdc.gov/foodnet/pus.htm; P. S. Mead and coauthors, Food-related illness and death in the United States, *Emerging Infectious Diseases* 5 (2000), available at www.cdc.gov/ncidod/eid/vol5no5/mead.htm.

Table 14•2

FOOD-BORNE ILLNESSES (continued)

Disease and Organism That Causes It	Yearly Occurrence and Deaths[a]	Most Frequent Food Source	Onset and General Symptoms	Prevention Methods
Food-Borne Infections				
Salmonellosis *Salmonella* bacteria	1–34 million; 600 deaths	Raw or undercooked eggs, meats, poultry, milk and other dairy products, shrimp, frog legs, yeast, coconut, pasta, produce, chocolate, and unpasteurized juices.	Onset: 6 to 48 hr. Nausea, fever, chills, vomiting, abdominal cramps, diarrhea, headache; can be fatal.	Use sanitary food-handling methods; use pasteurized milk; cook foods thoroughly; refrigerate foods promptly and properly.
Shigellosis *Shigella* bacteria varieties	90,000; 14 deaths	Contaminated food (may look and smell normal); produce and other foods contaminated by poor sanitation practices of infected farm workers or food handlers, sewage fertilizer in growing fields, or exposure to flies or other insects. Contaminated drinking or swimming water.	Onset: 1 to 2 days. Diarrhea, fever, stomach cramps. The diarrhea is often bloody. In young children, high fever, seizures. Symptomless sufferers can spread the bacteria to others.	Frequent and careful hand washing with soap. Those with shigellosis should not prepare food or beverages for others. In areas of uncertain sanitation, drink only treated or boiled water, and eat only cooked hot foods or fruits you peel yourself.
"Stomach flu"[b] (mistakenly called) Norwalk-type viruses	9.2 million; 124 deaths	Foods, such as sandwiches and salads, contaminated by infected food handlers. Contaminated produce. Oysters from waters contaminated with human sewage, such as boat bilge.	Onset: 18 to 72 hr. Acute digestive illness, with pain, vomiting, possibly diarrhea, headache, and low-grade fever.	Choose restaurants that pass health department inspections and enforce worker sanitation. If uncertain, order cooked foods served steaming hot. Avoid raw oysters.
"Stomach flu"[c] (mistakenly called) *Vibrio parahaemolyticus* and other *Vibrio* bacteria	5,000; 31 deaths	Raw or undercooked shellfish, often oysters. Less commonly, skin infection when an open wound is exposed to warm seawater.	Onset: 24 hr. Watery diarrhea, abdominal cramping, nausea, vomiting, fever and chills.	Cook shellfish well, especially oysters. Purchase shellfish from reputable dealer. Avoid exposing wounds to warm seawater.
Traveler's diarrhea A variety of microorganisms including *Giardia* and other protozoa.	10 million international travelers affected	Contaminated water, undercooked ground beef, raw foods, imported unpasteurized soft cheeses.	Onset: 12 hr. to several days. Loose and watery stools, nausea, vomiting bloating, abdominal cramps.	Cook foods thoroughly; use safe, treated water and pasteurized milk; wash raw fruits and vegetables or avoid them in areas of uncertain sanitation.
Trichinosis *Trichinella spiralis* parasite	50; no deaths	Raw or undercooked pork or wild game (bear). Worms burrow through the body tissues to reach muscle tissue where they remain alive.	Onset: 24 hr. Abdominal pain, nausea, vomiting, diarrhea, and fever. One to two weeks later, muscle pain, low-grade fever, pain on breathing, edema (swelling), skin eruptions, loss of appetite, and weight loss. Drug therapy kills the worms, and deaths are rare.	Cook foods thoroughly.

[b]Though popularly called "stomach flu," the digestive disturbances caused by Norwalk-type viruses are unrelated to influenza.
[c]Though popularly called "stomach flu," the digestive disturbances caused by *Vibrio* organisms are unrelated to influenza.

Continued on next page

Table 14•2

FOOD-BORNE ILLNESSES (continued)

Disease and Organism That Causes It	Yearly Occurrence and Deaths[a]	Most Frequent Food Source	Onset and General Symptoms	Prevention Methods
Food Intoxications				
Botulism Botulinum toxin (produced by the *Clostridium botulinum* bacterium)	60; 4 deaths	Anaerobic environment of low acidity (canned corn, peppers, green beans, soups, beets, asparagus, mushrooms, ripe olives, spinach, tuna, chicken, chicken liver, liver paté, luncheon meats, ham, sausage, stuffed eggplant, herb-flavored oils, lobster, and smoked and salted fish).	Onset: 4 to 36 hr. Nervous system symptoms, including double vision, inability to swallow, speech difficulty, and progressive paralysis of the respiratory system; often fatal; leaves prolonged symptoms in survivors.	Use proper canning methods for low-acid foods; avoid commercially prepared foods with leaky seals or with bent, bulging, or broken cans.
Staphylococcal food poisoning Staphylococcal toxin (produced by the *Staphylococcus aureus* bacterium)	185,100; 2 deaths	Toxin produced in meats, poultry, egg products, tuna, potato, and macaroni salads, and cream-filled pastries.	Onset: ½ to 8 hr. Diarrhea, nausea, vomiting, abdominal cramps, fatigue; mimics flu; lasts 24 to 48 hr; rarely fatal.	Use sanitary food-handling methods; cook food thoroughly; refrigerate foods promptly and properly.

pasteurization the treatment of milk with heat sufficient to kill certain pathogens (disease-causing microbes) commonly transmitted through milk; not a sterilization process. Pasteurized milk retains bacteria that cause milk spoilage. Raw milk, even if labeled "certified," transmits many food-borne diseases to people each year and should be avoided.

microbe grows only in the absence of oxygen, in low-acid conditions, and at temperatures that support the growth of most bacteria—40° to 120° Fahrenheit.

Botulism occurs rarely, but its symptoms constitute an immediate medical emergency (see the table above). Even with medical assistance, survivors can suffer symptoms for months, years, or a lifetime. So potent is the botulinum toxin that an amount as tiny as a single grain of salt can kill several people within an hour. The botulinum toxin is destroyed by heat, so canned foods that contain the toxin can be rendered harmless by boiling them for ten minutes. Food can be canned safely at home if proper canning techniques are followed to the letter.*

Each year in the United States, many millions of people suffer from mild to life-threatening symptoms caused by food-borne illness.

For safety, when making flavored oils, wash and dry the herbs before adding them to the oil and keep the oil refrigerated.

Food Safety from Farm to Table

Figure 14-1 shows that careful food handling is required to prevent microbes from becoming a problem—on the farm, in processing plants, during transportation, and at supermarkets and restaurants. Equally critical to the chain of food safety, however, is the final handling of food by people who purchase it and consume it at home.

The overwhelming majority of food-poisoning cases result from errors consumers make in handling foods *after* purchase. Commercially prepared food is usually safe but rare accidents do occur, and they often affect many people at once. Dairy farmers, for example, rely on **pasteurization,** a process of heating milk to kill many disease-causing organisms and make milk safe for consumption. When, on occasion, a major dairy develops flaws in its pasteurization system, tens of thousands of cases of food-borne illness may result. Other types of farming require other safeguards. Growing food usually involves soil, and soil is made up partly of bacterial colonies, making contamination of food likely.

*Complete, up-to-date, home canning instructions are available in the USDA's 172-page *Complete Guide to Home Canning*, available from the Superintendent of Documents, Government Printing Office, Washington, DC 20402.

Figure 14•1

FLOW OF FOOD SAFETY: FARM TO TABLE

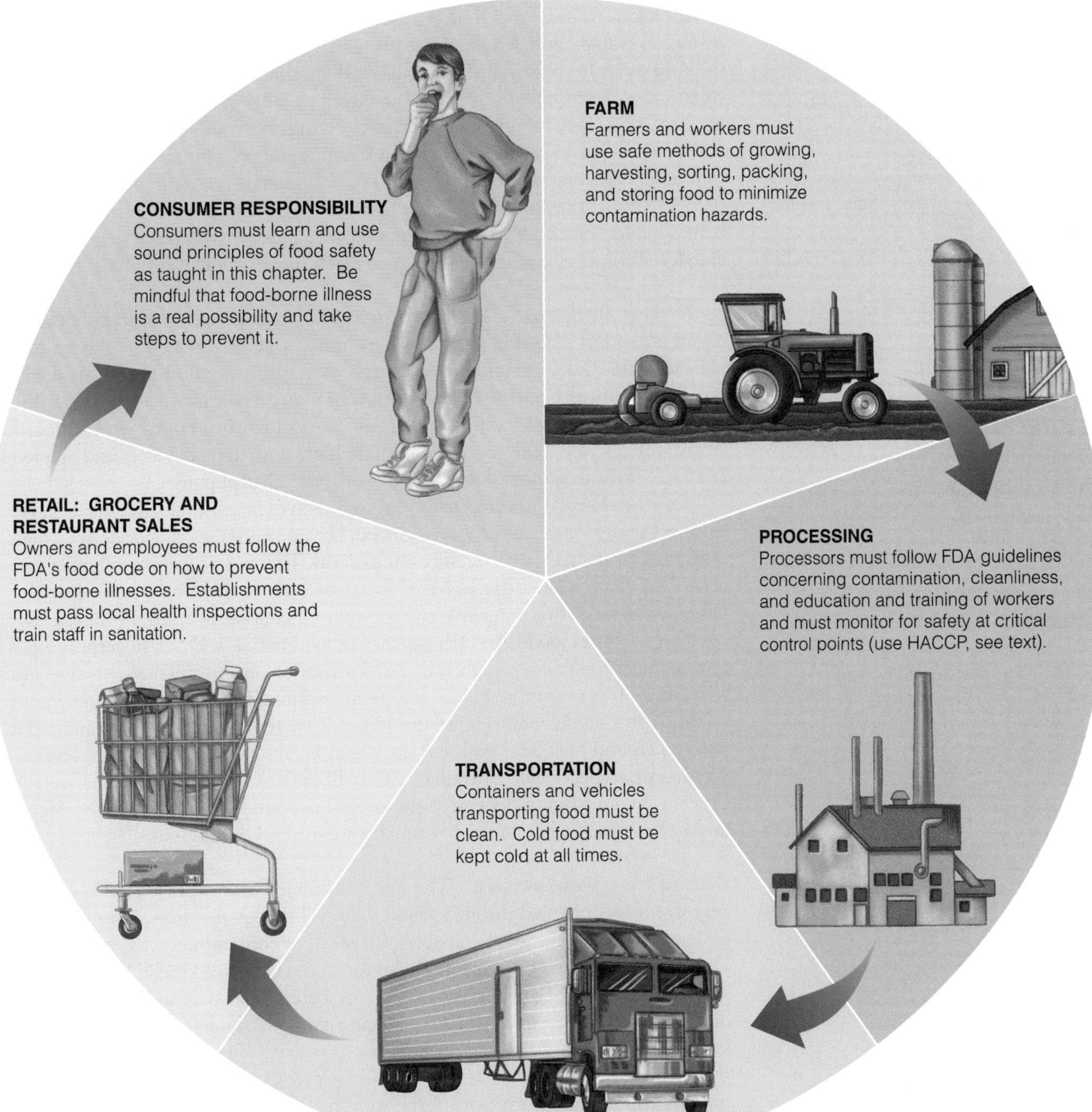

FARM
Farmers and workers must use safe methods of growing, harvesting, sorting, packing, and storing food to minimize contamination hazards.
PROCESSING
Processors must follow FDA guidelines concerning contamination, cleanliness, and education and training of workers and must monitor for safety at critical control points (use HACCP, see text).
TRANSPORTATION
Containers and vehicles transporting food must be clean. Cold food must be kept cold at all times.
RETAIL: GROCERY AND RESTAURANT SALES
Owners and employees must follow the FDA's food code on how to prevent food-borne illnesses. Establishments must pass local health inspections and train staff in sanitation.
CONSUMER RESPONSIBILITY
Consumers must learn and use sound principles of food safety as taught in this chapter. Be mindful that food-borne illness is a real possibility and take steps to prevent it.

Hazard Analysis Critical Control Point (HACCP) a systematic plan to identify and correct potential microbial hazards in the manufacturing, distribution, and commercial use of food products.

biotechnology the science that manipulates biological systems or organisms to modify their products or components or create new products (see the chapter Controversy).

biosensor a genetically altered microbe that provides a rapid, low-cost, and accurate test for toxic products of microbial agents in foods.

DNA fingerprinting a laboratory analysis of DNA that reveals the genetic pattern unique to an organism. The pattern is then compared to others (similar to grocery bar codes) to find a match, thus identifying the organism.

Attention on *E. coli* In the mid-1990s, a fast-food restaurant chain in the Northwest served undercooked hamburgers from meat contaminated with the dangerous bacterium *E. coli* 0157:H7 during processing. Four people died, and hundreds of other patrons were stricken with serious illness. News coverage focused the national spotlight on two important food safety issues: live, disease-causing organisms of many types are routinely found in raw meats, and thorough cooking is necessary to make animal-derived foods safe. Revelations about *E. coli* 0157:H7 have led to a much needed overhaul of the country's mechanisms for ensuring meat safety.

Infections from *E. coli* 0157:H7 cause severe illness, with bloody diarrhea, severe intestinal cramps, and dehydration setting in a few days after eating tainted meat, raw milk, or even fresh berries or organic produce that has been contaminated. In the worst cases, *hemolytic-uremic syndrome* develops with abnormal blood clotting that can lead to kidney failure, damage to the central nervous system and other organs, and death.[5] Obtain medical help when symptoms appear (see margin), and be aware that antibiotics may worsen the condition.[6] Children given antibiotics to treat *E. coli* 0157:H7 infection were seven times more likely than untreated children to develop hemolytic-uremic syndrome. The drugs seem to make the bacterial toxin more available for absorption and thus enable it to reach higher concentrations in body tissue.

Obtain medical help for these symptoms *E. coli* 0157:H7:

- Bloody diarrhea.
- Severe intestinal cramps.
- Dehydration.

More on biotechnology in this chapter's Controversy.

Industry Controls Media reports and public concern about food-borne illness resulted in a law requiring that producers of meat, poultry, seafood, fresh fruit juices, and vegetable juices employ an effective prevention method known as a **Hazard Analysis Critical Control Point (HACCP)** plan.[7] Each producer, packer, distributor, and transporter of food must identify "critical control points" in its procedures where the risk of food contamination is high and then develop and implement HACCP plans to prevent loss of control at those critical points.

Meat and seafood inspectors had always relied on their senses of sight, smell, and touch to detect tainted meat, poultry, and seafood, but human senses cannot detect dangerous organisms until after the food has begun to decay. Thanks to advances in **biotechnology,** geneticists have altered the DNA of living bacteria to yield **biosensor** organisms. A biosensor detects tiny amounts of metabolic chemicals released by disease-causing microorganisms in foods. Another advance in food testing is **DNA fingerprinting.** This test can quickly reveal the identity of a contamination organism. These sensitive, specific tests can help to verify that HACCP plans are working to reduce food contamination.

Since the implementation of the HACCP system, *Salmonella* contamination of poultry, ground beef, and pork has decreased by almost 50 percent, 40 percent, and 25 percent, respectively.[8] Encouraging data from the Centers for Disease Control and Prevention also show a decline in the overall incidence of *Campylobacter* infections, another common cause of food-borne disease in the United States.[9]

Consumer Protection The safety of canned and packaged foods sold in grocery stores is controlled through sound food technology practices, but rare accidents do happen. Batch numbering enables the recall of contaminated foods through public announcements via newspapers, television, and radio, and the FDA monitors large suppliers. You can help protect yourself, too. Carefully inspect the seals and wrappers of packages. Reject open, leaking, or bulging cans, jars, and packages. Many jars have safety "buttons" on the lid, designed to pop up once the jar is opened; make sure that they are firmly sealed. If a package on the shelf looks ragged, soiled, or punctured, do not buy the product; turn it in to the store manager. A badly dented can or a mangled package is useless in protecting food from microorganisms, insects, spoilage, or even vandals. Frozen foods should be solidly frozen, and those in a chest-type freezer case should be stored below the frost line. See the margin for some food safety myths than often make consumers sick.

Myths that often make consumers sick:

- "If it tastes okay, it's safe to eat."
- "We have always handled our food this way and nothing has ever happened."
- "I sampled it a couple of hours ago and didn't get sick, so it should be safe to eat."

key point *Industry employs sound practices to safeguard the commercial food supply from microbial threats. Still, incidents of commercial food-borne illness have caused widespread harm to health.*

Food Safety in the Kitchen

Large-scale commercial incidents make up only a fraction of the nation's total food-poisoning cases each year. Most cases arise from one person's error in a small setting and affect just a few victims. Some people have come to accept a yearly bout or two of intestinal illness as inevitable, but these illnesses can and should be prevented. Just for fun, take the food safety quiz in Table 14-3 to see how well you follow food safety rules. Then read on to learn how you can do your part to make meals at home as safe as they can be.

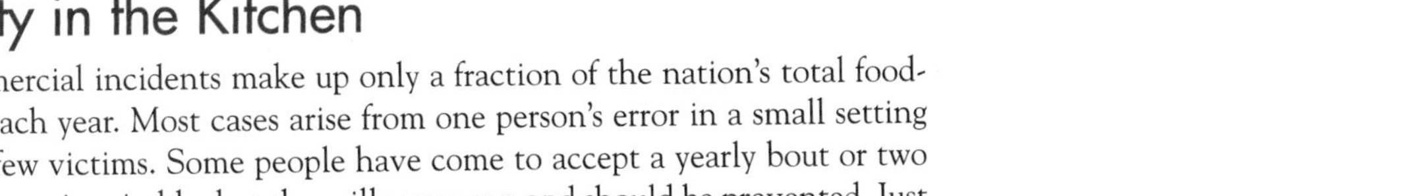

Table 14•3

CAN YOU PASS THE KITCHEN FOOD SAFETY QUIZ?

1. The temperature of the refrigerator in my home is:
   A. 50°F (10° Celsius).
   B. 40°F (5°C).
   C. I don't know; I don't own a refrigerator thermometer.
2. The last time we had leftover cooked stew or other meaty food, the food was:
   A. cooled to room temperature, then put in the refrigerator.
   B. put in the refrigerator immediately after the food was served.
   C. left at room temperature overnight or longer.
3. If a cutting board is used in my home to cut raw meat, poultry, or fish and it is going to be used to chop another food, the board is:
   A. reused as is.
   B. wiped with a damp cloth or sponge.
   C. washed with soap and water.
   D. washed with soap and hot water and then sanitized.
4. The last time I had a hamburger, I ate it:
   A. rare.
   B. medium.
   C. well-done.
5. The last time there was cookie dough where I live, the dough was:
   A. made with raw eggs, and I sampled some of it.
   B. store-bought, and I sampled some of it.
   C. not sampled until baked.
6. I clean my kitchen counters and food preparation areas with:
   A. a damp sponge that I rinse and reuse.
   B. a clean sponge or cloth and water.
   C. a clean cloth with hot water and soap.
   D. the same as above, then a bleach solution or other sanitizer.
7. When dishes are washed in my home, they are:
   A. cleaned by an automatic dishwasher and then air-dried.
   B. left to soak in the sink for several hours and then washed with soap in the same water.
   C. washed right away with hot water and soap in the sink and then air-dried.
   D. washed right away with hot water and soap in the sink and immediately towel-dried.
8. The last time I handled raw meat, poultry, or fish, I cleaned my hands afterward by:
   A. wiping them on a towel.
   B. rinsing them under warm tap water.
   C. washing with soap and water.
9. Meat, poultry, and fish products are defrosted in my home by:
   A. setting them on the counter.
   B. placing them in the refrigerator.
   C. microwaving and cooking promptly when thawed.
   D. soaking them in warm water.
10. I realize that eating raw seafood poses special problems for people with:
    A. diabetes.
    B. HIV infection.
    C. cancer.
    D. liver disease.

Answers

1. Refrigerators should stay at 40°F or less, so if you chose answer B, give yourself two points; zero for other answers.
2. Answer B is the best practice; give yourself two points if you picked it; zero for other answers.
3. If answer D best describes your household's practice, give yourself two points; if C, one point.
4. Give yourself two points if you picked answer C; zero for other answers.
5. If you answered A, you may be putting yourself at risk for infection from bacteria in raw shell eggs. Answer C—eating the baked product—will earn you two points and so will answer B. Commercial products are made with pasteurized eggs.
6. Answers C or D will earn you two points each; answer B, one point; answer A, zero.
7. Answers A and C are worth two points each; other answers, zero.
8. The only correct practice is answer C. Give yourself two points if you picked it; zero for others.
9. Give yourself two points if you picked B or C; zero for others.
10. This is a trick question: all of the answers apply. Give yourself two points for knowing one or more of the risky conditions.

Rating Your Home's Food Practices

20 points: Feel confident about the safety of foods served in your home.
12 to 19 points: Re-examine food safety practices in your home. Some key rules are being violated.
11 points or below: Take steps immediately to correct food-handling, storage, and cooking techniques used in your home. Current practices are putting you and other members of your household in danger of food-borne illness.

SOURCE: Adapted from U.S. Food and Drug Administration, Can your kitchen pass the food safety test? *FDA Consumer*, October 1998.

Figure 14•2

FIGHT BAC!

Four Ways to Keep Food Safe

Food can provide ideal conditions for bacteria to thrive or to produce toxins. Disease-causing bacteria require warmth (40° to 140°F), moisture, and nutrients. To defeat bacteria, deprive them of one of these conditions—usually, their temperature range. Remember these four "keepers": keep hot foods hot, keep cold foods cold, keep raw foods separate, and keep your hands and the kitchen clean (see Figure 14-2).

Keep Hot Foods Hot Keeping hot foods hot includes cooking foods long enough to reach an internal temperature that will kill microbes. To alert consumers to this fact, the USDA invented "Thermy," the cartoon character in the margin who urges the use of a thermometer to test the temperatures of cooked foods. Figure 14-3 illustrates the safe internal temperatures of cooked foods and various types of thermometers. Table 14-4 provides a glossary of thermometer terms.

After cooking, foods must be held at 140°F or higher until served because cooking does not destroy all bacterial toxins. Even hot cooked foods, if handled improperly prior to serving, can cause illness. Delicious looking meatballs on a buffet may harbor bacteria unless they have been kept steaming hot. Food at 140°F feels hot, not just warm. After the meal, cooked foods should be refrigerated immediately or definitely before two hours have passed (one hour if room temperature approaches 90°F). If food has been left out longer than this, toss it out.

Keep Cold Foods Cold Keeping cold foods cold starts when you leave the grocery store. If you are running errands, shop last so that the groceries do not stay in the car too long. (If ice cream begins to melt, it has been too long.) Upon arrival home, load foods into the refrigerator or freezer immediately. Keeping foods cold applies to defrosting foods, too. Thaw meats or poultry in the refrigerator, not at room temperature, and marinate meats in the refrigerator, too. Table 14-5 on page 520 lists some safe keeping times for foods stored at or below 40°F.

Any food with an "off" appearance or odor should not be used or even tasted. Most hazards are not detectable by odor, taste, or appearance, however, so you cannot rely on your senses of smell and sight alone to warn you.

cross-contamination the contamination of a food through exposure to utensils, hands, or other surfaces that were previously in contact with a contaminated food.

Requirements of disease-causing bacteria: warmth, moisture, and nutrients.

Keep Raw Foods Separate Keeping raw foods separate means preventing **cross-contamination** of foods. Raw foods, especially meats, eggs, and seafood, are likely to contain bacteria. To prevent them from spreading, keep the raw foods and their juices away from ready-to-eat foods. For example, if you take burgers out to the grill on a plate, wash that plate in hot, soapy water before using it to hold the cooked burgers. If you use a cutting board to cut raw meat, wash the board, the knife, and your hands thoroughly before using the utensils to make a salad or other foods that are eaten raw.

The USDA's "Thermy" character is intended to encourage consumers to use a food thermometer to test for doneness.

Keep Your Hands and the Kitchen Clean Keeping your hands and the kitchen clean requires using freshly washed utensils and laundered towels and washing your hands with warm water and soap for a minimum of 20 seconds before food handling. Wash your hands again after handling raw meat or poultry. If you are ill or have open sores, stay away from food. Clean equipment frequently and effectively. Be especially careful to wash surfaces and utensils in hot, soapy water after contact with raw meats. Microbes love to nestle down in small, damp spaces such as the inner cells of sponges or the pores between the fibers of wooden cutting boards. Antibacterial sponges, cloths, boards, and utensils possess a chemical additive intended to prevent rapid bacterial growth, but the protection is not perfect, so these products still need special handling to keep them safe. You can ensure the safety of cutting boards and reduce the microbes in sponges by washing them in a dishwasher or by treating them as suggested below. Alternatively, save the sponges for car washing and other heavy cleaning chores, and clean the kitchen with washable dishcloths that can be laundered often.

Figure 14•3

FOOD SAFETY TEMPERATURES (FAHRENHEIT) AND HOUSEHOLD THERMOMETERS

Different thermometers do different jobs. To choose the right one, pay attention to its temperature range: some have high temperature ranges intended to test the doneness of meats and other hot foods. Others have lower ranges for testing temperatures of refrigerators and freezers.

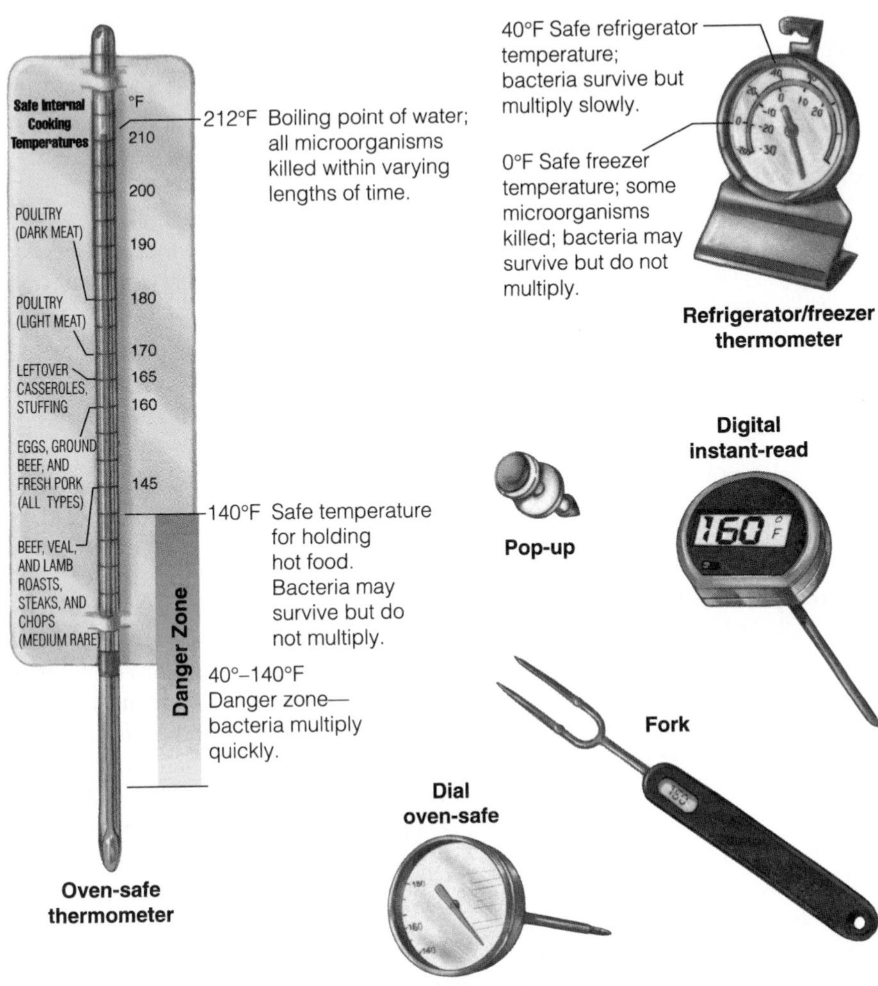

Table 14•4

GLOSSARY OF THERMOMENTER TERMS

- **appliance thermometer** a thermometer that verifies the temperature of an appliance. An *oven thermometer* verifies that the oven is heating properly; a *refrigerator/freezer thermometer* tests for proper refrigerator (<40°F) or freezer temperature (0°F).
- **fork thermometer** a utensil combing a meat fork and an instant-read food thermometer.
- **instant-read thermometer** a thermometer that, when inserted into food, measures its temperature within seconds; designed to test temperature of food at intervals, and not to be left in food during cooking.
- **oven-safe thermometer** a thermometer designed to remain in the food to give constant readings during cooking.
- **pop-up thermometer** a disposable timing device commonly used in turkeys. The center of the device contains a stainless steel spring that "pops up" when food reaches the right temperature.
- **single-use temperature indicator** a type of instant-read thermometer that changes color to indicate that the food has reached the desired temperature. Discarded after one use, they are often used in retail food markets to eliminate cross-contamination.

Wash your hands with warm water and soap before preparing food to reduce the chance of microbial contamination.

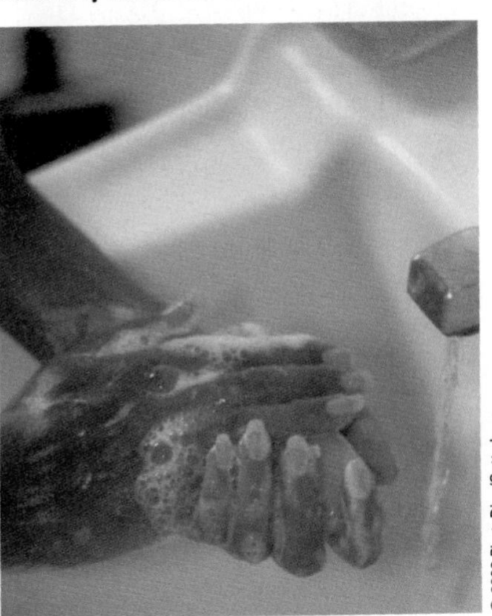

mad cow disease an often-fatal illness of cattle affecting the nerves and brain. Also called *bovine spongiform encephalopathy (BSE)*.

foot-and-mouth disease a severely disabling contagious viral disease that threatens domestic and wild animals. Foot-and-mouth disease rarely affects human beings and produces only mild symptoms in people.

To eliminate microbes in your kitchen, you have three choices, each with benefits and drawbacks. One is to poison the microbes on cutting boards, sponges, and other equipment with toxic chemicals such as bleach (one teaspoon per quart of water). The benefit is that chlorine can kill even the hardiest organism. The drawback is that chlorine is toxic to handle, can ruin clothing, and washes down household drains into the water supply and forms chemicals that can harm waterways and fish.

A second option is to treat kitchen equipment with heat. Soapy water heated to 140°F kills most harmful organisms and washes most others away. This method takes effort, though, since you have to use truly scalding water heated well beyond the temperature of the tap. Third, an automatic dishwasher can combine both methods: it washes in water hotter than hands can tolerate, and most dishwasher detergents contain chlorine but, of course, with the environmental disadvantage that chlorine entails. Pick one of these strategies to ensure safe implements for food preparation.

key point *To prevent food-borne illness, always remember that it can happen. Keep hot foods hot, keep cold foods cold, keep raw foods separate, and keep your hands and the kitchen clean.*

Which Foods Are Most Likely to Make People Sick?

Some foods are more hospitable to microbial growth than others. Foods that are high in moisture and nutrients and those that are chopped or ground are especially favorable hosts.

Meats and Poultry Raw meats and poultry require special handling, and packages bear labels to instruct consumers on meat safety (see Figure 14-4).* Meats in the grocery cooler often contain all sorts of bacteria, and they provide a moist, nutritious environment that is just right for microbial growth. Ground meat or poultry is handled more than meats left whole and exposes much more surface area for bacteria to land on, so experts advise cooking it well-done. Use a thermometer to test the internal temperature of poultry and meats, even hamburgers, before declaring them done. Burgers often turn brown and appear cooked before their internal temperature is high enough to kill harmful bacteria.[10]

Consumers may worry about animal diseases from overseas threatening their health. In past decades, an illness affecting the nervous systems of cattle, *bovine spongiform encephalopathy (BSE)* or **mad cow disease,** was linked with a rare but fatal brain disorder in a few people who consumed products from infected cattle.† The infective agent concentrates in the nervous system and digestive tissues of infected animals, making the consumption of foods such as sausages that may contain such parts especially risky.‡[11] The infection swept through herds across Great Britain and into Europe fueled by the now-banned practice of enriching cattle feed with bone meal and protein derived from cattle brains and other cattle parts. Since the emergence of mad cow disease, many tens of thousands of BSE-infected cattle have been destroyed.

To date, no cases of BSE have been observed in the United States, and no-nonsense safeguards are in place to prevent its occurrence.[12] Likewise, the 2001 European outbreak of **foot-and-mouth disease,** an infection debilitating to animals but having little effect on people, has not touched U.S. cattle. Thus, for U.S. beef eaters, animal diseases pose almost no threat.

Eggs Over the past ten years, increased occurrences of *Salmonella* of a most virulent type have been detected in blood samples from food-poisoning victims and samples of illness-causing foods.[13] Raw, unpasteurized eggs are most likely to be

*The USDA's meat and poultry hotline answers questions about meat and poultry safety: (800) 535-4555.

†The human disease is variant Creutzfeldt-Jakob disease (vCJD).

‡The agent is believed to be a *prion*, a protein molecule that acts as a template to distrupt the structures of normal brain proteins. Prions are heat-resistant and are not destroyed by cooking.

Table 14•5

SAFE REFRIGERATOR STORAGE TIMES (≤40°F)

1 to 2 Days
Raw ground meats, breakfast or other raw sausages, raw fish or poultry; gravies
3 to 5 Days
Raw steaks, roasts, or chops; cooked meats, vegetables, and mixed dishes; ham slices; mayonnaise salads (chicken, egg, pasta, tuna)
1 Week
Hard-cooked eggs, bacon or hot dogs (opened packages); smoked sausages.
2 to 4 Weeks
Raw eggs (in shells); bacon or hot dogs (packages unopened); dry sausages (pepperoni, hard salami); most aged and processed cheeses (Swiss, brick)
2 Months
Mayonnaise (opened jar); most dry cheeses (parmesan, romano)

Figure 14•4

SAFE HANDLING INSTRUCTIONS FOR MEAT AND POULTRY

Safe Handling Instructions

THIS PRODUCT WAS PREPARED FROM INSPECTED AND PASSED MEAT AND/OR POULTRY. SOME FOOD PRODUCTS MAY CONTAIN BACTERIA THAT CAN CAUSE ILLNESS IF THE PRODUCT IS MISHANDLED OR COOKED IMPROPERLY. FOR YOUR PROTECTION, FOLLOW THESE SAFE HANDLING INSTRUCTIONS.

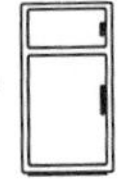

KEEP REFRIGERATED OR FROZEN. THAW IN REFRIGERATOR OR MICROWAVE.

KEEP RAW MEAT AND POULTRY SEPARATE FROM OTHER FOODS. WASH WORKING SURFACES (INCLUDING CUTTING BOARDS), UTENSILS, AND HANDS AFTER TOUCHING RAW MEAT OR POULTRY.

COOK THOROUGHLY.

KEEP HOT FOODS HOT. REFRIGERATE LEFTOVERS IMMEDIATELY OR DISCARD.

Never allow frozen meat to defrost at room temperature or in a bath of warm water. In both cases, meat thaws from outside in, and the outside meat layer can easily warm up to temperatures that permit bacterial growth before the core defrosts.

Microwave cooking of meats requires special care. Large, thick, dense foods such as roasts or meat loaves may register "cooked" on an internal meat thermometer, but may harbor cool spots in which dangerous microorganisms, such as the *Trichinella spiralis* parasite, sometimes present in pork, can survive. Such foods are best cooked by another method or divided into thin individual portions to be microwaved.

Properly cooked food hot from the oven or stove is relatively free of bacteria, but as soon as it is taken out to serve, it is reinoculated. Kitchen utensils recontaminate the food, or bacteria from the air land on its surface. Promptly after serving, even while the food is still hot, refrigerate leftovers in shallow containers for quick, even chilling. Large amounts of food refrigerated in deep containers may take hours to cool though, allowing bacteria time to multiply in the warm internal portions.

Take care when preparing meats along with foods intended to be served raw, such as chopped salads or lettuce and tomato toppers for hamburgers. A grave error is to prepare raw foods on the same board or with the same utensils as were used to prepare raw meats for cooking.

contaminated. The FDA has proposed a label for all egg cartons instructing consumers to keep eggs refrigerated, cook eggs until yolks are firm, and cook foods containing eggs thoroughly. No longer is it safe to drop a raw egg into a food or beverage that will not be cooked before consumption. Healthy people can still safely enjoy classic foods that call for raw or undercooked eggs, such as Caesar salad dressing and hollandaise sauce, by preparing them with pasteurized egg substitute. People who are elderly or very young, or suffer immune dysfunction, should avoid all uncooked or undercooked eggs, including pasteurized egg substitutes, which may contain a few bacteria that escape the pasteurization process.[14]

Seafood For adults and children alike, eating raw or lightly steamed seafood is a risky proposition even when it is prepared by a master chef. The microorganisms that lurk there are undetectable, even to an expert.*

As population density increases along the seashores, the offshore waters are becoming polluted, contaminating the seafood living there. Viruses that cause human diseases have been detected in some 90 percent of the waters off U.S. coasts.[15] Watchdog agencies monitor commercial fishing areas to keep harvesters out of unsafe waters, but unwholesome food can still reach the market. In one season alone, black-market dealers may sell millions of dollars worth of clams and oysters taken illegally from closed harvesting areas.

The food-borne infections that lurk in normal-appearing seafood can be even worse than those of spoilage: viral hepatitis; worms, flukes, and other parasites; severe

A safe hamburger is cooked well-done, has juices that run clear, and has reached an internal temperature of 160°F. Place it on a clean plate when it's done.

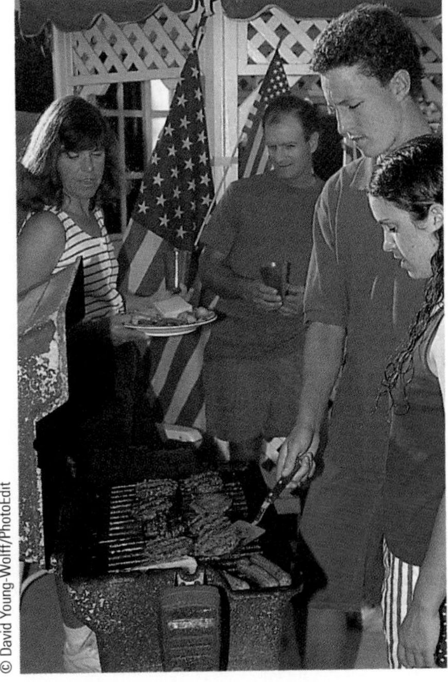

© David Young-Wolff/PhotoEdit

*To speak with an expert on seafood safety, call the FDA's seafood hotline: (800) FDA-4010.

sushi a Japanese dish that consists of vinegar-flavored rice, seafood, and colorful vegetables, typically wrapped in seaweed. Some sushi is wrapped in raw fish; other sushi contains only cooked ingredients.

In many states, containers of raw oysters must bear this warning: "There is a risk associated with consuming raw oysters or any raw animal protein. If you have chronic illness of the liver, stomach or blood or have immune disorders, you are at greater risk of serious illness from raw oysters and should eat oysters fully cooked. If unsure of your risk, consult a physician."

Unpasteurized or untreated juice must have the following warning on the label:

WARNING: This product has not been pasteurized and therefore may contain harmful bacteria that can cause serious illness in children, the elderly, and persons with weakened immune systems.

viral intestinal disorders; and poisoning by naturally occurring toxins. Hepatitis infection causes prolonged illness that persists for months or years, severely damages the liver, greatly increases the risk of developing liver cancer, and, once in the body, is transmissible to others. Many types of worms depend on the blood of their host for food and reproduction; they attack digestive membranes, sometimes causing life-threatening perforations. Flukes attack and damage the liver. The rumor that freezing fish will make it safe to eat raw is only partly true. Freezing fish will kill mature parasitic worms, but only cooking can kill all worm eggs and other microorganisms that can cause illness.

People who have enjoyed raw oysters and other raw seafood for years may be tempted to ignore these threats because they have never experienced serious illness. Some have heard that drinking alcoholic beverages when eating raw seafood eliminates risks or that hot sauce kills the bacteria, but these assertions are not true. One study, long ago, did find a correlation between taking one drink of whiskey or wine and a reduced risk of disease after eating contaminated seafood, but this evidence is no guarantee of protection. Hot sauce is useless against the infectious agents that contaminate oysters. Experts unanimously agree that the risks of eating raw or lightly cooked seafood today are unacceptably high due to environmental contamination. People who are fond of **sushi** should remember that not all varieties are made from raw fish; sushi made with cooked crabmeat and vegetables, avocado, and similar delicacies are safe.

Fruits and Vegetables This book champions a diet with abundant fresh fruits and vegetables for its health benefits, but these foods can also present a microbial threat unless they are thoroughly rinsed in cold running water to remove microbes before peeling, chopping, or eating. Salad ingredients, such as lettuce, tomatoes, and spinach, for example, grow close to the ground, making bacterial contamination from the soil and organic fertilizers likely. Also, much produce is imported from countries where fields may be irrigated with contaminated water or produce picked by infected farm workers with poor hygiene practices.

Rough skins of melons such as cantaloupes provide crevices that harbor bacteria and so should be scrubbed with a brush under running water before cutting. Raspberries, other berries, and, in fact, all produce should be rinsed thoroughly under running water for at least 10 seconds (see Table 14-6). Sprays and "wash" products are sold to help remove the waxes and residues on produce, but no evidence indicates that such products outperform plain water for removing microbes. Unpasteurized, or raw, juices and ciders are not safe because microbes on the original fruit may multiply in the product. The margin shows a mandatory label that announces their danger.

Table 14•6

PRODUCE SAFETY

Cleaning Fresh Fruits and Vegetables

1. Thoroughly rinse raw fruits and vegetables under running water for at least 10 seconds before peeling, cutting, or eating. Don't use soap, detergents, or bleach solutions; commercial vegetable washing products are safe to use.
2. To clean rough-skinned produce, scrub while rinsing using a small vegetable brush to loosen dirt and dislodge microbes.
3. Cut away damaged or bruised areas that may contain bacteria. Toss out moldly fruit or vegetables.

Juice Safety

1. Choose chilled pasteurized juices or shelf-stable juices (canned or boxed) that have been treated with high temperature to kill microbes and check their seals to be sure no microbes have entered after processing.
2. Especially, infants, children, the elderly, and people with weakened immune systems should never be given commercial raw or unpasteurized juice products.

Sprouts (including alfalfa, clover, and radish) are a special case. Although they are often preferred raw in salads, wraps, and other sandwiches, there seems to be no sure way to make them safe except to cook them. Sprout seeds often harbor the dangerous *E. coli* 0157:H7 bacteria even before the sprouts are grown, making even homegrown raw sprouts a risky food. Rinsing the seeds or mature sprouts is no guarantee that *E. coli* or *Salmonella* organisms have been removed. Some experts are calling for all consumers to avoid raw sprouts for safety's sake.

Picnics and Lunch Bags For safe picnics and safe packed lunches, keep these precautions in mind. Choose foods that remain safe to eat without refrigeration, such as fresh uncut fruits and vegetables, breads and crackers, and canned spreads and cheeses that you can open and use on the spot. Aged cheeses, such as cheddar and Swiss, do well at environmental temperatures for an hour or two, but for longer periods, carry them in a cooler or thermal lunch bag. Mayonnaise is somewhat resistant to spoilage because of its acid content, but when it is mixed with chopped ingredients in pasta, meat, or vegetable salads, the mixtures spoil easily. The chopped ingredients have extensive surface areas for bacteria to invade, and foods that have been in contact with cutting boards, hands, and kitchen utensils have picked up at least a few bacteria. Start with chilled ingredients, then chill chopped salads in shallow containers before, during, and after a picnic, and keep salad sandwiches cold until eaten. To keep lunch bag foods chilled, choose a thermal lunch bag and freeze beverages to pack in with the foods. As the beverages thaw in the hours before lunch, they keep the foods cold.

Honey Honey can contain dormant spores of *Clostridium botulinum* that can awaken (germinate) in the human body to produce the deadly botulinum toxin. Mature adults are usually protected against this threat, but infants under one year of age should never be fed honey, which can also be contaminated with environmental pollutants picked up by the bees. Honey has been implicated in several cases of sudden infant death.

Today, consumers bear much of the responsibility for staying safe from food-borne illnesses. They must cook meats and eggs to the well-done stage to kill dangerous microorganisms lurking in the raw products. They must scrub vegetables and fruits to remove pesticide residues applied to kill molds and insects that attack food during storage. They must avoid foods like raw sprouts and others that may harbor disease-causing microorganisms. One proposal to help reduce the consumer's burden of responsibility involves exposing foods to ionizing radiation in doses that kill microorganisms. The Consumer Corner provides some details.

key point *Some foods pose special microbial threats and so require special handling. Raw seafood is especially likely to be contaminated. Almost all types of food poisoning can be prevented by safe food preparation, storage, and cleanliness. Honey is unsafe for infants.*

How Can I Avoid Illness When Traveling?

About half of the people who travel to places where cleanliness standards are lacking suffer from food-borne illnesses—commonly known as traveler's diarrhea (see Table 14-2). A bout of this illness can ruin a trip. To avoid food-borne illness while traveling:

- Before you travel, ask your physician which medicines to take with you in case you get sick.
- Wash your hands often with soap and water, especially before handling food or eating.
- Eat only cooked and canned foods. Eat raw fruits or vegetables only if you have washed them with your own clean hands in boiled water and peeled them yourself. Skip salads.

- Be aware that water, and ice made from it, may be unsafe. Take along disinfecting tablets or an element that boils water in a cup. Drink only treated, boiled, canned, or bottled beverages, and drink them without ice, even if they are not chilled to your liking.
- Avoid using the local water supply, even if you are just brushing your teeth, unless you boil or disinfect it first.
- Be aware that mad cow disease poses an extremely small risk (1 in 10 billion servings) to travelers to countries where BSE is a problem.[16] To err on the safe side when traveling to these countries, avoid eating beef altogether or select solid pieces of muscle meat that may have less contamination. Avoid variety meats such as sweetbreads, brains, or tripe, as well as sausages or ground meats that may contain them.

In general, remember these rules: boil it, cook it, peel it, or forget it. If you follow these recommendations, chances are excellent that you will remain well.

Some special food safety concerns arise when traveling. To avoid food-borne illnesses, remember to boil it, cook it, peel it, or forget it.

Irradiation and Food Safety

The American Dietetic Association makes this statement concerning **irradiation:** "Food irradiation enhances the safety and quality of the food supply and helps protect consumers from food-borne illness."[1] The FDA has approved irradiation for controlling microbial contamination of many foods (see Table 14-7).[2] Can food irradiation solve our food safety and supply problems? What, if any, are the risks to human health?

irradiation the application of ionizing radiation to foods to reduce insect infestation or microbial contamination or to slow the ripening or sprouting process. Also called *cold pasteurization*.

Potential Benefits

Food irradiation's greatest potential benefit is its ability to kill almost all disease-producing microorganisms present in food, a characteristic that earns it the alternative name of *cold pasteurization*. Each year, the many millions of children, elderly, and susceptible people worldwide who sicken or die from food-borne illnesses could be spared by the power of irradiation. Raw poultry, for example, emerges from irradiation treatment 99.9 percent free of disease-causing microorganisms. Chicken is thus safer when purchased, and the threat of cross-contamination in home or industrial kitchens is reduced. Undercooking becomes less of a health threat, too.

Growers and marketers of fresh produce also stand to benefit. Irradiation eliminates the need to quarantine and spray fresh produce with pesticides before shipping to prevent the spread of plant diseases or harmful insects and their eggs. Irradiation kills mold spores, replacing the fungicides now sprayed on harvested foods. Irradiation also slows decay in fruits and vegetables, making them last longer and appear fresher.

Up to half of the world's food bounty is lost each year to pests and decay; irradiation could greatly diminish these losses. Irradiation facilities are expensive to build, however, particularly in areas where people most need them—see Chapter 15 for more information on world hunger.

The Irradiation Process

Irradiation works by exposing foods to controlled doses of gamma rays from the radioactive compound cobalt 60. As radiation passes through a living cell, it disrupts the internal structures and so kills or deactivates the cell. Low doses can kill the growth cells in the "eyes" of potatoes and ends of onions, preventing them from sprouting. Low doses also delay ripening of bananas, avocados, and other fruits. High doses can penetrate tough insect exoskeletons and mold or bacterial cell walls to destroy their life-maintaining DNA, proteins, and other molecules. Irradiation can even kill microbes while food is in a frozen state, making irradiation uniquely useful in protecting foods such as whole turkeys that are ordinarily marketed frozen. However, the spores of one dangerous bacterium are resistant to radiation—those of *Clostridium botulinum*, the bacterium responsible for the lethal food-poisoning agent, botulinum toxin. This problem is being addressed

These groups agree that irradiation can make the food supply safer:

- American Council on Science and Health.
- American Dietetic Association.
- American Medical Association.
- American Veterinary Medical Association.
- Centers for Disease Control and Prevention.
- Council for Agricultural Science and Technology.
- Institute of Food Technologists.
- International Atomic Energy Agency.
- Scientific Committee of the European Union.
- United Nations Food and Agricultural Organization.
- World Health Organization.
- Rarely do so many scientists agree on a single issue.

SOURCE: J. Farkas, Irradiation as a method of decontaminating food: A review, *International Journal of Food Microbiology* 44 (1998): 189–204.

Table 14•7

FOODS APPROVED FOR IRRADIATION BY THE FDA

- Wheat
- Flour
- Spices
- Citrus fruits
- Tropical fruits
- Strawberries
- Tomatoes
- Mushrooms
- Potatoes
- Onions
- Poultry
- Fresh and frozen red meats, such as lamb, beef, and pork

SOURCE: Meat irradiation can boost food safety, *FDA Consumer*, March/April 2000, p. 4.

by manufacturers hoping to produce irradiated low-acid, ready-to-serve food, such as pasta dishes and stews, intended to be stored without refrigeration.

For perspective, compare the doses of radiation used on foods with the lethal human dose. The lowest doses of radiation needed to delay ripening and sprouting of fragile fruits and vegetables are 10 to 20 times higher than the doses that would kill human beings. The dose required to sterilize foods is many times higher still, but most foods are not completely sterilized because doses that high would destroy the food. Dried herbs and spices are notable exceptions—they can withstand sterilizing doses and are commonly irradiated before being marketed to consumers in this country.

Labeling of Irradiated Foods

The FDA requires that irradiated foods bear a label, in letters as large as those of the ingredient list, stating that the foods have been treated with radiation and displaying the irradiation symbol, also called the radura logo, shown here.[3] No label is required for foods containing irradiated ingredients, including spices that are mixed with processed foods, and for irradiated foods served in restaurants.

If Irradiation Is So Great, Why Aren't More Foods Irradiated?

Companies that process foods follow consumer preferences—some consumers seem willing to purchase irradiated foods, but others vigorously challenge the idea of food irradiation. Among the most common reasons people give for fearing irradiation are these:

- Fear that the foods will become radioactive.
- Fear that the foods will lose substantial nutrients during irradiation.
- Fear that irradiated foods are not safe to eat.
- Fear of unique, untested chemicals that form in foods during irradiation.
- Fear that the radioactive substances used to irradiate foods will endanger plant workers, the general population, and the environment.

The first concern can be immediately put to rest—properly irradiated food does not become radioactive any more than teeth become radioactive after dental X-ray procedures. The use of radioactivity demands great care, though. Foods exposed to extremely high doses of the wrong sort of radiation can indeed become radioactive—however, such foods would also be rendered inedible.

Irradiation's Effects on Nutrients

Most nutrients, such as proteins, fats, carbohydrates, and minerals, survive irradiation intact or sustain only insignificant losses. Nutrients sensitive to heat treatment, such as the B vitamins and ascorbic acid, are sensitive to irradiation. Even so, substantial losses occur only with high doses of radiation well beyond those permitted in the treatment of food. In general the nutrient losses sustained during properly administered irradiation are similar to those caused by canning or other common processes. The FDA deems nutrient losses of less than 2 percent of the total as insignificant, and most losses incurred through irradiation fall within this limit.

Irradiation Safety

More than 40 years of research on animals have shown no toxic effects from eating irradiated foods.[4] Human volunteers who ate a diet composed entirely of irradiated food have shown no ill effect. The World Health Organization (WHO) has concluded that the use of irradiation for the purpose of reducing microbial contamination of food is safe and is analogous to cooking in its effects on the composition and nutrients of foods.

Two previous safety questions have been resolved to WHO's satisfaction. The first concerns chemicals formed in foods during irradiation. These *radiolytic products* have been identified as chemicals commonly formed in foods during many forms of processing, including cooking. Among them are minuscule amounts of highly reactive free radicals that could, theoretically, be damaging to health. However, the tiny amounts are unlikely to pose a hazard to the body, whose antioxidant systems are equipped to neutralize free radicals from many sources. (Chapter 7 and its Controversy discuss the body's free-radical defenses.)

This radura logo is the symbol for foods treated with radiation.

The second question concerns potential genetic damage from consuming irradiated food.[5] Early studies indicated that rats developed chromosomal abnormalities, impaired fertility, and depressed immune responses in response to a diet of irradiated chow.[6] In another study performed in the days before research ethics would have prevented it, malnourished children who were fed freshly irradiated wheat seemed to exhibit increased chromosomal abnormalities.[7] Since that time, despite many years of scientific investigation, no other evidence has emerged to support the idea that irradiated food can damage the chromosomes or cause any other adverse effect on health. Indeed, much research supports its safety.

Irradiation does require extremely cautious handling. The process necessitates transporting radioactive materials, exposing workers to them, and then disposing of the spent wastes, which remain radioactive for many years. Opponents of irradiation point out that birth defects are common in children who were exposed to nonlethal low doses of radiation during their fetal development and in children whose parents were exposed to radiation *before* the children were conceived.

These concerns are echoed by the food industry, which is working to safeguard both workers and future generations through strict operating standards and enforcement of regulations limiting radiation exposure.

Finally, some consumers worry that food manufacturers might use the technology unethically. Current law prohibits sale of old or tainted food found to have high bacterial counts, usually from insect or rodent droppings. Food condemned by the FDA represents lost profits, and unscrupulous manufacturers could irradiate the old or tainted food, killing the telltale bacteria and fooling FDA inspectors into judging the food wholesome and fit for sale. This concern raises an important point: irradiation is intended to complement other traditional food safety methods, not to replace them. Even irradiation cannot protect people from faulty food safety practices and contamination due to poor sanitation in the marketplace or at home.

Consumers: The Final Authority

Just a few irradiated products are now available for sale. Whether more irradiated foods appear in markets depends largely upon whether consumers choose to buy them. A national survey found that almost 80 percent of those interviewed were willing to buy packages of food with labels stating "irradiated to destroy harmful bacteria." People must be willing to pay a premium for irradiated meats, poultry, and seafood, however. The enormous costs of building and operating irradiation facilities and the expensive packaging required for the irradiation process will increase the price of the food.

In the end, it may come down to taste. Most consumers consider flavor to be important, and irradiation changes the taste of food slightly. Whether this change will influence consumer purchases of irradiated foods is for the future to tell.

contaminant any substance occurring in food by accident; any food constituent that is not normally present.

persistent of a stubborn or enduring nature; with respect to food contaminants, the quality of remaining unaltered and unexcreted in plant foods or in the bodies of animals and human beings.

Read more about the association between cancer and foods in Chapter 11.

Chemical contaminants of concern in foods:

- Heavy metals:
  - Lead.
  - Mercury.
  - Cadmium.
  - Selenium.
  - Arsenic.
- Halogens and organic halogens:
  - Chlorine.
  - Iodine.
  - Vinyl chloride.
  - Ethylene dichloride.
  - Trichloroethylene (TCE).
  - Polybrominated biphenyl (PBB).
  - Polychlorinated biphenyls (PCBs).
- Others:
  - Asbestos.
  - Dioxins.
  - Acrylonitrile.
  - Lysinoalanine.
  - Diethylstilbestrol (DES).
  - Heat-induced mutagens.
  - Antibiotics (in animal feed).

Natural Toxins in Foods

Some people think they can eliminate all poisons from their diets by eating only "natural" foods. On the contrary, nature has provided many plants used for food with natural poisons to fend off diseases, insects, and other predators. Humans rarely suffer actual harm from such poisons, but the potential for harm does exist.

Although the herbs belladonna and hemlock have reputations as deadly poisons, few people know that the herb sassafras contains a cancer-causing agent and is banned from use in commercially produced foods and beverages (see Table 11-9 page 421, for more on potentially harmful herbs). Cabbage, turnips, mustard greens, and radishes all contain small quantities of harmful goitrogens, compounds that can enlarge the thyroid gland and aggravate thyroid problems. These effects show up only under extreme conditions when people have little but cabbage to eat. Ordinarily, cabbages and their relatives are celebrated for their phytochemicals—compounds associated with low cancer rates.

Other natural poisons in *raw* lima and fava beans and in fruit seeds such as apricot pits are members of a group called cyanogens, precursors to the deadly poison cyanide. Many countries restrict commercially grown lima beans to those varieties with the lowest cyanogen contents. Fruit seeds are seldom deliberately eaten; an occasional swallowed seed or two presents no danger, but a couple of dozen seeds could be fatal to a small child. An infamous cyanogen is laetrile, a compound erroneously represented as a cancer cure in the 1970s. True, the poison laetrile kills cancer cells, but only at doses that kill the person, too. Research over the past century has proved that laetrile is an ineffective cancer treatment and dangerous to the taker.

Potatoes contain many natural poisons, including solanine, a powerful, bitter, narcotic-like substance. The small amounts of solanine normally found in potatoes are harmless, but solanine can build up to toxic levels when potatoes are exposed to light during storage. Cooking does not destroy solanine, but because most of a potato's solanine is in a green layer that develops just beneath the skin, it can be peeled off, making the potato safe to eat. If the potato tastes bitter, however, throw it out.

At certain times of the year, seafood may become contaminated with the so-called red tide toxin that occurs during algae blooms. Eating seafood contaminated with red tide causes a form of food poisoning that paralyzes the eater. The FDA monitors fishing waters and closes these waters to fishing when red tide algae appear.

These examples of naturally occurring toxins serve as a reminder of three principles. First, any substance can be toxic when consumed in excess. Practice moderation in the use of all foods. Second, poisons are poisons, whether made by people or by nature. It is not the source of a chemical that makes it hazardous, but its chemical structure. Third, by including a variety of foods in the diet, consumers ensure that toxins in foods are diluted by the volume of the other foods eaten.

key point *Natural foods contain natural toxins that can be hazardous if consumed in excess. To avoid poisoning by toxins, eat all foods in moderation, treat chemicals from all sources with respect, and choose a variety of foods.*

Environmental Contaminants

As populations increase worldwide and nations become more industrialized, concerns grow about the environmental contamination of foods.[17] A food **contaminant** is anything that does not belong there.

Harmfulness of Contaminants The potential harmfulness of a contaminant depends in part on the extent to which it lingers in the environment or in the human body—that is, on how **persistent** it is. Some contaminants are short-lived because microorganisms or agents such as sunlight or oxygen can break them down. Some contaminants linger in the body for only a short time because the body can rapidly excrete them or metabolize them to harmless compounds. These contaminants present little cause for concern. Some contaminants resist breakdown, however, and interact with the body's systems without being metabolized or excreted. These contaminants can pass from one species to the next and accumulate at higher concentrations in each level of the food chain, a process called **bioaccumulation**—see Figure 14-5.

How much of a threat do environmental contaminants pose to the food supply? It depends on the contaminant. In general, the threat remains small because the FDA monitors the presence of contaminants in foods and issues warnings when contaminated foods appear in the market. In the event of an industrial spill or a natural event, such as a volcano, however, the hazard can suddenly become great.

Other contaminants build in the food supply more insidiously. For example, increasing levels of the **heavy metal** mercury expelled from industrial sites have been detected in U.S. lakes, rivers, and ocean fisheries. The FDA has detected unacceptably high mercury levels in fish and other wildlife, prompting an advisory in 2001 to all pregnant women, women who may become pregnant, nursing mothers, and young children against eating large predatory fish, such as king mackerel,

bioaccumulation the accumulation of a contaminant in the tissues of living things at higher and higher concentrations along the food chain.

heavy metal any of a number of mineral ions such as mercury and lead; so called because they are of relatively high atomic weight. Many heavy metals are poisonous.

Figure 14•5

BIOACCUMULATION OF TOXINS IN THE FOOD CHAIN

If none of the chemicals are lost along the way, one person ultimately receives all of the toxic chemicals that were present in the original several tons of producer organisms.

4. A person whose principal animal-protein source is fish may consume about 100 pounds of fish in a year.

3. Larger fish consume a few tons of plankton-eating fish in the course of their lifetimes—and the toxic chemicals from the small fish become more concentrated in the flesh of the larger species.

2. The toxic chemicals become more concentrated in the plankton-eating fish that consume several tons of producer organisms in their lifetimes.

1. Producer organisms may become contaminated with toxic chemicals.

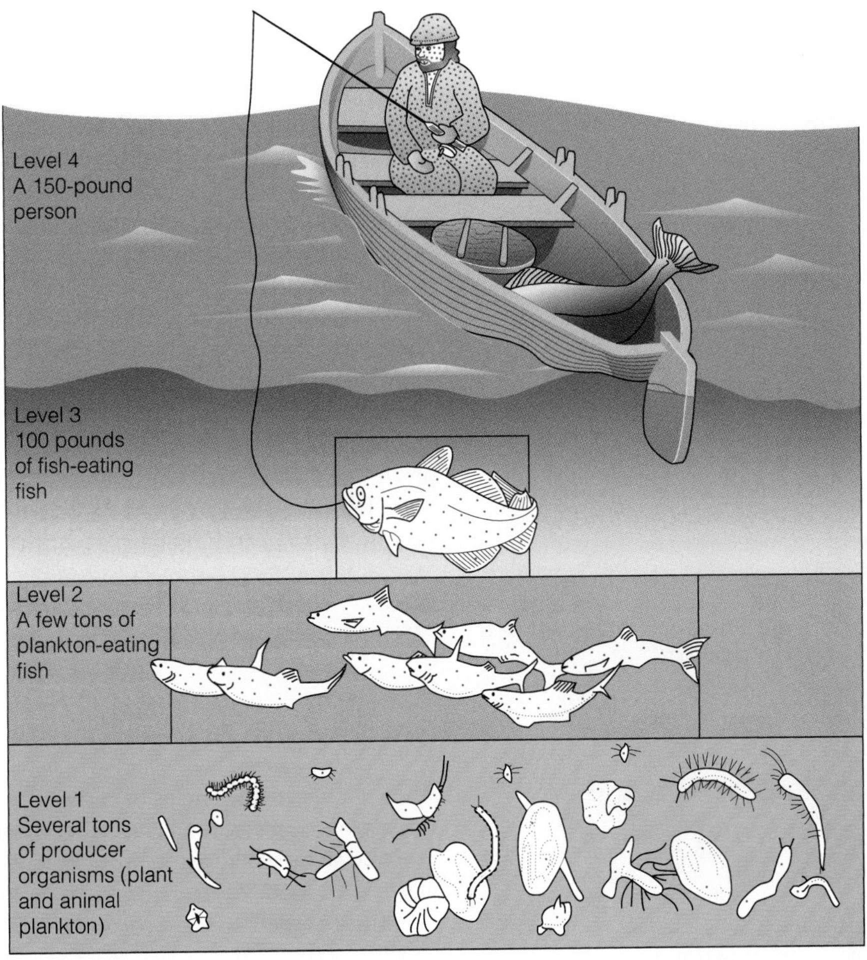

organic halogen an organic compound containing one or more atoms of a halogen—fluorine, chlorine, iodine, or bromine.

pesticides chemicals used to control insects, diseases, weeds, fungi, and other pests on crops and around animals. Used broadly, the term includes *herbicides* (to kill weeds), *insecticides* (to kill insects), and *fungicides* (to kill fungi).

swordfish, shark, and tilefish.[18] Further, the Environmental Protection Agency (EPA) has warned the same groups to limit their intakes of freshwater fish caught by family and friends in lakes, rivers, and streams to one fish meal per week.[19] Fish with mercury levels below the danger zone and thus safe to consume include canned and farm-raised fish, shellfish, and small ocean fish. Mercury is persistent in the environment, so efforts begun today to clean up U.S. waters will take years to diminish the threat to health.

The following paragraphs describe the results when contaminants found their way into the food supply in the past. In the first case, vast amounts of mercury were released into waterways by industry and accumulated in fish that people ate. In the second, an **organic halogen** (polybrominated biphenyl or PBB) was accidentally spilled into livestock feed and eaten by animals whose meat people eventually ate.

Mercury A classic example of acute contamination occurred in 1953 when a number of people in Minamata, Japan, became ill with a disease no one had seen before. By 1960, 121 cases had been reported, including 23 in infants. Mortality was high; 46 died, and the survivors suffered progressive, irreversible blindness, deafness, loss of coordination, and severely impaired mental function.* The cause of this misery was ultimately revealed: manufacturing plants in the region were discharging mercury into the waters of the bay, the mercury was turning to methylmercury upon leaving the factories, and the fish in the bay were accumulating this poison in their bodies. Some of the people who were poisoned had been eating fish from the bay every day. The infants who contracted the disease had not eaten any fish, but their mothers had, and the mothers were spared damage during their pregnancies because the poison concentrated in the tissues of their unborn babies.

Pesticides:

- Kill pests' natural predators.
- Accumulate in the food chain.
- Pollute the water, soil, and air.

PBB In 1973, half a ton of the toxic compound PBB was accidentally mixed into livestock feed that was distributed throughout the state of Michigan. The chemical found its way into millions of animals and then into people who ate their meat. The seriousness of the accident came to light when dairy farmers reported that their cows were going dry, aborting their calves, and developing abnormal growths on their hooves. More than 30,000 cattle, sheep, and swine and more than a million chickens were destroyed, but the effects on people were not prevented. An estimated 97 percent of Michigan's residents had been exposed to PBB. Some of the exposed farm residents suffered nervous system aberrations and liver disorders.

Heavy metals and organic halogens are among the most toxic of chemicals and are still being liberated into our environment daily. Much more information is available about other contaminants, but a discussion of them is far beyond the scope of this text. Table 14-8 selects a few contaminants of great concern in foods to show how pervasively a contaminant can affect the body.

Minamata disease. The effects of mercury contamination can be severe.

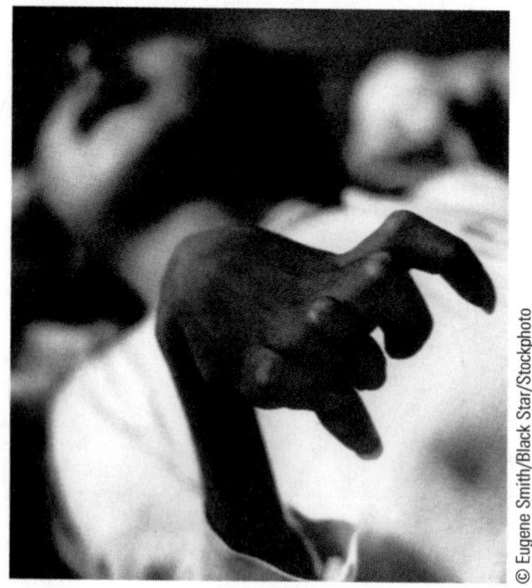

© Eugene Smith/Black Star/Stockphoto

Persistent environmental contaminants pose a significant, but generally small, threat to the safety of food. An accidental spill can create an extreme hazard.

Pesticides

The use of **pesticides** helps to ensure the survival of food crops, but the damage pesticides do to the environment is considerable and increasing. Moreover, there is some question about whether the widespread use of pesticides has really improved the overall yield of food. Even with extensive pesticide use, the world's farmers lose large quantities of their crops to pests every year.

**Minamata disease* was named for the location of the disaster.

Table 14•8

EXAMPLES OF CONTAMINANTS IN FOODS

Name and Description	Sources	Toxic Effects	Typical Route to Food Chain
Cadmium (heavy metal)	Used in industrial processes including electroplating, plastics, batteries, alloys, pigments, smelters, and burning fuels. Present in cigarette smoke and in smoke and ash from volcanic eruptions.	No immediately detectable symptoms; slowly and irreversibly damages kidneys and liver.	Enters air in smokestack emissions, settles on ground, absorbed into food plants, consumed by farm animals, and eaten in vegetables and meat by people. Sewage sludge and fertilizers leave large amounts in soil; runoff contaminates shellfish.
Lead[a] (heavy metal)	Lead crystal decanters and glassware, painted china, old house paint, batteries, pesticides, old plumbing, and some food-processing chemicals.	Displaces calcium, iron, zinc, and other minerals from their sites of action in the nervous system, bone marrow, kidneys, and liver, causing failure to function.	Originates from industrial plants and pollutes air, water, and soil. Still present in soil from many years of leaded gasoline use.
Mercury (heavy metal)	Widely dispersed in gases from earth's crust; local high concentrations from industry, electrical equipment, paints, and agriculture.	Poisons the nervous system, especially in fetuses.	Inorganic mercury released into waterways by industry and acid rain is converted to methylmercury by bacteria and ingested by food species of fish (tuna, swordfish, and others).
Polychlorinated biphenyls (PCBs) (organic compounds)	No natural source; produced for use in electrical equipment (transformers, capacitors).	Long-lasting skin eruptions, eye irritations, growth retardation in children of exposed mothers, anorexia, fatigue, others.	Discarded electrical equipment; accidental industrial leakage, or reuse of PCB containers for food.

[a]For answers to questions concerning lead, call the National Lead Information Center at (800) 424-LEAD.

Do Pesticides on Foods Pose a Hazard to Consumers?

Many pesticides are broad-spectrum poisons that damage all living cells, not just those of pests. Their use poses hazards to the plants and animals in natural systems, and especially to workers involved with pesticide production, transport, and application. High doses of pesticides applied to laboratory animals cause birth defects, sterility, tumors, organ damage, and central nervous system impairment. At one time, the law stated that no traces of pesticides found to cause cancer in animals would be allowed in foods. In 1996, however, this provision was eliminated.

Ironically, pesticides also promote the survival of the very pests they are intended to wipe out. A pesticide aimed at certain insects may kill *almost* 100 percent of them, but thanks to the genetic variability of large populations, some insects are likely to survive exposure. The resistant insects can then multiply free of competition and soon will produce many offspring—offspring that have inherited resistance to the pesticide and can attack the crop with enhanced vigor. To control these resistant insects requires application of a new and more powerful pesticide, which leads to the emergence of a population of still more resistant insects. The same effects arise from use of herbicides and fungicides. One alternative to this destructive series of events is to manage pests using a combination of natural and biological controls, as discussed in Controversy 15.

Pesticides are not produced only in laboratories; they also occur in nature. The nicotine in tobacco and psoralens in celery are examples. A bacterium from soil yields a pesticide often used in organic gardening; the genetic blueprint for producing this

In some small gardens, handwork can take the place of pesticides.

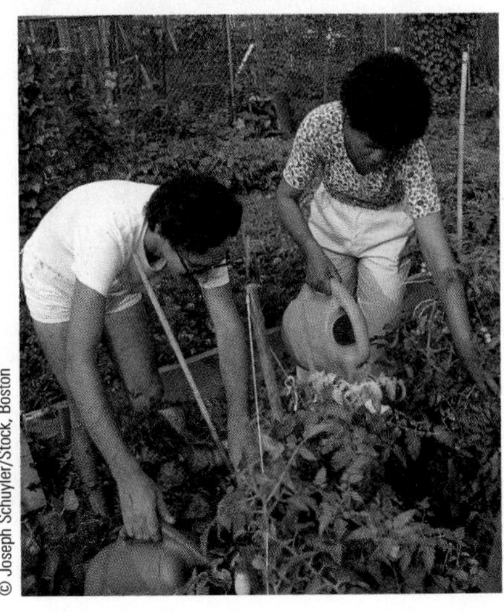

residues whatever remains. In the case of pesticides, those amounts that remain on or in foods when people buy and use them.

tolerance limit the maximum amount of a residue permitted in a food when a pesticide is used according to label directions.

bacterial pesticide has been transferred to vegetables, resulting in plants that grow their own pesticides in the field. See this chapter's Controversy for details. Natural pesticides are less damaging to other living things and leave less persistent **residues** in the environment than most human-made ones. An ideal pesticide would destroy pests in the field but vanish long before consumers ate the food; no such "perfect" pesticide yet exists, however.

Figure 14-6 demonstrates that pesticide residues on agricultural products can survive processing and may be present in and on foods served to people. Chemical companies are working to develop safer pesticides, and government agencies monitor the new products that appear. If a pesticide is deemed to pose a danger, its use is disallowed.

Pesticides can be part of a safe food protection program, but can also be hazardous when handled or used inappropriately.

Regulation of Pesticides

Pesticide residues in foods are subject to a low legal **tolerance limit,** generally 1/100 to 1/1,000 of the level found to cause no effect in laboratory animals. Over 10,000 tolerance regulations state maximum levels for the more than 300 pesticide chemicals allowed for use on various specific crops in the United States. If a pesticide is misused, growers risk fines, lawsuits, and destruction of their crops. In 25 years of testing, the FDA has seldom found crop residues above tolerance levels, so it appears that pesticides are generally used according to regulations. This makes sense because growers are not anxious to spend extra capital on unneeded chemicals.

A loophole in federal regulations, however, allows companies in the United States to make banned pesticides and export them to other countries. The banned pesticides can then slip back into the United States on imported foods, a circuitous route that has been called the "circle of poison." The FDA is stepping up surveillance of imported foods and can deny entry to any imports found to contain illegal residues. The FDA collects samples of both domestic and imported foods and analyzes them using methods that can detect residues well below tolerances. If any residues exceed permitted limits, the FDA can seize the foods or order them destroyed. The overwhelming majority of both imported and domestically produced foods tested by the FDA are found to contain either no residues or residues within federally permitted limits.[20]

A problem is that budget constraints limit the FDA's testing capacity. The FDA does not sample *all* food shipments or test for *all* pesticides. Fewer than 700 inspectors and scientists test food samples from the multitude of farms, groves, docks, airports, warehouses, and processing plants the agency oversees. The FDA cannot (nor can it be expected to) guarantee 100 percent safety in the food supply. Instead, it sets conditions so that substances do not become a hazard and acts promptly when problems or suspicions arise.

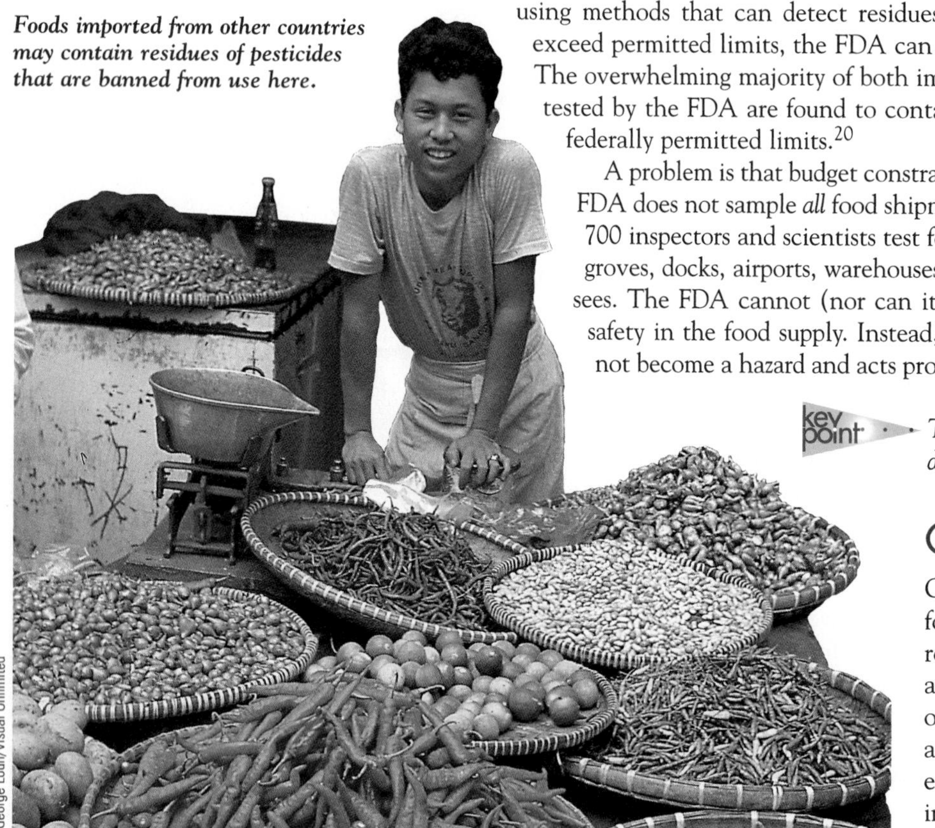
Foods imported from other countries may contain residues of pesticides that are banned from use here.

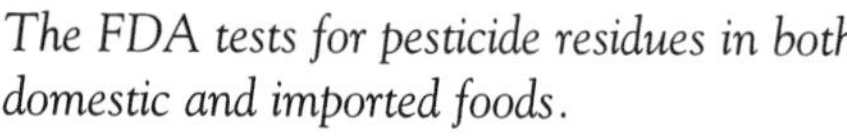
The FDA tests for pesticide residues in both domestic and imported foods.

Consumer Concerns

Consumers also bear some responsibility for their own health and safety with respect to pesticides. They can learn about the potential benefits and dangers of pesticide use, discuss regulations and alternatives with others, advise their government representatives about their findings, and apply pressure wherever it will help change inappropriate procedures.

Figure 14•6

POSSIBLE PATHWAYS OF PESTICIDE RESIDUES TO A FAST-FOOD MEAL

The red dots in the figure represent pesticide residues left on foods from field spraying or postharvest application. Notice that most pesticides follow fats in foods and that some processing methods, such as washing and peeling vegetables, reduce pesticide concentrations while others tend to concentrate them.

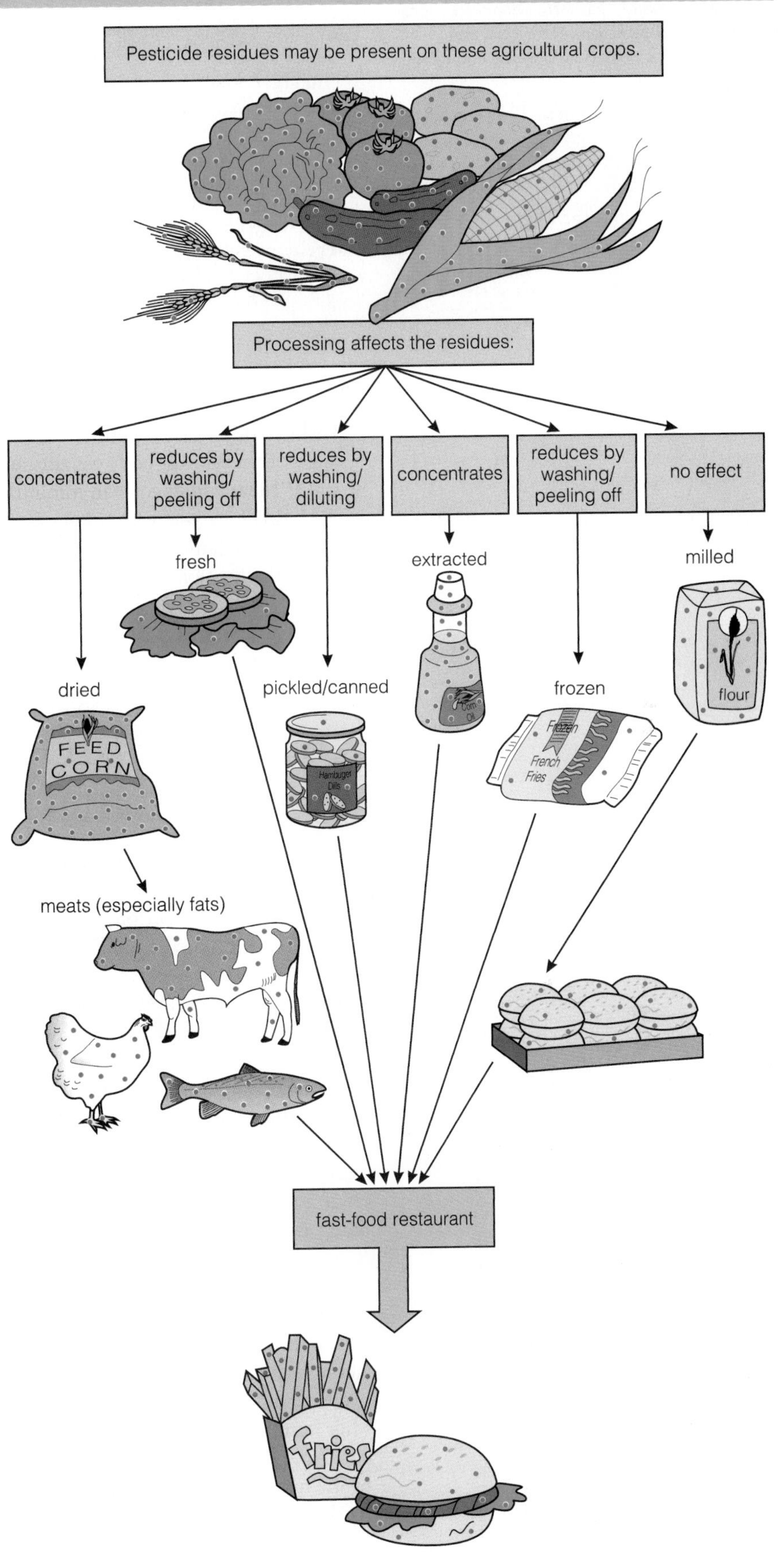

organic foods products grown and processed without the use of synthetic chemicals such as pesticides, herbicides, fertilizers, and preservatives and without genetic engineering or irradiation.

growth hormone a hormone (somatotropin) that promotes growth and is produced naturally in the pituitary gland of the brain.

bovine somatotropin (bST) growth hormone of cattle, which can be produced for agricultural use by genetic engineering. Also called *bovine growth hormone (bGH)*.

Table 14•9

WAYS TO REDUCE PESTICIDE RESIDUE INTAKE

- Trim fat from meat and remove skin from poultry and fish; discard fats and oils in broths and pan drippings. (Pesticide residues concentrate in the animal's fat.)
- Vary meat, poultry, and fish choices from day to day and do not take fish oil capsules.
- Wash fresh produce in water, use a scrub brush, and rinse thoroughly.
- Use a knife to peel an orange or grapefruit; do not bite into the peel.
- Discard the outer leaves of leafy vegetables such as cabbage and lettuce.
- Peel waxed fruit and vegetables. (Waxes don't wash off and can seal in pesticide residues.)
- Peel vegetables such as carrots and fruits such as apples when appropriate. (Peeling removes pesticides that remain in or on the peel, but also removes fibers, vitamins, and minerals.)

Risks to health from pesticide exposure are small for healthy adults, but children, because of their lower body weights and immature detoxifying systems, may be more at risk.[21] Meanwhile, you can minimize your risks by following the guidelines offered in Table 14-9.* If you want fewer pesticides in your produce and meat, weigh the pros and cons of **organic foods.** The Controversy in this chapter discusses organic foods and genetically modified foods.

Consumers can take steps to minimize their ingestion of pesticide residues in foods.

Growth Hormone in Meat and Milk

Hormones are sometimes administered to livestock that produce food. The FDA has deemed the practice safe and does not require testing of food products for traces of the drugs. This section provides the details about the genetically engineered hormones that prompted the FDA's decision. Some consumer groups oppose the common U.S. practice of injecting the cattle form of **growth hormone, bovine somatotropin (bST),** into meat animals or dairy herds. The hormone is produced by genetically engineered bacteria and was engineered to be identical to growth hormone made naturally in the pituitary gland of the animal's brain.

Ranchers advocate the use of bST because it makes meat animals develop more meat and less fat. bST may also increase milk production in dairy cows by up to 25 percent, while requiring less feed. To the farmers, these changes mean higher profits without incurring the costs of more cattle, more farmhands, or more equipment. The environment may profit as well. Smaller herds can live on smaller plots of cleared land, and less feed means the use of less resources to produce and transport it (Controversy 15 gives details). Consumer groups counter that while ranchers may benefit from the use of bST, consumers do not, and so its use is not justified.

The European Union and Canada have prohibited the use of bST for milk cows. The reason is that bST stimulates release of another hormone, insulin-like growth factor I (IGF-I), and some questions have been raised about its effects on human health.[22] IGF-I is produced naturally in people, with levels declining as they age. Milk from bST-treated cows, is somewhat higher in IGF-I than milk from untreated cows, and the hormone appears to survive pasteurization and digestion to be absorbed by the human digestive tract. After a thorough review of the literature, the FDA contends that IGF-I levels in milk from bST-treated cows are within the normal range of

*For answers to questions about any sort of pesticides, call the EPA's 24-hour national pesticide hotline: (800) 858-PEST.

variation seen in milk from untreated cows and that bST therefore presents no additional risk from this source.[23] The FDA's position has been reviewed and its conclusions supported by other agencies including the World Health Organization's expert committee on food additives.

As for bST itself, it differs from steroid hormones, such as estrogen, which can be taken orally because they survive digestion. bST is one of the peptide hormones, which are destroyed by digestive enzymes. About 90 percent of the bST in milk, regardless of its source, is destroyed by pasteurization. The rest is destroyed during digestion.

Even if some bST were to survive and enter the bloodstream, it would have no effect on the body because the chemical structures of animal growth hormones differ widely from the structure of human growth hormone, **human somatotropin (hST).** When bST was first discovered, scientists hoped to use it to treat growth hormone–deficient children. Tests proved disappointing, for bST failed to stimulate receptors for human growth hormone and thus had no effect on the children's growth.

Cows treated with bST suffer more udder infections (mastitis) and so are given more antibiotics; these drugs then show up in the cows' milk and meat. Eating the meat could thus pose a hazard to those allergic to the drugs, but milk and meat are tested for drug residues, and contaminated products are not sold. Overall, antibiotic residues appear in about 2 percent of milk tested in the United States. The National Institutes of Health concludes that meat and milk from hormone-treated U.S. cattle are as safe as those from untreated animals.

human somatotropin (hST) human growth hormone.

ultrahigh temperature (UHT) a process of sterilizing food by exposing it for a short time to temperatures above those normally used in processing.

key point *Bovine somatotropin causes cattle to produce more meat and milk on less feed than untreated cattle. The FDA has deemed the practice safe, but consumer groups oppose it.*

Effects of Food Processing on the Nutrients in Foods

Many consumers rely on packaged and processed foods for convenience and speed, and so lose some control over exactly what their foods contain. What does processing do to foods and to their nutritional value? Food processing involves tradeoffs. It makes food safer, or it gives food a longer usable lifetime, or it cuts preparation time—but at the cost of some vitamin and mineral losses. A process such as pasteurization, which makes milk safe to drink, is clearly worth that cost. Boxes of milk that can be kept at room temperature have been treated with a process called **ultrahigh temperature (UHT).** The milk is exposed to temperatures above those of pasteurization for just long enough to sterilize it. Irradiation, already discussed in this chapter's Consumer Corner, is a similar case: the price in terms of nutrient loss for gaining a safer food supply is probably trivial. Sometimes processed foods even gain a nutritional edge over their unprocessed counterparts, such as when fat is removed from milk or other foods.

Many forms of processing aim to extend the usable life of a food—that is, to preserve it. To preserve food, a process must prevent three kinds of events: (1) microbial growth, (2) oxidative changes, and (3) enzymatic destruction of food molecules. The first two have already been discussed—microbial growth earlier in this chapter and oxidative damage in Chapter 5. Enzymatic destruction occurs as active enzymes in food cells break down their internal molecular structures and cell membranes and walls. Processes involving heat denature the enzymes, and those applying cold slow enzyme activity. The next sections describe some of the most important preservation or processing techniques—modified atmosphere packaging (MAP), canning, freezing, drying, and extrusion—and their effects on nutrients.

key point *Some nutrients are lost in food processing. Processing aims to protect food from microbial, oxidative, and enzymatic spoilage.*

modified atmosphere packaging (MAP) a preservation technique in which a perishable food is packaged in a gas-impermeable container from which air has been removed, or to which another gas mixture has been added.

canning a method of preserving food by killing all microorganisms present in the food and then sealing out air. The food, container, and lid are heated until sterile; as the food cools, the lid makes an airtight seal, preventing contamination.

high-temperature–short-time (HTST) principle the rule that every 18°F (10°C) rise in processing temperature brings about an approximately tenfold increase in microbial destruction, while only doubling nutrient losses.

Modified Atmosphere Packaging

Today, in most produce departments, shoppers can choose bags of washed, trimmed, fresh, chilled salads and chopped vegetables. These convenient products are more expensive than comparable loose vegetables, but, unopened, they last much longer and so save on waste. The secret to these vegetables' long shelf life is a technique called **modified atmosphere packaging (MAP).** The method also preserves freshness in soft pasta noodles, baked goods, prepared foods, fresh and cured meats, seafoods, dry beans and other dry products, ground and whole-bean coffee, and other foods.

Food manufacturers using MAP first package foods in plastic film or other wraps that oxygen cannot penetrate. Then they remove the air inside the package, creating a vacuum, or they replace the air with a mixture of oxygen-free gases, such as carbon dioxide and nitrogen. By excluding oxygen, MAP:

- Slows ripening of fruits and vegetables.
- Reduces spoilage by mold and bacterial growth.
- Prevents discoloration of cut vegetables and fruits.
- Prevents spoilage of fats by rancidity.
- Slows development of "off" flavors from accelerated enzyme action that breaks down flavor and aroma molecules.
- Slows enzyme-induced breakdown of vitamins.

Chilling of all foods packaged this way is imperative to keep them fresh and safe.

MAP foods retain their vitamins much longer than the same foods exposed to the air. MAP foods also taste fresh, making them especially popular with consumers. One concern is that MAP may permit growth of the *Clostridium botulinum* bacterium in moist, low-acid foods, such as lunch meats or cooked dishes, when they are kept for long periods at too-warm temperatures. Properly stored, however, MAP foods present very little hazard and can be considered as safe and nutritious as fresh foods. An important exception is food that has become spoiled—such foods pose a serious threat of food-borne illness.[24] Consumers must read labels to distinguish between shelf-stable foods and those that must be refrigerated.

key point *Modified atmosphere packaging makes many fresh packaged foods available to consumers. MAP foods compare well to fresh foods in terms of nutrient quality. MAP foods may pose a threat of food-borne illness if not properly stored.*

Do Canned Foods Contain Any Nutrients?

Canning is one of the more effective methods of protecting food against the growth of microbes (bacteria, fungi, and yeasts) that might otherwise spoil it, but canned foods do have fewer nutrients. Like other heat treatments, the canning process is based on time and temperature. Each small increase in temperature has a major killing effect on microbes and only a minor effect on nutrients. In contrast, long heating times are costly in terms of nutrient losses. Therefore, industry chooses canning treatments that employ the **high-temperature–short-time (HTST) principle.**

Which nutrients does canning affect, and how? The fat-soluble vitamins and most minerals are relatively stable and are not affected much by canning. Food scientists have thus paid particular attention to three vulnerable water-soluble vitamins: thiamin, riboflavin, and vitamin C.

Acid stabilizes thiamin, but heat rapidly destroys it; therefore, the foods that lose the most thiamin during canning are the low-acid foods such as lima beans, corn, and meat. Up to half, or even more, of the thiamin in these foods can be lost during canning. Unlike thiamin, riboflavin is stable to heat but sensitive to light, so glass-packed, not canned, foods are most likely to lose riboflavin. Vitamin C's special enemy is an enzyme (ascorbic acid oxidase) present in fruits and vegetables as well as in microorganisms. By destroying this enzyme, HTST processes such as canning actually help to preserve at least some of the product's vitamin C.

Minerals are unaffected by heat, so they cannot be destroyed as vitamins can be. Both minerals and water-soluble vitamins can be lost, however, when they leach into canning or cooking water that the consumer then throws away. Losses are closely related to the extent to which a food's tissues have been broken, cut, or chopped and to the length of time the food is in the water.

Some minerals are added when foods are canned. Important in this respect is sodium chloride, table salt, which is added for flavoring. Many food companies have begun making low-salt versions of their products, which may cost more because fewer low-salt batches are made.

freezing a method of preserving food by lowering the food's temperature to a point that halts life processes. Microorganisms do not die but remain dormant until the food is thawed.

drying a method of preserving food by removing sufficient water from the food to inhibit microbial growth.

key point *Some water-soluble vitamins are destroyed by canning, but many more diffuse into the canning liquid. Fat-soluble vitamins and minerals are not affected by canning, but minerals also leach into canning liquid.*

Freezing

Freezing preserves foods because it stops bacterial reproduction and dramatically slows enzymatic reactions. The nutrient contents of frozen foods are similar to those of fresh foods—losses are minimal. The freezing process itself does not destroy any nutrients, but some losses can occur during the steps before freezing, such as the quick dunking into boiling water (blanching), washing, trimming, or grinding. Vitamin C losses are especially likely because they occur whenever tissues are broken and exposed to air (oxygen destroys vitamin C). Uncut fruits, especially if they are acidic, do not lose their vitamin C; strawberries, for example, can be kept frozen for over a year without losing any vitamin C. Mineral contents of frozen foods are much the same as for fresh.

The Food Feature later in this chapter gives tips on preserving nutrients during cooking.

Frozen foods may even have a nutrient advantage over fresh. Fresh foods are often shipped long distances, and to ensure that they make the trip without bruising or spoiling, they are often harvested unripe. Frozen foods are shipped frozen, so produce is allowed to ripen in the field and to develop nutrients to their fullest potential. Foods frozen and stored under proper conditions will often contain more nutrients when served at the table than fresh fruits and vegetables that have stayed in the produce department of the grocery store for even a day.

Frozen foods have to be kept solidly frozen at below 32°F or 0°C, if they are to retain their nutrients. Vitamin C converts to its inactive forms rapidly at warmer temperatures. Food may seem frozen at 36°F or 2°C, but much of it is actually unfrozen, and enzyme-mediated changes can occur fast enough to completely destroy the vitamin C in only two months. If you want to maximize the nutritive value of the foods you store at home, invest in a freezer thermometer and keep your freezer below 0°F.

 *Foods frozen promptly and kept frozen lose few nutrients.*

Drying

Dried or dehydrated foods offer several advantages. **Drying** eliminates microbial spoilage (because microbes need water to grow), and it greatly reduces the weight and volume of foods (because foods are mostly water). Commercial drying does not cause major nutrient losses. Foods dried in heated ovens at home, however, may sustain dramatic nutrient losses. Vacuum puff drying and freeze drying, which take place at cold temperatures, conserve nutrients especially well.

Sulfite additives are added during the drying of fruits such as peaches, grapes (raisins), and plums (prunes) to prevent browning. Some people suffer allergic reactions when they consume sulfites.[25] Sulfur dioxide helps to preserve vitamin C as well, but it destroys thiamin. This is of small concern, however, because most dehydrated products with added sulfur dioxide were not major sources of thiamin before processing.

extrusion a process by which the form of a food is changed, such as changing corn to corn chips; not a preservation measure.

additives substances that are added to foods, but are not normally consumed by themselves as foods.

Commercially dried foods retain most of their nutrients, but home-dried foods often sustain dramatic losses.

Extrusion

Some food products, particularly cereals and snack foods, have undergone a process known as **extrusion.** In this process, the food is heated, ground, and pushed through various kinds of screens to yield different shapes, such as breakfast "puffs," potato "tots" and snack products, the "bits" you sprinkle on salad, and so-called food novelties. Considerable nutrient losses occur during extrusion, and nutrients are usually added to compensate. But foods this far removed from the original fresh state are still lacking significant nutrients (notably, vitamin E) and fiber, and consumers should not rely on them as staple foods. Enjoy them as occasional snacks and as additions to enhance the appearance, taste, and variety of meals.

Extrusion involves heat and destroys nutrients.

Food Additives

What are **additives,** why are they there, and are they dangerous in any way? In the FDA's list of concerns at the start of this chapter, food additives were not a high priority. Compared with unregulated and untested "dietary supplements" sold directly to consumers, the 3,000 food additives used in this country are strictly controlled and pose little cause for concern.[26]

Manufacturers use food additives to give foods desirable characteristics: color, flavor, texture, stability, enhanced nutrient composition, or resistance to spoilage. Additives, classed by their functions, are listed with their definitions in Table 14-10, and some are discussed in the following sections.

Without additives, bread would quickly mold and salad dressing would go rancid.

Table 14•10

FOOD ADDITIVES BY FUNCTION

- **antimicrobial agents** preservatives that prevent spoilage by mold or bacterial growth. Familiar examples are acetic acid (vinegar) and sodium chloride (salt). Others are benzoic, propionic, and sorbic acids; nitrites and nitrates; and sulfur dioxide.
- **antioxidants** preservatives that prevent rancidity of fats in foods and other damage to food caused by oxygen. Examples are vitamins E and and C, BHA, BHT, propyl gallate, and sulfites.
- **artificial colors** certified food colors, added to enhance appearance. (Certified means approved by the FDA.) Vegetable dyes are extracted from vegetables such as beta-carotene from carrots. Food colors are a mix of vegetable dyes and synthetic dyes approved by the FDA for use in food.
- **artificial flavors, flavor enhancers** chemicals that mimic natural flavors and those that enhance flavor.
- **bleaching agents** substances used to whiten foods such as flour and cheese. Peroxides are examples.
- **chelating agents** defined in Chapter 4 as molecules that bind other molecules. As additives, they prevent discoloration, flavor changes, and rancidity that might occur because of processing. Examples are citric acid, malic acid, and tartaric acid (cream of tartar).
- **nutrient additives** vitamins and minerals added to improve nutritive value.
- **preservatives** antimicrobial agents, antioxidants, chelating agents, radiation, and other additives that retard spoilage or preserve desired qualities, such as softness in baked goods.
- **thickening and stabilizing agents** ingredients that maintain emulsions, foams, or suspensions or lend a desirable thick consistency to foods. Dextrins (short chains of glucose formed as a breakdown product of starch), starch, and pectin are examples. (Gums such as carrageenan, guar, locust bean, agar, and gum arabic are others.)

Regulations Governing Additives

The FDA has the responsibility for deciding what additives shall be in foods. To obtain permission to use a new additive in food products, a manufacturer must test the additive and satisfy the FDA that:

- It is effective (it does what it is supposed to do).
- It can be detected and measured in the final food product.

Then the manufacturer must study the effects of the additive when fed in large doses to animals under strictly controlled conditions to prove that:

- It is safe for consumption (it causes no birth defects or other injury).

Finally, the manufacturer must submit all test results to the FDA. The whole process may take many years.

The FDA then schedules a public hearing, inviting consumers to participate, where experts present testimony for and against granting permission to use the additive. FDA approval of an additive does not give manufacturers free license to add it to all foods. The FDA regulation states in what amounts, for what purposes, and in what foods the additive may be used. No additives are permanently approved; all are periodically reviewed.

generally recognized as safe (GRAS) list a list, established by the FDA, of food additives long in use and believed safe.

toxicity the ability of a substance to harm living organisms. All substances are toxic if the concentration is high enough.

margin of safety in reference to food additives, a zone between the concentration normally used and that at which a hazard exists. For common table salt, for example, the margin of safety is ⅕ (five times the concentration normally used would be hazardous).

The GRAS List Many substances were exempted from complying with this procedure when it was first instituted because they had been used for a long time and their use entailed no known hazards. Some 700 substances in all were put on the **generally recognized as safe (GRAS) list.** When substantial scientific evidence or public outcry has questioned the safety of a GRAS list additive, however, its safety has been reevaluated. All substances about which any legitimate question was raised have been removed or reclassified.

The Margin of Safety Decisions about an additive's safety are governed by the important distinction between **toxicity** and hazard associated with substances. Toxicity is a general property of all substances; hazard is the capacity of a substance to produce injury *under conditions of its use*.* All substances can be toxic at some level of consumption, but they are called hazardous only if they are toxic in the amounts ordinarily consumed. Thus, an additive is not a hazard if it proves toxic only in an immense amount that people never consume; an additive is a hazard only if it is toxic as actually used.

A food additive is supposed to have a wide **margin of safety.** Most additives that involve risk are allowed in foods only at levels 100 times below those at which the risk is still known to be zero. Experiments to determine the extent of risk involve feeding test animals the substance at different concentrations throughout their lifetimes. The additive is then permitted in foods at 1/100 the level that causes no harmful effect whatever in the animals. In many foods, *naturally* occurring toxins appear at levels that bring their margins of safety close to 1/10. Even nutrients, as you have seen, involve risks at high dosage levels. The margin of safety for vitamins A and D is 1/25 to 1/40; it may be less than 1/10 in infants. For some trace elements, it is about 1/5. People consume common table salt daily in amounts only three to five times less than those that cause serious toxicity.

The margin-of-safety concept also applies to nutrients used to fortify foods. Iodine, added in minute amounts to salt to prevent iodine deficiency, is a deadly poison in excess. Iron added to grain products has doubtless helped prevent many cases of iron-deficiency anemia in women and children, but iron in excess can cause iron overload.

Most additives used in foods offer benefits that outweigh their risks or that make the risks worth taking. In the case of color additives that only enhance the appearance of foods without improving their health value or safety, no amount of risk may

*The Delaney Clause, a legal requirement of zero cancer risk for additives, was eliminated in 1996.

be deemed worth taking. Only 10 of an original 80 synthetic color additives are still approved by the FDA for use in foods, and screening of these substances continues.

Manufacturers must comply with other regulations as well. Additives must not be used:

- In quantities larger than those necessary to achieve the needed effects.
- To disguise faulty or inferior products.
- To deceive the consumer.
- Where they significantly destroy nutrients.
- Where their effects can be achieved by economical, sound manufacturing processes.

The regulations governing the management of intentional additives are well conceived and, on the whole, have been effective. Funding shortages limit the capabilities of watchdog agencies such as the FDA, however, and some mistakes and false reports do slip by.

The following sections focus on the food additives that receive the most publicity because people ask questions about them most often. The order is alphabetical, not in order of importance.

key point *The FDA regulates the use of intentional additives. Additives must be safe, effective, and measurable in the final product. Additives on the GRAS list are assumed to be safe because they have long been used. Additives used must have wide margins of safety.*

Antimicrobial Agents

Preservatives known as *antimicrobial agents* protect food from the growth of microbes that can spoil the food and cause food-borne illnesses. Three of these preservatives—salt, sugar, and nitrites—are commonly used.

Examples of common antimicrobial additives:

- Salt
- Sugar
- Nitrites

Salt and Suger The best-known, most widely used antimicrobial agents are two common substances—salt and sugar. Salt has been used since before recorded history to preserve meat and fish; sugar serves the same purpose in jams, jellies, and canned and frozen fruits. (Any jam or jelly that toots its "no preservatives" horn is exaggerating. There is no need to add extra preservatives, so most makers do not.) Both salt and sugar work by withdrawing water from the food; microbes cannot grow without water. Today, other additives such as potassium sorbate and sodium propionate are also used to extend the shelf life of baked goods, cheese, beverages, mayonnaise, margarine, and many other products.

Nitrities The *nitrites* are added to meats and meat products for three main purposes: to preserve their color (especially the pink color of hot dogs and other cured meats); to enhance their flavor by inhibiting rancidity (in cured meats); and to protect against bacterial growth. In particular, in amounts much smaller than needed to confer color, nitrites prevent the growth of the bacterium that produces the deadly botulinum toxin.

Two long-used preservatives.

Nitrites perform important jobs, but they have been the object of controversy because they can be converted in the human body to nitrosamines, which cause cancer in animals. Some cured meats are available without nitrites. However, reducing nitrites consumed in meats would hardly make a difference in a person's overall exposure to nitrosamine-related compounds. For example, an average cigarette smoker inhales 100 times the nitrosamines that the average bacon eater ingests. Likewise, a beer drinker imbibes up to roughly five times the amount that the bacon eater receives. Even the air inside automobiles delivers measurable nitrites.

key point *Microbial food spoilage can be prevented by antimicrobial additives. Of these, sugar and salt have a long history of use. Nitrites added to meats have been associated with cancer in laboratory animals.*

Antioxidants Food can also go bad when it undergoes changes in color and flavor caused by exposure to oxygen in the air (oxidation). Often these changes involve little hazard to health, but they damage the food's appearance, taste, and nutritional quality. Antioxidants are often added to vulnerable foods to prevent the damage caused by oxidation. Familiar examples of oxidative changes are sliced apples or potatoes turning brown and oil going rancid. Antioxidant preservatives protect food from this kind of spoilage. Some 27 antioxidants, including vitamin C (ascorbate) and vitamin E (tocopherol), are approved for use in food.

Controversy 7 describes how antioxidants break the destructive chain reactions of oxidation.

Sulfites The sulfites are another group of antioxidants. They are used to prevent oxidation in many processed foods, in alcoholic beverages (especially wine), and in drugs. Sulfites were used to keep the raw fruits and vegetables in salad bars looking fresh, but this practice was banned after a few people experienced dangerous allergic reactions to the sulfites. The FDA now prohibits sulfite use on food meant to be eaten raw, with the exception of grapes, and it requires foods and drugs to list on their labels any sulfites that are present. For most people, sulfites do not pose a hazard in the amounts used in products, but they have one other drawback. Because sulfites can destroy a lot of thiamin in foods, you can't count on a food that contains sulfites to contribute to your daily thiamin intake.

The ban on sulfites has stimulated a search for alternatives. Some producers now use honey to clarify browned apple juice. Agriculturists have created a hybrid apple that does not turn brown. A combination of four GRAS additives can also substitute for sulfites.*

BHA and BHT Two other antioxidants in wide use are BHA and BHT, which prevent rancidity in baked goods and snack foods. BHT provides a refreshing change from the many tales of woe and cancer scares associated with other additives. Among the many tests performed on BHT were several showing that animals fed large amounts of this substance developed *less* cancer when exposed to carcinogens and lived longer than controls. BHT apparently protects against cancer through an antioxidant effect similar to that of vitamin E. To obtain this effect, though, a much larger amount of BHT must be present in the diet than the U.S. average. A caution: used experimentally at very high levels of intake, the substance has *produced* cancer.

Examples of common antioxidant additives:

- Vitamin C.
- Vitamin E (tocopherol).
- Sulfites.
- BHA and BHT.

This discussion provides the opportunity to mention an important point about additives. No two additives are alike, so generalizations about them are meaningless. No single valid statement can apply to all of the 3,000-odd substances commonly added to foods. Questions about which additives are safe and under what conditions of use must be asked and answered item by item.

Raw grapes may be treated with sulfites. Wash them thoroughly before eating them.

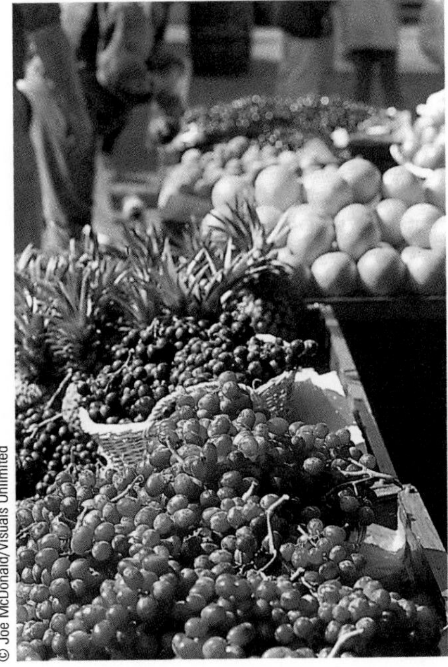

© Joe McDonald/Visuals Unlimited

key point *Antioxidants prevent oxidative changes in foods that would lead to unacceptable discoloration and texture changes in the food. Ingestion of the antioxidant sulfites can cause problems for some people; BHT may offer antioxidant effects in the body.*

Artificial Colors

As mentioned, only about ten artificial colors are still on the GRAS list; this select group has survived considerable screening. Among the most intensively investigated of all additives, artificial colors are much better known than the *natural* pigments of plants, and the limits on the safety of their use can be stated with greater certainty. Examples of natural pigments in common use are the caramel that tints cola beverages and baked goods and the carotenoids that color margarine, cheeses, and pastas. Nevertheless, the food colors have been criticized more than almost any other group of additives. Simply stated, they only make foods pretty, whereas other additives, such

*The four GRAS additives are citric acid, ascorbic acid, sodium acid pyrophosphate, and calcium chloride.

MSG symptom complex the acute, temporary, and self-limiting reactions experienced by sensitive people upon ingesting a large dose of MSG. The name MSG *symptom complex*, given by the FDA, replaces the former *Chinese restaurant syndrome*.

incidental additives substances that can get into food not through intentional introduction but as a result of contact with the food during growing, processing, packaging, storing, or some other stage before the food is consumed. Also called *accidental* or *indirect additives*.

as preservatives, make foods safe. Hence, with food colors, we can afford to require that their use entail no risk. With other food additives, we must weigh the risks of using them against the risks of *not* using them.

The food color tartrazine (yellow number 5) causes an allergic reaction in susceptible people. Symptoms include hives, itching, and nasal congestion, sometimes severe enough to require medical treatment. It is not a common problem; only 1 or 2 in 10,000 individuals may experience the reaction. In addition, some hyperactive children may be sensitive to tartrazine, as Chapter 13 explained. U.S. law requires that tartrazine be listed on all labels of foods that contain it so that people with allergy, parents of children who react to tartrazine, and consumers can avoid it if they wish. They cannot just avoid yellow-colored foods because tartrazine is used to confer turquoise, green, and maroon colors on foods and drugs as well.

The addition of artificial colors is tightly controlled. Some people react adversely to the colorant tartrazine.

Foods containing tartrazine:

- Orange drinks.
- Sports drinks.
- Gelatin desserts.
- Salad dressing.
- Some cake mixes and icings.
- Imitation banana or pineapple extract.
- Seasoning salt.
- Macaroni and cheese dinner.
- Fried or baked cheesy snacks, such as curls and balls.
- Fruit chews and candies.
- Butterscotch squares and candy corn.

Q Are Artificial Flavors and the Flavor Enhancer MSG Safe to Eat?

Although only a few artificial colors are currently permitted in foods, close to 2,000 artificial flavors and flavor enhancers are approved, making them the largest single group of food additives. The safety evaluation of flavoring agents is somewhat problematic because so many flavoring agents are already in use, the flavors are strong and so are used in tiny amounts unlikely to impose risks, and they occur naturally in a wide variety of foods.

A well-known flavor enhancer is monosodium glutamate, or MSG (trade name Accent), the sodium salt of the common amino acid glutamic acid. MSG is used widely in restaurants, especially Asian restaurants. In addition to enhancing other flavors, MSG itself possesses a basic taste (termed *umami*) independent of the well-known sweet, salty, bitter, and sour tastes.

In a few sensitive individuals, MSG produces adverse reactions known as the **MSG symptom complex.** Symptoms may include burning sensations, chest and facial flushing or pain, and throbbing headaches. A probable link lies in elevated blood levels of the MSG component glutamate, a compound known to stimulate the release of some types of pituitary hormones in experimental animals. Meals containing carbohydrate seem less likely to induce adverse effects from MSG than meals of broth, so when dining on Asian-style foods, potentially sensitive people should try ordering dishes such as soups that contain noodles and eat plenty of plain rice with main dishes to provide carbohydrate, as do Asians themselves.

MSG has been investigated extensively enough to be deemed safe for adults to use (except people who react adversely to it, of course), but it is kept out of foods for infants because very large doses have been shown to destroy brain cells in developing mice. Infants have not yet developed the capacity to fully exclude such substances from their brains. For other foods, the FDA requires that food label ingredient lists itemize each additive by its full name, including MSG as monosodium glutamate.[27]

Among flavorings added to foods, the flavor enhancer MSG has been determined to cause reactions in people with sensitivities to it.

Color additives not only make foods attractive, but identify flavors as well. Everyone agrees that yellow jellybeans should taste lemony and black ones like licorice.

Incidental Food Additives

Indirect or **incidental additives** are called *additives*, but they are really contaminants from some phase of production, processing, storage, or packaging. Examples of incidental additives include tiny bits of plastic, glass, paper, tin, and the like from packages and chemicals from processing, such as the solvent used to decaffeinate some coffees.

Some microwave products are sold in "active packaging" that participates in cooking the food. Pizza, for example, may rest on a cardboard pan coated with a thin film of metal that absorbs microwave energy and may heat up to 500°F. During the intense heat, some particles of the packaging components migrate into the food. Regular plastic packages heat up less, but particles still migrate. Materials from such packaging may not be entirely safe for consumption. A wise choice is to use only glass or ceramic containers or those plastics labeled as safe for microwaving. Avoid reusing disposable containers, such as margarine tubs or single-use trays from frozen microwavable meals, for microwaving.

MSG symptom complex:

- Breathing problems in people with asthma.
- Burning sensations on forearms, chest, and back of neck.
- Chest pain.
- Drowsiness.
- Facial pressure or tightness of skin.
- Headache.
- Nausea.
- Numbness of the neck that spreads to arms and back.
- Palpitations.
- Tingling, warmth, and weakness in face, upper back, neck, and arms.
- Weakness.

Coffee filters, paper milk cartons, paper plates, and frozen food boxes can all be made of bleached paper and so can contaminate foods with trace amounts of compounds known as dioxins. Dioxins form during the chlorination step in making bleached paper. Dioxins can migrate into foods that come in contact with bleached paper, but the amounts entering food are infinitesimally small—one part per trillion, or the equivalent of one second in 32,000 years. Such amounts do not appear to present a health risk to people, and drinking milk from bleached cartons appears to be safe. Dioxins are persistent, however, and they leach into the environment by way of both paper mill effluent and discarded paper products in landfills. Like heavy metals and organic halogens, dioxins accumulate, becoming more and more concentrated in land, water, and animals until they build up to hazardous levels.

Incidental additives sometimes find their way into foods, but adverse effects are rare. These additives are well regulated—all food packagers are required to test whether materials from packages are migrating into foods. If they are, their safety must be confirmed by strict procedures like those governing intentional additives.

Incidental additives are substances that get into food during processing. They are well regulated, and most present no hazard.

Nutrient Additives

Nutrients added to improve or to maintain the nutritional value of foods make up another class of additives. Among them are the enrichment nutrients added to refined grains, the iodine added to salt, vitamins A and D added to dairy products, and the nutrients used to fortify breakfast cereals. When nutrients are added to a nutrient-poor food, it may appear from its label to be nutrient-rich. It is, but only in those nutrients chosen for addition. Nutrients are sometimes also added for other purposes. Vitamins C and E used as antioxidants and beta-carotene as a colorant are examples already mentioned.

Examples of common nutrient additives:

- Thiamin, niacin, riboflavin, folate, and iron in grain products.
- Iodine in salt.
- Vitamins A and D in milk.
- Vitamin C in fruit drinks.
- Beta-carotene in cheeses.

key point *Nutrients are added to foods to enrich or to fortify them. These additives do not necessarily make the foods nutritious; they are rich only in the vitamins and minerals that have been added.*

To sum up the messages of this chapter, the U.S. food supply is safe and hazards are rare. Precautions against food-borne microbial illnesses are the most urgent measures for people to take to avoid food-related diseases. For optimal nutrition, though, people can do more. The Food Feature that follows offers pointers on the selection and cooking of foods for the healthiest possible diet.

Making Wise Food Choices and Cooking to Preserve Nutrients

IN GENERAL, THE more heavily processed foods are, the less nutritious they become. Does that mean that you should avoid all processed food? The answer is not simple: in each case, it depends on the food and on the process. Consider the case of orange juice and vitamin C.

The Choice of Orange Juice

Orange juice is available in several forms, each processed a different way. Fresh juice is squeezed from the orange, a process that extracts the fluid juice from the fibrous structures that contain it. Each 100 calories of fresh-squeezed juice contains 111 milligrams of vitamin C. When this juice is condensed by heat, frozen, and then reconstituted, as is the juice from the freezer case of the grocery store, 100 calories of the reconstituted juice contain just 88 milligrams of vitamin C because vitamin C is destroyed in the condensing process. Canning is even harder on vitamin C: 100 calories of canned orange juice have 82 milligrams of vitamin C.

These figures seem to indicate that fresh juice is the superior food, but consider this: most people's recommended intake of vitamin C (75 milligrams for women or 90 milligrams for men) is approximately the same as the vitamin C in a single serving of any of the above choices. Thus, for vitamin C, the losses due to processing are not a problem. Besides, processing confers enormous convenience and distribution advantages. Fresh orange juice spoils. Shipping fresh juice to distant places in refrigerated trucks costs much more than shipping frozen juice (which takes up less space) or canned juice (which requires no refrigeration). The fresh product still contains active enzymes that continue to degrade its compounds (including vitamin C) and so cannot be stored indefinitely without compromising nutrient quality. The savings gained from shipping and storing canned and frozen juices are passed on to consumers. Without canned or frozen juice, people with limited incomes or those with no access to fresh juice would be deprived of this excellent food.

Processing Mischief

Some processing stories are not so rosy. In Chapter 8, for instance, you saw how processed foods are often loaded with sodium as their potassium is leached away, exactly the wrong effect for people with hypertension. A related mischief of processing is the addition of sugar and fat—palatable, high-calorie additives that reduce nutrient density. For example, nuts and raisins covered with "natural yogurt" may sound like one healthy food being added to another, but the ingredient panel shows that generous amounts of sugar and fat accompany the yogurt. About 75 percent of the weight of the product is sugar and fat; only 8 percent is yogurt. To pick one nutrient for an example, look at what happens to the iron density of the raisins: 100 calories of raisins have 0.71 milligram of iron; 100 calories of "yogurt" raisins have 0.26 milligram of iron. These foods taste so good that wishful thinking can take hold, but the reality is that sugar- and fat-coated food is candy. The word *yogurt* on the label means only that one of the ingredients of the candy coating is some small amount of yogurt.

Best Nutrient Buys

Here are two good general rules for making food choices:

- Choose whole foods to the greatest extent possible.
- Seek out among processed foods only the ones that processing has improved nutritionally. For example, consumers benefit from processing that removes fat, as in fat-free milk, and pasteurization that improves the safety of juices.

Commercially prepared whole-grain breads, frozen cuts of meats, bags of frozen vegetables, and canned or frozen fruit juices do little disservice to nutrition and enable the consumer to eat a wide variety of foods at great savings in time and human energy. The nutrient density of processed foods exists on a continuum:

- Whole-grain bread > refined white bread > sugared doughnuts.
- Milk > fruit-flavored yogurt > canned chocolate pudding.

In terms of nutrient density, canned juice is almost as nutritious as fresh, but yogurt-covered raisins are not as nutritious as plain raisins.

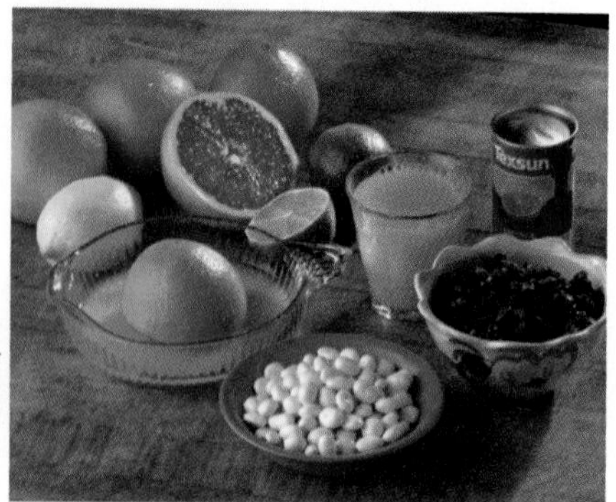

- Corn on the cob > canned creamed corn > caramel popcorn.
- Oranges > orange juice > orange-flavored drink.
- Baked ham > deviled ham > fried bacon.

The nutrient continuum is paralleled by another continuum—the nutrition status of the consumer. The closer to the farm the foods you eat, the better nourished you are, but that doesn't mean you have to live in the fields.

Conserving Nutrients at Home

Wise food choices are half the story of smart nutrition self-care; skillful food preparation is the other half. In modern commercial processing, losses of vitamins seldom exceed 25 percent. In contrast, losses in the 60 to 75 percent range are not unusual in food preparation at home, and they can be close to 100 percent. The kinds of foods you buy make a difference, but what you do with them in your kitchen can make an even greater difference.

Preventing Enzymatic Destruction

Vitamins are organic compounds synthesized and broken down by enzymes found in the foods that contain them. Like all enzymes, the enzymes that break down nutrients in fruits and vegetables have a temperature optimum. They work best at the temperatures at which the plants grow, normally about 70°F (25°C), which is also the room temperature in most homes. Chilling fresh produce slows down enzymatic destruction of nutrients. To protect the vitamin content, most fruits and vegetables should be vine ripened (if possible), chilled immediately after picking, and kept cold until use.

Protecting from Light and Air

Besides being vulnerable to enzyme-mediated spoilage, the vitamin riboflavin is light sensitive. It can be destroyed by the ultraviolet rays of the sun or by fluorescent light. For this reason, milk is not sold (and should not be stored) in transparent glass containers. Cardboard or opaque plastic containers screen out light, protecting the riboflavin. Since grain products such as macaroni and rice are also important sources of riboflavin, cooks who store them in glass jars should stow the jars in closed cupboards.

Some vitamins are acids or antioxidants and so are most stable in an acid solution away from air. Citrus fruits, tomatoes, and many juices are acid—as long as the skin is uncut or the can unopened, their vitamins are protected from air. If you store a cut vegetable or fruit, cover it with an airtight wrapper; close an opened carton of juice tightly and store it in the refrigerator.

Refreezing

Labels on frozen foods tell you "Do not refreeze." As food freezes, the cellular water expands into long, spiky ice crystals that puncture cell membranes and disrupt tissue structures, changing the texture of the food. There is usually no danger in eating a twice-frozen food although some nutrients are lost upon thawing and refreezing. Provided that the food hasn't spoiled while it was thawed or wasn't thawed at warm temperature, the main problem with a twice-frozen food is that it may be unappealing.

Preventing Nutrient Losses in Water

Minerals and water-soluble vitamins in fresh-cut vegetables readily dissolve into the water in which the vegetables are washed, boiled, or canned. If the water is discarded, as much as half of the vitamins and minerals in foods go down the drain with it. A bit of southern folk wisdom is to serve the cooking liquid with the vegetable rather than throwing it away; this liquid is known as the "pot liquor" and may be used to moisten cornbread or to make gravies or soups. Other ways to minimize cooking losses: steam vegetables over water rather than in it, stir-fry them in small amounts of oil, or microwave them. Wash the intact food vigorously and briefly—don't soak it. Cut vegetables after washing except for those such as broccoli that you have to cut to wash adequately. For peeled vegetables, such as potatoes, add them to water that is vigorously boiling, not

Purchase mostly whole foods or those that processing has benefited nutritionally.

Steam vegetables or cook them in a microwave oven.

Wrap foods tightly and refrigerate them. Space foods to allow chilled air to circulate around them.

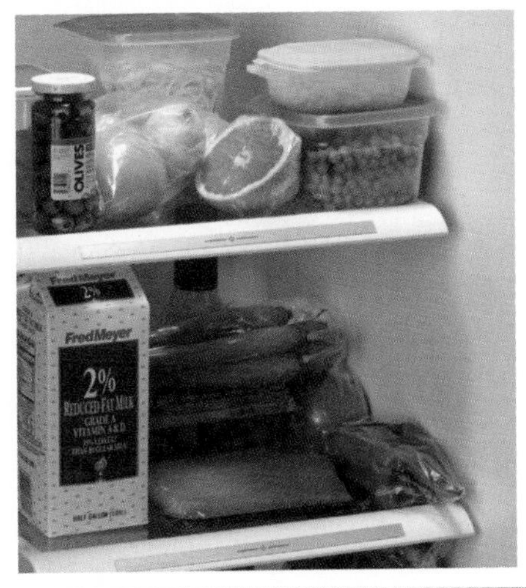

Take care when cooking in a microwave oven. Food can become extraordinarily hot or build up steam that can scald unprotected hands or face. Before cooking eggs, sausages, potatoes, or any food encased in a membrane, pierce the membrane to prevent explosion of the food. Never warm baby formula or food in a microwave oven because hot spots can form that can scald the baby.

Here's a way to tell if glass or other containers are made of microwave-safe materials. Microwave the empty container for one minute and carefully touch it.

- Warm = unsafe for microwave.
- Lukewarm = safe for short reheating use.
- Cool = safe for long microwave cooking times.

to cold water, to minimize the length of time the vegetables are exposed to nutrient-leaching water. Microwave ovens are excellent for conserving nutrients. They cook fast without requiring the addition of fats or excess liquid. Some special microwaving concerns appear in the margin.

During other types of cooking, minimize the destruction of vitamins by avoiding high temperatures and long cooking times. Iron destroys vitamin C by catalyzing its oxidation, but perhaps the benefit of increasing the iron content of foods by cooking in iron utensils outweighs this disadvantage. Each of these tactics is small by itself, but saving a small percentage of the vitamins in foods each day can mean saving significant amounts in a year's time.

Meanwhile, however, a law of diminishing returns operates. Most vitamin losses under reasonable conditions are not catastrophic. You need not fret over small vitamin losses that occur in your kitchen; you may waste energy or time that is valuable to you in other ways. If you start with fresh, whole foods containing ample amounts of vitamins and are reasonably careful in their preparation, you will receive a bounty of the nutrients that they contain.

Answers to these Self Check questions are in Appendix G.

1. Which of the following food hazards has the FDA prioritized as its number one concern?
   a. pesticides in food
   b. microbial food-borne illnesses
   c. intentional food additives
   d. environmental contaminants
2. In order to prevent food-borne illnesses, temperatures for holding cooked foods should be higher than:
   a. 85°F
   b. 100°F
   c. 140°F
   d. 212°F
3. Which of the following may be contracted from normal-appearing seafood?
   a. hepatitis
   b. worms and flukes
   c. viral intestinal disorders
   d. all of the above
4. Which of the following is correct concerning fruits that have been irradiated?
   a. They decay and ripen more slowly.
   b. They lose substantial nutrients.
   c. They are not safe to eat.
   d. They become radioactive.
5. Which of the following are likely sources of nitrates?
   a. beer
   b. cigarette smoke
   c. bacon and hot dogs
   d. all of the above
6. It is possible to eliminate all poisons from your diet by eating only "natural" foods. T F
7. Pregnant women are advised not to eat king mackerel, swordfish and shark because the FDA found unacceptably high levels of mercury in these fish. T F
8. The canning industry chooses treatments that employ the low-temperature-long-time (LTLT) principles for canning. T F
9. The artificial flavors and flavor enhancers are the largest single group of food additives. T F
10. Infants under one year of age should never be fed honey, which can contain spores of *Clostridium botulinum*. T F

nutrition on the net

For further study of the topics of this chapter, access these websites:

1. Find updates and quick links to these and other nutrition-related sites at our website: www.wadsworth.com/nutrition
2. Learn more about food-borne illnesses and livestock diseases such as BSE from the National Center for Infectious Diseases at the Centers for Disease Control and Prevention: www.cdc.gov/ncidod
3. Learn about the various types of food thermometers and how and when to use them from the USDA Thermy Campaign: www.fsis.usda.gov/thermy
4. Get commonsense health tips for travelers from the Centers for Disease Control and Prevention: www.cdc.gov/travel
5. Learn more about food irradiation from the Foundation for Food Irradiation Education and the International Consultative Group on Food Irradiation: www.food-irradiation.com and www.iaea.org/icgfi
6. Visit the Environmental Protection Agency to review tips for food buying and preparation that will help minimize pesticide exposure: www.epa.gov/pesticides/food
7. Visit the Canadian Food Inspection Agency: www.cfia-acia.agr.ca
8. Learn more about food safety in the marketplace from the Food Safety Inspection Service: www.usda.gov/fsis
9. Learn more about organic foods and national organic food standards: www.ams.usda.gov/nop
10. Visit the USDA Biotechnology Information Center: www.agnic.org

INTERNET ACTIVITY

Disorders of the digestive system are sometimes due to food spoilage. The Partnership for Food Safety Education is a coalition of organizations devoted to educating consumers how to buy, store, prepare and eat foods safely.

1. Go to the site of The Partnership for Food Safety Education http://www.fightbac.org
2. Click "Four Steps to Fight Bac."
3. For each step, click on the word (Clean, Separate, Cook, and Chill).

From the page on each step, list two ways to reduce the risk of food poisoning (or to increase food safety) that you would use. This means selecting a total of eight strategies to maintain food safety.

ORGANIC FOODS OR PRODUCTS OF BIOTECHNOLOGY: WHICH ARE BEST FOR OUR FUTURE?

IN 2000, THE USDA issued some long-awaited regulations for the production of **certified organic foods.** The ruling specifies that plant foods labeled *organic* must be produced without the use of most synthetic fertilizers and pesticides, using farming methods that are environmentally benign; animals must be provided with low-stress habitats that are natural to them. Organic foods must be free of ingredients produced by way of specified technologies, such as irradiation and biotechnology, including **genetic engineering** or **recombinant DNA (rDNA) technology.**

Proponents of organic foods claim that producing foods this way yields natural, safe foods that are low in pesticides and high in nutrients. This farming method, they say, will produce food harmlessly and indefinitely into the future. Meanwhile, people who develop **genetically modified foods** (or **GM foods**) by way of rDNA technology contend that GM food products are safe and wholesome. They claim that GM foods often present less pesticide contamination to consumers and the environment than conventional foods and that they are the most nutritious foods available today. Supporters hail biotechnology as nothing short of a revolutionary means of solving many of the planet's food problems today and in the future.

Who is correct? This Controversy strives to clarify the issues. First, it offers a brief glimpse into methods of organic food production; next, it outlines the workings of selective breeding, a conventional technique, and examines the basics of biotechnology. Then we take a scientific look at nutrition and health questions to see whether organic foods or those produced through biotechnology are better—or whether they are even substantially different from each other. Some terms in this section are defined in Table C14-1.

Organic Foods

A shopper picks up two fragrant, orange-yellow mangoes, one from a bin marked "organic" and another from a regular bin; both bear stickers identifying them as the "Hayden" variety. Both may be sweet and succulent. Both may have been kept in storage or shipped from faraway destinations. In fact, both may have been harvested from the same grove, separated by only a thin strip of land. The only obvious differences are the organic label and the cost: the organic mango costs more. The not-so-obvious differences are the methods used to produce them.

To be sold or labeled as organic, or to bear the USDA organic seal (see Figure C14-1), a food must be produced according to the procedures outlined in Table C14-2. A farmer wishing to grow and market organic foods must receive certification by USDA inspectors who check everything from the seed sown in the ground to the methods of making compost for fertilizer and the manufacturing of the final product.

In the 1990s, cropland used to grow certified organic foods doubled, making organic farming the fastest-growing segment of American agriculture. Organic foods are grown by using the techniques of *sustainable* agriculture (see Chapter 15). Organically grown vegetables and fruits are given organic fertilizer made

Table C14•1

ORGANIC FOOD AND FOOD TECHNOLOGY TERMS

- **anthrax** a potentially fatal disease caused by inhalation of or skin contact with spores of the toxin-producing anthrax bacterium; historically caused by contact with infected livestock, but currently a concern because it has been used as a biological weapon of terrorists.
- **antisense gene** a gene's chemical opposite, which interferes with the native working gene and prevents it from producing proteins.
- **certified organic foods** foods meeting strict USDA production regulations, including prohibition of most synthetic pesticides, herbicides, fertilizers, drugs, and preservatives, as well as genetic engineering and irradiation.
- **clone** an individual created asexually from a single ancestor, such as a plant grown from a single stem cell; a group of genetically identical individuals descended from a single common ancestor, such as a colony of bacteria arising from a single bacterial cell; in genetics, a replica of a segment of DNA, such as a gene, produced by genetic engineering.
- **genetic engineering** a field within biotechnology that involves the direct, intentional manipulation of the genetic material of living things in order to obtain some desirable trait not present in the original organism. Also called *recombinant DNA technology.*
- **genetically modified foods (GM foods)** transgenic foods produced through rDNA technology.
- **outcrossing** the unintended breeding of a domestic crop with a related wild species.
- **plant pesticides** substances produced within plant tissues that kill or repel attacking organisms.
- **recombinant DNA technology (rDNA technology)** a form of biotechnology that changes the characteristics of living things by manipulating the genes; includes methods of removing genes, doubling genes, introducing foreign genes, and changing gene positions to influence the growth and development of organisms.
- **stem cell** an undifferentiated cell that can mature into any number of specific specialized cell types. A stem cell of bone marrow may mature into one of many kinds of blood cells, for example.
- **transgenic organism** an organism resulting from the growth of an embryonic, stem, or germ cell into which a new gene has been inserted. Also called *genetically modified organism.*

Table C14•2

USDA CRITERIA FOR ORGANIC FOODS

All organic foods must be:

1. Processed or preserved with only agricultural substances or inorganic substances approved by the USDA for use in organic foods.
2. Produced without genetic modification that would not be possible under natural conditions, including rDNA technology.
3. Processed without ionizing radiation.

Food crops and feed for food-producing animals must be:

4. Produced without the use of most synthetic chemicals such as pesticides, herbicides, or fertilizers; a few low-toxicity exceptions are specified.
5. Fertilized without sewage sludge.
6. Managed in ways that preserve the integrity of farmland and surrounding environments, including reducing erosion, maintaining soil quality, and protecting drinking water sources from contamination.

Food-producing animals must be:

7. Grown without medications such as antibiotics or hormones, except appropriate vaccines to prevent diseases in healthy animals.[a]
8. Provided with living conditions similar to the animals' natural habitat, such as access to the outdoors, and maintained in conditions that reduce animal stress.
9. Grown in control-size herds, managed in ways that reduce the impact of manure and other aspects of animal ranching on the environment.
10. Fed only 100% organic feed.

[a]Under the requirements, sick animals may be treated with medications such as antibiotics, but products derived from treated animals, such as eggs, milk, or meat, may not be sold as organic.

from animal manure or vegetable compost and no synthetic fertilizers. No synthetic pesticides or disease-fighting agents are applied. Pests and diseases are battled by rotating crops each season, introducing predatory insects to kill off pests, and picking off large insects or diseased plant parts by hand. Only pesticides derived from natural sources, such as a pesticidal peptide toxin extracted from a bacterium that lives in the soil, are allowed for use on organic foods. Such technologies are familiar and fairly predictable.

To produce organic eggs, dairy products, and meats, farmers and ranchers raise food-producing animals in spacious, low-stress surroundings with access to the outdoors. Animals raised this way can grow large and stay healthy without growth hormones, daily antibiotics, and other drugs that are required when animals are stressed in overcrowded pens that make diseases likely.

For the most part, these methods are environmentally benevolent. People feel good about purchasing foods that they believe were produced in environment-friendly ways and are free of pesticides.

Now consider two other painstaking methods of food production: selective breeding and biotechnology.

Selective Breeding

For centuries, season after season, organic and conventional farmers have been changing the genetic makeup of their crop plants and farm animals by selectively breeding them for desirable traits. Today's lush, hefty, healthy agricultural crops and animals, from cabbage and squash to pigs and cattle, all demonstrate the results of those efforts.

Take corn, for example. Today's large, full, sweet ears and high yields bear little resemblance to the original wild, native corn with its sparse two or three kernels to a stalk. Breeders have even trained corn to "stay sweet" by breeding out an enzyme that normally turns sugar to starch within days after harvest. Selective breeding works, but slowly and imprecisely.

The Basics of Biotechnology

A shopper picks up two tomatoes from two adjacent bins. Both are red and ripe. Both may be delicious and nutrient-rich. Both may have been shipped a long distance. There are no obvious differences. Yet one may be a genetically altered tomato—the consumer may not be able to detect which one because no label is required for GM foods.

In contrast to the lengthy process of selective breeding, rDNA technology can change the genetic makeup of an organism in a year or two of work. A desirable gene from another species is inserted into the receiving DNA using a carrier that can deliver the gene to the

Figure C14•1

USDA SEAL AND ORGANIC FOOD LABELS

The USDA organic seal on a food means that it was produced in accordance with the USDA standards for organic foods and that it is 100 percent organic.[a]

Other organic labeling regulations include:

- Labels on foods containing *only* organically produced ingredients may state that the food is "100 percent organic," or simply "organic," and they may bear the USDA organic seal.
- Labels on foods containing *70 percent or more* organic ingredients may claim "made with organic ingredients" and may bear a different mark or seal of the certifying agent, but *not* the USDA organic seal.
- Labels of foods containing *less than 70 percent* organic ingredients may not make any organic claims but may identify organic ingredients on the ingredient list of the information panel.

[a]Organic labeling information is available at www.ams.usda.gov/nop/facts/labeling.htm.

The original wild corn from which today's corn was developed over centuries of selective breeding.

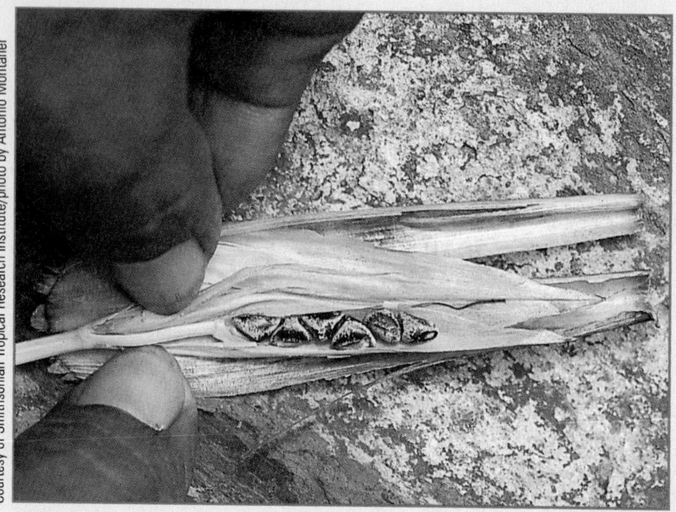

Courtesy of Smithsonian Tropical Research Institute/photo by Antonio Montaner

DNA. Marker genes that test the success of the method are also inserted. With great economy and precision, biotechnology can thus change one or more characteristics of a food. Figure C14-2 compares the genetic results of selective breeding and rDNA technology.

Plant cells make likely candidates for genetic engineering because a single plant cell can often be coaxed into producing an entire new plant. Each cell of the resulting plant contains an exact replica of the genetic information contained in the original cell. If scientists introduce any DNA fragments into that first single cell, those fragments will be faithfully reproduced in all of the cell's offspring. All of the resulting cells are **clone** cells—exact genetic replicas of the original.

For example, scientists can start with an immature cell, known as a **stem cell,** from the "eye" of a potato plant. Into that cell they can implant a gene for a protein from a virus that attacks potato plants—not the infective part but one of the other viral proteins. Then they can stimulate the stem cell to grow into a **transgenic organism,** in this case, a potato plant with a piece of viral protein in each of its cells. The viral protein stimulates the plant to develop resistance to an attack from the real virus in the growing field.

Like plants, animals have stem cells and can therefore be cloned. Twin calves have been cloned from a stem cell of a fetal calf; and a famous barnyard cousin, Dolly the sheep, was cloned from a stem cell harvested from a ewe's udder. Thus far, however, most cloned mammals have exhibited severe health problems, demonstrating the experimental nature of this research. Much more must be known before it can yield practical results.

The breadth of rDNA technology goes beyond simple improvements in selective breeding. Its techniques wield the awesome power to change the most basic patterns of life in ways never before possible. Its potential progeny reach far beyond nutrition into medicine, forestry, international trade policies, and even weapons of bioterrorism, such as treatment-resistant **anthrax** or other disease-causing organisms, and biodefenses against them.[1] Such uses and abuses, beyond the scope of this nutrition text, lie in the domain of ethical minds to ponder and control.

Food Biotechnology

Three areas of research in genetic engineering relate to the food supply:

- New strains of agricultural crops and animals offer desired traits, such as improved resistance to diseases or insect pests.
- Strains of microorganisms have been engineered to produce substances that, in nature, occur in only small amounts or not at all.

Figure C14•2

COMPARING SELECTIVE BREEDING AND rDNA TECHNOLOGY

Traditional Breeding

DNA is a strand of genes, much like a strand of pearls. Traditional selective breeding combines many genes from two individuals of the same species.

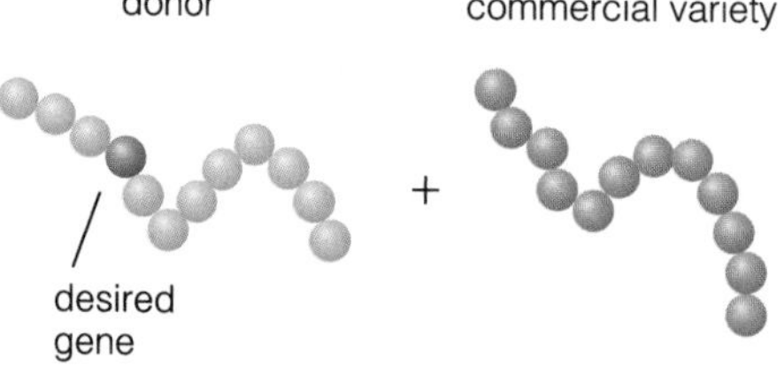

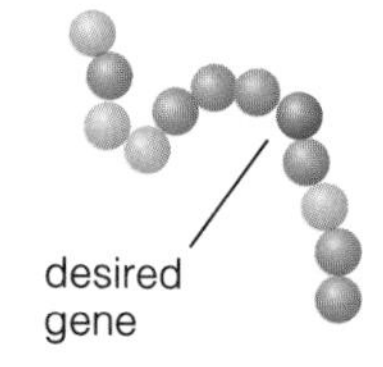

rDNA Technology

Through rDNA technology, a single gene or several may be transferred to the receiving DNA from the same species or others.

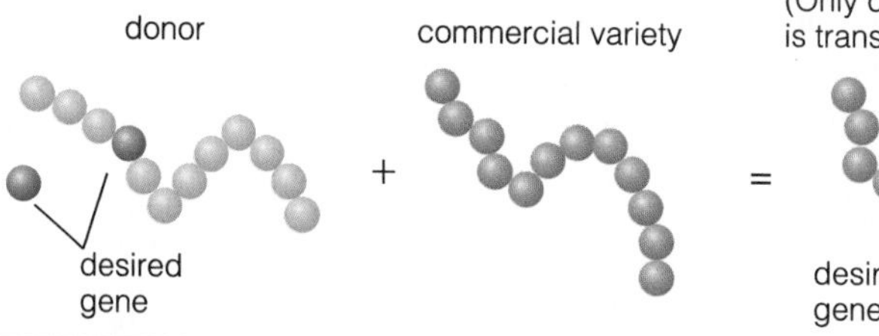

= new variety (Only desired gene is transferred.) desired gene

SOURCE: © 1995 Monsanto Company.

- Agricultural crops have been developed that withstand applications of herbicides.

The technique just described allows organisms to make proteins native to other living things. Another option is to block or suppress production of unwanted cell products. Tomatoes produce a protein that softens them after picking. Scientists introduced into a tomato plant an **antisense gene,** a mirror image of the native gene that coded for the "softening" enzyme. The antisense gene blocks production of the softener and produces otherwise normal tomatoes (see Figure C14-3). A vine-ripe tomato with the antisense gene can be harvested at its most flavorful and nutritious red-ripe stage and still last long enough to go to market.

Genetically modified microorganisms are currently at work. One bacterium, for example, was given the ability to make the enzyme rennin, necessary to produce cheese. Historically, rennin was harvested from the stomachs of calves, an expensive process. Through rDNA technology, the gene responsible for making rennin was snipped (enzymes do the snipping) from some calf DNA and transferred to a single bacterial cell. The resulting transgenic colony of bacteria became a factory mass-producing rennin. In the same way, another transgenic bacterium produces human growth hormone, so more children with growth hormone deficiency can grow normally.

Similarly, a group of crops is intended to benefit farmers by easing the task of controlling weeds or insect pests. Fields of plants genetically engineered to withstand potent herbicides can be sprayed with weed-killers, leaving only the desired crop in the field. A cotton plant that produces its own insecticide has rendered pesticide sprays unnecessary for its harvest.[2] The plants produce what the Environmental Protection Agency (EPA) calls **plant pesticides,** a group of pesticides made by the plants themselves. The technology is not yet a panacea for farmers, however. Limitations include a lack of control of pests *not* killed by the plant pesticide, the high cost of producing the transgenic plants, and the development of resistance to the pesticide among the target population of insects.[3] Researchers are working to find solutions to the problems as they arise.

These salmon are all of the same age and type. The largest one received a growth-enhancing gene, greatly accelerating its growth rate.

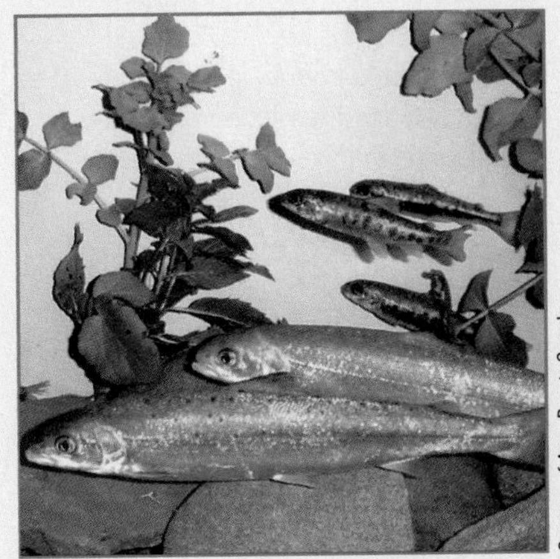

Courtesy of Aqua Bounty Canada

Animals, too, may be modified for the benefit of human beings.[4] The salmon in the photo grows at an astonishing rate and can be ready for market in far less time than it takes to grow an ordinary salmon.

The Promises of Biotechnology

Researchers are working toward creating animals and plants with the ability to produce needed pharmaceutical products as well as food. For example, a cow cloned with the genetic equipment to make milk containing a vaccine against a human disease could provide immunization as well as nourishment to a whole village of people who now lack medical resources. In the same way, researchers hope to create bananas or potatoes or other common foods that produce needed pharmaceutical products in their tissues, a process whimsically called "biopharming."

Recently, the genome of rice was sequenced, leading to the hope that the staple food for half the world's population may be improved in meaningful ways. Some rice has already been implanted with the genetic equipment to produce beta-carotene and so can potentially fight vitamin A deficiency in places where it causes blindness. Other rice varieties may become more productive; disease, pest, and drought resistant; or more complete sources of protein. Likewise, shrimp may fight diseases with genetic ammunition contributed by sea urchins. Some GM

Figure C14•3

HOW AN ANTISENSE GENE CAN BLOCK FORMATION OF A PROTEIN

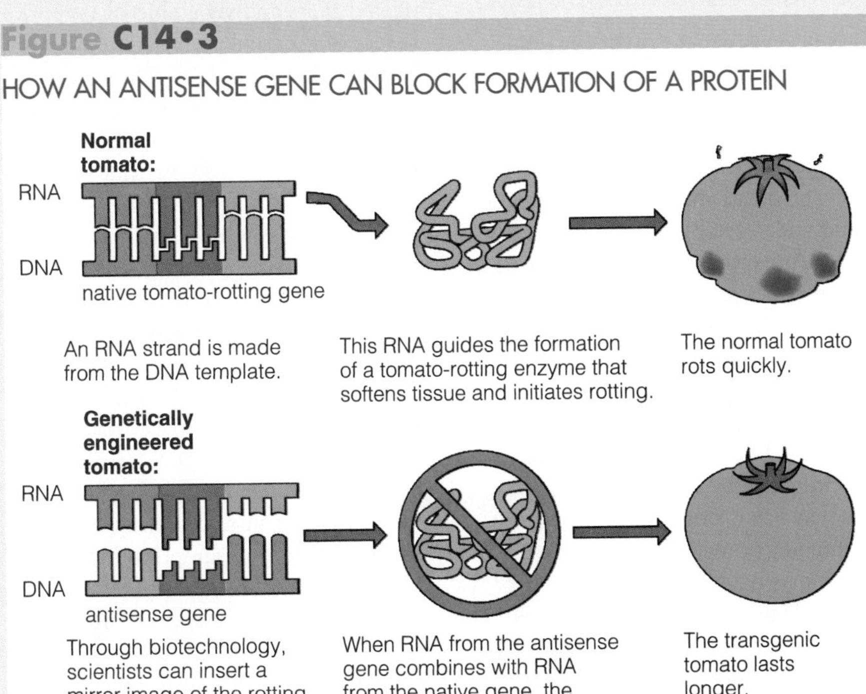

Normal tomato:
RNA
DNA
native tomato-rotting gene
An RNA strand is made from the DNA template.
This RNA guides the formation of a tomato-rotting enzyme that softens tissue and initiates rotting.
The normal tomato rots quickly.
Genetically engineered tomato:
RNA
DNA
antisense gene
Through biotechnology, scientists can insert a mirror image of the rotting gene (an antisense gene) into tomatoes.
When RNA from the antisense gene combines with RNA from the native gene, the native RNA is blocked from producing the rotting enzyme.
The transgenic tomato lasts longer.

plants may grow food in soil so polluted, or dry, or salty that all other plants wither and die. In 1998, genetically engineered crops accounted for about 40 percent of total planted acreage in this country, and about 40 genetically engineered agricultural products were approved for sale.[5]

Issues of Nutrients and Safety

Do organic foods and products of biotechnology differ substantially from each other in their nutrient contents or safety to consumers and the environment? The next sections address these questions.

Nutrient Composition of Organic Foods and GM Foods

Supporters claim that foods produced the organic way have more vitamins and minerals than conventional foods because organic compost fertilizes plants with more vitamins and trace minerals than those provided by synthetic chemical fertilizers. However, plants do not take up vitamins from either compost or other fertilizers; through their roots, they absorb only simple elements and water, which they use in making complex organic molecules like vitamins in their tissues.

Trace minerals, on the other hand, do get into plants, and they *are* often more abundant in organic compost than in synthetic fertilizers, which contain only the minerals that are added to the mix. A reviewer who studied the available research on the nutritional quality of organic foods deemed the data suggestive, but not conclusive, of their superiority.[6] Some findings showed slightly greater vitamin C content, lower nitrate, and somewhat improved protein quality of organic foods, but the differences were small and not significant to human health. Some people claim that organic fruits and vegetables taste better, and if people who choose organic food choose fruits and vegetables more often, then so much the better.

The notion that any traditional food, whether organic, conventional, or a GM food, has superior nutrient composition is invalid. The nutrient contents of traditional fruits and vegetables are limited to ranges set by nature. The *potential* ranges for nutrients in GM foods are virtually without limit, but except for these intentional variations, nutrients in GM foods are identical to those of comparable traditional foods. The rice enriched with beta-carotene, mentioned earlier, is an example of such an intentionally altered food. Practically speaking, eating the rice is the same as eating plain rice and taking a beta-carotene supplement.

One day the additional vitamins and minerals that now must be added to enrich the grain products and dairy products for sale in the United States may be grown in GM wheat, rice, and other grains or be produced in the milk of GM cows, eliminating the possibility of human error. Soon consumers may see common foods such as potatoes turned into "functional foods," sporting heavy doses of disease-fighting phytochemicals that occur naturally in less familiar foods such as flaxseed or ginger. Reduced nutrient contents of GM foods is not a concern. Instead, informed consumers may more rightly worry about nutrient or phytochemical overdoses from all sources.

Pesticide Residues in Organic Foods

Consumers who choose organic foods often wish to avoid pesticide contamination. Can consumers who pay more for organic foods trust that such foods are free of pesticides? Recently, a testing group purchased and tested 1,000 pounds of fruit and vegetables from across the nation. As expected, their tests revealed residues of pesticide in about three-quarters of the conventional food samples. The researchers also detected pesticide contamination in one-quarter of the organic food samples.[7] Pesticides may have "drifted" onto the organic foods from nearby sprayed fields; pesticides may have contaminated them during shipping or marketing; or a few farmers may have used pesticides on foods and later sold them, illegally, as organic. Based on these results, the word *organic* on the label is no guarantee that a food is pesticide-free.

Even if organic foods are pesticide-free, it is unlikely that this would constitute a health advantage for consumers. Federal standards for pesticide residues are set far below the threshold of any known threat to human health. An old adage, "the dose makes the poison," is applicable: the human body is well equipped to handle poisonous substances without apparent harm—as long as the doses are small enough.

In the study of pesticides mentioned above, all food samples but one fell within the federal guideline for safety. That one exception involved traditionally grown green peppers containing a pesticide not approved for use on that food. Good news: When the researchers washed the fruits and vegetables for 10 seconds under running water, the pesticide residues were dramatically reduced in both the conventional foods and the organic foods.

Pesticide Residues in GM Foods

Industry scientists contend that rDNA technology could eliminate problematic pesticide residues. This is because both the nature of the pesticide and the amount present in the food are predetermined by the genetics of the plant and are not left open to human error or misuse.

Pesticides produced by bioengineered fruits and vegetables exist in the tissues of the food and therefore cannot be washed or peeled off. The nature of the pesticides produced by plants differs from that of synthetic ones, however. From Chapter 6 we know that DNA governs the synthesis of protein, and plant pesticides are products of the activities of DNA—that is, they are peptide molecules. In the human body, peptide molecules are denatured and digested by the digestive system and so are rendered harmless. It is unlikely, then, that peptide plant pesticides pose a danger to the body. Besides, all plant pesticides are regulated as food additives by the FDA and so must be proved safe for consumption before being marketed to consumers.

The same pesticide internally produced by bioengineered fruits and vegetables may also appear in organic foods. As already mentioned, one of the few pesticides approved for use on organic foods is derived from a bacterium in soil. The extracted bacterial product is a peptide chain that is lethal to some insects that feed on corn. Recently, food technologists inserted into a corn plant the genetic information for producing this peptide, resulting in transgenic corn that produces the bacterial pesticide—the same one that organic farmers spray on their crops to kill off the same invading insects. The safety of this peptide, however, was established through extensive studies performed by the company producing the new corn approved by the FDA for animal feed.* The FDA had not yet approved the corn for the human food supply because of concerns that it might contain an allergen, however, when an unforeseen turn of events derailed it, as explained later on. This event seemed to confirm many people's worst fears of biotechnology—that unforeseen events, unpredictable outcomes, and unwanted consequences of rDNA technology threaten human health.

Potential Health Risks from GM Foods

The case for requiring rigorous safety testing of new GM foods is strong. For example, if a disease-producing microorganism has donated genetic material to make an rDNA food, scientists must prove that no dangerous characteristic from the microorganism has also entered the food. If the inserted genetic material comes from a food to which people develop allergies, such as nuts, unless the new food can be certified as allergen-free, product labeling must alert those with allergies.[8] Furthermore, newly altered genetic material may create unique proteins never before encountered by the human body. Their effects should be studied and their presence regulated to ensure that people can eat them safely.

To help determine the safety of GM foods, the FDA has established a National Center for Food Safety and Technology (NCFST) in Illinois. Studies performed at the NCFST guide the FDA in setting regulations governing food processes and products.

*Trade name "StarLink."

Potential Health Risks from Organic Foods

Although proponents of organic foods believe that these foods are safer for consumers than conventional foods, this may not be the case. For example, the application of improperly composted animal manure fertilizer may expose consumers to dangerous microbial diseases, such as *E. coli* 0157:H7. Organic sprouts and greens have caused many cases of serious food poisoning in recent years. Unpasteurized organic juices, milk, or cheeses may also constitute microbial hazards because pasteurization is required to kill disease-causing microorganisms in these foods. Furthermore, organic foods contain no preservatives and tend to spoil faster than other foods. Those purchasing organic foods are urged to buy only the amounts that can be consumed within a few days, to store and cook the food properly, to wash raw produce vigorously, and to buy only pasteurized organic dairy products and juices.

Environmental Effects of GM Crops

Advocates say that advances in rDNA technology will enable farmers to produce bumper crops of food on far fewer acres of land, with less loss of water and topsoil, and far less use of toxic pesticides and herbicides to end up in foods and drinking water. By one estimate, biotechnology has already led to an 80 percent reduction in insecticide use among U.S. transgenic cotton crops.[9] Some other potential undesirable effects are worthy of study, however.

One concern is the possibility of **outcrossing,** accidental cross-pollination of plant-pesticide crops with related wild weeds. Opponents of biotechnology refer to outcrossing as "genetic pollution"; as yet there is no way to clean up this pollution once it occurs. If a weed by chance were to inherit a plant-pesticide trait from a GM parent, it would have an enormous survival advantage over other wild species and could crowd them out. A way to prevent outcrossing exists, but it has spawned problems of its own. Scientists can use so-called *terminator technology*, altering a transgenic plant's genetic material to ensure the destruction of all of the plant's offspring. Terminator technology keeps all transgenic plants from passing their genes to unwanted weeds, but makes it impossible for the world's farmers to save fertile seeds from their own harvests from year to year. Thus, all farmers would be forced to buy expensive new seeds from the biotechnology companies each year. Poor subsistence farmers would suffer the most economic hardship.[10] An alternative possible answer to genetic pollution: modify only the genetics of structures (chloroplasts) whose DNA passes through female lines; only male genes are carried by pollen.[11]

Outcrossing is what probably happened to the pesticide-producing corn mentioned earlier, causing chaos in the U.S. corn market. After winning FDA approval to grow the corn for animal feed but not yet for human consumption, the company sold the GM corn seed to farmers who grew and harvested it successfully. But some of the corn turned up in the human food supply in tortillas, taco shells, tostadas, and chips. It is suspected that the pesticide-producing gene had outcrossed to nearby fields of ordinary corn destined for people's dinner tables. The company that developed the corn gave up its license to produce the corn and is helping to pay for the costs incurred by those who grew the transgenic corn, either intentionally or accidentally.

In addition to threats of outcrossing, the possibility also exists that the new crops may damage wildlife. In the laboratory, monarch butterfly larvae die when they feed on pollen from pesticide-producing corn. In real life, wild butterflies do not seem to consume enough of the toxic pollen to be harmed, at least in the short term.[12] In this case, it seems that the new technology may even protect countless numbers of monarchs and other harmless or beneficial insects that now die annually when they dine on conventionally sprayed fields.[13] Still, long-term effects on butterflies are unknown, and whether wildlife of all kinds may be adversely affected by transgenic plants or animals requires scientific vigilance. Organic farming, in contrast, poses no threat to benign species such as butterflies.

Environmental Benefits or Organic Foods

Organic farming methods bring benefits to the environment. Crop rotation and natural fertilizers not only minimize the chemical impact on wildlife and human beings but are beneficial to the soil. Such techniques curtail erosion and prevent contamination of drinking water from manure runoff, which is common on commercial farms. Chapter 15 and Controversy 15 revisit issues of agriculture and the environment.

Opposition to Biotechnology

Some consumers fear what they call "Frankenfoods," and Table C14-3 out-

Table C14•3

FOOD BIOTECHNOLOGY: POINT, COUNTERPOINT

Arguments in Opposition to Genetic Engineering	Arguments in Support of Genetic Engineering
1. **Ethical and moral issues.** It's immoral to "play God" by mixing genes from organisms unable to do so naturally. Religious and vegetarian groups object to genes from prohibited species occurring in their allowable foods.	1. **Ethical and moral issues.** Scientists throughout history have been persecuted and even put to death by fearful people who accuse them of playing God. Yet, today many of the world's citizens enjoy a long and healthy life of comfort and convenience due to once-feared scientific advances put to practical use.
2. **Imperfect technology.** The technology is young and imperfect—genes rarely function in just one way, their placement is imprecise ("shotgun"), and all of their potential effects are impossible to predict. Toxins are as likely to be produced as the desired trait. Over 95 percent of DNA is called "junk" because scientists have not yet determined its function.	2. **Advanced technology.** Recombinant DNA technology is precise and reliable. Many of the most exciting recent advances in medicine, agriculture, and technology were made possible by the application of this technology.
3. **Environmental concerns.** Environmental side effects are unknown. The power of a genetically modified organism to change the world's environments is unknown until such changes actually occur—then the "genie is out of the bottle." Once out, the genie cannot be put back in the bottle because insects, birds, and the wind distribute genetically altered seed and pollen to points unknown.	3. **Environmental protection.** Genetic engineering may be the only hope of saving rain forest and other habitats from destruction by impoverished people desperate for arable land. Through genetic engineering, farmers can make use of previously unproductive lands such as salt-rich soils and arid areas.
4. **"Genetic pollution."** Other kinds of pollution can often be cleaned up with money, time, and effort. Once genes are spliced into living things, those genes forever bear the imprint of human tampering.	4. **Genetic improvements.** Genetic side effects are more likely to benefit the environment than to harm it.
5. **Crop vulnerability.** Pests and disease can quickly adapt to overtake genetically identical plants or animals around the world. Diversity is key to defense.	5. **Improved crop resistance.** Pests and diseases can be specifically fought on a case-by-case basis. Biotechnology is the key to defense.
6. **Loss of gene pool.** Loss of genetic diversity threatens to deplete valuable gene banks from which scientists can develop new agricultural crops.	6. **Gene pool preserved.** Thanks to advances in genetics, laboratories around the world are able to stockpile the genetic material of millions of species that, without such advances, would have been lost forever.
7. **Profit motive.** Genetic engineering will profit industry more than the world's poor and hungry.	7. **Everyone profits.** Industries benefit from genetic engineering, and a thriving food industry benefits the nation and its people, as witnessed by countries lacking such industries. Genetic engineering promises to provide adequate nutritious food for millions who lack such food today. Developed nations gain cheaper, more attractive, more delicious foods with greater variety and availability year round.
8. **Unproven safety for people.** Human safety testing of genetically altered products is generally lacking. The population is an unwitting experimental group in a nationwide laboratory study for the benefit of industry.	8. **Safe for people.** Human safety testing of genetically altered products is unneeded because the products are essentially the same as the original foodstuffs.
9. **Increased allergens.** Allergens can unwittingly be transferred into foods.	9. **Control of allergens.** A few allergens can be transferred into foods, but these are known, and foods likely to contain them are clearly labeled to warn consumers.
10. **Decreased nutrients.** A fresh-looking tomato or other produce held for several weeks may have lost substantial nutrients.	10. **Increased nutrients.** Genetic modifications can easily enhance the nutrients in foods.
11. **No product tracking.** Without labeling, the food industry cannot track problems to the source.	11. **Excellent product tracking.** The identity and location of genetically altered foodstuffs are known, and they can be tracked should problems arise.
12. **Overuse of herbicides.** Farmers, knowing that their crops resist herbicide effects, will use them liberally.	12. **Conservative use of herbicides.** Farmers will not waste expensive herbicides in second or third applications when the prescribed amount gets the job done the first time.
13. **Increased consumption of pesticides.** When a pesticide is produced by the flesh of produce, consumers cannot wash it off the skin of the produce with running water as they can with ordinary sprays.	13. **Reduced pesticides on foods.** Pesticides produced by produce in tiny amounts known to be safe for consumption are more predictable than applications by agricultural workers who make mistakes. Because other genetic manipulations will eliminate the need for postharvest spraying, fewer pesticides will reach the dinner table.
14. **Lack of oversight.** Government oversight is run by industry people for the benefit of industry—no one is watching out for the consumer.	14. **Sufficient regulation and rapid response.** Government agencies are efficient in identifying and correcting problems as they occur in the industry.

lines the issues. Some say that genetic tampering produces effects that are not fully understood and that the industry is driven solely by potential profits without moral judgment or laws to harness its effects.[14] Consumer benefit has not been demonstrated, they argue, and consumer risk is not defined. Genetic decisions, they say, are best left to the powers of nature. Those given to flights of imagination envision a biotechnology run amok, used for frivolous, greedy purposes such as cloning dinosaurs for entertainment. More serious concerns for many people center on the prospect of cloning human beings for certain traits and genetic "improvements." Federal funds are withheld from any laboratory involved in human genetic engineering experiments, but independent laboratories have stated their intent to move ahead in such experiments.

Some people erroneously accuse rDNA technology of causing the illness and death associated with the consumption of manufactured tryptophan supplements (EMS, described in the Consumer Corner of Chapter 6). In reality, a "natural" fermentation step used by one Japanese company in the production of manufactured L-tryptophan introduced impurities suspected of causing the disease in susceptible people, although the mechanism remains unknown.[15]

Other concerns about biotechnology focus on "foreign" proteins produced by organisms that receive new genes. Most proteins are degraded by digestive enzymes and rendered nontoxic to the body. Several dangerous exceptions exist, though, including the botulinum toxin and certain other peptide toxins, which are absorbed into the bloodstream intact. Some scientists worry that such peptides may accidentally form in products of genetic engineering and not be detected promptly enough to prevent harm to those who eat the foods. Concerns also focus on lack of independent long-term testing and lack of consumer labeling information.

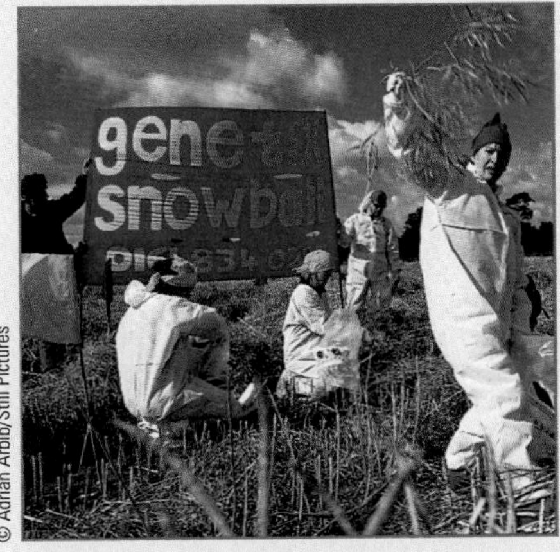

Protesters uprooting genetically altered plants.

© Adrian Arbib/Still Pictures

The FDA's Position on GM Foods

The FDA has taken the position that GM foods require no special safety testing or labeling unless they are substantially different from foods already in use.[16] The FDA holds the developers of the new foods responsible for testing those that differ significantly from traditional foods. The FDA assumes that any product with an antisense gene, including the tomato described earlier, is safe since the antisense gene merely prevents the synthesis of a protein and adds nothing but a tiny fragment of genetic material. The FDA does require that any extra substances produced in the food, such as new enzymes, hormones, or resistance traits, meet the same safety standards applied to all additives.

Besides calling for independent testing of GM products, consumer advocacy groups want labels to clearly identify all genetically altered products so that consumers can make informed decisions. They claim that by not requiring such labeling, the FDA forces millions of consumers to be the unwitting testers of GM foods. Without labeling, people with religious objections to particular foods may be unable to avoid consuming genes of prohibited organisms that have been added to permitted foods. For example, someone keeping a kosher kitchen may purchase a food product containing genes normally found in pork. One group filed suit against the FDA, hoping that the courts would mandate labeling of GM foods.[17] So far, the FDA has held to its position that no labeling is needed.

Speaking in defense of the FDA's position are the agency itself, recognized as the nation's leading expert and advocate for food safety, and the American Dietetic Association, which represents current scientific thinking in nutrition.[18] Many other scientific organizations agree, contending that rDNA technology can deliver on promises for an improved food supply if we give it a fair chance to do so.

According to the FDA, these transgenic tomatoes are essentially the same as regular tomatoes.

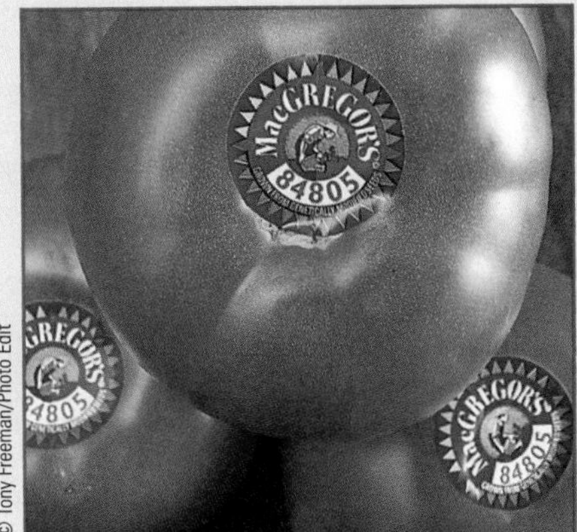

© Tony Freeman/Photo Edit

The Final Word

For those who would worry themselves into a diet of crackers and water, rest assured that eating more fruits and vegetables brings health advantages that far outweigh any slight risks that might arise from eating conventionally grown, organic, or genetically modified foods.[19] Table C14-4 sums up some pros and cons of food production methods and Table C14-5 provides Internet sources for information about biotechnology and associated issues. As presented in Controversy 6 and Chapter 11 of this book, the evidence is overwhelming: those who eat five or more servings of fruits and vegetables each day remain healthier than those who do not, regardless of the source of those foods. Furthermore, research has shown that the great majority of all fruits and vegetables, both domestic and imported, test well within safety margins set for pesticide contamination.[20] Consumers can best serve their health needs, then, by washing all produce thoroughly and by handling it properly before consumption but, most of all, by eating five or more servings every day.

Table C14•4

FOOD PRODUCTION METHODS COMPARED: ORGANIC, CONVENTIONAL, AND rDNA TECHNOLOGY

Soil Condition and Environment

- *Organic*: Improves soil condition through crop rotation and the addition of complex fertilizers such as manure; controls erosion; highly protective of waterways and wildlife. Uses sustainable agriculture techniques.
- *Conventional*: Depletes soil; adds synthetic chemical fertilizers containing only a few key elements; can create soil erosion problems. Runoff pollutes waterways, and sprays poison wildlife such as birds and beneficial insect predators.
- *rDNA technology*: No direct effect on soil or erosion; may require fewer pesticide sprays, thus protecting waterways and wildlife, but may harm wildlife by exposing wild species to altered genes or plant pesticides; may soon make use of salty, dry, or other currently unusable lands. May produce "genetic pollution."

Nutrients in Foods

- *Organic*: Suggestive evidence of slightly increased content of trace minerals, vitamin C, and improved amino acid balance in produce over conventionally farmed produce.
- *Conventional*: Standards for nutrient composition of foods are set by analysis of conventionally produced foods.
- *rDNA technology*: Virtually unlimited potential for nutrient and phytochemical content, at the will of the producer.

Benefits to Consumers

- *Organic*: Reduced exposure to pesticides and other sprays and animal medications and hormones. New standards define organic techniques, with regulatory oversight. Long history of safety for human consumption of food varieties. Ethical comfort of knowing that food-producing animals are well-treated.
- *Conventional*: General safety and pesticide residues monitored regularly; many varieties of foods available at low cost.
- *rDNA technology*: Greater food production at low cost, keeping consumer prices low and availability high. Particular products may meet particular consumer demands, such as better flavor, increased vitamin or phytochemical content, or improved freshness of foods. Potential exists for helping to ease world hunger. Crops may produce medicines needed in impoverished areas of the world.

Consumer Safety Issues

- *Organic*: Consumer must wash produce well to remove possible dangerous microbial contamination and pesticides that may have "drifted" onto produce.
- *Conventional*: Consumer must wash produce well to remove possible dangerous microbial contamination and pesticides that are applied to produce.
- *rDNA technology*: Consumer must wash produce well to remove possible dangerous microbial contamination and pesticides (especially herbicides) that are applied to produce. Internally produced plant pesticides do not wash off. Other dangers include introduction of allergens from other species and unproven safety of consuming rDNA products over a lifetime. Unknown dangers may also exist.

Table C14•5

BIOTECHNOLOGY INTERNET SITES

Much more information (and misinformation) is available on the Internet. Here are some sites to explore.

1. For policies and applications, search for "biotechnology" at the USDA Biotechnology Information Center: www.agnic.org
2. Get a "pro" biotechnology perspective from the Council for Biotechnology Information: www.whybiotech.com
3. A scientific view is available by searching for "biotechnology" at the International Food Information Council: http://ific.org
4. Consider the perspectives of those who oppose biotechnology at the Genetic Engineering section of Greenpeace, USA: www.greenpeaceusa.org
5. Another opposition view is available at the Transgenic Café of the Union of Concerned Scientists: www.ucsusa.org

Jacob Lawrence (1917–2000), *Street to M'bari*, 1964

Hunger and the Global Environment

Frequently Asked Questions

What U.S. food programs are directed at stopping domestic hunger? p. 561

What is the state of world hunger? p. 563

How can people engage in activism and simpler lifestyles at home? p. 568

Contents

ONE PERSON IN every five worldwide experiences persistent hunger—not the healthy appetite triggered by anticipation of a hearty meal, but the painful sensation caused by a lack of food. Tens of thousands die of starvation each day—one every two seconds. Table 15-1 presents the current best estimate of the number of hungry people worldwide. For perspective, the table also includes world incidences of vitamin and mineral deficiencies and overnutrition (obesity). The presence of any one of these conditions—hunger, nutrient deficiencies, or overnutrition—does not exclude the coexistence of any of the others within the same country, the same area, or even the same family.[1] Up to 80 percent of hungry children live in countries that produce surplus food—the decisions of policymakers in those areas largely determine who among the population has access to the bounty.

In the United States, where most people enjoy a life of relative abundance, well over two and a half million children are hungry at least some of the time.[2] Under the broad definition of **food insecurity,** over nine million U.S. children do not know where their next meal is coming from or when it will come (see Figure 15-1). Terms relating to hunger are defined in Table 15-2.

The chronically hungry people of this world suffer from undernutrition, a condition of energy and nutrient deficiency that causes general weakness and fatigue. The greater the caloric deficit, the greater the susceptibility to nutrition-related health risks. Undernutrition stymies mental and physical development in children and makes people susceptible to potentially fatal diseases such as dysentery, whooping cough, and tuberculosis. Consequences of chronic hunger in children include infant mortality, stunted growth, iron-deficiency anemia, poor learning, extreme weakness, clinical signs of protein-energy malnutrition (PEM), increased susceptibility to disease, loss of the ability to stand or walk, and premature death.[3]

The tragedy described on these pages may seem at first to be beyond the influence of the ordinary person. What possible difference can one person make? Can one person's choice to recycle a bottle or to serve a meal to the homeless or to join a hunger-relief organization make a difference? In truth, such choices produce several benefits. For one, a single person's awareness and example, shared with others, can influence many people over time. For another, an action repeated becomes a habit. For still another, making choices with awareness of their impacts lends a sense of control over those impacts. That sense of personal control, in turn, helps people to take effective action in many areas.

Students can play a powerful role in bringing about change. Students everywhere are helping to change governments, human predicaments, and environmental problems for the better. Student movements persuaded 127 universities and many institutions, corporations, and government agencies to put pressure on South Africa and succeeded in ending apartheid. Student pressure opened the way for the first deaf president at a university for the deaf. Students offer major services to communities through soup kitchens, home repair programs, and child education. The young people of today are the best hope of millions for a better tomorrow.

Table 15•1

GLOBAL UNDERNUTRITION AND OVERNUTRITION, 2000

Condition	Global Incidence
Hunger, protein-energy malnutrition	≥1.2 billion people
Vitamin and mineral deficiencies[a]	2.0–3.5 billion people (about half the world's population)
Overnutrition, obesity	≥1.2 billion people

[a]Vitamin and mineral deficiencies occur in both underfed and overfed populations.
SOURCE: Adapted from G. Gardner and B. Halweil, *Underfed and Overfed: The Global Epidemic of Malnutrition*, World Watch Paper 150 (Washington, D.C.: Worldwatch Institute, 2000), p. 7.

The Challenge to Change

Creating an adequate food supply for all of the world's citizens poses two challenges. The first is to provide enough food to meet the needs of the earth's expanding population, without destroying natural resources needed to continue producing food. The second challenge is to ensure food security—that is, to make sure all people have access to enough food to live active, healthy lives. By all accounts, today's total **world food supply** can abundantly feed the entire current population, but this does not guarantee that people in need are able to get adequate food. If people do not have enough money to buy food or to buy the land, seeds, and tools to grow food or if natural or human-made disasters such as drought or war prevent them from getting food, hunger remains a problem.

Worries for the future also exist. Many forces compound to threaten world food production and distribution in the next decade, and we may face more global food insecurity because of them:

- *Hunger, poverty, and population growth*. Millions of the world's people are starving. Fifteen children die of malnutrition every 30 seconds, but 125 children are born during that same 30 seconds. Every day, the earth gains another 220,000 new residents to feed, most of them born in impoverished areas.[4]
- *Loss of food-producing land*. Food-producing land is becoming saltier, eroding, and being paved over. Each year, the world's farmers try to feed some 85 million additional people with 24 billion fewer tons of topsoil. This loss threatens overall food security.
- *Accelerating fossil fuel use*. Fossil fuel use is growing, with attendant pollution of air, soil, and water; ozone depletion; and global climate changes.
- *Increasing air pollution*. As populations increase, air quality diminishes in many areas around the globe.*
- *Atmosphere and climate changes, droughts, and floods*. Climbing atmospheric levels of heat-trapping carbon dioxide are a concern. The concentration of carbon dioxide is now 26 percent higher than 200 years ago. As a result, a warming trend seems to be taking place.[5] Climate changes cause both droughts and floods, which destroy crops and people's homelands.
- *Ozone loss from the outer atmosphere*. The outer atmosphere's protective ozone layer is growing thinner, permitting harmful radiation from the sun to damage crops and ecosystems and increasing the likelihood of skin cancers and cataracts in people and animals.
- *Water shortages*. The world's supplies of fresh water are dwindling and becoming polluted.

*Many older references have been removed to save space, but they are available in older editions of this book.

world food supply the quantity of food, including stores from previous harvests, available to the world's people at a given time.

Figure 15•1

U.S. FOOD SECURITY AND HUNGER, 1995 AND 1999

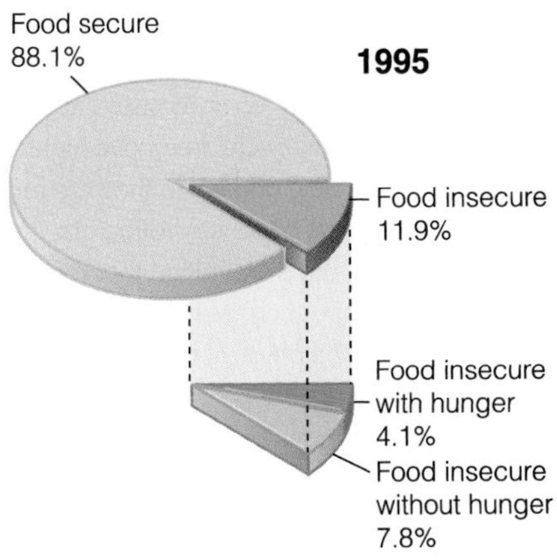

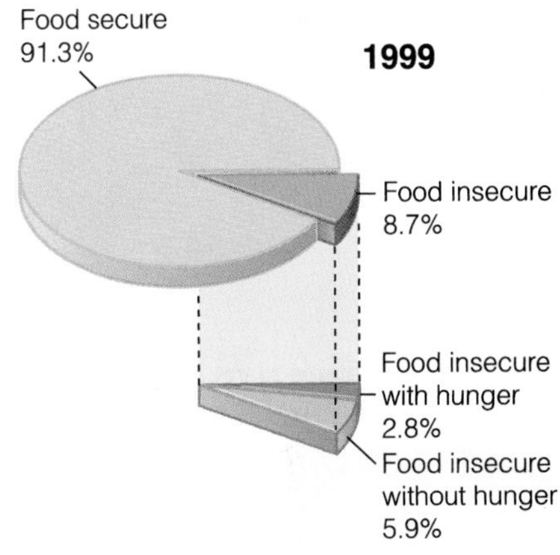

SOURCE: USDA.

Table 15•2

HUNGER TERMS

- **famine** widespread scarcity of food in an area that causes starvation and death in a large portion of the population.
- **food insecurity** the condition of uncertain access to food of sufficient quality or quantity.
- **food poverty** hunger occurring when enough food exists in an area but some of the people cannot obtain it because they lack money, are being deprived for political reasons, live in a country at war, or suffer from other problems such as lack of transportation.
- **food shortage** hunger occurring when an area of the world lacks enough total food to feed its people.
- **hunger** lack or shortage of basic foods needed to provide the energy and nutrients that support health.

- *Deforestation and desertification*. Forests are shrinking and deserts are growing.
- *Ocean pollution*. Ocean pollution is killing fish; overfishing is depleting the numbers of those that remain.[6]
- *Extinctions of species*. More than 140 species of animals and plants are going extinct each day. Another 20 percent of all species are expected to die out in the next ten years. Thousands of animals and plants, including many kinds of whales, birds, giant mammals, and colorful butterflies, will never again be seen in the universe.

Food poverty is the prevailing form of hunger in the United States.

"Hunger, particularly childhood hunger, [is] not only a moral issue [but] a competitiveness issue. . . . Hunger compromises the ability to learn. Hungry children have higher school absence rates, and when they are in school, their powers of concentration are greatly reduced. The malnutrition that results from chronic hunger can even slow or permanently inhibit the physical development of the brain. So hunger is . . . an issue of failed beginnings for millions of American children—the future of our country."

—M. Mudd, vice president, Kraft General Foods

These global problems are all related. The causes overlap, and so do the solutions. To think positively, this means that any initiative a person takes to help solve one problem will help solve many others. In particular, control of the earth's population is urgent, as a later section spells out. This chapter's Controversy shows how U.S. consumers and agricultural practices affect the world's resources.

key point *The world's chronically hungry people suffer the effects of undernutrition, and many in the United States live with food insecurity. Many forces combine to threaten the world's future food supply and its distribution.*

Hunger

The hunger of concern is a chronic, painful **hunger** people feel when no food or too little food is available. Severe deficiencies of vitamins and minerals accompany this hunger, afflicting more than 40 percent of the world's people to some degree. An estimated two billion people, mostly women and children, suffer the effects of iron-deficiency anemia, and many more suffer milder degrees of insufficiency.[7] Iodine deficiency remains the single greatest cause of preventable brain damage and mental retardation; 750 million adults suffer from goiter. Deficiency of vitamin A takes a terrible toll on the world's children—it stands out as the world's leading cause of blindness in young children and robs many millions of the ability to fight off infections. Worldwide, three-fourths of those who die each year from starvation and related illnesses are children.

In developed countries, the primary cause of hunger is **food poverty.** People are hungry not because there is no food nearby to purchase, but because they lack money with which to buy the food. Contributing to food poverty are problems such as abuse of alcohol and other drugs, mental or physical illness, depression, lack of awareness of or access to available food programs, and the reluctance of people, particularly the elderly, to accept what they perceive as "welfare" or "charity." Lack of resources remains the major cause of food poverty, and solving this problem would do much to relieve hunger.

Each person's choice to get involved and be heard can help lead to needed change.

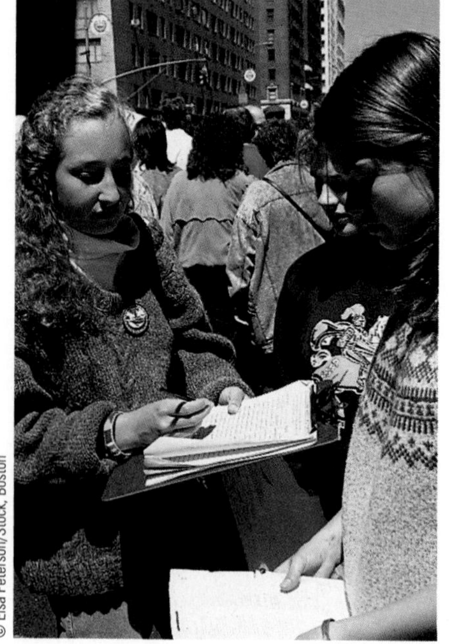

© Elsa Peterson/Stock, Boston

In the United States, food poverty reaches into many segments of society, affecting not only the chronic poor (migrant workers, the unskilled and unemployed, the homeless, and some elderly) but also the so-called working poor. Some are displaced farm families. Some are former blue-collar and white-collar workers forced out of their trades and professions into minimum-wage jobs. These people outnumber the chronic poor, and they are not on welfare—they have jobs, but the pay is too low to meet their needs. Families with incomes below a certain level are simply unable to buy sufficient amounts of nourishing foods, even if they are skilled in food shopping. Their worry about food security leads them to skip meals or cut their portions. Children in such families sometimes go hungry for an entire day until the adults find money for food.

Hunger is not always easy to recognize. Table 15-3 shows how national surveys identify it in the United States. A family that answers yes to these questions is suffering from the hunger caused by food poverty. The American Dietetic Association holds that "aggressive action is needed to bring an end to domestic hunger and to achieve food and nutrition security for all residents of the United States."[8] Such

Table 15•3

HOW TO DIAGNOSE FOOD INSECURITY IN A U.S. HOUSEHOLD

Questions like these are asked on surveys to determine the extent of food insecurity in a household. The more questions that receive a "Yes" answer, the more intense the hunger the household is experiencing.

- Do you often go hungry?
- Do you often have too little food to eat because you have no money, transportation, or kitchen appliances that work?
- Do you ever rely on nutritionally inferior foods to feed yourself or your children because you lack any of these resources?
- Do you ever eat less than you feel you should because you lack any of these resources?
- Do you ever skip meals or cut the size of meals because you lack any of these resources?
- Do you ever rely on neighbors, friends, relatives, or schools to feed any of your children because there is not enough food in the house?
- Do your children ever say they are hungry because there is not enough food in the house?
- Do you or any of your children ever go to bed hungry because there is not enough food in the house?

SOURCE: Adapted from C. A. Wehler, R. I. Scott, and J. J. Anderson, The Community Childhood Hunger Identification Project: A model of domestic hunger—Demonstration project in Seattle, Washington, *Journal of Nutrition Education* (1 supplement), January/February 1992, pp. 29S–35S; and R. R. Briefel and C. E. Woteki, Development of food sufficiency questions for the Third National Health and Nutrition Examination Survey, *Journal of Nutrition Education* (1 supplement), January/February 1992, pp. S24–S28.

These people and many others like them in the United States face food insecurity daily.

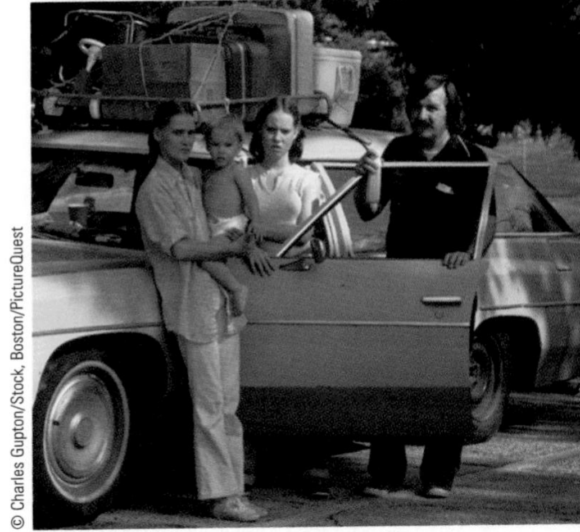

© Charles Gupton/Stock, Boston/PictureQuest

action is in everyone's interest because the hunger of individual families affects the nation as a whole, as the quotation in the margin on page 560 makes clear.

key point *Chronic hunger causes many deaths worldwide, especially among children. Intermittent hunger is frequently seen in U.S. children. The immediate cause of hunger is poverty.*

What U.S. Food Programs Are Directed at Stopping Domestic Hunger?

An extensive network of food assistance programs delivers life-giving food daily to millions of U.S. citizens.[9] One of every six Americans receives food assistance of some kind, at a total cost of almost $33 billion per year.[10] Even so, the programs are not fully successful in preventing hunger, even among those who receive their benefits.

Programs described in earlier chapters include children's school lunch and school breakfast programs, child-care and elder-care food programs, programs to supply low-income pregnant women and mothers with nourishing food (WIC), and food assistance programs for older adults such as congregate meals and Meals on Wheels. In particular, the WIC program is effective in improving the health of mothers and their infants. A 1990 study of WIC and Medicaid costs in five states showed that women who participate in WIC during pregnancy have far lower Medicaid costs for themselves and their babies in the first weeks after birth than eligible women who do not participate. Participation during pregnancy was associated with increased weight and longer gestation, positive occurrences in a low-income population. In a more recent study, children of WIC families had increased intakes of iron, folate, and vitamin B_6 when compared with children in nonparticipating but eligible families.[11]

The centerpiece of U.S. food programs for low-income people is the Food Stamp Program, administered by the U.S. Department of Agriculture (USDA). Eligible households (defined as people who live and purchase food together) receive food stamp coupons or debit cards similar to credit cards through state social services or welfare agencies. Recipients can use the coupons or cards like cash to purchase food and food-bearing plants and seeds, but not to buy tobacco, cleaning items, alcohol, or other nonfood items. About 17 million people in the United States receive food

Job training, when partnered with food programs can often help people out of poverty permanently.

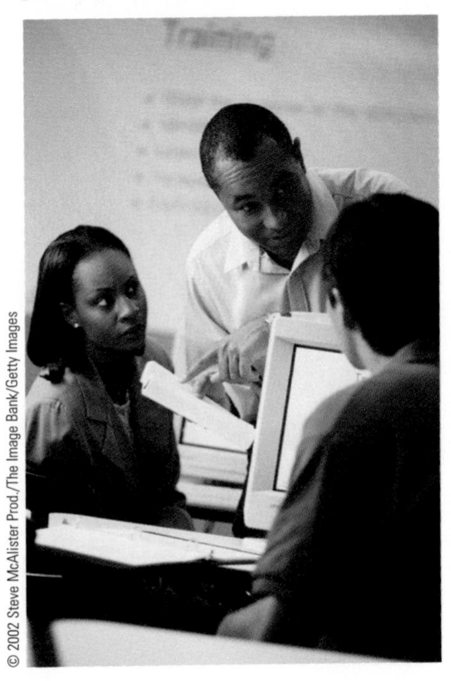

© 2002 Steve McAlister Prod./The Image Bank/Getty Images

gleaning traditionally, the practice of gathering crops left in the field after a harvest; today, refers to the recovery of excess food from various sources including restaurants, hotels, corporations, farms, wholesalers, farmers' markets, and supermarkets.

U.S. federal food programs include:

- Special Supplemental Food Program for Women, Infants, and Children (WIC, see Chapter 12).
- WIC Farmers' Market Nutrition Program.
- Food Stamp Program.
- National School Lunch and Breakfast Programs (see Chapter 13).
- Emergency Food Assistance Program.
- Commodity Supplemental Food Program.
- Commodity Distribution to Charitable Institutions.
- Senior Nutrition Program (see Chapter 13).
- Food Distribution Program on Indian Reservations.
- Nutrition Assistance to Puerto Rico.
- Temporary Assistance to Needy Families (often called "welfare").

For more information about gleaning, call the USDA's food gleaning hotline: (800) GLEAN-IT.

School lunches provide low-income children with nourishment at little or no cost.

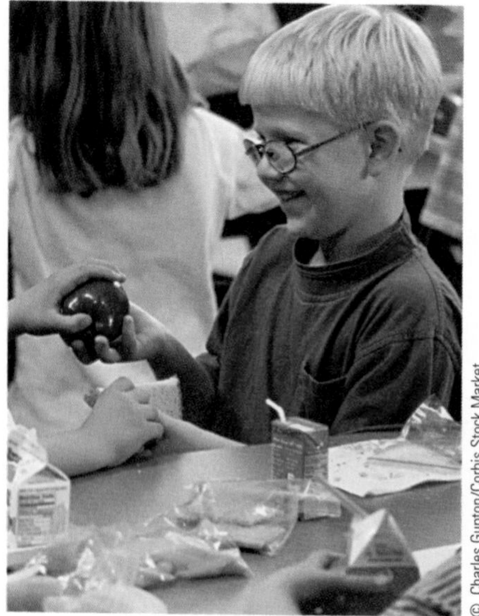

© Charles Gupton/Corbis Stock Market

stamps, and many millions more are thought to be eligible to receive them.[12] Of the homeless people in the United States who are eligible for food assistance, only 15 percent of single adults and 50 percent of families receive food stamps.

The Temporary Assistance to Needy Families (TANF) program also helps to stretch many families' budgets. An adult recipient can receive up to four years of cash assistance with the requirement of working toward self-sufficiency.

To assist where federal programs fall short, concerned citizens in many communities work through local agencies and churches to help deliver food to hungry people. Food recovery, or **gleaning,** from private industry has become a national priority, and federal funds are available as part of the Community Food Security Initiative, which assists communities in grassroots efforts to reduce hunger and improve nutrition.[13] Instead of being discarded, surplus produce from grocery stores and farms, excess food prepared for banquets and restaurant meals, and nonperishable foods from any source can be put to good use feeding those in need. Industries donating the food often qualify for tax deductions in proportion to their donations. Other initiatives to supply the hungry include community-based soup kitchens and shelters. Table 15-4 presents a 14-step program for developing a hunger-free community. Although these efforts provide emergency relief to hungry people, they leave unsolved the greater problems of low wages and poverty among people who lack higher education or training.

key point *Poverty and hunger coexist with affluence and bounty in the United States, not only among the unemployed, but also among working people. Government programs to relieve poverty and hunger are tremendously helpful, if not fully successful.*

Table 15•4

FOURTEEN WAYS COMMUNITIES CAN ADDRESS THEIR LOCAL HUNGER PROBLEMS

1. Establish a community-based emergency food-delivery network.
2. Assess food-insecurity problems and evaluate community services. Create strategies for responding to unmet needs.
3. Establish a group of individuals, including low-income participants, to develop and to implement policies and programs to combat food insecurity; monitor responsiveness of existing services; and address underlying causes of hunger.
4. Participate in federally assisted nutrition programs that are easily assessible to targeted populations.
5. Integrate public and private resources, including local businesses, to relieve food insecurity.
6. Establish an education program that addresses the food needs of the community and the need for increased local citizen participation in activities to alleviate food insecurity.
7. Provide information and referral services for accessing both public and private programs and services.
8. Support programs to provide transportation and assistance in food shopping, where needed.
9. Identify high-risk populations, and target services to meet their needs.
10. Provide adequate transportation and distribution of food from all resources.
11. Coordinate food services with parks and recreation programs and other community-based outlets to which area residents have easy access.
12. Improve public transportation to human services agencies and food resources.
13. Establish nutrition education programs for low-income citizens to enhance their food purchasing and preparation skills and to make them aware of the connections between diet and health.
14. Establish a program for collecting and distributing nutritious foods, either agricultural commodities in farmers' fields or prepared foods that would have been wasted.

SOURCE: House Select Committee on Hunger, legislation introduced by Tony P. Hall, excerpted in *Seeds*, Sprouts edition, January 1992, p. 3 with permission (SEEDS Magazine; P.O. Box 6170; Waco, TX 76706). For more on developing a hunger-free community, write: Hunger Free, House Select Committee on Hunger, 505 Ford House Office Building, Washington, DC 20515.

What Is the State of World Hunger?

In the developing world, hunger and poverty are even more intense, and the causes more diverse. The primary form of hunger is still food poverty, but the poverty is more extreme. Many of the world's people face hunger every day. Grasping the severity of poverty in the developing world can be difficult, but some statistics may help. One-fifth of the world's six billion people have no land and no possessions *at all*. They survive on less than one dollar a day each, they lack water that is safe to drink, and they cannot read or write. Many spend about 80 percent of all they earn on food, but still cannot meet their needs. The average U.S. housecat eats twice as much protein every day as one of these people, and the cost of keeping that cat is greater than that person's annual income.

The majority of today's undernourished people are concentrated in developing countries such as China, India, and nations of Africa and Latin America. Recent advances in agricultural technology, economic development, and commitment to eradicating hunger have begun to make inroads into the problem in some areas (see Figure 15-2 on the next page). In other areas, however, the number of hungry people continues to increase, and hunger remains an enormous challenge. In the year 2000, the United Nations World Food Programme fed 83 million people across the globe. Hunger and poverty, population growth, political strife, armed conflicts, and environmental degradation are all linked, and they tend to worsen each other.

Food Shortage and Armed Conflict The most visible form of hunger is **famine,** a true **food shortage** in an area that causes multitudes of people to starve and die (for definitions, see Table 15-1). The natural causes of famine—drought, flood, and pests—have, in recent years, taken second place behind the social causes. Between 1959 and 1961, for example, 15 to 30 million people died in China in the worst famine of the twentieth century. The famine was caused mainly by government policies that devastated Chinese agriculture. In parts of Africa, killer famines recur whenever human conflict converges with drought on a country such as Sudan that has little food in reserve even in a peaceful year.

Since the 1990s, the violence of armed conflict has been a dominant cause of all the famines reported worldwide. Farmers become warriors and agricultural fields become battlegrounds while citizens go hungry. Warring factions often repel famine relief efforts in hopes of starving their opponents before they succumb to starvation themselves. The world continues to struggle to find a middle ground between respecting the sovereignty of nations and insisting that all nations allow humanitarian assistance to reach their people. The recent war on terrorism has shone a spotlight on the estimated six million hungry, oppressed people of Afghanistan and also has unintentionally placed them more at risk by disrupting law enforcement, disturbing precarious food lines to remote areas, and sending people fleeing as refugees to countries that cannot feed them without help from other nations.[14]

During natural disasters without war, food aid from other countries has provided a safety net for countries whose crops fail. But food aid now does more than just offset poor harvests; it also delivers food relief to countries, such as Ethiopia, that are chronically short of food and without resources to buy it. Some people are concerned that as many nations cut their foreign aid, this food aid backup may become insufficient.

Chronic Hunger Though we usually associate world hunger with famine, the numbers affected by famine are relatively small compared with those suffering from less severe but chronic hunger. Nearly 800 million people, mostly women and children, in developing countries suffer from chronic malnutrition. The ravages to the body of nutrient deficiencies were spelled out in earlier chapters of this book.

Tens of thousands die of malnutrition every day. Most children who die of malnutrition do not starve to death—they die because their health has been compromised by dehydration from infections that cause diarrhea. Currently, **oral rehydration therapy (ORT)** is saving an estimated one million lives each year by helping to stop

oral rehydration therapy (ORT) oral fluid replacement for children with severe diarrhea caused by infectious disease. ORT enables parents to mix a simple solution for their child from substances that they have at home.

In 1999, Congressman Tony P. Hall conducted a survey of private U.S. food banks and reported that:

- Requests for food assistance were increasing at a rapid rate.
- People who need help are often working but not earning a living wage, are elderly, or have recently lost federal welfare or food stamp benefits.
- Fewer food banks are meeting the needs of the hungry on private food donations alone; they are turning to federal agencies for access to bulk foods, but this source is not steadily available to food banks.
- Gleaning from restaurants and other food industries is a potential partial solution, if problems of distribution can be solved.

SOURCE: T. P. Hall, Empty shelves: 1999 survey of U.S. food banks, a report available on the Internet at www.house.gov/tonyhall/pr49.htm.

The symptoms of malnutrition vary according to the nutrients lacking and the individual's stage of life. See Chapter 6 for effects of protein and energy deficiency; Chapters 7 and 8 for vitamin and mineral deficiencies; Chapter 11 for effects on immunity; Chapter 12 for effects on newborns and pregnant women; and Chapter 13 for effects on children, teens, and the elderly.

The primary cause of hunger is poverty.

Figure 15•2

CHRONIC HUNGER[a] AND UNDERWEIGHT AMONG THE WORLD'S CHILDREN

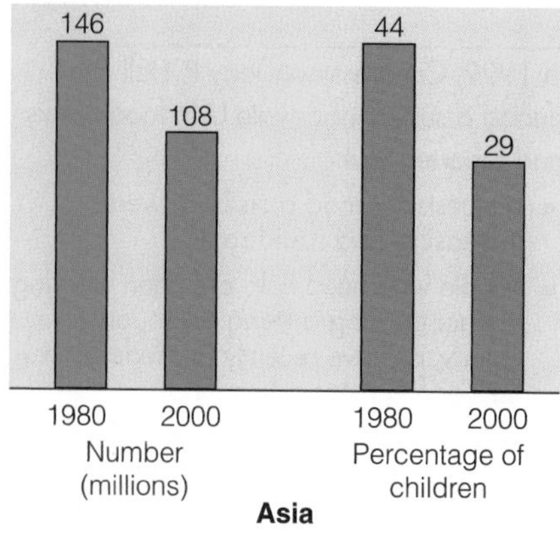

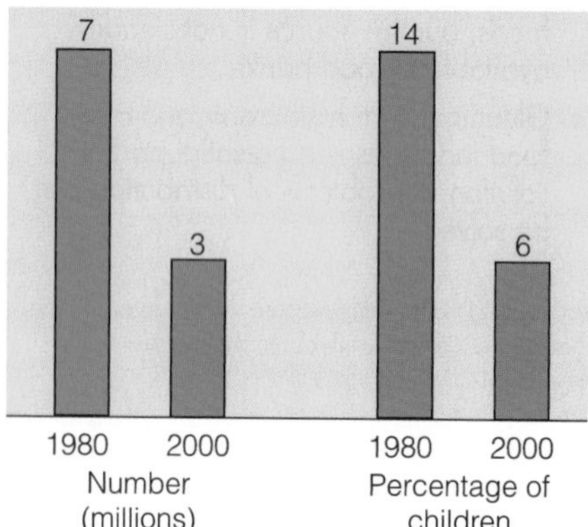

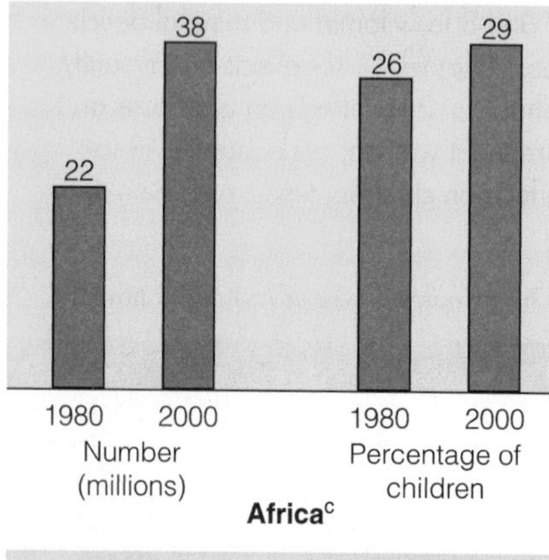

Decreased Increased

[a]Underweight is a proxy measure for chronic hunger.
[b]Haiti and some Central American areas have much higher levels of hunger.
[c]In sub-Saharan Africa, 36 percent of children are underweight; in Somalia, Ethiopia, and Niger, underweight among children approaches 50 percent.
SOURCE: Adapted from G. Gardner and B. Halweil, *Underfed and Overfed: The Global Epidemic of Malnutrition*, World Watch Paper 150 (Washington, D.C.: Worldwatch Institute, 2000) p. 12.

the destructive spiral in which infection worsens diarrhea and diarrhea causes dehydration. The ORT solution increases a body's ability to absorb fluids 25-fold. Clean or boiled drinking water is essential, however, because contaminated water will reinfect the child.

World Food Supply Most disturbingly, such misery and starvation exist side by side with adequate food supplies. In the decades after 1960, world food production grew faster than the world's population, mainly because of agricultural advances known as the green revolution and efforts to increase and diversify crop yields in agriculturally less advanced regions of the world. Today, the world's supply of grain, an index of the sufficiency of the world food supply, can still feed the world for several months. Wheat and corn, for example, the staple foods of many nations, are abundant and are now priced at less than half their cost of 40 years ago.

The future may not be so bright, however. At its present rate of growth, the world's population will soon outstrip the current rate of food production.[15] The green revolution has produced dramatically higher yields per acre of farmland over past decades, but progress has slowed and may not generate the greater crop yields needed to keep pace with the increasing numbers of people being born. A 2001 United Nations document states: "Only by doubling food production, improving distribution, and protecting the environment can we ensure food security for the 8 billion people that will inhabit the planet in 2025. Research suggests that the world's farmers will have to produce 40 per cent more grain by 2020 to meet rising demand." Environmental degradation and dwindling water supplies may ultimately prevent further growth in the world's food output in many agricultural areas. No part of the world is safely insulated against future food shortages. Developed countries may be the last to feel the effects, but they will ultimately go as the world goes.

Focus on Women Malnourished women in poverty bear sickly infants who cannot fend off the diseases of poverty, and many succumb within the first years of life. One child in six in the world is born underweight, and 10 million die by age five, half from malnutrition-related causes.[16] Breastfeeding helps prolong an infant's life, but eventually the child must be weaned to thin gruels of scant quantity made with unclean water. All too often, children sicken and die soon after weaning. Because of poverty, infection, and malnutrition, the life expectancy in some African countries averages 50 years; in Uganda it is only 42 years, little more than half of the U.S. life expectancy.

When crops fail or war and violence erupt in an already impoverished area, women are first to suffer. Seven out of ten of the world's hungry people are women and girls, yet they receive only about half of the available food aid and must use it to feed their children as well as themselves. These facts are offered by the World Food Programme (WFP):

- Even when women are starving, they are likely to give food received to their hungry children. When food is delivered to some government agencies, much of it is often diverted from its intended recipients.
- In Asia and Africa, 60 to 80 percent of women are engaged in farming.
- In a third of households worldwide, women are the sole breadwinners.

Therefore, the WFP has targeted women as the recipients of 80 percent of its food relief. Education for girls and women is also a high priority.

key point *Natural causes such as drought, flood, and pests and social causes such as armed conflicts and overpopulation all contribute to hunger and poverty in developing countries. To meet future demands for food, technology must continue to improve food production, food must be fairly distributed, and birthrates need to decline. The world's women and girls are major allies in the effort to fight hunger.*

Environmental Degradation and Hunger

Hunger and poverty interact with a third force—environmental degradation. Poor people often destroy the very resources they need for survival. Desperate to obtain money for food, they sell everything they own—even the seeds that would have provided next year's crops. They cut their trees for firewood or timber to sell, then lose the soil to erosion. Without these resources, they become still poorer. Thus, poverty causes environmental ruin, and the ruin leads to hunger.

Soil Erosion Soil erosion affects agriculture in every nation. Deforestation of the world's rain forests dramatically adds to land loss. In Sierra Leone, 60 percent of the land was primary rain forest in 1961; this had plummeted to 6 percent by 1994. Without the forest covering to hold the soil in place, it washes off the rocks beneath, drastically reducing the land's productivity.[17]

Around the world, irrigation and fertilizer can no longer compensate for these losses by improving crop yields because all the land that can benefit from these measures is already receiving them. Compounding the problem, continuous irrigation leaves deposits of salt in the soil, and rising salt concentrations are lowering yields on close to a quarter of the world's irrigated cropland.

Grazing Lands and Fisheries Meat and fish outputs are also endangered. Grasslands for growing beef are already being fully used or overused on every continent. Despite persistent expansion of the world's fishing industry, the yield of fish from the oceans has been declining in recent years due to overfishing and pollution.[18] Big fish, such as tuna, swordfish, and shark, are being overfished. According to the Food and Agriculture Organization (FAO), an agency of the United Nations that monitors the world's food supplies, about 47 to 50 percent of major marine fish stocks are currently fully exploited, with no room for further expansion of fishing. Another 28 percent are overexploited or depleted and in danger of extinction unless given relief from overfishing.[19] Almost every major bay in Japan suffers serious pollution sufficient to interrupt the normal breeding cycles of food species in the region.[20]

Inland fisheries have also suffered tremendous drops in yield as a result of environmental damage. In the early 1990s, 14,000 Canadian lakes were declared biologically dead as a result of acid rain. The FAO predicts that, unless something changes, the world demand for fish will outstrip the supply in about ten years.[21] Preventing this outcome will require commitments from the world's fishing nations to refrain from overfishing and to protect the environments of ocean fisheries. Developing greater production of fish from aquaculture ("fish farms") and alterations of wild species through biotechnology may also help meet human demand.

Climate, Air, and Water Both air pollution and the resulting climate change also reduce food outputs. According to the United Nations Intergovernmental Panel on Climate Change, a major international collaboration involving more than 2,500 scientists from around the world, changes in climate are expected to result from a buildup of so-called greenhouse gases, such as carbon dioxide, methane, and nitrous oxide, and airborne particles.[22] These pollutants are produced by human industry, agriculture, and transportation activities. A rise of only a degree or so in average global temperature may reduce soil moisture, impair pollination of major food crops such as rice and corn, slow growth, weaken disease resistance, and disrupt many other factors affecting crop yields.

Supplies of fresh water have shrunk to the point where they are limiting the numbers of people who can survive in some areas. In fact, lack of water may limit human

The hunger of conflict—people desperate for food in Somalia.

© Christopher Morris/Black Star/PictureQuest

To prevent death from diarrheal disease, provide:

- Adequate sanitation.
- Safe water.
- Oral rehydration therapy (ORT). A simple recipe for ORT calls for 1 cup boiled water, 2 teaspoons sugar, and a pinch of salt.

More about overgrazing appears in Controversy 15.

Worldwide 1.2 billion people live without access to clean, safe water.

As groundwater is used up, deserts spread.

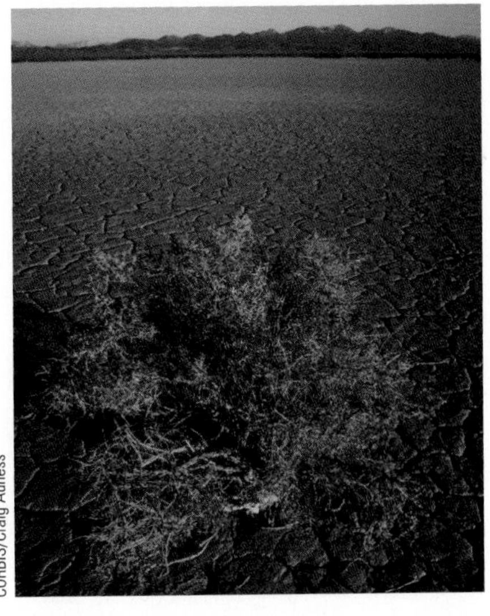

CORBIS/Craig Aurness

carrying capacity the total number of living organisms that a given environment can support without deteriorating in quality.

population growth even before lack of food does. If present consumption patterns continue, two of every three persons on earth will live in water-stressed conditions by the year 2025.[23]

Overpopulation The world's population reached six billion in 1999, but the rate of growth has begun to taper off somewhat.[24] Still, by 2033 the human population will exceed the earth's estimated **carrying capacity.** Many authorities in many fields—and more every year—are calling for a reduction in the rate at which the world's population increases. Overpopulation may well be the most serious threat that humankind faces today.

The sheer magnitude of our annual population increase is difficult to comprehend. Each month the world adds the equivalent of another New York City. During six months of the terrible 1992 famine in Somalia, an estimated 300,000 people starved to death. Yet it took the world only 29 *hours* to replace their numbers.

Years needed for the world's population to reach . . .

Its 1st billion	2,000,000 years
2nd billion	105 years
3rd billion	30 years
4th billion	15 years
5th billion	12 years
6th billion	11 years

Is it any wonder that food supplies may one day fall behind?

Population stabilization is one of the most pressing needs of our time because it appears to be the only way to enable the world's food output to keep up with demands. Without population stabilization, the world can neither support the lives of people already born nor halt environmental deterioration around the globe. And before the population problem can be resolved, it may be necessary to remedy the poverty problem. Of the many millions added to the population each year, 98 percent are born in the most poverty-stricken areas of the world.

Poverty and hunger exert an ironic effect on people, driving them to bear more children. Figure 15-3 shows the high correlation of income and high birthrate. Poverty and hunger are also correlated with lack of education (Figure 15-4), which includes lack of knowledge about controlling family size. A family in poverty also depends on its children to farm the land, haul water, and care for the adults in their old age. If a family faces ongoing poverty, and its young children are among the most likely to die from disease and other causes, the parents will choose to have many children as a form of "insurance" that some will survive to adulthood.

Figure 15•3

INCOME AND BIRTHRATE

Greater wealth means lower rates of birth. Each dot represents a country.

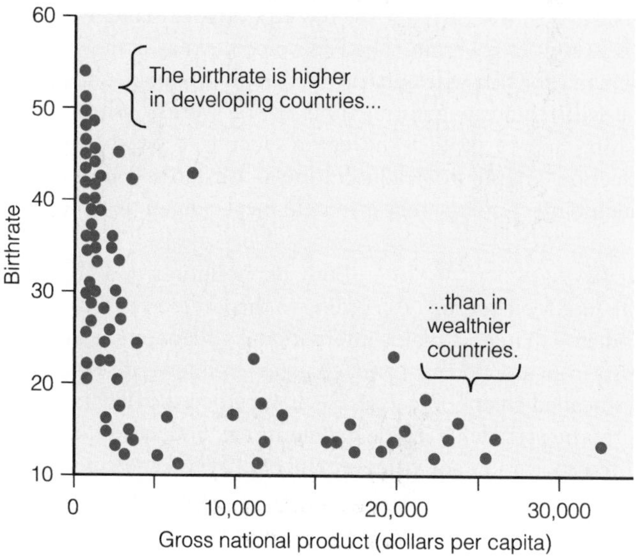

SOURCE: H. R. Pulliam and N. M. Haddad, Human population growth and the carrying capacity concept. *Bulletin of the Ecological Society of America*, September 1994, pp. 141–157.

Figure 15•4

EDUCATION AND BIRTHRATE

Higher education means lower rates of birth. Each dot represents a country.

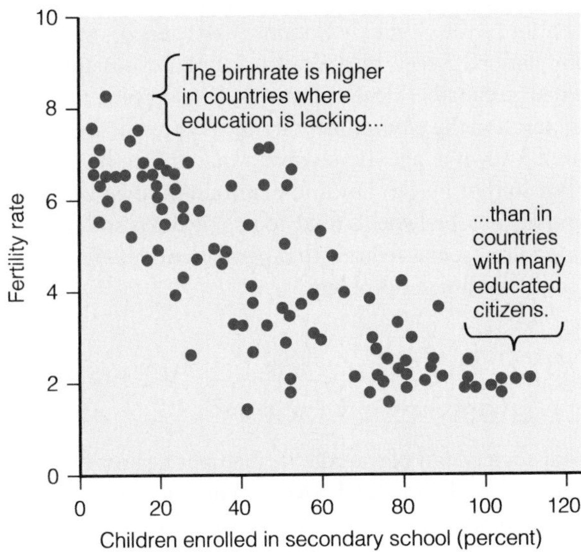

SOURCE: H. R. Pulliam and N. M. Haddad, Human population growth and the carrying capacity concept, *Bulletin of the Ecological Society of America*, September 1994, pp. 141–157.

sustainable able to continue indefinitely. Here the term refers to the use of resources at such a rate that the earth can keep on replacing them; for example, cutting trees no faster than new ones grow and producing pollutants at a rate with which the environment and human cleanup efforts can keep pace. In a sustainable economy, resources do not become depleted, and pollution does not accumulate.

Relieving poverty and hunger, then, may be a necessary first step in curbing population growth. When people attain better access to health care, education, and family planning, the death rate falls. After a time, the birthrate follows suit. Thus, improvements in living standards help stabilize the population. Wealth distribution matters, too. In countries where economic growth has benefited only the rich, population growth has remained high. Examples include Brazil, Mexico, the Philippines, and Thailand, where large families continue to be a major economic asset for the poor.

October 16 is World Food Day—visit the website listed in this chapter's Nutrition on the Net for details.

key point *Environmental degradation caused by the impacts of growing numbers of people is threatening the world's future ability to feed all of its citizens. Improvements in agriculture can no longer keep up with people's growing numbers. Human population growth is an urgent concern. Controlling population growth requires improving people's economic status and providing them with health care, education, and family planning.*

Moving toward Solutions

Slowly but surely, improvements are becoming evident in developing nations. For example, most nations have seen a rise in their gross domestic product, a key measure of economic well-being. Adult literacy rates have increased by more than 50 percent in some areas since 1970, and the proportion of children being sent to school has risen, while the proportion of chronically undernourished people has declined. Today, optimism abounds, and keys to solving the world's environmental, poverty, and hunger problems are within the reach of both the poor and the rich nations—if they will make the effort required.

The poor nations need to make contraceptive technology and information more widely available, educate their citizens, assist the poor, and adopt **sustainable** development practices that slow and reverse the destruction of their forests, waterways, and soil. The rich nations need to stem their wasteful and polluting uses of resources and energy, which are contributing to global environmental degradation.

In some countries, every pair of little hands is needed to help feed the family.

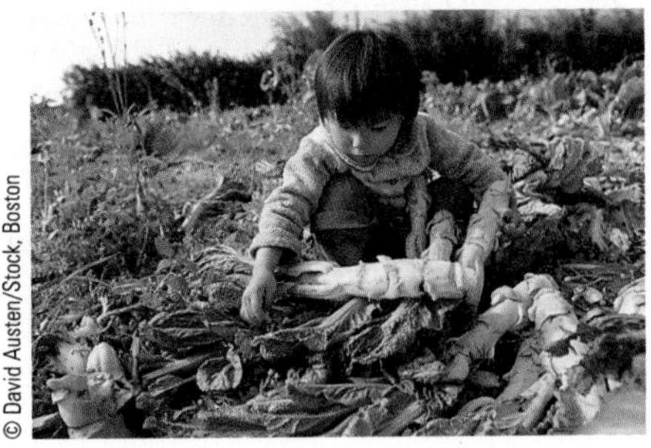

Sustainable Development Worldwide

Many nations now agree that improvement of all nations' economies is a prerequisite to meeting the world's other urgent needs: population stabilization, arrest of environmental degradation, sustainable treatment of resources, and relief of hunger. At a summit of over a hundred nations, the United Nations Conference on Environment and Development, many nations agreed to a set of principles of sustainable development. The conferees defined sustainable development as development that would equitably meet both the economic and the environmental needs of present and future generations.

To rephrase a well-known adage: If you give a man a fish, he will eat for a day. If you teach him to fish so that he can buy and maintain his own gear and bait, he will eat for a lifetime and help to feed you. Unlike food giveaways and money doles, which are only stop-gap measures, social reforms that permanently better the lot of the poor can permanently solve the hunger problem.

"For every person in the world to reach present U.S. levels of consumption with existing technology would require four more planet earths."—E. O. Wilson, 2002

How Can People Engage in Activism and Simpler Lifestyles at Home?

Every segment of our society can play a role in the fight against poverty, hunger, and environmental degradation. The federal government, the states, local communities, big business and small companies, educators, and all individuals, including dietitians and foodservice managers, have many opportunities to forward the effort.[25]

Government Action Government policies can change to promote sustainability. For example, the government can devote tax dollars and other resources to development of energy-conservation services and crop protection and to national and international education on sustainable development techniques.

Private and Community Enterprises Businesses can take initiatives to help; some already have—AT&T, Prudential, and Kraft General Foods are major supporters of antihunger programs. Restaurants and other food facilities can participate in the nation's gleaning effort by giving their fresh leftover foods to community distribution centers. As mentioned earlier, the Community Food Security Initiative aids grassroots efforts to reduce hunger and improve nutrition within communities.

Educators Educators, including nutrition educators, have a crucial role to play. They can teach others about the underlying social and political causes of poverty, the root cause of hunger. At the college level, they can teach the relationships between hunger and birthrate, hunger and environmental degradation, hunger and the status of women, and hunger and global economics.

Food and Nutrition Professionals Dietitians and foodservice managers are being urged by their professional organization, the American Dietetic Association (ADA), to promote the saving of resources by reuse, recycling (including composting), energy conservation, and water conservation in both their professional and their personal lives. In addition, the ADA urges its members to work for policy changes in private and government food assistance programs, to intensify education about hunger, and to be advocates on the local, state, and national levels to help end hunger in the United States.

"Never doubt that a small group of thoughtful, committed people can change the world. Indeed, it is the only thing that ever has."

—Margaret Mead

Individuals All individuals can become involved in these large trends. Many small decisions each day add up to large impacts on the environment. The Consumer Corner that follows sums up some of these decisions and actions.

Government, business, educators, and all individuals have many opportunities to promote sustainability worldwide and wise resource use at home.

Consumer Corner

Saving Money and Protecting the Environment

Consumers can "tread lightly on the earth" through their daily choices. Consider this list:

- Shop "carless" and plan to make fewer shopping trips. Motor vehicles constitute the single largest source of air pollution, causing lung problems, reduced crop yields, and acid rain that damages forests.
- Ride a bike to work or classes.
- Choose foods that are low on the food chain more often (see the chapter Controversy).
- Limit use of imported canned beef products, including stews, chili, corned beef, and pet foods. Many of these foods come at the expense of cleared rain forest land: 200 square feet of rain forest are lost *permanently* for every pound of beef produced.
- Choose small fish more often. Small fish eat tiny aquatic animals and plants—that is, they eat low on the food chain.
- Choose chicken from local farms.
- Buy more local foods grown close to home. Locally grown foods require less transportation, packaging, and refrigeration than shipped foods.
- Avoid overly packaged items; buy bulk items with minimal packaging or reusable or recyclable packaging. Each can, foam tray, waxed or clay-coated cardboard container, plastic bottle, or glass jar requires land and many other resources to produce, and its disposal pollutes and costs more land.
- Use reusable pans and dishes, rather than disposable items that are used once and thrown away. Use pumps instead of spray cans, which are hard to recycle because they are made of many materials.
- Carry reusable string or cloth grocery sacks or bring plastic sacks back to the store and refill them. Production of paper and plastic grocery bags represents a huge drain on resources. Paper factories use chemicals such as toxic forms of chlorine bleach, which are released into waterways in quantities so large that the chemicals can destroy whole bays and fisheries.
- Use fast cooking methods. Stir-frying, pressure cooking, and microwaving all use less energy than conventional stovetop or oven cooking methods.
- Reduce use of aluminum foil, paper towels, plastic wraps, plastic storage bags, and other disposable items. Find permanent reusable replacements for each, such as reusable storage containers and washable cloths.
- Use fewer electric gadgets. Mix batters, chop vegetables, and open cans by hand.
- Purchase the most efficient large appliances possible—look for the Energy Star logo (see Figure 15-5). Products that rank highest in their category for energy efficiency earn this logo from the Environmental Protection Agency. By purchasing Energy Star products, consumers can save many energy dollars each year.
- Insulate the home.
- Consider using solar power, especially to heat water.
- Reduce, reuse, recycle.

The personal rewards of all these behaviors are many, from saving money to the satisfaction of knowing that you are enjoying and preserving the earth (see Figure 15-6 on the next page for other suggestions). But do they really help? They do, if enough people join in. To make the greatest impact, people can also support organizations that lobby for changes in economic policies toward developing countries. Another way to help solve these problems is to join with others to work for international hunger-relief organizations. Table 15-5 (page 571) lists some of the major ones.

Figure 15•5

ENERGY STAR

Products bearing the U.S. government's Energy Star logo rank highest for energy efficiency. For example, a ten-year-old refrigerator uses as much energy as two refrigerators with the Energy Star label. Energy Star products range from large appliances to light bulbs and building materials.

By choosing products with the Energy Star logo when replacing old equipment, the typical household would save almost $400 per year in energy costs; if everyone chose nothing but Energy Star products the next ten years, the national energy bill would be reduced by about $100 billion. The reduction in greenhouse gas emissions (carbon dioxide) would be equivalent to taking 17 million cars off the road for each of those ten years.

Courtesy of NASA

"We do not inherit the earth from our ancestors, we borrow it from our children." Ascribed to Chief Seattle, a nineteenth-century Native American leader.

Figure 15•6

INDIVIDUAL RESPONSIBILITY AND RESPECT FOR THE ENVIRONMENT

Reduce resource use, reduce fuel use, and reduce pollution with these individual actions.

Eat lower on the food chain—more grains, local chicken, and small fish, and less meat from large animals.

Eat local foods—visit your farmers' market.

© 2002 PhotoDisc/Getty Images

Buy bulk items to save packaging.

© 2002 PhotoDisc/Getty Images

Recycle glass, cans, paper, and plastic.

Buy fruits of differing ripeness.

Eat most perishable foods first.

Shop carless. It can be a pleasure and a great source of exercise.

© Corbis Images/PictureQuest

Use reusable bags instead of throwaway bags.

© Burke/Triolo/Brand X Pictures/PictureQuest

Buy recycled goods to close the loop.

Use reusable items . . .

. . . instead of nonreusable items.

Use items that don't use energy . . .

. . . instead of those that do use energy (even small appliances).

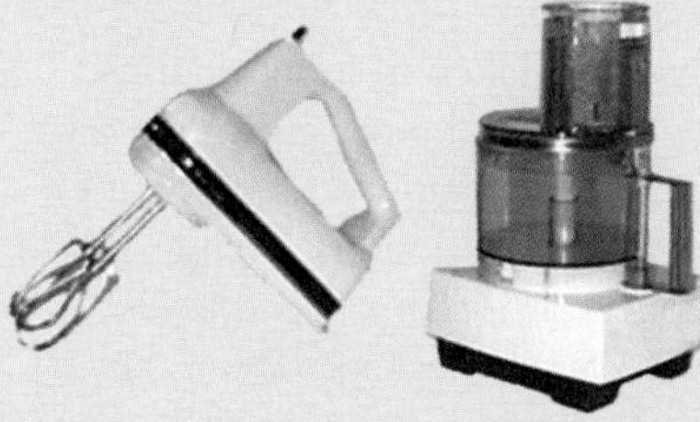

Use appliances that take less energy.

Run your refrigerator efficiently.

Even today's efficient refrigerators use substantial energy because they run day and night. Consumers can take several steps to minimize the energy a refrigerator uses:

- Set it at 37° to 40°F; set the freezer at 0°F.
- Clean the coils and the insulating gaskets around the doors regularly.
- Keep it in good repair.

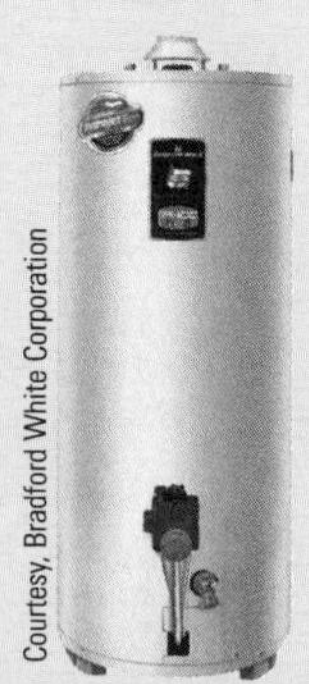

Courtesy, Bradford White Corporation

The water heater can also waste a lot of energy. Keep the water heater set at 120° to 130°F (no hotter) to save energy. For safe household dishes, sterilization is not necessary. Water of 120° to 130°F enhances the action of dishwashing detergents, making microorganisms slippery and removing them from the dishes. These measures will keep food fresh and clean while keeping energy use low.

Table 15•5

HUNGER-RELIEF ORGANIZATIONS PEOPLE CAN JOIN

This chapter's Nutrition on the Net feature lists many others.

Action without Borders
350 Fifth Ave., Suite 6614
New York, NY 10118
(212) 843-3973
www.idealist.org

America's Second Harvest
35 E. Wacker Dr. #2000
Chicago, IL 60601
(800) 771-2303
www.secondharvest.org

Bread for the World
50 F St. NW, Suite 500
Washington, DC 20010
(800) 82-BREAD or
(800) 822-7323
(202) 639-9400;
fax (202) 639-9401
www.bread.org

Children's Hunger Relief Fund
182 Farmer's Lane, Suite 200
Santa Rosa, CA 95405
(888) 781-1585
www.childrenshungerrelief.org

Congressional Hunger Center
229½ Pennsylvania Ave.
Washington, DC 20003
(202) 547-7022
www.hungercenter.org

Food Research and Action Center
1875 Connecticut Ave.
Suite 540
Washington, D.C. 20009
www.frac.org

Foodchain
912 Baltimore, #300
Kansas City, MO 64105
(800) 845-3008
www.foodchain.org

OXFAM America
26 West St.
Boston, MA 02111-1206
(800) 77-OXFAM or
(800) 776-9326
www.oxfam.america.org

Pan American Health Organization
525 23 St. NW
Washington, DC 20037
(202) 974-3000
www.paho.org

Society of St. Andrew
3383 Sweet Hollow Rd.
Big Island, VA 24526
(800) 333-4597
www.endhunger.org

United Nations Food and Agriculture Organization (FAO)
1001 22nd St. NW, Suite 300
Washington, DC 20437
(202) 653-2400
www.fao.org

United Nations International Children's Emergency Fund (UNICEF)
3 United Nations Plaza
New York, NY 10017-4414
(212) 326-7035
www.unicef.org

United Nations World Food Program
Via Cesare Giulio
Viola, 68
Parco dé Medici
Rome, Italy 00148
www.wfp.org

World Health Organization (WHO)
525 23rd St. NW
Washington, DC 20037
(202) 861-3200
www.who.org

World Hunger Program
Brown University
Box 1831
Providence, RI 02912
(401) 863-2700
www.brown.edu/Departments/World_Hunger_Program/hungerweb/WHP/overview.html

World Hunger Year
505 Eighth Ave., 21st Floor
New York, NY 10018-6582
(800) GleanIt
www.worldhungeryear.org

Answers to these Self Check questions are in Appendix G.

1. Which of the following is a symptom of food insecurity?
   a. You worry about gaining weight.
   b. You rely on neighbors to feed your children because there is not enough food in the house.
   c. You shop daily to get the best prices.
   d. You buy fresh rather than frozen foods.
2. Which of the items can be purchased with Food Stamps?
   a. cigarettes
   b. bread
   c. dishwashing liquid
   d. alcohol
3. What is the primary cause of famine in the world?
   a. poor agricultural practices
   b. drought
   c. social causes such as war
   d. flood
4. Which of the following is an example of environmental degradation?
   a. soil erosion
   b. diminished grazing lands
   c. air pollution
   d. all of the above
5. Which of the following activities are recommended due to the small impact they have on the environment?
   a. Use the oven whenever possible.
   b. Line pans with aluminum foil to reduce use of cleanup resources.
   c. Use a pressure cooker or microwave to cook foods.
   d. Carry groceries home in paper bags rather than plastic.
6. At least 140 species of animals and plants are becoming extinct every day in the world. T F
7. Most children who die of malnutrition starve to death. T F
8. More people in the world suffer from famine than from chronic hunger. T F
9. The higher a nation's economic status, the faster its population grows over the long run. T F
10. Deficiency of vitamin A is the world's leading cause of blindness in young children. T F

nutrition on the net

For further study of the topics of this chapter, access these websites:

1. Find updates and quick links to these and other nutrition-related sites at our website: www.wadsworth.com/nutrition
2. Learn about constructive, community-based solutions to the problems of poverty and hunger within and between the public and private sectors from the National Hunger Clearinghouse: www.worldhungeryear.org/nhc
3. Visit the USDA Food Stamp Program: www.fns.usda.gov/fsp
4. Visit the Gleaning section of the USDA Food and Nutrition Service: www.fns.usda.gov/fns
5. Find information on feeding the hungry from the Emergency Food and Shelter Program: www.efsp.unitedway.org
6. Donate free food at The Hunger Site: www.thehungersite.com
7. Details about World Food Day are available at: www.worldfooddayusa.org
8. Read about the worldwide hunger-relief efforts of the United Nations World Food Programme: www.wfp.org
9. Make a difference by joining many who work to distribute surplus food to those who need it: www.resourcelink.org
10. Learn more about the government's Energy Star program: www.energystar.gov
11. Many ideas and organizations for fighting hunger are listed with the Kitchen Link: www.kitchenlink.com
12. For information about farmers' markets, visit: www.ams.usda.gov/farmersmarkets/
13. See Table 15-5 (on p. 571) for additional websites.

INTERNET ACTIVITY

Hunger and global environmental issues are always evolving. The best way to understand the changing factors that influence such issues is to plug into Internet sites such as that of the World Watch Institute, an organization providing cross-disciplinary, global environmental information.

1. Go to World Watch Institute at: http://www.worldwatch.org
2. Explore the website for the latest hunger and global environmental issues.

In the Updates section of the site, select two topics. Read and write summaries of the updates of the two topics chosen and relate to issues discussed in this chapter.

CONTROVERSY 15

AGRIBUSINESS AND FOOD PRODUCTION: HOW TO GO FORWARD?

WHILE SOME individuals are making their own personal lifestyles more environmentally benign, as suggested in the chapter, others are seeking ways to improve whole sectors of human enterprise, such as agriculture. To date, large agricultural enterprises have been among the world's biggest polluters and resource users. Is it possible for agriculture to become sustainable? Do our new technologies hold promise for advancing sustainability? How are small farmers faring? This Controversy addresses these questions.

Costs of Producing Food

The environmental and social costs of agriculture and the food industry take many forms. Among them are resource waste and pollution, energy overuse, and tolls on life in farm communities. Table C15-1 offers some terms important to these concepts.

Impacts on Land and Water

Producing food has always cost the earth dearly. To grow food, we clear land—prairie, wetland, or forest—causing losses of native ecosystems and wildlife. Then we plant crops or graze animals on the land. The soil loses nutrients as each crop is taken from it, so fertilizer is applied. Some fertilizer runs off and pollutes the waterways. Some plowed soil runs off, clouds the water, and interferes with the growth of aquatic plants and animals.

Then, to protect crops against weeds and pests, we apply herbicides and pesticides. Most herbicides and pesticides kill not only weeds and pests, but also native plants, native insects, and animals that eat those plants and insects. Widespread use of pesticides and herbicides also causes resistant pests and weeds to evolve. Pesticides pose hazards for farm workers who handle and apply them, and pesticide residues can become a problem for people who consume them along with foods.

Agricultural pesticides and herbicides, if not used conservatively, also pollute rivers, lakes, and groundwater. Pollution from "point sources," such as sewage plants or factories, is relatively easy to control, but runoff from fields and pastures enters waterways across broad regions and is nearly impossible to control.

Finally, we irrigate, a practice that adds salts to the soil in many areas. The water evaporates, but the salts do not. As soils become salty, plant growth fails. Irrigation can also deplete the water supply over time because water is pulled from surface waters or from underground and then evaporates or runs off. This process, carried to an extreme, can dry up whole rivers and lakes and lower the water table of entire regions. The lower the water table, the more farmers must irrigate; and the more they irrigate, the more groundwater they use up.

Soil Depletion and Losses of Species

The soil can also be depleted by some agricultural practices, particularly indiscriminate land clearing (deforestation) and overuse by cattle (overgrazing). In just the past 40 years, human agricultural activities have ruined more than 10 percent of the earth's fertile land, an area the size of China and India combined. Over 20 million acres have been so damaged that they may be impossible to reclaim. With soil erosion proceeding unchecked, people in the year 2025 may see a 40 percent reduction in food-producing land per person, along with many more people to feed.

Unsustainable agriculture has already destroyed many once-fertile regions, where high civilizations formerly flourished. The dry, salty deserts of North Africa were once plowed and irrigated wheat fields, the breadbasket of the Roman Empire. Mistreatment of soil and water is now causing destruction on a scale never known before.

Agriculture is also weakening its own underpinnings by failing to conserve species diversity. By the year 2050, some 40,000 more plant species, existing in the 1990s, may go extinct. The United Nations Food and Agriculture Organization attributes many of the losses, which are occurring daily, to modern farming practices, as well as to population growth. The growing uniformity of global eating habits also contributes. As people everywhere eat the same limited array of foods, demand for local, genetically diverse, native plants is insufficient to make them seem worth preserving. Yet, in the future, as the climate warms, those very plants may be needed as food sources. A wild species of corn that grows in a dry climate, for example, might contain just the genetic information necessary to help make the domestic corn crop resistant to drought.

Energy

Massive fossil fuel use is threatening our planet by causing ozone depletion, water pollution, ocean pollution, and other ills

Table C15•1

AGRICULTURAL AND ENVIRONMENTAL TERMS

- **agribusiness** agriculture practiced on a massive scale by large corporations owning vast acreages and employing intensive technological, fuel, and chemical inputs.
- **alternative (low-input, or sustainable) agriculture** agriculture practiced on a small scale using individualized approaches that vary with local conditions so as to minimize technological, fuel, and chemical inputs.
- **integrated pest management (IPM)** management of pests using a combination of natural and biological controls and minimal or no application of pesticides.

© Paul Stover/Stone/Getty Images

and by making global warming likely. In the United States, the food industry consumes about 20 percent of all the energy the nation uses. Each year we spend 1,500 liters (over 350 gallons) of oil per person to produce, process, distribute, and prepare our food. Energy is used to run farm machinery and to produce fertilizers and pesticides. Energy is also used to prepare, package, transport, refrigerate, and otherwise store, cook, and wash our foods.

The Problems of Livestock and Fishing

Raising livestock also takes a toll. Like plant crops, herds of livestock occupy land that once maintained itself in a natural state. The land pays a price in losses of native plants and animals, soil erosion, water depletion, and desert formation. If animals are raised in concentrated areas such as cattle feedlots or giant hog "farms" instead, huge masses of animal wastes produced in these overcrowded, factory-style farms leach into local soils and water supplies, polluting them.[1] In an effort to control this source of pollution, the Environmental Protection Agency offers incentives to livestock farmers who agree to clean up their wastes and allow their operations to be monitored for pollution.[2] The U.S. Senate Agricultural Committee has concluded that current regulations for handling animal wastes are inadequate and should be changed to protect against this source of pollution.[3] In addition to the waste problem, animals in such feedlots still have to be fed; grain is grown for them on other land (Figure C15-1, on p. 575, compares the grain required to produce various foods). That grain may require fertilizers, herbicides, pesticides, and irrigation, too. In the United States, one-fifth of all cropland is used to produce grain for livestock—more land than is used to produce grain for people.

Other environmental costs attend fishing. Fishing easily becomes overfishing and depletes stocks of the very fish that people need to eat. Most nets also collect many nonfood species that are killed during harvest but returned to the sea instead of being put to use. Other aquatic animals are also vulnerable to injury and death, and populations of ecologically important nonfood animals, such as dolphins, are diminished. In short, our ways of producing foods are, for the most part, not sustainable.

Vast areas under plow are exposed to erosion, and those that must be irrigated can, over time, become salty and unusable.

© Stone/Getty Images

Pure rivers represent irreplaceable water resources.

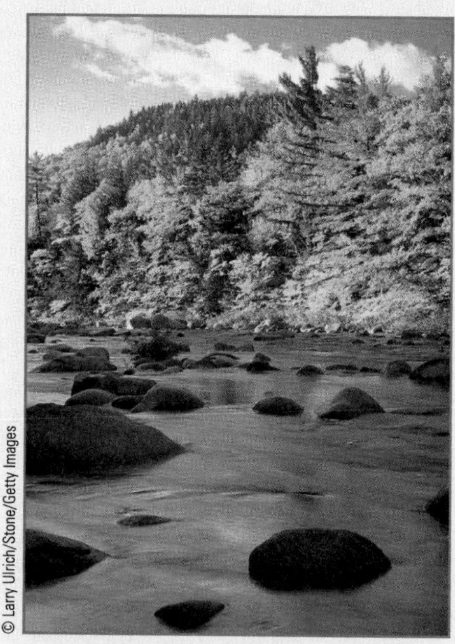
© Larry Ulrich/Stone/Getty Images

From Family Farms To Agribusiness

Beginning in the early 1980s and continuing today, U.S. agriculture has encountered serious economic problems: declining markets for U.S. farm products abroad as other countries increased their agricultural production and exports, reduced international economic ability to import grain, and an increase in energy and other costs of producing food domestically.[4] Many U.S. farmers, particularly those who specialize in export crops, have suffered heavy financial losses. Some have been unable to pay their debts and have had to leave farming. Between 1980 and 1994, more than 15 percent of the farms in the United States disappeared.[5] Tens of thousands of farms are still struggling, especially medium-size family farms, because of competition from foreign producers and a trend toward large food-producing operations.[6]

Taking the place of family farms are huge farms and ranches, many of which are being operated in Mexico or other developing nations; collectively, they are part of the massive food-producing enterprise called **agribusiness.** These huge operations in the United States tend to use little local labor, and the profits they make tend not to stay in

Figure C15•1

POUNDS OF GRAIN NEEDED TO PRODUCE ONE POUND OF BREAD AND ONE POUND OF ANIMAL WEIGHT GAIN

SOURCE: Idea and data from T. R. Reid, Feeding the planet, *National Geographic*, October 1998, pp. 58–74.

local communities. Those in foreign countries tend to hire local laborers willing to work for much less than laborers in the United States.

Agribusinesses also tend to place a higher priority on producing abundant, inexpensive food than on protecting soil, water, and local biodiversity. When these large operations overuse fertilizers and pesticides, overuse land at the cost of soil erosion, use excessive irrigation water, and promote intensive forms of livestock production, their impacts can be enormous. Then, in an effort to compete, small farmers may be driven to adopt similar unsustainable practices.

Because of economies of scale, agribusinesses can price their products so low that consumers tend to buy more products from them than from smaller, local farms. Thus, local U.S. grocers offer broccoli from Mexico, carrots from California, pineapples from Hawaii, and bananas from Central America at prices no local farmers can match, even if they could grow those products. Roadside stands and farmers' markets offer bundles of local green vegetables and baskets of local fruits and tomatoes, but less conveniently and sometimes at higher prices than many shoppers are willing to pay.

If food prices had to include a "tax" to pay for pollution cleanup, water protection, and land restoration, the prices of the products produced unsustainably would be higher. If they included a living wage, education, and benefits for the migrant farm workers, they would be higher still.

Proposed Solutions

For each of the problems described above, solutions are being devised, and indeed, some are being put into practice. To fully exploit these new sustainable agriculture techniques across the country will require some new learning. Sustainable agriculture is not one system but a set of practices that can be matched to particular needs in local areas. The first of these ideas, **alternative,** or **low-input, agriculture,** emphasizes careful use of natural processes wherever possible, rather than chemically intensive methods.

Low-Input Agriculture

One form of low-input agriculture is **integrated pest management.** Farmers using this system employ many techniques, such as crop rotation and natural predators, to control pests rather than depending on heavy use of pesticides alone. Not all crops can grow reliably without pesticides, but many can. Table C15-2 on page 576 contrasts low-input agriculture methods with unsustainable methods. Many sustainable techniques are not really new—they would be familiar to our great-grandparents. Many farmers today are rediscovering the benefits of old techniques as they adapt and experiment with them in the search for sustainable methods.

Low-input agriculture has some apparent disadvantages, but advantages offset them. For example, as chemical use falls, yields per acre also fall somewhat, but costs per acre also fall, so the return per acre may be the same as or greater than before. More money goes to farmers and less to the fuels, fertilizers, pesticides, and irrigation. The end result of such farming is to make both farmers and consumers better off financially and environmentally.

Low-input agriculture works. As the world's population grows, and its land and water dwindle, the need to adopt sustainable agriculture and development around the globe grows urgent.[7] More than 30,000 U.S. farmers are successfully using sustainable techniques such as those described in Table C15-2 on the next page. They see it as a system that can indefinitely sustain a healthy food supply, restore soil and water resources, and revitalize farming communities, while reducing reliance on fossil fuels.

Precision Agriculture

An exciting development in agriculture is the application of powerful new

Industrial farms generate huge masses of wastes that can contaminate local soil and water.

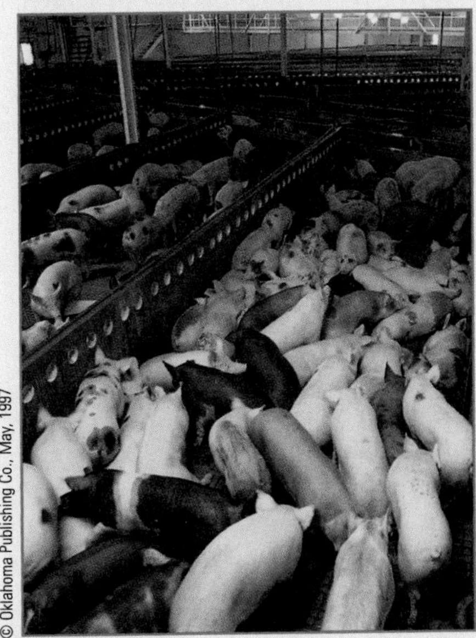

© Oklahoma Publishing Co., May, 1997

Table C15•2

HIGH-INPUT AND LOW-INPUT AGRICULTURAL TECHNIQUES COMPARED

Unsustainable Practice	Sustainable Practice
• Growing the same crop repeatedly on the same patch of land. This takes more and more nutrients out of the soil, makes fertilizer use necessary; favors soil erosion; and invites weeds and pests to become established, making pesticide use necessary.	• Rotate crops. This increases nitrogen in the soil so there is less need to buy fertilizers. If used with appropriate plowing methods, crop rotation reduces soil erosion. Crop rotation also reduces weeds and pests.
• Using fertilizers generously. Excess fertilizer pollutes ground and surface water and costs both farmers' household money and consumers' tax money.	• Reduce the use of fertilizers and use livestock manure more effectively. Store manure during the nongrowing season and apply it during the growing season. • Alternate nutrient-devouring crops with nutrient-restoring crops, such as legumes. • Compost on a large scale, including all plant residues not harvested. Plow the compost into the soil to improve its water-holding capacity.
• Feeding livestock in feedlots where their manure produces major water and soil pollutant problems. Piled in heaps, manure also releases methane, a global-warming gas.	• Feed livestock or buffalo on the open range where their manure will fertilize the ground on which plants grow and will release no methane. Alternatively, at least collect feedlot animals' manure and use it as fertilizer, or, at the very least, treat it before release.
• Spraying herbicides and pesticides over large areas to wipe out weeds and pests.	• Apply technology in weed and pest control. Use precision agriculture techniques if affordable or use rotary hoes twice instead of herbicides once. Spot treat weeds by hand. • Rotate crops to foil pests that lay their eggs in the soil where last year's crop was grown. • Use genetically resistant crops. • Use biological controls such as predators that destroy the pests.
• Plowing the same way everywhere, allowing unsustainable water runoff and erosion.	• Plow in ways tailored to different areas. Conserve both soil and water by using cover crops, crop rotation, no-till planting, and contour plowing.
• Injecting animals with antibiotics to prevent disease in livestock.	• Maintain animals' health so that they can resist disease.
• Irrigating on a large scale.	• Irrigate only during dry spells and only where needed.

computer technologies to food production. Through techniques collectively known as *precision agriculture*, farmers can adjust soil and crop management to meet the precise needs of various areas of the farm. For example, a farmer growing crops in a field with hills, which tend to stay drier, and with low-lying areas, which tend to stay wetter, can adjust irrigation water to meet the specific needs of each part of the field. Similarly, if one section of a field needs nitrogen fertilizer while another needs a different mix, the farmer can preprogram a computer to apply fertilizer of just the right type and amount for each area. Likewise, pesticide application can be programmed to prescribed applications, thus avoiding areas too close to streams or other water sources. The preprogrammed system turns off the pesticide flow when it comes to a designated safety zone.

The *global positioning satellite* (GPS) system is at the heart of precision farming. In the GPS system, satellites beam accurate information about land positions and elevations of an area, such as a field, to receivers placed on farm equipment here on earth. The GPS system delivers a grid map, pinpointing locations on a farm. Farmers can use the GPS information grid to target, within a meter's accuracy, areas that need treatments. They can then program computerized farm equipment to apply chemicals or other treatments accordingly. Farmers can also use the information to adjust the depths to which they till the soil. The goal is to till deeply enough to prepare seedbeds properly and control weeds, but to avoid excessive tilling that wastes fuel and worsens erosion. Finally, at harvest, a GPS system produces an accurate accounting of crop yield, acre by acre, so that spot adjustments can be made in the next planting season.

The future of precision agriculture seems bright, and the potential savings to farmers in terms of water, fertilizers, and pesticides are enormous. The accompanying reduction in polluting chemicals introduced into the environment means that everyone benefits.

Agricultural Biotechnology

Although not every farmer worldwide may be in a position to reap benefits from the technologies of precision agriculture, the advances of biotechnology may prove to be an essential part of a worldwide move toward sustainable agriculture. If health and safety issues are addressed and resolved, rDNA technology promises economic, environmental, and agricultural benefits by shrinking the acreage needed for crops, reducing soil losses, minimizing use of chemical insecticides, and bettering crop protection (see Controversy 14).

Genetically modified microbes offer benefits to sustainable agriculture if research and testing can ensure that rDNA microorganisms released into the environment will not turn out to be more harmful than the products they are intended to replace. Bioengineered

Table C15•3

SUSTAINABLE ENERGY-SAVING AGRICULTURAL TECHNIQUES

- Use machinery scaled to the job at hand and operate it at efficient speeds.
- Combine operations. Harrow, plant, and fertilize in the same operation.
- Use diesel fuel. Use solar and wind energy on farms. Use methane from manure. Be open-minded to alternative energy sources.
- Use new disease- and pest-resistant plant varieties developed through genetic engineering.
- Save on technological and chemical inputs and spend some of the savings paying people to do manual jobs. Increasing labor inputs has been considered inefficient. Reverse this thinking: creating more jobs is preferable to using more machinery and fuel.
- Partially return to the techniques of using animal manure and crop rotation. This would save energy because chemical fertilizers require large energy inputs to produce.
- Choose crops that require low energy inputs (fertilizer, pesticides, irrigation).
- Educate people to cook food efficiently and to eat low on the food chain.

microbes could contribute to continuous renewal of soil structure and fertility by fixing nitrogen and releasing other nutrients into the soil, lessening the need for chemical fertilizers and easing the environmental burden.[8] Scientific laboratories are also working to engineer microbes that can recycle agricultural, industrial, and household wastes into fertilizers, an obvious boon to the environment. Bacterial and fungal herbicides, fungicides, and insecticides are in advanced experimental stages and promise to augment other integrated pest management systems of low-input agriculture such as crop rotation.

Energy Efficiency

Some 6,560 calories of fuel are used to produce a can of corn (including the can and transportation), and 7,980 calories are needed to produce a package of frozen corn (including packaging, freezing, and transportation). Much of this energy input could be reduced, as Table C15-3 shows. The last item in the table suggests that consumers should center their diets on foods that require low energy inputs, a choice that is described next.

Eating Lower on the Food Chain

Studies of energy use in the U.S. food system have revealed which foods require the most and least energy to produce. The least energy is needed for grain: about one-third calorie of fuel is burned to grow each calorie of grain. Fruits and vegetables are intermediate, and most animal protein requires from 10 to 90 calories of fossil energy per calorie of usable food. An exception is livestock raised on the open range; these animals eat grass and require low energy inputs as do most plant foods.[9] So much of our beef is grain fed, rather than range fed, however, that the average energy requirement for beef production is high.

To support our meat intake, we maintain several billion livestock, about four times our own weight in animals. Livestock consume ten times as much grain each day as we do. We could use much of that grain to make grain products for ourselves and share them. The shift could free up enough grain to feed 400 million people, would necessitate burning less fuel and using less water, and could also free up much more land. According to the United Nations, the "ecological footprint"—the productive land area needed to support a person's lifestyle—of each individual is four times larger in an industrialized country than in a developing one (see Figure C15-2).

Some individuals are taking action to do their part to solve these problems. Some meat eaters are choosing to cut down on their meat portions or to eat range-fed beef or buffalo only. "Rangeburger" buffalo also offers nutrition advantages over grain-fed beef because it is lower in fat, and the fat has more polyunsaturated fatty acids, including the omega-3 type. Some people are switching to nonmeat, and even pure vegan, diets. The fish farming industry shows promise of being able to feed large numbers of people in the

Figure C15•2

HOW BIG IS YOUR ECOLOGICAL FOOTPRINT?

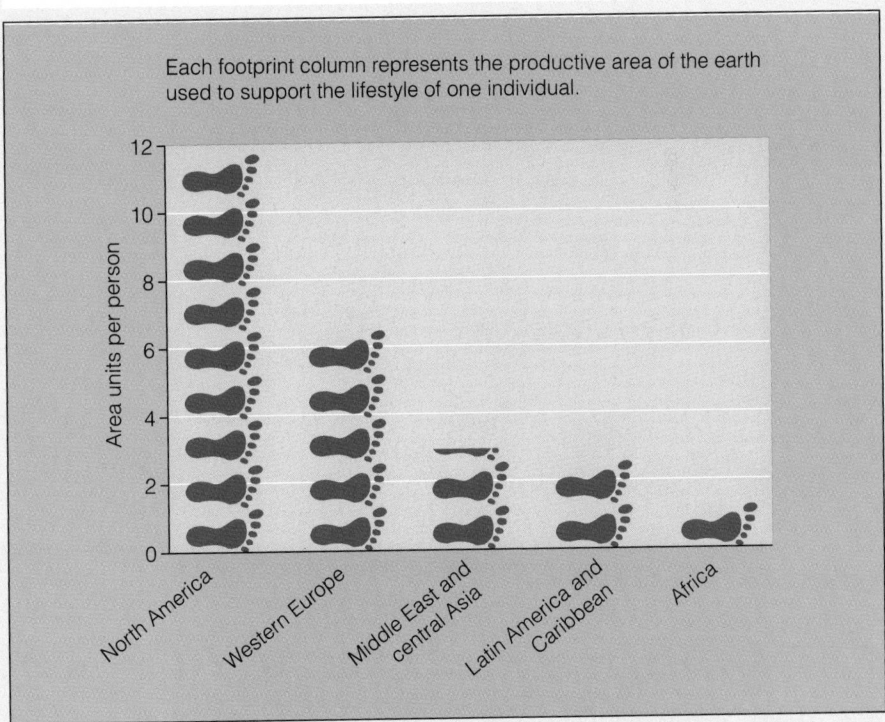

SOURCE: Adapted from United Nations Population Fund, *The State of the World Population 2001*, available at www.unfpa.org.

future and could help greatly to provide nutritious meat at a price people and the environment could afford.

Conclusion

Although many problems are global in scope, the actions of individual people lie at the heart of their solutions. Do what you can. Concerned people should not take a perfectionist attitude, believing that they "should" be doing more than they realistically can, and so feel defeated. Striving for perfection, even while falling short, is a way to achieve progress well worth celebrating. A positive attitude can bring about improvement, and improvement is enough to be proud of. Celebrate the changes that are possible today by making them a permanent part of your life; do the same with changes that become possible tomorrow and every day thereafter. The results may add up to more than you dared to hope for.

appendixes

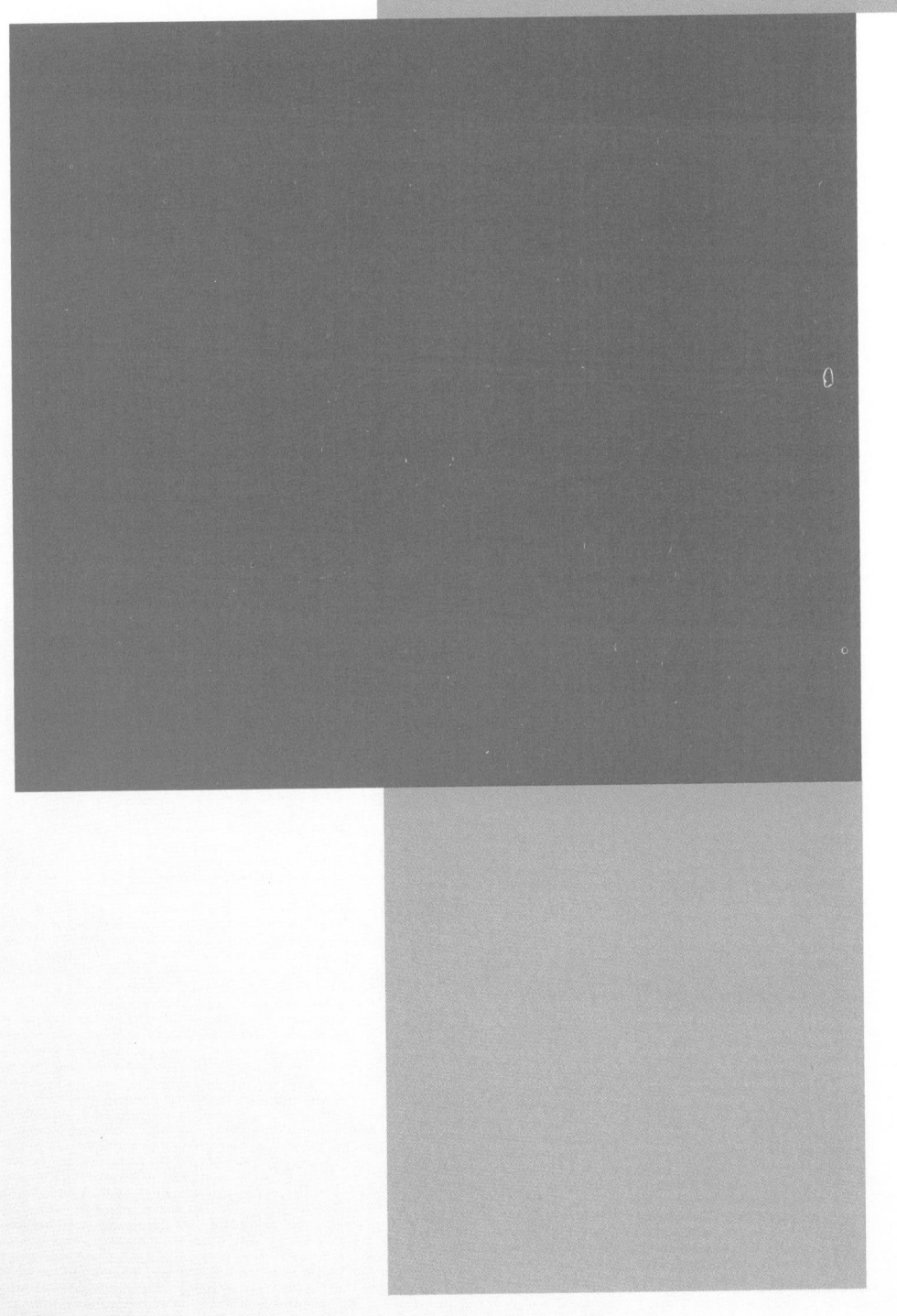

Table of Food Composition

This edition of the table of food composition includes a wide variety of foods from all food groups. It is updated yearly to reflect nutrient changes for current foods, remove outdated foods, and add foods that are new to the marketplace.*

The nutrient database for this appendix is compiled from a variety of sources, including the USDA Standard Release database (Release 14), literature sources, and manufacturers' data. The USDA database provides data for a wider variety of foods and nutrients than other sources. Because laboratory analysis for each nutrient can be quite costly, manufacturers tend to provide data only for those nutrients mandated on food labels. Consequently, data for their foods are often incomplete; any missing information is designated in this table as a blank space. Keep in mind that a blank space means only that the information is unknown and should not be interpreted as a zero.

Whenever using nutrient data, remember that many factors influence the nutrient contents of foods, including the mineral content of the soil, the diet of the animal or the fertilizer of the plant, the season of harvest, the method of processing, the length and method of storage, the method of cooking, the method of analysis, and the moisture content of the sample analyzed. With so many factors involved, users must view nutrient data as a close approximation of the actual amount.

For updates, corrections, and a list of 3000 additional foods and codes found in the diet analysis software that accompanies this text, visit www.wadsworth.com/nutrition and click on *Diet Analysis*.

Fats Total fats, as well as the breakdown to saturated, monounsaturated, and polyunsaturated fats, are listed in the table. The fatty acids seldom add up to the total due to rounding and to other fatty acid components that are not included in these basic categories, such as *trans*-fatty acids and glycerol. *Trans*-fatty acids can comprise a large share of the total fat in margarine and shortening (hydrogenated oils) and in any foods that include them as ingredients.

Vitamin A The vitamin A data in this table are reported in micrograms retinol activity equivalents, if available, otherwise the data are reported in micrograms retinol equivalents. (An asterisk is used to designate retinol equivalents.) In 2001, the DRI committee established retinol activity equivalents as the preferred unit of measure for vitamin A. This unit reflects recent research suggesting a new, lower rate of conversion of carotenoids to vitamin A in the body.

Vitamin E Databases, including this one, currently report vitamin E in milligrams α-tocopherol equivalents, a measure of vitamin E activity. This measure derives from all eight naturally occurring forms of the vitamin, but recent evidence has determined that the body derives vitamin E activity only from the α-tocopherol form. The 2000 DRI values for vitamin E are based only on the α-tocopherol form. Future editions of this table will include new vitamin E values as they become available.

Bioavailability Keep in mind that the availability of nutrients from foods depends not only on the quantity provided by a food, but also on the amount absorbed and used by the body—the bioavailability. The bioavailability of folate from fortified foods, for example, is greater than from naturally occurring sources. (Note that this appendix has been updated with data to reflect the folate fortification of grain products.) Similarly, the body can make niacin from the amino acid tryptophan, but niacin values in this table (and most databases) report preformed niacin only. Chapter 7 provides conversion factors and additional details.

Using the Table The items in this table have been organized into several categories, which are listed at the head of each right-hand page. Page numbers have been provided, and each group has been color-coded to make it easier to find individual items.

In an effort to conserve space, the following abbreviations have been used in the food descriptions and nutrient breakdowns:

- diam = diameter
- ea = each
- enr = enriched
- f/ = from
- frzn = frozen
- g = grams
- liq = liquid
- pce = piece
- pkg = package
- w/ = with
- w/o = without
- t = trace
- 0 = zero (no nutrient value)
- blank space = information not available

*This food composition table has been prepared for Wadsworth Publishing Company and is copyrighted by ESHA Research in Salem, Oregon—the developer and publisher of the Food Processor and Genesis nutritional software programs. The nutritional data are supported by over 1300 references. Because the list of sources is so extensive, it is not provided here, but is available from the publisher.

Table A-1

Food Composition (Computer code number is for Wadsworth Diet Analysis program) (For purposes of calculations, use "0" for t, <1, <.1, <.01, etc.)

Computer Code Number	Food Description	Measure	Wt (g)	H_2O (%)	Ener (cal)	Prot (g)	Carb (g)	Dietary Fiber (g)	Fat (g)	Fat Breakdown (g)		
										Sat	Mono	Poly
	BEVERAGES											
	Alcoholic:											
	Beer:											
1	Regular (12 fl oz)	1½ c	356	92	146	1	13	1	0	0	0	0
2	Light (12 fl oz)	1½ c	354	95	99	1	5	0	0	0	0	0
1506	Non alcohol beer (12 fl oz)	1 ea	360	98	32	1	5	0	0	0	0	0
	Gin, rum, vodka, whiskey:											
3	80 proof	1½ fl oz	42	67	97	0	0	0	0	0	0	0
4	86 proof	1½ fl oz	42	64	105	0	<1	0	0	0	0	0
5	90 proof	1½ fl oz	42	62	110	0	0	0	0	0	0	0
	Liqueur:											
1359	Coffee liqueur, 53 proof	1½ fl oz	52	31	175	<1	24	0	<1	.1	t	.1
1360	Coffee & cream liqueur, 34 proof	1½ fl oz	47	46	154	1	10	0	7	4.5	2.1	.3
1361	Creme de menthe, 72 proof	1½ fl oz	50	28	186	0	21	0	<1	t	t	.1
	Wine, 4 fl oz:											
6	Dessert, sweet	½ c	118	72	181	<1	14	0	0	0	0	0
7	Red	½ c	118	88	85	<1	2	0	0	0	0	0
8	Rose'	½ c	118	89	84	<1	2	0	0	0	0	0
9	White medium	½ c	118	90	80	<1	1	0	0	0	0	0
1592	Nonalcoholic	1 c	232	98	14	1	3	0	0	0	0	0
1593	Nonalcoholic light	1 c	232	98	14	1	3	0	0	0	0	0
1409	Wine cooler, bottle (12 fl oz)	1½ c	340	90	170	<1	20	<1	<1	t	t	t
1595	Wine cooler, cup	1 c	227	90	113	<1	13	<1	<1	t	t	t
	Carbonated:											
10	Club soda (12 fl oz)	1½ c	355	100	0	0	0	0	0	0	0	0
11	Cola beverage (12 fl oz)	1½ c	372	89	153	0	39	0	0	0	0	0
12	Diet cola w/aspartame (12 fl oz)	1½ c	355	100	4	<1	<1	0	0	0	0	0
13	Diet soda pop w/saccharin (12 fl oz)	1½ c	355	100	0	0	<1	0	0	0	0	0
14	Ginger ale (12 fl oz)	1½ c	366	91	124	0	32	0	0	0	0	0
15	Grape soda (12 fl oz)	1½ c	372	89	160	0	42	0	0	0	0	0
16	Lemon-lime (12 fl oz)	1½ c	368	89	147	0	38	0	0	0	0	0
17	Orange (12 fl oz)	1½ c	372	88	179	0	46	0	0	0	0	0
18	Pepper-type soda (12 fl oz)	1½ c	368	89	151	0	38	0	<1	.3	0	0
19	Root beer (12 fl oz)	1½ c	370	89	152	0	39	0	0	0	0	0
20	Coffee, brewed	1 c	237	99	5	<1	1	0	<1	t	0	t
20592	Coffee, cappuccino w/lowfat milk	1½ c	244		110	8	11	0	3	2.5		
20639	Coffee, cappuccino w/whole milk	1½ c	244		140	7	11	0	7	4.5		
20668	Coffee, latte w/lowfat milk	1½ c	366		170	12	17	0	6	4		
21	Coffee, prepared from instant	1 c	238	99	5	<1	1	0	<1	t	0	t
	Fruit drinks, noncarbonated:											
22	Fruit punch drink, canned	1 c	248	88	117	0	29	<1	0	0	0	0
1358	Gatorade	1 c	241	93	60	0	15	0	0	0	0	0
23	Grape drink, canned	1 c	250	87	125	<1	32	<1	0	0	0	0
1304	Koolade sweetened with sugar	1 c	262	90	97	0	25	0	0	0	0	0
1356	Koolade sweetened with nutrasweet	1 c	240	95	43	0	11	0	0	0	0	0
26	Lemonade,frzn concentrate (6-oz can)	¾ c	219	52	396	1	103	1	<1	.1	t	.1
27	Lemonade, from concentrate	1 c	248	89	99	<1	26	<1	0	0	0	0
28	Limeade, frzn concentrate (6-oz can)	¾ c	218	50	408	<1	108	1	<1	t	t	.1
29	Limeade, from concentrate	1 c	247	89	101	0	27	<1	<1	t	t	t
24	Pineapple grapefruit, canned	1 c	250	88	118	<1	29	<1	<1	t	t	.1
25	Pineapple orange, canned	1 c	250	87	125	3	29	<1	0	0	0	0
20559	Powerade	1 c	247	92	72	0	19	0	0	0	0	0
20737	Snapple, fruit punch	1 c	252	88	110	0	29		0	0	0	0
20761	Snapple, tropical	1 c	252	89	110	0	27		0	0	0	0
	Fruit and vegetable juices: see Fruit and Vegetable sections											
	Ultra Slim Fast, ready to drink, can:											
	Chocolate Royale	1 ea	350	83	220	10	40	5	3	1	1.5	.5
	French Vanilla	1 ea	350	84	220	10	40	5	3	.5	1.5	.5
	Strawberries n' cream	1 ea	350	84	220	10	40	5	2	.5	1.5	.5
2427	Water, bottled: La Croix	1 c	236	100	0	0	0	0	0	0	0	0

Chol (mg)	Calc (mg)	Iron (mg)	Magn (mg)	Pota (mg)	Sodi (mg)	Zinc (mg)	VT-A (µg)	Thia (mg)	VT-E (mg)	Ribo (mg)	Niac (mg)	V-B6 (mg)	Fola (µg)	VT-C (mg)
0	18	.11	21	89	18	.07	0	.02	0	.09	1.61	.18	21	0
0	18	.14	18	64	11	.11	0	.03	0	.11	1.39	.12	14	0
0	25	.04	32	90	18	.04	0	.02	0	.09	1.63	.18	22	0
0	0	.02	0	1	<1	.02	0	<.01	0	<.01	<.01	0	0	0
0	0	.02	0	1	<1	.02	0	<.01	0	<.01	<.01	0	0	0
0	0	.02	0	1	<1	.02	0	<.01	0	<.01	<.01	0	0	0
0	1	.03	2	16	4	.02	0	<.01	0	.01	.07	0	0	0
7	8	.06	1	15	43	.07	20*	0	.12	.03	.04	.01	0	0
0	0	.03	0	0	2	.02	0	0	0	0	<.01	0	0	0
0	9	.28	11	109	11	.08	0	.02	0	.02	.25	0	0	0
0	9	.51	15	132	6	.11	0	.01	0	.03	.1	.04	2	0
0	9	.45	12	117	6	.07	0	<.01	0	.02	.09	.03	1	0
0	11	.38	12	94	6	.08	0	<.01	0	.01	.08	.02	0	0
0	21	.93	23	204	16	.19	0	0	0	.02	.23	.05	2	0
0	21	.93	23	204	16	.19	0	0	0	.02	.23	.05	2	0
0	19	.92	18	153	29	.2	<1	.02	.02	.02	.15	.04	4	6
0	13	.62	12	102	19	.13	<1	.01	.01	.02	.1	.03	3	4
0	18	.04	4	7	75	.35	0	0	0	0	0	0	0	0
0	11	.11	4	4	15	.04	0	0	0	0	0	0	0	0
0	14	.11	4	0	21	.28	0	.02	0	.08	0	0	0	0
0	14	.14	4	7	57	.18	0	0	0	0	0	0	0	0
0	11	.66	4	4	26	.18	0	0	0	0	0	0	0	0
0	11	.3	4	4	56	.26	0	0	0	0	0	0	0	0
0	7	.26	4	4	40	.18	0	0	0	0	.05	0	0	0
0	19	.22	4	7	45	.37	0	0	0	0	0	0	0	0
0	11	.15	0	4	37	.15	0	0	0	0	0	0	0	0
0	18	.18	4	4	48	.26	0	0	0	0	0	0	0	0
0	5	.12	12	128	5	.05	0	0	0	0	.53	0	0	0
15	250	0			110		98							2
30	250	0			105		73							2
25	400	0			170		122							4
0	7	.12	10	86	7	.07	0	0	0	<.01	.67	0	0	0
0	20	.52	5	62	55	.3	2	.05	0	.06	.05	0	2	73
0	0	.12	2	26	96	.05	0	.01	0	0	0	0	0	0
0	7	.25	10	87	2	.07	<1	.02	0	.02	.25	.05	2	40
0	42	.13	3	3	37	.08	0	0	0	<.01	<.01	0	0	31
0	17	.65	5	50	50	.26	1	.02	0	.05	.05	0	5	77
0	15	1.58	11	147	9	.17	11	.06	0	.21	.16	.05	22	39
0	7	.4	5	37	7	.1	2	.01	0	.05	.04	.01	5	10
0	11	.22	9	129	0	.09	0	.02	0	.02	.22	0	9	26
0	7	.07	2	32	5	.05	0	<.01	0	<.01	.05	0	2	7
0	17	.77	15	153	35	.15	5	.07	0	.04	.67	.1	27	115
0	12	.67	15	115	7	.15	66	.07	0	.05	.52	.12	27	56
0	0	0		32	28		0							0
0					10									0
0					10									
5	400	2.7	140	600	220	2.25	350*	.52	20	.59	7	.7	120	60
5	400	2.7	140	600	220	2.25	350*	.52	20	.59	7	.7	120	60
5	400	2.7	140	600	220	2.25	350*	.52	20	.59	7	.7	120	60
0					5									

*This value is expressed in retinol equivalents (RE). All other values are in retinol activity equivalents (RAE).

Table A–1

Food Composition (Computer code number is for Wadsworth Diet Analysis program) (For purposes of calculations, use "0" for t, <1, <.1, <.01, etc.)

Computer Code Number	Food Description	Measure	Wt (g)	H_2O (%)	Ener (cal)	Prot (g)	Carb (g)	Dietary Fiber (g)	Fat (g)	Fat Breakdown (g) Sat	Mono	Poly
	BEVERAGES—Continued											
1357	Water, bottled: Perrier (6½ fl oz)	1 ea	192	100	0	0	0	0	0	0	0	0
1594	Water, bottled: Tonic water	1½ c	366	91	124	0	32	0	0	0	0	0
	Tea:											
30	Brewed, regular	1 c	237	100	2	0	1	0	0	0	0	0
1662	Brewed, herbal	1 c	237	100	2	0	<1	0	<1	t	t	t
32	From instant, sweetened	1 c	259	91	88	<1	22	0	<1	t	t	t
31	From instant, unsweetened	1 c	237	100	2	0	<1	0	0	0	0	0
DAIRY												
	Butter: see Fats and Oils, #158,159,160											
	Cheese, natural:											
33	Blue	1 oz	28	42	99	6	1	0	8	5.2	2.2	.2
34	Brick	1 oz	28	41	104	7	1	0	8	5.2	2.4	.2
35	Brie	1 oz	28	48	93	6	<1	0	8	4.9	2.2	.2
36	Camembert	1 oz	28	52	84	6	<1	0	7	4.3	2	.2
37	Cheddar:	1 oz	28	37	113	7	<1	0	9	5.9	2.6	.3
38	1" cube	1 ea	17	37	68	4	<1	0	6	3.6	1.6	.2
39	Shredded	1 c	113	37	455	28	1	0	37	23.8	10.6	1.1
1406	Low fat, low sodium	1 oz	28	65	48	7	1	0	2	1.2	.6	.1
	Cottage:											
2425	Fat Free	1 c	230	83	160	26	12	0	0	0	0	0
984	Low Sodium, low fat	1 c	225	83	162	28	6	0	2	1.4	.6	.1
40	Creamed, large curd	1 c	225	79	232	28	6	0	10	6.4	2.9	.3
41	Creamed, small curd	1 c	210	79	216	26	6	0	9	6	2.7	.3
42	With fruit	1 c	226	72	280	22	30	0	8	4.9	2.2	.2
43	Low fat 2%	1 c	226	79	203	31	8	0	4	2.8	1.2	.1
44	Low fat 1%	1 c	226	82	163	28	6	0	2	1.5	.7	.1
46	Cream	1 tbs	15	54	52	1	<1	0	5	3.3	1.5	.2
983	low fat	1 tbs	15	64	35	2	1	0	3	1.7	.7	.1
47	Edam	1 oz	28	42	100	7	<1	0	8	4.9	2.3	.2
48	Feta	1 oz	28	55	74	4	1	0	6	4.2	1.3	.2
49	Gouda	1 oz	28	41	100	7	1	0	8	4.9	2.2	.2
50	Gruyere	1 oz	28	33	116	8	<1	0	9	5.3	2.8	.5
51	Gorgonzola	1 oz	28	43	97	6	1	0	8	5		
1676	Limburger	1 oz	28	48	92	6	<1	0	8	4.7	2.4	.1
53	Monterey Jack	1 oz	28	41	104	7	<1	0	8	5.3	2.4	.3
54	Mozzarella, whole milk	1 oz	28	54	79	5	1	0	6	3.7	1.8	.2
55	Mozzarella, part-skim milk, low moisture	1 oz	28	49	78	8	1	0	5	3	1.4	.1
56	Muenster	1 oz	28	42	103	7	<1	0	8	5.3	2.4	.2
2422	Neufchatel	1 oz	28	62	73	3	1	0	7	4.1	1.9	.2
1399	Nonfat cheese (Kraft Singles)	1 oz	28	61	44	6	4	0	0	0	0	0
59	Parmesan, grated:	1 oz	28	18	128	12	1	0	8	5.3	2.4	.2
57	Cup, not pressed down	1 c	100	18	456	42	4	0	30	19.1	8.7	.7
58	Tablespoon	1 tbs	6	18	27	2	<1	0	2	1.1	.5	t
60	Provolone	1 oz	28	41	98	7	1	0	7	4.8	2.1	.2
61	Ricotta, whole milk	1 c	246	72	428	28	7	0	32	20.4	8.9	.9
62	Ricotta, part-skim milk	1 c	246	74	339	28	13	0	19	12.1	5.7	.6
63	Romano	1 oz	28	31	108	9	1	0	8	4.8	2.2	.2
64	Swiss	1 oz	28	37	105	8	1	0	8	5	2	.3
976	low fat	1 oz	28	60	50	8	1	0	1	.9	.4	t
	Pasteurized processed cheese products:											
65	American	1 oz	28	39	105	6	<1	0	9	5.5	2.5	.3
66	Swiss	1 oz	28	42	93	7	1	0	7	4.5	2	.2
67	American cheese food, jar	½ c	57	43	187	11	4	0	14	8.8	4.1	.4
68	American cheese spread	1 tbs	15	48	43	2	1	0	3	2	.9	.1
982	Velveeta cheese spread, low fat, low sodium, slice	1 pce	34	62	61	8	1	0	2	1.5	.7	.1
	Cream, sweet:											
69	Half & half (cream & milk)	1 c	242	81	315	7	10	0	28	17.3	8	1
70	Tablespoon	1 tbs	15	81	19	<1	1	0	2	1.1	.5	.1

Chol (mg)	Calc (mg)	Iron (mg)	Magn (mg)	Pota (mg)	Sodi (mg)	Zinc (mg)	VT-A (µg)	Thia (mg)	VT-E (mg)	Ribo (mg)	Niac (mg)	V-B6 (mg)	Fola (µg)	VT-C (mg)
0	27	0	0	0	2	0	0	0	0	0	0	0	0	0
0	4	.04	0	0	15	.37	0	0	0	0	0	0	0	0
0	0	.05	7	88	7	.05	0	0	0	.03	0	0	12	0
0	5	.19	2	21	2	.09	0	.02	0	.01	0	0	2	0
0	5	.05	5	49	8	.08	0	0	0	.05	.09	<.01	10	0
0	5	.05	5	47	7	.07	0	0	0	<.01	.09	<.01	0	0
21	148	.09	6	72	391	.74	64*	.01	.18	.11	.28	.05	10	0
26	189	.12	7	38	157	.73	85*	<.01	.14	.1	.03	.02	6	0
28	51	.14	6	43	176	.67	51*	.02	.18	.15	.11	.07	18	0
20	109	.09	6	52	236	.67	71*	.01	.18	.14	.18	.06	17	0
29	202	.19	8	27	174	.87	78*	.01	.1	.1	.02	.02	5	0
18	123	.12	5	17	106	.53	47*	<.01	.06	.06	.01	.01	3	0
119	815	.77	32	111	702	3.51	314*	.03	.41	.42	.09	.08	20	0
6	197	.2	8	31	6	.86	17*	.01	.05	.01	.02	.02	5	0
10	240	0		380	1000									0
9	137	.31	11	194	29	.85	25	.04	.25	.36	.29	.16	27	0
34	135	.31	11	189	911	.83	108*	.05	.27	.37	.28	.15	27	0
31	126	.29	10	176	851	.78	101*	.04	.25	.34	.26	.14	25	0
25	108	.25	9	151	915	.65	81*	.04	.2	.29	.23	.12	23	0
18	156	.36	14	217	918	.95	45*	.05	.14	.42	.32	.17	29	0
9	138	.32	11	194	918	.86	25*	.05	.25	.37	.29	.15	27	0
16	12	.18	1	18	44	.08	57*	<.01	.14	.03	.01	.01	2	0
8	17	.25	1	25	44	.11	33*	<.01	.07	.04	.02	.01	3	0
25	205	.12	8	53	270	1.05	71*	.01	.21	.11	.02	.02	4	0
25	138	.18	5	17	312	.81	36*	.04	.01	.24	.28	.12	9	0
32	196	.07	8	34	229	1.09	49*	.01	.1	.09	.02	.02	6	0
31	283	.05	10	23	94	1.09	84*	.02	.1	.08	.03	.02	3	0
30	170	.18			280		43*							0
25	139	.04	6	36	224	.59	88*	.02	.18	.14	.04	.02	16	0
25	209	.2	8	23	150	.84	71*	<.01	.09	.11	.03	.02	5	0
22	145	.05	5	19	104	.62	67*	<.01	.1	.07	.02	.02	2	0
15	205	.07	7	27	148	.88	53*	.01	.13	.1	.03	.02	3	0
27	201	.11	8	37	176	.79	88*	<.01	.13	.09	.03	.02	3	0
21	21	.08	2	32	112	.15	84*	<.01	.26	.05	.03	.01	3	0
7	221	0		88	398	.88	84*			.15				0
22	385	.27	14	30	521	.89	48*	.01	.22	.11	.09	.03	2	0
79	1376	.95	51	107	1862	3.19	173*	.04	.8	.39	.31	.1	8	0
5	83	.06	3	6	112	.19	10*	<.01	.05	.02	.02	.01	<1	0
19	212	.15	8	39	245	.9	74*	<.01	.1	.09	.04	.02	3	0
125	509	.93	27	258	207	2.85	330*	.03	.86	.48	.26	.11	29	0
76	669	1.08	37	308	308	3.3	278*	.05	.52	.45	.19	.05	32	0
29	298	.22	11	24	336	.72	39*	.01	.2	.1	.02	.02	2	0
26	269	.05	10	31	73	1.09	71*	.01	.14	.1	.03	.02	2	0
10	269	.05	10	31	73	1.09	18*	.01	.05	.1	.02	.02	2	0
26	172	.11	6	45	400	.84	81*	.01	.13	.1	.02	.02	2	0
24	216	.17	8	60	384	1.01	64*	<.01	.19	.08	.01	.01	2	0
36	327	.48	18	159	678	1.7	125*	.02	.4	.25	.08	.08	4	0
8	84	.05	4	36	202	.39	28*	.01	.11	.06	.02	.02	1	0
12	233	.15	8	61	2	1.13	22*	.01	.17	.13	.03	.03	3	0
89	254	.17	24	315	99	1.23	259*	.08	.27	.36	.19	.09	7	2
6	16	.01	1	19	6	.08	16*	<.01	.02	.02	.01	.01	<1	<1

*This value is expressed in retinol equivalents (RE). All other values are in retinol activity equivalents (RAE).

Table A–1

Food Composition

(Computer code number is for Wadsworth Diet Analysis program) (For purposes of calculations, use "0" for t, <1, <.1, <.01, etc.)

Computer Code Number	Food Description	Measure	Wt (g)	H_2O (%)	Ener (cal)	Prot (g)	Carb (g)	Dietary Fiber (g)	Fat (g)	Fat Breakdown (g) Sat	Mono	Poly
	DAIRY—Continued											
71	Light, coffee or table:	1 c	240	74	468	6	9	0	46	28.8	13.4	1.7
72	Tablespoon	1 tbs	15	74	29	<1	1	0	3	1.8	.8	.1
73	Light whipping cream, liquid:	1 c	239	63	698	5	7	0	74	46.2	21.7	2.1
74	Tablespoon	1 tbs	15	63	44	<1	<1	0	5	2.9	1.4	.1
75	Heavy whipping cream, liquid:	1 c	238	58	821	5	7	0	88	54.8	25.4	3.3
76	Tablespoon	1 tbs	15	58	52	<1	<1	0	6	3.4	1.6	.2
77	Whipped cream, pressurized:	1 c	60	61	154	2	7	0	13	8.3	3.8	.5
78	Tablespoon	1 tbs	4	61	10	<1	<1	0	1	.6	.3	t
79	Cream, sour, cultured:	1 c	230	71	492	7	10	0	48	30	13.9	1.8
80	Tablespoon	1 tbs	14	71	30	<1	1	0	3	1.8	.8	.1
2423	Fat free	1 tbs	15	79	12	1	2	0	0	0	0	0
	Cream products-imitation and part dairy:											
81	Coffee whitener, frozen or liquid	1 tbs	15	77	20	<1	2	0	1	1.4	t	0
82	Coffee whitener, powdered	1 tsp	2	2	11	<1	1	0	1	.6	t	0
83	Dessert topping, frozen, nondairy:	1 c	75	50	239	1	17	0	19	16.3	1.2	.4
84	Tablespoon	1 tbs	5	50	16	<1	1	0	1	1.1	.1	t
85	Dessert topping, mix with whole milk:	1 c	80	67	151	3	13	0	10	8.5	.7	.2
86	Tablespoon	1 tbs	5	67	9	<1	1	0	1	.5	t	t
88	Dessert topping, pressurized	1 c	70	60	185	1	11	0	16	13.2	1.3	.2
87	Tablespoon	1 tbs	4	60	11	<1	1	0	1	.8	.1	t
91	Sour cream, imitation:	1 c	230	71	478	6	15	0	45	40.9	1.3	.1
92	Tablespoon	1 tbs	14	71	29	<1	1	0	3	2.5	.1	t
89	Sour dressing, part dairy:	1 c	235	75	418	8	11	0	39	31.2	4.6	1.1
90	Tablespoon	1 tbs	15	75	27	<1	1	0	2	2	.3	.1
	Milk, fluid:											
93	Whole milk	1 c	244	88	149	8	11	0	8	5.1	2.3	.3
94	2% lowfat milk	1 c	244	89	122	8	12	0	5	2.9	1.3	.2
95	2% milk solids added	1 c	245	89	125	9	12	0	5	2.9	1.4	.2
96	1% lowfat milk	1 c	244	90	102	8	12	0	3	1.6	.7	.1
97	1% milk solids added	1 c	245	90	105	9	12	0	2	1.5	.7	.1
98	Nonfat milk, vitamin A added	1 c	245	91	86	8	12	0	<1	.3	.1	t
99	Nonfat milk solids added	1 c	245	90	91	9	12	0	1	.4	.2	t
100	Buttermilk, skim	1 c	245	90	98	8	12	0	2	1.3	.6	.1
	Milk, canned:											
101	Sweetened condensed	1 c	306	27	982	24	166	0	27	16.8	7.4	1
103	Evaporated, nonfat	1 c	256	79	200	19	29	0	1	.3	.2	t
	Milk, dried:											
104	Buttermilk, sweet	1 c	120	3	464	41	59	0	7	4.3	2	.3
105	Instant, nonfat, vit A added-makes 1 qt	1 ea	91	4	326	32	47	0	1	.4	.2	t
106	Instant nonfat, vit A added, cup	1 c	68	4	243	24	35	0	<1	.3	.1	t
107	Goat milk	1 c	244	87	168	9	11	0	10	6.5	2.7	.4
108	Kefir	1 c	233	88	149	8	11	0	8			
	Milk beverages and powdered mixes:											
	Chocolate:											
109	Whole	1 c	250	82	208	8	26	2	8	5.3	2.5	.3
110	2% fat	1 c	250	84	180	8	26	1	5	3.1	1.5	.2
111	1% fat	1 c	250	84	158	8	26	1	2	1.5	.7	.1
	Chocolate-flavored beverages:											
112	Powder containing nonfat dry milk:	1 oz	28	1	101	3	22	<1	1	.7	.4	t
113	Prepared with water	1 c	275	86	138	4	30	3	2	.9	.5	t
114	Powder without nonfat dry milk:	1 oz	28	1	98	1	25	2	1	.5	.3	t
115	Prepared with whole milk	1 c	266	81	226	9	31	1	9	5.5	2.6	.3
116	Eggnog, commercial	1 c	254	74	343	10	34	0	19	11.3	5.7	.9
974	2% low-fat eggnog	1 c	254	85	191	12	17	0	8	3.7	2.7	.7
1027	Instant Breakfast, envelope,powder only:	1 ea	37	7	131	7	24	<1	1	.2	.1	.1
1028	Prepared with whole milk	1 c	281	77	280	15	36	<1	9	5.3		
1029	Prepared with 2% milk	1 c	281	78	252	15	36	<1	5	3.1		
1283	Prepared with 1% milk	1 c	281	79	233	15	36	<1	3	1.8		
1284	Prepared with nonfat milk	1 c	282	80	216	16	36	<1	1	.7		
117	Malted milk, chocolate, powder:	3 tsp	21	1	79	1	18	<1	1	.5	.2	.1
118	Prepared with whole milk	1 c	265	81	228	9	30	<1	9	5.5	2.6	.4
1661	Ovaltine with whole milk	1 c	265	81	225	9	29	<1	9	5.5	2.5	.4

Chol (mg)	Calc (mg)	Iron (mg)	Magn (mg)	Pota (mg)	Sodi (mg)	Zinc (mg)	VT-A (µg)	Thia (mg)	VT-E (mg)	Ribo (mg)	Niac (mg)	V-B6 (mg)	Fola (µg)	VT-C (mg)
158	230	.1	22	293	96	.65	437*	.08	.36	.35	.14	.08	5	2
10	14	.01	1	18	6	.04	27*	<.01	.02	.02	.01	<.01	<1	<1
265	165	.07	17	232	81	.6	705*	.06	1.43	.3	.1	.07	10	1
17	10	<.01	1	15	5	.04	44*	<.01	.09	.02	.01	<.01	1	<1
326	155	.07	17	179	90	.55	1001*	.05	1.5	.26	.09	.06	10	1
21	10	<.01	1	11	6	.03	63*	<.01	.09	.02	.01	<.01	1	<1
46	61	.03	7	88	78	.22	124*	.02	.36	.04	.04	.02	2	0
3	4	<.01		6	5	.01	8*	<.01	.02	<.01	<.01	<.01	<1	0
101	267	.14	25	331	122	.62	449*	.08	1.31	.34	.15	.04	25	2
6	16	.01	2	20	7	.04	27*	<.01	.08	.02	.01	<.01	2	<1
0	27	0		42	15									0
0	1	<.01	0	29	12	<.01	1	0	.24	0	0	0	0	0
0		.02		16	4	.01	<1	0	<.01	<.01	0	0	0	0
0	4	.09	1	13	19	.02	32	0	.14	0	0	0	0	0
0		.01		1	1	<.01	2	0	.01	0	0	0	0	0
8	72	.03	8	121	53	.22	39*	.02	.11	.09	.05	.02	3	1
<1	4	<.01		8	3	.01	2*	<.01	.01	.01	<.01	<.01	<1	<1
0	3	.01	1	13	43	.01	33*	0	.12	0	0	0	0	0
0		<.01		1	2	0	2*	0	.01	0	0	0	0	0
0	7	.9	14	370	235	2.71	0	0	.34	0	0	0	0	0
0		.05	1	22	14	.16	0	0	.02	0	0	0	0	0
12	266	.07	23	381	113	.87	5*	.09	.28	.38	.17	.04	28	2
1	17	<.01	1	24	7	.06	<1*	.01	.02	.02	.01	<.01	2	<1
34	290	.12	32	371	120	.93	76*	.09	.24	.39	.2	.1	12	2
19	298	.12	34	376	122	.95	139*	.09	.17	.4	.21	.1	12	2
20	314	.12	34	397	127	.98	140*	.1	.17	.42	.22	.11	12	2
10	300	.12	34	381	124	.95	144*	.09	.1	.41	.21	.1	12	2
10	314	.12	34	397	127	.98	145*	.1	.1	.42	.22	.11	12	2
5	301	.1	27	407	127	.98	149*	.09	.1	.34	.22	.1	12	2
5	316	.12	37	419	130	1	149*	.1	.1	.43	.22	.11	12	2
10	284	.12	27	370	257	1.03	20*	.08	.15	.38	.14	.08	12	2
104	869	.58	80	1135	389	2.88	248*	.27	.64	1.27	.64	.16	34	8
10	742	.74	69	850	294	2.3	300	.11	0	.79	.44	.14	23	3
83	1420	.36	132	1910	620	4.82	65*	.47	.48	1.89	1.05	.41	56	7
16	1120	.28	106	1551	500	4.01	646	.38	.02	1.59	.81	.31	45	5
12	837	.21	80	1159	373	3	483	.28	.01	1.19	.61	.23	34	4
27	327	.12	34	498	122	.73	137	.12	.22	.34	.68	.11	2	3
		.3	33	373	107									
30	280	.6	32	418	150	1.03	72*	.09	.22	.4	.31	.1	12	2
17	285	.6	32	423	150	1.03	143*	.09	.12	.41	.31	.1	12	2
7	288	.6	32	425	153	1.03	148*	.09	.07	.41	.32	.1	12	2
1	91	.33	23	199	141	.41	1*	.03	.04	.16	.16	.03	0	<1
3	129	.47	33	270	198	.6	<1	.04	.06	.21	.22	.04	0	1
0	10	.88	27	165	59	.43	<1	.01	.11	.04	.14	<.01	2	<1
32	301	.8	53	497	165	1.28	77*	.1	.21	.43	.32	.1	13	2
150	330	.51	48	419	137	1.17	203*	.09	.58	.48	.27	.13	3	4
194	270	.71	32	368	155	1.26	197*	.11	.58	.55	.21	.15	30	2
4	105	4.74	84	350	142	3.16	554*	.31	5.31	.07	5.27	.42	105	28
38	396	4.87	117	721	262	4.09	430*	.41	5.55	.47	5.47	.52	118	31
23	402	4.87	118	727	264	4.12	469*	.41	5.48	.48	5.48	.53	118	31
14	406	4.87	118	731	266	4.12	469*	.41	5.4	.48	5.48	.53	118	31
9	407	4.83	112	755	268	4.14	469*	.4	5.3	.42	5.47	.52	118	31
1	13	.48	15	130	53	.17	4*	.04	.08	.04	.42	.03	4	<1
34	305	.61	48	498	172	1.09	79*	.13	.26	.44	.62	.13	16	3
34	384	3.76	53	620	244	1.17	901*	.73	.32	1.26	10.9	1.02	32	34

*This value is expressed in retinol equivalents (RE). All other values are in retinol activity equivalents (RAE).

Table A-1

Food Composition (Computer code number is for Wadsworth Diet Analysis program) (For purposes of calculations, use "0" for t, <1, <.1, <.01, etc.)

Computer Code Number	Food Description	Measure	Wt (g)	H_2O (%)	Ener (cal)	Prot (g)	Carb (g)	Dietary Fiber (g)	Fat (g)	Fat Breakdown (g) Sat	Mono	Poly
	DAIRY—Continued											
119	Malted mix powder, natural:	3 tsp	21	2	87	2	16	<1	2	.9	.4	.3
120	Prepared with whole milk	1 c	265	81	236	10	27	0	10	5.9	2.8	.6
121	Milk shakes, chocolate	1 c	166	71	211	6	34	1	6	3.8	1.8	.2
122	Milk shakes, vanilla	1 c	166	75	184	6	30	1	5	3.1	1.4	.2
	Milk desserts:											
134	Custard, baked	1 c	282	79	296	14	30	0	13	6.6	4.3	1
1548	Low-fat frozen dessert bars	1 ea	81	72	88	2	19	0	1	.2	.1	.4
	Ice cream, vanilla (about 10% fat):											
124	Hardened	1 c	132	61	265	5	31	0	14	9	4.2	.5
126	Soft serve	1 c	172	60	370	7	38	0	22	12.9	6	.8
	Ice cream, rich vanilla (16% fat):											
128	Hardened	1 c	148	57	357	5	33	0	24	14.8	6.9	.9
1724	Ben & Jerry's	½ c	108	60	250	4	22	0	16	11		
	Ice milk, vanilla (about 4% fat):											
130	Hardened	1 c	132	68	183	5	30	0	6	3.5	1.6	.2
131	Soft serve (about 3.3% fat)	1 c	176	70	222	9	38	0	5	2.9	1.3	.2
	Pudding, canned (5 oz can = .55 cup):											
135	Chocolate	1 ea	142	69	189	4	32	1	6	1	2.4	2
136	Tapioca	1 ea	142	74	169	3	27	<1	5	.9	2.2	1.9
137	Vanilla	1 ea	142	71	185	3	31	<1	5	.8	2.2	1.9
	Puddings, dry mix with whole milk:											
138	Chocolate, instant	1 c	294	74	326	9	55	3	9	5.4	2.7	.5
139	Chocolate, regular, cooked	1 c	284	74	315	9	51	3	10	5.9	2.8	.4
140	Rice, cooked	1 c	288	72	351	9	60	<1	8	5.1	2.3	.3
141	Tapioca, cooked	1 c	282	74	321	8	55	0	8	5.1	2.3	.3
142	Vanilla, instant	1 c	284	73	324	8	56	0	8	4.9	2.4	.4
143	Vanilla, regular, cooked	1 c	280	75	311	8	52	0	8	5.1	2.4	.4
133	Sherbet (2% fat)	1 c	198	66	273	2	60	0	4	2.3	1	.2
20440	Rice Milk	1 c	245	89	120	<1	25	0	2	.2	1.3	.3
20590	Rice/Soy Milk, blend	1 c	241	88	120	7	18	0	3	.5		
144	Soy Milk	1 c	245	93	81	7	4	3	5	.5	.8	2
2301	Soy Milk, fortified, fat free	1 c	240	88	110	6	22	1	0	0	0	0
	Yogurt, fat free:											
2851	Strawberry, container	1 ea	227	86	120	8	22	0	0	0	0	0
2424	Vanilla	1 c	245	76	223	12	43	0	<1	.3	.1	t
	Yogurt, frozen, low-fat											
1584	Cup 1 c	144	65	229	6	35	0	8	4.9	2.3	.3	
1512	Scoop	1 ea	79	74	78	4	15	0	<1	.1	t	t
	Yogurt, lowfat:											
1172	Fruit added with low-calorie sweetener	1 c	241	86	122	11	19	1	<1	.2	.1	t
145	Fruit added	1 c	245	74	250	11	47	0	3	1.7	.7	.1
146	Plain	1 c	245	85	154	13	17	0	4	2.4	1	.1
147	Vanilla or coffee flavor	1 c	245	79	208	12	34	0	3	2	.8	.1
148	Yogurt, made with nonfat milk	1 c	245	85	137	14	19	0	<1	.3	.1	t
149	Yogurt, made with whole milk	1 c	245	88	149	8	11	0	8	5.1	2.2	.2
	EGGS											
	Raw, large:											
150	Whole, without shell	1 ea	50	75	74	6	1	0	5	1.5	1.9	.7
151	White	1 ea	33	88	16	3	<1	0	0	0	0	0
152	Yolk 1 ea	17	49	61	3	<1	0	5	1.6	2	.7	
	Cooked:											
153	Fried in margarine	1 ea	46	69	91	6	1	0	7	1.9	2.7	1.3
154	Hard-cooked, shell removed	1 ea	50	75	77	6	1	0	5	1.6	2	.7
155	Hard-cooked, chopped	1 c	136	75	211	17	2	0	14	4.4	5.5	1.9
156	Poached, no added salt	1 ea	50	75	74	6	1	0	5	1.5	1.9	.7
157	Scrambled with milk & margarine	1 ea	61	73	101	7	1	0	7	2.2	2.9	1.3
1681	Egg substitute, liquid:	½ c	126	83	106	15	1	0	4	.8	1.1	2
1254	Egg Beaters, Fleischmann's	½ c	122		60	12	2	0	0	0	0	0
1262	Egg substitute, liquid, prepared	½ c	105	80	107	11	2	0	6	1.1	2.1	2.1

Chol (mg)	Calc (mg)	Iron (mg)	Magn (mg)	Pota (mg)	Sodi (mg)	Zinc (mg)	VT-A (µg)	Thia (mg)	VT-E (mg)	Ribo (mg)	Niac (mg)	V-B6 (mg)	Fola (µg)	VT-C (mg)
4	63	.15	19	159	104	.21	18*	.11	.08	.19	1.1	.09	10	1
37	355	.26	53	530	223	1.14	95*	.2	.32	.59	1.31	.19	21	3
22	188	.51	28	332	161	.68	38*	.1	.12	.41	.27	.08	7	1
18	203	.15	20	289	136	.6	53*	.07	.1	.3	.31	.09	5	1
245	316	.85	39	431	217	1.49	169*	.09	.68	.64	.24	.14	28	1
1	81	.04	9	107	44	.26	38	.03	.07	.11	.06	.03	3	1
58	169	.12	18	263	106	.91	154*	.05	0	.32	.15	.06	7	1
157	225	.36	21	304	105	.89	265*	.08	.64	.31	.16	.08	15	1
90	173	.07	16	235	83	.59	272*	.06	0	.24	.12	.06	7	1
75	100	.36			60		150*							0
18	183	.13	20	279	112	.58	62*	.08	0	.35	.12	.09	8	1
21	276	.11	25	389	123	.93	51*	.09	0	.35	.21	.08	11	2
4	128	.72	30	256	183	.6	16*	.04	.17	.22	.49	.04	4	3
1	119	.33	11	138	226	.38	0	.03	.13	.14	.44	.03	4	1
10	125	.18	11	160	192	.35	9*	.03	.17	.2	.36	.02	0	0
32	300	.85	53	488	835	1.23	62*	.1	.18	.41	.28	.11	12	3
34	315	1.02	43	463	293	1.28	74*	.09	.17	.49	.29	.1	11	2
32	297	1.09	37	372	314	1.09	58*	.22	.17	.4	1.28	.1	11	2
34	293	.17	34	372	341	.96	76*	.08	.23	.4	.21	.11	11	2
31	287	.2	34	364	812	.94	71*	.09	.17	.39	.21	.1	11	2
34	300	.14	36	381	448	.98	76*	.08	.17	.4	.21	.09	11	2
12	107	.28	16	190	91	.95	28*	.05	.16	.15	.12	.05	10	6
0	20	.2	10	69	86	.24	<1	.08	1.76	.01	1.91	.04	91	1
0	13	1.08	40	270	85	.9	0	.09		.1	.4	.08	26	
0	10	1.42	47	345	29	.56	4	.39	.02	.17	.36	.1	5	0
0	400	1.44		20	60		0	.07		.1	3			0
5	350	0		380	160		0							5
4	436	.21	42	559	168	2.13	4	.1	.01	.52	.27	.11	27	2
3	206	.43	20	304	125	.6	82*	.05	.07	.32	.41	.11	9	1
1	137	.07	13	175	53	.67	1	.03	<.01	.16	.08	.04	8	1
3	369	.61	41	550	139	1.83	6*	.1	.17	.45	.5	.11	32	26
10	372	.17	37	478	142	1.81	27*	.09	.07	.44	.23	.1	22	2
15	448	.2	42	573	172	2.18	39*	.11	.1	.52	.28	.12	27	2
12	419	.17	39	537	162	2.03	32*	.1	.07	.49	.26	.11	27	2
5	488	.22	47	625	189	2.38	5*	.12	0	.57	.3	.13	29	2
32	296	.12	29	380	113	1.45	73*	.07	.22	.35	.18	.08	17	1
213	24	.72	5	60	63	.55	95	.03	.52	.25	.04	.07	23	0
0	2	.01	4	47	54	<.01	0	<.01	0	.15	.03	<.01	1	0
218	23	.6	2	16	7	.53	99	.03	.54	.11	<.01	.07	25	0
211	25	.72	5	61	162	.55	114	.03	.75	.24	.03	.07	17	0
212	25	.59	5	63	62	.52	84	.03	.52	.26	.03	.06	22	0
577	68	1.62	14	171	169	1.43	228	.09	1.43	.7	.09	.16	60	0
212	24	.72	5	60	140	.55	95	.02	.52	.21	.03	.06	17	0
215	43	.73	7	84	171	.61	119*	.03	.8	.27	.05	.07	18	<1
1	67	2.65	11	416	223	1.64	272	.14	.62	.38	.14	<.01	19	0
0	40	2.16		170	250	1.2	120*		1.61	1.7		.16	64	0
1	82	1.85	11	337	201	1.25	233*	.09	.83	.29	.12	.01	11	<1

*This value is expressed in retinol equivalents (RE). All other values are in retinol activity equivalents (RAE).

Table A–1

Food Composition (Computer code number is for Wadsworth Diet Analysis program) (For purposes of calculations, use "0" for t, <1, <.1, <.01, etc.)

Computer Code Number	Food Description	Measure	Wt (g)	H_2O (%)	Ener (cal)	Prot (g)	Carb (g)	Dietary Fiber (g)	Fat (g)	Fat Breakdown (g) Sat	Mono	Poly
	FATS AND OILS											
158	Butter: Stick	½ c	114	16	817	1	<1	0	92	57.5	26.7	3.4
159	Tablespoon:	1 tbs	14	16	100	<1	<1	0	11	7.1	3.3	.4
8025	Unsalted	1 tbs	14	18	100	<1	<1	0	11	7.1	3.3	.4
160	Pat (about 1 tsp)	1 ea	5	16	36	<1	<1	0	4	2.5	1.2	.2
1682	Whipped	1 tsp	3	16	21	<1	<1	0	2	1.5	.7	.1
	Fats, cooking:											
1363	Bacon fat	1 tbs	14		125	0	0	0	14	6.3	5.9	1.1
1362	Beef fat/tallow	1 c	205	0	1849	0	0	0	205	102	85.7	8.2
1364	Chicken fat	1 c	205		1845	0	0	0	205	61.1	91.6	42.8
161	Vegetable shortening:	1 c	205	0	1812	0	0	0	205	51.3	91.2	53.5
162	Tablespoon	1 tbs	13	0	115	0	0	0	13	3.2	5.8	3.4
163	Lard:	1 c	205	0	1849	0	0	0	205	81.1	87	28.3
164	Tablespoon	1 tbs	13	0	117	0	0	0	13	5.1	5.5	1.8
	Margarine:											
165	Imitation (about 40% fat), soft:	1 c	232	58	800	1	1	0	90	17.9	36.4	32
166	Tablespoon	1 tbs	14	58	48	<1	<1	0	5	1.1	2.2	1.9
167	Regular, hard (about 80% fat):	½ c	114	16	820	1	1	0	92	18	40.8	29
168	Tablespoon	1 tbs	14	16	101	<1	<1	0	11	2.2	5	3.6
169	Pat	1 ea	5	16	36	<1	<1	0	4	.8	1.8	1.3
170	Regular, soft (about 80% fat):	1 c	227	16	1625	2	1	0	183	31.3	64.7	78.5
171	Tablespoon	1 tbs	14	16	100	<1	<1	0	11	1.9	4	4.8
2056	Saffola, unsalted	1 tbs	14	20	100	0	0	0	11	2	3	4.5
2057	Saffola, reduced fat	1 tbs	14	37	60	0	0	0	8	1.3	2.7	4.4
172	Spread (about 60% fat), hard:	1 c	227	37	1225	1	0	0	138	32	59	41.1
173	Tablespoon	1 tbs	14	37	76	<1	0	0	9	2	3.6	2.5
174	Pat	1 ea	5	37	27	<1	0	0	3	.7	1.2	1
175	Spread (about 60% fat), soft:	1 c	227	37	1225	1	0	0	138	29.1	71.5	31.3
176	Tablespoon	1 tbs	14	37	76	<1	0	0	9	1.8	4.4	1.9
2160	Touch of Butter (47% fat)	1 tbs	14		60	0	0	0	7	1.5	3.1	1.5
	Oils:											
1585	Canola:	1 c	218	0	1927	0	0	0	218	15.5	128	64.5
1586	Tablespoon	1 tbs	14	0	124	0	0	0	14	1	8.2	4.1
177	Corn:	1 c	218	0	1927	0	0	0	218	27.7	52.8	128
178	Tablespoon	1 tbs	14	0	124	0	0	0	14	1.8	3.4	8.2
179	Olive:	1 c	216	0	1909	0	0	0	216	29.2	159	18.1
180	Tablespoon	1 tbs	14	0	124	0	0	0	14	1.9	10.3	1.2
1683	Olive, extra virgin	1 tbs	14		126	0	0	0	14	2	10.8	1.3
181	Peanut:	1 c	216	0	1909	0	0	0	216	36.5	99.8	69.1
182	Tablespoon	1 tbs	14	0	124	0	0	0	14	2.4	6.5	4.5
183	Safflower:	1 c	218	0	1927	0	0	0	218	13.5	31.3	163
184	Tablespoon	1 tbs	14	0	124	0	0	0	14	.9	2	10.4
185	Soybean:	1 c	218	0	1927	0	0	0	218	31.4	50.8	126
186	Tablespoon	1 tbs	14	0	124	0	0	0	14	2	3.3	8.1
187	Soybean/cottonseed:	1 c	218	0	1927	0	0	0	218	39.2	64.3	105
188	Tablespoon	1 tbs	14	0	124	0	0	0	14	2.5	4.1	6.7
189	Sunflower:	1 c	218	0	1927	0	0	0	218	22.5	42.5	143
190	Tablespoon	1 tbs	14	0	124	0	0	0	14	1.4	2.7	9.2
	Salad dressings/sandwich spreads:											
191	Blue cheese, regular	1 tbs	15	32	76	1	1	0	8	1.5	1.8	4.2
1040	Low calorie	1 tbs	15	80	15	1	<1	0	1	.4	.3	.4
1684	Caesar's	1 tbs	12		55	1	<1	<1	5	.9		
192	French, regular	1 tbs	16	38	69	<1	3	0	7	1.5	1.3	3.5
193	Low calorie	1 tbs	16	69	21	<1	3	0	1	.1	.2	.5
194	Italian, regular	1 tbs	15	38	70	<1	2	0	7	1	1.7	4.2
195	Low calorie	1 tbs	15	82	16	<1	1	<1	1	.2	.3	.9
	Kraft, Deliciously Right:											
2150	1000 Island	1 tbs	16	64	34	0	3	0	2	.2		
2153	Bacon & tomato	1 tbs	16		31	1	2	0	3	.5		
2154	Cucumber ranch	1 tbs	16	76	31	0	1	0	3	.5		
2151	French	1 tbs	16		25	0	3	0	1	.2		
2152	Ranch	1 tbs	16		52	0	3	0	5	.8		

Chol (mg)	Calc (mg)	Iron (mg)	Magn (mg)	Pota (mg)	Sodi (mg)	Zinc (mg)	VT-A (µg)	Thia (mg)	VT-E (mg)	Ribo (mg)	Niac (mg)	V-B6 (mg)	Fola (µg)	VT-C (mg)
250	27	.18	2	30	942	.06	860*	.01	1.8	.04	.05	<.01	3	0
31	3	.02		4	116	.01	106*	<.01	.22	<.01	.01	0	<1	0
31	3	.02		4	2	.01	106*	<.01	.22	<.01	.01	0	<1	0
11	1	.01		1	41	<.01	38*	0	.08	<.01	<.01	0	<1	0
7	1	<.01		1	25	<.01	23*	0	.05	<.01	<.01	0	<1	0
14		<.01			76	<.01	0	0	.31	0	0	0	0	0
223	0	0	0	0	0	0	0	0	5.54	0	0	0	0	0
174	0	0	0	0	0	0	0	0	5.54	0	0	0	0	0
0	0	0	0	0	0	0	0	0	17	0	0	0	0	0
0	0	0	0	0	0	0	0	0	1.08	0	0	0	0	0
195		0			<1	.23	0*	0	2.46	0	0	0	0	0
12		0			<1	.01	0*	0	.16	0	0	0	0	0
0	42	0	5	58	2227	0	1853*	.01	5.41	.05	.03	.01	2	<1
0	3	0		3	134	0	112*	<.01	.33	<.01	<.01	<.01	<1	<1
0	34	.07	3	48	1075	0	911*	.01	14.6	.04	.03	.01	1	<1
0	4	.01		6	132	0	112*	<.01	1.79	<.01	<.01	<.01	<1	<1
0	1	<.01		2	47	0	40*	<.01	.64	<.01	<.01	0	<1	<1
0	61	0	5	86	2449	0	1813*	.02	27.2	.07	.04	.02	2	<1
0	4	0		5	151	0	112*	<.01	1.68	<.01	<.01	<.01	<1	<1
	0	0			0		51*							0
	0	0			115		51*							0
0	48	0	5	68	2256	0	1813*	.02	11.4	.06	.04	.01	2	<1
0	3	0		4	139	0	112*	<.01	.7	<.01	<.01	<.01	<1	<1
0	1	0		1	50	0	40*	0	.25	<.01	<.01	0	<1	<1
0	48	0	5	68	2256	0	1813*	.02	20.5	.06	.04	.01	2	<1
0	3	0		4	139	0	112*	<.01	1.26	<.01	<.01	<.01	<1	<1
0	0	0		0	110		100*		1.27					0
0	0	0	0	0	0	0	0	0	45.7	0	0	0	0	0
0	0	0	0	0	0	0	0	0	2.93	0	0	0	0	0
0	0	0	0	0	0	0	0	0	46	0	0	0	0	0
0	0	0	0	0	0	0	0	0	2.96	0	0	0	0	0
0	0	.82	0	0	0	.13	0	0	26.8	0	0	0	0	0
0	0	.05	0	0	0	.01	0	0	1.74	0	0	0	0	0
									1.74					
0	0	.06	0	0	0	.02	0	0	27.9	0	0	0	0	0
0	0	<.01	0	0	0	<.01	0	0	1.81	0	0	0	0	0
0	0	0	0	0	0	0	0	0	93.9	0	0	0	0	0
0	0	0	0	0	0	0	0	0	6.03	0	0	0	0	0
0	0	.04	0	0	0	0	0	0	39.7	0	0	0	0	0
0	0	<.01	0	0	0	0	0	0	2.55	0	0	0	0	0
0	0	0	0	0	0	0	0	0	61.5	0	0	0	0	0
0	0	0	0	0	0	0	0	0	3.95	0	0	0	0	0
0	0	0	0	0	0	0	0	0	110	0	0	0	0	0
0	0	0	0	0	0	0	0	0	7.08	0	0	0	0	0
3	12	.03	0	6	164	.04	10*	<.01	1.4	.01	.01	.01	1	<1
<1	13	.07	1	1	180	.04	<1	<.01	.13	.01	.01	<.01	<1	<1
12	22	.2	3	20	210	.12	5*	<.01	.72	.02	.49	.01	2	1
0	2	.06	0	13	219	.01	10	<.01	1.35	<.01	0	<.01	1	0
0	2	.06	0	13	126	.03	10	0	.19	0	0	0	0	0
0	1	.03		2	118	.02	4*	<.01	1.55	<.01	0	<.01	1	0
1		.03	0	2	118	.02	0	0	.22	0	0	0	0	0
5	0	0		29	165		0							0
1	0	0		21	155		0		.75					0
0	0	0		10	248		0							0
0	0	0		7	130		50*		.42					0
0	0	0		5	165		0*		1.31					0

*This value is expressed in retinol equivalents (RE). All other values are in retinol activity equivalents (RAE).

Table A-1

Food Composition (Computer code number is for Wadsworth Diet Analysis program) (For purposes of calculations, use "0" for t, <1, <.1, <.01, etc.)

Computer Code Number	Food Description	Measure	Wt (g)	H_2O (%)	Ener (cal)	Prot (g)	Carb (g)	Dietary Fiber (g)	Fat (g)	Fat Breakdown (g) Sat	Mono	Poly
	FATS AND OILS—Continued											
199	Mayo type, regular	1 tbs	15	40	58	<1	4	0	5	.7	1.3	2.7
1030	Low calorie	1 tbs	14	54	36	<1	3	0	3	.4	.6	1.4
	Mayonnaise:											
197	Imitation, low calorie	1 tbs	15	63	35	<1	2	0	3	.5	.7	1.6
196	Regular (soybean)	1 tbs	14	15	100	<1	<1	0	11	1.6	3.2	5.8
1488	Regular, low calorie, low sodium	1 tbs	14	63	32	<1	2	0	3	.5	.6	1.5
1493	Regular, low calorie	1 tbs	15	63	35	<1	2	0	3	.5	.7	1.6
198	Ranch, regular	1 tbs	15	39	80	0	<1	0	8	1.2		
2251	Low calorie	1 tbs	14	72	29	<1	1	<1	3	.4	.3	.8
1685	Russian	1 tbs	15	34	74	<1	2	0	8	1.1	1.8	4.4
1502	Salad dressing, low calorie, oil free	1 tbs	15	88	4	<1	1	<1	<1	0	0	0
	Salad dressing, no cholesterol											
1605	Miracle Whip	1 tbs	15	57	48	0	2	0	4	.6	1	2.5
203	Salad dressing, from recipe, cooked	1 tbs	16	69	25	1	2	0	2	.5	.6	.3
200	Tartar sauce, regular	1 tbs	14	34	74	<1	1	<1	8	1.5	2.6	4.1
1503	Low calorie	1 tbs	14	63	31	<1	2	<1	3	.4	.6	1.4
201	Thousand island, regular	1 tbs	16	46	60	<1	2	0	6	1	1.3	3.2
202	Low calorie	1 tbs	15	69	24	<1	2	<1	2	.2	.4	.9
204	Vinegar and oil	1 tbs	16	47	72	0	<1	0	8	1.5	2.4	3.9
	Wishbone:											
2180	Creamy Italian, lite	1 tbs	15	72	26	<1	2		2	.4	.9	.7
2166	Italian, lite	1 tbs	16	90	6	0	1		<1	0	.2	.1
8427	Ranch, lite	1 tbs	15	56	50	0	2	0	4	.7		
	FRUITS and FRUIT JUICES											
	Apples:											
	Fresh, raw, with peel:											
205	2¾" diam (about 3 per lb w/cores)	1 ea	138	84	81	<1	21	4	<1	.1	t	.1
206	3¼" diam (about 2 per lb w/cores)	1 ea	212	84	125	<1	32	6	1	.1	t	.2
207	Raw, peeled slices	1 c	110	84	63	<1	16	2	<1	.1	t	.1
208	Dried, sulfured	10 ea	64	32	156	1	42	6	<1	t	t	.1
209	Apple juice, bottled or canned	1 c	248	88	117	<1	29	<1	<1	t	t	.1
210	Applesauce, sweetened	1 c	255	80	194	<1	51	3	<1	.1	t	.1
211	Applesauce, unsweetened	1 c	244	88	105	<1	27	3	<1	t	t	t
	Apricots:											
212	Raw, w/o pits (about 12 per lb w/pits)	3 ea	105	86	50	1	12	3	<1	t	.2	.1
	Canned (fruit and liquid):											
213	Heavy syrup	1 c	240	78	199	1	51	4	<1	t	.1	t
214	Halves	3 ea	120	78	100	1	26	2	<1	t	t	t
215	Juice pack	1 c	244	87	117	2	30	4	<1	t	t	t
216	Halves	3 ea	108	87	52	1	13	2	<1	t	t	t
217	Dried, halves	10 ea	35	31	83	1	22	3	<1	t	.1	t
218	Dried, cooked, unsweetened, w/liquid	1 c	250	76	213	3	55	8	<1	t	.2	.1
219	Apricot nectar, canned	1 c	251	85	141	1	36	2	<1	t	.1	t
	Avocados, raw, edible part only:											
220	California	1 ea	173	73	306	4	12	8	30	4.5	19.4	3.5
221	Florida	1 ea	304	80	340	5	27	16	27	5.3	14.8	4.5
222	Mashed, fresh, average	1 c	230	74	370	5	17	11	35	5.6	22.1	4.5
	Bananas, raw, without peel:											
223	Whole, 8¾" long (175g w/peel)	1 ea	118	74	109	1	28	3	1	.2	t	.1
224	Slices	1 c	150	74	138	2	35	4	1	.3	.1	.1
1285	Bananas, dehydrated slices	½ c	50	3	173	2	44	4	1	.3	.1	.2
225	Blackberries, raw	1 c	144	86	75	1	18	8	1	t	.1	.3
	Blueberries:											
226	Fresh	1 c	145	85	81	1	20	4	1	t	.1	.2
227	Frozen, sweetened	10 oz	284	77	230	1	62	6	<1	t	.1	.2
228	Frozen, thawed	1 c	230	77	186	1	50	5	<1	t	t	.1
3239	Breadfruit	1 c	220	71	227	2	60	11	<1	.1	.1	.1
	Cherries:											
229	Sour, red pitted, canned water pack	1 c	244	90	88	2	22	3	<1	.1	.1	.1
230	Sweet, red pitted, raw	10 ea	68	81	49	1	11	2	1	.1	.2	.2
231	Cranberry juice cocktail,vitamin C added	1 c	253	85	144	0	36	<1	<1	t	t	.1
1411	Cranberry juice, low calorie	1 c	237	95	45	0	11	0	0	0	0	0
232	Cranberry-apple juice,vitamin C added	1 c	245	83	164	<1	42	<1	0	0	0	0

Chol (mg)	Calc (mg)	Iron (mg)	Magn (mg)	Pota (mg)	Sodi (mg)	Zinc (mg)	VT-A (µg)	Thia (mg)	VT-E (mg)	Ribo (mg)	Niac (mg)	V-B6 (mg)	Fola (µg)	VT-C (mg)
4	2	.03		1	107	.03	13*	<.01	.6	<.01	<.01	<.01	1	0
4	2	.03		1	99	.02	9	<.01	.6	<.01	0	<.01	1	0
4	0	0	0	1	75	.02	0	0	.96	0	0	0	0	0
8	3	.07		5	79	.02	12*	0	.32	0	<.01	.08	1	0
3	0	0	0	1	15	.01	1	0	.53	<.01	0	0	<1	0
4	0	0	0	1	75	.02	0	0	.96	0	0	0	0	0
5	0	0			105		0							0
5	5	.03		5	118		1*		.7					<1
3	3	.09		24	130	.06	31*	.01	1.53	.01	.09	<.01	1	1
0	1	.04	2	7	256	<.01	<1	0	0	0	<.01	<.01	<1	<1
0	0	<.01	0	0	102	0	2	0	.64	0	0	0	0	0
9	13	.08	1	19	117	.06	20*	.01	.3	.02	.04	<.01	1	<1
7	3	.13		11	99	.02	9*	<.01	2.24	<.01	0	<.01	1	<1
3	1	.06		4	82	.02	1	0	.84	<.01	.01	<.01	<1	<1
4	2	.1		18	112	.02	15*	<.01	.18	<.01	<.01	<.01	1	0
2	2	.09		17	150	.02	14	<.01	.18	<.01	0	<.01	1	0
0	0	0	0	1	<1	0	0	0	1.41	0	0	0	0	0
<1	0	0			148		0	0	.56	0	0			0
0	1	0			255		2*	0	.24	0	0			<1
2	0	0			120		0							0
0	10	.25	7	159	0	.05	3	.02	.44	.02	.11	.07	4	8
0	15	.38	11	244	0	.08	5	.04	.68	.03	.16	.1	6	12
0	4	.08	3	124	0	.04	2	.02	.09	.01	.1	.05	0	4
0	9	.9	10	288	56	.13	0	0	.35	.1	.59	.08	0	2
0	17	.92	7	295	7	.07	<1	.05	.02	.04	.25	.07	0	2
0	10	.89	8	156	8	.1	1	.03	.03	.07	.48	.07	3	4
0	7	.29	7	183	5	.07	4	.03	.02	.06	.46	.06	2	3
0	15	.57	8	311	1	.27	137	.03	.93	.04	.63	.06	9	10
0	22	.72	17	336	10	.26	148	.05	2.14	.05	.9	.13	5	7
0	11	.36	8	168	5	.13	74	.02	1.07	.03	.45	.06	2	4
0	29	.73	24	403	10	.27	206	.04	2.17	.05	.84	.13	5	12
0	13	.32	11	178	4	.12	91	.02	.96	.02	.37	.06	2	5
0	16	1.65	16	482	3	.26	127	<.01	.52	.05	1.05	.05	3	1
0	40	4.18	42	1222	7	.65	295	.01	1.25	.07	2.36	.28	0	4
0	18	.95	13	286	8	.23	166	.02	.2	.03	.65	.05	3	2
0	19	2.04	71	1096	21	.73	53	.19	2.32	.21	3.32	.48	114	14
0	33	1.61	103	1483	15	1.28	93	.33	2.37	.37	5.84	.85	161	24
0	25	2.35	90	1377	23	.97	70	.25	3.08	.28	4.42	.64	143	18
0	7	.37	34	467	1	.19	5	.05	.32	.12	.64	.68	22	11
0	9	.46	43	594	1	.24	6	.07	.4	.15	.81	.87	28	14
0	11	.57	54	746	1	.3	8	.09	0	.12	1.4	.22	7	3
0	46	.82	29	282	0	.39	11	.04	1.02	.06	.58	.08	49	30
0	9	.25	7	129	9	.16	7	.07	1.45	.07	.52	.05	9	19
0	17	1.11	6	170	3	.17	6	.06	2.02	.15	.72	.17	20	3
0	14	.9	5	138	2	.14	5	.05	1.63	.12	.58	.14	16	2
0	37	1.19	55	1078	4	.26	4	.24	2.46	.07	1.98	.22	31	64
0	27	3.34	15	239	17	.17	91	.04	.32	.1	.43	.11	19	5
0	10	.26	7	152	0	.04	7	.03	.09	.04	.27	.02	3	5
0	8	.38	5	45	5	.18	<1	.02	0	.02	.09	.05	0	90
0	21	.09	5	52	7	.05	<1	.02	0	.02	.08	.04	0	76
0	17	.15	5	66	5	.1	<1	.01	0	.05	.15	.05	0	78

*This value is expressed in retinol equivalents (RE). All other values are in retinol activity equivalents (RAE).

Table A–1

Food Composition (Computer code number is for Wadsworth Diet Analysis program) (For purposes of calculations, use "0" for t, <1, <.1, <.01, etc.)

Computer Code Number	Food Description	Measure	Wt (g)	H_2O (%)	Ener (cal)	Prot (g)	Carb (g)	Dietary Fiber (g)	Fat (g)	Fat Breakdown (g) Sat	Mono	Poly
	FRUITS and FRUIT JUICES—Continued											
233	Cranberry sauce, canned, strained	1 c	277	61	418	1	108	3	<1	t	.1	.2
234	Dates, whole, without pits	10 ea	83	22	228	2	61	6	<1	.2	.1	t
235	Dates, chopped	1 c	178	22	490	4	131	13	1	.3	.3	.1
236	Figs, dried	10 ea	190	28	485	6	124	23	2	.4	.5	1.1
	Fruit cocktail, canned, fruit and liq:											
237	Heavy syrup pack	1 c	248	80	181	1	47	2	<1	t	t	.1
238	Juice pack	1 c	237	87	109	1	28	2	<1	t	t	t
	Grapefruit:											
	Raw 3¾" diam (half w/rind = 241g)											
239	Pink/red, half fruit, edible part	1 ea	123	91	37	1	9	2	<1	t	t	t
240	White, half fruit, edible part	1 ea	118	90	39	1	10	1	<1	t	t	t
241	Canned sections with light syrup	1 c	254	84	152	1	39	1	<1	t	t	.1
	Grapefruit juice:											
242	Fresh, white,raw	1 c	247	90	96	1	23	<1	<1	t	t	.1
243	Canned, unsweetened	1 c	247	90	94	1	22	<1	<1	t	t	.1
244	Sweetened	1 c	250	87	115	1	28	<1	<1	t	t	.1
	Frozen concentrate, unsweetened:											
246	Diluted with 3 cans water	1 c	247	89	101	1	24	<1	<1	t	t	.1
	Grapes, raw European (adherent skin):											
247	Thompson seedless	10 ea	50	81	35	<1	9	<1	<1	.1	t	.1
248	Tokay/Emperor, seeded types	10 ea	50	81	35	<1	9	<1	<1	.1	t	.1
	Grape juice:											
249	Bottled or canned	1 c	253	84	154	1	38	<1	<1	.1	t	.1
	Frozen concentrate, sweetened:											
251	Diluted with 3 cans water,vit C added	1 c	250	87	128	<1	32	<1	<1	.1	t	.1
1410	Low calorie	1 c	253	84	154	1	38	<1	<1	.1	t	.1
3636	Jackfruit, fresh, sliced	1 c	165	73	155	2	40	3	<1	.1	.1	.1
252	Kiwi fruit, raw, peeled (88g with peel)	1 ea	76	83	46	1	11	3	<1	t	t	.2
253	Lemons, raw, without peel and seeds (about 4 per lb whole)	1 ea	58	89	17	1	5	2	<1	t	t	.1
	Lemon juice:											
254	Fresh:	1 c	244	91	61	1	21	1	0	0	0	0
255	Tablespoon	1 tbs	15	91	4	<1	1	<1	0	0	0	0
256	Canned or bottled, unsweetened:	1 c	244	92	51	1	16	1	1	.1	t	.2
257	Tablespoon	1 tbs	15	92	3	<1	1	<1	<1	t	t	t
258	Frozen, single strength, unsweetened:	1 c	244	92	54	1	16	1	1	.1	t	.2
2298	Tablespoon	1 tbs	15	92	3	<1	1	<1	<1	t	t	t
	Lime juice:											
260	Fresh:	1 c	246	90	66	1	22	1	<1	t	t	.1
261	Tablespoon	1 tbs	15	90	4	<1	1	<1	<1	t	t	t
262	Canned or bottled, unsweetened	1 c	246	93	52	1	16	1	1	.1	.1	.2
3758	Pomelos, raw	1 ea	609	89	231	5	59	6	<1			
263	Mangos, raw, edible part (300g w/skin & seeds)	1 ea	207	82	135	1	35	4	1	.1	.2	.1
	Melons, raw, without rind and contents:											
264	Cantaloupe, 5" diam (2⅓ lb whole with refuse), orange flesh	½ ea	276	90	97	2	23	2	1	.2	t	.3
265	Honeydew, 6½ diam (5¼ lb whole with refuse), slice = ⅒ melon	1 pce	160	90	56	1	15	1	<1	t	t	.1
266	Nectarines, raw, w/o pits, 2" diam	1 ea	136	86	67	1	16	2	1	.1	.2	.3
	Oranges, raw:											
267	Whole w/o peel and seeds, 2⅝" diam (180g with peel and seeds)	1 ea	131	87	62	1	15	3	<1	t	t	t
268	Sections, without membranes	1 c	180	87	85	2	21	4	<1	t	t	t
	Orange juice:											
269	Fresh, all varieties	1 c	248	88	112	2	26	<1	<1	.1	.1	.1
270	Canned, unsweetened	1 c	249	89	105	1	24	<1	<1	t	.1	.1
3480	Calcium fortified	1 c	247	89	110	1	27	0	0	0	0	0
271	Chilled	1 c	249	88	110	2	25	<1	1	.1	.1	.2
	Frozen concentrate:											
273	Diluted w/3 parts water by volume	1 c	249	88	112	2	27	<1	<1	t	t	t
1345	Orange juice, from dry crystals	1 c	248	88	114	0	29	0	0	0	0	0
274	Orange and grapefruit juice, canned	1 c	247	89	106	1	25	<1	<1	t	t	t

Chol (mg)	Calc (mg)	Iron (mg)	Magn (mg)	Pota (mg)	Sodi (mg)	Zinc (mg)	VT-A (µg)	Thia (mg)	VT-E (mg)	Ribo (mg)	Niac (mg)	V-B6 (mg)	Fola (µg)	VT-C (mg)
0	11	.61	8	72	80	.14	3	.04	.28	.06	.28	.04	3	6
0	27	.95	29	541	2	.24	2	.07	.08	.08	1.83	.16	11	0
0	57	2.05	62	1160	5	.52	4	.16	.18	.18	3.92	.34	23	0
0	274	4.24	112	1352	21	.97	12	.13	0	.17	1.32	.43	15	2
0	15	.72	12	218	15	.2	25	.04	.72	.05	.93	.12	7	5
0	19	.5	17	225	9	.21	37	.03	.47	.04	.95	.12	7	6
0	13	.15	10	159	0	.09	16	.04	.31	.02	.23	.05	15	47
0	14	.07	11	175	0	.08	1	.04	.29	.02	.32	.05	12	39
0	36	1.02	25	328	5	.2	0	.1	.63	.05	.62	.05	23	54
0	22	.49	30	400	2	.12	1	.1	.12	.05	.49	.11	25	94
0	17	.49	25	378	2	.22	1	.1	.12	.05	.57	.05	25	72
0	20	.9	25	405	5	.15	0	.1	.12	.06	.8	.05	25	67
0	20	.35	27	336	2	.12	1	.1	.12	.05	.54	.11	10	83
0	5	.13	3	92	1	.02	2	.05	.35	.03	.15	.05	2	5
0	5	.13	3	92	1	.02	2	.05	.35	.03	.15	.05	2	5
0	23	.61	25	334	8	.13	1	.07	0	.09	.66	.16	8	<1
0	10	.25	10	52	5	.1	1	.04	.12	.06	.31	.1	2	60
0	23	.61	25	334	8	.13	1	.07	0	.09	.66	.16	8	<1
0	56	.99	61	500	5	.69	25	.05	.25	.18	.66	.18	23	11
0	20	.31	23	252	4	.13	7	.01	.85	.04	.38	.07	29	74
0	15	.35	5	80	1	.03	1	.02	.14	.01	.06	.05	6	31
0	17	.07	15	303	2	.12	2	.07	.22	.02	.24	.12	32	112
0	1	<.01	1	19	<1	.01	<1	<.01	.01	<.01	.01	.01	2	7
0	27	.32	19	249	51	.15	2	.1	.22	.02	.48	.1	24	60
0	2	.02	1	15	3	.01	<1	.01	.01	<.01	.03	.01	1	4
0	19	.29	19	217	2	.12	1	.14	.22	.03	.33	.15	24	77
0	1	.02	1	13	<1	.01	<1	.01	.01	<.01	.02	.01	1	5
0	22	.07	15	268	2	.15	1	.05	.22	.02	.25	.11	20	72
0	1	<.01	1	16	<1	.01	<1	<.01	.01	<.01	.01	.01	1	4
0	29	.57	17	185	39	.15	2	.08	.17	.01	.4	.07	20	16
0	24	.67	36	1315	6	.49	0	.21	.55	.16	1.34	.22	158	371
0	21	.27	19	323	4	.08	403	.12	2.32	.12	1.21	.28	29	57
0	30	.58	30	853	25	.44	444	.1	.41	.06	1.58	.32	47	116
0	10	.11	11	434	16	.11	3	.12	.24	.03	.96	.09	10	40
0	7	.2	11	288	0	.12	50	.02	1.21	.06	1.35	.03	5	7
0	52	.13	13	237	0	.09	14	.11	.31	.05	.37	.08	39	70
0	72	.18	18	326	0	.13	19	.16	.43	.07	.51	.11	54	96
0	27	.5	27	496	2	.12	25	.22	.22	.07	.99	.1	74	124
0	20	1.1	27	436	5	.17	22	.15	.22	.07	.78	.22	45	86
0	300	0		430	15		0	0	0		0	0	40	78
0	25	.42	27	473	2	.1	10	.28	.47	.05	.7	.13	45	82
0	22	.25	25	473	2	.12	10	.2	.47	.04	.5	.11	110	97
0	62	.2	2	50	12	.1	275	<.01	0	.04	0	0	144	121
0	20	1.14	25	390	7	.17	15	.14	.17	.07	.83	.06	35	72

*This value is expressed in retinol equivalents (RE). All other values are in retinol activity equivalents (RAE).

Table A–1

Food Composition (Computer code number is for Wadsworth Diet Analysis program) (For purposes of calculations, use "0" for t, <1, <.1, <.01, etc.)

Computer Code Number	Food Description	Measure	Wt (g)	H_2O (%)	Ener (cal)	Prot (g)	Carb (g)	Dietary Fiber (g)	Fat (g)	Fat Breakdown (g) Sat	Mono	Poly
	FRUITS and FRUIT JUICES—Continued											
	Papayas, raw:											
275	½" slices	1 c	140	89	55	1	14	3	<1	.1	.1	t
276	Whole, 3" diam by 5⅛" w/o seeds and skin (1 lb w/refuse)	1 ea	304	89	119	2	30	5	<1	.1	.1	.1
1031	Papaya nectar, canned	1 c	250	85	143	<1	36	1	<1	.1	.1	.1
	Peaches:											
277	Raw, whole, 2½" diam, peeled, pitted (about 4 per lb whole)	1 ea	98	88	42	1	11	2	<1	t	t	t
278	Raw, sliced	1 c	170	88	73	1	19	3	<1	t	.1	.1
	Canned, fruit and liquid:											
279	Heavy syrup pack:	1 c	262	79	194	1	52	3	<1	t	.1	.1
280	Half	1 ea	98	79	72	<1	19	1	<1	t	t	t
281	Juice pack:	1 c	248	87	109	2	29	3	<1	t	t	t
282	Half	1 ea	98	87	43	1	11	1	<1	t	t	t
283	Dried, uncooked	10 ea	130	32	311	5	80	11	1	.1	.4	.5
284	Dried, cooked, fruit and liquid	1 c	258	78	199	3	51	7	1	.1	.2	.3
	Frozen, slice, sweetened:											
285	10-oz package,vitamin C added	1 ea	284	75	267	2	68	5	<1	t	.1	.2
286	Cup, thawed measure, vitamin C added	1 c	250	75	235	2	60	4	<1	t	.1	.2
1032	Peach nectar, canned	1 c	249	86	134	1	35	1	<1	4.7	19	2.6
	Pears:											
	Fresh, with skin, cored:											
287	Bartlett, 2½" diam (about 2½ per lb)	1 ea	166	84	98	1	25	4	1	t	.1	.2
288	Bosc, 2⅕" diam (about 3 per lb)	1 ea	139	84	82	1	21	3	1	t	.1	.1
289	D'Anjou, 3" diam (about 2 per lb)	1 ea	209	84	123	1	32	5	1	t	.2	.2
	Canned, fruit and liquid:											
290	Heavy syrup pack:	1 c	266	80	197	1	51	4	<1	t	.1	.1
291	Half	1 ea	76	80	56	<1	15	1	<1	t	t	t
292	Juice pack:	1 c	248	86	124	1	32	4	<1	t	t	t
293	Half	1 ea	76	86	38	<1	10	1	<1	t	t	t
294	Dried halves	10 ea	175	27	459	3	122	13	1	.1	.2	.3
1033	Pear nectar, canned	1 c	250	84	150	<1	39	1	<1	t	t	t
	Pineapple:											
295	Fresh chunks, diced	1 c	155	86	76	1	19	2	1	t	.1	.2
	Canned, fruit and liquid:											
	Heavy syrup pack:											
296	Crushed, chunks, tidbits	½ c	127	79	99	<1	26	1	<1	t	t	.1
297	Slices	1 ea	49	79	38	<1	10	<1	<1	t	t	t
298	Juice pack, crushed, chunks, tidbits	1 c	250	84	150	1	39	2	<1	t	t	.1
299	Juice pack, slices	1 ea	47	84	28	<1	7	<1	<1	t	t	t
300	Pineapple juice, canned, unsweetened	1 c	250	86	140	1	34	<1	<1	t	t	.1
	Plantains, yellow fleshed, without peel:											
301	Raw slices (whole=179g w/o peel)	1 c	148	65	181	2	47	3	1	.2	t	.1
302	Cooked, boiled, sliced	1 c	154	67	179	1	48	4	<1	.1	t	.1
	Plums:											
303	Fresh, medium, 2⅛" diam	1 ea	66	85	36	1	9	1	<1	t	.3	.1
304	Fresh, small, 1½" diam	1 ea	28	85	15	<1	4	<1	<1	t	.1	t
	Canned, purple, with liquid:											
305	Heavy syrup pack:	1 c	258	76	230	1	60	3	<1	t	.2	.1
306	Plums	3 ea	138	76	123	<1	32	1	<1	t	.1	t
307	Juice pack:	1 c	252	84	146	1	38	3	<1	4.8	3.4	12
308	Plums	3 ea	138	84	80	1	21	1	<1	2.6	1.8	6.6
1698	Pomegranate, fresh	1 ea	154	81	105	1	26	1	<1	.1	.1	.1
	Prunes, dried, pitted:											
309	Uncooked (10 = 97g w/pits, 84g w/o pits)	10 ea	84	32	201	2	53	6	<1	t	.3	.1
310	Cooked, unsweetened, fruit & liq (250g w/pits)	1 c	248	70	265	3	70	16	1	t	.4	.1
311	Prune juice, bottled or canned	1 c	256	81	182	2	45	3	<1	7.4	5.2	17.3
	Raisins, seedless:											
312	Cup, not pressed down	1 c	145	15	435	5	115	6	1	.2	t	.2
313	One packet, ½ oz	½ oz	14	15	42	<1	11	1	<1	t	t	t
	Raspberries:											

Chol (mg)	Calc (mg)	Iron (mg)	Magn (mg)	Pota (mg)	Sodi (mg)	Zinc (mg)	VT-A (µg)	Thia (mg)	VT-E (mg)	Ribo (mg)	Niac (mg)	V-B6 (mg)	Fola (µg)	VT-C (mg)
0	34	.14	14	360	4	.1	20	.04	1.57	.04	.47	.03	53	86
0	73	.3	30	781	9	.21	43	.08	3.4	.1	1.03	.06	116	188
0	25	.85	7	77	12	.37	14	.01	.05	.01	.37	.02	5	7
0	5	.11	7	193	0	.14	26	.02	.69	.04	.97	.02	3	6
0	8	.19	12	335	0	.24	46	.03	1.19	.07	1.68	.03	5	11
0	8	.71	13	241	16	.24	43	.03	2.33	.06	1.61	.05	8	7
0	3	.26	5	90	6	.09	16	.01	.87	.02	.6	.02	3	3
0	15	.67	17	317	10	.27	47	.02	3.72	.04	1.44	.05	7	9
0	6	.26	7	125	4	.11	19	.01	1.47	.02	.57	.02	3	4
0	36	5.28	55	1294	9	.74	140	<.01	0	.28	5.69	.09	0	6
0	23	3.38	33	826	5	.46	26	.01	0	.05	3.92	.1	0	10
0	9	1.05	14	369	17	.14	40	.04	2.53	.1	1.85	.05	9	268
0	7	.92	12	325	15	.12	35	.03	2.23	.09	1.63	.04	7	236
0	12	.47	10	100	17	.2	32	.01	.02	.03	.72	.02	2	13
0	18	.41	10	208	0	.2	2	.03	.83	.07	.17	.03	12	7
0	15	.35	8	174	0	.17	1	.03	.69	.06	.14	.02	10	6
0	23	.52	12	261	0	.25	2	.04	1.05	.08	.21	.04	15	8
0	13	.58	11	173	13	.21	0	.03	1.33	.06	.64	.04	3	3
0	4	.17	3	49	4	.06	0	.01	.38	.02	.18	.01	1	1
0	22	.72	17	238	10	.22	1	.03	1.24	.03	.5	.03	2	4
0	7	.22	5	73	3	.07	<1	.01	.38	.01	.15	.01	1	1
0	59	3.68	58	933	10	.68	<1	.01	0	.25	2.4	.13	0	12
0	12	.65	7	32	10	.17	<1	<.01	.25	.03	.32	.03	2	3
0	11	.57	22	175	2	.12	2	.14	.15	.06	.65	.13	17	24
0	18	.48	20	132	1	.15	1	.11	.13	.03	.36	.09	6	9
0	7	.19	8	51	<1	.06	<1	.04	.05	.01	.14	.04	2	4
0	35	.7	35	305	2	.25	5	.24	.25	.05	.71	.18	12	24
0	7	.13	7	57	<1	.05	1	.04	.05	.01	.13	.03	2	4
0	42	.65	32	335	2	.27	1	.14	.05	.05	.64	.24	57	27
0	4	.89	55	739	6	.21	84	.08	.4	.08	1.02	.44	33	27
0	3	.89	49	716	8	.2	70	.07	.22	.08	1.16	.37	40	17
0	3	.07	5	114	0	.07	11	.03	.4	.06	.33	.05	1	6
0	1	.03	2	48	0	.03	4	.01	.17	.03	.14	.02	1	3
0	23	2.17	13	235	49	.18	33	.04	1.81	.1	.75	.07	8	1
0	12	1.16	7	126	26	.1	18	.02	.97	.05	.4	.04	4	1
0	25	.86	20	388	3	.28	127	.06	1.76	.15	1.19	.07	8	7
0	14	.47	11	213	1	.15	70	.03	.97	.08	.65	.04	4	4
0	5	.46	5	399	5	.18	0	.05	.85	.05	.46	.16	9	9
0	43	2.08	38	626	3	.44	84	.07	1.22	.14	1.65	.22	3	3
0	57	2.75	50	828	5	.59	38	.06	0	.25	1.79	.54	0	7
0	31	3.02	36	707	10	.54	<1	.04	.03	.18	2.01	.56	0	10
0	71	3.02	48	1088	17	.39	1	.23	1.02	.13	1.19	.36	4	5
0	7	.29	5	105	2	.04	<1	.02	.1	.01	.11	.03	<1	<1

*This value is expressed in retinol equivalents (RE). All other values are in retinol activity equivalents (RAE).

Table A-1

Food Composition (Computer code number is for Wadsworth Diet Analysis program) (For purposes of calculations, use "0" for t, <1, <.1, <.01, etc.)

Computer Code Number	Food Description	Measure	Wt (g)	H_2O (%)	Ener (cal)	Prot (g)	Carb (g)	Dietary Fiber (g)	Fat (g)	Fat Breakdown (g) Sat	Mono	Poly
	FRUITS and FRUIT JUICES—Continued											
314	Fresh	1 c	123	87	60	1	14	8	1	t	.1	.4
315	Frozen, sweetened	10 oz	284	73	293	2	74	12	<1	t	t	.3
316	Cup, thawed measure	1 c	250	73	258	2	65	11	<1	t	t	.2
317	Rhubarb, cooked, added sugar	1 c	240	68	278	1	75	5	<1	t	t	.1
	Strawberries:											
318	Fresh, whole, capped	1 c	144	92	43	1	10	3	1	t	.1	.3
	Frozen, sliced, sweetened:											
319	10-oz container	10 oz	284	73	273	2	74	5	<1	t	.1	.2
320	Cup, thawed measure	1 c	255	73	245	1	66	5	<1	t	t	.2
	Tangerines, without peel and seeds:											
321	Fresh (2⅜" whole) 116g w/refuse	1 ea	84	88	37	1	9	2	<1	t	t	t
322	Canned, light syrup, fruit and liquid	1 c	252	83	154	1	41	2	<1	t	t	t
323	Tangerine juice, canned, sweetened	1 c	249	87	125	1	30	<1	<1	t	t	.1
	Watermelon, raw, without rind and seeds:											
324	Piece, 1⁄16th wedge	1 pce	286	92	91	2	20	1	1	.1	.3	.4
325	Diced	1 c	152	92	49	1	11	1	1	.1	.2	.2
	BAKED GOODS: BREADS, CAKES, COOKIES, CRACKERS, PIES											
42100	Bagel, cinnamon raisin, 3½" diam.	1 ea	71	32	195	7	39	2	1	.2	.1	.5
326	Bagel, plain, enriched, 3½" diam.	1 ea	71	33	195	7	38	2	1	.2	.1	.5
1663	Bagel, oat bran	1 ea	110	33	281	12	59	4	1	.2	.3	.5
42617	Bagel, whole wheat	1 ea	110	28	291	12	62	10	2	.3	.2	.6
	Biscuits:											
327	From home recipe	1 ea	60	29	212	4	27	1	10	2.6	4.2	2.5
328	From mix	1 ea	57	29	191	4	28	1	7	1.6	2.4	2.4
329	From refrigerated dough	1 ea	74	27	276	4	34	1	13	8.7	3.4	.5
330	Bread crumbs, dry, grated (see 364, 365 for soft crumbs)	1 c	108	6	427	13	78	3	6	1.3	2.6	1.2
2087	Bread sticks, brown & serve	1 ea	57	34	150	7	28	1	1	.5	.5	.5
	Breads:											
331	Boston brown, canned, 3¼" slice	1 pce	45	47	88	2	19	2	1	.1	.1	.3
	Cracked wheat (¼ cracked-wheat & ¾ enriched wheat flour):											
333	Slice (18 per loaf)	1 pce	25	36	65	2	12	1	1	.2	.5	.2
334	Slice, toasted	1 pce	23	30	65	2	12	1	1	.2	.5	.2
	French/Vienna, enriched:											
337	Slice, 4¾ x 4½"	1 pce	25	34	68	2	13	1	1	.2	.3	.2
336	French, slice, 5 x 2½"	1 pce	25	34	68	2	13	1	1	.2	.3	.2
	French toast: see Mixed Dishes, and Fast Foods, #691											
2083	Honey wheatberry	1 pce	38	38	100	3	18	2	1	0	.5	1
	Italian, enriched:											
339	Slice, 4½ x 3¼ x ¾"	1 pce	30	36	81	3	15	1	1	.3	.2	.4
	Mixed grain, enriched:											
341	Slice (18 per loaf)	1 pce	26	38	65	3	12	2	1	.2	.4	.2
342	Slice, toasted	1 pce	24	32	65	3	12	2	1	.2	.4	.2
	Oatmeal, enriched:											
344	Slice (18 per loaf)	1 pce	27	37	73	2	13	1	1	.2	.4	.5
345	Slice, toasted	1 pce	25	31	73	2	13	1	1	.2	.4	.5
346	Pita pocket bread, enr, 6½" round	1 ea	60	32	165	5	33	1	1	.1	.1	.3
	Pumpernickel(⅔ rye & ⅓ enr wheat flr):											
348	Slice, 5 x 4 x ⅜"	1 pce	26	38	65	2	12	2	1	.1	.2	.3
349	Slice, toasted	1 pce	29	32	80	3	15	2	1	.1	.3	.4
	Raisin, enriched:											
351	Slice (18 per loaf)	1 pce	26	34	71	2	14	1	1	.3	.6	.2
352	Slice, toasted	1 pce	24	28	71	2	14	1	1	.3	.6	.2
353	Rye, light (⅓ rye & ⅔ enr wheat flr):											
	1-lb loaf	1 ea	454	37	1175	39	219	26	15	2.8	5.9	3.6
354	Slice, 4¾ x 3¾ x 7⁄16"	1 pce	32	37	83	3	15	2	1	.2	.4	.3
355	Slice, toasted	1 pce	24	31	68	2	13	2	1	.2	.3	.2

Chol (mg)	Calc (mg)	Iron (mg)	Magn (mg)	Pota (mg)	Sodi (mg)	Zinc (mg)	VT-A (μg)	Thia (mg)	VT-E (mg)	Ribo (mg)	Niac (mg)	V-B6 (mg)	Fola (μg)	VT-C (mg)
0	27	.7	22	187	0	.57	8	.04	.55	.11	1.11	.07	32	31
0	43	1.85	37	324	3	.51	9	.05	1.28	.13	.65	.1	74	47
0	37	1.63	32	285	2	.45	7	.05	1.13	.11	.57	.08	65	41
0	348	.5	29	230	2	.19	8	.04	.48	.05	.48	.05	12	8
0	20	.55	14	239	1	.19	2	.03	.2	.09	.33	.08	26	82
0	31	1.68	20	278	9	.17	3	.04	.4	.14	1.14	.08	43	118
0	28	1.5	18	250	8	.15	3	.04	.36	.13	1.02	.08	38	106
0	12	.08	10	132	1	.2	39	.09	.2	.02	.13	.06	17	26
0	18	.93	20	197	15	.6	106	.13	.86	.11	1.12	.11	13	50
0	45	.5	20	443	2	.07	52	.15	.22	.05	.25	.08	12	55
0	23	.49	31	332	6	.2	53	.23	.43	.06	.57	.41	6	27
0	12	.26	17	176	3	.11	28	.12	.23	.03	.3	.22	3	15
0	13	2.7	20	105	229	.8	0*	.27	.11	.2	2.19	.04	64	<1
0	52	2.53	21	72	379	.62	0	.38	.03	.22	3.24	.04	62	0
0	13	3.39	34	127	558	.99	<1	.36	.15	.37	3.26	.05	89	<1
0	32	3.52	116	379	592	2.52	0	.34	.99	.28	5.75	.3	66	<1
2	141	1.74	11	73	348	.32	14*	.21	.78	.19	1.77	.02	37	<1
2	105	1.17	14	107	544	.35	15*	.2	.23	.2	1.72	.04	30	<1
5	89	1.64	9	87	584	.29	24*	.27	.44	.18	1.63	.03	6	0
0	245	6.61	50	239	931	1.32	<1	.83	.62	.47	7.4	.11	118	0
0	60	2.7			290		0*	.22		.1	1.6			0
<1	31	.94	28	143	284	.22	5*	.01	.25	.05	.5	.04	5	0
0	11	.7	13	44	135	.31	0	.09	.15	.06	.92	.08	15	0
0	11	.7	13	44	135	.31	0*	.07	.14	.05	.83	.07	7	0
0	19	.63	7	28	152	.22	0	.13	.07	.08	1.19	.01	24	0
0	19	.63	7	28	152	.22	0	.13	.07	.08	1.19	.01	24	0
0	20	.72			200		0	.12	.24	.07	.8			0
0	23	.88	8	33	175	.26	0	.14	.11	.09	1.31	.01	28	0
0	24	.9	14	53	127	.33	0	.11	.17	.09	1.13	.09	21	<1
0	24	.9	14	53	127	.33	0	.08	.16	.08	1.02	.08	16	<1
0	18	.73	10	38	162	.27	1*	.11	.16	.06	.85	.02	17	0
0	18	.73	10	38	163	.28	<1*	.09	.09	.06	.77	.02	13	<1
0	52	1.57	16	72	322	.5	0	.36	.02	.2	2.78	.02	57	0
0	18	.75	14	54	174	.38	0	.08	.11	.08	.8	.03	21	0
0	21	.91	17	66	214	.47	0	.08	.17	.09	.89	.04	20	0
0	17	.75	7	59	101	.19	0	.09	.13	.1	.9	.02	23	<1
0	17	.76	7	59	102	.19	<1	.07	.2	.09	.81	.02	18	<1
0	331	12.8	182	754	2996	5.18	5*	1.97	1.68	1.52	17.3	.34	390	2
0	23	.91	13	53	211	.36	<1*	.14	.12	.11	1.22	.02	27	<1
0	19	.74	10	44	174	.3	<1	.09	.15	.08	.9	.02	17	<1

*This value is expressed in retinol equivalents (RE). All other values are in retinol activity equivalents (RAE).

Table A-1

Food Composition (Computer code number is for Wadsworth Diet Analysis program) (For purposes of calculations, use "0" for t, <1, <.1, <.01, etc.)

Computer Code Number	Food Description	Measure	Wt (g)	H_2O (%)	Ener (cal)	Prot (g)	Carb (g)	Dietary Fiber (g)	Fat (g)	Fat Breakdown (g) Sat	Mono	Poly
	BAKED GOODS: BREADS, CAKES, COOKIES, CRACKERS, PIES—Continued											
	Wheat (enr wheat & whole-wheat flour):											
357	Slice (18 per loaf)	1 pce	25	37	65	2	12	1	1	.2	.4	.2
358	Slice, toasted	1 pce	23	32	65	2	12	1	1	.2	.4	.2
	White, enriched:											
360	Slice	1 pce	42	35	120	3	21	1	2	.5	.5	1.2
361	Slice, toasted	1 pce	38	29	119	3	21	1	2	.5	.5	1.2
	Whole Wheat:											
367	Slice (16 per loaf)	1 pce	28	38	69	3	13	2	1	.3	.5	.3
368	Slice, toasted	1 pce	25	30	69	3	13	2	1	.3	.5	.3
	Bread stuffing, prepared from mix:											
369	Dry type	1 c	200	65	356	6	43	6	17	3.5	7.6	5.2
370	Moist type, with egg and margarine	1 c	232	65	390	9	51	5	17	3.4	7.4	4.9
	Cakes, prepared from mixes using enrich flour and veg shortening, w/frostings made from margarine:											
372	Angel Food, 1/12 of cake	1 pce	28	33	72	2	16	<1	<1	t	t	.1
373	Boston cream pie, 1/8 of cake	1 pce	92	45	232	2	39	1	8	2.2	4.2	.9
375	Coffee Cake, 1/6 of cake	1 pce	56	30	178	3	30	1	5	1	2.2	1.8
	Devil's food, chocolate frosting:											
377	Piece, 1/16 of cake	1 pce	64	23	235	3	35	2	10	3	5.6	1.2
378	Cupcake, 2½" diam	1 ea	42	23	154	2	23	1	7	2	3.7	.8
380	Gingerbread, 1/9 of cake	1 pce	67	33	207	3	34	1	7	1.8	3.8	.9
	Yellow, chocolate frosting, 2 layer:											
382	Piece, 1/16 of cake	1 pce	64	22	243	2	35	1	11	3	6.1	1.3
	Cakes from home recipes w/enr flour:											
	Carrot cake, made with veg oil, cream cheese frosting:											
384	Piece, 1/16 of cake, 2¼ x 3¼" slice	1 pce	111	21	484	5	52	1	29	5.4	7.2	15.1
	Fruitcake, dark:											
386	Piece, 1/32 of cake, ⅔" arc	1 pce	43	25	139	1	26	2	4	.5	1.8	1.4
	Sheet, plain, made w/veg shortening,											
388	no frosting, 1/9 of cake	1 pce	86	24	313	4	48	<1	12	3.3	5.7	2.8
	Sheet, plain, made w/margarine,											
390	uncooked white frosting, 1/9 of cake	1 pce	64	22	239	2	38	<1	9	1.5	3.9	3.3
	Cakes, commerical:											
402	Cheesecake, 1/12 of cake	1 pce	80	46	257	4	20	<1	18	7.9	6.9	1.3
394	Pound cake, 1/17 of loaf, 2" slice	1 pce	28	25	109	2	14	<1	6	3.2	1.6	.3
	Snack cakes:											
395	Chocolate w/creme filling,Ding Dong	1 ea	50	20	188	2	30	<1	7	1.4	2.8	2.6
396	Sponge cake w/creme filling,Twinkie	1 ea	43	20	157	1	27	<1	5	1.1	1.7	1.4
1677	Sponge cake, 1/12 of 12" cake	1 pce	38	30	110	2	23	<1	1	.3	.4	.2
398	White, white frosting, 2 layer, 1/16	1 pce	71	20	266	2	45	1	10	4.3	3.8	1
	Yellow, chocolate frosting, 2 layer:											
400	Slice, 1/16 of cake	1 pce	64	22	243	2	35	1	11	3	6.1	1.3
1332	Bagel Chips	5 pce	70	3	298	6	52	4	7	1.3	2.1	3.4
2225	Bagel chips, onion garlic, toasted	1 oz	28		181	5	30	3	7	1.6	4.9	0
1035	Cheese puffs/Cheetos	1 c	20	1	111	2	11	<1	7	1.3	4.1	1
	Cookies made with enriched flour:											
	Brownies with nuts:											
403	Commercial w/frosting, 1½ x 1¾ x ⅞"	1 ea	61	14	247	3	39	1	10	2.6	5.5	1.4
1902	Fat free fudge, Entenmann's	1 pce	40	24	110	2	27	1	0	0	0	0
	Chocolate chip cookies:											
405	Commercial, 2¼" diam	4 ea	60	12	275	2	35	2	15	4.4	7.8	2.1
406	Home recipe, 2¼" diam	4 ea	64	6	312	4	37	2	18	5.2	6.6	5.4
407	From refrigerated dough, 2¼" diam	4 ea	64	13	284	3	39	1	13	4.3	6.7	1.4
408	Fig bars	4 ea	64	16	223	2	45	3	5	.7	1.9	1.8
2052	Fruit bar, no fat	1 ea	28		90	2	21	0	0	0	0	0
2162	Fudge, fat free, Snackwell	1 ea	16	14	53	1	12	<1	<1	.1	.1	t
409	Oatmeal raisin, 2⅝" diam	4 ea	60	6	261	4	41	2	10	1.9	4.1	3
410	Peanut butter, home recipe, 2⅝"diam	4 ea	80	6	380	7	47	2	19	3.5	8.7	5.8

Chol (mg)	Calc (mg)	Iron (mg)	Magn (mg)	Pota (mg)	Sodi (mg)	Zinc (mg)	VT-A (µg)	Thia (mg)	VT-E (mg)	Ribo (mg)	Niac (mg)	V-B6 (mg)	Fola (µg)	VT-C (mg)
0	26	.83	11	50	133	.26	0	.1	.13	.07	1.03	.02	19	0
0	26	.83	11	50	132	.26	0	.08	.14	.06	.93	.02	15	0
1	24	1.25	8	61	151	.27	9*	.17	.36	.16	1.51	.02	38	<1
1	24	1.24	8	61	150	.27	8*	.13	.46	.14	1.35	.02	12	<1
0	20	.92	24	71	148	.54	0	.1	.24	.06	1.07	.05	14	0
0	20	.93	24	71	148	.54	0	.08	.29	.05	.97	.04	10	0
0	64	2.18	24	148	1086	.56	162	.27	2.8	.21	2.95	.08	202	0
0	148	3.8	35	304	1069	.74	160*	.39	2.78	.33	3.69	.12	39	4
0	39	.15	3	26	210	.02	0	.03	.03	.14	.25	.01	10	0
34	21	.35	6	36	132	.15	21*	.37	.97	.25	.18	.02	14	<1
27	76	.8	10	63	236	.25	22*	.09	.93	.1	.85	.03	27	<1
27	27	1.41	22	128	214	.44	16*	.02	1.08	.08	.37	.03	11	<1
18	18	.92	14	84	140	.29	10*	.01	.71	.06	.24	.02	7	<1
23	46	2.22	11	161	307	.27	11*	.13	.92	.12	1.05	.02	7	<1
35	24	1.33	19	114	216	.4	21	.08	1.45	.1	.8	.02	14	0
60	28	1.39	20	124	273	.54	426*	.15	4.68	.17	1.12	.08	13	1
2	14	.89	7	66	116	.12	2*	.02	.71	.04	.34	.02	8	<1
56	55	1.3	12	68	258	.3	41*	.14	1.22	.15	1.12	.03	6	<1
35	40	.68	4	34	220	.16	12*	.06	1.22	.04	.32	.02	17	0
44	41	.5	9	72	166	.41	117*	.02	1.26	.15	.16	.04	14	<1
62	10	.39	3	33	111	.13	44*	.04	.18	.06	.37	.01	11	0
8	36	1.68	20	61	213	.25	2*	.11	1.68	.15	1.21	.01	14	0
7	19	.55	3	37	157	.12	2*	.07	.87	.06	.53	.01	12	<1
39	27	1.03	4	38	93	.19	17*	.09	.1	.1	.73	.02	15	0
6	34	.57	4	41	166	.11	23*	.07	1.28	.09	.64	.01	4	<1
35	24	1.33	19	114	216	.4	21	.08	1.45	.1	.8	.02	14	0
0	9	1.39	39	167	419	.87	0	.13	1.71	.12	1.62	.15	46	<1
0	0	2.37			461		0	.37	<.01	.22	3.29			0
1	12	.47	4	33	210	.08	7*	.05	1.02	.07	.65	.03	24	<1
10	18	1.37	19	91	190	.44	4*	.16	1.27	.13	1.05	.02	13	0
0	0	1.08		90	140		0		.01					0
0	9	1.45	21	56	196	.28	0*	.07	1.74	.12	.97	.1	23	0
20	25	1.57	35	143	231	.59	105*	.12	1.84	.11	.87	.05	21	<1
15	16	1.44	15	115	134	.32	11*	.12	1.48	.12	1.26	.02	36	0
0	41	1.86	17	132	224	.25	3*	.1	.8	.14	1.2	.05	17	<1
0	0	.36			95		0		.01					0
0	3	.29	5	26	71	.08	<1*	.02	<.01	.02	.26	<.01		0
20	60	1.59	25	143	323	.52	98*	.15	1.5	.1	.75	.04	18	<1
25	31	1.78	31	185	414	.66	125*	.18	3.04	.17	2.81	.07	44	<1

*This value is expressed in retinol equivalents (RE). All other values are in retinol activity equivalents (RAE).

Table A-1

Food Composition (Computer code number is for Wadsworth Diet Analysis program) (For purposes of calculations, use "0" for t, <1, <.1, <.01, etc.)

Computer Code Number	Food Description	Measure	Wt (g)	H₂O (%)	Ener (cal)	Prot (g)	Carb (g)	Dietary Fiber (g)	Fat (g)	Fat Breakdown (g) Sat	Mono	Poly
	BAKED GOODS: BREADS, CAKES, COOKIES, CRACKERS, PIES—Continued											
411	Sandwich-type, all	4 ea	40	2	189	2	28	1	8	1.5	3.4	2.9
412	Shortbread, commercial, small	4 ea	32	4	161	2	21	1	8	1.9	4.3	1
413	Shortbread, home recipe, large	2 ea	22	3	120	1	12	<1	7	4.5	2.1	.3
414	Sugar from refrigerated dough,2" diam	4 ea	48	5	232	2	31	<1	11	2.8	6.2	1.4
1874	Vanilla sandwich, Snackwell's	2 ea	26	4	109	1	21	1	2	.5	.8	.2
415	Vanilla wafers	10 ea	40	5	176	2	29	1	6	1.5	2.6	1.5
42672	Cornbread, 2.5 x 2.5 x 1.5" piece	1 pce	65	49	152	4	23	2	5	1.6	2.4	.5
416	Corn chips	1 c	26	1	140	2	15	1	9	1.2	2.5	4.3
	Crackers:											
417	Cheese-enriched	10 ea	10	3	50	1	6	<1	3	.9	1.2	.2
418	Cheese with peanut butter-enriched	4 ea	28	4	135	4	16	1	6	1.5	3.3	1.3
	Fat free-enriched:											
2161	Cracked pepper, Snackwell	1 ea	14	2	61	1	10	<1	2	.3	.6	.2
2159	Wheat, Snackwell	7 ea	15	1	60	2	12	1	<1	.1	.1	.1
2075	Whole wheat, herb seasoned	5 ea	14	5	50	2	11	2	0	0	0	0
2077	Whole wheat, onion	5 ea	14	4	50	2	11	2	0	0	0	0
419	Graham-enriched	2 ea	14	4	59	1	11	<1	1	.2	.6	.5
420	Melba toast, plain-enriched	1 pce	5	5	19	1	4	<1	<1	t	t	.1
1514	Rice cakes, unsalted-enriched	2 ea	18	6	70	1	15	1	<1	.1	.2	.2
421	Rye wafer, whole grain	2 ea	22	5	73	2	18	5	<1	t	t	.1
422	Saltine-enriched	4 ea	12	4	52	1	9	<1	1	.4	.8	.2
1971	Saltine, unsalted tops-enriched	2 ea	6		25	1	4	0	<1			
423	Snack-type, round like Ritz-enriched	3 ea	9	3	45	1	5	<1	2	.3	1	.9
424	Wheat, thin-enriched	4 ea	8	3	38	1	5	<1	2	.3	.9	.2
425	Whole-wheat wafers	2 ea	8	3	35	1	5	1	1	.3	.5	.5
426	Croissants, 4½ x 4 x 1¾"	1 ea	57	23	231	5	26	1	12	6.6	3.1	.6
1699	Croutons, seasoned	½ c	20	4	93	2	13	1	4	1	1.9	.5
	Danish pastry:											
428	Round piece, plain, 4¼" diam, 1" high	1 ea	88	21	349	5	47	<1	17	3.5	10.6	1.6
429	Ounce, plain	1 oz	28	21	111	2	15	<1	5	1.1	3.4	.5
430	Round piece with fruit	1 ea	94	29	335	5	45		16	3.3	10.1	1.6
	Desserts, 3 x 3" piece:											
1348	Apple crisp	1 pce	78	61	127	1	25	1	3	.6	1.2	.9
1353	Apple cobbler	1 pce	104	57	199	2	35	2	6	1.2		
1349	Cherry crisp	1 pce	138	75	157	2	27	2	5	.9		
1352	Cherry cobbler	1 pce	129	66	197	2	34	1	6	1.2		
1350	Peach crisp	1 pce	139	73	166	1	30	2	5	.8		
1351	Peach cobbler	1 pce	130	64	203	2	36	2	6	1.2		
	Doughnuts:											
431	Cake type, plain, 3¼" diam	1 ea	47	21	198	2	23	1	11	1.7	4.4	3.7
432	Yeast-leavened, glazed, 3¾" diam	1 ea	60	25	242	4	27	1	14	3.5	7.7	1.7
	English muffins:											
433	Plain, enriched	1 ea	57	42	134	4	26	2	1	.1	.2	.5
434	Toasted	1 ea	52	37	133	4	26	2	1	.1	.2	.5
1504	Whole wheat	1 ea	66	46	134	6	27	4	1	.2	.3	.6
1414	Granola bar, soft	1 ea	28	6	124	2	19	1	5	2	1.1	1.5
1415	Granola bar, hard	1 ea	25	4	118	3	16	1	5	.6	1.1	3
1985	Granola bar, fat free, all flavors	1 ea	42	10	140	2	35	3	0	0	0	0
	Muffins, 2½" diam, 1½" high:											
	From home recipe:											
435	Blueberry	1 ea	57	39	165	4	23	1	6	1.4	1.6	3.1
436	Bran, wheat	1 ea	57	35	164	4	24	4	7	1.5	1.8	3.6
437	Cornmeal	1 ea	57	32	183	4	25	2	7	1.6	1.8	3.5
	From commercial mix:											
438	Blueberry	1 ea	50	36	150	3	24	1	4	.7	1.8	1.5
439	Bran, wheat	1 ea	50	35	138	3	23	2	5	1.2	2.3	.7
440	Cornmeal	1 ea	50	30	161	4	25	1	5	1.4	2.6	.6
1864	Nabisco Newtons, fat free, all flavors	1 ea	23		69	1	16		0	0	0	0
	Pancakes, 4" diam:											
441	Buckwheat, from mix w/ egg and milk	1 ea	30	54	62	2	8	1	2	.6	.6	.8
442	Plain, from home recipe	1 ea	38	53	86	2	11	1	4	.8	.9	1.7
443	Plain, from mix; egg, milk, oil added	1 ea	38	53	74	2	14	<1	1	.2	.3	.3

Chol (mg)	Calc (mg)	Iron (mg)	Magn (mg)	Pota (mg)	Sodi (mg)	Zinc (mg)	VT-A (µg)	Thia (mg)	VT-E (mg)	Ribo (mg)	Niac (mg)	V-B6 (mg)	Fola (µg)	VT-C (mg)
0	10	1.55	18	70	242	.32	0*	.03	1.38	.07	.83	.01	17	0
6	11	.88	5	32	146	.17	4*	.11	1.03	.1	1.07	.03	19	0
20	4	.58	3	15	102	.09	67*	.08	.18	.06	.64	<.01	2	0
15	43	.88	4	78	225	.13	5*	.09	1.54	.06	1.16	.01	25	0
<1	17	.61	5	28	95	.16	<1*	.05		.07	.69	.01		0
20	19	.95	6	39	125	.14	3	.11	.51	.13	1.24	.03	20	0
22	70	.82	13	105	356	.38	32*	.12	.64	.16	.93	.05	6	<1
0	33	.34	20	37	164	.33	1	.01	.35	.04	.31	.06	5	0
1	15	.48	4	14	99	.11	3*	.06	.26	.04	.47	.05	8	0
1	22	.82	16	69	278	.3	10*	.11	1.05	.1	1.83	.42	25	<1
0	24	.51	4	16	117	.11	<1*	.04	0	.05	.75	.01	11	<1
<1	28	.58	7	43	169	.21	<1*	.04		.07	.73	.02		0
0	0				80		100*							2
0	0	0			80		100*							2
0	3	.52	4	19	85	.11	0	.03	.29	.04	.58	.01	8	0
0	5	.18	3	10	41	.1	0	.02	<.01	.01	.21	<.01	6	0
0	2	.27	24	52	5	.54	<1	.01	.02	.03	1.41	.03	4	0
0	9	1.31	27	109	175	.62	<1	.09	.31	.06	.35	.06	10	<1
0	14	.65	3	15	156	.09	0	.07	.19	.05	.63	<.01	15	0
0		.36		5	50				.1					
0	11	.32	2	12	76	.06	0	.04	.41	.03	.36	<.01	7	0
0	4	.35	5	15	64	.13	0	.04	.32	.03	.4	.01	1	0
0	4	.25	8	24	53	.17	0	.02	.09	.01	.36	.01	2	0
38	21	1.16	9	67	424	.43	106*	.22	.24	.14	1.25	.03	35	<1
1	19	.56	8	36	248	.19	2*	.1	.44	.08	.93	.02	18	0
27	37	1.8	14	96	326	.48	5*	.25	.79	.19	2.2	.05	55	3
9	12	.57	4	30	104	.15	2*	.08	.25	.06	.7	.02	17	1
19	22	1.4	14	110	333	.48	24*	.29	.85	.21	1.8	.06	31	2
0	22	.58	5	76	142	.12	24*	.07		.06	.6	.03	4	2
1	66	.87	6	86	345	.16	57*	.09	.92	.08	.77	.03	11	<1
0	29	2.15	12	164	73	.15	209*	.06	.95	.08	.6	.05	15	3
1	73	1.89	10	113	352	.19	175*	.09	1.02	.11	.88	.04	17	2
0	23	.95	13	198	69	.2	129*	.05	2.46	.05	1.05	.03	10	5
1	68	1	10	139	349	.23	116*	.09	2.15	.09	1.22	.02	13	3
17	21	.92	9	60	257	.26	8*	.1	1.8	.11	.87	.03	22	<1
4	26	1.22	13	65	205	.46	2*	.22	1.84	.13	1.71	.03	26	<1
0	99	1.43	12	75	264	.4	0	.25	.1	.16	2.21	.02	46	0
0	98	1.41	11	74	262	.39	0	.2	.09	.14	1.98	.02	38	<1
0	175	1.62	47	139	420	1.06	0	.2	.46	.09	2.25	.11	32	0
<1	29	.72	21	91	78	.42	0	.08	.34	.05	.14	.03	7	0
0	15	.74	24	84	73	.51	4*	.07	.33	.03	.39	.02	6	<1
0	0	3.6			5		100*							0
22	107	1.29	9	69	251	.31	16*	.15	1.03	.16	1.26	.02	7	1
20	106	2.39	44	181	335	1.57	136*	.19	1.31	.25	2.29	.18	30	4
26	147	1.49	13	82	333	.35	23*	.17	1.08	.18	1.36	.05	10	<1
23	12	.56	5	39	219	.19	11*	.07	.7	.16	1.12	.04	5	<1
34	16	1.27	28	73	234	.57	15*	.1	.75	.12	1.44	.09	8	0
31	37	.97	10	65	398	.32	22*	.12	.75	.14	1.05	.05	28	<1
					77									
20	77	.56	17	70	160	.35	20*	.05	.62	.08	.4	.04	5	<1
22	83	.68	6	50	167	.21	20*	.08	.36	.11	.59	.02	14	<1
5	48	.59	8	66	239	.15	3*	.08	.32	.08	.65	.03	14	<1

*This value is expressed in retinol equivalents (RE). All other values are in retinol activity equivalents (RAE).

Table A-1

Food Composition (Computer code number is for Wadsworth Diet Analysis program) (For purposes of calculations, use "0" for t, <1, <.1, <.01, etc.)

Computer Code Number	Food Description	Measure	Wt (g)	H_2O (%)	Ener (cal)	Prot (g)	Carb (g)	Dietary Fiber (g)	Fat (g)	Fat Breakdown (g) Sat	Mono	Poly
	BAKED GOODS: BREADS, CAKES, COOKIES, CRACKERS, PIES—Continued											
1468	Pan dulce, sweet roll w/topping	1 ea	79	21	291	5	48	1	9	1.8	4	2.5
	Piecrust,with enriched flour, vegetable shortening, baked:											
444	Home recipe, 9" shell	1 ea	180	10	949	11	85	3	62	15.5	27.3	16.4
	From mix:											
445	Piecrust for 2-crust pie	1 ea	320	10	1686	20	152	5	111	27.6	48.5	29.2
446	1 pie shell	1 ea	160	11	802	11	81	3	49	12.3	27.6	6.2
	Pies, 9" diam; pie crust made with vegetable shortening, enriched flour:											
448	Apple, ⅛ of pie	1 pce	117	52	277	2	40	2	13	4.4	5.1	2.6
450	Banana cream, ⅛ of pie	1 pce	144	48	387	6	47	1	20	5.4	8.2	4.7
452	Blueberry, ⅛ of pie	1 pce	147	51	360	4	49	2	17	4.3	7.5	4.5
454	Cherry, ⅛ of pie	1 pce	180	46	486	5	69	3	22	5.4	9.6	5.8
456	Chocolate cream, ⅛ of pie	1 pce	199	63	358	8	47	2	16	5.9		
458	Custard, ⅛ of pie	1 pce	105	61	221	6	22	2	12	2.5	5	3.9
460	Lemon meringue, ⅛ of pie	1 pce	113	42	303	2	53	1	10	2	3	4.1
462	Peach, ⅛ of pie	1 pce	139	47	354	4	53	3	15			
464	Pecan, ⅛ of pie	1 pce	113	19	452	5	65	4	21	4	12.1	3.6
466	Pumpkin, ⅛ of pie	1 pce	109	58	229	4	30	3	10	1.9	4.4	3.4
467	Pies, fried, commercial: Apple	1 ea	85	40	266	2	33	1	14	6.5	5.8	1.2
468	Pies, fried, commercial: Cherry	1 ea	128	38	404	4	54	3	21	3.1	9.5	6.9
	Pretzels, made with enriched flour:											
469	Thin sticks, 2¼" long	1 oz	28	3	107	3	22	1	1	.2	.4	.3
470	Dutch twists	10 pce	60	3	229	5	47	2	2	.4	.8	.7
471	Thin twists, 3¼ x 2¼ x ¼"	10 pce	60	3	229	5	47	2	2	.4	.8	.7
	Rolls & buns, enriched, commercial:											
472	Cloverleaf rolls, 2½" diam, 2" high	1 ea	28	32	84	2	14	1	2	.5	1	.3
473	Hot dog buns	1 ea	40	34	114	3	20	1	2	.5	.3	1
474	Hamburger buns	1 ea	43	34	123	4	22	1	2	.5	.4	1.1
475	Hard roll, white, 3¾" diam, 2" high	1 ea	57	31	167	6	30	1	2	.3	.6	1
476	Submarine rolls/hoagies, 11¼ x 3 x 2½"	1 ea	135	34	386	11	68	4	7	1.6	3.4	1.2
	Rolls & buns, enriched, home recipe:											
477	Dinner rolls 2½" diam, 2" high	1 ea	35	29	112	3	19	1	3	.7	1.1	.7
	Sports/fitness bar:											
2043	Forza energy bar	1 ea	70	18	231	10	45	4	1			
2042	Power bar	1 ea	65		230	10	45	3	2			
2041	Tiger sports bar	1 ea	65	14	260	11	33	2	9	1.9		
478	Toaster pastries, fortified (Poptarts)	1 ea	52	12	204	2	37	1	5	.8	2.1	2
2132	Toaster strudel pastry—cream cheese	1 ea	54	30	200	3	24	<1	10	3		
2134	Toaster strudel pastry—french toast	1 ea	54	30	200	3	24	<1	10	3		
	Tortilla chips:											
1271	Plain	10 pce	18	2	90	1	11	1	5	.9	2.8	.7
1036	Nacho flavor	1 c	26	2	129	2	16	1	7	1.3	3.9	.9
1037	Taco flavor	1 pce	18	2	86	1	11	1	4	.8	2.6	.6
	Tortillas:											
479	Corn, enriched, 6" diam	1 ea	26	44	58	1	12	1	1	.1	.2	.3
480	Flour, 8" diam	1 ea	49	27	159	4	27	2	3	.9	1.8	.5
1301	Flour, 10" diam	1 ea	72	27	234	6	40	2	5	1.3	2.7	.8
481	Taco shells	1 ea	14	6	65	1	9	1	3	.5	1.3	1.2
	Waffles, 7" diam:											
482	From home recipe	1 ea	75	42	218	6	25	1	11	2.1	2.6	5.1
483	From mix, egg/milk added	1 ea	75	42	218	5	26	1	10	1.7	2.7	5.2
1510	Whole grain, prepared from frozen	1 ea	39	43	105	4	13	1	4	1.2	1.7	1.1
	GRAIN PRODUCTS: CEREAL, FLOUR, GRAIN, PASTA and NOODLES, POPCORN											
38070	Amaranth	1 c	195	10	729	28	129	30	13	3.2	2.8	5.6
484	Barley, pearled, dry, uncooked	1 c	200	10	704	20	155	31	2	.5	.3	1.1
485	Barley, pearled, cooked	1 c	157	69	193	4	44	6	1	.1	.1	.3

Chol (mg)	Calc (mg)	Iron (mg)	Magn (mg)	Pota (mg)	Sodi (mg)	Zinc (mg)	VT-A (μg)	Thia (mg)	VT-E (mg)	Ribo (mg)	Niac (mg)	V-B6 (mg)	Fola (μg)	VT-C (mg)
26	13	1.84	9	57	75	.35	67*	.23	1.22	.21	2.02	.03	19	<1
0	18	5.2	25	121	976	.79	0	.7	9.94	.5	5.95	.04	121	0
0	32	9.25	45	214	1734	1.41	0	1.25	17.7	.89	10.6	.08	214	0
0	96	3.44	24	99	1166	.62	0	.48	8.83	.3	3.8	.09	112	0
0	13	.53	8	76	311	.19	35*	.03	.08	.03	.31	.04	26	4
73	108	1.5	23	238	346	.69	101*	.2	2.12	.3	1.52	.19	39	2
0	10	1.81	12	73	272	.29	6*	.22	3.09	.19	1.76	.05	34	1
0	18	3.33	16	139	344	.36	86*	.27	3.42	.22	2.3	.06	49	2
18	171	1.47	28	284	347	.82	33*	.17	1.85	.4	1.22	.06	28	1
35	84	.61	12	111	252	.55	70*	.04	1.96	.22	.31	.05	21	1
51	63	.69	17	101	165	.55	59*	.07	2.45	.24	.73	.03	15	4
4	12	2.22	16	280	228	.33	42*	.16	2.19	.13	2.55	.04	28	3
36	19	1.18	20	84	479	.64	53*	.1	2.09	.14	.28	.02	30	1
22	65	.86	16	168	307	.49	405*	.06	1.85	.17	.2	.06	22	1
13	13	.88	8	51	325	.17	33*	.1	.37	.08	.98	.03	4	1
0	28	1.56	13	83	479	.29	22*	.18	.55	.14	1.82	.04	23	2
0	10	1.21	10	41	480	.24	0	.13	.06	.17	1.47	.03	48	0
0	22	2.59	21	88	1029	.51	0	.28	.13	.37	3.15	.07	103	0
0	22	2.59	21	88	1029	.51	0	.28	.13	.37	3.15	.07	103	0
<1	33	.88	6	37	146	.22	0	.14	.25	.09	1.13	.01	27	<1
0	56	1.27	8	56	224	.25	0	.19	.62	.12	1.57	.02	38	<1
0	60	1.36	9	61	241	.27	0	.21	.67	.13	1.69	.02	41	<1
0	54	1.87	15	62	310	.54	0	.27	.19	.19	2.42	.02	54	0
0	188	4.28	27	190	756	.84	0	.65	.62	.42	5.31	.06	36	0
13	21	1.04	7	53	145	.24	28*	.14	.35	.14	1.21	.02	15	<1
0	300	6.3	160	220	65	5.25		1.5	27.1	1.7	20	2	400	60
0	300	5.4	140	150	110	5.25		1.5		1.7	20	2	400	60
0	557	5.01	186		139		279*	2.37		1.11	5.57	1.11		11
0	13	1.81	9	58	218	.34	100*	.15	1.19	.19	2.05	.2	34	<1
10	0	1.08			220		0							0
10	0	1.08			220		0							0
0	28	.27	16	35	95	.27	2	.01	.24	.03	.23	.05	2	0
1	38	.37	21	56	184	.31	5	.03	.35	.05	.37	.07	4	<1
1	28	.36	16	39	142	.23	8	.04	.24	.04	.36	.05	4	<1
0	45	.36	17	40	42	.24	0	.03	.04	.02	.39	.06	30	0
0	61	1.62	13	64	234	.35	0	.26	.45	.14	1.75	.02	60	0
0	90	2.38	19	94	344	.51	0	.38	.66	.21	2.57	.04	89	0
0	22	.35	15	25	51	.2	5	.03	.42	.01	.19	.05	1	0
52	191	1.73	14	119	383	.51	49*	.2	1.73	.26	1.55	.04	34	<1
38	93	1.22	15	134	458	.35	19*	.15	1.5	.19	1.23	.07	9	<1
37	102	.81	15	90	132	.44	30*	.08	.55	.13	.77	.04	7	<1
0	298	14.8	519	714	41	6.2	0	.16	2.01	.41	2.51	.43	96	8
0	58	5	158	560	18	4.26	2	.38	.26	.23	9.21	.52	46	0
0	17	2.09	34	146	5	1.29	1	.13	.08	.1	3.24	.18	25	0

*This value is expressed in retinol equivalents (RE). All other values are in retinol activity equivalents (RAE).

Table A–1

Food Composition (Computer code number is for Wadsworth Diet Analysis program) (For purposes of calculations, use "0" for t, <1, <.1, <.01, etc.)

Computer Code Number	Food Description	Measure	Wt (g)	H_2O (%)	Ener (cal)	Prot (g)	Carb (g)	Dietary Fiber (g)	Fat (g)	Fat Breakdown (g) Sat	Mono	Poly
	GRAIN PRODUCTS: CEREAL, FLOUR, GRAIN, PASTA and NOODLES, POPCORN—Continued											
2009	Breakfast bars, fat free, all flavors	1 ea	38	25	110	2	26	3	0	0	0	0
	Breakfast bar, Snackwell:											
2165	Apple-cinnamon	1 ea	37	16	119	1	29	1	<1	.1	t	.1
2164	Blueberry	1 ea	37	16	121	1	29	1	<1	t	t	.1
2163	Strawberry	1 ea	37	16	120	1	29	1	<1	t	t	.1
	Breakfast cereals, hot, cooked:w/o salt added											
	Corn grits (hominy) enriched:											
486	Regular/quick prep w/o salt, yellow:	1 c	242	85	145	3	31	<1	<1	.1	.1	.2
487	Instant, prepared from packet, white	1 ea	137	82	89	2	21	1	<1	t	t	.1
	Cream of wheat:											
488	Regular, quick, instant	1 c	239	87	129	4	27	1	<1	.1	.1	.3
489	Mix and eat, plain, packet	1 ea	142	82	102	3	21	<1	<1	t	t	.2
1664	Farina cereal, cooked w/o salt	1 c	233	88	117	3	25	3	<1	t	t	.1
490	Malt-O-Meal, cooked w/o salt	1 c	240	88	122	4	26	1	<1	.1	.1	t
494	Maypo	1 c	216	83	153	5	29	5	2	.4	.7	.8
	Oatmeal or rolled oats:											
491	Regular, quick, instant,nonfortified cooked w/o salt	1 c	234	85	145	6	25	4	2	.4	.7	.9
	Instant, fortified:											
492	Plain, from packet	½ c	118	85	70	3	12	2	1	.2	.4	.4
493	Flavored, from packet	½ c	109	76	106	3	21	2	1	.2	.5	.5
	Breakfast cereals, ready to eat:											
495	All-Bran	1 c	62	3	164	8	47	20	2	.4	.4	1.1
1306	Alpha Bits	1 c	28	1	110	2	24	1	1	.1	.2	.2
1307	Apple Jacks	1 c	33	3	127	2	29	1	<1	.1	.1	.2
1308	Bran Buds	1 c	90	3	248	8	72	36	2	.4	.4	1.3
1305	Bran Chex	1 c	49	2	156	5	39	8	1	.2	.3	.7
1309	Honey BucWheat Crisp	1 c	38	5	147	4	31	3	1	.2	.2	.5
1310	C.W. Post, plain	1 c	97	2	421	9	73	7	13	1.7	6	4.7
1311	C.W. Post, with raisins	1 c	103	4	446	9	74	14	15	11	1.7	1.4
496	Cap'n Crunch	1 c	37	2	147	2	32	1	2	.5	.4	.3
1312	Cap'n Crunchberries	1 c	35	2	140	2	30	1	2	.5	.3	.3
1313	Cap'n Crunch, peanut butter	1 c	35	2	146	3	28	1	3	.7	1.1	.7
497	Cheerios	1 c	23	3	84	2	17	2	1	.3	.5	.2
1314	Cocoa Krispies	1 c	41	2	159	2	36	1	1	.8	.1	.2
1316	Cocoa Pebbles	1 c	32	3	127	1	28	1	1	1.2	.1	t
1315	Corn Bran	1 c	36	3	120	2	30	6	1	.3	.3	.4
1317	Corn Chex	1 c	28	2	105	2	24	<1	<1	.1	.1	.2
498	Corn Flakes, Kellogg's	1 c	28	3	102	2	24	1	<1	.1	t	.1
499	Corn Flakes, Post Toasties	1 c	28	3	101	2	24	1	<1	0	t	t
1340	Corn Pops	1 c	31	3	118	1	28	<1	<1	.1	.1	t
1318	Cracklin' Oat Bran	1 c	65	4	266	5	47	8	8	3.4	3.8	.9
1038	Crispy Wheat 'N Raisins	1 c	43	7	150	3	35	3	1	.1	.1	.1
1319	Fortified Oat Flakes	1 c	48	3	180	8	36	1	1	.2	.3	.4
500	40% Bran Flakes, Kellogg's	1 c	39	4	128	4	31	6	1	.2	.2	.5
501	40% Bran Flakes, Post	1 c	47	4	150	4	38	8	1	.2	.2	.7
502	Froot Loops	1 c	32	2	125	2	28	1	1	.4	.2	.3
518	Frosted Flakes	1 c	41	3	158	2	37	1	<1	.1	3.4	.1
1320	Frosted Mini-Wheats	1 c	55	5	186	5	45	6	1	.2	.1	.6
1321	Frosted Rice Krispies	1 c	35	2	132	2	32	<1	<1	.1	.1	.1
1324	Fruit & Fibre w/dates	1 c	57	9	193	5	43	8	3	.4	.5	1.5
1322	Fruity Pebbles	1 c	32	3	128	1	28	<1	1	.3	.6	.4
503	Golden Grahams	1 c	39	3	150	2	33	1	1	.2	.4	.2
504	Granola, homemade	½ c	61	5	285	9	32	6	15	2.9	4.8	6.4
	Granola, low fat	½ c	47	3	181	5	38	3	3	0		
1670	Granola, low fat, commercial	½ c	45	5	165	4	36	3	2	.8	.5	.9
505	Grape Nuts	½ c	55	3	197	6	45	5	1	.2	.2	.6
1326	Grape Nuts Flakes	1 c	39	3	142	4	32	3	1	.2	.3	.6
1665	Heartland Natural with raisins	1 c	110	5	468	11	76	6	16	4	4.2	6.2
1327	Honey & Nut Corn Flakes	1 c	37	2	150	3	31	1	2	.3	.8	.6
506	Honey Nut Cheerios	1 c	33	2	126	3	27	2	1	.3	.5	.2

PAGE KEY: A–2 = Beverages A–4 = Dairy A–8 = Eggs A–10 = Fat/Oil A–12 = Fruit A–18 = Bakery A–24 = Grain A–30 = Fish A–32 = Meats A–36 = Poultry A–38 = Sausage A–38 = Mixed/Fast A–44 = Nuts/Seeds A–46 = Sweets A–50 = Vegetables/Legumes A–60 = Vegetarian A–62 = Misc A–64 = Soups/Sauces A–66 = Fast A–80 = Convenience A–86 = Baby foods

Chol (mg)	Calc (mg)	Iron (mg)	Magn (mg)	Pota (mg)	Sodi (mg)	Zinc (mg)	VT-A (µg)	Thia (mg)	VT-E (mg)	Ribo (mg)	Niac (mg)	V-B6 (mg)	Fola (µg)	VT-C (mg)
0	20	.72			25		20*							1
<1	17	5	6	68	103	3.88	260*	.39		.44	5.2	.52		<1
<1	14	4.83	5	43	107	3.85	260*	.39		.44	5.2	.52		<1
<1	14	4.82	6	47	102	3.83	260*	.39		.44	5.2	.52		2
0	0	1.55	10	53	0	.17	7	.24	.05	.14	1.96	.06	75	0
0	8	8.19	11	38	289	.21	0	.15	.03	.08	1.38	.05	47	0
0	50	10.3	12	45	139	.33	0	.24	.02	0	1.43	.03	108	0
0	20	8.09	7	38	241	.24	376	.43	.01	.28	4.97	.57	101	0
0	5	1.17	5	30	0	.16	0	.19	.02	.12	1.28	.02	54	0
0	5	9.6	5	31	2	.17	0	.48	.02	.24	5.76	.02	5	0
0	112	7.56	45	190	233	1.34	633	.65	1.51	.65	8.42	.86	9	26
0	19	1.59	56	131	2	1.15	2	.26	.23	.05	.3	.05	9	0
0	109	4.2	28	66	190	.58	302*	.35	.14	.19	3.65	.49	65	0
0	112	4.45	34	91	169	.66	306*	.35	.14	.25	3.92	.51	100	<1
0	219	9.3	266	706	126	7.75	466	.81	1.14	.87	10.4	1.05	186	31
0	8	2.66	16	54	178	1.48	371	.36	.02	.42	4.93	.5	99	0
0	4	4.95	10	35	148	4.13	248	.43	.05	.46	5.51	.56	116	16
0	60	13.5	250	809	599	19.4	676	1.17	1.42	1.26	15	1.53	270	45
0	29	14	69	216	345	6.48	5	.64	.56	.26	8.62	.88	173	26
0	54	10.9	43	142	361	.68	914	.9	8.99	1.03	12.1	1.88	11	36
<1	47	15.4	67	198	167	1.64	1284*	1.26	.68	1.46	17.1	1.75	342	0
<1	50	16.4	74	261	161	1.64	1363	1.34	.72	1.55	18.1	1.85	364	0
0	7	6.17	13	47	286	5.14	2	.51	.18	.58	6.86	.68	137	0
0	9	6.06	13	49	256	5.4	6*	.5	.25	.57	6.73	.67	135	<1
0	3	5.83	24	80	264	4.86	2	.49	.19	.55	6.48	.65	130	0
0	42	6.21	25	68	218	2.88	288	.29	.16	.33	3.83	.38	153	11
0	5	2.38	15	79	278	1.97	298	.49	.19	.57	6.6	.66	123	20
0	4	1.99	12	47	173	1.65	248	.41		.47	5.52	.55	110	0
0	27	10.1	19	75	338	5	3	.1	.19	.56	6.67	.67	134	0
0	93	8.4	8	30	270	.35	0	.35	.09	0	4.67	.47	186	6
0	1	8.68	3	25	298	.17	210	.36	.03	.39	4.68	.48	99	14
0	1	5.4	4	33	266	.13	225	.37	.67	.43	5	.5	100	0
0	2	1.86	2	23	123	1.55	233	.4	.03	.43	5.18	.53	109	15
0	29	2.41	90	301	231	1.95	299	.5	.43	.56	6.63	.66	181	20
0	54	3.52	33	180	223	.85	293	.29	.45	.33	3.91	.39	157	0
0	68	13.7	58	228	220	2.54	636	.62	.34	.72	8.45	.86	169	0
0	19	10.9	81	236	304	5.04	488	.51	7.22	.58	6.72	.66	138	20
0	26	12.7	101	290	344	2.35	353	.59		.67	7.83	.78	157	0
0	4	4.51	9	34	150	4	225	.42	.12	.45	5.34	.54	96	15
0	1	5.95	4	27	264	.2	298	.49	.05	.57	6.6	.66	123	20
0	20	15.4	56	183	2	1.6	0	.38	.49	.44	5.39	.49	110	0
0	2	2.42	8	27	256	.42	303	.49	.03	.56	6.72	.66	140	20
0	30	10.1	81	335	270	3.02	725	.75	1.32	.85	10.1	1	201	0
0	2	2.13	6	35	187	1.78	267	.44		.5	5.93	.59	118	0
0	19	5.85	12	69	357	4.88	293	.49	.29	.55	6.5	.65	130	19
0	49	2.56	109	328	15	2.48	1	.45	7.86	.17	1.25	.19	52	1
0		2.71	36	143	90	5.64	226*	.56	7.57	.64	7.52	.75	151	
0	20	1.58	37	127	101	2.84	169	.27	4.53	.31	3.74	.36	90	1
0	19	15.4	55	169	336	1.14	214	.36		.4	4.74	.47	95	0
0	15	10.9	40	133	188	1.61	303	.5	.1	.57	6.72	.67	135	0
0	66	4.02	141	415	226	2.83	3	.32	.77	.14	1.54	.2	44	1
0	4	3.03	3	40	249	.26	152	.26	.09	.3	3.37	.33	74	10
0	22	4.95	32	94	285	4.13	248	.41	.34	.47	5.5	.55	220	16

*This value is expressed in retinol equivalents (RE). All other values are in retinol activity equivalents (RAE).

Table A-1

Food Composition (Computer code number is for Wadsworth Diet Analysis program) (For purposes of calculations, use "0" for t, <1, <.1, <.01, etc.)

Computer Code Number	Food Description	Measure	Wt (g)	H_2O (%)	Ener (cal)	Prot (g)	Carb (g)	Dietary Fiber (g)	Fat (g)	Fat Breakdown (g) Sat	Mono	Poly
	GRAIN PRODUCTS: CEREAL, FLOUR, GRAIN, PASTA and NOODLES, POPCORN—Continued											
1328	HoneyBran	1 c	35	2	119	3	29	4	1	.3	.1	.3
1329	HoneyComb	1 c	22	1	87	1	20	1	<1	.1	.1	.2
1330	King Vitaman	1 c	21	2	81	2	18	1	1	.2	.3	.2
1039	Kix	1 c	19	2	72	1	16	1	<1	.1	.1	t
1331	Life	1 c	44	4	167	4	35	3	2	.3	.6	.8
507	Lucky Charms	1 c	32	2	124	2	27	1	1	.2	.4	.2
1323	Mueslix Five Grain	1 c	82	8	289	6	63	6	5	.7	2	1.8
508	Nature Valley Granola	1 c	113	4	510	12	74	7	20	2.6	13.3	3.8
1666	Nutri Grain Almond Raisin	1 c	40	6	147	3	31	3	2	.1	1	1.2
1336	100% Bran	1 c	66	3	178	8	48	19	3	.6	.6	1.9
509	100% Natural cereal, plain	1 c	104	2	462	11	71	8	17	7.4	7.4	2.2
1337	100% Natural with apples & cinnamon	1 c	104	2	477	11	70	7	20	15.5	1.8	1.3
1338	100% Natural with raisins & dates	1 c	110	3	496	12	72	7	20	13.6	3.7	1.7
510	Product 19	1 c	30	3	110	3	25	1	<1	t	.1	.2
1339	Quisp	1 c	30	3	121	1	25	1	2	.5	.4	.2
511	Raisin Bran, Kellogg's	1 c	61	8	186	6	47	8	1	.1	.2	.8
512	Raisin Bran, Post	1 c	59	9	187	5	46	8	1	.2	.2	.7
1667	Raisin Squares	1 c	71	9	241	6	55	7	2	.2	.1	.6
1041	Rice Chex	1 c	33	2	125	2	29	<1	<1	t	t	.1
513	Rice Krispies, Kellogg's	1 c	28	2	111	2	25	<1	<1	t	t	t
514	Rice, puffed	1 c	14	4	54	1	12	<1	<1	t	t	t
515	Shredded Wheat	1 c	43	5	154	5	35	4	1	.1	.1	.4
516	Special K	1 c	31	3	115	6	22	1	<1	t	0	.2
517	Super Golden Crisp	1 c	33	1	123	2	30	<1	<1	.1	.1	.1
519	Honey Smacks	1 c	36	3	137	2	31	1	1	.4	.1	.3
1341	Tasteeos	1 c	24	2	94	3	19	3	1	.2	.2	.2
1342	Team	1 c	42	4	164	3	36	1	1	.1	.2	.3
520	Total, wheat, with added calcium	1 c	40	3	140	4	32	4	1	.2	.2	.1
521	Trix	1 c	28	2	114	1	24	1	2	.4	.8	.3
1344	Wheat Chex	1 c	46	2	159	5	37	5	1	.2	.2	.4
1043	Wheat cereal, puffed, fortified	1 c	12	4	44	2	9	1	<1	t	t	.1
522	Wheaties	1 c	29	3	106	3	23	2	1	.2	.2	.1
523	Buckwheat flour, dark	1 c	120	11	402	15	85	12	4	.8	1.1	1.1
525	Buckwheat, whole grain, dry	1 c	170	10	583	22	122	17	6	1.3	1.8	1.8
526	Bulgar, dry, uncooked	1 c	140	9	479	17	106	26	2	.3	.2	.8
527	Bulgar, cooked	1 c	182	78	151	6	34	8	<1	.1	.1	.2
	Cornmeal:											
528	Whole-ground, unbolted, dry	1 c	122	10	442	10	94	9	4	.6	1.2	2
530	Degermed, enriched, dry	1 c	138	12	505	12	107	10	2	.3	.6	1
38041	Degermed, enriched, baked	1 c	138	12	505	12	107	10	2	.3	.6	1
38076	Couscous, cooked	1 c	157	73	176	6	36	2	<1	t	t	.1
38329	Cracked wheat	1 c	120	10	407	16	87	14	2	.4	.3	.9
	Macaroni, cooked:											
532	Enriched	1 c	140	66	197	7	40	2	1	.1	.1	.4
533	Whole wheat	1 c	140	67	174	7	37	4	1	.1	.1	.3
534	Vegetable, enriched	1 c	134	68	172	6	36	6	<1	t	t	.1
535	Millet, cooked	1 c	240	71	286	8	57	3	2	.4	.4	1.2
7508	Natto	1 c	175	55	371	31	25	9	19	2.8	4.2	10.9
	Noodles (see also Pasta and Spaghetti):											
1507	Cellophane noodles, cooked	1 c	190	79	160	<1	39	<1	<1	t	t	t
1995	Cellophane noodles, dry	1 c	140	13	491	<1	121	1	<1	t	t	t
537	Chow Mein, dry	1 c	45	1	237	4	26	2	14	2	3.5	7.8
536	Egg noodles, cooked, enriched	1 c	160	69	213	8	40	2	2	.5	.7	.7
538	Spinach noodles, dry	3½ oz	100	8	372	13	75	11	2	.2	.2	.6
1343	Oat bran, dry	¼ c	24	7	59	4	16	4	2	.3	.6	.7
	Pasta, cooked:											
1418	Fresh	2 oz	57	69	75	3	14	1	1	.1	.1	.2
1417	Linguini/Rotini	1 c	140	66	197	7	40	2	1	.1	.1	.4
	Popcorn:											
539	Air popped, plain	1 c	8	4	31	1	6	1	<1	t	.1	.2
1042	Microwaved, low fat, low sodium	1 c	6	3	25	1	4	1	1	.1	.2	.2

Chol (mg)	Calc (mg)	Iron (mg)	Magn (mg)	Pota (mg)	Sodi (mg)	Zinc (mg)	VT-A (µg)	Thia (mg)	VT-E (mg)	Ribo (mg)	Niac (mg)	V-B6 (mg)	Fola (µg)	VT-C (mg)
0	16	5.57	46	151	202	.9	463	.45	.81	.52	6.16	.63	23	19
0	4	2.05	8	26	163	1.14	171	.28		.32	3.79	.38	76	0
0	3	5.92	18	58	176	2.65	212	.26	1.42	.3	3.53	.35	71	8
0	28	5.13	6	26	167	2.38	238	.24	.05	.27	3.17	.32	127	9
0	134	12.3	43	109	240	5.5	1	.55	.22	.62	7.33	.73	147	0
0	35	4.8	21	58	217	4	240	.4	.14	.45	5.33	.53	213	16
0	67	8.94	82	369	107	7.46	747	.75	8.94	.84	9.84	.99	197	1
0	85	3.53	107	375	183	2.27	0	.35	7.97	.12	1.25	.16	17	0
0	122	1	9	143	142	2.72	0	.28	4	.32	3.64	.36	80	0
0	46	8.12	312	652	457	5.74	0	1.58	1.53	1.78	20.9	2.11	47	63
1	100	3.11	109	457	28	2.5	1*	.36	1.19	.17	1.84	.19	26	<1
0	157	2.89	72	514	52	2	3	.33	.73	.57	1.87	.11	17	1
0	160	3.12	124	538	47	2.11	3	.31	.77	.65	2.09	.16	45	0
0	3	18	12	40	216	15	225	1.5	22.2	1.71	20	2.01	390	60
0	6	5.09	15	40	216	4.25	2	.42	.16	.48	5.66	.56	113	0
0	35	5	89	437	354	4.15	250	.43	.55	.49	5.55	.55	122	0
0	27	10.8	88	357	360	2.25	225	.38		.42	5	.5	100	0
0	24	21.7	62	335	4	1.99	0	.5	.38	.57	6.67	.64	142	0
0	110	9.58	10	38	310	0	0	.4	0	.02	5.32	.53	213	6
0	5	.7	12	27	206	.46	371	.52	.03	.59	6.92	.69	138	15
0	1	.41	4	16	1	.15	0	.06	.01	.01	.87	0	1	0
0	16	1.81	57	155	4	1.42	0	.11	.23	.12	2.26	.11	21	0
0	5	8.71	18	55	250	3.75	225	.53	.08	.59	7.01	.71	93	15
0	7	2.08	20	48	51	1.75	437	.43	.12	.49	5.81	.59	116	0
0	4	2.4	21	56	68	.47	300	.5	.18	.58	6.66	.68	133	20
0	11	6.86	26	71	183	.69	318	.31	.17	.36	4.22	.43	85	13
0	6	12	12	71	260	.58	556*	.55	.1	.63	7.39	.76	7	22
0	344	24	43	129	265	20	500	2	31.3	2.27	26.8	2.67	533	80
0	30	4.2	3	16	184	3.5	210	.35	.56	.4	4.67	.47	93	14
0	92	13.8	51	178	412	1.13	0	.34	.68	.06	4.6	.46	368	6
0	3	.56	16	44	1	.37	<1	.05	.08	.03	1.43	.02	4	0
0	53	7.83	31	101	215	.68	218	.36	.36	.41	4.83	.48	97	14
0	49	4.87	301	692	13	3.74	0	.5	1.24	.23	7.38	.7	65	0
0	31	3.74	393	782	2	4.08	0	.17	1.75	.72	11.9	.36	51	0
0	49	3.44	230	574	24	2.7	0	.32	.22	.16	7.16	.48	38	0
0	18	1.75	58	124	9	1.04	0	.1	.05	.05	1.82	.15	33	0
0	7	4.21	155	350	43	2.22	29	.47	.82	.24	4.43	.37	30	0
0	7	5.7	55	224	4	.99	28	.99	.45	.56	6.95	.35	258	0
0	7	5.7	55	224	4	.99	28	.74	.45	.48	6.6	.27	52	0
0	13	.6	13	91	8	.41	0	.1	.02	.04	1.54	.08	24	0
0	41	4.67	166	486	6	3.53	0	.54	.5	.26	7.64	.41	53	0
0	10	1.96	25	43	1	.74	0	.29	.04	.14	2.34	.05	98	0
0	21	1.48	42	62	4	1.13	0	.15	.14	.06	.99	.11	7	0
0	15	.66	25	41	8	.59	3	.15	.05	.08	1.44	.03	87	0
0	7	1.51	106	149	5	2.18	0	.25	.14	.2	3.19	.26	46	0
0	380	15.1	201	1275	12	5.3	0	.28	.02	.33	0	.23	14	23
0	13	.85	3	3	8	.2	0	.04	.06	0	.06	.01	1	0
0	35	3.04	4	14	14	.57	0	.21	.18	0	.28	.07	3	0
0	9	2.13	23	54	198	.63	2	.26	.07	.19	2.68	.05	40	0
53	19	2.54	30	45	11	.99	10	.3	.08	.13	2.38	.06	102	0
0	58	2.13	174	376	36	2.76	23	.37	.04	.2	4.55	.32	48	0
0	14	1.3	56	136	1	.75	0	.28	.41	.05	.22	.04	12	0
19	3	.65	10	14	3	.32	3	.12	.09	.09	.56	.02	36	0
0	10	1.96	25	43	1	.74	0	.29	.08	.14	2.34	.05	98	0
0	1	.21	10	24	<1	.27	1	.02	.01	.02	.16	.02	2	0
0	1	.14	9	14	29	.23	<1	.02	.06	.01	.12	.01	1	0

*This value is expressed in retinol equivalents (RE). All other values are in retinol activity equivalents (RAE).

Table A-1

Food Composition

(Computer code number is for Wadsworth Diet Analysis program) (For purposes of calculations, use "0" for t, <1, <.1, <.01, etc.)

Computer Code Number	Food Description	Measure	Wt (g)	H_2O (%)	Ener (cal)	Prot (g)	Carb (g)	Dietary Fiber (g)	Fat (g)	Fat Breakdown (g) Sat	Mono	Poly
	GRAIN PRODUCTS: CEREAL, FLOUR, GRAIN, PASTA and NOODLES, POPCORN—Continued											
540	Popped in vegetable oil/salted	1 c	11	3	55	1	6	1	3	.5	.9	1.5
541	Sugar-syrup coated	1 c	35	3	151	1	28	2	4	1.3	1	1.6
38079	Quinoa, dry	1 c	170	9	636	22	117	10	10	1	2.6	4
	Rice:											
542	Brown rice, cooked	1 c	195	73	216	5	45	4	2	.4	.6	.6
8858	Mexican rice, cooked	1 c	250		349	17	66	2	4		2	
2216	Spanish rice, cooked	1 c	246	85	130	3	28	2	1	0		
	White, enriched, all types:											
543	Regular/long grain, dry	1 c	185	12	675	13	148	2	1	.3	.4	.3
544	Regular/long grain, cooked	1 c	158	68	205	4	44	1	<1	.1	.1	.1
545	Instant, prepared without salt	1 c	165	76	162	3	35	1	<1	.1	.1	.1
	Parboiled/converted rice:											
546	Raw, dry	1 c	185	10	686	13	151	3	1	.3	.3	.3
547	Cooked	1 c	175	72	200	4	43	1	<1	.1	.1	.1
1486	Sticky Rice (Glutinous), cooked	1 c	174	77	169	4	37	2	<1	.1	.1	.1
548	Wild rice, cooked	1 c	164	74	166	7	35	3	1	.1	.1	.3
1700	Rice and pasta (Rice-a-Roni), cooked	1 c	202	72	246	5	43	5	6	1.1	2.3	1.9
549	Rye flour, medium	1 c	102	10	361	10	79	15	2	.2	.2	.8
1044	Soy flour, low-fat	1 c	88	3	325	45	30	9	6	.9	1.3	3.3
	Spaghetti pasta:											
550	Without salt, enriched	1 c	140	66	197	7	40	2	1	.1	.1	.4
551	With salt, enriched	1 c	140	66	197	7	40	2	1	.1	.1	.4
552	Whole-wheat spaghetti, cooked	1 c	140	67	174	7	37	6	1	.1	.1	.3
1302	Tapioca-pearl, dry	1 c	152	11	544	<1	135	1	<1	t	t	t
553	Wheat bran, crude	1 c	58	10	125	9	37	25	2	.4	.4	1.3
554	Wheat germ, raw	1 c	115	11	414	27	60	15	11	1.9	1.6	6.9
555	Wheat germ, toasted	1 c	113	6	432	33	56	15	12	2.1	1.7	7.5
1669	Wheat germ, with brown sugar & honey	1 c	113	3	420	30	66	11	9	1.5	1.2	5.5
556	Rolled wheat, cooked	1 c	240	84	149	5	33	4	1	.1	.1	.5
557	Whole-grain wheat, cooked	1 c	150	86	84	4	20	3	<1	.1	.1	.2
	Wheat flour (unbleached):											
	All-purpose white flour, enriched:											
558	Sifted	1 c	115	12	419	12	88	3	1	.2	.1	.5
559	Unsifted	1 c	125	12	455	13	95	3	1	.2	.1	.5
560	Cake or pastry, enriched, sifted	1 c	96	13	348	8	75	2	1	.1	.1	.4
561	Self-rising, enriched, unsifted	1 c	125	11	443	12	93	3	1	.2	.1	.5
562	Whole wheat, from hard wheats	1 c	120	10	407	16	87	15	2	.4	.3	.9
	MEATS: FISH and SHELLFISH											
1045	Bass, baked or broiled	4 oz	113	69	165	27	0	0	5	1.1	2.1	1.5
1046	Bluefish, baked or broiled	4 oz	113	63	180	29	0	0	6	1.3	2.6	1.5
1686	Catfish, breaded/flour fried	4 oz	113	49	329	21	14	<1	20	4.5	9.1	5.2
	Clams:											
563	Raw meat only	1 ea	145	82	107	18	4	0	1	.1	.1	.4
564	Canned, drained	1 c	160	64	237	41	8	0	3	.3	.3	.9
1290	Steamed, meat only	10 ea	95	64	141	24	5	0	2	.2	.2	.5
	Cod:											
565	Baked	4 oz	113	76	119	26	0	0	1	.2	.1	.3
566	Batter fried	4 oz	113	67	197	20	8	<1	9	1.8	3.6	3
567	Poached, no added fat	4 oz	113	77	116	25	0	0	1	.1	.1	.3
	Crab, meat only:											
1048	Blue crab, cooked	1 c	118	77	120	24	0	0	2	.3	.3	.8
1049	Dungeness crab, cooked	1 c	118	73	130	26	1	0	1	.2	.3	.5
568	Blue crab, canned	1 c	135	76	134	28	0	0	2	.3	.3	.6
1587	Crab, imitation, from surimi	4 oz	113	74	115	14	11	0	1	.3	.2	.8
569	Fish sticks, breaded pollock	2 ea	56	46	152	9	13	0	7	1.8	2.8	1.8
572	Flounder/sole, baked	4 oz	113	73	132	27	0	0	2	.4	.3	.7
1599	Grouper, baked or broiled	4 oz	113	73	133	28	0	0	1	.3	.3	.5
573	Haddock, breaded, fried	4 oz	113	60	247	23	10	<1	12	2.6	5.3	3.7
1050	Haddock, smoked	4 oz	113	71	131	28	0	0	1	.2	.2	.4

Chol (mg)	Calc (mg)	Iron (mg)	Magn (mg)	Pota (mg)	Sodi (mg)	Zinc (mg)	VT-A (µg)	Thia (mg)	VT-E (mg)	Ribo (mg)	Niac (mg)	V-B6 (mg)	Fola (µg)	VT-C (mg)
0	1	.31	12	25	97	.29	1	.01	.01	.01	.17	.02	2	<1
2	15	.61	12	38	72	.2	3*	.02	.42	.02	.77	.01	1	0
0	102	15.7	357	1258	36	5.61	0	.34	8.28	.67	4.98	.38	83	0
0	19	.82	84	84	10	1.23	0	.19	.43	.05	2.98	.28	8	0
0					1162		120*							
0	0	0			1340		0							0
0	52	7.97	46	213	9	2.02	0	1.07	.24	.09	7.76	.3	427	0
0	16	1.9	19	55	2	.77	0	.26	.08	.02	2.33	.15	92	0
0	13	1.04	8	7	5	.4	0	.12	.08	.08	1.45	.02	68	0
0	111	6.59	57	222	9	1.78	0	1.1	.24	.13	6.72	.65	427	0
0	33	1.98	21	65	5	.54	0	.44	.09	.03	2.45	.03	87	0
0	3	.24	9	17	9	.71	0	.03	.05	.02	.5	.04	2	0
0	5	.98	52	166	5	2.2	0	.08	.38	.14	2.11	.22	43	0
2	16	1.9	24	85	1147	.57	0	.25	.27	.16	3.6	.2	89	<1
0	24	2.16	76	347	3	2.03	0	.29	1.36	.12	1.76	.27	19	0
0	165	5.27	202	2261	16	1.04	2	.33	.17	.25	1.9	.46	361	0
0	10	1.96	25	43	1	.74	0	.29	.08	.14	2.34	.05	98	0
0	10	1.96	25	43	140	.74	0	.29	.38	.14	2.34	.05	98	0
0	21	1.48	42	62	4	1.13	0	.15	.07	.06	.99	.11	7	0
0	30	2.4	2	17	2	.18	0	.01	0	0	0	.01	6	0
0	42	6.13	354	686	1	4.22	0	.3	1.35	.33	7.87	.76	46	0
0	45	7.2	275	1025	14	14.1	0	2.16	20.7	.57	7.83	1.5	323	0
0	51	10.3	362	1070	5	18.8	0	1.89	20.5	.93	6.32	1.11	398	7
0	56	9.1	307	1089	12	15.7	6	1.51	24.9	.78	5.34	.56	376	0
0	17	1.49	53	170	0	1.15	0	.17	.48	.12	2.14	.17	26	0
0	9	.88	35	99	1	.73	0*	.12	.3	.03	1.5	.08	12	0
0	17	5.34	25	123	2	.8	0	.9	.07	.57	6.79	.05	177	0
0	19	5.8	27	134	2	.87	0	.98	.07	.62	7.38	.05	193	0
0	13	7.03	15	101	2	.59	0	.86	.06	.41	6.52	.03	148	0
0	423	5.84	24	155	1587	.77	0	.84	.07	.52	7.29	.06	193	0
0	41	4.66	166	486	6	3.52	0	.54	1.48	.26	7.64	.41	53	0
98	116	2.16	43	515	102	.94	40	.1	.84	.1	1.72	.16	19	2
86	10	.7	47	539	87	1.18	156	.08	.71	.11	8.19	.52	2	0
91	62	1.88	36	391	240	1.17	33*	.46	2.87	.21	3.78	.22	16	1
49	67	20.3	13	455	81	1.99	131	.12	1.45	.31	2.56	.09	23	19
107	147	44.7	29	1004	179	4.37	274	.24	1.6	.68	5.37	.18	46	35
64	87	26.6	17	597	106	2.59	162	.14	1.86	.4	3.19	.1	28	21
62	16	.55	47	276	88	.65	16	.1	.34	.09	2.84	.32	9	1
56	33	.81	28	437	104	.57	10*	.08	1.49	.11	2.58	.37	10	2
52	10	.37	30	484	90	.56	9	.02	.32	.05	2.45	.45	7	3
118	123	1.07	39	382	329	4.98	2	.12	1.18	.06	3.89	.21	60	4
90	70	.51	68	481	446	6.45	37	.07	1.33	.24	4.28	.2	50	4
120	136	1.13	53	505	450	5.43	3*	.11	1.35	.11	1.85	.2	58	4
23	15	.44	49	102	950	.37	23	.04	.11	.03	.2	.03	2	0
63	11	.41	14	146	326	.37	17	.07	.77	.1	1.19	.03	25	0
77	20	.38	65	389	119	.71	12	.09	2.14	.13	2.46	.27	10	0
53	24	1.29	42	537	60	.58	56	.09	.71	.01	.43	.4	11	0
87	70	2.02	49	376	194	.63	28*	.11	1.93	.12	4.94	.31	15	<1
87	55	1.58	61	469	862	.56	25	.05	.45	.05	5.73	.45	17	0

*This value is expressed in retinol equivalents (RE). All other values are in retinol activity equivalents (RAE).

Table A-1

Food Composition (Computer code number is for Wadsworth Diet Analysis program) (For purposes of calculations, use "0" for t, <1, <.1, <.01, etc.)

Computer Code Number	Food Description	Measure	Wt (g)	H_2O (%)	Ener (cal)	Prot (g)	Carb (g)	Dietary Fiber (g)	Fat (g)	Fat Breakdown (g) Sat	Mono	Poly
	MEATS: FISH and SHELLFISH—Continued											
	Halibut:											
17291	Baked	4 oz	113	72	158	30	0	0	3	.5	1.1	1.1
1051	Smoked	4 oz	113	64	203	34	0	0	4	.6	1.2	1.5
1054	Raw	4 oz	113	78	124	23	0	0	3	.4	.8	.8
575	Herring, pickled	4 oz	113	55	296	16	11	0	20	2.7	13.5	1.9
1052	Lobster meat, cooked w/moist heat	1 c	145	76	142	30	2	0	1	.2	.2	.1
1687	Ocean perch, baked/broiled	4 oz	113	73	137	27	0	0	2	.4	.9	.6
576	Ocean perch, breaded/fried	4 oz	113	58	255	23	10	<1	13	2.7	5.8	3.9
1056	Octopus, raw	4 oz	113	80	93	17	2	0	1	.3	.2	.3
	Oysters:											
577	Raw, Eastern	1 c	248	85	169	17	10	0	6	1.9	.8	2.4
578	Raw, Pacific	1 c	248	82	201	23	12	0	6	1.3	.9	2.2
	Cooked:											
579	Eastern, breaded, fried, medium	5 ea	73	65	144	6	8	<1	9	2.3	3.4	2.4
580	Western, simmered	5 ea	125	64	204	24	12	0	6	1.3	.9	2.2
581	Pollock, baked, broiled, or poached	4 oz	113	74	128	27	0	0	1	.3	.2	.6
	Salmon:											
582	Canned pink, solids and liquid	4 oz	113	69	157	22	0	0	7	1.7	2	2.3
583	Broiled or baked	4 oz	113	62	244	31	0	0	12	2.2	6	2.7
584	Smoked	4 oz	113	72	132	21	0	0	5	1	2.3	1.1
585	Atlantic sardines, canned, drained, 2 = 24g	4 oz	113	60	235	28	0	0	13	1.7	4.4	5.8
586	Scallops, breaded, cooked from frozen	6 ea	93	58	200	17	9	<1	10	2.5	4.2	2.7
1588	Scallops, imitation, from surimi	4 oz	113	74	112	14	12	0	<1	.1	.1	.2
1688	Scallops, steamed/boiled	½ c	60	77	65	10	1	0	2	.3	.7	.6
	Shrimp:											
587	Cooked, boiled, 2 large=11g	16 ea	88	77	87	18	0	0	1	.3	.2	.4
588	Canned, drained	½ c	64	73	77	15	1	0	1	.2	.2	.5
589	Fried, 2 large=15g,breaded	12 ea	90	53	218	19	10	<1	11	1.9	3.4	4.6
1057	Raw, large, about 7g each	14 ea	98	76	104	20	1	0	2	.3	.2	.7
1589	Shrimp, imitation, from surimi	4 oz	113	75	114	14	10	0	2	.3	.2	.8
1053	Snapper, baked or broiled	4 oz	113	70	145	30	0	0	2	.4	.4	.7
1060	Squid, fried in flour	4 oz	113	65	198	20	9	0	8	2.1	3.1	2.4
1590	Surimi	4 oz	113	76	112	17	8	0	1	.2	.2	.5
1058	Swordfish, raw	4 oz	113	76	137	22	0	0	5	1.2	1.7	1
1059	Swordfish, baked or broiled	4 oz	113	69	175	29	0	0	6	1.6	2.2	1.3
590	Trout, baked or broiled	4 oz	113	70	170	26	0	0	7	1.8	2	2.1
	Tuna, light, canned, drained solids:											
591	Oil pack	1 c	145	60	287	42	0	0	12	2.2	4.3	4.2
592	Water pack	1 c	154	75	179	39	0	0	1	.4	.2	.5
1061	Bluefin tuna, fresh	4 oz	113	68	163	26	0	0	6	1.4	1.8	1.6
	MEATS: BEEF, LAMB, PORK and others											
	BEEF, cooked, trimmed to ½" outer fat:											
	Braised, simmered, pot roasted:											
	Relatively fat, choice chuck blade:											
593	Lean and fat, piece 2½ x 2½ x ¾"	4 oz	113	47	393	30	0	0	29	11.5	12.5	1.1
594	Lean only	4 oz	113	55	297	35	0	0	16	6.3	7	.5
	Relatively lean, like choice round:											
595	Lean and fat, pce 4 ⅛ x 2½ x ¾"	4 oz	113	52	311	32	0	0	19	7.2	8.3	.7
596	Lean only	4 oz	113	57	249	36	0	0	11	3.6	4.7	.4
	Ground beef, broiled, patty 3 x ⅝":											
597	Extra lean, about 16% fat	4 oz	113	54	299	32	0	0	18	7	7.8	.7
598	Lean, 21% fat	4 oz	113	53	316	32	0	0	20	7.8	8.7	.7
	Roasts, oven cooked, no added liquid:											
	Relatively fat, prime rib:											
601	Lean and fat, piece 4⅛ x 2¼ x ½"	4 oz	113	46	425	25	0	0	35	14.2	15.2	1.2
602	Lean only	4 oz	113	58	271	31	0	0	16	7	7.9	.7
	Relatively lean, choice round:											
603	Lean and fat, piece 2½ x 2½ x ¾"	4 oz	113	59	272	30	0	0	16	6.2	6.8	.6
604	Lean only	4 oz	113	65	198	33	0	0	6	2.3	2.7	.2
1701	Steak, rib, broiled, lean	4 oz	113	58	250	32	0	0	13	5.1	5.3	.4

Chol (mg)	Calc (mg)	Iron (mg)	Magn (mg)	Pota (mg)	Sodi (mg)	Zinc (mg)	VT-A (µg)	Thia (mg)	VT-E (mg)	Ribo (mg)	Niac (mg)	V-B6 (mg)	Fola (µg)	VT-C (mg)
46	68	1.21	121	651	78	.6	61	.08	1.23	.1	8.05	.45	16	0
59	87	1.56	154	833	2260	.78	86	.11	1.11	.14	10.8	.64	22	0
36	53	.95	94	509	61	.47	53	.07	.96	.08	6.61	.39	14	0
15	87	1.38	9	78	983	.6	292	.04	1.13	.16	3.73	.19	2	0
104	88	.57	51	510	551	4.23	38	.01	1.45	.1	1.55	.11	16	0
61	155	1.33	44	396	108	.69	16	.15	1.84	.15	2.75	.3	11	1
71	151	1.88	39	335	201	.75	23*	.18	2.87	.2	2.98	.24	13	1
54	60	5.99	34	396	260	1.9	51	.03	1.36	.04	2.37	.41	18	6
131	112	16.5	117	387	523	225	74	.25	2.11	.24	3.42	.15	25	9
124	20	12.7	55	417	263	41.2	201	.17	2.11	.58	4.98	.12	25	20
59	45	5.07	42	178	304	63.6	66	.11	1.66	.15	1.2	.05	23	3
125	20	11.5	55	378	265	41.6	183	.16	2.21	.55	4.52	.11	19	16
108	7	.32	82	437	131	.68	26	.08	.23	.09	1.86	.08	5	0
62	241	.95	38	368	626	1.04	19	.03	1.53	.21	7.39	.34	17	0
98	8	.62	35	424	75	.58	71	.24	1.42	.19	7.54	.25	6	0
26	12	.96	20	198	886	.35	29*	.03	1.53	.11	5.33	.31	2	0
160	432	3.3	44	449	571	1.48	76	.09	.34	.26	5.93	.19	14	0
57	39	.76	55	310	432	.99	20	.04	1.77	.1	1.4	.13	34	2
25	9	.35	49	116	898	.37	23	.01	.12	.02	.35	.03	2	0
19	15	.15	33	171	112	.56	22*	.01	.8	.04	.61	.08	7	1
172	34	2.72	30	160	197	1.37	58	.03	.45	.03	2.28	.11	4	2
111	38	1.75	26	134	108	.81	11	.02	.59	.02	1.76	.07	1	1
159	60	1.13	36	203	310	1.24	50	.12	1.35	.12	2.76	.09	16	1
149	51	2.36	36	181	145	1.09	53	.03	.8	.03	2.5	.1	3	2
41	21	.68	49	101	797	.37	23	.03	.12	.04	.19	.03	2	0
53	45	.27	42	590	64	.5	40	.06	.71	<.01	.39	.52	7	2
294	44	1.14	43	315	346	1.97	12	.06	2.09	.52	2.94	.07	16	5
34	10	.29	49	127	162	.37	23	.02	.28	.02	.25	.03	2	0
44	5	.91	30	325	102	1.3	41	.04	.56	.11	10.9	.37	2	1
56	7	1.18	38	417	130	1.66	46	.05	.71	.13	13.3	.43	2	1
78	97	.43	35	506	63	.58	17	.17	.57	.11	6.52	.39	21	2
26	19	2.02	45	300	513	1.31	33	.05	1.74	.17	18	.16	7	0
46	17	2.36	42	365	521	1.19	26	.05	.82	.11	20.5	.54	6	0
43	9	1.15	56	285	44	.68	740	.27	1.13	.28	9.78	.51	2	0
112	11	3.45	21	275	67	7.57	0	.08	.26	.27	3.54	.32	10	0
120	15	4.16	26	297	80	11.6	0	.09	.16	.32	3.02	.33	7	0
108	7	3.53	25	319	56	5.55	0	.08	.21	.27	4.21	.37	11	0
108	6	3.91	28	348	58	6.19	0	.08	.2	.29	4.61	.41	12	0
112	10	3.13	28	417	93	7.27	0	.08	.2	.36	6.61	.36	12	0
114	14	2.77	27	394	101	7.01	0	.07	.23	.27	6.75	.34	12	0
96	12	2.61	21	334	71	5.92	0	.08	.27	.19	3.8	.26	8	0
91	11	2.95	28	425	84	7.84	0*	.09	.14	.24	4.64	.34	9	0
81	7	2.07	27	406	67	4.87	0	.09	.23	.18	3.92	.4	7	0
78	6	2.2	30	446	70	5.36	0	.1	.12	.19	4.24	.43	8	0
90	15	2.9	30	445	78	7.9	0	.11	.16	.25	5.42	.45	9	0

*This value is expressed in retinol equivalents (RE). All other values are in retinol activity equivalents (RAE).

Table A-1

Food Composition (Computer code number is for Wadsworth Diet Analysis program) (For purposes of calculations, use "0" for t, <1, <.1, <.01, etc.)

Computer Code Number	Food Description	Measure	Wt (g)	H_2O (%)	Ener (cal)	Prot (g)	Carb (g)	Dietary Fiber (g)	Fat (g)	Fat Breakdown (g) Sat	Mono	Poly
	MEATS: BEEF, LAMB, PORK and others—Continued											
	Steak, broiled, relatively lean,											
606	choice sirloin, lean only	4 oz	113	62	228	34	0	0	9	3.5	3.8	.3
	Steak, broiled, relatively fat, choice T-bone:											
1063	Lean and fat	4 oz	113	52	349	26	0	0	26	10.3	11.6	.9
1064	Lean only	4 oz	113	61	232	30	0	0	11	4.1	5.1	.3
	Variety meats:											
1086	Brains, panfried	4 oz	113	71	221	14	0	0	18	4.2	4.5	2.6
599	Heart, simmered	4 oz	113	64	198	32	<1	0	6	1.9	1.4	1.5
600	Liver, fried	4 oz	113	56	245	30	9	0	9	3	1.8	1.9
1062	Tongue, cooked	4 oz	113	56	320	25	<1	0	23	10.1	10.7	.9
607	Beef, canned, corned	4 oz	113	58	283	31	0	0	17	7	6.7	.7
608	Beef, dried, cured	1 oz	28	56	46	8	<1	0	1	.4	.5	.1
	LAMB, domestic, cooked:											
	Chop, arm, braised (5.6 oz raw w/bone):											
609	Lean and fat	1 ea	70	44	242	21	0	0	17	6.9	7.1	1.2
610	Lean only	1 ea	55	49	153	19	0	0	8	2.8	3.4	.5
	Chop, loin, broiled (4.2oz raw w/bone):											
611	Lean and fat	1 ea	64	52	202	16	0	0	15	6.3	6.2	1.1
612	Lean only	1 ea	46	61	99	14	0	0	4	1.6	2	.3
1067	Cutlet, avg of lean cuts, cooked	4 oz	113	54	330	28	0	0	23	9.9	9.8	1.7
	Leg, roasted, 3 oz = 4 ⅛ x 2¼ x ½":											
613	Lean and fat	4 oz	113	57	292	29	0	0	19	7.8	7.9	1.3
614	Lean only	4 oz	113	64	216	32	0	0	9	3.1	3.8	.6
615	Rib, roasted, lean and fat	4 oz	113	48	406	24	0	0	34	14.4	14.1	2.4
616	Rib, roasted, lean only	4 oz	113	60	262	30	0	0	15	5.4	6.6	1
1065	Shoulder, roasted, lean and fat	4 oz	113	56	312	25	0	0	23	9.5	9.2	1.8
1066	Shoulder, roasted, lean only	4 oz	113	63	231	28	0	0	12	4.6	4.9	1.1
	Variety meats:											
1069	Brains, pan-fried	4 oz	113	76	164	14	0	0	11	2.9	2.1	1.2
1068	Heart, braised	4 oz	113	64	209	28	2	0	9	3.5	2.5	.9
1070	Sweetbreads, cooked	4 oz	113	60	264	26	0	0	17	7.7	6.2	.8
1071	Tongue, cooked	4 oz	113	58	311	24	0	0	23	8.8	11.3	1.4
	PORK, cured, cooked (see also Sausages and Lunch Meats)											
617	Bacon, medium slices	3 pce	19	13	109	6	<1	0	9	3.3	4.5	1.1
1087	Breakfast strips, cooked	2 pce	23	27	106	7	<1	0	8	2.9	3.8	1.3
618	Canadian-style bacon	2 pce	47	62	87	11	1	0	4	1.3	1.9	.4
	Ham, roasted:											
619	Lean and fat, 2 pces 4⅛ x 2¼ x ¼"	4 oz	113	65	201	26	0	0	10	3.5	5	1.6
620	Lean only	4 oz	113	68	164	24	2	0	6	2	3	.6
621	Ham, canned, roasted, 8% fat	4 oz	113	69	154	24	1	0	6	1.8	2.8	.5
	PORK, fresh, cooked:											
	Chops, loin (cut 3 per lb with bone):											
1291	Braised, lean and fat	1 ea	89	58	213	24	0	0	12	4.5	5.4	1
1292	Lean only	1 ea	80	61	163	23	0	0	7	2.7	3.3	.6
622	Broiled, lean and fat	1 ea	82	58	197	23	0	0	11	3.9	4.8	.8
623	Broiled, lean only	1 ea	74	61	149	22	0	0	6	2.2	2.7	.4
624	Panfried, lean and fat	1 ea	78	53	216	23	0	0	13	4.7	5.5	1.5
625	Panfried, lean only	1 ea	63	59	152	16	0	0	9	3.2	3.9	1.2
626	Leg, roasted, lean and fat	4 oz	113	55	308	30	0	0	20	7.3	8.9	1.9
627	Leg, roasted, lean only	4 oz	113	61	233	35	0	0	9	3.2	4.3	.9
628	Rib, roasted, lean and fat	4 oz	113	56	288	31	0	0	17	6.7	7.9	1.4
629	Rib, roasted, lean only	4 oz	113	59	252	32	0	0	13	4.9	5.9	1
630	Shoulder, braised, lean and fat	4 oz	113	48	372	32	0	0	26	9.6	11.7	2.6
631	Shoulder, braised, lean only	4 oz	113	54	280	36	0	0	14	4.7	6.5	1.3
1088	Spareribs, cooked, yield from 1 lb raw with bone	4 oz	113	40	449	33	0	0	34	12.6	15.2	3.1
1095	Rabbit, roasted (1 cup meat=140g)	4 oz	113	61	223	33	0	0	9	2.7	2.4	1.8
	VEAL, cooked:											
632	Cutlet, braised or broiled, 4⅛ x 2¼ x ½"	4 oz	113	52	321	34	0	0	19	7.6	7.6	1.3

Chol (mg)	Calc (mg)	Iron (mg)	Magn (mg)	Pota (mg)	Sodi (mg)	Zinc (mg)	VT-A (µg)	Thia (mg)	VT-E (mg)	Ribo (mg)	Niac (mg)	V-B6 (mg)	Fola (µg)	VT-C (mg)
101	12	3.8	36	455	75	7.37	0	.15	.16	.33	4.84	.51	11	0
76	9	3.06	26	363	72	5.03	0	.1	.15	.24	4.46	.37	8	0
67	7	3.58	32	427	80	6	0	.12	.16	.28	5.23	.44	9	0
2254	10	2.51	17	400	179	1.53	0	.15	2.37	.29	4.27	.44	7	4
218	7	8.49	28	263	71	3.54	0	.16	.81	1.74	4.6	.24	2	2
545	12	7.1	26	411	120	6.16	12123	.24	.72	4.68	16.3	1.62	249	26
121	8	3.83	19	203	68	5.42	0	.03	.4	.4	2.43	.18	6	1
97	14	2.35	16	154	1136	4.03	0	.02	.17	.17	2.75	.15	10	0
12	2	1.26	9	124	972	1.47	0	.02	.04	.06	1.53	.1	3	0
84	17	1.67	18	214	50	4.26	0	.05	.1	.17	4.66	.08	13	0
67	14	1.49	16	186	42	4.02	0	.04	.1	.15	3.48	.07	12	0
64	13	1.16	15	209	49	2.23	0	.06	.08	.16	4.54	.08	11	0
44	9	.92	13	173	39	1.9	0	.05	.07	.13	3.15	.07	11	0
110	12	2.26	25	340	77	4.67	0	.12	.15	.32	7.48	.16	19	0
105	12	2.24	27	354	75	4.97	0	.11	.17	.3	7.45	.17	23	0
101	9	2.4	29	382	77	5.58	0	.12	.2	.33	7.16	.19	26	0
110	25	1.81	23	306	82	3.94	0	.1	.11	.24	7.63	.12	17	0
99	24	2	26	356	91	5.05	0	.1	.17	.26	6.96	.17	25	0
104	23	2.23	26	284	75	5.91	0	.1	.16	.27	6.95	.15	24	0
98	21	2.41	28	299	77	6.83	0	.1	.2	.29	6.51	.17	28	0
2308	14	1.9	16	232	151	1.54	0	.12	1.73	.27	2.79	.12	6	14
281	16	6.24	27	212	71	4.16	0	.19	.79	1.34	4.93	.34	2	8
452	14	2.4	21	329	59	3.03	0	.02	.78	.24	2.89	.06	15	23
214	11	2.97	18	179	76	3.38	0	.09	.36	.47	4.17	.19	3	8
16	2	.31	5	92	303	.62	0	.13	.1	.05	1.39	.05	1	0
24	3	.45	6	107	483	.85	0	.17	.07	.08	1.75	.08	1	0
27	5	.38	10	183	727	.8	0	.39	.12	.09	3.25	.21	2	0
67	9	1.51	25	462	1695	2.79	0	.82	.29	.37	6.95	.35	3	0
60	9	1.67	16	324	1359	3.25	0	.85	.29	.23	4.55	.45	3	0
34	7	1.04	24	393	1282	2.52	0	1.17	.29	.28	5.53	.51	6	0
71	19	.95	17	333	43	2.12	2	.56	.23	.23	3.93	.33	3	1
63	14	.9	16	310	40	1.98	2	.53	.21	.21	3.67	.31	3	<1
67	27	.66	20	294	48	1.85	2	.87	.27	.24	4.3	.35	5	<1
61	23	.63	20	278	44	1.76	1	.85	.31	.23	4.1	.35	4	<1
72	21	.71	23	332	62	1.8	2	.89	.2	.24	4.37	.37	5	1
52	14	.67	16	230	49	2.44	1	.46	.16	.23	2.8	.26	3	<1
106	16	1.14	25	398	68	3.34	3	.72	.29	.35	5.17	.45	11	<1
108	8	1.29	33	442	73	3.4	3	.91	.46	.4	5.56	.38	3	<1
82	32	1.06	24	476	52	2.33	2	.82	.41	.34	6.91	.37	3	<1
80	29	1.11	25	494	53	2.41	2	.86	.55	.36	7.25	.38	3	<1
123	20	1.82	21	417	99	4.72	3	.61	.29	.35	5.89	.4	5	<1
129	9	2.2	25	458	115	5.62	2	.68	.29	.41	6.71	.46	6	<1
137	53	2.09	27	362	105	5.2	3	.46	.29	.43	6.19	.4	5	0
93	21	2.57	24	433	53	2.57	0	.1	.96	.24	9.53	.53	12	0
133	32	1.23	27	316	90	4.1	0	.04	.45	.34	10.2	.29	16	0

*This value is expressed in retinol equivalents (RE). All other values are in retinol activity equivalents (RAE).

Table A-1

Food Composition (Computer code number is for Wadsworth Diet Analysis program) (For purposes of calculations, use "0" for t, <1, <.1, <.01, etc.)

Computer Code Number	Food Description	Measure	Wt (g)	H_2O (%)	Ener (cal)	Prot (g)	Carb (g)	Dietary Fiber (g)	Fat (g)	Fat Breakdown (g) Sat	Mono	Poly
	MEATS: BEEF, LAMB, PORK and others—Continued											
633	Rib roasted, lean, 2 pieces 4⅛ x 2¼ x ¼"	4 oz	113	60	258	27	0	0	16	6.1	6.1	1.1
634	Liver, panfried	4 oz	113	67	186	24	3	0	8	2.9	1.7	1.2
1096	Venison (deer meat), roasted	4 oz	113	65	179	34	0	0	4	1.4	1	.7
	MEATS: POULTRY and POULTRY PRODUCTS											
	CHICKEN, cooked:											
	Fried, batter dipped:											
635	Breast	1 ea	280	52	728	70	25	1	37	9.9	15.3	8.6
636	Drumstick	1 ea	72	53	193	16	6	<1	11	3	4.6	2.7
637	Thigh	1 ea	86	51	238	19	8	<1	14	3.8	5.8	3.3
638	Wing	1 ea	49	46	159	10	5	<1	11	2.9	4.4	2.5
	Fried, flour coated:											
639	Breast	1 ea	196	57	435	62	3	<1	17	4.8	6.9	3.8
1212	Breast, without skin	1 ea	172	60	322	57	1	0	8	2.2	3	1.8
640	Drumstick	1 ea	49	57	120	13	1	<1	7	1.8	2.7	1.6
641	Thigh	1 ea	62	54	162	17	2	<1	9	2.5	3.6	2.1
1099	Thigh, without skin	1 ea	52	59	113	15	1	0	5	1.4	2	1.3
642	Wing	1 ea	32	49	103	8	1	<1	7	1.9	2.8	1.6
	Roasted:											
643	All types of meat	1 c	140	64	266	40	0	0	10	2.9	3.7	2.4
644	Dark meat	1 c	140	63	287	38	0	0	14	3.7	5	3.2
645	Light meat	1 c	140	65	242	43	0	0	6	1.8	2.2	1.4
646	Breast, without skin	1 ea	172	65	284	53	0	0	6	1.7	2.1	1.3
647	Drumstick, without skin	1 ea	44	67	76	12	0	0	2	.7	.8	.6
1703	Leg, without skin	1 ea	95	65	181	26	0	0	8	2.2	2.9	1.9
648	Thigh	1 ea	62	59	153	15	0	0	10	2.7	3.8	2.1
1100	Thigh, without skin	1 ea	52	63	109	13	0	0	6	1.6	2.2	1.3
649	Stewed, all types	1 c	140	67	248	38	0	0	9	2.6	3.3	2.2
656	Canned, boneless chicken	4 oz	113	69	186	25	0	0	9	2.5	3.6	2
1102	Gizzards, simmered	1 c	145	67	222	39	2	0	5	1.5	1.3	1.5
1101	Hearts, simmered	1 c	145	65	268	38	<1	0	11	3.3	2.9	3.3
2300	Liver, simmered: Ounce	3 oz	85	68	133	21	1	0	5	1.6	1.1	.8
1098	Liver, simmered: Piece = 20g	6 ea	120	68	188	29	1	0	7	2.2	1.6	1.1
	DUCK, roasted:											
1293	Meat with skin, about 2.7 cups	½ ea	382	52	1287	72	0	0	108	36.9	49.3	13.9
651	Meat only, about 1.5 cups	½ ea	221	64	444	52	0	0	25	9.2	8.2	3.2
	GOOSE, domesticated, roasted:											
1294	Meat only, about 4.2 cups	½ ea	591	57	1406	171	0	0	75	26.9	25.6	9.1
1295	Meat with skin, about 5.5 cups	½ ea	774	52	2360	195	0	0	170	53.2	79.3	19.5
	TURKEY:											
	Roasted, meat only:											
652	Dark meat	4 oz	113	63	211	32	0	0	8	2.7	1.8	2.4
653	Light meat	4 oz	113	66	177	34	0	0	4	1.2	.6	1
654	All types, chopped or diced	1 c	140	65	238	41	0	0	7	2.3	1.4	2
1103	Ground, cooked	4 oz	113	59	266	31	0	0	15	3.8	5.5	3.6
1106	Gizzard, cooked	2 ea	134	65	218	39	1	0	5	1.5	1	1.5
1107	Heart, cooked	4 ea	64	64	113	17	1	0	4	1.1	.8	1.1
1108	Liver, cooked	1 ea	75	66	127	18	3	0	4	1.4	1.1	.8
	POULTRY FOOD PRODUCTS (see also items in Sausage & Lunchmeats section):											
1567	Chicken patty, breaded, cooked	1 ea	75	49	213	12	11	<1	13	4.1	6.4	1.6
659	Turkey and gravy, frozen package	3 oz	85	85	57	5	4	0	2	.7	.8	.4
	Turkey breast, Louis Rich:											
1104	Barbecued	2 oz	56	72	57	11	2	0	<1	.2	.2	.1
1943	Hickory smoked	1 pce	80	73	80	16	2	0	1	0		
1947	Honey roasted	1 pce	80	73	80	16	3	0	1	.5		
1945	Oven roasted	1 pce	80		70	16	0	0	1	0		
661	Turkey patty, breaded, fried	2 oz	57	50	161	8	9	<1	10	2.7	4.3	2.7
662	Turkey, frozen, roasted, seasoned	4 oz	113	68	175	24	3	0	7	2.1	1.4	1.9
1704	Turkey roll, light meat	1 pce	28	72	41	5	<1	0	2	.6	.7	.5

Chol (mg)	Calc (mg)	Iron (mg)	Magn (mg)	Pota (mg)	Sodi (mg)	Zinc (mg)	VT-A (µg)	Thia (mg)	VT-E (mg)	Ribo (mg)	Niac (mg)	V-B6 (mg)	Fola (µg)	VT-C (mg)
124	12	1.1	25	333	104	4.62	0	.06	.4	.3	7.89	.28	15	0
634	8	2.96	21	232	60	10.8	9095	.15	.38	2.19	9.58	.55	858	35
127	8	5.05	27	379	61	3.11	0	.2	.28	.68	7.58	.42	5	0
238	56	3.5	67	563	770	2.66	56	.32	2.97	.41	29.5	1.2	42	0
62	12	.97	14	134	194	1.68	19	.08	.88	.15	3.67	.19	13	0
80	15	1.25	18	165	248	1.75	25	.1	1.05	.19	4.91	.22	16	0
39	10	.63	8	68	157	.68	17	.05	.52	.07	2.58	.15	9	0
174	31	2.33	59	508	149	2.16	29	.16	1.12	.26	26.9	1.14	12	0
157	27	1.96	53	475	136	1.86	12	.14	.72	.21	25.4	1.1	7	0
44	6	.66	11	112	44	1.42	12	.04	.41	.11	2.96	.17	5	0
60	9	.92	15	147	55	1.56	18	.06	.52	.15	4.31	.2	7	0
53	7	.76	13	135	49	1.45	11	.05	.3	.13	3.7	.2	5	0
26	5	.4	6	57	25	.56	12	.02	.18	.04	2.14	.13	2	0
125	21	1.69	35	340	120	2.94	22	.1	.36	.25	12.8	.66	8	0
130	21	1.86	32	336	130	3.92	31	.1	.36	.32	9.17	.5	11	0
119	21	1.48	38	346	108	1.72	13	.09	.36	.16	17.4	.84	6	0
146	26	1.79	50	440	127	1.72	10	.12	.45	.2	23.6	1.03	7	0
41	5	.57	11	108	42	1.4	8	.03	.11	.1	2.67	.17	4	0
89	11	1.24	23	230	86	2.72	18	.07	.25	.22	6	.35	8	0
58	7	.83	14	138	52	1.46	30	.04	.16	.13	3.95	.19	4	0
49	6	.68	12	124	46	1.34	10	.04	.13	.12	3.39	.18	4	0
116	20	1.64	29	252	98	2.79	21	.07	.36	.23	8.56	.36	8	0
70	16	1.79	14	156	568	1.59	38	.02	.24	.15	7.15	.4	5	2
281	14	6.02	29	260	97	6.35	81	.04	1.73	.35	5.76	.17	77	2
351	28	13.1	29	191	70	10.6	13	.1	2.32	1.07	4.06	.46	116	3
536	12	7.2	18	119	43	3.69	4176	.13	1.22	1.48	3.78	.49	655	13
757	17	10.2	25	168	61	5.21	5895	.18	1.73	2.1	5.34	.7	924	19
321	42	10.3	61	779	225	7.11	241	.66	2.67	1.03	18.4	.69	23	0
197	26	5.97	44	557	144	5.75	51	.57	1.55	1.04	11.3	.55	22	0
567	83	17	148	2293	449	18.7	71	.54	9.16	2.3	24.1	2.78	71	0
704	101	21.9	170	2546	542	20.3	163	.6	13.5	2.5	32.3	2.86	15	0
96	36	2.63	27	328	89	5.04	0	.07	.72	.28	4.12	.41	10	0
78	21	1.53	32	345	72	2.31	0	.07	.1	.15	7.73	.61	7	0
106	35	2.49	36	417	98	4.34	0	.09	.46	.25	7.62	.64	10	0
115	28	2.18	27	305	121	3.23	0	.06	.38	.19	5.45	.44	8	0
311	20	7.29	25	283	72	5.57	74	.04	.21	.44	4.12	.16	70	2
145	8	4.41	14	117	35	3.37	5	.04	.1	.56	2.08	.2	51	1
470	8	5.85	11	146	48	2.32	2805	.04	2.18	1.07	4.46	.39	500	1
45	12	.94	15	185	399	.78	11	.07	1.46	.1	5.04	.23	8	<1
15	12	.79	7	52	471	.59	6	.02	.3	.11	1.53	.08	3	0
25	14	.61	16	173	592	.58	0	.02		.06	5.28	.22	2	0
35	0	.72			1060		0							0
35	0	.72			940		0							0
35	0				910		0							0
35	8	1.25	9	157	456	.82	6	.06	1.36	.11	1.31	.11	16	0
60	6	1.84	25	337	768	2.87	0	.05	.43	.18	7.09	.3	6	0
12	11	.36	4	70	137	.44	0	.02	.04	.06	1.96	.09	1	0

*This value is expressed in retinol equivalents (RE). All other values are in retinol activity equivalents (RAE).

Table A-1

Food Composition (Computer code number is for Wadsworth Diet Analysis program) (For purposes of calculations, use "0" for t, <1, <.1, <.01, etc.)

Computer Code Number	Food Description	Measure	Wt (g)	H_2O (%)	Ener (cal)	Prot (g)	Carb (g)	Dietary Fiber (g)	Fat (g)	Fat Breakdown (g) Sat	Mono	Poly
	MEATS: SAUSAGES and LUNCHMEATS (see also Poultry Food Products)											
1072	Beerwurst/beer salami, beef	1 oz	28	53	92	3	<1	0	8	3.6	3.9	.3
1074	Beerwurst/beer salami, pork	1 oz	28	61	67	4	1	0	5	1.8	2.5	.7
1075	Berliner sausage	1 oz	28	61	64	4	1	0	5	1.7	2.2	.4
	Bologna:											
1297	Beef	1 pce	23	55	72	3	<1	0	7	2.8	3.2	.3
2115	Beef, light, Oscar Mayer	1 pce	28	65	55	3	2	0	4	1.6	2	.1
663	Beef & pork	1 pce	28	54	88	3	1	0	8	3	3.7	.7
2155	Healthy Favorites	1 pce	23		22	3	1	0	<1	0		
1298	Pork	1 pce	23	61	57	4	<1	0	5	1.6	2.2	.5
2114	Regular, light, Oscar Mayer	1 pce	28	65	56	3	2	0	4	1.6	2	.4
664	Turkey	1 pce	28	65	56	4	<1	0	4	1.4	1.3	1.2
1970	Turkey, Louis Rich	1 pce	56	67	115	6	1	0	10	2.9	3.6	2.6
665	Braunschweiger sausage	2 pce	57	48	205	8	2	0	18	6.2	8.5	2.1
1073	Bratwurst-link	1 ea	70	51	226	10	2	0	19	6.9	9.3	2
666	Brown & serve sausage links, cooked	2 ea	26	45	102	4	1	0	10	3.4	4.4	1
1089	Cheesefurter/cheese smokie	2 ea	86	52	281	12	1	0	25	9	11.7	2.6
2157	Chicken breast, Healthy Favorites	4 pce	52		40	9	1	0	0	0	0	0
1556	Chorizo, pork & beef	1 ea	60	32	273	14	1	0	23	8.6	11	2.1
1090	Corned beef loaf, jellied	1 pce	28	69	43	6	0	0	2	.7	.7	.1
	Frankfurters:											
1077	Beef, large link, 8/package	1 ea	57	55	180	7	1	0	16	6.9	7.8	.8
1078	Beef and pork, large link, 8/package	1 ea	45	54	144	5	1	0	13	4.8	6.1	1.2
667	Beef and pork, small link, 10/pkg	1 ea	45	54	144	5	1	0	13	4.8	6.1	1.2
668	Turkey frankfurter, 10/package	1 ea	45	63	102	6	1	0	8	2.6	2.5	2.2
1968	Turkey/chicken frank 8/pkg	1 ea	43	67	81	5	2	0	6	1.6	2.4	1.4
	Ham:											
669	Ham lunchmeat, canned, 3 x 2 x ½"	1 pce	21	52	70	3	<1	0	6	2.3	3	.7
670	Chopped ham, packaged	2 pce	42	64	96	7	0	0	7	2.4	3.4	.9
2156	Honey ham, Healthy Favorites	4 pce	52	73	55	9	2	0	1	.4	.8	.1
2113	Oscar Mayer lower sodium ham	1 pce	21	73	23	3	1	0	1	.3	.4	.1
673	Turkey ham lunchmeat	2 pce	57	71	73	11	<1	0	3	1	.7	.9
1091	Kielbasa sausage	1 pce	26	54	81	3	1	0	7	2.6	3.4	.8
1092	Knockwurst sausage, link	1 ea	68	55	209	8	1	0	19	6.9	8.7	2
1093	Mortadella lunchmeat	2 pce	30	52	93	5	1	0	8	2.8	3.4	.9
1097	Olive loaf lunchmeat	2 pce	57	58	134	7	5	0	9	3.3	4.5	1.1
1952	Turkey breast, fat free	1 pce	28	76	23	4	1	0	<1	.1	.1	t
1080	Turkey pastrami	2 pce	57	71	80	10	1	0	4	1	1.2	.9
1969	Turkey salami	1 pce	28	72	41	4	<1	0	3	.8	.9	.7
1081	Pepperoni sausage	2 pce	11	27	55	2	<1	0	5	1.8	2.3	.5
1094	Pickle & pimento loaf	2 pce	57	57	149	7	3	0	12	4.5	5.5	1.5
1082	Polish sausage	1 oz	28	53	91	4	<1	0	8	2.9	3.8	.9
674	Pork sausage, cooked, link, small	2 ea	26	45	96	5	<1	0	8	2.8	4.1	.8
1079	Pork sausage, cooked, patty	4 oz	113	45	417	22	1	0	35	12.1	17.7	3.3
675	Salami, pork and beef	2 pce	57	60	143	8	1	0	11	4.6	5.2	1.1
677	Salami, pork and beef, dry	3 pce	30	35	125	7	1	0	10	3.7	5.1	1
676	Salami, turkey	2 pce	57	66	112	9	<1	0	8	2.3	2.6	2
	Sandwich spreads:											
1300	Ham salad spread	2 tbsp	30	63	65	3	3	0	5	1.5	2.2	.8
678	Pork and beef	2 tbsp	30	60	70	2	4	<1	5	1.8	2.3	.8
1296	Chicken/turkey	2 tbsp	26	66	52	3	2	0	4	.9	.8	1.6
1084	Smoked link sausage, beef and pork	1 ea	68	52	228	9	1	0	21	7.2	9.6	2.2
1083	Smoked link sausage, pork	1 ea	68	39	265	15	1	0	22	7.7	10	2.6
1085	Summer sausage	2 pce	46	51	154	7	<1	0	14	5.5	6	.6
1076	Turkey breakfast sausage	1 pce	28	60	64	6	0	0	5	1.6	1.8	1.2
679	Vienna sausage, canned	2 ea	32	60	89	3	1	0	8	3	4	.5
	MIXED DISHES and FAST FOODS											
	MIXED DISHES:											
1445	Almond Chicken	1 c	242	77	280	22	16	3	14	1.9	6.1	5.6
1981	Baked beans, fat free, honey	½ c	120	73	110	7	24	7	0	0	0	0
1454	Bean cake	1 ea	32	23	130	2	16	1	7	1	2.9	2.6
680	Beef stew w/vegetables, homemade	1 c	245	82	218	16	15	2	10	4.9	4.5	.5
1109	Beef stew w/vegetables, canned	1 c	245	82	194	14	17	2	8	2.4	3.1	.3

Chol (mg)	Calc (mg)	Iron (mg)	Magn (mg)	Pota (mg)	Sodi (mg)	Zinc (mg)	VT-A (µg)	Thia (mg)	VT-E (mg)	Ribo (mg)	Niac (mg)	V-B6 (mg)	Fola (µg)	VT-C (mg)
17	3	.42	3	49	288	.68	0	.02	.05	.03	.95	.05	1	0
16	2	.21	4	71	347	.48	0	.15	.06	.05	.91	.1	1	0
13	3	.32	4	79	363	.69	0	.11	.06	.06	.87	.06	1	0
13	3	.38	3	36	226	.5	0	.01	.04	.02	.55	.03	1	0
13	4	.34	4	44	314	.53	0							0
15	3	.42	3	50	285	.54	0	.05	.06	.04	.72	.05	1	0
7		.18			255									
14	3	.18	3	65	272	.47	0	.12	.06	.04	.9	.06	1	0
15	14	.39	6	46	312	.45	0							0
28	23	.43	4	56	246	.49	0	.01	.15	.05	.99	.06	2	0
44	68	.9	10	103	484	1.14	0	.03		.1	2.15	.1		0
89	5	5.34	6	113	652	1.6	2405	.14	.2	.87	4.77	.19	25	0
44	34	.72	11	197	778	1.47	0	.17	.19	.16	2.31	.09	3	0
16	2	.62	4	70	248	.3	0*	.21	.06	.09	.96	.06	1	0
58	50	.93	11	177	931	1.94	33*	.21	.27	.14	2.49	.11	3	0
25		.72			620									
53	5	.95	11	239	741	2.05	0	.38	.13	.18	3.08	.32	1	0
13	3	.57	3	28	267	1.15	0	0	.05	.03	.49	.03	2	0
35	11	.81	2	95	585	1.24	0	.03	.11	.06	1.38	.07	2	0
22	5	.52	4	75	504	.83	0	.09	.11	.05	1.19	.06	2	0
22	5	.52	4	75	504	.83	0	.09	.11	.05	1.19	.06	2	0
48	48	.83	6	81	642	1.4	0	.02	.28	.08	1.86	.1	4	0
40	56	.94	10	69	488	.8	0							0
13	1	.15	2	45	271	.31	0*	.08	.05	.04	.66	.04	1	<1
21	3	.35	7	134	576	.81	0	.26	.11	.09	1.63	.15	<1	0
24	6	.7	18	144	635	1.02	0							0
9	1	.3	5	197	174	.42	0							0
32	6	1.57	9	185	568	1.68	0	.03	.36	.14	2.01	.14	3	0
17	11	.38	4	70	280	.52	0	.06	.06	.06	.75	.05	1	0
39	7	.62	7	135	687	1.13	0	.23	.39	.09	1.86	.12	1	0
17	5	.42	3	49	374	.63	0	.04	.07	.05	.8	.04	1	0
22	62	.31	11	169	846	.79	11*	.17	.14	.15	1.05	.13	1	0
9	3	.31	8	57	334	.24	0							0
31	5	.95	8	148	596	1.23	0	.03	.12	.14	2.01	.15	3	0
21	11	.35	6	60	281	.65	0							0
9	1	.15	2	38	224	.27	0	.03	.02	.03	.55	.03	<1	0
21	54	.58	10	194	792	.8	2	.17	.14	.14	1.17	.11	3	0
20	3	.4	4	66	245	.54	0	.14	.06	.04	.96	.05	1	<1
22	8	.33	4	94	336	.65	0*	.19	.07	.07	1.18	.09	1	<1
94	36	1.42	19	408	1462	2.84	0*	.84	.29	.29	5.11	.37	2	2
37	7	1.52	9	113	607	1.22	0	.14	.12	.21	2.03	.12	1	0
24	2	.45	5	113	558	.97	0	.18	.08	.09	1.46	.15	1	0
47	11	.92	9	139	572	1.03	0	.04	.32	.1	2.01	.14	2	0
11	2	.18	3	45	274	.33	0	.13	.52	.04	.63	.04	<1	0
11	4	.24	2	33	304	.31	3*	.05	.52	.04	.52	.04	1	0
8	3	.16	3	48	98	.27	11*	.01	.57	.02	.43	.03	1	<1
48	7	.99	8	129	643	1.43	0	.18	.15	.12	2.19	.12	1	0
46	20	.79	13	228	1020	1.92	0	.48	.17	.17	3.08	.24	3	1
34	6	1.17	6	125	571	1.18	0	.07	.1	.15	1.98	.12	1	0
23	5	.51	6	75	188	.96	0	.03	.14	.08	1.4	.08	1	0
17	3	.28	2	32	305	.51	0	.03	.07	.03	.52	.04	1	0
40	69	1.97	60	549	526	1.62	37*	.08	3.8	.2	9.48	.44	26	7
0	40	2.7			135		225							12
0	3	.67	6	58	1	.16	0	.07	1.24	.05	.55	.02	9	0
64	29	2.94	40	613	292	5.29	568*	.15	.49	.17	4.66	.28	37	17
34	29	2.21	39	426	1006	4.24	262*	.07	.34	.12	2.45	.2	31	7

*This value is expressed in retinol equivalents (RE). All other values are in retinol activity equivalents (RAE).

Table A-1

Food Composition (Computer code number is for Wadsworth Diet Analysis program) (For purposes of calculations, use "0" for t, <1, <.1, <.01, etc.)

Computer Code Number	Food Description	Measure	Wt (g)	H_2O (%)	Ener (cal)	Prot (g)	Carb (g)	Dietary Fiber (g)	Fat (g)	Fat Breakdown (g) Sat	Mono	Poly
	MIXED DISHES and FAST FOODS MIXED DISHES:—Continued											
1116	Beef, macaroni, tomato sauce casserole	1 c	226	76	255	16	26	2	10			
2295	Beef fajita	1 ea	223	65	399	23	36	3	18	5.5	7.6	3.5
1265	Beef flauta	1 ea	113	51	354	14	13	2	28	4.8	11.8	9.4
681	Beef pot pie, homemade	1 pce	210	55	517	21	39	3	30	8.4	14.7	7.3
1898	Broccoli, batter fried	1 c	85	74	122	3	9	2	9	1.3	2.1	4.9
1462	Buffalo wings/spicy chicken wings	2 pce	32	53	98	8	<1	<1	7	1.8	2.6	1.8
1675	Carrot raisin salad	½ c	88	58	203	1	21	2	14	2.1		
2248	Cheeseburger deluxe	1 ea	219	52	563	28	38		33	15	12.6	2
682	Chicken a la king, homemade	1 c	245	68	468	27	12	1	34	12.7	14.3	6.2
683	Chicken & noodles, homemade	1 c	240	71	367	22	26	2	18	5.9	7.1	3.5
684	Chicken chow mein, canned	1 c	250	89	95	6	18	2	1	0	.1	.8
685	Chicken chow mein, homemade	1 c	250	78	255	31	10	1	10	2.4	4.3	3.1
1266	Chicken fajita	1 ea	223	65	363	20	44	5	12	2.2	5.5	3.1
1264	Chicken flauta	1 ea	113	55	330	13	12	2	26	4.2	10.7	9.2
686	Chicken pot pie, homemade (⅓)	1 pce	232	57	545	23	42	3	31	10.9	14.5	5.8
1672	Chili con carne	½ c	127	77	128	12	11	2	4	1.7	1.7	.3
1112	Chicken salad with celery	½ c	78	53	268	11	1	<1	25	3.1		
1382	Chicken teriyaki, breast	1 ea	128	67	178	27	7	<1	4	.9	1.1	.9
687	Chili with beans, canned	1 c	256	76	287	15	30	11	14	6	6	.9
1479	Chinese Pastry	1 oz	28	46	67	1	13	<1	2	.2	.5	.8
688	Chop suey with beef & pork	1 c	220	63	421	22	31	4	24	4.7	8.3	9.2
690	Coleslaw	1 c	132	74	195	2	17	2	15	2.1	3.2	8.5
689	Corn pudding	1 c	250	76	273	11	32	4	13	6.3	4.3	1.7
1110	Corned beef hash, canned	1 c	220	67	398	19	23	1	25	11.9	10.9	.9
1255	Deviled egg (½ egg + filling)	1 ea	31	70	63	4	<1	0	5	1.2	1.7	1.5
	Egg Foo Yung Patty:											
1467	Meatless	1 ea	86	77	113	6	3	1	8	2	3.4	2.1
1458	With beef	1 ea	86	76	119	8	3	<1	8	2	2.9	2.2
1465	With chicken	1 ea	86	76	121	8	4	<1	8	1.9	2.8	2.3
1602	Egg roll, meatless	1 ea	64	70	101	3	10	1	6	1.2	2.9	1.3
1550	Egg roll, with meat	1 ea	64	66	113	5	9	1	6	1.4	3	1.3
1113	Egg salad	1 c	183	57	584	17	3	0	56	10.5	17.4	23.9
56102	Falafel	1 ea	17	35	57	2	5	1	3	.4	1.7	.7
	French toast w/wheat bread,											
691	homemade	1 pce	65	54	151	5	16	<1	7	2	3	1.7
1355	Green Pepper, stuffed	1 ea	172	75	229	12	18	1	12	5.3		
1487	Hot & Sour Soup (Chinese)	1 c	244	87	162	15	5	1	8	2.7	3.4	1.2
2242	Hamburger deluxe	1 ea	110	49	279	13	27		13	4.1	5.3	2.6
1997	Hummous/hummus	¼ c	62	65	106	3	12	3	5	.8	2.2	2
16335	Kung Pao Chicken	1 c	162	54	431	29	11	2	31	5.2	13.9	9.7
	Lasagna:											
1346	With meat, homemade	1 pce	245	67	392	23	40	3	16	8	5.2	.8
1111	Without meat, homemade	1 pce	218	69	306	16	40	3	10	5.6	2.5	.6
1117	Frozen entree	1 ea	340	75	389	24	41	4	14	6.7	5.5	.8
1606	Lo mein, meatless	1 c	200	82	135	6	27	4	1	.1	.1	.3
1607	Lo mein, with meat	1 c	200	72	283	20	21	3	14	2.6	3.9	6
692	Macaroni & cheese, canned	1 c	240	80	228	9	26	1	10	4.2	3.1	1.4
693	Macaroni & cheese, homemade	1 c	200	70	302	15	30	1	14	8.5		
1115	Macaroni salad, no cheese	1 c	177	60	460	5	28	2	37	4		
1120	Meat loaf, beef	1 pce	87	63	182	16	4	<1	11	4.3		
1119	Meat loaf, beef and pork (⅓)	1 pce	87	60	210	14	5	<1	15	5.5		
1303	Moussaka (lamb & eggplant)	1 c	250	82	238	16	13	4	13	4.5		
1899	Mushrooms, batter fried	5 ea	70	63	155	2	11	1	12	1.5	3.6	6
715	Potato salad with mayonnaise											
	and eggs	½ c	125	76	179	3	14	2	10	1.8	3.1	4.7
1674	Pizza, combination, ¹⁄₁₂ of 12″ round	1 pce	79	48	184	13	21		5	1.5	2.5	.9
1673	Pizza, pepperoni, ¹⁄₁₂ of 12″ round	1 pce	71	47	181	10	20		7	2.2	3.1	1.2
694	Quiche Lorraine ⅛ of 8″ quiche	1 pce	176	53	526	15	25	1	41	18.8	14.3	5.2
1449	Ramen noodles-cooked	1 c	227	86	153	3	20	1	6	1.6	1.2	3.3
1671	Ravioli, meat	½ c	125	69	197	11	18	1	9	3	3.7	1
1597	Fried rice (meatless)	1 c	166	68	271	5	34	1	12	1.8	3.2	6.7
2142	Roast beef hash	½ c	117	66	230	9	11	1	16	7	5.8	3.2
	Spaghetti (enriched) in tomato sauce											
	With cheese:											

Chol (mg)	Calc (mg)	Iron (mg)	Magn (mg)	Pota (mg)	Sodi (mg)	Zinc (mg)	VT-A (µg)	Thia (mg)	VT-E (mg)	Ribo (mg)	Niac (mg)	V-B6 (mg)	Fola (µg)	VT-C (mg)
39	26	2.69	40	522	882	3.13	49	.22	1.55	.21	4.31	.29	59	14
45	84	3.76	38	479	316	3.52	21	.39	1.74	.3	5.4	.38	23	27
37	51	1.87	28	313	68	3.45	21*	.06	4.65	.13	1.88	.23	10	19
44	29	3.78	6	334	596	3.17	519*	.29	3.78	.29	4.83	.24	29	6
15	66	.98	20	242	64	.38	98*	.08	2.85	.13	.73	.11	36	53
26	5	.4	6	59	25	.57	17*	.01	.27	.04	2.06	.13	1	<1
10	26	.74	14	317	118	.18	2905*	.08	2.4	.05	.64	.22	9	5
88	206	4.66	44	445	1108	4.6	129*	.39	1.18	.46	7.38	.28	81	8
186	127	2.45	20	404	760	1.8	272*	.1	.98	.42	5.39	.23	11	12
96	26	2.16	26	149	600	1.53	10*	.05		.17	4.32	.19	10	0
7	45	1.25	14	418	725	1.3	28*	.05	.05	.1	1	.09	12	12
77	57	2.5	28	473	718	2.12	50*	.07	.75	.22	4.25	.41	19	10
39	101	3.32	48	533	343	1.65	65*	.43	1.71	.33	6.12	.38	42	37
35	50	.95	27	268	71	1.13	26*	.05	4.36	.09	3.1	.22	8	18
72	70	3.02	25	343	594	2	735*	.32	3.25	.32	4.87	.46	29	5
67	34	2.6	23	347	505	1.79	42	.06	.81	.57	1.24	.16	23	1
48	16	.62	11	138	201	.79	31*	.03	6.27	.07	3.28	.34	8	1
82	27	1.71	35	309	1683	1.96	16*	.08	.35	.19	8.75	.47	12	3
43	120	8.78	115	934	1336	5.12	43	.12	1.89	.27	.92	.34	59	4
0	6	.18	7	25	3	.16	<1	.02	.26	<.01	.27	.04	1	0
43	39	4.19	54	519	950	3.48	103*	.36	1.8	.36	5.73	.39	44	20
7	45	.96	12	236	356	.26	66*	.05	5.28	.04	.11	.14	51	11
250	100	1.4	37	403	138	1.25	90*	1.03	.52	.32	2.47	.29	62	7
73	29	4.4	36	440	1188	3.3	0*	.02	.48	.2	4.62	.43	20	0
122	15	.35	3	37	50	.3	50	.02	.6	.15	.02	.05	13	0
185	31	1.04	12	117	317	.7	86*	.04	1.22	.26	.43	.09	29	5
166	25	1.01	11	139	131	1.01	86*	.05	1.06	.23	.65	.13	22	3
167	27	.82	11	136	132	.76	87*	.05	1.1	.23	.89	.12	22	3
30	14	.81	9	97	274	.25	16*	.08	.85	.11	.8	.05	13	3
37	15	.83	10	124	273	.46	16*	.16	.8	.12	1.28	.09	10	2
581	74	1.81	13	181	464	1.45	262	.09	7.66	.66	.09	.46	61	0
0	9	.58	14	99	50	.25	<1	.02	.19	.03	.18	.02	16	<1
76	64	1.09	11	86	311	.44	81*	.13	.31	.21	1.06	.05	15	<1
37	16	1.73	20	245	233	2.38	23	.14	.78	.11	2.75	.31	42	59
34	29	1.9	29	384	1010	1.51	2*	.27	.15	.25	5	.2	13	1
26	63	2.63	22	227	504	2.06	4	.23	.82	.2	3.69	.12	52	2
0	31	.97	18	108	151	.68	1	.06	.62	.03	.25	.25	37	5
64	49	1.96	63	428	907	1.5	58*	.15	3.9	.15	13.2	.59	43	8
58	270	3.07	50	460	391	3.33	158*	.24	1.16	.34	4.2	.25	20	14
32	265	2.35	44	373	364	1.82	157*	.23	1.09	.28	2.51	.17	17	14
55	265	3.82	64	759	839	3.7	360*	.29	3.45	.39	5.07	.32	28	12
0	46	2.03	33	386	564	.92	65	.23	.35	.24	2.82	.19	48	12
42	29	2.07	42	332	142	1.83	8*	.41	2.06	.28	5.02	.36	53	11
24	199	.96	31	139	730	1.2	73*	.12	.14	.24	.96	.02	8	<1
41	257	.82	41	204	493	1.36	153*	.28	.7	.35	1.57	.11	74	1
27	30	1.55	20	155	360	.54	37*	.18	10.3	.1	1.45	.33	70	3
84	29	1.6	14	187	145	3.22	23*	.05	.32	.22	2.62	.15	14	1
82	29	1.39	13	181	289	2.56	23*	.11	.33	.2	2.46	.12	13	1
96	75	1.74	40	565	460	2.57	109*	.16	.98	.31	4.14	.23	46	6
2	15	1.22	7	154	112	.42	6*	.11	2.34	.26	2.25	.04	8	1
85	24	.81	19	318	661	.39	41*	.1	2.33	.07	1.11	.18	9	12
20	101	1.53	18	179	382	1.11	101*	.21		.17	1.96	.09	32	2
14	65	.94	9	153	267	.52	55*	.13		.23	3.05	.06	37	2
221	231	1.88	24	239	221	1.5	279*	.26	2.02	.49	2.01	.1	19	1
<1	13	.39	10	49	802	.18	2	.02	2.34	.01	.25	.01	3	<1
85	35	2.15	20	264	90	1.7	87*	.16	1.32	.22	2.99	.14	14	4
43	28	1.94	23	128	261	.92	20*	.21	2.51	.11	2.24	.15	22	4
40	10	.9	22	362	695	2.99	0*	.09		.12	2.33	.3	12	0

*This value is expressed in retinol equivalents (RE). All other values are in retinol activity equivalents (RAE).

Table A-1

Food Composition

(Computer code number is for Wadsworth Diet Analysis program) (For purposes of calculations, use "0" for t, <1, <.1, <.01, etc.)

Computer Code Number	Food Description	Measure	Wt (g)	H_2O (%)	Ener (cal)	Prot (g)	Carb (g)	Dietary Fiber (g)	Fat (g)	Fat Breakdown (g) Sat	Mono	Poly
	MIXED DISHES and FAST FOODS MIXED DISHES:—Continued											
695	Canned	1 c	250	80	190	5	38	2	1	0	.4	.5
696	Home recipe	1 c	250	77	260	9	37	2	9	2		
	With meatballs:											
697	Canned	1 c	250	78	258	12	28	6	10	2.1	3.9	3.9
698	Home recipe	1 c	248	71	370	19	29	3	18	5		
716	Spinach souffle	1 c	136	74	219	11	3	3	18	7.1	6.8	3.1
2995	Spring roll, vegetable	1 ea	63	50	157	4	20	1	7	.9		
56313	Sushi, fish and vegetable	1 c	166	65	232	9	47	2	1	.2	.2	.2
56314	Sushi, vegetable seaweed	1 c	166	71	194	4	43	1	<1	.1	.1	.1
1553	Sweet & sour pork	1 c	226	77	231	15	25	2	8	2.1	3.1	2.3
1263	Sweet & sour chicken breast	1 ea	131	79	118	8	15	1	3	.5	.9	1.4
56916	Tabouli	1 c	160	77	199	3	16	4	15	2	10.8	1.4
1994	Thai lemongrass vegetables, svg	1 ea	187	75	238	8	19	4	16	2.8		
2426	Thai peanut chicken, svg	1 ea	309	81	271	19	26	3	10	2		
1515	Three bean salad	1 c	150	81	140	4	15	5	8	1.1	1.7	4.4
717	Tuna salad	1 c	205	63	383	33	19	0	19	3.2	5.9	8.4
1121	Tuna noodle casserole, homemade	1 c	202	75	238	17	25	1	7			
1270	Waldorf salad	1 c	137	58	411	4	12	3	41	4.2		
56111	Wonton, meat filled	1 ea	19	45	55	3	5	<1	3	.8	1.2	.3
	FAST FOODS and SANDWICHES (see end of this appendix for additional Fast Foods)											
699	Burrito, beef & bean	1 ea	116	52	255	11	33	3	9	4.2	3.5	.6
700	Burrito, bean	1 ea	109	53	225	7	36	4	7	3.5	2.4	.6
2106	Burrito, chicken con queso	1 ea	299	73	350	14	60	6	6	2.5		
701	Cheeseburger with bun, regular	1 ea	154	55	359	18	28		20	9.2	7.2	1.5
702	Cheeseburger with bun, 4-oz patty	1 ea	166	51	417	21	35		21	8.7	7.8	2.7
703	Chicken patty sandwich	1 ea	182	47	515	24	39	1	29	8.5	10.4	8.4
704	Corndog	1 ea	175	47	460	17	56		19	5.2	9.1	3.5
1922	Corndog, chicken	1 ea	113	52	271	13	26		13			
705	Enchilada	1 ea	163	63	319	10	28		19	10.6	6.3	.8
706	English muffin with egg, cheese, bacon	1 ea	146	57	308	18	28	2	13	5	5	1.7
	Fish sandwich:											
707	Regular, with cheese	1 ea	183	45	523	21	48	<1	29	8.1	8.9	9.4
708	Large, no cheese	1 ea	158	47	431	17	41	<1	23	5.2	7.7	8.2
709	Hamburger with bun, regular	1 ea	107	45	275	12	35	2	10	3.6	3.4	1
710	Hamburger with bun, 4-oz patty	1 ea	215	51	576	32	39		32	12	14.1	2.8
711	Hotdog/frankfurter with bun	1 ea	98	54	242	10	18		14	5.1	6.8	1.7
	Lunchables:											
2129	Bologna & American cheese	1 ea	128		450	18	19	0	34	15		
2130	Ham & cheese	1 ea	128		320	22	19	0	17	8		
2117	Honey ham & Amer. w/choc pudding	1 ea	176		390	18	34	<1	20	9		
2118	Honey turkey & cheddar w/Jello	1 ea	163		320	17	27	<1	16	9		
2131	Pepperoni & American cheese	1 ea	128		480	20	19	0	36	17		
2125	Salami & American cheese	1 ea	128		430	18	18	0	32	15		
2127	Turkey & cheddar cheese	1 ea	128		360	20	20	1	22	11		
712	Pizza, cheese, ⅛ of 15" round	1 pce	63	48	140	8	20	1	3	1.5	1	.5
	SANDWICHES:											
	Avocado, chesse, tomato & lettuce:											
1276	On white bread, firm	1 ea	210	62	429	14	35	5	27	7.7		
1278	On part whole wheat	1 ea	201	63	402	14	30	6	27	7.8		
1277	On whole wheat	1 ea	214	63	424	15	33	8	28	8.2		
	Bacon, lettuce & tomato sandwich:											
1137	On white bread, soft	1 ea	124	52	318	10	29	2	17	4.1		
1139	On part whole wheat	1 ea	124	53	314	11	26	3	19	4.6		
1138	On whole wheat	1 ea	137	52	339	12	29	5	20	4.9		
	Cheese, grilled:											
1140	On white bread, soft	1 ea	119	37	399	17	30	1	23	11.9		
1142	On part whole wheat	1 ea	119	37	402	18	26	2	25	13.1		
1141	On whole wheat	1 ea	132	38	432	20	30	4	27	13.8		
1596	Chicken fillet	1 ea	182	47	515	24	39	1	29	8.5	10.4	8.4

Chol (mg)	Calc (mg)	Iron (mg)	Magn (mg)	Pota (mg)	Sodi (mg)	Zinc (mg)	VT-A (µg)	Thia (mg)	VT-E (mg)	Ribo (mg)	Niac (mg)	V-B6 (mg)	Fola (µg)	VT-C (mg)
7	40	2.75	21	303	955	1.12	120*	.35	2.13	.27	4.5	.13	6	10
7	80	2.25	26	408	955	1.3	215*	.25	2.75	.17	2.25	.2	8	12
22	52	3.25	20	245	1220	2.39	100*	.15	1.5	.17	2.25	.12	5	5
66	99	3.38	44	469	1108	3.5	90*	.26	2.32	.31	4.47	.28	67	14
184	230	1.35	38	201	763	1.29	675*	.09	1.22	.3	.48	.12	80	3
3	58	1.77	12	75	261	.38	70*	.18	1.29	.14	1.87	.03	31	1
11	25	2.33	27	218	93	.84	136*	.28	.62	.07	2.96	.16	15	4
0	21	1.65	21	106	5	.75	33	.21	.13	.04	1.99	.15	11	3
38	28	1.44	34	386	838	1.47	31*	.55	1.09	.21	3.63	.41	10	20
23	15	.84	21	185	506	.67	22*	.06	.67	.08	3.09	.18	6	12
0	29	1.25	36	245	799	.48	34	.07	2.16	.05	1.14	.11	31	28
0	144	2.85	48	457	724	1.08	523	.18	2.27	.12	1.45	.29	47	90
36	43	2.69	49	374	882	1.24	183*	.22	1.59	.12	8.24	.48	72	68
0	35	1.48	27	246	520	.58	19*	.08	1.74	.1	.44	.04	56	4
27	35	2.05	39	365	824	1.15	55*	.06	1.95	.14	13.7	.17	16	5
41	34	2.3	31	182	686	1.21	8*	.18	1.07	.15	7.8	.2	55	1
21	45	.97	37	258	234	.7	38*	.09	8.6	.05	.53	.36	33	5
20	4	.4	4	51	10	.32	6*	.09	.18	.06	.63	.04	3	<1
24	53	2.46	42	329	670	1.93	16	.27	.7	.42	2.71	.19	58	1
2	57	2.27	44	328	495	.76	8	.32	.87	.3	2.04	.15	44	1
35	40	1.8			590		300*							6
52	182	2.65	26	229	976	2.62	71*	.32	1.34	.23	6.38	.15	65	2
60	171	3.42	30	335	1050	3.49	65*	.35		.28	8.05	.18	61	2
60	60	4.68	35	353	957	1.87	31	.33	.55	.24	6.81	.2	100	9
79	102	6.18	17	263	973	1.31	18	.28	.7	.7	4.17	.09	103	0
64					668									
44	324	1.32	50	240	784	2.51	186*	.08	1.47	.42	1.91	.39	65	1
250	161	2.6	25	212	777	1.66	166*	.53	.9	.48	3.55	.16	73	2
68	185	3.5	37	353	939	1.17	97*	.46	1.83	.42	4.23	.11	91	3
55	84	2.61	33	340	615	.99	30	.33	.87	.22	3.4	.11	85	3
30	127	2.74	23	254	539	2.27	5	.29	.01	.24	3.95	.12	52	2
103	92	5.55	45	527	742	5.81	2	.34	1.61	.41	6.73	.37	84	1
44	23	2.31	13	143	670	1.98	0	.23	.27	.27	3.65	.05	48	<1
85	300	2.7			1620		60*							0
60	300	1.8			1770		80*							
55	250	2.7			1540		40*							
50	20	6			1360		80*							
95	250	2.7			1840		60*							
80	250	2.7			1740		60*							
70	300	1.8			1650		60*							
9	117	.58	16	110	336	.81	74*	.18		.16	2.48	.04	35	1
29	283	6.1	48	548	507	1.45	175*	.37	3.15	.43	3.76	.3	102	12
29	272	5.96	56	576	454	1.59	175*	.3	3.19	.37	3.47	.31	85	12
30	270	6.36	83	636	499	2.21	181*	.3	3.51	.36	3.67	.37	77	13
20	68	2.21	22	240	632	.98	49*	.41	2.32	.26	3.7	.16	66	6
22	61	2.22	32	288	625	1.21	53*	.36	2.55	.21	3.68	.18	53	6
23	51	2.54	61	346	690	1.87	56*	.37	2.91	.2	3.97	.25	45	7
53	407	2	27	162	1154	2.03	167*	.29	1.01	.4	2.37	.08	60	<1
57	430	2	38	208	1196	2.37	182*	.24	1.15	.36	2.25	.09	46	<1
60	439	2.33	68	264	1293	3.13	192*	.24	1.44	.35	2.46	.16	36	<1
60	60	4.68	35	353	957	1.87	31	.33	.55	.24	6.81	.2	100	9

*This value is expressed in retinol equivalents (RE). All other values are in retinol activity equivalents (RAE).

Table A-1

Food Composition (Computer code number is for Wadsworth Diet Analysis program) (For purposes of calculations, use "0" for t, <1, <.1, <.01, etc.)

Computer Code Number	Food Description	Measure	Wt (g)	H_2O (%)	Ener (cal)	Prot (g)	Carb (g)	Dietary Fiber (g)	Fat (g)	Fat Breakdown (g) Sat	Mono	Poly
	MIXED DISHES and FAST FOODS MIXED DISHES:—Continued											
	Chicken salad:											
1143	On white bread, soft	1 ea	110	41	366	10	31	2	22	2.7		
1145	On part whole wheat	1 ea	110	41	369	11	27	3	24	3.1		
1144	On whole wheat	1 ea	123	41	398	13	31	5	26	3.4		
1146	Corned beef & swiss on rye	1 ea	156	47	427	28	22	6	26	9.5		
	Egg salad:											
1147	On white bread, soft	1 ea	117	43	379	9	31	1	24	3.8		
1149	On part whole wheat	1 ea	116	44	378	10	27	2	26	4.3		
1148	On whole wheat	1 ea	130	44	410	11	31	5	28	4.6		
	Ham:											
1279	On rye bread	1 ea	150	57	312	24	20	6	16	3.3		
1151	On white bread, soft	1 ea	157	52	364	24	30	2	16	3.3		
1153	On part whole wheat	1 ea	156	54	354	25	26	2	17	3.6		
1152	On whole wheat	1 ea	169	53	378	27	29	4	18	3.8		
	Ham & cheese:											
1280	On white bread, soft	1 ea	157	48	423	24	31	2	22	8.1		
1282	On part whole wheat	1 ea	156	49	417	25	26	2	24	8.7		
1281	On whole wheat	1 ea	170	48	445	27	30	4	25	9.2		
1150	Ham & swiss on rye	1 ea	150	53	359	24	21	6	21	7.1		
	Ham salad:											
1154	On white bread, soft	1 ea	131	47	361	10	37	1	19	4.2		
1156	On part whole wheat	1 ea	131	48	358	11	33	2	20	4.7		
1155	On whole wheat	1 ea	144	48	383	12	37	4	22	5		
1157	Patty melt: Ground beef & cheese on rye	1 ea	182	46	561	37	22	6	37	12.7		
	Peanut butter & jelly:											
1158	On white bread, soft	1 ea	101	27	348	11	47	3	14	2.7		
1160	On part whole wheat	1 ea	101	27	352	11	45	4	15	3.1		
1159	On whole wheat	1 ea	114	27	383	13	50	6	17	3.4		
1161	Reuben, grilled: Corned beef, swiss cheese, sauerkraut on rye	1 ea	239	60	555	22	27	5	40	17.3		
	Roast beef:											
713	On a bun	1 ea	139	49	346	21	33		14	3.6	6.8	1.7
1162	On white bread, soft	1 ea	157	46	405	29	34	1	16	2.9		
1164	On part whole wheat	1 ea	156	47	397	30	30	2	17	3.2		
1163	On whole wheat	1 ea	169	47	422	32	34	4	18	3.4		
	Tuna salad:											
1165	On white bread, soft	1 ea	122	46	326	13	35	1	14	1.9		
1167	On part whole wheat	1 ea	122	47	322	14	32	2	15	2.2		
1166	On whole wheat	1 ea	135	47	347	16	36	4	17	2.4		
	Turkey:											
1168	On white bread, soft	1 ea	156	54	346	24	29	1	14	1.9		
1170	On part whole wheat	1 ea	155	55	336	25	25	2	15	2.1		
1169	On whole wheat	1 ea	169	54	360	27	29	4	16	2.3		
	Turkey ham:											
1272	On rye bread	1 ea	150	60	280	21	20	6	13	2.5		
1273	On white bread, soft	1 ea	156	55	331	21	30	2	14	2.5		
1275	On part whole wheat	1 ea	156	52	363	23	28	4	19	3.8		
1274	On whole wheat	1 ea	169	56	343	24	29	4	15	3		
714	Taco	1 ea	171	58	369	21	27		21	11.4	6.6	1
	Tostada:											
1114	With refried beans	1 ea	144	66	223	10	26	7	10	5.4	3	.7
1118	With beans & beef	1 ea	225	70	333	16	30	4	17	11.5	3.5	.6
1354	With beans & chicken	1 ea	156	70	242	19	16	3	11	4.5		
	NUTS, SEEDS and PRODUCTS											
	Almonds:											
1365	Dry roasted, salted	1 c	138	3	824	30	27	16	73	5.6	46.4	17.4
718	Slivered, packed, unsalted	1 c	108	5	624	23	21	13	55	4.2	34.7	13.2
720	Whole, dried, unsalted	1 oz	28	5	162	6	6	3	14	1.1	9	3.4
721	Almond butter:	1 tbs	16	1	101	2	3	1	9	.9	6.1	2
4572	Salted	1 tbs	16	1	101	2	3	1	9	.9	6.1	2
722	Brazil nuts, dry (about 7)	1 c	140	3	918	20	18	8	93	22.6	32.2	33.8

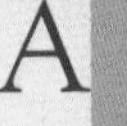

Chol (mg)	Calc (mg)	Iron (mg)	Magn (mg)	Pota (mg)	Sodi (mg)	Zinc (mg)	VT-A (µg)	Thia (mg)	VT-E (mg)	Ribo (mg)	Niac (mg)	V-B6 (mg)	Fola (µg)	VT-C (mg)
30	76	2.21	20	146	483	.79	21*	.3	5.47	.24	4.09	.27	63	1
33	70	2.25	32	194	468	1.04	23*	.25	6.07	.2	4.15	.3	49	1
34	59	2.6	63	252	528	1.76	24*	.25	6.65	.18	4.47	.38	39	1
83	267	3.03	28	232	1469	3.59	58*	.2	2.59	.33	2.76	.18	32	<1
154	87	2.36	18	122	500	.76	50*	.31	4.24	.38	2.45	.21	74	0
167	80	2.38	29	165	479	.99	54*	.25	4.65	.34	2.3	.24	60	0
177	71	2.75	60	222	543	1.71	57*	.26	5.17	.33	2.54	.31	51	0
56	49	2.84	29	375	1155	2.66	7*	.82	2.52	.36	5.89	.4	27	<1
54	76	3.1	32	384	1237	2.61	6*	.9	2.52	.44	6.82	.39	60	<1
56	68	3.12	43	438	1251	2.92	7*	.88	2.71	.4	6.9	.42	44	<1
59	58	3.46	72	500	1336	3.66	7*	.9	3.05	.38	7.28	.49	34	<1
64	246	2.82	33	328	1366	2.71	74*	.7	2.57	.46	5.36	.31	61	<1
67	248	2.82	44	379	1388	3.03	78*	.67	2.77	.42	5.35	.34	45	<1
70	246	3.17	74	441	1487	3.79	82*	.68	3.13	.41	5.7	.41	36	<1
63	258	2.61	32	353	1326	2.93	57*	.61	2.64	.39	4.37	.31	27	<1
29	72	2.25	21	167	935	1.06	5*	.55	3.33	.28	3.69	.18	59	0
30	65	2.26	32	213	950	1.31	6*	.52	3.66	.23	3.65	.21	44	0
32	54	2.59	62	269	1029	2.02	6*	.53	4.07	.21	3.92	.28	34	0
113	222	4.17	39	391	715	7.04	95*	.25	3.51	.46	6.14	.37	37	<1
1	76	2.29	52	240	428	1.05	<1	.29	2.55	.23	5.44	.15	79	2
0	70	2.36	67	297	412	1.32	<1	.24	2.87	.18	5.68	.17	67	2
0	61	2.72	99	361	472	2.05	<1	.25	3.28	.16	6.12	.24	60	2
106	418	3.54	46	381	2166	3.89	175*	.18	2.59	.33	2.64	.33	56	14
51	54	4.23	31	316	792	3.39	21	.37	.19	.31	5.87	.26	57	2
43	76	4.14	30	436	1605	3.73	8*	.35	3.38	.36	6.79	.4	67	0
45	67	4.21	41	493	1642	4.11	8*	.29	3.62	.32	6.86	.44	51	0
47	57	4.6	70	557	1743	4.9	8*	.29	4	.3	7.24	.51	42	0
13	76	2.41	24	168	589	.68	15*	.3	2.77	.24	5.89	.13	63	1
13	69	2.45	36	215	578	.9	17*	.25	3.06	.19	6.07	.16	48	1
14	59	2.79	67	273	642	1.61	17*	.25	3.46	.17	6.47	.22	38	1
43	72	2.19	31	307	1589	1.33	8*	.31	3.27	.29	9.29	.42	60	0
45	63	2.15	42	356	1625	1.56	8*	.25	3.51	.24	9.52	.45	45	0
47	53	2.46	71	417	1735	2.25	8*	.25	3.91	.22	10.1	.53	35	0
55	51	4.04	27	343	1178	2.94	7*	.22	2.86	.33	4.32	.3	29	<1
52	77	4.21	29	351	1252	2.85	6*	.32	2.82	.41	5.29	.29	62	<1
50	49	2.53	65	408	1553	2.52	8*	.58	2.28	.27	7.98	.48	32	0
58	59	4.72	69	466	1361	3.95	7*	.26	3.4	.35	5.62	.39	37	<1
56	221	2.41	70	474	802	3.93	147*	.15	1.88	.44	3.21	.24	68	2
30	210	1.89	59	403	543	1.9	85*	.1	1.15	.33	1.32	.16	43	1
74	189	2.45	67	491	871	3.17	173*	.09	1.8	.49	2.86	.25	85	4
55	146	1.57	41	263	386	1.93	63*	.08	.66	.15	4.26	.31	25	5
0	367	6.22	395	1029	468	4.89	<1	.1	36.3	1.19	5.31	.17	45	0
0	268	4.64	297	786	1	3.63	1	.26	28.3	.88	4.24	.14	31	0
0	69	1.2	77	204	<1	.94	<1	.07	7.33	.23	1.1	.04	8	0
0	43	.59	48	121	2	.49	0	.02	3.25	.1	.46	.01	10	<1
0	43	.59	48	121	72	.49	0	.02	3.24	.1	.46	.01	10	<1
0	246	4.76	315	840	3	6.43	0	1.4	10.6	.17	2.27	.35	6	1

*This value is expressed in retinol equivalents (RE). All other values are in retinol activity equivalents (RAE).

Table A–1

Food Composition (Computer code number is for Wadsworth Diet Analysis program) (For purposes of calculations, use "0" for t, <1, <.1, <.01, etc.)

Computer Code Number	Food Description	Measure	Wt (g)	H_2O (%)	Ener (cal)	Prot (g)	Carb (g)	Dietary Fiber (g)	Fat (g)	Fat Breakdown (g) Sat	Mono	Poly
	NUTS, SEEDS and PRODUCTS—Continued											
	Cashew nuts, dry roasted:											
724	Salted	1 oz	28	2	161	4	9	1	13	2.6	7.6	2.2
4621	Unsalted	1 oz	28	2	161	4	9	1	13	2.6	7.6	2.2
725	Oil roasted:	1 c	130	4	749	21	37	5	63	12.4	36.9	10.6
726	Ounce	1 oz	28	4	161	5	8	1	13	2.7	7.9	2.3
4622	Unsalted:	1 c	130	4	749	21	37	5	63	12.4	36.9	10.6
4622	Ounce	1 oz	28	4	161	5	8	1	13	2.7	7.9	2.3
727	Cashew butter, unsalted	1 tbs	16	3	94	3	4	<1	8	1.6	4.7	1.3
4662	Cashew butter, salted	1 tbs	16	3	94	3	4	<1	8	1.6	4.7	1.3
728	Chestnuts, European, roasted (1 cup = approx 17 kernels)	1 c	143	40	350	5	76	7	3	.6	1.1	1.2
	Coconut, raw:											
729	Piece 2 x 2 x ½"	1 pce	45	47	159	1	7	4	15	13.4	.6	.2
730	Shredded/grated, unpacked	½ c	40	47	142	1	6	4	13	11.9	.6	.1
	Coconut, dried, shredded/grated:											
731	Unsweetened	1 c	78	3	515	5	19	13	50	44.6	2.1	.6
732	Sweetened	1 c	93	13	466	3	44	4	33	29.3	1.4	.4
4559	Coconut milk, canned	1 c	226	73	445	5	6	3	48	42.7	2	.5
734	Filberts/hazelnuts, chopped	1 oz	28	5	176	4	5	3	17	1.2	12.8	2.2
735	Macadamias, oil roasted, salted:	1 c	134	2	962	10	17	12	103	15.4	80.9	1.8
736	Ounce	1 oz	28	2	201	2	4	3	21	3.2	16.9	.4
1368	Macadamias, oil roasted, unsalted	1 c	134	2	962	10	17	12	103	15.4	80.9	1.8
	Mixed Nuts:											
737	Dry roasted, salted	1 c	137	2	814	24	35	12	70	9.4	43	14.7
738	Oil roasted, salted	1 c	142	2	876	24	30	13	80	12.4	45	18.9
1369	Oil roasted, unsalted	1 c	142	2	876	24	30	14	80	12.4	45	18.9
	Peanuts:											
740	Oil roasted, salted	1 oz	28	2	163	7	5	3	14	1.9	6.8	4.4
742	Dried, salted	1 oz	28	2	164	7	6	2	14	1.9	6.9	4.4
743	Peanut butter:	½ c	128	1	759	32	25	8	65	13.2	31.1	17.6
1371	Tablespoon	2 tbs	32	1	190	8	6	2	16	3.3	7.8	4.4
745	Pecan halves, dried, unsalted	1 oz	28	4	193	3	4	3	20	1.7	11.4	6
1372	Pecan halves, dry roasted, salted	¼ c	28	1	199	3	4	3	21	1.8	12.3	5.8
746	Pine nuts/pinons, dried	1 oz	28	6	176	3	5	3	17	2.6	6.4	7.2
747	Pistachios, dried, shelled	1 oz	28	4	156	6	8	3	12	1.5	6.5	3.8
1373	Pistachios,dry roasted,salted,shelled	1 c	128	2	727	27	34	13	59	7.1	31	17.8
748	Pumpkin kernels, dried, unsalted	1 oz	28	7	151	7	5	1	13	2.4	4	5.8
1374	Pumpkin kernels, roasted, salted	1 c	227	7	1184	75	30	9	96	18.1	29.7	43.6
749	Sesame seeds, hulled, dried	¼ c	38	5	223	10	4	4	21	2.9	7.9	9.1
8878	Soy nuts, BBQ	5 pce	28		119	12	9	4	4	1		
8877	Soy nuts, salted	5 pce	28		119	12	9	5	4	1		
	Sunflower seed kernels:											
750	Dry	¼ c	36	5	205	8	7	4	18	1.9	3.4	11.8
751	Oil roasted	¼ c	34	3	209	7	5	2	19	2	3.7	12.9
752	Tahini (sesame butter)	1 tbs	15	3	91	3	3	1	8	1.2	3.2	3.7
1334	Trail Mix w/chocolate chips	1 c	146	7	707	21	66	8	47	8.9	19.8	16.5
754	Black walnuts, chopped	1 oz	28	4	170	7	3	1	16	1	3.6	10.5
756	English walnuts, chopped	1 oz	28	4	183	4	4	2	18	1.7	2.5	13.2
	SWEETENERS and SWEETS (see also Dairy (milk desserts) and Baked Goods)											
757	Apple butter	2 tbs	36	56	62	<1	15	1	0	0	0	0
1124	Butterscotch topping	2 tbs	41	32	103	1	27	<1	<1	3.5	.6	0
1125	Caramel topping	2 tbs	41	32	103	1	27	<1	<1	3.5	.6	0
	Cake frosting, creamy vanilla:											
1127	Canned	2 tbs	39	13	163	<1	27	<1	7	1.9	3.4	.9
1123	From mix	2 tbs	39	12	165	<1	28	<1	6	1.3	2.6	2.2
	Cake frosting, lite:											
2061	Milk chocolate	1 tbs	16	18	58	<1	11	<1	1		.9	
2062	Vanilla	1 tbs	16	15	60	0	12	<1	1	0	1	
	Candy:											

Chol (mg)	Calc (mg)	Iron (mg)	Magn (mg)	Pota (mg)	Sodi (mg)	Zinc (mg)	VT-A (μg)	Thia (mg)	VT-E (mg)	Ribo (mg)	Niac (mg)	V-B6 (mg)	Fola (μg)	VT-C (mg)
0	13	1.68	73	158	179	1.57	0	.06	.16	.06	.39	.07	19	0
0	13	1.68	73	158	4	1.57	0	.06	.16	.06	.39	.07	19	0
0	53	5.33	332	689	814	6.18	0	.55	2.03	.23	2.34	.32	88	0
0	11	1.15	71	148	175	1.33	0	.12	.44	.05	.5	.07	19	0
0	53	5.33	332	689	22	6.18	0	.55	2.03	.23	2.34	.32	88	0
0	11	1.15	71	148	5	1.33	0	.12	.44	.05	.5	.07	19	0
0	7	.8	41	87	2	.83	0	.05	.25	.03	.26	.04	11	0
0	7	.8	41	87	98	.83	0	.05	.25	.03	.26	.04	11	0
0	41	1.3	47	847	3	.81	1	.35	1.72	.25	1.92	.71	100	37
0	6	1.09	14	160	9	.49	0	.03	.33	.01	.24	.02	12	1
0	6	.97	13	142	8	.44	0	.03	.29	.01	.22	.02	10	1
0	20	2.59	70	424	29	1.57	0	.05	1.05	.08	.47	.23	7	1
0	14	1.79	46	313	244	1.69	0	.03	1.26	.02	.44	.25	7	1
0	41	7.46	104	497	29	1.27	0	.05	1.47	0	1.44	.06	32	2
0	32	1.32	46	190	0	.69	1	.18	4.25	.03	.5	.16	32	2
0	60	2.41	157	441	348	1.47	1*	.28	.55	.15	2.71	.26	21	0
0	13	.5	33	92	73	.31	<1*	.06	.11	.03	.57	.05	4	0
0	60	2.41	157	441	9	1.47	1*	.28	.55	.15	2.71	.26	21	0
0	96	5.07	308	818	917	5.21	1	.27	8.22	.27	6.44	.41	68	1
0	153	4.56	334	825	926	7.21	1	.71	8.52	.31	7.19	.34	118	1
0	153	4.56	334	825	16	7.21	1	.71	8.52	.31	7.19	.34	118	1
0	25	.51	52	191	121	1.86	0	.07	2.07	.03	4	.07	35	0
0	15	.63	49	184	228	.93	0	.12	2.07	.03	3.79	.07	41	0
0	49	2.36	204	856	598	3.74	0	.11	12.8	.13	17.2	.58	95	0
0	12	.59	51	214	149	.93	0	.03	3.2	.03	4.29	.14	24	0
0	20	.71	34	115	0	1.27	1	.18	1.13	.04	.33	.06	6	<1
0	20	.78	37	119	107	1.42	2	.13	1.05	.03	.33	.05	4	<1
0	2	.86	65	176	20	1.2	<1	.35	.98	.06	1.22	.03	16	1
0	30	1.16	34	287	<1	.62	8	.24	1.28	.04	.36	.48	14	1
0	141	5.38	154	1333	518	2.94	34	1.08	5.45	.2	1.82	2.18	64	3
0	12	4.19	150	226	5	2.09	5	.06	.28	.09	.49	.06	16	1
0	98	33.9	1212	1829	1305	16.9	43	.48	2.27	.72	3.95	.2	129	4
0	50	2.96	132	155	15	3.9	1	.27	.86	.03	1.78	.05	36	0
0	59	1.07			415		0							0
0	59	1.07			148		0							0
0	42	2.44	127	248	1	1.82	1	.82	18.1	.09	1.62	.28	82	<1
0	19	2.28	43	164	1	1.77	1	.11	17.1	.09	1.4	.27	80	<1
0	21	.95	53	69	<1	1.57	1	.24	.34	.02	.85	.02	15	0
6	159	4.95	235	946	177	4.58	7*	.6	15.6	.33	6.43	.38	95	2
0	16	.86	57	147	<1	.96	4	.06	.73	.03	.19	.15	18	1
0	27	.81	44	123	1	.86	1	.09	.82	.04	.56	.15	27	<1
0	5	.11	2	33	1	.02	4*	.01	<.01	.01	.04	.02	<1	<1
<1	22	.08	3	34	143	.08	11*	<.01	0	.04	.02	.01	1	<1
<1	22	.08	3	34	143	.08	11*	<.01	0	.04	.02	.01	1	<1
0	1	.04		14	35	0	88*	0	1.84	<.01	<.01	0	0	0
0	4	.09	1	9	87	.04	42*	.01	.79	.01	.13	<.01	0	0
0							<1*							
0							0*	0				0		

*This value is expressed in retinol equivalents (RE). All other values are in retinol activity equivalents (RAE).

Table A–1

Food Composition (Computer code number is for Wadsworth Diet Analysis program) (For purposes of calculations, use "0" for t, <1, <.1, <.01, etc.)

Computer Code Number	Food Description	Measure	Wt (g)	H_2O (%)	Ener (cal)	Prot (g)	Carb (g)	Dietary Fiber (g)	Fat (g)	Fat Breakdown (g) Sat	Mono	Poly
	SWEETENERS and SWEETS (see also Dair (milk desserts) and Baked Goods)—Continued											
1128	Almond Joy candy bar	1 oz	28	10	131	1	16	1	7	4.8	1.8	.4
2069	Butterscotch morsels	¼ c	43	8	243	0	27	0	12	10.6		
758	Caramel, plain or chocolate	1 pce	10	8	38	<1	8	<1	1	.7	.1	t
1961	Chewing gum, sugarless	1 pce	3		6	0	2		0	0	0	0
	Chocolate (see also #784, 785, 971):											
	Milk chocolate:											
759	Plain	1 oz	28	1	144	2	17	1	9	5.2	2.8	.3
760	With almonds	1 oz	28	1	147	3	15	2	10	4.7	3.8	.6
761	With peanuts	1 oz	28	1	155	5	11	2	11	3.4	5.1	2.5
762	With rice cereal	1 oz	28	2	139	2	18	1	7	4.4	2.4	.2
763	Semisweet chocolate chips	1 c	168	1	805	7	106	10	50	29.8	16.7	1.6
764	Sweet Dark chocolate (candy bar)	1 ea	41	1	226	2	25	2	13	8.3	4.6	.4
765	Fondant candy, uncoated (mints, candy corn, other)	1 pce	16	7	57	0	15	0	0	0	0	0
1697	Fruit Roll-Up (small)	1 ea	14	11	49	<1	12	<1	<1	.1	.2	.1
766	Fudge, chocolate	1 pce	17	10	65	<1	13	<1	1	.9	.4	.1
767	Gumdrops	1 c	182	1	703	0	180	0	0	0	0	0
768	Hard candy-all flavors	1 pce	6	1	24	0	6	0	<1			
769	Jellybeans	10 pce	11	6	40	0	10	0	<1	t	t	t
1134	M&M's plain chocolate candy	10 pce	7	2	34	<1	5	<1	1	.9	.5	t
1135	M&M's peanut chocolate candy	10 pce	20	2	103	2	12	1	5	2.1	2.2	.8
1130	MARS almond bar	1 ea	50	4	234	4	31	1	11	3.6	5.3	2
1129	MILKY WAY candy bar	1 ea	60	6	254	3	43	1	10	4.7	3.6	.4
1708	Milk chocolate-coated peanuts	1 c	149	2	773	19	74	7	50	21.8	19.3	6.4
1709	Peanut brittle, recipe	1 c	147	2	666	11	102	3	28	7.4	12.5	6.9
1132	Reese's peanut butter cup	2 ea	50	2	271	5	27	2	16	5.5	6.5	2.8
1133	Skor English toffee candy bar	1 ea	39	3	217	2	22	1	13	8.5	4.3	.5
1131	Snickers candy bar (2.2oz)	1 ea	62	5	297	5	37	2	15	5.6	6.5	3
23082	Chewing gum	1 pce	3	3	10	0	3	0	<1	t	t	t
1482	Fruit juice bar (2.5 fl oz)	1 ea	77	78	63	1	16	0	<1	t	0	t
771	Gelatin dessert/Jello, prepared	½ c	135	85	80	2	19	0	0	0	0	0
1702	SugarFree	½ c	117	98	8	1	1	0	0	0	0	0
772	Honey:	1 c	339	17	1030	1	279	1	0	0	0	0
773	Tablespoon	1 tbs	21	17	64	<1	17	<1	0	0	0	0
774	Jams or preserves:	1 tbs	20	29	54	<1	14	<1	<1	0	t	t
775	Packet	1 ea	14	30	39	<1	10	<1	<1	t	t	0
776	Jellies:	1 tbs	19	29	54	<1	13	<1	<1	.2	1.3	.2
777	Packet	1 ea	14	29	40	<1	10	<1	<1	.2	1	.1
1136	Marmalade	1 tbs	20	33	49	<1	13	<1	0	0	0	0
770	Marshmallows	1 ea	7	16	22	<1	6	<1	<1	t	t	t
1126	Marshmallow creme topping	2 tbs	38	20	122	<1	30	<1	<1	t	t	t
778	Popsicle/ice pops	1 ea	128	80	92	0	24	0	0	0	0	0
23171	Rice crispie bar	1 ea	28	13	107	1	20	<1	3	.6	1.3	.8
	Sugars:											
779	Brown sugar	1 c	220	2	827	0	214	0	0	0	0	0
780	White sugar, granulated:	1 c	200	0	774	0	200	0	0	0	0	0
781	Tablespoon	1 tbs	12	0	46	0	12	0	0	0	0	0
782	Packet	1 ea	6	0	23	0	6	0	0	0	0	0
783	White sugar, powdered, sifted	1 c	100		389	0	99	0	<1	t	t	t
	Sweeteners:											
1711	Equal, packet	1 ea	1	12	4	<1	1	0	<1	0	t	t
1712	Sweet 'N Low, packet	1 ea	1		4	0	1	0	0	0	0	0
	Syrups:											
	Chocolate:											
785	Hot fudge type	2 tbs	43	22	151	2	27	1	4	1.7	1.7	.1
784	Thin type	2 tbs	38	29	93	1	25	1	<1	.3	.2	t
25003	Molasses	2 tbs	41	26	109	0	28	0	0	0	0	0
1710	Light cane	2 tbs	41	24	103	0	27	0	0	0	0	0
787	Pancake table syrup (corn and maple)	2 tbs	40	24	115	0	30	0	0	0	0	0

Chol (mg)	Calc (mg)	Iron (mg)	Magn (mg)	Pota (mg)	Sodi (mg)	Zinc (mg)	VT-A (µg)	Thia (mg)	VT-E (mg)	Ribo (mg)	Niac (mg)	V-B6 (mg)	Fola (µg)	VT-C (mg)
1	17	.39	18	69	41	.22	1*	.01	.63	.04	.13	.02		<1
0	0	0		79	45		0	.03		.03	.02			0
1	14	.01	2	21	24	.04	1*	<.01	.05	.02	.02	<.01	<1	<1
				0	0									
6	53	.39	17	108	23	.39	8	.02	.35	.08	.09	.01	2	<1
5	63	.46	25	124	21	.37	4*	.02	.53	.12	.21	.01	3	<1
3	32	.52	34	150	11	.68	5*	.08	1.3	.05	2.12	.04	23	0
5	48	.21	14	96	41	.31	3*	.02	.3	.08	.13	.02	3	<1
0	54	5.26	193	613	18	2.72	2	.09	2	.15	.72	.06	5	0
<1	11	.98	45	123	3	.59	2*	.01	.18	.03	.16	.01	1	0
0		.01		3	6	.01	0*	0	0	<.01	0	0	0	0
0	4	.14	3	41	9	.03	1	.01	.04	<.01	.01	.04	1	1
2	7	.08	4	17	10	.07	8*	<.01	.02	.01	.02	<.01	<1	<1
0	5	.73	2	9	80	0	0	0	0	<.01	<.01	0	0	0
0		.02			2	<.01	0	0	0	0	0	0	0	0
0		.12		4	3	.01	0	0	0	0	0	0	0	0
1	7	.08	3	19	4	.07	4*	<.01	.06	.01	.02	<.01	<1	<1
2	20	.23	15	69	10	.46	5*	.02	.49	.03	.75	.02	7	<1
8	84	.55	36	163	85	.55	25*	.02	2.33	.16	.47	.03	9	<1
8	78	.46	20	145	144	.43	19*	.02	.39	.13	.21	.03	6	1
13	155	1.95	140	748	61	2.89	0	.17	3.8	.26	6.33	.31	12	0
19	44	2.03	73	306	664	1.43	69*	.28	2.41	.08	5.14	.15	103	0
2	39	.6	44	176	159	.91	9*	.12	2.04	.08	2.31	.07	27	<1
20	51	.19	13	93	108	.3	27*	.01	.53	.13	.03	.01		<1
8	58	.47	45	201	165	1.46	24*	.06	.95	.09	2.6	.05	25	<1
0	0	0	0		<1	0	0	0	0	0	0	0	0	0
0	4	.15	3	41	3	.04	1	.01	0	.01	.12	.02	5	7
0	3	.04	1	1	57	.04	0	0	0	<.01	<.01	<.01	0	0
0	2	.01	1	0	56	.03	0	0	0	<.01	<.01	<.01	0	0
0	20	1.42	7	176	14	.75	0	0	0	.13	.41	.08	7	2
0	1	.09		11	1	.05	0	0	0	.01	.02	<.01	<1	<1
0	4	.2	1	18	2	.01	<1*	<.01	.02	.01	.04	<.01	2	<1
0	3	.07	1	11	4	.01	<1	0	0	<.01	<.01	<.01	5	1
0	2	.04	1	12	5	.01	<1	0	0	<.01	.01	<.01	<1	<1
0	1	.03	1	9	4	.01	<1	0	0	<.01	<.01	<.01	<1	<1
0	8	.03		7	11	.01	<1	<.01	0	<.01	.01	<.01	7	1
0		.02			3	<.01	<1	0	0	0	<.01	0	<1	0
0	1	.08	1	2	19	.01	<1	0	0	0	.03	<.01	<1	0
0	0	0	1	5	15	.03	0	0	0	0	0	0	0	0
0	2	.51	4	12	123	.15	85*	.1	.41	.11	1.31	.13	27	4
0	187	4.2	64	761	86	.4	0	.02	0	.01	.18	.06	2	0
0	2	.12	0	4	2	.06	0	0	0	.04	0	0	0	0
0		.01	0		<1	<.01	0	0	0	<.01	0	0	0	0
0		<.01	0		<1	<.01	0	0	0	<.01	0	0	0	0
0	1	.06	0	2	1	.03	0	0	0	0	0	0	0	0
0		<.01			<1	0	0	0	0	0	0	0	0	0
0	0	0			0		0							0
1	35	.56	22	156	149	.29	2*	.02	1.26	.09	.13	.03	2	<1
0	5	5.15	25	183	58	.28	494*	<.01	.01	.31	12.8	.01	2	<1
0	84	1.94	99	600	15	.12	0	.02	0	0	.38	.27	0	0
0	68	1.76	100	376	6	.12	0*	.03	0	.02	.08	.27	0	0
0		.04	1	1	33	.02	0	<.01	0	<.01	.01	0	0	0

*This value is expressed in retinol equivalents (RE). All other values are in retinol activity equivalents (RAE).

Table A–1

Food Composition (Computer code number is for Wadsworth Diet Analysis program) (For purposes of calculations, use "0" for t, <1, <.1, <.01, etc.)

Computer Code Number	Food Description	Measure	Wt (g)	H_2O (%)	Ener (cal)	Prot (g)	Carb (g)	Dietary Fiber (g)	Fat (g)	Fat Breakdown (g) Sat	Mono	Poly
	VEGETABLES AND LEGUMES											
788	Alfalfa sprouts	1 c	33	91	10	1	1	1	<1	t	t	.1
1815	Amaranth leaves, raw, chopped	1 c	28	92	6	1	1	<1	<1	t	t	t
1816	Amaranth leaves, raw, each	1 ea	14	92	3	<1	1	<1	<1	t	t	t
1817	Amaranth leaves, cooked	1 c	132	91	28	3	5	2	<1	.1	.1	.1
1987	Arugula, raw, chopped	½ c	10	92	2	<1	<1	<1	<1	t	t	t
789	Artichokes, cooked globe (300g with refuse)	1 ea	120	84	60	4	13	6	<1	t	t	.1
1177	Artichoke hearts, cooked from frozen	1 c	168	86	76	5	15	8	1	.2	t	.4
1176	Artichoke hearts, marinated	1 c	130	80	116	5	14	5	7	0		
2021	Artichoke hearts, in water	½ c	100	91	37	2	6	0	0	0	0	0
	Asparagus, green, cooked:											
	From fresh:											
790	Cuts and tips	½ c	90	92	22	2	4	1	<1	.1	t	.1
791	Spears, ½" diam at base	4 ea	60	92	14	2	3	1	<1	t	t	.1
	From frozen:											
792	Cuts and tips	½ c	90	91	25	3	4	1	<1	.1	t	.2
793	Spears, ½" diam at base	4 ea	60	91	17	2	3	1	<1	.1	t	.1
794	Canned, spears, ½" diam at base	4 pce	72	94	14	2	2	1	<1	.1	t	.2
795	Bamboo shoots, canned, drained slices	1 c	131	94	25	2	4	2	1	.1	t	.2
1795	Bamboo shoots, raw slices	1 c	151	91	41	4	8	3	<1	.1	t	.2
1798	Bamboo shoots, cooked slices	1 c	120	96	14	2	2	1	<1	.1	t	.1
	Beans (see also alphabetical listing this section):											
1990	Adzuki beans, cooked	½ c	115	66	147	9	28	8	<1	t	t	t
796	Black beans, cooked	½ c	86	66	114	8	20	7	<1	.1	t	.2
	Canned beans (white/navy):											
803	With pork and tomato sauce	½ c	127	73	124	7	25	6	1	.5	.6	.2
804	With sweet sauce	½ c	130	71	144	7	27	7	2	.7	.8	.2
805	With frankfurters	½ c	130	69	185	9	20	9	9	3.1	3.7	1.1
	Lima beans:											
797	Thick seeded (Fordhooks), cooked from frozen	½ c	85	73	85	5	16	5	<1	.1	t	.1
798	Thin seeded (Baby), cooked from frozen	½ c	90	72	94	6	17	5	<1	.1	t	.1
799	Cooked from dry, drained	½ c	94	70	108	7	20	7	<1	.1	t	.2
1998	Red Mexican, cooked f/dry	½ c	112	70	127	8	24	9	<1	.1	.1	.2
	Snap bean/green string beans cuts and french style:											
800	Cooked from fresh	½ c	63	89	22	1	5	2	<1	t	t	.1
801	Cooked from frozen	½ c	68	91	19	1	4	2	<1	t	t	.1
802	Canned, drained	½ c	68	93	14	1	3	1	<1	t	t	t
1713	Snap bean, yellow, cooked f/fresh	½ c	63	89	22	1	5	2	<1	t	t	.1
	Bean sprouts (mung):											
806	Raw	½ c	52	90	16	2	3	1	<1	t	t	t
807	Cooked, stir fried	½ c	62	84	31	3	7	1	<1	t	t	t
808	Cooked, boiled, drained	½ c	62	93	13	1	3	<1	<1	t	t	t
1788	Canned, drained	½ c	63	96	8	1	1	<1	<1	t	t	t
	Beets, cooked from fresh:											
809	Sliced or diced	½ c	85	87	37	1	8	2	<1	t	t	.1
810	Whole beets, 2" diam	2 ea	100	87	44	2	10	2	<1	t	t	.1
	Beets, canned:											
811	Sliced or diced	½ c	79	91	24	1	6	1	<1	t	t	t
812	Pickled slices	½ c	114	82	74	1	19	3	<1	t	t	t
813	Beet greens, cooked, drained	½ c	72	89	19	2	4	2	<1	t	t	t
	Broccoli, raw:											
817	Chopped	½ c	44	91	12	1	2	1	<1	t	t	.1
818	Spears	1 ea	31	91	9	1	2	1	<1	t	t	.1
	Brocoli, cooked from fresh:											
819	Spears	1 ea	180	91	50	5	9	5	1	.1	t	.3
820	Chopped	½ c	78	91	22	2	4	2	<1	t	t	.1
	Broccoli, cooked from frozen:											
821	Spear, small piece	½ c	92	91	26	3	5	3	<1	t	t	.1

Chol (mg)	Calc (mg)	Iron (mg)	Magn (mg)	Pota (mg)	Sodi (mg)	Zinc (mg)	VT-A (µg)	Thia (mg)	VT-E (mg)	Ribo (mg)	Niac (mg)	V-B6 (mg)	Fola (µg)	VT-C (mg)
0	11	.32	9	26	2	.3	3	.02	.01	.04	.16	.01	12	3
0	60	.65	15	171	6	.25	41	.01	.22	.04	.18	.05	24	12
0	30	.32	8	85	3	.13	20	<.01	.11	.02	.09	.03	12	6
0	276	2.98	73	846	28	1.16	183	.03	.66	.18	.74	.23	75	54
0	16	.15	5	37	3	.05	12	<.01	.04	.01	.03	.01	10	1
0	54	1.55	72	425	114	.59	11	.08	.23	.08	1.2	.13	61	12
0	35	.94	52	444	89	.6	13	.1	.32	.26	1.54	.15	200	8
0	0	0			488		0							46
0	0	1.35		0	250		6							4
0	18	.66	9	144	10	.38	24	.11	.34	.11	.97	.11	131	10
0	12	.44	6	96	7	.25	16	.07	.23	.08	.65	.07	88	6
0	21	.58	12	196	4	.5	37	.06	1.13	.09	.93	.02	122	22
0	14	.38	8	131	2	.34	25	.04	.75	.06	.62	.01	81	15
0	11	1.32	7	124	207	.29	19	.04	.31	.07	.69	.08	69	13
0	10	.42	5	105	9	.85	1	.03	.5	.03	.18	.18	4	1
0	20	.75	5	805	6	1.66	2	.23	1.51	.11	.91	.36	11	6
0	14	.29	4	640	5	.56	0	.02	.8	.06	.36	.12	2	0
0	32	2.3	60	612	9	2.04	1	.13	.11	.07	.82	.11	139	0
0	23	1.81	60	305	1	.96	<1	.21	.07	.05	.43	.06	128	0
9	71	4.17	44	381	559	7.44	8	.07	.69	.06	.63	.09	29	4
9	79	2.16	44	346	437	1.95	7	.06	.7	.08	.46	.11	48	4
8	62	2.25	36	306	559	2.43	10	.07	.61	.07	1.17	.06	39	3
0	19	1.16	29	347	45	.37	8	.06	.25	.05	.91	.1	18	11
0	25	1.76	50	370	26	.49	8	.06	.58	.05	.69	.1	14	5
0	16	2.25	40	478	2	.89	0	.15	.17	.05	.4	.15	78	0
0	42	1.87	48	371	6	.87	<1	.13	.08	.07	.38	.11	94	2
0	29	.81	16	188	2	.23	21	.05	.09	.06	.39	.03	21	6
0	33	.6	16	86	6	.33	14	.02	.09	.06	.26	.04	16	3
0	18	.61	9	74	178	.2	12	.01	.09	.04	.14	.02	22	3
0	29	.81	16	188	2	.23	3	.05	.18	.06	.39	.03	21	6
0	7	.47	11	77	3	.21	1	.04	<.01	.06	.39	.05	32	7
0	8	1.18	20	136	6	.56	1	.09	.01	.11	.74	.08	43	10
0	7	.4	9	63	6	.29	<1	.03	.01	.06	.51	.03	18	7
0	9	.27	6	17	88	.18	1	.02	.01	.04	.14	.02	6	<1
0	14	.67	20	259	65	.3	2	.02	.25	.03	.28	.06	68	3
0	16	.79	23	305	77	.35	2	.03	.3	.04	.33	.07	80	4
0	12	1.44	13	117	153	.17	<1	.01	.24	.03	.12	.04	24	3
0	12	.47	17	169	301	.3	1	.01	.15	.05	.29	.06	31	3
0	82	1.37	49	654	174	.36	184	.08	.22	.21	.36	.09	10	18
0	21	.39	11	143	12	.18	34	.03	.73	.05	.28	.07	31	41
0	15	.27	8	101	8	.12	24	.02	.51	.04	.2	.05	22	29
0	83	1.51	43	526	47	.68	125	.1	3.04	.2	1.03	.26	90	134
0	36	.65	19	228	20	.3	54	.04	1.32	.09	.45	.11	39	58
0	47	.56	18	166	22	.28	87	.05	.95	.07	.42	.12	28	37

*This value is expressed in retinol equivalents (RE). All other values are in retinol activity equivalents (RAE).

Table A-1

Food Composition (Computer code number is for Wadsworth Diet Analysis program) (For purposes of calculations, use "0" for t, <1, <.1, <.01, etc.)

Computer Code Number	Food Description	Measure	Wt (g)	H_2O (%)	Ener (cal)	Prot (g)	Carb (g)	Dietary Fiber (g)	Fat (g)	Fat Breakdown (g) Sat	Mono	Poly
	VEGETABLES AND LEGUMES—Continued											
822	Chopped	½ c	92	91	26	3	5	3	<1	t	t	.1
1603	Broccoflower-steamed	½ c	78	90	25	2	5	2	<1	t	t	.1
823	Brussels sprouts, cooked from fresh	½ c	78	87	30	2	7	2	<1	.1	t	.2
824	Brussels sprouts, cooked from frozen	½ c	78	87	33	3	6	3	<1	.1	t	.2
	Cabbage, common varieties:											
825	Raw, shredded or chopped	1 c	70	92	17	1	4	2	<1	t	t	.1
826	Cooked, drained	1 c	150	94	33	2	7	3	1	.1	t	.3
	Cabbage, Chinese:											
1178	Bok Choy, raw, shredded	1 c	70	95	9	1	2	1	<1	t	t	.1
827	Bok Choy, cooked, drained	1 c	170	96	20	3	3	3	<1	t	t	.1
1937	Kim chee style	1 c	150	92	31	2	6	2	<1	t	t	.1
828	Pe Tsai, raw, chopped	1 c	76	94	12	1	2	2	<1	t	t	.1
1796	Pe Tsai, cooked	1 c	119	95	17	2	3	3	<1	t	t	.1
	Cabbage, red, coarsely chopped:											
829	Raw	1 c	89	92	24	1	5	2	<1	t	t	.1
830	Cooked, drained	1 c	150	94	31	2	7	3	<1	t	t	.1
831	Cabbage, savoy, coarsely chopped, raw	1 c	70	91	19	1	4	2	<1	t	t	t
1785	Cabbage, savoy, cooked	1 c	145	92	35	3	8	4	<1	t	t	.1
1896	Capers	1 ea	5			<1	<1		<1			
	Carrots, raw:											
832	Whole, 7½ x 1⅛"	1 ea	72	88	31	1	7	2	<1	t	t	.1
833	Grated	½ c	55	88	24	1	6	2	<1	t	t	t
	Carrots, cooked, sliced, drained:											
834	From raw	½ c	78	87	35	1	8	3	<1	t	t	.1
835	From frozen	½ c	73	90	26	1	6	3	<1	t	t	t
836	Carrots, canned, sliced, drained	½ c	73	93	18	<1	4	1	<1	t	t	.1
837	Carrot juice,canned	1 c	236	89	94	2	22	2	<1	.1	t	.2
5625	Cassava, cooked	1 c	137	59	221	2	53	2	<1	.1	.1	.1
	Cauliflower, flowerets:											
838	Raw	½ c	50	92	12	1	3	1	<1	t	t	t
839	Cooked from fresh, drained	½ c	62	93	14	1	3	2	<1	t	t	.1
840	Cooked, from frozen, drained	½ c	90	94	17	1	3	2	<1	t	t	.1
	Celery, pascal type, raw:											
841	Large outer stalk,8 x 1 ½"(root end)	1 ea	40	95	6	<1	1	1	<1	t	t	t
842	Diced	1 c	120	95	19	1	4	2	<1	t	t	.1
1789	Celeriac/celery root, cooked	1 c	155	92	42	1	9	2	<1	.1	.1	.2
1179	Chard, swiss, raw, chopped	1 c	36	93	7	1	1	1	<1	t	t	t
1180	Chard, swiss, cooked	1 c	175	93	35	3	7	4	<1	t	t	t
1855	Chayote fruit, raw	1 ea	203	94	39	2	9	3	<1	.1	t	.1
1856	Chayote fruit, cooked	1 c	160	93	38	1	8	4	1	.1	.1	.3
	Chickpeas (see Garbanzo Beans #854)											
	Collards, cooked, drained:											
843	From raw	½ c	95	92	25	2	5	3	<1	t	t	.2
844	From frozen	½ c	85	88	31	3	6	2	<1	.1	t	.2
	Corn, yellow, cooked, drained:											
845	From raw, on cob, 5" long	1 ea	77	73	72	2	17	2	1	.1	.2	.3
846	From frozen, on cob, 3½" long	1 ea	63	73	59	2	14	2	<1	.1	.1	.2
847	Kernels, cooked from frozen	½ c	82	77	66	2	16	2	<1	.1	.1	.2
	Corn, canned:											
848	Cream style	½ c	128	79	92	2	23	2	1	.1	.2	.3
849	Whole kernel, vacuum pack	½ c	105	77	83	3	20	2	1	.1	.2	.2
	Cowpeas (see Black-eyed peas #814-816)											
850	Cucumber slices with peel	7 pce	28	96	4	<1	1	<1	<1	t	t	t
1948	Cucumber, kim chee style	1 c	150	91	31	2	7	2	<1	t	t	.1
	Dandelion Greens:											
851	Raw	1 c	55	86	25	1	5	2	<1	.1	t	.2
852	Chopped, cooked, drained	1 c	105	90	35	2	7	3	1	.2	t	.3
853	Eggplant, cooked	1 c	99	92	28	1	7	2	<1	t	t	.1
1714	Endive, fresh, chopped	1 c	50	94	8	1	2	2	<1	t	t	t
856	Escarole/curly endive-chopped	1 c	50	94	8	1	2	2	<1	t	t	t
854	Garbanzo beans (Chickpeas), cooked	1 c	164	60	269	14	45	12	4	.4	1	1.9

PAGE KEY: A–2 = Beverages A–4 = Dairy A–8 = Eggs A–10 = Fat/Oil A–12 = Fruit A–18 = Bakery A–24 = Grain A–30 = Fish A–32 = Meats A–36 = Poultry A–38 = Sausage A–38 = Mixed/Fast A–44 = Nuts/Seeds A–46 = Sweets A–50 = Vegetables/Legumes A–60 = Vegetarian A–62 = Misc A–64 = Soups/Sauces A–66 = Fast A–80 = Convenience A–86 = Baby foods

Chol (mg)	Calc (mg)	Iron (mg)	Magn (mg)	Pota (mg)	Sodi (mg)	Zinc (mg)	VT-A (μg)	Thia (mg)	VT-E (mg)	Ribo (mg)	Niac (mg)	V-B6 (mg)	Fola (μg)	VT-C (mg)
0	47	.56	18	166	22	.28	87	.05	1.52	.07	.42	.12	51	37
0	25	.55	16	251	18	.39	3	.06	.23	.07	.59	.14	38	49
0	28	.94	16	247	16	.26	28	.08	.66	.06	.47	.14	47	48
0	19	.58	19	254	18	.28	23	.08	.45	.09	.42	.22	79	36
0	33	.41	10	172	13	.13	5	.03	.07	.03	.21	.07	30	22
0	46	.25	12	146	12	.13	10	.09	.15	.08	.42	.17	30	30
0	73	.56	13	176	45	.13	105	.03	.08	.05	.35	.14	46	31
0	158	1.77	19	631	58	.29	218	.05	.2	.11	.73	.28	70	44
0	145	1.28	27	375	995	.36	213	.07	.24	.1	.75	.34	88	80
0	58	.24	10	181	7	.17	46	.03	.09	.04	.3	.18	60	20
0	38	.36	12	268	11	.21	58	.05	.14	.05	.59	.21	63	19
0	45	.44	13	183	10	.19	2	.04	.09	.03	.27	.19	19	51
0	55	.52	16	210	12	.22	2	.05	.18	.03	.3	.21	19	52
0	24	.28	20	161	20	.19	35	.05	.07	.02	.21	.13	56	22
0	43	.55	35	267	35	.33	64	.07	.15	.03	.03	.22	67	25
0	2	.05			105		1							0
0	19	.36	11	233	25	.14	1012	.07	.33	.04	.67	.11	10	7
0	15	.27	8	178	19	.11	773	.05	.25	.03	.51	.08	8	5
0	24	.48	10	177	51	.23	957	.03	.33	.04	.39	.19	11	2
0	20	.34	7	115	43	.17	646	.02	.31	.03	.32	.09	8	2
0	18	.47	6	131	177	.19	503	.01	.31	.02	.4	.08	7	2
0	57	1.09	33	689	68	.42	1292	.22	.02	.13	.91	.51	9	20
0	21	.35	28	337	18	.45	1	.1	.26	.06	1.06	.11	24	18
0	11	.22	7	152	15	.14	<1	.03	.02	.03	.26	.11	28	23
0	10	.2	6	88	9	.11	1	.03	.02	.03	.25	.11	27	27
0	15	.37	8	125	16	.12	1	.03	.04	.05	.28	.08	37	28
0	16	.16	4	115	35	.05	3	.02	.14	.02	.13	.03	11	3
0	48	.48	13	344	104	.16	8	.05	.43	.05	.39	.1	34	8
0	40	.67	19	268	95	.31	0	.04	.31	.06	.66	.16	5	6
0	18	.65	29	136	77	.13	59	.01	.68	.03	.14	.04	5	11
0	102	3.96	151	961	313	.58	275	.06	3.31	.15	.63	.15	16	31
0	34	.69	24	254	4	1.5	6	.05	.24	.06	.95	.15	189	16
0	21	.35	19	277	2	.5	4	.04	.19	.06	.67	.19	29	13
0	113	.44	16	247	9	.4	149	.04	.84	.1	.55	.12	88	17
0	179	.95	25	213	42	.23	254	.04	.42	.1	.54	.1	65	22
0	2	.47	22	193	3	.48	8	.13	.07	.05	1.17	.17	24	4
0	2	.38	18	158	3	.4	7	.11	.06	.04	.96	.14	19	3
0	3	.29	16	121	4	.33	9	.07	.07	.06	1.07	.11	25	3
0	4	.49	22	172	365	.68	6	.03	.11	.07	1.23	.08	58	6
0	5	.44	24	195	286	.48	13	.04	.09	.08	1.23	.06	51	9
0	4	.07	3	40	1	.06	3	.01	.02	.01	.06	.01	4	1
0	13	7.23	12	176	1531	.76	25	.04	.24	.04	.69	.16	34	5
0	103	1.71	20	218	42	.23	385	.1	1.38	.14	.44	.14	15	19
0	147	1.89	25	244	46	.29	614	.14	2.63	.18	.54	.17	14	19
0	6	.35	13	246	3	.15	3	.07	.03	.02	.59	.08	14	1
0	26	.41	7	157	11	.39	51	.04	.22	.04	.2	.01	71	3
0	26	.41	7	157	11	.39	51	.04	.22	.04	.2	.01	71	3
0	80	4.74	79	477	11	2.51	2	.19	.57	.1	.86	.23	282	2

*This value is expressed in retinol equivalents (RE). All other values are in retinol activity equivalents (RAE).

Table A–1

Food Composition (Computer code number is for Wadsworth Diet Analysis program) (For purposes of calculations, use "0" for t, <1, <.1, <.01, etc.)

Computer Code Number	Food Description	Measure	Wt (g)	H_2O (%)	Ener (cal)	Prot (g)	Carb (g)	Dietary Fiber (g)	Fat (g)	Fat Breakdown (g) Sat	Mono	Poly
	VEGETABLES AND LEGUMES—Continued											
1939	Grape leaf, raw:	1 ea	3	73	3	<1	1	<1	<1	t	t	t
7914	Cup	1 c	14	73	13	1	2	2	<1	t	t	.1
855	Great northern beans, cooked	1 c	177	69	209	15	37	12	1	.2	t	.3
857	Jerusalem artichoke, raw slices	1 c	150	78	114	3	26	2	<1	0	t	t
1794	Jicama	1 c	120	90	46	1	11	6	<1	t	t	.1
	Kale, cooked, drained:											
858	From raw	1 c	130	91	36	2	7	3	1	.1	t	.3
859	From frozen	1 c	130	90	39	4	7	3	1	.1	t	.3
860	Kidney beans, canned	1 c	256	77	218	13	40	16	1	.1	.1	.5
1181	Kohlrabi, raw slices	1 c	135	91	36	2	8	5	<1	t	t	.1
861	Kohlrabi, cooked	1 c	165	90	48	3	11	2	<1	t	t	.1
1183	Leeks, raw, chopped	1 c	89	83	54	1	13	2	<1	t	t	.1
1182	Leeks, cooked, chopped	1 c	104	91	32	1	8	1	<1	t	t	.1
862	Lentils, cooked from dry	1 c	198	70	230	18	40	16	1	.1	.1	.3
1288	Lentils, sprouted, stir fried	1 c	124	69	125	11	26	5	1	.1	.1	.2
1289	Lentils, sprouted, raw	1 c	77	67	82	7	17	3	<1	t	.1	.2
	Lettuce:											
	Butterhead/Boston types:											
863	Head, 5" diameter	¼ ea	41	96	5	1	1	<1	<1	t	t	t
864	Leaves, inner or outer	4 ea	30	96	4	<1	1	<1	<1	t	t	t
	Iceberg/crisphead:											
867	Chopped or shredded	1 c	55	96	7	1	1	1	<1	t	t	.1
865	Head, 6" diameter	1 ea	539	96	65	5	11	8	1	.1	t	.5
866	Wedge, ¼ head	1 ea	135	96	16	1	3	2	<1	t	t	.1
868	Looseleaf, chopped	½ c	28	94	5	<1	1	1	<1	t	t	t
869	Romaine, chopped	½ c	28	95	4	<1	1	<1	<1	t	t	t
870	Romaine, inner leaf	3 pce	30	95	4	<1	1	1	<1	t	t	t
1930	Luffa, cooked (Chinese okra)	1 c	178	90	57	3	13	4	<1	.1	t	.1
6777	Manioc, raw	1 c	206	60	330	3	78	4	1	.2	.2	.1
	Mushrooms:											
871	Raw, sliced	½ c	35	92	9	1	1	<1	<1	t	t	t
872	Cooked from fresh, pieces	½ c	78	91	21	2	4	2	<1	t	t	.1
1962	Stir fried, shitake slices	½ c	73	83	40	1	10	2	<1	t	t	t
873	Canned, drained	½ c	78	91	19	1	4	2	<1	t	t	.1
1951	Mushroom caps, pickled	8 ea	47	92	11	1	2	<1	<1	t	t	.1
	Mustard greens:											
874	Cooked from raw	½ c	70	94	10	2	1	1	<1	t	.1	t
875	Cooked from frozen	½ c	75	94	14	2	2	2	<1	t	.1	t
876	Navy beans, cooked from dry	1 c	182	63	258	16	48	12	1	.3	.1	.4
	Okra, cooked:											
877	From fresh pods	8 ea	85	90	27	2	6	2	<1	t	t	t
878	From frozen slices	1 c	184	91	51	4	11	5	1	.1	.1	.1
1236	Batter Fried from fresh	1 c	92	67	175	2	14	2	12	1.7	3.1	7.1
1930	Chinese, (Luffa), cooked	1 c	178	90	57	3	13	4	<1	.1	t	.1
	Onions:											
879	Raw, chopped	½ c	80	90	30	1	7	1	<1	t	t	t
880	Raw, sliced	½ c	58	90	22	1	5	1	<1	t	t	t
881	Cooked, drained, chopped	½ c	105	88	46	1	11	1	<1	t	t	.1
882	Dehydrated flakes	¼ c	14	4	49	1	12	1	<1	t	t	t
1934	Onions, pearl, cooked	½ c	93	88	41	1	9	1	<1	t	t	.1
	Spring/green onions, bulb and top,											
883	chopped	½ c	50	90	16	1	4	1	<1	t	t	t
884	Onion rings, breaded, heated f/frozen	2 ea	20	28	81	1	8	<1	5	1.7	2.2	1
1917	Palm hearts, cooked slices	1 c	146	69	150	4	39	2	<1	.1	t	.1
885	Parsley, raw, chopped	½ c	30	88	11	1	2	1	<1	t	.1	t
888	Parsnips, sliced, cooked	½ c	78	78	63	1	15	3	<1	t	.1	t
	Peas:											
	Black-eyed, cooked:											
814	From dry, drained	½ c	86	70	100	7	18	6	<1	.1	t	.2
815	From fresh, drained	½ c	82	75	79	3	17	4	<1	.1	t	.1
816	From frozen, drained	½ c	85	66	112	7	20	5	1	.1	.1	.2
889	Edible pod peas, cooked	½ c	80	89	34	3	6	2	<1	t	t	.1
890	Green, canned, drained:	½ c	85	82	59	4	11	3	<1	.1	t	.1

Chol (mg)	Calc (mg)	Iron (mg)	Magn (mg)	Pota (mg)	Sodi (mg)	Zinc (mg)	VT-A (μg)	Thia (mg)	VT-E (mg)	Ribo (mg)	Niac (mg)	V-B6 (mg)	Fola (μg)	VT-C (mg)
0	11	.08	3	8	<1	.02	40	<.01	.06	.01	.07	.01	2	<1
0	51	.37	13	38	1	.09	189	.01	.28	.05	.33	.06	12	2
0	120	3.77	88	692	4	1.56	<1	.28	.53	.1	1.21	.21	181	2
0	21	5.1	25	644	6	.18	1	.3	.28	.09	1.95	.12	19	6
0	14	.72	14	180	5	.19	1	.02	.55	.03	.24	.05	14	24
0	94	1.17	23	296	30	.31	481	.07	1.11	.09	.65	.18	17	53
0	179	1.22	23	417	19	.23	413	.06	.23	.15	.87	.11	18	33
0	61	3.23	72	658	873	1.41	0	.27	.13	.22	1.17	.06	131	3
0	32	.54	26	473	27	.04	3	.07	.65	.03	.54	.2	22	84
0	41	.66	31	561	35	.51	3	.07	2.76	.03	.64	.25	20	89
0	52	1.87	25	160	18	.11	4	.05	.82	.03	.36	.21	57	11
0	31	1.14	15	90	10	.06	3	.03	.63	.02	.21	.12	25	4
0	38	6.59	71	731	4	2.51	1	.33	.22	.14	2.1	.35	358	3
0	17	3.84	43	352	12	1.98	2	.27	.11	.11	1.49	.2	83	16
0	19	2.47	28	248	8	1.16	2	.18	.07	.1	.87	.15	77	13
0	13	.12	5	105	2	.07	20	.02	.18	.02	.12	.02	30	3
0	10	.09	4	77	1	.05	15	.02	.13	.02	.09	.01	22	2
0	10	.27	5	87	5	.12	9	.02	.15	.02	.1	.02	31	2
0	102	2.7	48	852	48	1.19	89	.25	1.51	.16	1.01	.22	302	21
0	26	.67	12	213	12	.3	22	.06	.38	.04	.25	.05	76	5
0	19	.39	3	74	3	.08	27	.01	.12	.02	.11	.01	14	5
0	10	.31	2	81	2	.07	36	.03	.12	.03	.14	.01	38	7
0	11	.33	2	87	2	.07	39	.03	.13	.03	.15	.01	41	7
0	112	.8	101	573	9	.98	52	.23	1.23	.1	1.55	.33	81	29
0	33	.56	43	558	29	.7	3	.18	.39	.1	1.76	.18	56	42
0	2	.36	3	130	1	.26	0	.03	.04	.15	1.41	.04	4	1
0	5	1.36	9	278	2	.68	0	.06	.09	.23	3.48	.07	14	3
0	2	.32	10	85	3	.97	0	.03	.09	.12	1.1	.12	15	<1
0	9	.62	12	101	332	.56	0	.07	.09	.02	1.24	.05	9	0
0	2	.51	5	140	2	.28	0	.03	.05	.16	1.42	.03	6	1
0	52	.49	10	141	11	.08	106	.03	1.41	.04	.3	.07	51	18
0	76	.84	10	104	19	.15	168	.03	1.31	.04	.19	.08	52	10
0	127	4.51	107	670	2	1.93	<1	.37	.73	.11	.97	.3	255	2
0	54	.38	48	274	4	.47	25	.11	.59	.05	.74	.16	39	14
0	177	1.23	94	431	6	1.14	47	.18	1.27	.23	1.44	.09	269	22
2	61	1.26	36	190	122	.5	39*	.18	3.04	.14	1.44	.12	38	10
0	112	.8	101	573	9	.98	52	.23	1.23	.1	1.55	.33	81	29
0	16	.18	8	126	2	.15	0	.03	.1	.02	.12	.09	15	5
0	12	.13	6	91	2	.11	0	.02	.07	.01	.09	.07	11	4
0	23	.25	12	174	3	.22	0	.04	.14	.02	.17	.13	16	5
0	36	.22	13	227	3	.26	0	.07	.19	.01	.14	.22	23	10
0	20	.22	10	154	3	.19	0	.04	.12	.02	.15	.12	14	5
0	36	.74	10	138	8	.19	10	.03	.06	.04	.26	.03	32	9
0	6	.34	4	26	75	.08	2	.06	.14	.03	.72	.01	13	<1
0	26	2.47	15	2636	20	5.45	5	.07	.73	.25	1.25	1.06	30	10
0	41	1.86	15	166	17	.32	78	.03	.54	.03	.39	.03	46	40
0	29	.45	23	286	8	.2	0	.06	.78	.04	.56	.07	45	10
0	21	2.16	46	239	3	1.11	1	.17	.24	.05	.43	.09	179	<1
0	105	.92	43	343	3	.84	32	.08	.18	.12	1.15	.05	104	2
0	20	1.8	42	319	4	1.21	3	.22	.33	.05	.62	.08	120	2
0	34	1.58	21	192	3	.3	5	.1	.31	.06	.43	.11	23	38
0	17	.81	14	147	214	.6	33	.1	.32	.07	.62	.05	37	8

*This value is expressed in retinol equivalents (RE). All other values are in retinol activity equivalents (RAE).

Table A–1

Food Composition (Computer code number is for Wadsworth Diet Analysis program) (For purposes of calculations, use "0" for t, <1, <.1, <.01, etc.)

Computer Code Number	Food Description	Measure	Wt (g)	H_2O (%)	Ener (cal)	Prot (g)	Carb (g)	Dietary Fiber (g)	Fat (g)	Fat Breakdown (g) Sat	Mono	Poly
	VEGETABLES AND LEGUMES—Continued											
5267	Unsalted	½ c	124	86	66	4	12	4	<1	.1	t	.2
891	Green, cooked from frozen	½ c	80	80	62	4	11	4	<1	t	t	.1
1786	Snow peas, raw	½ c	49	89	21	1	4	1	<1	t	t	t
1787	Snow peas, raw	10 ea	34	89	14	1	3	1	<1	t	t	t
892	Split, green, cooked from dry	½ c	98	69	116	8	21	8	<1	.1	.1	.2
1187	Peas & carrots, cooked from frozen	½ c	80	86	38	2	8	2	<1	.1	t	.2
1186	Peas & carrots, canned w/liquid	½ c	128	88	49	3	11	3	<1	.1	t	.2
	Peppers, hot:											
893	Hot green chili, canned	½ c	68	92	14	1	3	1	<1	t	t	t
894	Hot green chili, raw	1 ea	45	88	18	1	4	1	<1	t	t	t
1715	Hot red chili, raw, diced	1 tbs	9	88	4	<1	1	<1	<1	t	t	t
1988	Jalapeno, raw	1 ea	45	90	11	<1	2		<1			
895	Jalapeno, chopped, canned	½ c	68	89	18	1	3	2	1	.1	t	.3
1918	Jalapeno wheels, in brine (Ortega)	2 tbs	29		10	0	2		0	0	0	0
	Peppers, sweet, green:											
896	Whole pod (90g with refuse), raw	1 ea	119	92	32	1	8	2	<1	t	t	.1
897	Cooked, chopped (1 pod cooked = 73g)	½ c	68	92	19	1	5	1	<1	t	t	.1
	Peppers, sweet, red:											
1286	Raw, chopped	½ c	75	92	20	1	5	1	<1	t	t	.1
1807	Raw, each	1 ea	74	92	20	1	5	1	<1	t	t	.1
1287	Cooked, chopped	½ c	68	92	19	1	5	1	<1	t	t	.1
	Peppers, sweet, yellow:											
1872	Raw, large	1 ea	186	92	50	2	12	2	<1	.1	t	.2
1873	Strips	10 pce	52	92	14	1	3	<1	<1	t	t	.1
898	Pinto beans, cooked from dry	½ c	85	64	116	7	22	7	<1	.1	.1	.2
1191	Poi - two finger	½ c	120	72	134	<1	33	<1	<1	t	t	.1
	Potatoes:											
	Baked in oven, 4¾"x2⅓" diam											
899	With skin	1 ea	202	71	220	5	51	5	<1	.1	t	.1
900	Flesh only	1 ea	156	75	145	3	34	2	<1	t	t	.1
901	Skin only	1 ea	58	47	115	2	27	5	<1	t	t	t
	Baked in microwave, 4¾"x 2⅓"dm:											
902	With skin	1 ea	202	72	212	5	49	5	<1	.1	t	.1
903	Flesh only	1 ea	156	74	156	3	36	2	<1	t	t	.1
904	Skin only	1 ea	58	63	77	3	17	3	<1	t	t	t
	Boiled, about 2½ diam:											
905	Peeled after boiling	1 ea	136	77	118	3	27	2	<1	t	t	.1
906	Peeled before boiling	1 ea	135	77	116	2	27	2	<1	t	t	.1
	French fried, strips 2-3½" long:											
907	Oven heated	10 ea	50	35	167	2	20	2	9	3	5.7	.7
908	Fried in vegetable oil	10 ea	50	38	158	2	20	2	8	1.9	4.7	.7
1188	Fried in veg and animal oil	10 ea	50	38	158	2	20	2	8	1.9	4.7	.7
909	Hashed browns from frozen	1 c	156	56	340	5	44	3	18	7	8	2.1
	Mashed:											
910	Home recipe with whole milk	½ c	105	78	81	2	18	2	1	.3	.2	.1
911	Home recipe with milk and marg	½ c	105	76	111	2	17	2	4	1.1	1.9	1.3
912	Prepared from flakes; water, milk, margarine, salt added	½ c	110	76	124	2	16	3	6	1.6	2.5	1.7
	Potato products, prepared:											
	Au gratin:											
913	From dry mix	½ c	123	79	114	3	16	1	5	3.2	1.4	.2
914	From home recipe,using butter	½ c	122	74	161	6	14	2	9	4.3	3.2	1.3
	Scalloped:											
915	From dry mix	½ c	122	79	113	3	16	1	5	3.2	1.5	.2
916	From home recipe,using butter	½ c	123	81	106	4	13	2	5	1.7	1.7	.9
	Potato Salad (see Mixed Dishes #715)											
1192	Potato Puffs, cooked from frozen	½ c	64	53	142	2	19	2	7	3.3	2.8	.5
918	Pumpkin, cooked from fresh, mashed	½ c	123	94	25	1	6	1	<1	t	t	t
919	Pumpkin, canned	½ c	123	90	42	1	10	4	<1	.2	t	t
1891	Radicchio, raw, shredded	½ c	20	93	5	<1	1	<1	<1	t	t	t
1894	Radicchio, raw, leaf	10 ea	80	93	18	1	4	1	<1	t	t	.1
920	Red radishes	10 ea	45	95	9	<1	2	1	<1	t	t	t

Chol (mg)	Calc (mg)	Iron (mg)	Magn (mg)	Pota (mg)	Sodi (mg)	Zinc (mg)	VT-A (µg)	Thia (mg)	VT-E (mg)	Ribo (mg)	Niac (mg)	V-B6 (mg)	Fola (µg)	VT-C (mg)
0	22	1.26	21	124	11	.87	24	.14	.47	.09	1.04	.08	36	12
0	19	1.26	23	134	70	.75	27	.23	.14	.08	1.18	.09	47	8
0	21	1.02	12	98	2	.13	3	.07	.19	.04	.29	.08	21	29
0	15	.71	8	68	1	.09	2	.05	.13	.03	.2	.05	14	20
0	14	1.26	35	355	2	.98	<1	.19	.38	.05	.87	.05	64	<1
0	18	.75	13	126	54	.36	310	.18	.26	.05	.92	.07	21	6
0	29	.96	18	128	333	.74	369	.09	.24	.07	.74	.11	23	8
0	5	.34	10	127	798	.12	21	.01	.47	.03	.54	.1	7	46
0	8	.54	11	153	3	.13	17	.04	.31	.04	.43	.12	10	109
0	2	.11	2	31	1	.03	48	.01	.06	.01	.09	.02	2	22
				2	2		30*		.37					53
0	16	1.28	10	131	1136	.23	58	.03	.47	.03	.27	.13	10	7
0				55	390		10*		.2					21
0	11	.55	12	211	2	.14	37	.08	.82	.04	.61	.29	26	106
0	6	.31	7	113	1	.08	20	.04	.47	.02	.32	.16	11	51
0	7	.34	7	133	1	.09	214	.05	.52	.02	.38	.19	16	143
0	7	.34	7	131	1	.09	211	.05	.51	.02	.38	.18	16	141
0	6	.31	7	113	1	.08	128	.04	.47	.02	.32	.16	11	116
0	20	.86	22	394	4	.32	22	.05	1.28	.05	1.66	.31	48	341
0	6	.24	6	110	1	.09	6	.01	.36	.01	.46	.09	13	95
0	41	2.22	47	398	2	.92	<1	.16	.8	.08	.34	.13	146	2
0	19	1.06	29	220	14	.26	1	.16	.22	.05	1.32	.33	25	5
0	20	2.75	54	844	16	.65	0	.22	.1	.07	3.32	.7	22	26
0	8	.55	39	610	8	.45	0	.16	.06	.03	2.18	.47	14	20
0	20	4.08	25	332	12	.28	0	.07	.02	.06	1.78	.36	13	8
0	22	2.5	54	903	16	.73	0	.24	.1	.06	3.46	.69	24	30
0	8	.64	39	641	11	.51	0	.2	.06	.04	2.54	.5	19	24
0	27	3.45	21	377	9	.3	0	.04	.02	.04	1.29	.28	10	9
0	7	.42	30	515	5	.41	0	.14	.07	.03	1.96	.41	14	18
0	11	.42	27	443	7	.36	0	.13	.07	.03	1.77	.36	12	10
0	6	.83	11	270	307	.2	0	.04	.25	.02	1.33	.11	11	3
0	9	.38	17	366	108	.19	0*	.09	.25	.01	1.63	.12	14	5
6	9	.38	17	366	108	.19	0*	.09	.25	.01	1.63	.12	14	5
0	23	2.36	26	680	53	.5	0	.17	.3	.03	3.78	.2	11	10
2	27	.28	19	314	318	.3	6*	.09	.05	.04	1.17	.24	8	7
2	27	.27	19	303	310	.28	21*	.09	.31	.04	1.13	.23	8	6
4	54	.24	20	256	365	.2	23*	.12	.77	.05	.74	.01	8	11
18	102	.39	18	269	540	.29	38*	.02	1.48	.1	1.15	.05	9	4
18	145	.78	24	483	528	.84	46*	.08	.64	.14	1.21	.21	13	12
13	44	.46	17	248	416	.3	26*	.02	.18	.07	1.26	.05	12	4
7	70	.7	23	465	412	.49	23*	.08	.4	.11	1.3	.22	13	13
0	19	1	12	243	477	.19	1	.12	.03	.05	1.38	.15	11	4
0	18	.7	11	283	1	.28	66	.04	1.3	.1	.51	.05	11	6
0	32	1.71	28	253	6	.21	1356	.03	1.3	.07	.45	.07	15	5
0	4	.11	3	60	4	.12	<1	<.01	.45	.01	.05	.01	12	2
0	15	.46	10	242	18	.5	1	.01	1.81	.02	.2	.05	48	6
0	9	.13	4	104	11	.13	<1	<.01	0	.02	.13	.03	12	10

*This value is expressed in retinol equivalents (RE). All other values are in retinol activity equivalents (RAE).

Table A-1

Food Composition (Computer code number is for Wadsworth Diet Analysis program) (For purposes of calculations, use "0" for t, <1, <.1, <.01, etc.)

Computer Code Number	Food Description	Measure	Wt (g)	H_2O (%)	Ener (cal)	Prot (g)	Carb (g)	Dietary Fiber (g)	Fat (g)	Fat Breakdown (g) Sat	Mono	Poly
	VEGETABLES AND LEGUMES—Continued											
1793	Daikon radishes (Chinese) raw	½ c	44	95	8	<1	2	1	<1	t	t	t
921	Refried beans, canned	½ c	126	76	118	7	20	7	2	.6	.7	.2
1375	Rutabaga, cooked cubes	½ c	85	89	33	1	7	2	<1	t	t	.1
922	Sauerkraut, canned with liquid	½ c	118	93	22	1	5	3	<1	t	t	.1
923	Seaweed, kelp, raw	½ c	40	82	17	1	4	1	<1	.1	t	t
924	Seaweed, spirulina, dried	½ c	8	5	23	5	2	<1	1	.2	.1	.2
1866	Shallots, raw, chopped	1 tbs	10	80	7	<1	2	<1	<1	t	t	t
1557	Snow Peas, stir fried	½ c	83	89	35	2	6	2	<1	t	t	.1
925	Soybeans, cooked from dry	½ c	86	63	149	14	9	5	8	1.1	1.7	4.4
1996	Soybeans, dry roasted	½ c	86	1	387	34	28	7	19	2.7	4.1	10.5
	Soybean products:											
926	Miso	½ c	138	41	284	16	39	7	8	1.2	1.8	4.7
	Soy milk (see #144 and #2301 under Dairy)											
	Tofu (soybean curd)											
7540	Extra firm, silken	½ c	126	88	69	9	3	<1	2	.4	.4	1.3
7542	Firm, silken	½ c	126	87	78	9	3	<1	3	.5	.7	1.9
927	Regular	½ c	124	87	76	8	2	<1	5	.7	1	2.6
7541	Soft, silken	½ c	124	89	68	6	4	<1	3	.4	.6	1.9
	Spinach:											
928	Raw, chopped	½ c	15	92	3	<1	1	<1	<1	t	t	t
929	Cooked, from fresh, drained	½ c	90	91	21	3	3	2	<1	t	t	.1
930	Cooked from frozen (leaf)	½ c	95	90	27	3	5	3	<1	t	t	.1
931	Canned, drained solids:	½ c	107	92	25	3	4	3	1	.1	t	.2
5149	Unsalted	½ c	107	92	25	3	4	3	1	.1	t	.2
	Spinach souffle (see Mixed Dishes)											
	Squash, summer varieties,cooked w/skin:											
932	Varieties averaged	½ c	90	94	18	1	4	1	<1	.1	t	.1
933	Crookneck	½ c	90	94	18	1	4	1	<1	.1	t	.1
934	Zucchini	½ c	90	95	14	1	4	1	<1	t	t	t
	Squash, winter varieties, cooked:											
	Average of all varieties, baked:											
935	Mashed	1 c	245	89	96	2	21	7	2	.3	.1	.6
936	Cubes	1 c	205	89	80	2	18	6	1	.3	.1	.5
937	Acorn, baked, mashed	½ c	123	83	69	1	18	5	<1	t	t	.1
1218	Acorn, boiled, mashed	½ c	122	90	41	1	11	3	<1	t	t	t
	Butternut squash:											
938	Baked cubes	½ c	103	88	41	1	11	3	<1	t	t	t
1219	Baked, mashed	½ c	103	88	41	1	11	3	<1	t	t	t
1193	Cooked from frozen	½ c	120	88	47	1	12	3	<1	t	t	t
1194	Hubbard, baked, mashed	½ c	120	85	60	3	13	3	1	.2	.1	.3
1195	Hubbard, boiled, mashed	½ c	118	91	35	2	8	3	<1	.1	t	.2
1196	Spaghetti, baked or boiled	½ c	77	92	21	<1	5	1	<1	t	t	.1
1189	Succotash, cooked from frozen	½ c	85	74	79	4	17	3	1	.1	.1	.4
	Sweet potatoes:											
939	Baked in skin, peeled, 5 x 2" diam	1 ea	114	73	117	2	28	3	<1	t	t	.1
940	Boiled without skin, 5 x 2" diam	1 ea	151	73	159	2	37	3	<1	.1	t	.2
941	Candied, 2½ x 2"	1 pce	105	67	144	1	29	3	3	1.4	.7	.2
	Canned:											
942	Solid pack	½ c	128	74	129	3	30	2	<1	.1	t	.1
943	Vacuum pack, mashed	½ c	127	76	116	2	27	2	<1	.1	t	.1
944	Vacuum pack, 3¾ x 1"	2 pce	80	76	73	1	17	1	<1	t	t	.1
1940	Taro shoots, cooked slices	1 c	140	95	20	1	4	1	<1	t	t	t
1941	Taro, tahitian, cooked slices	1 c	137	86	60	6	9	1	1	.2	.1	.4
	Tomatillos:											
1877	Raw, each	1 ea	34	92	11	<1	2	1	<1	t	.1	.1
1875	Raw, chopped	1 c	132	92	42	1	8	3	1	.2	.2	.5
	Tomatoes:											
945	Raw, whole, 2⅗" diam	1 ea	123	94	26	1	6	1	<1	.1	.1	.2
946	Raw, chopped	1 c	180	94	38	2	8	2	1	.1	.1	.2
947	Cooked from raw	1 c	240	92	65	3	14	2	1	.1	.2	.4
948	Canned, solids and liquid:	1 c	240	94	46	2	10	2	1	t	t	.1
5741	Unsalted	1 c	240	94	46	2	10	2	<1	t	t	.1
1879	Tomatoes, sundried:	1 c	54	15	139	8	30	7	2	.2	.3	.6

PAGE KEY: A–2 = Beverages A–4 = Dairy A–8 = Eggs A–10 = Fat/Oil A–12 = Fruit A–18 = Bakery A–24 = Grain A–30 = Fish A–32 = Meats A–36 = Poultry A–38 = Sausage A–38 = Mixed/Fast A–44 = Nuts/Seeds A–46 = Sweets A–50 = Vegetables/Legumes A–60 = Vegetarian A–62 = Misc A–64 = Soups/Sauces A–66 = Fast A–80 = Convenience A–86 = Baby foods

Chol (mg)	Calc (mg)	Iron (mg)	Magn (mg)	Pota (mg)	Sodi (mg)	Zinc (mg)	VT-A (μg)	Thia (mg)	VT-E (mg)	Ribo (mg)	Niac (mg)	V-B6 (mg)	Fola (μg)	VT-C (mg)
0	12	.18	7	100	9	.07	0	.01	0	.01	.09	.02	12	10
10	44	2.09	42	336	377	1.47	0	.03	0	.02	.4	.18	14	8
0	41	.45	20	277	17	.3	24	.07	.13	.03	.61	.09	13	16
0	35	1.73	15	201	780	.22	1	.02	.12	.03	.17	.15	28	17
0	67	1.14	48	36	93	.49	2	.02	.35	.06	.19	<.01	72	1
0	10	2.28	16	109	84	.16	2	.19	.4	.29	1.03	.03	8	1
0	4	.12	2	33	1	.04	6	.01	.01	<.01	.02	.03	3	1
0	36	1.73	20	166	3	.22	5	.11	.32	.06	.47	.13	28	42
0	88	4.42	74	443	1	.99	<1	.13	1.68	.24	.34	.2	46	1
0	120	3.4	196	1173	2	4.1	1	.37	3.96	.65	.91	.19	176	4
0	91	3.78	58	226	5032	4.58	6	.13	.01	.34	1.19	.3	45	0
0	39	1.5	34	194	79	.76	0	.1	.18	.04	.3	.01		0
0	40	1.3	34	244	45	.77	0	.13	.24	.05	.31	.01		0
0	138	1.38	33	149	10	.79	<1	.06	.01	.05	.66	.06	55	<1
0	38	1.02	36	223	6	.64	0	.12	.25	.05	.37	.01		0
0	15	.41	12	84	12	.08	50	.01	.28	.03	.11	.03	29	4
0	122	3.21	78	419	63	.68	369	.09	.86	.21	.44	.22	131	9
0	139	1.44	66	283	82	.66	370	.06	.91	.16	.4	.14	103	12
0	136	2.46	81	370	29	.49	470	.02	1.39	.15	.41	.11	105	15
0	136	2.46	81	370	29	.49	470	.02	1.39	.15	.41	.11	105	15
0	24	.32	22	173	1	.35	13	.04	.11	.04	.46	.06	18	5
0	24	.32	22	173	1	.35	13	.04	.11	.04	.46	.08	18	5
0	12	.31	20	228	3	.16	11	.04	.11	.04	.38	.07	15	4
0	34	.81	20	1070	2	.64	436	.21	.29	.06	1.72	.18	69	23
0	29	.68	16	896	2	.53	365	.17	.25	.05	1.44	.15	57	20
0	54	1.14	53	538	5	.21	26	.2	.15	.02	1.08	.24	23	13
0	32	.68	32	321	4	.13	16	.12	.15	.01	.65	.14	13	8
0	42	.62	30	293	4	.13	361	.07	.17	.02	1	.13	20	16
0	42	.62	30	293	4	.13	361	.07	.17	.02	1	.13	20	16
0	23	.7	11	160	2	.14	200	.06	.16	.05	.56	.08	19	4
0	20	.56	26	430	10	.18	362	.09	.14	.06	.67	.21	19	11
0	12	.33	15	253	6	.12	237	.05	.14	.03	.39	.12	12	8
0	16	.26	8	90	14	.15	4	.03	.09	.02	.62	.08	6	3
0	13	.76	20	225	38	.38	10	.06	.31	.06	1.11	.08	28	5
0	32	.51	23	397	11	.33	1243	.08	.32	.14	.69	.27	26	28
0	32	.85	15	278	20	.41	1287	.08	.42	.21	.97	.37	17	26
8	27	1.19	12	198	73	.16	220	.02	3.99	.04	.41	.04	12	7
0	38	1.7	31	269	96	.27	968	.03	.35	.11	1.22	.3	14	7
0	28	1.13	28	396	67	.23	507	.05	.32	.07	.94	.24	22	33
0	18	.71	18	250	42	.14	319	.03	.2	.05	.59	.15	14	21
0	20	.57	11	482	3	.76	3	.05	1.4	.07	1.13	.16	4	26
0	204	2.14	70	854	74	.14	121	.06	3.7	.27	.66	.16	10	52
0	2	.21	7	91	<1	.07	2	.01	.13	.01	.63	.02	2	4
0	9	.82	26	354	1	.29	7	.06	.5	.05	2.44	.07	9	15
0	6	.55	13	273	11	.11	38	.07	.47	.06	.77	.1	18	23
0	9	.81	20	400	16	.16	56	.11	.68	.09	1.13	.14	27	34
0	14	1.34	34	670	26	.26	89	.17	.91	.14	1.8	.23	31	55
0	72	1.32	29	530	355	.38	72	.11	.77	.07	1.76	.22	19	34
0	72	1.32	29	545	24	.38	72	.11	.91	.07	1.76	.22	19	34
0	59	4.91	105	1850	1131	1.07	23	.28	<.01	.26	4.89	.18	37	21

*This value is expressed in retinol equivalents (RE). All other values are in retinol activity equivalents (RAE).

Table A–1

Food Composition

(Computer code number is for Wadsworth Diet Analysis program) (For purposes of calculations, use "0" for t, <1, <.1, <.01, etc.)

Computer Code Number	Food Description	Measure	Wt (g)	H_2O (%)	Ener (cal)	Prot (g)	Carb (g)	Dietary Fiber (g)	Fat (g)	Fat Breakdown (g) Sat	Mono	Poly
	VEGETABLES AND LEGUMES—Continued											
1881	Pieces	10 pce	20	15	52	3	11	2	1	.1	.1	.2
1885	Oil pack, drained	10 pce	30	54	64	2	7	2	4	.6	2.6	.6
2020	Tomato, raw	1 ea	62	94	13	1	3	1	<1	t	t	.1
949	Tomato juice, canned:	1 c	243	94	41	2	10	1	<1	t	t	.1
5397	Unsalted	1 c	243	94	41	2	10	2	<1	t	t	.1
	Tomato products, canned:											
950	Paste-no added salt	1 c	262	74	215	10	51	11	1	.2	.2	.6
951	Puree-no added salt	1 c	250	87	100	4	24	5	<1	.1	.1	.2
952	Sauce-with salt	1 c	245	89	73	3	18	3	<1	.1	.1	.2
953	Turnips, cubes, cooked from fresh	1 c	156	94	33	1	8	3	<1	t	t	.1
	Turnip greens, cooked:											
954	From fresh, leaves and stems	1 c	144	93	29	2	6	5	<1	.1	t	.1
955	From frozen, chopped	1 c	164	90	49	5	8	6	1	.2	t	.3
956	Vegetable juice cocktail, canned	1 c	242	94	46	2	11	2	<1	t	t	.1
	Vegetables, Mixed:											
957	Canned, drained	½ c	81	87	38	2	7	2	<1	t	t	.1
958	Frozen, cooked, drained	½ c	91	83	54	3	12	4	<1	t	t	.1
1818	Water chestnuts, Chinese, raw	½ c	62	73	60	1	15	2	<1	t	t	t
	Water chestnuts, canned:											
959	Slices	½ c	70	86	35	1	9	2	<1	t	t	t
960	Whole	4 ea	28	86	14	<1	3	1	<1	t	0	t
1190	Watercress, fresh, chopped	½ c	17	95	2	<1	<1	<1	<1	t	t	t
	VEGETARIAN FOODS:											
7509	Bacon strips, meatless	3 ea	15	49	46	2	1	<1	4	.7	1.1	2.3
1511	Baked beans, canned	½ c	127	73	118	6	26	6	1	.1	t	.2
7526	Bakon Crumbles	¼ c	7	8	31	2	2	1	2	0		
7548	Chicken, breaded, fried, meatless	1 pce	57	70	97	6	3	3	7	1	1.6	3.9
7547	Chicken slices, meatless	2 ea	60	59	132	10	4	3	8	1.3	2	4.3
7557	Chili w/meat substitute	½ c	107	65	141	19	15	5	2	.3	.6	.9
7549	Fish stick, meatless	2 ea	57	45	165	13	5	3	10	1.6	2.5	5.3
7550	Frankfurter, meatless	1 ea	51	58	102	10	4	2	5	.8	1.2	2.6
7504	GardenBurger, patty	1 ea	71	58	130	8	18	5	3	1	1.5	.5
7505	GardenSausage, patty	1 ea	71	59	130	7	18	4	3	2	.7	.3
7551	Luncheon slice, meatless	1 sl	67	46	188	17	6	3	11	1.7	2.6	5.6
7560	Meatloaf, meatless	1 ea	71	58	142	15	6	3	6	1	1.5	3.3
1171	Nuteena	1 ea	55	58	162	6	6	2	13	5.1	5.8	1.7
7556	Pot pie, meatless	1 ea	227	60	510	14	41	5	32	8.6	12.4	9.6
7554	Soyburger, patty	1 ea	71	58	142	15	6	3	6	1	1.5	3.3
7562	Soyburger w/cheese, patty	1 ea	135	51	308	20	30	4	12	3.6	3.6	3.6
8832	Soyburger, veggie, patty	1 ea	85	76	70	11	7	2	0	0	0	0
7517	Soy protein isolate	1 oz	28.35	5	96	23	2	2	1	.1	.2	.5
7564	Tempeh	1 c	166	60	320	31	16	9	18	3.7	5	6.3
7670	Vegan burger, patty	1 ea	78	71	83	13	7	4	<1	.1	.3	.1
8842	Veggie slices, soy	1 pce	15	68	17	4	1	0	0	0	0	0
8830	Veggie ground soy	⅓ c	55	70	60	12	3	3	0	0	0	0
	Vegetarian foods, Green Giant:											
7677	Breakfast links	3 ea	68	65	114	12	5	4	5		4.3	
7676	Breakfast patties	2 ea	57	65	95	10	5	3	4		3.6	
	Burger, harvest, patty:											
7673	Italian	1 ea	90	67	140	17	8	5	4	1.5	.5	.5
7674	Original	1 ea	90	65	138	18	7	6	4	1	2.1	.3
7675	Southwestern	1 ea	90	68	140	16	9	5	4	1.5	0	.5
	Vegetarian Foods, Loma Linda											
7727	Chik nuggets, frozen	5 pce	85	47	245	12	13	5	16	2.5	4	8.8
7753	Chik-fried, frozen	1 pce	57	51	178	11	1	1	15	1.9	3.7	8.7
7744	Franks, big, canned	1 ea	51	58	118	12	2	1	7	.8	1.5	3.7
7747	Linketts, canned	1 ea	35	60	72	7	1		4	.7	1.2	2.5
1173	Redi-burger, patty	1 ea	85	59	172	16	5		10	1.5	2.4	5.8
7755	Swiss stake w/gravy, canned	1 pce	92	71	120	9	8	4	6	.8	1.5	3.3
1174	Vege-Burger, patty	1 ea	55	71	66	10	2	2	2	.4	.6	.5
	Vegetarian foods, Morningstar Farms:											
7672	Better-n-burgers, svg	1 ea	78	71	83	13	7	4	<1	.1	.3	.1

Chol (mg)	Calc (mg)	Iron (mg)	Magn (mg)	Pota (mg)	Sodi (mg)	Zinc (mg)	VT-A (µg)	Thia (mg)	VT-E (mg)	Ribo (mg)	Niac (mg)	V-B6 (mg)	Fola (µg)	VT-C (mg)
0	22	1.82	39	685	419	.4	9	.11	<.01	.1	1.81	.07	14	8
0	14	.8	24	470	80	.23	19	.06	.16	.11	1.09	.1	7	30
0	3	.28	7	138	6	.06	19	.04	.24	.03	.39	.05	9	12
0	22	1.41	27	535	877	.34	68	.11	2.21	.07	1.64	.27	49	44
0	22	1.41	27	535	24	.34	68	.11	2.21	.07	1.64	.27	49	44
0	92	5.08	134	2454	231	2.1	320	.41	11.3	.5	8.44	1	58	111
0	42	3.1	60	1065	85	.55	160	.18	6.3	.13	4.29	.38	27	26
0	34	1.89	47	909	1482	.61	120	.16	3.43	.14	2.82	.38	22	32
0	34	.34	12	211	78	.31	0	.04	.05	.04	.47	.1	14	18
0	197	1.15	32	292	42	.2	396	.06	2.48	.1	.59	.26	170	39
0	249	3.18	43	367	25	.67	654	.09	4.79	.12	.77	.11	64	36
0	27	1.02	27	467	653	.48	142	.1	.77	.07	1.76	.34	51	67
0	22	.85	13	236	121	.33	472	.04	.49	.04	.47	.06	19	4
0	23	.75	20	154	32	.45	195	.06	.33	.11	.77	.07	17	3
0	7	.04	14	362	9	.31	0	.09	.74	.12	.62	.2	10	2
0	3	.61	3	83	6	.27	0	.01	.35	.02	.25	.11	4	1
0	1	.24	1	33	2	.11	0	<.01	.14	.01	.1	.04	2	<1
0	20	.03	4	56	7	.02	40	.01	.17	.02	.03	.02	2	7
0	3	.36	3	25	220	.06	1	.66	1.04	.07	1.13	.07	6	0
0	63	.37	41	376	504	1.78	11	.19	.67	.08	.54	.17	30	4
0	16	.28			163		0							0
0	13	.97	7	171	228	.37	0	.4	1.11	.27	2.68	.28	32	0
0	21	.78	10	198	474	.42	0	.66	1.61	.24	3.18	.42	46	0
0	54	4.38	36	366	355	1.27	37	.12	1.28	.07	1.22	.15	82	6
0	54	1.14	13	342	279	.8	0	.63	2.25	.51	6.84	.85	58	0
0	17	.92	9	76	219	.61	0	.56	.98	.61	8.16	.5	40	0
11	84	0	30	193	290	.89	10*	.11	.2	.15	1.08	.08	10	0
10	80	.5	28	143	300	.84	24*	.1	.2	.15	1.01	.07	10	<1
0	27	1.54	15	188	576	1.07	0	.64	2.01	.37	7.37	.74	67	0
0	21	1.49	13	128	391	1.28	0	.64	1.23	.43	7.1	.85	55	0
0	9	.27	33	166	119	.46	0	.1		.35	1.04	.45	49	0
19	68	2.96	32	378	486	1.08	785*	.82	4.23	.44	5.17	.41	58	10
0	21	1.49	13	128	391	1.28	0	.64	1.23	.43	7.1	.85	55	0
9	158	2.94	27	242	922	1.91	36*	.8	1.58	.59	8.32	.84	64	1
0	60	1.8		300	520		0							0
0	50	4.11	11	23	285	1.14	0	.05	0	.03	.41	.03	50	0
0	184	4.48	134	684	15	1.89	0	.13	.03	.59	4.38	.36	40	0
0	80	2.66	15	398	351	.69	0	.23	.01	.51	3.77	.18	225	0
0				37	133									
0	40	9		220	250	3.75	0	.22		.17	.4	.2		0
0							0*	0		.18	.09	.27		<1
0							0*	0		.15	.07	2.28		<1
0	80	2.7			370	6.75	0	.3		.14	4	.3		0
0	102	3.85	70	432	411	8.07	0	.31	1.56	.2	6.3	.39	22	0
0	80	2.7			370	6.75	0	.3		.14	4	.3		0
2	40	1.4		153	709	.43	0	.67		.3	2.89	.45		0
4	2	.63		76	503	.2	0	.98		.46	2.1	.35		0
0	10	.99		61	224	1.2	0	.28		.68	5.78	.67		
1	4	.39		29	160	.46	0	.13		.22	.64	.29		0
1	12	1.06	16	121	455	1.11	0	.14		.3	1.9	.51	21	0
2	24	.31		225	433	.41	0	1.25		.65	5.41	1		0
0	8	.5	12	30	114	.58	0	.2		.25	.78	.31	15	0
0	80	2.66	15	398	351	.69	0	.23	.01	.51	3.77	.18	225	0

*This value is expressed in retinol equivalents (RE). All other values are in retinol activity equivalents (RAE).

Table A-1

Food Composition (Computer code number is for Wadsworth Diet Analysis program) (For purposes of calculations, use "0" for t, <1, <.1, <.01, etc.)

Computer Code Number	Food Description	Measure	Wt (g)	H_2O (%)	Ener (cal)	Prot (g)	Carb (g)	Dietary Fiber (g)	Fat (g)	Fat Breakdown (g) Sat	Mono	Poly
	VEGETARIAN FOODS—Continued											
7766	Better-n-eggs	¼ c	57	88	26	5	<1	0	<1	.1	.1	.1
57436	Breakfast links	2 pce	45	67	64	9	2	1	2	.4	.5	1
7752	Breakfast strips	2 pce	16	43	56	2	2	1	4	.7	.9	2.6
7725	Burger crumbles, svg	1 ea	55	60	116	11	3	3	6	1.6	2.3	2.5
7726	Burger, spicy black bean	1 ea	78	60	115	12	15	5	1	.2	.2	.4
7665	Chik pattie	1 ea	71	54	153	9	14	3	6	.9	1.6	3.6
7724	Frank, deli	1 ea	57	52	141	13	5	3	8	1.1	2.5	4.2
7722	Garden vege pattie	1 ea	67	60	119	11	10	4	4	.5	1.1	2.2
7746	Grillers	1 ea	64	56	139	15	5	2	6	1.1	1.5	3.1
7664	Prime pattie	1 ea	64	64	94	16	4	3	2	.2	.4	.6
	Vegetarian foods, Worthington:											
7634	Beef style, meatless, frzn	3 pce	55	58	113	9	4	3	7	1.2	2.7	2.6
7732	Burger, meatless, patty	¼ c	55	71	60	9	2	1	2	.3	.5	1.1
1846	Chik slices, canned	2 pce	60	78	62	6	1	1	4	.6	.9	2.3
1833	Chili, canned	½ c	106	73	136	9	10	4	7	1.1	1.7	4.1
1835	Choplets, slices, canned	2 pce	92	72	93	17	3	2	2	.9	.3	.3
7608	Corned beef style, meatless, frzn	4 pce	57	55	138	10	5	2	9	1.9	4.1	3.1
1831	Country stew, canned	1 c	240	81	208	13	20	5	9	1.6	2.3	4.8
7632	Egg Roll, meatless, frzn	1 ea	85	53	181	6	20	2	8	1.7	4.5	2.3
1838	Numete, slices, canned	1 pce	55	58	132	6	5	3	10	2.4	4.4	2.7
1839	Prime stakes, slices, canned	1 pce	92	71	136	9	4	4	9	1.4	2.9	4.9
1840	Protose, slices, canned	1 pce	55	53	131	13	5	3	7	1	3	2.4
7606	Roast, dinner, meatless, frzn	1 ea	85	63	180	12	5	3	12	2.2	5	5.2
1842	Saucette links, canned	1 pce	38	62	86	6	1		6	1.1	1.6	3.8
1844	Savory slices, canned	1 pce	28	66	48	3	2	1	3	1.2	1.3	.6
7735	Stakelets, frzn	1 pce	71	58	145	12	6	2	8	1.4	2.7	3.9
1847	Turkee slices, canned	1 pce	33	64	68	5	1	1	5	.8	1.9	2.1
	MISCELLANEOUS											
	Baking Powders for home use:											
	Sodium aluminum sulfate:											
962	With monocalcium phosphate monohydrate	1 tsp	5	2	6	<1	2	0	0	0	0	0
963	With monocalcium phosphate monohydrate, calcium sulfate	1 tsp	5	5	3	0	1	<1	0	0	0	0
964	Straight Phosphate	1 tsp	5	4	3	<1	1	<1	0	0	0	0
965	Low sodium	1 tsp	5	6	5	<1	2	<1	<1	t	0	t
1204	Baking soda	1 tsp	5		0	0	0	0	0	0	0	0
966	Basil, dried	1 tbsp	5	6	13	1	3	2	<1	t	t	.1
2068	Cajun seasoning	1 tsp	3	5	6	<1	1	<1	<1			
961	Carob flour	1 c	103	4	229	5	91	41	1	.1	.2	.2
967	Catsup:	¼ c	61	67	63	1	17	1	<1	t	t	.1
968	Tablespoon	1 tbsp	15	67	16	<1	4	<1	<1	t	t	t
1200	Cayenne/red pepper	1 tbsp	5	8	16	1	3	1	1	.2	.1	.4
969	Celery seed	1 tsp	2	6	8	<1	1	<1	<1	t	.3	.1
1203	Chili powder:	1 tbsp	8	8	25	1	4	3	1	.2	.3	.6
970	Teaspoon	1 tsp	3	8	9	<1	2	1	<1	.1	.1	.2
	Chocolate:											
971	Baking, unsweetened, square	1 oz	28	1	146	3	8	4	15	9.1	5.2	.5
	For other chocolate items, see Sweeteners & Sweets											
972	Cilantro/Coriander, fresh	1 tbsp	1	92		<1	<1	<1	<1	0	t	0
2287	Cinnamon	1 tsp	2	10	5	<1	2	1	<1	t	t	t
1197	Cornstarch	1 tbsp	8	8	30	<1	7	<1	<1	.7	t	t
2239	Curry powder	1 tsp	2	10	6	<1	1	1	<1	t	.1	.1
1202	Dill weed, dried	1 tbsp	3	7	8	1	2	<1	<1	t	.1	t
975	Garlic cloves	1 ea	3	59	4	<1	1	<1	<1	t	0	t
2238	Garlic powder	1 tsp	3	6	10	<1	2	<1	<1	t	t	t
977	Gelatin, dry, unsweetened: Envelope	1 ea	7	13	23	6	0	0	<1	t	.3	.5
978	Ginger root, slices, raw	2 pce	5	82	3	<1	1	<1	<1	t	t	t
1198	Horseradish, prepared	1 tbsp	15	85	7	<1	2	<1	<1	t	t	.1
1997	Hummous/hummus	1 c	246	65	421	12	50	12	21	3.1	8.7	7.8
1909	Mustard, country dijon	1 tsp	5		5	<1	<1	0	0	0	0	0

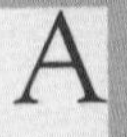

Chol (mg)	Calc (mg)	Iron (mg)	Magn (mg)	Pota (mg)	Sodi (mg)	Zinc (mg)	VT-A (µg)	Thia (mg)	VT-E (mg)	Ribo (mg)	Niac (mg)	V-B6 (mg)	Fola (µg)	VT-C (mg)
1	24	.83		60	98	.6	32	.05		.36	0	.11		0
1	9	1.77	16	46	355	.35	0	5.43		.15	2.5	.38	12	0
<1	3	.33		16	228	.06	0	.67		.05	.75	.08		0
0	40	3.2	1	89	238	.82	0	4.96	.35	.18	1.49	.27		0
1	56	1.84	44	269	499	.93	7	8.06	.36	.14	0	.21		0
2	14	1.31		202	581	.35	0	.7		.15	2.26	.18		0
1	22	.77	5	63	545	.48	0	.18	1.59	.03	0	.01		0
1	48	1.21	29	180	382	.58	38	6.47	.98	.1	0	0	29	0
2	22	2.5		122	269	.67	0	11.8		.2	4.86	.48		0
1	46	2.14		142	247	.74	0*	.51		.25	.92	.41		2
0	4	2.63		44	624	.22	0	.89		.34	6.46	.56		0
0	4	1.73		25	269	.38	0	.13		.1	1.96	.24		0
1	9	.73		111	257	.26	0	.06		.05	.37	.08		0
0	20	1.49		195	523	.57	0	.02		.03	1.04	.31		0
0	6	.37		40	500	.65	0	.05		.05	0	.05		0
1	6	1.17		58	524	.26	0	10.6		.07	1.36	.3		0
2	51	5.09		270	826	1.03	108	1.85		.29	4.22	.86		0
1	15	.57		96	384	.31	0	1.22		.19	0	.03		0
0	10	1.12		155	272	.56	0	.08		.06	.54	.2		0
2	12	.38		82	445	.38	0	.12		.13	1.98	.38		0
<1	1	1.84		50	283	.7	0	.18		.13	1.34	.24		0
2	36	2.87		38	566	.64	0	2.13		.25	6.02	.6		0
1	9	1.15		25	205	.26	0	.59		.08	.09	.13		0
<1		.47		14	179	.08	0	.08		.06	.48	.1		0
2	49	.99		95	484	.5	0	1.51		.12	3.1	.26		0
1	3	.47		16	203	.11	0	1.13		.05	.39	.09		0
0	97	0		7	547	0	0*	0	0	0	0	0	0	0
0	294	.55	1	1	530	<.01	0	0	0	0	0	0	0	0
0	368	.56	2		395	<.01	0	0	0	0	0	0	0	0
0	217	.41	1	505	4	.04	0	0	<.01	0	0	0	0	0
0	0	0	0	0	1368	0	0	0	0	0	0	0	0	0
0	106	2.1	21	172	2	.29	23	.01	.08	.02	.35	.06	14	3
				29	474									
0	358	3.03	56	852	36	.95	1	.05	.65	.47	1.95	.38	30	<1
0	12	.43	13	293	723	.14	31	.05	.89	.04	.83	.11	9	9
0	3	.1	3	72	178	.03	8	.01	.22	.01	.2	.03	2	2
0	7	.39	8	101	1	.12	104	.02	.24	.05	.43	.1	5	4
0	35	.9	9	28	3	.14	<1	.01	.02	.01	.06	.01	<1	<1
0	22	1.14	14	153	81	.22	140	.03	.08	.06	.63	.15	8	5
0	8	.43	5	57	30	.08	52	.01	.03	.02	.24	.06	3	2
0	21	1.77	87	233	4	1.12	1	.02	.34	.05	.31	.03	2	0
0	1	.02		5	<1	<.01	3	<.01	.02	<.01	.01	<.01	1	<1
0	25	.76	1	10	1	.04	<1	<.01	0	<.01	.03	<.01	1	1
0		.04			1	<.01	0	0	0	0	0	0	0	0
0	10	.59	5	31	1	.08	1	<.01	.01	.01	.07	.01	3	<1
0	53	1.46	13	99	6	.1	9	.01		.01	.08	.04		1
0	5	.05	1	12	1	.03	0	.01	0	<.01	.02	.04	<1	1
0	2	.08	2	33	1	.08	0	.01	0	<.01	.02	.08	<1	1
0	4	.08	2	1	14	.01	0	<.01	0	.02	.01	0	2	0
0	1	.02	2	21	1	.02	0	<.01	.01	<.01	.03	.01	1	<1
0	8	.06	4	37	47	.12	<1	<.01	<.01	<.01	.06	.01	9	4
0	123	3.86	71	428	600	2.71	2	.23	2.46	.13	1.01	.98	145	19
0				10	120									

*This value is expressed in retinol equivalents (RE). All other values are in retinol activity equivalents (RAE).

Table A-1

Food Composition (Computer code number is for Wadsworth Diet Analysis program) (For purposes of calculations, use "0" for t, <1, <.1, <.01, etc.)

Computer Code Number	Food Description	Measure	Wt (g)	H_2O (%)	Ener (cal)	Prot (g)	Carb (g)	Dietary Fiber (g)	Fat (g)	Fat Breakdown (g) Sat	Mono	Poly
	MISCELLANEOUS—Continued											
2019	Mustard, gai choy chinese	1 tbsp	16	94	3	<1	1		<1			
979	Mustard, prepared (1 packet = 1 tsp)	1 tsp	5	80	4	<1	<1	<1	<1	t	.2	t
	Miso (see #926 under Vegetables and Legumes, Soybean products)											
980	Olives, green	5 ea	20	78	23	<1	<1	<1	3	.3	1.9	.2
981	Olives, ripe, pitted	5 ea	22	80	25	<1	1	1	2	.3	1.7	.2
26008	Onion powder	1 tsp	2	5	7	<1	2	<1	<1	t	t	t
2237	Oregano, ground	1 tsp	2	7	6	<1	1	1	<1	.1	t	.1
2236	Paprika	1 tsp	2	10	6	<1	1	<1	<1	t	t	.2
887	Parsley, freeze dried	¼ c	1	2	3	<1	<1	<1	<1	t	t	t
	Parsley, fresh (see #885 and #886)											
985	Pepper, black	1 tsp	2	11	5	<1	1	1	<1	t	t	t
	Pickles:											
986	Dill, medium, 3¾ x 1¼" diam	1 ea	65	92	12	<1	3	1	<1	t	t	t
987	Fresh pack, slices, 1½" diam x ¼"	2 pce	15	79	11	<1	3	<1	<1	0	0	t
988	Sweet, medium	1 ea	35	65	41	<1	11	<1	<1	t	t	t
989	Pickle relish, sweet	1 tbsp	15	63	21	<1	5	<1	<1	t	t	t
	Popcorn (see Grain Products #539-541)											
917	Potato chips:	10 pce	20	2	107	1	11	1	7	2.2	2	2.4
44076	Unsalted	1 oz	28	2	150	2	15	1	10	3.1	2.8	3.4
1201	Sage, ground	1 tsp	1	8	3	<1	1	<1	<1	.1	t	t
1347	Salsa, from recipe	1 tbsp	15	95	3	<1	1	<1	<1	t	t	t
2218	Salsa, pico de gallo, medium	1 tbsp	15	92	2	0	1	<1	0	0	0	0
990	Salt	1 tsp	6		0	0	0	0	0	0	0	0
	Salt Substitutes:											
1205	Morton, salt substitute	1 tsp	6			0	<1		0	0	0	0
1207	Morton, light salt	1 tsp	6			0	<1		0	0	0	0
2067	Seasoned salt, no MSG	1 tsp	5	5	0	0	0	0	0	0	0	0
991	Vinegar, cider	½ c	120	94	17	0	7	0	0	0	0	0
2172	Balsamic	1 tbsp	15	64	21	0	5	0	0	0	0	0
2176	Malt	1 tbsp	15	90	5	0	1	0	0	0	0	0
2182	Tarragon	1 tbsp	15	95	3	0	<1	0	0	0	0	0
2181	White wine	1 tbsp	15	89	5	0	1	0	0	0	0	0
	Yeast:											
992	Baker's, dry, active, package	1 ea	7	8	21	3	3	1	<1	t	.2	t
993	Brewer's, dry	1 tbsp	8	5	23	3	3	3	<1	t	t	0
	SOUPS, SAUCES, and GRAVIES											
	SOUPS, canned, condensed:											
	Unprepared, condensed:											
1210	Cream of celery	1 c	251	85	181	3	18	2	11	2.8	2.6	5
1215	Cream of chicken	1 c	251	82	233	7	18	<1	15	4.2	6.5	3
1216	Cream of mushroom	1 c	251	81	259	4	19	1	19	5.1	3.6	8.9
1220	Onion	1 c	246	86	113	8	16	2	3	.5	1.5	1.3
	Prepared w/equal volume of whole milk:											
994	Clam chowder, New England	1 c	248	85	164	9	17	1	7	2.9	2.3	1.1
1209	Cream of celery	1 c	248	86	164	6	14	1	10	3.9	2.5	2.6
995	Cream of chicken	1 c	248	85	191	7	15	<1	11	4.6	4.5	1.6
996	Cream of mushroom	1 c	248	85	203	6	15	<1	14	5.1	3	4.6
1214	Cream of potato	1 c	248	87	149	6	17	<1	6	3.8	1.7	.6
1213	Oyster stew	1 c	245	89	135	6	10	0	8	5	2.1	.3
997	Tomato	1 c	248	85	161	6	22	3	6	2.9	1.6	1.1
	Prepared with equal volume of water:											
998	Bean with bacon	1 c	253	84	172	8	23	9	6	1.5	2.2	1.8
999	Beef broth/bouillon/consomme'	1 c	240	98	17	3	<1	0	1	.3	.2	t
1000	Beef noodle	1 c	244	92	83	5	9	1	3	1.1	1.2	.5
1001	Chicken noodle	1 c	241	92	75	4	9	1	2	.7	1.1	.6
1002	Chicken rice	1 c	241	94	60	4	7	1	2	.5	.9	.4
1208	Chili beef	1 c	250	85	170	7	21	9	7	3.3	2.8	.3
1003	Clam chowder, Manhattan	1 c	244	92	78	2	12	1	2	.4	.4	1.3
1004	Cream of chicken	1 c	244	91	117	3	9	<1	7	2.1	3.3	1.5
1005	Cream of mushroom	1 c	244	90	129	2	9	<1	9	2.4	1.7	4.2
1006	Minestrone	1 c	241	91	82	4	11	1	3	.6	.7	1.1

Chol (mg)	Calc (mg)	Iron (mg)	Magn (mg)	Pota (mg)	Sodi (mg)	Zinc (mg)	VT-A (µg)	Thia (mg)	VT-E (mg)	Ribo (mg)	Niac (mg)	V-B6 (mg)	Fola (µg)	VT-C (mg)
0	4	.1	2	6	63	.03	0*	0	.09	0	0	<.01	0	0
0	12	.32	4	11	480	.01	6*	0	.6	0	0	<.01	<1	0
0	19	.73	1	2	192	.05	4	<.01	.66	0	.01	<.01	0	<1
0	7	.05	2	19	1	.05	0	.01	<.01	<.01	.01	.03	3	<1
0	31	.88	5	33	<1	.09	7	.01	.03	.01	.12	.02	5	1
0	4	.47	4	47	1	.08	61	.01	.01	.03	.31	.04	2	1
0	2	.54	4	63	4	.06	32	.01	.06	.02	.1	.01	15	1
0	9	.58	4	25	1	.03	<1	<.01	.02	<.01	.02	.01	<1	<1
0	6	.34	7	75	833	.09	11	.01	.1	.02	.04	.01	1	1
0	5	.27	1	30	101	0	2*	0	.02	<.01	0	<.01	0	1
0	1	.21	1	11	329	.03	2	<.01	.06	.01	.06	<.01	<1	<1
0	3	.12	1	30	107	.01	1*	0	.02	<.01	0	0	0	1
0	5	.33	13	255	119	.22	0	.03	.98	.04	.76	.13	9	6
0	7	.46	19	357	2	.3	0	.05	1.37	.05	1.07	.18	13	9
0	16	.28	4	11	<1	.05	3	.01	.02	<.01	.06	.01	3	<1
0	1	.05	1	23	1	.01	3	.01	.04	<.01	.06	.01	2	2
0												0		
0	1	.02			2325	.01	0	0	0	0	0	0	0	0
	33			3018	<1									
	2		4	1560	1170									
				15	1583									
0	7	.72	26	120	1	0	0	0	0	0	0	0	0	0
	2	.07		10	3		<1*	.07		.07	.07			<1
	2	.07		13	4		<1*	.07		.07	.07			1
		.07		2	1		<1*	.07		.07	.07			<1
	1	.07		12	1		<1*	.07		.07	.07			<1
0	4	1.16	7	140	3	.45	<1	.16	.01	.38	2.78	.11	164	<1
0	17	1.38	18	151	10	.63	0*	1.25		.34	3.03	.4	313	0
28	80	1.26	13	246	1900	.3	60*	.06	.38	.1	.66	.02	5	<1
20	68	1.2	5	176	1972	1.26	113*	.06	.33	.12	1.64	.03	3	<1
3	65	1.05	10	168	1736	1.18	0	.06	2.61	.17	1.62	.02	8	2
0	54	1.35	5	138	2115	1.23	0	.07	.57	.05	1.21	.1	29	2
22	186	1.49	22	300	992	.79	40*	.07	.15	.24	1.03	.13	10	3
32	186	.69	22	310	1009	.2	67*	.07	.97	.25	.44	.06	7	1
27	181	.67	17	273	1046	.67	94*	.07	.25	.26	.92	.07	7	1
20	179	.59	20	270	918	.64	37*	.08	1.34	.28	.91	.06	10	2
22	166	.55	17	322	1061	.67	67*	.08	.1	.24	.64	.09	10	1
32	167	1.05	20	235	1041	10.3	44*	.07	.49	.23	.34	.06	10	4
17	159	1.81	22	449	744	.3	109*	.13	2.6	.25	1.52	.16	20	68
3	81	2.05	45	402	951	1.04	44	.09	.08	.03	.57	.04	33	2
0	14	.41	5	130	782	0	0	<.01	0	.05	1.87	.02	5	0
5	15	1.1	5	100	952	1.54	63*	.07	0	.06	1.07	.04	19	<1
7	17	.77	5	55	1106	.39	72*	.05	.07	.06	1.39	.03	22	<1
7	17	.75	0	101	815	.26	65*	.02	.05	.02	1.13	.02	0	<1
12	42	2.13	30	525	1035	1.4	150*	.06	.17	.07	1.07	.16	17	4
2	27	1.63	12	188	578	.98	98*	.03	.73	.04	.82	.1	10	4
10	34	.61	2	88	986	.63	56*	.03	.19	.06	.82	.02	2	<1
2	46	.51	5	100	881	.59	0	.05	1.24	.09	.72	.01	5	1
2	34	.92	7	313	911	.75	117	.05	.07	.04	.94	.1	36	1

*This value is expressed in retinol equivalents (RE). All other values are in retinol activity equivalents (RAE).

Table A-1

Food Composition

(Computer code number is for Wadsworth Diet Analysis program) (For purposes of calculations, use "0" for t, <1, <.1, <.01, etc.)

Computer Code Number	Food Description	Measure	Wt (g)	H_2O (%)	Ener (cal)	Prot (g)	Carb (g)	Dietary Fiber (g)	Fat (g)	Fat Breakdown (g) Sat	Mono	Poly
	SOUPS, SAUCES, and GRAVIES—Continued											
1211	Onion	1 c	241	93	58	4	8	1	2	.3	.7	.7
1007	Split pea & ham	1 c	253	82	190	10	28	2	4	1.8	1.8	.6
1008	Tomato	1 c	244	90	85	2	17	<1	2	.4	.4	1
1009	Vegetable beef	1 c	244	92	78	6	10	<1	2	.9	.8	.1
1010	Vegetarian vegetable	1 c	241	92	72	2	12	<1	2	.3	.8	.7
	Ready to serve:											
1707	Chunky chicken soup	1 c	251	84	178	13	17	2	7	2	3	1.4
	SOUPS, dehydrated:											
	Prepared with water:											
1299	Beef broth/bouillon	1 c	244	97	19	1	2	0	1	.3	.3	t
1376	Chicken broth	1 c	244	97	22	1	1	0	1	.3	.4	.4
1013	Chicken noodle	1 c	252	94	58	2	9	<1	1	.3	.5	.4
1122	Cream of chicken	1 c	261	91	107	2	13	<1	5	3.4	1.2	.4
1014	Onion	1 c	246	96	27	1	5	1	1	.1	.3	.1
1217	Split pea	1 c	255	87	125	7	21	3	1	.4	.7	.3
1015	Tomato vegetable	1 c	253	94	56	2	10	<1	1	.4	.3	.1
	Unprepared, dry products:											
1011	Beef bouillon, packet	1 ea	6	3	14	1	1	0	1	.3	.2	t
1012	Onion soup, packet	1 ea	39	4	115	5	21	4	2	.5	1.4	.3
	SAUCES											
	From dry mixes, prepared with milk:											
1016	Cheese sauce	1 c	279	77	307	17	23	1	17	9.3	5.3	1.6
1017	Hollandaise	1 c	259	84	240	5	14	<1	20	11.6	5.9	.9
1018	White sauce	1 c	264	81	240	10	21	<1	13	6.4	4.7	1.7
	From home recipe:											
1206	Lowfat cheese sauce	¼ c	61	74	81	6	4	<1	5	1.8	1.8	.8
1019	White sauce, medium	¼ c	72	77	102	2	6	<1	8	2.3		
	Ready to serve:											
2202	Alfredo sauce, reduced fat	¼ c	69	64	144	5	9	0	9	6.2		
1020	Barbeque sauce	1 tbsp	16	81	12	<1	2	<1	<1	t	.1	.1
1706	Chili sauce, tomato base	1 tbsp	17	68	18	<1	4	<1	<1	t	t	t
2126	Creole sauce	¼ c	62	89	25	1	4	1	1	.1	.2	.3
2124	Hoisin sauce	1 tbsp	17	47	35	<1	7	0	1	0		
2199	Pesto sauce	2 tbsp	16	34	83	2	1	<1	8	1.5		
1021	Soy sauce	1 tbsp	16	71	8	1	1	<1	<1	t	t	t
2123	Szechuan sauce	1 tbsp	16	71	21	<1	3	<1	1	.1	.3	.4
1380	Teriyaki sauce	1 tbsp	18	68	15	1	3	<1	0	0	0	0
	Spaghetti sauce, canned:											
1377	Plain	1 c	249	75	271	5	40	8	12	1.7	6.1	3.3
1378	With meat	1 c	250	85	178	7	19	4	8	1.8	3.3	1.8
1379	With mushrooms	½ c	123	84	108	2	13	1	3	.4	1.5	.8
	GRAVIES											
	Canned:											
1022	Beef	1 c	233	87	123	9	11	1	5	2.7	2.2	.2
1023	Chicken	1 c	238	85	188	5	13	1	14	3.4	6.1	3.6
1024	Mushroom	1 c	238	89	119	3	13	1	6	1	2.8	2.4
1025	From dry mix, brown	1 c	258	92	75	2	13	<1	2	.8	.7	.1
1026	From dry mix, chicken	1 c	260	91	83	3	14	<1	2	.5	.9	.4
	FAST FOOD RESTAURANTS											
	ARBY'S											
1402	Bac'n cheddar deluxe	1 ea	231	59	512	21	39	<1	31	8.7	12.7	10.1
	Roast beef sandwiches:											
1403	Regular	1 ea	155	54	326	21	35	2	14	6.9		
1404	Junior	1 ea	89	50	200	11	23	1	8	3.4	3.3	1.6
1405	Super	1 ea	254	61	467	23	50	3	22	8.3		
1407	Beef 'n cheddar	1 ea	194	53	451	22	42	2	22	8.8	8.8	5
1408	Chicken breast sandwich	1 ea	204	49	539	23	46	2	29	4.9	12	12.5
1412	Ham'n cheese sandwich	1 ea	169	57	338	23	35	1	13	4.5	5.1	3.3
1726	Italian sub sandwich	1 ea	297	57	743	28	47	3	50	14.3	23.5	12.7
1413	Turkey sandwich, deluxe	1 ea	218	68	292	26	37	3	6	.6	2.5	2.6
1680	Turkey sub sandwich	1 ea	277	62	570	23	46	2	33	8.1	11.7	13.7
	Milkshakes:											

Chol (mg)	Calc (mg)	Iron (mg)	Magn (mg)	Pota (mg)	Sodi (mg)	Zinc (mg)	VT-A (µg)	Thia (mg)	VT-E (mg)	Ribo (mg)	Niac (mg)	V-B6 (mg)	Fola (µg)	VT-C (mg)
0	26	.67	2	67	1053	.6	0	.03	.29	.02	.6	.05	14	1
8	23	2.28	48	400	1006	1.32	45*	.15	.15	.08	1.47	.07	3	2
0	12	1.76	7	264	695	.24	34	.09	2.49	.05	1.42	.11	15	66
5	17	1.12	5	173	791	1.54	95	.04	.32	.05	1.03	.08	10	2
0	22	1.08	7	210	822	.46	301*	.05	.79	.05	.92	.05	10	1
30	25	1.73	8	176	889	1	131*	.08	.18	.17	4.42	.05	5	1
0	10	.02	7	37	1361	.07	1	<.01	.02	.02	.36	0	0	0
0	15	.07	5	24	1483	0	12*	.01	.02	.03	.19	0	2	0
10	5	.5	8	33	577	.2	5*	.2	.1	.08	1.09	.02	18	0
3	76	.26	5	214	1184	1.57	123*	.1	.16	.2	2.61	.05	5	1
0	12	.15	5	64	849	.05	<1	.03	.1	.06	.48	0	2	<1
3	20	.94	43	224	1147	.56	5*	.21	.13	.14	1.26	.05	41	0
0	8	.63	20	104	1146	.18	10	.06	.81	.05	.79	.05	10	6
1	4	.06	3	27	1018	0	<1*	<.01	.01	.01	.27	.01	2	0
2	55	.58	25	260	3493	.23	<1	.11	.42	.24	1.99	.04	6	1
53	569	.28	47	552	1565	.97	117*	.15	.33	.56	.32	.14	13	2
52	124	.9	8	124	1564	.7	220*	.04	.26	.18	.06	.5	22	<1
34	425	.26	264	444	797	.55	92*	.08	1.58	.45	.53	.07	16	3
9	167	.24	10	101	307	.74	56*	.03	.51	.14	.16	.03	4	<1
8	75	.21	9	100	82	.26	62*	.05	.98	.12	.28	.03	4	1
31	103	0	8	93	618		154*	0		.1	0		0	0
0	3	.14	3	28	130	.03	14*	<.01	.18	<.01	.14	.01	1	1
0	3	.14	2	63	227	.05	24*	.01	.05	.01	.27	.02	1	3
0	35	.31	9	187	339	.1	12	.03	.61	.02	.53	.07	9	0
0	0	0			250		0*							0
4	64	.09		26	137		26*	0		.02	0		2	0
0	3	.32	5	29	914	.06	0	.01	0	.02	.54	.03	3	0
0	2	.12	2	13	218	.02	10*	<.01	.07	<.01	.1	.01	1	<1
0	4	.31	11	40	690	.02	0	<.01	0	.01	.23	.02	4	0
0	70	1.62	60	956	1235	.52	306*	.14	4.98	.15	3.76	.88	54	28
15	53	2.1	43	742	982	1.27	176*	.13	2.98	.12	3.47	.31	25	19
0	15	1	15	332	494	.34	120	.08	1.35	.08	.93	.16	12	9
7	14	1.63	5	189	1304	2.33	0	.07	.14	.08	1.54	.02	5	0
5	48	1.12	5	259	1373	1.9	264*	.04	.38	.1	1.05	.02	5	0
0	17	1.57	5	252	1356	1.67	0	.08	.19	.15	1.6	.05	29	0
3	67	.23	10	57	1075	.31	0*	.04	.05	.08	.81	0	0	0
3	39	.26	10	62	1133	.32	0*	.05	.05	.15	.78	.03	3	3
38	110	4.32		491	1094	3	40*	.34		.46	9.6			11
44	59	3.55	16	422	879	3.75	0	.28		.48	11	.2	14	0
28	41	1.86	8	201	483	1.5		.18		.25	6.6	.1	7	
47	62	3.73	25	533	1098	3.73	30*	.39		.58	12.4	.3	21	1
49	98	3.53		321	1146	2.94		.42		.63	9.8			1
88	78	1.77	30	330	1137	.15		.22		.54	8.99	.38	18	4
89	149	2.68	31	380	1441	.89	40*	.82		.37	7.75	.31	26	1
114	238	2.57		565	2322		97*	.91		.49	8.19			2
45	90	2.02	33	394	1157	1.69	45*	.09		.46	17.2	.58	22	1
90	181	.33		500	1964		20*	13.2		.54	18.8			2

*This value is expressed in retinol equivalents (RE). All other values are in retinol activity equivalents (RAE).

Table A–1

Food Composition (Computer code number is for Wadsworth Diet Analysis program) (For purposes of calculations, use "0" for t, <1, <.1, <.01, etc.)

Computer Code Number	Food Description	Measure	Wt (g)	H_2O (%)	Ener (cal)	Prot (g)	Carb (g)	Dietary Fiber (g)	Fat (g)	Fat Breakdown (g) Sat	Mono	Poly
	FAST FOOD RESTAURANTS—Continued											
1419	Chocolate	1 ea	340	71	411	9	72	0	14	6.8	5.5	1.3
1420	Jamocha	1 ea	326	72	386	8	67	0	12	5.7	5.3	1.3
1421	Vanilla	1 ea	312	72	369	8	65	0	12	5.5	4.4	1.9
1728	Salad, roast chicken	1 ea	400	90	152	19	14	6	2	0		
1729	Sports drink, Upper Ten	1 ea	358	88	169	0	42		0	0	0	0
	Source: Arby's											
	BURGER KING											
1423	Croissant sandwich, egg,sausage&cheese	1 ea	176	46	600	22	25	1	46	16		
	Whopper sandwiches:											
1425	Whopper	1 ea	270	58	640	27	45	3	39	11		
1426	Whopper with cheese	1 ea	294	57	730	33	46	3	46	16		
	Sandwiches:											
1629	BK broiler chicken sandwich	1 ea	248	59	550	30	41	2	29	6		
1432	Cheeseburger	1 ea	138	48	380	23	28	1	19	9		
1434	Chicken sandwich	1 ea	229	45	710	26	54	2	43	9		
1427	Double beef	1 ea	351	57	870	46	45	3	56	19		
1428	Double beef & cheese	1 ea	375	56	960	52	46	3	63	24		
1433	Double cheeseburger with bacon	1 ea	218	48	640	44	28	1	39	18		
1431	Hamburger	1 ea	126	48	330	20	28	1	15	6		
1437	Ocean catch fish fillet	1 ea	255	51	700	26	56	3	41	6		
1435	Chicken tenders	1 ea	88	50	230	16	14	2	12	3		
1439	French fries (salted)	1 svg	116	40	370	5	43	3	20	5		
1630	French toast sticks	1 svg	141	33	500	4	60	1	27	7		
1440	Onion rings	1 svg	124	51	310	4	41	6	14	2	8	4
1441	Milk shakes, chocolate	1 ea	284	75	320	9	54	3	7	4		
1442	Milk shakes, vanilla	1 ea	284	75	300	9	53	1	6	4		
1443	Fried apple pie	1 ea	113	47	300	3	39	2	15	3		
	Source: Burger King Corporation											
	CHICK-FIL-A											
	Sandwiches:											
69153	Chargrilled chicken	1 ea	150	54	280	27	36	1	3	1		
69152	Chicken	1 ea	167	61	290	24	29	1	9	2		
69155	Chicken salad	1 ea	167	55	320	25	42	1	5	2		
69154	Chicken salad club	1 ea	232	62	390	33	38	2	12	5		
	Salads:											
52139	Carrot and raisin	1 ea	76	53	150	5	28	2	2	0		
52136	Chicken plate	1 ea	468	85	290	21	40	6	5	0		
52134	Chicken garden, charbroiled	1 ea	397	89	170	26	10	5	3	1		
52135	Chick-n-strips	1 ea	451	86	290	32	21	5	9	2		
52138	Cole slaw	1 ea	79	70	130	6	11	1	6	1		
52137	Tossed salad	1 ea	130	85	70	5	13	1	0	0	0	0
15263	Chicken nuggets, svg	1 ea	110	51	290	28	12	0	14	3		
15262	Chicken-n- strips, svg	1 ea	119	59	230	29	10	0	8	2		
50885	Hearty breast of chicken soup, svg	1 ea	215	86	110	16	10	1	1	0		
7973	Waffle potato fries, svg	1 ea	85	28	290	1	49	0	10	4		
46489	Cheesecake, svg	1 ea	88	52	270	13	7	0	21	9		
49134	Fudge nut brownie, svg	1 ea	74	8	350	10	41	0	16	3		
20601	Icedream, svg	1 ea	127	74	140	11	16	0	4	1		
48214	Lemon pie, svg	1 ea	99	56	280	1	19	0	22	6		
	Source: Chick-Fil-A											
	DAIRY QUEEN											
	Ice cream cones:											
1446	Small vanilla	1 ea	142	63	230	6	38	0	7	4.5		
1447	Regular vanilla	1 ea	213	64	355	9	57	0	10	6.4		
1448	Large vanilla	1 ea	253	65	410	10	65	0	12	8		
1450	Chocolate dipped	1 ea	220	58	490	8	59	1	24	12.5		
1453	Chocolate sundae	1 ea	234	62	400	8	71	0	10	6		

Chol (mg)	Calc (mg)	Iron (mg)	Magn (mg)	Pota (mg)	Sodi (mg)	Zinc (mg)	VT-A (µg)	Thia (mg)	VT-E (mg)	Ribo (mg)	Niac (mg)	V-B6 (mg)	Fola (µg)	VT-C (mg)
38	428	.62	48	410	317	1.5	60*	.12		.68	.8	.14	14	2
37	411	.59	36	525	320	1.48	60*	.12		.68	.8	.14	14	2
35	393	.85	36	686	283	1.49	60*	.12		.68	4	.14	37	2
38	76	1.71		877	667		970*	.31		.54	5.6			<1
0				0	40									
260	150	3.6			1140		80*							0
90	80	4.5			870		100*	.33		.41	7	.35		9
115	250	4.5			1350		150*	.34		.48	7	.33		9
80	60	5.4			480		60*							6
65	100	2.7			770		60*							0
60	100	3.6			1400		0							0
170	80	7.2			940		100*	.34		.56	10			9
195	250	7.2			1420		150*	.35		.63	10			9
145	200	4.5			1240		80*	.31		.42	6			0
55	40	1.8			530		20*	.28		.31	4.89			0
90	60	2.7			980		20*							1
35	0	.72			530		0							0
0	0	1.08			240		0							4
0	60	2.7			490		0							0
0	100	1.44			810		0							0
20	200	1.8			230		60*	.13		.55	.13			0
20	300	0			230		60*	.11		.57	.13			4
0	0	1.44			230		0							6
40	0	1.8			640		80*							1
50	0	1.8			870		80*							0
10	0	1.44			810		80*							0
70	80	1.8			980		140*							8
6	40	3.6			650		200*							6
35	60	3.24			570		160*							13
25	60	2.52			650		240*							13
20		2.52			430		200*							13
15	20	3.6			430		140*							6
0	0	3.6			0		100*							8
60	0	1.44			770		80*							0
20	0	1.08			380		60*							8
45	0	2.52			760		140*							2
5	0	0			960		0							0
10	20	.72			510		60*	.03		.19	.19		18	0
30	0	.72			650		80*	.12		.21	.97		29	0
40	80	1.08			240		100*							0
5	150	1.08			550		150*							5
20	200	1.08		250	115		100*	.05		.28				1
32	269	1.94		390	172		161*	.09		.38	.16	.13		3
40	350	1.8		451	200		200*	.11		.4	.2			2
30	250	1.8		409	190		150*	.08		.36	.15	.13		2
30	250	1.44		383	210		150*	.08		.34	.39	.18		0

*This value is expressed in retinol equivalents (RE). All other values are in retinol activity equivalents (RAE).

Table A–1

Food Composition (Computer code number is for Wadsworth Diet Analysis program) (For purposes of calculations, use "0" for t, <1, <.1, <.01, etc.)

Computer Code Number	Food Description	Measure	Wt (g)	H_2O (%)	Ener (cal)	Prot (g)	Carb (g)	Dietary Fiber (g)	Fat (g)	Fat Breakdown (g) Sat	Mono	Poly
	FAST FOOD RESTAURANTS—Continued											
1455	Banana split	1 ea	369	67	510	8	96	3	12	8		
1456	Peanut buster parfait	1 ea	305	51	730	16	99	2	31	17		
1457	Hot fudge brownie delight	1 ea	305	52	710	11	102	1	29	14	12	2
1459	Buster bar	1 ea	149	45	450	10	41	2	28	12		
1645	Breeze, strawberry, regular	1 ea	383	70	460	13	99	1	1	1	0	0
1460	Dilly bar	1 ea	85	55	210	3	21	0	13	7	3	3
1461	DQ ice cream sandwich	1 ea	61	46	150	3	24	1	5	2		
1463	Milk shakes, regular	1 ea	397	71	520	12	88	<1	14	8	2	2
1464	Milk shakes, large	1 ea	461	71	600	13	101	<1	16	10	2	2
1466	Milk shakes, malted	1 ea	418	68	610	13	106	<1	14	8	2	2
1470	Misty slush, small	1 ea	454	88	220	0	56	0	0	0	0	0
2250	Starkiss	1 ea	85	75	80	0	21	0	0	0	0	0
	Yogurt:											
1641	Yogurt cone, regular	1 ea	198	65	260	9	56	0	1	.5		
1643	Yogurt sundae, strawberry	1 ea	234	69	280	8	61	1	<1	0		
	Sandwiches:											
1481	Cheeseburger, double	1 ea	219	55	540	35	30	2	31	16		
1480	Cheeseburger, single	1 ea	152	55	340	20	29	2	17	8		
1474	Chicken	1 ea	191	56	430	24	37	2	20	4		
1647	Chicken fillet, grilled	1 ea	184	64	310	24	30	3	10	2.5		
1475	Fish fillet sandwich	1 ea	170	57	370	16	39	2	16	3.5		
1476	Fish fillet with cheese	1 ea	184	56	420	19	40	2	21	6	7	8
1477	Hamburger, single	1 ea	138	56	290	17	29	2	12	5	6	1
1478	Hamburger, double	1 ea	212	62	440	30	29	2	22	10		
	Hotdog:											
1483	Regular	1 ea	99	57	240	9	19	1	14	5		
1484	With cheese	1 ea	113	55	290	12	20	1	18	8	8	2
1485	With chili	1 ea	128	61	280	12	21	2	16	6		
1489	French fries, small	1 ea	112	41	350	4	42	3	18	3.5		
1490	French fries, large	1 ea	128	40	390	5	52	6	18	4	8	6
1491	Onion rings	1 ea	113	46	320	5	39	3	16	4		
	Source: International Dairy Queen											
	HARDEES'S											
	Sandwiches:											
56414	Cheeseburger	1 ea	120	47	300	15	34		13	6.5	4.6	1.9
1734	Frisco burger hamburger	1 ea	242	46	760	36	43		50	18		
56412	Hamburger	1 ea	107	49	260	11	33		9	3.6	3.6	1.8
56415	Quarter pound cheeseburger	1 ea	184	51	490	27	37		25	11.5	11.5	1.9
56422	Chicken sandwich	1 ea	187	56	400	19	48		14	3	4	6
6146	French fries, svg	1 ea	96	48	240	4	33		10	2.5	4.2	3.3
	JACK IN THE BOX											
	Breakfast items:											
1492	Breakfast jack sandwich	1 ea	126	54	280	17	28	1	12	5	4.7	2.3
1494	Sausage crescent	1 ea	181	41	660	20	37	0	48	15		
1495	Supreme crescent	1 ea	164	43	530	21	37	0	34	10	17	7
1496	Pancake platter	1 ea	231	45	610	15	87	0	22	9	7.6	3.5
1497	Scrambled egg platter	1 ea	213	52	560	18	50	0	32	9	16.6	4.4
	Sandwiches:											
1654	Bacon cheeseburger	1 ea	274	51	760	39	39	2	50	17	21.2	11.8
1499	Cheeseburger	1 ea	116	48	300	14	31	2	13	6	5	2
1739	Chicken caesar pita sandwich	1 ea	237	59	520	27	44	4	26	6		
1655	Chicken sandwich	1 ea	164	53	400	15	38	3	21	3		
1656	Chicken sandwich, sourdough ranch	1 ea	225	57	490	29	45	1	21	6		
1505	Chicken supreme	1 ea	305	49	830	33	66	3	49	7		
1583	Double cheeseburger	1 ea	158	48	440	24	31	2	24	11	10.3	2.7
1651	Grilled sourdough burger	1 ea	233	49	690	34	37	2	45	15	20.8	9.2
1498	Hamburger	1 ea	104	50	250	12	30	2	9	3.5	3.9	1.6
1500	Jumbo jack burger	1 ea	271	62	550	27	43	2	30	10	12.4	7.6
1501	Jumbo jack burger with cheese	1 ea	296	60	640	31	44	2	38	15	14.4	8.6
1740	Monterey roast beef sandwich	1 ea	238	57	540	30	40	3	30	9		

Chol (mg)	Calc (mg)	Iron (mg)	Magn (mg)	Pota (mg)	Sodi (mg)	Zinc (mg)	VT-A (µg)	Thia (mg)	VT-E (mg)	Ribo (mg)	Niac (mg)	V-B6 (mg)	Fola (µg)	VT-C (mg)
30	250	1.8		860	180		200*	.15		.25	.4	.2		15
35	300	1.8		660	400		150*	.15		.51	3	.22		1
35	300	5.4		510	340		80*	.15		.68	.3	.18		1
15	150	1.08		400	280		80*	.09		.17	3	.08		0
10	450	2.7		530	270		0	.13		.73				9
10	100	.36		170	75		60*	.03		.14		.06		0
5	60	.72		105	115		40*	.03		.25	.4	.05		0
45	400	1.44		570	230		80*	.12		.59	.8	.19		<1
50	450	1.44		660	260		200*	.15		.68	.8			<1
45	400	1.44		570	230		80*	.12		.59	.8	.19		<1
0	0	0			20		0							0
0	0	0			10		0							0
5	250	1.8		265	160		0	.08		.35				2
5	300	1.44		323	160		0	.08		.45				6
115	250	4.5		426	1130		150*	.29		.49	6.78			4
55	150	3.6		263	850		100*	.29		.33	3.89			4
55	40	1.8		350	760		0	.37		.34	11			0
50	200	2.7		330	1040		0	.3		1.02	12			0
45	40	1.8		280	630		0	.3		.22	3			0
60	100	1.8		290	850		80*	.3		.25	5			0
45	60	2.7		252	630		40*	.29		.25	3.88			4
90	60	4.5		444	680		60*	.32		.45	7.49			6
25	60	1.8		170	730		20*	.22		.14	2			4
40	150	1.8		180	950		60*	.22		.17	2			4
35	60	1.8		262	870		80*	.23		.14	3			4
0	20	.72		678	630		0	.14		.05	3.15			4
0	0	1.44		780	200		0*	.15		.07	3			9
0	20	1.44		120	180		0	.12		.07	.53			0
25	178	3		210	690									
70					1280									
20	111	3		200	460									
35	248	5		350	980									
55	123	3		290	1100									
0	12	1		350	100		0							
190	150	3.6		120	750		80*	.49		.43	3.12			10
240	100	1.8		160	860		80*	.7		.59	5.3			0
225	100	1.8		165	1060		80*	.7		.58	4.5			4
100	100	1.8		310	890		80*	.03		.85	7			6
380	150	4.5		450	1060		150*			.66	5			9
135	250	4.5		530	1570		150*	.27		.54	9.96	.44		9
40	150	3.6		180	840		40*	.24		.24	3.16			0
55	250	2.7		490	1050		80*							2
40	100	2.7		200	770		40*							5
65	150	1.8		340	1060									0
65	200	3.6		250	2140		100*	.49		.4	13.7			9
80	250	4.5		290	1100		80*	.16		.35	6.23			1
105	200	4.5		480	1180		150*	.68		.5	8.36	.34		9
30	100	3.6		155	610		0	.16		.28	2.14			0
75	150	4.5		490	880		100*	.43		.34	2.1			9
105	250	4.5		530	1340		150*	.44		.54	1.96			9
75	300	3.6		500	1270		80*							5

*This value is expressed in retinol equivalents (RE). All other values are in retinol activity equivalents (RAE).

Table A-1

Food Composition (Computer code number is for Wadsworth Diet Analysis program) (For purposes of calculations, use "0" for t, <1, <.1, <.01, etc.)

Computer Code Number	Food Description	Measure	Wt (g)	H_2O (%)	Ener (cal)	Prot (g)	Carb (g)	Dietary Fiber (g)	Fat (g)	Fat Breakdown (g) Sat	Mono	Poly
	FAST FOOD RESTAURANTS—Continued											
1508	Tacos, regular	1 ea	90	66	170	7	12	2	10	3.5		
1509	Tacos, super	1 ea	138	63	270	12	19	4	17	6		
	Teriyaki bowl:											
1679	Beef	1 ea	440	62	640	28	124	7	3	1		
1668	Chicken	1 ea	502	68	670	26	128	3	4	1		
1516	French fries	1 ea	113	40	350	4	46	3	16	4		
1517	Hash browns	1 ea	57	52	170	1	14	1	12	2	9.6	.4
1518	Onion rings	1 ea	120	30	450	7	50	3	25	5	18.9	1.1
	Milkshakes:											
1519	Chocolate	1 ea	332	62	630	11	85	1	27	16		
1520	Strawberry	1 ea	382	67	640	10	85	0	28	15		
1521	Vanilla	1 ea	332	64	610	12	73	0	31	18		
1522	Apple turnover	1 ea	107	40	340	4	41	2	18	4	12.3	1.7
	Source: Jack in the Box Restaurant, Inc											
	KENTUCKY FRIED CHICKEN											
	Rotisserie gold:											
1472	Dark qtr, no skin	1 ea	117	66	217	27	0	0	12	3.5		
1473	Dark qtr, w/skin	1 ea	146	62	333	30	1		24	6.6		
1513	White qtr with wing, w/skin	1 ea	176	65	335	40	1		19	5.4		
1525	White qtr with wing, no skin	1 ea	117	63	199	37	0	0	6	1.7		
	Original Recipe:											
1253	Center breast	1 ea	103	52	260	25	9	<1	14	3.8	7.8	2
1251	Side breast	1 ea	153	53	400	29	16	1	24	6	14.4	3.6
1250	Drumstick	1 ea	61	56	140	13	4	0	9	2	5.3	1.7
1252	Thigh	1 ea	91	54	250	16	6	1	18	4.5	10.2	3.3
1249	Wing	1 ea	47	48	140	9	5	0	10	2.5	5.8	1.7
	Hot & spicy:											
1451	Center breast	1 ea	180	48	505	38	23	1	29	8		
1452	Side breast	1 ea	120	43	400	22	16		28	6		
1430	Thigh	1 ea	107	45	355	19	13	1	26	7		
1471	Wing	1 ea	55	37	210	10	9	1	15	4		
	Extra Crispy Recipe:											
1261	Center breast	1 ea	168	48	470	39	17	1	28	8	16.7	3.3
1259	Side breast	1 ea	116	40	400	21	19	<1	27	5.5	12.9	2.3
1258	Drumstick	1 ea	67	48	195	15	7	1	12	3	7.4	1.6
1260	Thigh	1 ea	118	45	380	21	14	1	27	7	15.8	4.2
1257	Wing	1 ea	55	35	220	10	10	1	15	4	8.9	2.1
1390	Baked beans	½ c	167	71	203	6	35	6	3	1.1	1.4	.7
1526	Breadstick	1 ea	33	30	110	3	17	0	3	0		
1388	Buttermilk biscuit	1 ea	56	37	180	4	20	1	10	2.5	5.4	2.1
1391	Chicken little sandwich	1 ea	47	35	169	6	14	<1	10	2	4.7	3.4
1269	Coleslaw	1 svg	142	70	232	2	26	3	13	2	3.5	7
1527	Cornbread	1 ea	56	26	228	3	25	1	13	2		
1268	Corn-on-the-cob	1 ea	162	73	150	5	35	2	1	0	.6	.9
1429	Chicken, hot wings	1 svg	135	41	471	27	18	2	33	8		
1386	Kentucky fries	1 svg	77	42	228	3	26	3	12	3.2		
1381	Kentucky nuggets	6 ea	95	48	284	16	15	<1	18	4		
1534	Macaroni & cheese	1 svg	153	74	180	7	21	2	8	3		
1387	Mashed potatoes & gravy	1 svg	136	82	120	1	17	2	6	1	3.6	1.4
1530	Pasta salad	1 svg	108	78	135	2	14	1	8	1		
1389	Potato salad	½ c	160	73	230	4	23	3	14	1.7	4.5	7.8
1383	Potato wedges	1 svg	135	65	278	5	28	5	13	4		
1535	Red beans & rice	1 svg	112	76	114	4	18	3	3	1		
1529	Vegetable medley salad	1 ea	114	77	126	1	21	3	4	1		
	Source: Kentucky Fried Chicken Corp											
	LONG JOHN SILVER'S											
1528	Chicken plank dinner, 3 piece	1 ea	399	56	890	32	101		44	9.5	24.8	9.4
1531	Clam chowder	1 ea	198	86	140	11	10	1	6	1.8	2.5	1.7
1532	Clam dinner	1 ea	361	46	990	24	114		52	10.9	31.3	9.9

Chol (mg)	Calc (mg)	Iron (mg)	Magn (mg)	Pota (mg)	Sodi (mg)	Zinc (mg)	VT-A (μg)	Thia (mg)	VT-E (mg)	Ribo (mg)	Niac (mg)	V-B6 (mg)	Fola (μg)	VT-C (mg)
15	100	1.08	40	235	390	1.38	60*	.08		.2	1.15	.15		<1
30	200	1.44	49	365	630	1.8	100*	.13		.09	1.53	.2		2
25	150	4.5		430	930		1000*							6
15	100	4.5		620	1730		1300*							24
0	10	.72		590	710		0	.19		.03	3.94			6
0	10	.18		100	250		0	.05			1			0
0	40	2.7		150	780		40*	.34		.2	3.03			18
85	350	.36		720	330		150*							0
85	350	0		620	300		150*							0
95	400	0		730	320		150*							0
0	10	1.8		85	510		20*	.19		.12	1.75			10
128	10	.18			772		15*							1
163	10	.18			980		15*							1
157	10	.18			1104		15*							1
97	10	.18			667		15*							1
92	30	.72			609		15*	.09		.17	11.5			
135	40	1.08			1116		20*							1
75	20	.72			422		20*							1
95	20	.72			747		20*							1
55	20	.36			414		20*							1
162	60	1.08			1170		20*							1
80	40	1.08			850		15*							6
126	20	.72			630		20*							1
55	20	.72			350		20*							1
160	20	1.08			874		20*							1
75	20	.72			710		15*	.09		.1	8.5			
77	20	.72			375		20*							1
118	20	1.08			625		20*							1
55	20	.36			415		20*							1
5	86	1.93			814		21							1
0	30	.18			15		0*							0
0	20	1.08			560		20*							1
18	23	1.7			331		5*	.16		.12	2.2			
8	30	.36			285		90*							34
42	60	.72			194		10*							
0	20	.36			20		5							4
150	40	1.44			1230		20*							1
4	11	.98			535		0*							0
66	2	.1			865		15*	.02		.02	1	.05		<1
10	150	.36			860		200*							1
1	20	.36			440		20*							1
1	20	1.08			663		110*							7
15	20	2.7			540		90*							1
5	20	1.79			744		5							1
4	10	.72			315									
0	20	.36			240		375*							5
55	200	4.5		1170	2000	3	40*	.52		.51	16			9
20	200	1.8		380	590	.6	150*	.09		.25	2			
75	200	4.5		910	1830	3	40*	.75		.42	12			12

*This value is expressed in retinol equivalents (RE). All other values are in retinol activity equivalents (RAE).

Table A–1

Food Composition (Computer code number is for Wadsworth Diet Analysis program) (For purposes of calculations, use "0" for t, <1, <.1, <.01, etc.)

Computer Code Number	Food Description	Measure	Wt (g)	H_2O (%)	Ener (cal)	Prot (g)	Carb (g)	Dietary Fiber (g)	Fat (g)	Fat Breakdown (g) Sat	Mono	Poly
	FAST FOOD RESTAURANTS—Continued											
	Fish, batter fried:											
1523	Fish & fryes (fries), 3 piece	1 ea	384	54	980	31	92		50	11.3	28.4	9.7
1524	Fish & fryes, 2 piece	1 ea	261	54	610	27	52		37	7.9	23.5	5.3
2240	Fish and lemon crumb dinner, 3 piece	1 ea	493	71	610	39	86		13	2.2	3.9	5.3
2241	Fish and lemon crumb dinner, 2 piece	1 ea	334	77	330	24	46		5	.9	1.6	1.2
1533	Fish & chicken dinner	1 ea	431	55	950	36	102		49	10.6	28.8	9.5
1537	Shrimp dinner, batter fried	1 ea	331	54	840	18	88		47	9.7	27.2	9.1
	Salads:											
1541	Cole slaw	1 ea	98	70	140	1	20	1	6	1	1.5	3.5
1539	Ocean chef salad	1 ea	234	89	110	12	13	2	1	.4	.4	.2
1540	Seafood salad	1 ea	278	79	380	15	12	2	31	5.1	8.2	17.5
1542	Fryes (fries) serving	1 ea	85	43	250	3	28	1	15	2.5	7.4	5.1
1543	Hush puppies	1 ea	24	40	70	2	10	<1	2	.4	1.3	.2
	Source: Long John Silver's, Lexington KY											
	McDONALD'S											
	Sandwiches:											
1221	Big mac	1 ea	216	51	560	26	45	3	31	10		
1226	Cheeseburger	1 ea	122	46	323	15	35	2	13	6		
1224	Filet-o-fish	1 ea	156	45	450	16	42	2	25	4.5		
1225	Hamburger	1 ea	108	49	262	13	34	2	9	3.5		
1444	McChicken	1 ea	189	52	491	17	42	2	29	5.4	8.5	10.2
1591	McLean deluxe	1 ea	214	64	345	23	37	2	12	4.4	3.6	1.2
1438	McLean deluxe with cheese	1 ea	228	63	398	26	38	2	16	6.8	4.6	1.3
1222	Quarter-pounder	1 ea	171	52	418	23	37	2	21	7.9		
1223	Quarter-pounder with cheese	1 ea	199	50	527	28	38	2	30	12.9		
1227	French fries, small serving	1 ea	68	40	210	3	26	2	10	1.5		
1228	Chicken McNuggets	4 pce	71	51	190	12	10	0	11	2.5		
	Sauces (packet):											
1229	Hot mustard	1 ea	28	60	60	1	7	1	3	0		
1230	Barbecue	1 ea	28	60	45	0	10	0	0	0	0	0
1231	Sweet & sour	1 ea	28	57	50	0	11	0	0	0	0	0
	Low-fat (frozen yogurt) milk shakes:											
1232	Chocolate	1 ea	295	72	360	11	60	1	9	6		
1233	Strawberry	1 ea	294	72	360	11	60	0	9	6		
1234	Vanilla	1 ea	293	72	360	11	59	0	9	6		
	Low-fat (frozen yogurt) sundaes:											
1237	Hot caramel	1 ea	182	56	360	7	61	0	10	6		
1235	Hot fudge	1 ea	179	59	340	8	52	1	12	9		
1267	Strawberry	1 ea	178	63	290	7	50	1	7	5		
1238	Vanilla	1 ea	90	64	150	4	23	0	4	3		
1241	Cookies, McDonaldland	1 ea	42	3	180	3	32	1	5	1		
1242	Cookies, chocolaty chip	1 ea	35	1	170	2	22	1	10	6		
1240	Muffin, apple bran, fat-free	1 ea	114	37	300	6	61	3	3	.5		
1239	Pie, apple	1 ea	77	34	260	3	34	1	13	3.5		
	Breakfast items:											
1243	English muffin with spread	1 ea	63	33	189	5	30	2	6	2.4	1.5	1.3
1244	Egg McMuffin	1 ea	137	57	292	17	27	1	12	4.5		
1245	Hotcakes with marg & syrup	1 ea	228	41	610	9	104	2	18	3.5		
1246	Scrambled eggs	1 ea	102	73	160	13	1	0	11	3.5		
1247	Pork sausage	1 ea	43	45	170	6	0	0	16	5		
1248	Hashbrown potatoes	1 ea	53	55	130	1	14	1	8	1.5		
1392	Sausage McMuffin	1 ea	112	42	360	13	26	1	23	8		
1393	Sausage McMuffin with egg	1 ea	163	52	443	19	27	1	28	10.1		
1394	Biscuit with biscuit spread	1 ea	84	32	290	5	34	1	15	3		
1395	Biscuit with sausage	1 ea	127	37	470	11	35	1	31	9		
1396	Biscuit with sausage & egg	1 ea	178	48	550	18	35	1	37	10		
1397	Biscuit with bacon, egg, cheese	1 ea	157	46	470	18	36	1	28	8		
	Salads:											
1398	Chef salad	1 ea	313	86	206	19	9	3	11	4.2	3	1.2
1400	Garden salad	1 ea	149	88	100	7	4	2	6	3		

Chol (mg)	Calc (mg)	Iron (mg)	Magn (mg)	Pota (mg)	Sodi (mg)	Zinc (mg)	VT-A (µg)	Thia (mg)	VT-E (mg)	Ribo (mg)	Niac (mg)	V-B6 (mg)	Fola (µg)	VT-C (mg)
70	200	4.5		1120	1530	3	40*	.45		.42	8			15
60	40	1.8		900	1480	1.2		.37		.34	8			9
125	200	5.4		990	1420	2.25	700*	.75		.59	24			6
75	80	1.8		440	640	.9	1000*	.3		.25	14			18
75	200	4.5		1280	2090	3	40*	.6		.59	14			9
100	200	3.6		840	1630	3	40*	.45		.42	9			9
15	60	.72		190	260	.6	40*	.06		.07	2			
40	100	3.6		95	730	.3	500*	.12		.14	3			21
55	150	4.5		130	980	.9	200*	.15		.25	3			21
0	200	.72		370	500	.3	0	.09			1.6			6
	40	.72		65	25	.3		.06		.03	.8			
85	250	4.5	45	455	1070	4.8	60*	.49	1.01	.44	6.07	.25	49	4
40	202	2.72	27	281	827	2.62	60*	.33	.46	.31	3.81	.15	24	2
50	150	1.8	34	286	870	.76	40*	.34	1.64	.24	2.78	.07	32	0
30	151	2.73	24	260	585	2.25	22*	.33	.23	.26	3.81	.14	21	2
52	128	2.5	32	319	797	1.06	29*	.91	6.16	.24	7.74	.38	37	1
59	131	4.29	40	537	811	4.9	74*	.42	.63	.34	7.16	.28	44	8
73	139	4.29	43	559	1046	5.26	115*	.42	.85	.39	7.16	.29	47	8
70	149	4.47	33	405	815	4.66	20*	.39	.36	.32	6.78	.24	27	2
94	299	4.48			1283		99*	.39	.81	.43	6.78	.27	33	2
0	9	.36	26	469	135	.32	0	.05	.83	0	1.94	.24	26	9
40	9	.72	17	204	340	.67	0	.08	.94	.11	5.01	.2		0
5	7	.72		27	240		4*	.01		.01	.14			0
0	3	0		45	250		0	.01		.01	.15			4
0	2	.14		7	140		60*	0		.01	.07			0
40	350	.72		543	250		73*	.12		.51	.4	.1		1
40	350	.72		542	180		73*	.12		.51	.4	.11		6
40	350	.36		533	250		73*	.12		.51	.31			1
35	250				180		100*							1
30	250	.72			170		100*							1
30	200	.36			95		100*							1
20	100	.36			75		60*							1
0	20	1.8	8	46	190	.29	0	.18	.74	.12	1.5	.02		0
20	20	1.08	15	89	120	.25	40*	.09	.58	.1	.92			0
0	100	1.44	20	117	380	.5	0	.22	0	.22	2.01	.04	8	1
0	20	1.08	7	63	200	.21		.18	1.38	.11	1.42	.03	8	24
13	103	1.59	13	69	386	.42	33*	.25	.13	.31	2.61	.04	57	1
237	201	2.72	24	199	796	1.56	101*	.49	.85	.44	3.32	.15	33	1
25	150	1.08	28	293	680	.54	80*	.25	1.23	.26	1.91	.09	<1	<1
425	40	1.08	10	126	170	1.06	135	.07	.92	.51	.06	.12	44	0
35	7	.36	7	102	290	.78	0	.18	.26	.06	1.7	.09		0
0	7	.36	11	213	330	.15	0	.08	.58	.02	.9	.08	8	2
45	200	1.8	22	191	740	1.51	40*	.56	.66	.27	3.76	.13	16	0
257	252	2.72	26	251	895	2.06	101*	.59	1.11	.49	3.79	.19	30	0
0	60	1.8	10	116	780	.33	2*	.32	.89	.26	2.46	.03	5	0
35	80	2.7	16	221	1080	1.08	2*	.51	1.14	.31	4.19	.13	5	0
245	100	2.7	21	283	1160	1.69	60*	.53	1.6	.57	4.14	.19	28	0
235	150	2.7	21	253	1250	1.7	100*	.4	1.54	.59	3.43	.13	31	0
179	157	1.81	40	605	727	2.16	1179*	.33	1.45	.37	4.32	.36	100	22

*This value is expressed in retinol equivalents (RE). All other values are in retinol activity equivalents (RAE).

Table A–1

Food Composition

(Computer code number is for Wadsworth Diet Analysis program) (For purposes of calculations, use "0" for t, <1, <.1, <.01, etc.)

Computer Code Number	Food Description	Measure	Wt (g)	H_2O (%)	Ener (cal)	Prot (g)	Carb (g)	Dietary Fiber (g)	Fat (g)	Fat Breakdown (g) Sat	Mono	Poly
	FAST FOOD RESTAURANTS—Continued											
1401	Chunky chicken salad	1 ea	296	87	164	23	8	3	5	1.3	1.6	1
	Source: McDonald's Corporation.											
	PIZZA HUT											
	Pan pizza:											
1657	Cheese	2 pce	216	50	569	24	55	4	27	11.8		
1658	Pepperoni	2 pce	208	48	549	22	55	4	27	9.8		
1659	Supreme	2 pce	273	54	657	27	59	6	35	12.3		
1660	Super supreme	2 pce	286	55	680	28	60	6	36	12		
	Thin 'n crispy pizza:											
1649	Cheese	2 pce	174	49	409	20	45	4	18	10.2		
1623	Pepperoni	2 pce	168	50	394	19	44	4	19	8.3		
1622	Supreme	2 pce	232	57	496	24	46	4	26	11.9		
1620	Super supreme	2 pce	247	58	532	25	44	4	28	11.4		
	Hand tossed pizza:											
1619	Cheese	2 pce	216	51	489	24	57	4	20	10.2		
1618	Pepperoni	2 pce	208	52	502	23	50	4	23	10.8		
1648	Supreme	2 pce	273	56	568	32	60	6	24	10		
1617	Super supreme	2 pce	286	56	599	29	64	7	27	11.1		
	Personal pan pizza:											
1610	Pepperoni	1 ea	255	49	615	26	69	5	28	10.9		
1609	Supreme	1 ea	327	57	721	33	70	6	34	12	14.7	5.6
	Source: Pizza Hut.											
	SUBWAY											
	Deli style sandwich:											
69104	Bologna	1 ea	171	64	292	10	38	2	12	4		
69102	Ham	1 ea	171	69	234	11	37	2	4	1		
69103	Roast beef	1 ea	180	69	245	13	38	2	4	1		
69105	Seafood and crab:	1 ea	178	66	298	12	37	2	11	2		
69106	With light mayo	1 ea	178	68	256	12	37	2	7	2		
69108	Tuna:	1 ea	178	63	354	11	37	2	18	3		
69107	With light mayo	1 ea	178	67	279	11	38	2	9	2		
69101	Turkey	1 ea	180	69	235	12	38	2	4	1		
	Sandwiches, 6 inch:											
	B.L.T.:											
69135	On white bread	1 ea	191	67	311	14	38	3	10	3		
69136	On wheat bread	1 ea	198	65	327	14	44	3	10	3		
	Chicken taco sub:											
69131	On white bread	1 ea	286	70	421	24	43	3	16	5		
69132	On wheat bread	1 ea	293	69	436	25	49	4	16	5		
	Club :											
69117	On white bread	1 ea	246	73	297	21	40	3	5	1		
69118	On wheat bread	1 ea	253	71	312	21	46	3	5	1		
	Cold cut trio:											
69113	On white bread	1 ea	246	71	362	19	39	3	13	4		
69114	On wheat bread	1 ea	253	68	378	20	46	3	13	4		
	Ham:											
69115	On white bread	1 ea	232	73	287	18	39	3	5	1		
69116	On wheat bread	1 ea	232	71	293	18	44	3	5	1		
	Italian B.M.T.											
69139	On white bread	1 ea	246	66	445	21	39	3	21	8		
69140	On wheat bread	1 ea	253	64	460	21	45	3	22	7		
	Meatball:											
69129	On white bread	1 ea	260	70	404	18	44	3	16	6		
69130	On wheat bread	1 ea	267	67	419	19	51	3	16	6		
	Melt with turkey, ham, bacon, cheese:											
69127	On white bread	1 ea	251	70	366	22	40	3	12	5		
69128	On wheat bread	1 ea	258	68	382	23	46	3	12	5		

Chol (mg)	Calc (mg)	Iron (mg)	Magn (mg)	Pota (mg)	Sodi (mg)	Zinc (mg)	VT-A (µg)	Thia (mg)	VT-E (mg)	Ribo (mg)	Niac (mg)	V-B6 (mg)	Fola (µg)	VT-C (mg)
75	150	1.08			120		300*							15
76	54	1.62	44	673	318	1.52	1973*	.51	1.28	.2	8.46	.52	83	30
20	393	3.53			1158		295*							5
29	196	3.53			1196		196*							5
41	308	3.69			1375		205*							12
50	300	3.6			1560		200*							12
20	409	2.95			1207		307*							5
31	207	2.99			1265		207*							5
40	297	3.57			1407		198*							18
47	285	3.42			1596		190*							17
20	408	2.93			1324		306*							5
36	359	3.23			1416		269*							43
60	232	4.6	87	589	1769	5.48	192*	.82		.66	8.45			14
44	333	3.99			1618		222*							13
30	298	4.46			1418		248*							6
66	276	5.19	74	603	1757	4.69	240*	.73		.82	9.91	.4		14
20	39	3			744		113*							14
14	24	3			773		113*							14
13	23	3			638		113*							14
17	24	3			544		113*							14
16	24	3			556		118*							14
18	26	3			557		116*							14
16	26	3			583		126*							14
12	26	3			944		113*							14
16	27	3			945		120*							15
16	33	3			957		120*							15
52	118	4			1264		209*		3.08					18
52	124	4			1275		209*		2.83					18
26	29	4			1341		120*							15
26	35	4			1352		120*							15
64	49	4			1401		130*							16
64	55	4			1412		130*							16
28	28	3			1308		120*							15
27	34	2.91			1280		117*							15
56	44	4			1652		151*							15
56	50	4			1664		151*							15
33	32	4			1035		142*							16
33	39	4			1046		142*							16
42	93	4			1735		155*							15
42	100	3			1746		156*							15

*This value is expressed in retinol equivalents (RE). All other values are in retinol activity equivalents (RAE).

Table A–1

Food Composition (Computer code number is for Wadsworth Diet Analysis program) (For purposes of calculations, use "0" for t, <1, <.1, <.01, etc.)

Computer Code Number	Food Description	Measure	Wt (g)	H_2O (%)	Ener (cal)	Prot (g)	Carb (g)	Dietary Fiber (g)	Fat (g)	Fat Breakdown (g) Sat	Mono	Poly
	FAST FOOD RESTAURANTS—Continued											
	Pizza sub:											
69133	On white bread	1 ea	250	66	448	19	41	3	22	9		
69134	On wheat bread	1 ea	257	65	464	19	48	3	22	9		
	Roast beef:											
69121	On white bread	1 ea	232	72	288	19	39	3	5	1		
69122	On wheat bread	1 ea	239	70	303	20	45	3	5	1		
	Roasted chicken breast:											
69125	On white bread	1 ea	246	70	332	26	41	3	6	1		
69126	On wheat bread	1 ea	253	68	348	27	47	3	6	1		
	Seafood and crab:											
69145	On white bread:	1 ea	246	69	415	19	38	3	19	3		
69147	With light mayo	1 ea	246	72	332	19	39	3	10	2		
69146	On wheat bread:	1 ea	253	67	430	20	44	3	19	3		
69148	With light mayo	1 ea	253	70	347	20	45	3	10	2		
	Spicy italian:											
69123	On white bread	1 ea	232	64	467	20	38	3	24	9		
69124	On wheat bread	1 ea	239	62	482	21	44	3	25	9		
	Steak and cheese:											
69119	On white bread	1 ea	257	68	383	29	41	3	10	6		
69120	On wheat bread	1 ea	264	67	398	30	47	3	10	6		
	Tuna:											
69141	On white bread:	1 ea	246	62	527	18	38	3	32	5		
69143	with light mayo	1 ea	246	70	376	18	39	3	15	2		
69142	On wheat bread:	1 ea	253	62	542	19	44	3	32	5		
69144	With light mayo	1 ea	253	68	391	19	46	3	15	2		
	Turkey:											
69111	On white bread	1 ea	232	73	273	17	40	3	4	1		
69112	On wheat bread	1 ea	239	71	289	18	46	3	4	1		
	Turkey breast and ham:											
69137	On white bread	1 ea	232	73	280	18	39	3	5	1		
69138	On wheat bread	1 ea	239	71	295	18	46	3	5	1		
	Veggie delite:											
69109	On white bread	1 ea	175	71	222	9	38	3	3	0		
69110	On wheat bread	1 ea	182	69	237	9	44	3	3	0		
	Salads:											
52128	B.L.T.	1 ea	276	91	140	7	10	2	8	3		
52124	B.M.T., classic italian	1 ea	331	86	274	14	11	1	20	7		
52127	Chicken taco	1 ea	370	87	250	18	15	2	14	5		
52115	Club	1 ea	331	91	126	14	12	1	3	1		
52120	Cold cut trio	1 ea	330	89	191	13	11	1	11	3		
52123	Ham	1 ea	316	91	116	12	11	1	3	1		
52129	Meatball	1 ea	345	88	233	12	16	2	14	5		
52131	Melt	1 ea	336	88	195	16	12	1	10	4		
52121	Pizza	1 ea	335	86	277	12	13	2	20	8		
52126	Roast beef	1 ea	316	92	117	12	11	1	3	1		
52119	Roasted chicken breast	1 ea	331	89	162	20	13	1	4	1		
52117	Seafood and crab:	1 5	331	88	244	13	10	2	17	3		
52116	With light mayo	1 5	331	90	161	13	11	2	8	1		
52130	Steak and cheese	1 ea	342	87	212	22	13	1	8	5		
52122	Tuna:	1 ea	331	84	356	12	10	1	30	5		
52118	With light mayo	1 ea	331	89	205	12	11	1	13	2		
52114	Turkey breast	1 ea	316	92	102	11	12	1	2	1		
52125	With ham	1 ea	316	92	109	11	11	1	3	1		
52113	Veggie delite	1 ea	260	94	51	2	10	1	1	0		
	Cookies:											
47662	Brazil nut and chocolate chip	1 ea	48	12	229	3	27	1	12	3.5		
47655	Chocolate chip:	1 ea	48	14	209	2	29	1	10	3.5		
47658	With M&M's	1 ea	48	14	209	2	29	1	10	3		
47659	Chocolate chunk	1 ea	48	14	209	2	29	1	10	3.5		
47656	Oatmeal raisin	1 ea	48	15	199	3	29	1	8	2		
47657	Peanut butter	1 ea	48	13	219	3	26	1	12	2.5		
47660	Sugar	1 ea	48	11	229	2	28	0	12	3		

Chol (mg)	Calc (mg)	Iron (mg)	Magn (mg)	Pota (mg)	Sodi (mg)	Zinc (mg)	VT-A (µg)	Thia (mg)	VT-E (mg)	Ribo (mg)	Niac (mg)	V-B6 (mg)	Fola (µg)	VT-C (mg)
50	103	4			1609		238*							16
50	110	3			1621		238*							16
20	25	4			928		120*							15
20	32	3			939		120*							15
48	35	3			967		123*							15
48	42	3			978		123*							15
34	28	3			849		121*							15
32	28	3			873		131*							15
34	34	3			860		121*							15
32	34	3			884		131*							15
57	40	4			1592		169*							15
57	47	4			1604		169*							15
70	88	5			1106		175*							18
70	95	5			1117		176*							18
36	32	3			875		125*							15
32	32	3			928		146*							15
36	38	3			886		126*							15
32	38	3			940		146*							15
19	30	4			1391		120*							15
19	37	3			1403		120*							15
24	29	3			1350		120*							15
24	36	3			1361		120*							15
0	25	3			582		120*							15
0	32	3			593		120*							15
16	24	1			672		273*							32
56	41	2			1379		303*							32
52	115	3			990		361*							35
26	26	2			1067		273*							32
64	46	2			1127		282*							33
28	25	2			1034		273*							32
33	30	2			761		295*							33
42	90	2			1461		308*							32
50	100	2			1336		390*							33
20	23	2			654		273*							32
48	32	2			693		276*							32
34	25	2			575		273*							32
32	25	2			599		284*							32
70	86	3			832		328*							35
36	29	2			601		278*							32
32	29	2			654		298*							32
19	28	2			1117		273*							32
24	27	2			1076		273*							32
0	23	1			308		136*							32
10	32	1.99			115		0							0
10	16	1.99			139		0							0
15	16	1			139		0							0
10	16	1			139		0							0
15	32	1			159		0							0
0	16	1			179		0							0
20	0	.72			179		0							0

*This value is expressed in retinol equivalents (RE). All other values are in retinol activity equivalents (RAE).

Table A–1

Food Composition (Computer code number is for Wadsworth Diet Analysis program) (For purposes of calculations, use "0" for t, <1, <.1, <.01, etc.)

Computer Code Number	Food Description	Measure	Wt (g)	H_2O (%)	Ener (cal)	Prot (g)	Carb (g)	Dietary Fiber (g)	Fat (g)	Fat Breakdown (g) Sat	Mono	Poly
	FAST FOOD RESTAURANTS—Continued											
47661	White chip macademia	1 ea	48	12	229	2	28	1	12	2.5		
	Source: Subway International											
	TACO BELL											
	Breakfast burrito:											
1601	Bacon breakfast burrito	1 ea	99	48	291	11	23		17	4		
1627	Country breakfast burrito	1 ea	113	55	220	8	26	2	14	5		
1626	Fiesta breakfast burrito	1 ea	92	44	280	9	25	2	16	6		
1625	Grande breakfast burrito	1 ea	177	56	420	13	43	3	22	7		
1604	Sausage breakfast burrito	1 ea	106	49	303	11	23		19	6		
	Burritos:											
1544	Bean with red sauce	1 ea	198	58	380	13	55	13	12	4		
1545	Beef with red sauce	1 ea	198	57	432	22	42	4	19	8	6.7	.7
1546	Beef & bean with red sauce	1 ea	198	57	412	17	50	5	16	6	6.1	2.1
1569	Big beef supreme	1 ea	298	64	520	24	54	11	23	10		
1552	Chicken burrito	1 ea	171	58	345	17	41		13	5		
1547	Supreme with red sauce	1 ea	255	64	440	17	51	10	19	8		
1571	7 layer burrito	1 ea	283	61	530	16	66	13	23	7		
1538	Chilito	1 ea	156	49	391	17	41		18	9		
1549	Chilito, steak	1 ea	257	62	496	26	47		23	10		
	Tacos:											
1551	Taco	1 ea	78	58	180	9	12	3	10	4		
1554	Soft taco	1 ea	90	63	220	9	12	3	10	4		
1536	Soft taco supreme	1 ea	142	64	260	12	23	3	14	7		
1568	Soft taco, chicken	1 ea	121	63	200	14	21	2	7	2.5		
1572	Soft taco, steak	1 ea	128	63	230	15	20	2	10	2.5		
1555	Tostada with red sauce	1 ea	177	67	300	10	31	12	15	5		
1558	Mexican pizza	1 ea	220	53	570	21	42	8	35	10		
1559	Taco salad with salsa	1 ea	539	71	850	30	65	16	52	15		
1560	Nachos, regular	1 ea	99	40	320	5	34	3	18	4		
1561	Nachos, bellgrande	1 ea	312	51	770	21	84	17	39	11		
1562	Pintos & cheese with red sauce	1 ea	120	68	190	9	18	10	9	4		
1563	Taco sauce, packet	1 ea	11	94	3	<1	<1	<1	<1			
1564	Salsa	1 ea	10	28	27	1	6		<1	0	0	0
1565	Cinnamon twists	1 ea	28	6	140	1	19	0	6	0		
1628	Caramel roll	1 ea	85	19	353	6	46		16	4		
	Source: Taco Bell Corporation											
	WENDY'S											
	Hamburgers:											
1566	Single on white bun, no toppings	1 ea	133	44	358	24	31	1	15	6.1		
1570	Cheeseburger, bacon	1 ea	166	55	384	20	34	2	19	7.3	6.7	2.6
1730	Chicken sandwich, grilled	1 ea	189	64	303	24	36	2	7	1.6	.8	1.9
	Baked potatoes:											
1573	Plain	1 ea	284	71	310	7	71	7	0	0	0	0
1574	With bacon & cheese	1 ea	380	69	525	16	78	7	17	4	5.9	7.2
1575	With broccoli & cheese	1 ea	411	74	478	8	80	7	14	2.7	4.3	6.8
1576	With cheese	1 ea	383	68	571	14	78	7	23	8.4	7.1	7.2
1577	With chili & cheese	1 ea	439	69	625	20	83	9	24	8.9	6.3	7.1
1578	With sour cream & chives	1 ea	439	71	515	10	102	10	8	5.2	2	.4
1579	Chili	1 ea	227	79	206	15	21	5	7	2.4		
1582	Chocolate chip cookies	1 ea	57	6	270	3	36	1	13	6		
1580	French fries	1 ea	130	41	390	5	50	5	19	3	11.9	2.4
1581	Frosty dairy dessert	1 ea	227	67	333	8	56	0	8	5.4	2.1	.3
	Source: Wendy's International											
	CONVENIENCE FOODS and MEALS											
	BUDGET GOURMET											
1695	Chicken cacciatore	1 ea	312	80	300	20	27		13			

Chol (mg)	Calc (mg)	Iron (mg)	Magn (mg)	Pota (mg)	Sodi (mg)	Zinc (mg)	VT-A (µg)	Thia (mg)	VT-E (mg)	Ribo (mg)	Niac (mg)	V-B6 (mg)	Fola (µg)	VT-C (mg)
10	16	1			139		0							0
181	80	1.8			652		310*							
195	80	1.08			690		250*							0
25	80	.72			580		150*							0
205	100	1.8			1050		500*							0
183	80	1.8			661		320*							
10	150	2.7		495	1100		450*	.04		2.02	1.98	.31		0
57	160	3.96		380	1303		530*	.4		2.14	3.44	.32		1
32	170	3.78	50	442	1221	2.67	450*	.49		.41	3.09	.59	38	1
55	150	2.7			1520		600*							5
57	140	2.52			854		440*							1
35	150	9	50	422	1230		500*	.4		2.1	2.89	.35		5
25	200	3.6			1280		300*							6
47	300	3.06			980		950*							
78	200	2.7			1313		970*							2
25	80	1.08		159	330		100*	.05		.14	1.2	.12		0
25	80	1.08		192	330		100*	.38		.22	2.68	.98		0
35	100	1.8			590		150*							4
35	80	.72			540		60*							1
25	80	1.44			1020		40*							0
15	150	1.8		455	650		500*	.06		.19	.71	.29		1
45	250	3.6	79	403	1040	5.3	400*	.32		.33	2.92	1.1	59	5
60	300	6.3		966	1780	1.54	1600*	.47		.7	4.42	.51	9	24
5	100	.72		149	570	1.57	60*	.16		.15	.64	.18	9	0
35	200	3.6		733	1310		150*	.11		.37	2.36			4
15	150	1.8	103	360	650	2.03	250*	.05		.14	.4	.2	64	0
0	0	.08		11	92		37*	0			.02			<1
0	50	.6		376	709		168*	.02		.14	0			10
0	0	.36		22	190		40*	.08		.03	.57	.03		0
15	60	1.44			312		330*							4
65	110	4.13		296	580		0	.43		.38	6.7			3
57	167	3.49	38	307	874	5.94	76*	.3		.31	6.43		28	9
65	89	2.92		428	746		43*							10
0	20	3.6	75	1187	25	.74	0	.31	.14	.12	4.3	.8	31	36
26	116	4.36	87	1385	821	2.75	103*	.24		.19	5.04	.94	36	39
5	112	6.24	93	1266	471	.97	105*	.34		.29	4.5	.97	74	79
33	312	4.06	85	1293	641	.67	172*	.25		.28	3.6	.88	36	39
41	301	5.07	122	1435	776	4.15	201*	.33		.29	4.5	.99	55	39
20	89	5.5	99	1723	108	1.28	49*	.32		.19	4.25	1.11	45	54
29	83	2.86		488	797		82*	.11		.15	2.66			4
30	10	1.8	13	89	120	.41	0*	.05		.06	.36	.03	5	0
0	20	1.08	55	845	120	.62	0*	.18		.04	3.6	.33	40	6
37	311	1.11	46	585	197	.97	163*	.11		.47	.32	.13	17	0
60	150	1.8			810		40*	.23		.51	5			21

*This value is expressed in retinol equivalents (RE). All other values are in retinol activity equivalents (RAE).

Table A-1

Food Composition (Computer code number is for Wadsworth Diet Analysis program) (For purposes of calculations, use "0" for t, <1, <.1, <.01, etc.)

Computer Code Number	Food Description	Measure	Wt (g)	H_2O (%)	Ener (cal)	Prot (g)	Carb (g)	Dietary Fiber (g)	Fat (g)	Fat Breakdown (g) Sat	Mono	Poly
	CONVENIENCE FOODS and MEALS—Continued											
1692	Linguini & shrimp	1 ea	284		364	14	55	3	10	3.1		
1691	Scallops & shrimp	1 ea	326	79	320	16	43		9			
2245	Seafood newburg	1 ea	284	74	350	17	43		12			
1693	Sirloin tips with country gravy	1 ea	284	80	310	16	21		18			
1694	Sweet & sour chicken with rice	1 ea	284		354	14	64	2	4	1.2		
1689	Teriyaki chicken	1 ea	340	77	360	20	44		12			
1690	Veal parmigiana	1 ea	340	75	440	26	39		20			
1696	Yankee pot roast	1 ea	312	77	380	27	22		21			
	Source: The All American Gourmet Co.											
	HAAGEN DAZS											
1755	Ice cream bar, vanilla almond	1 ea	87	42	304	5	21	1	22	11.5	8.2	2.5
	Sorbet:											
1758	Lemon	½ c	113	72	120	0	31	<1	0	0	0	0
1760	Orange	½ c	113	68	140	0	36		0	0	0	0
1759	Raspberry	½ c	105	71	120	0	30	2	0	0	0	0
	Yogurt, frozen:											
1753	Chocolate	½ c	98		171	8	26		4		2	
1754	Strawberry	½ c	98		171	6	27		4		2	
	Yogurt extra, frozen:											
1752	Brownie nut	½ c	101		220	8	29		9		5	
1751	Raspberry rendezvous	½ c	101		132	4	26		2		1	
	Source: Pillsbury											
	HEALTHY CHOICE											
	Entrees:											
2112	Fish, lemon pepper	1 ea	303	78	320	14	50	5	7	2		
1624	Lasagna	1 ea	383	75	420	26	59	6	9	3		
2111	Meatloaf, traditional	1 ea	340	78	316	15	52	6	5	2.5	1.9	.6
2104	Zucchini lasagna	1 ea	396	83	290	13	49	5	4	2.6		
2110	Dinner, pasta shells marinara	1 ea	340	74	380	25	55	5	6	3.5		
	Low-Fat ice cream:											
973	Brownie	½ c	71	60	120	3	22	1	2	1	.3	.7
259	Butter pecan	½ c	71	60	120	3	22	1	2	1	.3	.7
650	Chocolate chip	½ c	71	62	120	3	21	1	2	1	1	0
1608	Cookie & cream	½ c	71	62	120	3	21	1	2	1	1	0
45	Rocky road	½ c	71	52	140	3	28	1	2	1	1	0
1621	Vanilla	½ c	71	66	100	3	18	1	2	1	1	0
391	Vanilla fudge	½ c	71	62	120	3	21	1	2	1.5		
	Source: ConAgra Frozen Foods, Omaha, NE											
	HEALTH VALLEY											
	Soups, fat-free:											
2001	Beef broth, no salt added	1 c	240	98	18	5	0	0	0	0	0	0
2073	Beef broth, w/salt	1 c	240	98	30	5	2	0	0	0	0	0
2016	Black bean & vegetable	1 c	240	85	110	11	24	12	0	0	0	0
2017	Chicken broth	1 c	240	97	30	6	0	0	0	0	0	0
2018	14 garden vegetable	1 c	240	90	80	6	17	4	0	0	0	0
2015	Lentil & carrot	1 c	240	85	90	10	25	14	0	0	0	0
2014	Split pea & carrot	1 c	240	89	110	8	17	4	0	0	0	0
2013	Tomato vegetable	1 c	240	90	80	6	17	5	0	0	0	0
	Source: Health Valley											
	LA CHOY											
2100	Egg rolls, mini, chicken	1 svg	106		108	3	13	1	5	1.3		
2099	Egg rolls, mini, shrimp	1 svg	106		98	3	14	1	3	.8		
	Source: Beatrice/Hunt-Wesson											

Chol (mg)	Calc (mg)	Iron (mg)	Magn (mg)	Pota (mg)	Sodi (mg)	Zinc (mg)	VT-A (µg)	Thia (mg)	VT-E (mg)	Ribo (mg)	Niac (mg)	V-B6 (mg)	Fola (µg)	VT-C (mg)
31	50	2.26			565		75*							5
70	150	.72			690		150*			.26	3			12
70	100	.72			660		40*	.23		.26	2			
40	60	.36			570		150*	.15		.17	4	.28		2
24	71	3.18			519		354*							25
55	80	1.4			610		300*	.15		.34	6			12
165	30	4.5			1160		1000*	.45		.6	6			6
70	150	1.8			690		600*	.15		.43	7			6
74	123	.59		180	66		82*			.15				0
0	0	0		30	5		0							4
0												0		
0	0	0		56	0		0							2
40				147			20*				.17			
50				147			20*			.03	.17			
55				152			20*				.14			
20				61			0*	0			.1			
30	20	1.08			480		100*							30
35	150	3.6		500	580		100*	.3		.26	2			6
37	48	2.24			459		149*							55
10	207	1.86			321		258*							0
25	400	1.8			390		100*							0
5	100	0		268	55		40*							0
2	100	0		211	60		40*							0
5	100	0		240	50		40*							0
5	100			254	90		40*	.03		.15				0
5	100	0		168	60		40*	.03		.15				0
5	100	0		254	50		60*	.05		.22				0
2	100	0		296	50		40*							0
0				196	74						.98			
0	0	0		196	160		0				.98			5
0	40	3.6		676	280		2000*	.34		.11	1.35	.22	135	9
0	20	1.8		147	170		0			.03	2.45			1
0	40	1.8		406	250		2000*	.26		.08	2.25	.18	27	15
0	60	5.4		439	220		2000*	.1		.16	5.63	.45	27	2
0	40	5.4		439	230		2000*	.1		.16	5.63	.45		9
0	40	5.4		609	240		2000*	.1		.08	2.25	.13	<1	9
8	10	.56			335		10*							0
5	10	.56			377		52*							0

*This value is expressed in retinol equivalents (RE). All other values are in retinol activity equivalents (RAE).

Table A–1

Food Composition (Computer code number is for Wadsworth Diet Analysis program) (For purposes of calculations, use "0" for t, <1, <.1, <.01, etc.)

Computer Code Number	Food Description	Measure	Wt (g)	H_2O (%)	Ener (cal)	Prot (g)	Carb (g)	Dietary Fiber (g)	Fat (g)	Fat Breakdown (g) Sat	Mono	Poly
	CONVENIENCE FOODS and MEALS—Continued											
	LEAN CUISINE											
	Dinners:											
1639	Baked cheese ravioli	1 ea	241	76	260	12	38	4	7	3.5	1.5	.5
1632	Chicken chow mein	1 ea	255	78	240	14	37	3	3	1	1.5	.5
1633	Lasagna	1 ea	291	78	293	19	37	4	8	4.4	1.9	1
1634	Macaroni & cheese	1 ea	255	77	261	13	38	2	6	3.6	1.3	.4
1631	Spaghetti w/meatballs	1 ea	269	75	299	18	39	5	8	2.1	2.7	1.3
	Pizza:											
1636	French bread sausage pizza	1 ea	170	75	210	8	24	1	9	3.5		
	Source: Stouffer's Foods Corp, Solon OH											
	TASTE ADVENTURE SOUPS											
1905	Black bean	1 c	242	8	807	51	148	36	4	.9	.3	1.5
1904	Curry lentil	1 c	241	4	795	66	135	71	3	.4	.5	1.2
1906	Lentil chili	1 c	242		411	24	75	14	1			
1903	Split pea	1 c	244	4	807	58	143	60	3	.4	.6	1.2
	Source: Taste Adventure Soups											
	WEIGHT WATCHERS											
	Cheese, fat-free slices:											
1978	Cheddar, sharp	2 pce	21	65	30	5	2	0	0	0	0	0
1980	Swiss	2 pce	21	65	30	5	2	0	0	0	0	0
1977	White	2 pce	21	65	30	5	2	0	0	0	0	0
1979	Yellow	2 pce	21	65	30	5	2	0	0	0	0	0
	Dinners:											
2029	Chicken chow mein	1 ea	255	81	200	12	34	3	2	.5		
1646	Oven fried fish	1 ea	218	78	230	15	25	2	8	2.1	4.2	1.7
1972	Margarine, reduced fat	1 tbsp	14	49	59	0	0	0	7	1.5		
	Pizza:											
1653	Cheese	1 ea	163	48	390	23	49	6	12	4	3	1
1650	Deluxe combination pizza	1 ea	186	56	380	23	47	6	11	3.5	5	2
1652	Pepperoni pizza	1 ea	158	48	390	23	46	4	12	4	5	2
	Desserts:											
1644	Chocolate brownie	1 ea	91	75	95	3	17	2	2	.5	1	.5
2024	Chocolate eclair	1 ea	60	48	143	2	24	1	4	.8	1.1	1.9
2247	Chocolate mousse	1 ea	78	44	190	6	33	3	4	1.5		
1642	Strawberry cheesecake	1 ea	111	62	180	7	28	2	5	2	1	2
2027	Triple chocolate cheesecake	1 ea	89	52	199	7	32	1	5	2.5		
	Source: Weight Watchers											
	SWEET SUCCESS:											
	Drinks, prepared:											
1776	Chocolate chip	1 c	265	81	180	15	30	6	3	1.6		
1777	Chocolate fudge	1 c	265	81	180	15	30	6	2			
1774	Chocolate mocha	1 c	265	81	180	15	30	6	1	1		
1778	Milk chocolate	1 c	265	81	180	15	30	6	2	1		
1775	Vanilla	1 c	265	81	180	15	33	6	1	.6		
	Drinks, ready to drink:											
2147	Chocolate mint	1 c	265	82	167	10	32	5	3	0		
2148	Strawberry	1 c	265	82	167	10	32	5	3	0		
	Shakes:											
1771	Chocolate almond	1 c	250	82	158	9	30	5	2	0	2.1	.2
1773	Chocolate fudge	1 c	250	82	158	9	30	5	2	0	2.1	.2
1768	Chocolate mocha	1 c	250	82	158	9	30	5	2	0	.6	1.8
1769	Chocolate raspberry truffle	1 c	250	82	158	9	30	5	2	0	2.2	.2
1770	Vanilla creme	1 c	250	82	158	9	30	5	2	0	2.1	.3
	Snack bars:											
1767	Chocolate brownie	1 ea	33	9	120	2	23	3	4	2	.5	.6

Chol (mg)	Calc (mg)	Iron (mg)	Magn (mg)	Pota (mg)	Sodi (mg)	Zinc (mg)	VT-A (µg)	Thia (mg)	VT-E (mg)	Ribo (mg)	Niac (mg)	V-B6 (mg)	Fola (µg)	VT-C (mg)
35	150	1.44	42	450	590	1.5	100*	.06		.25	1.2	.2	48	5
35	40	.72	30	300	590	1.1	20*	.15		.17	5			0
29	244	1.41	44	596	576	2.9	98*	.15		.25	3	.32		6
18	180	.65		423	567		0	.12		.25	1.2			0
5	94	2.37		539	465		0							6
10	100	2.7	39	165	630	2.2	30*	.45		.51	.05	.07		1
0	296	12.2	405	3521	1978	8.67	39	2.12	.12	.47	4.8	.84	1042	1
0	140	22	256	2169	2182	8.51	48	1.12	.71	.59	6.33	1.25	1005	16
				1476	1016									
0	140	10.7	272	2324	1728	7.14	19	1.71	2.75	.51	6.84	.42	646	5
0	99	0		64	306		56*							0
0	99	0		74	276		56*							0
0	99	0		64	306		56*							0
0	99	0		64	306		56*							0
25	40	.72		360	430		300*							36
25	20	1.44		370	450		40*	.09		.14	1.6			0
0	0	0		5	128		49*							0
35	700	1.8		290	590		80*	.3		.51	3	.06		6
40	500	3.6		370	550		150*	.3		.51	3	.2		5
45	450	1.8		320	650		80*	.23		.51	3			5
2	40	.54		115	80		0	.03		.01	.1	.01		0
31					189		0							
5	60	1.8		320	150		0							0
15	80	.36		115	230		40*	.06		.07	1.6			2
10	80	1.08		169	199		0							0
6	500	6.3	140	600	288	5.25	350*	.52	7.05	.59	7	.7	140	21
6	500	6.3	140	750	336	5.25	350*	.52	7.05	.59	7	.7	140	21
6	500	6.3	140	800	336	5.25	350*	.52	7.05	.59	7	.7	140	21
6	500	6.3	140	750	336	5.25	350*	.52	7.05	.59	7	.7	140	21
6	500	6.3	140	830	312	5.25	250*	.52	7.05	.59	7	.7	140	21
5	419	5.3	117	469	201	4.51	294*	.45	5.86	.5	5.83	.58	117	17
5	419	5.3	117	310	175	4.51	294*	.45	5.86	.5	5.83	.58	117	17
5	396	5	110	443	190	4.25	277*	.42	7.5	.47	5.5	.55	110	16
5	396	5	110	443	175	4.25	277*	.42	7.5	.47	5.5	.55	110	16
5	396	5	110	403	175	4.25	277*	.42	7.5	.47	5.5	.55	110	16
5	383	5	110	443	175	4.25	277*	.42	7.5	.47	5.5	.55	110	16
5	396	5	110	293	175	4.25	277*	.42	7.5	.47	5.5	.55	110	16
3	150	2.71	60	140	45	.59	150*	.22	3.01	.25	3	.3	60	9
3	150	2.71	60	110	40	.59	150*	.22	3.01	.25	3	.3	60	9

*This value is expressed in retinol equivalents (RE). All other values are in retinol activity equivalents (RAE).

Table A-1

Food Composition (Computer code number is for Wadsworth Diet Analysis program) (For purposes of calculations, use "0" for t, <1, <.1, <.01, etc.)

Computer Code Number	Food Description	Measure	Wt (g)	H_2O (%)	Ener (cal)	Prot (g)	Carb (g)	Dietary Fiber (g)	Fat (g)	Fat Breakdown (g) Sat	Mono	Poly
	CONVENIENCE FOODS and MEALS—Continued											
1766	Chocolate chip	1 ea	33	9	120	2	23	3	4	2	.4	.5
1921	Oatmeal raisin	1 ea	33	9	120	2	23	3	4	2		
1765	Peanut butter	1 ea	33	9	120	2	23	3	4	2	.6	.6
	Source: Foodway National Inc, Boise, ID											
	BABY FOODS											
1720	Apple juice	½ c	125	88	59	0	15	<1	<1	t	t	t
1721	Applesauce, strained	1 tbsp	16	89	7	<1	2	<1	<1	t	t	t
1716	Carrots, strained	1 tbsp	14	92	4	<1	1	<1	<1	t	t	t
1718	Cereal, mixed, milk added	1 tbsp	15	75	17	1	2	<1	1	.3		
1719	Cereal, rice, milk added	1 tbsp	15	75	17	1	3	<1	1	.3		
1723	Chicken and noodles, strained	1 tbsp	16	86	11	<1	1	<1	<1	.1	.1	.1
1722	Peas, strained	1 tbsp	15	87	6	1	1	<1	<1	t	t	t
1717	Teething biscuits	1 ea	11	6	43	1	8	<1	<1	.2	.2	.1

Chol (mg)	Calc (mg)	Iron (mg)	Magn (mg)	Pota (mg)	Sodi (mg)	Zinc (mg)	VT-A (µg)	Thia (mg)	VT-E (mg)	Ribo (mg)	Niac (mg)	V-B6 (mg)	Fola (µg)	VT-C (mg)
3	150	2.71	60		30	.59	150*	.22	3.01	.25	3	.3	60	9
3	150	2.71	60	125	35	.59	150*	.22	3.01	.25	3	.3	60	9
0	5	.71	4	114	4	.04	1	.01	.75	.02	.1	.04	0	72
0	1	.03		11	<1	<.01	<1*	<.01	.1	<.01	.01	<.01	<1	6
0	3	.05	1	27	5	.02	160*	<.01	.07	.01	.06	.01	2	1
2	33	1.56	4	30	7	.11	4*	.06		.09	.87	.01	2	<1
2	36	1.83	7	28	7	.1	4*	.07		.07	.78	.02	1	<1
3	4	.1	2	22	4	.09	17	.01	.03	.01	.12	.01	2	<1
0	3	.14	2	17	1	.05	8*	.01	.08	.01	.15	.01	4	1
0	29	.39	4	35	40	.1	1*	.03	.05	.06	.48	.01	5	1

*This value is expressed in retinol equivalents (RE). All other values are in retinol activity equivalents (RAE).

B Canadiana: Recommendations, Choice System, and Labels

CONTENTS

This appendix presents the 1991 Recommended Nutrient Intakes (RNI) values for energy. It also offers the Choice System for Meal Planning. The U.S. Exchange System is found in Appendix D, and its Canadian equivalent, the Canadian Choice System, is presented here. Canada's new food label appears on page B-13 along with a brief explanation of how the label has changed. Visit the Internet website of Health Canada at www.hc-sc.gc.ca/ for more details. Appendix E includes addresses of Canadian governmental agencies and professional organizations that may provide additional information.

RNI

As Chapter 2 mentioned, the Dietary Reference Intakes (DRI) have replaced the 1991 RNI in Canada. Recommendations from the DRI reports are presented on the inside front cover. For energy intake recommendations, the RNI will continue to serve health professionals in Canada (Table B-1).

Table B•1

AVERAGE ENERGY REQUIREMENTS FOR CANADIANS

Age	Sex	Average Height (cm)	Average Weight (kg)	Requirements[a]					
				(kcal/kg)[b]	(MJ/kg)[b]	(kcal/day)	(MJ/day)	(kcal/cm)	(MJ/cm)
Infants (months)									
0–2	Both	55	4.5	120–100	0.50–0.42	500	2.0	9	0.04
3–5	Both	63	7.0	100–95	0.42–0.40	700	2.8	11	0.05
6–8	Both	69	8.5	95–97	0.40–0.41	800	3.4	11.5	0.05
9–11	Both	73	9.5	97–99	0.41	950	3.8	12.5	0.05
Children and Adults (years)									
1	Both	82	11	101	0.42	1100	4.8	13.5	0.06
2–3	Both	95	14	94	0.39	1300	5.6	13.5	0.06
4–6	Both	107	18	100	0.42	1800	7.6	17	0.07
7–9	M	126	25	88	0.37	2200	9.2	17.5	0.07
	F	125	25	76	0.32	1900	8.0	15	0.06
10–12	M	141	34	73	0.30	2500	10.4	17.5	0.07
	F	143	36	61	0.25	2200	9.2	15.5	0.06
13–15	M	159	50	57	0.24	2800	12.0	17.5	0.07
	F	157	48	46	0.19	2200	9.2	14	0.06
16–18	M	172	62	51	0.21	3200	13.2	18.5	0.08
	F	160	53	40	0.17	2100	8.8	13	0.05
19–24	M	175	71	42	0.18	3000	12.6		
	F	160	58	36	0.15	2100	8.8		
25–49	M	172	74	36	0.15	2700	11.3		
	F	160	59	32	0.13	1900	8.0		
50–74	M	170	73	31	0.13	2300	9.7		
	F	158	63	29	0.12	1800	7.6		
75+	M	168	69	29	0.12	2000	8.4		
	F	155	64	23	0.10	1500	6.3		

[a]Requirements can be expected to vary within a range of ±30 percent.
[b]First and last figures are averages at the beginning and end of the three-month period.
SOURCE: Health and Welfare Canada, *Nutrition Recommendations: The Report of the Scientific Review Committee* (Ottawa: Canadian Government Publishing Centre, 1990), Tables 5 and 6, pp. 25, 27.

B

Choice System for Meal Planning

The *Good Health Eating Guide* is the Canadian choice system of meal planning.[a] It is similar to the U.S. exchange system in the following ways:

- Foods are divided into lists according to carbohydrate, protein, and fat content.
- Foods are interchangeable within a group.
- Most foods are eaten in measured amounts.
- An energy value is given for each food group.

Tables B-2 through B-9 present the Canadian choice system.

[a]The tables for the Canadian choice system are adapted from the *Good Health Eating Guide Resource*, copyright 1994, with permission of the Canadian Diabetes Association.

Table B•2

CANADIAN CHOICE SYSTEM: STARCH FOODS

1 starch choice = 15 g carbohydrate (starch), 2 g protein, 290 kJ (68 kcal)

Food	Measure	Mass (Weight)
Breads		
Bagels	½	30 g
Bread crumbs	50 mL (¼ c)	30 g
Bread cubes	250 mL (1 c)	30 g
Bread sticks	2	20 g
Brewis, cooked	50 mL (¼ c)	45 g
Chapati	1	20 g
Cookies, plain	2	20 g
English muffins, crumpets	½	30 g
Flour	40 mL (2½ tbs)	20 g
Hamburger buns	½	30 g
Hot dog buns	½	30 g
Kaiser rolls	½	30 g
Matzo, 15 cm	1	20 g
Melba toast, rectangular	4	15 g
Melba toast, rounds	7	15 g
Pita, 20 cm (8″) diameter	¼	30 g
Pita, 15 cm (6″) diameter	½	30 g
Plain rolls	1 small	30 g
Pretzels	7	20 g
Raisin bread	1 slice	30 g
Rice cakes	2	30 g
Roti	1	20 g
Rusks	2	20 g
Rye, coarse or pumpernickel	½ slice	30 g
Soda crackers	6	20 g
Tortilla, corn (taco shell)	1	10 g
Tortilla, flour (9″ diameter)	1	30 g
White (French and Italian)	1 slice	25 g
Whole-wheat, cracked-wheat, rye, white enriched	1 slice	30 g
Cereals		
Bran flakes, 100% bran	125 mL (½ c)	30 g
Cooked cereals, cooked	125 mL (½ c)	125 g
Dry	30 mL (2 tbs)	20 g
Cornmeal, cooked	125 mL (½ c)	125 g
Dry	30 mL (2 tbs)	20 g
Ready-to-eat unsweetened cereals	125 mL (½ c)	20 g
Shredded wheat biscuits, rectangular or round	1	20 g
Shredded wheat, bite size	125 mL (½ c)	20 g
Wheat germ	75 mL (⅓ c)	30 g
Cornflakes	300 mL (1¼ c)	30 g
Rice Krispies	250 mL (1 c)	30 g

(continued on the next page)

Table B•2

CANADIAN CHOICE SYSTEM: STARCH FOODS—(Continued)

1 starch choice = 15 g carbohydrate (starch), 2 g protein, 290 kJ (68 kcal)

Food	Measure	Mass (Weight)
Cereals—continued		
Cheerios	200 mL (¾ c)	20 g
Muffets	1	20 g
Puffed rice	300 mL (1¼ c)	15 g
Puffed wheat	425 mL (1⅔ c)	20 g
Grains		
Barley, cooked	125 mL (½ c)	120 g
Dry	30 mL (2 tbs)	20 g
Bulgur, kasha, cooked, moist	125 mL (½ c)	70 g
Cooked, crumbly	75 mL (⅓ c)	40 g
Dry	30 mL (2 tbs)	20 g
Rice, cooked, brown and white (short and long grain)	75 mL (⅓ c)	60 g
Rice, cooked, wild	75 mL (⅓ c)	60 g
Tapioca, pearl and granulated, quick cooking, dry	30 mL (2 tbs)	15 g
Couscous, cooked moist	125 mL (½ c)	70 g
Dry	30 mL (2 tbs)	20 g
Quinoa, cooked moist	125 mL (½ c)	70 g
Dry	30 mL (2 tbs)	20 g
Pastas		
Macaroni, cooked	125 mL (½ c)	70 g
Noodles, cooked	125 mL (½ c)	80 g
Spaghetti, cooked	125 mL (½ c)	70 g
Starchy Vegetables		
Beans and peas, dried, cooked	125 mL (½ c)	98 g
Breadfruit	1 slice	75 g
Corn, canned, whole kernel	125 mL (½ c)	85 g
Corn on the cob	½ medium cob	140 g
Cornstarch	30 mL (2 tbs)	15 g
Plantains	⅓ small	50 g
Popcorn, air-popped, unbuttered	750 mL (3 c)	20 g
Potatoes, whole (with or without skin)	½ medium	95 g
Yams, sweet potatoes, (with or without skin)	½	75 g

Food	Choices per Serving	Measure	Mass (Weight)
NOTE: Food items found in this category provide more than 1 starch choice:			
Bran flakes	1 starch + ½ sugar	150 mL (⅔ c)	24 g
Croissant, small	1 starch + 1½ fats	1 small	35 g
Large	1 starch + 1½ fats	1/2 large	30 g
Corn, canned creamed	1 starch + ½ sugar	12 mL (½ c)	113 g
Potato chips	1 starch + 2 fats	15 chips	30 g
Tortilla chips (nachos)	1 starch + 1½ fats	13 chips	30 g
Corn chips	1 starch + 2 fats	30 chips	30 g
Cheese twists	1 starch + 1½ fats	30 chips	30 g
Cheese puffs	1 starch + 2 fats	27 chips	30 g
Tea biscuit	1 starch + 2 fats	1	30 g
Pancakes, homemade using 50 mL (¼ c) batter (6″ diameter)	1½ starches + 1 fat	1 medium	50 g
Potatoes, french fried (homemade or frozen)	1 starch + 1 fat	10 regular size	35 g
Soup, canned* (prepared with equal volume of water)	1 starch	250 mL (1 c)	260 g
Waffles, packaged	1 starch + 1 fat	1	35 g

*Soup can vary according to brand and type. Check the label for Food Choice Values and Symbols or the core nutrient listing.

B

Table B•3

CANADIAN CHOICE SYSTEM: FRUITS AND VEGETABLES

1 fruits and vegetables choice = 10 g carbohydrate, 1 g protein, 190 kJ (44 kcal)

Food	Measure	Mass (Weight)
Fruits (fresh, frozen, without sugar, canned in water)		
Apples, raw (with or without skin)	½ medium	75 g
Sauce unsweetened	125 mL (½ c)	120 g
Sweetened	*see Combined Food Choices*	
Apple butter	20 mL (4 tsp)	20 g
Apricots, raw	2 medium	115 g
Canned, in water	4 halves, plus 30 mL (2 tbs) liquid	110 g
Bake-apples (cloudberries), raw	125 mL (½ c)	120 g
Bananas, with peel	½ small	75 g
Peeled	½ small	50 g
Berries (blackberries, boysenberries, raspberries)	250 mL (1 c)	150 g
(blueberries, loganberries, huckleberries)	125 mL (½ c)	70 g
Raw	125 mL (½ c)	70 g
Cantaloupe, wedge with rind	¼	160 g
Cubed or diced	175 mL (⅔ c)	135 g
Cherries, raw, with pits	10	75 g
Raw, without pits	10	70 g
Canned, in water, with pits	75 mL (⅓ c), plus 30 mL (2 tbs) liquid	90 g
Canned, in water, without pits	75 mL (⅓ c), plus 30 mL (2 tbs) liquid	85 g
Crabapples, raw	1 small	55 g
Cranberries, raw	250 mL (1 c)	100 g
Figs, raw	1 medium	50 g
Canned, in water	3 medium, plus 30 mL (2 tbs) liquid	100 g
Foxberries, raw	250 mL (1 c)	100 g
Fruit cocktail, canned, in water	125 mL (½ c), plus 30 mL (2 tbs) liquid	120 g
Fruit, mixed, cut-up	125 mL (½ c)	120 g
Gooseberries, raw	250 mL (1 c)	150 g
Canned, in water	250 mL (1 c), plus 30 mL (2 tbs) liquid	230 g
Grapefruit, raw, with rind	½ small	185 g
Raw, sectioned	125 mL (½ c)	100 g
Canned, in water	125 mL (½ c), plus 30 mL (2 tbs) liquid	120 g
Grapes, raw, slip skin	125 mL (½ c)	75 g
Raw, seedless	125 mL (½ c)	75 g
Canned, in water	75 mL (⅓ c), plus 30 mL (2 tbs) liquid	115 g
Guavas, raw	½	50 g
Honeydew melon, raw, with rind	⅒ wedge	130 g
Cubed or diced	175 mL (⅔ c)	115 g
Kiwis, raw, with skin	1 medium	76 g
Kumquats, raw	3	60 g
Loquats, raw	8	130 g
Lychee fruit, raw	8	120 g
Mandarin oranges, raw, with rind	1	135 g
Raw, sectioned	125 mL (½ c)	100 g
Canned, in water	125 mL (½ c), plus 30 mL (2 tbs) liquid	100 g
Mangoes, raw, without skin and seed, Diced	75 mL (⅓ c)	65 g
Nectarines	½ medium	75 g
Oranges, raw, with rind	1 small	130 g
Raw, sectioned	125 mL (½ c)	95 g
Papayas, raw, with skin and seeds	¼ medium	150 g
Raw, without skin and seeds	¼ medium	100 g
Cubed or diced	125 mL (½ c)	100 g
Peaches, raw, with seed and skin	1 large	100 g
Raw, sliced or diced	125 mL (½ c)	85 g
Canned in water, halves or slices	125 mL (½ c), plus 30 mL (2 tbs) liquid	120 g

(continued on the next page)

B

Table B•3

CANADIAN CHOICE SYSTEM: FRUITS AND VEGETABLES—(Continued)

1 fruits and vegetables choice = 10 g carbohydrate, 1 g protein, 190 kJ (44 kcal)

Food	Measure	Mass (Weight)
Pears, raw, with skin and core	½	90 g
Raw, without skin and core	½	85 g
Canned, in water, halves	1 half plus 30 mL (2 tbs) liquid	90 g
Persimmons, raw, native	1	30 g
Raw, Japanese	¼	50 g
Pineapple, raw	1 slice	75 g
Raw, diced	125 mL (½ c)	75 g
Canned, in juice, diced	75 mL (⅓ c), plus 15 mL (1 tbs) liquid	55 g
Canned, in juice, sliced	1 slice, plus 15 mL (1 tbs) liquid	55 g
Canned, in water, diced	125 mL (½ c), plus 30 mL (2 tbs) liquid	100 g
Canned, in water, sliced	2 slices, plus 15 mL (1 tbs) liquid	100 g
Plums, raw	2 small	60 g
Damson	6	65 g
Japanese	1	70 g
Canned, in apple juice	2, plus 30 mL (2 tbs) liquid	70 g
Canned, in water	3, plus 30 mL (2 tbs) liquid	100 g
Pomegranates, raw	½	140 g
Strawberries, raw	250 mL (1 c)	150 g
Frozen/canned, in water	250 mL (1 c), plus 30 mL (2 tbs) liquid	240 g
Rhubarb	250 mL (1 c)	150 g
Tangelos, raw	1	205 g
Tangerines, raw	1 medium	115 g
Raw, sectioned	125 mL (½ c)	100 g
Watermelon, raw, with rind	1 wedge	310 g
Cubed or diced	250 mL (1 c)	160 g
Dried Fruit		
Apples	5 pieces	15 g
Apricots	4 halves	15 g
Banana flakes	30 mL (2 tbs)	15 g
Currants	30 mL (2 tbs)	15 g
Dates, without pits	2	15 g
Peaches	½	15 g
Pears	½	15 g
Prunes, raw, with pits	2	15 g
Raw, without pits	2	10 g
Stewed, no liquid	2	20 g
Stewed, with liquid	2, plus 15 mL (1 tbs) liquid	35 g
Raisins	30 mL (2 tbs)	15 g
Juices (no sugar added or unsweetened)		
Apricot, grape, guava, mango, prune	50 mL (¼ c)	55 g
Apple, carrot, papaya, pear, pineapple, pomegranate	75 mL (⅓ c)	80 g
Cranberry (see Sugars Section)		
Clamato (see Sugars Section)		
Grapefruit, loganberry, orange, raspberry, tangelo, tangerine	125 mL (½ c)	130 g
Tomato, tomato-based mixed vegetables	250 mL (1 c)	255 g
Vegetables (fresh, frozen, or canned)		
Artichokes, French, globe	2 small	50 g
Beets, diced or sliced	125 mL (½ c)	85 g
Carrots, diced, cooked or uncooked	125 mL (½ c)	75 g
Chestnuts, fresh	5	20 g
Parsnips, mashed	125 mL (½ c)	80 g
Peas, fresh or frozen	125 mL (½ c)	80 g
Canned	75 mL (⅓ c)	55 g
Pumpkin, mashed	125 mL (½ c)	45 g
Rutabagas, mashed	125 mL (½ c)	85 g
Sauerkraut	250 mL (1 c)	235 g
Snow peas	250 mL (1 c)	135 g
Squash, yellow or winter, mashed	125 mL (½ c)	115 g

(continued on the next page)

B

Table B•3

CANADIAN CHOICE SYSTEM: FRUITS AND VEGETABLES—(Continued)

1 fruits and vegetables choice = 10 g carbohydrate, 1 g protein, 190 kJ (44 kcal)

Food	Measure	Mass (Weight)
Succotash	75 mL (⅓ c)	55 g
Tomatoes, canned	250 mL (1 c)	240 g
Tomato paste	50 mL (¼ c)	55 g
Tomato sauce*	75 mL (⅓ c)	100 g
Turnips, mashed	125 mL (½ c)	115 g
Vegetables, mixed	125 mL (½ c)	90 g
Water chestnuts	8 medium	50 g

*Tomato sauce varies according to brand name. Check the label or discuss with your dietitian.

Table B•4

CANADIAN CHOICE SYSTEM: MILK

Type of Milk	Carbohydrate (g)	Protein (g)	Fat (g)	Energy
Nonfat (0%)	6	4	0	170 kJ (40 kcal)
1%	6	4	1	206 kJ (49 kcal)
2%	6	4	2	244 kJ (58 kcal)
Whole (4%)	6	4	4	319 kJ (76 kcal)

Food	Measure	Mass (Weight)
Buttermilk (higher in salt)	125 mL (½ c)	125 g
Evaporated milk	50 mL (¼ c)	50 g
Milk	125 mL (½ c)	125 g
Powdered milk, regular	30 mL (2 tbs)	15 g
Instant	50 mL (¼ c)	15 g
Plain yogurt	125 mL (½ c)	125 g

Food	Choices per Serving	Measure	Mass (Weight)
Note: Food items found in this category provide more than 1 milk choice:			
Milkshake	1 milk + 3 sugars + ½ protein	250 mL (1 c)	300 g
Chocolate milk, 2%	2 milks 2% + 1 sugar	250 mL (1 c)	300 g
Frozen yogurt	1 milk + 1 sugar	125 mL (½ c)	125 g

Table B•5

CANADIAN CHOICE SYSTEM: SUGARS

1 sugar choice = 10 g carbohydrate (sugar), 167 kJ (40 kcal)

Food	Measure	Mass (Weight)
Beverages		
Condensed milk	15 mL (1 tbs)	
Flavoured fruit crystals*	75 mL (⅓ c)	
Iced tea mixes*	75 mL (⅓ c)	
Regular soft drinks	125 mL (½ c)	
Sweet drink mixes*	75 mL (⅓ c)	
Tonic water	125 mL (½ c)	
*These beverages have been made with water.		
Miscellaneous		
Bubble gum (large square)	1 piece	5 g
Cranberry cocktail	75 mL (⅓ c)	80 g
Cranberry cocktail, light	350 mL (1⅓ c)	260 g
Cranberry sauce	30 mL (2 tbs)	
Hard candy mints	2	5 g
Honey, molasses, corn and cane syrup	10 mL (2 tsp)	15 g
Jelly bean	4	10 g
Licorice	1 short stick	10 g
Marshmallows	2 large	15 g
Popsicle	1 stick (½ popsicle)	
Powdered gelatin mix (Jello®) (reconstituted)	50 mL (¼ c)	
Regular jam, jelly, marmalade	15 mL (1 tbs)	
Sugar, white, brown, icing, maple	10 mL (2 tsp)	10 g
Sweet pickles	2 small	100 g
Sweet relish	30 mL (2 tbs)	

Food	Choices per Serving	Measures	Mass (Weight)
The following food items provide more than 1 sugar choice:			
Brownie	1 sugar + 1 fat	1	20 g
Clamato juice	1½ sugars	175 mL (⅔ c)	
Fruit salad, light syrup	1 sugar + 1 fruits & vegetables	125 mL (½ c)	130 g
Aero® bar	2½ sugars + 2½ fats	1 bar	43 g
Smarties®	4½ sugars + 2 fats	1 box	60 g
Sherbet	3 sugars + ½ fat	125 mL (½ c)	95 g

Table B•6

CANADIAN CHOICE SYSTEM: PROTEIN FOODS

1 protein choice = 7 g protein, 3 g fat, 230 kJ (55 kcal)

Food	Measure	Mass (Weight)
Cheese		
Low-fat cheese, about 7% milk fat	1 slice	30 g
Cottage cheese, 2% milkfat or less	50 mL (¼ c)	55 g
Ricotta, about 7% milkfat	50 mL (¼ c)	60 g
Fish		
Anchovies (see Extras, Table B-9)		
Canned, drained (e.g., mackerel, salmon, tuna packed in water)	50 mL (¼ c)	30 g
Cod tongues, cheeks	75 mL (⅓ c)	50 g
Fillet or steak (e.g., Boston blue, cod, flounder, haddock, halibut, mackerel, orange roughy, perch, pickerel, pike, salmon, shad, snapper, sole, swordfish, trout, tuna, whitefish)	1 piece	30 g
Herring	⅓ fish	30 g
Sardines, smelts	2 medium or 3 small	30 g
Squid, octopus	50 mL (¼ c)	40 g
Shellfish		
Clams, mussels, oysters, scallops, snails	3 medium	30 g
Crab, lobster, flaked	50 mL (¼ c)	30 g
Shrimp, fresh	5 large	30 g
Frozen	10 medium	30 g
Canned	18 small	30 g
Dry pack	50 mL (¼ c)	30 g
Meat and Poultry (e.g., beef, chicken, goat, ham, lamb, pork, turkey, veal, wild game)		
Back, peameal bacon	3 slices, thin	30 g
Chop	½ chop, with bone	40 g
Minced or ground, lean or extra-lean	30 mL (2 tbs)	30 g
Sliced, lean	1 slice	30 g
Steak, lean	1 piece	30 g
Organ Meats		
Hearts, liver	1 slice	30 g
Kidneys, sweetbreads, chopped	50 mL (¼ c)	30 g
Tongue	1 slice	30 g
Tripe	5 pieces	60 g
Soyabean		
Bean curd or tofu	½ block	70 g
Eggs		
In shell, raw or cooked	1 medium	50 g
Without shell, cooked or poached in water	1 medium	45 g
Scrambled	50 mL (¼ c)	55 g

Food	Choices per Serving	Measures	Mass (Weight)
Note: The following choices provide more than 1 protein exchange:			
Cheese			
Cheeses	1 protein + 1 fat	1 piece	25 g
Cheese, coarsely grated (e.g., cheddar)	1 protein + 1 fat	50 mL (¼ c)	25 g
Cheese, dry, finely grated (e.g., parmesan)	1 protein + 1 fat	45 mL	15 g
Cheese, ricotta, high fat	1 protein + 1 fat	50 mL (¼ c)	55 g
Fish			
Eel	1 protein + 1 fat	1 slice	50 g
Meat			
Bologna	1 protein + 1 fat	1 slice	20 g
Canned lunch meats	1 protein + 1 fat	1 slice	20 g
Corned beef, canned	1 protein + 1 fat	1 slice	25 g
Corned beef, fresh	1 protein + 1 fat	1 slice	25 g
Ground beef, medium-fat	1 protein + 1 fat	30 mL (2 tbs)	25 g

(continued on the next page)

Table B•6

CANADIAN CHOICE SYSTEM: PROTEIN FOODS—(Continued)

1 protein choice = 7 g protein, 3 g fat, 230 kJ (55 kcal)

Food	Choices per Serving	Measures	Mass (Weight)
Meat—continued			
Meat spreads, canned	1 protein + 1 fat	45 mL	35 g
Mutton chop	1 protein + 1 fat	½ chop, with bone	35 g
Paté (*see Fats and Oils group*, Table B-8)			
Sausages, garlic, Polish or knockwurst	1 protein + 1 fat	1 slice	50 g
Sausages, pork, links	1 protein + 1 fat	1 link	25 g
Spareribs or shortribs, with bone	1 protein + 1 fat	1 large	65 g
Stewing beef	1 protein + 1 fat	1 cube	25 g
Summer sausage or salami	1 protein + 1 fat	1 slice	40 g
Weiners, hot dog	1 protein + 1 fat	½ medium	25 g
Miscellaneous			
Blood pudding	1 protein + 1 fat	1 slice	25 g
Peanut butter	1 protein + 1 fat	15 mL (1 tbs)	15 g

Table B•7

CANADIAN CHOICE SYSTEM: FATS AND OILS

1 fat choice = 5 g fat, 190 kJ (45 kcal)

Food	Measure	Mass (Weight)
Avocado*	⅛	30 g
Bacon, side, crisp*	1 slice	5 g
Butter*	5 mL (1 tsp)	5 g
Cheese spread	15 mL (1 tbs)	15 g
Coconut, fresh*	45 mL (3 tbs)	15 g
Coconut, dried*	15 mL (1 tbs)	10 g
Cream, Half and half (cereal), 10%*	30 mL (2 tbs)	30 g
Light (coffee), 20%*	15 mL (1 tbs)	15 g
Whipping, 32 to 37%*	15 mL (1 tbs)	15 g
Cream cheese*	15 mL (1 tbs)	15 g
Gravy*	30 mL (2 tbs)	30 g
Lard*	5 mL (1 tsp)	5 g
Margarine	5 mL (1 tsp)	5 g
Nuts, shelled:		
Almonds	8	5 g
Brazil nuts	2	10 g
Cashews	5	10 g
Filberts, hazelnuts	5	10 g
Macadamia	3	5 g
Peanuts	10	10g
Pecans	5 halves	5 g
Pignolias, pine nuts	25 mL (5 tsp)	10 g
Pistachios, shelled	20	10 g
Pistachios, in shell	20	20 g
Pumpkin and squash seeds	20 mL (4 tsp)	10 g
Nuts (continued):		
Sesame seeds	15 mL (1 tbs)	10 g
Sunflower seeds		
Shelled	15 mL (1 tbs)	10 g
In shell	45 mL (3 tbs)	15 g
Walnuts	4 halves	10 g
Oil, cooking and salad	5 mL (1 tsp)	5 g
Olives, green	10	45 g
Ripe black	7	57 g
Pâté, liverwurst, meat spreads	15 mL (1 tbs)	15 g
Salad dressing: blue, French, Italian,	10 mL (2 tsp)	10 g
mayonnaise, Thousand Island	5 mL (1 tsp)	5 g
Salad dressing, low-calorie	30 mL (2 tbs)	30 g
Salt pork, raw or cooked*	5 mL (1 tsp)	5 g
Sesame oil	5 mL (1 tsp)	5 g
Sour cream		
12% milkfat	30 mL (2 tbs)	30 g
7% milkfat	60 mL (4 tbs)	60 g
Shortening*	5 mL (1 tsp)	

*These items contain higher amounts of saturated fat.

Table B•8

CANADIAN CHOICE SYSTEM: EXTRAS

Extras have no more than 2.5 g carbohydrate, 60 kJ (14 kcal)

Vegetables 125 mL (½ c)
Artichokes
Asparagus
Bamboo shoots
Bean sprouts, mung or soya
Beans, string, green, or yellow
Bitter melon (balsam pear)
Bok choy
Broccoli
Brussels sprouts
Cabbage
Cauliflower
Celery
Chard
Cucumbers
Eggplant
Endive
Fiddleheads
Greens: beet, collard, dandelion, mustard, turnip, etc.
Kale
Kohlrabi
Leeks
Lettuce
Mushrooms
Okra
Onions, green or mature
Parsley
Peppers, green, yellow or red
Radishes
Rapini
Rhubarb
Sauerkraut
Shallots
Spinach
Sprouts: alfalfa, radish, etc.
Tomato wedges
Watercress
Zucchini

Free Foods (may be used without measuring)
Artificial sweetener, such as cyclamate or aspartame
Baking powder, baking soda
Bouillon from cube, powder, or liquid
Bouillon or clear broth
Chowchow, unsweetened
Coffee, clear
Consommé
Dulse
Flavorings and extracts
Garlic
Gelatin, unsweetened
Ginger root
Herbal teas, unsweetened
Horseradish, uncreamed
Lemon juice or lemon wedges
Lime juice or lime wedges
Marjoram, cinnamon, etc.
Mineral water
Mustard
Parsley
Pimentos
Salt, pepper, thyme
Soda water, club soda
Soya sauce
Sugar-free Crystal Drink
Sugar-free Jelly Powder
Sugar-free soft drinks
Tea, clear
Vinegar
Water
Worcestershire sauce

Condiments

Food	**Measure**
Anchovies	2 fillets
Barbecue sauce	15 mL (1 tbs)
Bran, natural	30 mL (2 tbs)
Brewer's yeast	5 mL (1 tsp)
Carob powder	5 mL (1 tsp)
Catsup	5 mL (1 tsp)
Chili sauce	5 mL (1 tsp)
Cocoa powder	5 mL (1 tsp)
Cranberry sauce, unsweetened	15 mL (1 tbs)
Dietetic fruit spreads	5 mL (1 tsp)
Maraschino cherries	1
Nondairy coffee whitener	5 mL (1 tsp)
Nuts, chopped pieces	5 mL (1 tsp)
Pickles	
Unsweetened dill	2
Sour mixed	11
Sugar substitutes, granular	5 mL (1 tsp)
Whipped toppings	15 mL (1 tbs)

B

Table B•9

CANADIAN CHOICE SYSTEM: COMBINED FOOD CHOICES

Food	Choices per Serving	Measure	Mass (Weight)
Angel food cake	½ starch + 2½ sugars	1/12 cake	50 g
Apple crisp	½ starch + 1½ fruits & vegetables + 1 sugar + 1–2 fats	125 mL (½ c)	
Applesauce, sweetened	1 fruits & vegetables + 1 sugar	125 mL (½ c)	
Beans and pork in tomato sauce	1 starch + ½ fruits & vegetables + ½ sugar + 1 protein	125 mL (½ c)	135 g
Beef burrito	2 starches + 3 proteins + 3 fats		110 g
Brownie	1 sugar + 1 fat	1	20 g
Cabbage rolls*	1 starch + 2 proteins	3	310 g
Caesar salad	2–4 fats	20 mL dressing (4 tsp)	
Cheesecake	½ starch + 2 sugars + ½ protein + 5 fats	1 piece	80 g
Chicken fingers	1 starch + 2 proteins + 2 fats	6 small	100 g
Chicken and snow pea Oriental	2 starches + ½ fruits & vegetables + 3 proteins + 1 fat	500 mL (2 c)	
Chili	1½ starches + ½ fruits & vegetables + 3½ protein	300 mL (1¼ c)	325 g
Chips			
Potato chips	1 starch + 2 fats	15 chips	30 g
Corn chips	1 starch + 2 fats	30 chips	30 g
Tortilla chips	1 starch + 1½ fats	13 chips	
Cheese twist	1 starch + 1½ fats	30 chips	30 g
Chocolate bar			
Aero®	2½ sugars + 2½ fats	bar	43 g
Smarties®	4½ sugars + 2 fats	package	60 g
Chocolate cake (without icing)	1 starch + 2 sugars + 3 fats	1/10 of a 8″ pan	
Chocolate devil's food cake (without icing)	2 starches + 2 sugars + 3 fats	1/12 of a 9″ pan	
Chocolate milk	2 milks 2% + 1 sugar	250 mL (1 c)	300 g
Clubhouse (triple-decker) sandwich	3 starches + 3 proteins + 4 fats		
Cookies			
Chocolate chip	½ starch + ½ sugar + 1½ fats	2	22 g
Oatmeal	1 starch + 1 sugar + 1 fat	2	40 g
Donut (chocolate glazed)	1 starch + 1½ sugars + 2 fats	1	65 g
Egg roll	1 starch + ½ protein + 1 fat		75 g
Four bean salad	1 starch + ½ protein + 1 fat	125 mL (½ c)	
French toast	1 starch + ½ protein + 2 fats	1 slice	65 g
Fruit in heavy syrup	1 fruits & vegetables + 1½ sugars	125 mL (½ c)	
Granola bar	½ starch + 1 sugar + 1–2 fats		30 g
Granola cereal	1 starch + 1 sugar + 2 fats	125 mL (½ c)	45 g
Hamburger	2 starches + 3 proteins + 2 fats	junior size	
Ice cream and cone, plain flavour			
Ice cream	½ milk + 2–3 sugars + 1–2 fats		100 g
Cone	½ sugar		4 g
Lasagna			
Regular cheese	1 starch + 1 fruits & vegetables + 3 proteins + 2 fats	3″ x 4″ piece	
Low-fat cheese	1 starch + 1 fruits & vegetables + 3 proteins	3″ x 4″ piece	
Legumes			
Dried beans (kidney, navy, pinto, fava, chick peas)	2 starches + 2 protein	250 mL (1 c)	175 g
Dried peas	2 starches + 2 protein	250 mL (1 c)	196 g
Lentils	2 starches + 2 protein	250 mL (1 c)	196 g
Macaroni and cheese	2 starches + 2 proteins + 2 fats	250 mL (1 c)	210 g
Minestrone soup	1½ starches + ½ fruits & vegetables + ½ fat	250 mL (1 c)	
Muffin	1 starch + ½ sugar + 1 fat	1 small	45 g

*If eaten with sauce, add ½ fruits & vegetables exchange.

(continued on the next page)

B

Table B•9

CANADIAN CHOICE SYSTEM: COMBINED FOOD CHOICES—(Continued)

Food	Choices per Serving	Measure	Mass (Weight)
Nuts (dry or roasted without any oil added).			
Almonds, dried sliced	½ protein + 2 fats	50 mL (¼ c)	22 g
Brazil nuts, dried unblanched	½ protein + 2½ fats	5 large	23 g
Cashew nuts, dry roasted	½ starch + ½ protein + 2 fats	50 mL (¼ c)	28 g
Filbert hazelnut, dry	½ protein + 3½ fats	50 mL (¼ c)	30 g
Macadamia nuts, dried	½ protein + 4 fats	50 mL (¼ c)	28 g
Peanuts, raw	1 protein + 2 fats	50 mL (¼ c)	30 g
Pecans, dry roasted	½ fruits & vegetables + 3 fats	50 mL (¼ c)	22 g
Pine nuts, pignolia dried	1 protein + 3 fats	50 mL (¼ c)	34 g
Pistachio nuts, dried	½ fruits & vegetables + ½ protein + 2½ fats	50 mL (¼ c)	27 g
Pumpkin seeds, roasted	2 proteins + 2½ fats	50 mL (¼ c)	47 g
Sesame seeds, whole dried	½ fruits & vegetables 1 ½ protein 1 2½ fats	50 mL (¼ c)	30 g
Sunflower kernel, dried	½ protein + 1½ fats	50 mL (¼ c)	17 g
Walnuts, dried chopped	½ protein + 3 fats	50 mL (¼ c)	26 g
Perogies	2 starches + 1 protein + 1 fat	3	
Pie, fruit	1 starch + 1 fruits & vegetables + 2 sugars + 3 fats	1 piece	120 g
Pizza, cheese	1 starch + 1 protein + 1 fat	1 slice (⅛ of a 12″)	50 g
Pork stir fry	½ to + fruits & vegetables + 3 proteins	200 mL (¾ c)	
Potato salad	1 starch + 1 fat	125 mL (½ c)	130 g
Potatoes, scalloped	2 starches + 1 milk + 1–2 fats	200 mL (¾ c)	210 g
Pudding, bread or rice	1 starch + 1 sugar + 1 fat	125 mL (½ c)	
Pudding, vanilla	1 milk + 2 sugars	125 mL (½ c)	
Raisin bran cereal	1 starch + ½ fruits & vegetables + ½ sugar	175 mL (⅔ c)	40 g
Rice krispie squares	½ starch + 1½ sugars 1 ½ fat	1 square	30 g
Shepherd's pie	2 starches + 1 fruits & vegetables + 3 proteins	325 mL (1⅓ c)	
Sherbet, orange	3 sugars + ½ fat	125 mL (½ c)	
Spaghetti and meat sauce	2 starches + 1 fruits & vegetables + 2 proteins + 3 fats	250 mL (1 c)	
Stew	2 starches + 2 fruits & vegetables + 3 proteins + ½ fat	200 mL (¾ c)	
Sundae	4 sugars + 3 fats	125 mL (½ c)	
Tuna casserole	1 starch + 2 proteins + ½ fat	125 mL (½ c)	
Yogurt, fruit bottom	1 fruits & vegetables + 1 milk + 1 sugar	125 mL (½ c)	125 g
Yogurt, frozen	1 milk + 1 sugar	125 mL (½ c)	125 g

B

Food Labels

Consumers can gather a lot of information from a nutrition label. Figure B-1 points out changes to the Canadian food label and Table B-10 (below) and Table B-11 (next page) define terms.

Figure B•1

EXAMPLES OF CANADA'S NEW FOOD LABELS

English Label

Nutrition Facts Per 1 cup (264g)	
Amount	**% Daily Value**
Calories 260	
Fat 13g	20%
Saturated Fat 3g + Trans Fat 2g	25%
Cholesterol 30mg	
Sodium 660mg	28%
Carbohydrate 31g	10%
Fibre 0g	0%
Sugars 5g	
Protein 5g	
Vitamin A 4% • Vitamin C 2%	
Calcium 15% • Iron 4%	

Bilingual Label

Nutrition Facts Valeur nutritive Per 1 cup (264g) pour 1 tasse (264g)	
Amount Quantité	**% Daily Value % valeur quotidienne**
Calories / Calories 260	
Fat / Lipides 13g	20%
Saturated / saturés 3g + Trans / trans 2g	25%
Cholesterol / Cholestérol 30mg	
Sodium / Sodium 660mg	28%
Carbohydrate / Glucides 31g	10%
Fibre / Fibres 0g	0%
Sugars / Sucres 5g	
Protein / Protéines 5g	
Vitamin A / Vitamine A	4%
Vitamin C / Vitamine C	2%
Calcium / Calcium	15%
Iron / Fer	4%

French Label

Valeur nutritive pour 1 tasse (264g)	
Quantité	**% valeur quotidienne**
Calories 260	
Lipides 13g	20%
Saturés 3g + Trans 2g	25%
Cholestérol 30mg	
Sodium 660mg	28%
Glucides 31g	10%
Fibres 0g	0%
Sucres 5g	
Protéines 5g	
Vitamine A 4% • Vitamine C 2%	
Calcium 15% • Fer 4%	

The new and improved nutrition labelling includes:

- A new title: *Nutrition Facts.*
- Consistent serving sizes on which nutrient information is based.
- An expanded list of nutrients.
- A standardized format that is bold, clear, and easy to read.
- Appearance consistent from product to product.
- More clearly identified nutrient information.
- The Daily Value gives a context to the nutrient values.

For more information, visit the Internet website:

http://www.hc-sc.gc.ca/hppb/nutrition/labels/e_before.html

Table B•10

TERMS ON FOOD LABELS

Energy

kcalorie reduced 50% or fewer kcalories than the regular version.
light term may be used to describe anything (for example, light in colour, texture, flavour, taste, or kcalories); read the label to find out what is "light" about the product.
low kcalorie kcalorie-reduced and no more than 15 kcalories per serving.

Fat and Cholesterol

low cholesterol no more than 3 mg of cholesterol per 100 g of the food and low in saturated fat; does not always mean low in total fat.
low fat no more than 3 g of fat per serving; does not always mean low in kcalories.
lower fat at least 25% less fat than the comparison food; be aware that 80% fat-free still means the food is 20% fat.

Carbohydrates: Fibre and Sugar

carbohydrate reduced not more than 50% of the carbohydrate found in the regular version; does not always mean the product is lower in kcalories because other ingredients such as fat may have increased.
source of dietary fibre a product that provides 2–4 g of fibre.
high source of dietary fibre a product that provides 4–6 g of fibre.
very high source of fibre a product that provides 6 g (or more) of fibre.
sugar free low in carbohydrates and kcalories; can be used as an extra food in the exchange system.
unsweetened or no sugar added no sugar was added to the product; sugar may be found naturally in the food (for example, fruit canned in its own juice).

B

Table B•11

NUTRIENT CONTENT CLAIMS

Certain nutrient content claims are allowable on labels if the food meets certain criteria compared with a reference amount and serving of stated size. For example, a food claiming to be "high fibre" would have to contain 4 grams or more of fibre per reference amount and serving of stated size. For prepackaged meals or entrées for which no reference amount exists, the criteria for most claims would be based per 100 grams food for most claims.

Nutrient Content Claim	Compositional Criteria
	(per reference amount and serving of stated size)[1]
Calories	
Calorie-free	less than 5 calories
Low in calories	40 calories or less
Reduced or lower in calories	at least 25% less energy[2]
Source of calories	at least 100 calories
Protein	
Source of protein	protein rating of 20 or more[3]
Excellent source of protein	protein rating of 40 or more[3]
More protein	protein rating of 20 or more[3], at least 25% or more protein[2], and 7 g or more protein
Fat	
Fat-free	less than 0.5 g fat
Low in fat	3 g or less fat
Reduced or lower in fat	at least 25% less fat[2]
100% fat-free	less than 0.5 g fat per 100 g, no added fat, and "free of fat"
(naming the %) fat-free	"low in fat"
Saturated Fatty Acids	
Saturated fatty acid-free	less than 0.2 g saturated fatty acids and less than 0.2 g *trans* fatty acids
Low in saturated fatty acids	2 g or less saturated fatty acids and *trans* fatty acids combined and 15% or less energy from saturated fatty acids plus *trans* fatty acids
Reduced or lower in saturated fatty acids	at least 25% less saturated fatty acids and *trans* fatty acids not increased[2]
***Trans* Fatty Acids**	
Free of *trans* fatty acids	less than 0.2 g *trans* fatty acids and "low in saturated fatty acids"
Reduced or lower in *trans* fatty acids	at least 25% less *trans* fatty acids and saturated fatty acids not increased[2]
Polyunsaturated fatty acids	
Source of omega-3 polyunsaturated fatty acids	0.3 g or more omega-3 polyunsaturated fatty acids
Source of omega-6 polyunsaturated fatty acids	2 g or more omega-6 polyunsaturated fatty acids
Cholesterol	
Cholesterol-free	less than 2 mg cholesterol and "low in saturated fatty acids"
Low in cholesterol	20 mg or less cholesterol and "low in saturated fatty acids"
Reduced or lower in cholesterol	at least 25% less cholesterol[2] and "low in saturated fatty acids"
Sodium	
Sodium-free or salt-free	less than 5 mg sodium or salt
Low in sodium or salt	140 mg or less sodium or salt
Reduced or lower in sodium or salt	at least 25% less sodium or salt[2]
No added sodium or salt	no salt or other sodium salts added during processing
Lightly salted	at least 50% less added sodium or salt[2]
Sugars	
Sugar-free	less than 0.5 g sugars and (except chewing gum) "free of energy"
Reduced or lower in sugar	at least 25% less sugars[2]
No added sugar	no sugars added in processing or packaging, including ingredients that contain added sugars or ingredients that functionally substitute for added sugars (e.g. concentrated fruit juice), and sugars not increased through some other means
Fibre	
Source of fibre	2 g or more fibre or of each identified fibre
High source of fibre	4 g or more fibre or of each identified fibre
Very high source of fibre	6 g or more fibre or of each identified fibre
More fibre	at least 25% more fibre[2] and at least 2 g fibre
Light	
Light	"reduced in energy" or "reduced in fat"

[1]Criteria for "reduced" or "lower in" claims are based on the reference amount of the food, and for protein claims, criteria are given per reasonable daily intake of food (Schedule K of the Regulations) or, in the case of breakfast cereals, per 30 g of breakfast cereal combined with 125 mL of milk.
[2]As compared with the reference amount of a reference food. Reference food must not be "low" in subject nutrient.
[3]As determined by official method FO-1, Determination of Protein Rating, Oct. 15, 1981.

B

Figure B•2

CANADA'S FOOD GUIDE TO HEALTHY EATING

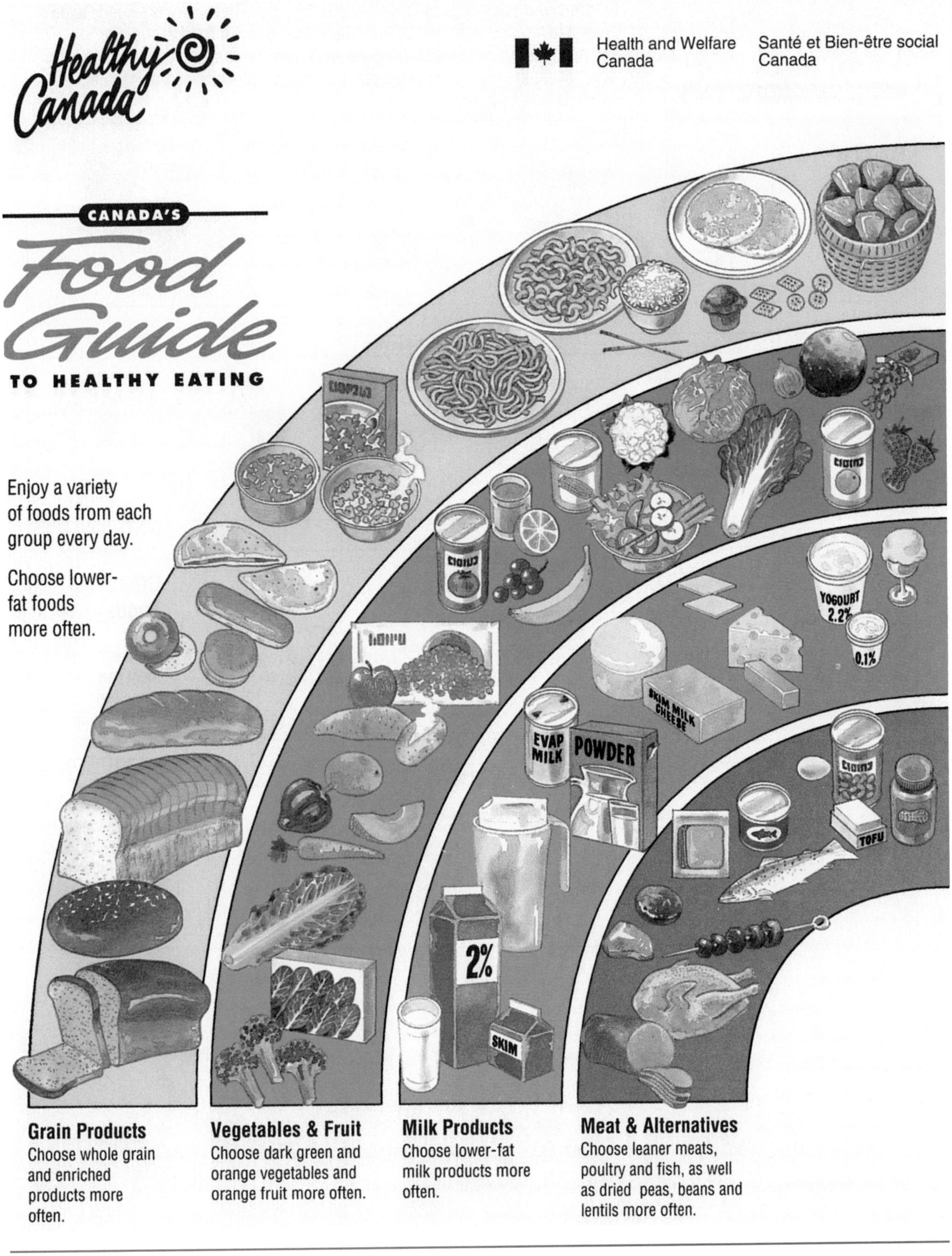

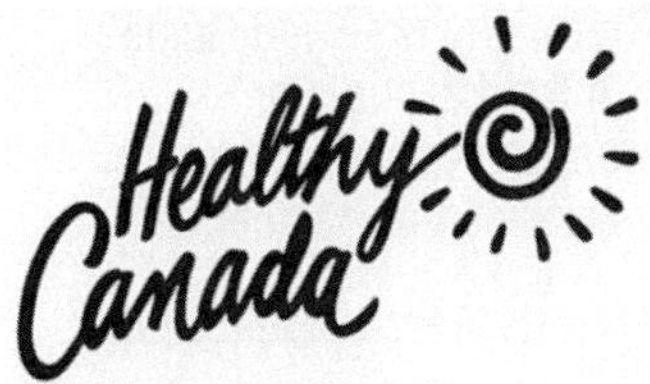

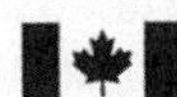

Healthy Canada
Health and Welfare Canada
Santé et Bien-être social Canada
CANADA'S
Food Guide
TO HEALTHY EATING
Enjoy a variety of foods from each group every day.
Choose lower-fat foods more often.
EVAP MILK
POWDER
SKIM MILK CHEESE
YOGOURT 2.2%
0.1%
2%
SKIM
TOFU
Grain Products
Choose whole grain and enriched products more often.
Vegetables & Fruit
Choose dark green and orange vegetables and orange fruit more often.
Milk Products
Choose lower-fat milk products more often.
Meat & Alternatives
Choose leaner meats, poultry and fish, as well as dried peas, beans and lentils more often.

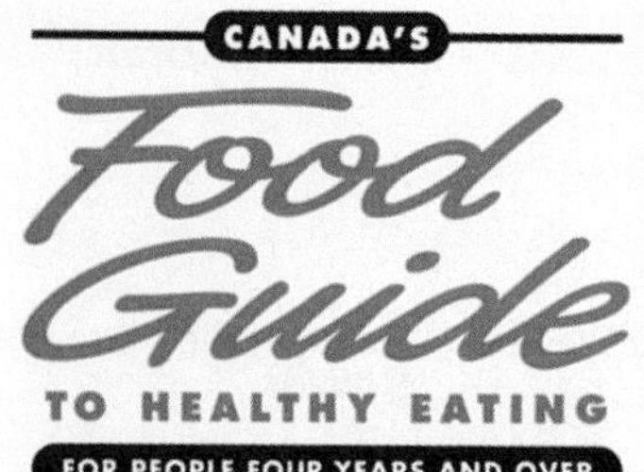

Different People Need Different Amounts of Food

The amount of food you need every day from the 4 food groups and other foods depends on your age, body size, activity level, whether you are male or female and if you are pregnant or breast-feeding. That's why the Food Guide gives a lower and higher number of servings for each food group. For example, young children can choose the lower number of servings, while male teenagers can go to the higher number. Most other people can choose servings somewhere in between.

Grain Products
5–12
SERVINGS PER DAY

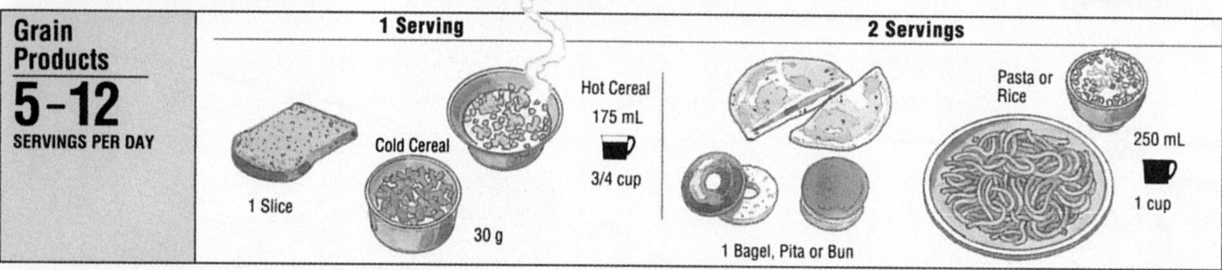

Vegetables & Fruit
5–10
SERVINGS PER DAY

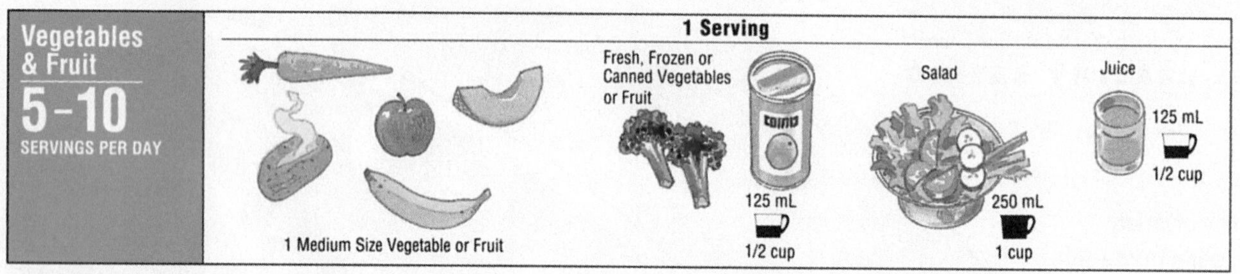

Milk Products
SERVINGS PER DAY
Children 4–9 years: 2–3
Youth 10–16 years: 3–4
Adults: 2–4
Pregnant & Breast-feeding Women: 3–4

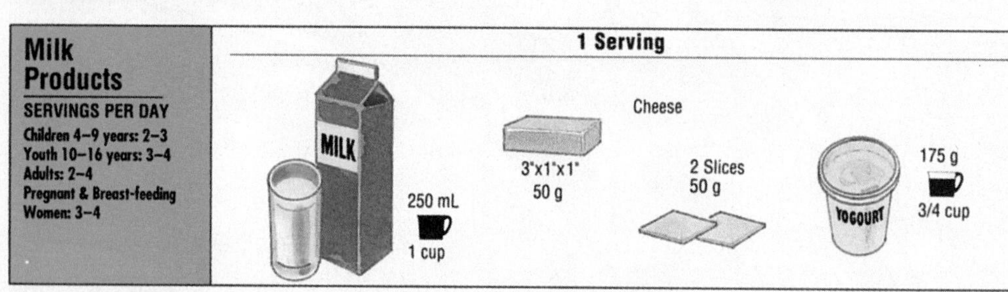

Meat & Alternatives
2–3
SERVINGS PER DAY

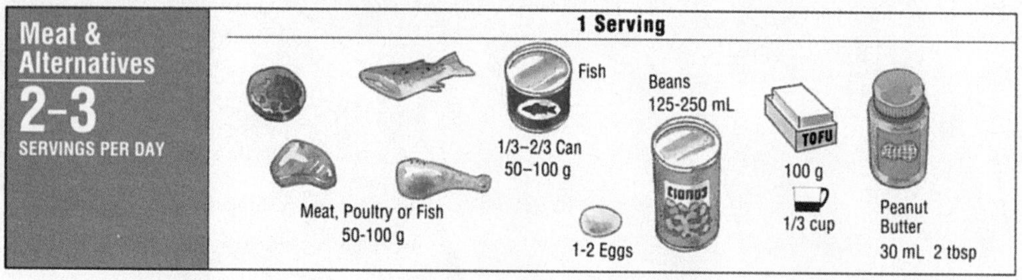

Other Foods

Taste and enjoyment can also come from other foods and beverages that are not part of the 4 food groups. Some of these foods are higher in fat or Calories, so use these foods in moderation.

Enjoy eating well, being active and feeling good about yourself. That's VITALIT**E**

Aids to Calculation

CONTENTS

C

Mathematical problems have been worked out for you as examples at appropriate places in the text. This appendix aims to help with the use of the metric system and with those problems not fully explained elsewhere.

Conversion Factors

Conversion factors are useful mathematical tools in everyday calculations, like the ones encountered in the study of nutrition. A conversion factor is a fraction in which the numerator (top) and the denominator (bottom) express the same quantity in different units. For example, 2.2 pounds (lb) and 1 kilogram (kg) are equivalent; they express the same weight. The conversion factor used to change pounds to kilograms or vice versa is:

$$\frac{2.2 \text{ lb}}{1 \text{ kg}} \text{ or } \frac{1 \text{ kg}}{2.2 \text{ lb}}$$

Because both factors equal 1, measurements can be multiplied by the factor without changing the value of the measurement. Thus the units can be changed.

The correct factor to use in a problem is the one with the unit you are seeking in the numerator (top) of the fraction. Following are some examples of problems commonly encountered in nutrition study; they illustrate the usefulness of conversion factors.

Example 1

Convert the weight of 130 pounds to kilograms.

1. Choose the conversion factor in which the unit you are seeking is on top:

$$\frac{1 \text{ kg}}{2.2 \text{ lb}}$$

2. Multiply 130 pounds by the factor:

$$130 \text{ lb} \times \frac{1 \text{ kg}}{2.2 \text{ lb}} = \frac{130 \text{ kg}}{2.2}$$

= 59 kg (rounded off to the nearest whole number)

Example 2

How many grams (g) of saturated fat are contained in a 3-ounce (oz) hamburger?

1. Appendix A shows that a 4-ounce hamburger contains 7 grams of saturated fat. You are seeking grams of saturated fat; therefore, the conversion factor is:

$$\frac{7 \text{ g saturated fat}}{4 \text{ oz hamburger}}$$

2. Multiply 3 ounces of hamburger by the conversion factor:

$$3 \text{ oz hamburger} \times \frac{7 \text{ g saturated fat}}{4 \text{ oz hamburger}} = \frac{3 \times 7}{4} = \frac{21}{4}$$

= 5 g saturated fat (rounded off to the nearest whole number

Energy Units

1 calorie[a] (cal) = 4.2 kilojoules
1 millijoule (MJ) = 240 cal
1 kilojoule (kJ) = 0.24 cal
1 gram (g) carbohydrate = 4 cal = 17 kJ
1 g fat = 9 cal = 37 kJ
1 g protein = 4 cal = 17 kJ
1 g alcohol = 7 cal = 29 kJ

Percentages

A percentage is a comparison between a number of items (perhaps your intake of energy) and a standard number (perhaps the number of calories recommended for your age and sex—your energy RDA). The standard number is the number you divide by. The answer you get after the division must be multiplied by 100 to be stated as a percentage (*percent* means "per 100").

Example 3

What percentage of the 1989 RDA for energy is your energy intake?

1. Find your energy RDA (see 1989 RDA, inside front cover). We'll use 2,100 calories to demonstrate.
2. Total your energy intake for a day—for example, 1,200 calories.
3. Divide your calorie intake by the RDA calories:

1,200 cal (your intake) ÷ 2,100 cal (RDA) = 0.571

4. Multiply your answer by 100 to state it as a percentage:

0.571 × 100 = 57.1 = 57% (rounded off to the nearest whole number)

In some problems in nutrition, the percentage may be more than 100. For example, suppose your daily intake of vitamin A is 3,200 RE and your RDA (male) is 1,000 RE. Your intake as a percentage of the RDA is more than 100 percent (that is, you consume more than 100 percent of your vitamin A RDA). The following calculations show your vitamin A intake as a percentage of the RDA:

3,200 ÷ 1,000 = 3.2
3.2 × 100 = 320% of RDA

[a]NOTE: Throughout this book and in the appendixes, the term *calorie* is used to mean kilocalorie. Thus, when converting calories to kilojoules, do not enlarge the calorie values—they are kilocalorie values.

C

Example 4

Food labels express nutrients and energy contents of foods as percentages of the Daily Values. If a serving of a food contains 200 milligrams of calcium, for example, what percentage of the calcium Daily Value does the food provide?

1. Find the calcium Daily Value on the inside front cover, page c.
2. Divide the milligrams of calcium in the food by the Daily Value standard:

$$\frac{200}{1,000} = 0.2$$

3. Multiply by 100:

$$0.2 \times 100 = 20\% \text{ of the Daily Value}$$

Example 5

This example demonstrates how to calculate the percentage of fat in a day's meals.

1. Recall the general formula for finding percentages of calories from a nutrient:

(one nutrient's calories ÷ total calories) × 100 = the percentage of calories from that nutrient

2. Say a day's meals provide 1,754 calories and 54 grams of fat. First, convert fat grams to fat calories:

54 g × 9 cal per g = 486 cal from fat

3. Then apply the general formula for finding percentage of calories from fat:

(fat calories ÷ total calories) × 100 = percentage of calories from fat

(486 ÷ 1,754) × 100 = 27. 7 (28%, rounded)

Weights and Measures

Length
1 inch (in) = 2.54 centimeters (cm)
1 foot (ft) = 30.48 cm
1 meter (m) = 39.37 in

Temperature
Steam = 100° Celsius[a] (C) or 212° Fahrenheit (F)
Body temperature = 37°C or 98.6°F
Ice = 0°C or 32°F

To convert Fahrenheit (t_F) to Celsius:

$$t_C = \frac{5}{9}(t_F - 32)$$

To convert Celsius (t_C) to Fahrenheit:

$$t_F = \frac{9}{5}(t_C + 32)$$

Volume
Used to measure fluids or pourable dry substances such as cereal.
1 milliliter (ml) = ⅕ teaspoon or 0.034 fluid ounce or 1⁄1000 liter
1 deciliter (dl) = 1⁄10th liter
1 teaspoon (tsp or t) = 5 ml or about 5 grams (weight) salt
1 tablespoon (tbs or T) = 3 tsp or 15 ml
1 ounce, fluid (fl oz) = 2 tbs or 30 ml
1 cup (c) = 8 fl oz or 16 tbs or 250 ml
1 quart (qt) = 32 fl oz or 4 c or 0.95 liter
1 liter (1) = 1.06 qt or 1,000 ml
1 gallon (gal) = 16 c or 4 qt or 128 fl oz or 3.79 l

Weight
1 microgram (µg or mcg) = 1⁄1000 milligram
1 milligram (mg) = 1,000 mg or 1⁄1,000 gram
1 gram (g) = 1,000 mg or 1⁄1,000 kilogram
1 ounce, weight (oz) = about 28 g or 1⁄16 pound
1 pound (lb) = 16 oz (wt) or about 454 g
1 kilogram (kg) = 1,000 g or 2.2 lb

International Units (IU)
To convert IU to:
- µg vitamin D: divide by 40 or multiply by 0.025.
- mg α-TE:[b] divide by 1.5.
- vitamin A, see below.

Sodium
To convert milligrams of sodium to grams of salt:

mg sodium ÷ 400 = g of salt

The reverse is also true:

g salt × 400 = mg sodium

Folate
To convert micrograms of synthetic folate in supplements and enriched foods to Dietary Folate Equivalents (µg DFE):

µg synthetic folate × 1.7 = µg DFE

For naturally occurring folate, assign each microgram folate a value of 1 µg DFE:

µg folate = µg DFE

Vitamin A
Equivalencies for vitamin A:
1 RAE = 1 µg retinol
= 12 µg beta-carotene
= 24 µg other vitamin A carotenoids

1 IU = 0.3 µg retinol
= 3.6 µg beta-carotene
= 7.2 µg other vitamin A carotenoids

To convert older RE values to micrograms RAE:

1 µg RE retinol = 1 µg RAE retinol
6 µg RE beta-carotene = 12 µg RAE beta-carotene
12 µg RE other vitamin A carotenoids = 24 µg RAE other vitamin A carotenoids

[a]Also known as centigrade.

[b]Alpha-tocopherol equivalents (vitamin E).

U.S. Food Exchange System

D

Contents

The U.S. food exchange system is intended to help people with diabetes control the levels of glucose and lipids in the blood by controlling the grams of carbohydrate and fat they consume. Other diet planners have found the system invaluable for achieving calorie control and moderation.

Planning a Diet

Unlike the Daily Food Guide of Chapter 2, which sorts foods primarily by their protein, vitamin, and mineral contents, the exchange system sorts foods into three main groups by their proportions of carbohydrate, fat, and protein. These three groups—the carbohydrate group, the fat group, and the meat and meat substitute group (protein)—are each subdivided into several exchange lists of foods (Table D-1 below).

Portion Sizes

All of the food portions in a given list provide approximately the same amounts of energy nutrients (carbohydrate, fat, and protein) and the same number of calories. Portion sizes are strictly defined so that every item on a given list provides roughly the same amount of energy. Any food on a list can then be exchanged, or traded, for any other food on that same list without affecting a plan's balance or total calories.

To apply the system successfully, users must become familiar with portion sizes. A convenient way to remember the portion sizes and energy values is to keep in mind a typical item from each list. Table D-1 below includes some representative portion sizes; Figure D-1 shows the foods on each of the exchange lists and their accurate portion sizes.

The Foods on the Lists

Foods are not always on the exchange list where you might first expect them to be because they are grouped according to their energy-nutrient contents rather than by their source (such as milks), their outward appearance, or their vitamin and mineral contents. For example, cheeses are grouped with meats in the exchange system because, like meats, cheeses contribute energy from protein and fat but provide negligible carbohydrate. (In the food group plans presented earlier, cheeses are classed with milk because they are milk products with a similar calcium content.)

For similar reasons, starchy vegetables such as corn, green peas, and potatoes are listed on the starch list in the exchange system, rather than with the vegetables. Likewise, olives are not classed as a "fruit" as a botanist would claim; they are classified as a "fat" because their fat content makes them more similar to butter than to berries. Bacon is also on the fat list to remind users of its high fat content. These groupings permit you to see the characteristics of foods that are significant to energy intake.

Users of the exchange lists learn to view mixtures of foods, such as casseroles and

(text continues on page D–4)

Table D•1

EXCHANGE GROUPS AND LISTS

	List	Portion Size	Carbohydrate (g)	Protein (g)	Fat (g)	Energy (cal)
Carbohydrate Group						
	Starch	1 slice; ½ c	15	3	1 or less	80
	Fruit	varies	15	—	—	60
	Milk	1 c				
	Nonfat[a]		12	8	0–3	90
	Low-fat		12	8	5	120
	Whole		12	8	8	150
	Other carbohydrates	varies	15	varies	varies	varies
	Vegetable	½ c	5	2	—	25
Meat and Meat Substitute Group		1 oz				
	Very lean		—	7	0–1	35
	Lean		—	7	3	55
	Medium-fat		—	7	5	75
	High-fat		—	7	8	100
	Fat Group	1 tsp pure fat	—	—	5	45

[a]Nonfat is the same as fat-free or skim.

Figure D•1

THE EXCHANGE SYSTEM: EXAMPLE FOODS, PORTION SIZES, AND ENERGY-NUTRIENT CONTRIBUTIONS

THE CARBOHYDRATE GROUP

Starch
1 starch exchange is like:
1 slice bread
¾ c ready-to-eat cereal
½ c cooked pasta
⅓ c cooked rice
½ c cooked beans[a]
½ c corn, peas, or yams
1 small (3 oz) potato
½ bagel, English muffin, or bun
1 tortilla, waffle, or roll
(1 starch = 15 g carbohydrate, 3 g protein, 0–1 g fat, and 80 cal)

Vegetables
1 vegetable exchange is like:
½ c cooked carrots, greens, green beans, brussels sprouts, beets, broccoli, cauliflower, or spinach
1 c raw carrots, radishes, or salad greens
1 lg tomato
(1 vegetable = 5 g carbohydrate, 2 g protein, and 25 cal)

Fruits
1 fruit exchange is like:
1 small banana, nectarine, apple, or orange
½ large grapefruit or pear
½ c orange, apple, or grapefruit juice
17 small grapes
⅓ cantaloupe (or 1 c cubes)
2 tbs raisins
(1 fruit = 15 g carbohydrate and 60 cal)

[a]½ c cooked beans = 1 very lean meat exchange plus 1 starch exchange.

THE MEAT AND MEAT SUBSTITUTES GROUP (PROTEIN)

Meat and substitutes (very lean)
1 very lean meat exchange is like:
1 oz chicken (white meat, no skin)
1 oz cod, flounder, or trout
1 oz tuna (canned in water)
1 oz clams, crab, lobster, scallops, shrimp, or imitation seafood
1 oz fat-free cheese
½ c cooked beans, peas, or lentils
¼ c nonfat or low-fat cream cheese
2 egg whites (or ¼ c egg substitute)
(1 very lean meat = 7 g protein, 0–1 g fat, and 35 cal)

Meats and substitutes (lean)
1 lean meat exchange is like:
1 oz beef or pork tenderloin
1 oz chicken (dark meat, no skin)
1 oz herring or salmon
1 oz tuna (canned in oil, drained)
1 oz low-fat cheese or luncheon meats
(1 lean meat = 7 g protein, 3 g fat, and 55 cal)

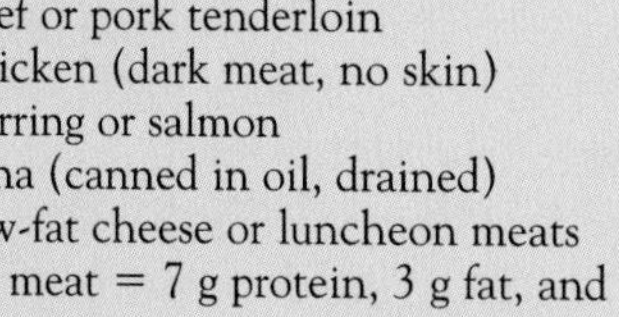

Meats and substitutes (medium-fat)
1 medium-fat meat exchange is like:
1 oz ground beef
1 oz pork chop
1 egg
¼ c ricotta
4 oz tofu
(1 medium-fat meat = 7 g protein, 5 g fat, and 75 cal)

D

Other carbohydrates
1 other carbohydrates exchange is like:
2 small cookies
1 small brownie or cake
5 vanilla wafers
1 granola bar
½ c ice cream
(1 other carbohydrate = 15 g carbohydrate and may be exchanged for 1 starch, 1 fruit, or 1 milk. Because many items on this list contain added sugar and fat, their fat and calorie values vary and their portion sizes are small.)

Milks (nonfat and very-low-fat)
1 nonfat milk exchange is like:
1 c nonfat milk
¾ c nonfat yogurt, plain
1 c nonfat or lowfat buttermilk
½ c evaporated nonfat milk
⅓ c dry nonfat milk
(1 nonfat milk = 12 g carbohydrate, 8 g protein, 0–3 g fat, and 90 cal)

Milks (low-fat)
1 low-fat milk exchange is like:
1 c 2% milk
¾ c low-fat yogurt, plain
(1 low-fat milk = 12 g carbohydrate, 8 g protein, 5 g fat, and 120 cal)

Milks (whole)
1 whole milk exchange is like:
1 c whole milk
½ c evaporated whole milk
(1 whole milk = 12 g carbohydrate, 8 g protein, 8 g fat, and 150 cal)

Meats and substitutes (high-fat)
1 high-fat meat exchange is like:
1 oz pork sausage
1 oz luncheon meat (such as bologna)
1 oz regular cheese (such as cheddar or Swiss)
1 small hot dog (turkey or chicken)[b]
2 tbs peanut butter[c]
(1 high-fat meat = 7 g protein, 8 g fat, and 100 cal)

[b]A beef or pork hot dog counts as 1 high-fat meat exchange *plus* 1 fat exchange.
[c]Peanut butter counts as 1 high-fat meat exchange *plus* 1 fat exchange.

THE FAT GROUP

Fats
1 fat exchange is like:
1 tsp butter
1 tsp margarine or mayonnaise (1 tbs reduced fat)
1 tsp any oil
1 tbs salad dressing (2 tbs reduced fat)
8 large black olives
10 large peanuts
⅛ medium avocado
1 slice bacon
2 tbs shredded coconut
1 tbs cream cheese (2 tbs reduced fat)
(1 fat = 5 g fat and 45 cal)

D

soups, as combinations of foods from different exchange lists. They also learn to interpret food labels with the exchange system in mind. Knowing that foods on the starch list provide 15 grams of carbohydrate and those on the vegetable list provide 5, you can interpret the label of a lasagna dinner that lists 37 grams of carbohydrate as "2 starches" (mostly noodles) and "1 vegetable" (the sauce).

Controlling Energy, Fat, and Sodium

The exchange system helps people control their energy intakes by paying close attention to portion sizes. A portion of any food on a given list provides roughly the same amount of energy nutrients and total calories. The portion sizes have been adjusted so that all portions have the same energy value. For example, 17 grapes count as one fruit portion, as does ½ grapefruit. A whole grapefruit counts as two portions.

A *portion* in the exchange system is not the same as a *serving* in the Daily Food Guide, especially when it comes to meats. The exchange system lists meats and most cheeses in single ounces; that is, 1 *portion* (or *exchange*) of meat is 1 ounce, whereas one *serving* is 2 to 3 ounces. Calculating meat by the ounce encourages the planner to keep close track of the exact amounts eaten. This, in turn, helps control energy and fat intakes.

By allocating items like bacon and avocados to the fat list, the exchange system alerts consumers to foods that are unexpectedly high in fat. Even the starch list specifies which grain products contain added fat (such as biscuits, muffins, and waffles). In addition, the exchange system encourages users to think of nonfat milk as milk and of whole milk as milk with added fat, and to think of very lean meats as meats and of lean, medium-fat, and high-fat meats as meats with added fat. To that end, foods on the milk and meat lists are separated into categories based on their fat contents.

Control of food energy and fat intake can be highly successful with the exchange system. Exchange plans do not, however, guarantee adequate intakes of vitamins and minerals. Food group plans work better from that standpoint because the food groupings are based on similarities in vitamin-mineral content. In the exchange system, for example, meats are grouped with cheeses, yet the meats are iron-rich and calcium-poor, whereas the cheeses are iron-poor and calcium-rich. To take advantage of the strengths of both food group plans and exchange patterns, and to compensate for their weaknesses, diet planners often combine these two diet planning tools, as the following section shows.

People wishing to control the sodium in their diets can begin by eliminating any foods bearing this symbol [🧂] found on any exchange list. The symbol identifies each food that, in one exchange, provides 400 milligrams or more of sodium. Other foods may also contribute substantially to sodium, however (consult Chapter 8 for details).

Combining Food Group Plans and Exchange Lists

A diet planner may find that using a food group plan together with the exchange lists eases the task of choosing foods that will provide all the nutrients. The food group plan ensures that all classes of nutritious foods are included, thus promoting adequacy, balance, and variety. The exchange system classifies the food selections by their energy-yielding nutrients, thus controlling energy and fat intakes.

Table D-2 shows how to use the Daily Food Guide plan together with the exchange lists to plan a diet. The Daily Food Guide ensures that a certain number of servings is chosen from each of the five food groups (see the first column of the table). The second column translates the number of servings (using the midpoint) into exchanges. With the addition of a small amount of fat, this sample diet plan provides about 1,750 calories. Most people can meet their needs for all the nutrients within this reasonable energy allowance. (Table D-3 shows patterns for other energy intakes.) The next step in diet planning is to assign the exchanges to meals and snacks. The final plan might look like the one in Table D-4. To aid you in the development of your own diet plan, Tables D-5 through D-13 present the U.S. exchange system in detail.

Next, a person could begin to fill in the plan with real foods to create a menu. For example, the breakfast plan calls for 2 starches, 1 fruit, and 1 nonfat milk. A person might select a bowl of shredded wheat with banana slices and milk (1 cup shredded wheat = 2 starches, 1 small banana = 1 fruit, and 1 cup nonfat milk = 1 milk); or a bagel and a bowl of cantaloupe pieces topped with yogurt (1 bagel = 2 starches, ⅓ cantaloupe melon = 1 fruit, and ¾ cup nonfat plain yogurt = 1 milk). A person who

Table D•2

DIET PLANNING WITH THE EXCHANGE SYSTEM USING THE DAILY FOOD GUIDE PATTERN

Pattern from Daily Food Guide Plan	Selections Made Using the Exchange System	Energy Cost (cal)
Grains (breads and cereals)—6 to 11 servings	Starch list—select 9 exchanges	720
Vegetables—3 to 5 servings	Vegetable list—select 4 exchanges	100
Fruits—2 to 4 servings	Fruit list—select 3 exchanges	180
Meat—2 to 3 servings[a]	Meat list—select 6 lean exchanges	330
Milk—2 servings	Milk list—select 2 nonfat exchanges	180
	Fat list—select 5 exchanges	225
Total		1,735

[a]In the food group plan, 1 serving is 2 to 3 ounces; in the exchange system, 1 exchange is 1 ounce. The Daily Food Guide suggests that amounts should total 5 to 7 ounces of meat daily.

wanted butter on the bagel could move a fat exchange or two from dinner to breakfast. If willing to use two fat exchanges at breakfast, the person could have pancakes with strawberries and milk (4 small pancakes = 2 starches plus 2 fats, 1¼ cup strawberries = 1 fruit, and a cup of nonfat milk = 1 milk). Then the person could move on to complete the menu for lunch, dinner and snacks.

U.S. Exchange Lists for Meal Planning[a]

[a]SOURCE: The Exchange Lists are the basis of a meal planning system designed by a committee of the American Dietetic Association. © 1995 American Diabetes Association, Inc.

D

Table D•3

DIET PATTERNS FOR DIFFERENT ENERGY INTAKES

	Energy Level (cal)						
EXCHANGE	**1,200**	**1,500**	**1,800**	**2,000**	**2,200**	**2,600**	**3,000**
Starch	6	7	8	9	11	13	15
Meat (lean)	4	5	6	6	6	7	8
Vegetable	3	4	5	5	5	6	6
Fruit	2	3	4	4	4	5	6
Milk (nonfat)	2	2	2	3	3	3	3
Fat	3	5	6	7	8	10	12

NOTE: These patterns follow the Daily Food Guide plan and supply less than 30 percent of calories as fat.

Table D•4

A SAMPLE DIET PLAN

Exchange	Breakfast	Lunch	Snack	Dinner	Evening Snack
9 starch	2	2	1	3	1
4 vegetable				4	
3 fruit	1	1	1		
6 lean meat		2		4	
2 nonfat milk	1	1			1
5 fat		1		4	

NOTE: This diet plan is one of many possibilities. It follows the number of servings suggested by the Daily Food Guide and meets dietary recommendations to provide 55 to 60 percent of its calories from carbohydrate, 15 to 20 percent from protein, and less than 30 percent from fat.

Table D•5

U.S. EXCHANGE SYSTEM: STARCH LIST

1 starch exchange = 15 g carbohydrate, 3 g protein, 0–1 g fat, and 80 cal
Note: In general, a starch serving is ½ c cereal, grain, pasta, or starchy vegetable; 1 oz of bread; ¾ to 1 oz snack food.

Serving Size	Food
Bread	
½ (1 oz)	Bagels
2 slices (1½ oz)	Bread, reduced-calorie
1 slice (1 oz)	Bread, white (including French and Italian), whole-wheat, pumpernickel, rye
2 (⅔ oz)	Bread sticks, crisp, 4″ × ½″
½	English muffins
½ (1 oz)	Hot dog or hamburger buns
½	Pita, 6″ across
1 (1 oz)	Plain rolls, small
1 slice (1 oz)	Raisin bread, unfrosted
1	Tortillas, corn, 6″ across
1	Tortillas, flour, 7–8″ across
1	Waffles, 4½″ square, reduced-fat
Cereals and Grains	
½ c	Bran cereals
½ c	Bulgur, cooked
½ c	Cereals, cooked
¾ c	Cereals, unsweetened, ready-to-eat
3 tbs	Cornmeal (dry)
⅓ c	Couscous
3 tbs	Flour (dry)
¼ c	Granola, low-fat
¼ c	Grape nuts
½ c	Grits, cooked
½ c	Kasha
¼ c	Millet
¼ c	Muesli
½ c	Oats
½ c	Pasta, cooked
1½ c	Puffed cereals
½ c	Rice milk
⅓ c	Rice, white or brown, cooked
½ c	Shredded wheat
½ c	Sugar-frosted cereal
3 tbs	Wheat germ
Starchy Vegetables	
⅓ c	Baked beans
½ c	Corn
1 (5 oz)	Corn on cob, medium
1 c	Mixed vegetables with corn, peas, or pasta
½ c	Peas, green
½ c	Plantains
1 small (3 oz)	Potatoes, baked or boiled
½ c	Potatoes, mashed
1 c	Squash, winter (acorn, butternut)
½ c	Yams, sweet potatoes, plain
Crackers and Snacks	
8	Animal crackers
3	Graham crackers, 2½″ square
¾ oz	Matzoh
4 slices	Melba toast
24	Oyster crackers
3 c	Popcorn (popped, no fat added or low-fat microwave)
¾ oz	Pretzels
2	Rice cakes, 4″ across
6	Saltine-type crackers
15–20 (¾ oz)	Snack chips, fat-free (tortilla, potato)
2–5 (¾ oz)	Whole-wheat crackers, no fat added
Dried Beans, Peas, and Lentils	
½ c	Beans and peas, cooked (garbanzo, lentils, pinto, kidney, white, split, black-eyed)
⅔ c	Lima beans
3 tbs	Miso
Starchy Foods Prepared with Fat	
Count as 1 starch + 1 fat exchange.	
1	Biscuit, 2½″ across
½ c	Chow mein noodles
1 (2 oz)	Corn bread, 2″ cube
6	Crackers, round butter type
1 c	Croutons
16–25 (3 oz)	French-fried potatoes
¼ c	Granola
1 (1½ oz)	Muffin, small
2	Pancake, 4″ across
3 c	Popcorn, microwave
3	Sandwich crackers, cheese or peanut butter filling
⅓ c	Stuffing, bread (prepared)
2	Taco shell, 6″ across
1	Waffle, 4½″ square
4–6 (1 oz)	Whole-wheat crackers, fat added

= 400 mg or more of sodium per serving.

D

Table D•6

U.S. EXCHANGE SYSTEM: FRUIT LIST

1 fruit exchange = 15 g carbohydrate and 60 cal
Note: In general, a fruit serving is 1 small to medium fresh fruit; ½ c canned or fresh fruit or fruit juice; ¼ c dried fruit. The weights given include skin, core, seeds, and rind.

Serving Size	Food
1 (4 oz)	Apples, unpeeled, small
½ c	Applesauce, unsweetened
4 rings	Apples, dried
4 whole (5½ oz)	Apricots, fresh
8 halves	Apricots, dried
½ c	Apricots, canned
1 (4 oz)	Bananas, small
¾ c	Blackberries
¾ c	Blueberries
⅓ melon (11 oz) or 1 c cubes	Cantaloupe, small
12 (3 oz)	Cherries, sweet, fresh
½ c	Cherries, sweet, canned
3	Dates
1½ large or 2 medium (3½ oz)	Figs, fresh
1½	Figs, dried
½ c	Fruit cocktail
½ (11 oz)	Grapefruit, large
¾ c	Grapefruit sections, canned
17 (3 oz)	Grapes, small
1 slice (10 oz) or 1 c cubes	Honeydew melon
1 (3½ oz)	Kiwi
¾ c	Mandarin oranges, canned
½ (5½ oz) or ½ c	Mangoes, small
1 (5 oz)	Nectarines, small
1 (6½ oz)	Oranges, small
½ (8 oz) or 1 c cubes	Papayas
1 (6 oz)	Peaches, medium, fresh
½ c	Peaches, canned
½ (4 oz)	Pears, large, fresh
½ c	Pears, canned
¾ c	Pineapple, fresh
½ c	Pineapple, canned
2 (5 oz)	Plums, small
½ c	Plums, canned
3	Prunes, dried
2 tbs	Raisins
1 c	Raspberries
1¼ c whole berries	Strawberries
2 (8 oz)	Tangerines, small
1 slice (13½ oz) or 1¼ c cubes	Watermelon
Fruit Juice	
½ c	Apple juice/cider
⅓ c	Cranberry juice cocktail
1 c	Cranberry juice cocktail, reduced-calorie
⅓ c	Fruit juice blends, 100% juice
⅓ c	Grape juice
½ c	Grapefruit juice
½ c	Orange juice
½ c	Pineapple juice
⅓ c	Prune juice

Table D•7

U.S. EXCHANGE SYSTEM: MILK LIST

Serving Size	Food
Nonfat and Very-Low-Fat Milk	
1 nonfat/low-fat milk exchange = 12 g carbohydrate, 8 g protein, 0–3 g fat, 90 cal	
1 c	Nonfat milk
1 c	½% milk
1 c	1% milk
1 c	Nonfat or low-fat buttermilk
½ c	Evaporated nonfat milk
⅓ c dry	Dry nonfat milk
¾ c	Plain nonfat yogurt
1 c	Nonfat or low-fat fruit-flavored yogurt sweetened with aspartame or with a nonnutritive sweetener
Low-Fat Milk	
1 low-fat milk exchange = 12 g carbohydrate, 8 g protein, 5 g fat, 120 cal	
1 c	2% milk
¾ c	Plain low-fat yogurt
1 c	Sweet acidophilus milk
Whole Milk	
1 whole milk exchange = 12 g carbohydrate, 8 g protein, 8 g fat, 150 cal	
1 c	Whole milk
½ c	Evaporated whole milk
1 c	Goat's milk
1 c	Kefir

D

Table D•8

U.S. EXCHANGE SYSTEM: OTHER CARBOHYDRATES LIST

1 other carbohydrate exchange = 15 g carbohydrate, or 1 starch, or 1 fruit, or 1 milk exchange

Food	Serving Size	Exchanges per Serving
Angel food cake, unfrosted	1⁄12 cake	2 carbohydrates
Brownies, small, unfrosted	2″ square	1 carbohydrate, 1 fat
Cake, unfrosted	2″ square	1 carbohydrate, 1 fat
Cake, frosted	2″ square	2 carbohydrates, 1 fat
Cookie, fat-free	2 small	1 carbohydrate
Cookies or sandwich cookies	2 small	1 carbohydrate, 1 fat
Cupcakes, frosted	1 small	2 carbohydrates, 1 fat
Cranberry sauce, jellied	¼ c	2 carbohydrates
Doughnuts, plain cake	1 medium, (1½ oz)	1½ carbohydrates, 2 fats
Doughnuts, glazed	3¾″ across (2 oz)	2 carbohydrates, 2 fats
Fruit juice bars, frozen, 100% juice	1 bar (3 oz)	1 carbohydrate
Fruit snacks, chewy (pureed fruit concentrate)	1 roll (¾ oz)	1 carbohydrate
Fruit spreads, 100% fruit	1 tbs	1 carbohydrate
Gelatin, regular	½ c	1 carbohydrate
Gingersnaps	3	1 carbohydrate
Granola bars	1 bar	1 carbohydrate, 1 fat
Granola bars, fat-free	1 bar	2 carbohydrates
Hummus	⅓ c	1 carbohydrate, 1 fat
Ice cream	½ c	1 carbohydrate, 2 fats
Ice cream, light	½ c	1 carbohydrate, 1 fat
Ice cream, fat-free, no sugar added	½ c	1 carbohydrate
Jam or jelly, regular	1 tbs	1 carbohydrate
Milk, chocolate, whole	1 c	2 carbohydrates, 1 fat
Pie, fruit, 2 crusts	⅙ pie	3 carbohydrates, 2 fats
Pie, pumpkin or custard	⅛ pie	1 carbohydrate, 2 fats
Potato chips	12–18 (1 oz)	1 carbohydrate, 2 fats
Pudding, regular (made with low-fat milk)	½ c	2 carbohydrates
Pudding, sugar-free (made with low-fat milk)	½ c	1 carbohydrate
Salad dressing, fat-free 🖉	¼ c	1 carbohydrate
Sherbet, sorbet	½ c	2 carbohydrates
Spaghetti or pasta sauce, canned 🖉	½ c	1 carbohydrate, 1 fat
Sweet roll or danish	1 (2½ oz)	2½ carbohydrates, 2 fats
Syrup, light	2 tbs	1 carbohydrate
Syrup, regular	1 tbs	1 carbohydrate
Syrup, regular	¼ c	4 carbohydrates
Tortilla chips	6–12 (1 oz)	1 carbohydrate, 2 fats
Vanilla wafers	5	1 carbohydrate, 1 fat
Yogurt, frozen, low-fat, fat-free	⅓ c	1 carbohydrate, 0–1 fat
Yogurt, frozen, fat-free, no sugar added	½ c	1 carbohydrate
Yogurt, low-fat with fruit	1 c	3 carbohydrates, 0–1 fat

🖉 = 400 mg or more sodium per exchange.

Table D•9

U.S. EXCHANGE SYSTEM: VEGETABLE LIST

1 vegetable exchange = 5 g carbohydrate, 2 g protein, 25 cal

Note: In general, a vegetable serving is ½ c cooked vegetables or vegetable juice; 1 c raw vegetables. Starchy vegetables such as corn, peas, and potatoes are on the starch list.

Artichokes
Artichoke hearts
Asparagus
Beans (green, wax, Italian)
Bean sprouts
Beets
Broccoli
Brussels sprouts
Cabbage
Carrots
Cauliflower
Celery
Cucumbers
Eggplant
Green onions or scallions
Greens (collard, kale, mustard, turnip)
Kohlrabi
Leeks
Mixed vegetables (without corn, peas, or pasta)
Mushrooms
Okra
Onions
Pea pods
Peppers (all varieties)
Radishes
Salad greens (endive, escarole, lettuce, romaine, spinach)
Sauerkraut 🧂
Spinach
Summer squash (crookneck)
Tomatoes
Tomatoes, canned
Tomato sauce 🧂
Tomato/vegetable juice 🧂
Turnips
Water chestnuts
Watercress
Zucchini

🧂 = 400 mg or more sodium per exchange.

D

Table D•10

U.S. EXCHANGE SYSTEM: MEAT AND MEAT SUBSTITUTES LIST

Note: In general, a meat serving is 1 oz meat, poultry, or cheese; ½ c dried beans (weigh meat and poultry and measure beans after cooking).

Serving Size	Food
Very Lean Meat and Substitutes	
1 very lean meat exchange = 7 g protein, 0–1 g fat, 35 cal	
1 oz	Poultry: Chicken or turkey (white meat, no skin), Cornish hen (no skin)
1 oz	Fish: Fresh or frozen cod, flounder, haddock, halibut, trout; tuna, fresh or canned in water
1 oz	Shellfish: Clams, crab, lobster, scallops, shrimp, imitation shellfish
1 oz	Game: Duck or pheasant (no skin), venison, buffalo, ostrich
	Cheese with ≤ 1 g fat/oz:
¼ c	Nonfat or low-fat cottage cheese
1 oz	Fat-free cheese
	Other:
1 oz	Processed sandwich meats with ≤ 1 g fat/oz (such as deli thin, shaved meats, chipped beef 🧂, turkey ham)
2	Egg whites
¼ c	Egg substitutes, plain
1 oz	Hot dogs with ≤ 1 g fat/oz
1 oz	Kidney (high in cholesterol)
1 oz	Sausage with ≤ 1 g fat/oz 🧂
Count as one very lean meat and one starch exchange:	
½ c	Dried beans, peas, lentils (cooked)
Lean Meat and Substitutes	
1 lean meat exchange = 7 g protein, 3 g fat, 55 cal	
1 oz	Beef: USDA Select or Choice grades of lean beef trimmed of fat (round, sirloin, and flank steak); tenderloin; roast (rib, chuck, rump); steak (T-bone, porterhouse, cubed), ground round
1 oz	Pork: Lean pork (fresh ham); canned, cured, or boiled ham; Canadian bacon 🧂; tenderloin, center loin chop
1 oz	Lamb: Roast, chop, leg
1 oz	Veal: Lean chop, roast
1 oz	Poultry: Chicken, turkey (dark meat, no skin), chicken white meat (with skin), domestic duck or goose (well drained of fat, no skin)
	Fish:
1 oz	Herring (uncreamed or smoked)
6 medium	Oysters
1 oz	Salmon (fresh or canned), catfish
2 medium	Sardines (canned)
1 oz	Tuna (canned in oil, drained)
1 oz	Game: Goose (no skin), rabbit
	Cheese:
¼ c	4.5%-fat cottage cheese
2 tbs	Grated Parmesan
1 oz	Cheeses with ≤ 3 g fat/oz
	Other:
1½ oz	Hot dogs with ≤ 3 g fat/oz 🧂
1 oz	Processed sandwich meat with ≤ 3 g fat/oz (turkey pastrami or kielbasa)
1 oz	Liver, heart (high in cholesterol)
Medium-Fat Meat and Substitutes	
1 medium-fat meat exchange = 7 g protein, 5 g fat, and 75 cal	
1 oz	Beef: Most beef products (ground beef, meat loaf, corned beef, short ribs, Prime grades of meat trimmed of fat, such as prime rib)
1 oz	Pork: Top loin, chop, Boston butt, cutlet
1 oz	Lamb: Rib roast, ground
1 oz	Veal: Cutlet (ground or cubed, unbreaded)
1 oz	Poultry: Chicken dark meat (with skin), ground turkey or ground chicken, fried chicken (with skin)
1 oz	Fish: Any fried fish product
	Cheese with ≤ 5 g fat/oz:
1 oz	Feta
1 oz	Mozzarella
¼ c (2 oz)	Ricotta
	Other:
1	Egg (high in cholesterol, limit to 3/week)
1 oz	Sausage with ≤ 5 g fat/oz
1 c	Soy milk
¼ c	Tempeh
4 oz or ½ c	Tofu
High-Fat Meat and Substitutes	
1 high-fat meat exchange = 7 g protein, 8 g fat, 100 cal	
1 oz	Pork: Spareribs, ground pork, pork sausage
1 oz	Cheese: All regular cheeses (American 🧂, cheddar, Monterey Jack, Swiss)
	Other:
1 oz	Processed sandwich meats with ≤ 8 g fat/oz (bologna, pimento loaf, salami)
1 oz	Sausage (bratwurst, Italian, knockwurst, Polish, smoked)
1 (10/lb)	Hot dog (turkey or chicken) 🧂
3 slices (20 slices/lb)	Bacon
Count as one high-fat meat plus one fat exchange:	
1 (10/lb)	Hot dog (beef, pork, or combination) 🧂
2 tbs	Peanut butter (contains unsaturated fat)

🧂 = 400 mg or more of sodium per serving.

Table D•11

U.S. EXCHANGE SYSTEM: FAT LIST

1 fat exchange = 5 g fat, 45 cal

Note: In general, a fat serving is 1 tsp regular butter, margarine, or vegetable oil; 1 tbs regular salad dressing. Many fat-free and reduced fat foods are on the Free Foods List.

Serving Size	Food
Monounsaturated Fats	
⅛ medium (1 oz)	Avocados
1 tsp	Oil (canola, olive, peanut)
8 large	Olives, ripe (black)
10 large	Olives, green, stuffed 🧂
6 nuts	Almonds, cashews
6 nuts	Mixed nuts (50% peanuts)
10 nuts	Peanuts
4 halves	Pecans
2 tsp	Peanut butter, smooth or crunchy
1 tbs	Sesame seeds
2 tsp	Tahini paste
Polyunsaturated Fats	
1 tsp	Margarine, stick, tub, or squeeze
1 tbs	Margarine, lower-fat (30% to 50% vegetable oil)
1 tsp	Mayonnaise, regular
1 tbs	Mayonnaise, reduced-fat
4 halves	Nuts, walnuts, English
1 tsp	Oil (corn, safflower, soybean)
1 tbs	Salad dressing, regular
2 tbs	Salad dressing, reduced-fat
2 tsp	Mayonnaise-type salad dressing, regular 🧂
1 tbs	Mayonnaise-type salad dressing, reduced-fat
1 tbs	Seeds: pumpkin, sunflower
Saturated Fats[a]	
1 slice (2″ slices/lb)	Bacon, cooked
1 tsp	Bacon, grease
1 tsp	Butter, stick
2 tsp	Butter, whipped
1 tbs	Butter, reduced-fat
2 tbs (½ oz)	Chitterlings, boiled
2 tbs	Coconut, sweetened, shredded
2 tbs	Cream, half and half
1 tbs (½ oz)	Cream cheese, regular
2 tbs (1 oz)	Cream cheese, reduced-fat
	Fatback or salt pork[b]
1 tsp	Shortening or lard
2 tbs	Sour cream, regular
3 tbs	Sour cream, reduced-fat

🧂 = 400 mg or more sodium per exchange.

[a]Saturated fats can raise blood cholesterol levels.

[b]Use a piece 1″ × 1″ × ¼″ if you plan to eat the fatback cooked with vegetables. Use a piece 2″ × 1″ × ½″ when eating only the vegetables with the fatback removed.

Table D•12

U.S. EXCHANGE SYSTEM: FREE FOODS LIST

Note: A serving of free food contains fewer than 20 calories; those with serving sizes should be limited to three servings a day whereas those without serving sizes can be eaten freely.

Serving Size	Food
Fat-Free or Reduced-Fat Foods	
1 tbs	Cream cheese, fat-free
1 tbs	Creamers, nondairy, liquid
2 tsp	Creamers, nondairy, powdered
1 tbs	Mayonnaise, fat-free
1 tsp	Mayonnaise, reduced-fat
4 tbs	Margarine, fat-free
1 tsp	Margarine, reduced-fat
1 tbs	Mayonnaise type salad dressing, nonfat
1 tsp	Mayonnaise type salad dressing, reduced-fat
	Nonstick cooking spray
1 tbs	Salad dressing, fat-free
2 tbs	Salad dressing, fat-free, Italian
¼ c	Salsa
1 tbs	Sour cream, fat-free, reduced-fat
2 tbs	Whipped topping, regular or light
Sugar-Free or Low-Sugar Foods	
1 piece	Candy, hard, sugar-free
	Gelatin dessert, sugar-free
	Gelatin, unflavored
	Gum, sugar-free
2 tsp	Jam or jelly, low-sugar or light
	Sugar substitutes
2 tbs	Syrup, sugar-free
Drinks	
	Bouillon, broth, consommé 🧂
	Bouillon or broth, low-sodium
	Carbonated or mineral water
1 tbs	Cocoa powder, unsweetened
	Coffee
	Club soda
	Diet soft drinks, sugar-free
	Drink mixes, sugar-free
	Tea
	Tonic water, sugar-free
Condiments	
1 tbs	Catsup
	Horseradish
	Lemon juice
	Lime juice
	Mustard
1½ large	Pickles, dill 🧂
	Soy sauce, regular or light 🧂
1 tbs	Taco sauce
	Vinegar
Seasonings	
Flavoring extracts	Spices
Garlic	Hot pepper sauces
Herbs, fresh or dried	Wine, used in cooking
Pimento	Worcestershire sauce

🧂 = 400 mg or more sodium per exchange.

Table D•13

U.S. EXCHANGE SYSTEM: COMBINATION FOODS LIST

Food	Serving Size	Exchanges per Serving
Entrées		
Tuna noodle casserole, lasagna, spaghetti with meatballs, chili with beans, macaroni and cheese 🧂	1 c (8 oz)	2 carbohydrates, 2 medium-fat meats
Chow mein (without noodles or rice)	2 c (16 oz)	1 carbohydrate, 2 lean meats
Pizza, cheese, thin crust 🧂 (5 oz)	¼ of 10″	2 carbohydrates, 2 medium-fat meats, 1 fat
Pizza, meat topping, thin crust 🧂 (5 oz)	¼ of 10″	2 carbohydrates, 2 medium-fat meats, 2 fats
Potpie 🧂	1 (7 oz)	2 carbohydrates, 1 medium-fat meat, 4 fats
Frozen Entrées		
Salisbury steak with gravy, mashed potato	1 (11 oz)	2 carbohydrates, 3 medium-fat meats, 3–4 fats
Turkey with gravy, mashed potato, dressing 🧂	1 (11 oz)	2 carbohydrates, 2 medium-fat meats, 2 fats
Entrée with less than 300 calories 🧂	1 (8 oz)	2 carbohydrates, 3 lean meats
Soups		
Bean 🧂	1 c 1 very lean meat	1 carbohydrate,
Cream (made with water) 🧂	1 c (8 oz)	1 carbohydrate, 1 fat
Split pea (made with water) 🧂	½ c (4 oz)	1 carbohydrate
Tomato (made with water) 🧂	1 c (8 oz)	1 carbohydrate
Vegetable beef, chicken noodle, or other broth-type 🧂	1 c (8 oz)	1 carbohydrate
Fast Foods		
Burritos with beef 🧂	2	4 carbohydrates, 2 medium-fat meats, 2 fats
Chicken nuggets 🧂	6	1 carbohydrate, 2 medium-fat meats, 1 fat
Chicken breast and wing, breaded and fried 🧂	1	1 carbohydrate, 4 medium-fat meats, 2 fats
Fish sandwich/tartar sauce 🧂	1	3 carbohydrates, 1 medium-fat meat, 3 fats
French fries, thin	20–25	2 carbohydrates, 2 fats
Hamburger, regular	1	2 carbohydrates, 2 medium-fat meats
Hamburger, large 🧂	1	2 carbohydrates, 3 medium-fat meats, 1 fat
Hot dog with bun 🧂	1	1 carbohydrate, 1 high-fat meat, 1 fat
Individual pan pizza 🧂	1	5 carbohydrates, 3 medium-fat meats, 3 fats
Soft serve cone	1 medium	2 carbohydrates, 1 fat
Submarine sandwich 🧂	1 (6″)	3 carbohydrates, 1 vegetable, 2 medium-fat meats, 1 fat
Taco, hard shell 🧂	1 (6 oz)	2 carbohydrates, 2 medium-fat meats, 2 fats
Taco, soft shell 🧂	1 (3 oz)	1 carbohydrate, 1 medium-fat meat, 1 fat

🧂 = 400 mg or more of sodium per serving.

Nutrition Resources

Contents

E

People interested in nutrition often want to know where they can find reliable nutrition information. Wherever you live, there are several sources you can turn to:

- The Department of Health may have a nutrition expert.
- The local extension agent is often an expert.
- The food editor of your local paper may be well informed.
- The dietitian at the local hospital had to fulfill a set of qualifications before he or she became an RD.
- There may be knowledgeable professors of nutrition or biochemistry at a nearby college or university.

In addition, you may be interested in building a nutrition library of your own. Books you can buy, journals you can subscribe to, and addresses you can contact for general information are given below.

Books

For students seeking to establish a personal library of nutrition references, the authors of this text recommend the following books:

- *Present Knowledge in Nutrition*, 7th ed. (Washington, D.C.: International Life Sciences Institute—Nutrition Foundation, 1996).

This 646-page paperback has a chapter on each of 64 topics, including energy, obesity, each of the nutrients, several diseases, malnutrition, growth and its assessment, immunity, alcohol, fiber, exercise, drugs, and toxins. Watch for an update; new editions come out every few years.

- M. E. Shils and coeditors, *Modern Nutrition in Health and Disease*, 9th ed. (Baltimore: Williams & Wilkins, 1999).

This reference book contains encyclopedic articles on the nutrients, foods, diet, metabolism, malnutrition, age-related needs, and nutrition in disease.

- Committee on Dietary Reference Intakes, *Dietary Reference Intakes for Calcium, Phosphorus, Magnesium, Vitamin D, and Fluoride* (Washington, D.C.: National Academy Press, 1997).
- Committee on Dietary Reference Intakes, *Dietary Reference Intakes for Thiamin, Riboflavin, Niacin, Vitamin B_6, Folate, Vitamin B_{12}, Pantothenic Acid, Biotin, and Choline* (Washington, D.C.: National Academy Press, 1998).
- Committee on Dietary Reference Intakes, *Dietary Reference Intakes for Vitamin C, Vitamin E, Selenium, and Carotenoids* (Washington, D.C.: National Academy Press, 2000).
- Committee on Dietary Reference Intakes, *Dietary Reference Intakes for Vitamin A, Vitamin K, Arsenic, Boron, Chromium, Copper, Iodine, Iron, Manganese, Molybdenum, Nickel, Silicon, Vanadium, and Zinc* (Washington, D.C.: National Academy Press, 2001).

These reports review the function of each nutrient, dietary sources, and deficiency and toxicity symptoms as well as provide recommendations for intakes. Watch for additional reports on the Dietary Reference Intakes for the remaining nutrients and other food components.

- Committee on Diet and Health, *Diet and Health Implications for Reducing Chronic Disease Risk* (Washington, D.C.: National Academy Press, 1989).

This 749-page book presents the integral relationship between diet and chronic disease prevention. Its nutrient chapters provide evidence on how diet influences disease development, and its disease chapters review the dietary patterns implicated in each chronic disease.

- S. S. Gropper, *The Biochemistry of Human Nutrition: A Desk Reference*, 2nd ed. (Belmont, Calif.: Wadsworth/ Thomson Learning, 2000).

This 263-page paperback presents the biochemical concepts necessary for an understanding of nutrition. It is a handy reference book for those who need a refresher in the basics of biochemistry or for those who are learning biochemistry for the first time.

We also recommend two of our own books that explore current topics in nutrition, health, and the life span:

- S. R. Rolfes, L. K. DeBruyne, and E. N. Whitney, *Life Span Nutrition: Conception through Life*, 2nd ed. (Belmont, Calif.: West/Wadsworth, 1998).
- E. N. Whitney, C. B. Cataldo, L. K. DeBruyne, and S. R. Rolfes, *Nutrition for Health and Health Care*, 2nd ed. (Belmont, Calif.: Wadsworth/ Thomson Learning, 2001).

Journals

Nutrition Today is an excellent magazine for the interested layperson. It makes a point of raising controversial issues and providing a forum for conflicting opinions. Six issues per year are published. Order from Williams & Wilkins, 12107 Insurance Way, Hagerstown, MD 21740.

The *Journal of the American Dietetic Association*, the official publication of the ADA, contains articles of interest to dietitians and nutritionists, news of legislative action on food and nutrition, and a very useful section of abstracts of articles from many other journals of nutrition and related areas. There are 12 issues per year, available from the American Dietetic Association (see “Addresses,” later).

Nutrition Reviews, a publication of the International Life Sciences Institute, does much of the work for the library researcher, compiling recent evidence on current topics and presenting extensive bibliographies. Twelve issues per year are available from Nutrition Reviews, P.O. Box 1897, Lawrence, KS 66044-8897.

Nutrition and the M.D. is a monthly newsletter that provides up-to-date, easy-to-read, practical information on nutrition for health care providers. It is available from Lippincott-Williams & Wilkens, 16522 Hunters Green Parkway, Hagerstown, MD 21740.

Other journals that deserve mention here are *Food Technology*, *Journal of Nutrition*, *American Journal of Clinical*

E

Nutrition, *Nutrition Research*, and *Journal of Nutrition Education*. *FDA Consumer*, a government publication with many articles of interest to the consumer, is available from the Food and Drug Administration (see "Addresses," below). Many other journals of value are referred to throughout this book.

Addresses

Many of the organizations listed below will provide publication lists free on request. Government and international agencies and professional nutrition organizations are listed first, followed by organizations in the following areas: aging, alcohol and drug abuse, consumer organizations, fitness, food safety, health and disease, infancy and childhood, pregnancy and lactation, trade and industry organizations, weight control and eating disorders, and world hunger.

U.S. Government

- Federal Trade Commission (FTC)
  Public Reference Branch
  600 Pennsylvania Avenue NW
  Washington, DC 20580
  (202) 326-2222
  www.ftc.gov
- Food and Drug Administration (FDA)
  Office of Consumer Affairs, HFE 1
  Room 16-85
  5600 Fishers Lane
  Rockville, MD 20857
  (301) 443-1726
  www.fda.gov
- FDA Consumer Information Line
  (888) INFO-FDA
  (888) 463-6332
- FDA Center for Food Safety & Applied Nutrition, HFS 150
  200 C Street SW
  Washington, DC 20204
  (202) 205-4561; fax (202) 205-4564
  www.cfsan.fda.gov
- Food and Nutrition Information Center
  National Agricultural Library, Room 304
  10301 Baltimore Avenue
  Beltsville, MD 20705-2351
  fax (301) 504-6409
  www.nal.usda.gov/fnic
- National Institutes of Health (NIH)
  9000 Rockville Pike
  Bethesda, MD 20892
  (301) 496-2433
  www.nih.gov
- Superintendent of Documents
  U.S. Government Printing Office
  Washington, DC 20402
  (202) 512-1530
  www.access.gpo.gov/su_docs
- U.S. Department of Agriculture (USDA)
  14th Street and Independence Avenue SW
  Washington, DC 20250
  (202) 720-2791
  www.fns.usda.gov/fncs
- USDA Center for Nutrition Policy and Promotion (Dietary Guidelines and Food Guide Pyramid)
  1120 20th Street NW, Suite 200
  North Lobby
  Washington, DC 20036
  (800) 687-2258 or (202) 418-2312
  www.cnpp.usda.gov
- USDA Food Safety and Inspection Service
  Food Safety Education Office
  1400 Independence Avenue SW, Room 2932-S
  Washington, DC 20250
  (202) 690-0351
  www.usda.gov/fsis
- U.S. Department of Education (DOE)
  Accreditation and State Liaison
  Accrediting Agency Evaluation
  1990 K Street NW, #7105
  Washington, DC 20006-8509
  (202) 219-7011
  www.ed.gov/offices/OPE/accreditation
- U.S. Department of Health and Human Services
  200 Independence Avenue SW
  Washington, DC 20201
  (202) 619-0257
  www.os.dhhs.gov
- U.S. Environmental Protection Agency (EPA)
  1200 Pennsylvania Avenue NW
  Washington, DC 20460
  (202) 260-2090
  www.epa.gov
- Assistant Secretary of Health
  Office of Public Health and Science
  Department of Health and Human Sciences
  200 Independence Avenue SW, Room 725-H
  Washington, DC 20201
  (202) 690-7694

Canadian Government

Federal

- Bureau of Nutritional Sciences
  Food Directorate Health Protection Branch, 3-West
  Sir Frederick Banting Research Centre, AL0904A Tunney's Pasture, Ottawa, Ontario K1A 0K9
  (613) 957-2991 fax (613) 941-5366
  www.hc-sc.gc.ca
- Canadian Food Inspection Agency
  Agriculture and Agri-Food Canada
  59 Camelot Drive
  Nepean, Ontario K1A 0Y9
  (613) 225-CFIA or (613) 225-2342
  fax (613) 228-6653
  www.agr.ca
- Office of Nutrition Policy and Promotion, Health Products and Food Branch AL 90761, 7th Floor—Jeanne Mance Bldg., Tunney's Pasture, Ottawa, Ontario K1A 1B4
  www.hc-sc.gc.ca/hppb/nutrition

Provincial and Territorial

- Population Health Strategies Branch
  Alberta Health 23rd Floor,
  TELUS Plaza,
  North Tower
  10025 Jasper Ave
  Edmonton, AB T5J 2N3
  www.health.gov.ab.ca
- Preventive Services Branch
  Ministry of Health
  1520 Blanshard Street
  Victoria, BC V8W 3C8
  www.gov.bc.ca/healthservices
- Manitoba Health
  Health Programs Branch
  300 Carlton Street
  Winnipeg, MB R3B 3M9
  www.gov.mb.ca/health
- Public Health Mgmt. Services
  Health and Community Services
  P.O. Box 5100 520 King Street
  Fredericton, NB E3B 5G8
  www.gnb.ca/0051/index-e.asp

  Department of Health
  Newfoundland and Labrador
  P.O. Box 8700
  Confederation Building, West Bank
  St. John's, NF A1B 4J6
  www.gov.nf.ca/health
- Health Promotion Unit
  Population Health Division
  Dept. of Health and Social Services
  Gov't of the Northwest Territories

Centre Square Tower, 6th Floor
P.O. Box 1320
Yellowknife, NT X1A 2L9
www.gov.nt.ca/agendas/health

- Nova Scotia Dept. of Health
P.O. Box 488
Halifax, NS B3J 2R8
www.gov.ns.ca/health

- Ontario Ministry of Health and Long-Term Care
8th FLR
5700 Yonge Street
North York, ON M2M 4K5
www.gov/on.ca/MOH

- Health Promotion
Government of Nunavut
Dept. of Health and Social Services
P.O. Box 800
Iqaluit, NT X0A 0H0
www.gov.nu.ca/hss.htm

PEI, Dept. of Health and Social Services
11 Kent Street, 2nd floor
Jones Building, Box 2000
Charlottetown, PEI C1A 7N8
www.gov/pe.ca/hss

- Ministère de la Santé et des Services Sociaux,
Service de la promotion des saines habitudes de vie et dépistage
1075, Chemin Ste-Foy, 3e étage
Québec (Québec) G1S 2M1
www.msss.gouv.qc.ca

- Health Promotion Unit
Population Health Branch
Saskatchewan Health
3475 Albert Street
Regina, SK S4S 6X6
www.health.gov.sk.ca

- Government of Yukon
Dept. of Health and Social Services
#5 Hospital Road
Whitehorse, YT Y1A 3H7
www.hss.gov.yk.ca

International Agencies

- Food and Agriculture Organization of the United Nations (FAO)
Liaison Office for North America
2175 K Street, Suite 300
Washington, DC 20437
(202) 653-2400; fax (202) 653-5760
www.fao.org

- International Food Information Council Foundation
1100 Connecticut Avenue NW,
Suite 430
Washington, DC 20036
(202) 296-6540
ificinfo.health.org

- UNICEF
3 United Nations Plaza
New York, NY 10017
(212) 326-7000
www.unicef.org

- World Health Organization (WHO)
Regional Office
525 23rd Street NW
Washington, DC 20037
(202) 974-3000
www.who.org

Professional Nutrition Organizations

- American Society of Nutritional Sciences
9650 Rockville Pike
Bethesda, MD 20814
(301) 530-7050; fax (301) 571-1892
www.nutrition.org

- American Dietetic Association (ADA)
216 West Jackson Boulevard, Suite 800
Chicago, IL 60606-6995
(800) 877-1600; (312) 899-0040
www.eatright.org

- ADA, Consumer Nutrition Hotline
(800) 366-1655

- American Society for Clinical Nutrition
9650 Rockville Pike
Bethesda, MD 20814-3998
(301) 530-7110; fax (301) 571-1863
www.faseb.org/ascn

- Dietitians of Canada
480 University Avenue, Suite 604
Toronto, Ontario, Canada M5G 1V2
(416) 596-0857; fax (416) 596-0603
www.dietitians.ca

- International Life Sciences Institute
1126 Sixteenth Street NW
Washington, DC 20036
(202) 659-0074
www.ilsi.org

- National Academy of Sciences/National Research Council (NAS/NRC)
2101 Constitution Avenue, NW
Washington, DC 20418
(202) 334-2000
www.nas.edu

- National Institute of Nutrition
265 Carling Avenue, Suite 302
Ottawa, Ontario K1S 2E1
(613) 235-3355; fax (613) 235-7032
www.nin.ca

- Society for Nutrition Education
1001 Connecticut Avenue NW,
Suite 528
Washington, DC 20036
(202) 452-8534; fax (202) 452-8536
www.sne.org

Aging

- Administration on Aging
330 Independence Avenue SW
Washington, DC 20201
(202) 619-0724
www.aoa.dhhs.gov

- American Association of Retired Persons (AARP)
601 E Street NW
Washington, DC 20049
(800) 424-3410
www.aarp.org

- National Aging Information Center
330 Independence Avenue SW
Washington, DC 20201
(202) 619-7501
www.aoa.dhhs.gov/naic

- National Institute on Aging
Public Information Office
31 Center Drive, MSC 2292
Bethesda, MD 20892
(800) 222-2225
www.nih.gov/nia

Alcohol and Drug Abuse

- Al-Anon Family Groups, Inc.
1600 Corporate Landing Parkway
Virginia Beach, VA 23454-5617
(888) 4AL-ANON or (888) 425-2666
(757) 563-1600; fax (757) 563-1655
www.al-anon.alateen.org

- Alcohol & Drug Abuse Information Line
Adcare Hospital
(800) 252-6465

- Alcoholics Anonymous (AA)
Grand Central Station
P.O. Box 459
New York, NY 10163
(212) 870-3400
www.alcoholics-anonymous.org

- Narcotics Anonymous (NA)
P.O. Box 9999
Van Nuys, CA 91409
(818) 773-9999; fax (818) 700-0700
www.wsoinc.com

- National Clearinghouse for Alcohol and Drug Information (NCADI)
P.O. Box 2345
Rockville, MD 20847-2345
(800) 729-6686
www.health.org

- National Council on Alcoholism and Drug Dependence (NCADD)
12 West 21st Street
New York, NY 10010
(800) NCA-CALL or (800) 622-2255
(212) 206-6770; fax (212) 645-1690
www.ncadd.org
- U.S. Center for Substance Abuse Prevention
1010 Wayne Avenue, Suite 850
Silver Spring, MD 20910
(301) 459-1591 ext. 244
fax (301) 495-2919
www.covesoft.com/csap.html

Consumer Organizations

- Center for Science in the Public Interest (CSPI)
1875 Connecticut Avenue NW, Suite 300
Washington, DC 20009-5728
(202) 332-9110; fax (202) 265-4954
www.cspinet.org
- Choice in Dying, Inc.
1035 30th Street NW
Washington, DC 20007
(800) 989-WILL or (800) 989-9455
(202) 338-9790; fax (202) 338-0242
www.choices.org
- Consumer Information Center
Pueblo, CO 81009
(888) 8 PUEBLO or (888) 878-3256
www.pueblo.gsa.gov
- Consumers Union of US Inc.
101 Truman Avenue
Yonkers, NY 10703-1057
(914) 378-2000
www.consumersunion.org
- National Council Against Health Fraud, Inc. (NCAHF)
P.O. Box 141
Fort Lee, NJ 07024
(212) 723-2955
www.ncahf.org

Fitness

- American College of Sports Medicine
401 West Michigan Street
Indianapolis, IN 46206-1440
(317) 637-9200
www.acsm.org
- American Council on Exercise (ACE)
5820 Oberlin Drive, Suite 102
San Diego, CA 92121
(800) 825-3636
www.acefitness.org
- Shape Up America!
6707 Democracy Boulevard, Suite 306
Bethesda, MD 20817
(301) 493-5368
www.shapeup.org

Food Safety

- Alliance for Food & Fiber Food Safety Hotline
(800) 266-0200
- FDA Center for Food Safety and Applied Nutrition Outreach and Information
200 C Street SW
Washington, DC 20204
www.cfsan.fda.gov
- National Lead Information Center
(800) LEAD-FYI or (800) 532-3394
(800) 424-LEAD or (800) 424-5323
- National Pesticide Telecommunications Network (NPTN)
Oregon State University
333 Weniger Hall
Corvallis, OR 97331-6502
(800) 858-7378
www.ace.orst.edu/info/nptn
- USDA Meat and Poultry Hotline
(800) 535-4555
- U.S. EPA Safe Drinking Water Hotline
(800) 426-4791

Health and Disease

- Alzheimer's Disease Education and Referral Center
P. O. Box 8250
Silver Spring, MD 20907-8250
(800) 438-4380
www.alzheimers.org
- Alzheimer's Disease Information and Referral Service
919 North Michigan Avenue, Suite 1100
Chicago, IL 60611
(800) 272-3900
www.alz.org
- American Academy of Allergy, Asthma, and Immunology
611 East Wells Street
Milwaukee, WI 53202
(414) 272-6071; fax (414) 272-6070
www.aaaai.org
- American Cancer Society
National Home Office
1599 Clifton Road NE
Atlanta, GA 30329-4251
(800) ACS-2345 or (800) 227-2345
www.cancer.org
- American Council on Science and Health
1995 Broadway, 2nd Floor
New York, NY 10023-5860
(212) 362-7044; fax (212) 362-4919
www.acsh.org
- American Dental Association
211 East Chicago Avenue
Chicago, IL 60611
(312) 440-2500; fax (312) 440-2800
www.ada.org
- American Diabetes Association
1701 North Beauregard Street
Alexandria, VA 22311
(800) 232-3472 or (703) 549-1500
www.diabetes.org
- American Heart Association
Box BHG, National Center
7272 Greenville Avenue
Dallas, TX 75231
(800) 242-8721
www.americanheart.org
- American Institute for Cancer Research
1759 R Street NW
Washington, DC 20009
(800) 843-8114 or (202) 328-7744
fax (202) 328-7226
www.aicr.org
- American Medical Association
515 North State Street
Chicago, IL 60610
(312) 464-5000
www.ama-assn.org
- American Public Health Association (APHA)
800 I Street NW
Washington, DC 20001-3710
(282) 777-2742
www.apha.org
- American Red Cross
National Headquarters
8111 Gatehouse Road
Falls Church, VA 22042
(703) 206-8143
www.redcross.org
- Arthritis Foundation
(800) 283-7800
www.arthritis.org
- Canadian Diabetes Association
15 Toronto Street, Suite 800
Toronto, ON M5C 2E3
(800) BANTING or (800) 226-8464
(416) 363-3373
www.diabetes.ca
- Canadian Public Health Association
400-1565 Carling Avenue
Ottawa, Ontario K1Z 8R1
(613) 725-3769; fax (613) 725-9826
www.cpha.ca
- Centers for Disease Control and Prevention (CDC)
1600 Clifton Road NE
Atlanta, GA 30333
(404) 639-3311
www.cdc.gov
- The Food Allergy Network
10400 Eaton Place, Suite 107

Fairfax, VA 22030-2208
(800) 929-4040 or (703) 691-2713
www.foodallergy.org

- Internet Health Resources
www.ihr.com
- Mayo Clinic Health Oasis
www.mayohealth.org
- National AIDS Hotline (CDC)
(800) 342-AIDS (English)
(800) 344-SIDA (Spanish)
(800) 2437-TTY (Deaf)
(900) 820-2437
- National Cancer Institute
Office of Cancer Communications
Building 31, Room 10A31
31 Center Drive MSC 2580
Bethesda, MD 20892
(800) 4-CANCER or
(800) 422-6237
www.nci.nih.gov
- National Diabetes Information Clearinghouse
31 Center Drive MSC 2560
Bethesda, MD 20892-2560
(301) 654-3327
www.niddk.nih.gov
- National Digestive Disease Information Clearinghouse (NDDIC)
31 Center Drive MSC 2560
Bethesda, MD 20892-2560
(301) 654-3810
www.niddk.nih.gov
- National Health Information Center (NHIC)
P.O. Box 1133
Washington, DC 20013
(800) 336-4797
nhic-nt.health.org
- National Heart, Lung, and Blood Institute Information Center
P.O. Box 30105
Bethesda, MD 20824-0105
(301) 592-8573
www.nhlbi.nih.gov
- National Institute of Allergy and Infectious Diseases
Office of Communications
Building 31, Room 7A50
31 Center Drive, MSC 2520
Bethesda, MD 20892-2520
(301) 496-5717
www.niaid.nih.gov
- National Institute of Dental Research (NIDR)
National Institute of Health
Bethesda, MD 20892-2190
www.nidr.nih.gov
- National Osteoporosis Foundation
1232 22 Street NW
Washington, DC 20037
(202) 223-2226
www.nof.org
- Office of Aids Research
www.nih.gov/od/oar
- Office of Disease Prevention and Health Promotion
odphp.osophs.dhhs.gov
- Osteoporosis and Related Bone Diseases
(800) 624-BONE or (800) 624-2663
www.osteo.org

Infancy and Childhood

- American Academy of Pediatrics
141 Northwest Point Boulevard
Elk Grove Village, IL 60007-1098
(847) 434-4000; fax (847) 434-8000
www.aap.org
- Birth Defect Research for Children
930 Woodcock Road, Suite 225
Orlando, FL 32803
(407) 895-0802; fax (407) 895-0824
www.birthdefects.org
- Canadian Paediatric Society
100-2204 Walkley Road
Ottawa, ON K1G 4G8
(613) 526-9397; fax (613) 526-3332
www.cps.ca
- National Center for Education in Maternal & Child Health
2000 15th Street North, Suite 701
Arlington, VA 22201-2617
(703) 524-7802
www.ncemch.org

Pregnancy and Lactation

- American College of Obstetricians and Gynecologists Resource Center
409 12th Street SW
Washington, DC 20090
(202) 638-5577
www.acog.org
- La Leche International, Inc.
1400 N. Meacham Road
Schaumburg, IL 60173
(847) 519-7730
www.lalecheleague.org
- March of Dimes Birth Defects Foundation
1275 Mamaroneck Avenue
White Plains, NY 10605
(888) MoDimes or (888) 663-4637
www.modimes.org

Trade and Industry Organizations

- Beech-Nut Nutrition Corporation
100 South 4th Street
St. Louis, MO 63102
(800) 523-6633
www.beechnut.com
- Borden Foods Nutrition Department
180 East Broad Street
Columbus, OH 43215
(800) 426-7336
- Campbell Soup Company
Consumer Response Center
Campbell Place, Box 26B
Camden, NJ 08103-1701
(800) 257-8443
www.campbellssoup.com
- General Mills, Inc.
Number One General Mills Boulevard
Minneapolis, MN 55440
(800) 328-6787
www.generalmills.com
- Kellogg Company
P.O. Box 3599
Battle Creek, MI 49016-3599
(800) 962-1413
www.kelloggs.com
- Kraft Foods
Consumer Response and Information Center
One Kraft Court
Glenview, IL 60025
(800) 323-0768
www.kraftfoods.com
- Mead Johnson Nutritionals
2400 West Lloyd Expressway
Evansville, IN 47721
(800) 247-7893
www.meadjohnson.com
- Nabisco Consumer Affairs
100 DeForest Avenue
East Hanover, NJ 07936
(800) NABISCO or (800) 932-7800
www.nabisco.com
- National Dairy Council
10255 West Higgins Road, Suite 900
Rosemond, IL 60018-5616
(847) 803-2000
www.dairyinfo.com
- NutraSweet/KELCO
P.O. Box 2986
Chicago, IL 60654-0986
www.equal.com
- Pillsbury Company
2866 Pillsbury Center
Minneapolis, MN 55402
(800) 767-4466
www.pillsbury.com
- Procter and Gamble Company
One Procter and Gamble Plaza
Cincinnati, OH 45202
(513) 983-1100
www.pg.com/info
- Ross Laboratories, Abbot Laboratory
625 Cleveland Avenue
Columbus, OH 43215

(800) 227-5767
www.abbot.com

- Sunkist Growers
  Consumer Affairs
  Fresh Fruit Division
  14130 Riverside Drive
  Sherman Oaks, CA 91423
  www.sunkist.com
- United Fresh Fruit and Vegetable Association
  727 North Washington Street
  Alexandria, VA 22314
  (703) 836-3410
- USA Rice Federation
  4301 North Fairfax Drive,
  Suite 305
  Arlington, VA 22203
  Phone: (703) 351-8161
  www.usarice.com

Weight Control and Eating Disorders

- American Anorexia & Bulimia Association, Inc.
  165 West 46th Street #1108
  New York, NY 10036
  (212) 575-6200
  www.aabaine.org
- American Obesity Association
  1250 24 Street NW, Suite 300
  Washington, DC 20037
  (800) 98-OBESE or (800) 986-2373
  www.obesity.org
- Anorexia Nervosa and Related Eating Disorders (ANRED)
  P.O. Box 5102
  Eugene, OR 97405
  (541) 344-1144
  www.anred.com
- National Association of Anorexia Nervosa and Associated Disorders, Inc. (ANAD)
  P.O. Box 7
  Highland Park, IL 60035
  (847) 831-3438; fax (847) 433-4632
  www.anad.org
- National Eating Disorder Information Centre
  Toronto General Hospital
  200 Elizabeth Street, CW 1-211
  Toronto, Ontario M5G 2C4
  (416) 340-4156; fax (416) 340-4736
  www.nedic.ca
- National Institute of Diabetes and Digestive Diseases Weight-Control Information Network (WIN)
  31 Center Drive MSC 2560
  Bethesda, MD 20892
  (800) WIN-8098
  www.niddk.nih.gov/health/nutrit/win.htm
- Overeaters Anonymous (OA)
  World Service Office
  6075 Zenith Court NE
  Rio Rancho, NM 87124
  (505) 891-2664; fax (505) 891-4320
  www.overeatersanonymous.org
- TOPS (Take Off Pounds Sensibly)
  4575 South Fifth Street
  P.O. Box 07360
  Milwaukee, WI 53207-0360
  (800) 932-8677 or (414) 482-4620
  www.tops.org
- Weight Watchers International, Inc.
  Consumer Affairs Department/IN
  175 Crossways Park West
  Woodbury, NY 11797
  (800) 651-6000; fax (516) 390-1632
  www.weightwatchers.com

World Hunger

- Bread for the World
  50 F Street NW, Suite 500
  Washington, DC 20001
  (800) 82-BREAD or (800) 822-7323
  (202) 639-9400; fax (202) 639-9401
  www.bread.org
- Center on Hunger, Poverty and Nutrition Policy
  Tufts University School of Nutrition
  11 Curtis Avenue
  Medford, MA 02155
  (617) 627-3956
- Freedom from Hunger
  P.O. Box 2000
  1644 DaVinci Court
  Davis, CA 95616
  (530) 758-6241
  www.freefromhunger.org
- Oxfam America
  26 West Street
  Boston, MA 02111-1206
  (800) 77-OXFAM or (800) 776-9326
  www.oxfamamerica.org
- Worldwatch Institute
  1776 Massachusetts Avenue NW
  Washington, DC 20036
  (202) 452-1999
  www.worldwatch.org

F

Notes

Chapter 1

1. M. J. Stampfer and coauthors, Primary prevention of coronary heart disease in women through diet and lifestyle, *New England Journal of Medicine* 343 (2000): 16–22; A. K. Kant and coauthors, A prospective study of diet quality and mortality in women, *Journal of the American Medical Assocaition* 283 (2000): 2109–2115.

2. N. S. Scrimshaw, Nutrition and health from womb to tomb, *Nutrition Today*, March/April 1996, pp. 55–67.

3. M. Merti, Bridging genomics and genetics, in *BioMedNet Conference Reporter*, American Society of Human Genetics 2000, (http://news.bmn.com/conferences).

4. Y. Okada and coauthors, Small volumes of enteral feedings normalise immune function in infants receiving parenteral nutrition, *Journal of Pediatric Surgery* 33 (1998): 16–19.

5. Position of The American Dietetic Association: Food fortification and dietary supplements, *Journal of the American Dietetic Association* 101 (2001): 115–125.

6. A. S. Levine and C. J. Billington, Why do we eat? A neural systems approach, *Annual Review of Nutrition* 17 (1997): 597–619; R. D. Mattes, Physiologic responses to sensory stimulation by food: Nutritional implications, *Journal of the American Dietetic Association* 97 (1997): 406–410, 413.

7. W. W. Souba, Nutritional support, *New England Journal of Medicine* 336 (1997): 41–48.

8. Belsville Human Nutrition Research Center, Pyramid servings data: Results from USDA's 1994 Continuing Survey of Food Intakes by Individuals, as cited in research and evaluation activities in USDA, *Family Economics and Nutrition Review* 10 (1997): 66–67.

9. F. B. Hu Dietary pattern analysis: a new direction in nutritional epidemiology, *Current Opinion in Lipidology* 13 (2002): 3–9.

10. F. M. Clydesdale, A proposal for the establishment of scientific criteria for health claims for functional foods, *Nutrition Reviews* 55 (1997): 413–422.

11. R. M. Krauss and coauthors, AHA Dietary Guidelines: Revision 2000: A statement for healthcare professionals from the nutrition committee of the American Heart Association, *Circulation* 102 (2000): 2284–2299.

12. R. K. Johnson and E. Kennedy, The 2000 Dietary Guidelines for Americans: What are the changes and why were they made? *Journal of the American Dietetic Association* 100 (2000): 769–774.

13. M. Nestle and coauthors, Behavioral and social influences on food choices, *Nutrition Reviews* 56 (1998): 550–574.

14. F. Katz, "How nutritious?" meets "How convenient?" *Food Technology* 53 (1999): 44–50.

15. R. S. Gradgenett, Watching the trends helps focus nutrition education strategies, *Journal of the American Dietetic Association* 98 (1998): 1000.

16. L. L. Birch, Development of food preferences, *Annual Review of Nutrition* 19 (1999): 41–62; M. B. M. van den Bree, L. J. Eaves, and J. T. Dwyer, Genetic and environmental influences on eating patterns of twins aged ≥50 y, *American Journal of Clinical Nutrition* 70 (1999): 456–465; A. Drewnowski, Taste preferences and food intake, *Annual Review of Nutrition* 17 (1997): 237–253.

17. S. Haberman and D. Luffey, Weighing in college students' diet and exercise behaviors, *Journal of American College Health* 46 (1998): 189–191.

18. U.S. Department of Agriculture and U.S. Department of Health and Human Services, *Nutrition and Your Health: Dietary Guidelines for Americans, 5th edition*, 2000, Home and Garden Bulletin no. 232, available online at www.usda.gov/cnpp or call (888) 878-3256.

19. J. L. Dodd, Incorporating genetics into dietary guidance, *Food Technology* 51 (1997): 80–82.

20. U.S. Department of Health and Human Services, *Healthy People 2010*, 2nd ed. (Washington, D.C.: Government Printing Office, 2000), available online at www.health.gov/healthypeople or call (800) 367-4725.

21. D. Satcher, U.S. Surgeon General, as cited in C. Marwick, Healthy People 2010 initiative launched, *Journal of the American Medical Association* 283 (2000): 989–990.

22. D. S. Satcher, Healthy People at 2000, *Public Health Reports* 114 (1999): 563–564.

23. S. S. Smith, NCHS launches latest National Health and Nutrition Examination Survey, *Public Health Reports* 114 (1999): 190–192.

24. N. S. Wellman and coauthors, Do we facilitate the scientific process and the development of dietary guidance when findings from single studies are publicized? An American Society for Nutritional Sciences Controversy Session Report, *American Journal of Clinical Nutrition* 70 (1999): 802–805.

25. C. W. Enns, J. D. Goldman, and A. Cook, Trends in food and nutrient intakes by adults: NFCS 1977–78, CSFII 1989–91, and CSFII 1994–95, *Family Economics and Nutrition Review* 10 (1997): 2–19.

26. S. M. Krebs-Smith and coauthors, Characterizing food intake patterns of American adults, *American Journal of Clinical Nutrition* 65 (1997): 1264S–1268S.

27. J. D. Anding, R. R. Suminski, and L. Boss, Dietary intake, body mass index, exercise, and alcohol: Are college women following the *Dietary Guidelines for Americans? Journal of American College Health* 49 (2001): 167–171.

Controversy 1

1. W. M. Silberg, G. D. Lundberg, and R. A. Musacchio, Assessing, controlling, and assuring the quality of medical information on the Internet, *Journal of the American Medical Association* 277 (1997): 1244–1245.

2. Top health frauds, *FDA Backgrounder*, November 6, 1996 (available from www.fda.gov/opacom/backgrounders).

3. E. Cunningham and W. Marcason, Internet hoaxes: How to spot them and how to debunk them, *Journal of the American Dietetic Association* 101 (2001): 460.

4. J. A. Schulman, Nutrition education in medical schools: Trends and implications for health educators, *Med Ed Online* www.med-ed-Online.org/f0000015.htm (accessed October 24, 2000).

5. Position of The American Dietetic Association: Nutrition education of health professionals. *Journal of the American Dietetic Association* 98 (1998): 343–346.

6. C. Gopalan, Dietetics and nutrition: Impact of scientific advances and development, *Journal of the American Dietetic Association* 97 (1997): 737–741.

7. J. Arena and P. Walters, Do you know what a dietetic technician can do? A focus on clinical technicians and their expanded roles and responsibilities, *Journal of the American Dietetic Association* 97 (1997): S139–S141.

Chapter 2

1. USDA Center for Nutrition Policy and Promotion, *The Healthy Eating Index, 1994–1996* (Washington, D.C.: Government Printing Office, 1998), or available from www.usda.gov/cnpp.

2. Standing Committee on the Scientific Evaluation of Dietary Reference Intakes, Food and Nutrition Board, Institute of Medicine, *Dietary Reference Intakes: Applications in Dietary Assessment* (Washington, D.C.: National Academy Press, 2000), pp. 5–7.

3. Standing Committee on the Scientific Evaluation of Dietary Reference Intakes, Food and Nutrition Board, Institute of Medicine, *Dietary Reference Intakes for Calcium, Phosphorus, Magnesium, Vitamin D, and Fluoride* (Washington, D.C.: National Academy Press, 1997), p. S-5.

4. Standing Committee on the Scientific Evaluation of Dietary Reference Intakes, 1997, p. S-3.
5. Dietary Reference Intakes, *Nutrition Reviews* 55 (1997): 319–326.
6. M. B. Hogbin and M. A. Hess, Public confusion over food portions and servings, *Journal of the American Dietetic Association* 99 (1999): 1209–1211.
7. M. Hogbin, A. Shaw, and R. S. Anand, Food portions and servings: How do they differ? *Family and Economics Review* 13 (2001): 92–94.
8. L. R. Young and M. Nestle, Variation in perceptions of a "medium" food portion: Implications for dietary guidance, *Journal of the American Dietetic Association* 98 (1998): 458–459.

Consumer Corner 2

1. USDA, Away-from-home foods increasingly important to quality of American diet, 1999, available from www.econ.ag.gov.
2. Foods in menu claims must meet FDA rule, *FDA Consumer*, October 1996, p. 5.
3. FDA authorizes new coronary heart disease health claim for plant sterol and plant stanol esters, *FDA Talk Paper*, 2000 available at http://vm.ctsan.tda.gov/_lrd/.

Controversy 2

1. L. Hilakivi-Clarke and coauthors, Maternal and prepubertal diet, mammary development and breast cancer risk, *Journal of Nutrition* 131 (2001): S154–S157.
2. L. Le Marchand and coauthors, Intake of flavonoids and lung cancer, *Journal of the National Cancer Institute* 92 (2000): 154–160; P. Knekt and coauthors, Quercetin intake and the incidence of cerebrovascular disease, *European Journal of Clinical Nutrition* 54 (2000): 415–417; L. F. Macrae, Wheat bran fiber and development of adenomatous polyps: Evidence from randomized, controlled, clinical trials, *American Journal of Medicine* 106 (1999): S38–S42; S. V. Nigdikar and coauthors, Consumption of red wine polyphenols reduces the susceptibility of low-density lipoproteins to oxidation in vivo, *American Journal of Clinical Nutrition* 68 (1998): 258–265; L. Chatenoud and coauthors, Whole grain food intake and cancer risk, *International Journal of Cancer* 77 (1998): 24–28.
3. A. Drewnowski and C. Gomez-Carneros, Bitter taste, phytonutrients, and the consumer: A review, *American Journal of Clinical Nutrition* 72 (2000): 1424–1435.
4. J. H. M. de Vries and coauthors, Red wine is a poor source of bioavailable flavonols in men, *Journal of Nutrition* 131 (2001): 745–748; Y. Scheider and coauthors, Anti-proliferative effect of resveratrol, a natural component of grapes and wine, on human colonic cancer cells, *Cancer Letter* 158 (2000): 85–91.
5. U. Wenzel and coauthors, Dietary flavone is a potent apoptosis inducer in human colon carcinoma cells, *Cancer Research* 60 (2000): 3823–3831.
6. D. Rein and coauthors, Epicatechin in human plasma: In vivo determination and effect of chocolate consumption on plasma oxidation status, *Journal of Nutrition* 130 (2000): S2109–S2114.
7. J. F. Wang and coauthors, A dose-response effect from chocolate consumption on plasma epicatechin and oxidative damage, *Journal of Nutrition* 130 (2000): S2115–S2119.
8. D. Rein and coauthors, Cocoa and wine polyphenols modulate platelet activation and function, *Journal of Nutrition* 130 (2000): S2120–S2126.
9. T. L. Dillinger and coauthors, Food of the gods: Cure for humanity? A cultural history of the medicinal and ritual use of chocolate, *Journal of Nutrition* 130 (2000): S2057–S2072.
10. K. E. Wangen and coauthors, Soy isoflavones improve plasma lipids in normocholesterolemic and mildly hypercholesterolemic postmenopausal women, *American Journal of Clinical Nutrition* 73 (2001): 225–231; S. M. Potter, Soy protein and cardiovascular disease: The impact of bioactive components in soy, *Nutrition Reviews* 56 (1998): 231–235.
11. C. A. Lamartiniere, Protection against breast cancer with genistein: A component of soy, *American Journal of Clinical Nutrition* 71 (2000): S1705–S1707; Potter, 1998; D. Ingram and coauthors, Case-control study of phyto-estrogens and breast cancer, *Lancet* 350 (1997): 990–994; H. Adlercreutz and W. Mazur, Phyto-estrogens and western diets, *Annals of Medicine* 29 (1997): 95–120; M. S. Morton and coauthors, Lignans and isoflavonoids in plasma and prostatic fluid in men: Samples from Portugal, Hong Kong, and the United Kingdom, *Prostate* 32 (1997): 122–128.
12. A Seow and coauthors, Diet, reproductive factors and lung cancer risk among Chinese women in Singapore: evidence for a protective effect of soy in nonsmokers, *International Journal of Cancer* 20 (2002): 365–371; P. L. Horn-Ross, K. J. Hoggatt, and M. M. Lee, Phytoestrogens and thyroid cancer risk: the San Francisco Bay Area thyroid cancer study, *Cancer Epidemiology, Biomarkers, and Prevention* 11 (2002): 43–49.
13. C. Nagata and coauthors, Soy product intake and hot flashes in Japanese women: Results from a community-based prospective study, *American Journal of Epidemiology* 153 (2001): 790–793.
14. A. St. Germain and coauthors, Isoflavone-rich or isoflavone-poor soy protein does not reduce menopausal symptoms during 24 weeks of treatment, *Menopause* 8 (2001): 17–26; S. R. Davis, Phytoestrogen therapy for menopausal symptoms? (editorial), *British Medical Journal* 323 (2001): 354–355.
15. C. D. Allred and coauthors, Soy diets containing varying amounts of genistein stimulate growth of estrogen-dependent (MCF-7) tumors in a dose-dependent manner, *Cancer Research* 61 (2001): 5045–5050.
16. R. R. Newbold and coauthors, Uterine adenocarcinoma in mice treated neonatally with genistein, *Cancer Research* 61 (2001): 4325–4328.
17. P. E. Bowen, Evaluating the health claim of flaxseed and cancer prevention, *Nutrition Today* 36 (2001): 144–158.
18. S. K. Clinton, Lycopene: Chemistry, biology, and implications for human health and disease, *Nutrition Reviews* 56 (1998): 35–51.
19. Fat-soluble vitamins, in C. D. Berdanier, *Advanced Nutrition: Micronutrients* (Boca Raton, Fla.: CRC Press, 1998), pp. 21–22; G. R. Beecher, Nutrient contents of tomatoes and tomato products, *Proceedings of the Society for Experimental and Biological Medicine* 218 (1998): 98–100; H. Nishino, Cancer prevention by natural carotenoids, *Journal of Cellular Biochemistry* (supplement) 37 (1997): 86–91.
20. P. Prakash, R. M. Russell, and N. I. Krinsky, In vitro inhibition of proliferation of estrogen-dependent and estrogen-independent human breast cancer cells treated with carotenoids or retinoids, *Journal of Nutrition* 131 (2001): 1574–1580; H. Gerster, The potential role of lycopene for human health, *Journal of the American College of Nutrition* 16 (1997): 109–126.
21. P. A. Kantesky and coauthors, Dietary intake and blood levels of lycopene: Association with cervical dysplasia among non-Hispanic black women, *Nutrition and Cancer* 31 (1998): 31–40.
22. P. Toniolo and coauthors, Serum carotenoids and breast cancer, *American Journal of Epidemiology* 153 (2001): 1142–1147.
23. T. H. Rissanen and coauthors, Low serum lycopene concentration is associated with an excess incidence of acute coronary events and stroke: The Kupio Ischaemic Heart Disease Risk Factor Study, *British Journal of Nutrition* 85 (2001): 749–754.
24. J. D. Ribaya-Mercado and coauthors, Skin lycopene is destroyed preferentially over β-carotene during ultraviolet irradiation in humans, *Journal of Nutrition* 125 (1995): 1854—1859.
25. J. A. Milner, A historical perspective on garlic and cancer, *Journal of Nutrition* 131 (2001): S1027–S1031.
26. S. V. Singh and coauthors, Differential induction of NAD(P)H: quinone oxidoreductase by anticarcinogenic organosulfides from garlic, *Biochemical and Biophysical Research Communications* 27 (1998): 917–920; S. Fukushima and coauthors, Cancer prevention by organosulfur compounds from garlic and onion, *Journal of Cell Biochemistry* 27 (1997): 100–105.
27. J. L. Issaacsohn and coauthors, Garlic powder and plasma lipids and lipoproteins: A multicenter, randomized placebo-controlled trial, *Archives of Internal Medicine* 158 (1998): 1189–1194.
28. J. T. Pinto and coauthors, Effects of garlic thioallyl derivatives on growth, glutathione concentration, and polyamine formation of human prostate carcinoma cells in culture, *American Journal of Clinical Nutrition* 66 (1997): 398–405.
29. L. O. Dragsted, M. Strube, and T. Leth, Dietary levels of plant phenols and other non-nutritive components: Could they prevent cancer? *European Journal of Cancer Prevention* 6 (1997): 522–528.
30. Position of The American Dietetic Association: Functional foods, *Journal of the American Dietetic Association* 99 (1999): 1278–1285.
31. N. B. Cater, Plant stanol ester foods: New tools in the dietary management of cholesterol, *Nutrition and the MD*, November 1999, pp. 1–4.
32. P. J. H. Jones and F. Ntanios, Comparable efficacy of hydrogenated versus nonhydrogenated plant sterol esters on circulating cholesterol levels in humans, *Nutrition Reviews* 56 (1998): 245–252; T. A. Miettinen and coau-

thors, Reduction of serum cholesterol with sitostanol-ester margarine in a mildly hypercholesterolemic population, *New England Journal of Medicine* 333 (1995): 1308–1312.

33. H. Gylling and coauthors, Retinol, vitamin D, carotenes and alpha-tocopherol in serum of a moderately hypercholesterolemic population consuming sitostanol ester margarine, *Atherosclerosis* 145 (1999): 279–285; H. Gylling and T. A. Miettinen, Cholesterol reduction by different plant stanol mixtures and with variable fat intake, *Metabolism* 48 (1999): 575–580.

34. J. M. Saavedra, Clinical applications of probiotic agents, *American Journal of Clinical Nutrition* 73 (2001): S1147–S1151; J. Schrezenmeir and M. de Vrese, Probiotics, prebiotics, and synbiotics—approaching a definition, *American Journal of Clinical Nutrition* 73 (2001): S361–S364.

35. L. Sherwood and M. D. Gorbach, Probiotics and gastrointestinal health, *American Journal of Gastroenterology* 95 (2000): S2–S4; J. A. Vanderhoof and R. J. Young, Use of probiotics in childhood gastrointestinal disorders, *Journal of Pediatric Gastroenterology and Nutrition* 27 (1998): 323–332.

36. M. de Vrese and coauthors, Probiotics—compensation for lactase insufficiency, *American Journal of Clinical Nutrition* 73 (2001): S421–S429; M. B. Roberfroid, Prebiotics and probiotics: Are they functional foods? *American Journal of Clinical Nutrition* 71 (2000): S1682–S1687; P. Michetti and coauthors, Effect of whey-based culture supernatant of *Lactobacillus acidophilus (johnsonii) La1* on *helicobacter pylori* infection in humans, *Digestion* 60 (1999): 203–209; Y. Delneste, A. Dounet-Hughes, and E. J. Schiffin, Functional foods: Mechanisms of action on immunocompetent cells, *Nutrition Reviews* 56 (1998): 593–598; M. H. Coconnier and coauthors. Antagonistic activity against *Helicobacter* infection in vitro and in vivo by the human *Lactobacillus acidophilus* strain LB, *Applied and Environmental Microbiology* 64 (1998): 4573–4580; M. M. Velraeds and coauthors, Interference in initial adhesion of uropathogenic bacteria and yeasts to silicone rubber by *Lactobacillus acidophilus* biosurfactant, *Journal of Medical Microbiology* 47 (1998): 1081–1085; K. Gupta and coauthors, Inverse association of H202-producing lactobacilli and vaginal *Escherichia coli* colonization in women with recurrent urinary tract infections, *Journal of Infectious Diseases* 178 (1998): 446–450.

37. M. Nestle, Broccoli sprouts in cancer prevention, *Nutrition Reviews* 56 (1998): 127–130.

38. T. Kontiokari and coauthors, Randomised trial of cranberry-lingonberry juice and *Lactobacillus* GG drink for the prevention of urinary tract infections in women, *British Medical Journal* 322 (2001): 1571, available at http://bmj.com/cgi/content/full/322/7302/1571?view=full&pmid=11431298.

39. M. K. Terris, M. M. Issa, and J. R. Tacker, Dietary supplementation with cranberry concentrate tablets may increase the risk of nephrolithiasis, *Urology* 57 (2001): 26–29.

40. P. F. Jacques and coauthors, Long-term nutrient intake and early age-related nuclear lens opacities, *Archives of Ophthalmology* 119 (2001): 1009–1019; J. T. Landrum and R. A. Bone, Lutein, zeaxanthin, and the macular pigment, *Archives of Biochemistry and Biophysics* 385 (2001): 28–40.

41. L. D. Whigham, M. E. Cook, and R. L. Atkinson, Conjugated linoleic acid: Implications for human health, *Pharmacological Research* 42 (2000): 503–150.

42. Position of The American Dietetic Association: The role of nutrition in health promotion and disease prevention programs, *Journal of the American Dietetic Association* 98 (1998): 205–208.

43. D. R. Farr, Functional foods, *Cancer Letters* 114 (1997): 59–63.

44. B. Liebman and D. Schardt, Wild claims, weak evidence, *Nutrition Action Healthletter*, July/August 1998, pp. 8–9.

45. Adapted from C. Hasler and coauthors, How to evaluate the safety, efficacy, and quality of functional foods and their ingredients, *Journal of the American Dietetic Association* 101 (2001): 733–736; B. Brophy and D. Schardt, Functional foods, *Nutrition Action Healthletter*, April 1999, pp. 3–7.

46. D. Josefson, US moves to tighten law on health supplements, *British Medical Journal* 323 (2001): 654.

47. S. Liu and coauthors, Intake of vegetables rich in carotenoids and risk of coronary heart disease in men: The Physicians' Health Study, *International Journal of Epidemiology* 30 (2001): 130–135.

48. J. W. Lampe, Health effects of vegetables and fruit: Assessing mechanism of action in human experimental studies, *American Journal of Clinical Nutrition* 70 (1999): S475–S490.

Chapter 3

1. A. P. Simopoulos and coauthors, Impact of diet and genetic interactions on chronic disease risk, A symposium in *Food Technology* 51 (1997): 65–82.

2. D. P. Huston, The biology of the immune system, *Journal of the American Medical Association* 278 (1997): 1804–1814.

3. L. L. Birch, Development of food preferences, *Annual Reviews of Nutrition* 19 (1999): 41–62; A Drewnowski, Why do we like fat? *Journal of the American Dietetic Association* 97 (1997): S58–S62.

4. G. A. Falciglia and P. A. Norton, Evidence for a genetic influence on preference for some foods, *Journal of the American Dietetic Association* 94 (1994): 154–158.

5. J. L. Jeraci, B. A. Lewis, and P. J. Van Soest, Interaction between human gut bacteria and fiberous structures, in G. A. Spiller, ed., *Dietary Fiber in Human Nutrition* (Boca Raton, Fla.: CRC Press, 1993), pp. 371–376.

6. M. Robinson and coauthors, Heartburn requiring frequent antacid use may indicate significant illness, *Archives of Internal Medicine* 158 (1998): 2373–2376.

7. N. Uemura and coauthors, *Helicobacter pylori* infection and the development of gastric cancer, *New England Journal of Medicine* 345 (2001): 784–789; J. G. Fox and T. C. Wang, *Helicobacter pylori*—Not a good bug after all, *New England Journal of Medicine* 345 (2001): 829–832.

8. T. Gislason and coauthors, Respiratory symptoms and nocturnal gastroesophageal reflux: a population-based study of young adults in three European countries, *Chest* 121 (2002): 158–163.

Controversy 3

1. K. A. Douglas and coauthors, Results from the 1995 National College Health Risk Behavior Survey, *Journal of College Health*, September 1997, pp. 55–66.

2. M. W. Westerterp-Plantenga and C. R. Verwegen, The appetizing effect of an aperitif in overweight and normal-weight humans, *American Journal of Clinical Nutrition* 69 (1999): 205–212; P. M. Suter, E. Häsler, and W. Vetter, Effects of alcohol on energy metabolism and body weight regulation: Is alcohol a risk factor for obesity? *Nutrition Reviews* 55 (1997): 157–171.

3. H. Wechsler and coauthors, College binge drinking in the 1990s: A continuing problem—Results of the Harvard School of Public Health 1999 College Alcohol Study, *Journal of American College Health* 48 (2000): 199–210; H. Wechsler and coauthors, Health and behavioral consequences of binge drinking in college: A national survey of students at 140 campuses, *Journal of the American Medical Association* 272 (1994): 1672–1677.

4. H. Wechsler and coauthors, College alcohol use: A full or empty glass? *Journal of American College Health* 47 (1999): 247–252.

5. K. A. Bradley and coauthors, Medical risks for women who drink alcohol, *Journal of General Internal Medicine* 13 (1998): 627–639.

6. M. A. Emanuele and coauthors, Reversal of ethanol-induced testosterone suppression in prepubertal male rats by opiate blockage, *Alcoholism: Clinical and Experimental Research* 22 (1998): 1199–1204.

7. Suter, Häsler, and Vetter, 1997.

8. B. B. Duncan and coauthors, Association of the waist-to-hip ratio is different with wine than with beer or hard liquor consumption, *American Journal of Epidemiology* 142 (1995): 1034–1038.

9. C. S. Leiber, Alcohol: Its metabolism and interaction with nutrients, *Annual Review of Nutrition* 20 (2000): 395–430.

10. K. W. Singletary and S. M. Gapstur, Alcohol and breast cancer: review of epidemiologic and experimental evidence and potential mechanisms, *Journal of the American Medical Association* 286 (2001): 2143–2151; S. A. Smith-Warner and coauthors, Alcohol and breast cancer in women, *Journal of the American Medical Association* 279 (1998): 535–540.

11. J. D. Piette, P. G. Barnett, and R. H. Moos, First-time admissions with alcohol-related medical problems: A 10-year follow-up of a national sample of alcoholic patients, *Journal of Studies on Alcohol* 59 (1998): 89–96.

12. Substance Abuse and Mental Health Services Administration (SAMHSA), 1998, pp. 170–717.

13. X. Wang, Chronic alcohol intake interferes with retinoid metabolism and signaling, *Nutrition Reviews* 57 (1999): 51–59.

14. M. L. Ambrose, S. C. Bowden, and G. Whelan, Thiamin treatment and working memory function of alcohol-dependent people: Preliminary

F

findings, *Alcoholism: Clinical and Experimental Research* 25 (2001): 112–116.

15. J. W. Miller, Does lowering plasma homocysteine reduce vascular disease risk? *Nutrition Reviews* 59 (2001): 242–244; N. M. van der Put and coauthors, Folate, homocysteine and neural tube defects: An overview, *Experimental Biology and Medicine* 226 (2001): 243–270.

16. J. M. Gaziano and J. E. Buring, Alcohol intake, lipids and risks of myocardial infarction, *Novartis Foundation Symposium* 216 (1998): 86–95.

17. E. V. Leino and coauthors, Alcohol consumption and mortality: II Studies of male populations, *Addiction* 93 (1998): 205–218.

18. Smith-Warner and coauthors, 1998.

19. M. J. Thun and coauthors, Alcohol consumption and mortality among middle-aged and elderly U.S. adults, *New England Journal of Medicine* 337 (1997): 1705–1714; E. B. Rimm and coauthors, Review of moderate alcohol consumption and reduced risk of coronary heart disease: The effect due to beer, wine, or spirits? *British Medical Journal* 312 (1996): 731–736.

20. H. D. Sesso and coauthors, Seven-year changes in alcohol consumption and subsequent risk of cardiovascular disease in men, *Archives of Internal Medicine* 160 (2000): 2605–2612.

21. M. Bobak, A. Skodova, and M. Marmot, Effect of beer drinking on risk of myocardial infarction: Population based case-control study, *British Medical Journal* 320 (2000): 1378–1379.

22. C. L. Hart and coauthors, Alcohol consumption and mortality from all causes, coronary heart disease, and stroke: Results from a prospective cohort study of Scottish men with 21 years of follow up, *British Medical Journal* 318 (1999): 1725–1729.

23. G. A. Rosito, F. D. Fuchs, and B. B. Duncan, Dose-dependent biphasic effect of ethanol on 24-h blood pressure in normotensive subjects, *American Journal of Hypertension* 12 (1999): 236–240.

24. R. Brouillard, F. George, and A. Fougerousse, Polyphenols produced during red wine aging, *Biofactors* 6 (1997): 403–410.

25. J. H. M. de Vries and coauthors, Red wine is a poor source of bioavailable flavonols in men, *Journal of Nutrition* 131 (2001): 745–748.

26. S. I. Oh and coauthors, Chronic ethanol consumption affects glutathione status in rat liver, *Journal of Nutrition* 128 (1998): 758–763; I. D. Norton and coauthors, Chronic ethanol administration causes oxidative stress in the rat pancreas, *Journal of Laboratory and Clinical Medicine* 131 (1998): 442–446.

27. M. Serafini, G. Maiani, and A. Ferro-Luzzi, Alcohol-free red wine enhances plasma antioxidant capacity in humans, *Journal of Nutrition* 128 (1998): 1003–1007.

Chapter 4

1. J. H. Cummings and coauthors, A new look at dietary carbohydrate: Chemistry, physiology and health (Paris Carbohydrate Group), *European Journal of Clinical Nutrition* 51 (1997): 417–423.

2. L. Van Horn and N. Ernst, A summary of the science supporting the new National Cholesterol Education program dietary recommendations: What dietitians should know, *Journal of the American Dietetic Association* 101 (2001): 1148–1154; L. Brown and coauthors, Cholesterol-lowering effects of dietary fiber: A meta-analysis, *American Journal of Clinical Nutrition* 69 (1999): 30–42.

3. M. L. Fernandez, Soluble fiber and nondigestible carbohydrate effects on plasma lipids and cardiovascular risk, *Current Opinion in Lipidology* 12 (2001): 35–40; Brown and coauthors, 1999; D. J. Jenkins and coauthors, Effect of dietary fiber on plasma lipoproteins, in G. A. Spiller, ed., *Handbook of Lipids in Human Nutrition* (Boca Raton, Fla.: CRC Press, 1996), pp. 173–198.

4. Jenkins and coauthors, 1996.

5. J. Salmerón and coauthors, Dietary fiber, glycemic load, and risk of non-insulin-dependent diabetes mellitus in women, *Journal of the American Medical Association* 277 (1997): 472–477; J. Salmerón and coauthors, Dietary fiber, glycemic load, and risk of NIDDM in men, *Diabetes Care* 20 (1997): 545–550; T. M. S. Wolever and coauthors, Low dietary fiber and high protein intakes associated with newly diagnosed diabetes in a remote aboriginal community, *American Journal of Clinical Nutrition* 66 (1997): 1470–1474.

6. A. Sparti and coauthors, Effect of diet high or low in unavailable and slowly digestible carbohydrates on the pattern of 24-h substrate oxidation and feelings of hunger in humans, *American Journal of Clinical Nutrition* 72 (2000): 1461–1468.

7. Z. Yu and coauthors, A comparison of whole wheat, refined wheat, and wheat bran as inhibitors of heterocyclic amines in the Salmonell mutagenicity assay in the rat colonic aberrant crypt focus assay, *Food and Chemical Toxicology* 39 (2001): 655–665; G. H. McIntosh, P. J. Royle, and G. Pointing, Wheat aleurone flour increases cecal beta-glucuronidase activity and butyrate concentration and reduces colon adenoma burden in azoxymethane-treated rats, *Journal of Nutrition* 131 (2001): 127–131; P. Perrin and coauthors, Only fibres promoting a stable butyrate producing colonic ecosystem decrease the rate of aberrant crypt foci in rats, *Gut* 48 (2001): 53–61; C. Avivi-Green, Z. Madar, and B. Schwartz, Pectin-enriched diet affects distribution and expression of apoptosis-cascade proteins in colonic crypts of dimethlhydrazine-treated rats, *International Journal of Molecular Medicine* 6 (2000): 689–698.

8. A. Schatzkin and coauthors, Lack of effect of a low-fat, high-fiber diet on the recurrence of colorectal adenomas, *New England Journal of Medicine* 342 (2000): 1149–1155; D. S. Alberts and coauthors, Lack of effect of a high-fiber cereal supplement on the recurrence of colorectal adenomas, *New England Journal of Medicine* 342 (2000): 1156–1162; P. Boyle and J. S. Langman, ABC of colorectal cancer: Epidemiology, *British Medical Journal* 321 (2000): 805–808; B. S. Reddy, Role of dietary fiber in colon cancer: An overview, *American Journal of Medicine* 106 (1999): S16–S19; D. Kritchevsky, Protective role of wheat bran fiber: Preclinical data, *American Journal of Medicine* 106 (1999): S28–S31; C. S. Fuchs and coauthors, Dietary fiber and the risk of colorectal cancer and adenoma in women, *New England Journal of Medicine* 340 (1999): 169–176.

9. J. A. Story and D. A. Savaiano, Dietary fiber and colorectal cancer: What is appropriate advice? *Nutrition Reviews* 59 (2001): 84–86.

10. J. L. Slavin, Mechanisms for the impact of whole grain foods on cancer risk, *Journal of the American College of Nutrition* 19 (2000): S300–S307.

11. K. Alaimo and coauthors, Dietary intake of vitamins, minerals, and fiber of persons ages 2 months and over in the United States: Third National Health and Nutrition Examination Survey, Phase 1, 1988–1991, as cited in Position of The American Dietetic Association: Health implications of dietary fiber, *Journal of the American Dietetic Association* 97 (1997): 1157–1159.

12. D. D. Gallaher and B. O. Schneeman, Dietary fiber, in E. E. Ziegler and L. J. Filer, Jr., eds., *Present Knowledge in Nutrition*, 7th ed. (Washington, D.C.: International Life Sciences Institute Press, 1996), pp. 87–97.

13. K. N. Englyst and coauthors, Rapidly available glucose in foods: An in vitro measurement that reflects the glycemic response, *American Journal of Clinical Nutrition* 69 (1999): 448–454.

14. N.-G. Asp, J. M. M. van Amelsvoort, and J. G. A. J. Hautvast, Nutritional implications of resistant starch, *Nutrition Research Reviews* 9 (1996): 1–31.

15. F. L. Suarez and D. A. Savaiano, Diet, genetics, and lactose intolerance, *Food Technology* 51 (1997): 74–76; B. Levine, Most frequently asked questions about lactose intolerance, *Nutrition Today* 31 (1996): 78–79.

16. F. L. Suarez and coauthors, Tolerance to the daily ingestion of two cups of milk by individuals claiming lactose intolerance, *American Journal of Clinical Nutrition* 65 (1997): 1502–1506.

17. F. L. Suarez and M. D. Levitt, Abdominal symptoms and lactose: The discrepancy between patients' claims and the results of blinded trials, *American Journal of Clinical Nutrition* 64 (1996): 251–252.

18. J. L. Rosado, Lactose digestion and maldigestion: Implications for dietary habits in developing countries, *Nutrition Research Reviews* (1997): 137–149.

19. F. L. Suarez and coauthors, Lactose maldigestion is not an impediment to the intake of 1500 mg calcium daily as dairy products, *American Journal of Clinical Nutrition* 68 (1998): 1118–1122.

20. M. de Vrese and coauthors, Probiotics—Compensation for lactase insufficiency, *American Journal of Clinical Nutrition* 73 (2001): S421–S429; L. Kopp-Hoolihan, Prophylactic and therapeutic uses of probiotics: A review, *Journal of the American Dietetic Association* 101 (2001): 229–238, 241; J. M. Saavedra, Clinical applications of probiotic agents, *American Journal of Clinical Nutrition* 73 (2001): S1147–S1151.

21. A. E. Buyken and coauthors, Glycemic index in the diet of European outpatients with type 1 diabetes: Relations to glycated hemoglobin and serum lipids, *American Journal of Clinical Nutrition* 73 (2001): 574–581.

22. D. S. Ludwig and coauthors, Dietary fiber, weight gain, and cardiovascular disease risk factors in young adults, *Journal of the American Medical*

Association 282 (1999): 1539–1546; D. S. Ludwig and coauthors, High glycemic index foods, overeating, and obesity, *Pediatrics* 103 (1999): e26 (www.pediatrics.org).

23. T. M. S. Wolever, Dietary recommendations for diabetes: High carbohydrate or high monounsaturated fat? *Nutrition Today* 34 (1999): 73–77.

24. J. Brand-Miller and K. Foster-Powell, Diets with a low glycemic index: From theory to practice, *Nutrition Today* 34 (1999): 64–72; H. Katanas, Diets with a low glycemic index are ready for practice, *Nutrition Today* 34 (1999): 87–88.

25. C. Beebe, Diets with a low glycemic index: Not ready for practice yet! *Nutrition Today* 34 (1999): 82–86.

26. Salmerón and coauthors, Dietary fiber, glycemic load, and risk of NIDDM in men, 1997.

27. A. H. Mokdad and coauthors, Diabetes trends in the U.S.: 1990–1998, *Diabetes Care* 23 (2000): 1278–1283; K. M. V. Narayan and coauthors, Diabetes—A common, growing, serious, costly, and potentially preventable public health problem, *Diabetes Research and Clinical Practice* 50 (2000): S77–S84.

28. National diabetes month—November 1996, *Mortality and Morbidity Weekly Report* 45 (1996): 937; Blindness caused by diabetes—Massachusetts, 1987–1994, *Mortality and Morbidity Weekly Report* 45 (1996): 937–941.

29. Report of the Expert Committee on the Diagnosis and Classification of Diabetes Mellitus, *Clinical Diabetes*, July/August 1997, pp. 158–174; C. R. Scott and coauthors, Characteristics of youth-onset noninsulin-dependent diabetes mellitus and insulin-dependent diabetes mellitus at diagnosis, *Pediatrics* 100 (1997): 84–91.

30. A. A. Skolnick, First type 1 diabetes prevention trials, *Journal of the American Medical Association* 277 (1997): 1101–1102.

31. C. D. Berdanier, Mitochondrial gene expression in diabetes mellitus: Effect of nutrition, *Nutrition Reviews* 59 (2001): 61–70; F. A. Darlsson and coauthors, Beta-cell activity and destruction in type 1 diabetes, *Upsala Journal of Medical Sciences* 105 (2000): 85–95; J. R. Baker, Autoimmune endocrine disease, *Journal of the American Medical Association* 278 (1997): 1931–1937.

32. Evidence supporting the link between cow's milk formula and diabetes is found in American Academy of Pediatrics Work Group on Cow's Milk Protein and Diabetes, Infant feeding practices and their possible relationship to the etiology of diabetes mellitus, *Pediatrics* 94 (1994): 752–754; evidence against the link is in J. M. Norris and coauthors, Lack of association between early exposure to cow's milk protein and ß-cell autoimmunity, *Journal of the American Medical Association* 276 (1996): 609–614; see also Letters, *Journal of the American Medical Association* 276 (1996): 1799–1801.

33. M. Atkinson and G. Eisenbarth, Type 1 diabetes: New perspectives on disease pathogenesis and treatment, *Lancet* 358 (2001): 221–229; J. Hopkins, Treating diabetes with aerosolized insulin, *Chest* 120 (2001): 99S–106S; A. M. Davalli and coauthors, *Journal of Clinical Endocrinology and Metabolism* 85 (2000): 3847–3852.

34. A. Rubenstein, as quoted by Skolnick, 1997.

35. New type II diabetes drug may reduce insulin needs, *FDA Consumer*, April 1997, p. 4.

36. A. Fagot-Campagna, K. M. Narayan, and G. Imperatore, Type 2 diabetes in children exemplifies the growing problem of chronic diseases, *British Medical Journal* 322 (2001): 377–378; American Diabetes Association Consensus Statement, Type 2 diabetes in children and adolescents, *Diabetes Care* 23 (2000): 381–389, accessed June 3, 2001, at www.diabetes.org/ada/Consensus/pg381.htm; American Diabetes Association, Type 2 diabetes in children and adolescents, *Pediatrics* 105 (2000): 671–680.

37. C. M. Steppan and coauthors, The hormone resistin links obesity to diabetes, *Nature* 409 (2001): 292–293.

38. R. M. van Dam and coauthors, Dietary patterns and risk for type 2 diabetes mellitus in U.S. men, *Annals of Internal Medicine* 136 (2002): 201–209; F. B. Hu and coauthors, Diet, lifestyle, and the risk of type 2 diabetes mellitus in women, *New England Journal of Medicine* 345 (2001): 790–797.

39. R. Kikkawa, Chronic complications in diabetes mellitus, *British Journal of Nutrition* 84 (2000): S183–S185.

40. A. Ceriello and coauthors, Antioxidant defenses are reduced during the oral glucose tolerance test in normal and non-insulin-dependent diabetic subjects, *European Journal of Clinical Investigation* 28 (1998): 329–333.

41. Position of the American Dietetic Association: Total diet approach to communicating food and nutrition information, *Journal of the American Dietetic Association* 102 (2002): 100–108.

42. American Diabetes Association, Evidence-based nutrition principles and recommendations for the treatment and prevention of diabetes and related complications, *Diabetes Care* 25 (2002): S50–S60.

43. H. T. Pigman, D. X. Gan, and M. A. Krousel-Wood, Role of exercise for type 2 diabetic patient management, *Southern Medical Journal* 95 (2002): 72–77; M. Wei and coauthors, The association between cardiorespiratory fitness and impaired fasting glucose and type 2 diabetes mellitus in men, *Annals of Internal Medicine* 130 (1999): 89–96; Joint position statement of the American College of Sports Medicine and the American Diabetes Association, Diabetes and exercise, *Medicine and Science in Sports and Exercise* 29 (1997): i–vi.

44. G. Perseghin and coauthors, Increased glucose transport-phosphorylation and muscle glycogen synthesis after exercise training in insulin-resistant subjects, *New England Journal of Medicine* 335 (1996): 1357–1362; A. Kriska, Physical activity and the prevention of type II (non-insulin dependent) diabetes, *President's Council on Physical Fitness and Sports Research Digest*, June 1997, available from the President's Council on Physical Fitness and Sports, HHH Building, Room 738H, 200 Independence Avenue, W.W., Washington, DC 20201.

45. L. Dye, A. Lluch, and J. E. Blundell, Macronutrients and mental performance, *Nutrition* 16 (2000): 1021–1034.

46. C. V. Ford, Somatization and fashionable diagnoses: Illness as a way of life, *Scandinavian Journal of Work Environment and Health* (supplement 3) 23 (1997): 7–16.

47. J. F. Brun, C. Fedou, and J. Mercier, Postprandial reactive hypoglycemia, *Diabetes and Metabolism* 26 (2000): 337–351.

48. G. Pourmotabbed and A. E. Kitabchi, Hypoglycemia, *Obstetrics and Gynecology Clinics of North America* 28 (2001): 383–400.

49. D. Flanagan and coauthors, Gin and tonic and reactive hypoglycemia: What is important—The gin, the tonic, or both? *Journal of Clinical Endocrinology and Metabolism* 83 (1998): 796–800.

Controversy 4

1. A. Drewnowski, Taste preferences and food intake, *Annual Review of Nutrition* 17 (1997): 237–253.

2. U.S. Department of Agriculture, Economic Research Services, *A Dietary Assessment of the U.S. Food Supply: Comparing Per Capita Food Consumption with the Food Guide Pyramid Serving Recommendations*, AER no. 772 (Washington, D.C.: Government Printing Office, 1998).

3. Executive summary, from *Third Report on Nutrition Monitoring in the United States* (Washington, D.C.: Government Printing Office, 1996).

4. Economic Research Service, Food consumption: Food supply and use, *Briefing Room*, 2000, available at www.ers.usda.gov/briefing/consumption/Supply.htm.

5. U.S. Department of Agriculture, *Sugar and Sweetener Situation and Outlook Report*, March 1996, pp. 3–4.

6. Executive summary, 1996.

7. L. E. Spieth and coauthors, A low-glycemic index diet in the treatment of pediatric obesity, *Archives of Pediatrics and Adolescent Medicine* 154 (2000): 947–951.

8. D. S. Ludwig, Dietary glycemic index and obesity, *Journal of Nutrition* 130 (2000): S280–S283.

9. J. Salmerón and coauthors, Dietary fiber, glycemic load, and risk of non-insulin-dependent diabetes mellitus in women, *Journal of the American Medical Association* 277 (1997): 472–477; J. Salmerón and coauthors, Dietary fiber, glycemic load, and risk of NIDDM in men, *Diabetes Care* 20 (1997): 545–550.

10. L. C. Hudgins and coauthors, Human fatty acid synthesis is reduced after the substitution of dietary starch for sugar, *American Journal of Clinical Nutrition* 67 (1998): 631–639.

11. A. Raben and A. Astrup, Ad libitum intake of low-fat diets rich in either starchy foods or sucrose: Effects on blood lipids, factor VII coagulant activity, and fibrinogen, *Metabolism: Clinical and Experimental* 49 (2000): 731–735; J. P. Bantle and coauthors, Effects of dietary fructose on plasma lipids in healthy subjects, *American Journal of Clinical Nutrition* 72 (2000): 1128–1134; F. Abbasi and coauthors, High carbohydrate diets, triglyceride-rich lipoproteins, and coronary heart disease risk, *American Journal of Cardiology* 85 (2000): 45–48.

F

12. A. T. Erkkilä and coauthors, *APOE* polymorphism and the hypertriglyceridemic effect of dietary sucrose, *American Journal of Clinical Nutrition* 73 (2001): 746–752.

13. E. J. Parks and M. K. Hellerstein, Carbohydrate-induced hypertriacylglycerolemia: Historical perspective and review of biological mechanisms, *American Journal of Clinical Nutrition* 71 (2000): 412–433.

14. J. W. White and M. Wolraich, Effect of sugar on behavior and mental performance, *American Journal of Clinical Nutrition* 62 (1995): S242–S249; M. L. Wolraich, D. B. Wilson, and J. W. White, The effect of sugar on behavior or cognition of children: A meta-analysis, *Journal of the American Medical Association* 274 (1995): 1617–1621.

15. S. Gibson and S. Williams, Dental caries in pre-school children: Associations with social class, toothbrushing habit and consumption of sugars and sugar-containing foods. Further analysis of data from the National Diet and Nutrition Survey of children aged 1.5–4.5 years, *Caries Research* 33 (1999): 101–113.

16. K. McNutt, Why some consumers don't believe some nutrition claims, *Nutrition Today* 32 (1997): 252–256.

17. K. McNutt, What clients need to know about sugar replacers, *Journal of the American Dietetic Association* 100 (2000): 466–469.

18. Diarrhea from sugarless candy, *Health Gazette,* April 1998, p. 2.

19. K. McNutt and A. Sentko, Isomalt: The sugar that isn't, a 1997 fact sheet available from Consumer Choices, Inc., 28W176 Belleau Drive, Winfield, IL 60190.

20. S. S. Natah and coauthors, Metabolic response to lactitol and xylitol in healthy men, *American Journal of Clinical Nutrition* 65 (1997): 947–950.

21. Fact Sheet: The Report on Carcinogens, 9th ed., National Institutes of Health News Release available at www.nih.gov/news/pr/may2000/niehs-15.htm.

22. Position of The American Dietetic Association: Use of nutritive and nonnutritive sweeteners, *Journal of the American Dietetic Association* 98 (1998): 581–587.

23. S. M. Cohen and coauthors, Calcium phosphate–containing precipitate and the carcinogenicity of sodium salt in rats, *Carcinogenesis* 21 (2000): 783–792.

24. A bibliography of 167 research articles on aspartame can be found in J. Van de Kamp, Adverse effects of aspartame, *Current Bibliographies in Medicine* (Washington, D.C.: Government Printing Office, 1991).

25. Notebook, *FDA Consumer*, September 1996, p. 29.

26. MediaScams, *Priorities*, no. 4, 1996, p. 8.

27. P. A. Spiers and coauthors, Aspartame: Neuropsychologic and neurophysiologic evaluation of acute and chronic effects, *American Journal of Clinical Nutrition* 68 (1998): 531–537; J. G. Gurney and coauthors, Aspartame consumption in relation to childhood brain tumor risk: Results from a case-control study, *Journal of the National Cancer Institute* 89 (1997): 1072–1074.

28. Position of The American Dietetic Association, 1998.

Chapter 5

1. A. H. Lichenstein and coauthors, Dietary fat consumption and health, *Nutrition Reviews* 56 (1998): S3–S19.

2. E. Faloia and coauthors, Adipose tissue as an endocrine organ? A review of some recent data, *Eating and Weight Disorders* 5 (2000): 116–123.

3. A. Drewnowski, Why do we like fat? *Journal of the American Dietetic Association* 97 (1997): S58–S62.

4. A. Drewnowski and B. M. Popkin, The nutrition transition: New trends in the global diet, *Nutrition Reviews* 55 (1997): 31–43.

5. P. Tso and coauthors, The role of apolipoprotein A-IV in the regulation of food intake, *Annual Review of Nutrition* 21 (2001): 231–254; G. A. Bray, Afferent signals regulating food intake, *Proceedings of the Nutrition Society* 59 (2000): 373–384; T. H. Moran, Cholecystokinin and satiety: current perspectives, *Nutrition* 16 (2000): 858–865.

6. M. T. Montoya and coauthors, Fatty acid saturation of the diet and plasma lipid concentrations, lipoprotein particle concentrations, and cholesterol efflux capacity, *American Journal of Clinical Nutrition* 75 (2002): 484–491; J. Zhang and H. Kesteloot, differences in all-cause, cardiovascular and cancer mortality between Hong Kong and Singapore: role of nutrition, *European Journal of Epidemiology* 17 (2001): 469–477.

7. S. A. Morgan, K. O'Dea, and A. J. Sinclair, A low-fat diet supplemented with monounsaturated fat results in less HDL-C lowering than a very-low-fat diet, *Journal of the American Dietetic Association* 97 (1997): 151–156.

8. N. R. Simonsen and coauthors, Tissue stores of individuals' monounsaturated fatty acids and breast cancer: The EURAMIC study, European Community Multicenter Study on Antioxidants, Myocardial Infarction, and Breast Cancer, *American Journal of Clinical Nutrition* 68 (1998): 134–141; R. McPherson and G. A. Spiller, Effects of dietary fatty acids and cholesterol on cardiovascular disease risk factors in man, in G. A. Spiller, ed., *Handbook of Lipids in Human Nutrition* (Boca Raton, Fla.: CRC Press, 1996), pp. 41–49.

9. C. T. Phan and P. Tso, Intestinal lipid absorption and transport, *Frontiers in Bioscience* 6 (2001): d299–d319.

10. A. L. Lichtenstein and P. J. H. Jones, Lipids: Absorption and transport, in B. A. Bowman and R. M. Russell, eds., *Present Knowledge in Nutrition* (Washington, D.C.: International Life Sciences Institute Press, 2001), p. 99; C. F. Semenkovich, Nutrient and genetic regulation of lipoprotein metabolism, in M. E. Shils and coeditors, *Modern Nutrition in Health and Disease* (Baltimore: Williams & Wilkins, 1999), p. 1196.

11. P. M. Kris-Etherton and S. Yu, Individual fatty acid effects on plasma lipids and lipoproteins: Human studies, *American Journal of Clinical Nutrition* 65 (1997): S1628–S1644.

12. Executive Summary of the Third Report of the National Cholesterol Education Program (NCEP) Expert Panel on Detection, Evaluation, and Treatment of High Blood Cholesterol in Adults (Adult Treatment Panel III), *Journal of the American Medical Association* 285 (2001): 2486–2497.

13. E. Sarkkinen and coauthors, Effect of apolipoprotein E polymorphism on serum lipid response to the separate modification of dietary fat and dietary cholesterol, *American Journal of Clinical Nutrition* 68 (1998): 1215–1222; McPherson and Spiller, 1996.

14. M. Rantala and coauthors, Apolipoprotein B gene polymorphisms and serum lipids: Meta-analysis of the role of genetic variation in responsiveness to diet, *American Journal of Clinical Nutrition* 71 (2000): 713–724.

15. S. M. Grundy, Nutrition and diet in the management of hyperlipidemia and atherosclerosis, in M. E. Shils and coauthors, eds., *Modern Nutrition in Health and Disease* (Baltimore: Williams & Wilkins, 1999), p. 1203.

16. W. E. Boden and T. A. Pearson, Raising low levels of high-density lipoprotein cholesterol is an important target of therapy, *American Journal of Cardiology* 85 (2000): 645–650, A10.

17. Executive Summary of the Third Report of the National Cholesterol Education Program (NCEP) Expert Panel on Detection, Evaluation, and Treatment of High Blood Cholesterol in Adults (Adult Treatment Panel III), 2001.

18. S. M. Grundy, The optimal ratio of fat-to-carbohydrate in the diet, *Annual Review of Nutrition* 19 (1999): 325–341; F. B. Hu and coauthors, Dietary fat intakes and the risk of coronary heart disease in women, *New England Journal of Medicine* 337 (1997): 1491–1499.

19. D. Hwang, Fatty acids and immune responses—A new perspective in searching for clues to mechanism, *Annual Review of Nutrition* 20 (2000): 431–456; A. P. Simopoulos, Omega-3 fatty acids, part II: Epidemiological aspects of omega-3 fatty acids in disease states, in G. A. Spiller, ed., *Handbook of Lipids in Human Nutrition* (Boca Raton, Fla.: CRC Press, 1996), pp. 75–89.

20. J. P. SanGiovanni and coauthors, Meta-analysis of dietary essential fatty acids and long-chain polyunsaturated fatty acids as they relate to visual resolution acuity in healthy preterm infants, *Pediatrics* 105 (2000): 1292–1298.

21. J. P. Middaugh, Cardiovascular deaths among Alaskan Natives, 1980–1986, *American Journal of Public Health* 80 (1990): 282–285; J. Dyerberg, Linolenate-derived polyunsaturated fatty acids and prevention of atherosclerosis, *Nutrition Reviews* 44 (1986): 125–134.

22. R. Uauy and P. Mena, Lipids and neurodevelopment, *Nutrition Reviews* 59 (2001): S34–S48.

23. I. Hiroyasu and coauthors, Intake of fish and omega-3 fatty acids and risk of stroke in women, *Journal of the American Medical Association* 285 (2001): 304–312; R. De Caterina, J. K. Liao, and P. Libby, Fatty acid modulation of endothelial activation, *American Journal of Clinical Nutrition* 71 (2000): S213–S223; K. D. Stark and coauthors, Effect of a fish-oil concentrate on serum lipids in postmenopausal women receiving and not receiving hormone replacement therapy in a placebo-controlled, double-blind trial, *American Journal of Clinical Nutrition* 72 (2000): 389–394;

R. Marchioli and coauthors, Dietary supplementation with n-3 polyunsaturated fatty acids and vitamin E after myocardial infarction: Results of the GISSI-Prevenzione trial, *Lancet* 354 (1999): 447–455; T. Tashiro and coauthors, n-3 versus n-6 polyunsaturated fatty acids in critical illness, *Nutrition* 14 (1998): 551–553.

24. A. P. Simopoulos, Omega-3 fatty acids, part I: Metabolic effects of omega-3 fatty acids and essentiality, in G. A. Spiller, ed., *Handbook of Lipids in Human Nutrition* (Boca Raton, Fla.: CRC Press, 1996), pp. 51–73.

25. Fish oil, American Heart Association Recommendation, 2000, available at www.americanheart.org; S. L. Connor and W. E. Connor, Are fish oils beneficial in the prevention and treatment of coronary artery disease? *American Journal of Clinical Nutrition* 66 (1997): S1020–S1031.

26. D. S. Siscovick and coauthors, Dietary intake of long-chain n-3 polyunsaturated fatty acids and the risk of primary cardiac arrest, *American Journal of Clinical Nutrition* 72 (2000): S208–S212; N. F. Sheard, Fish consumption and risks of sudden cardiac death, *Nutrition Reviews* 56 (1998): 177–179; C. M. Albert and coauthors, Fish consumption and risk of sudden cardiac death, *Journal of the American Medical Association* 279 (1998): 23–28; R. F. Gillum, M. E. Mussolino, and J. H. Madans, The relationship between fish consumption and stroke incidence: The NHANES I Epidemiologic Follow-up Study, *Archives of Internal Medicine* 156 (1996): 537–542.

27. V. M. Montori and coauthors, Fish oil supplementation in type 2 diabetes: A quantitative systematic review, *Diabetes Care* 23 (2000): 1407–1415.

28. S. M. Innis, Essential fatty acids in infant nutrition: Lessons and limitations from animal studies in relation to studies on infant fatty acid requirements, *American Journal of Clinical Nutrition* 71 (2000): S238–S244.

29. R. C. Wander and S-H. Du, Oxidation of plasma proteins is not increased after supplementation with eicosapentaenoic and docosahexaenoic acids, *American Journal of Clinical Nutrition* 71 (2000): 731–737; J. V. Higdon and coauthors, Supplementation of postmenopausal women with fish oil rich in eicosapentaenoic acid and docosahexaenoic acid is not associated with greater in vivo lipid peroxidation compared with oils rich in oleate and linoleate as assessed by plasma malondialehyde and F_2-isoprostanes, *American Journal of Clinical Nutrition* 72 (2000): 714–722; M. Meydani and J. Mayer, Omega-3 fatty acids alter soluble markers of endothelial function in coronary heart disease patients, *Nutrition Reviews* 58 (2000): 56–58.

30. K. Ando and coauthors, Effect of n-3 fatty acid supplementation on lipid peroxidation and protein aggregation in rat erythrocyte membranes, *Lipids* 33 (1998): 505–512.

31. L. D. Yam, A. Peled, and M. Shinitzky, Suppression of tumor growth and metastasis by dietary fish oil combined with vitamins E and C and cisplatin, *Cancer Chemotherapy and Pharmacology* 47 (2001): 34–40; L. Klieveri and coauthors, Promotion of colon cancer metastases in rat liver by fish oil diet is not due to reduced stroma formation, *Clinical and Experimental Metastasis* 18 (2000): 371–377; P. Griffini and coauthors, Dietary omega-3 polyunsaturated fatty acids promote colon carcinoma metastasis in rat liver, *Cancer Research* 58 (1998): 3312–3319.

32. Fish: Weighing the risks and benefits, *Consumer Reports on Health*, April 2001, pp. 1, 4.

33. T. A. Mori and L. J. Beilin, Long-chain omega 3 fatty acids, blood lipids and cardiovascular risk reduction, *Current Opinion in Lipidology* 12 (2001): 11–17.

34. B. M. Popkin and coauthors, Where's the fat? Trends in U.S. diets 1965–1996, *Preventive Medicine* 32 (2001): 245–254.

35. L. Hansen and M. S. Rose, Sensory acceptability is inversely related to development of fat rancidity in bread made from stored flour, *Journal of the American Dietetic Association* 96 (1996): 792–793.

36. M. B. Katan, *Trans* fatty acids and plasma lipoproteins, *Nutrition Reviews* 58 (2000): 188–191; P. O. Kwiterovich, The effect of dietary fat, antioxidants, and pro-oxidants on blood lipids, lipoproteins, and atherosclerosis, *Journal of the American Dietetic Association* 97 (1997): S31–S41.

37. G. J. Nelson, Dietary fat, *trans* fatty acids, and risk of coronary heart diseases, *Nutrition Reviews* 56 (1998): 250–252; S. Shapiro, Do *trans* fatty acids increase the risk of coronary artery disease? A critique of the epidemiologic evidence, *American Journal of Clinical Nutrition* 66 (1997): S1011–S1017.

38. L. Kohlmeier and coauthors, Adipose tissue *trans*-fatty acids and breast cancer in the European Community Multicenter Study on Antioxidants, Myocardial Infarction, and Breast Cancer, *Cancer, Epidemiology, Biomarkers and Prevention* 6 (1997): 705–710; C. Ip, Review of the effects of *trans*-fatty acids, oleic acid, n-3 polyunsaturated fatty acids, and conjugated linoleic acid on mammary carcinogenesis in animals, *American Journal of Clinical Nutrition* 66 (1997): S1523–S1529; M. L. Gurr, Dietary fatty acids with *trans* unsaturation, *Nutrition Research Reviews* 9 (1996): 259–279.

39. M. A. Hallikainen and M. I. Uusitupa, Effects of 2 low-fat stanol ester–containing margarines on serum cholesterol concentrations as part of a low-fat diet in hypercholesterolemic subjects, *American Journal of Clinical Nutrition* 69 (1999): 403–410.

40. D. B. Allison and coauthors, Estimated intakes of *trans* fatty and other fatty acids in the US population, *Journal of the American Dietetic Association* 99 (1999): 166–174; A. P. Simopoulos, Trans fatty acids, in G. A. Spiller, ed., *Handbook of Lipids in Human Nutrition* (Boca Raton, Fla.: CRC Press, 1996), pp. 91–99.

41. A. H. Lichtenstein and coauthors, Effects of different forms of dietary hydrogenated fats on serum lipoprotein cholesterol levels, *New England Journal of Medicine* 340 (1999): 1933–1940; F. B. Hu and coauthors, Dietary fat intake and the risk of coronary heart disease in women, *New England Journal of Medicine* 337 (1997): 1491–1499; M. B. Katan, High-oil compared with low-fat, high-carbohydrate diets in the prevention of ischemic heart disease, *American Journal of Clinical Nutrition* 66 (1997): S974–S979.

42. Is total fat consumption really decreasing? *Nutrition Insights*, April 1998.

43. Popkin and coauthors, 2001.

44. B. Hardin, Genetic testing helps single out leanness gene, *Agricultural Research Service News*, June 1999, available from www.ars.usda.gov/is/pr/1999/990622.htm.

Consumer Corner 5

1. R. S. Sandler and coauthors, Gastrointestinal symptoms in 3181 volunteers ingesting snack foods containing olestra or triglycerides: A 6-week randomized, placebo-controlled trial, *Annals of Internal Medicine* 130 (1999): 253–261.

2. G. S. Allgood and coauthors, Postmarketing surveillance of new food ingredients: Results from the program with the fat replacer olestra, *Regulatory Toxicology and Pharmacology* 33 (2001): 224–233.

3. J. A. Westrate and K. H. van het Hof, Sucrose polyester and plasma carotenoid concentrations in healthy subjects, *American Journal of Clinical Nutrition* 62 (1995): 591–597.

4. E. Kennedy and S. Bowman, Assessment of the effect of fat-modified foods on diet quality in adults, 19 to 50 years, using data from the Continuing Survey of Food Intake by Individuals, *Journal of the American Dietetic Association* 101 (2001): 455–460.

5. R. E. Patterson and coauthors, Changes in diet, weight, and serum lipid levels associated with olestra consumption, *Archives of Internal Medicine* 160 (2000): 2600–2604.

6. Position of The American Dietetic Association: Fat replacements, *Journal of the American Dietetic Association* 98 (1998): 463–468.

Controversy 5

1. A. Trichopoulou and coauthors, Cancer and Mediterranean dietary traditions, *Cancer Epidemiology, Biomarkers, and Prevention* 9 (2000): 869–873.

2. C. Lasheras, S. Fernandez, and A. M. Patterson, Mediterranean diet and age with respect to overall survival in institutionalized, nonsmoking elderly people, *American Journal of Clinical Nutrition* 71 (2000): 987–992; A. Trichopoulou and E. Vasilopoulou, Mediterranean diet and longevity, *British Journal of Nutrition* 84 (2000): 205–209.

3. M. de Lorgeril and coauthors, Mediterranean dietary pattern in a randomized trial: Prolonged survival and possible reduced cancer rate, *Archives of Internal Medicine* 158 (1998): 1181–1187. For a debate on issues surrounding Mediterranean diets, see articles by P. Crothy and K. D. Gifford in *Nutrition Today* 33 (1998).

4. A. Trichopoulou and P. Lagiou, Healthy traditional Mediterranean diet: An expression of culture, history, and lifestyle, *Nutrition Reviews* 55 (1997): 383–389.

5. M. de Lorgeril and coauthors, Mediterranean diet, traditional risk factors, and the rate of cardiovascular complications after myocardial

F

infarction: Final report of the Lyon Diet Heart Study, *Circulation*, 99 (1999): 779–785; B. Haber, The Mediterranean diet: A view from history, *American Journal of Clinical Nutrition* 66 (1997): S1053–S1057.

6. J. Hurley and B. Liebman, Greek food: A Mediterranean mixed bag, *Nutrition Action Health Letter*, November 2000, pp. 1–6.

7. A. Drewnowski and B. M. Popkin, The nutrition transition: New trends in the global diet, *Nutrition Reviews* 55 (1997): 31–43.

8. L. Serra-Majem and coauthors, Nutrition policies in Mediterranean Europe, *Nutrition Reviews* 55 (1997): S42–S57.

9. A. Trichopoulou, E. Vasilopoulou, and A. Lagiou, Mediterranean diet and coronary heart disease: Are antioxidants critical? *Nutrition Reviews* 57 (1999): 253–255.

10. A. Naska and coauthors, Fruit and vegetable availability among ten European countries: How does it compare with the "five-a-day" recommendation? DAFNE I and II projects of the European Commission, *British Journal of Nutrition* 84 (2000): 549–556.

11. E. Giovannucci and coauthors, A prospective study of tomato products, lyocpene, and prostate cancer risk, *Journal of the National Cancer Institute* 94 (2002): 391–398; M. Peluso and coauthors, White blood cell DNA adducts and fruit and vegetable consumption in bladder cancer, *Carcinogenesis* 21 (2000): 183–187; Trichopoulou, Vasilopoulou, and Lagiou, 1999.

12. R. W. Owen and coauthors, The antioxidant/anticancer potential of phenolic compounds isolated from olive oil, *European Journal of Cancer* 36 (2000): 1235–1247; J. P. De La Cruz and coauthors, Antithrombotic potential of olive oil administration in rabbits with elevated cholesterol, *Thrombosis Research* 100 (2000): 305–315; L. F. Larsen, J. Jespersen, and P. Marckmann, Are olive oil diets antithrombotic? Diets enriched with olive, rapeseed, or sunflower oil affect postprandial factor VII differently, *American Journal of Clinical Nutrition* 70 (1999): 976–982.

13. F. Visioli, A. Poli, and C. Gall, Antioxidant and other biological activities of phenols from olives and olive oil, *Medicinal Research Reviews* 22 (2002): 65–75.

14. L. Van Horn and N. Ernst, A summary of the science supporting the new National Cholesterol Education program dietary recommendations: What dietitians should know, *Journal of the American Dietetic Association* 101 (2001): 1148–1154.

15. C. von Schacky, n-3 fatty acids and the prevention of coronary atherosclerosis, *American Journal of Clinical Nutrition* 71 (2000): S224–S227.

16. J. H. M. de Vries and coauthors, Red wine is a poor source of bioavailable flavonols in men, *Journal of Nutrition* 131 (2001): 745–748.

17. J. Ruidavets and coauthors, Catechin in the Mediterranean diet: Vegetable, fruit or wine? *Atherosclerosis* 153 (2000): 107–117; A. Tjønneland and coauthors, Wine intake and diet in a random sample of 48,763 Danish men and women, *American Journal of Clinical Nutrition* 69 (1999): 49–54.

18. M. de Lorgeril and P. Salen, Diet as preventive medicine in cardiology, *Current Opinions in Cardiology* 15 (2000): 364–370; Trichopoulou and coauthors, 2000.

19. F. B. Hu, Dietary pattern analysis: a new direction in nutritional epidemiology, *Current Opinion in Lipidology* 13 (2002): 3–9.

Chapter 6

1. C. Wu, Understanding how proteins fold, *Science News* 152 (1997): 270; E. Strauss, How proteins take shape: Guardians give a new twist to protein folding, *Science News* 152 (1997): 155.

2. A. Tomer and coauthors, Thrombogenesis in sickle cell disease, *Journal of Laboratory and Clinical Medicine* 137 (2001): 398–407; S. T. Miller and coauthors, Prediction of adverse outcomes in children with sickle cell disease, *New England Journal of Medicine* 342 (2000): 83–89.

3. B. R. Stevens, Digestion and absorption of protein, in M. H. Stipanuk, ed., *Biochemical and Physiological Aspect of Human Nutrition* (Philadelphia: Saunders, 2000), pp. 121–122.

4. M. C. Crim and H. N. Munro, Proteins and amino acids, in M. E. Shils, J. A. Olson, and M. Shike, eds., *Modern Nutrition in Health and Disease* (Philadelphia: Lea & Febiger, 1994), p. 31.

5. Position of The American Dietetic Association: Vegetarian diets, *Journal of the American Dietetic Association* 97 (1997): 1317–1321; V. R. Young and P. L. Pellett, Plant proteins in relation to human protein and amino acid nutrition, *American Journal of Clinical Nutrition* 59 (1994): S1203–S1212.

6. *Protein Quality Evaluation: Report of a Joint FAO/WHO Expert Consultation*, Food and Nutrition paper no. 51 (Rome, Italy: FAO and WHO, 1990).

7. A thorough review of methodology in whole-body protein turnover is provided by J. C. Waterlow, Whole-body protein turnover in humans—Past, present, and future, *Annual Review of Nutrition* 15 (1995): 57–92.

8. S. M. Smith and coauthors, Nutrition in space, *Nutrition Today* 32 (1997): 6–12.

9. A. Miyamoto and coauthors, Medical baseline data collection on bone and muscle change with space flight, *Bone* 22 (1998): S79–S82.

10. Position of The American Dietetic Association, Dietitians of Canada, and the American College of Sports Medicine: Nutrition and athletic performance, *Journal of the American Dietetic Association* 100 (2000); 1543–1556.

11. D. G. Schroeder and R. Martorell, Enhancing child survival by preventing malnutrition, *American Journal of Clinical Nutrition* 65 (1997): 1080–1081.

12. A full discussion of PEM appears in B. Torun and F. Chew, Protein-energy malnutrition, in M. E. Shils, J. A. Olson, and M. Shike, eds., *Modern Nutrition in Health and Disease* (Philadelphia: Lea & Febiger, 1999), pp. 963–988.

13. B. Woodward, Protein, calories, and immune defenses, *Nutrition Reviews* 56 (1998): S84–S92.

14. J. C. Waterlow, Childhood malnutrition in developing nations: Looking back and forward, *Annual Review of Nutrition* 14 (1994): 1–19.

15. L. Combaret, D. Taillandier, and D. Attaix, Nutritional and hormonal control of protein breakdown, *American Journal of Kidney Diseases* 37 (2001): S108–S111.

16. S. M. Day and D. H. DeHeer, Reversal of the detrimental effects of chronic protein malnutrition on long bone fracture healing, *Journal of Orthopaedic Trauma* 15 (2001): 47–53.

17. Woodward, 1998; Protein deficiency abets tuberculosis, *Science News* 150 (1996): 374.

18. V. Scherbaum and P. Furst, New concepts on nutritional management of severe malnutrition: The role of protein, *Current Opinion in Clinical Nutrition and Metabolic Care* 3 (2000): 31–38.

19. S. M. Grantham-McGregor, S. P. Walker, and S. Chang, Nutritional deficiencies and later behavioural development, *Proceedings of the Nutrition Society* 59 (2000): 47–54.

20. T. Liu and coauthors, Kwashiorkor in the United States: Fad diets, perceived and true milk allergy, and nutritional ignorance, *Archives of Dermatology* 137 (2001): 630–636; N. F. Carvalho and coauthors, Severe nutritional deficiencies in toddlers resulting from health food milk alternatives (electronic article), *Pediatrics* 107 (2001): e46.

21. Nutritional status of poor children in the United States, *Journal of the American Dietetic Association* 95 (1995): 248–250.

22. T. Liu and coauthors, 2001; Carvalho and coauthors, 2001.

23. D. H. Sullivan, S. Sun, and R. C. Walls, Protein-energy undernutrition among elderly hospitalized patients: A prospective study, *Journal of the American Medical Association* 281 (1999): 2013–2019; Distinguishing malnutrition from the effects of stress and disease, *Nutrition and the M.D.*, April 1998, p. 6.

24. Federal Interagency Forum on Child and Family Statistics, *America's Children: Key National Indicators of Well-Being*, 2001 available at www.childstats.gov.

25. B. A. Underwood and S. Smitasiri, Micronutrient malnutrition: Policies and programs for control and their implications, *Annual Review of Nutrition* 19 (1999): 303–324; C. G. Victora and coauthors, Potential interventions for the prevention of childhood pneumonia in developing countries: Improving nutrition, *American Journal of Clinical Nutrition* 70 (1999): 309–320.

26. M. Daenzer and coauthors, Prenatal high protein exposure decreases energy expenditure and increases adiposity in young rats, *Journal of Nutrition* 132 (2002): 142–144.

27. E. Brändle, H. G. Sieberth, and R. E. Hautmann, Effect of chronic dietary protein intake on the renal function in healthy subjects, *European Journal of Clinical Nutrition* 50 (1996): 734–740.

28. L. Borghi and coauthors, Comparison of two diets for the prevention of recurrent stones in idopathic hypercalciuria, *New England Journal of Medicine*

346 (2002): 77–84; M. T. Pedrini and coauthors, The effects of dietary protein restriction on the progression of diabetic and nondiabetic renal diseases: A meta-analysis, *Annals of Internal Medicine* 124 (1996): 627–632.
29. Two points of view concerning protein and bone loss are found in U. S. Barzel and L. K. Massey, Excess dietary protein can adversely affect bone, *Journal of Nutrition* 128 (1998): 1051–1053; and R. P. Heaney, Excess dietary protein may not adversely affect bone, *Journal of Nutrition* 128 (1998): 1054–1057.
30. L. K. Massey, Does excess dietary protein adversely affect bone? Symposium overview, *Journal of Nutrition* 128 (1998): 1048–1050.
31. B. J. Abelow, T. R. Holford, and K. L. Insogna, Cross-cultural associations between dietary animal protein and hip fracture: A hypothesis, *Calcified Tissue International* 50 (1996): 14–18. D. Feskanich and coauthors, Protein consumption and bone fractures in women, *American Journal of Epidemiology* 143 (1996): 472–479.
32. R. G. Munger, J. R. Cerhan, and B. C-H. Chiu, Prospective study of dietary protein intake and risk of hip fracture in postmenopausal women, *American Journal of Clinical Nutrition* 69 (1999): 147–152; J. P. Bonjour, M. A. Schurch, and R. Rizzoli, Nutritional aspects of hip fractures, *Bone* 18 (1996): S139–S144.

Consumer Corner 6

1. N. W. Flodin, The metabolic roles, pharmacology and toxicology of lysine, *Journal of the American College of Nutrition* 16 (1997): 7–21.
2. Food and Drug Administration, Impurities confirmed in dietary supplement 5-hydroxy-L-tryptophan, FDA Talk Paper, August 1998, available from Food and Drug Administration, U.S. Department of Health and Human Services, Public Health Service, 5600 Fishers Lane, Rockville, MD 20857.
3. P. Garlick, Assessment of the safety of glutamine and other amino acids, *Journal of Nutrition* 131 (2001): S2556–S2561.

Controversy 6

1. L. B. Szabo, The health risks of new-wave vegetarianism, *Canadian Medical Association Journal* 156 (1997): 1454–1455.
2. U.S. Department of Agriculture, Dietary Guidelines Advisory Committee, *Nutrition and Your Health: Dietary Guidelines for Americans, 5th ed.*, 2000, Home and Garden Bulletin no. 232, available online at www.usda.gov/cnpp or call (888) 878-3256.
3. J. Raloff, Soya-nara, heart disease: The United States' top-selling legume gains heartfelt respect, *Science News* 153 (1998): 348–349; Position of The American Dietetic Association: Vegetarian diets, *Journal of the American Dietetic Association* 97 (1997): 1317–1321; P. Walter, Effects of vegetarian diets on aging and longevity, *Nutrition Reviews* 55 (1997): S61–S68.
4. P. N. Appleby and coauthors, Low body mass index in non-meat eaters: The possible roles of animal fat, dietary fiber and alcohol, *International Journal of Obesity and Related Metabolic Disorders* 22 (1998): 454–460; T. Key, Prevalence of obesity is low in people who do not eat meat, *British Medical Journal* 313 (1996): 816–817.
5. A. Golay and E. Bobbioni, The role of dietary fat in obesity, *International Journal of Obesity and Related Metabolic Disorders* 21 (1997): S2–S11.
6. R. R. Wolfe, Metabolic interactions between glucose and fatty acids in humans, *American Journal of Clinical Nutrition* 67 (1998): S519–S526.
7. J. I. Mann and coauthors, Dietary determinants of ischaemic heart disease in health conscious individuals, *Heart* 78 (1997): 450–455; M. Krajcovicova-Kudlackova and coauthors, Plasma fatty acid profile and alternative nutrition, *Annals of Nutrition and Metabolism* 41 (1997): 365–370.
8. J. W. Anderson, B. M. Smith, and C. S. Washnock, Cardiovascular and renal benefits of dry bean and soybean intake, *American Journal of Clinical Nutrition* 70 (1999): S464–S467.
9. D. J. Jenkins and coauthors, High-protein diets in hyperlipidemia: Effect of wheat gluten on serum lipids, uric acid, and renal function, *American Journal of Clinical Nutrition* 74 (2001): 57–63.
10. C. D. Gardner and coauthors, The effect of soy protein with or without isoflavones relative to milk protein on plasma lipids in hypercholesterolemic postmenopausal women, *American Journal of Clinical Nutrition* 73 (2001): 728–735; M. Pfeuffer and J. Schrezenmeir, Bioactive substances in milk with properties decreasing risk of cardiovascular diseases, *British Journal of Nutrition* 84 (2000): S155–S159.
11. T. J. A. Key and coauthors, Dietary habits and mortality in 11,000 vegetarians and health conscious people: Results of a 17 year follow up, *British Medical Journal* 313 (1996): 775–779.
12. J. Sabaté, Nut consumption, vegetarian diets, ischemic heart disease risk, and all-cause mortality: Evidence from epidemiologic studies, *American Journal of Clinical Nutrition* 71 (2000): 1077–1084; Raloff, 1998.
13. S. R. Teixeira and coauthors, Effects of feeding 4 levels of soy protein for 3 and 6 wk on blood lipids and apolipoproteins in moderately hyercholesterolemic men, *American Journal of Clinical Nutrition* 71 (2000): 1077–1084; Raloff, 1998.
14. S. K. Harman and W. R. Parnell, The nutritional health of New Zealand vegetarian and nonvegetarian Seventh-Day Adventists: Selected vitamin, mineral, and lipid levels, *New Zealand Medical Journal* 27 (1998): 91–94; D. C. Knight and J. A. Eden, A review of the clinical effects of phytoestrogens, *Obstetrics and Gynecology* 87 (1996): 897–904.
15. Harman and Parnell, 1998.
16. G. E. Fraser, Associations between diet and cancer, ischemic heart disease, and all-cause mortality in non-Hispanic white California Seventh-Day Adventists, *American Journal of Clinical Nutrition* 70 (1999): S532–S538.
17. B.C.-H. Chiu and coauthors, Diet and risk of non-Hodgkins lymphoma in older women, *Journal of the American Medical Association* 275 (1996): 1315–1321.
18. J. G. Holmen and coauthors, Long-term effects of a change from a mixed diet to a lacto-vegetarian diet on human urinary and fecal mutagenic activity, *Mutagenesis* 13 (1998): 167–171; J. G. Holmen and coauthors, Dietary influence on some proposed risk factors for colon cancer: Fecal and urinary mutagenic activity and the activity of some intestinal bacterial enzymes, *Cancer Detection and Prevention* 21 (1997): 258–266.
19. B. C. Pence and coauthors, Feeding of a well-cooked beef diet containing a high heterocyclic amine content enhances colon and stomach carcinogenesis in 1,2-dimethylhydrazine-treated rats, *Nutrition and Cancer* 30 (1998): 220–226.
20. P. K. Johnston, Nutritional implications of vegetarian diets, in M. E. Shils and coeditors, *Modern Nutrition in Health and Disease* (Baltimore: Williams & Wilkins, 1999), p. 1755.
21. A. Hackett, I. Nathan, and L. Burgess, Is a vegetarian diet adequate for children? *Nutrition and Health* 12 (1998): 189–195.
22. T. A. B. Sanders and S. Reddy, Vegetarian diets and children, *American Journal of Clinical Nutrition* 59 (1994): S1176–S1181.
23. T. Liu, coauthors, Kwashiorkor in the United States: Fad diets, perceived and true milk allergy, and nutritional ignorance, *Archives of Dermatology* 137 (2001): 630–636; N. F. Carvalho and coauthors, Severe nutritional deficiencies in toddlers resulting from health food milk alternatives (electronic article), *Pediatrics* 107 (2001): e46. Hackett, Nathan, and Burgess, 1998.
24. I. Nathan, A. F. Hackett, and S. Kirby, A longitudinal study of the growth of matched pairs of vegetarian and omnivorous children, aged 7–11 years, in the north-west of England, *European Journal of Clinical Nutrition* 51 (1997): 20–25.
25. W. Herrmann and coauthors, Total homocysteine, vitamin B(12), and total antioxidant status of vegetarians, *Clinical Chemistry* 47 (2001): 1094–1101.
26. P. J. Grattan-Smith and coauthors, The neurological syndrome of infantile cobalamin deficiency: Developmental regression and involuntary movements, *Movement Disorders* 12 (1997): 39–46.
27. K. Lovblad and coauthors, Retardation of myelination due to dietary vitamin B_{12} deficiency: Cranial MRI findings, *Pediatric Radiology* 27 (1997): 155–158.
28. U. von Schenck, C. Bender-Gotze, and B. Kolezko, Persistence of neurological damage induced by dietary vitamin B-12 deficiency in infancy, *Archives of Diseases in Childhood* 77 (1997): 137–139.
29. G. Massa and coauthors, Protein malnutrition due to replacement of milk by rice drink, *European Journal of Pediatrics* 160 (2001): 382–384.
30. Liv, 2001; Cavallo, 2001.
31. V. Messina and A. R. Mangels, Considerations in planning vegan diets: Children, *Journal of the American Dietetic Association* 101 (2001): 661–669.
32. T. J. Parsons and coauthors, Reduced bone mass in Dutch adolescents fed a macrobiotic diet in early life, *Journal of Bone Mineral Research* 12 (1997): 1486–1494.
33. T. A. B. Sanders, Essential fatty acid requirements of vegetarians in

pregnancy, lactation, and infancy, *American Journal of Clinical Nutrition* 70 (1999): S555–S559.

34. A. R. Mangels and V. Messina, Considerations in planning vegan diets: Infants, *Journal of the American Dietetic Association* 101 (2001): 670–677.

35. J. R. Hunt and Z. K. Roughead, Nonheme-iron absorption, fecal ferritin excretion, and blood indexes of iron status in women consuming controlled lactoovovegetarian diet for 8 wk, *American Journal of Clinical Nutrition* 69 (1999): 944–952.

36. J. A. Conquer and B. J. Holub, Supplementation with an algae source of docosahexaenoic acid increases (n-3) fatty acid status and alters selected risk factors for heart disease in vegetarian subjects, *Journal of Nutrition* 126 (1996): 3032–3039.

Chapter 7

1. B. N. Ames, Micronutrient deficiencies: A major cause of DNA damage, *Annals of the New York Academy of Sciences* 889 (1999): 87–106.

2. C. D. Berdanier, Vitamin A, in *Advanced Nutrition: Micronutrients* (Boca Raton, Fla.: CRC Press, 1998), pp. 22–37.

3. S. Kato, Molecular mechanism of transcriptional control by nuclear vitamin receptors, *British Journal of Nutrition* 84 (2000): S229–S233; S. Nagpal and R. A. Chandraratna, Vitamin A and regulation of gene expression, *Current Opinion in Clinical Nutrition and Metabolic Care* 1 (1998): 341–346.

4. M. H. Zile, Vitamin A and embryonic development: An overview, *Journal of Nutrition* 128 (1998): S455–S458; J. A. Olson, Vitamin A, in E. E. Ziegler and L. J. Filer, Jr., eds., *Present Knowledge in Nutrition*, 7th ed. (Washington, D.C.: International Life Sciences Institute Press, 1996), pp. 109–119.

5. P. Christian and coauthors, Working after the sun goes down: Exploring how night blindness impairs women's work activities in rural Nepal, *European Journal of Clinical Nutrition* 52 (1998): 519–524.

6. A. Sommer, Xerophthalmia and vitamin A status, *Progress in Retinal and Eye Research* 17 (1998): 9–31.

7. R. D. Semba, The role of vitamin A and related retinoids in immune function, *Nutrition Reviews* 56 (1998): S38–S48.

8. J. Sakar, Vitamin A is required for regulation of polymeric immunoglobulin receptor (pIgR) expression by interleukin-4 and interferon-gamma in a human intestinal epithelial cell line, *Journal of Nutrition* 128 (1998): 1063–1069; M. Cippitelli and coauthors, Retinoic acid–induced transcriptional modulation of the human interferon-gamma promoter, *Journal of Biological Chemistry* 43 (1996): 26783–26793.

9. W. W. Fawzi and coauthors, Dietary vitamin A intake in relation to child growth, *Epidemiology* 8 (1997): 402–407.

10. W. W. Fawzi and coauthors, The effect of vitamin A supplementation on the growth of preschool children in the Sudan, *American Journal of Public Health* 87 (1997): 1359–1362.

11. C. Ballew and coauthors, Serum retinol distributions in residents of the United States: Third National Health and Nutrition Examination Survey, 1988–1994, *American Journal of Clinical Nutrition* 73 (2001): 586–593.

12. Standing Committee on the Scientific Evaluation of Dietary Reference Intakes, Food and Nutrition Board, Institute of Medicine, *Dietary Reference Intakes for Vitamin A, Vitamin K, Arsenic, Boron, Chromium, Copper, Iodine, Iron, Manganese, Molybdenum, Nickel, Silicon, Vanadium, and Zinc* (Washington, D.C.: National Academy Press, 2001), pp. 4-9–4-10.

13. M. A. Dijkhuizen and coauthors, Concurrent micronutrient deficiencies in lactating mothers and their infants in Indonesia, *American Journal of Clinical Nutrition* 73 (2001): 786–791.

14. D. Feskanich and coauthors, Vitamin A intake and hip fractures among postmenopausal women, *Journal of the American Medical Association* 287 (2002): 47–54.

15. J. W. Coates and coauthors, Gastric ulceration and suspected vitamin A toxicosis in grower pigs fed fish silage, *Canadian Veterinary Journal* 39 (1998): 167–170.

16. E. R. Greenberg and coauthors, Mortality associated with low plasma concentration of beta carotene and effect of oral supplementation, *Journal of the American Medical Association* 275 (1996): 699–703.

17. Kato, 2000; I. Nemere and M. C. Farach-Carson, Membrane receptors for steroid hormones: A case for specific cell surface binding sites for vitamin D metabolites and estrogens, *Biochemical and Biophysical Research Communications* 248 (1998): 443–449; C. Carlberg and P. Polly, Gene regulation by vitamin D3, *Critical Reviews in Eukaryotic Gene Expression* 8 (1998): 19–42.

18. S. R. Kreiter and coauthors, Nutritional rickets in African American breast-fed infants, *Journal of Pediatrics* 137 (2000): 143–145.

19. Standing Committee on the Scientific Evaluation of Dietary Reference Intakes, Food and Nutrition Board, Institute of Medicine, *Dietary Reference Intakes for Calcium, Phosphorus, Magnesium, Vitamin D, and Fluoride* (Washington, D.C.: National Academy Press, 1997), pp. 279–280.

20. J. B. Randlov and coauthors, Acute cardiovascular effect of 1,25-dihydroxycholecalciferol in essential hypertension, *American Journal of Hypertension* 11 (1998): 659–666; N. Niederhoffer and coauthors, Calcification of medial elastic fibers and aortic elasticity, *Hypertension* 29 (1997): 999–1006.

21. J. L. Giunta, Dental changes in hypervitaminosis D, *Oral Surgery, Oral Medicine, Oral Pathology, Oral Radiology and Endodontics* 85 (1998): 410–413.

22. S. S. Harris and B. Dawson-Hughes, Seasonal changes in plasma 25-hydroxyvitamin D concentrations of young American black and white women, *American Journal of Clinical Nutrition* 67 (1998): 1232–1236.

23. S. N. Meydani and A. A. Beharka, Recent developments in vitamin E and immune response, *Nutrition Reviews* 56 (1998): S49–S58.

24. M. J. Fryer, The possible role of nitric oxide and impaired mitochondrial function in ataxia due to severe vitamin E deficiency, *Medical Hypotheses* 50 (1998): 353–354.

25. M. A. Beck, Nutritionally induced oxidative stress: Effect on vital disease, *American Journal of Clinical Nutrition* 71 (2000): S1676–S1679.

26. S. N. Meydani and coauthors, Assessment of the safety of supplementation with different amounts of vitamin E in healthy older adults, *American Journal of Clinical Nutrition* 68 (1998): 311–318.

27. C. T. Taylor and coauthors, Vitamin K to reverse excessive anticoagulation: A review of the literature, *Phamacotherapy* 19 (1999): 1415–1425.

28. G. Ferland, The vitamin K–dependent proteins: An update, *Nutrition Reviews* 56 (1998): 223–230.

29. D. Feskanich and coauthors, Vitamin K and hip fractures in women: A prospective study, *American Journal of Clinical Nutrition* 69 (1999): 77–79.

30. J. B. Hack and R. S. Hoffman, Thiamine before glucose to prevent Wernicke encephalopathy: Examining the conventional wisdom (letter), *Journal of the American Medical Association* 279 (1998): 583–584.

31. C. M. Chesney and coauthors, Effect of niacin, warfarin, and antioxidant therapy on coagulation parameters in patients with peripheral arterial disease in the Arterial Disease Multiple Intervention Trial (ADMIT), *American Heart Journal* 140 (2000): 631–636; M. B. Elam and coauthors, Effect of niacin on lipid and lipoprotein levels and glycemic control in patients with diabetes and peripheral arterial disease: The ADMIT study: A randomized trial. Arterial Disease Multiple Intervention Trial, *Journal of the American Medical Association* 284 (2000): 1263–1270.

32. D. Callanan, B. A. Blodi, and D. F. Martin, Macular edema associated with nicotinic acid, *Journal of the American Medical Association* 279 (1998): 1702.

33. P. W. Jungnickel and coauthors, Effect of two aspirin pretreatment regimens on niacin-induced cutaneous reactions, *Journal of General Internal Medicine* 12 (1997): 591–596.

34. T. A. Jacobson, Combination lipid-altering therapy: an emerging treatment paradigm for the 21st century, *Current Atherosclerosis Reports* 3 (2001): 373–382.

35. Standing Committee on the Scientific Evaluation of Dietary Reference Intakes, Food and Nutrition Board, Institute of Medicine, *Dietary Reference Intakes for Thiamin, Riboflavin, Niacin, Vitamin B_6, Folate, Vitamin B_{12}, Pantothenic Acid, Biotin, and Choline* (Washington, D.C.: National Academy Press, 1998), p. 8-1.

36. A. Fleming, The role of folate in the prevention of neural tube defects: Human and animal studies, *Nutrition Reviews* 59 (2001): S13–S23.

37. R. J. Hine, What practitioners need to know about folic acid, *Journal of the American Dietetic Association* 96 (1996): 451–452. A full discussion of folate and neural tube defects is found in C. E. Butterworth and A. Bendich, Folic acid and the prevention of birth defects, *Annual Review of Nutrition* 16 (1996): 73–97.

38. M. A. Honein and coauthors, Impact of folic acid fortification of the U.S. food supply on the occurrence of neural tube defects, *Journal of the American Medical Association* 285 (2001): 2981–2986.

39. I. A. Brouwer and coauthors, Low-dose folic acid supplementation

decreases plasma homocysteine concentrations: A randomized trial, *American Journal of Clinical Nutrition* 69 (1999): 99–104; T. K. A. B. Eskes, Open or closed? A world of difference: A history of homocysteine research, *Nutrition Reviews* 56 (1998): 236–244.

40. L. B. Bailey, Dietary Reference Intakes for folate: The debut of Dietary Folate Equivalents, *Nutrition Reviews* 56 (1998): 294–299.

41. S. P. Rothenberg, Increasing the dietary intake of folate: Pros and cons, *Seminars in Hematology* 36 (1999): 65–74.

42. Standing Committee on the Scientific Evaluation of Dietary Reference Intakes, 1998, p. 331.

43. D. S. Inagaki and coauthors, Effects of vitamin B_6 deficiency on cytokine levels and lymphocytes in mice, *Bioscience, Biotechnology and Biochemistry* 62 (1998): 1008–1010.

44. J. E. Leklem, Vitamin B-6, in E. E. Ziegler and L. J. Filer, Jr., eds., *Present Knowledge in Nutrition*, 7th ed. (Washington, D.C.: International Life Sciences Institute Press, 1996), pp. 174–183.

45. E. B. Rimm and coauthors, Folate and vitamin B_6 from diet and supplements in relation to risk of coronary heart disease among women, *Journal of the American Medical Association* 279 (1998): 359–364.

46. Y. C. Huang and coauthors, Vitamin B-6 requirement and status assessment of young women fed a high-protein diet with various levels of vitamin B-6, *American Journal of Clinical Nutrition* 67 (1998): 208–220; C. M. Hansen, J. E. Leklem, and L. T. Miller, Changes in vitamin B-6 status indicators of women fed a constant protein diet with varying levels of vitamin B-6, *American Journal of Clinical Nutrition* 66 (1997): 1379–1387.

47. Standing Committee on the Scientific Evaluation of Dietary Reference Intakes, 1998, pp. 7-1–7-27.

48. J. W. Miller, Does lowering plasma homocysteine reduce vascular disease risk? *Nutrition Reviews* 59 (2001): 242–244; P. F. Jacques and coauthors, Determinants of plasma total homocysteine concentration in the Framingham Offspring cohort, *American Journal of Clinical Nutrition* 73 (2001): 613–621.

49. C. J. Schorah and coauthors, The responsiveness of plasma homocysteine to small increases in dietary folic acid: A primary care study, *European Journal of Clinical Nutrition* 52 (1998): 407–411; Rimm and coauthors, 1998; A. R. Folsom and coauthors, Prospective study of coronary heart disease incidence in relation to fasting total homocysteine, related genetic polymorphisms, and B vitamins: The Atherosclerosis Risk in Communities (ARIC) study, *Circulation* 98 (1998): 204–210; J. B. Ubbink, P. J. Becker, and W. J. H. Vermaak, Will an increased dietary folate intake reduce the incidence of cardiovascular disease? *Nutrition Reviews* 54 (1996): 213–216.

50. J. V. Woodside and coauthors, Effect of B-group vitamins and antioxidant vitamins on hyperhomocysteinemia: A double-blind, randomized, factorial design, controlled trial, *American Journal of Clinical Nutrition* 67 (1998): 858–866.

51. N. M. van der Put and coauthors, Folate, homocysteine and neural tube defects: An overview, *Experimental Biology and Medicine* 226 (2001): 243–270; P. F. Jacques and coauthors, The effect of folic acid fortification on plasma folate and total homocysteine concentrations, *New England Journal of Medicine* 340 (1999): 1449–1454.

52. Woodside and coauthors, 1998.

53. J. Selhum and A. D'Angelo, Relationship between homocysteine and thrombotic disease, *American Journal of the Medical Sciences* 316 (1998): 129–141.

54. S. Mendiratta, Z. C. Qu, and J. M. May, Erythrocyte ascorbate recycling: Antioxidant effects in blood, *Free Radical Biology and Medicine* 24 (1998): 789–797; J. M. May, Ascorbate function and metabolism in the human erythrocyte, *Frontiers in Bioscience* 3 (1998): D1–D10.

55. K. Nyyssonen and coauthors, Effect of supplementation of smoking men with plain or slow release ascorbic acid on lipoprotein oxidation, *European Journal of Clinical Nutrition* 51 (1997): 154–163.

56. M. Levine and coauthors, Criteria and recommendations for vitamin C intake, *Journal of the American Medical Association* 281 (1999): 1415–1423.

57. M. C. Nesheim, Regulation of dietary supplements, *Nutrition Today* 33 (1998): 62–68.

Consumer Corner 7

1. S. B. Mossad, Treatment of the common cold, *British Medical Journal* 317 (1998): 33–36.

2. R. M. Douglas, E. B. Chalker, and B. Tracey, Vitamin C for preventing and treating the common cold (Cochrane Review), *The Cochrane Library* 2 (2001): available at www.update-software.com/abstracts/ab000980.htm; H. Hemiliä, Does vitamin C alleviate the symptoms of the common cold? A review of the current evidence, *Scandinavian Journal of Infectious Diseases* 26 (1994): 1–6; L. Pauling, *Vitamin C and the Common Cold* (San Francisco: Freeman, 1970); T. C. Chalmers, Effects of ascorbic acid on the common cold: An evaluation of the evidence, *American Journal of Medicine* 58 (1975): 532–536, as cited by H. Hemiliä and Z. S. Herman, Vitamin C and the common cold: A retrospective analysis of Chalmers' review, *Journal of the American College of Nutrition* 14 (1995): 116–123.

3. Hemiliä and Herman, 1995.

4. H. Hemiliä, Vitamin C supplementation and common cold symptoms: Factors affecting the magnitude of the benefit, *Medical Hypotheses* 52 (1999): 171–178.

5. M. Del Rio and coauthors, Improvement by several antioxidants of macrophage function in vito, *Life Sciences* 63 (1998): 871–881.

6. B. Takkouche and coauthors, Intake of vitamin C and zinc and risk of common cold: a cohort study, *Epidemiology* 13 (2002): 38–44; H. H. Kaprio and coauthors, Vitamin C, vitamin E, and beta-carotene in relation to common cold incidence in male smokers, *Epidemiology* 13 (2002): 32–37.

7. M. Levine and coauthors, Criteria and recommendations for vitamin C intake, *Journal of the American Medical Association* 281 (1999): 1415–1423.

Controversy 7

1. B. Halliwell, Antioxidants in human health and disease, *Annual Review of Nutrition* 16 (1996): 33–50.

2. Standing Committee on the Scientific Evaluation of Dietary Reference Intakes, Food and Nutrition Boards, Institute of Medicine, *Dietary Reference Intakes for Vitamin C, Vitamin E, Selenium, and Carotenoids* (Washington, D.C.: National Academy Press, 2000), pp. 45–46.

3. C. L. Rock, R. A. Jacob, and P. E. Bowen, Update on the biological characteristics of the antioxidant micronutrient: Vitamin C, vitamin E, and the carotenoids, *Journal of the American Dietetic Association* 96 (1996): 693–702.

4. H. Hu and coauthors, Antioxidants may contribute in fight against ageing: An in vitro model, *Mechanisms of Ageing and Development* 121 (2001): 217–230.

5. G. Cao and coauthors, Increases in human plasma antioxidant capacity after consumption of controlled diets high in fruit and vegetables, *American Journal of Clinical Nutrition* 68 (1998): 1081–1087.

6. J. L. Groff and S. S. Gropper, *Advanced Nutrition and Human Metabolism* (Belmont, Calif.: Wadsworth Thompson Learning, 2000), pp. 366–367.

7. M. P. Longnecker and coauthors, Intake of carrots, spinach, and supplements containing vitamin A in relation to breast cancer, *Cancer, Epidemiology, Biomarkers and Prevention* 6 (1997): 887–892; World Cancer Research Fund/American Institute for Cancer Research, *Food, Nutrition and the Prevention of Cancer: A Global Perspective* (American Institute for Cancer Research, 1997), pp. 436–446.

8. E. R. Berton and coauthors, A population-based case-control study of carotenoid and vitamin A intake and ovarian cancer (United States), *Cancer Causes and Control* 12 (2001): 83–90; Standing Committee on the Scientific Evaluation of Dietary Reference Intakes, 2000, p. 346; Y. M. Peng and coauthors, Concentrations of carotenoids, tocopherols, and retinol in paired plasma and cervical tissue of patients with cervical cancer, precancer, and noncancerous diseases, *Cancer, Epidemiology, Biomarkers and Prevention* 7 (1998): 347–350; Y. Kumagai and coauthors, Serum antioxidant vitamins and risk of lung and stomach cancers in Shenyang, China, *Cancer Letters* 129 (1998): 145–149; A. R. Giuliano and coauthors, Antioxidant nutrients: Associations with persistent human papillomavirus infection, *Cancer, Epidemiology, Biomarkers and Prevention* 6 (1997): 917–923.

9. S. T. Mayne, Beta-carotene, carotenoids, and disease prevention in humans, *FASEB Journal* 10 (1996): 690–701; E. R. Greenberg and coauthors, A clinical trial of antioxidant vitamins to prevent colorectal adenoma, *New England Journal of Medicine* 33 (1994): 141–147.

10. K. Smigel, Beta-carotene fails to prevent cancer in two major studies; CARET intervention stopped, *Journal of the National Cancer Institute* 88 (1996): 145; G. S. Omenn and coauthors, Effects of a combination of beta-carotene and vitamin A on lung cancer and cardiovascular disease, *New England Journal of Medicine* 334 (1996): 1150–1155.

F

11. I. Min Lee and coauthors, ß-carotene supplementation and incidence of cancer and cardiovascular disease: The Women's Health Study, *Journal of the National Cancer Institute* 91 (1999): 2102–2106.

12. Smigel, 1996; Omenn and coauthors, 1996; O. P. Heinonen, J. K. Huttunen, and D. Albanes (and other participants in the alpha-tocopherol, beta-carotene cancer prevention study group), The effect of vitamin E and beta carotene on the incidence of lung cancer and other cancers in male smokers, *New England Journal of Medicine* 330 (1994): 1029–1035.

13. Standing Committee on the Scientific Evaluation of Dietary Reference Intakes, 2000, pp. 342–343.

14. O. P. Heinonen and coauthors, Prostate cancer and supplementation with alpha-tocopherol and beta-carotene: Incidence and mortality in a controlled trial, *Journal of the National Cancer Institute* 90 (1998): 440–446.

15. J. C. Fleet, Dietary selenium repletion may reduce cancer incidence in people at high risk who live in areas with low soil selenium, *Nutrition Reviews* 55 (1997): 277–286.

16. L. C. Clark and coauthors, Effects of selenium supplementation for cancer prevention in patients with carcinoma of the skin, *Journal of the American Medical Association* 276 (1996): 1957–1963.

17. V. Herbert, Selenium supplementation and cancer rates, *Journal of the American Medical Association* 277 (1997): 880.

18. K. Yoshizawa and coauthors, Study of prediagnostic selenium level in toenails and the risk of advanced prostate cancer, *Journal of the National Cancer Institute* 90 (1998): 1219–1224.

19. Fleet, 1997.

20. J. W. Finley and C. D. Davis, Selenium (Se) from high-selenium broccoli is utilized differently than selenite, selenate and selenomethionine, but is more effective in inhibiting colon carcinogensis, *Biofactors* 14 (2001): 191–196.

21. B. R. Hammond, Jr., B. R. Wooten, and J. Curran-Celentano, Carotenoids in the retina and lens: Possible acute and chronic effects on human visual performance, *Archives of Biochemistry and Biophysics* 385 (2001): 41–46.

22. P. F. Jacques and coauthors, Long-term nutrient intake and early age-related nuclear lens opacities, *Archives of Ophthalmology* 119 (2001): 1009–1019; J. T. Landrum and R. A. Bone, Lutein, zeaxanthin, and the macular pigment, *Archives of Biochemistry and Biophysics* 385 (2001): 28–40.

23. R. C. Rose, S. P. Richer, and A. M. Bode, Ocular oxidants and antioxidant protection, *Proceedings of the Society for Experimental Biology and Medicine* 217 (1998): 397–407.

24. J. E. Buring and C. H. Hennekens, Antioxidant vitamins and cardiovascular disease, *Nutrition Reviews* 55 (1997): S53–S60.

25. Standing Committee on the Scientific Evaluation of Dietary Reference Intakes, 2000, pp. 211–212.

26. F. Carrasquedo, M. Glanc, and C. G. Fraga, Tissue damage in acute myocardial infarction: Selective protection by vitamin E, *Free Radical Biology & Medicine* 26 (1999): 1587–1590.

27. E. K. Parkkala-Sarataho and coauthors, A randomized, single-blind, placebo-controlled trial of the effects of 200 mg α-tocopherol on the oxidation resistance of atherogenic lipoproteins, *American Journal of Clinical Nutrition* 68 (1998): 1034–1041.

28. J. Regnström and coauthors, Inverse relation between the concentration of low-density-lipoprotein, vitamin E and severity of coronary artery disease, *American Journal of Clinical Nutrition* 63 (1996): 377–385.

29. L. H. Kushi and coauthors, Dietary antioxidant vitamins and death from coronary heart disease in postmenopausal women, *New England Journal of Medicine* 334 (1996): 1156–1162.

30. M. J. Stampfer and coauthors, Vitamin E consumption and the risk of coronary disease in women, *New England Journal of Medicine* 328 (1993): 1444–1449; E. B. Rimm and coauthors, Vitamin E consumption and the risk of coronary disease in men, *New England Journal of Medicine* 328 (1993): 1450–1456.

31. M. Meydani, Effect of functional food ingredients: Vitamin E modulation of cardiovascular diseases and immune status in the elderly, *American Journal of Clinical Nutrition* 71 (2000): S1665–S1668; K. G. Losonczy, T. B. Harris, and R. J. Havlik, Vitamin E and vitamin C supplementation use and risk of all-cause and coronary heart disease mortality in older persons: The Established Populations for Epidemiologic Studies of the Elderly, *American Journal of Clinical Nutrition* 64 (1996): 190–196.

32. N. G. Stephens and coauthors, Randomized controlled trial of vitamin E in patients with coronary disease: Cambridge Heart Antioxidant Study (CHAOS), *Lancet* 347 (1996): 781–786; A. Ness and G. D. Smith, Mortality in the CHAOS trial (correspondence), *Lancet* 353 (1999): 1017–1018.

33. The Heart Outcomes Prevention Evaluation Study Investigators, Vitamin E supplementation an cardiovascular events in high-risk patients, *New England Journal of Medicine* 342 (2000): 154–160.

34. Collaborative Group of the Primary Prevention Project, Low-dose aspirin and vitamin E in people at cardiovascular risk: A randomized trial in general practice, *Lancet* 357 (2001): 89–95; E. A. Meagher and coauthors, Effects of vitamin E on lipid peroxidation in healthy persons, *Journal of the American Medical Association* 285 (2001): 1178–1182; N. J. Stone, The Gruppo Italiano per lo Studio della Sopravvivenza nell'Infarto Miocardio (GISSI)-Prevenzione Trial on fish oil and vitamin E supplementation in myocardial infarction survivors, *Current Cardiology Reports* 2 (2000); 445–451; S. Yusuf and coauthors, Vitamin E supplementation and cardiovascular events in high-risk patients. The Heart Outcomes Prevention Evaluation Study Investigators, *New England Journal of Medicine* 342 (2000): 154–160.

35. R. A. Riemersma and coauthors, Vitamin C and the risk of acute myocardial infarction, *American Journal of Clinical Nutrition* 71 (2000): 1181–1186.

36. D. Harats, Citrus fruit supplementation reduces lipoprotein oxidation in young men ingesting a diet high in saturated fat: Presumptive evidence for an interaction between vitamins C and E in vivo, *American Journal of Clinical Nutrition* 67 (1998): 240–245.

37. D. Kritchevsky, Antioxidant vitamins in the prevention of cardiovascular disease, *Nutrition Today*, January/February 1992, pp. 30–33.

38. M. M. Mahfouz, H. Kawano, and F. A. Kummerow, Effect of cholesterol-rich diets with and without added vitamins E and C on the severity of atherosclerosis in rabbits, *American Journal of Clinical Nutrition* 66 (1997): 1240–1249.

39. K. Nyyssönen and coauthors, Vitamin C deficiency and risk of myocardial infarction: Prospective population study of men from eastern Finland, *British Medical Journal* 314 (1997): 634–638.

40. S. Liu and coauthors, Intake of vegetables rich in carotenoids and risk of coronary heart disease in men: The Physican's Heart Study, *International Journal of Epidemiology* 30 (2001): 130–135.

Chapter 8

1. S. M. Kleiner, Water: An essential but overlooked nutrient, *Journal of the American Dietetic Association* 99 (1999): 200–206.

2. A. C. Grandjean and coauthors, The effect of caffeinated, non-caffeinated, caloric and noncaloric beverages on hydration, *Journal of the American College of Nutrition* 19 (2000): 591–600.

3. M. P. Sauvant and D. Pepin, Geographic variation of the mortality from cardiovascular disease and drinking water in a French small area (Puy de Dome), *Environmental Research* 84 (2000): 219–227.

4. H. Bohmer, H. Muller, and K. L. Resch, Calcium supplementation with calcium-rich mineral waters: A systematic review and meta-analysis of its bioavailability, *Osteoporosis International* 11 (2000): 938–943; R. Maheswaran and coauthors, Magnesium in drinking water supplies and mortality from acute myocardial infarction in north west England, *Heart* 82 (1999): 455–460.

5. J. Chmielnicka and B. Sowa, Cadmium interaction with essential metals (Zn, Cu, Fe), metabolism metallothionein, and ceruloplasmin in pregnant rats and fetuses, *Ecotoxicology and Environmental Safety* 36 (1996): 277–281.

6. Standing Committee on the Scientific Evaluation of Dietary Reference Intakes, Food and Nutrition Board, Institute of Medicine, *Dietary Reference Intakes for Calcium, Phosphorus, Magnesium, Vitamin D, and Fluoride* (Washington, D.C.: National Academy Press, 1997), pp. 71–145.

7. M. E. Wastney and coauthors, Changes in calcium kinetics in adolescent girls induced by high calcium intake, *Journal of Clinical Endocrinology and Metabolism* 85 (2000): 4470–4475.

8. Standing Committee on the Scientific Evaluation of Dietary Reference Intakes, 1997, p. 72.

9. Standing Committee on the Scientific Evaluation of Dietary Reference Intakes, 1997, pp. 106–107.

10. J. Tuomilehto and coauthors, Urinary sodium excretion and cardiovascular mortality in Finland: A prospective study, *Lancet* 357 (2001): 848–851; L. Liu and coauthors, Comparative studies of diet-related factors

and blood pressure among Chinese and Japanese: Results from the China-Japan Cooperative Research of the WHO-CARDIAC Study. Cardiovascular Disease and Alimentary Comparison, *Hypertension Research* 23 (2000): 413–420; The sixth report of the Joint National Committee on Prevention, Detection, Evaluation, and Treatment of High Blood Pressure, *Archives of Internal Medicine* 157 (1997): 2413–2446.

11. J. Stamler, The INTERSALT Study: Background, methods, findings, and implications, *American Journal of Clinical Nutrition* 65 (1997): S626–S642.

12. F. M. Sacks and coauthors, Effects on blood pressure of reduced dietary sodium and the Dietary Approaches to Stop Hypertension (DASH) diet, *New England Journal of Medicine* 344 (2001): 3–10.

13. S. M. Groziak and G. D. Miller, Natural bioactive substances in milk and colostrum: Effects on the arterial blood pressure system, *British Journal of Nutrition* 84 (2000): S119–S125; L. M. Resnick and coauthors, Factors affecting blood pressure responses to diet: The Vanguard study, *American Journal of Hypertension* 13 (2000): 956–965.

14. N. M. Karanja and coauthors, Descriptive characteristics of the dietary patterns used in the Dietary Approaches to Stop Hypertension Trial, *Journal of the American Dietetic Association* 99 (1999): S19–S27; L. P. Svetkey and coauthors, Effects of dietary patterns on blood pressure: Subgroup analysis of the Dietary Approaches to Stop Hypertension (DASH) randomized clinical trial, *Archives of Internal Medicine* 159 (1999): 285–293.

15. E. Saltos and S. Bowman, Dietary guidance on sodium: Should we take it with a grain of salt? *Family Economics and Nutrition Review, USDA* 11 (1998): 49–51.

16. N. A. Graudal and coauthors, Effects of sodium restriction on blood pressure, renin, aldosterone, catecholamines, cholesterol, and triglycerides, *Journal of the American Medical Association* 279 (1998): 1383–1391.

17. F. P. Cappuccio and coauthors, Double-blind randomized trial of modest salt restriction in older people, *Lancet* 350 (1997): 850–854.

18. J. B. Neily and coauthors, Potential contributing factors to noncompliance with dietary sodium restriction in patients with heart failure, *American Heart Journal* 143 (2002): 29–33.

19. F. J. He and G. A. MacGregor, Beneficial effects of potassium, *British Medical Journal* 323 (2001): 497–501.

20. N. Bleichrodt and coauthors, The benefits of adequate iodine intake, *Nutrition Reviews* 54 (1996): S72–S78.

21. S. C. Das, U. P. Isichei, and P. O. Obekpa, Iodine deficiency disorders in pre-adolescent and adolescent children in Nigeria, West Africa, *West African Journal of Medicine* 17 (1998): 113–120.

22. Standing Committee on the Scientific Evaluation of Dietary Reference Intakes, Food and Nutrition Board, Institute of Medicine, *Dietary Reference Intakes for Vitamin A, Vitamin K, Arsenic, Boron, Chromium, Copper, Iodine, Iron, Manganese, Molybdenum, Nickel, Silicon, Vanadium, and Zinc* (Washington, D.C.: National Academy Press, 2001), p. 8-15.

23. Food and Drug Administration Center for Drug Evaluation and Research, Guidance: Potassium iodide as a thyroid blocking agent in radiation emergencies, November 2001, available at www.fda.gov/cder/guidance/index.htm.

24. Standing Committee on the Scientific Evaluation of Dietary Reference Intakes, 2001, p. 9-14.

25. N. C. Andrews, Disorders of iron metabolism, *New England Journal of Medicine* 341 (1999): 1986–1995; E. Beutler, How little we know about the absorption of iron, *American Journal of Clinical Nutrition* 66 (1997): 419–420.

26. J. D. Haas and T. Brownlie IV, Iron deficiency and reduced work capacity: A critical review of the research to determine a causal relationship, *Journal of Nutrition* 131 (2001): 676S–690S; S. Horton and C. Levin, Commentary on "evidence that iron deficiency anemia causes reduced work capacity," *Journal of Nutrition* 131 (2001): S691–S696.

27. D. Pinero, B. Jones, and J. Beard, Variations in dietary iron alter behavior in developing rats, *Journal of Nutrition* 131 (2001): 311–318; Standing Committee on the Scientific Evaluation of Dietary Reference Intakes, 2001, pp. 9-5–9-6.

28. C. I. Obialo and coauthors, Clay pica has no hematologic or metabolic correlate in chronic hemodialysis patients, *Journal of Renal Nutrition* 11 (2001): 32–36.

29. Centers for Disease Control and Prevention, Recommendations to prevent and control iron deficiency in the United States, *Morbidity and Mortality Weekly Report* (April supplement) 47 (1998), 3; R. Yip and P. R. Dallman, Iron, in E. E. Ziegler and L. J. Filer, Jr., eds., *Present Knowledge in Nutrition*, 7th ed. (Washington, D.C.: International Life Sciences Institute Press, 1996), pp. 277–292.

30. B. Annibale and coauthors, Reversal of iron deficiency anemia after *Helicobacter pylori* eradication in patients with asymptomatic gastritis, *Annals of Internal Medicine* 131 (1999): 668–672.

31. M. Wessling-Resnick, Iron transport, *Annual Review of Nutrition* 20 (2000): 129–151.

32. J. R. Hunt and Z. K. Roughead, Adaptation of iron absorption in men consuming diets with high or low iron bioavailability, *American Journal of Clinical Nutrition* 71 (2000): 94–102.

33. R. E. Fleming and W. S. Sly, Mechanisms of iron accumulation in hereditary hemochromatosis, *Annual Review of Physiology* 64 (2002): 663–680; A. S. Tavill, Clinical implications of the hemochromatosis gene, *New England Journal of Medicine* 341 (1999): 755–757.

34. J. M. McCord, Effects of positive iron status at a cellular level, *Nutrition Reviews* 54 (1996): 85–88.

35. K. Klipstein-Grobusch and coauthors, Serum ferritin and risk of myocardial infarction in the elderly: The Rotterdam Study, *American Journal of Clinical Nutrition* 69 (1999): 1231–1236.

36. H. van Jaarsveld, G. F. Pool, and H. C. Barnard, Dietary iron concentration alters LDL oxidatively: The effect of antioxidants, *Research Communications in Molecular Pathology and Pharmacology* 99 (1998): 69–80.

37. B. de Valk and J. J. Marx, Iron, atherosclerosis, and ischemic heart disease, *Archives of Internal Medicine* 159 (1999): 241–242 (positive); C. T. Sempos and coauthors, Serum ferritin and death from all causes and cardiovascular disease: The NHANES II Mortality Study. National Health and Nutrition Examination Study, *Annals of Epidemiology* 10 (2000): 441–448 (negative); J. Danesh and P. Appleby, Coronary heart disease and iron status: Meta-analyses of prospective studies, *Circulation* 99 (1999): 852–854 (negative); K. Klipstein-Grobusch and coauthors, Dietary iron and risk of myocardial infarction in the Rotterdam Study, *American Journal of Epidemiology* 149 (1999): 421–428.

38. J. D. Cook and M. B. Reddy, Effect of ascorbic acid intake on nonheme-iron absorption from a complete diet, *American Journal of Clinical Nutrition* 73 (2001): 93–98.

39. M. Shannon, Ingestion of toxic substances by children, *New England Journal of Medicine* 342 (2000): 186–191; C. C. Morris, Pediatric iron poisonings in the United States, *Southern Medicine Journal* 93 (2000): 352–358.

40. Standing Committee on the Scientific Evaluation of Dietary Reference Intakes, 2001, p. 9-48.

41. M. B. Reddy, R. F. Hurrell, and J. D. Cook, Estimation of nonheme-iron bioavailability from meal composition, *American Journal of Clinical Nutrition* 71 (2000): 937–943.

42. I. M. Zijp, O. Korver, and L. B. Tijburg, Effect of tea and other dietary factors on iron absorption, *Critical Reviews in Food Science and Nutrition* 40 (2000): 371–398.

43. R. J. Cousins, Zinc, in E. E. Ziegler and L. J. Filer, Jr., eds., *Present Knowledge in Nutrition*, 7th ed. (Washington, D.C.: International Life Sciences Institute Press, 1996), pp. 293–306.

44. A. H. Shankar and A. S. Prasad, Zinc and immune function: The biological basis of altered resistance to infection, *American Journal of Clinical Nutrition* 69 (1998): S447–S463.

45. N. W. Solomons, Mild human zinc deficiency produces an imbalance between cell-mediated and humoral immunity, *Nutrition Reviews* 56 (1998): 27–28.

46. R. B. Costello and J. Grumstrup-Scott, Zinc: What role might supplements play? *Journal of the American Dietetic Association* 100 (2000): 371–375; R. E. Black, Therapeutic and preventive effects of zinc on serious childhood infectious diseases in developing countries, *American Journal of Clinical Nutrition* 68 (1998): S476–S479; G. J. Fuchs, Possibilities for zinc in the treatment of acute diarrhea, *American Journal of Clinical Nutrition* 68 (1998): S480–S483; J. L. Rosado and coauthors, Zinc supplementation reduced morbidity, but neither zinc nor iron supplementation affected growth or body composition of Mexican preschoolers, *American Journal of Clinical Nutrition* 65 (1997): 13–19.

47. K. A. Keller, A. Grider, and J. A. Coffield, Age-dependent influence of dietary zinc restriction on short-term memory in male rats, *Physiology and Behavior* 72 (2001): 339–348; M. S. Golub and coauthors, Activity and attention in zinc-deprived adolescent monkeys, *American Journal of Clinical Nutrition* 64 (1996): 908–915.

F

48. Cousins, 1996.

49. M. L. Macknin and coauthors, Zinc gluconate lozenges for treating the common cold in children: A randomized controlled trial, *Journal of the American Medical Association* 279 (1998): 1962, 1967; S. B. Mossad and coauthors, Zinc gluconate lozenges for treating the common cold: A randomized, double-blind, placebo-controlled study, *Annals of Internal Medicine* 125 (1996): 81–88.

50. Standing Committee on the Scientific Evaluation of Dietary Reference Intakes, 2001, p. 12-29.

51. D. H. Holben and A. M. Smith, The diverse role of selenium within selenoproteins: A review, *Journal of the American Dietetic Association* 99 (1999): 836–843.

52. K. Yoshizawa and coauthors, Study of prediagnostic selenium level in toenails and the risk of advanced prostate cancer, *Journal of the National Cancer Institute* 90 (1998): 1219–1224; O. A. Levander and R. F. Burk, Selenium, in E. E. Ziegler and L. J. Filer, Jr., eds., *Present Knowledge in Nutrition,* 7th ed. (Washington, D.C.: International Life Sciences Institute Press, 1996), pp. 320–328.

53. Standing Committee on the Scientific Evaluation of Dietary Reference Intakes, 1997, p. 290.

54. Fluorides and fluorosis, *Nutrition and the M.D.*, February 1997, pp. 4–5.

55. Achievements in public health, 1900–1999: Fluoridation of drinking water to prevent dental caries, *Morbidity and Mortality Weekly Report* 48 (1999): 933–940.

56. Centers for Disease Control and Prevention, Recommendations for using fluoride to prevent and control dental caries in the United States, *Morbidity and Mortality Weekly Report* 50 (2001): 1–42; Position of The American Dietetic Association: The impact of fluoride on health, *Journal of the American Dietetic Association* 100 (2000): 1208–1213.

57. NIH plans chromium study for type 2 diabetes, Federal Update, *Journal of the American Dietetic Association* 101 (2001): 1136; J. B. Vincent, Quest for the molecular mechanism of chromium action and its relationship to diabetes, *Nutrition Reviews* 58 (2000): 67–72.

58. R. A. Anderson, Effects of chromium on body composition and weight loss, *Nutrition Reviews* 56 (1998): 266–270; W. J. Pasman, M. S. Weterterp-Plantenga, and W. H. Saris, The effectiveness of long-term supplementation of carbohydrate, chromium, fiber and caffeine on weight maintenance, *International Journal of Obesity Related Metabolic Disorders* 21 (1997): 1143–1151; H. C. Lukaski and coauthors, Chromium supplementation and resistance training: Effects on body composition, strength, and trace element status of men, *American Journal of Clinical Nutrition* 63 (1996): 954–965.

59. V. A. Dubrovskaya and K. E. Wetterhahn, Effects of Cr(IV) on the expression of the oxidative stress genes in human lung cells, *Carcinogenesis* (1998): 1401–1407.

60. R. A. Anderson, Chromium supplements safe, *USDA Food and Nutrition Research Brief,* October 1997, pp. 5–6, available from www.nal.usda.gov/fnic/usda/fnrb/fnrb1097.html.

61. J. F. Fowler, Jr., Systemic contact dermatitis caused by oral chromium picolinate, *Cutis* 65 (2000): 116.

62. W. Mertz, Interaction of chromium with insulin: A progress report, *Nutrition Reviews* 56 (1998): 174–177.

63. R. Uauy, M. Olivares, and M. Gonzalez, Essentiality of copper in humans, *American Journal of Clinical Nutrition* 67 (1998): S952–S959.

64. A. Cordano, Clinical manifestations of nutritional copper deficiency in infants and children, *American Journal of Clinical Nutrition* 67 (1998): S1012–S1016.

65. D. J. Fitzgerald, Safety guidelines for copper in water, *American Journal of Clinical Nutrition* 67 (1998): S1098–S1102.

66. Standing Committee on the Scientific Evaluation of Dietary Reference Intakes, 2001, p. 7-17.

67. National Dairy Council, Calcium Summit, *Breaking News*, available on the Internet at www.nationaldairycouncil.org/nwbdbrek.html.

68. A. Azoulay, P. Garzon, and M. J. Eisenberg, Comparison of the mineral content of tap water and bottled waters, *Journal of General Internal Medicine* 16 (2001): 168–175; Bohmer, Muller, and Resch, 2000.

Consumer Corner 8

1. Another example is found in Outbreaks of cyclosporiasis—United States, 1997, *Morbidity and Mortality Weekly Report* 46 (1997): 451–452; Position of The American Dietetic Association: Food and water safety, *Journal of the American Dietetic Association* 97 (1997): 184–189.

2. R. S. Barwick and coauthors, Surveillance for waterborne-disease outbreaks—United States, 1997–1998, *Morbidity and Mortality Weekly Report* 49 (2000): 1–35.

3. N. W. Woodruff and coauthors, Human cell mutagenicity of chlorinated and unchlorinated water and the disinfection byproduct 3-chloro-4-(dichloromethyl)-5-hydroxy-2(5H)-furanone (MX), *Mutation Research* 495 (2001): 157–168; S. F. Thai and coauthors, Detection of early gene expression changes by differential display in the livers of mice exposed to dichloroacetic acid, *Carcinogenesis* 22 (2001): 1317–1322; J. H. Miller and coauthors, In vivo MRI measurements of tumor growth induced by dichloroacetate: Implications for mode of action, *Toxicology* 145 (2000): 115–125.

4. K. P. Cantor and coauthors, Drinking water source and chlorination byproducts in Iowa. III. Risk of brain cancer, *American Journal of Epidemiology* 150 (1999): 552–560; T. J. Doyle and coauthors, The association of drinking water source and chlorination by-products with cancer incidence among postmenopausal women in Iowa: A prospective cohort study, *American Journal of Public Health* 87 (1997): 1168–1176.

5. D. S. Michaud and coauthors, Fluid intake and the risk of bladder cancer in men, *New England Journal of Medicine* 340 (1999): 1390–1397.

6. Natural Resources Defense Council, *Bottled Water: Pure Drink or Pure Hype?* 1999, a report available at www.nrdc.org/water/drinking/nbw.asp.

7. Water sold state to state safe to drink, agency says, *FDA Consumer,* July/August 1999, p. 6.

8. National Sanitation Foundation, Consumer guide to bottled water, *Information for Consumers*, March 1998.

9. P. Garzon and M. J. Eisenberg, Variation in the mineral content of commercially available bottled waters: Implications for health and disease, *American Journal of Medicine* 105 (1998): 125–130.

Controversy 8

1. NIH Consensus Development Panel, Osteoporosis prevention, diagnosis, and therapy, *Journal of the American Medical Association* 285 (2001): 785–795.

2. F. H. Anderson, Osteoporosis in men, *International Journal of Clinical Practice* 52 (1998): 176–180.

3. National Center for Injury Prevention and Control, Falls and hip fractures among the elderly, *Unintentional Injury Fact Sheet*, 1998 (available from www.cdc.gov); Standing Committee on the Scientific Evaluation of Dietary Reference Intakes, Food and Nutrition Board, Institute of Medicine, *Dietary Reference Intakes for Calcium, Phosphorus, Magnesium, Vitamin D, and Fluoride* (Washington, D.C.: National Academy Press, 1997), pp. 4-1–4-57.

4. T. D. Galsworthy and P. L. Wilson, Osteoporosis: It steals more than bone, *American Journal of Nursing* 96 (1996): 27–32.

5. M. K. Jeffcoat, Osteoporosis: A possible modifying factor in oral bone loss, *Annals of Periodontology* 3 (1998): 312–321.

6. S. H. Ralston, Science, medicine, and the future: Osteoporosis, *British Medical Journal* 315 (1997): 469–472.

7. R. J. Wood and J. C. Fleet, The genetics of osteoporosis: Vitamin D receptor polymorphisms, *Annual Review of Nutrition* 18 (1998): 233–258.

8. A. M. Parfitt, Genetic effects on bone mass and turnover—Relevance to black/white differences, *Journal of the American College of Nutrition* 16 (1997): 325–333.

9. W. C. Annie and coauthors, Age-related osteoporosis in Chinese: An evaluation of the response of intestinal calcium absorption and calcitropic hormones to dietary calcium deprivation, *American Journal of Clinical Nutrition* 68 (1998): 1291–1297.

10. D. Hartman and coauthors, Bone density of women who have recovered from anorexia nervosa, *International Journal of Eating Disorders* 28 (2000): 107–112; E. R. Brooks, B. W. Ogden, and D. S. Cavalier, Compromised bone density 11.4 years after diagnosis of anorexia nervosa, *Journal of Women's Health* 7 (1998): 567–574.

11. T. V. Nguyen, P. N. Sambrook, and J. A. Eisman, Bone loss, physical activity, and weight change in elderly women: The Dubbo Osteoporosis Epidemiology Study, *Journal of Bone Mineral Research* 13 (1998): 1458–1467.

12. S. Suleiman and coauthors, Effect of calcium intake and physical activity level protect bone mass and turnover in healthy white postmenopausal women, *American Journal of Clinical Nutrition* 66 (1997): 937–943.

13. N. K. Henderson, C. P. White, and J. A. Eisman, The roles of exercise and fall risk reduction in the prevention of osteoporosis, *Endocrinology and Metabolism Clinics of North America* 27 (1998): 369–387.

14. G. F. Maddalozzo and C. M. Snow, High intensity resistance training: Effects on bone in older men and women, *Calcified Tissue International* 66 (2000): 399–404; R. D. Lewis and C. M. Modlesky, Nutrition, physical activity, and bone health in women, *International Journal of Sports Nutrition* 8 (1998): 250–284; E. Ernst, Exercise for female osteoporosis: A systematic review of randomized clinical trials, *Sports Medicine* 25 (1998): 359–368.

15. L. M. Salamone and coauthors, Effect of a lifestyle intervention on bone mineral density in premenopausal women: A randomized trial, *American Journal of Clinical Nutrition* 70 (1999): 97–103; L. W. Turner, M. Q. Wang, and Q. Fu, Risk factors for hip fracture among southern older women, *Southern Medical Journal* 91 (1998): 533–540; M. E. Mussolino and coauthors, Risk factors for hip fracture in white men: The NHANES I Epidemiologic follow-up study, *Journal of Bone Mineral Research* 13 (1998): 918–924; Nguyen, Sambrook, and Eisman, 1998.

16. S. Gillette-Guyonnet and coauthors, Body composition and osteoporosis in elderly women, *Gerontology* 46 (2000): 189–193.

17. J. C. Fleet, Leptin and bone: Does the brain control bone biology? *Nutrition Reviews* 58 (2000): 209–211.

18. M. N. Hadley and S. V. Reddy, Smoking and the human vertebral column: A review of the impact of cigarette use on vertebral bone metabolism and spinal fusion, *Neurosurgery* 41 (1997): 116–124.

19. S. W. Sampson, Alcohol, osteoporosis, and bone regulating hormones, *Alcoholism: Clinical and Experimental Research* 21 (1997): 400–403; R. F. Klein, Alcohol-induced bone disease: Impact of ethanol on osteoblast proliferation, *Alcoholism: Clinical and Experimental Research* 21 (1997): 392–399.

20. D. E. Sellmeyer and coauthors, A high ratio of dietary animal to vegetable protein increases the rate of bone loss and the risk of fracture in postmenopausal women, *American Journal of Clinical Nutrition* 73 (2001): 118–122; U. S. Barzel and L. K. Massey, Excess dietary protein can adversely affect bone, *Journal of Nutrition* 128 (1998): 1051–1053.

21. R. Itoh, N. Nishiyama, and Y. Suyama, Dietary protein intake and urinary excretion of calcium: A cross-sectional study in a healthy Japanese population, *American Journal of Clinical Nutrition* 67 (1998): 438–444.

22. L. A. Frassetto and coauthors, Worldwide incidence of hip fracture in elderly women: Relation to consumption of animal to vegetable foods, *Journals of Gerontology Series A: Biological Sciences and Medical Sciences* 55 (2000): M585–M592.

23. R. G. Munger, J. R. Cerhan, and B. C-H. Chiu, Prospective study of dietary protein intake and risk of hip fracture in postmenopausal women, *American Journal of Clinical Nutrition* 69 (1999): 147–152; J. P. Bonjour, M. A. Schurch, and R. Rizzoli, Nutritional aspects of hip fractures, *Bone* 18 (1996): S139–S144.

24. R. P. Heaney, Excess dietary protein may not adversely affect bone, *Journal of Nutrition* 128 (1998): 1054–1057.

25. J. Raloff, Do meat and dairy harm aging bones? *Science News* 159 (2001): 20.

26. F. Ginty, A. Flynn, and K. D. Cashman, The effect of dietary sodium intake on biochemical markers of bone metabolism in young women, *British Journal of Nutrition* 79 (1998): 343–350; Standing Committee on the Scientific Evaluation of Dietary Reference Intakes, 1997.

27. A. J. Cohen and F. J. Roe, Review of the risk factors for osteoporosis with particular reference to a possible aetiological role of dietary salt, *Food and Chemical Toxicology* 38 (2000): 237–253.

28. P. T. Packard and R. P. Heaney, Medical nutrition therapy for patients with osteoporosis, *Journal of the American Dietetic Association* 97 (1997): 414–417.

29. T. Lloyd and coauthors, Dietary caffeine intake and bone status of postmenopausal women, *American Journal of Clinical Nutrition* 65 (1997): 1826–1830.

30. A. D. Maravilla and coauthors, Acute effects of soft drink intake on calcium and phosphate metabolism in immature and adult rats (abstract), *Revista de Investigacion Clinica* 50 (1998): 185–189; K. Lau and coauthors, Differing effects of acid versus neutral phosphate therapy of hypercalciuria, *Kidney International* (1979), as cited by Barzel and Massey, 1998.

31. F. Garcia-Contreras and coauthors, Cola beverage consumption induces bone mineralization reduction in ovariectomized rats, *Archives of Medical Research* 31 (2000): 360–365.

32. E. Petridou and coauthors, The role of dairy products and non alcoholic beverages in bone fractures among schoolage children, *Scandinavian Journal of Social Medicine* 25 (1997): 119–125.

33. G. Ferland, The vitamin K–dependent proteins: An update, *Nutrition Reviews* 56 (1998): 223–230.

34. D. Feskanich and coauthors, Vitamin K intake and hip fractures in women: A prospective study, *American Journal of Clinical Nutrition* 69 (1999): 74–79.

35. A. M. Craciun and coauthors, Improved bone metabolism in female elite athletes after vitamin K supplementation, *International Journal of Sports Medicine* 19 (1998): 479–484.

36. Feskanich and coauthors, 1999; K. L. Tucker and coauthors, Potassium, magnesium, and fruit and vegetable intakes are associated with greater bone mineral density in elderly men and women, *American Journal of Clinical Nutrition* 69 (1999): 727–736; R. K. Rude and coauthors, Magnesium deficiency induces bone loss in the rat, *Mineral and Electrolyte Metabolism* 24 (1998): 314–320.

37. D. Feskanich and coauthors, Vitamin A intake and hip fractures among postmenopausal women, *Journal of the American Medical Association* 287 (2002): 47–54; S. J. Whiting and B. Lemke, Excess retinol intake may explain the high incidence of osteoporosis in northern Europe, *Nutrition Reviews* 57 (1999): 192–198; J. Melhus and coauthors, Excessive dietary intake of vitamin A is associated with reduced bone mineral density and increased risk for hip fracture, *Annals of Internal Medicine* 129 (1998): 770–778.

38. S. A. New and coauthors, Dietary influences on bone mass and bone metabolism: Further evidence of a positive link between fruit and vegetable consumption and bone health? *American Journal of Clinical Nutrition* 71 (2000): 142–151; J. J. B. Anderson, Plant-based diets and bone health: Nutritional implications, *American Journal of Clinical Nutrition* 70 (1999) S539–S542; Tucker and coauthors, 1999.

39. D. J. van der Voort and coauthors, Screening for osteoporosis using easily obtainable biometrical data: Diagnostic accuracy of measured, self-reported and recalled BMI, and related costs of bone mineral density measurements, *Osteoporosis International* 11 (2000): 233–239; L. W. Turner, P. A. Faile, and R. Tomlinson, Jr., Osteoporosis diagnosis and fracture, *Orthopaedic Nursing*, September/October 1999, pp. 21–27.

40. P. A. Rochon and J. H. Gurwitz, Prescribing for seniors: Neither too much nor too little, *Journal of the American Medical Association* 282 (1999): 113–115; D. L. Schneider, E. L. Barrett-Connor, and D. J. Morton, Timing of postmenopausal estrogen for optimal bone mineral density, *Journal of the American Medical Association* 277 (1997): 543–547; L. Speroff and coauthors, The comparative effect on bone density, endometrium, and lipids of continuous hormones as replacement therapy (CHART Study): A randomized controlled trial, *Journal of the American Medical Association* 276 (1996): 1397–1403; The Writing Group for the PEPI Trial, Effects of hormone therapy on bone mineral density: Results from the Postmenopausal Estrogen/Progestin Interventions (PEPI) Trial, *Journal of the American Medical Association* 276 (1996): 1389–1396.

41. B. H. Arjmandi and coauthors, Role of soy protein with normal or reduced isoflavone content in reversing bone loss induced by ovarian hormone deficiency in rats, *American Journal of Clinical Nutrition* 68 (1998): S1358–S1363; B. H. Arjmandi and coauthors, Bone-sparing effect of soy protein in ovarian-hormone-deficient rats is related to its isoflavone content, *American Journal of Clinical Nutrition* 68 (1998): S1364–S1368; J. P. Williams and coauthors, Tyrosine kinase inhibitor effects on avian osteoclastic and transport, *American Journal of Clinical Nutrition* 68 (1998): S1369–S1374.

42. Anderson, 1998; E. Velazquez and G. Bellabarba Arata, Testosterone replacement therapy, *Archives of Andrology* 41 (1998): 79–90.

43. D. Teegarden and coauthors, Previous milk consumption is associated with greater bone density in young women, *American Journal of Clinical Nutrition* 69 (1999): 1014–1017.

44. NIH Consensus Development Panel, 2001.

45. I. R. Reid, The roles of calcium and vitamin D in the prevention of osteoporosis, *Endocrinology and Metabolism Clinics of North America* 27 (1998): 389–398; K. O. O'Brien, Combined calcium and vitamin D supplementation reduces bone loss and fracture incidence in older men and women, *Nutrition Reviews* 56 (1998): 148–150.

46. S. J. Whiting and R. J. Wood, Adverse effects of high-calcium diets in humans, *Nutrition Reviews* 55 (1997): 1–9.

47. J. Guillemant and coauthors, Mineral water as a source of dietary

calcium: Acute effects on parathyroid function and bone resorption in young men, *American Journal of Clinical Nutrition* 71 (2000): 999–1002.

Chapter 9

1. A. H. Mokdad and coauthors, Centers for Disease Control and Prevention, The continuing epidemic of obesity in the United States (letter), *Journal of the American Medical Association* 284 (2000): 1650–1651; World Health Organization, as quoted in J. M. Rippe, S. Crossley, and R. Ringer, Obesity as a chronic disease: Modern medical and lifestyle management, *Journal of the American Dietetic Association* 98 (1998): S9–S15.
2. Centers for Disease Control and Prevention, Obesity and overweight: a public health epidemic, 2001, available at www.cdc.gov/nccdphp/dnpa/obesity/epidemic.htm.
3. Centers for Disease Control and Prevention, 2001.
4. U.S. Department of Health and Human Services, *Healthy People 2010*, 2nd ed. (Washington, D.C.: Government Printing Office, 2000), available online at www.health.gov/healthypeople or call (800) 367-4725.
5. S. E. Gariballa and coauthors, Nutritional status of hospitalized acute stroke patients, *British Journal of Nutrition* 79 (1998): 481–487.
6. J. K. Alexander, Obesity and coronary heart disease, *American Journal of the Medical Sciences* 321 (2001): 215–224; R. H. Eckel and R. M. Krauss, American Heart Association call to action: Obesity as a major risk factor for coronary heart disease, *Circulation* 97 (1998): 2099–2100.
7. S. W. Farrell and coauthors, Influences of cardiorespiratory fitness levels and other predictors of cardiovascular disease mortality in men, *Medicine and Science in Sports and Exercise* 30 (1998): 899–904.
8. 1997 Heart and Stroke Statistical Update, Dallas, Texas, as cited in Rippe, Crossley, and Ringer, 1998.
9. D. T. Felson, Weight and osteoarthritis, *American Journal of Clinical Nutrition* 63 (1996): S430–S432.
10. D. B. Allison and coauthors, Annual deaths attributable to obesity in the United States, *Journal of the American Medical Association* 282 (1999): 1530–1538.
11. E. E. Calle and coauthors, Body mass index and mortality in a prospective cohort of U.S. adults, *New England Journal of Medicine* 341 (1999): 1140–1141.
12. K. N. Frayn, Regulation of fatty acid delivery in vivo, *Advances in Experimental Medicine and Biology* 441 (1998): 171–179.
13. P. R. Jones and D. A. Edwards, Areas of fat loss in overweight young females following an 8-week period of energy intake reduction, *Annals of Human Biology* 26 (1999): 151–162; M. D. Jensen, Lipolysis: Contribution from regional fat, *Annual Review of Nutrition* 17 (1997): 127–139.
14. National Heart, Lung, and Blood Institute Expert Panel, 1998.
15. J. Stevens and coauthors, Evaluation of WHO and NHANES II standards for overweight using mortality rates, *Journal of the American Dietetic Association* 100 (2000): 825–827; A. Must and coauthors, The disease burden associated with overweight and obesity, *Journal of the American Medical Association* 282 (1999): 1523–1529.
16. J. C. Seidell, Societal and personal costs of obesity, *Experimental and Clinical Endocrinology and Diabetes* 106 (1998): S7–S9.
17. M. R. Lowe, K. Miller-Kovach, and S. Phelan, Weight-loss maintenance in overweight individuals one to five years following successful completion of a commercial weight-loss program, *International Journal of Obesity and Related Metabolic Disorders* 25 (2001): 325–331.
18. J. P. Flatt, How NOT to approach the obesity problem (comment), *Obesity Research* 5 (1997): 632–633.
19. J. Calles-Escandon and E. T. Poehlman, Aging, fat oxidation, and exercise, *Aging* 9 (1997): 57–63.
20. M. Gilliat-Wimberely and coauthors, Effects of habitual physical activity on the resting metabolic rates and body compositions of women aged 35 to 50 years, *Journal of the American Dietetic Association* 101 (2001): 1181–1188.
21. National Heart, Lung, and Blood Institute Expert Panel, 1998.
22. Are fat calipers becoming obsolete? *ACSM'S Health and Fitness Journal*, May/June 2001, p. 26.
23. G. Panotopoulos and coauthors, Dual X-ray absorptiometry, bioelectrical impedance, and near infrared interactance in obese women, *Medicine and Science in Sports and Exercise* 33 (2001): 665–670.
24. L. J. Harnack, R. W. Jeffrey, and K. N. Boutelle, Temporal trends in energy intake in the United States: An ecologic perspective, *American Journal of Clinical Nutrition* 71 (2000): 1478–1484.
25. A. S. Livine and C. J. Billington, Why do we eat? A neural systems approach, *Annual Review of Nutrition* 17 (1997): 597–619.
26. A. Geliebter and coauthors, Reduced stomach capacity in obese subjects after dieting, *American Journal of Clinical Nutrition* 63 (1996): 170–173.
27. T. A. Spiegel and coauthors, Contribution of gastric and postgastric feedback to satiation and satiety in women, *Physiology and Behavior* 62 (1997): 1125–1136.
28. A. L. Hirschberg, Hormonal regulation of appetite and food intake, *Annals of Medicine* 30 (1998): 7–20.
29. M. Yao and S. B. Roberts, Dietary energy density and weight regulation, *Nutrition Reviews* (2001): 247–258; N. C. Howarth, E. Saltzman, and S. B. Roberts, Dietary fiber and weight regulation, *Nutrition Reviews* 59 (2001): 129–139.
30. D. S. Ludwig, Dietary glycemic index and obesity, *Journal of Nutrition* 130 (2000): S280–S283.
31. L. E. Spieth and coauthors, A low-glycemic index diet in the treatment of pediatric obesity, *Archives of Pediatrics and Adolescent Medicine* 154 (2000): 947–951.
32. B. J. Rolls, E. A. Bell, and M. L. Thorwart, Water incorporated into a food but not served with a food decreases energy intake in lean women, *American Journal of Clinical Nutrition* 70 (1999): 448–455.
33. B. J. Rolls, E. A. Bell, and B. A. Waugh, Increasing the volume of food by incorporating air increases satiety in men, *American Journal of Clinical Nutrition* 72 (2000): 361–368.
34. M. S. Treuth and coauthors, Familial resemblance of body composition in prepubertal girls and their biological parents, *American Journal of Clinical Nutrition* 74 (2001): 529–533; L. Pérusse and C. Bouchard, Genotype-environment interaction in human obesity, *Nutrition Reviews* 57 (1999): S31–S38.
35. P. Hakala and coauthors, Environmental factors in the development of obesity in identical twins, *International Journal of Obesity and Related Metabolic Disorders* 23 (1999): 746–753.
36. B. E. Levin and A. A. Dunn-Meynell, Defense of body weight depends on dietary composition and palatability in rats with diet-induced obesity, *American Journal of Physiology: Regulatory, Integrative and Comparative Physiology* 282 (2002): R46–54; L. Pérusse and C. Bouchard, Gene-diet interactions in obesity, *American Journal of Clinical Nutrition* 72 (2000): S1285–S1290; D. B. Allison and M. S. Faith, Genetic and environmental influences on human body weight: Implications for the behavior therapist, *Nutrition Today* 35 (2000): 18–21; J. P. Foreyt and W. S. C. Poston II, Diet, genetics, and obesity, *Food Technology* 51 (1997): 70–73.
37. C. Bouchard and A. Tremblay, Genetic influences on the response of body fat and fat distribution to positive and negative energy balances in human identical twins, *Journal of Nutrition* 127 (1997): S943–S947.
38. W. Siffert, G protein beta 3 subunit 825T allele, hypertension, obesity, and diabetic nephropathy, *Nephrology, Dialysis, Transplantation* 15 (2000): 1298–1306.
39. J. S. Flier and B. B. Lowell, Obesity research springs a proton leak, *Nature Genetics* 15 (1997): 223–224; C. Fleury and coauthors, Uncoupling protein-2: A novel gene linked to obesity and hyperinsulinemia, *Nature Genetics* 15 (1997): 269–272.
40. G. Wolf, A new uncoupling protein: A potential component of the human body weight regulation system, *Nutrition Reviews* 55 (1997): 178–179.
41. J. A. Yanovski and coauthors, Associations between uncoupling protein 2, body composition, and resting energy expenditure in lean and obese African American, white, and Asian children, *American Journal of Clinical Nutrition* 71 (2000): 1405–1412.
42. L. P. Kozak and M. E. Harper, Mitochondrial uncoupling proteins in energy expenditure, *Annual Review of Nutrition* 20 (2000): 339–363.
43. R. B. Ceddia, W. N. William, Jr., and R. Curi, The response of skeletal muscle to leptin, *Frontiers in Bioscience* 6 (2001): D90–D97.
44. E. Faloia and coauthors, Adipose tissue as an endocrine organ? A review of some recent data, *Eating and Weight Disorders* 5 (2000): 116–123.
45. Ceddia, William, Jr., and Curi, 2001; M. S. Westerterp-Plantenga and coauthors, Effects of weekly administration of pegylated recombinant human OB protein on appetite profile and energy metabolism in obese men, *American Journal of Clinical Nutrition* 74 (2001): 426–434.
46. W. A. Banks, Leptin transport across the blood-brain barrier: Implications for the cause and treatment of obesity, *Current Pharmaceutical Design* 7 (2001): 125–133; N. D. Quinton and coauthors, A single

nucleotide polymorphism (SNP) in the leptin receptor is associated with BMI, fat mass and leptin levels in postmenopausal Caucasian women, *Human Genetics* 108 (2001): 233–236.

47. N. F. Chu and coauthors, Plasma leptin concentrations and four-year weight gain among U.S. men, *International Journal of Obesity and Related Metabolic Disorders* 25 (2001): 346–353.

48. C. A. Baile and M. A. Della-Fera, Regulation of metabolism and body fat mass by leptin, *Annual Review of Nutrition* 20 (2000): 105–127.

49. R. B. S. Harris, Leptin—Much more than a satiety signal, *Annual Review of Nutrition* 20 (2000): 45–75; R. L. Bradley, K. A. Cleveland, and B. Cheatham, The adipocyte as a secretory organ: Mechanisms of vesicle transport and secretory pathways, *Recent Progress in Hormone Research* 56 (2001): 329–358; Faloia and coauthors, 2000.

50. S. Moschos, J. L. Chan, and C. S. Mantzoros, Leptin and reproduction: a review, *Fertility and Sterility* 77 (2002): 433–444; T. Thomas and coauthors, Leptin acts on human marrow stromal cells to enhance differentiation to osteoblasts and to inhibit differentiation to adipocytes, *Endocrinology* 140 (1999): 1630–1638; P. Trayhurn and coauthors, Hormonal and neuroendocrine regulation of energy balance—The role of leptin, *Arch Tierernahr* 51 (1998): 177–185; G. Frunbeck, S. A. Jebb, and A. M. Prentice, Leptin: Physiology and pathophysiology, *Clinical Physiology* 18 (1998): 399–419; M. R. Sierra-Honigmann and coauthors, Biological action of leptin as an angiogenic factor, *Science* 281 (1998): 1683–1686.

51. J. P. Koplan and W. H. Dietz, Caloric imbalance and public health policy (editorial), *Journal of the American Medical Association* 282 (1999): 1579.

52. M. A. McCrory and coauthors, Dietary variety within food groups: Association with energy intake and body fatness in men and women, *American Journal of Clinical Nutrition* 69 (1999): 440–447.

53. G. A. Bray and B. M. Popkin, Dietary fat intake does affect obesity! *American Journal of Clinical Nutrition* 68 (1998): 1157–1173.

54. J. E. Blundell and J. I. Macdiarmid, Fat as a risk factor for overconsumption: Satiation, satiety, and patterns of eating, *Journal of the American Dietetic Association* 97 (1997): S63–S69; L. H. Nelson and L. A. Tucker, Diet composition related to body fat in a multivariate study of 203 men, *Journal of the American Dietetic Association* 96 (1996): 771–777.

55. K. Brownell, The pressure to eat—Why we're getting fatter, *Nutrition Action Healthletter*, July/August 1998, pp. 3–6.

56. J. K. Binkley, J. Eales, and M. Jekanowski, The relation between dietary change and rising U.S. obesity, *International Journal of Obesity and Related Metabolic Disorders* 24 (2000): 1032–1039.

57. Blundell and Macdiarmid, 1997.

58. M. Wei and coauthors, The association between cardiorespiratory fitness and impaired glucose and type 2 diabetes mellitus in men, *Annals of Internal Medicine* 130 (1999): 89–96; U.S. Department of Health and Human Services, *Physical Activity and Health—A Report of the Surgeon General Executive Summary*, 1996.

59. A. H. C. Goris, M. S. Westerterp-Plantenga, and K. R. Westerterp, Undereating and underrecording of habitual food intake in obese men: Selective underreporting of fat intake, *American Journal of Clinical Nutrition* 71 (2000): 130–134; L. A. Braam and coauthors, Determinants of obesity-related underreporting of energy intake, *American Journal of Epidemiology* 147 (1998): 1081–1086; L. Johansson and coauthors, Under- and overreporting of energy intake related to weight status and lifestyle, *American Journal of Clinical Nutrition* 68 (1998): 266–274; L. M. Carter and S. J. Whiting, Underreporting of energy intake, socioeconomic status, and expression of nutrient intake, *Nutrition Reviews* 56 (1998): 179–182.

60. D. A. Schoeller, Balancing energy expenditure and body weight, *American Journal of Clinical Nutrition* 68 (1998): S956–S961.

61. J. M. Rippe and S. Hess, The role of physical activity in the prevention and management of obesity, *Journal of the American Dietetic Association* 98 (1998): S31–S38.

62. National Heart, Lung, and Blood Institute Expert Panel, 1998, p. 78.

63. M. M. McGrane, Carbohydrate metabolism—Synthesis and oxidation, in M. H. Stipanuk, ed., *Biochemical and Physiological Aspects of Human Nutrition* (Philadelphia: W. B. Saunders, 2000), p. 192.

64. R. P. Ferraris and H. V. Carey, Intestinal transport during fasting and malnutrition, *Annual Review of Nutrition* 20 (2000): 195–219.

65. J. Polivy, Psychological consequences of food restriction, *Journal of the American Dietetic Association* 96 (1996): 589–592.

66. National Task Force on the Prevention and Treatment of Obesity, Dieting and the development of eating disorders in overweight and obese adults, *Archives of Internal Medicine* 260 (2000): 2581–2589.

67. S. M. Shick and coauthors, Persons successful at long-term weight loss and maintenance continue to consume a low-energy, low-fat diet, *Journal of the American Dietetic Association* 98 (1998): 408–413.

68. S. D. Poppitt and coauthors, Long-term effects of ad libitum low-fat, high-carbohydrate diets on body weight and serum lipids in overweight subjects with metabolic syndrome, *American Journal of Clinical Nutrition* 75 (2002): 11–20.

69. E. T. Kennedy and coauthors, Popular diets: Correlation to health, nutrition, and obesity, *Journal of the American Dietetic Association* 101 (2001): 411–420.

70. USDA coordinated nutrition research program on health and nutrition effects of popular weight-loss diets, *USDA Backgrounder*, 2001, available at www.usda.gov/news/releases/2001/01/whitebac.htm.

71. P. M. Suter, E. Häsler, and W. Vetter, Effects of alcohol on energy metabolism and body weight regulation: Is alcohol a risk factor for obesity? *Nutrition Reviews* 55 (1997): 157–171; Y. Sakurai and coauthors, Relation of total and beverage-specific alcohol intake to body mass index and waist-to-hip ratio: A study of self-defense officials in Japan, *European Journal of Epidemiology* 13 (1997): 893–898.

72. D. Festi and coauthors, Gallbladder motility and gallstone formation in obese patients following very low calorie diets: Use it (fat) to lose it (well), *International Journal of Obesity and Related Metabolic Disorders* 22 (1998): 592–600.

73. National Heart, Lung, and Blood Institute, National Institutes of Health, *The Practical Guide: Identification, Evaluation, and Treatment of Overweight and Obesity in Adults*, NIH publication no. 00-4084 (Washington, D.C.: Government Printing Office, 2000), p. 26.

74. National Heart, Lung, and Blood Institute Expert Panel, 1998, p. 75.

75. National Heart, Lung, and Blood Institute Expert Panel, 1998, p. 74.

76. A. Golay and E. Bobbioni, The role of dietary fat in obesity, *International Journal of Obesity and Related Metabolic Disorders* 21 (1997): S2–S11.

77. W. J. McCarthy, Strategies for achieving long-term weight maintenance (letter), *Journal of the American Dietetic Association* 98 (1998): 1273.

78. L. K. Khan and coauthors, Use of prescription weight loss pills among U.S. adults in 1996–1998, *Annals of Internal Medicine* 134 (2001): 282–286.

79. FDA approves implanted stomach band to treat severe obesity, *FDA Talk Paper*, June 5, 2001, available at www.fda.gov.

80. E. C. Mun, G. L. Blackburn, and J. B. Matthews, Current status of medical and surgical therapy for obesity, *Gastroenterology* 120 (2001): 669–681.

81. J. G. Bruner and R. H. de Jong, Lipoplasty claims experience of U.S. insurance companies, *Plastic and Reconstructive Surgery* 107 (2001): 1285–1291.

82. R. B. Rao, S. F. Ely, and R. S. Hoffman, Deaths related to liposuction, *New England Journal of Medicine* 340 (1999): 1471–1475.

83. FDA warns against drug promotion or "herbal fen-phen," *FDA Talk Paper*, 1997, available from the Food and Drug Administration, 5600 Fishers Lane, Rockville, MD 20857.

84. W. K. Jones, Safety of dietary supplements containing ephedrine alkaloids, FDA Public Meeting Summary, August 2000, available at http://vm.cfsan.fda.gov/~dms/.

85. P. Kurtzweil, Dieter's brews make tea time a dangerous affair, *FDA Consumer*, July/August 1997, pp. 6–11.

86. USDA Coordinated nutrition research program on health and nutrition effects of popular weight-loss diets, 2001.

87. M. T. McGuire and coauthors, Long-term maintenance of weight loss: Do people who lose weight through various weight loss methods use different behaviors to maintain their weight? *International Journal of Obesity and Related Metabolic Disorders* 22 (1998): 572–577.

88. W. J. Pasman, M. S. Westerterp-Plantenga, and W. H. Saris, The effectiveness of long-term supplementation of carbohydrate, chromium, fiber and caffeine on weight maintenance, *International Journal of Obesity and Related Metabolic Disorders* 21 (1997): 1143–1151.

89. P. D. Wood, Clinical applications of diet and physical activity in weight loss, *Nutrition Reviews* 54 (1998): S131–S135; M. L. Klem and coauthors, A descriptive study of individuals successful at long-term maintenance of substantial weight loss, *American Journal of Clinical Nutrition* 66 (1997): 239–246.

90. McCarthy, 1998.
91. Howarth, Saltzman, and Roberts, 2001.
92. J. A. Yanovski and coauthors, A prospective study of holiday weight gain, *New England Journal of Medicine* 342 (2000): 861–867; S. B. Roberts, Holiday weight gain: Fact or fiction? *Nutrition Reviews* 58 (2000): 378–379.
93. K. A. Petersmark and coauthors, The effect of weight cycling on blood lipids and blood pressure in the Multiple Risk Factor intervention Trial Special Intervention group, *International Journal of Obesity and Related Metabolic Disorders* 23 (1999): 1246–1255.
94. Klem and coauthors, 1997.

Consumer Corner 9

1. S. L. Booth and coauthors, Environmental and societal factors affect food choice and physical activity: Rationale, influences, and leverage points, *Nutrition Reviews* 59 (2001): S21–S39.
2. L. R. Young and M. Nestle, The contribution of expanding portion sizes to the US obesity epidemic, *American Journal of Public Health* 92 (2002): 246–249.
3. M. Nestle and M. F. Jacobson, Halting the obesity epidemic: A public health policy approach, *Public Health Reports* 115 (2000): 12–24.

Controversy 9

1. C. A. Arriaza and T. Mann, Ethnic differences in eating disorder symptoms among college students: The confounding role of body mass index, *Journal of American College Health* 49 (2001): 309–315.
2. Position of The American Dietetic Association: Nutrition intervention in the treatment of anorexia nervosa, bulimia nervosa, and eating disorders not otherwise specified (EDNOS), *Journal of the American Dietetic Association* 101 (2001): 810–819.
3. D. Neumark-Sztainer and P. J. Hannan, Weight-related behaviors among adolescent girls and boys: Results from a national survey, *Archives of Pediatrics and Adolescent Medicine* 154 (2000): 569–577.
4. S. Haberman and D. Luffey, Weighing in college students' diet and exercise behaviors, *Journal of American College Health* 46 (1998): 189–191.
5. J. Polivy and C. P. Herman, Causes of eating disorders, *Annual Review of Psychology* 53 (2002): 187–213; G. Tsai, Eating disorders in the Far East, *Eating and Weight Disorders* 5 (2000): 183–197.
6. J. A. McLean, S. I. Barr, and J. C. Prior, Cognitive dietary restraint is associated with higher urinary cortisol excretion in healthy premenopausal women, *American Journal of Clinical Nutrition* 73 (2001): 7–12; G. C. Patton and coauthors, Onset of adolescent eating disorders: Population based cohort study over 3 years, *British Journal of Medicine* 318 (1999): 765–768.
7. Girls in the 90's: Working to undo stereotypes, *Eating Disorders Review*, September/October 1997, pp. 6–7.
8. E. Cooley and T. Toray, Disordered eating in college freshman women: A prospective study, *Journal of American College Health* 49 (2001): 229–235.
9. R. V. West, The female athlete: The triad of disordered eating, amenorrhea and osteoporosis, *Sports Medicine* 26 (1998): 63–71.
10. N. Bettle and coauthors, Adolescent ballet school students: Their quest for body weight change, *Psychopathology* 31 (1998): 153–159.
11. N. A. Armsey, Stress injury to bone in the female athlete, *Clinical Sports Medicine* 16 (1997): 197–224.
12. M. Hotta and coauthors, The importance of body weight history in the occurrence and recovery of osteoporosis in patients with anorexia nervosa: Evaluation by dual X-ray absorptiometry and bone metabolic markers, *European Journal of Endocrinology* 139 (1998): 276–283.
13. E. R. Brooks, B. W. Ogden, and D. S. Cavalier, Compromised bone density 11.4 years after diagnosis of anorexia nervosa, *Journal of Women's Health* 7 (1998): 567–574.
14. M. L. Shoene and C. Nelson, An ominous addition to the female athletic triad? *Sports Medicine Digest*, 23 (2001): 73, 80–81.
15. Are you finding and treating males with eating disorders at your school? *School Health Professional*, August 2000, pp. 1–2.
16. Hyperthermia and dehydration-related deaths associated with intentional rapid weight loss in three collegiate wrestlers—North Carolina, Wisconsin, and Michigan, November-December 1997, *Morbidity and Mortality Weekly Report* 47 (1998): 105–108.
17. American College of Sports Medicine, Position stand: Weight loss in wrestlers, *Medicine and Science in Sports and Exercise* 28 (1996): ix–xii.
18. H. G. Pope, Jr., and coauthors, Muscle dysmorphic—An underrecognized form of body dysmorphic disorder, *Psychosomatics* 38 (1997): 548–557.
19. A. Gila and coauthors, Subjective body-image dimensions in normal and anorexic adolescents, *British Journal of Medical Psychology* 71 (1998): 175–184.
20. G. Waller, Perceived control in eating disorders: Relationship with reported sexual abuse, *International Journal of Eating Disorders* 23 (1998): 213–216.
21. T. Pryor and M. W. Weiderman, Personality features and expressed concerns of adolescents with eating disorders, *Adolescence* 33 (1998): 291–300.
22. P. Santonastaso, A. Sala, and A. Favaro, Water intoxication in anorexia nervosa: A case report, *International Journal of Eating Disorders* 24 (1998): 439–442.
23. D. M. McLoughlin and coauthors, Structural and functional changes in skeletal muscle in anorexia nervosa, *Acta Neuropathologica* 95 (1998): 632–640.
24. C. Panagiotopoulos and coauthors, Electrocardiographic findings in adolescents with eating disorders, *Pediatrics* 105 (2000): 1100–1105.
25. G. Addolorato and coauthors, A case of marked cerebellar atrophy in a woman with anorexia nervosa and cerebral atrophy and a review of the literature, *International Journal of Eating Disorders* 24 (1998): 443–447; L. M. Allende and coauthors, Immunodeficiency associated with anorexia nervosa is secondary and improves after refeeding, *Immunology* 94 (1998): 543–551; V. W. Swayze, Brain imaging and eating disorders, *Eating Disorders Review*, May/June 1997, pp. 1–4.
26. G. F. Russell, J. Treasure, and I. Eisler, Mothers with anorexia nervosa who underfeed their children: Their recognition and management, *Psychological Medicine* 28 (1998): 93–108.
27. Position of The American Dietetic Association, 2001.
28. A. E. Becker and coauthors, Eating disorders, *New England Journal of Medicine* 340 (1999): 1092–1098.
29. J. T. Dwyer, Adolescence, in E. E. Ziegler and L. J. Filer, Jr., eds., *Present Knowledge in Nutrition*, 7th ed. (Washington, D.C.: International Life Sciences Institute Press, 1996), pp. 404–413.
30. M. Moukaddem and coauthors, Increase in diet-induced thermogenesis at the start of refeeding in severely malnourished anorexia nervosa patients, *American Journal of Clinical Nutrition* 66 (1997): 133–140.
31. E. D. Eckert and coauthors, Leptin in anorexia nervosa, *Journal of Clinical Endocrinology and Metabolism* 83 (1998): 791–795.
32. L. Scalfi and coauthors, Body composition changes in patients with anorexia nervosa after complete weight recovery, *European Journal of Clinical Nutrition* 56 (2002): 15–20; L. Mayer, Body composition and anorexia nervosa: Does physiology explain psychology? *American Journal of Clinical Nutrition* 73 (2001): 851–852; S. Grinspoon and coauthors, Changes in regional fat distribution and the effects of estrogen during spontaneous weight gain in women with anorexia nervosa, *American Journal of Clinical Nutrition* 73 (2001): 865–869; K. M. Pike, Long-term course of anorexia nervosa: Response, relapse, remission, and recovery, *Clinical Psychology Reviews* 18 (1998): 447–475.
33. D. B. Herzog and coauthors, Mortality in eating disorders: A descriptive study, *International Journal of Eating Disorders* 28 (2000): 20–26.
34. M. L. Granner, D. A. Abood, and D. R. Black, Racial differences in eating disorder attitudes, cigarette, and acohol use, *American Journal of Health Behavior* 25 (2001): 83–99; N. A. Troop, A. Holbrey, and J. L. Treasure, Stress, coping, and crisis support in eating disorders, *International Journal of Eating Disorders* 24 (1998): 157–166.
35. C. G. Fairburn and coauthors, Risk factors for bulimia nervosa: A community-based case-control study, *Archives of General Psychiatry* 54 (1997): 509–517.
36. D. Neumark-Sztainer and coauthors, Disordered eating among adolescents: Associations with sexual/physical abuse and other familial/psychosocial factors, *International Journal of Eating Disorders* 28 (2000): 249–258.
37. Waller, 1998.
38. K. A. Gendall and coauthors, The nutrient intake of women with bulimia nervosa, *International Journal of Eating Disorders* 21 (1997): 115–127.

39. J. Sundgot-Borgen and coauthors, The effect of exercise, cognitive therapy, and nutritional counseling in treating bulimia nervosa, *Medicine and Science in Sports and Exercise* 34 (2002): 190–195.

40. P. K. Keel and J. E. Mitchell, Outcome in bulimia nervosa, *American Journal of Psychiatry* 154 (1997): 313–321.

41. A. F. Schatzberg, New indications for antidepressants, *Journal of Clinical Psychiatry* 61 (2000): S9–S17.

42. S. Rubinstein and B. Caballero, Is Miss America an undernourished role model? *Journal of the American Medical Association* 283 (2000): 1569.

43. Federal Interagency Forum on Child and Family Statistics, *America's Children: Key National Indicators of Well-Being, 1999,* a report from the National Institutes of Health, available from National Maternal and Child Health Clearinghouse, 2070 Chain Bridge Road, Suite 450, Vienna, VA 22182 or on the Internet at http://childstats.gov.

44. W. Kaye, K. Gendall, and M. Strober, Serotonin neuronal function and selective serotonin reuptake inhibitor treatment in anorexia and bulimia nervosa, *Biological Psychiatry* 44 (1998): 825–838; B. Baranowska and coauthors, Neuropeptide Y, galanin, and leptin release in obese women and in women with anorexia nervosa, *Metabolism* 46 (1997): 1384–1389; B. E. Wolfe, E. Metzger, and D. C. Jimerson, Research update on serotonin function in bulimia nervosa and anorexia nervosa, *Psychopharmacology Bulletin* 33 (1997): 345–354.

Chapter 10

1. A. L. Dunn and coauthors, Comparison of lifestyle and structured interventions to increase physical activity and cardiorespiratory fitness, *Journal of the American Medical Association* 281 (1999): 327–334.

2. Y. A. Kesaniemi and coauthors, Dose-response issues concerning physical activity and health: An evidence-based symposium, *Medicine and Science in Sports and Exercise* 33 (2001): 351S–358S; U. M. Kujala and coauthors, Relationship of leisure-time physical activity and mortality, *Journal of the American Medical Association* 279 (1998): 440–444.

3. F. B. Hu and coauthors, Physical activity and risk of stroke in women, *Journal of the American Medical Association* 283 (2000): 2961–2967; J. E. Manson and coauthors, A prospective study of walking as compared with vigorous exercise in the prevention of coronary heart disease in women, *New England Journal of Medicine* 341 (1999): 650–658; M. L. Slattery and coauthors, Lifestyle and colon cancer: An assessment of factors associated with risk, *American Journal of Epidemiology* 150 (1999): 869–877; S. A. Oliveria and P. J. Christos, The epidemiology of physical activity and cancer, *Annals of the New York Academy of Sciences* 833 (1997): 79–90; G. Perseghin and coauthors, Increased glucose transport-phosphorylation and muscle glycogen synthesis after exercise training in insulin-resistant subjects, *New England Journal of Medicine* 335 (1996): 1357–1362; S. N. Blair, Physical inactivity and cardiovascular disease risk in women, *Medicine and Science in Sports and Exercise* 28 (1996): 9–10; NIH Consensus Development Panel on Physical Activity and Cardiovascular Health, Physical activity and cardiovascular health, *Journal of the American Medical Association* 276 (1996): 241–246.

4. American College of Sports Medicine, Position stand: The recommended quantity and quality of exercise for developing and maintaining cardiorespiratory and muscular fitness, and flexibility in healthy adults, *Medicine and Science in Sports and Exercise* 30 (1998): 975–991.

5. D. A. Leaf, D. L. Parker, and D. Schaad, Changes in VO_2 max, physical activity, and body fat with chronic exercise: Effects on plasma lipids, *Medicine and Science in Sports and Exercise* 29 (1997): 1152–1159; P. T. Williams, Relationship of distance run per week to coronary heart disease risk factors in 8283 male runners, *Archives of Internal Medicine* 157 (1997): 191–198.

6. American College of Sports Medicine, Position Stand: Progression models in resistance training for healthy adults, *Medicine and Science in Sports and Exercise,* 34 (2002): 364–380; M. L. Pollock and coauthors, AHA Science Advisory: Resistance exercise in individuals with and without cardiovascular disease: Benefits, rationale safety, and prescription, *Circulation* 101 (2000): 828–833.

7. American College of Sports Medicine, Position stand: Exercise and physical activity for older adults, *Medicine and Science in Sports and Exercise* 30 (1998): 992–1008; P. A. Ades and coauthors, Weight training improves walking endurance in healthy elderly persons, *Annals of Internal Medicine* 124 (1996): 568–572.

8. J. E. Layne and M. E. Nelson, The effects of progressive resistance training on bone density: A review, *Medicine and Science in Sports and Exercise* 31 (1999): 25–30.

9. L. Metcalfe and coauthors, Postmenopausal women and exercise for prevention of osteoporosis: The bone, estrogen, strength training (BEST) study, *ACSM's Health and Fitness Journal,* May/June 2001, pp. 6–14.

10. P. T. Williams, High-density lipoprotein cholesterol and other risk factors for coronary heart disease in female runners, *New England Journal of Medicine* 334 (1996): 1298–1303.

11. C. M. Donovan and K. D. Sumida, Training enhanced hepatic gluconeogenesis: The importance for glucose homeostasis during exercise, *Medicine and Science in Sports and Exercise* 29 (1997): 628–634.

12. C. Williams and C. Chryssanthopoulos, Pre-exercise food intake and performance, in A. P. Simopoulos and K. N. Pavlou, eds., *Nutrition and Fitness: Metabolic and Behavioral Aspects in Health and Disease* (New York: Karger, 1997), pp. 33–45.

13. E. Coleman, Carbohydrate and exercise, in C. A. Rosenbloom, ed., *Sports Nutrition: A Guide for the Professional Working with Active People,* 3rd ed. (Chicago: The American Dietetic Association, 2000), pp. 13–31.

14. J. A. M. Parkin and coauthors, Muscle glycogen storage following prolonged exercise: Effect of timing of ingestion of high glycemic index food, *Medicine and Science in Sports and Exercise* 29 (1997): 220–224.

15. Coleman, 2000; L. M. Burke, G. R. Collier, and M. Hargreaves, Glycemic index—A new tool in sports nutrition? *International Journal of Sports Nutrition* 8 (1998): 401–415.

16. A. R. Coggan, Plasma glucose metabolism during exercise: Effect of endurance training in humans, *Medicine and Science in Sports and Exercise* 29 (1997): 620–627.

17. Diabetes in the elite athlete, *Sports Medicine Digest* 18 (1998): 109–111.

18. P. J. Horvath and coauthors, The effects of varying dietary fat on performance and metabolism in trained male and female runners, *Journal of the American College of Nutrition* 19 (2000): 52–60.

19. E. Coleman, Does a low-fat diet impair nutrition and performance? *Sports Medicine Digest* 22 (2000): 41; Position of The American Dietetic Association, Dietitians of Canada, and the American College of Sports Medicine: Nutrition and athletic performance, *Journal of the American Dietetic Association* 100 (2000): 1543–1556.

20. P. W. R. Lemon, Is increased dietary protein necessary or beneficial for individuals with a physically active lifestyle? *Nutrition Reviews* 54 (1996): S169–S175.

21. Lemon, 1996.

22. Position of The American Dietetic Association, Dietitians of Canada, and the American College of Sports Medicine, 2000.

23. M. J. Webster, Physiological and performance responses to supplementation with thiamin and pantothenic acid derivatives, *European Journal of Applied Physiology* 77 (1998): 486–491.

24. R. S. Virk and coauthors, Effect of vitamin B-6 supplementation on fuels, catecholamines, and amino acids during exercise in men, *Medicine and Science in Sports and Exercise* 31 (1999): 400–408.

25. S. K. Powers, L. L. Ji, and C. Leeuwenburgh, Exercise training-induced alterations in skeletal muscle antioxidant capacity: A brief review, *Medicine and Science in Sports and Exercise* 31 (1999): 987–997; J. M. McBride and coauthors, Effect of resistance exercise on free radical production, *Medicine and Science in Sports and Exercise* 30 (1998): 67–72; D. A. Leaf and coauthors, The effect of exercise intensity on lipid peroxidation, *Medicine and Science in Sports and Exercise* 29 (1997): 1036–1039; R. A. Fielding and M. Meydani, Exercise, free radical generation, and aging, *Aging: Clinical and Experimental Research* 9 (1997): 12–18.

26. P. M. Clarkson and H. S. Thompson, Antioxidants: What role do they play in physical activity and health? *American Journal of Clinical Nutrition* 72 (2000): S637–S646; K. V. Reddy and coauthors, Pulmonary lipid peroxidation and antioxidant defenses during exhaustive physical exercise: The role of vitamin E and selenium, *Nutrition* 14 (1998): 448–451; McBride and coauthors, 1998; L. Grievink and coauthors, Acute effects of ozone on pulmonary function of cyclists receiving antioxidant supplements, *Occupational and Environmental Medicine* 55 (1998): 13–17; M. Kanter, Free radicals, exercise and antioxidant supplementation, *Proceedings of the Nutrition Society* 57 (1998): 9–13.

27. W. J. Evans, Vitamin E, vitamin C, and exercise, *American Journal of Clinical Nutrition* 72 (2000): S647–S652; L. Packer, Oxidants, antioxidant nutrients, and the athlete, *Journal of Sports Science* 15 (1997): 353–363.

28. L. E. Armstrong and C. M. Maresh, Vitamin and mineral supplements as nutritional aids to exercise performance and health, *Nutrition Reviews* 54 (1996): S149–S158.

29. R. V. West, The female athlete: The triad of disordered eating, amenorrhea, and osteoporosis, *Sports Medicine* 26 (1998): 63–71.

30. J. Beard and B. Tobin, Iron status and exercise, *American Journal of Clinical Nutrition* 72 (2000): S594–S597.

31. M. F. Waller and E. M. Haymes, The effects of heat and exercise on sweat iron loss, *Medicine and Science in Sports and Exercise* 28 (1996): 197–203.

32. Beard and Tobin, 2000; E. Coleman, Nutritional concerns of vegetarian athletes, *Sports Medicine Digest* 20 (1998): 22–23.

33. D. C. Nieman, Physical fitness and vegetarian diets: Is there a relation? *American Journal of Clinical Nutrition* 70 (1999): S570–S575.

34. E. R. Eichner, Anemia in female athletes, *Sports Medicine Digest* 22 (2000): 42–43; Z. Y. Haas, Iron depletion without anemia and physical performance in young women, *American Journal of Clinical Nutrition* 66 (1997): 334–341.

35. H. C. Lukaski, Magnesium, zinc, and chromium nutriture and physical activity, *American Journal of Clinical Nutrition* 72 (2000): S585–S593.

36. Lukaski, 2000.

37. M. N. Sawka and S. J. Montain, Fluid and electrolyte supplementation for exercise heat stress, *American Journal of Clinical Nutrition* 72 (2000): S564–S572; C. V. Gisolfi, Fluid balance for optimal performance, *Nutrition Reviews* 54 (1996): S159–S168.

38. American College of Sports Medicine, Position stand: Heat and cold illness during distance running, *Medicine and Science in Sports and Exercise* 28 (1996): i–x.

Consumer Corner 10

1. X. Shi and C. V. Gisolfi, Fluid and carbohydrate replacement during intermittent exercise, *Sports Medicine* 25 (1998): 157–172.

2. R. Murray, Fluid and electrolytes, in C. A. Rosenbloom, ed., *Sports Nutrition: A Guide for the Professional Working with Active People*, 3rd ed. (Chicago: The American Dietetic Association, 2000), pp. 95–106.

3. American College of Sports Medicine, Position stand: Heat and cold illness during distance running, *Medicine and Science in Sports and Exercise* 28 (1996): i–x.

4. J. H. Wilmore and coauthors, Role of taste preference on fluid intake during and after 90 minutes of running at 60% of VO_2 max in the heat, *Medicine and Science in Sports and Exercise* 30 (1998): 587–595.

Controversy 10

1. E. Blomstrand, Amino acids and central fatigue, *Amino Acids* 20 (2001): 25–34; K. D. Mittleman, M. R. Ricci, and S. P. Bailey, Branched-chain amino acids prolong exercise during heat stress in men and women, *Medicine and Science in Sports and Exercise* 30 (1998): 83–91; P. Calders and coauthors, Pre-exercise branched-chain amino acid administration increases endurance performance in rats, *Medicine and Science in Sports and Exercise* 29 (1997): 1182–1186; M. D. Vukovich and coauthors, Effects of a low-dose amino acid supplement on adaptations to cycling training in untrained individuals, *International Journal of Sports Nutrition* 7 (1997): 298–309.

2. M. J. Gibala, Regulation of skeletal muscle amino acid metabolism during exercise, *International Journal of Sports Nutrition and Exercise Metabolism* 11 (2001): 87–108; Mittleman, Ricci, and Bailey, 1998; E. Coleman, Branched-amino acids and fatigue, *Sports Medicine Digest* 18 (1996): 44.

3. M. J. Rennie and K. D. Tipton, Protein and amino acid metabolism during and after exercise and the effects of nutrition, *Annual Review of Nutrition* 20 (2000): 457–483; R. R. Wolfe, Protein supplements and exercise, *American Journal of Clinical Nutrition* 72 (2000): S551–S557.

4. C. D. Paton, W. G. Hopkins, and L. Vollebregt, Little effect of caffeine ingestion on repeated sprints in team-sport athletes, *Medicine and Science in Sports and Exercise* 33 (2001): 822–825; C. J. Sinclair and J. D. Geiger, Caffeine use in sports: A pharmacological review, *Journal of Sports Medicine and Physical Fitness* 40 (2000): 71–79; L. R. Bucci, Dietary supplements as ergogenic aids, in I. Wolinsky, ed., *Nutrition in Exercise and Sport*, 3rd ed. (New York: CRC Press, 1998), pp. 315–368; T. E. Graham and L. L. Spriet, Caffeine and exercise performance, *Sport Science Exchange* 1 (1996): 9.

5. C. R. Bruce and coauthors, Enhancement of 2000-m rowing performance after caffeine ingestion, *Medicine and Science in Sports and Exercise* 32 (2000): 1958–1963; M. E. Anderson and coauthors, Improved 2000-meter rowing performance in competitive oarswomen after caffeine ingestion, *International Journal of Sports Nutrition and Exercise Metabolism* 10 (2000): 464–475.

6. Paton, Hopkins, and Vollebregt, 2001.

7. T. E. Graham and coauthors, Caffeine ingestion does not alter carbohydrate or fat metabolism in human skeletal muscle during exercise, *Journal of Physiology* 529 (2000): 837–847.

8. D. Laurent and coauthors, Effects of caffeine on muscle glycogen utilization and the neuroendocrine axis during exercise, *Journal of Clinical Endocrinology and Metabolism* 85 (2000): 2170–2175.

9. P. R. Sticker, Other ergogenic agents, *Clinics in Sports Medicine* 17 (1998): 283–297.

10. G. W. Evans, The effect of chromium picolinate on insulin-controlled parameters in humans, *Journal of Biosocial Medicine Research* 11 (1989): 163–180.

11. J. M. Davis, R. S. Welsh, and N. A. Alerson, Effects of carbohydrate and chromium ingestion during intermittent high-intensity exercise to fatigue, *International Journal of Sports Nutrition and Exercise Metabolism* 10 (2000): 476–485; H. C. Lukaski and coauthors, Chromium supplementation and resistance training: Effects on body composition, strength, and trace element status of men, *American Journal of Clinical Nutrition* 63 (1996): 954–965; M. A. Hallmark and coauthors, Effects of chromium and resistance training on muscle strength and body composition, *Medicine and Science in Sports and Exercise* 28 (1996): 139–144.

12. W. R. Fuller, Suspected chromium picolinate–induced rhabdomyolysis, *Pharmacotherapy* 18 (1998): 860–862.

13. J. F. Fowler, Systemic contact dermatitis caused by oral chromium picolinate, *Cutis* 65 (2000): 116.

14. J. B. Vincent, The biochemistry of chromium, *Journal of Nutrition* 130 (2000): 715–718.

15. E. A. Applegate and L. E. Grivetti, Search for the competitive edge: A history of dietary fads and supplements, *Journal of Nutrition* 127 (1997): S869–S873.

16. D. Preen and coauthors, Effect of creatine loading on long-term sprint exercise performance and metabolism, *Medicine and Science in Sports and Exercise* 33 (2001): 814–821; T. Ziegenfuss and coauthors, Performance benefits following a five-day creatine loading procedure persist for at least four weeks, *Medicine and Science in Sports and Exercise* 30 (1998): S265; R. B. Kreider and coauthors, Effects of creatine supplementation on body composition, strength, and sprint performance, *Medicine and Science in Sports and Exercise* 30 (1998): 73–82.

17. J. D. Gilliam, C. Hohzom, and A. D. Martin, Effect of oral creatine supplementation on isokinetic force production, *Medicine and Science in Sports and Exercise* 30 (1998): S140; L. M. Odland and coauthors, Effect of oral creatine supplementation on muscle [PCr] and short-term maximum power output, *Medicine and Science in Sports and Exercise* 29 (1997): 216–219.

18. E. B. Feldman, Creatine: A dietary supplement and ergogenic aid, *Nutrition Reviews* 57 (1999): 45–50.

19. M. Greenwood and coauthors, Creatine supplementation patterns and perceived effects in select division I collegiate athletes, *Clinical Journal of Sports Medicine* 10 (2000): 191–194.

20. T. Noakes, as quoted in M. Gaie, Olympic athletes face heat, other health hurdles, *Journal of the American Medical Association* 276 (1996): 231–237.

21. Applegate and Grivetti, 1997.

22. D. H. Catlin and T. H. Murray, Performance-enhancing drugs, fair competition and Olympic sport, *Journal of the American Medical Association* 276 (1996): 231–237.

23. R. C. Daly and coauthors, Cerebrospinal fluid and behavioral changes after methyltestosterone administration, *Archives of General Psychiatry* 58 (2001): 172–177.

24. I. Thiblin, O. Lindquist, and J. Rajs, Cause and manner of death among users of anabolic androgenic steroids, *Journal of Forensic Science* 45 (2000): 16–23.

25. Andrew Taber, 'Roid rage, *Salon.com*, November 18, 1999, www.salon.com/health/feature/1999/11/18/steroids/index.html.

26. B. B. Rasmussen and coauthors, Androstenedione does not stimulate muscle protein anabolism in young healthy men, *Journal of Clinical Endocrinology and Metabolism* 85 (2000): 55–59.

27. G. A. Brown and coauthors, Effects of anabolic precursors on serum testosterone concentration and adaptations to resistance training in young men, *International Journal of Sports Nutrition and Exercise Metabolism* 10 (2000): 340–359.
28. R. Skinner, E. Coleman, and C. A. Rosenbloom, Ergogenic aids, in C. A. Rosenbloom, ed., *Sports Nutrition: A Guide for the Professional Working with Active People*, 3rd ed. (Chicago: The American Dietetic Association, 2000), pp.107–146; E. Coleman, DHEA—An anabolic aid? *Sports Medicine Digest* 18 (1996): 140–141.
29. D. H. Catlin and coauthors, Trace contamination of over-the-counter androstenedione and positive urine test results for a nandrolone metabolite, *Journal of the American Medical Association* 284 (2000): 2618–2621.
30. G. Lombardi and coauthors, Is growth hormone bad for your heart? Cardiovascular impact of GH deficiency and of acromegaly, *Journal of Endocrinology* 155 (1997): S33–S37.
31. FDA warns against consuming dietary supplements containing tiratricol, *FDA Talk Paper*, November 2001, available from http://vm.cfsan.fda.gov.
32. R. M. Philen and coauthors, Survey of advertising for nutritional supplements in health and bodybuilding magazines, *Journal of the American Medical Association* 268 (1992): 1008–1011.

Chapter 11

1. S. H. Gillespie, Antibiotic resistance in the absence of selective pressure, *International Journal of Antimicrobial Agents* 17 (2001): 171–176; D. K. Warren and V. J. Fraser, Infection control measures to limit antimicrobial resistance, *Critical Care Medicine* 29 (2001): N128–N134; Centers for Disease Control and Prevention, The global HIV and AIDS epidemic, 2001, *Morbidity and Mortality Weekly Report* 50 (2001): 434–439; A. E. Platt, Confronting infectious diseases, in L. R. Brown, ed., *State of the World 1996: A Worldwatch Institute Report on Progress toward a Sustainable Society* (New York: Norton, 1996), pp. 114–132.
2. M. J. Stampfer and coauthors, Primary prevention of coronary heart disease in women through diet and lifestyle, *New England Journal of Medicine* 343 (2000): 16–22; A. K. Kant and coauthors, A prospective study of diet quality and mortality in women, *Journal of the American Medical Association* 283 (2000): 2109–2115.
3. R. K. Chandra, Nutrition and the immune system: An introduction, *American Journal of Clinical Nutrition* 66 (1997): S460–S463.
4. C. J. Field, I. R. Johnson, P. D. Schley, Nutrients and their role in host resistance to infection, *Journal of Leukocyte Biology* 71 (2002): 16–32; N. S. Scrimshaw and J. P. SanGiovanni, Synergism of nutrition, infection, and immunity: An overview, *American Journal of Clinical Nutrition* 66 (1997): S464–S477.
5. A. Nimmagadda and coauthors, The significance of vitamin A and carotenoid status in persons infected by the human immunodeficiency virus, *Clinical Infectious Diseases* 26 (1998): 711–718; M. K. Baum and coauthors, High risk of HIV-related mortality is associated with selenium deficiency, *Journal of Acquired Immune Deficiency Syndromes and Human Retrovirology* 15 (1997): 370–374.
6. J. S. Young, HIV and medical nutrition therapy, *Journal of the American Dietetic Association* 97 (1997): S161–S166.
7. W. J. Evans, R. Roubenoff, and A. Shevitz, Exercise and the treatment of wasting: Aging and human immunodeficiency virus infection, *Seminars in Oncology* 25 (1998): 112–122.
8. M. Serafini, Dietary vitamin E and T cell-mediated function in the elderly: Effectiveness and mechanism of action, *International Journal of Developmental Neuroscience* 18 (2000): 401–410; A. Valenti and coauthors, Effects of vitamin E and prostaglandin E2 on expression of CREB1 and CREB2 proteins by human T lymphocytes, *Physiological Research* 49 (2000): 363–368; C. Y. Lee and J. Man-Fan Wan, Vitamin E supplementation improves cell-mediated immunity and oxidative stress of Asian men and women, *Journal of Nutrition* 130 (2000): 2932–2937; S. N. Meydani and A. A. Beharka, Recent developments in vitamin E and immune response, *Nutrition Reviews* 56 (1998): S49–S58.
9. J. C. Fleet, Zinc, copper, and manganese, in M. H. Stipanuk, ed., *Biochemical and Physiological Aspects of Human Nutrition* (Philadelphia: Saunders, 2000), p. 753; Standing Committee on the Scientific Evaluation of Dietary Reference Intakes, Food and Nutrition Board, Institute of Medicine, *Dietary Reference Intakes for Vitamin A, Vitamin K, Arsenic, Boron, Chromium, Copper, Iodine, Iron, Manganese, Molybdenum, Nickel, Silicon, Vanadium, and Zinc* (Washington, D.C.: National Academy Press, 2001), pp. 12–30; Scrimshaw and SanGiovanni, 1997.
10. Position of The American Dietetic Association: The role of nutrition in health promotion and disease prevention programs, *Journal of the American Dietetic Association* 98 (1998): 205–208.
11. A. J. Blumenfeld and coauthors, Nutritional aspects of prostate cancer: A review, *Canadian Journal of Urology* 7 (2000): 927–935; W. R. Fair, N. E. Fleshner, and W. Heston, Cancer of the prostate: A nutritional disease? *Urology* 50 (1997): 840–848.
12. A. P. Simopoulos, Diet and gene interactions, *Food Technology* 51 (1997): 66–69.
13. Prepare for the future: Know your ancestors, *Consumer Reports on Health*, September 1999, pp. 1, 3–6.
14. Age-adjusted death rates for 1997, percentage of change in age-adjusted death rates for the 15 leading causes of death, 1996–1997 and 1979–1997, and ratio of age-adjusted death rates, by sex and race of decedent, 1997—United States, *Morbidity and Mortality Weekly Report* 48 (1999): 664.
15. P. D. Reaven and J. L. Witztum, Oxidized low density lipoproteins in atherogenesis: Role of dietary modification, *Annual Review of Nutrition* 16 (1996): 51–71.
16. H. C. McGill and coauthors, Origin of atherosclerosis in childhood and adolescence, *American Journal of Clinical Nutrition* 72 (2000): S1307–S1325.
17. K. K. Griendling and W. Alexander, Oxidative stress and cardiovascular disease (editorial), *Circulation* 96 (1997): 3264–3265; H. Sies, Oxidative stress: Oxidants and antioxidants, *Experimental Physiology* 82 (1997): 291–295.
18. McGill and coauthors, 2000.
19. Third report of the National Cholesterol Education Program (NCEP) Expert Panel on Detection, Evaluation, and Treatment of High Blood Cholesterol in Adults (Adult Treatment Panel III), *Journal of the American Medical Association* 285 (2001): 2486–2497.
20. J. W. Miller, Does lowering plasma homocysteine reduce vascular disease risk? *Nutrition Reviews* 59 (2001): 242–244
21. American Heart Association, www.americanheart.org, accessed December 19, 2001.
22. J. P. Despres and coauthors, Hyperinsulinemia as an independent risk factor for ischemic heart disease, *New England Journal of Medicine* 334 (1996): 952–957.
23. D. L. Sprecher and G. L. Pearce, How deadly is the "deadly quartet"? A post-CABG evaluation, *Journal of the American College of Cardiology* 36 (2000): 1159–1165; M. E. Daly and coauthors, Dietary carbohydrates and insulin sensitivity: A review of the evidence and clinical implications, *American Journal of Clinical Nutrition* 66 (1997): 1072–1085.
24. National Institutes of Health, National Heart, Lung, and Blood Institute, NCEP issues major new cholesterol guidelines (press release), 2001, available at www.nhlbi.nih.gov.
25. R. A. Hegele, Premature atherosclerosis associated with monogenic insulin resistance, *Circulation* 103 (2001): 2225–2229.
26. C. K. Roberts and coauthors, Reversibility of chronic experimental syndrome X by diet modification, *Hypertension* 37 (2001): 1323–1328.
27. J. Jeppesen and coauthors, Triglyceride concentration and ischemic heart disease: An eight-year follow-up in the Copenhagen Male Study, *Circulation* 97 (1998): 1029–1036.
28. Third Report of the National Cholesterol Education Program (NCEP) Expert Panel on Detection, Evaluation, and Treatment of High Blood Cholesterol in Adults (Adult Treatment Panel III), 2001.
29. P. L. Zock and M. B. Katan, Diet, LDL oxidation, and coronary artery disease, *American Journal of Clinical Nutrition* 68 (1998): 759–760.
30. A. Ascherio and coauthors, *Trans*-fatty acids and coronary heart disease, *New England Journal of Medicine* 340 (1999): 1994–1998.
31. A dietary intervention trial for nutritional management of cardiovascular risk factors, *Nutrition Reviews* 55 (1997): 54–60.
32. F. B. Hu and coauthors, Dietary fat intake and the risk of coronary heart disease in women, *New England Journal of Medicine* 337 (1997): 1491–1499.
33. R. M. Krauss and coauthors, AHA Dietary Guidelines: Revision 2000: A statement for healthcare professionals from the nutrition committee of the American Heart Association, *Circulation* 102 (2000): 2284–2299.

34. M. de Longeril and coauthors, Mediterranean diet, traditional risk factors and the rate of cardiovascular complications after myocardial infarction: Final report of the Lyon Diet Heart Study, *Circulation* 99 (1999): 779–785.

35. R. Cooper and coauthors, Trends and disparities in coronary heart disease, stroke, and other cardiovascular diseases in the United States: Findings of the national conference on cardiovascular disease prevention, *Circulation* 102 (1999): 3137–3147.

36. M. L. Pollock and coauthors, Resistance exercise in individuals with and without cardiovascular disease, *Circulation* 101 (2000): 828–833.

37. S. W. Farrell and coauthors, Influences of cardiorespiratory fitness levels and other predictors on cardiovascular disease mortality in men, *Medicine and Science in Sports and Exercise* 30 (1998): 899–905.

38. S. N. Blair and coauthors, Influences of cardiorespiratory fitness and other precursors on cardiovascular disease and all-cause mortality in men and women, *Journal of the American Medical Association* 276 (1996): 205–210.

39. G. A. Bray, Obesity, in E. E. Ziegler and L. J. Filer, Jr., eds., *Present Knowledge in Nutrition*, 7th ed. (Washington, D.C.: International Life Sciences Institute Press, 1996), pp. 19–32.

40. D. Ornish, Avoiding revascularization with lifestyle changes: The Multicenter Lifestyle Demonstration Project, *American Journal of Cardiology* 82 (1998): T72–T76.

41. J. L. Abramson and coauthors, Moderate alcohol consumption and risk of heart failure among older persons, *Journal of the American Medical Association* 285 (2001): 1971–1977; E. Rimm, Alcohol and cardiovascular disease, *Current Atherosclerosis Reports* 2 (2002): 529–535; C. T. Valmadrid and coauthors, Alcohol intake and the risk of coronary heart disease mortality in persons with older-onset diabetes mellitus, *Journal of the American Medical Association* 282 (1999): 239–246.

42. M. J. Williams, N. J. Restieaux, and C. J. Low, Myocardial infarction in young people with normal coronary arteries, *Heart* 79 (1998): 191–194.

43. Third Report of the National Cholesterol Education Program (NCEP) Expert Panel on Detection, Evaluation, and Treatment of High Blood Cholesterol in Adults (Adult Treatment Panel III), 2001.

44. P. Bjorntorp, Stress and cardiovascular disease, *Acta Physiologica Scandinavica Supplementum* 640 (1997): 144–148.

45. S. Ramachandran and M. D. Vasan, Residual lifetime risk for developing hypertension in middle-aged women and men: the Framingham Heart Study, *Journal of the American Medical Association* 287 (2002): 1003–1010; U.S. Department of Agriculture, Third Report on Nutrition Monitoring in the United States: Executive summary, *Journal of Nutrition* 126 (1996): S1907–S1936.

46. W. B. Kannel, Blood pressure as a cardiovascular risk factor, *Journal of the American Medical Association* 275 (1996): 1571–1575.

47. Third Report of the National Cholesterol Education Program (NCEP) Expert Panel on Detection, Evaluation, and Treatment of High Blood Cholesterol in Adults (Adult Treatment Panel III), 2001.

48. R. F. Gillum, M. E. Mussolino, and J. H. Madans, Body fat distribution and hypertension incidence in women and men: The NHANES I Epidemiologic Follow-up Study, *International Journal of Obesity and Related Metabolic Disorders* 22 (1998): 127–134.

49. M. R. Sierra-Honigmann and coauthors, Biological action of leptin as an angiogenic factor, *Science* 281 (1998): 1683–1686.

50. J. M. Lalouel and A. Rohrwasser, Development of genetic hypotheses in essential hypertension, *Journal of Human Genetics* 46 (2001): 299–306; K. W. Sellers and coauthors, Gene therapy to control hypertension: Current studies and future perspectives, *American Journal of the Medical Sciences* 322 (2001): 1–6.

51. M. L. Nurminen, R. Korpela, and H. Vapaatalo, Dietary factors in the pathogenesis and treatment of hypertension, *Annals of Medicine* 30 (1998): 143–150; D. A. McCarron and coauthors, Comprehensive nutrition plan improves cardiovascular risk factors in essential hypertension, *American Journal of Hypertension* 11 (1998): 31–40.

52. A. W. Cowley, Genetic and nongenetic determinants of salt sensitivity and blood pressure, *American Journal of Clinical Nutrition* 65 (1997): S587–S593; D. L. Ely, Overview of dietary sodium effects on and interactions with cardiovascular and neuroendocrine functions, *American Journal of Clinical Nutrition* 65 (1997): S594–S605.

53. J. Stamler, The INTERSALT Study: Background, methods, findings, and implications, *American Journal of Clinical Nutrition* 65 (1997): S626–S642.

54. J. A. Staessen and coauthors, Salt and blood pressure in community-based intervention trials, *American Journal of Clinical Nutrition* 65 (1997): S661–S670.

55. Stamler, 1997; J. A. Cutler, D. Follmann, and P. S. Allender, Randomized trials of sodium reduction: An overview, *American Journal of Clinical Nutrition* 65 (1997): S643–S651.

56. F. M. Sacks and coauthors, Effects on blood pressure of reduced dietary sodium and the Dietary Approaches to Stop Hypertension (DASH) diet, *New England Journal of Medicine* 344 (2001): 3–10; E. Saltos and S. Bowman, Dietary guidance on sodium: Should we take it with a grain of salt? *USDA Family Economics and Nutrition Review* vol. 11, no. 4, 1998, pp. 49–51.

57. V. J. Stevens and coauthors, Long-term weight loss and changes in blood pressure: Results of the Trials of Hypertension Prevention, phase II, *Annals of Internal Medicine* 134 (2001): 1–11.

58. D. A. McCarron and M. E. Reusser, Body weight and blood pressure regulation, *American Journal of Clinical Nutrition* 63 (1996): S423–S425.

59. P. D. Thompson and coauthors, The acute versus the chronic response to exercise, *Medicine and Science in Sports and Exercise* 33 (2001): S452–S453.

60. Thompson and coauthors, 2001.

61. M. Iwane and coauthors, Walking 10,000 steps/day or more reduces blood pressure and sympathetic nerve activity in mild essential hypertension, *Hypertension Research* 23 (2000): 573–580.

62. M. Hillbom, S. Juvela, and V. Karttunen, Mechanisms of alcohol-related strokes, in *Alcohol and Cardiovascular Diseases: Novartis Foundation Symposium 216* (New York: John Wiley & Sons, 1998).

63. M. E. Reusser and D. A. McCarron, Micronutrient effects on blood pressure regulation, *Nutrition Reviews* 52 (1994): 367–375.

64. D. A. McCarron, Dietary calcium and lower blood pressure: We can all benefit, *Journal of the American Medical Association* 275 (1996): 1128–1129.

65. D. A. McCarron, Role of adequate dietary calcium intake in the prevention and management of salt-sensitive hypertension, *American Journal of Clinical Nutrition* 65 (1997): S712–S716; C. G. Osborne and coauthors, Evidence for the relationship of calcium to blood pressure, *Nutrition Reviews* 54 (1996): 365–381; McCarron, 1996.

66. P. K. Whelton and coauthors, Effects of oral potassium on blood pressure: Meta-analysis of randomized controlled clinical trials, *Journal of the American Medical Association* 277 (1997): 1624–1632.

67. C. J. Bates and coauthors, Does vitamin C reduce blood pressure? Results of a large study of people aged 65 or older, *Journal of Hypertension* 16 (1998): 925–932.

68. L. J. Appel and coauthors, A clinical trial of the effects of dietary patterns on blood pressure, *New England Journal of Medicine* 336 (1997): 1117–1124.

69. E. Obarzanek and coauthors, Effects on blood lipids of a blood pressure–lowering diet: The Dietary Approaches to Stop Hypertension (DASH) trial, *American Journal of Clinical Nutrition* 74 (2001): 80–89.

70. R. J. Kreitman and coauthors, Efficacy of the anti-CD22 recombinant immunotoxin BL22 in chemotherapy-resistant hairy-cell leukemia, *New England Journal of Medicine* 345 (2001): 241–247.

71. P. Lichtenstein and coauthors, Environmental and heritable factors in the causation of cancer: Analyses of cohorts of twins from Sweden, Denmark, and Finland, *New England Journal of Medicine* 343 (2000): 78–85.

72. M. L. Slattery and coauthors, Lifestyle and colon cancer: An assessment of factors associated with risk, *American Journal of Epidemiology* 150 (1999): 869–877; S. A. Oliveria and P. J. Christos, The epidemiology of physical activity and cancer, *Annals of the New York Academy of Sciences* 833 (1997): 79–90.

73. D. M. DeMarini, Dietary interventions of human carcinogenesis, *Mutation Research* 400 (1998): 457–465.

74. Committee on Comparative Toxicity of Naturally Occurring Carcinogens, *Carcinogens and Anticarcinogens in the Human Diet* (Washington, D.C.: National Academy Press, 1996), pp. 1–18.

75. T. Sugimura, Cancer prevention: Past, present, future, *Mutation*

Research 402 (1998): 7–14; M. J. Hill, Nutrition and human cancer, *Annals of the New York Academy of Sciences* 883 (1997): 68–78.

76. L. F. Macrae, Wheat bran fiber and development of adenomatous polyps: Evidence from randomized, controlled, clinical trials, *American Journal of Medicine* 106 (1999): S38–S42; L. Chatenoud and coauthors, Whole grain food intake and cancer risk, *International Journal of Cancer* 77 (1998): 24–28.

77. P. L. Zock, Dietary fats and cancer, *Current Opinions in Lipidology* 12 (2001): 5–10.

78. Zock, 2001.

79. Zock, 2001; M. D. Holmes and coauthors, Association of dietary intake of fat and fatty acids with risk of breast cancer, *Journal of the American Medical Association* 281 (1999): 914–920; D. J. Hunter and coauthors, Cohort studies of fat intake and the risk of breast cancer—A pooled analysis, *New England Journal of Medicine* 334 (1996): 356–361.

80. P. Terry and coauthors, Fatty fish consumption and risk of prostate cancer, *Lancet* 357 (2001): 1764–1766; Zock, 2001; W. C. Willet, Diet and cancer, *Oncologist* 5 (2000): 393–404.

81. E. Giovannucci and B. Goldin, The role of fat, fatty acids, and total energy intake in the etiology of human colon cancer, *American Journal of Clinical Nutrition* 66 (1997): S1564–S1571.

82. C. M. Yang and coauthors, Thermally oxidized dietary fat and colon carcinogenesis in rodents, *Nutrition and Cancer* 30 (1998): 69–73; J. G. Ernhardt and coauthors, A diet rich in fat and poor in dietary fiber increases the in vitro formation of reactive oxygen species in human feces, *Journal of Nutrition* 127 (1997): 706–709.

83. G. E. Cowing and K. E. Saker, Polyunsaturated fatty acids and epidermal growth factor receptor/mitogen-activated protein kinase signaling in mammary cancer, *Journal of Nutrition* 131 (2001): 1125–1128; E. D. Collett and coauthors, n-6 and n-3 polyunsaturated fatty acids differentially modulate oncogenic Ras activation in colonocytes, *American Journal of Physiology: Cell Physiology* 280 (2001): C1066–C1075; G. K. Ogilvie and coauthors, Effect of fish oil, arginine, and doxorubicin chemotherapy on remission and survival time for dogs with lymphoma: a double-blind, randomized placebo-controlled study, *Cancer* 88 (2000): 1916–1928.

84. L. Klieveri and coauthors, Promotion of colon cancer metastases in rat liver by fish oil diet is not due to reduced stroma formation, *Clinical and Experimental Metastasis* 18 (2000): 371–377.

85. Giovannucci and Goldin, 1997.

86. D. Kritchevsky, Caloric restriction and experimental mammary carcinogenesis, *Breast Cancer Research and Treatment* 46 (1997): 161–167; Committee on Comparative Toxicity of Naturally Occurring Carcinogens, 1996, pp. 35–126.

87. B. J. Caan and coauthors, Body size and the risk of colon cancer in a large case-control study, *International Journal of Obesity and Related Metabolic Disorders* 22 (1998): 178–184; C. La Vecchia and coauthors, Diabetes mellitus and colorectal cancer risk, *Cancer, Epidemiology, Biomarkers, and Prevention* 6 (1997): 1007–1010.

88. M. L. Slattery, Diet, lifestyle, and colon cancer, *Seminars in Gastrointestinal Disease* 11 (2000): 142–146.

89. B. C. Pence and coauthors, Feeding of a well-cooked beef diet containing a high heterocyclic amine content enhances colon and stomach carcinogenesis in 1,2-dimethylhydrazine-treated rats, *Nutrition and Cancer* 30 (1998): 220–226.

90. T. Norat and E. Riboli, Meat consumption and colorectal cancer: A review of epidemiologic evidence, *Nutrition Reviews* 59 (2001): 37–47; Giovannucci and Goldin, 1997.

91. A. Schatzkin and coauthors, Lack of effect of a low-fat, high-fiber diet on the recurrence of colorectal adenomas, *New England Journal of Medicine* 342 (2000): 1149–1155; C. S. Fuchs and coauthors, Dietary fiber and the risk of colorectal cancer and adenoma in women, *New England Journal of Medicine* 340 (1999): 169–176.

92. American Gastroenterological Association, Medical position statement: Impact of dietary fiber on colon cancer occurrence, *Gastroenterology* 118 (2000): 1233–1234; B. S. Reddy, Role of dietary fiber in colon cancer: An overview, *American Journal of Medicine* 106 (1999): S16–S19; D. Kritchevsky, Protective role of wheat bran fiber: Preclinical data, *American Journal of Medicine* 106 (1999): S28–S31; M. C. Jansen and coauthors, Dietary fiber and plant foods in relation to colorectal cancer mortality: The Seven Countries Study, *International Journal of Cancer* 81 (1999): 174–179; M. J. Hill, Cereals, cereal fibre, and colorectal cancer risk: A review of the epidemiological literature, *European Journal of Cancer Prevention* 6 (1997): 219–225; L. Le Marchand and coauthors, Dietary fiber and colorectal cancer risk, *Epidemiology* 8 (1997): 658–665.

93. J. L. Slavin, Mechanisms for the impact of whole grain foods on cancer risk, *Journal of the American College of Nutrition* 19 (2000): S300–S307.

94. D. S. Michaud and coauthors, Fluid intake and the risk of bladder cancer in men, *New England Journal of Medicine* 340 (1999): 1390–1397.

95. S. W. Choi and J. B. Mason, Folate and carcinogenesis: An integrated scheme, *Journal of Nutrition* 130 (2000): 129–132; K. Young-In, Folate and cancer prevention: A new medical application for folate beyond hyperhomocysteinemia and neural tube defects, *Nutrition Reviews* 57 (1999): 314–324; A. Kwasniewska, A. Tukendorf, and M. Semczuk, Folate deficiency and cervical intraepithelial neoplasia, *European Journal of Gynaecological Oncology* 18 (1997): 526–530.

96. B. N. Ames, Micronutrient deficiencies: A major cause of DNA damage, *Annals of the New York Academy of Sciences* 889 (1999): 87–106.

97. A. S. Vrablic and coauthors, Altered mitochondrial function and overgeneration of reactive oxygen species precede the induction of apoptosis by 1-O-octadecyl-2-methyl-rac-glycero-3-phophocholine in p53-defective hepatocytes, *FASEB Journal* 15 (2001): 1739–1744.

98. R. I. Salganik and coauthors, Dietary antioxidant depletion: Enhancement of tumor apoptosis and inhibition of brain tumor growth in transgenic mice, *Carcinogenesis* 21 (2000): 909–914.

99. E. Kallay and coauthors, Dietary calcium and growth modulation of human colon cancer cells: Role of the extracellular calcium-sensing receptor, *Cancer Detection and Prevention* 24 (2000): 127–136.

100. E. Kampman and coauthors, Calcium, vitamin D, sunshine exposure, dairy products and colon cancer risk (United States), *Cancer Causes and Control* 11 (2000): 459–466; S. Mobarhan, Calcium and the colon: Recent findings, *Nutrition Reviews* 57 (1999): 124–129.

101. R. L. Nelson, Iron and colorectal cancer risk: Human studies, *Nutrition Reviews* 59 (2001): 140–148.

102. American Institute for Cancer Research, *Food, Nutrition and the Prevention of Cancer: A Global Perspective*, 1997, available at www.aicr.org/report2.htm.

103. Norat and Riboli, 2001; Nelson, 2001.

104. M. Peluso and coauthors, White blood cell DNA adducts and fruit and vegetable consumption in bladder cancer, *Carcinogenesis* 21 (2000): 183–187.

105. K. A. Steinmetz and J. D. Potter, Vegetables, fruit, and cancer prevention: A review, *Journal of the American Dietetic Association* 96 (1996): 1027–1039.

106. National Academy of Sciences, quoted in First International Conference on East-West Perspectives on Functional Foods, *Nutrition Today*, March/April 1996, pp. 70–73.

107. *The Surgeon General's Report on Nutrition and Health: Summary and Recommendations* (Washington, D.C.: Department of Health and Human Services—Public Health Service publication no. 88-50211, 1988).

108. Kant and coauthors, 2000.

Consumer Corner 11

1. R. J. Lamarine, Alternative medicine: More than a harmless option, *Journal of School Health* 71 (2001): 114–116.

2. M. M. Lipman, The power of placebos, *Consumer Reports on Health*, February 1996, p. 23.

3. A. White and coauthors, Adverse events following acupuncture: Prospective survey of 32,000 consultations with doctors and physiotherapists, *British Medical Journal* 323 (2001): 485–486; H. MacPherson and coauthors, The York acupuncture safety study: Prospective survey of 34,000 treatments by traditional acupuncturists, *British Medical Journal* 323 (2001): 486–487.

4. P. Lipkin, An ancient salve dampens pain, *Science News* 149 (1996): 20.

5. J. A. Bakerlink and coauthors, Multiple organ failure after ingestion of pennyroyal oil from herbal tea in two infants, *Pediatrics* 98 (1996): 944–947.

6. U.S. Food and Drug Administration, Letter to health professionals regarding safety concerns related to the use of botanical products containing aristolochic acid, April 2001, available at www.cfsan.fda.gov.

7. E. Meseguer and coauthors, Life-threatening parkinsonism induced by kava-kava, *Movement Disorders* 17 (2002): 195–196.

Controversy 11

1. E. B. Feldman, How grapefruit juice potentiates drug bioavailability, *Nutrition Reviews* 55 (1997): 398–400.

2. K. Sasaki and coauthors, Bilobalide, a constituent of ginkgo biloba L., potentiates drug metabolizing enzyme activities in mice: Possible mechanism for anticonvulsant activity against 4-O-methylpyridoxine induced convulsions, *Research Communications in Molecular Pathology and Pharmacology* 96 (1997): 45–56.

3. T. J. Green and coauthors, Oral contraceptives did not affect biochemical folate indexes and homocysteine concentrations in adolescent females, *Journal of the American Dietetic Association* 98 (1998): 49–55.

4. G. Berg, L. Kohlmeier, and H. Brenner, Use of oral contraceptives and serum beta-carotene, *European Journal of Clinical Nutrition* 51 (1997): 181–187.

5. S. S. Harris and B. Dawson-Hughes, The association of oral contraceptive use with plasma 25-hydroxyvitamin D levels, *Journal of the American College of Nutrition* 17 (1998): 282–284.

6. S. G. Stoney and coauthors, Oral contraceptive use is associated with increased cardiovascular reactivity in nonsmokers, *Annals of Behavioral Medicine* 23 (2001): 149–157.

7. J. R. Huges and coauthors, Endorsement of DSM-IV dependence criteria among caffeine users, *Drug and Alcohol Dependence* 52 (1998): 99–107.

8. B. J. Fine and coauthors, Effects of caffeine or diphenhydramine on visual vigilance, *Psychopharmacology (Berl)* 114 (1994): 233–238 as cited by H. R. Lieberman, The effects of ginseng, ephedrine, and caffeine on cognitive performance, mood, and energy, *Nutrition Reviews* 59 (2001): 91–102.

9. P. J. Durlach, The effects of a low dose of caffeine on cognitive performance, *Psychopharmacology* 140 (1998): 116–119.

10. T. R. Hartley and coauthors, Hypertension risk status and effect of caffeine on blood pressure, *Hypertension* 36 (2000): 137–141.

11. M. R. Olthof and coauthors, Consumption of high doses of chlorogenic acid, present in coffee, or of black tea increases plasma total homocysteine concentrations in humans, *American Journal of Clinical Nutrition* 73 (2001): 532–538.

12. M. F. Leitzmann and coauthors, A prospective study of coffee consumption and the risk of symptomatic gallstone disease in men, *Journal of the American Medical Association* 281 (1999): 2106–2112.

13. A. Rodgers, Effect of cola consumption on urinary biochemical and physiochemical risk factors associated with calcium oxalate urolithianis, *Urological Research* 27 (1999): 77–81.

14. K. Neilann and coauthors, Potential mechanisms of diet therapy for fibrocystic breast conditions show inadequate evidence of effectiveness, *Journal of the American Dietetic Association* 100 (2000): 1368–1380.

15. Standing Committee on the Scientific Evaluation of Dietary Reference Intakes, Food and Nutrition Board, Institute of Medicine, *Dietary Reference Intakes for Vitamin C, Vitamin E, Selenium, and Carotenoids* (Washington, D.C.: National Academy Press, 2000), pp. 152–153.

Chapter 12

1. M. Hack and coauthors, Outcomes in young adulthood for very low-birth-weight infants, *New England Journal of Medicine* 346 (2002): 149–157.

2. K. M. Godfrey and D. J. P. Barker, Fetal nutrition and adult disease, *American Journal of Clinical Nutrition* 71 (2000): S1344–S1352.

3. B. Guyer and coauthors, Annual summary of vital statistics: Trends in the health of Americans during the 20th century, *Pediatrics* 106 (2000): 1307–1317.

4. F. Galtier-Dereure, C. Boegner, and J. Bringer, Obesity and pregnancy: Complications and cost, *American Journal of Clinical Nutrition* 71 (2000): S1242–S1248; M. M. Werler and coauthors, Prepregnant weight in relation to risk of neural tube defects, *Journal of the American Medical Association* 275 (1996): 1089–1092; G. M. Shaw, E. M. Velie, and D. Schaffer, Risk of neural tube defect—Affected pregnancies among obese women, *Journal of the American Medical Association* 275 (1996): 1093–1096.

5. Galtier-Dereure, Boegner, and Bringer, 2000; R. L. Goldenberg and T. Tamura, Prepregnancy weight and pregnancy outcome, *Journal of the American Medical Association* 275 (1996): 1127–1128.

6. D. J. Barker and P. M. Clark, Fetal undernutrition and disease in later life, *Reviews of Reproduction* 2 (1997): 105–112.

7. Godfrey and Barker, 2000; J. Newnham, Consequences of fetal growth restriction, *Current Opinion in Obstetrics and Gynecology* 10 (1998): 145–149; W. P. T. James, Long-term fetal programming of body composition and longevity, *Nutrition Reviews* 55 (1997): S31–S43.

8. R. M. Pitkin, Energy in pregnancy, *American Journal of Clinical Nutrition* 69 (1999): 583; L. E. Kopp-Hoolihan and coauthors, Longitudinal assessment of energy balance in well-nourished, pregnant women, *American Journal of Clinical Nutrition* 69 (1999): 697–704.

9. S. J. Otto and coauthors, Changes in the maternal essential fatty acid profile during early pregnancy and the relation of the profile to diet, *American Journal of Clinical Nutrition* 73 (2001): 302–307; G. Hornstra, Essential fatty acids in mothers and their neonates, *American Journal of Clinical Nutrition* 71 (2000): S1262–S1269; R. Uauy and coauthors, Role of essential fatty acids in the function of the developing nervous system, *Lipids* 31 (1996): S167–S176.

10. Committee on Genetics, American Academy of Pediatrics, Folic acid for the prevention of neural tube defects, *Pediatrics* 104 (1999): 325–327; B. Burke and coauthors, *Preventing Neural Tube Birth Defects: A Prevention Model and Resource Guide* (Atlanta, Ga.: Centers for Disease Control and Prevention, 1998).

11. Standing Committee on the Scientific Evaluation of Dietary Reference Intakes, Food and Nutrition Board, Institute of Medicine, *Dietary Reference Intakes for Thiamin, Riboflavin, Niacin, Vitamin B_6, Folate, Vitamin B_{12}, Pantothenic Acid, Biotin, and Choline* (Washington, D.C.: National Academy Press, 1998), pp. 196–305; G. J. Locksmith and P. Duff, Preventing neural tube defects: The importance of periconceptional folic acid supplements, *Obstetrics and Gynecology* 91 (1998): 1027–1034.

12. Burke and coauthors, 1998.

13. J. E. Brown and coauthors, Predictors of red cell folate level in women attempting pregnancy, *Journal of the American Medical Association* 277 (1997): 548–552.

14. H. McNulty, G. J. Cuskelly, and M. Ward, Response of red blood cell folate to intervention: Implications for folate recommendations for the prevention of neural tube defects, *American Journal of Clinical Nutrition* 71 (2000): S1308–S1311.

15. N. M. van der Put and coauthors, Folate, homocysteine and neural tube defects: An overview, *Experimental Biology and Medicine* 226 (2001): 243–270; P. F. Jacques and coauthors, The effect of folic acid fortification on plasma folate and total homocysteine concentrations, *New England Journal of Medicine* 340 (1999): 1449–1454.

16. Standing Committee on the Scientific Evaluation of Dietary Reference Intakes, 1998.

17. Standing Committee on the Scientific Evaluation of Dietary Reference Intakes, Food and Nutrition Board, Institute of Medicine, *Dietary Reference Intakes for Calcium, Phosphorus, Magnesium, Vitamin D, and Fluoride* (Washington, D.C: National Academy Press, 1997), p. 4-38.

18. A. Prentice, Maternal calcium metabolism and bone mineral status, *American Journal of Clinical Nutrition* 71 (2000): S1312–S1316.

19. L. H. Allen, Pregnancy and iron deficiency: Unresolved issues, *Nutrition Reviews* 55 (1997): 91–101.

20. J. C. King, Determinants of maternal zinc status during pregnancy, *American Journal of Clinical Nutrition* 71 (2000): S1334–S1343.

21. King, 2000.

22. M. M. Werler and coauthors, Multivitamin supplementation and risk of birth defects, *American Journal of Epidemiology* 150 (1999): 675–682; T. O. Scholl and coauthors, Use of multivitamin/mineral prenatal supplements: Influence on the outcome of pregnancy, *American Journal of Epidemiology* 146 (1997): 134–141.

23. A. R. Swensen, L. J. Harnack, and J. A. Ross, Nutritional assessment of pregnant women enrolled in the Special Supplemental Program for Women, Infants, and Children (WIC), *Journal of the American Dietetic Association* 101 (2001): 903–908.

24. A. L. Owen and G. M. Owen, Twenty years of WIC: A review of some effects of the program, *Journal of the American Dietetic Association* 97 (1997): 777–782.

25. J. F. Clapp, The effect of continuing regular endurance exercise on the

physiologic adaptations to pregnancy and pregnancy outcome, *American Journal of Sports Medicine* 24 (1996): S28–S29.

26. Committee on Adolescence, American Academy of Pediatrics, Adolescent pregnancy—Current trends and issues: 1998, *Pediatrics* 103 (1999): 516–520.

27. R. J. Trissler, The child within: A guide to nutrition counseling for pregnant teens, *Journal of the American Dietetic Association* 99 (1999): 916–917.

28. C. D. Drews and coauthors, The relationship between idiopathic mental retardation and maternal smoking during pregnancy, *Pediatrics* 97 (1996): 547–553.

29. J. M. Lightwood, C. S. Phibbs, and S. A. Glantz, Short-term health and economic benefits of smoking cessation: Low birth weight, *Pediatrics* 104 (1999): 1312–1320.

30. American Academy of Pediatrics, Task Force on Infant Sleep Position and Sudden Infant Death Syndrome, Changing concepts of sudden infant death syndrome: Implication for infant sleeping environment and sleep position, *Pediatrics* 105 (2000): 650–656.

31. American Academy of Pediatrics, 2000; Environmental tobacco smoke affects birth weight, *Journal of the American Medical Association* 279 (1998): 739; E. Cutz and coauthors, Maternal smoking and pulmonary neuroendocrine cells in sudden infant death syndrome, *Pediatrics* 98 (1996): 668–672.

32. M. S. Scher, G. A. Richardson, and N. L. Day, Effects of prenatal cocaine/crack and other drug exposure on electroencephalographic sleep studies at birth and one year, *Pediatrics* 105 (2000): 39–48; F. D. Eyler and coauthors, Birth outcome from a prospective, matched study of prenatal crack/cocaine use: II. Interactive and dose effects on neurobehavioral assessment, *Pediatrics* 101 (1998): 237–241.

33. FDA announces advisory on methylmercury in fish, *FDA Talk Paper*, 2001, available from www.cfsan.fda.gov.

34. S. Cnattingius and coauthors, Caffeine intake and the risk of first-trimester spontaneous abortion, *New England Journal of Medicine* 343 (2000): 1839–1845; M. A. Klebanoff and coauthors, Maternal serum paraxanthine, a caffeine metabolite, and the risk of spontaneous abortion, *New England Journal of Medicine* 341 (1999): 1639–1644.

35. T. S. Hinds and coauthors, The effect of caffeine on pregnancy variables, *Nutrition Reviews* 54 (1996): 203–207.

36. I. S. Santos and coauthors, Caffeine intake and low birth weight: A population-based, case-control study, *American Journal of Epidemiology* 147 (1998): 620–627.

37. National Institutes of Health News Release, Long-chain alcohol found to block mechanism of fetal alcohol syndrome, www.nih.gov/news/pr/may2001/niaaa-18.htm, website visited May 23, 2001; K. Strömand and A. Hellström, Fetal alcohol syndrome: An ophthalmological and socioeducational prospective study, *Pediatrics* 97 (1996): 845–850.

38. C. Ikonomidou and coauthors, Ethanol-induced apoptotic neurodegeneration and fetal alcohol syndrome, *Science* 287 (5455) (2000): 947–948.

39. Centers for Disease Control, Division of Birth Defects, Child Development, and Disability and Health, Fetal alcohol syndrome, www.cdc.gov/nceh/cddh/fas/fasfact.htm, website visited July 13, 2000.

40. Committee on Substance Abuse and Committee on Children with Disabilities, American Academy of Pediatrics, Fetal alcohol syndrome and alcohol-related neurodevelopmental disorders, *Pediatrics* 106 (2000): 358–361.

41. National Institutes of Health New Release, 2001.

42. Committee on Substance Abuse and Committee on Children with Disabilities, 2000.

43. D. B. Carr and S. Gabbe, Gestational diabetes: Detection, management, and implications, *Clinical Diabetes* 16 (1998): 4–11.

44. C. D. Naylor and coauthors, Cesarean delivery in relation to birth weight and gestational glucose tolerance: Pathophysiology or practice style? *Journal of the American Medical Association* 275 (1996): 1165–1170.

45. The Expert Committee on the Diagnosis and Classification of Diabetes Mellitus, Report of the Expert Committee on the diagnosis and classification of diabetes mellitus, *Diabetes Care* (supplement 1) 21 (1998): 5–19.

46. D. Maine, Role of nutrition in the prevention of toxemia, *American Journal of Clinical Nutrition* 72 (2000): S298–S300.

47. K. G. Dewey, Energy and protein requirements during lactation, *Annual Review of Nutrition* 17 (1997): 19–36.

48. M. J. Heinig and K. G. Dewey, Health effects of breast feeding for mothers: A critical review, *Nutrition Research Reviews* 10 (1997): 35–56.

49. M. A. McCrory, Does dieting during lactation put infant growth at risk? *Nutrition Reviews* 59 (2001): 18–27; C. A. Lovelady and coauthors, The effect of weight loss in overweight, lactating women on the growth of their infants, *New England Journal of Medicine* 342 (2000): 449–453.

50. J. L. B. Pharm, Breastfeeding and the use of recreational drugs—Alcohol, caffeine, nicotine, and marijuana, *Breastfeeding Reviews* 2 (1998): 27–30.

51. Pharm, 1998.

52. R. Nduati and coauthors, Effect of breastfeeding and formula feeding on transmission of HIV-1: A randomized clinical trial, *Journal of the American Medical Association* 283 (2000): 1167–1174; P. G. Miotti and coauthors, HIV transmission through breastfeeding: A study in Malawi, *Journal of the American Medical Association* 282 (1999): 744–749.

53. R. F. Black, Transmission of HIV-1 in the breast-feeding process, *Journal of the American Dietetic Association* 96 (1996): 267–274.

54. Black, 1996.

55. J. Humphrey and P. Iliff, Is breast not best? Feeding babies born to HIV-positive mothers: Bringing balance to a complex issue, *Nutrition Reviews* 59 (2001): 119–127.

56. Committee on Nutrition, American Academy of Pediatrics, *Pediatric Nutrition Handbook*, 4th ed., ed. R. E. Kleinman (Elk Grove Village, Ill.: American Academy of Pediatrics, 1998), pp. 277–278.

57. Position of The American Dietetic Association: Promotion of breast feeding, *Journal of the American Dietetic Association* 97 (1997): 662–666.

58. Work Group on Breastfeeding, American Academy of Pediatrics, Breastfeeding and the use of human milk, *Pediatrics* 100 (1997): 1035–1039.

59. A. L. Wright and coauthors, Increasing breastfeeding rates to reduce infant illness at the community level, *Pediatrics* 101 (1998): 837–844.

60. J. Raisler and coauthors, Breast-feeding and infant illness: A dose-response relationship? *American Journal of Public Health* 89 (1999): 25–30.

61. D. S. Newburg and coauthors, Role of human-milk lactadherin in protection against symptomatic rotavirus infection, *Lancet* 351 (1998): 1160–1164.

62. W. H. Oddy and coauthors, Association between breastfeeding and asthma in 6 year old children: Findings of a prospective birth cohort study, *British Medical Journal* 319 (1999): 815–819; C. G. Victora and coauthors, Potential interventions for the prevention of childhood pneumonia in developing countries: Improving nutrition, *American Journal of Clinical Nutrition* 70 (1999): 309–320; A. H. Cushing and coauthors, Breastfeeding reduces risk of respiratory illness in infants, *American Journal of Epidemiology* 147 (1998): 863–870; D. S. Newburg and J. M. Street, Bioactive materials in human milk, *Nutrition Today* 32 (1997): 191–201.

63. Newburg and Street, 1997.

64. M. A. Atkinson and T. M. Ellis, Infants' diets and insulin-dependent diabetes: Evaluating the "cows' milk hypothesis" and a role for anti-bovine serum albumin immunity, *Journal of the American College of Nutrition* 16 (1997): 334–340.

65. R. A. Gibson and M. Makrides, n-3 Polyunsaturated fatty acid requirements of term infants, *American Journal of Clinical Nutrition* 71 (2000): S251–S255; M. Neuringer, Infant vision and retinal function in studies of dietary long-chain polyunsaturated fatty acids: Methods, results, and implications, *American Journal of Clinical Nutrition* 71 (2000): S256–S267.

66. C. Williams and coauthors, Stereoacuity at age 3.5 y in children born full-term is associated with prenatal and postnatal dietary factors: A report from a population-based cohort study, *American Journal of Clinical Nutrition* 73 (2001): 316–322.

67. Committee on Nutrition, American Academy of Pediatrics, Hypoallergenic infant formulas, *Pediatrics* 106 (2000): 346–349.

68. Committee on Nutrition, American Academy of Pediatrics, Iron fortification of infant formulas, *Pediatrics* 104 (1999): 119–123.

69. Committee on Nutrition, 1999.

70. B. A. Dennison, H. L. Rockwell, and S. L. Baker, Excess fruit juice consumption by preschool-aged children is associated with short stature and obesity, *Pediatrics* 99 (1997): 15–22.

71. J. D. Skinner and B. R. Carruth, A longitudinal study of children's juice intake and growth: The juice controversy revisited, *Journal of the American Dietetic Association* 101 (2001): 432–437; Committee on Nutrition, American Academy of Pediatrics, The use and misuse of fruit juice in pediatrics, *Pediatrics* 107 (2001): 1210–1213.

F

Consumer Corner 12

1. I. Ozturk, N. Votto, and J. M. Leventhal, The timing and predictors of the early termination of breast feeding, *Pediatrics* 107 (2001): 543–548.
2. B. L. Phillipp and coauthors, Baby-friendly hospital initiative improves breastfeeding initiation rates in a U.S. hospital setting, *Pediatrics* 108 (2001): 677–681.
3. Position of The American Dietetic Association: Promotion of breastfeeding, *Journal of the American Dietetic Association* 97 (1997): 662–666.
4. Position of The American Dietetic Association: Breaking the barriers to breastfeeding, *Journal of the American Dietetic Association* 101 (2001): 1213–1220.
5. I. B. Stehlin, Infant formula: Second best but good enough, *FDA Consumer*, June 1996, pp. 17–20.
6. A. N. J. Malik and W. A. M. Cutting, Breast feeding: The baby friendly initiative, *British Medical Journal* 316 (1998): 1548–1549.

Controversy 12

1. K. M. Godfrey and D. J. P. Barker, Fetal nutrition and adult disease, *American Journal of Clinical Nutrition* 71 (2000): S1344–S1352.
2. American Diabetes Association Consensus Statement, Type 2 diabetes in children and adolescents, *Diabetes Care* 23 (2000): 381–389, accessed June 3, 2001, at www.diabetes.org/ada/Consensus/pg381.htm.
3. A. Fagot-Campagna, K. M. Narayan, and G. Imperatore, Type 2 diabetes in children exemplifies the growing problem of chronic diseases, *British Medical Journal* 322 (2001): 377–378.
4. American Diabetes Association, Type 2 diabetes in children and adolescents, *Pediatrics* 105 (2000): 671–680.
5. D. McNamara, Overcoming juvenile diabetes with a little planning and high-tech tools, *FDA Consumer*, July/August 2000, pp. 28–32.
6. M. I. Goran, Energy expenditure, body composition, and disease risk in children and adolescents, *Proceedings of the Nutrition Society* 56 (1997): 195–209.
7. Update: Prevalence of overweight among children, adolescents, and adults—United States, 1988–1994, *Morbidity and Mortality Weekly Report* 46 (1997): 199–202.
8. D. S. Hardin and coauthors, Treatment of childhood syndrome X (abstract), *Pediatrics* 100 (1997): E5.
9. M. I. Goran, Metabolic precursors and effects of obesity in children: A decade of progress, 1990–1999, *American Journal of Clinical Nutrition* 73 (2001): 158–171.
10. R. Martorell, A. D. Stein, and D. G. Schroeder, Early nutrition and later adiposity, *Journal of Nutrition* 131 (2001): S874–S880.
11. M. W. Gillman and coauthors, Risk of overweight among adolescents who were breastfed as infants, *Journal of the American Medical Association* 285 (2001): 2461–2467.
12. M. L. Hediger and coauthors, Association between infant breastfeeding and overweight in young children, *Journal of the American Medical Association* 285 (2001): 2453–2460.
13. T. M. Cutting and coauthors, Like mother, like daughter: Familiar patterns of overweight are mediated by mothers' dietary disinhibition, *American Journal of Clinical Nutrition* 69 (1999): 608–613.
14. R. E. Anderson and coauthors, Relationship of physical activity and television watching with body weight and levels of fatness among children: Results from the third National Health and Nutrition Examination Survey, *Journal of the American Medical Association* 279 (1998): 938–942.
15. S. J. Marshall and coauthors, Tracking of health-related fitness components in youth age 9 to 12, *Medicine and Science in Sports & Exercise* 30 (1998): 910–916; S. B. Craig and coauthors, The impact of physical activity on lipids, lipoproteins, and blood pressure in preadolescent girls, *Pediatrics* 98 (1996): 389–395.
16. N. F. Chu and coauthors, Clustering of cardiovascular disease risk factors among obese schoolchildren: The Taipei Children Heart Study, *American Journal of Clinical Nutrition* 67 (1998): 1141–1146; F. J. van Lenthe and coauthors, Association of a central pattern of body fat with blood pressure and lipoproteins from adolescence into adulthood: The Amsterdam Growth and Health Study, *American Journal of Epidemiology* 147 (1998): 686–693.
17. H. C. McGill and coauthors, Origin of atherosclerosis in childhood and adolescence, *American Journal of Clinical Nutrition* 72 (2000): 1307S–1315S.
18. W. Bao and coauthors, Longitudinal changes in cardiovascular risk from childhood to young adulthood in offspring of parents with coronary artery disease: The Bogalusa Heart Study, *Journal of the American Medical Association* 278 (1997): 1749–1754.
19. R. M. Lauer and coauthors, Efficacy and safety of lowering dietary intake of total fat, saturated fat, and cholesterol in children with elevated LDL cholesterol: The Dietary Intervention Study in Children, *American Journal of Clinical Nutrition* 72 (2000): 1333S–1342S.
20. Committee on Nutrition, American Academy of Pediatrics, Cholesterol in childhood, *Pediatrics* 101 (1998): 141–147.
21. Chu and coauthors, 1998; A. R. Sinaiko, Hypertension in children, *New England Journal of Medicine* 335 (1996): 1968–1973.
22. Committee on Sports Medicine and Fitness, American Academy of Pediatrics, Athletic participation by children and adolescents who have systemic hypertension, *Pediatrics* 99 (1997): 637–638.
23. A. Singhal, T. J. Cole, and A. Lucas, Early nutrition in preterm infants and later blood pressure: Two cohorts after randomised trials, *Lancet* 357 (2001): 413–419; C. P. M. Leeson and coauthors, Duration of breast feeding and arterial distensibility in early adult life: Population based study, *British Medical Journal* 322 (2001): 643–647.
24. Fagot-Campagna, Narayan, and Imperatore, 2001; R. P. Troiano and coauthors, Energy and fat intakes of children and adolescents in the United States: Data from the National Health and Nutrition Examination Surveys, *American Journal of Clinical Nutrition* 72 (2000): 1343S–1453S.
25. American Diabetes Association, 2000.
26. K. M. Rasmussen, The "fetal origins" hypothesis: Challenges and opportunities for maternal and child nutrition, *Annual Reviews of Nutrition* 21 (2001): 73–95; R. Robinson, The fetal origins of adult disease (editorial), *British Medical Journal* 322 (2001): 375–376.
27. Rasmussen, 2001.
28. D. J. P. Barker, as cited in Robinson, 2001.
29. J. W. Rich-Edwards and coauthors, Birthweight and the risk for type 2 diabetes mellitus in adult women, *Annals of Internal Medicine* 130 (1999): 322–324.
30. R. S. Lindsay and coauthors, Secular trends in birthweight, BMI, and diabetes in the offspring of diabetic mothers, *Diabetes Care* 23 (2000): 1249–1254; G. M. Egeland, R. Skjærven, and L. M. Irgens, Birth characteristics of women who develop gestational diabetes: Population based study, *British Medical Journal* 321 (2000): 546–547.
31. Rasmussen, 2001.
32. C. S. Yajnik, The insulin resistance epidemic in India: Fetal origins, later lifestyle, or both? *Nutrition Reviews* 59 (2001): 1–9.
33. J. V. Neel, The "thrifty genotype" in 1998, *Nutrition Reviews* 57 (1999): S2–S9.
34. A. Lucas, M. S. Fewtrell, and T. J. Cole, Fetal origins of adult disease—The hypothesis revisited, *British Medical Journal* 319 (1999): 245–249.
35. J. G. Eriksson and coauthors, Early growth and coronary heart disease in later life: Longitudinal study, *British Medical Journal* 322 (2001): 949–953.
36. J. Tuomilehto and coauthors, Prevention of type 2 diabetes mellitus by changes in lifestyle among subjects with impaired glucose tolerance, *New England Journal of Medicine* 344 (2001): 1343–1350.
37. J. Paronen and coauthors, Effect of cow's milk exposure and maternal type 1 diabetes on cellular and humoral immunization to dietary insulin in infants at genetic risk for type 1 diabetes. Finnish Trial to Reduce IDDM in the Genetically at Risk Study Group, *Diabetes* 49 (2000): 1657–1665.
38. T. Kimpimaki and coauthors, Short-term exclusive breastfeeding predisposes young children with increased genetic risk of type I diabetes to progressive beta-cell autoimmunity, *Diabetologia* 44 (2001): 63–69; E. Hypponen and coauthors, Infant feeding, early weight gain, and risk of type 1 diabetes: Childhood Diabetes in Finland (DiMe) Study Group, *Diabetes Care* 22 (1999): 1961–1965.
39. Paronen and coauthors, 2000.
40. L. Monetini and coauthors, Bovine beta-casein antibodies in breast- and bottle-fed infants: Their relevance in type 1 diabetes, *Diabetes/Metabolism Research and Reviews* 17 (2001): 51–54.
41. W. Karges and coauthor, Immunological aspects of nutritional diabetes prevention in NOD mice: a pilot study for the cow's milk-based IDDM prevention trial, *Diabetes* 46 (1997): 557–564.
42. C. D. Bernadier, Diabetes mellitus: Is there a connection with infant feeding practices? *Nutrition Today* 36 (2001): 241–248; M. Hummel and coau-

thors, No major association of breast-feeding, vaccinations, and child viral diseases with early islet autoimmunity in the German BABYDIAB study, *Diabetes Care* 23 (2000): 969–974; J. J. Couper and coauthors, Lack of association between duration of breast-feeding or introduction of cow's milk and development of islet autoimmunity, *Diabetes* 48 (1999): 2145–2149.

43. F. Esfarjani, M. R. Azar, and M. Gafarpour, IDDM and early exposure of infant to cow's milk and solid food, *Indian Journal of Pediatrics* 68 (2001): 107–110.

44. J. J. Couper and coauthors, Lack of association between duration of breast feeding or introduction of cow's milk and the development of islet immunity, *Diabetes* 48 (1999): 2145–2149.

45. I. Thorsdottir and coauthors, Different beta-casein fractions in Icelandic versus Scandinavian cow's milk may influence diabetogenicity of cow's milk in infants and explain low incidence of insulin-dependent diabetes mellitus in Iceland, *Pediatrics* 106 (2000): 719–724.

46. P. A. McKinney and coauthors, Perinatal and neonatal determinants of childhood type 1 diabetes: A case-control study in Yorkshire, U.K., *Diabetes Care* 22 (1999): 928–932.

47. Fagot-Campagna, Narayan, and Imperatore, 2001.

48. R. Lowry and coauthors, Recent trends in participation in physical education among U.S. high school students, *Journal of School Health* 71 (2001): 145–152; Committee on Sports Medicine and Fitness and Committee on School Health, American Academy of Pediatrics, Physical fitness and activity in schools, *Pediatrics* 105 (2000): 1156–1157.

Chapter 13

1. S. B. Roberts and M. B. Heyman, Micronutrient shortfalls in young children's diets: Common, and owing to inadequate intakes both at home and at child care centers, *Nutrition Reviews* 58 (2000): 27–29.

2. T. A. Nicklas and coauthors, Breakfast consumption with and without vitamin-mineral supplement use favorably impacts daily nutrient intake of ninth-grade students, *Journal of Adolescent Health* 27 (2000): 314–321.

3. R. E. Olson, Is it wise to restrict fat in the diets of children? *Journal of the American Dietetic Association* 100 (2000): 28–32; E. Satter, A moderate view on fat restriction, *Journal of the American Dietetic Association* 100 (2000): 32–36; L. A. Lytle, In defense of a low-fat diet for healthy children, *Journal of the American Dietetic Association* 100 (2000): 39–41; N. F. Butte, Fat intake of children in relation to energy requirements, *American Journal of Clinical Nutrition* 72 (2000): S1246–S1252.

4. L. Rask-Nissilä, Neurological development of 5-year-old children receiving a low-saturated fat, low-cholesterol diet since infancy: A randomized controlled trial, *Journal of the American Medical Association* 284 (2000): 993–1000.

5. U.S. Department of Agriculture, Dietary Guidelines Advisory Committee, *Nutrition and Your Health: Dietary Guidelines for Americans, 5th edition, 2000*, Home and Garden Bulletin no. 232, available at www.usda.gov/cnpp or call (888) 878-3256.

6. L. L. Birch, Development of food preferences, *Annual Review of Nutrition* 19 (1999): 41–62.

7. J. O. Fisher and L. L. Birch, Restricting access to palatable foods affects children's behavioral response, food selection, and intake, *American Journal of Clinical Nutrition* 69 (1999): 1264–1272.

8. J. O. Fisher and coauthors, Parental influences on young girls' fruit and vegetable, micronutrient, and fat intakes, *Journal of the American Dietetic Association* 102 (2002): 58–64; J. Skinner and coauthors, Toddlers' food preferences: Concordance with family members' preferences, *Journal of Nutrition Education* 30 (1998): 17–22.

9. S. J. Schoenthaler and I. D. Bier, The effect of vitamin-mineral supplementation on juvenile delinquency among American schoolchildren: A randomized, double-blind placebo-controlled trial, *Journal of Alternative and Complementary Medicine* 6 (2000): 7–17.

10. A. C. Looker and coauthors, Prevalence of iron deficiency in the United States, *Journal of the American Medical Association* 277 (1997): 973–976.

11. E. Pollitt, Iron deficiency and educational deficiency, *Nutrition Reviews* 55 (1997): 133–141.

12. E. K. Hurtado, A. H. Claussen, and K. G. Scott, Early childhood anemia and mild or moderate mental retardation, *American Journal of Clinical Nutrition Society* 59 (2000): 47–54.

13. D. Pinero, B. Jones, and J. Beard, Variations in dietary iron alter behavior in developing rats, *Journal of Nutrition* 131 (2001): 311–318; S. M. Grantham-McGregor, S. P. Walker, and S. Chang, Nutritional deficiencies and later behavioural development, *Proceedings of the Nutrition Society* 59 (2000): 47–54.

14. W. J. Rogan and coauthors, The effect of chelation therapy with succimer on neuropsychological development in children exposed to lead, *New England Journal of Medicine* 344 (2001): 1421–1426; J. F. Rosen and P. Mushak, Primary prevention of childhood lead poisoning—The only solution, *New England Journal of Medicine* 344 (2001): 1470–1471; National Center for Environment Health, CDC Childhood Lead Poisoning Prevention Program, www.cdc.gov/nceh/lead/lead.htm.

15. Fatal pediatric lead poisoning—New Hampshire, 2000, *Morbidity and Mortality Weekly Report* 50 (2001): 457–459.

16. H. L. Needleman and coauthors, Bone lead levels and delinquent behavior, *Journal of the American Medical Association* 275 (1996): 363–369.

17. T. D. Matte, Reducing blood lead levels: Benefits and strategies, *Journal of the American Medical Association* 281 (1999): 2340–2342; J. A. Simon and E. S. Hudes, Relationship of ascorbic acid to blood lead levels, *Journal of the American Medical Association* 281 (1999): 2289–2293; Y. Cheng and coauthors, Relation of nutrition to bone lead and blood lead levels in middle-aged to elderly men: The Normative Aging Study, *American Journal of Epidemiology* 147 (1998): 1162–1174; R. A. Goyer, Toxic and essential metal interactions, *Annual Review of Nutrition* 17 (1997): 37–50.

18. M. J. Brown and coauthors, The effectiveness of housing policies in reducing children's lead exposure, *American Journal of Public Health* 91 (2001): 621–624; Federal Interagency Forum on Child and Family Statistics, *America's Children: Key National Indicators of Well-Being, 1999* (Washington, D.C.: Government Printing Office, 1999). Yearly updates available from www.childstats.gov.

19. Advisory Committee on Childhood Lead Poisoning Prevention, Recommendations for blood lead screening of young children enrolled in Medicaid: Targeting a high-risk group, *Morbidity and Mortality Weekly Report* 49 (2000): 1–13; C. Campbell and K. C. Osterhoudt, Prevention of childhood lead poisoning, *Current Opinion in Pediatrics* 12 (2000): 428–437.

20. D. Farley, Dangers of lead still linger, *FDA Consumer*, January/February 1998, pp. 16–21.

21. Food Allergy and Intolerances, National Institutes of Health Fact Sheet, available at www.niaid.hih.gov/factsheets/food.htm (accessed May 25, 2001).

22. H. S. Skolnick and coauthors, The natural history of peanut allergy, *Journal of Allergy and Clinical Immunology* 107 (2001): 367–374; U.S. Department of Health and Human Services, *Healthy People 2010*, 2nd ed, (Washington, D.C.: Government Printing Office, 2000), available online at www.health.gov/healthypeople or from (800) 367-4725.

23. K. J. Falci, K. L. Gombas, and E. L. Elliot, Food allergen awareness: An FDA priority, *Food Safety Magazine*, February/March 2001, available at www.cfsan.fda.gov/~dms/.

24. B. Wuthrich, Lethal or life-threatening allergic reactions to food, *Journal of Investigational Allergology and Clinical Immunology* 10 (2000): 59–65.

25. G. Iacono, Intolerance of cow's milk and chronic constipation in children, *New England Journal of Medicine* 339 (1998): 1100–1104; B. Wuthrich, Food-induced cutaneous adverse reactions, *Allergy* 53 (1998): 131–135.

26. G. S. Rhim and M. S. McMorris, School readiness for children with food allergies, *Annals of Allergy, Asthma and Immunololgy* 86 (2001): 172–176.

27. S. A. Bock, A. Munoz-Furlong, and H. A. Sampson, Fatalities due to anaphylactic reactions to foods, *Journal of Allergy and Clinical Immunology* 107 (2001): 191–193.

28. J. M. Yeung, R. S. Applebaum, and R. Hildwine, Criteria to determine food allergen priority, *Journal of Food Protection* 63 (2000): 982–986.

29. R. Formanek Jr., Food allergies: When food becomes the enemy, *FDA Consumer*, July/August 2001, pp. 10–16.

30. K. Deibel and coauthors, A comprehensive approach to reducing the risk of allergens in foods, *Journal of Food Protection* 60 (1997): 436–441.

31. R. Sporik, D. J. Hill, and C. S. Hosking, Specificity of allergen skin testing in predicting positive open food challenges to milk, egg and peanut

F

in children, *Clinical and Experimental Allergy* 30 (2000): 1495–1498; S. L. Taylor, S. L. Hefle, and A. Munoz-Furlong, Food allergies and avoidance diets, *Nutrition Today* 34 (1999): 15–22; A. W. Burks and coauthors, Atopic dermatitis and food hypersensitivity reactions, *Journal of Pediatrics* 132 (1998): 132–136.

32. T. Spencer, J. Biederman, and T. Wilens, Growth deficits in children with attention deficit hyperactivity disorder, *Pediatrics* 102 (1998): 501–506.

33. J. Breakey, The role of diet and behaviour in childhood, *Journal of Paediatrics and Child Health* 33 (1997): 190–194.

34. J. M. Murphy and coauthors, Relationship between hunger and psychosocial functioning in low-income American children, *Journal of the American Academy of Child and Adolescent Psychiatry* 37 (1998): 163–170.

35. R. E. Andersen and coauthors, Relationship of physical activity and television watching with body weight and level of fatness among children, *Journal of the American Medical Association* 279 (1998): 938–942.

36. K. A. Coon and coauthors, Relationships between use of television during meals and children's food consumption patterns, *Pediatrics* 107 (2001): 167 [www.pediatrics.org/cgi/content/full/107/1/e7]; C. J. Crespo and coauthors, Television watching, energy intake, and obesity in U.S. children: Results from the third National Health and Nutrition Examination Survey, 1988–1994, *Archives of Pediatrics and Adolescent Medicine* 155 (2001): 360–365.

37. M. I. Goran, Metabolic precursors and effects of obesity in children: A decade of progress, 1990–1999, *American Journal of Clinical Nutrition* 73 (2001):158–171.

38. E. Luder, T. A. Melnik, and M. DiMaio, Association of being overweight with greater asthma symptoms in inner city black and Hispanic children, *Journal of Pediatrics* 132 (1998): 699–703.

39. B. J. Rolls, D. Engell, and L. L. Birch, Serving portion size influences 5-year-old but not 3-year-old children's food intakes, *Journal of the American Dietetic Association* 100 (2000): 232–234.

40. R. Lowry and coauthors, Recent trends in participation in physical education among U.S. high school students, *Journal of School Health* 71 (2001): 145–152.

41. T. N. Robinson, Reducing children's television viewing to prevent obesity: A randomized controlled trial, *Journal of the American Medical Association* 282 (1999): 1561–1567.

42. Coon and coauthors, 2001.

43. M. Golan and coauthors, Parents as the exclusive agents of change in the treatment of childhood obesity, *American Journal of Clinical Nutrition* 67 (1998): 1130–1135.

44. A. E. Field and coauthors, Peer, parent, and media influences on the development of weight concerns and frequent dieting among preadolescent and adolescent girls and boys, *Pediatrics* 107 (2001): 54–60.

45. A. F. Subar and coauthors, Dietary sources of nutrients among US children, 1989–1991, *Pediatrics* 102 (1998): 913–923; C. H. Ruxton and T. R. Kirk, Breakfast: A review of associations with measures of dietary intake, physiology, and biochemistry, *British Journal of Nutrition* 78 (1997): 199–214.

46. Murphy and coauthors, 1998.

47. Position of The American Dietetic Association: Dietary guidance for healthy children aged 2 to 11 years, *Journal of the American Dietetic Association* 99 (1999): 93–101.

48. A. J. Rainville, Nutritional quality of reimbursable school lunches compared with lunches brought from home in elementary schools in two southeastern Michigan districts, *Journal of Child Nutrition and Management* 1 (2000): 13–18.

49. E. A. Bergman and coauthors, Time spent by schoolchildren to eat lunch, *Journal of the American Dietetic Association* 100 (2000): 696–698.

50. K. Schuster, Feds put schools on a lowfat diet, *Food Management*, August 1994, pp. 78–84. Evidence that children grow well on a low-fat diet was provided by L. Van Horn, the principal investigator of the Dietary Intervention Study in Children (DISC), which follows children up to age 18, as cited in Kids grow well on low-fat diet, *Nutrition and the M.D.*, January 1996, pp. 6–7.

51. Position of The American Dietetic Association: Local support for nutrition integrity in schools, *Journal of the American Dietetic Association* 100 (2000): 108–111; K. W. Cullen and coauthors, Effect of a la carte and snack bar foods at school on children's lunchtime intake of fruits and vegetables, *Journal of the American Dietetic Association* 100 (2000): 1482–1486.

52. M. B. Wildey and coauthors, Fat and sugar levels are high in snacks purchased from student stores in middle schools, *Journal of the American Dietetic Association* 100 (2000): 319–322; L. Harnack and coauthors, Availability of a la carte food items in junior and senior high schools: A needs assessment, *Journal of the American Dietetic Association* 100 (2000): 701–703.

53. The Writing Group for the DISC Collaborative Research Group, Efficacy and safety of lowering dietary intake of fat and cholesterol in children with elevated low-density lipoprotein cholesterol: The Dietary Intervention Study in Children (DISC), *Journal of the American Medical Association* 273 (1995): 1429–1435.

54. M. Story, M. Hayes, and B. Kalina, Availability of foods in high schools: Is there cause for concern? *Journal of the American Dietetic Association* 96 (1996): 123–126.

55. Committee on Nutrition, *American Academy of Pediatrics*, The use and misuse of fruit juice in pediatrics, *Pediatrics* 107 (2001): 1210–1213.

56. J. D. Skinner and B. R. Carruth, A longitudinal study of children's juice intake and growth: The juice controversy revisited, *Journal of the American Dietetic Association* 101 (2001): 432–437.

57. D. S. Ludwig, K. E. Peterson, and S. L. Gortmaker, Relation between consumption of sugar-sweetened drinks and childhood obesity: A prospective, observational analysis, *Lancet* 357 (2001): 505–508.

58. V. C. Lysen and R. Walker, Osteoporosis risk factors in eighth grade students, *Journal of School Health* 67 (1997): 317–321.

59. L. Harnack, J. Stang, and M. Story, Soft drink consumption among U.S. children and adolescents: Nutritional consequences, *Journal of the American Dietetic Association* 99 (1999): 436–441; J. Cadogan and coauthors, Milk intake and bone mineral acquisition in adolescent girls: Randomised, controlled intervention trial, *British Medical Journal* 315 (1997): 1255–1260.

60. D. Neumark-Sztainer and coauthors, Adolescents engaging in unhealthy weight control behaviors: Are they at risk for other health-compromising behaviors? *American Journal of Public Health* 88 (1998): 952–955.

61. Field and coauthors, 2001; K. K. Davison and L. L. Birch, Weight status, parent reaction, and self-concept in five-year-old girls, *Pediatrics* 107 (2001): 46–53.

62. A. M. Siega-Riz, C. Cavadini, and B. M. Popkin, U.S. Teens and the nutrient contribution and differences of their selected meal patterns, *Family Economics and Nutrition Review* 13 (2001): 15–26; L. A. Lytle and coauthors, How do children's eating patterns and food choices change over time? Results from a cohort study, *American Journal of Health Promotion* 14 (2000): 222–228.

63. A. M. Siega-Riz, T. Carson, and B. Popkin, Three squares or mostly snacks—What do teens really eat? A sociodemographic study of meal patterns, *Journal of Adolescent Health* 22 (1998): 29–36.

64. Siega-Riz, Carson, and Popkin, 1998.

65. E. J. van Hoogdalem, I. J. Terpstra, and A. L. Baven, Evaluation of the effect of zinc acetate on the stratum corneum penetration kinetics of erythromycin in healthy male volunteers, *Skin Pharmacology* 9 (1996): 104–110.

66. H. Kerschner and J. M. Pegues, Productive aging: A quality of life agenda, *Journal of the American Dietetic Association* 98 (1998): 1445–1448.

67. B. Guyer and coauthors, Annual summary of vital statistics—1998, *Pediatrics* 104 (1999): 1229–1246.

68. K. G. Manton and J. W. Vaupel, Survival after the age of 80 in the United States, Sweden, France, England, and Japan, *New England Journal of Medicine* 333 (1995): 1232–1235.

69. K. G. Manton and E. Stallard, Longevity in the United States: Age and sex-specific evidence on life span limits from mortality patterns 1960–1990, *Journal of Gerontology* 51A (1996): B362–B375.

70. D. A. Banks and M. Fossel, Telomeres, cancer, and aging: Altering the human life span, *Journal of the American Medical Association* 278 (1997): 1345–1348.

71. D. B. Allison and coauthors, Body mass index and all-cause mortality among people age 70 and over: The Longitudinal Study of Aging, *International Journal of Obesity and Related Metabolic Disorders* 21 (1997): 424–431.

72. L. Di Pietro, The epidemiology of physical activity and physical function in older people, *Medicine and Science in Sports and Exercise* 28 (1996): 596–660; K. Buzina-Suboticanec and coauthors, Aging, nutritional status and immune response, *International Journal of Vitamin and Nutrition Research* 68 (1998): 133–141.

73. A. M. Egbert, The dwindles: Failure to thrive in older patients, *Nutrition Reviews* 54 (1996): S25–S30.

74. A. S. Nicolas and coauthors, Successful aging and nutrition, *Nutrition Reviews* 59 (2001): S88–S92.

75. W. J. Evans, Exercise training guidelines for the elderly, *Medicine and Science in Sports and Exercise* 31 (1999): 12–17.

76. R. Marcus and A. R. Hoffman, Growth hormone as therapy for older men and women, *Annual Review of Pharmacology and Toxicology* 38 (1998): 45–61; S. A. Lieberman and A. R. Hoffman, The somatopause: Should growth hormone deficiency in older people be treated? *Clinical Geriatric Medicine* 13 (1997): 671–684.

77. Targeting arthritis: The nation's leading cause of disability, *At-A-Glance*, 1998, available from Centers for Disease Control and Prevention, Mail Stop K-13, 4770 Buford Highway NE, Atlanta, GA 30341-3724.

78. T. McAlindon and D. T. Felson, Nutrition: Risk factors for osteoarthritis, *Annals of the Rheumatic Diseases* 56 (1997): 397–400; S. E. Edmonds and coauthors, Putative analgesic activity of repeated oral doses of vitamin E in the treatment of rheumatoid arthritis: Results of a prospective placebo controlled double blind trial, *Annals of the Rheumatic Diseases* 56 (1997): 649–655.

79. McAlindon and Felson, 1997.

80. J. J. Shrander and coauthors, Does food intolerance play a role in juvenile chronic arthritis? *British Journal of Rheumatology* 36 (1997): 905–908; J. A. Shapiro and coauthors, Diet and rheumatoid arthritis in women: A possible protective effect of fish consumption, *Epidemiology* 7 (1996): 256–263.

81. J. M. Kremer, n-3 Fatty acid supplements in rheumatoid arthritis, *American Journal of Clinical Nutrition* 71 (2000): S349–S351; Shapiro and coauthors, 1996.

82. National Institute of Arthritis and Musculoskeletal and Skin Diseases, National Institutes of Health, Glucosamine/Chondroitin Arthritis Intervention Trial (GAIT) begins patient recruitment, Press releases, 2000 available at www.niams.nih.gov/ne/press/2000/12_11a.httm; T. E. McAlindon and coauthors, Glucosamine and chondroitin for treatment of osteoarthritis: A systematic quality assessment and meta-analysis, *Journal of the American Medical Association* 283 (2000): 1469–1475.

83. Standing Committee on the Scientific Evaluation of Dietary Reference Intakes, Food and Nutrition Board, Institute of Medicine, *Dietary Reference Intakes for Calcium, Phosphorus, Magnesium, Vitamin D, and Fluoride* (Washington, D.C.: National Academy Press, 1997).

84. C. Ho, G. P. A. Kauwell, and L. B. Bailey, Practitioners' guide to meeting the vitamin B12 Recommended Dietary Allowance for people aged 51 years and older, *Journal of the American Dietetic Association* 99 (1999): 725–727; Standing Committee on the Scientific Evaluation of Dietary Reference Intakes, *Dietary Reference Intakes for Thiamin, Riboflavin, Niacin, Vitamin B_6, Folate, Vitamin B_{12}, Pantothenic Acid, Biotin, and Choline* (Washington, D.C.: National Academy Press, 1998), pp. 7-1–7-27.

85. H. W. Baik and R. M. Russell, Vitamin B_{12} deficiency in the elderly, *Annual Review of Nutrition* 19 (1999): 357–377.

86. Ho, Kauwell, and Bailey, 1999; Baik and Russell, 1999.

87. I. H. Rosenberg, B vitamins, homocysteine, and neurocognitive function, *Nutrition Reviews* 59 (2001): S69–S74.

88. M. Meydani, Antioxidants and cognitive function, *Nutrition Reviews* 59 (2001): S75–S82.

89. B. R. Hammond and coauthors, Dietary modification of human macular pigment density, *Investigative Ophthalmology and Visual Science* 38 (1997): 1795–1801.

90. E. Cho and coauthors, Prospective study of dietary fat and the risk of age-related macular degeneration, *American Journal of Clinical Nutrition* 73 (2000): 209–218.

91. S. T. Mayne, Beta-carotene, carotenoids, and disease prevention in humans, *FASEB Journal* 10 (1996): 690–701.

92. M. C. Leske and coauthors, Antioxidant vitamins and nuclear opacities: The longitudinal study of cataract, *Ophthalmology* 105 (1998): 831–836; P. F. Jacques and coauthors, Long-term vitamin C supplement use and prevalence of early age-related lens opacities, *American Journal of Clinical Nutrition* 66 (1997): 911–916.

93. J. C. Chidester and A. A. Spangler, Fluid intake in the institutionalized elderly, *Journal of the American Dietetic Association* 97 (1997): 23–28.

94. D. H. Holben and coauthors, Fluid intake compared with established standards and symptoms of dehydration among elderly residents of a long-term-care facility, *Journal of the American Dietetic Association* 99 (1999): 1447–1450.

95. F. Girodon and coauthors, Impact of trace elements and vitamin supplementation on immunity and infections in institutionalized elderly patients: A randomized controlled trial, *Archives of Internal Medicine* 159 (1999): 748–754; C. Fortes and coauthors, The effect of zinc and vitamin A supplementation on immune response in an older population, *Journal of the American Geriatrics Society* 46 (1998): 19–26; M. A. Johnson and K. H. Porter, Micronutrient supplementation and infection in institutionalized elders, *Nutrition Reviews* 55 (1997): 400–404.

96. R. D. Chandra, Graying of the immune system: Can nutrient supplements improve immunity in the elderly? *Journal of the American Medical Association* 277 (1997): 1398–1399.

97. Johnson and Porter, 1997.

98. S. N. Meydani and coauthors, Vitamin E supplementation and in vivo immune response in healthy elderly subjects, *Journal of the American Medical Association* 277 (1997): 1380–1386.

99. N. B. Belloc and L. Breslow, Relationship of physical health status and health practices, *Preventive Medicine* 1 (1972): 409–421.

100. G. E. Vaillant and K. Mukamal, Successful aging, *American Journal of Psychiatry* 158 (2001): 839–847.

101. R. Weindruch, Caloric restriction and aging, *New England Journal of Medicine* 337 (1997): 986–994.

102. T. A. Gresl and coauthors, Dietary restriction and glucose regulation in aging rhesus monkeys: A follow-up report at 8.5 yr, *American Journal of Physiology: Endocrinology and Metabolism* 281 (2001): E757–E765; J. J. Ramsey and coauthors, Dietary restriction and aging in rhesus monkeys: The University of Wisconsin study, *Experimental Gerontology* 35 (2000): 1131–1149.

103. P. R. Johnson and coauthors, Longevity in obese and lean male and female rats of the Zucker strain: Prevention of hyperphagia, *American Journal of Clinical Nutrition* 66 (1997): 890–903.

104. J. M. Dhahbi and coauthors, Caloric restriction alters the feeding response of key metabolic enzyme genes, *Mechanics of Aging and Development* 122 (2001): 1033–1048; C. K. Lee and coauthors, Gene expression profile of aging and its retardation by calorie restriction, *Science* 285 (1999): 1390–1393.

105. S. X. Cao and coauthors, Genomic profiling of short- and long-term caloric restriction effects in the liver of aging mice, *Proceedings of the National Academy of Sciences* 98 (2001): 10630–10635.

106. L. E. Rikans and K. R. Hornbrook, Lipid peroxidation, antioxidant protection and aging, *Biochimica et Biophysica Acta* 1362 (1997): 116–127.

107. J. Wanagat, D. B. Allison, and R. Weindruch, Calorie intake and aging: Mechanisms in rodents and a study in nonhuman primates, *Toxicological Sciences* 52 (1999): 35–40; G. Paolisso and coauthors, Oxidative stress and advancing age: Results in healthy centenarians, *Journal of the American Geriatrics Society* 46 (1998): 833–838.

108. H. C. Hendrie, Epidemiology of dementia and Alzheimer's disease, *American Journal of Psychiatry* 6 (1998): S3–S18.

109. D. Schenk and coauthors, Immunization with amyloid-beta attenuates Alzheimer-disease-like pathology in the PDAPP mouse, *Nature* 400 (1999): 173–177; R. Vassar and coauthors, Beta-secretase cleavage of Alzheimer's amyloid precursor protein by the transmembrane aspartic protease BACE, *Science* 286 (1999): 735; I. Hussain and coauthors, Identification of novel aspartic protease (Asp 2) as beta-secretase, *Molecular and Cellular Neuroscience* 14 (1999): 419–427.

110. Y. Christen, Oxidative stress and Alzheimer disease, *American Journal of Clinical Nutrition* 71 (2000): S621–S629; M. A. Lovely and coauthors, Copper, iron and zinc in Alzheimer's disease senile plaques, *Journal of the Neurological Sciences* 158 (1998): 47–52; C. R. Cornett, W. R. Markesbery, and W. D. Ehmann, Imbalances of trace elements related to oxidative damage in Alzheimer's disease brain, *Neurotoxicology* 19 (1998): 339–345.

111. M. Grundman, Vitamin E and Alzheimer disease: The basis for additional clinical trials, *American Journal of Clinical Nutrition* 71 (2000): S630–S636.

112. L. R. White and coauthors, Brain aging and midlife tofu consumption, *Journal of the American College of Nutrition* 19 (2000): 207–209.

113. M. A. Lovely, C. Xie, and W. R. Markesbery, Protections against amyloid beta peptide toxicity by zinc, *Brain Research* 823 (1999): 88–95; Cornett, Markesbery, and Ehmann, 1998; Lovely and coauthors, 1998; M. P. Cuajungco and G. J. Lees, Zinc metabolism in the brain, *Neurobiology of Disease* 4 (1997): 137–169; F. C. Potocnik and coauthors, Zinc and

F

platelet membrane microviscosity in Alzheimer's disease: The in vivo effect of zinc on platelet membranes and cognition, *South African Medical Journal* 87 (1997): 1116–1119.

114. M. Sano and coauthors, A controlled trial of selegiline, alpha-tocopherol, or both as treatment for Alzheimer's disease, *New England Journal of Medicine* 336 (1997): 1216–1222.

115. H. X. Wang and coauthors, Vitamin B_{12} and folate in relation to the development of Alzheimer's disease, *Neurology* 56 (2001): 1188–1194D; A. Snowdon and coauthors, Serum folate and the severity of atrophy of the neocortex in Alzheimer disease: Findings from the Nun Study, *American Journal of Clinical Nutrition* 71 (2000): 993–998.

116. P. L. LeBars and coauthors, A placebo-controlled, double-blind, randomized trial of an extract of Ginkgo biloba for dementia: North American EGb study group, *Journal of the American Medical Association* 278 (1997): 1327–1332.

117. G. W. Small, Treatment of Alzheimer's disease: Current approaches and promising developments, *American Journal of Medicine* 27 (1998): S32–S38.

118. D. Schenk and coauthors, 1999.

119. S. Gillette-Guyonnet and coauthors, Weight loss in Alzheimer disease, *American Journal of Clinical Nutrition* 71 (2000): S637–S642; E. T. Poehlman and R. V. Dvorak, Energy expenditure, energy intake, and weight loss in Alzheimer disease, *American Journal of Clinical Nutrition* 71 (2000): S650–S655; S. Rivière and coauthors, Nutrition and Alzheimer's disease, *Nutrition Reviews* 57 (1999): 363–367.

120. B. Finley, Nutritional needs of the person with Alzheimer's disease: Practical approaches to quality care, *Journal of the American Dietetic Association* 97 (1997): S177–S180.

121. Y. Liao and coauthors, Quality of the last year of life of older adults: 1986 vs 1993, *Journal of the American Medical Association* 283 (2000): 512–518; K. G. Manton, L. Corder, and E. Stallard, Chronic disability trends in elderly United States population: 1982–1994, *Proceedings of the National Academy of Sciences of the USA* 94 (1997): 2593–2598.

122. K. J. Joshipura, W. C. Willett, and C. W. Douglass, The impact of edentulousness on food and nutrient intake, *Journal of the American Dental Association* 127 (1996): 459–467.

123. S. B. Roberts, Energy regulation and aging: Recent findings and their implications, *Nutrition Reviews* 58 (2000): 91–97; M. J. Toth and E. T. Poehlman, Energetic adaptation to chronic disease in the elderly, *Nutrition Reviews* 58 (2000): 61–66; F. Landi and coauthors, Body mass index and mortality among older people living in the community, *Journal of the American Geriatrics Society* 47 (1999): 1072–1076.

124. An excellent review of the many problems associated with alcoholism in the elderly is found in AMA Council on Scientific Affairs, Alcoholism in the elderly, *Journal of the American Medical Association* 275 (1996): 797–801.

125. Position of The American Dietetic Association: Nutrition, aging, and the continuum of care, *Journal of the American Dietetic Association* 100 (2000): 580–595.

126. B. E. Millen and coauthors, The Elderly Nutrition Program: an effective national framework for preventive nutrition interventions, *Journal of the American Dietetic Association* 102 (2002): 234–240.

127. P. Kurtzweil, Growing older, eating better, *FDA Consumer*, March 1996, pp. 12–16.

128. The takeout-food trend: Who carries out their meals and why, *Journal of the American Dietetic Association* 98 (1998): 820.

129. Economic Research Service of the U.S. Department of Agriculture, USDA report encourages Americans to remember nutritional needs when eating out, USDA News Release no. 0060.99 (1999) (available at www.econ.ag.gov/whatsnew/ news/diet.htm).

130. L. H. Clemens, D. L. Slawson, and R. C. Klesges, The effect of eating out on quality of diet in premenopausal women, *Journal of the American Dietetic Association* 99 (1999): 442–444.

Consumer Corner 13

1. ACOG News Release, ACOG issues guidelines on diagnosis and treatment of PMS, March 31, 2001, available at www.acog.org.

2. G. R. Kraemer and R. R. Kraemer, Premenstrual syndrome: Diagnosis and treatment experiences, *Journal of Women's Health* 7 (1998): 893–907.

3. P. J. Schmidt and coauthors, Differential behavioral effects of gonadal steroids in women with and in those without premenstrual syndrome, *New England Journal of Medicine* 338 (1998): 209–216.

4. D. R. Rubinow, P. J. Schmidt, and C. A. Roca, Estrogen-serotonin interactions: Implications for affective regulation, *Biological Psychiatry* 44 (1998): 839–850.

5. G. B. Cross and coauthors, Changes in nutrient intake during the menstrual cycle of overweight women with premenstrual syndrome, *British Journal of Nutrition* 85 (2001): 475–482.

6. A. F. Walker and coauthors, Magnesium supplementation alleviates premenstrual symptoms of fluid retention, *Journal of Women's Health* 7 (1998): 1157–1165.

7. S. Thys-Jacobs and coauthors, Calcium carbonate and the premenstrual syndrome: Effects on premenstrual and menstrual symptoms, *American Journal of Obstetrics and Gynecology* 179 (1998): 444–452.

8. S. S. Girdler and coauthors, Dysregulation of cardiovascular and neuroendocrine responses to stress in premenstrual dysphoric disorder, *Psychiatry Research* 81 (1998): 163–178.

Controversy 13

1. F. Bellisle and coauthors, Functional food science and behaviour and psychological functions, *British Journal of Nutrition* 80 (1998): S173–S193.

2. J. D. Fernstrom and M. H. Fernstrom, Monoamines and protein intake: Are control mechanisms designed to monitor a threshold intake or a set point? *Nutrition Reviews* 59 (2001): S60–S68.

3. D. Benton and R. T. Donohoe, The effects of nutrients on mood, *Public Health Nutrition* 2 (1999): 403–409.

4. T. Hamazaki and coauthors, The effect of docosa hexaenoic acid on aggression in young adults: A placebo-controlled double-blind study, *Journal of Clinical Investigation* 97 (1996): 1129–1133.

5. K. A. Bruinsma and D. L. Taren, Dieting, essential fatty acid intake, and depression, *Nutrition Reviews* 58 (2000): 98–108.

6. J. R. Hibbeln and coauthors, Plasma total cholesterol concentrations do not predict cerebrospinal fluid neurotransmitter metabolites: Implications for the biophysical role of highly unsaturated fatty acids, *American Journal of Clinical Nutrition* 71 (2000): S331–S338; J. Wardle and coauthors, Randomized trial of the effects of cholesterol-lowering dietary treatments on psychological function, *American Journal of Medicine* 108 (2000): 547–553.

7. D. Schardt, Memory pills—Mostly forgettable, *Nutrition Action Healthletter*, September 2001, pp. 9–11; P. Hollingsworth, Beverages: Redefining new age, *Food Technology* 51 (1997): 44–51.

8. M. C. J. M. van Dongen and coauthors, The efficacy of ginkgo for elderly people with dementia and age-associated memory impairment: New results of a randomized clinical trial, *Journal of the American Geriatrics Society* 48 (2000): 1183–1194; J. A. Mix and W. D. Crews Jr., An examination of the efficacy of Ginkgo biloba extract EGb761 on the neuropsychologic functioning of cognitively intact older adults, *Journal of Alternative and Complementary Medicine* 6 (2000): 219–229.

9. D. O. Kennedy, A. B. Scholey, and K. A. Wesnes, The dose-dependent cognitive effects of acute administration of Ginkgo biloba to healthy young volunteers, *Psychopharmacology* 151 (2000): 416–423; K. A. Wesnes and coauthors, The memory enhancing effects of a Ginkgo biloba/Panax ginseng combination in healthy middle-aged volunteers, *Psychopharmacology* 152 (2000): 353–361.

10. H. Miwa and coauthors, Generalized convulsions after consuming a large amount of ginkgo nuts, *Epilepsia* 42 (2001): 280–281; J. Benjamin and coauthors, A case of cerebral haemorrhage—Can Ginkgo biloba be implicated? *Postgraduate Medicine* 77 (2001): 112–113; J. M. Fessenden, W. Wittenborn, and L. Clarke, Ginkgo biloba: A case report of herbal medicine and bleeding postoperatively from a laparoscopic cholecystectomy, *American Surgeon* 67 (2001): 33–35.

11. H. R. Lieberman, The effects of ginseng, ephedrine, and caffeine on cognitive performance, mood and energy, *Nutrition Reviews* 59 (2001): 91–102.

12. B. J. Cardinal and H. J. Engels, Ginseng does not enhance psychological well-being in healthy, young adults: Results of a double-blind, placebo-controlled, randomized clinical trial, *Journal of the American Dietetic Association* 101 (2001): 655–660.

13. Lieberman, 2001.

14. D. Benton, The impact of the supply of glucose to the brain on mood and memory, *Nutrition Reviews* 59 (2001): S20–S21.

15. P. J. Rogers and H. J. Smit, Food craving and food "addiction": A critical review of the evidence from a biopsychosocial perspective, *Pharmacology, Biochemistry, and Behavior* 66 (2000): 3–14.

16. Lieberman, 2001.

17. E. diTomaso, M. Beltramo, and D. Piomelli, Brain cannabinoids in chocolate (scientific correspondence), *Nature* 382 (1996): 677–678.

18. M. S. Gevaerd and coauthors, Caffeine reverses the memory disruption induced by intra-nigral MPTP-injection in rats, *Brain Research Bulletin* 55 (2001): 101–106; K. P. Corodimas, J. C. Pruitt, and J. M. Stieg, Acute exposure to caffeine selectively disrupts context conditioning in rats, *Psychopharmacology* 152 (2000): 376–382.

19. Bellisle and coauthors, 1998.

20. C. R. Markus and coauthors, Does carbohydrate-rich, protein-poor food prevent a deterioration of mood and cognitive performance of stress-prone subjects when subjected to a stressful task? *Appetite* 31 (1998): 49–65.

21. R. C. Markus and coauthors, The bovine protein α-lactalbumin increases the plasma ratio of tryptophan to the other large neutral amino acids, and in vulnerable subjects raises brain serotonin activity, reduces cortisol concentration, and improves mood under stress, *American Journal of Clinical Nutrition* 71 (2000): 1536–1544.

22. J. D. Fernstrom, Can nutrient supplements modify brain function? *American Journal of Clinical Nutrition* 71 (2000): S1669–S1673; J. Selhub and coauthors, B vitamins, homocysteine, and neurocognitive function in the elderly, *American Journal of Clinical Nutrition* 71 (2000): S614–S620.

Chapter 14

1. Food and Drug Administration, Food security guidance: Availability, *Federal Register* 67 (2002): 1224–1225; J. Sobel, A. S. Khan, and D. L. Swerdlow, Threat of a biological terrorist attack on the US food supply: the CDC perspective, *The Lancet* 359 (2002): 874–880.

2. Centers for Disease Control and Prevention, Preliminary FoodNet data on the incidence of foodborne illnesses—Selected sites, United States, *Morbidity and Mortality Weekly Report* 50 (2001): 241–246; R. G. Villar and coauthors, Investigation of multidrug-resistant *Salmonella* serotype *typhimurium* DT104 infections linked to raw-milk cheese in Washington state, *Journal of the American Medical Association* 281 (1999): 1811–1816.

3. P. S. Mead and coauthors, Food-related illness and death in the United States, *Emerging Infectious Diseases*, September/October 2001, available at www.cdc.gov/ncidod/eid/vol5no5/mead.htm; Surgeon General, Food safety: A growing global health problem, *Journal of the American Medical Association* 283 (2000): 1817.

4. *Salmonella* and *Campylobacter* illnesses on the decline, Centers for Disease Control and Prevention News Release, March 11, 1999, available at www.hhs.gov/news/press/1999pres/990311b.html. Data from Foodborne Diseases Active Surveillance Network (Food Net) indicate a 14 percent decline in illnesses from both *Salmonella* and *Campylobacter* bacteria.

5. W. L. Chandler and coauthors, Prothrombotic coagulation abnormalities preceding the hemolytic-uremic syndrome, *New England Journal of Medicine* 346 (2002): 23–32; E. F. Grabowski, The hemolytic-uremic syndrome: Toxin, thrombin, and thrombosis, *New England Journal of Medicine* 346 (2002): 58–61.

6. L. B. Zimmerhackl, *E. coli*, antibiotics, and the hemolytic-uremic syndrome, *New England Journal of Medicine* 342 (2000): 1990–1991.

7. National Advisory Committee on Microbiological Criteria for Foods, Hazard analysis and critical control point principles and application guidelines, *Journal of Food Protection* 61 (1998): 762–775; HHS News, www.fda.gov/bbs/topics/NEWS/2001.

8. U.S. Department of Agriculture, Second progress report on *Salmonella* testing for raw meat and poultry products, *FSIS Backgrounder*, January 21, 1999.

9. Centers for Disease Control and Prevention, *Salmonella* and *campylobacter* illnesses on the decline, CDC Press Release, March 11, 1999, available from www.hhs.gov/news/press/1999pres/990311b.html.

10. J. Henkel, "Thermy" promotes thorough food cooking, *FDA Consumer*, September/October 2000, p. 35.

11. Position statement of the Institute of Food Science and Technology, Public Affairs and Technical and Legislative Committees, Bovine spongiform encephalopathy (BSE): Parts 1 / 2, June 16, 1999, available at http://ific.org.

12. Food and Drug Administration Action Plan, Transmissible spongiform encephalopathies including bovine spongiform encephalopathy and chronic wasting disease, April 2001, available at www.fda.gov/oc/oca/roundtable/bse/FDA_actionplan.html.

13. Outbreaks of *Salmonella* serotype enteritidis infection associated with consumption of raw shell eggs—United States, 1994–1995, *Morbidity and Mortality Weekly Report* 45 (1996): 737–742.

14. Even pasteurized eggs may contain harmful bacteria, *Tufts University Health and Nutrition Letter*, July 1998, p. 6.

15. Viruses—Just a flush away? *Science News* 155 (1999): 107.

16. Centers for Disease Control and Prevention, National Center for Infectious Diseases, Travelers' health: Bovine spongiform encephalopathy (mad cow disease) and new variant Creutzfeldt-Jakob disease (nvCJD), January 29, 2001, available at www.cdc.gov/travel/\madcow.htm.

17. C. P. Dougherty and coauthors, Dietary exposures to food contaminants across the United States, *Environmental Research* 84 (2000): 170–185.

18. FDA announces advisory on methylmercury in fish, *FDA Talk Paper*, 2001, available from www.cfsan.fda.gov.

19. EPA national advice on mercury in fish caught by family and friends for women who are or may become pregnant, nursing mothers, and young children, *EPA National Advisory*, 2001, available from www.epa.gov/ost/fish.

20. FDA, *Food and Drug Administration Pesticide Program—Residue Monitoring*, a report obtainable from FDA, HFF-420, 200 C Street S.W., Washington, DC 20204.

21. "Provocative" report issued on use of pesticides, *Journal of the American Medical Association* 275 (1996): 899.

22. J. Ma and coauthors, Milk intake, circulating levels of insulin-like growth factor-I, and risk of colorectal cancer in men, *Journal of the National Cancer Institute* 93 (2001): 1330–1336; J. L. Outwater, A. Nicholson, and N. Barnard, Dairy products and breast cancer: The IGF-I, estrogen, and bGH hypothesis, *Medical Hypotheses* 48 (1997): 453–461.

23. FDA Center for Veterinary Medicine, Report on the Food and Drug Administration's review of the safety of recombinant bovine somatotropin, available at www.fda.gov/cvm/index/bst/RBRPTFNL.htm.

24. A. E. Larson and coauthors, Evaluation of the botulism hazard from vegetables in modified atmosphere packaging, *Journal of Food Protection* 60 (1997): 1208–1214.

25. R. Papazian, Sulfites: Safe for most, dangerous for some, *FDA Consumer*, December 1996, pp. 11–14.

26. International Food Information Council, What's this doing in my food? A guide to food ingredients, *Food Insight*, November/December 2000, available at http://ific.org.

27. International Food Information Council, Glutamate and monosodium glutamate: Examining the myths, *IFIC Review*, November 2001, available at http://ific.org/relatives/17660.pdf.

Consumer Corner 14

1. Position of The American Dietetic Association: Food irradiation, *Journal of the American Dietetic Association* 100 (2000): 246–253.

2. Meat irradiation can boost food safety, *FDA Consumer*, March/April 2000, p. 4.

3. FDA clarifies radiation labeling, *Journal of the American Dietetic Association* 98 (1998): 1403.

4. Food and Drug Administration, Irradiation in the production, processing and handling of food, 62 *Federal Register* 232 (1997): 64107–64121.

5. World Health Organization, *Weekly Epidemiological Record*, January 16, 1998, pp. 9–11.

6. Vijayalaxmi and S. G. Srikantia, A preview of the studies on the wholesomeness of irradiated wheat, conducted at the National Institute of Nutrition, India, *Radiation, Physics, and Chemistry* 34 (1989): 941–952.

7. Vijayalaxmi and Srikantia, 1989.

Controversy 14

1. C. Robinson, Making forest biotechnology a commercial reality: Do we need a tree genome project, or will Arabidopsis point the way? *Nature Biotechnology* 17 (1999): 27–30; S. M. Pepa, Research and trade in genetics: How countries should structure for the future, *Medicine and Law* 17 (1998): 437–454; R. A. Dixon and coauthors, Metabolic engineering: Prospects for crop improvement through genetic manipulation of phenylpropanoid biosynthesis and defense responses—a review, *Gene* 179 (1996): 61–71.

F

F

2. J. Fernandez-Cornejo and W. D. McBride, Genetically engineered crops for pest management in U.S. agriculture: Farm-level effects, Economic Research Service, USDA, 2001, available from www.aphis.usda.gov/biotechnology/research.html.

3. H. C. Sharma and coauthors, Prospects for using transgenic resistance to insects in crop improvement, *EJB Electronic Journal of Biotechnology* (online) vol. 3, no. 2 (2000), available from www.ejb.org/contrent/vol3/issue2/full/3/index/html.

4. C. Lewis, A new kind of fish story: The coming of biotech animals, *FDA Consumer*, January/February 2001, pp. 15–20.

5. U.S. Department of Agriculture, Agricultural biotechnology: Frequently asked questions, 2001, available from www.aphis.usda.gov/biotechnology/faqs.html.

6. V. Worthington, Effect of agricultural methods on nutritional quality: A comparison of organic with conventional crops, *Alternative Therapies in Health and Medicine* 4 (1998): 58–69.

7. Organic food test, *Consumer Reports*, January 1998, pp. 13–18.

8. International Food Information Council, Food biotechnology, *Backgrounder*, 1998, available from http://www.starpass.net/winific.html; J. A. Nordlee and coauthors, Identification of a brazil-nut allergen in transgenic soybeans, *New England Journal of Medicine* 334 (1996): 688–692.

9. Amy Ridenour, director of the National Center for Public Policy Research in Washington, D.C.

10. H. Shand and P. Mooney, Terminator seeds threaten an end to farming, *Earth Island Journal*, Fall 1998, pp. 30–31.

11. Sarma and coauthors, 2000.

12. M. K. Sears and coauthors, Impact of BT corn pollen on monarch butterfly populations: A risk assessment, *Proceedings of the National Academy of Sciences* 98 (2001): 11937–11942.

13. D. E. Stanley-Horn and coauthors, Assessing the impact of Cry1Ab-expressing corn pollen on monarch butterfly larvae in field studies, *Proceedings of the National Academy of Sciences* 98 (2001): 11931–11936.

14. J. Walsh, Brave new farm, *Time*, January 11, 1999, pp. 86–88.

15. FDA Center for Food Safety and Applied Nutrition, Information paper on L-tryptophan and 5-hydroxy-L-tryptophan, February 2001, available at www.cfsan.fda.gov/~dms/.

16. R. Formanek, Jr., Proposed rule issued for bioengineered foods, *FDA Consumer*, March/April 2001, pp. 9–11.

17. J. Puzzanghera, Coalition sues over genetically altered foods, *Tallahassee Democrat*, May 28, 1998, p. 3A.

18. Position of The American Dietetic Association: Biotechnology and the future of food, *Journal of the American Dietetic Association* 95 (1995): 1429–1432.

19. Notebook, *FDA Consumer*, March/April 1998, p. 36.

20. Pesticides not pesty, *Priorities* (1996): 6.

Chapter 15

1. G. Gardner and B. Halweil, *Underfed and Overfed: The Global Epidemic of Malnutrition*, World Watch Paper 150 (Washington, D.C.: Worldwatch Institute, 2000).

2. Federal Interagency Forum on Child and Family Statistics, *America's Children: Key National Indicators of Well-Being*, 2001, available at www.childstats.gov.

3. S. Collins and M. Myatt, Short-term prognosis in severe adult and adolescent malnutrition during famine, *Journal of the American Medical Association* 284 (2000): 621–626.

4. World Health Organization, Executive summary, World Health Report 1998: Life in the 21st century—A vision for all, available from www.ch/whr/1998/exsum98e.htm.

5. C. Flavin and O. Tunali, *Climate of Hope: New Strategies for Stabilizing the World's Atmosphere*, World Watch Paper 130 (Washington, D.C.: Worldwatch Institute, 1996).

6. United Nations Food and Agriculture Organization, *The State of Food and Agriculture 2000*, available at www.fao.org.

7. The sweeping toll of malnutrition, *Public Health Reports* 112 (1997): 186–187.

8. Position of The American Dietetic Association: Domestic food and nutrition security, *Journal of the American Dietetic Association* 98 (1998): 337–342.

9. C. S. Kramer-LeBlanc and K. McMurry, Discussion paper on domestic food security, *Family Economics and Nutrition Review* 11 (1998): 49–78.

10. Economic Research Service, U.S. Department of Agriculture, Food and assistance programs, Briefing Room 2000, available at www.ers.usda.gov/briefing/foodnutritionassistance/.

11. V. Oliveira and C. Gundersen, WIC increases the nutrient intakes of children, *Food Review*, January/April 2001, pp. 27–30.

12. USDA Office of Analysis, The decline in food stamp participation: A report to Congress, 2001, Nutrition and Evaluation, Nutrition Assistance Research Report Series no. FSP-01-WEL, available from USDA Food and Nutrition Service.

13. L. Trivers and J. Borland, Glickman announces new community anti-hunger initiative, USDA press release, February 11, 1999, available from www.usda.gov/news/releases/; Kramer-LeBlanc and McMurry, 1998.

14. United Nations World Food Programme, accessible at www.wfp.org.

15. United Nations Population Fund, Food and water: Can we meet increasing demand? *The State of World Population*, 2001 available from www.unfpa.org.

16. World Health Organization, Meeting of Interested Parties, Nutrition risk factors, June 2001, available from www.who.int//mipfiles/2233/NHDbrochurecentrefold.pdf.

17. J. W. Clay and coauthors, *The Spoils of Famine: Ethiopian Famine Policy and Peasant Agriculture* (Cambridge, Mass.: Cultural Survival, 1988), as cited in L. R. Brown and H. Kane, *Full House* (New York: W. W. Norton, 1994), pp. 146–157.

18. United Nations Food and Agriculture Organization, Elements for a plan of action on fishing capacity, Preparatory meeting for the FAO consultation on the management of fishing capacity, shark fisheries, and incidental catch of seabirds in longline fisheries, July 22–24, 1998, available from www.nmfs.noaa.gov.

19. New international plan of action targets illegal, unregulated and unreported fishing, Press release, 2001, available at www.fao.org/WAICENT/OIS/PRESS_NE/PRESSENG/2001/pren0111.htm; United Nations Food and Agricultural Organization, Understanding the cultures of fishing communities: A key to fisheries management and food security, *The State of World Fisheries and Aquaculture*, 2000, available at www.fao.org/docrep/003/x8002e/x8002e00.htm.

20. T. Suzuki, Oxygen-deficient waters along the Japanese coast and their effects upon the estuarine ecosystem, *Journal of Environmental Quality* 30 (2001): 291–302.

21. United Nations Food and Agriculture Organization, 1998.

22. Intergovernmental Panel on Climate Change, Climate Change 2001: IPCC Third Assessment Report, available at www.ipcc.ch/pub/spm2201.pdf.

23. United Nations Environmental Programme, Major global trends: The state of the environment, *Global Environmental Outlook 2000*, 2000, overview available at www.unep.org.

24. United Nations Environmental Programme, 2000.

25. Position of The American Dietetic Association: Dietetic professionals can implement practices to conserve natural resources and protect the environment, *Journal of the American Dietetic Association* 101 (2001): 1221–1227.

Controversy 15

1. P. Smith and J. Warrick, Boss hog: North Carolina's pork revolution, *Amicus Journal*, Spring 1996, pp. 36–42.

2. EPA and hog farmers agree, *Progressive Farmer*, January 1999, pp. 6–7.

3. A. Beers, Animal waste report gives traction to reform bill, *Food Chemical News* 39 (1998): 24–26.

4. USDA Interagency Agricultural Projections Committee, *USDA Agricultural Baseline Projections to 2008*, 1999, available at usda.mannlib.cornell.edu.

5. U.S. Department of Commerce, *Statistical Abstract of the United States, 1994* (Washington, D.C.: Bureau of the Census, 1994), p. 668.

6. USDA National Agricultural Statistics Service, *Farms and Land in Farms, 1998* (Washington, D.C.: USDA Interagency Agricultural Projections Committee, 1999).

7. H. Carsalade, Food and Agriculture Organization, Sustainable Development Department, Sustainable food security, March 20, 1998, available from www.fao.org.

8. R. P. Tengerdy and G. Szakacs, Perspectives in agrobiotechnology, *Journal of Biotechnology* 66 (1998): 91–99.

9. A. T. Durning and H. B. Brough, Reforming the livestock economy, in L. R. Brown and coauthors, *State of the World 1992* (New York: W. W. Norton, 1992), pp. 66–82.

Answers to Self Check Questions

G

Chapter 1

1. a
2. a
3. c
4. b
5. d
6. b
7. False. Heart disease and cancer are influenced by many factors with genetics and diet among them.
8. True
9. True
10. False. Only when a finding has been repeatedly confirmed by science is it wise to change your diet accordingly.

Chapter 2

1. b
2. d
3. b
4. d
5. a
6. True
7. False. The DRI are estimates of the needs of healthy persons only. Medical problems alter nutrient needs.
8. False. People who choose to eat no meats or products taken from animals can still use the Food Guide Pyramid to make their diets adequate.
9. False. By law, food labels must state as a percentage of the Daily Values the amounts of vitamin A, vitamin C, calcium, and iron present in a food.
10. True

Chapter 3

1. c
2. d
3. c
4. c
5. d
6. False. Vitamin and other deficiencies easily damage the tissues of the immune system.
7. False. Hydrochloric acid initiates protein digestion and activates a protein-digesting enzyme in the stomach.
8. False. The digestive tract works efficiently to digest all foods simultaneously, regardless of composition.
9. True
10. False. Absorption of the majority of nutrients takes place across the specialized cells of the small intestine.

Chapter 4

1. b
2. a
3. c
4. b
5. a
6. True
7. False. Type 1 diabetes is most often controlled with insulin injections.
8. False. Whole-grain bread remains more nutritious despite the enrichment of white flour.
9. True
10. True

Chapter 5

1. c
2. a
3. c
4. b
5. d
6. True
7. False. The fatty acids of coconut oil are more saturated than those of cream and are of a type that seems to add to heart disease risk.
8. False. Consuming large amounts of *trans*-fatty acids elevates serum LDL cholesterol and thus raises the risk of heart disease and heart attack.
9. False. When olestra is present in the digestive tract, fat-soluble vitamins, including vitamin E, become unavailable for absorption.
10. True

Chapter 6

1. b
2. b
3. c
4. a
5. d
6. True
7. False. Excess protein in the diet may have adverse effects such as obesity, enlarged liver or kidneys, worsened kidney disease, and accelerated bone loss.
8. False. Impoverished people living on U.S. Indian reservations, in inner cities, and in rural areas of the United States, as well as some elderly, homeless, and ill people in hospitals, have been diagnosed with PEM.
9. True
10. False. Astronauts in space are in negative nitrogen balance.

Chapter 7

1. d
2. c
3. d
4. a
5. d
6. d
7. False. No study to date has conclusively demonstrated that vitamin C can prevent colds or reduce their severity.
8. True

9. False. With a few exceptions, nutrients are absorbed best from foods where they are dispersed among other ingredients that facilitate their absorption.
10. True

Chapter 8

1. d
2. c
3. b
4. d
5. a
6. False. You can survive a deficiency of any of the other nutrients for a long time, in some cases even for months or years, but you can survive only a few days without water.
7. True.
8. False. Calcium is the most abundant mineral in the body.
9. False. Butter, cream, and cream cheese contain negligible calcium, being almost pure fat. Many vegetables, such as broccoli are good sources of available calcium.
10. False. The standards for bottled water are substantially less rigorous than those applied to U.S. tap water.

Chapter 9

1. d
2. b
3. c
4. a
5. a
6. d
7. False. The thermic effect of food is believed to have negligible effects on total energy expenditure.
8. True
9. False. The BMI is unsuitable for use with athletes and adults over age 65.
10. True

Chapter 10

1. c
2. b
3. a
4. d
5. d
6. False. Weight training to improve muscle strength and endurance also helps maximize and maintain bone mass.
7. False. The average resting pulse for adults is around 70 beats per minute, but the rate is lower for active people.
8. True
9. True
10. True

Chapter 11

1. b
2. d
3. c
4. d
5. d
6. False. The best way to plan a diet to support the immune system is to meet the recommended intake for each nutrient while not ingesting a dose from a supplement that would cause harm.
7. True
8. False. Hypertension is more severe and occurs earlier in life among African Americans than those of European or Asian descent.
9. True
10. True

Chapter 12

1. d
2. a
3. b
4. d
5. d
6. a
7. True
8. True
9. False. In general, the effect of nutritional deprivation of the mother is to reduce the quantity, not the quality, of her milk.
10. False. There is no proof for the theory that "stuffing the baby" at bedtime will promote sleeping through the night.

Chapter 13

1. c
2. d
3. c
4. d
5. b
6. d
7. False. A food intolerance is an adverse effect of a food or food additive not involving the immune response, as in a food allergy.
8. False. Research to date does not support the idea that food allergies or intolerances cause hyperactivity in children, but studies continue.
9. True
10. False. Vitamin A absorption appears to increase with aging.

Chapter 14

1. b
2. c
3. d
4. a
5. d
6. False. Nature has provided many plants used for food with natural poisons to fend off diseases, insects, and other predators.
7. True
8. False. The canning industry chooses treatments that employ the high-temperature-short-time (HTST) principle for canning.
9. True
10. True

Chapter 15

1. b
2. b
3. c
4. d
5. c
6. True
7. False. Most children who die of malnutrition do not starve to death—they die because their health has been compromised by dehydration from infections that cause diarrhea.
8. False. The number of people affected by famine is relatively small compared with the number suffering from less severe but chronic hunger.
9. False. The link between improved economic status and slowed population growth has been demonstrated in country after country.
10. True

Glossary

A

absorb to take in, as nutrients are taken into the intestinal cells after digestion; the main function of the digestive tract with respect to nutrients.

acceptable daily intake (ADI) the estimated amount of sweetener that can be consumed daily over a person's lifetime without any adverse effects.

accredited approved; in the case of medical centers or universities, certified by an agency recognized by the U.S. Department of Education.

acesulfame (AY-sul-fame) **potassium,** also called **acesulfame-K** a zero-calorie sweetener approved by the FDA and Health Canada.

acetaldehyde (ass-et-AL-deh-hide) a substance to which ethanol is metabolized on its way to becoming harmless waste products that can be excreted.

acid reducers and **acid controllers** drugs that reduce the acid output of the stomach. They are most suitable for treating severe, persistent forms of heartburn but are useless for neutralizing acid already present in the stomach. These drugs are now sold without prescription, but the packages bear warnings of side effects; some types interfere with the stomach's ability to destroy alcohol, so more of the alcohol in a drink enters the bloodstream.

acid-base balance equilibrium between acid and base concentrations in the body fluids.

acidosis (acid-DOH-sis) blood acidity above normal, indicating excess acid (*osis* means "too much in the blood").

acids compounds that release hydrogens in a watery solution.

acne chronic inflammation of the skin's follicles and oil-producing glands, which leads to an accumulation of oils inside the ducts that surround hairs; usually associated with the maturation of young adults.

added sugars sugars added to a food for any purpose, such as to add sweetness or bulk or to aid in browning (baked goods).

additives substances that are added to foods, but are not normally consumed by themselves as foods.

adequacy the dietary characteristic of providing all of the essential nutrients, fiber, and energy in amounts sufficient to maintain health and body weight.

adipose tissue the body's fat tissue, consisting of masses of fat-storing cells and blood vessels to nourish them. Adipose tissue produces and releases hormones—among them, the hormone leptin involved in appetite regulation.

advertorials lengthy advertisements in newspapers and magazines that read like feature articles but are written for the purpose of touting the virtues of products.

aerobic (air-ROE-bic) requiring oxygen. Aerobic activity strengthens the heart and lungs by requiring them to work harder than normal to deliver oxygen to the tissues.

agribusiness agriculture practiced on a massive scale by large corporations owning vast acreages and employing intensive technological, fuel, and chemical inputs.

AIDS acquired immune deficiency syndrome; caused by infection with human immunodeficiency virus (HIV), which is transmitted primarily by sexual contact, contact with infected blood, needles shared among drug users, or fluids transferred from an infected mother to her fetus or infant.

alcohol dehydrogenase (dee-high-DRAH-gen-ace) **(ADH)** an enzyme system that breaks down alcohol. The antidiuretic hormone listed below is also abbreviated ADH.

alcoholism a dependency on alcohol marked by compulsive uncontrollable drinking with negative effects on physical health, family relationships, and social health.

alcohol-related birth defects (ARBD) a condition caused by prenatal alcohol exposure. ARBD is diagnosed when there is a history of substantial regular maternal alcohol intake or heavy episodic drinking, combined with birth defects known to be associated with alcohol exposure.

alcohol-related neurodevelopmental disorder (ARND) a condition caused by prenatal alcohol exposure. ARND is diagnosed when there is a confirmed history of substantial regular maternal alcohol intake or heavy episodic drinking, combined with behavioral, cognitive, or central nervous system abnormalities in the child that are known to be associated with alcohol exposure.

alitame a noncaloric sweetener formed from the amino acids L-aspartic acid and L-alanine. In the United States, the FDA is considering its approval.

alkalosis (al-kah-LOH-sis) blood alkalinity above normal (*alka* means "base"; *osis* means "too much in the blood").

allergy an immune reaction to a foreign substance, such as a component of food. Also called *hypersensitivity* by researchers.

aloe a tropical plant with widely claimed value as a topical treatment for minor skin injury. Some scientific evidence supports this claim; evidence against its use in severe wounds also exists.

alpha-lactalbumin (lact-AL-byoo-min) the chief protein in human breast milk. The chief protein in cow's milk is *casein* (CAY-seen).

alternative (low-input, or **sustainable) agriculture** agriculture practiced on a small scale using individualized approaches that vary with local conditions so as to minimize technological, fuel, and chemical inputs.

amenorrhea the absence or cessation of menstruation.

American Dietetic Association (ADA) the professional organization of dietitians in the United States. The Canadian equivalent is the Dietitians of Canada (DC), which operates similarly.

amine (a-MEEN) **group** the nitrogen-containing portion of an amino acid.

amino (a-MEEN-o) **acids** the building blocks of protein. Each has an amine group at one end, an acid group at the other, and a distinctive side chain.

amino acid chelates (KEY-lates) compounds of minerals (such as calcium) combined with amino acids in a form that favors their absorption. A *chelating agent* is a molecule that surrounds another molecule and can then either promote or prevent its movement from place to place (*chele* means "claw").

amino acid pools amino acids dissolved in cellular fluid that provide cells with ready raw materials from which to build new proteins or other molecules.

amniotic (am-nee-OTT-ic) **sac** the "bag of waters" in the uterus in which the fetus floats.

anabolic steroid hormones chemical messengers related to the male sex hormone testosterone that stimulate building up of body tissues (*anabolic* means "promoting growth"; *sterol* refers to compounds chemically related to cholesterol).

anaerobic (AN-air-ROE-bic) not requiring oxygen. Anaerobic activity may require strength but does not work the heart and lungs very hard for a sustained period.

anaphylactic (an-AFF-ill-LAC-tic) **shock** a life-threatening whole-body allergic reaction to an offending substance.

androstenedione (AN-droh-STEEN-dee-own) a precursor of testosterone that elevates both testosterone and estrogen in the blood of both males and females. Often called *andro,* it is sold with claims of producing increased muscle strength, but controlled studies disprove such claims.

anecdotal evidence information based on interesting and entertaining, but not scientific, personal accounts of events.

anemia the condition of inadequate or impaired red blood cells; a reduced number or volume of red blood cells along with too little hemoglobin in the blood. The red blood cells may be immature and therefore too large or too small to function properly. Anemia can result from blood loss, excessive red blood cell destruction, defective red blood cell formation, and many nutrient deficiencies. Anemia is not a disease, but a symptom of another problem; its name literally means "too little blood."

anencephaly (an-en-SEFF-ah-lee) a severe neural tube defect in which the brain fails to form. Anencephaly leads to death soon after birth.

aneurysm (AN-you-rism) the ballooning out of an artery wall at a point that is weakened by deterioration.

anorexia nervosa an eating disorder characterized by a refusal to maintain a minimally normal body weight, self-starvation to the extreme, and a disturbed perception of body weight and shape; seen (usually) in teenage girls and young women (*anorexia* means "without appetite"; *nervos* means "of nervous origin").

antacids acid-buffering agents used to counter excess acidity in the stomach. Calcium-containing preparations (such as Tums) contain available calcium. Antacids with aluminum or magnesium hydroxides (such as Rolaids) can accelerate calcium losses.

antacids medications that react directly and immediately with the acid of the stomach, neutralizing it. Antacids are most suitable for treating occasional heartburn.

anthrax a potentially fatal disease caused by inhalation of or skin contact with spores of the toxin-producing anthrax bacterium; historically caused by contact with infected livestock, but currently a concern because it has been used as a biological weapon of terrorists.

antibodies (AN-te-bod-ees) large proteins of the blood, produced by the immune system in response to an invasion of the body by foreign substances (antigens). Antibodies combine with and inactivate the antigens.

anticarcinogens compounds in foods that act in any of several ways to oppose the formation of cancer.

antidiuretic (AN-tee-dye-you-RET-ick) **hormone (ADH)** a hormone produced by the pituitary gland in response to dehydration (or a high sodium concentration in the blood). It stimulates the kidneys to reabsorb more water and so to excrete less. (This hormone should not be confused with the enzyme alcohol dehydrogenase, which is also abbreviated ADH.)

antigen a substance foreign to the body that elicits the formation of antibodies or an inflammation reaction from immune system cells. Food antigens are usually glycoproteins (large proteins with glucose molecules attached). Inflammation consists of local swelling and irritation and attracts white blood cells to the site.

antimicrobial agents preservatives that prevent spoilage by mold or bacterial growth. Familiar examples are acetic acid (vinegar) and sodium chloride (salt). Others are benzoic, propionic, and sorbic acids; nitrites and nitrates; and sulfur dioxide.

antioxidant (anti-OX-ih-dant) a compound that protects other compounds from damaging reactions involving oxygen by itself reacting with oxygen (*anti* means "against"; *oxy* means "oxygen"). *Oxidation* is a potentially damaging effect of normal cell chemistry involving oxygen.

antioxidants preservatives that prevent rancidity of fats in foods and other damage to food caused by oxygen. Examples are vitamins E and C, BHA, BHT, propyl gallate, and sulfites.

antisense gene a gene's chemical opposite, which interferes with the native working gene and prevents it from producing proteins.

aorta (ay-OR-tuh) the large, primary artery that conducts blood from the heart to the body's smaller arteries.

Apgar score a system of scoring an infant's physical condition right after birth. Heart rate, respiration, muscle tone, response to stimuli, and color are ranked 0, 1, or 2. A low score indicates that medical attention is required to facilitate survival.

appendicitis inflammation and/or infection of the appendix, a sac protruding from the intestine.

appetite the psychological desire to eat; a learned motivation and a positive sensation that accompanies the sight, smell, or thought of appealing foods.

appliance thermometer a thermometer that verifies the temperature of an appliance. An *oven thermometer* verifies that the oven is heating properly; a *refrigerator/freezer thermometer* tests for proper refrigerator (<40°F) or freezer temperature (0°F).

aquifers underground rock formations containing water that can be drawn to the surface for use.

arginine a nonessential amino acid falsely promoted as enhancing the secretion of human growth hormone, the breakdown of fat, and the development of muscle.

arousal heightened activity of certain brain centers associated with attention, excitement, and anxiety.

arteries blood vessels that carry blood containing fresh oxygen supplies from the heart to the tissues.

artesian water water drawn from a well that taps a confined aquifer in which the water is under pressure.

arthritis a usually painful inflammation of the joints caused by many conditions, including infections, metabolic disturbances, or injury; usually results in altered joint structure and loss of function.

artificial colors certified food colors, added to enhance appearance. (Certified means approved by the FDA.) Vegetable dyes are extracted from vegetables such as beta-carotene from carrots. Food colors are a mix of vegetable dyes and synthetic dyes approved by the FDA for use in food.

artificial fats zero-energy fat replacers that are chemically synthesized to mimic the sensory and cooking qualities of naturally occurring fats, but are totally or partially resistant to digestion. Also called *fat analogues*.

artificial flavors, flavor enhancers chemicals that mimic natural flavors and those that enhance flavor.

ascorbic acid one of the active forms of vitamin C (the other is *dehydroascorbic* acid); an antioxidant nutrient.

aspartame a compound of phenylalanine and aspartic acid that tastes like the sugar sucrose but is much sweeter. It is used in both the United States and Canada.

atherosclerosis (ath-er-oh-scler-OH-sis) the most common form of cardiovascular disease; characterized by plaques along the inner walls of the arteries (*scleros* means "hard"; *osis* means "too much"). The term *arteriosclerosis* refers to all forms of hardening of the arteries and includes some rare diseases.

atrophy (AT-tro-fee) a decrease in size (for example, of a muscle) because of disuse.

B

balance study a laboratory study in which a person is fed a controlled diet and the intake and excretion of a nutrient are measured. Balance studies are valid only for nutrients like calcium (chemical elements) that do not change while they are in the body.

balance the dietary characteristic of providing foods of a number of types in proportion to each other, such that foods rich in some nutrients do not crowd out of the diet foods that are rich in other nutrients. Also called *proportionality*.

basal metabolic rate (BMR) the rate at which the body uses energy to support its basal metabolism.

basal metabolism the sum total of all the involuntary activities that are necessary to sustain life, including circulation, respiration, temperature maintenance, hormone secretion, nerve activity, and new tissue synthesis, but excluding digestion and voluntary activities. Basal metabolism is the largest component of the average person's daily energy expenditure.

bases compounds that accept hydrogens from solutions.

basic foods milk and milk products; meats and similar foods such as fish and poultry; vegetables, including dried beans and peas; fruits; and grains. These foods are generally considered to form the basis of a nutritious diet. Also called *whole foods*.

B-cells lymphocytes that produce antibodies. B stands for bursa, an organ in the chicken where B-cells were first identified.

bee pollen a product consisting of bee saliva, plant nectar, and pollen that confers no benefit on athletes and may cause an allergic reaction in individuals sensitive to it.

beer belly central-body fatness associated with alcohol consumption.

behavior modification alteration of behavior using methods based on the theory that actions can be controlled by manipulating the environmental factors that cue, or trigger, the actions.

belladonna any part of the deadly nightshade plant; a fatal poison.

beriberi the thiamin-deficiency disease; characterized by loss of sensation in the hands and feet, muscular weakness, advancing paralysis, and abnormal heart action.

beta-carotene an orange pigment with antioxidant activity; a vitamin A precursor made by plants and stored in human fat tissue.

bicarbonate a common alkaline chemical; a secretion of the pancreas; also, the active ingredient of baking soda.

bile an emulsifier made by the liver from cholesterol and stored in the gallbladder. Bile does not digest fat as enzymes do but emulsifies it so that enzymes in the watery fluids may contact it and split the fatty acids from their glycerol for absorption.

binge drinkers people who drink 4 or more drinks in a short period.

binge eating disorder an eating disorder whose criteria are similar to those of bulimia nervosa, excluding purging or other compensatory behaviors.

bioaccumulation the accumulation of a contaminant in the tissues of living things at higher and higher concentrations along the food chain.

bioavailability absorbability; the individual differences in the proportion of a nutrient that is available for absorption from various sources.

bioelectrical impedance (im-PEE-dense) a technique for measuring body fatness by measuring the body's electrical conductivity.

biosensor a genetically altered microbe that provides a rapid, low-cost, and accurate test for toxic products of microbial agents in foods.

biotechnology the science that manipulates biological systems or organisms to modify their products or components or create new products.

bioterrorism the intentional spreading of disease-causing organisms or agricultural pests by terrorists.

biotin (BY-o-tin) a B vitamin; a coenzyme necessary for fat synthesis and other metabolic reactions.

bladder the sac that holds urine until time for elimination.

bleaching agents substances used to whiten foods such as flour and cheese. Peroxides are examples.

blind experiment an experiment in which the subjects do not know whether they are members of the experimental group or the control group. In a *double-blind experiment,* neither the subjects nor the researchers know to which group the members belong until the end of the experiment.

blood clotting Proteins provide the netting on which blood clots are built.

blood the fluid of the cardiovascular system; composed of water, red and white blood cells, other formed particles, nutrients, oxygen, and other constituents.

blood-brain barrier a barrier composed of the cells lining the blood vessels in the brain. These cells are so tightly glued to each other that blood-borne substances cannot get into the brain between the cells, but only by crossing the cell bodies themselves. Thus, the cells, using all their sophisticated equipment, can screen substances for entry.

body composition the proportions of muscle, bone, fat, and other tissue that make up a person's total body weight.

body mass index (BMI) an indicator of obesity, calculated by dividing the weight of a person by the square of the person's height.

body system a group of related organs that work together to perform a function. Examples are the circulatory system, respiratory system, and nervous system.

bone density a measure of bone strength; the degree of mineralization of the bone matrix.

bone meal or **powdered bone** crushed or ground bone preparations intended to supply calcium to the diet. Calcium from bone is not well absorbed and is often contaminated with toxic materials such as arsenic, mercury, lead, and cadmium.

boron a nonessential mineral that is promoted as a "natural" steroid replacement.

bottled water drinking water sold in bottles.

botulism an often-fatal food poisoning caused by botulinum toxin, a toxin produced by the *Clostridium botulinum* bacterium that grows without oxygen in nonacidic canned foods.

bovine somatotropin (bST) growth hormone of cattle, which can be produced for agricultural use by genetic engineering. Also called *bovine growth hormone (bGH)*.

bran the protective fibrous coating around a grain; the chief fiber donator of a grain.

branched-chain amino acids (BCAA) the amino acids leucine, isoleucine, and valine, which are present in large amounts in skeletal muscle tissue; promoted often by supplement makers as fuel for exercising muscles.

brewer's yeast a preparation of yeast cells, containing a concentrated amount of B vitamins and some minerals; falsely promoted as an energy booster.

broccoli sprouts the sprouted seed of *Brassica italica,* or the common broccoli plant; believed to be a functional food by virtue of its high phytochemical content.

brown bread bread containing ingredients such as molasses that lend a brown color; may be made with any kind of flour, including white flour.

brown sugar white sugar with molasses added, 95% pure sucrose.

buffers compounds that help keep a solution's acidity or alkalinity constant.

bulimia (byoo-LEEM-ee-uh) **nervosa** recurring episodes of binge eating combined with a morbid fear of becoming fat; usually followed by self-induced vomiting or purging.

C

caffeine a stimulant that in small amounts may produce alertness and reduced reaction time in some people, but that also creates fluid losses. Overdoses cause headaches, trembling, an abnormally fast heart rate, and other undesirable effects.

calcium compounds the simplest forms of purified calcium. They include calcium carbonate, citrate, gluconate, lactate, malate, and phosphate. These supplements vary in the amount of calcium they contain, so read the labels carefully. A 500-milligram tablet of calcium gluconate may provide only 45 milligrams of calcium, for example.

caloric effect the drop in cancer incidence seen whenever intake of food energy (calories) is restricted.

calorie control control of energy intake; a feature of a sound diet plan.

calories units of energy. Strictly speaking, the unit used to measure the energy in foods is a kilocalorie (*kcalorie* or *Calorie*): it is the amount of heat energy necessary to raise the temperature of a kilogram (a liter) of water 1 degree **Celsius.** This book follows the common practice of using the lowercase term *calorie* (abbreviated *cal*) to mean the same thing.

cancer a disease in which cells multiply out of control and disrupt normal functioning of one or more organs.

canning a method of preserving food by killing all microorganisms present in the food and then sealing out air. The food, container, and lid are heated until sterile; as the food cools, the lid makes an airtight seal, preventing contamination.

capillaries minute, weblike blood vessels that connect arteries to veins and permit transfer of materials between blood and tissues.

carbohydrate loading a regimen of moderate exercise, followed by eating a high-carbohydrate diet, that enables muscles to temporarily store glycogen beyond their normal capacity; also called *glycogen loading* or *glycogen supercompensation*.

carbohydrate sweeteners carbohydrates containing sugars that are used as ingredients for sweetening food products. They include glucose, fructose, corn syrup, concentrated grape juice, and other sweet carbohydrates.

carbohydrates compounds composed of single or multiple sugars. The name means "carbon and water," and a chemical shorthand for carbohydrate is CHO, signifying carbon (C), hydrogen (H), and oxygen (O).

carbonated water water that contains carbon dioxide gas, either naturally occurring or added, that causes bubbles to form in it; also called *bubbling* or *sparkling water*. Seltzer, soda, and tonic waters are legally soft drinks and are not regulated as water.

carcinogen (car-SIN-oh-jen) a cancer-causing substance (*carcin* means "cancer"; *gen* means "gives rise to").

carcinogenesis the origination or beginning of cancer.

cardiac output the volume of blood discharged by the heart each minute.

cardiorespiratory endurance the ability to perform large-muscle dynamic exercise of moderate-to-high intensity for prolonged periods.

cardiovascular disease (CVD) disease of the heart and blood vessels; also called *coronary heart disease (CHD)*. The two most common forms of CVD are atherosclerosis and hypertension.

carnitine a nitrogen-containing compound, formed in the body from lysine and methionine, that helps transport fatty acids across the mitochondrial membrane. Carnitine is claimed to "burn" fat and spare glycogen during endurance events, but it does neither.

carotenoid (CARE-oh-ten-oyd) any of a group of pigments in foods ranging from light yellow to reddish orange. Carotenoids are chemical relatives of beta-carotene; many have a degree of vitamin A activity in the body.

carrying capacity the total number of living organisms that a given environment can support without deteriorating in quality.

case studies studies of individuals. In clinical settings, researchers can observe treatments and their apparent effects. To prove that a treatment has produced an effect requires simultaneous observation of an untreated similar subject (a *case control*).

casein or **sodium caseinate** the principal protein of cow's milk. Another milk protein found in human milk's whey is **lactalbumin.**

cat's claw an herb from the rain forests of Brazil and Peru; claimed, but not proved, to be an "all-purpose" remedy.

cataracts (CAT-uh-racts) thickening of the lens of the eye that can lead to blindness. Cataracts can be caused by injury, viral infection, toxic substances, genetic disorders, and, possibly, some nutrient deficiencies or imbalances.

catecholamines (CAT-eh-COAL-ah-meenz) neurotransmitters made from the amino acid tyrosine: dopamine, epinephrine, and norepinephrine.

cathartic a strong laxative.

cell differentiation (dih-fer-en-she-AY-shun) the process by which immature cells are stimulated to mature and gain the ability to perform functions characteristic of their cell type.

cell salts a mineral preparation supposedly prepared from living cells. No scientific evidence supports benefits from such preparations.

cells the smallest units in which independent life can exist. All living things are single cells or organisms made of cells.

cellulite a term popularly used to describe dimpled fat tissue on the thighs and buttocks; not recognized in science.

central obesity excess fat in the abdomen (visceral fat) and around the trunk.

certified lactation consultant a health-care provider, often a registered nurse, with specialized training in breast and infant anatomy and physiology who teaches the mechanics of breastfeeding to new mothers. Certification is granted after passing a standardized post-training examination.

certified organic foods foods meeting strict USDA production regulations, including prohibition of most synthetic pesticides, herbicides, fertilizers, drugs, and preservatives, as well as genetic engineering and irradiation.

cesarean (see-ZAIR-ee-un) **section** surgical childbirth, in which the infant is taken through an incision in the woman's abdomen.

chamomile flowers that may provide some limited medical value in soothing menstrual, intestinal, and stomach discomforts.

chaparral an herbal product made from ground leaves of the creosote bush and sold in tea or capsule form; supposedly, this herb has antioxidant effects, delays aging, "cleanses" the bloodstream, and treats skin conditions—all unproven claims. Chaparral has been found to cause acute toxic hepatitis, a severe liver illness.

chelating (KEE-late-ing) **agents** molecules that surround other molecules and are therefore useful in either preventing or promoting movement of substances from place to place.

chitin an indigestible polysaccharide forming the hard shells of insects, lobsters, and shrimp; used in medicine to slow the absorption of drugs and sold in pill form as an unproven weight-loss aid.

chlorophyll the green pigment of plants that captures energy from sunlight for use in photosynthesis.

cholesterol (koh-LESS-ter-all) a member of the group of lipids known as sterols; a soft, waxy substance made in the body for a variety of purposes and also found in animal-derived foods.

choline (KOH-leen) a nonessential nutrient used to make the phospholipid lecithin and other molecules.

chromium picolinate a trace element supplement; falsely promoted to increase lean body mass, enhance energy, and burn fat.

chronic diseases long-duration degenerative diseases characterized by deterioration of the body organs; examples include heart disease, cancer, and diabetes.

chylomicrons (KYE-low-MY-krons) clusters formed when lipids from a meal are combined with carrier proteins in the intestinal lining. Chylomicrons transport food fats through the watery body fluids to the liver and other tissues.

chyme (KIME) the fluid resulting from the actions of the stomach upon a meal.

cirrhosis (seer-OH-sis) advanced liver disease, often associated with alcoholism, in which liver cells have died, hardened, turned an orange color, and permanently lost their function.

clone an individual created asexually from a single ancestor, such as a plant grown from a single stem cell; a group of genetically identical individuals descended from a single common ancestor, such as a colony of bacteria arising from a single bacterial cell; in genetics, a replica of a segment of DNA, such as a gene, produced by genetic engineering.

coenzyme (co-EN-zime) a small molecule that works with an enzyme to promote the enzyme's activity. Many coenzymes have B vitamins as part of their structure (*co* means "with").

coenzyme Q10 a lipid found in cells (mitochondria) that has been shown to improve exercise performance in heart disease patients, but is not effective in improving performance of healthy athletes.

cognitive therapy psychological therapy aimed at changing undesirable behaviors by changing underlying thought processes contributing to these behaviors; in anorexia, a goal is to replace false beliefs about body weight, eating, and self-worth with health-promoting beliefs.

collagen (COLL-a-jen) the chief protein of most connective tissues, including scars, ligaments, and tendons, and the underlying matrix on which bones and teeth are built.

colon the large intestine.

colostrum (co-LAHS-trum) a milklike secretion from the breasts during the first day or so after delivery before milk appears; rich in protective factors.

comfrey leaves and roots of the comfrey plant; believed, but not proved, to have drug effects. Comfrey contains cancer-causing chemicals.

complementary proteins two or more proteins whose amino acid assortments complement each other in such a way that the essential amino acids missing from one are supplied by the other.

complex carbohydrates long chains of sugar units arranged to form starch or fiber; also called *polysaccharides*.

concentrated fruit juice sweetener a concentrated sugar syrup made from dehydrated, deflavored fruit juice, commonly grape juice; used to sweeten products that can then claim to be "all fruit."

conditionally essential amino acid an amino acid that is normally nonessential, but must be supplied by the diet in special circumstances when the need for it exceeds the body's ability to produce it.

confectioner's sugar finely powdered sucrose, 99.9% pure.

congeners (CON-jen-ers) chemical substances other than alcohol that account for some of the physiological effects of alcoholic beverages, such as appetite, taste, and aftereffects.

conjugated linoleic acid (CLA) a type of fat in butter, milk, and other dairy products believed by some to have biological activity in the body. A biologically active chemical produced by animals.

constipation hardness and dryness of bowel movements, associated with discomfort in passing them from the body.

contaminant any substance occurring in food by accident; any food constituent that is not normally present.

control group a group of individuals who are similar in all possible respects to the group being treated in an experiment but who receive a sham treatment instead of the real one. Also called *control subjects*. See also *experimental group* and *intervention studies*.

corn sweeteners corn syrup and sugar solutions derived from corn.

corn syrup a syrup, mostly glucose, partly maltose, produced by the action of enzymes on cornstarch. *High-fructose corn syrup (HFCS)* is mostly fructose; glucose (dextrose) and maltose make up the balance.

cornea (KOR-nee-uh) the hard, transparent membrane covering the outside of the eye.

correlation the simultaneous change of two factors, such as the increase of weight with increasing height (a *direct* or *positive* correlation) or the decrease of cancer incidence with increasing fiber intake (an *inverse* or *negative* correlation). A correlation between two factors suggests that one may cause the other, but does not rule out the possibility that both may be caused by chance or by a third factor.

correspondence school a school that offers courses and degrees by mail. Some correspondence schools are accredited; others are *diploma mills*.

cortex the outermost layer of something. The brain's cortex is the part of the brain where conscious thought takes place.

cortical bone the ivorylike outer bone layer that forms a shell surrounding trabecular bone and that comprises the shaft of a long bone.

creatine a nitrogen-containing compound that combines with phosphate to burn a high-energy compound stored in muscle. Claims that creatine safely enhances energy and stimulates muscle growth are unconfirmed.

cretinism (CREE-tin-ism) severe mental and physical retardation of an infant caused by the mother's iodine deficiency during pregnancy.

critical period a finite period during development in which certain events may occur that will have irreversible effects on later developmental stages. A critical period is usually a period of cell division in a body organ.

cross-contamination the contamination of a food through exposure to utensils, hands, or other surfaces that were previously in contact with a contaminated food.

cruciferous vegetables vegetables with cross-shaped blossoms—the cabbage family. Their intake is associated with low cancer rates in human populations. Examples include broccoli, brussels sprouts, cabbage, cauliflower, rutabagas, and turnips.

cuisines styles of cooking.

cyclamate a zero-calorie sweetener under consideration for use in the United States and used with restrictions in Canada.

D

degenerative diseases chronic, irreversible diseases characterized by degeneration of body organs due in part to such personal lifestyle elements as poor food choices, smoking, alcohol use, and lack of physical activity. Also called *lifestyle diseases*, *chronic diseases*, or the *diseases of old age*.

dehydration loss of water. The symptoms progress rapidly, from thirst to weakness to exhaustion and delirium, and end in death.

denaturation the change in a protein's shape brought about by heat, acids, bases, alcohol, salts of heavy metals, or other agents.

dental caries decay of the teeth (*caries* means "rottenness").

desiccated liver a powder sold in health-food stores and supposed to contain in concentrated form all the nutrients found in liver (*desiccated* means "totally dried").

dextrose an older name for glucose.

DHEA (dehydroepiandrosterone) a hormone made in the adrenal glands that serves as a precursor to the male hormone testosterone; falsely promoted as burning fat, building muscle, and slowing aging.

diabetes (dye-uh-BEET-eez) a disease (technically termed *diabetes mellitus*) characterized by elevated blood glucose and inadequate or ineffective insulin, which renders a person unable to regulate blood glucose normally.

diarrhea frequent, watery bowel movements usually caused by diet, stress, or irritation of the colon. Severe, prolonged diarrhea robs the body of fluid and certain minerals, causing dehydration and imbalances that can be dangerous if left untreated.

diastolic (dye-as-TOL-ik) **pressure** the second figure in a blood pressure reading (the "lub" of the heartbeat), which reflects the arterial pressure when the heart is between beats.

diet the foods (including beverages) a person usually eats and drinks.

dietary antioxidants (anti-OX-ih-dants) substances in food that significantly decrease the damaging effects of reactive compounds, such as reactive forms of oxygen and nitrogen on tissue functioning (*anti* means "against"; *oxy* means "oxygen").

Dietary Folate Equivalent (DFE) a unit of measure expressing the amount of folate available to the body from naturally occurring sources. The measure mathematically accounts for the greater absorption of synthetic folate added to enriched foods and supplements.

Dietary Reference Intakes (DRI) a set of four lists of values for the dietary nutrient intakes of healthy people in the United States and Canada. The values include Estimated Average Requirements (EAR), Recommended Dietary Allowances (RDA), Adequate Intakes (AI), and Tolerable Upper Intake Levels (UL).

dietary supplement a product, other than tobacco, that is added to the diet and contains one of the following ingredients: a vitamin, mineral, herb, botanical (plant extract), amino acid, metabolite, constituent, or extract, or a combination of any of these ingredients.

dietetic technician a person who has completed a two-year acadmic degree from an accredited college or university and an approved dietetic technician program. A **dietetic technician, registered** (DTR) has also passed a national examination and maintains registration through continuing professional education.

dietitian a person trained in nutrition, food science, and diet planning. See also *registered dietitian*.

digest to break molecules into smaller molecules; a main function of the digestive tract with respect to food.

digestive system the body system composed of organs that break down complex food particles into smaller, absorbable products. The *digestive tract* and *alimentary canal* are names for the tubular organs that extend from the mouth to the anus. The whole system, including the pancreas, liver, and gallbladder, is sometimes called the *gastrointestinal*, or *GI*, system.

dipeptides (dye-PEP-tides) protein fragments that are two amino acids long. A peptide is a strand of amino acids (*di* means "two").

diploma mill an organization that awards meaningless degrees without requiring its students to meet educational standards.

disaccharides pairs of single sugars linked together (*di* means "two").

distilled water water that has been vaporized and recondensed, leaving it free of dissolved minerals.

diuretics (dye-you-RET-ics) compounds, usually medications, causing increased urinary water excretion; "water pills."

diverticulosis (dye-ver-tic-you-LOH-sis) outpocketing or ballooning out of areas of the intestinal wall, caused by weakening of the muscle layers that encase the intestine.

DNA and **RNA (deoxyribonucleic acid** and **ribonucleic acid)** the genetic materials of cells necessary in protein synthesis; falsely promoted as ergogenic aids.

DNA fingerprinting a laboratory analysis of DNA that reveals the genetic pattern unique to an organism. The pattern is then compared to others (similar to grocery bar codes) to find a match, thus identifying the organism.

dolomite a compound of minerals (calcium magnesium carbonate) found in limestone and marble. Dolomite is powdered and is sold as a calcium-magnesium supplement, but may be contaminated with toxic minerals, is not well absorbed, and interacts adversely with absorption of other essential minerals.

drink a dose of any alcoholic beverage that delivers ½ ounce of pure ethanol.

drug any substance that when taken into a living organism may modify one or more of its functions.

drying a method of preserving food by removing sufficient water from the food to inhibit microbial growth.

dual energy X-ray absorptiometry (ab-sorp-tee-OM-eh-tree) **(DEXA)** a noninvasive method of determining total body fat, fat distribution, and bone density by passing two low-dose X-ray beams through the body. Also used in evaluation of osteoporosis.

dysentery (DISS-en-terry) an infection of the digestive tract that causes diarrhea.

E

eating disorder a disturbance in eating behavior that jeopardizes a person's physical or psychological health.

echinacea an herb popular before the advent of antibiotics for its "anti-infectious" properties and as an all-purpose remedy, especially for colds and allergy and for healing of wounds. A small body of research seems to lend preliminary support for some of the claims, but also points to an insecticidal property, leading to questions about safety. Also called *cone-flower*.

edema (eh-DEEM-uh) swelling of body tissue caused by leakage of fluid from the blood vessels; seen in protein deficiency (among other conditions).

eicosanoids (eye-COSS-ah-noyds) biologically active compounds that regulate body functions.

electrolytes compounds that partly dissociate in water to form ions, such as the potassium ion (K+) and the chloride ion (Cl^-).

electrons parts of an atom; negatively charged particles. Stable atoms (and molecules, which are made of atoms) have even numbers of electrons in pairs. An atom or molecule with an unpaired electron is an unstable *free radical*.

elemental diets diets composed of purified ingredients of known chemical composition; intended to supply all essential nutrients to people who cannot eat foods.

embolism an embolus that causes sudden closure of a blood vessel.

embolus (EM-boh-luss) a thrombus that breaks loose (*embol* means "to insert").

embryo (EM-bree-oh) the stage of human gestation from the third to the eighth week after conception.

emetic (em-ETT-ic) an agent that causes vomiting.

emulsification the process of mixing lipid with water by adding an emulsifier.

emulsifier (ee-MULL-sih-fire) a compound with both water-soluble and fat-soluble portions that can attract fats and oils into water to form an emulsion.

endorphins (en-DORE-fins), **endogenous opiates** compounds of the brain whose actions mimic those of opiate drugs (morphine, heroin) in reducing pain and producing pleasure. In appetite control, endorphins are released on seeing, smelling, or tasting delicious food and are believed to enhance the drive to eat or continue eating.

endosperm the bulk of the edible part of a grain, the starchy part.

energy the capacity to do work. The energy in food is chemical energy; it can be converted to mechanical, electrical, heat, or other forms of energy in the body. Food energy is measured in calories.

energy-yielding nutrients the nutrients the body can use for energy. They may also supply building blocks for body structures.

enriched, fortified refers to the addition of nutrients to a refined food product. As defined by U.S. law, these terms mean that specified levels of thiamin, riboflavin, niacin, folate, and iron have been added to refined grains and grain products. The terms *enriched* and *fortified* can refer to the addition of more nutrients than just these five; read the label.

enterotoxins poisons that act upon mucous membranes, such as those of the digestive tract.

environmental tobacco smoke (ETS) the combination of exhaled smoke (mainstream smoke) and smoke from lighted cigarettes, pipes, or cigars (sidestream smoke) that enters the air and may be inhaled by other people.

enzymes (EN-zimes) protein catalysts. A catalyst is a compound that facilitates a chemical reaction without itself being altered in the process.

EPA, DHA eicosapentaenoic (EYE-cossa-PENTA-ee-NO-ick) acid, docosahexaenoic (DOE-cossa-HEXA-ee-NO-ick) acid; omega-3 fatty acids made from linolenic acid in the tissues of fish.

ephedrine One of a group of compounds with dangerous amphetamine-like stimulant effects; commonly added to herbal preparations such as Ma huang, to weight-loss products, and to products claiming to imitate the effects of illegal drugs of abuse. The most severe reported side effects of ephedrine include sudden death, heart attack, and stroke; other reported effects include chest pain, dizziness, fatigue, headache, insomnia, nausea, psychosis, seizure, tremor, and vomiting. The World Health Organization has called for worldwide controls on ephedrine-containing products.

epidemiological studies studies of populations; often used in nutrition to search for correlations between dietary habits and disease incidence; a first step in seeking nutrition-related causes of diseases.

epinephrine (EP-ih-NEFF-rin) a hormone of the adrenal gland administered by injection to counteract anaphylactic shock by opening the airways and maintaining heartbeat and blood pressure.

epiphyseal (eh-PIFF-ih-seal) **plate** a thick, cartilage-like layer at the ends of bones that forms new cells that are eventually calcified, lengthening the bone (*epiphysis* means "growing" in Greek).

epithelial (ep-ith-THEE-lee-ull) **tissue** the layers of the body that serve as selective barriers to environmental factors. Examples are the cornea, the skin, the respiratory tract lining, and the lining of the digestive tract.

epoetin a drug derived from the human hormone erythropoietin and marketed under the trade name Epogen; illegally used to increase oxygen capacity.

ergogenic (ER-go-JEN-ic) **aids** products that supposedly enhance performance, although none actually do so; the term *ergogenic* implies "energy giving" (*ergo* means "work"; *genic* means "give rise to").

erythrocyte (eh-REETH-ro-sight) **hemolysis** (HE-moh-LIE-sis, he-MOLL-ih-sis) rupture of the red blood cells, caused by vitamin E deficiency (*erythro* means "red"; *cyte* means "cell"; *hemo* means "blood"; *lysis* means "breaking").

essential amino acids amino acids that either cannot be synthesized at all by the body or cannot be synthesized in amounts sufficient to meet physiological needs. Also called *indispensable amino acids*.

essential fatty acids fatty acids that the body needs but cannot make in amounts sufficient to meet physiological needs.

essential nutrients the nutrients the body cannot make for itself (or cannot make fast enough) from other raw materials; nutrients that must be obtained from food to prevent deficiencies.

ethanol the alcohol of alcoholic beverages, produced by the action of microorganisms on the carbohydrates of grape juice or other carbohydrate-containing fluids.

ethnic foods foods associated with particular cultural subgroups within a population.

euphoria (you-FOR-ee-uh) an inflated sense of well-being and pleasure brought on by normal body functions such as the consumption of a meal or sexual activity or by a moderate dose of alcohol and some other drugs.

evaporated milk milk concentrated to half volume by evaporation. Adding water reconstitutes the milk; the taste is altered by the processing, however.

exchange system a diet-planning tool that organizes foods with respect to their nutrient contents and calorie amounts. Foods on any single exchange list can be used interchangeably.

experimental group the people or animals participating in an experiment who receive the treatment under investigation. Also called *experimental subjects*.

extracellular fluid fluid residing outside the cells that transport materials to and from the cells.

extreme obesity clinically severe overweight, presenting very high risks to health; the condition of having a BMI of 40 or above; also called *morbid obesity*.

extrusion a process by which the form of a food is changed, such as changing corn to corn chips; not a preservation measure.

F

famine widespread scarcity of food in an area that causes starvation and death in a large portion of the population.

fast foods restaurant foods that are available within minutes after customers order them—traditionally, hamburgers, french fries, and milkshakes; more recently, salads and other vegetable dishes as well. These foods may or may not meet people's nutrient needs well, depending on the selections made and on the energy allowances and nutrient needs of the eaters.

fasting hypoglycemia hypoglycemia that occurs after 8 to 14 hours of fasting.

fat cells cells that specialize in the storage of fat and form the fat tissue.

fat replacers ingredients that replace some or all of the functions of fat and may or may not provide energy. Often used interchangeably with *fat substitutes*, but the latter technically applies only to ingredients that replace all of the functions of fat and provide no energy.

fatfold test measurement of the thickness of a fold of skin on the back of the arm (over the triceps muscle), below the shoulder blade (subscapular), or in other places, using a caliper; also called *skinfold test*.

fats lipids that are solid at room temperature (70°F or 25°C).

fatty acids organic acids composed of carbon chains of various lengths. Each fatty acid has an acid end and hydrogens attached to all of the carbon atoms of the chain.

fatty liver an early stage of liver deterioration seen in several diseases, including kwashiorkor and alcoholic liver disease in which fat accumulates in the liver cells.

feces waste material remaining after digestion and absorption are complete; eventually discharged from the body.

female athlete triad a potentially fatal triad of medical problems seen in female athletes: disordered eating, amenorrhea, and osteoporosis.

fetal alcohol syndrome (FAS) the cluster of symptoms seen in an infant or child whose mother consumed excessive alcohol during her pregnancy. FAS includes, but is not limited to, brain damage, growth retardation, mental retardation, and facial abnormalities.

fetus (FEET-us) the stage of human gestation from eight weeks after conception until the birth of an infant.

feverfew an herb sold as a migraine headache preventive. Some evidence exists to support this claim.

fibers the indigestible polysaccharides in food, consisting mostly of cellulose, hemicellulose, and pectin; also called *nonstarch polysaccharides*.

fibrosis (fye-BROH-sis) an intermediate stage of alcoholic liver deterioration. Liver cells lose their function and assume the characteristics of connective tissue cells (fibers).

fight-or-flight reaction the body's instinctive hormone- and nerve-mediated reaction to danger. Also known as the *stress response*.

filtered water water treated by filtration, usually through *activated carbon filters* that reduce the lead in tap water, or by *reverse osmosis* units that force pressurized water across a membrane removing lead, arsenic, and some microorganisms from tap water.

flavonoids (FLAY-von-oyds) yellow pigments in foods; phytochemicals that may exert physiological effects on the body. *Flavus* means "yellow."

flaxseed the small brown seed of the flax plant; used in baking, cereals, or other foods and valued by industry as a source of linseed oil and fiber.

flexibility the capacity of the joints to move through a full range of motion; the ability to bend and recover without injury.

fluid and electrolyte balance maintenance of the proper amounts and kinds of fluids and minerals in each compartment of the body.

fluid and electrolyte imbalance failure to maintain the proper amount and kind of fluid in every body compartment; a medical emergency.

fluorapatite (floor-APP-uh-tight) a crystal of bones and teeth, formed when fluoride displaces the hydroxy portion of hydroxyapatite. Fluorapatite resists being dissolved back into body fluid.

fluorosis (floor-OH-sis) discoloration of the teeth due to ingestion of too much fluoride during tooth development.

folate (FOH-late) a B vitamin that acts as part of a coenzyme important in the manufacture of new cells. The form added to foods and supplements is *folic acid*.

food medically, any substance that the body can take in and assimilate that will enable it to stay alive and to grow; the carrier of nourishment; socially, a more limited number of such substances defined as acceptable by each culture.

food aversion an intense dislike of a food, possibly biological in nature, resulting from an illness or other negative experience associated with that food.

food group plans diet-planning tools that sort foods into groups based on origin and nutrient content and then specify that people should eat certain minimum numbers of servings of foods from each group.

food insecurity the condition of uncertain access to food of sufficient quality or quantity.

food intolerance an adverse effect of a food or food additive not involving an immune response.

food poverty hunger occurring when enough food exists in an area but some of the people cannot obtain it because they lack money, are being deprived for political reasons, live in a country at war, or suffer from other problems such as lack of transportation.

food shortage hunger occurring when an area of the world lacks enough total food to feed its people.

food-borne illness illness transmitted to human beings through food and water; caused by a poisonous substance (*food intoxication*) or an infectious agent (*food-borne infection*). Also called *food poisoning*.

foodways the sum of a culture's habits, customs, beliefs, and preferences concerning food.

foot-and-mouth disease a severely disabling contagious viral disease that threatens domestic and wild animals. Foot-and-mouth disease rarely affects human beings and produces only mild symptoms in people.

fork thermometer a utensil combining a meat fork and an instant-read food thermometer.

formaldehyde a substance to which methanol is metabolized on the way to being converted to harmless waste products that can be excreted.

foxglove a plant that contains a substance used in the heart medicine digoxin.

fraud or **quackery** the promotion, for financial gain, of devices, treatments, services, plans, or products (including diets and supplements) that alter or claim to alter a human condition without proof of safety or effectiveness. (The word *quackery* comes from the term *quacksalver,* meaning a person who quacks loudly about a miracle product—a lotion or a salve.)

free radicals atoms or molecules with one or more unpaired electrons that make the atom or molecule unstable and highly reactive.

freezing a method of preserving food by lowering the food's temperature to a point that halts life processes. Microorganisms do not die but remain dormant until the food is thawed.

fructose (FROOK-tose) a monosaccharide; sometimes known as fruit sugar (*fruct* means "fruit"; *ose* means "sugar").

fructose, galactose, glucose the monosaccharides.

fruitarian includes only raw or dried fruits, seeds, and nuts in the diet.

functional foods a general term for foods with beneficial physiological or psychological effects beyond providing essential nutrients. May also be referred to as *medical foods, foods for medical purposes,* or other terms with no legal or scientific definitions.

G

galactose (ga-LACK-tose) a monosaccharide; part of the disaccharide lactose (milk sugar).

garlic oil an extract of garlic; may or may not contain the chemicals associated with garlic; claims for health benefits unproved.

gastric banding a surgical means of producing weight loss by restricting stomach size with a contricting band or pouch; used in people whose severe obesity brings extreme health risks.

gastric bypass surgery that reroutes food from the stomach to the lower part of the small intestine; creates a chronic, lifelong state of malabsorption by preventing normal digestion and absorption of nutrients.

gastric juice the digestive secretion of the stomach.

gastro-esophageal reflux disease (GERD) a severe and chronic splashing of stomach acid and enzymes into the esophagus, throat, mouth, or airway that causes inflammation and injury to those organs. Untreated GERD may increase the risk of esophageal cancer; treatment may require surgery or management with medication.

gastroplasty surgery that partitions the stomach by stapling off a "pouch" or otherwise modifying the stomach, thereby reducing total food intake.

gatekeeper with respect to nutrition, a key person who controls other people's access to foods and thereby affects their nutrition profoundly. Examples are the spouse who buys and cooks the food, the parent who feeds the children, and the caretaker in a day-care center.

gelatin a soluble form of the protein collagen, used to thicken foods; sometimes falsely promoted as a strength enhancer.

generally recognized as safe (GRAS) list a list, established by the FDA, of food additives long in use and believed safe.

genes units of a cell's inheritance, made of the chemical DNA (deoxyribonucleic acid). Each gene directs the making of a protein to do the body's work.

genetic engineering a field within biotechnology that involves the direct, intentional manipulation of the genetic material of living things in order to obtain some desirable trait not present in the original organism. Also called *recombinant DNA technology.*

genetically modified foods (GM foods) transgenic foods produced through rDNA technology.

genistein (GEN-ih-steen) a phytosterol found primarily in soybeans that both mimics and blocks the action of estrogen in the body.

genome the complete set of chromosomes that comprises the entirety of an organism's genetic information. The *human genome* is the genetic map of a human being, revealing the location of hereditary information carried on the chromosomes. The map may be used to identify disease-causing genetic variations, allowing advances in disease prevention and treatment.

germ the nutrient-rich inner part of a grain.

gestation the period of about 40 weeks (three trimesters) from conception to birth; the term of a pregnancy.

gestational diabetes abnormal glucose tolerance appearing during pregnancy, with subsequent return to normal after the end of pregnancy.

ginkgo biloba an extract of a tree of the same name, claimed to enhance mental alertness, but not proved to be effective or safe.

ginseng (JIN-seng) a plant root containing chemicals that have stimulant drug effects. *Ginseng abuse syndrome* is a group of symptoms associated with the overuse of ginseng, including high blood pressure, insomnia, nervousness, confusion, and depression.

glandular products extracts or preparations of raw animal glands and organs; sold with the false claim of boosting athletic performance, but may present disease hazards if collected from infected animals.

gleaning traditionally, the practice of gathering crops left in the field after a harvest; today, refers to the recovery of excess food from various sources including restaurants, hotels, corporations, farms, wholesalers, farmers' markets, and supermarkets.

glucagon a hormone secreted by the pancreas that stimulates the liver to release glucose into the blood when blood glucose concentration dips.

glucose (GLOO-cose) a single sugar used in both plant and animal tissues for energy; sometimes known as blood sugar or *dextrose*.

glucose polymers compounds that supply glucose, not as single molecules, but linked in chains somewhat like starch. The objective is to attract less water from the body into the digestive tract.

glycemic (gligh-SEEM-ic) **effect** the extent to which a food raises the blood glucose concentration and elicits an insulin response as compared with pure glucose.

glycemic index (GI) a ranked measure of the glycemic effect of foods.

glycerol (GLISS-er-all) an organic compound, three carbons long, of interest here because it serves as the backbone for triglycerides.

glycine a nonessential amino acid, promoted as an ergogenic aid because it is a precursor of creatine.

glycogen (GLY-co-gen) a polysaccharide composed of glucose that is made and stored by liver and muscle tissues of human beings and animals as a storage form of glucose. Glycogen is not a significant food source of carbohydrate and is not counted as one of the complex carbohydrates in foods.

goiter (GOY-ter) enlargement of the thyroid gland due to iodine deficiency is *simple goiter;* enlargement due to an iodine excess is *toxic goiter.*

gout (GOWT) a painful form of arthritis caused by the abnormal buildup of the waste product uric acid in the blood, with uric acid salt deposited as crystals in the joints.

grams units of weight. A gram (g) is the weight of a cubic centimeter (cc) or milliliter (ml) of water under defined conditions of temperature and pressure. About 28 grams equal an ounce.

granulated sugar common table sugar, crystalline sucrose, 99.9% pure.

granules small grains. Starch granules are packages of starch molecules. Various plant species make starch granules of varying shapes.

green pills, fruit pills pills containing dehydrated, crushed vegetable or fruit matter. An advertisement may claim that each pill equals a *pound* of fresh produce, but in reality a pill may equal one small forkful—minus nutrient losses incurred in processing.

ground water water that comes from underground aquifers.

growth and maintenance Proteins serve as building materials for growth and repair of body tissues.

growth hormone a hormone (somatotropin) that promotes growth and is produced naturally in the pituitary gland of the brain.

growth hormone releasers herbs or pills that supposedly regulate hormones; falsely promoted as enhancing athletic performance.

guarana a reddish berry found in Brazil's Amazon basin that contains seven times as much caffeine as its relative the coffee bean. It is used as an ingredient in carbonated sodas and, taken in powder or tablet form, it supposedly enhances speed and endurance and serves as an aphrodisiac, a "cardiac tonic," an "intestinal disinfectant," and a "smart drug" touted to improve mental functions. High doses may stress the heart and can cause panic attacks.

H

hard water water with high calcium and magnesium concentrations.

hazard a state of danger; used to refer to any circumstance in which harm is possible under normal conditions of use.

Hazard Analysis Critical Control Point (HACCP) a systematic plan to identify and correct potential microbial hazards in the manufacturing, distribution, and commercial use of food products.

heart attack the event in which the vessels that feed the heart muscle become closed off by an embolism, thrombus, or other cause with resulting sudden tissue death. A heart attack is also called a *myocardial infarction* (*myo* means "muscle"; *cardial* means "of the heart"; *infarct* means "tissue death").

heartburn a burning sensation in the chest (in the area of the heart) area caused by backflow of stomach acid into the esophagus.

heat stroke an acute and life-threatening reaction to heat buildup in the body.

heavy metal any of a number of mineral ions such as mercury and lead; so called because they are of relatively high atomic weight. Many heavy metals are poisonous.

heme (HEEM) the iron-containing portion of the hemoglobin and myoglobin molecules.

hemlock any part of the hemlock plant, which causes severe pain, convulsions, and death within 15 minutes.

hemoglobin (HEEM-oh-globe-in) the oxygen-carrying protein of the blood; found in the red blood cells (*hemo* means "blood"; *globin* means "spherical protein").

hemorrhoids (HEM-or-oids) swollen, hardened (varicose) veins in the rectum, usually caused by the pressure resulting from constipation.

herbal steroids or **plant sterols** mixtures of compounds from herbs that supposedly enhance human hormone activity. Products marketed as herbal steroids include astragalus, damiana, dong quai, fo ti teng, ginseng root, licorice root, palmetto berries, sarsaparilla, schizardra, unicorn root, yohimbe bark, and yucca.

hernia a protrusion of an organ or part of an organ through the wall of the body chamber that normally contains the organ. An example is a *hiatal* (high-AY-tal) *hernia,* in which part of the stomach protrudes up through the diaphragm into the chest cavity, which contains the esophagus, heart, and lungs.

hiccups spasms of both the vocal cords and the diaphragm, causing periodic, audible, short, inhaled coughs. Can be caused by irritation of the diaphragm, indigestion, or other causes. Hiccups usually resolve in a few minutes, but can have serious effects if prolonged. Breathing into a paper bag (inhaling carbon dioxide) or dissolving a teaspoon of sugar in the mouth may stop them.

high-density lipoproteins (HDL) lipoproteins that are critical in the process of carrying cholesterol from body cells to the liver for dismantling and disposal; contain a large proportion of protein.

high-temperature–short-time (HTST) principle the rule that every 18°F (10°C) rise in processing temperature brings about an approximately tenfold increase in microbial destruction, while only doubling nutrient losses.

histamine a substance that participates in causing inflammation; produced by cells of the immune system as part of a local immune reaction to an antigen.

HMB (beta-hydroxy-beta-methylbutyrate) a metabolite of the branched-chain amino acid leucine. Claims that HMB increases muscle mass and strength stem from "evidence" from the company that developed HMB as a supplement.

homocysteine (home-oh-SIS-teen) an amino acid produced as an intermediate compound during amino acid metabolism. A buildup of homocysteine in the blood is associated with deficiencies of folate and other B vitamins and may increase the risk of diseases.

homogenized milk milk treated to mix the fat evenly with the watery part (fat ordinarily floats to the top as cream). Heated milk is forced under high pressure through small openings to emulsify the fat.

honey a concentrated solution primarily composed of glucose and fructose produced by enzymatic digestion of the sucrose in nectar by bees.

hormones chemicals that are secreted by glands into the blood in response to conditions in the body that require regulation. These chemicals serve as messengers, acting on other organs to maintain constant conditions.

human growth hormone (HGH) a hormone produced by the brain's pituitary gland that regulates normal growth and development (see text discussion); also called *somatotropin*.

hunger the physiological need to eat, experienced as a drive for obtaining food; an unpleasant sensation that demands relief. Also, a lack or shortage of basic foods needed to provide the energy and nutrients that support health.

husk the outer, inedible part of a grain.

hydrogenation (high-dro-gen-AY-shun) the process of adding hydrogen to unsaturated fatty acids to make fat more solid and resistant to the chemical change of oxidation.

hydroxyapatite (hi-DROX-ee-APP-uh-tight) the chief crystal of bone, formed from calcium and phosphorus.

hyperactivity (in children) a syndrome characterized by inattention, impulsiveness, and excess motor activity; usually diagnosed before age seven, lasts six months or more, and usually does not entail mental illness or mental retardation. Properly called *attention-deficit/hyperactivity disorder (ADHD)* and may be associated with minimal brain damage.

hypertension high blood pressure.

hypertrophy (high-PURR-tro-fee) an increase in size (for example, of a muscle) in response to use.

hypoallergenic formulas clinically tested infant formulas that do not provoke reactions in 90% of infants or children with confirmed cow's milk allergy. Like all infant formulas, hypoallergenic formulas must demonstrate nutritional suitability to support infant growth and development. Extensively hydrolyzed and free amino acid–based formulas are examples.

hypoglycemia (HIGH-poh-gly-see-mee-ah) a blood glucose concentration below normal, a symptom that may indicate any of several diseases, including impending diabetes.

hypothalamus (high-poh-THAL-uh-mus) a part of the brain that senses a variety of conditions in the blood, such as temperature, glucose content, salt content, and others. It signals other parts of the brain or body to adjust those conditions when necessary.

hypothermia a below-normal body temperature.

I

immune system a system of tissues and organs that defend the body against antigens, foreign materials that have penetrated the skin or body linings.

immunity specific disease resistance, derived from the immune system's memory of prior exposure to specific disease agents and its ability to mount a swift defense against them.

implantation the stage of development, during the first two weeks after conception, in which the fertilized egg (fertilized ovum or zygote) embeds itself in the wall of the uterus and begins to develop.

incidental additives substances that can get into food not through intentional introduction but as a result of contact with the food during growing, processing, packaging, storing, or some other stage before the food is consumed. Also called *accidental* or *indirect additives*.

infectious diseases diseases that are caused by bacteria, viruses, parasites, and other microbes and can be transmitted from one person to another through air, water, or food; by contact; or through vector organisms such as mosquitoes or fleas.

infomercials feature-length television commercials that follow the format of regular programs but are intended to convince viewers to buy products, and not to educate or entertain them.

initiation an event, probably occurring in a cell's genetic material, caused by radiation or by a chemical carcinogen that can give rise to cancer.

inosine an organic chemical that is falsely said to "activate cells, produce energy, and facilitate exercise." Studies have shown that it actually reduces the endurance of runners.

inositol (in-OSS-ih-tall) a nonessential nutrient found in cell membranes.

insoluble fibers the tough, fibrous structures of fruits, vegetables, and grains; indigestible food components that do not dissolve in water.

instant-read thermometer a thermometer that, when inserted into food, measures its temperature within seconds; designed to test temperature of food at intervals, and not to be left in food during cooking.

insulin a hormone secreted by the pancreas in response to a high blood glucose concentration. It assists cells in drawing glucose from the blood.

integrated pest management (IPM) management of pests using a combination of natural and biological controls and minimal or no application of pesticides.

intervention studies studies of populations in which observation is accompanied by experimental manipulation of some population members—for example, a study in which half of the subjects (the *experimental subjects*) follow diet advice to reduce fat intakes while the other half (the *control subjects*) do not, and both groups' heart health is monitored.

intestine the body's long, tubular organ of digestion and the site of nutrient absorption.

intracellular fluid fluid residing inside the cells that provides the medium for cellular reactions.

intrinsic factor a factor found inside a system. The intrinsic factor necessary to prevent pernicious anemia is now known to be a compound that helps in the absorption of vitamin B_{12}.

invert sugar a mixture of glucose and fructose formed by the splitting of sucrose in an industrial process. Sold only in liquid form and sweeter than sucrose, invert sugar forms during certain cooking procedures and works to prevent crystallization of sucrose in soft candies and sweets.

ions (EYE-ons) electrically charged particles, such as sodium (positively charged) or chloride (negatively charged).

iron deficiency the condition of having depleted iron stores, which, at the extreme, causes iron-deficiency anemia.

iron overload the state of having more iron in the body than it needs or can handle, usually arising from a hereditary defect. Also called *hemochromatosis*.

iron-deficiency anemia a form of anemia caused by a lack of iron and characterized by red blood cell shrinkage and color loss. Accompanying symptoms are weakness, apathy, headaches, pallor, intolerance to cold, and inability to pay attention. (For other anemias, see the index.)

irradiation the application of ionizing radiation to foods to reduce insect infestation or microbial contamination or to slow the ripening or sprouting process. Also called *cold pasteurization*.

irritable bowel syndrome intermittent disturbance of bowel function, especially diarrhea or alternating diarrhea and constipation; associated with diet, lack of physical activity, or psychological stress.

isomalt, mannitol, sorbitol, xylitol sugar alcohols that can be derived from fruits or commercially produced from dextrose; absorbed more slowly and metabolized differently than other sugars in the human body and not readily used by ordinary mouth bacteria.

IU (international unit) a measure of fat-soluble vitamin activity sometimes used on supplement labels.

K

kava-kava the root of a tropical pepper plant, often brewed as a tea consumed for its calming effects. Limited scientific research supports the effectiveness of kava-kava for treating anxiety. Adverse effects include skin rash, metabolic abnormalities, elevated blood cholesterol, lethargy, mental disorientation, and possible life-threatening Parkinson's disease.

kefir a yogurt-based beverage.

kelp tablets tablets made from dehydrated kelp, a kind of seaweed used by the Japanese as a foodstuff.

keratin (KERR-uh-tin) the normal protein of hair and nails.

keratinization accumulation of keratin in a tissue; a sign of vitamin A deficiency.

ketone (kee-tone) **bodies** acidic compounds derived from fat and certain amino acids. Normally rare in the blood, they help to feed the brain during times when too little carbohydrate is available.

ketosis (kee-TOE-sis) an undesirable high concentration of ketone bodies, such as acetone, in the blood or urine.

kidneys a pair of organs that filter wastes from the blood, make urine, and release it to the bladder for excretion from the body.

kombucha a product of fermentation of sugar-sweetened tea by various yeasts and bacteria. Proclaimed as a treatment for everything from AIDS to cancer but lacking scientific evidence. Microorganisms in home-brewed teas have caused serious illnesses in people with weakened immunity. Also known as *Manchurian tea*, *mushroom tea*, or *Kargasok tea*.

kudzu a weedy vine, whose roots are harvested and used by Chinese herbalists as a treatment for alcoholism. Kudzu reportedly reduces alcohol absorption by up to 50 percent in rats.

kwashiorkor (kwash-ee-OR-core, kwashee-or-CORE) a disease related to protein malnutrition, with a set of recognizable symptoms, such as edema.

L

laboratory studies studies that are performed under tightly controlled conditions and are designed to pinpoint causes and effects. Such studies often use animals as subjects.

lactase the intestinal enzyme that splits the disaccharide lactose to monosaccharides during digestion.

lactation production and secretion of breast milk for the purpose of nourishing an infant.

lactic acid a product of the incomplete breakdown of glucose during anaerobic metabolism. When oxygen becomes available, lactic acid can be completely broken down for energy or converted back to glucose.

lactoferrin (lack-toe-FERR-in) a factor in breast milk that binds iron and keeps it from supporting the growth of the infant's intestinal bacteria.

lacto-ovo vegetarian includes dairy products, eggs, vegetables, grains, legumes, fruits, and nuts; excludes flesh and seafood.

lactose intolerance inability to digest lactose due to a lack of the enzyme lactase.

lactose a disaccharide composed of glucose and galactose; sometimes known as milk sugar (*lact* means "milk"; *ose* means "sugar").

lactose, maltose, sucrose the disaccharides.

lacto-vegetarian includes dairy products, vegetables, grains, legumes, fruits, and nuts; excludes flesh, seafood, and eggs.

large intestine the portion of the intestine that completes the absorption process.

learning disability any of a group of conditions resulting in an altered ability to learn basic cognitive skills such as reading, writing, and mathematics.

leavened (LEV-end) literally, "lightened" by yeast cells, which digest some carbohydrate components of the dough and leave behind bubbles of gas that make the bread rise.

lecithin (LESS-ih-thin) a phospholipid manufactured by the liver and also found in many foods; a major constituent of cell membranes.

legumes (leg-GOOMS, LEG-yooms) plants of the bean, pea, and lentil family that have roots with nodules containing special bacteria. These bacteria can trap nitrogen from the air in the soil and make it into compounds that become part of the plant's seeds. The seeds are rich in protein compared with those of most other plant foods.

leptin an appetite-suppressing hormone produced in the fat cells that conveys information about body fatness to the brain; believed to be involved in the maintenance of body composition (*leptos* means "slender").

levulose an older name for fructose.

license to practice permission under state or federal law, granted on meeting specified criteria, to use a certain title (such as *dietitian*) and to offer certain services. Licensed dietitians may use the initials LD after their names.

life expectancy the average number of years lived by people in a given society.

life span the maximum number of years of life attainable by a member of a species.

lignans phytochemicals present in flaxseed, but not in flax oil, that are converted to phytosterols by intestinal bacteria and are under study as possible anticancer agents.

limiting amino acid an essential amino acid that is present in dietary protein in an insufficient amount, thereby limiting the body's ability to build protein.

linoleic (lin-oh-LAY-ic) **acid** and **linolenic** (lin-oh-LEN-ic) **acid** polyunsaturated fatty acids that are essential nutrients for human beings.

lipid (LIP-id) a family of organic (carbon-containing) compounds soluble in organic solvents but not in water. Lipids include triglycerides (fats and oils), phospholipids, and sterols.

lipoic (lip-OH-ic) **acid** a nonessential nutrient.

lipoprotein lipase (LPL) an enzyme mounted on the surfaces of fat cells that splits triglycerides in the blood into fatty acids and glycerol to be absorbed into the cells for reassembly and storage.

lipoproteins (LYE-poh-PRO-teens, LIH-poh-PRO-teens) clusters of lipids associated with protein, which serve as transport vehicles for lipids in blood and lymph. Major lipoprotein classes are the chylomicrons, the VLDL, the LDL, and the HDL.

liver a large, lobed organ that lies just under the ribs. It filters the blood, removes and processes nutrients, manufactures materials for export to other parts of the body, and destroys toxins or stores them to keep them out of the circulation.

longevity long duration of life.

low birthweight a birthweight of less than 5½ pounds (2,500 grams); used as a predictor of probable health problems in the newborn and as a probable indicator of poor nutrition status of the mother before and/or during pregnancy. Low-birthweight infants are of two different types. Some are premature; they are born early and are the right size for their gestational age. Others have suffered growth failure in the uterus; they may or may not be born early, but they are small for gestational age (small for date).

low-density lipoproteins (LDL) lipoproteins that transport lipids from the liver to other tissues such as muscle and fat; contain a large proportion of cholesterol.

lungs the body's organs of gas exchange. Blood circulating through the lungs releases its carbon dioxide and picks up fresh oxygen to carry to the tissues.

lutein (LOO-teen) a plant pigment of yellow hue; a phytochemical believed to play roles in eye functioning and health.

lycopene (LYE-koh-peen) a pigment responsible for the red color of tomatoes and other red-hued vegetables; a phytochemical that may act as an antioxidant in the body.

lymph (LIMF) the fluid that moves from the bloodstream into tissue spaces and then travels in its own vessels, which eventually drain back into the bloodstream.

lymphocytes (LIM-foe-sites) white blood cells that participate in the immune response; B-cells and T-cells.

M

ma huang an evergreen plant derivative that supposedly boosts energy and helps with weight control; Ma huang contains ephedrine, especially dangerous in combination with kola nut or other caffeine-containing substances.

macrobiotic diet a vegan diet that progressively eliminates more and more foods. Ultimately, only brown rice and small amounts of water or herbal tea are consumed; taken to extremes, macrobiotic diets have resulted in malnutrition and even death.

macular degeneration a common, progressive loss of function of the part of the retina that is most crucial to focused vision. This degeneration often leads to blindness.

mad cow disease an often-fatal illness of cattle affecting the nerves and brain. Also called *bovine spongiform encephalopathy (BSE)*.

major minerals essential mineral nutrients found in the human body in amounts larger than 5 grams.

malnutrition any condition caused by excess or deficient food energy or nutrient intake or by an imbalance of nutrients. Nutrient or energy deficiencies are classed as forms of undernutrition; nutrient or energy excesses are classed as forms of overnutrition.

maltose a disaccharide composed of two glucose units; sometimes known as malt sugar.

maple sugar a concentrated solution of sucrose derived from the sap of the sugar maple tree, mostly sucrose. This sugar was once common but is now usually replaced by sucrose and artificial maple flavoring.

marasmus (ma-RAZ-mus) the calorie-deficiency disease; starvation.

margin of safety in reference to food additives, a zone between the concentration normally used and that at which a hazard exists. For common table salt, for example, the margin of safety is ⅕ (five times the concentration normally used would be hazardous).

medical nutrition therapy nutrition services used in the treatment of injury, illness, or other conditions; includes assessment of nutrition status and dietary intake, and corrective applications of diet, counseling, and other nutrition services.

medicinal herbs nonwoody plants, plant parts, or extracts valued by some people for their medicinal qualities, both proved and unproved.

melatonin a hormone of the pineal gland believed to help regulate the body's daily rhythms, to reverse the effects of jet lag, and to promote sleep. Claims for life extension or enhancement of sexual prowess are without merit.

MEOS (microsomal ethanol oxidizing system) a system of enzymes in the liver that oxidize not only alcohol but also several classes of drugs.

metabolic syndrome a combination of four risk factors—diabetes, obesity (especially central obesity), hypertension, and high blood cholesterol—that greatly increase a person's risk of developing CVD. Also called *insulin resistance syndrome* or *syndrome X*.

metabolism the sum of all physical and chemical changes taking place in living cells; including all reactions by which the body obtains and spends the energy from food.

metastasis (meh-TASS-ta-sis) movement of cancer cells from one body part to another, usually by way of the body fluids.

methanol an alcohol produced in the body continually by all cells.

MFP factor a factor (identity unknown) present in meat, fish, and poultry that enhances the absorption of nonheme iron present in the same foods or in other foods eaten at the same time.

microbes bacteria, viruses, or other organisms invisible to the naked eye, some of which cause diseases. Also called *microorganisms*.

microvilli (MY-croh-VILL-ee, MY-croh-VILL-eye) tiny, hairlike projections on each cell of every villus that can trap nutrient particles and transport them into the cells (singular: *microvillus*).

milk anemia iron-deficiency anemia caused by drinking so much milk that iron-rich foods are displaced from the diet.

mineral water water from a spring or well that typically contains 250 to 500 parts per million (ppm) of minerals. Minerals give water a distinctive flavor. Many mineral waters are high in sodium.

minerals naturally occurring, inorganic, homogeneous substances; chemical elements.

miso fermented soybean paste used in Japanese cooking. Soy products are considered to be functional foods.

moderate drinkers people who do not drink excessively and do not behave inappropriately because of alcohol. A moderate drinker's health is not harmed by alcohol over the long term.

moderation the dietary characteristic of providing constituents within set limits, not to excess.

modified atmosphere packaging (MAP) a preservation technique in which a perishable food is packaged in a gas-impermeable container from which air has been removed, or to which another gas mixture has been added.

molasses a syrup left over from the refining of sucrose from sugarcane; a thick, brown syrup. The major nutrient in molasses is iron, a contaminant from the machinery used in processing it.

monoglycerides (mon-oh-GLISS-er-ides) products of the digestion of lipids; consist of glycerol molecules with one fatty acid attached (*mono* means "one"; *glyceride* means "a compound of glycerol").

monosaccharides (MON-oh-SACK-ah-rides) single sugar units (*mono* means "one"; *saccharide* means "sugar unit").

monounsaturated fats triglycerides in which most of the fatty acids have one point of unsaturation (are monounsaturated).

monounsaturated fatty acid a fatty acid containing one point of unsaturation.

MSG symptom complex the acute, temporary, and self-limiting reactions experienced by sensitive people upon ingesting a large dose of MSG. The name *MSG symptom complex*, given by the FDA, replaces the former *Chinese restaurant syndrome*.

mucus (MYOO-cus) a slippery coating of the digestive tract lining (and other body linings) that protects the cells from exposure to digestive juices (and other destructive agents). The adjective form is mucous (same pronunciation). The digestive tract lining is a *mucous membrane*.

muscle endurance the ability of a muscle to contract repeatedly within a given time without becoming exhausted.

muscle strength the ability of muscles to work against resistance.

mutual supplementation the strategy of combining two incomplete protein sources so that the amino acids in one food make up for those lacking in the other food. Such protein combinations are sometimes called *complementary proteins*.

myoglobin (MYE-oh-globe-in) the oxygen-holding protein of the muscles (*myo* means "muscle").

N

natural foods a term that has no legal definition.

natural water water obtained from a spring or well that is certified to be safe and sanitary. The mineral content may not be changed, but the water may be treated in other ways such as with ozone or by filtration.

naturally occurring sugars sugars that are not added to a food but are present as its original constituents, such as the sugars of fruit or milk.

nephrons the working units in the kidneys, consisting of intermeshed blood vessels and tubules.

neural tube defects (NTD) any of a group of nervous system abnormalities caused by interruption of the normal early development of the neural tube.

neural tube the embryonic tissue that later forms the brain and spinal cord.

neurotoxins poisons that act upon the cells of the nervous system.

neurotransmitters chemicals that are released at the end of a nerve cell when a nerve impulse arrives there. They diffuse across the gap to the next cell and alter the membrane of that second cell to either inhibit or excite it.

niacin (NYE-ah-sin) a B vitamin needed in energy metabolism. Niacin can be eaten preformed or can be made in the body from tryptophan, one of the amino acids. Other forms of niacin are *nicotinic acid*, *niacinamide*, and *nicotinamide*.

niacin equivalents the amount of niacin present in food, including the niacin that can theoretically be made from its precursor tryptophan that is present in the food.

night blindness slow recovery of vision after exposure to flashes of bright light at night; an early symptom of vitamin A deficiency.

nitrogen balance the amount of nitrogen consumed compared with the amount excreted in a given time period.

nonnutrients a term used in this book to mean compounds other than the six nutrients that are present in foods and that have biological activity in the body.

norepinephrine (NOR-EP-ih-NEFF-rin) a compound related to epinephrine that helps to elicit the stress response.

nori a type of seaweed popular in Asian, particularly Japanese, cooking.

nutraceutical a term with no legal or scientific meaning, but sometimes used to refer to foods, nutrients, or dietary supplements believed to have medicinal effects. Often used to sell unnecessary or unproved supplements.

nutrient additives vitamins and minerals added to improve nutritive value.

nutrient density a measure of nutrients provided per calorie of food.

nutrients components of food that are indispensable to the body's functioning. They provide energy, serve as building material, help maintain or repair body parts, and support growth. The nutrients include water, carbohydrate, fat, protein, vitamins, and minerals.

nutrition the study of the nutrients in foods and in the body; sometimes also the study of human behaviors related to food.

nutritional yeast a preparation of yeast cells, often praised for its high nutrient content. Yeast is a source of B vitamins, as are many other foods. Also called *brewer's yeast;* not the yeast used in baking.

nutritionist someone who engages in the study of nutrition. Some nutritionists are RDs, whereas others are self-described experts whose training is questionable and who are not qualified to give advice. In states with responsible legislation, the term applies only to people who have masters of science (MS) or doctor of philosophy (PhD) degrees from properly accredited institutions.

O

obesity overfatness with adverse health effects, as determined by reliable measures and interpreted with good medical judgment. Obesity is officially defined as a body mass index of 30 or higher.

octacosanol an alcohol extracted from wheat germ, often falsely promoted as enhancing athletic performance.

oils lipids that are liquid at room temperature (70°F or 25°C).

olestra a noncaloric artificial fat made from sucrose and fatty acids; formerly called *sucrose polyester.*

omega-3 fatty acid a polyunsaturated fatty acid with its endmost double bond three carbons from the end of the carbon chain; also called *n-3 fatty acid.* Linolenic acid is an example.

omega-6 fatty acid a polyunsaturated fatty acid with its endmost double bond six carbons from the end of the carbon chain; also called *n-6 fatty acid.* Linoleic acid is an example.

omnivores people who eat foods of both plant and animal origin, including animal flesh.

oral rehydration therapy (ORT) oral fluid replacement for children with severe diarrhea caused by infectious disease. ORT enables parents to mix a simple solution for their child from substances that they have at home.

organic carbon containing. Four of the six classes of nutrients are organic: carbohydrate, fat, protein, and vitamins. Strictly speaking, organic compounds include only those made by living things and do not include carbon dioxide and a few carbon salts.

organic foods products grown and processed without the use of synthetic chemicals such as pesticides, herbicides, fertilizers, and preservatives and without genetic engineering or irradiation.

organic halogen an organic compound containing one or more atoms of a halogen—fluorine, chlorine, iodine, or bromine.

organosulfur compounds a large group of phytochemicals containing the mineral sulfur. Organosulfur phytochemicals are responsible for the pungent flavors and aromas of foods belonging to the onion, leek, chive, shallot, and garlic family and are thought to stimulate cancer defenses in the body.

organs discrete structural units made of tissues that perform specific jobs. Examples are the heart, liver, and brain.

ornithine a nonessential amino acid falsely promoted as enhancing the secretion of human growth hormone, the breakdown of fat, and the development of muscle.

oryzanol a plant sterol that supposedly provides the same physical responses as anabolic steroids without the adverse side effects; also known as *ferulic acid, ferulate,* or FRAC.

osteomalacia (OS-tee-o-mal-AY-shuh) the vitamin D–deficiency disease in adults (*osteo* means "bone"; *mal* means "bad"). Symptoms include bending of the spine and bowing of the legs.

osteoporosis (OSS-tee-oh-pore-OH-sis) a reduction of the bone mass of older persons in which the bones become porous and fragile (*osteo* means "bones"; *poros* means "porous"); also known as **adult bone loss.**

outcrossing the unintended breeding of a domestic crop with a related wild species.

oven-safe thermometer a thermometer designed to remain in the food to give constant readings during cooking.

overload an extra physical demand placed on the body; an increase in the frequency, duration, or intensity of an activity. A principle of training is that for a body system to improve, it must be worked at frequencies, durations, or intensities that increase by increments.

overweight overfatness of a moderate degree; defined as a body mass index (BMI) of 25.0 through 29.9. BMI is defined later.

ovo-vegetarian includes eggs, vegetables, grains, legumes, fruits, and nuts; excludes flesh, seafood, and milk products.

ovum the egg, produced by the mother, that unites with a sperm from the father to produce a new individual.

oxidants compounds (such as oxygen itself) that oxidize other compounds. Compounds that prevent oxidation are called *anti*oxidants, whereas those that promote it are called *pro*oxidants (*anti* means "against"; *pro* means "for").

oxidation interaction of a compound with oxygen; in this case, a damaging effect by a chemically reactive form of oxygen.

oxidative stress damage inflicted on living systems by free radicals.

oyster shell a product made from the powdered shells of oysters that is sold as a calcium supplement, but is not well absorbed by the digestive system.

P

pancreas an organ with two main functions. One is an endocrine function—the making of hormones such as insulin, which it releases directly into the blood (*endo* means "into" the blood). The other is an exocrine function—the making of digestive enzymes, which it releases through a duct into the small intestine to assist in digestion (*exo* means "out" into a body cavity or onto the skin surface).

pancreatic juice fluid secreted by the pancreas that contains both enzymes to digest carbohydrate, fat, and protein and sodium bicarbonate, a neutralizing agent.

pangamic acid also called vitamin B_{15} (but not a vitamin, nor even a specific compound—it can be anything with that label); falsely claimed to speed oxygen delivery.

pantothenic (PAN-to-THEN-ic) **acid** a B vitamin.

partial vegetarian includes seafood, poultry, eggs, dairy products, vegetables, grains, legumes, fruits, and nuts; excludes or strictly limits certain meats, such as red meats. Also called *semivegetarian.*

partitioned foods foods composed of parts of whole foods, such as butter (from milk), sugar (from beets or cane), or corn oil (from corn). Partitioned foods are generally overused and provide few nutrients with many calories.

pasteurization the treatment of milk with heat sufficient to kill certain pathogens (disease-causing microbes) commonly transmitted through milk; not a sterilization process. Pasteurized milk retains bacteria that cause milk spoilage. Raw milk, even if labeled "certified," transmits many food-borne diseases to people each year and should be avoided.

pasteurized milk milk that is heat treated to eliminate disease-causing microbes and to reduce its total bacterial count to an acceptable level.

peak bone mass the highest attainable bone density for an individual; developed during the first three decades of life.

pellagra (pell-AY-gra) the niacin-deficiency disease (*pellis* means "skin"; *agra* means "rough"). Symptoms include the "4 Ds": diarrhea, dermatitis, dementia, and, ultimately, death.

peptide bond a bond that connects one amino acid with another, forming a link in a protein chain.

peristalsis (perri-STALL-sis) the wavelike muscular squeezing of the esophagus, stomach, and small intestine that pushes their contents along.

pernicious (per-NISH-us) **anemia** a vitamin B_{12}–deficiency disease, caused by lack of intrinsic factor and characterized by large, immature red blood cells and damage to the nervous system (*pernicious* means "highly injurious or destructive").

persistent of a stubborn or enduring nature; with respect to food contaminants, the quality of remaining unaltered and unexcreted in plant foods or in the bodies of animals and human beings.

pesco-vegetarian same as partial vegetarian, but eliminates poultry.

pesticides chemicals used to control insects, diseases, weeds, fungi, and other pests on crops and around animals. Used broadly, the term includes *herbicides* (to kill weeds), *insecticides* (to kill insects), and *fungicides* (to kill fungi).

pH a measure of acidity on a point scale. A solution with a pH of 1 is a strong acid; a solution with a pH of 7 is neutral; a solution with a pH of 14 is a strong base.

phagocytes (FAG-oh-sites) white blood cells that can ingest and destroy antigens. The process by which phagocytes engulf materials is called *phagocytosis*. The Greek word *phagein* means "to eat."

phosphate salt a product demonstrated to increase the levels of a metabolically important phosphate compound (diphosphoglycerate) in red blood cells and the potential of the cells to deliver oxygen to the body's muscle cells. However, it does not extend endurance or increase efficiency of aerobic metabolism, and it may cause calcium losses from the bones if taken in excess.

phospholipids (FOSS-foh-LIP-ids) one of the three main classes of dietary lipids. These lipids are similar to triglycerides, but each has a phosphorus-containing acid in place of one of the fatty acids. Phospholipids are present in all cell membranes.

photosynthesis the process by which green plants make carbohydrates from carbon dioxide and water using the green pigment chlorophyll to capture the sun's energy (*photo* means "light"; *synthesis* means "making").

phytates (FYE-tates) compounds present in plant foods (particularly whole grains) that bind iron and may prevent its absorption.

phytochemicals (FIGH-toe-CHEM-ih-cals) biologically active compounds of plants believed to confer resistance to diseases on the eater. *Phyto* means "plant."

phytosterols (FIGH-toe-STER-ols; figh-TOSS-ter-ols) phytochemicals structurally similar to mammalian steroid hormones, such as the female sex hormone estrogen. Phytosterols may or may not mimic hormone activity in the human body.

pica (PIE-ka) a craving for nonfood substances. Also known as *geophagia* (gee-oh-FAY-gee-uh) when referring to clay eating, and *pagophagia* (pag-oh-FAY-gee-uh) when referring to ice craving (*geo* means "earth"; *pago* means "frost"; *phagia* means "to eat").

placebo a sham treatment often used in scientific studies; an inert harmless medication. The *placebo effect* is the healing effect that the act of treatment, rather than the treatment itself, often has.

placenta (pla-SEN-tuh) the organ that develops inside the uterus in early pregnancy in which maternal and fetal blood circulate in close proximity and exchange materials. The fetus receives nutrients and oxygen across the placenta; the mother's blood picks up carbon dioxide and other waste materials to be excreted via her lungs and kidneys.

plant pesticides substances produced within plant tissues that kill or repel attacking organisms.

plant sterols lipid extracts of plants, called ferulic acid, oryzanol, phytosterols, or "adaptogens," marketed with false claims that they contain hormones or enhance hormonal activity.

plaque (PLACK) a mass of microorganisms and their deposits on the crowns and roots of the teeth, a forerunner of dental caries and gum disease.

plaques (PLACKS) mounds of lipid material mixed with smooth muscle cells and calcium that develop in the artery walls in atherosclerosis (*placken* means "patch"). The same word is also used to describe the accumulation of a different kind of deposits on teeth, which promote dental caries.

plasma the cell-free fluid part of blood and lymph.

platelets tiny cell-like fragments in the blood, important in blood clot formation (*platelet* means "little plate").

point of unsaturation a site in a molecule where the bonding is such that additional hydrogen atoms can easily be attached.

polypeptides protein fragments of many (more than ten) amino acids bonded together (*poly* means "many"). A chain of between four and ten amino acids is called an *oligopeptide*.

polysaccharides another term for complex carbohydrates; compounds composed of long strands of glucose units linked together (*poly* means "many"). Also called *complex carbohydrates*.

polyunsaturated fats triglycerides in which most of the fatty acids have two or more points of unsaturation (are polyunsaturated).

polyunsaturated fatty acid (PUFA) a fatty acid with two or more points of unsaturation.

pop-up thermometer a disposable timing device commonly used in turkeys. The center of the device contains a stainless steel spring that "pops up" when food reaches the right temperature.

postprandial hypoglycemia a drop in blood glucose that follows a meal and is accompanied by symptoms of the stress response; also called *reactive hypoglycemia*.

potassium iodide a medication approved by the Food and Drug Administration as safe and effective for the prevention of thyroid cancer caused by radioactive iodine known to be released during radiation emergencies.

powdered milk dehydrated milk solids. Some powdered milks rehydrate easily (instant milk); others require extensive blending. Both whole and fat-free milk can be powdered.

precursor control control of a compound's synthesis by the availability of that compound's precursor. The more precursor there is, the more of the compound is made.

precursors, provitamins compounds that can be converted into active vitamins.

preeclampsia a potentially dangerous condition during pregnancy characterized by edema, hypertension, and protein in the urine.

pregame meal a meal eaten three to four hours before athletic competition.

premenstrual syndrome (PMS) a cluster of symptoms that some women experience prior to and during menstruation. They include, among others, abdominal cramps, back pain, swelling, headache, painful breasts, and mood changes.

preservatives antimicrobial agents, antioxidants, chelating agents, radiation, and other additives that retard spoilage or preserve desired qualities, such as softness in baked goods.

probiotics consumable products containing live microorganisms in sufficient numbers to alter the bacterial colonies of the body in ways believed to benefit health. A *prebiotic* product is a substance that may not be digestible by the host, such as fiber, but serves as food for probiotic bacteria and thus promotes their growth.

problem drinkers or **alcohol abusers** people who suffer social, emotional, family, job-related, or other problems because of alcohol. A problem drinker is on the way to alcoholism.

processed foods foods subjected to any process, such as milling, alteration of texture, addition of additives, cooking, or others. Depending on the starting material and the process, a processed food may or may not be nutritious.

promoters factors that do not initiate cancer but speed up its development once initiation has taken place.

proof a statement of the percentage of alcohol in an alcoholic beverage. Liquor that is 100 proof is 50% alcohol, 90 proof is 45%, and so forth.

prooxidant (proh-OX-ih-dant) a compound that triggers reactions involving oxygen.

protein digestibility–corrected amino acid score (PDCAAS) a measuring tool used to determine protein quality. The PDCAAS reflects a protein's digestibility as well as the proportions of amino acids that it provides.

protein efficiency ratio (PER) a measure of protein quality assessed by determining how well a given protein supports weight gain in growing rats. The PER is used to judge the quality of protein in infant formulas and baby foods.

protein-energy malnutrition (PEM) the world's most widespread malnutrition problem, including both marasmus and kwashiorkor and states in which they overlap; also called *protein-calorie malnutrition (PCM)*.

proteins compounds composed of carbon, hydrogen, oxygen, and nitrogen and arranged as strands of amino acids. Some amino acids also contain the element sulfur.

protein-sparing action the action of carbohydrate and fat in providing energy that allows protein to be used for purposes it alone can serve.

puberty the period in life when a person develops sexual maturity and the ability to reproduce.

public health nutritionist a dietitian or other person with an advanced degree in nutrition who specializes in public health nutrition.

public water water from a municipal or county water system that has been treated and disinfected.

purified water water that has been treated by distillation or other physical or chemical processes that remove dissolved solids. Because purified water contains no minerals or contaminants, it is useful for medical and research purposes.

pyloric (pye-LORE-ick) **valve** the circular muscle of the lower stomach that regulates the flow of partly digested food into the small intestine. Also called *pyloric sphincter*.

pyruvate a 3-carbon compound derived during the metabolism of glucose, certain amino acids, and glycerol; falsely promoted as burning fat and enhancing endurance. Common side effects include intestinal gas and diarrhea.

R

raw sugar the first crop of crystals harvested during sugar processing. Raw sugar cannot be sold in the United States because it contains too much filth (dirt, insect fragments, and the like). Sugar sold as "raw sugar" domestically is not actually raw but has gone through more than half of the refining steps.

recombinant DNA technology (rDNA technology) a form of biotechnology that changes the characteristics of living things by manipulating the genes; includes methods of removing genes, doubling genes, introducing foreign genes, and changing gene positions to influence the growth and development of organisms.

reconstituted a concentrate that has been brought back to its original strength by the addition of water.

refined refers to the process by which the coarse parts of food products are removed. For example, the refining of wheat into flour involves removing three of the four parts of the kernel—the chaff, the bran, and the germ—leaving only the endosperm, composed mainly of starch and a little protein.

registered dietitian (RD) a dietitian who has graduated from a university or college after completing a program of dietetics. The program must be approved or accredited by the American Dietetic Association (or Dietitians of Canada). The dietitian must serve in an approved internship, coordinated program, or preprofessional practice program to practice the necessary skills; pass the five parts of the association's *registration* examination; and maintain competency through continuing education. Many states also require licensing for practicing dietitians.

registration listing with a professional organization that requires specific course work, experience, and passing of an examination.

requirement the amount of a nutrient that will just prevent the development of specific deficiency signs; distinguished from the DRI recommended intake value, which is a generous allowance with a margin of safety.

residues whatever remains. In the case of pesticides, those amounts that remain on or in foods when people buy and use them.

resistant starch the fraction of starch in a food that is digested slowly, or not at all, by human enzymes.

resistin a hormone made and released by fat cells under conditions of obesity and thought to increase tissue resistance to the effects of insulin.

retina (RET-in-uh) the layer of light-sensitive nerve cells lining the back of the inside of the eye.

retinol activity equivalents (RAE) a measure of vitamin A activity of beta-carotene and other vitamin A precursors that reflects the amount of retinol that the body will derive from a food containing the precursors.

retinol one of the active forms of vitamin A made from beta-carotene in animal and human bodies; an antioxidant nutrient. Other active forms are *retinal* and *retinoic acid*.

rhodopsin (roh-DOP-sin) the light-sensitive pigment of the cells in the retina, containing vitamin A (*rhod* refers to the rod-shaped cells; *opsin* means "visual protein").

riboflavin (RIBE-o-flay-vin) a B vitamin active in the body's energy-releasing mechanisms.

rickets the vitamin D–deficiency disease in children; characterized by abnormal growth of bone and manifested in bowed legs or knock-knees, outward-bowed chest, and knobs on the ribs.

risk factors factors known to be related to (or correlated with) diseases but not proved to be causal.

roughage (RUFF-idge) the rough parts of food; an imprecise term that has largely been replaced by the term *fiber*.

royal jelly a substance produced by worker bees and fed to the queen bee; often falsely promoted as enhancing athletic performance.

S

saccharin a zero-calorie sweetener used freely in the United States but restricted in Canada.

safety the practical certainty that injury will not result from the use of a substance.

salts compounds composed of charged particles (ions). An example is potassium chloride ($K+Cl-$).

SAM-e an amino acid derivative that may have an antidepressant effect on the brain in some people, but is not recommended as a substitute for standard antidepressant therapy.

sassafras root bark from the sassafras tree; once used in beverages but now banned as an ingredient in foods or beverages because it contains cancer-causing chemicals.

satiation (SAY-she-AY-shun) the perception of fullness that builds throughout a meal, eventually reaching the degree of fullness and satisfaction that halts eating. Satiation generally determines how much food is consumed at one sitting.

satiety (sah-TIE-eh-tee) the perception of fullness that lingers in the hours after a meal and inhibits eating until the next mealtime. Satiety generally determines the length of time between meals.

saturated fats triglycerides in which most of the fatty acids are saturated.

saturated fatty acid a fatty acid carrying the maximum possible number of hydrogen atoms (having no points of unsaturation). A saturated fat is a triglyceride that contains three saturated fatty acids.

saw palmetto the ripe fruit or extracts of the saw palmetto plant claimed to relieve symptoms associated with enlarged prostate; may act as a diuretic.

scurvy the vitamin C–deficiency disease.

self-efficacy a person's belief in his or her ability to succeed in an undertaking.

senile dementia the loss of brain function beyond the normal loss of physical adeptness and memory that occurs with aging.

serotonin (SER-eh-TONE-in) a compound related in structure to (and made from) the amino acid tryptophan. It serves as one of the brain's principal neurotransmitters and reduces appetite. Some weight-loss drugs aim to elevate the serotonin levels in the brain.

set-point theory the theory that the body tends to maintain a certain weight by means of its own internal controls.

side chain the unique chemical structure attached to the backbone of each amino acid that differentiates one amino acid from another.

simple carbohydrates sugars, including both single sugar units and linked pairs of sugar units. The basic sugar unit is a molecule containing six carbon atoms, together with oxygen and hydrogen atoms.

single-use temperature indicator a type of instant-read thermometer that changes color to indicate that the food has reached the desired temperature. Discarded after one use, they are often used in retail food markets to eliminate cross-contamination.

small intestine the 20-foot length of small-diameter intestine, below the stomach and above the large intestine, that is the major site of digestion of food and absorption of nutrients.

smoking point the temperature at which fat gives off an acrid blue gas.

social drinkers people who drink only on social occasions. Depending on how alcohol affects a social drinker's life, the person may be a moderate drinker or a problem drinker.

sodium bicarbonate baking soda; an alkaline salt believed to neutralize blood lactic acid and thereby reduce pain and enhance possible workload. "Soda loading" may cause intestinal bloating and diarrhea.

soft water water with a high sodium concentration.

soluble fibers food components that readily dissolve in water and often impart gummy or gel-like characteristics to foods. An example is pectin from fruit, which is used to thicken jellies. Soluble fibers are indigestible by human enzymes but may be broken down to absorbable products by bacteria in the digestive tract.

soy drink a milk-like beverage made from soybeans, claimed to be a functional food. Soy drink should be fortified with vitamin A, vitamin D, riboflavin, and calcium to approach nutritional equivalency with milk. Also called *soy milk*.

Special Supplemental Food Program for Women, Infants, and Children (WIC) a USDA program to provide nutrition support to low-income women who are pregnant or have infants or preschool children. WIC offers coupons redeemable for specific foods to supply the nutrients deemed most needed for growth and development.

sphincter (SFINK-ter) a circular muscle surrounding, and able to close, a body opening.

spina bifida (SPY-na BIFF-ih-duh) one of the most common types of neural tube defects. The infant is born with gaps in the bones of the spine, leaving the spinal cord protected only by a sheath of skin in those spots, or with no protection at all. The spinal cord may bulge and protrude through the gaps in the vertebral column.

spirulina a kind of alga ("blue-green manna") that supposedly contains large amounts of protein and vitamin B_{12}, suppresses appetite, and improves athletic performance. It does none of these things and is potentially toxic.

spring water water originating from an underground spring or well. It may be bubbly (carbonated), or "flat" or "still," meaning not carbonated. Brand names such as "Spring Pure" do not necessarily mean that the water comes from a spring.

St. John's wort an herb containing psychoactive substances that has been used for centuries to treat depression, insomnia, bedwetting, and "nervous conditions." Most scientific reports find St. John's wort equal in effectiveness to standard antidepressant medication for relief of depression. Long-term safety, however, has not been established.

stanol esters plant-derived compounds belonging to the sterol family of lipids that have been shown experimentally to reduce blood cholesterol when consumed in place of other fats in a low-fat diet.

stanol esters, sterol esters compounds derived from vegetable oils or wood pulp that lower blood cholesterol in human beings by competing with cholesterol for absorption from the digestive tract. The term *sterol esters* often refers to both stanol and sterol esters.

staple foods foods used frequently or daily, for example, rice (in East and Southeast Asia) or potatoes (in Ireland). If well chosen, these foods are nutritious; certainly, they should be.

starch a plant polysaccharide composed of glucose. After cooking, starch is highly digestible by human beings; raw starch often resists digestion.

stem cell an undifferentiated cell that can mature into any number of specific specialized cell types. A stem cell of bone marrow may mature into one of many kinds of blood cells, for example.

sterols (STEER-alls) one of the three main classes of dietary lipids. Sterols have a structure similar to that of cholesterol.

stomach a muscular, elastic, pouchlike organ of the digestive tract that grinds and churns swallowed food and mixes it with acid and enzymes, forming chyme.

stone ground refers to a milling process using limestone to grind any grain, including refined grains, into flour.

stone-ground flour flour made by grinding kernels of grain between heavy wheels made of limestone, a kind of rock derived from the shells and bones of marine animals. As the stones scrape together, bits of the limestone mix with the flour, enriching it with calcium.

stress fractures bone injuries or breaks caused by the stress of exercise on the bone surface.

stroke the sudden shutting off of the blood flow to the brain by a thrombus, embolism, or the bursting of a vessel (hemorrhage).

stroke volume the amount of oxygenated blood ejected from the heart toward body tissues at each beat.

structural components Proteins form integral parts of most body structures such as skin, tendons, ligaments, membranes, muscles, organs, and bones.

subclinical, or **marginal, deficiency** a nutrient deficiency that has no outward clinical symptoms. The term is often used to scare consumers into buying unneeded nutrient supplements.

subcutaneous fat fat stored directly under the skin (*sub* means "beneath"; *cutaneous* refers to the skin).

succinate a compound synthesized in the body and involved in the TCA cycle; falsely promoted as a metabolic enhancer.

sucralose a noncaloric sweetener derived from a chlorinated form of sugar that travels through the digestive tract unabsorbed. Approved by the FDA for use in the United States.

sucrose (SOO-crose) a disaccharide composed of glucose and fructose; sometimes known as table, beet, or cane sugar.

sucrose polyester any of a family of compounds in which fatty acids are bonded with sugars or sugar alcohols. Olestra is an example.

sugars simple carbohydrates, that is, molecules of either single sugar units or pairs of those sugar units bonded together.

superoxide dismutase (SOD) an enzyme that protects cells from oxidation. When it is taken orally, the body digests and inactivates this protein; it is useless to athletes.

supplements pills, liquids, or powders that contain purified nutrients or other ingredients.

surface water water that comes from lakes, rivers, and reservoirs.

sushi a Japanese dish that consists of vinegar-flavored rice, seafood, and colorful vegetables, typically wrapped in seaweed. Some sushi is wrapped in raw fish; other sushi contains only cooked ingredients.

sustainable able to continue indefinitely. Here the term refers to the use of resources at such a rate that the earth can keep on replacing them; for example, cutting trees no faster than new ones grow and producing pollutants at a rate with which the environment and human cleanup efforts can keep pace. In a sustainable economy, resources do not become depleted, and pollution does not accumulate.

systolic (sis-TOL-ik) **pressure** the first figure in a blood pressure reading (the "dub" of the heartbeat), which reflects arterial pressure caused by the contraction of the heart's left ventricle.

T

tannins compounds in tea (especially black tea) and coffee that bind iron. Tannins also denature proteins.

T-cells lymphocytes that attack antigens. *T* stands for the thymus gland of the neck, where the T-cells are stored and matured.

textured vegetable protein processed soybean protein used in products formulated to look and taste like meat, fish, or poultry.

thermic effect of food (TEF) the body's speeded-up metabolism in response to having eaten a meal; also called *diet-induced thermogenesis*.

thermogenesis the generation and release of body heat associated with the breakdown of body fuels. *Adaptive thermogenesis* describes adjustments in energy expenditure related to changes in environment such as cold and to physiological events such as underfeeding or trauma.

thiamin (THIGH-uh-min) a B vitamin involved in the body's use of fuels.

thickening and stabilizing agents ingredients that maintain emulsions, foams, or suspensions or lend a desirable thick consistency to foods. Dextrins (short chains of glucose formed as a breakdown product of starch), starch, and pectin are examples. (Gums such as carrageenan, guar, locust bean, agar, and gum arabic are others.)

thrombosis a thrombus that has grown enough to close off a blood vessel. A *coronary thrombosis* is the closing off of a vessel that feeds the heart muscle. A *cerebral thrombosis* is the closing off of a vessel that feeds the brain (*coronary* means "crowning" [the heart]; *thrombo* means "clot"; the cerebrum is part of the brain).

thrombus a stationary blood clot.

tissues systems of cells working together to perform specialized tasks. Examples are muscles, nerves, blood, and bone.

tocopherol (tuh-KOFF-er-all) a kind of alcohol. The active form of vitamin E is alpha-tocopherol.

tofu (TOE-foo) a curd made from soybeans that is rich in protein, often rich in calcium, and variable in fat content; used in many Asian and vegetarian dishes in place of meat.

tolerance limit the maximum amount of a residue permitted in a food when a pesticide is used according to label directions.

toxicity the ability of a substance to harm living organisms. All substances are toxic if the concentration is high enough.

trabecular (tra-BECK-you-lar) **bone** the weblike structure composed of calcium-containing crystals inside a bone's solid outer shell. It provides strength and acts as a calcium storage bank.

trace minerals essential mineral nutrients found in the human body in amounts less than 5 grams.

training regular practice of an activity, which leads to physical adaptations of the body with improvement in flexibility, strength, or endurance.

***trans*-fatty acids** fatty acids with unusual shapes that can arise when polyunsaturated oils are hydrogenated.

transgenic organism an organism resulting from the growth of an embryonic, stem, or germ cell into which a new gene has been inserted. Also called *genetically modified organism*.

transportation Proteins help transport needed substances, such as lipids, minerals, and oxygen, around the body.

triglycerides (try-GLISS-er-ides) one of the three main classes of dietary lipids and the chief form of fat in foods. A triglyceride is made up of three units of fatty acids and one unit of glycerol (fatty acids and glycerol are defined later). Triglycerides are also called *triacylglycerols*.

trimesters periods representing one-third of gestation. A trimester is about 13 to 14 weeks.

tripeptides (try-PEP-tides) protein fragments that are three amino acids long (*tri* means "three").

turbinado (ter-bih-NOD-oh) **sugar** raw sugar from which the filth has been washed; legal to sell in the United States.

type 1 diabetes the type of diabetes in which the person produces no or very little insulin; often diagnosed in childhood, although some cases arise in adulthood. Formerly called *juvenile-onset* or *insulin-dependent diabetes*.

type 2 diabetes the type of diabetes in which the person makes plenty of insulin, but the body cells resist insulin's action; often diagnosed in adulthood. Formerly called *adult-onset* or *noninsulin-dependent diabetes*.

U

ulcer an erosion in the topmost, and sometimes underlying, layers of cells that form a lining. Ulcers of the digestive tract commonly form in the esophagus, stomach, or upper small intestine.

ultrahigh temperature (UHT) a process of sterilizing food by exposing it for a short time to temperatures above those normally used in processing.

unbleached flour a beige-colored endosperm flour with texture and nutritive qualities that approximate those of regular white flour.

underwater weighing a measure of density and volume used to determine body fat content.

underweight too little body fat for health; defined as having a body mass index of less than 18.5.

unsaturated fatty acid a fatty acid that lacks some hydrogen atoms and has one or more points of unsaturation. An unsaturated fat is a triglyceride that contains one or more unsaturated fatty acids.

urban legend a story, usually false, that may travel rapidly throughout the world via the Internet gaining strength of conviction solely on the basis of repetition.

urea (yoo-REE-uh) the principal nitrogen-excretion product of metabolism; generated mostly by removal of amine groups from unneeded amino acids or from amino acids being sacrificed to a need for energy.

urethane a carcinogenic compound that commonly forms in alcoholic beverages.

uterus (YOO-ter-us) the womb; the muscular organ within which the infant develops before birth.

V

valerian a preparation of the root of an herb used as a sedative and sleep agent. Safety and effectiveness of valerian have not been scientifically established.

variety the dietary characteristic of providing a wide selection of foods—the opposite of monotony.

vegan includes only food from plant sources: vegetables, grains, legumes, fruits, seeds, and nuts; also called *strict vegetarian*.

vegetarian includes plant-based foods and eliminates some or all animal-derived foods.

vegetarians people who exclude from their diets animal flesh and possibly other animal products such as milk, cheese, and eggs.

veins blood vessels that carry blood, with the carbon dioxide it has collected, from the tissues back to the heart.

very-low-density lipoproteins (VLDL) lipoproteins that transport triglycerides and other lipids from the liver to various tissues in the body.

villi (VILL-ee, VILL-eye) fingerlike projections of the sheets of cells that line the intestinal tract. The villi make the surface area much greater than it would otherwise be (singular: *villus*).

visceral fat fat stored within the abdominal cavity in association with the internal abdominal organs; also called *intra-abdominal fat*.

vitamin B_6 a B vitamin needed in protein metabolism. Its three active forms are *pyridoxine*, *pyridoxal*, and *pyridoxamine*.

vitamin B_{12} a B vitamin that helps to convert folate to its active form and also helps maintain the sheaths around nerve cells. Vitamin B_{12}'s scientific name, not often used, is *cyanocobalamin*.

vitamins organic compounds that are vital to life and indispensable to body functions, but are needed only in minute amounts; noncaloric essential nutrients.

$VO_{2\,max}$ the maximum rate of oxygen consumption by an individual (measured at sea level).

voluntary activities intentional activities (such as walking, sitting, or running) conducted by voluntary muscles.

W

wasting the progressive, relentless loss of the body's tissues that accompanies certain diseases and shortens survival time.

water balance the balance between water intake and water excretion, which keeps the body's water content constant.

water intoxication the rare condition in which body water content is too high. Symptoms are headache, muscular weakness, lack of concentration, poor memory, and loss of appetite.

weight cycling repeated rounds of weight loss and subsequent regain, with reduced ability to lose weight with each attempt; also called *yo-yo dieting*.

weight training the use of free weights or weight machines to provide resistance for developing muscle strength and endurance. A person's own body weight may also be used to provide resistance as when a person does push-ups, pull-ups, or sit-ups. Also called *resistance training*.

well water water drawn from ground water by tapping into an aquifer.

Wernicke-Korsakoff (VER-nik-ee KOR-sah-koff) **syndrome** a cluster of symptoms involving nerve damage arising from a deficiency of the vitamin thiamin in alcoholism. Characterized by mental confusion, disorientation, memory loss, jerky eye movements, and staggering gait.

wheat flour any flour made from wheat, including white flour.

wheat germ oil the oil from the wheat kernel; often falsely promoted as an energy aid.

whey the liquid that remains after milk has coagulated (see also *casein*).

whey protein a by-product of cheese production; falsely promoted as increasing muscle mass. As for whey, it is the liquid left when most solids are removed from milk.

white flour an endosperm flour that has been refined and bleached for maximum softness and whiteness.

white sugar pure sucrose, produced by dissolving, concentrating, and recrystallizing raw sugar.

whole grain refers to a grain milled in its entirety (all but the husk), not refined.

whole milk full-fat cow's milk.

whole-wheat flour flour made from whole-wheat kernels; a whole-grain flour.

witch hazel leaves or bark of a witch hazel tree; not proved to have healing powers.

world food supply the quantity of food, including stores from previous harvests, available to the world's people at a given time.

X

xerophthalmia (ZEER-ahf-THALL-me-uh) hardening of the cornea of the eye in advanced vitamin A deficiency that can lead to blindness (*xero* means "dry"; *ophthalm* means "eye").

xerosis (zeer-OH-sis) drying of the cornea; a symptom of vitamin A deficiency.

Z

zygote (ZYE-goat) the term that describes the product of the union of ovum and sperm during the first two weeks after fertilization.

Credits

Chapter Opener Art Credits

CO 1: Collection of Frances O. Stem Babinsky

CO 2: The Barnes Foundation, Merion, PA. Photo: SuperStock

CO 3: Oil on canvas, 72.7 × 44.5 cm., The Baltimore Museum of Art: The Cone Collection, formed by Dr. Claribel Cone and Miss Etta Cone of Baltimore, Maryland. BMA 1950.213.

CO 4: Gouache on paper, 20 × 33 cm., Victoria & Albert Museum, London. Photo: Art Resource, NY.

CO 5: Oil on panel, 30.5 cm × 40.5 cm (12 × 16 in.), Private Collection. Photo: Courtesy of Christie's Images, New York.

CO 6: Collection of Siri von Reis, New York City

CO 7: Oil on canvas, 17 3/8 × 23 in., High Museum of Art, Atlanta, Georgia: Purchase, 75.37.

CO 8: Courtesy of Teresa Fasolino

CO 9: © 2002 Reproduccíon autorizada por el Instituto Nacional de Bellas Artes y Literatura, México D.F. & Banco de México Diego Rivera & Frida Kahlo Museum Trust. Av. Cinco de Mayo No. 2, Col. Centro, Del. Cuauhtémoc 06059, México, D.F. Photo: Bridgeman/SuperStock

CO 10: Oil on canvas. Musée National Fernand Leger, Biot, France. Photo: Erich Lessing/Art Resource, NY. © 2002 Artists Rights Society (ARS), NY/ADAGP, Paris

CO 11: Oil on canvas, 51 × 68 in. Acquired 1923, The Phillips Collection, Washington, DC

CO 12: Mixed media. Private Collection. Photo: Art Resource, NY. © 2002 Michael Escoffery/Artists Rights Society (ARS), NY

CO 13: Oil pastel, 30 × 22 in. Creative Growth Art Center, Oakland, CA.

CO 14: Silkscreen & acrylic on canvas. Private Collection/Courtesy of Curt Marcus Gallery, New York. Photo: SuperStock

CO 15: Gouache with graphite on paper, .565 × .784 mm. (22 1/4 × 30 7/8 in.). Gift of Mr. and Mrs. James T. Dyke, © 2002 Board of Trustees, National Gallery of Art, Washington, DC.

Photo Credits

Page 2 © 2001 PhotoDisc, Inc.; 6 © Bob Daemmrich/Stock, Boston; 7 © Richard Fukuhara/Corbis; 9 © David R. Frazier/Photolibrary; 9 © Polara Studios Inc.; 11 © Michael Newman/PhotoEdit; 12 © 1999 Daphne Hougard; 18 © Craig M. Moore; 21 © 2001 PhotoDisc, Inc.; 24 © 2002 Larry Bray/FPG/ Getty Images; 28 © Marilyn Herbert Photography (both); 32 © 1999 PhotoDisc, Inc.; 36–37 © Polara Studios, Inc. (all); 40 © Tony Freeman/ PhotoEdit; 40 © Felicia Martinez/PhotoEdit; 41 © Michael Newman/PhotoEdit; 41 © Bonnie Kamin/PhotoEdit; 43 Photo reprinted with permission of the Pillsbury Company, 2001; 44 © Tony Freeman/PhotoEdit (all); 45 © David Young-Wolff/PhotoEdit; 51 (top to bottom) © Polara Studios, Inc.; © Polara Studios, Inc.; © Polara Studios, Inc.; © Quest; © Quest; 52 (top to bottom) © Quest; Quest; © Polara Studios, Inc.; Quest; © Michael Newman/PhotoEdit; 63 (top) Brassica Protection Products; 63 © 2001 PhotoDisc, Inc.; © 2001 PhotoDisc, Inc.; © EyeWire, Inc.; 63 (center) © 2001 PhotoDisc, Inc.; © 2001 PhotoDisc, Inc.; © EyeWire, Inc.; © 2001 PhotoDisc, Inc.; 63 (bottom) © 2001 PhotoDisc, Inc.; Flax Council of Canada; © 2001 PhotoDisc; 64 © Craig M. Moore; 71 © Alan Oddie/PhotoEdit; 83 © 2001 PhotoDisc, Inc.; 84 (photo) © Don W. Fawcett; 92 © Polara Studies, Inc.; 95 © A. Glaubermann/ Photo Researchers, Inc.; 103 © Mary Kate Denny/ PhotoEdit; 109 © 2002 Steve Rothfeld/Stone/Getty Images; 117 © 2002 PhotoDisc/Getty Images; 119 © R. W. Jones/Corbis; 120 © Polara Studios, Inc.; 123 © Polara Studios, Inc.; 130 © Polara Studios, Inc.; 136 © Polara Studios, Inc.; 140 © David R. Fraizer/PhotoLibrary; 141 © Quest Photographic, Inc.; 147 © 2002 PhotoDisc/Getty Images; 150 © Polara Studios, Inc.; 153 © www.comstock.com; 155 © 2002, PhotoDisc, Inc.; 157 © Felicia Martinez/PhotoEdit; 159 © Polara Studios, Inc.; 160 © 2001 PhotoDisc, Inc.; 161 © Quest Photographic, Inc. (all); 163 © Polara Studios, Inc.; 164 © Polara Studios, Inc.; 172 © 2002 Robert Frerck/Stone/Getty Images; 174 © Corbis Images; 174 © 2001 PhotoDisc, Inc.; 179 © Gary Conners/ PhotoEdit; 192 © Japack Company/CORBIS; 193 (left to right) © Michael Newman/PhotoEdit; Felicia Martinez/PhotoEdit; © Michael Newman/ PhotoEdit; 196 United Nations Food and Agriculture Organization/Photo Library; 197 AP/Wide World Photos; 198 © Paul A. Souders/CORBIS; 199 © World Health Organization, Geneva; 202–203 © Polara Studios, Inc. (all); 205 © William H. Edwards/ The Image Bank/Getty Images; 205 © Polara Studios, Inc.; 207 © Polara Studios, Inc.; 214 © David Farr/Image Smythe; 215 © David Ulmer/Stock, Boston/PictureQuest; 215 H. Sanstead, University of Texas-Galveston; 218 © Fotografia/CORBIS; 219 © Biophoto Associates/ Science Source/ Photo Researchers; 219 © Photo Courtesy of Dr. Norman Carvalho at Children's Healthcare of Atlanta; 227 © L. V. Bergman & Associates/CORBIS; 229 © George L. Blackburn M.D., Ph.D., Harvard Medical School; 232 Martin M. Rotker (both); 234 © George L. Blackburn, M.D., Harvard Medical School; 237 © L. V. Bergman & Associates/CORBIS; Kronauer and Buhler, *The New England Journal of Medicine*, 6/16/95; 239 L. V. Bergman & Associates/CORBIS; 257 © 2002 J. Share/ Stone/Getty Images; 261 © 2002 Thomas Braise/Stone/Getty Images; 267 © Lawrence Migdale/Stock, Boston; 267 © Camera M. D. Studios, Inc.; 268 Felicia Martinez/© ; 272 © 2001 PhotoDisc, Inc.; 272 © Polara Studios, Inc.; 275 © Craig M. Moore; 283 © Polara Studios, Inc. (all); 286 L. V. Bergman & Associates/CORBIS; 288 © L. V. Bergman & Associates/CORBIS (both); 290 Michael Newman/ © ; 290 Courtesy of Ray Stanyard; 292 H. Sanstead, University of Texas-Galveston; 293 © Dr. P. Marazzi/Science Photo Library/ Photo Researchers; 294 © Ross Whitaker/ The Image Bank/ Getty Images; 298 National Dairy Council®—© 2000 Dairy Management, Inc.; 304 Courtesy of Gjon Mills; 305 Permission by David Dempster from J Bone Miner Res, 1986 (both); 307 © DigitalVision/ PictureQuest; 315 © George Sesota/ Newsmakers/ Getty Images; 316 © 2002 Jurgen Reisch/Stone/ Getty

Images; 319 © Rick Schaff; 320 (left to right) © Fitness & Wellness, Boise, Idaho; © Photo courtesy of Hologic, Inc.; © David Young-Wolff/PhotoEdit; 326 © R. Benali/Liaison Gamma/Getty Images; 330 © 2002 David Madison/Stone/ Getty Images; 333 © 1999 PhotoDisc, Inc.; 335 CORBIS; 341 © Bob Daemmrich Photography; 347 (top) © Michael Newman/PhotoEdit; (center and bottom) © Polara Studios, Inc.; 351 © Laura Wagner/ Index Stock Imagery; 353 © David Kelly Crow/PhotoEdit; 355 © Felicia Martinez/PhotoEdit; 361 (left to right) © Photo Disc, Inc.; © David Hanover Photography; © David Hanover Photography; 362 © 1999 PhotoDisc, Inc.; 365 © Bonnie Kamin/PhotoEdit; 368 © Kevin R. Morris/CORBIS; 371 © Rubber Ball Productions/PictureQuest; 374 CORBIS/Bill Ross; 375 © Polara Studios, Inc.; 377 © 2002 PhotoDisc, Inc.; 381 © Polara Studios, Inc. (all); 388 © Randy M. Ury/Corbis Stock Market; 400 (top) Courtesy of Zeneca Pharmaceutical Division, Cheshire, England (both); (bottom) © Science Photo Library/ Photo Researchers; 404 © Jacksonville Journal Courier/The Image Works; 408 © Michael Keller/Index Stock Imagery; 415 © Mark Andersen/ Rubber Ball Productions/ PictureQuest; 416 © Stanley Rowin/Index Stock Imagery; 420 © James Worrell/Time Pix (all); 424 © Tom McCarthy/PhotoEdit; 425 © Polara Studios, Inc. (all); 429 © Michael Newman/ PhotoEdit; 433 © 1999 PhotoDisc, Inc.; 436 © Masterfile; 438 © (Photos 1, 2, 3) Petit Format/Nestle/ Photo Researchers, © (Photo 4) Anthony M. Vanelli; 441 © Biophoto Associates/Photo Researchers, Inc.; 445 © 2002 Tracy Frankel/Image Bank/Getty Images; 450 © 1995 George Steinmetz; 455 © 2002 PhotoDisc, Inc.; 456 © 2002 David Young-Wolff/Stone/ Getty Images; 459 © Myrleen Cate/Index Stock Imagery; 461 © Michael Bridwell/PhotoEdit; 462 © Polara Studios, Inc.; 463 © Jose L. Pelaez/Corbis Stock Market; 464 Courtesy of Pamela R. Erickson (both); 465 © 2002 Stephanie Rausser/FPG/ Getty Images; 467 © Will Hart/PhotoEdit; 474 © Anthony Vanneli, (both); 477 © Mary Kate Denny/ PhotoEdit; 479 © Tony Freeman/PhotoEdit; 480 Polara Studios, Inc.; 486 © 2002 Henley and Savage/ Stone/Getty Images; © D. M. Phillips/Visuals Unlimited; 492 Courtesy of Dr. William Evans; 494 © Brian Bailey/Stone/Getty Images; 495 © Darama/Corbis Stock Market; 498 © D & I MacDonald/ PhotoEdit; 500 © Susan Van Etten/ PhotoEdit; 501 ©2002 Larry Dale Gordon/Image Bank/Getty Images; 503 © Bob Daemmrich/Stock, Boston/PictureQuest; 511 © Jeff Greenberg/PhotoEdit; 514 © Polara Studios, Inc.; 519 © 2002 PhotoDisc/Getty Images; 521 © David Young-Wolff/PhotoEdit; 530 © Eugene Smith/Black Star-Stockphoto; 531 © Joseph Schuyler/Stock, Boston; 532 © George Loun/Visual Unlimited; 538 © Polara Studios, Inc.; 540 © Polara Studios, Inc.; 541 © Joe McDonald/ Visuals Unlimited; 542 © 1999 Photo Disc, Inc.; 544 © Polara Studios, Inc.; 545 © Polara Studios, Inc. (all); 550 Courtesy of Smithsonian Tropical Research Institute/photo by Antonio Montaner; 551 Courtesy of Aqua Bounty Canada; 555 © Adrian Arbib/Still Pictures; 555 © Tony Freeman/Photo Edit;560 © Elsa Peterson/Stock, Boston; 561 © Charles Gupton/Stock, Boston/PictureQuest; © 2002 Steve McAlister Prod./The Image Bank/Getty Images; 562 © Charles Gupton/Corbis Stock Market; 563 © Robert Caputo/Aurora/PictureQuest; 565 © Christopher Morris/Black Star/PictureQuest; 565 © Craig Aurness/ CORBIS; 567 © David Austen/Stock, Boston; 569 Courtesy of NASA; 570 © 2002 PhotoDisc/Getty Images; © 2002 PhotoDisc/Getty Images; © Corbis Images/PictureQuest; © Burke/Triolo/Brand X Pictures/ PictureQuest; Courtesy, Bradford White Corporation; 574 (top-left) © Paul Stover/Stone/ Getty Images; (top-right) (© Larry Ulrich/Stone/ Getty Images; (bottom) © Stone/Getty Images; 575 © Oklahoma Publishing Co., May, 1997.

Index

The page letters A, B, and C that stand alone refer to the tables beginning on the inside front cover. The page letters Y and Z refer to the tables on the last pages of the book. The page numbers preceded by A through G are appendix page numbers. The boldfaced page numbers indicate definitions. Terms are also defined in the glossary. Page numbers followed by *n* indicate footnotes. Page numbers followed by *t* indicate tables. Page numbers followed by *f* indicate figures.

C

F

I

J

K

P

1989 RDA Median Heights and Weights and Recommended Energy Intakes (United States)

Age (years)	Weight		Height		Average Energy Allowance	
	kg	lb	cm	inches	cal per kg	cal per day[a]
Infants						
0.0–0.5	6	13	60	24	108	650
0.5–1.0	9	20	71	28	98	850
Children						
1–3	13	29	90	35	102	1,300
4–6	20	44	112	44	90	1,800
7–10	28	62	132	52	70	2,000
Males						
11–14	45	99	157	62	55	2,500
15–18	66	145	176	69	45	3,000
19–24	72	160	177	70	40	2,900
25–50	79	174	176	70	37	2,900
51+	77	170	173	68	30	2,300
Females						
11–14	46	101	157	62	47	2,200
15–18	55	120	163	64	40	2,200
19–24	58	128	164	65	38	2,200
25–50	63	138	163	64	36	2,200
51+	65	143	160	63	30	1,900
Pregnant (2nd and 3rd trimesters)						+ 300
Lactating						+ 500

[a]Average energy allowances have been rounded.

SOURCE: *Recommended Dietary Allowances,* © 1989 by the National Academy of Sciences, National Academy Press, Washington, D.C.

Body Mass Index (BMI)

Height	18	19	20	21	22	23	24	25	26	27	28	29	30	31	32	33	34	35	36	37	38	39	40
	Body Weight (pounds)																						
4'10"	86	91	96	100	105	110	115	119	124	129	134	138	143	148	153	158	162	167	172	177	181	186	191
4'11"	89	94	99	104	109	114	119	124	128	133	138	143	148	153	158	163	168	173	178	183	188	193	198
5'0"	92	97	102	107	112	118	123	128	133	138	143	148	153	158	163	168	174	179	184	189	194	199	204
5'1"	95	100	106	111	116	122	127	132	137	143	148	153	158	164	169	174	180	185	190	195	201	206	211
5'2"	98	104	109	115	120	126	131	136	142	147	153	158	164	169	175	180	186	191	196	202	207	213	218
5'3"	102	107	113	118	124	130	135	141	146	152	158	163	169	175	180	186	191	197	203	208	214	220	225
5'4"	105	110	116	122	128	134	140	145	151	157	163	169	174	180	186	192	197	204	209	215	221	227	232
5'5"	108	114	120	126	132	138	144	150	156	162	168	174	180	186	192	198	204	210	216	222	228	234	240
5'6"	112	118	124	130	136	142	148	155	161	167	173	179	186	192	198	204	210	216	223	229	235	241	247
5'7"	115	121	127	134	140	146	153	159	166	172	178	185	191	198	204	211	217	223	230	236	242	249	255
5'8"	118	125	131	138	144	151	158	164	171	177	184	190	197	203	210	216	223	230	236	243	249	256	262
5'9"	122	128	135	142	149	155	162	169	176	182	189	196	203	209	216	223	230	236	243	250	257	263	270
5'10"	126	132	139	146	153	160	167	174	181	188	195	202	209	216	222	229	236	243	250	257	264	271	278
5'11"	129	136	143	150	157	165	172	179	186	193	200	208	215	222	229	236	243	250	257	265	272	279	286
6'0"	132	140	147	154	162	169	177	184	191	199	206	213	221	228	235	242	250	258	265	272	279	287	294
6'1"	136	144	151	159	166	174	182	189	197	204	212	219	227	235	242	250	257	265	272	280	288	295	302
6'2"	141	148	155	163	171	179	186	194	202	210	218	225	233	241	249	256	264	272	280	287	295	303	311
6'3"	144	152	160	168	176	184	192	200	208	216	224	232	240	248	256	264	272	279	287	295	303	311	319
6'4"	148	156	164	172	180	189	197	205	213	221	230	238	246	254	263	271	279	287	295	304	312	320	328
6'5"	151	160	168	176	185	193	202	210	218	227	235	244	252	261	269	277	286	294	303	311	319	328	336
6'6"	155	164	172	181	190	198	207	216	224	233	241	250	259	267	276	284	293	302	310	319	328	336	345
	Under-weight (<18.5)	**Healthy Weight** (18.5–24.9)						**Overweight** (25–29.9)					**Obese** (≥30)										

Find your height along the left-hand column and look across the row until you find the number that is closest to your weight. The number at the top of that column identifies your BMI. Chapter 9 describes how BMI correlates with disease risks and defines obesity. The area shaded in blue represents healthy weight ranges.

Recommended BMI Cutoff Values for Adolescents

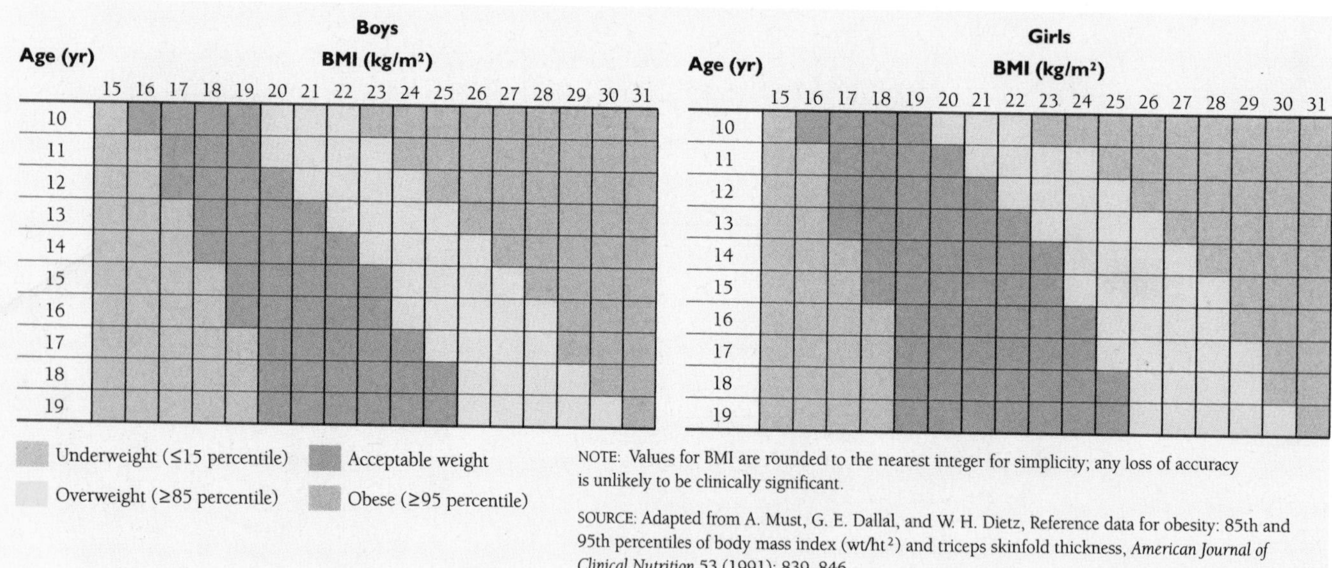

NOTE: Values for BMI are rounded to the nearest integer for simplicity; any loss of accuracy is unlikely to be clinically significant.

SOURCE: Adapted from A. Must, G. E. Dallal, and W. H. Dietz, Reference data for obesity: 85th and 95th percentiles of body mass index (wt/ht²) and triceps skinfold thickness, *American Journal of Clinical Nutrition* 53 (1991): 839–846.